FOURTH EDITION

THE ANALYSIS AND DESIGN OF LINEAR CIRCUITS: LAPLACE EARLY

ROLAND E. THOMAS
Professor Emeritus
United States Air Force Academy

ALBERT J. ROSA
Professor of Engineering
University of Denver

WILEY

JOHN WILEY & SONS, INC.

Executive Editor   Bill Zobrist
Senior Editorial Assistant   Angie Vennerstrom
Senior Marketing Manager   Katherine Hepburn
Senior Production Editor   Caroline Sieg
Design Manager   Madelyn Lesure
Illustration Editor   Sigmund Malinowski
Cover Designer   Madelyn Lesure
Production Management Services   Suzanne Ingrao/Ingrao Associates

This book was set in 10/12 Times Ten Roman by Pine Tree Composition and printed and bound by Von Hoffmann Corporation. The cover was printed by Von Hoffmann Corporation.

This book is printed on acid-free paper. ∞

TRADEMARK INFORMATION

Orcad release 9.2 lite Edition is a registered trademark of Cadence Design Systems, Inc.
Electronics Workbench is a registered trademark of Interactive Image Technologies Ltd.
MATLAB is a registered trademark of The Math Works, Inc.
Mathcad is a registered trademark of Mathsoft, Inc.
Excel is a registered trademark of Microsoft Corporation.

ISBN 0-471-43299-7
WIE ISBN: 0-471-45252-1

Printed in the United States of America

10   9   8   7   6   5   4   3   2   1

*To our wives
Juanita and Kathleen*

PREFACE

There are two versions of the fourth edition of *The Analysis and Design of Linear Circuits*. The standard version (ISBN 0-471-27213-2) is an incremental revision and updating of the third edition. The new *Laplace Early Version* (ISBN 0-471-43299-7) retains the core features of the third edition, but uses a new chapter sequencing that emphasizes transform methods. Both versions are aimed at introductory circuit analysis courses, and both assume the same student prerequisites. Although the sequencing is different, the two versions cover the same range of topics. John Wiley & Sons, Inc. offers two versions of this book as part of a continuing commitment to supplying a diversity of resources for teaching circuit courses.

The *Laplace Early Version* introduces Laplace transforms immediately after the classical treatment of circuit differential equations. This arrangement allows the concepts of sinusoidal steady-state response, network functions, frequency responses, impulse response, step responses, and convolution to be treated within a common framework. We have found that beginning students achieve a real understanding of transient and steady-state responses more rapidly by this method than the classical phasor-first approach. The Laplace early approach actually saves classroom time because students quickly master classical phasor analysis when it is presented as a logical outgrowth of an overall theme. This approach also better prepares students for the profusion of transforms they will encounter in subsequent signals and systems courses.

The fourth edition continues the authors' two-decade commitment to a modern approach to classical circuit analysis and design. Although this version stresses transform methods, it does so within a framework of continuing features that remains unchanged.

CONTINUING FEATURES

CIRCUIT DESIGN

Experience convinces us that students' grasp of circuit fundamentals is reinforced by an interweaving of analysis and design. Early involvement in design provides motivation as students apply their newly acquired knowledge in practical situations. The evaluation of alternative designs also introduces them to real-world engineering practices. Including design in the introductory course makes students aware that circuits perform useful functions and are not simply academic vehicles for teaching node-voltage and mesh-current analysis.

THE OP AMP

The ubiquitous OP AMP is the work horse of linear circuit analysis and design. Its early introduction and integration throughout the first circuit course offers the following advantages. The modular form of OP AMP circuits simplifies analog circuit analysis and design. The OP AMP plays an important role by relating the abstract concept of a dependent source to a real device. The close agreement between theory and laboratory results allows students to analyze, design, and successfully build meaningful OP AMP circuits in the laboratory.

SIGNAL PROCESSING

An emphasis on signal processing and systems is achieved through the use of block diagrams, input–output relationships, and transform methods. The study of dynamic circuits is preceded by a separate chapter on waveforms and signal characteristics. This chapter gives students an early familiarity with all of the important input and output signals encountered in linear circuits. The ultimate goal is for students to understand that time-domain waveforms and frequency-domain transforms are simply alternative ways to characterize signals and signal processing.

COMPUTER TOOLS

Three types of computer programs are used to illustrate computer-aided circuit analysis, namely spreadsheets (Excel®), math solvers (Mathcad® & MATLAB®), and circuit simulators (PSpice® and Electronics Workbench®). Examples of computer-aided circuit analysis are integrated into all chapters, beginning with Chapter 2. The purpose of these examples is to help students develop a problem-solving style that includes the productivity tools routinely used by practicing engineers.

EN ROUTE OBJECTIVES

The book is structured around a sequence of carefully defined learning objectives and related evaluation tools. Each objective is explicitly stated in terms of expected student proficiency, and each is supported by at least ten homework problems specifically designed to evaluate student mastery of the objective. This framework has been a standard feature of all four editions of this book and has enabled us to maintain a consistent level of expected student performance over the years.

NEW FEATURES

- New application examples have been added, emphasizing computer engineering, signal processing, and bioengineering.
- All PSpice examples are updated to Orcad Release 9.2.
- Examples of CMOS voltage-controlled switch models include digitally controlled analog circuits in Chapters 3 and 4 as well as the *RC* switching response of digital circuits in Chapter 7.

- Chapter 6 includes a new treatment of the sample-hold circuit and introduces the integrating A/D converter.

- A new section in Chapter 11 develops the equivalence of time-domain and *s*-domain convolution.

- The en route objectives and related homework problems are now cross-referenced to in-text worked examples and exercises. Cross-referencing allows students to review the solution of problems similar to those in their assigned homework.

- The fourth edition adds chapter length treatment of Fourier transforms and two-port networks as Web appendices. These appendices are fully integrated into the text with index references and answers to selected homework problems. These appendices are available at: *http://www.wiley.com/college/thomas.*

CHAPTER FEATURES

Chapter Openers provide a brief discussion of the historical context as well as an overview of the topical content of the chapter.

Examples provide students with a detailed discussion of the steps involved in solving an analysis or design problem. There are nearly 300 worked examples distributed throughout the text, including more than 30 *design examples*. Several of the design examples involve evaluating alternative solutions. Examples that use *computer tools* also show how to translate a problem statement into a form amenable to computer solution.

Exercises with answers are placed immediately after worked examples. These exercises give students an opportunity to practice the techniques illustrated in the preceding example. Providing answers allows students to test their proficiency before proceeding.

Chapter Summaries contain concise statements that help students review the chapter before tackling the homework problems that follow.

Homework Problems are directly related to and grouped with the En Route Objectives. The En route Objectives are cross-referenced to related in-text examples and exercises. About one-third of the nearly 900 homework problems have answers in the back of the book. Problems involving design are labeled with a **D** icon.

Integrating Problems are comprehensive problems that appear at the very end of a chapter. To solve these problems, students must combine several previously mastered skills. Integrating problems are labeled with **A**, **D**, and **E** icons to indicate whether analysis, design, and evaluation methods are required.

Computer Problems Homework problems for which computer tools offer a significant advantage are marked with a computer icon ▯ in the text. We do not specify a computer tool for each problem, but expect students to choose. To assist students in developing this ability, there is a separately bound **Student Solution Manual** (ISBN

0-471-46968-8) in which about 100 of the homework problems are solved using computer tools. Problems whose solutions are included in this manual are labeled with an 📖 icon in the text.

SUPPLEMENTS

INSTRUCTOR'S MANUAL

A complete set of detailed solutions to all of the end-of-chapter problems is available free to all adopting faculty. Solutions were double checked and documented using Mathcad.

eGRADE ONLINE ASSESSMENT

eGrade is an online problem-solving, quizzing, and testing tool. It also provides interactive self-scoring practice problems to help students learn the skills involved in solving circuits problems. For more information and to see a demonstration visit *www.wiley.com/college/egrade.*

CIRCUIT WORKS SOFTWARE

Circuit Works is a simulator based on a set library of 100 circuits, with adjustable parameters. Circuit Works is a tool for teaching and learning the principles and relationships that underlie basic first- and second-order circuits having resistors, capacitors, inductors, op-amps, dependent and independent sources, and transformers.

STUDENT SOLUTIONS MANUAL

Selected homework problems are solved using one or more of the computer tools demonstrated in the text. These worked-out solutions are available in a separately bound manual that is sold separately. Faculty can order a student bundle containing both the text and the Student Solutions Manual at a special reduced price.

CIRCUITS EXTRA

This Web site is designed to support excellence in teaching and student learning. Circuits Extra is a valuable tool for all instructors and students using Wiley circuits texts. This site contains student and instructor resources for all of the Wiley circuits texts, as well as circuits links and resources of general interest. It will be continually updated to provide enrichment for students and instructors alike. You can access this site at *www.wiley.com/college/circuits.*

ACKNOWLEDGMENTS

The authors wish thank the many people at John Wiley & Sons, Inc. who made both versions of this edition a reality, especially Senior Production Editor Caroline Sieg and Senior Marketing Manager Katherine Hepburn. We are deeply indebted to our Executive Editor Bill Zobrist whose guidance and support allowed us to continue to record our thoughts on teaching circuits. We cannot fail to mention our Production Manager Suzanne

Ingrao whose professionalism has made the task of converting manuscripts into books an experience in the vicinity of enjoyable.

Over the years the following reviewers have helped shape this work in many ways: Robert M. Anderson, Iowa State University; Doran J. Baker, Utah State University; James A. Barby, University of Waterloo; William E. Bennett, United States Naval Academy; Maqsood A. Chaudhry, California State University at Fullerton; Michael Chier, Milwaukee School of Engineering; Don E. Cottrell, University of Denver; Micheal L. Daley University of Memphis; Prasad Enjeti, Texas A&M University; James G. Gottling, Ohio State University; Robert Kotiuga, Boston University; Hans H. Kuehl, University of Southern California; K.S.P. Kumar, University of Minnesota; Michael Lightner, University of Colorado at Boulder; Jerry I. Lubell, Jaycor; Reinhold Ludwig, Worcester Polytechnic Institute; Lloyd W. Massengill, Vanderbilt University; Frank L. Merat, Case Western Reserve University; Richard L. Moat, Motorola; Anil Pahwa, Kansas State University; William Rison, New Mexico Institute of Mining and Technology; Martin S. Roden, California State University at Los Angeles; Pat Sannuti, the State University of New Jersey; Alan Schneider, University of Calfornia at San Diego; Ali O. Shaban, California Polytechnic State University; Jacob Shekel, Northeastern University; Kadagattur Srinidhi, Northeastern University; Peter J. Tabolt, University of Massachusetts at Boston; Len Trombetta, University of Houston; David Voltmer, Rose-Hulman Institute of Technology; and Bruce F. Wollenberg, University of Minnesota. We are also indebted to Ronald R. DeLyser of the University of Denver for developing the *Student Solutions Manual,* to John C. Getty for preparing the *Laboratory Manual,* and to James Kang of Cal Poly Pomona for his assistance in the preparation of the Instructor's Manual.

The first author wishes to express his indebtedness to his wife Juanita whose proofreading and constructive comments were invariably helpful.

CONTENTS

CHAPTER 1 • INTRODUCTION 1

  1-1 About This Book 2
  1-2 Symbols and Units 3
  1-3 Circuit Variables 5
      SUMMARY 11
      PROBLEMS 11
      INTEGRATING PROBLEMS 13

CHAPTER 2 • BASIC CIRCUIT ANALYSIS 14

  2-1 Element Constraints 15
  2-2 Connection Constraints 20
  2-3 Combined Constraints 28
  2-4 Equivalent Circuits 33
  2-5 Voltage and Current Division 39
  2-6 Circuit Reduction 46
  2-7 Computer-aided Circuit Analysis 52
      SUMMARY 57
      PROBLEMS 58
      INTEGRATING PROBLEMS 65

CHAPTER 3 • CIRCUIT ANALYSIS TECHNIQUES 68

  3-1 Node-voltage Analysis 69
  3-2 Mesh-current Analysis 84
  3-3 Linearity Properties 93
  3-4 Thévenin and Norton Equivalent Circuits 101
  3-5 Maximum Signal Transfer 114
  3-6 Interface Circuit Design 117
      SUMMARY 128
      PROBLEMS 129
      INTEGRATING PROBLEMS 138

CHAPTER 4 • ACTIVE CIRCUITS 140

  4-1 Linear Dependent Sources 141
  4-2 Analysis of Circuits with Dependent Sources 143

4-3   The Transistor   156
4-4   The Operational Amplifier   161
4-5   OP AMP Circuit Analysis   168
4-6   OP AMP Circuit Design   183
4-7   The Comparator   193
      SUMMARY   196
      PROBLEMS   197
      INTEGRATING PROBLEMS   204

CHAPTER 5 • SIGNAL WAVEFORMS   208
5-1   Introduction   209
5-2   The Step Waveform   211
5-3   The Exponential Waveform   215
5-4   The Sinusoidal Waveform   219
5-5   Composite Waveforms   227
5-6   Waveform Partial Descriptors   232
      SUMMARY   240
      PROBLEMS   241
      INTEGRATING PROBLEMS   243

CHAPTER 6 • CAPACITANCE AND INDUCTANCE   246
6-1   The Capacitor   247
6-2   The Inductor   254
6-3   Dynamic OP AMP Circuits   260
6-4   Equivalent Capacitance and Inductance   268
      SUMMARY   271
      PROBLEMS   272
      INTEGRATING PROBLEMS   276

CHAPTER 7 • FIRST-ORDER CIRCUITS   278
7-1   *RC* and *RL* Circuits   279
7-2   First-order Circuit Step Response   289
7-3   Initial and Final Conditions   298
7-4   First-order Circuit Sinusoidal Response   304
      SUMMARY   310
      PROBLEMS   311
      INTEGRATING PROBLEMS   314

CHAPTER 8 • SECOND-ORDER CIRCUITS   316
8-1   The Series *RLC* Circuit   317
8-2   The Parallel *RLC* Circuit   328
8-3   Second-order Circuit Step Response   333

8-4   Second-order OP AMP Circuits   339
      SUMMARY   343
      PROBLEMS   344
      INTEGRATING PROBLEMS   347

CHAPTER 9 • LAPLACE TRANSFORMS   348
9-1   Signal Waveforms and Transforms   349
9-2   Basic Properties and Pairs   353
9-3   Pole-zero Diagrams   362
9-4   Inverse Laplace Transforms   365
9-5   Some Special Cases   371
9-6   Circuit Response Using Laplace Transforms   376
9-7   Initial Value and Final Value Properties   384
      SUMMARY   389
      PROBLEMS   389
      INTEGRATING PROBLEMS   393

CHAPTER 10 • S-DOMAIN CIRCUIT ANALYSIS   394
10-1 Transformed Circuits   395
10-2 Basic Circuit Analysis in the s Domain   404
10-3 Circuit Theorems in the s Domain   409
10-4 Node-voltage Analysis in the s Domain   420
10-5 Mesh-current Analysis in the s Domain   428
10-6 Summary of s-Domain Circuit Analysis   435
      SUMMARY   440
      PROBLEMS   440
      INTEGRATING PROBLEMS   447

CHAPTER 11 • NETWORK FUNCTIONS   448
11-1 Definition of a Network Function   449
11-2 Network Functions of One- and Two-port Circuits   452
11-3 Network Functions and Impulse Response   463
11-4 Network Functions and Step Response   466
11-5 Network Functions and Sinusoidal Steady-state Response   471
11-6 Impulse Response and Convolution   476
11-7 Network Function Design   483
      SUMMARY   497
      PROBLEMS   498
      INTEGRATING PROBLEMS   503

CHAPTER 12 • FREQUENCY RESPONSE   505
12-1 Frequency-response Descriptors   506
12-2 First-order Circuit Frequency Response   511

12-3 Second-order Circuit Frequency Response   527

12-4 The Frequency Response of *RLC* Circuits   541

12-5 Bode Diagrams   548

12-6 Bode Diagrams with Complex Critical Frequencies   558

12-7 Frequency Response and Step Response   565

SUMMARY   572

PROBLEMS   573

INTEGRATING PROBLEMS   579

CHAPTER 13 • FOURIER SERIES   581

13-1 Overview of Fourier Analysis   582

13-2 Fourier Coefficients   584

13-3 Waveform Symmetries   593

13-4 Circuit Analysis Using Fourier Series   596

13-5 RMS Value and Average Power   602

SUMMARY   607

PROBLEMS   608

INTEGRATING PROBLEMS   612

CHAPTER 14 • ANALOG FILTER DESIGN   613

14-1 Frequency-domain Signal Processing   614

14-2 Design with First-order Circuits   616

14-3 Design with Second-order Circuits   624

14-4 Low-pass Filter Design   632

14-5 High-pass Filter Design   649

14-6 Bandpass and Bandstop Filter Design   654

SUMMARY   661

PROBLEMS   661

INTEGRATING PROBLEMS   666

CHAPTER 15 • SINUSOIDAL STEADY-STATE RESPONSE   668

15-1 Sinusoids and Phasors   669

15-2 Phasor Circuit Analysis   675

15-3 Basic Circuit Analysis with Phasors   678

15-4 Circuit Theorems with Phasors   690

15-5 General Circuit Analysis with Phasors   699

15-6 Energy and Power   709

SUMMARY   714

PROBLEMS   715

INTEGRATING PROBLEMS   720

CHAPTER 16 • MUTUAL INDUCTANCE   722

16-1 Coupled Inductors   723
16-2 The Dot Convention   726
16-3 Energy Analysis   729
16-4 The Ideal Transformer   731
16-5 Transformers in the Sinusoidal Steady State   737
16-6 Transformer Equivalent Circuits   742
    SUMMARY   744
    PROBLEMS   745
    INTEGRATING PROBLEMS   748

CHAPTER 17 • POWER IN THE SINUSOIDAL STEADY STATE   750

17-1 Average and Reactive Power   751
17-2 Complex Power   754
17-3 AC Power Analysis   758
17-4 Load-flow Analysis   761
17-5 Three-phase Circuits   768
17-6 Three-phase AC Power Analysis   772
    SUMMARY   783
    PROBLEMS   784
    INTEGRATING PROBLEMS   787

APPENDIX A — STANDARD VALUES   A-1
APPENDIX B — SOLUTION OF LINEAR EQUATIONS   A-2
APPENDIX C — COMPLEX NUMBERS   A-12

ANSWERS TO SELECTED PROBLEMS   A-16
INDEX   I-1

WEB APPENDICES   W-1
APPENDIX W1 — FOURIER TRANSFORMS   W-1

W1-1 Definition of Fourier Transforms   W-1
W1-2 Laplace Transforms and Fourier Transforms   W-7
W1-3 Basic Properties and Pairs   W-10
W1-4 Circuit Analysis Using Fourier Transforms   W-18
W1-5 Impulse Response and Convolution   W-22
W1-6 Parseval's Theorem   W-26

SUMMARY   W-32
PROBLEMS   W-33

APPENDIX W2—TWO-PORT CIRCUITS   W-37

W2-1 Introduction   W-37
W2-2 Impedance Parameters   W-39
W2-3 Admittance Parameters   W-41
W2-4 Hybrid Parameters   W-43
W2-5 Transmission Parameters   W-46
W2-6 Two-Port Conversion   W-49
W2-7 Two-Port Connections   W-50
     SUMMARY   W-53
     PROBLEMS   W-54

CHAPTER 1

INTRODUCTION

The electromotive action manifests itself in the form of two effects which I believe must be distinguished from the beginning by a precise definition. I will call the first of these "electric tension," the second "electric current."

André-Marie Ampère, 1820,
French Mathematician/Physicist

1–1 About This Book

1–2 Symbols and Units

1–3 Circuit Variables

Summary

Problems

Integrating Problems

This book deals with the analysis and design of linear electric circuits. A circuit is an interconnection of electric devices that processes energy or information. Understanding circuits is important because energy and information are the underlying technological commodities in electrical engineering. The study of circuits provides a foundation for areas of electrical engineering such as electronics, power systems, communication systems, and control systems.

This chapter describes the structure of this book, introduces basic notation, and defines the primary physical variables in electric circuits—voltage

1

and current. André Ampère (1775–1836) was the first to recognize the importance of distinguishing between the electrical effects now called voltage and current. He also invented the galvanometer, the forerunner of today's voltmeter and ammeter. A natural genius who had mastered all the then-known mathematics by age 12, he is best known for defining the mathematical relationship between electric current and magnetism. This relationship, now known as Ampère's law, is one of the basic concepts of modern electromagnetics.

The first section of this chapter describes the pedagogical framework and terminology that must be understood to use this book effectively. It describes how the learning objectives are structured to help the student develop the problem-solving abilities needed to analyze and design circuits. The second section provides some of the standard scientific notation and conventions used throughout the book. The last section introduces electric voltage, current, and power—the physical variables used throughout the book to describe the signal-processing and energy-transfer capabilities of linear circuits.

1–1 ABOUT THIS BOOK

The basic purpose of this book is to introduce the analysis and design of linear circuits. Circuits are important in electrical engineering because they process electrical signals that carry energy and information. For the present we can define a **circuit** as an interconnection of electrical devices, and a **signal** as a time-varying electrical entity. For example, the information stored on a compact disk is recovered in the CD-ROM player as electronic signals that are processed by circuits to generate audio and video outputs. In an electrical power system some form of stored energy is converted to electrical form and transferred to loads, where the energy is converted into the form required by the customer. The CD-ROM player and the electrical power system both involve circuits that process and transfer electrical signals carrying energy and information.

In this text we are primarily interested in **linear circuits**. An important feature of a linear circuit is that the amplitude of the output signal is proportional to the input signal amplitude. The proportionality property of linear circuits greatly simplifies the process of circuit analysis and design. Most circuits are only linear within a restricted range of signal levels. When driven outside this range they become nonlinear, and proportionality no longer applies. Although we will treat a few examples of nonlinear circuits, our attention is focused on circuits operating within their linear range.

Our study also deals with interface circuits. For the purposes of this book, we define an **interface** as a pair of accessible terminals at which signals may be observed or specified. The interface idea is particularly important with integrated circuit (IC) technology. Integrated circuits involve many thousands of interconnections, but only a small number are accessible to the user. Designing systems using integrated circuits involves interconnecting complex circuits that have only a few accessible terminals. This often involves relatively simple circuits whose purpose is to change signal levels or formats. Such interface circuits are intentionally introduced to ensure that the appropriate signal conditions exist at the connections between complex integrated circuits.

COURSE OBJECTIVES

This book is designed to help you develop the knowledge and application skills needed to solve three types of circuit problems: analysis, design, and evaluation. An **analysis** problem involves finding the output signals of a given circuit with known input signals. Circuit analysis is the foundation for understanding the interaction of signals and circuits. A **design** problem involves devising one or more circuits that perform a given signal-processing function. There usually are several possible solutions to a design problem. This leads to an **evaluation** problem which involves picking the best solution from among several candidates using factors such as cost, power consumption, and part counts. In real life the engineer's role is a blend of analysis, design, and evaluation, and in practice the boundaries between these categories are often blurred.

This text contains many worked examples to help you develop your problem solving skills. The **examples** include a problem statement and provide the intermediate steps needed to obtain the final answer. The examples often treat analysis problems, although design and evaluation examples are included. This text also contains a number of **exercises** that include only the problem statement and the final answer. You should use the exercises to test your understanding of the circuit concepts discussed in the preceding section. In addition, there is a **student solutions manual** available from the publisher in which 100 of the end-of-chapter problems are worked using computer tools. In the text these problems are marked with a 📖 icon. Other problems that lend themselves to computer solution are marked with a 💻 icon.

EN ROUTE OBJECTIVES

At the end of each chapter we provide a carefully structured set of enabling skills called **en route objectives** (EROs). Collectively, the ERO's define the basic knowledge and understanding needed to achieve the course objectives. They are a sequence of steps that must be mastered to progress to subsequent topics. Following each ERO there are a number of homework problems that test your ability to meet the objective and move on to subsequent topics. If you have difficulty, review the in-text examples and exercises listed in the ERO statement for guidance in problem solving. Once you understand the chapter ERO's you can move on to the integrating problems at the very end of the chapter. These problems require mastery of several en route objectives and offer you an opportunity to test your ability to deal with more comprehensive problems.

1-2 SYMBOLS AND UNITS

Throughout this text we will use the international system (SI) of units. The SI system includes six fundamental units: meter (m), kilogram (kg), second (s), ampere (A), kelvin (K), and candela (cd). All the other units of measure can be derived from these six.

Like all disciplines, electrical engineering has its own terminology and symbology. The symbols used to represent some of the more important physical quantities and their units are listed in Table 1–1. It is not our purpose to define these quantities here, nor to offer this list as an item for memorization. Rather, the purpose of this table is merely to list in one place all the electrical quantities used in this book.

TABLE 1–1   SOME IMPORTANT QUANTITIES, THEIR SYMBOLS,
AND UNIT ABBREVIATIONS

QUANTITY	SYMBOL	UNIT	UNIT ABBREVIATION
Time	t	second	s
Frequency	f	hertz	Hz
Radian frequency	ω	radian/second	rad/s
Phase angle	θ, ϕ	degree or radian	° or rad
Energy	w	joule	J
Power	p	watt	W
Charge	q	coulomb	C
Current	i	ampere	A
Electric field	$\mathscr{E}$	volt/meter	V/m
Voltage	v	volt	V
Impedance	Z	ohm	Ω
Admittance	Y	siemens	S
Resistance	R	ohm	Ω
Conductance	G	siemens	S
Reactance	X	ohm	Ω
Susceptance	B	siemens	S
Inductance, self	L	henry	H
Inductance, mutual	M	henry	H
Capacitance	C	farad	F
Magnetic flux	ϕ	weber	wb
Flux linkages	λ	weber-turns	wb-t
Power ratio	P	bel	B

Numerical values in engineering range over many orders of magnitude. Consequently, the system of standard decimal prefixes in Table 1–2 is used. These prefixes on a unit abbreviation symbol indicate the power of 10 that is applied to the numerical value of the quantity.

Exercise 1–1

Given the pattern in the statement 1 kΩ = 1 kilohm = 1×10^3 ohms, fill in the blanks in the following statements using the standard decimal prefixes.

(a) _____ = _____ = 5×10^{-3} watts
(b) 10.0 dB = _____ = _____
(c) 3.6 ps = _____ = _____
(d) _____ = 0.03 microfarads = _____
(e) _____ = _____ gigahertz = 6.6×10^9 Hertz

  Answers:
(a) 5.0 mW = 5 milliwatts
(b) 10.0 decibels = 1.0 bel
(c) 3.6 picoseconds = 3.6×10^{-12} seconds
(d) 30 nF or 0.03 μF = 30.0×10^{-9} Farads
(e) 6.6 GHz = 6.6 gigahertz

TABLE 1–2   STANDARD DECIMAL PREFIXES

MULTIPLIER	PREFIX	ABBREVIATION
10^{18}	exa	E
10^{15}	peta	P
10^{12}	tera	T
10^{9}	giga	G
10^{6}	mega	M
10^{3}	kilo	k
10^{-1}	deci	d
10^{-2}	centi	c
10^{-3}	milli	m
10^{-6}	micro	μ
10^{-9}	nano	n
10^{-12}	pico	p
10^{-15}	femto	f
10^{-18}	atto	a

1–3   CIRCUIT VARIABLES

The underlying physical variables in the study of electronic systems are **charge** and **energy**. The idea of electrical charge explains the very strong electrical forces that occur in nature. To explain both attraction and repulsion, we say that there are two kinds of charge—positive and negative. Like charges repel, while unlike charges attract one another. The symbol q is used to represent charge. If the amount of charge is varying with time, we emphasize the fact by writing $q(t)$. In the international system (SI), charge is measured in **coulombs** (abbreviated C). The smallest quantity of charge in nature is an electron's charge ($q_E = 1.6 \times 10^{-19}$C). Thus, there are $1/q_E = 6.25 \times 10^{18}$ electrons in 1 coulomb of charge.

Electrical charge is a rather cumbersome variable to measure in practice. Moreover, in most situations the charges are moving, so we find it more convenient to measure the amount of charge passing a given point per unit time. If $q(t)$ is the cumulative charge passing through a point, we define a signal variable i called **current** as follows:

$$i = \frac{dq}{dt} \tag{1–1}$$

Current is a measure of the flow of electrical charge. It is the time rate of change of charge passing a given point in a circuit. The physical dimensions of current are coulombs per second. In the SI system, the unit of current is the **ampere** (abbreviated A). That is,

$$1 \text{ coulomb/second} = 1 \text{ ampere}$$

Since there are two types of electrical charge, there is a bookkeeping problem associated with the direction assigned to the current. In engineering it is customary to define the direction of current as the direction of the net flow of positive charge.

A second signal variable called **voltage** is related to the change in energy that would be experienced by a charge as it passes through a circuit. The symbol w is commonly used to represent energy. In the SI system of units, energy carries the units of **joules** (abbreviated J). If a small charge dq was to experience a change in energy dw in passing from point A to point B in a circuit, then the voltage v between A and B is defined as the change in energy per unit charge. We can express this definition in differential form as

$$v = \frac{dw}{dq} \tag{1–2}$$

Voltage does not depend on the path followed by the charge dq in moving from point A to point B. Furthermore, there can be a voltage between two points even if there is no charge motion, since voltage is a measure of how much energy dw would be involved if a charge dq was moved. The dimensions of voltage are joules per coulomb. The unit of voltage in the SI system is the **volt** (abbreviated V). That is,

$$1 \text{ joule/coulomb} = 1 \text{ volt}$$

The general definition of physical variable called **power** is the time rate of change of energy:

$$p = \frac{dw}{dt} \tag{1–3}$$

The dimensions of power are joules per second, which in the SI system is called a **watt** (abbreviated W). In electrical circuits it is useful to relate power to the signal variables current and voltage. Using the chain rule, Eq. (1–3) can be written as

$$p = \left(\frac{dw}{dq} \right) \left(\frac{dq}{dt} \right) \tag{1–4}$$

Now using Eqs. (1–1) and (1–2), we obtain

$$p = vi \tag{1–5}$$

The electrical power associated with a situation is determined by the product of voltage and current. The total energy transferred during the period from t_1 to t_2 is found by solving for dw in Eq. (1–3) and then integrating

$$w_T = \int_{w_1}^{w_2} dw = \int_{t_1}^{t_2} p \, dt \tag{1–6}$$

EXAMPLE 1–1

The electron beam in the cathode-ray tube shown in Figure 1–1 carries 10^{14} electrons per second and is accelerated by a voltage of 50 kV. Find the power in the electron beam.

FIGURE 1 – 1

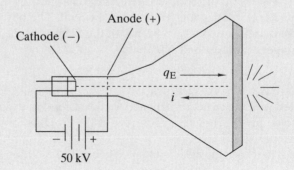

SOLUTION:

Since current is the rate of charge flow, we can find the net current by multiplying the charge of the electron q_E by the rate of electron flow dn_E/dt.

$$i = q_E \frac{dn_E}{dt} = (1.6 \times 10^{-19})(10^{14}) = 1.6 \times 10^{-5} \text{ A}$$

Therefore, the beam power is

$$p = vi = (50 \times 10^3)(1.6 \times 10^{-5}) = 0.8 \text{ W} \qquad \blacksquare$$

EXAMPLE 1-2

The current through a circuit element is 50 mA. Find the total charge and the number of electrons transferred during a period of 100 ns.

SOLUTION:

The relationship between current and charge is given in Eq. (1–1) as

$$i = \frac{dq}{dt}$$

Since the current i is given, we calculate the charge transferred by solving this equation for dq and then integrating

$$q_T = \int_{q_1}^{q_2} dq = \int_0^{10^{-7}} i \, dt$$

$$= \int_0^{10^{-7}} 50 \times 10^{-3} \, dt = 50 \times 10^{-10} \text{C} = 5 \text{ nC}$$

There are $1/q_E = 6.25 \times 10^{18}$ electrons/coulomb, so the number of electrons transferred is

$$n_E = (5 \times 10^{-9} \text{ C})(6.25 \times 10^{18} \text{ electrons/C}) = 31.2 \times 10^9 \text{ electrons} \quad \blacksquare$$

Exercise 1-2

A device dissipates 100 W of power. How much energy is delivered to it in 10 seconds?

Answer: 1 kJ

Exercise 1-3

The graph in Figure 1–2(a) shows the charge $q(t)$ flowing past a point in a wire as a function of time.
(a) Find the current $i(t)$ at $t = 1, 2.5, 3.5, 4.5,$ and 5.5 ms.
(b) Sketch the variation of $i(t)$ versus time.

Answers:
(a) -10 nA, $+40$ nA, 0 nA, -20 nA, 0 nA.
(b) The variations in $i(t)$ are shown in Figure 1–2(b).

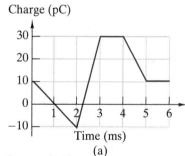
Charge (pC)

(a)

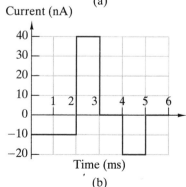
Current (nA)

(b)

FIGURE 1 - 2

THE PASSIVE SIGN CONVENTION

We have defined three circuit variables (current, voltage, and power) using two basic variables (charge and energy). Charge and energy, like mass, length, and time, are basic concepts of physics that provide the scientific foundation for electrical engineering. However, engineering problems rarely involve charge and energy directly, but are usually stated in terms of voltage, current, and power. The reason for this is simple: The circuit variables are much easier to measure and therefore are the most useful working variables in engineering practice.

At this point, it is important to stress the physical differences between current and voltage variables. Current is a measure of the time rate of charge passing a point in a circuit. We think of current as a *through variable,* since it describes the flow of electrical charge through a point in a circuit. On the other hand, voltage is not measured at a single point, but rather between two points or across an electrical device. Consequently, we think of voltage as an *across variable* that inherently involves two points.

The arrow and the plus and minus symbols in Figure 1–3 are *reference marks* that define the positive directions for the current and voltage associated with an electrical device. These reference marks do not represent an assertion about what is happening physically in the circuit. The response of an electrical circuit is determined by physical laws, and not by the reference marks assigned to the circuit variables.

The reference marks are benchmarks assigned at the beginning of the analysis. When the actual direction and reference direction agree, the answers found by circuit analysis will have positive algebraic signs. When they disagree, the algebraic signs of the answers will be negative. For example, if circuit analysis reveals that the current variable in Figure 1–3 is positive [i.e., $i(t) > 0$], then the sign of this answer, together with the assigned reference direction, indicates that the current passes through point A in Figure 1–3 from left to right. Conversely, when analysis reveals that the current variable is negative, then this result, combined with the assigned reference direction, tells us that the current passes through point A from right to left. In summary, the algebraic sign of the answer together with arbitrarily assigned reference marks tell us the actual directions of a voltage or current variable.

In Figure 1–3, the current reference arrow enters the device at the terminal marked with the plus voltage reference mark. This orientation is called the **passive sign convention**. Under this convention, the power $p(t)$ is positive when the device absorbs power and is negative when it delivers power to the rest of the circuit. Since $p(t) = v(t) \times i(t)$, a device absorbs power when the voltage and current variables have the same algebraic sign and delivers power when they have opposite signs. Certain devices, such as heaters (a toaster, for example), can only absorb power. The voltage and current variables associated with these devices must always have the same algebraic sign. On the other hand, a battery absorbs power [$p(t) > 0$] when it is being charged and delivers power [$p(t) < 0$] when it is discharging. Thus, the voltage and current variables for a battery can have the same or the opposite algebraic signs.

In a circuit some devices absorb power and others deliver power, but the sum of the power in all of the devices in the circuit is zero. This is more than just a conservation-of-energy concept. When electrical devices are interconnected to form a circuit, the only way that power can enter or leave the cir-

$i(t)$

Ⓐ

+

$v(t)$

Rest of the circuit

Device

−

Ⓑ

F I G U R E   1 – 3 *Voltage and current reference marks for a two-terminal device.*

cuit is via the currents and voltages at device terminals. The existence of a power balance in a circuit is one method of checking calculations.

The passive sign convention is used throughout this book. It is also the convention used by circuit simulation computer programs.[1] To interpret correctly the results of circuit analysis, it is important to remember that the reference marks (arrows and plus/minus signs) are reference directions, not indications of the circuit response. The actual direction of a response is determined by comparing its reference direction with the algebraic signs of the result predicted by circuit analysis based on physical laws.

GROUND

Since voltage is defined between two points, it is often useful to define a common voltage reference point called **ground**. The voltages at all other points in a circuit are then defined with respect to this common reference point. We indicate the voltage reference point using the ground symbol shown in Figure 1–4. Under this convention we sometimes refer to the variables $v_A(t)$, $v_B(t)$, and $v_C(t)$ as the voltages at points A, B, and C, respectively. This terminology appears to contradict the fact that voltage is an across variable that involves two points. However, the terminology means that the variables $v_A(t)$, $v_B(t)$, and $v_C(t)$ are the voltages defined between points A, B, and C, and the common voltage reference point at point G.

Using a common reference point for across variables is not an idea unique to electrical circuits. For example, the elevation of a mountain is the number of feet or meters between the top of the mountain and a common reference point at sea level. If a geographic point lies below sea level, then its elevation is assigned a negative algebraic sign. So it is with voltages. If circuit analysis reveals that the voltage variable at point A is negative [i.e., $v_A(t) < 0$], then this fact together with the reference marks in Figure 1–4 indicate that the potential at point A is less than the ground potential.

$$v_A(t) \qquad v_B(t) \qquad v_C(t)$$
$$+\,\circ \qquad\quad +\,\circ \qquad\quad +\,\circ$$
$$\text{A} \qquad\qquad \text{B} \qquad\qquad \text{C}$$

$$-\,\circ\,\text{G}$$

F I G U R E   1 – 4 *Ground symbol indicates a common voltage reference point.*

EXAMPLE 1–3

Figure 1–5 shows a circuit formed by interconnecting five devices, each of which has two terminals. A voltage and current variable has been assigned to each device using the passive sign convention. The working variables for each device are observed to be as follows:

	DEVICE 1	DEVICE 2	DEVICE 3	DEVICE 4	DEVICE 5
v	+100 V	?	+25 V	+75 V	−75 V
i	?	+5 mA	+5 mA	?	+5 mA
p	−1 W	+0.5 W	?	0.75 W	?

(a) Find the missing variable for each device and state whether the device is absorbing or delivering power.
(b) Check your work by showing that the sum of the device powers is zero.

---

1 We discuss computer-aided circuit analysis in subsequent chapters.

FIGURE 1–5

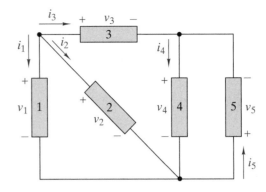

SOLUTION:

(a) We use $p = vi$ to solve for the missing variable since two of the three circuit variables are given for each device.

Device 1: $i_1 = p_1/v_1 = -1/100 = -10$ mA $[p(t) < 0$, delivering power$]$

Device 2: $v_2 = p_2/i_2 = 0.5/0.005 = 100$ V $[p(t) > 0$, absorbing power$]$

Device 3: $p_3 = v_3i_3 = 25 \times 0.005 = 0.125$ W $[p(t) > 0$, absorbing power$]$

Device 4: $i_4 = p_4/v_4 = 0.75/75 = 10$ mA $[p(t) > 0$, absorbing power$]$

Device 5: $p_5 = v_5i_5 = -75 \times 0.005 = -0.375$ W $[p(t) < 0$, delivering power$]$

(b) Summing the device powers yields

$$p_1 + p_2 + p_3 + p_4 + p_5 = -1 + 0.5 + 0.125 + 0.75 - 0.375$$

$$= +1.375 - 1.375 = 0$$

This example shows that the sum of the power absorbed by devices is equal in magnitude to the sum of the power supplied by devices. A power balance always exists in the types of circuits treated in this book and can be used as an overall check of circuit analysis calculations. ■

Exercise 1–4

The working variables of a set of two-terminal electrical devices are observed to be as follows:

	DEVICE 1	DEVICE 2	DEVICE 3	DEVICE 4	DEVICE 5
v	+10 V	?	−15 V	+5 V	?
i	−3 A	−3 A	+10 mA	?	−12 mA
p	?	+40 W	?	+10 mW	−120 mW

Using the passive sign convention, find the magnitude and sign of the unknown variable and state whether the device is absorbing or delivering power.

Answers:

Device 1: $p = -30$ W (delivering power)
Device 2: $v = -13.3$ V (absorbing power)
Device 3: $p = -150$ mW (delivering power)
Device 4: $i = +2$ mA (absorbing power)
Device 5: $v = +10$ V (delivering power)

SUMMARY

- Circuits are important in electrical engineering because they process signals that carry energy and information. A **circuit** is an interconnection of electrical devices. A **signal** is an electrical current or voltage that carries energy or information. An **interface** is a pair of accessible terminals at which signals may be observed or specified.

- This book defines overall course objectives at the analysis, design, and evaluation levels. In **circuit analysis** the circuit and input signals are given and the object is to find the output signals. The object of **circuit design** is to devise one or more circuits that produce prescribed output signals for given input signals. The **evaluation** problem involves appraising alternative circuit designs using criteria such as cost, power consumption, and parts count.

- Charge (q) and energy (w) are the basic physical variables involved in electrical phenomena. Current (i), voltage (v), and power (p) are the derived variables used in circuit analysis and design. In the SI system, charge is measured in coulombs (C), energy in joules (J), current in amperes (A), voltage in volts (V), and power in watts (W).

- **Current** is defined as dq/dt and is a measure of the flow of electrical charge. **Voltage** is defined as dw/dq and is a measure of the energy required to move a small charge from one point to another. **Power** is defined as dw/dt and is a measure of the rate at which energy is being transferred. Power is related to current and voltage as $p = vi$.

- The **reference marks** (arrows and plus/minus signs) assigned to a device are reference directions, not indications of the way a circuit responds. The actual direction of the response is determined by comparing the reference direction and the algebraic sign of the answer found by circuit analysis using physical laws.

- Under the **passive sign convention**, the current reference arrow is directed toward the terminal with the positive voltage reference mark. Under this convention, the device power is positive when it absorbs power and is negative when it delivers power. When current and voltage have the same (opposite) algebraic signs, the device is absorbing (delivering) power.

PROBLEMS

ERO 1–1 ELECTRICAL SYMBOLS AND UNITS (SECT. 1–2)

Given an electrical quantity described in terms of words, scientific notation, or decimal prefix notation, convert the quantity to an alternate description.
See Exercise 1–1

1–1 Write the following statements in symbolic form:
(a) twelve milliamps
(b) four hundred fifty five kilohertz
(c) two hundred picoseconds
(d) five megawatts

1–2 Express the following quantities using appropriate engineering prefixes (i.e., state the numeric to the nearest standard prefix).
(a) 0.022 volts
(b) 23×10^{-9} farads
(c) 56,000 ohms
(d) 7.52×10^5 joules
(e) 0.000235 henrys

1–3 An ampere-hour (Ah) meter measures the time-integral of the current in a conductor. During an 8-hour period a certain meter records 3300 Ah. Find the number of coulombs that flowed through the meter during the recording period.

1–4   Commercial electrical power companies measure energy consumption in kilowatt-hours, denoted kWh. One kilowatt-hour is the amount of energy transferred by 1 kW of power in a time period of 1 hour. During a one-month time period, a power company billing statement reports a user's total energy usage to be 2128 kWh. Determine the number of joules used during the billing period.

1–5   Fill in the blanks in the following statements.
 (a) To convert capacitance from microfarads to picofarads, multiply by _____.
 (b) To convert resistance from kilohms to megohms, multiply by _____.
 (c) To convert current from amperes to milliamperes, multiply by _____.
 (d) To convert power from watts to megawatts, multiply by _____.

ERO 1–2   CIRCUIT VARIABLES (SECT. 1–3)

Given any two of the three signal variables (i, v, p) or the two basic variables (q, w), find the magnitude and direction (sign) of the unspecified variables.
See Examples 1–1, 1–2, 1–3 and Exercises 1–1, 1–2, 1–3

1–6   The charge flowing through a device is $q(t) = 3t - 2$ mC. Find the current through the device.

1–7   The charge flowing through a device is $q(t) = 20e^{-3t}$ μC. Find the current through the device. .

1–8   Figure P1–8 shows a plot of the net positive charge flowing in a wire versus time. Sketch the corresponding current during the same time period.

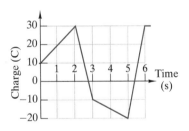

FIGURE   P1–8

1–9   The current through a device is $i(t) = 3t^2$ A. Find the total charge that flowed through the device between $t = 0$ and $t = 2$ s.

1–10   The current through a device is $i(t) = 3e^{-2t}$ A. Find the total charge that flowed through the device between $t = 0$ and $t = 0.1$ s.

1–11   For $t \leq 5$ s the current through a device is $i(t) = 10$ A. For $5 < t \leq 10$ s the current is $i(t) = 20 - 2t$ A, and it is zero for $t > 10$. Sketch $i(t)$ versus time and find the total charge through the device between $t = 0$ s and $t = 10$ s.

1–12   The charge flowing through a device is $q(t) = 1 - e^{-2000t}$ μC. Sketch the current through the device versus time for $t > 0$.

1–13   The 12-V automobile battery in Figure P1–13 has an output capacity of 200 ampere-hours (Ah) when connected to a headlamp that absorbs 50 watts of power. Assume that the battery voltage is constant.
 (a) Find the current supplied by the battery.
 (b) How long can the battery power the headlight?

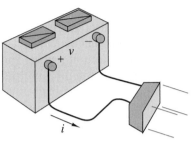

FIGURE   P1–13

1–14   An incandescent lamp absorbs 75 W when connected to a 120-V source.
 (a) Find the current through the lamp.
 (b) Find the cost of operating the lamp for 8 hours when electricity costs 6.8 cents/kW-hr.

1–15   A total charge of 50 ampere-hours is supplied to a 12-V battery during recharging. Determine the number of joules supplied to the battery. Assume that battery voltage is constant.

1–16   The voltage across a device is 25 V when the current through the device is $i(t) = 3e^{-2t}$ A. Find the total energy delivered to the device between $t = 0$ and $t = 0.5$ s.

1–17   When illuminated the i–v relationship for a photocell is $i = e^v - 10$ A. Calculate the device power and state whether the device is absorbing or delivering power when $v = -3$ V. Repeat for $v = 1.5$ V and 3 V.

1–18   Using the passive sign convention, the current through and voltage across a two-terminal device are $i = 1 - 2e^{-t}$ A and $v = 25$ V. Calculate the device power and state whether the device is absorbing or delivering power when $t = 0.5$ s, $t = 1$ s, and $t = 10$ s.

1–19   The maximum power the device can dissipate is 0.25 W. Determine the maximum current allowed by the device power rating when the voltage is 50 V.

1–20   A laser produces 5-kW bursts of power that last 20 ns. If the burst rate is 40 bursts per second, what is the average power in the laser's output?

1–21   Two electrical devices are connected as shown in Figure P1-21. Using the reference marks shown in the figure, find the power transferred and state whether the power is transferred from A to B or B to A when
 (a) $v = +33$ V and $i = -2.2$ A
 (b) $v = -12$ V and $i = -1.2$ mA
 (c) $v = +37.5$ V and $i = +40$ mA
 (d) $v = -15$ V and $i = -43$ mA

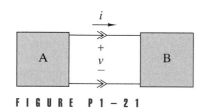

F I G U R E   P 1 – 2 1

1–22 Figure P1–22 shows an electric circuit with a voltage and a current variable assigned to each of the six devices. The device signal variables are observed to be as follows:

Device 1, $v = 15$ V, $i = -1$ A, and $p = ?$
Device 2, $v = 5$ V, $i = ?$, and $p = 5$ W
Device 3, $v = ?$ V, $i = 0.5$ A, and $p = 5$ W
Device 4, $v = 4$ V, $i = 0.5$ A, and $p = ?$
Device 5, $v = ?$ V, $i = 3$ A, and $p = 18$ W
Device 6, $v = ?$ V, $i = -2.5$ A, and $p = -15$ W

Find the unknown signal variable associated with each device and state whether the device is absorbing or delivering power. Use the power balance to check your work.

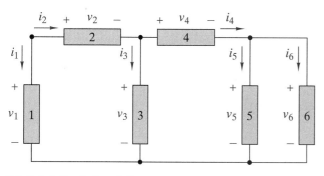

F I G U R E   P 1 – 2 2

1–23 Figure P1–22 shows an electric circuit with a voltage and a current variable assigned to each of the six devices. The device signal variables are observed to be as follows:

Device 1, $v = 30$ V, $i = -2$ A, and $p = ?$
Device 2, $v = 10$ V, $i = ?$, and $p = 20$ W
Device 3, $v = 20$ V, $i = ?$, and $p = 20$ W
Device 4, $v = 8$ V, $i = 1$ A, and $p = ?$
Device 5, $v = ?$ V, $i = -5$ A, and $p = -60$ W
Device 6, $v = 12$ V, $i = 6$ A, and $p = ?$ W

Find the unknown signal variable associated with each device and state whether the device is absorbing or delivering power. Use the power balance to check your work.

1–24 Using the passive sign convention, the voltage across a device is $v(t) = 5 \cos(10t)$ V and the current through the device $i(t) = 0.5 \sin(10t)$ A. Calculate the device power at $t = 0.2$ s and $t = 0.4$ s and state whether the device is absorbing or delivering power.

1–25 For $t \geq 0$ the voltage across and current through a device are $v(t) = 10(1 - e^{-25t})$ V and $i(t) = 0.5e^{-25t}$ A. Find the energy delivered to the device between $t = 0$ and $t = 1$ s.

INTEGRATING PROBLEMS

1–26 Power Ratio in dB

In complete darkness the voltage across and current through a two-terminal light detector are +5.6 V and +8 nA. In full sunlight the voltage and current are +0.9 V and +4 mA. Express the light/dark power ratio of the device in decibels (dB), where the power ratio in dB is

$$P_{dB} = 10 \log_{10}(p_2/p_1).$$

1–27 AC to DC Converter

A manufacturer's data sheet for the converter in Figure P1-27 states that the AC input voltage is 120 V, the DC output is 24 V, and the efficiency is 82% when the output power is 200 W. Find the input and output currents.

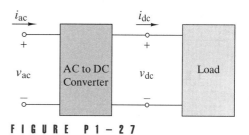

F I G U R E   P 1 – 2 7

1–28 Storage Battery Efficiency

The ampere-hour efficiency of a storage battery is the ratio of its ampere-hour output to the ampere-hour input required to recharge the battery. A certain 24-V battery has a rated output of 400 ampere-hours. When the battery is completely drained, a battery charger must deliver 75 A for 6 hours to recharge the battery. Assume that the battery voltage is constant.

(a) Determine the ampere-hour efficiency of the battery.

(b) Determine the number of joules required to recharge the battery.

CHAPTER 2

BASIC CIRCUIT ANALYSIS

The equation S = A/L shows that the current of a voltaic circuit is subject to a change, by each variation originating either in the magnitude of a tension or in the reduced length of a part, which latter is itself again determined, both by the actual length of the part as well as its conductivity and its section.

Georg Simon Ohm, 1827,
German Mathematician/Physicist

2–1 Element Constraints

2–2 Connection Constraints

2–3 Combined Constraints

2–4 Equivalent Circuits

2–5 Voltage and Current Division

2–6 Circuit Reduction

2–7 Computer-aided Circuit Analysis

Summary

Problems

Integrating Problems

This chapter begins our study of the ideal models used to describe the physical devices in electrical circuits. Foremost among these is the renowned Ohm's law, defining the model of a linear resistor. Georg Simon Ohm (1789–1854) originally discovered the law that now bears his name in 1827. His results drew heavy criticism and were not generally accepted for many years. Fortunately, the importance of his contribution was eventually recognized during his lifetime. He was honored by the Royal Society of England in 1841 and appointed a Professor of Physics at the University of Munich in 1849.

The first section of this chapter deals with the element constraints derived from the ideal models of resistors, voltage sources, and current sources. The models for other linear devices, such as amplifiers, inductors, capacitors, and transformers, are introduced in later chapters. When devices are interconnected to form a circuit, they are subject to connection constraints based on fundamental conservation principles known as Kirchhoff's laws. The remainder of the chapter uses the combined element and connection constraints to develop a basic set of circuit analysis tools that include equivalent circuits, voltage and current division, and circuit reduction. The basic circuit analysis tools developed in this chapter are used frequently in the rest of the book and by practicing engineers. These analysis tools promote basic understanding because they involve working directly with the circuit model.

2–1   ELEMENT CONSTRAINTS

A **circuit** is a collection of interconnected electrical devices. An electrical **device** is a component that is treated as a separate entity. The rectangular box in Figure 2–1 is used to represent any one of the two-terminal devices used to form circuits. A two-terminal device is described by its *i–v* **characteristic**; that is, by the relationship between the voltage across and current through the device. In most cases the relationship is complicated and nonlinear, so we use a linear model that approximates the dominant features of a device.

To distinguish between a device (the real thing) and its model (an approximate stand-in), we call the model a circuit **element**. Thus, a device is an article of hardware described in manufacturers' catalogs and parts specifications. An element is a model described in textbooks on circuit analysis. This book is no exception, and a catalog of circuit elements will be introduced as we go on. A discussion of real devices and their models is contained in Appendix A.

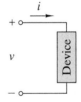

FIGURE 2–1 *Voltage and current reference marks for a two-terminal device.*

THE LINEAR RESISTOR

The first element in our catalog is a linear model of the device described in Figure 2–2. The actual *i–v* characteristic of this device is shown in Figure 2–2(b). To model this curve accurately across the full operating range shown in the figure would require at least a cubic equation. However, the graph in Figure 2–2(b) shows that a straight line is a good approximation to the *i–v* characteristic if we operate the device within its linear range. The power rating of the device limits the range over which the *i–v* characteristics can be represented by a straight line through the origin.

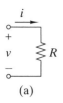

(a)

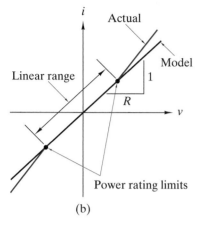

(b)

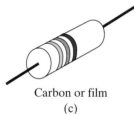

Wirewound

Carbon or film

(c)

FIGURE 2–2 *The resistor: (a) Circuit symbol. (b) i–v characteristics. (c) Some actual devices.*

For the passive sign convention used in Figure 2–2(a), the equations describing the *linear resistor* element are

$$v = Ri \quad \text{or} \quad i = Gv \qquad (2\text{–}1)$$

where R and G are positive constants that are reciprocally related.

$$G = \frac{1}{R} \qquad (2\text{–}2)$$

The relationships in Eq. (2–1) are collectively known as **Ohm's law**. The parameter R is called **resistance** and has the unit **ohms**, Ω. The parameter G is called **conductance**, with the unit **siemens**, S. In earlier times the unit of conductance was cleverly called the mho, with a unit abbreviation symbol ℧ (ohm spelled backward and the ohm symbol upside down). Note that Ohm's law presumes that the passive sign convention is used to assign the reference marks to voltage and current.

The Ohm's law model is represented graphically by the black straight line in Figure 2–2(b). The i–v characteristic for the Ohm's law model defines a circuit element that is said to be linear and bilateral. **Linear** means that the defining characteristic is a straight line through the origin. Elements whose characteristics do not pass through the origin or are not a straight line are said to be **nonlinear**. **Bilateral** means that the i–v characteristic curve has odd symmetry about the origin.[1] With a bilateral resistor, reversing the polarity of the applied voltage reverses the direction but not the magnitude of the current, and vice versa. The net result is that we can connect a bilateral resistor into a circuit without regard to which terminal is which. This is important because devices such as diodes and batteries are not bilateral, and we must carefully identify each terminal.

Figure 2–2(c) shows sketches of discrete resistor devices. Detailed device characteristics and fabrication techniques are discussed in Appendix A.

The power associated with the resistor can be found from $p = vi$. Using Eq. (2–1) to eliminate v from this relationship yields

$$p = i^2R \qquad (2\text{–}3)$$

or using the same equations to eliminate i yields

$$p = v^2G = \frac{v^2}{R} \qquad (2\text{–}4)$$

Since the parameter R is positive, these equations tell us that the power is always nonnegative. Under the passive sign convention, this means that the resistor **always absorbs power**.

EXAMPLE 2–1

A resistor operates as a linear element as long as the voltage and current are within the limits defined by its power rating. Suppose we have a 47-kΩ resistor with a power rating of 0.25 W. Determine the maximum current and voltage that can be applied to the resistor and remain within its linear operating range.

---

1   A curve $i = f(v)$ has odd symmetry if $f(-v) = -f(v)$.

SOLUTION:
Using Eq. (2–3) to relate power and current, we obtain

$$I_{MAX} = \sqrt{\frac{P_{MAX}}{R}} = \sqrt{\frac{0.25}{47 \times 10^3}} = 2.31 \text{ mA}$$

Similarly, using Eq. (2–4) to relate power and voltage, we obtain

$$V_{MAX} = \sqrt{RP_{MAX}} = \sqrt{47 \times 10^3 \times 0.25} = 108 \text{ V} \quad \blacksquare$$

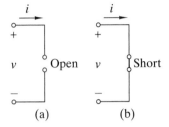

OPEN AND SHORT CIRCUITS

The next two circuit elements can be thought of as limiting cases of the linear resistor. Consider a resistor R with a voltage v applied across it. Let's calculate the current i through the resistor for different values of resistance. If $v = 10$ V and $R = 1$ Ω, using Ohm's law we readily find that $i = 10$ A. If we increase the resistance to 100 Ω, we find i has decreased to 0.1 A or 100 mA. If we continue to increase R to 1 MΩ, i becomes a very small 10 μA. Continuing this process, we arrive at a condition where R is very nearly infinite and i just about zero. When the current $i = 0$, we call the special value of resistance (i.e., $R = \infty$ Ω) an **open circuit**. Similarly, if we reduce R until it approaches zero, we find that the voltage is very nearly zero. When $v = 0$, we call the special value of resistance (i.e., $R = 0$ Ω), a **short circuit**. The circuit symbols for these two elements are shown in Figure 2–3. In circuit analysis the elements in a circuit model are assumed to be interconnected by zero-resistance wire (that is, by short circuits).

FIGURE 2-3 *Circuit symbols: (a) Open-circuit symbol. (b) Short-circuit symbol.*

THE IDEAL SWITCH

A switch is a familiar device with many applications in electrical engineering. The **ideal switch** can be modeled as a combination of an open- and a short-circuit element. Figure 2–4 shows the circuit symbol and the i–v characteristic of an ideal switch. When the switch is closed,

$$v = 0 \quad \text{and} \quad i = \text{any value} \qquad (2\text{–}5a)$$

and when it is open,

$$i = 0 \quad \text{and} \quad v = \text{any value} \qquad (2\text{–}5b)$$

When the switch is closed, the voltage across the element is zero and the element will pass any current that may result. When open, the current is zero and the element will withstand any voltage across its terminals. The power is always zero for the ideal switch, since the product $vi = 0$ when the switch is either open ($i = 0$) or closed ($v = 0$). Actual switch devices have limitations, such as the maximum current they can safely carry when closed and the maximum voltage they can withstand when open. The switch is operated (opened or closed) by some external influence, such as a mechanical motion, temperature, pressure, or an electrical signal.

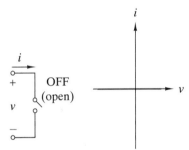

Circuit symbol i-v characteristics
(a)

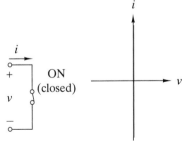

Circuit symbol i-v characteristics
(b)

FIGURE 2-4 *The circuit symbol and i–v characteristics of an ideal switch: (a) Switch OFF. (b) Switch ON.*

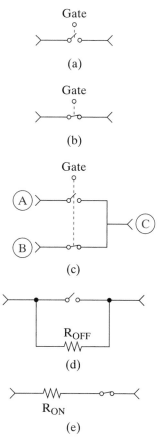

FIGURE 2–5 *The analog switch: (a) Normally open model. (b) Normally closed model. (c) Double throw model. (d) Model with finite OFF resistance. (e) Model with finite ON resistance.*

EXAMPLE 2–2

The **analog switch** is an important device found in analog-to-digital interfaces. Figures 2–5(a) and 2–5(b) show the two basic versions of the device. In either type the switch is actuated by applying a voltage to the terminal labeled "gate." The switch in Figure 2–5(a) is said to be *normally open* because it is open when no voltage is applied to the gate terminal and closes when voltage is applied. The switch in Figure 2–5(b) is said to be *normally closed* because it is closed when no voltage is applied to the controlling gate and opens when voltage is applied.

Figure 2–5(c) shows an application in which complementary analog switches are controlled by the same gate. When gate voltage is applied the upper switch closes and the lower opens so that point A is connected to point C. Conversely, when no gate voltage is applied the upper switch opens and the lower switch closes to connect point B to point C. In the analog world this arrangement is called a double throw switch since point C can be connected to two other points. In the digital world it is called a two-to-one multiplexer (or MUX) because it allows you to select the analog input at point A or point B under control of the digital signal applied to the gate.

In many applications an analog switch can be treated as an ideal switch. In other cases, it may be necessary to account for their nonideal characteristics. When the switch is open an analog switch acts like a very large resistance (R_{OFF}), as suggested in Figure 2–5(d). This resistance is usually negligible because it ranges from perhaps 10^9 to 10^{11} Ω. When the switch is closed it acts like a small resistor (R_{ON}), as suggested in Figure 2–5(e). Depending on other circuit resistances, it may be necessary to account for R_{ON} because it ranges from perhaps 20 to 200 Ω.

This example illustrates how ideal switches and resistors can be combined to model another electrical device. It also suggests that no single model can serve in all applications. It is up to the engineer to select a model that adequately represents the actual device in each application.

IDEAL SOURCES

The signal and power sources required to operate electronic circuits are modeled using two elements: **voltage sources** and **current sources**. These sources can produce either constant or time-varying signals. The circuit symbols and the *i–v* characteristic of an ideal voltage source are shown in Figure 2–6, while the circuit symbol and *i–v* characteristic of an ideal current source are shown in Figure 2–7. The symbol in Figure 2–6(a) represents either a time-varying or constant voltage source. The battery symbol in Figure 2–6(b) is used exclusively for a constant voltage source. There is no separate symbol for a constant current source.

The *i–v* characteristic of an **ideal voltage source** in Figure 2–6(c) is described by the following element equations:

$$v = v_S \quad \text{and} \quad i = \text{any value} \tag{2–6}$$

The element equations mean that the ideal voltage source produces v_S volts across its terminals and will supply whatever current may be required by the circuit to which it is connected.

The i–v characteristic of an **ideal current source** in Figure 2–7(b) is described by the following element equations:

$$i = i_S \quad \text{and} \quad v = \text{any value} \tag{2–7}$$

The ideal current source produces i_S amperes in the direction of its arrow symbol and will furnish whatever voltage is required by the circuit to which it is connected.

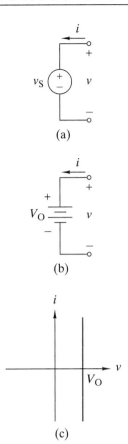

FIGURE 2 – 6 *Circuit symbols and i–v characteristic of an ideal independent voltage source: (a) Time-varying. (b) Constant (Battery). (c) Constant source i–v characteristics.*

FIGURE 2 – 7 *Circuit symbols and* i–v *characteristic of an ideal independent current source: (a) Time-varying or constant source. (b) Constant source* i–v *characteristics.*

The voltage or current produced by these ideal sources is called a **forcing function** or a **driving function** because it represents an input that causes a circuit response.

EXAMPLE 2–3

Given an ideal voltage source with the time-varying voltage shown in Figure 2–8(a), sketch its i–v characteristic at the times $t = 0, 1,$ and 2 ms.

FIGURE 2 – 8

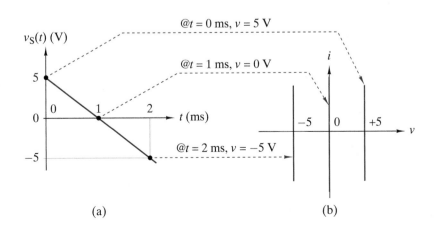

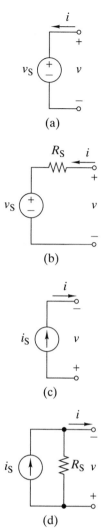

FIGURE 2 – 9 *Circuit symbols for ideal and practical independent sources: (a) Ideal voltage source. (b) Practical voltage source. (c) Ideal current source. (d) Practical current source.*

SOLUTION:

At any instant of time, the time-varying source voltage has only one value. We can treat the voltage and current at each instant of time as constants representing a snapshot of the source i–v characteristic. For example, at $t = 0$, the equations defining the i–v characteristic are $v_S = 5$ V and i = any value. Figure 2–8(b) shows the i–v relationship at the other instants of time. Curiously, the voltage source i–v characteristic at $t = 1$ ms ($v_S = 0$ and i = any value) is the same as that of a short circuit [see Eq. (2–5a) or Figure 2–3(b)]. ■

PRACTICAL SOURCES

The practical models for voltage and sources in Figure 2–9 may be more appropriate in some situations than the ideal models used up to this point. These circuits are called practical models because they more accurately represent the characteristics of real-world sources than do the ideal models. It is important to remember that models are interconnections of elements, not devices. For example, the resistance in a model does not always represent an actual resistor. As a case in point, the resistances R_S in the practical source models in Figure 2–9 do not represent physical resistors but are circuit elements used to account for resistive effects within the devices being modeled.

The linear resistor, open circuit, short circuit, ideal switch, ideal voltage source, and ideal current source are the initial entries in our catalog of circuit elements. In Chapter 4 we will develop models for active devices like the transistor and OP AMP. Models for dynamic elements like capacitors and inductors are introduced in Chapter 6.

2–2 CONNECTION CONSTRAINTS

In the previous section, we considered individual devices and models. In this section, we turn our attention to the constraints introduced by interconnections of devices to form circuits. The laws governing circuit behavior are based on the meticulous work of the German scientist Gustav Kirchhoff (1824–1887). **Kirchhoff's laws** are derived from conservation laws as applied to circuits. They tell us that element voltages and currents are forced to behave in certain ways when the devices are interconnected to form a circuit. These conditions are called **connection constraints** because they are based only on the circuit connections and not on the specific devices in the circuit.

In this book, we will indicate that crossing wires are connected (electrically tied together) using the dot symbol, as in Figure 2–10(a). Sometimes crossing wires are not connected (electrically insulated) but pass over or under each other. Since we are restricted to drawing wires on a planar surface, we will indicate unconnected crossovers by *not* placing a dot at their intersection, as indicated in the left of Figure 2–10(b). Other books sometimes show unconnected crossovers using the semicircular "hopover" shown on the right of Figure 2–10(b). In engineering systems two or more separate circuits are often tied together to form a larger circuit (for example, the interconnection of two integrated circuit packages). Interconnecting different circuits forms an interface between the circuits. The special jack or interface symbol in Figure 2–10(c) is used in this book because interface connections represent important points at which the interaction between two circuits can be observed or specified. On certain occasions a control line is required to show a mechanical or other nonelectrical dependency. Figure 2–10(d) shows how this dependency is indicated in this book. Figure 2–10(e) shows how power supply connections are often shown in electronic circuit diagrams. The implied power supply connection is indicated by an arrow pointing to the supply voltage, which may be given in numerical (+15 V) or symbolic form (+V_{CC}).

The treatment of Kirchhoff's laws uses the following definitions:

- A **circuit** is an interconnection of electrical devices.

- A **node** is an electrical juncture of two or more devices.

- A **loop** is a closed path formed by tracing through an ordered sequence of nodes without passing through any node more than once.

While it is customary to designate the juncture of two or more elements as a node, it is important to realize that a node is not confined to a point but includes all the zero-resistance wire from the point to each element. In the circuit of Figure 2–11, there are only three different nodes: A, B, and C. The points 2, 3, and 4, for example, are part of node B, while the points 5, 6, 7, and 8 are all part of node C.

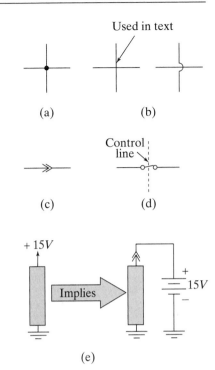

FIGURE 2-10 *Symbols used in circuit diagrams: (a) Electrical connection. (b) Crossover with no connection. (c) Jack connection. (d) Control line. (e) Power supply connection.*

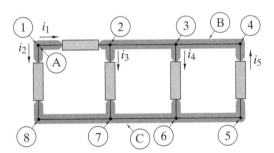

FIGURE 2-11 *Circuit for demonstrating Kirchhoff's current law.*

KIRCHHOFF'S CURRENT LAW

Kirchhoff's first law is based on the principle of conservation of charge. **Kirchhoff's current law (KCL)** states that

> *the algebraic sum of the currents entering a node is zero at every instant.*

In forming the algebraic sum of currents, we must take into account the current reference direction associated with each device. If the current reference direction is into the node, then we assign a positive sign to the corresponding current in the algebraic sum. If the reference direction is away from the node, we assign a negative sign. Applying this convention to the nodes in Figure 2–11, we obtain the following set of KCL connection equations:

$$\text{Node A: } - i_1 - i_2 \qquad = 0$$
$$\text{Node B: } i_1 - i_3 - i_4 + i_5 = 0 \qquad \text{(2–8)}$$
$$\text{Node C: } i_2 + i_3 + i_4 - i_5 = 0$$

The KCL equation at node A does not mean that the currents i_1 and i_2 are both negative. The minus signs in this equation simply mean that the reference direction for each current is directed away from node A. Likewise, the equation at node B could be written as

$$i_3 + i_4 = i_1 + i_5 \qquad \text{(2–9)}$$

This form illustrates an alternate statement of KCL:

> *The sum of the currents entering a node equals the sum of the currents leaving the node.*

There are two algebraic signs associated with each current in the application of KCL. First is the sign given to a current in writing a KCL connection equation. This sign is determined by the orientation of the current reference direction relative to a node. The second sign is determined by the actual direction of the current relative to the reference direction. The actual direction is found by solving the set of KCL equations, as illustrated in the following example.

EXAMPLE 2–4

Given $i_1 = +4$ A, $i_3 = +1$ A, $i_4 = +2$ A in the circuit shown in Figure 2–11, find i_2 and i_5.

SOLUTION:
Using the node A constraint in Eq. (2–8) yields

$$- i_1 - i_2 = - (+4) - i_2 = 0$$

The sign outside the parentheses comes from the node A KCL connection constraint in Eq. (2–8). The sign inside the parentheses comes from the actual direction of the current. Solving this equation for the unknown current, we find that $i_2 = -4$ A. In this case, the minus sign indicates that the actual direction of the current i_2 is directed upward in Figure 2–11, which is opposite to the reference direction assigned. Using the second KCL equation in Eq. (2–8), we can write

$$i_1 - i_3 - i_4 + i_5 = (+4) - (+1) - (+2) + i_5 = 0$$

which yields the result $i_5 = -1$ A.

Again, the signs inside the parentheses are associated with the actual direction of the current, and the signs outside come from the node B KCL connection constraint in Eq. (2–8). The minus sign in the final answer means that the current i_5 is directed in the opposite direction from its assigned reference direction. We can check our work by substituting the values found into the node C constraint in Eq. (2–8). These substitutions yield

$$+ i_2 + i_3 + i_4 - i_5 = (-4) + (+1) + (+2) - (-1) = 0$$

as required by KCL. Given three currents, we determined all the remaining currents in the circuit using only KCL without knowing the element constraints. ∎

In Example 2–4, the unknown currents were found using only the KCL constraints at nodes A and B. The node C equation was shown to be valid, but it did not add any new information. If we look back at Eq. (2–8), we see that the node C equation is the negative of the sum of the node A and B equations. In other words, the KCL connection constraint at node C is not independent of the two previous equations. This example illustrates the following general principle:

> *In a circuit containing a total of N nodes there are only N − 1 independent KCL connection equations.*

Current equations written at $N - 1$ nodes contain all the independent connection constraints that can be derived from KCL. To write these equations, we select one node as the reference or ground node and then write KCL equations at the remaining $N - 1$ nonreference nodes.

Exercise 2–1

Refer to Figure 2–12.

(a) Write KCL equations at nodes A, B, C, and D.
(b) Given $i_1 = -1$ mA, $i_3 = 0.5$ mA, $i_6 = 0.2$ mA, find i_2, i_4, and i_5.

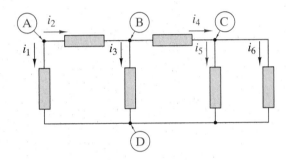

FIGURE   2 - 1 2

Answers:
(a) Node A: $-i_1 - i_2 = 0$; node B: $i_2 - i_3 - i_4 = 0$; node C: $i_4 - i_5 - i_6 = 0$; node D: $i_1 + i_3 + i_5 + i_6 = 0$
(b) $i_2 = 1$ mA, $i_4 = 0.5$ mA, $i_5 = 0.3$ mA

KIRCHHOFF'S VOLTAGE LAW

The second of Kirchhoff's circuit laws is based on the principle of conservation of energy. **Kirchhoff's voltage law** (abbreviated **KVL**) states that

> *the algebraic sum of all the voltages around a loop is zero at every instant.*

For example, three loops are shown in the circuit of Figure 2–13. In writing the algebraic sum of voltages, we must account for the assigned reference marks. As a loop is traversed, a positive sign is assigned to a voltage when we go from a "+" to "−" reference mark. When we go from "−" to "+", we use a minus sign. Traversing the three loops in Figure 2–13 in the indicated clockwise direction yields the following set of KVL connection equations:

$$\text{Loop 1: } -v_1 + v_2 + v_3 \qquad\quad = 0$$
$$\text{Loop 2: } -v_3 + v_4 + v_5 \qquad\quad = 0 \qquad\qquad (2–10)$$
$$\text{Loop 3: } -v_1 + v_2 + v_4 + v_5 = 0$$

FIGURE 2–13 *Circuit for demonstrating Kirchhoff's voltage law.*

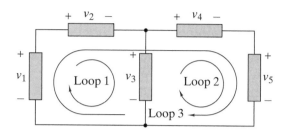

There are two signs associated with each voltage. The first is the sign given the voltage when writing the KVL connection equation. The second is the sign determined by the actual polarity of a voltage relative to its assigned reference polarity. The actual polarities are found by solving the set of KVL equations, as illustrated in the following example.

EXAMPLE 2–5

Given $v_1 = 5$ V, $v_2 = -3$ V, and $v_4 = 10$ V in the circuit shown in Figure 2–13, find v_3 and v_5.

SOLUTION:
Inserting the given numerical values into Eq. (2–10) yields the following KVL equation for loop 1:

$$-v_1 + v_2 + v_3 = -(+5) + (-3) + (v_3) = 0$$

The sign outside the parentheses comes from the loop 1 KVL constraint in Eq. (2–10). The sign inside comes from the actual polarity of the voltage. This equation yields $v_3 = +8$ V. Using this value in the loop 2, KVL constraint in Eq. (2–10) produces

$$-v_3 + v_4 + v_5 = -(+8) + (+10) + v_5 = 0$$

The result is $v_5 = -2$ V. The minus sign here means that the actual polarity of v_5 is the opposite of the assigned reference polarity indicated in Figure 2–13. The results can be checked by substituting all the aforementioned values into the loop 3 KVL constraint in Eq. (2–10). These substitutions yield

$$-(+5) + (-3) + (+10) + (-2) = 0$$

as required by KVL.   ■

In Example 2–5, the unknown voltages were found using only the KVL constraints for loops 1 and 2. The loop 3 equation was shown to be valid, but it did not add any new information. If we look back at Eq. (2–10), we see that the loop 3 equation is equal to the sum of the loop 1 and 2 equations. In other words, the KVL connection constraint around loop 3 is not independent of the previous two equations. This example illustrates the following general principle:

> *In a circuit containing a total of* **E** *two-terminal elements and* **N** *nodes, there are only* **E − N +1** *independent KVL connection equations.*

Writing voltage summations around a total of $E - N + 1$ *different* loops produces all the independent connection constraints that can be derived from KVL. A *sufficient condition* for loops to be different is that each contains at least one element that is not contained in any other loop. In simple circuits with no crossovers the open space between elements produces $E - N + 1$ independent loops. However, finding all the loops in a more complicated circuit can be a nontrivial problem.

Exercise 2–2

Find the voltages v_x and v_y in Figure 2–14.

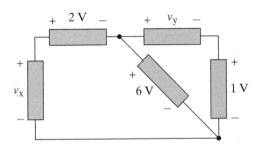

FIGURE 2 – 1 4

Answers: $v_x = + 8$ V, $v_y = + 5$ V

Exercise 2–3

Find the voltages v_x, v_y, and v_z in Figure 2–15.

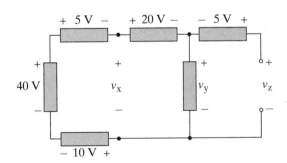

FIGURE 2 – 1 5

Answers: $v_x = + 25$ V; $v_y = + 5$ V; $v_z = + 10$ V. *Note:* KVL yields the voltage v_z even though it appears across an open circuit.

PARALLEL AND SERIES CONNECTIONS

Two types of connections occur so frequently in circuit analysis that they deserve special attention. Elements are said to be connected in **parallel** when they form a loop containing no other elements. For example, loop A in Figure 2–16 contains only elements 1 and 2. As a result, the KVL connection constraint around loop A is

$$-v_1 + v_2 = 0 \qquad (2\text{–}11)$$

which yields $v_1 = v_2$. In other words, in a parallel connection KVL requires equal voltages across the elements. The parallel connection is not restricted to two elements. For example, loop B in Figure 2–16 contains only elements 2 and 3; hence, by KVL $v_2 = v_3$. As a result, in this circuit we have $v_1 = v_2 = v_3$, and we say that elements 1, 2, and 3 are connected in parallel. In general, then, any number of elements connected between two common nodes are in parallel, and as a result, the same voltage appears across each of them. Existence of a parallel connection does not depend on the graphical position of the elements. For example, the position of elements 1 and 3 could be switched, and the three elements are still connected in parallel.

FIGURE 2–16 *A parallel connection.*

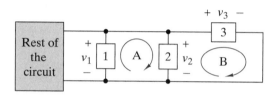

Two elements are said to be connected in **series** when they have one common node to which no other element is connected. In Figure 2–17 elements 1 and 2 are connected in series, since only these two elements are connected at node A. Applying KCL at node A yields

$$i_1 - i_2 = 0 \quad \text{or} \quad i_1 = i_2 \qquad (2\text{–}12)$$

In a series connection, KCL requires equal current through each element. Any number of elements can be connected in series. For example, element 3 in Figure 2–17 is connected in series with element 2 at node B, and KCL requires $i_2 = i_3$. Therefore, in this circuit $i_1 = i_2 = i_3$, we say that elements 1, 2, and 3 are connected in series, and the same current exists in each of the elements. In general, elements are connected in series when they form a single path between two nodes such that only elements in the path are connected to the intermediate nodes along the path.

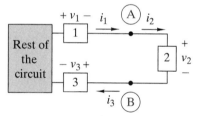

FIGURE 2–17 *A series connection.*

EXAMPLE 2–6

Identify the elements connected in parallel and in series in each of the circuits in Figure 2–18.

SOLUTION:
In Figure 2–18(a) elements 1 and 2 are connected in series at node A and elements 3 and 4 are connected in parallel between nodes B and C. In Figure 2–18(b) elements 1 and 2 are connected in series at node A, as are elements 4 and 5 at node D. There are no single elements connected in paral-

lel in this circuit. In Figure 2–18(c) there are no elements connected in either series or parallel. It is important to realize that in some circuits there are elements that are not connected in either series or in parallel. ∎

Exercise 2–4

Identify the elements connected in series or parallel when a short circuit is connected between nodes A and B in each of the circuits of Figure 2–18.

Answers:
Circuit in Figure 2–18(a): Elements 1, 3, and 4 are all in parallel.
Circuit in Figure 2–18(b): Elements 1 and 3 are in parallel; elements 4 and 5 are in series.
Circuit in Figure 2–18(c): Elements 1 and 3 are in parallel; elements 4 and 6 are in parallel.

Exercise 2–5

Identify the elements in Figure 2–19 that are connected in (a) parallel, (b) series, or (c) neither.

Answers:
(a) The following elements are in parallel: 1, 8, and 11; 3, 4, and 5.
(b) The following elements are in series: 9 and 10; 6 and 7.
(c) Only element 2 is not in series or parallel with any other element.

DISCUSSION: *The ground symbol indicates the reference node. When ground symbols are shown at several nodes, the nodes are effectively connected together by a short circuit to form a single node.*

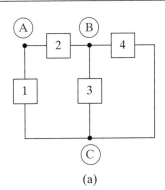

(a)

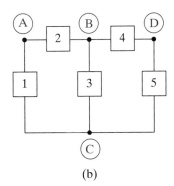

(b)

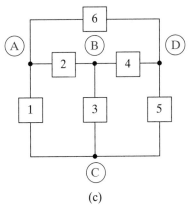

(c)

FIGURE 2–18

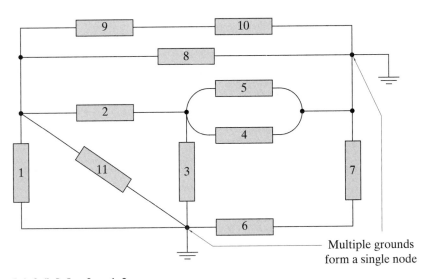

Multiple grounds
form a single node

FIGURE 2–19

2-3 COMBINED CONSTRAINTS

The usual goal of circuit analysis is to determine the currents or voltages at various places in a circuit. This analysis is based on constraints of two distinctly different types. The element constraints are based on the models of the specific devices connected in the circuit. The connection constraints are based on Kirchhoff's laws and the circuit connections. The element equations are independent of the circuit connections. Likewise, the connection equations are independent of the devices in the circuit. Taken together, however, the combination of the element and connection constraints supply the equations needed to describe a circuit.

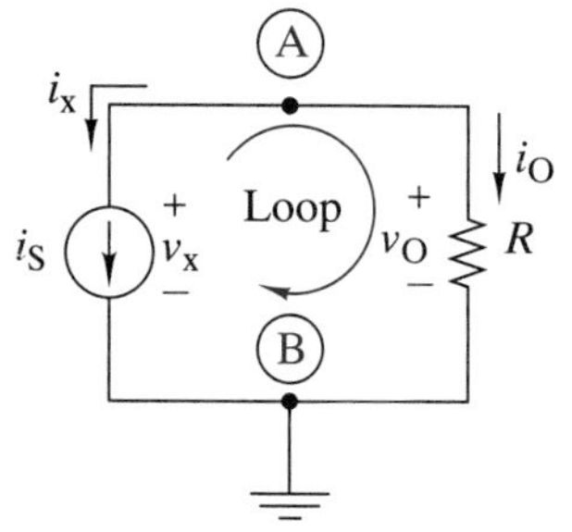

Our study of the combined constraints begins by considering the simple but important example in Figure 2–20. This circuit is driven by a current source i_S and the resulting responses are current/voltage pairs (i_x, v_x) and (i_O, v_O). The reference marks for the response pairs have been assigned using the passive sign convention.

To solve for all four responses, we must write four equations. The first two are the element equations

$$i_x = i_S$$

$$v_O = Ri_O \qquad (2\text{–}13)$$

The first element equation states that the response current i_x and the input driving force i_S are equal in magnitude and direction. The second element equation is Ohm's law relating v_O and i_O under the passive sign convention.

The connection equations are obtained by applying Kirchhoff's laws. The circuit in Figure 2–20 has two elements ($E = 2$) and two nodes ($N = 2$), so we need $E - N + 1 = 1$ KVL equation and $N - 1 = 1$ KCL equation. Selecting node B as the reference node, we apply KCL at node A and apply KVL around the loop to write

$$\text{KCL: } -i_x - i_O = 0$$

$$\text{KVL: } -v_x + v_O = 0 \qquad (2\text{–}14)$$

We now have two element constraints in Eq. (2–13) and two connection constraints in Eq. (2–14), so we can solve for all four responses in terms of the input driving force i_S. Combining the KCL connection equation and the first element equation yields $i_O = -i_x = -i_S$. Substituting this result into the second element equation (Ohm's law) produces

$$v_O = -Ri_S \qquad (2\text{–}15)$$

The minus sign in this equation does not mean that v_O is always negative. Nor does it mean the resistance is negative. It means that when the input driving force i_S is positive, then the response v_O is negative, and vice versa. This sign reversal is a result of the way we assigned reference marks at the beginning of our analysis. The reference marks defined the circuit input and outputs in such a way that i_S and v_O always have opposite algebraic signs. Put differently, Eq. (2–15) is an input-output relationship, not an element i–v relationship.

FIGURE   2 – 2 0 *Circuit used to demonstrate combined constraints.*

EXAMPLE 2–7

(a) Find the responses i_x, v_x, i_O, and v_O in the circuit in Figure 2–20 when $i_S = +2$ mA and $R = 2$ kΩ.

(b) Repeat for $i_S = -2$ mA.

SOLUTION:

(a) From Eq. (2–13) we have $i_x = i_S = +2$ mA and $v_O = 2000\, i_O$. From Eq. (2–14) we have $i_O = -i_x = -2$ mA and $v_x = v_O$. Combining these results, we obtain

$$v_x = v_O = 2000 i_O = 2000(-0.002) = -4 \text{ V}$$

(b) In this case $i_x = i_S = -2$ mA, $i_O = -i_x = -(-0.002) = +2$ mA, and

$$v_x = v_O = 2000 i_O = 2000(+0.002) = +4 \text{ V}$$

This example confirms that the algebraic signs of the outputs v_x, v_O, and i_O are always the opposite sign from that of the input driving force i_S. ∎

Analyzing the circuit in Figure 2–21 illustrates the formulation of combined constraints. We first assign reference marks for all the voltages and currents using the passive sign convention. Then, using these definitions we can write the element constraints as

$$v_A = V_O$$
$$v_1 = R_1 i_1 \qquad (2\text{–}16)$$
$$v_2 = R_2 i_2$$

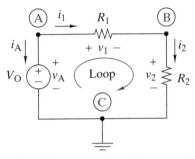

FIGURE   2 – 2 1 *Circuit used to demonstrate combined*

These equations describe the three devices and do not depend on how the devices are connected in the circuit.

The connection equations are obtained from Kirchhoff's laws. To apply these laws, we must first label the different loops and nodes. The circuit contains $E = 3$ elements and $N = 3$ nodes, so there are $E - N + 1 = 1$ independent KVL constraints and $N - 1 = 2$ independent KCL constraints. There is only one loop, but there are three nodes in this circuit. We will select one node as the reference point and write KCL equations at the other two nodes. Any node can be chosen as the reference, so we select node C as the reference node and indicate this choice by drawing the ground symbol there. The connection constraints are

$$\text{KCL: Node A } - i_A - i_1 \quad = 0$$
$$\text{KCL: Node B } i_1 - i_2 \quad\quad = 0 \qquad (2\text{–}17)$$
$$\text{KVL: Loop } - v_A + v_1 + v_2 = 0$$

These equations are independent of the specific devices in the circuit. They depend only on Kirchhoff's laws and the circuit connections.

This circuit has six unknowns: three element currents and three element voltages. Taken together, the element and connection equations give us six independent equations. For a network with (N) nodes and (E) two-terminal elements, we can write ($N - 1$) independent KCL connection equations, $(E - N + 1)$ independent KVL connection equations, and (E) element equations. The total number of equations generated is

Element equations	E
KCL equations	$N - 1$
KVL equations	$E - N + 1$
Total	$2E$

The grand total is then $(2E)$ combined connection and element equations, which is exactly the number of equations needed to solve for the voltage across and current through every element—a total of $(2E)$ unknowns.

EXAMPLE 2–8

Find all of the element currents and voltages in Figure 2–21 for $V_O = 10$ V, $R_1 = 2000$ Ω, and $R_2 = 3000$ Ω.

SOLUTION:
Substituting the element constraints from Eq. (2–16) into the KVL connection constraint in Eq. (2–17) produces

$$-V_O + R_1 i_1 + R_2 i_2 = 0$$

This equation can be used to solve for i_1 since the second KCL connection equation requires that $i_2 = i_1$.

$$i_1 = \frac{V_O}{R_1 + R_2} = \frac{10}{2000 + 3000} = 2 \text{ mA}$$

In effect, we have found all of the element currents since the elements are connected in series. Hence, collectively the KCL connection equations require that

$$-i_A = i_1 = i_2$$

Substituting all of the known values into the element equations gives

$$v_A = 10 \text{ V} \quad v_1 = R_1 i_1 = 4 \text{ V} \quad v_2 = R_2 i_2 = 6 \text{ V}$$

Every element voltage and current has been found. Note the analysis strategy used. We first found all the element currents and then used these values to find the element voltages. ∎

EXAMPLE 2–9

Consider the source-resistor-switch circuit of Figure 2–22 with $V_O = 10$ V and $R_1 = 2000$ Ω. Find all the element voltages and currents with the switch open and again with the switch closed.

SOLUTION:
The connection equations for the circuit are

$$\text{KCL: Node A} \quad -i_A - i_1 = 0$$
$$\text{KCL: Node B} \quad i_1 - i_2 = 0$$
$$\text{KVL: Loop} \quad -v_A + v_1 + v_2 = 0$$

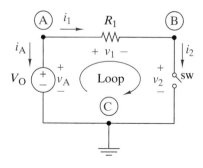

F I G U R E   2 – 2 2

These connection equations are the same as those in Eq. (2–17) for the circuit in Figure 2–21. The two circuits have the same connections but different devices. When the switch is open, the current through the switch is zero, so $i_2 = 0$. The KCL connection equation at node B requires $i_1 = i_2$, and therefore $i_1 = 0$. The element equation for R_1 yields $v_1 = R_1 i_1 = 0$, and the KVL connection equation yields $v_1 + v_2 = v_2 = v_A = 10$ V. Thus, when the switch is open, the current is zero and all the source voltage appears across the switch. When the switch is closed, its element equations require $v_2 = 0$. The KVL connection equation gives $v_1 + v_2 = v_1 = v_A = 10$ V. When the switch is closed, all the source voltage appears across R_1 rather than across the switch. So the current through the circuit is

$$i_1 = \frac{v_1}{R_1} = \frac{10}{2000} = 5 \text{ mA} \qquad \blacksquare$$

Exercise 2–6

For the circuit of Figure 2–23,

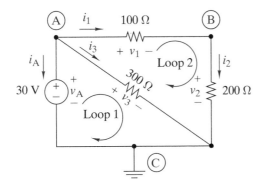

FIGURE 2–23

(a) Write a complete set of element equations.
(b) Write a complete set of connection equations.
(c) Solve the equations in (a) and (b) for all element currents and voltages.

Answers:

(a) $v_A = 30$ V; $v_3 = 300\, i_3$; $v_1 = 100\, i_1$; $v_2 = 200\, i_2$
(b) $-i_A - i_1 - i_3 = 0$; $+i_1 - i_2 = 0$; $-30 + v_3 = 0$; $v_1 + v_2 - v_3 = 0$
(c) $v_A = 30$ V; $v_1 = 10$ V; $v_2 = 20$ V; $v_3 = 30$ V; $i_A = -200$ mA; $i_1 = i_2 = 100$ mA; i_3
$=$
100 mA

ASSIGNING REFERENCE MARKS

In all of our previous examples and exercises, the reference marks for the element currents (arrows) and voltages (+ and –) were given. When reference marks are not shown on a circuit diagram, they must be assigned by the person solving the problem. Beginners sometimes wonder how to assign reference marks when the actual voltage polarities and current directions are unknown. It is important to remember that the reference marks do not indicate what is actually happening in the circuit. They are benchmarks assigned at the beginning of the analysis. If it turns out that the actual direction and reference direction agree, then the algebraic sign of the response will be positive. If they disagree, the algebraic sign will be negative. In other words, the sign of the answer together with assigned reference marks tell us the actual voltage polarity or current direction.

In this book the reference marks always follow the passive sign convention. This means that for any given two-terminal element we can arbitrarily assign either the + voltage reference mark or the current reference arrow, but not both. For example, we can arbitrarily assign the voltage reference marks to the terminals of a two-terminal device. Once the voltage reference is assigned, however, the passive sign convention requires that the current reference arrow be directed into the element at the terminal with the + mark. On the other hand, we could start by arbitrarily selecting the terminal at which the current reference arrow is directed into the device. Once the current reference is assigned, however, the passive sign convention requires that the + voltage reference be assigned to the selected terminal.

Following the passive sign convention avoids confusion about the direction of power flow in a device. In addition, the element constraints, such as Ohm's law, assume that the passive sign convention is used to assign the voltage and current reference marks to a device.

The next example illustrates the assignment of reference marks.

EXAMPLE 2–10

Find all of the element voltages and currents in the circuit shown in Figure 2–24(a).

FIGURE 2 – 24

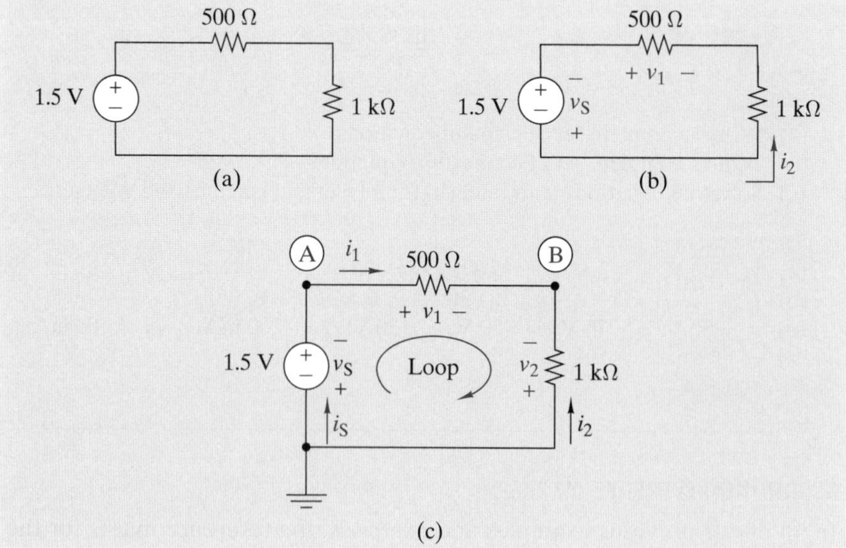

(a)

(b)

(c)

SOLUTION:

Since no reference marks are shown in Figure 2–24(a), we assign references to two voltages and one current, as shown in Figure 2–24(b). Other choices are possible, of course, but once these marks are selected, the passive sign convention dictates the reference marks for the remaining voltage and currents, as shown in Figure 2–24(c). Using all of these reference marks, we write the element equations as

$$v_1 = 500i_1$$
$$v_2 = 1000i_2$$
$$v_S = -1.5 \text{ V}$$

Using the indicated reference node, we write two KCL and one KVL equations:

$$\text{KCL: Node A } + i_S - i_1 \quad = 0$$
$$\text{KCL: Node B } + i_1 + i_2 \quad = 0$$
$$\text{KVL: Loop } + v_S + v_1 - v_2 = 0$$

Solving the combined element and connection equations yields

$$v_S = -1.5 \text{ V}, \quad i_S = +1.0 \text{ mA}$$
$$v_1 = +0.5 \text{ V}, \quad i_1 = +1.0 \text{ mA}$$
$$v_2 = -1.0 \text{ V}, \quad i_2 = -1.0 \text{ mA}$$

These results show that the reference marks for v_1, i_S, and i_1 agree with the actual voltage polarities and current directions, while the minus signs on the other responses indicate disagreement. It is important to realize that this disagreement does not mean that assigned reference marks are wrong.   ∎

2–4   EQUIVALENT CIRCUITS

The analysis of a circuit can often be made easier by replacing part of the circuit with one that is equivalent but simpler. The underlying basis for two circuits to be equivalent is contained in their *i–v* relationships.

> *Two circuits are said to be equivalent if they have identical **i–v** characteristics at a specified pair of terminals.*

In other words, when two circuits are equivalent the voltage and current at an interface do not depend on which circuit is connected to the interface.

EQUIVALENT RESISTANCE

The two resistors in Figure 2–25(a) are connected in series between a pair of terminals A and B. The objective is to simplify the circuit without altering the electrical behavior of the rest of the circuit.

The KVL equation around the loop from A to B is

$$v = v_1 + v_2 \qquad (2\text{–}18)$$

Since the two resistors are connected in series, the same current i exists in both. Applying Ohm's law, we get $v_1 = R_1 i$ and $v_2 = R_2 i$. Substituting these relationships into Eq. (2–18) and then simplifying yields

$$v = R_1 i + R_2 i = i(R_1 + R_2)$$

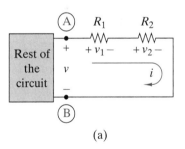

(a)

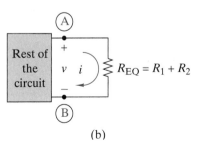

(b)

F I G U R E   2 – 2 5   *A series resistance circuit: (a) Original circuit. (b) Equivalent circuit.*

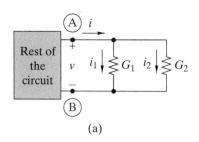

(a)

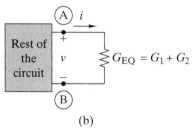

(b)

FIGURE 2-26 *A parallel resistance circuit: (a) Original circuit. (b) Equivalent circuit.*

We can write this equation in terms of an equivalent resistance R_{EQ} as

$$v = iR_{EQ} \quad \text{where} \quad R_{EQ} = R_1 + R_2 \qquad (2\text{--}19)$$

This result means the circuits in Figs. 2–25(a) and 2–25(b) have the same i–v characteristic at terminals A and B. As a result, the response of the rest of the circuit is unchanged when the series connection of R_1 and R_2 is replaced by a resistance R_{EQ}.

The parallel connection of two conductances in Figure 2–26(a) is the dual[2] of the series circuit in Figure 2–25(a). Again the objective is to replace the parallel connection by a simpler equivalent circuit without altering the response of the rest of the circuit.

A KCL equation at node A produces

$$i = i_1 + i_2 \qquad (2\text{--}20)$$

Since the conductances are connected in parallel, the voltage v appears across both. Applying Ohm's law, we obtain $i_1 = G_1 v$ and $i_2 = G_2 v$. Substituting these relationships into Eq. (2–20) and then simplifying yields

$$i = vG_1 + vG_2 = v(G_1 + G_2)$$

This result can be written in terms of an equivalent conductance G_{EQ} as follows:

$$i = vG_{EQ}, \quad \text{where} \quad G_{EQ} = G_1 + G_2 \qquad (2\text{--}21)$$

This result means the circuits in Figures 2–26(a) and 2–26(b) have the same i–v characteristic at terminals A and B. As a result, the response of the rest of the circuit is unchanged when the parallel connection of G_1 and G_2 is replaced by a conductance G_{EQ}.

Since conductance is not normally used to describe a resistor, it is sometimes useful to rewrite Eq. (2–21) as an equivalent resistance $R_{EQ} = 1/G_{EQ}$. That is,

$$R_1 \| R_2 = R_{EQ} = \frac{1}{G_{EQ}} = \frac{1}{G_1 + G_2} = \frac{1}{\dfrac{1}{R_1} + \dfrac{1}{R_2}} = \frac{R_1 R_2}{R_1 + R_2} \qquad (2\text{--}22)$$

where the symbol "∥" is shorthand for "in parallel." The expression on the far right in Eq. (2–22) is called the product over the sum rule for two resistors in parallel. This rule is useful in derivations in which the resistances are left in symbolic form (see Example 2–11), but it is not an efficient algorithm for calculating numerical values of equivalent resistance.

Caution: The product over sum rule only applies to two resistors connected in parallel. When more than two resistors are in parallel, we must use the following general result to obtain the equivalent resistance:

$$R_{EQ} = \frac{1}{G_{EQ}} = \frac{1}{\dfrac{1}{R_1} + \dfrac{1}{R_2} + \dfrac{1}{R_3} + \cdots} \qquad (2\text{--}23)$$

2 Dual circuits have identical behavior patterns when we interchange the roles of the following parameters: (1) voltage and current, (2) series and parallel, and (3) resistance and conductance. In later chapters we will see duality exhibited by other circuit parameters as well.

EXAMPLE 2−11

Given the circuit in Figure 2–27(a),

(a) Find the equivalent resistance R_{EQ1} connected between terminals A and B.

(b) Find the equivalent resistance R_{EQ2} connected between terminals C and D.

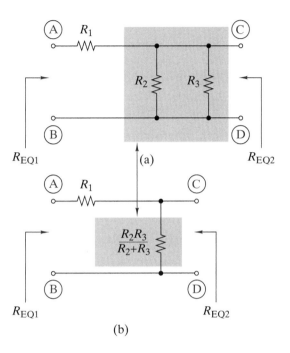

SOLUTION:

First we note that resistors R_2 and R_3 are connected in parallel. Applying the product over sum rule [Eq. (2–22)], we obtain

$$R_2\|R_3 = \frac{R_2 R_3}{R_2 + R_3}$$

As an interim step, we redraw the circuit, as shown in Figure 2–27(b).

(a) To find the equivalent resistance between terminals A and B, we note that R_1 and the equivalent resistance $R_2\|R_3$ are connected in series. The total equivalent resistance R_{EQ1} between terminals A and B is

$$R_{EQ1} = R_1 + (R_2\|R_3)$$

$$R_{EQ1} = R_1 + \frac{R_2 R_3}{R_2 + R_3}$$

$$R_{EQ1} = \frac{R_1 R_2 + R_1 R_3 + R_2 R_3}{R_2 + R_3}$$

(b) Looking between terminals C and D yields a different result. In this case R_1 is not involved, since there is an open circuit (an infinite resistance) between terminals A and B. Therefore, only $R_2 \parallel R_3$ affect the resistance between terminals C and D. As a result, R_{EQ2} is simply

$$R_{EQ2} = R_2 \| R_3 = \frac{R_2 R_3}{R_2 + R_3}$$

This example shows that equivalent resistance depends on the pair of terminals involved. ∎

One final note on checking numerical calculations of equivalent resistance. When several resistances are connected in parallel, the equivalent resistance must be smaller than the smallest resistance in the connection. Conversely, when several resistances are connected in series, the equivalent resistance must be larger than the largest resistance in the connection.

Exercise 2–7

Find the equivalent resistance between terminals A–C, B–D, A–D, and B–C in the circuit in Figure 2–27.

Answers: $R_{A-C} = R_1$, $R_{B-D} = 0\ \Omega$ (a short circuit), $R_{A-D} = R_1 + R_2 \| R_3$, and $R_{B-C} = R_2 \| R_3$.

Exercise 2–8

Find the equivalent resistance between terminals A–B, A–C, A–D, B–C, B–D, and C–D in the circuit of Figure 2–28.

Answers: $R_{A-B} = 100\ \Omega$, $R_{A-C} = 70\ \Omega$, $R_{A-D} = 65\ \Omega$, $R_{B-C} = 90\ \Omega$, $R_{B-D} = 85\ \Omega$, and $R_{C-D} = 55\ \Omega$.

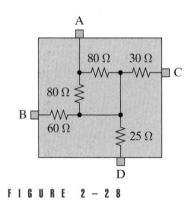

FIGURE 2–28

EQUIVALENT SOURCES

The practical source models introduced previously are shown in Figure 2–29. These models consist of an ideal voltage source in series with a resistance and an ideal current source in parallel with a resistance. We now determine the conditions under which the practical voltage source and the practical current sources are equivalent.

Figure 2–29 shows the two practical sources connected between terminals labeled A and B. A parallel analysis of these circuits yields the conditions for equivalency at terminals A and B. First, Kirchhoff's laws are applied as

Circuit A	Circuit B
KVL	KCL
$v_S = v_R + v$	$i_S = i_R + i$

Next, Ohm's law is used to obtain

Circuit A	Circuit B
$v_R = R_1 i$	$i_R = \dfrac{v}{R_2}$

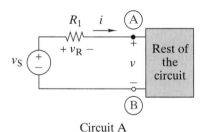

Circuit A

Circuit B

FIGURE 2–29 *Practical source models that are equivalent when Eq. (2–24) is satisfied.*

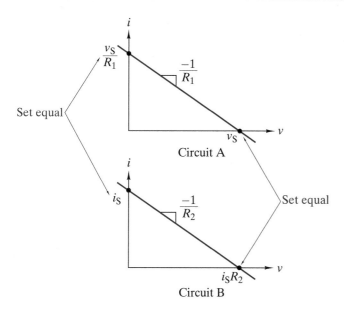

Combining these results, we find that the *i–v* relationships of each of the circuits at terminals A and B are

Circuit A   Circuit B

$$i = -\frac{v}{R_1} + \frac{v_S}{R_1} \qquad i = -\frac{v}{R_2} + i_S$$

These *i–v* characteristics take the form of the straight lines shown in Figure 2–30. The two lines are identical when the intercepts are equal. This requires that $v_S/R_1 = i_S$ and $v_S = i_S R_2$, which, in turn, requires that

$$R_1 = R_2 = R \quad \text{and} \quad v_S = i_S R \qquad (2\text{--}24)$$

When conditions in Eq. (2–24) are met, the response of the rest of the circuit is unaffected when we replace a practical voltage source by an equivalent practical current source, or vice versa. Exchanging one practical source model for an equivalent model is called *source transformation*.

Source transformation means that either model will deliver the same voltage and current to the rest of the circuit. It does not mean the two models are identical in every way. For example, when the practical voltage source drives an open circuit, there is no i^2R power loss since the current in the series resistance is zero. However, the current in the parallel resistance of a practical current source is not zero when the load is an open circuit. Thus, equivalent sources do not have the same internal power loss even though they deliver the same current and voltage to the rest of the circuit.

EXAMPLE 2-12

Convert the practical voltage source in Figure 2–31(a) into an equivalent current source.

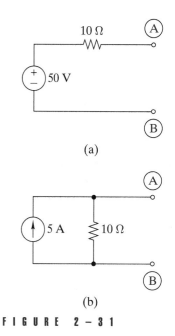

(a)

(b)

FIGURE 2-31

SOLUTION:
Using Eq. (2–24), we have

$$R_1 = R_2 = R = 10 \ \Omega$$

$$i_S = v_S/R = 5 \text{ A}$$

The equivalent practical current source is shown in Figure 2–31(b). ■

Exercise 2–9

A practical current source consists of a 2-mA ideal current source in parallel with a 0.002-S conductance. Find the equivalent practical voltage source.

Answer: The equivalent is a 1-V ideal voltage source in series with a 500-Ω resistance.

Figure 2–32 shows another source transformation in which a voltage source and resistor in parallel is replaced by a voltage source acting alone. The two circuits are equivalent because i–v constraint at the input to the rest of the circuit is $v = v_S$ in both circuits. In other words, the response of the rest of the circuit is unchanged if a resistor in parallel with a voltage source is removed from the circuit. However, removing the resistor does reduce the total current supplied by the voltage source by v_S/R. While the resistor does not affect the current and voltage delivered to the rest of the circuit, it does dissipate power that must be supplied by the source.

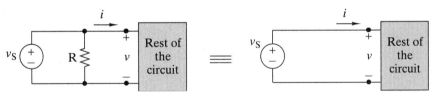

FIGURE 2–32 *Equivalent circuit of a voltage source and a resistor in parallel.*

The dual situation is shown in Figure 2–33. In this case a current source connected in series with a resistor can be replaced by a current source acting alone because the i–v constraint at the input to the rest of the circuit is $i = i_S$ for both circuits. In other words, the response of the rest of the circuit is unchanged if a resistor in series with a current source is removed from the circuit.

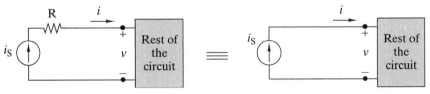

FIGURE 2–33 *Equivalent circuit of a current source and a resistor in series.*

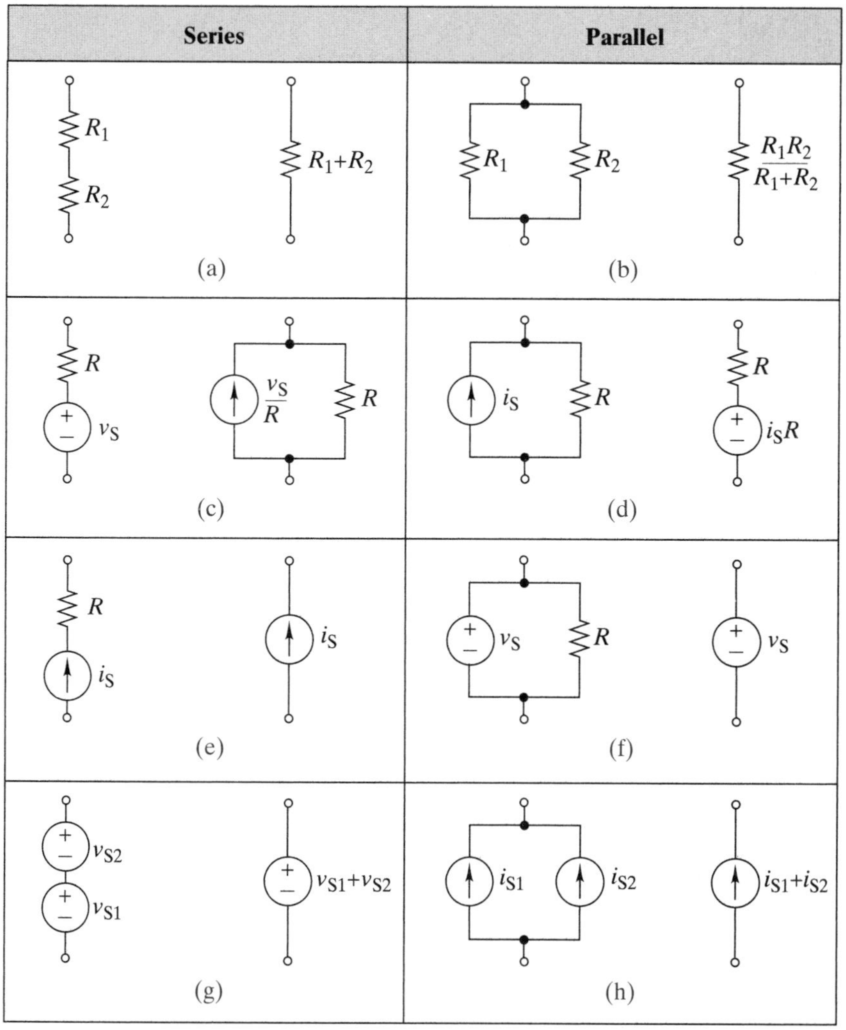

FIGURE 2-34 *Summary of two-terminal equivalent circuits.*

Series		Parallel	

(a)   (b)   (c)   (d)   (e)   (f)   (g)   (h)

SUMMARY OF EQUIVALENT CIRCUITS

Figure 2–34 is a summary of two-terminal equivalent circuits involving resistors and sources connected in series or parallel. The series and parallel equivalences in the first row and the source transformations in the second row are used regularly in subsequent discussions. The last row in Figure 2–34 presents additional source transformations that reduce series or parallel connections to a single ideal current or voltage source. Proof of these equivalences involves showing that the final single-source circuits have the same i–v characteristics as the original connections. The details of such a derivation are left as an exercise for the reader.

2-5 VOLTAGE AND CURRENT DIVISION

We complete our treatment of series and parallel circuits with a discussion of voltage and current division. These two analysis tools find wide application in circuit analysis and design.

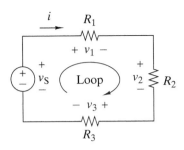

F I G U R E   2 – 3 5   *A voltage divider circuit.*

VOLTAGE DIVISION

Voltage division provides a simple way to find the voltage across each element in a series circuit. Figure 2–35 shows a circuit that lends itself to solution by voltage division. Applying KVL around the loop in Figure 2–35 yields

$$v_S = v_1 + v_2 + v_3 \qquad (2\text{--}25)$$

The elements in Figure 2–35 are connected in series, so the same current i exists in each of the resistors. Using Ohm's law, we find that

$$v_S = R_1 i + R_2 i + R_3 i \qquad (2\text{--}26)$$

Solving for i yields

$$i = \frac{v_S}{R_1 + R_2 + R_3} \qquad (2\text{--}27)$$

Once the current in the series circuit is found, the voltage across each resistor is computed using Ohm's law:

$$v_1 = R_1 i = \left(\frac{R_1}{R_1 + R_2 + R_3}\right) v_S \qquad (2\text{--}28)$$

$$v_2 = R_2 i = \left(\frac{R_2}{R_1 + R_2 + R_3}\right) v_S \qquad (2\text{--}29)$$

$$v_3 = R_3 i = \left(\frac{R_3}{R_1 + R_2 + R_3}\right) v_S \qquad (2\text{--}30)$$

Looking over these results, we see an interesting pattern. In a series connection, the voltage across each resistor is equal to its resistance divided by the equivalent series resistance of the connection times the voltage across the series circuit. Thus, the general expression of the *voltage division rule* is

$$v_k = \left(\frac{R_k}{R_{EQ}}\right) v_{TOTAL} \qquad (2\text{--}31)$$

In other words, the total voltage divides among the series resistors in proportion to their resistance over the equivalent resistance of the series connection. The following examples show several applications of this rule.

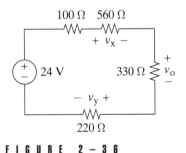

F I G U R E   2 – 3 6

EXAMPLE 2–13

Find the voltage across the 330-Ω resistor in the circuit of Figure 2–36.

SOLUTION:

Applying the voltage division rule, we find that

$$v_O = \left(\frac{330}{100 + 560 + 330 + 220}\right) 24 = 6.55 \text{ V} \qquad \blacksquare$$

Exercise 2–10

Find the voltages v_x and v_y in Figure 2–36.

   Answers: $v_x = 11.1$ V, $v_y = 4.36$ V

EXAMPLE 2–14

Select a value for the resistor R_x in Figure 2–37 so $v_O = 8$ V.

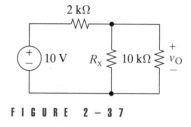

FIGURE 2–37

SOLUTION:

The unknown resistor is in parallel with the 10-kΩ resistor. Since voltages across parallel elements are equal, the voltage $v_O = 8$ V appears across both. We first define an equivalent resistance $R_{EQ} = R_x \,\|\, 10$ kΩ as

$$R_{EQ} = \frac{R_x \times 10000}{R_x + 10000}$$

We write the voltage division rule in terms of R_{EQ} as

$$v_O = 8 = \left(\frac{R_{EQ}}{R_{EQ} + 2000}\right)10$$

which yields $R_{EQ} = 8$ kΩ. Finally, we substitute this value into the equation defining R_{EQ} and solve for R_x to obtain $R_x = 40$ kΩ. ■

EXAMPLE 2–15

Use the voltage division rule to find the output voltage v_O of the circuit in Figure 2–38.

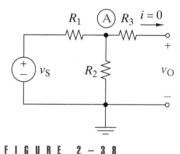

FIGURE 2–38

SOLUTION:

At first glance it appears that the voltage division rule does not apply, since the resistors are not connected in series. However, the current through R_3 is zero since the output of the circuit is an open circuit. Therefore, Ohm's law shows that $v_3 = R_3 i_3 = 0$. Applying KCL at node A shows that the same current exists in R_1 and R_2, since the current through R_3 is zero. Applying KVL around the output loop shows that the voltage across R_2 must be equal to v_O since the voltage across R_3 is zero. In essence, it is as if R_1 and R_2 were connected in series. Therefore, voltage division can be used and yields the output voltage as

$$v_O = \left(\frac{R_2}{R_1 + R_2}\right) v_S$$

The reader should carefully review the logic leading to this result because voltage division applications of this type occur frequently. ■

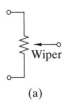

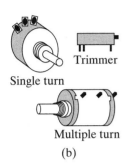

Single turn

Trimmer

Multiple turn

(b)

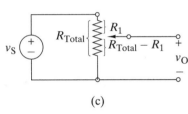

(c)

FIGURE 2–39 *The potentiometer: (a) Circuit symbol. (b) Actual device. (c) An application.*

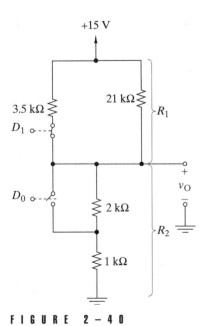

FIGURE 2–40

EXAMPLE 2–16

The operation of a potentiometer is based on the voltage division rule. The device is a three-terminal element that uses voltage (potential) division to meter out a fraction of the applied voltage. Simply stated, a **potentiometer** is an adjustable voltage divider. Figure 2–39 shows the circuit symbol of a potentiometer, sketches of three different types of actual potentiometers, and a typical application.

The voltage v_O in Figure 2–39(c) can be adjusted by turning the shaft on the potentiometer to move the wiper arm contact. Using the voltage division rule, the voltage v_O is found as

$$v_O = \left(\frac{R_{TOTAL} - R_1}{R_{TOTAL}}\right) v_S \qquad (2\text{–}32)$$

Adjusting the movable wiper arm all the way to the top makes R_1 zero, and voltage division yields

$$v_O = \left(\frac{R_{TOTAL} - 0}{R_{TOTAL}}\right) v_S = v_S \qquad (2\text{–}33)$$

In other words, 100% of the applied voltage is delivered to the rest of the circuit. Moving the wiper all the way to the bottom makes R_1 equal to R_{TOTAL}, and voltage division yields

$$v_O = \left(\frac{R_{TOTAL} - R_{TOTAL}}{R_{TOTAL}}\right) v_S = 0 \qquad (2\text{–}34)$$

This opposite extreme delivers zero voltage. By adjusting the wiper arm position, we can obtain an output voltage anywhere between zero and the applied voltage v_S. When the wiper is positioned halfway between the top and bottom, we naturally expect to obtain half of the applied voltage. Setting $R_1 = \frac{1}{2}R_{TOTAL}$ yields

$$v_O = \left(\frac{R_{TOTAL} - \dfrac{1}{2} R_{TOTAL}}{R_{TOTAL}}\right) v_S = \frac{v_S}{2} \qquad (2\text{–}35)$$

as expected. The many applications of the potentiometer include volume controls, voltage balancing, and fine-tuning adjustment.

EXAMPLE 2–17

Figure 2–40 shows a programmable voltage divider in which the digital signals D_1 and D_0 control the divider resistance R_1 and R_2 by opening and closing the analog switches shown. Determine the output v_O when $(D_1, D_0) = (0, 0)$, $(0, 1)$, $(1, 0)$, and $(1, 1)$. Assume that the analog switches are ideal switches.

SOLUTION:
When $(D_1, D_0) = (0, 0)$ the upper switch is closed and the lower switch is open. In this configuration the divider resistances are

$$R_1 = 3.5 \text{ k}\Omega \| 21 \text{ k}\Omega = 3 \text{ k}\Omega \qquad \text{and} \qquad R_2 = 2 \text{ k}\Omega + 1 \text{ k}\Omega = 3 \text{ k}\Omega$$

and the output voltage is

$$v_O = \frac{R_2}{R_1 + R_2} \times 15 = \frac{1}{2} \times 15 = 7.5 \text{ V}$$

When $(D_1, D_0) = (0, 1)$ both switches are closed and the divider resistances are

$$R_1 = 3.5 \text{ k}\Omega \| 21 \text{ k}\Omega = 3 \text{ k}\Omega \quad \text{and} \quad R_2 = 1 \text{ k}\Omega$$

and the output voltage is

$$v_O = \frac{R_2}{R_1 + R_2} \times 15 = \frac{1}{4} \times 15 = 3.75 \text{ V}$$

When $(D_1, D_0) = (1, 0)$ both switches are open and the divider resistances are

$$R_1 = 21 \text{ k}\Omega \quad \text{and} \quad R_2 = 2 \text{ k}\Omega + 1 \text{ k}\Omega = 3 \text{ k}\Omega$$

and the output voltage is

$$v_O = \frac{R_2}{R_1 + R_2} \times 15 = \frac{1}{8} \times 15 = 1.875 \text{ V}$$

Finally, when $(D_1, D_0) = (1, 1)$ the upper switch is open, the lower switch is closed, and the divider resistances are

$$R_1 = 21 \text{ k}\Omega \quad \text{and} \quad R_2 = 1 \text{ k}\Omega$$

and the output voltage is

$$v_O = \frac{R_2}{R_1 + R_2} \times 15 = \frac{1}{22} \times 15 = 0.682 \text{ V}$$

This example illustrates using digital signals to control an analog signal processor (the divider). ∎

CURRENT DIVISION

Current division provides a simple way to find the current through each element in a parallel circuit. Figure 2–41 shows a parallel circuit that lends itself to solution by current division. Applying KCL at node A yields

$$i_S = i_1 + i_2 + i_3$$

The voltage v appears across all three conductances since they are connected in parallel. Using Ohm's law, we can write

$$i_S = vG_1 + vG_2 + vG_3$$

Solving for v yields

$$v = \frac{i_S}{G_1 + G_2 + G_3}$$

Given the voltage v, the current through any element is found using Ohm's law as

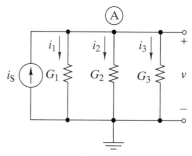

FIGURE   2 – 4 1 *A current divider circuit.*

$$i_1 = vG_1 = \left(\frac{G_1}{G_1 + G_2 + G_3}\right) i_S \qquad (2\text{--}36)$$

$$i_2 = vG_2 = \left(\frac{G_2}{G_1 + G_2 + G_3}\right) i_S \qquad (2\text{--}37)$$

$$i_3 = vG_3 = \left(\frac{G_3}{G_1 + G_2 + G_3}\right) i_S \qquad (2\text{--}38)$$

These results show that the source current divides among the parallel resistors in proportion to their conductances divided by the equivalent conductances in the parallel connection. Thus, the general expression for the *current division rule* is

$$i_k = \left(\frac{G_k}{G_{EQ}}\right) i_{TOTAL} \qquad (2\text{--}39)$$

Sometimes it is useful to express the current division rule in terms of resistance rather than conductance. For the two-resistor case in Figure 2–42, the current i_1 is found using current division as

$$i_1 = \left(\frac{G_1}{G_1 + G_2}\right) i_S = \frac{\dfrac{1}{R_1}}{\dfrac{1}{R_1} + \dfrac{1}{R_2}} i_S = \left(\frac{R_2}{R_1 + R_2}\right) i_S \qquad (2\text{--}40)$$

Similarly, the current i_2 in Figure 2–42 is found to be

$$i_2 = \left(\frac{G_2}{G_1 + G_2}\right) i_S = \frac{\dfrac{1}{R_2}}{\dfrac{1}{R_1} + \dfrac{1}{R_2}} i_S = \left(\frac{R_1}{R_1 + R_2}\right) i_S \qquad (2\text{--}41)$$

These two results lead to the following *two-path current division rule:* When a circuit can be reduced to two equivalent resistances in parallel, the current through one resistance is equal to the other resistance divided by the sum of the two resistances times the total current entering the parallel combination.

Caution: Equations (2–40) and (2–41) only apply when the circuit is reduced to two parallel paths in which one path contains the desired current and the other path is the equivalent resistance of all other paths.

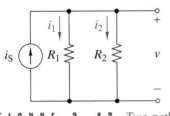

FIGURE 2–42 *Two-path current divider circuit.*

EXAMPLE 2–18

Find the current i_x in Figure 2–43(a).

SOLUTION:

To find i_x, we reduce the circuit to two paths, a path containing i_x and a path equivalent to all other paths, as shown in Figure 2–43(b). Now we can use the two-path current divider rule as

$$i_x = \frac{6.67}{20 + 6.67} \times 5 = 1.25 \text{ A} \qquad \blacksquare$$

FIGURE   2 - 4 3

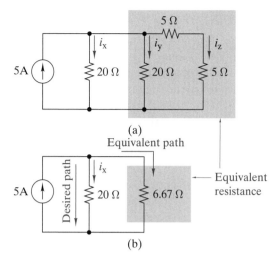

(a)

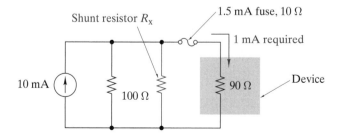

(b)

Exercise 2–11

(a) Find i_y and i_z in the circuit of Figure 2–43(a).
(b) Show that the sum of i_x, i_y, and i_z equals the source current.

Answers:
(a) $i_y = 1.25$ A, $i_z = 2.5$ A
(b) $i_x + i_y + i_z = 5$ A

Exercise 2–12

The circuit in Figure 2–44 shows a delicate device that is modeled by a 90-Ω equivalent resistance. The device requires a current of 1 mA to operate properly. A 1.5-mA fuse is inserted in series with the device to protect it from overheating. The resistance of the fuse is 10 Ω. Without the shunt resistance R_x, the source would deliver 5 mA to the device, causing the fuse to blow. Inserting a shunt resistor R_x diverts a portion of the available source current around the fuse and device. Select a value of R_x so only 1 mA is delivered to the device.

FIGURE   2 - 4 4

Answer: $R_x = 12.5$ Ω.

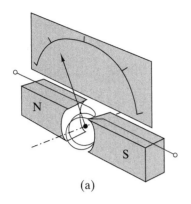

(a)

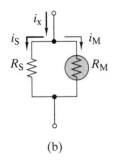

(b)

FIGURE 2–45 *The d'Arsonval meter: (a) Meter movement. (b) Equivalent circuit.*

APPLICATION NOTE: EXAMPLE 2–19

The **D'Arsonval meter** shown in Figure 2–45(a) is a device used to measure electrical currents or voltages. In simple terms, a coil of wire is mounted between the poles of a permanent magnet so it is free to rotate. The magnet produces a magnetic field that interacts with the coil current to produce a torque, which causes the coil to turn. The deflection of the pointer attached to the coil is linearly proportional to the current.

D'Arsonval movements are rated by a parameter I_{FS}, which is the electrical current required to produce full-scale deflection of the pointer. Ratings range from a few microamperes to several milliamperes depending on the structure of the device. To measure larger currents, a very precise shunt resistance R_S is connected in parallel with the meter movement, as shown in Figure 2–45(b). The current i_x divides between the shunt resistance and the meter. The shunt resistance diverts a precisely known fraction of i_x around the coil to keep the meter current i_M within the meter's full-scale deflection rating. Different current ranges can be measured with the same D'Arsonval movement by using different values of R_S.

For example, suppose currents up to 10 A are to be measured using a D'Arsonval movement with a full-scale rating of $I_{FS} = 10$ mA and a coil resistance of $R_M = 20\ \Omega$. Using the circuit model in Figure 2–45(b), the problem is to find the value of R_S that will shunt current around the meter so that $i_M = I_{FS} = 10$ mA when $i_x = 10$ A. From the model it is evident the problem can be solved using the two-path current division rule:

$$i_M = \frac{R_S i_x}{R_M + R_S}$$

$$10^{-2} = \frac{R_S(10)}{20 + R_S}$$

Solving the last equation yields $R_S = 20.02$ mΩ.

2–6 CIRCUIT REDUCTION

The concepts of series/parallel equivalence, voltage/current division, and source transformations can be used to analyze **ladder circuits** of the type shown in Figure 2–46. The basic analysis strategy is to reduce the circuit to a simpler equivalent in which the output is easily found by voltage or current division or Ohm's law. There is no fixed pattern to the reduction process, and much depends on the insight of the analyst. In any case, with

FIGURE 2–46 *A ladder circuit.*

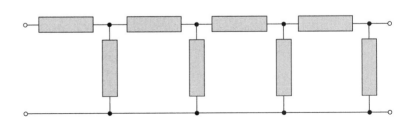

circuit reduction we work directly with the circuit model, and so the process gives us insight into circuit behavior.

With circuit reduction the desired unknowns are found by simplifying the circuit and, in the process, eliminating certain nodes and elements. However, we must be careful not to eliminate a node or element that includes the desired unknown voltage or current. The next three examples illustrate circuit reduction. The final example shows that rearranging the circuit can simplify the analysis.

EXAMPLE 2–20

Use series and parallel equivalence to find the output voltage v_O and the input current i_S in the ladder circuit shown in Figure 2–47(a).

SOLUTION:
One approach is to combine parallel resistors and use voltage division to find v_O, and then combine all resistances into a single equivalent to find the input current i_S. Figure 2–47(b) shows the step required to determine the equivalent resistance between the terminals B and ground. The equivalent resistance of the parallel $2R$ and R resistors is

$$R_{EQ1} = \frac{R \times 2R}{R + 2R} = \frac{2}{3}R$$

The reduced circuit in Figure 2–47(b) is a voltage divider. Notice that the two nodes needed to find the voltage v_O have been retained. The unknown voltage is found in terms of the source voltage as

$$v_O = \frac{\frac{2}{3}R}{\frac{2}{3}R + R}\, v_S = \frac{2}{5}v_S$$

The input current is found by combining the equivalent resistance found previously with the remaining resistor R to obtain

$$R_{EQ2} = R + R_{EQ1}$$

$$= R + \frac{2}{3}R = \frac{5}{3}R$$

Application of series/parallel equivalence has reduced the ladder circuit to the single equivalent resistance shown in Figure 2–47(c). Using Ohm's law, the input current is

$$i_S = \frac{v_S}{R_{EQ2}} = \frac{3}{5}\frac{v_S}{R}$$

Notice that the reduction step between Figures 2–47(b) and 2–47(c) eliminates node B, so the output voltage v_O must be calculated before this reduction step is taken.

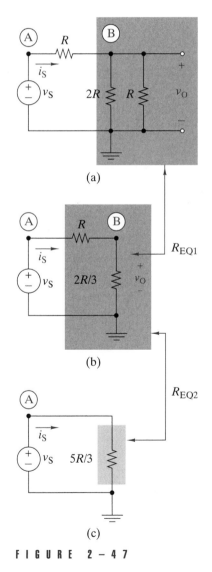

(a)

(b)

(c)

FIGURE 2–47

EXAMPLE 2–21

Use source transformations to find the output voltage v_O and the input current i_S in the ladder circuit shown in Figure 2–48(a).

FIGURE 2 – 48

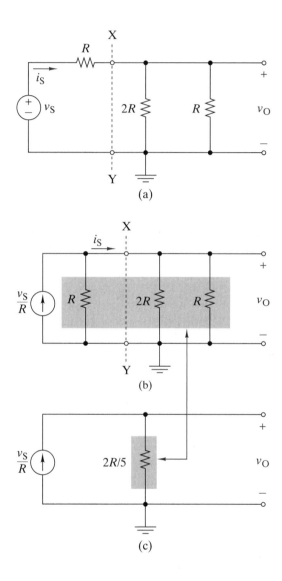

(a)

(b)

(c)

SOLUTION:

Figure 2–48 shows another way to reduce the circuit analyzed in Example 2–20. Breaking the circuit at points X and Y in Figure 2–48(a) produces a voltage source v_S in series with a resistor R. Using source transformation, this combination can be replaced by an equivalent current source in parallel with the same resistor, as shown in Figure 2–48(b).

Caution: The current source v_S/R is *not* the input current i_S, as is indicated in Figure 2–48(b). Applying the two-path current division rule to the circuit in Figure 2–48(b) yields the input current i_S as

$$i_S = \frac{R}{\frac{2}{3}R + R} \times \frac{v_S}{R} = \frac{v_S}{\frac{5}{3}R} = \frac{3}{5}\frac{v_S}{R}$$

The three parallel resistances in Figure 2–48(b) can be combined into a single equivalent conductance without eliminating the node pair used to define the output voltage v_O. Using parallel equivalence, we obtain

$$G_{EQ} = G_1 + G_2 + G_3$$

$$= \frac{1}{R} + \frac{1}{2R} + \frac{1}{R} = \frac{5}{2R}$$

which yields the equivalent circuit in Figure 2–48(c). The current source v_S/R determines the current through the equivalent resistance in Figure 2–48(c). The output voltage is found using Ohm's law.

$$v_O = \left(\frac{v_S}{R}\right) \times \left(\frac{2R}{5}\right) = \frac{2}{5}v_S$$

Of course, these results are the same as the result obtained in Example 2–20, except that here they were obtained using a different sequence of circuit reduction steps.   ∎

EXAMPLE 2−22

Find v_x in the circuit shown in Figure 2–49(a).

SOLUTION:
In the two previous examples the unknown responses were defined at the circuit input and output. In this example the unknown voltage appears across a 10-Ω resistor in the center of the network. The approach is to reduce the circuit at both ends while retaining the 10-Ω resistor defining v_x. Applying a source transformation to the left of terminals X–Y and a series reduction to the two 10-Ω resistors on the far right yields the reduced circuit shown in Figure 2–49(b). The two pairs of 20-Ω resistors connected in parallel can be combined to produce the circuit in Figure 2–49(c). At this point there are several ways to proceed. For example, a source transformation at the points W–Z in Figure 2–49(c) produces the circuit in Figure 2–49(d). Using voltage division in Figure 2–49(d) yields v_x.

$$v_x = \frac{10}{10 + 10 + 10} \times 7.5 = 2.5 \text{ V}$$

Yet another approach is to use the two-path current division rule in Figure 2–49(c) to find the current i_x.

$$i_x = \frac{10}{10 + 10 + 10} \times \frac{3}{4} = \frac{1}{4} \text{ A}$$

FIGURE 2–49

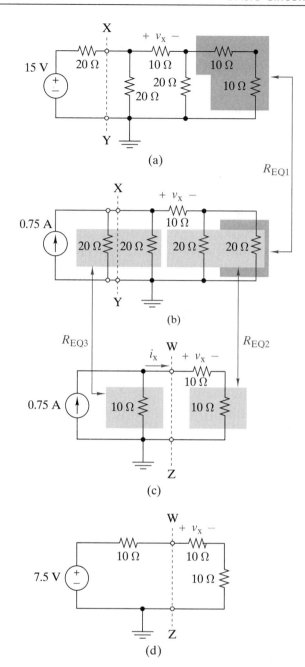

(a)

(b)

(c)

(d)

Then, applying Ohm's law to obtain v_x,

$$v_x = 10 \times i_x = 2.5 \text{ V}$$

Exercise 2–13

Find v_x and i_x using circuit reduction on the circuit in Figure 2–50.

Answers: $v_x = 3.33$ V, $i_x = 0.444$ A

FIGURE 2-50

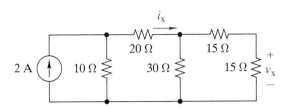

Exercise 2-14

Find v_x and v_y using circuit reduction on the circuit in Figure 2–51.

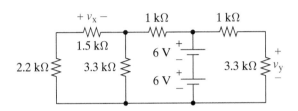

FIGURE 2-51

Answers: $v_x = 3.09$ V, $v_y = 9.21$ V

EXAMPLE 2-23

Using circuit reduction, find v_O in Figure 2–52(a).

SOLUTION:

One way to solve this problem is to notice that the source branch and the leftmost two-resistor branch are connected in parallel between node A and ground. Switching the order of these branches and replacing the two resistors by their series equivalent yields the circuit of Figure 2–52(b). A source transformation yields the circuit in Figure 2–52(c). This circuit contains a current source $v_S/2R$ in parallel with two $2R$ resistances whose equivalent resistance is

$$R_{EQ} = 2R \| 2R = \frac{2R \times 2R}{2R + 2R} = R$$

Applying a source transformation to the current source $v_S/2R$ in parallel with R_{EQ} results in the circuit of Figure 2–52(d), where

$$V_{EQ} = \left(\frac{v_S}{2R}\right) \times R_{EQ} = \left(\frac{v_S}{2R}\right)R = \frac{v_S}{2}$$

Finally, applying voltage division in Figure 2–52(d) yields

$$v_O = \left(\frac{2R}{R + R + 2R}\right)\frac{v_S}{2} = \frac{v_S}{4}$$

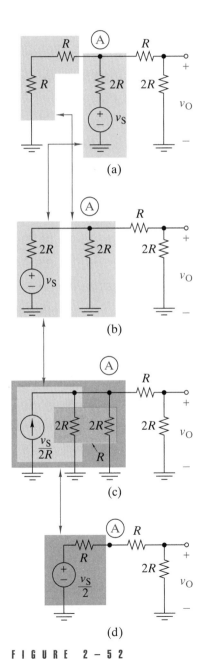

FIGURE 2-52

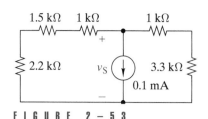

FIGURE 2 – 5 3

Find the voltage across the current source in Figure 2–53.

Answer: $v_S = -0.225$ V

2–7   COMPUTER-AIDED CIRCUIT ANALYSIS

In this book we use three types of computer programs to illustrate computer-aided circuit analysis, namely spreadsheets, math solvers, and circuit simulators. Practicing engineers routinely use these tools to analyze and design circuits, and so it is important to learn how to use them effectively. The purpose of having computer examples in this book is to help you develop an analysis style that includes the intelligent use of computer tools. As you develop your style, always keep in mind that computer tools are not problem solvers. *You* are the problem solver. Computer tools can be very useful, even essential once the problem is defined. But they do not substitute for an understanding of the fundamentals needed to formulate the problem, identify a practical approach, and interpret analysis results.

There are 36 worked examples in the text that use computer tools. The spreadsheet examples use Microsoft Excel. The math solver examples use Mathcad by MathSoft and MATHLAB by the MathWorks, Inc. The circuit simulation examples use Orcad Family Lite Edition by Cadence Design Systems, Inc. and Electronics WorkBench by Interactive Image Technologies, Inc. A student solutions manual supplement is available in which 100 of the end-of-chapter problems are solved using these five computer tools. In the text these problems are marked with a 📖 icon. Other problems suitable for these computer tools are marked with a 💻 icon.

However, keep in mind that our objective is to illustrate the effective use of computer tools rather than develop your ability to operate these specific software programs. Accordingly, this book does not emphasize details of how to operate any of these software tools. We assume that you learned how to operate computer tools in previous courses or have enough familiarity with the WINDOWS operating environment to learn how to do so using on-line tutorials or any of a number of commercially available paperback manuals.[1]

The following discussion gives a brief overview of circuit simulation, since you may be less familiar with this process than with the use of spreadsheets and math solvers.

CIRCUIT SIMULATION

Most circuit simulation programs are based on a circuit analysis package called SPICE, which is an acronym for **S**imulation **P**rogram with **I**ntegrated **C**ircuit **E**mphasis. The original SPICE program was developed in the 1970s at the University of California at Berkeley. Since then, various companies have added proprietary features to the basic SPICE program to produce an array of SPICE-based commercial products for personal computer and workstation platforms.

---

1 For example see, *PSpice for Linear Circuits*, by James A Svoboda, John Wiley & Sons, 2002 or *Simulation of Electric Circuits Using Electronic Workbench*, James L. Antonakos, Prentice Hall, 2001.

Figure 2–54 is a block diagram summarizing the major features of a SPICE-based circuit simulation program. The inputs are a circuit diagram and the type of analysis required. In contemporary programs the circuit diagram is drawn on the monitor screen using a graphical schematic editor. When the circuit diagram is complete, the input processor performs a *schematic capture*, a process that documents the circuit in what is called a *net list*. To initiate circuit simulation the input processor sends the netlist and analysis commands to the simulation processor. If the net list file is not properly prepared, the simulation will not run or (worse) will return erroneous results. Hence, it is important to check the net list to be sure that the circuit it defines is the one you want to analyze.

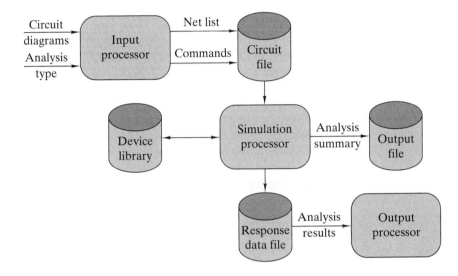

FIGURE 2–54 *Flow diagram for circuit simulation programs.*

The simulation processor uses the netlist together with data from the device library to formulate a set of equations that describes the circuit. The simulation processor then solves the equations, writes a dc analysis summary to a standard SPICE output file, and writes the other analysis results to a response data file. For simple dc analysis, the desired response data are accessible by examining the SPICE output file. For other types of analysis, the output processor can be used to generate graphical plots of the data in the response data file.

In the Orcad Family Lite Edition the input processor is called *Orcad Capture,* the simulation processor is a version of SPICE called *PSpice,* and the output processor is called *Probe.* These three programs allow the circuit diagram to be entered and analysis results viewed graphically on a computer monitor. Electronics WorkBench has three similar functions integrated into a single program called *MultiSIM.*

The following example illustrates a circuit simulation using Orcad.

EXAMPLE 2–24

Use Orcad to find the voltage v_x across the 50-Ω resistor in the circuit of Fig. 2–55

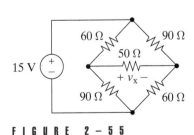

FIGURE 2–55

SOLUTION:

This example illustrates the steps involved in using Orcad to analyze a circuit. To begin, open Orcad Capture and the opening screen will come into view. Select **File\ New** from the main menu. Since this is an analog circuit, select **Analog or Mixed A/D**. Every new project must have a name, hence the circuit was named "Orcad Exercise." Drawing the circuit diagram on the monitor screen requires three actions:

(1) Selecting the desired element and placing it on the Orcad Capture workspace.

(2) Changing the element's default parameters to the desired values.

(3) Connecting the elements together using virtual wires.

Selecting the **Place** button and **Part** causes the "Place Part" dialog box to appear. The various elements needed to build the circuit are found by scrolling down the "Part List." For example, Figure 2–56 shows the "Place Part" dialog box with a resistor selected. Clicking **OK** causes a resistor to appear on the screen with a default label of "R1" and a default value of "1k." These default assignments can be changed by double-clicking on the element designator (R1) and the element value (1k). After the resistors and voltage source are selected and arranged in the capture workspace, the circuit is wired using the **Place** button and selecting **Wire**.

FIGURE 2–56

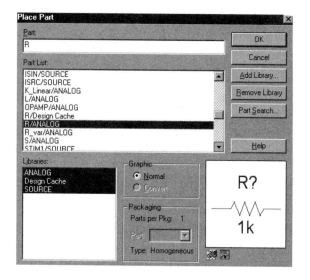

Figure 2–57 shows the circuit diagram as it might appear in Orcad Capture. Note that in an Orcad circuit diagram, connection points are indicated by a small dot and are assigned a node number. Orcad (as well as Electronic Workbench) requires that one of the nodes be chosen as ground. This is done by attaching the ground symbol found under the **Place\Ground** button to one of the nodes. The ground node is automatically assigned the number zero.

When the diagram is completed the circuit is documented using the **PSpice** button and then by selecting the **Create Netlist** command. If there are mistakes the program will issue error messages. Beginning users may receive an error message reporting that one or more nodes are "floating."

FIGURE 2-57

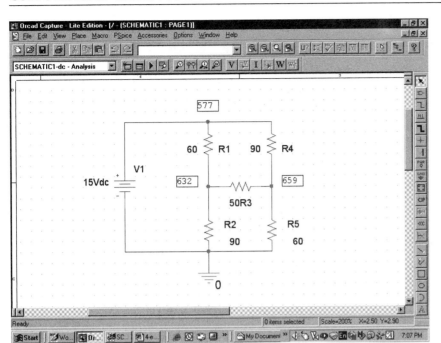

This message generally indicates a violation of one of two SPICE rules: (1) There must be at least two elements connected to every node; or (2) the circuit must include a reference (ground) node.

Using the **PSpice** and **View netlist** command produces the net list display shown in Figure 2–58. A net list is a sequence of statements that defines the circuit elements and connections. The first entry in each statement is the element type and name. The next two entries are the nodes to which the element is connected. The remaining entries define the element value(s). For example, the third statement in netlist says that the circuit contains a resistor named R4 connected between node 659 and node 577, and whose value is 90 Ω. All SPICE-based programs assign the positive reference mark for the element voltage to the first node in the element statement. For example, the plus reference mark for resistor R1 is at Node 577 and the minus mark is at Node 632. SPICE-based programs use the passive sign convention, so the reference direction for element current is from the first node toward the second node. For example, the reference direction for the current in resistor R2 is from Node 632 toward Node 0 (Ground).

Orcad analysis commands are found under **PSpice** on the main menu. Selecting **New Simulation Profile** produces the dialog box shown in Figure 2–59. In Orcad each different simulation must be given a name—"dc Analysis" in this example. Once a name is provided, a new dialog box appears called "Simulations Settings" containing the various types of analysis that PSpice performs. In the present case we select "Bias Point."[2] This option produces a dc analysis of the circuit and writes the results to the output file. This analysis type is the default option so PSpice always performs a dc analysis even when another analysis type is selected. Also select **General Settings.** Conclude by selecting **OK.**

```
1: * source ORCAD EXERCISE
2: R_R3        N00632 N00659  50
3: R_R1        N00577 N00632  90
4: R_R4        N00577 N00659  90
5: R_R2        N00632 0   60
6: R_R5        N00659 0   60
7: V_V1        N00577 0   15Vdc
8:
```

FIGURE 2-58

---

2 In later chapters we will use other analysis options on this menu such as Time Domain (transient), AC Sweep/Noise, and DC Sweep.

FIGURE 2–59

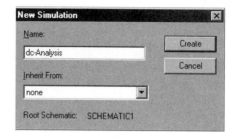

To perform the analysis select the **PSpice\Run** from the main menu. This causes PSpice to analyze the circuit defined by the netlist and to write the calculated dc responses in the output file. When the analysis is completed select **View\Output File** from the main menu. Paging down through the file we find the data shown in Figure 2–60. The output file lists the voltage between the numbered nodes and ground, namely $v_{577\text{-}0} = 15$ V, $v_{632\text{-}0} = 8.1148$ V, and $v_{639\text{-}0} = 6.8852$ V. The problem statement in this example asks for the voltage v_x across resistor R3 in Figure 2–57. To find this voltage we apply KVL to the loop formed by R2, R3, and R5.

$$-v_{632\text{-}0} + v_x + v_{659\text{-}0} = 0$$

Hence,

$$v_x = v_{632\text{-}0} - v_{659\text{-}0}$$
$$= 8.1148 - 6.8852 = 1.2296 \text{ V}$$

A nice feature of Orcad is that the calculated voltages and/or currents can be displayed directly on the circuit diagram. To do this, return to the diagram on the Orcad Capture screen and select the **V** and **I** buttons on the analysis bar. Selecting these buttons produces Figure 2–61 which displays the calculated values of current in each branch and voltage between each node and ground. ∎

FIGURE 2–60

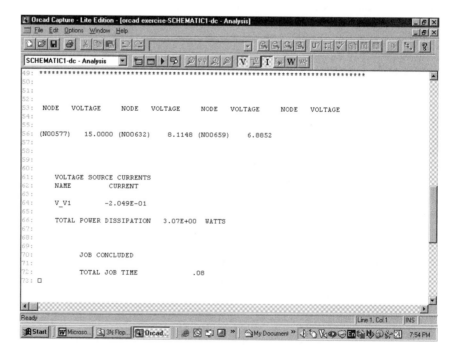

FIGURE 2 - 61

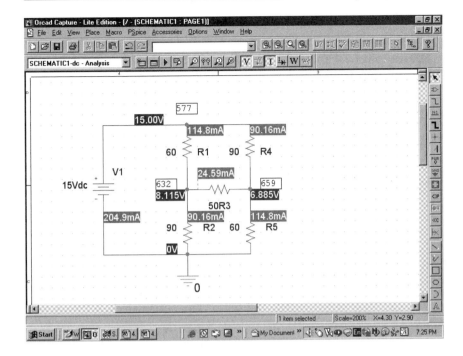

SUMMARY

- An **electrical device** is a real physical entity, while a **circuit element** is a mathematical or graphical model that approximates major features of the device.

- Two-terminal circuit elements are represented by a circuit symbol and are characterized by a single constraint imposed on the associated current and voltage variables.

- An **electrical circuit** is an interconnection of electrical devices. The interconnections form nodes and loops.

- A **node** is an electrical juncture of the terminals of two or more devices. A **loop** is a closed path formed by tracing through a sequence of devices without passing through any node more than once.

- Device interconnections in a circuit lead to two connection constraints: **Kirchhoff's current law (KCL)** states that the algebraic sum of currents at a node is zero at every instant; and **Kirchhoff's voltage law (KVL)** states that the algebraic sum of voltages around any loop is zero at every instant.

- A pair of two-terminal elements are connected in **parallel** if they form a loop containing no other elements. The same voltage appears across any two elements connected in parallel.

- A pair of two-terminal elements are connected in **series** if they are connected at a node to which no other elements are connected. The same current exists in any two elements connected in series.

- Two circuits are said to be **equivalent** if they each have the same i–v constraints at a specified pair of terminals.

- Series and parallel equivalence and voltage and current division are important tools in circuit analysis and design.

- **Source transformation** changes a voltage source in series with a resistor into an equivalent current source in parallel with a resistor, or vice versa.

- **Circuit reduction** is a method of solving for selected signal variables in ladder circuits. The method involves sequential application of the series/parallel equivalence rules, source transformations, and the voltage/current division rules. The reduction sequence used depends on the variables to be determined and the structure of the circuit and is not unique.

PROBLEMS

ERO 2–1   ELEMENT CONSTRAINTS (SECT. 2–1)

Given a two-terminal element with one or more electrical variables specified, use the element i–v constraint to find the magnitude and direction of the unknown variables.
See Example 2–1

2–1   Figure P2–1 shows a general two-terminal element with voltage and current reference marks assigned. Find the unknown electrical variables when the element is:
   (a) A linear 5-kΩ resistor with $v = 50$ V.
   (b) An ideal 10-mA current source with the arrow directed upward and $p = 30$ mW.

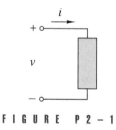

F I G U R E   P 2 – 1

2–2   Figure P2–1 shows a general two-terminal element. Determine as much as you can about the element value and signal variables when the element is:
   (a) An ideal 15-V voltage source with the plus terminal at the top and $i = 5$ mA.
   (b) An ideal switch with $i = 20$ A.

2–3   When the voltage across a linear resistor is 12 V the current is 3 mA. Find the conductance of the resistor.

2–4   The design guidelines for a circuit call for using only ¼-W resistors. The maximum voltage level in the circuit is known to be 12 V. Determine the smallest allowable value of resistance.

2–5   The i–v measurements for an unknown element are as follows: $i = 7.5$ mA @ $v = 15$ V, $i = 4.3$ mA @ $v = 8.6$ V, $i = -4$ mA @ $v = -8$ V. Use these measurements to estimate the current through the element when $v = -15$ V. Is the device linear?

2–6   The i–v relationship of a nonlinear resistor is $v = 75i + 0.2i^3$.
   (a) Calculate v and p for $i = \pm0.5, \pm1, \pm2, \pm5,$ and ±10 A.
   (b) If the operating range of the device is limited to $|i| < 0.5$ A, what is the maximum error in v when the device is approximated by a 75-Ω linear resistance?

2–7   A certain 10-kΩ resistor dissipates 12 mW. Find the current through the device.

2–8   A certain type of film resistor is available with resistance values between 10 Ω and 100 MΩ. The maximum ratings for all resistors of this type are 400 V and ½ W. Show that the voltage rating is the controlling limit when $R > 320$ kΩ, and that the power rating is the controlling limit when $R < 320$ kΩ.

2–9   Figure P2–9 shows the circuit symbol for a class of two-terminal devices called diodes. The i–v relationship for a p–n junction diode is

$$i = 2 \times 10^{-16}(e^{40\,v} - 1)$$

   (a) Use this equation to find i and p for $v = 0, \pm0.1, \pm0.2, \pm0.4,$ and ±0.8 V. Use these data to plot the i–v characteristic of the element.

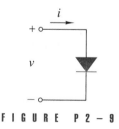

F I G U R E   P 2 – 9

(b) Is the diode linear or nonlinear, bilateral or nonbilateral, and active or passive?

(c) Use the diode model to predict i and p for $v = 5$ V. Do you think the model applies to voltages in this range? Explain.

(d) Repeat (c) for $v = -5$ V.

2–10 The resistance of a device is given by

$$R = 0.3T_C + 100$$

where T_C is the device temperature in degrees Celsius. Find the voltage across the device when the current is 1 mA and the temperature is 400°C.

ERO 2–2 CONNECTION CONSTRAINTS (SECT. 2–2)

Given a circuit composed of two-terminal elements:
(a) Identify nodes and loops in the circuit.
(b) Identify elements connected in series and in parallel.
(c) Use Kirchhoff's laws (KCL and KVL) to find selected signal variables.
See Examples 2–4, 2–5, 2–6 and Exercises 2–1, 2–2, 2–3, 2–4, 2–5

2–11 For the circuit in Figure P2–11,
(a) Identify the nodes and at least two loops.
(b) Identify any elements connected in series or in parallel.
(c) Write KCL and KVL connection equations for the circuit.
(d) If $i_1 = 6$ mA and $i_2 = -4$ mA, find the other element currents.

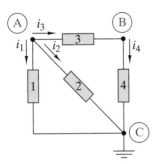

FIGURE P2-11

2–12 The currents in Figure P2–11 are observed to be $i_1 = 20$ mA and $i_3 = -30$ mA. Find the other element currents in the circuit.

2–13 For the circuit in Figure P2–13,
(a) Identify the nodes and at least three loops in the circuit.
(b) Identify any elements connected in series or in parallel.
(c) Write KCL and KVL connection equations for the circuit.

(d) If $v_3 = -8$ V, $v_4 = -8$ V, and $v_5 = 9$ V, find the other element voltages.

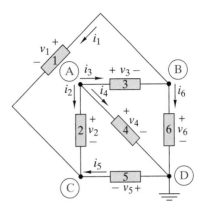

FIGURE P2-13

2–14 In Figure P2–13 $v_1 = -8$ V, $v_4 = 8$ V, and $v_6 = 6$ V. Find the other element voltages.

2–15 The circuit in Figure P2–15 is organized around the three signal lines A, B, and C.
(a) Identify the nodes and at least three loops in the circuit.
(b) Write KCL connection equations for the circuit.
(c) If $i_3 = 15$ mA, $i_4 = -12$ mA, and $i_5 = 5$ mA, find the other element currents.

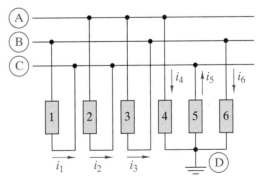

FIGURE P2-15

2–16 In Figure P2–16 $v_1 = 5$ V, $v_3 = -10$ V, and $v_4 = 10$ V. Find v_2 and v_5.

2–17 In the circuit in Figure P2–17 $i_1 = 2$ A, $i_2 = -5$ A, and $i_3 = 4$ A. Use KCL to find i_4 and i_5.

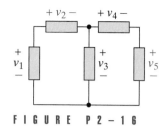

FIGURE P2-16

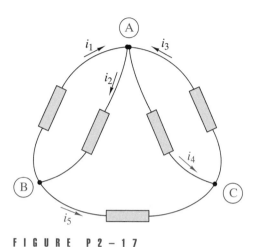

FIGURE P2-17

2–18 The connection equations for a certain circuit are

$$v_1 - v_2 - v_3 = 0 \qquad\qquad i_1 + i_2 = 0$$
$$v_3 + v_4 + v_5 = 0 \qquad -i_2 + i_3 - i_4 = 0$$
$$i_4 - i_5 = 0$$
$$-i_1 - i_3 + i_5 = 0$$

Draw the circuit diagram and indicate the reference marks for the element voltages and currents.

2–19 The incident matrix $A = [a_{ij}]$ of a circuit has one row for each node and one column for each element. If the jth element is not connected to the ith node, then $a_{ij} = 0$. If the jth element is connected to the ith node, then $a_{ij} = \pm 1$, where the plus sign applies if the current reference direction is into the node and the minus sign applies if it is away from the node. The incident matrix of a circuit is

$$A = \begin{bmatrix} -1 & +1 & 0 & -1 & 0 \\ 0 & -1 & -1 & 0 & +1 \\ +1 & 0 & +1 & +1 & -1 \end{bmatrix}$$

Draw the circuit diagram and indicate the reference directions for currents and voltages using the passive sign convention.

2–20 Given the definition in Problem 2–19, construct the incident matrix for the circuit corresponding to the connection equations in Problem 2–18.

ERO 2–3 COMBINED CONSTRAINTS (SECT. 2–3)

Given a circuit consisting of independent sources and linear resistors, use the element constraints and connection constraints to find selected signal variables.
See Examples 2–7, 2–8, 2–9, 2–10 and Exercise 2–6

2–21 Find v_x and i_x in Figure P2–21.

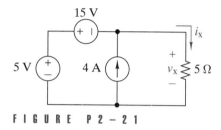

FIGURE P2-21

2–22 First use KVL to find the voltage across each resistor in Figure P2–22. Then use Ohm's law and KCL to find the current through every element, including the voltage sources.

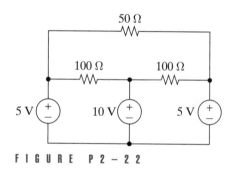

FIGURE P2-22

2–23 Find v_x in Figure P2–23.

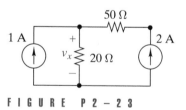

FIGURE P2-23

2–24 Figure P2–24 shows a subcircuit connected to the rest of the circuit at four points.
(a) Use element and connection constraints to find v_x and i_x.
(b) Show that the sum of the currents into the rest of the circuit is zero.

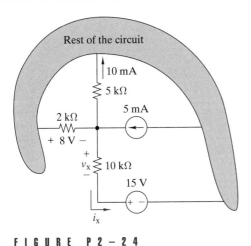

FIGURE   P 2 – 2 4

2–25 The circuit in Figure P2–25 is a model of the feedback path in an electronic amplifier circuit. The current i_x is known to be 4 mA. (a) Find the value of R. (b) Show that the sum of currents into the rest of the circuit is zero.

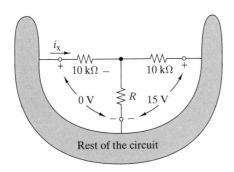

FIGURE   P 2 – 2 5

2–26 Figure P2–26 shows a resistor with one terminal connected to ground and the other connected to an arrow. The arrow symbol is used to indicate a connection to one terminal of a voltage source whose other terminal is connected to ground. The label next to the arrow indicates the source voltage at the ungrounded terminal. Find the voltage across, current through, and power dissipated in the resistor.

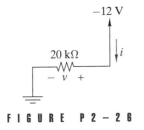

FIGURE   P 2 – 2 6

ERO 2–4 EQUIVALENT CIRCUITS (SECT. 2–4)

(a) Given a circuit consisting of linear resistors, find the equivalent resistance between a specified pair of terminals.
(b) Given a circuit consisting of a source-resistor combination, find an equivalent source-resistor circuit.
See Example 2–11, 2–12 and Exercises 2–7, 2–8, 2–9

2–27 Find the equivalent resistance R_{EQ} in Figure P2–27.

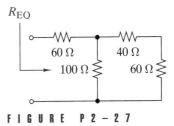

FIGURE   P 2 – 2 7

2–28 Find the equivalent resistance R_{EQ} in Figure P2–28.

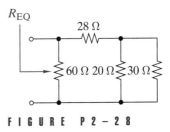

FIGURE   P 2 – 2 8

2–29 Find the equivalent resistance R_{EQ} in Figure P2–29 when the switch is open. Repeat when the switch closed.

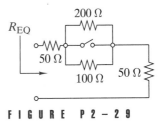

FIGURE   P 2 – 2 9

2–30 Find the equivalent resistance between terminals A–B, B–C, A–C, C–D, B–D, and A–D in the circuit of Figure P2–30.

2–31 Find the equivalent resistance between terminals A–B, A–C, A–D, B–C, B–D, and C–D in the circuit of Figure P2–31.

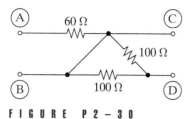

FIGURE P2–30

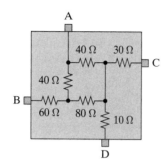

FIGURE P2–31

2–32 Find the equivalent resistance R_{EQ} in Figure P2–32 when $R_L = 10$ kΩ. Repeat when $R_L = 0$. Select the value of R_L so that $R_{EQ} = 22$ kΩ.

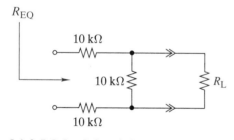

FIGURE P2–32

2–33 An ideal 15-V voltage source is connected in series with a 50-Ω resistor. Use source transformations to obtain an equivalent practical current source.

2–34 A 5-mA practical current source is connected in parallel with a 2-kΩ resistor. The voltage across the resistor is observed to be 5 V. Find the source resistance of the practical current source.

2–35 The circuit of Figure P2–35 is an R-2R resistance array package. All of the following equivalent resistances can be obtained by making proper connections of the array *except for one*: R/2, 2R/3, R, 8R/5, 2R, 3R, and 4R. Show how to interconnect the terminals of the array to produce the equivalent resistances, and identify the one that cannot be obtained using this array.

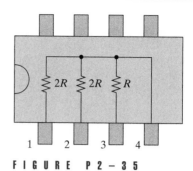

FIGURE P2–35

2–36 Select the value of R_x in Figure P2–36 so that $R_{EQ} = 49$ kΩ.

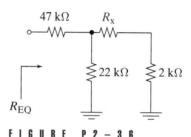

FIGURE P2–36

2–37 **D** Using no more than 14 resistors, show how to interconnect standard 4.3-kΩ resistors to obtain equivalent resistances of 1 kΩ $\pm$10%, 5 kΩ $\pm$10%, and 10 kΩ $\pm$10%.

2–38 Select the value of R in Figure P2–38 so that $R_{AB} = R_L$.

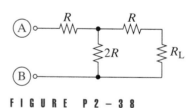

FIGURE P2–38

2–39 What is the range of R_{EQ} in Figure P2–39?

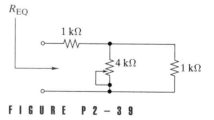

FIGURE P2–39

2–40 Find the equivalent resistance between terminals A and B in Figure P2–40.

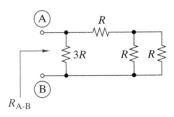

FIGURE P 2 - 4 0

ERO 2–5 VOLTAGE AND CURRENT DIVISION (SECT. 2–5)

(a) Given a circuit with elements connected in series or parallel, use voltage or current division to find specified voltages or currents.

(b) Design a voltage or current divider that delivers specified output signals within stated constraints.

See Examples 2–13, 2–14, 2–16, 2–17, 2–18 and Exercises 2–10, 2–11, 2–12

2–41 Use voltage division in Figure P2–41 to obtain an expression for v_L in terms of R, R_L, and v_S.

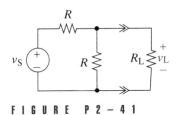

FIGURE P 2 - 4 1

2–42 Find v_x in Figure P2–42.

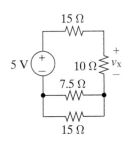

FIGURE P 2 - 4 2

2–43 Find i_x in Figure P2–43.

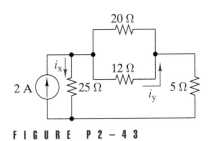

FIGURE P 2 - 4 3

2–44 Find i_y in Figure P2–43.

2–45 The electronic switch in Figure P2–45 is controlled by the voltage between gate G and ground. The switch is closed ($R_{ON} = 100\ \Omega$) when $v_G > 2$ V and open ($R_{OFF} = 500$ MΩ) when $v_G < 0.8$ V. Use voltage division to predict v_O when $v_G = 5$ V and $v_G = 0.5$ V.

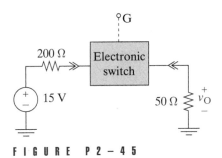

FIGURE P 2 - 4 5

2–46 Figure P2–46 shows a resistance divider connected in a general circuit.

(a) What is the relationship between v_1 and v_2 when $i_1 = 0$?

(b) What is the relationship between v_1 and v_2 when $i_2 = 0$?

(c) What is the relationship between i_1 and i_2 when $v_1 = 0$?

(d) What is the relationship between i_1 and i_2 when $v_2 = 0$?

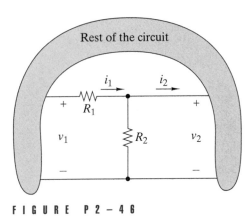

FIGURE P 2 - 4 6

2–47 Figure P2–47 shows an ammeter circuit consisting of a D'Arsonval meter, a two-position selector switch, and two shunt resistors. A current of 0.5 mA produces full-scale deflection of the D'Arsonval meter, whose internal resistance is $R_M = 50\ \Omega$. Select the shunt resistance R_1 and R_2 so that $i_x = 10$ mA produces full scale deflection when the switch is in position A, and i_x

= 50 mA produces full-scale deflection when the switch is in position B.

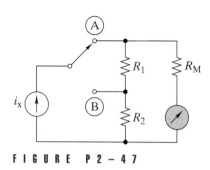

FIGURE P2–47

2–48 **D** Figure P2–48 shows the *R-2R* integrated circuit package connected as a voltage divider and across a 15-V source. The divider output is 5 V for the connections shown in the figure. Show how to interconnect the source and the R-2R package to obtain outputs of 3 V, 6 V, 9 V, and 12 V.

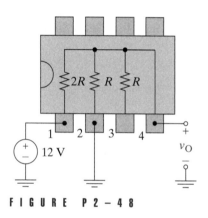

FIGURE P2–48

2–49 **D** Find the value of R_x in Figure P2–49 such that $v_L = 2$ V.

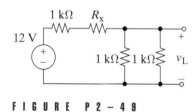

FIGURE P2–49

2–50 **D** Find the value of R_x in Figure P2–50 such that $v_L = 3$ V.

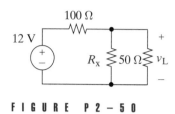

FIGURE P2–50

ERO 2-6 CIRCUIT REDUCTION (SECT. 2-6)

Given a circuit consisting of linear resistors and an independent source, find selected signal variables using successive application of series/parallel equivalence, source transformations, and voltage/current division.
See Example 2–20, 2–21, 2–22, 2–23 and Exercises 2–13, 2–14, 2–15

2–51 Use circuit reduction to determine v_x and i_x in the circuit shown in Figure P2–51.

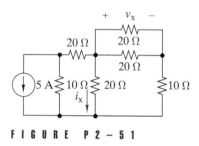

FIGURE P2–51

2–52 Use circuit reduction to find i_S and i_x in the circuit shown in Figure P2–52.

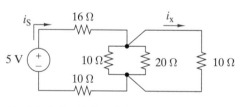

FIGURE P2–52

2–53 Use circuit reduction to find v_x and i_x in the circuit shown in Figure P2–53.

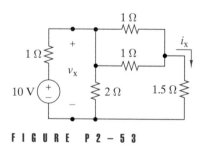

FIGURE P2–53

2–54 Use circuit reduction to find v_x and i_x in the circuit shown in Figure P2–54.

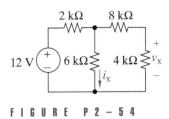

FIGURE P2 – 54

2–55 The resistance array circuit in Figure P2–55 has external terminals at pads A, B, C, and D. Connect the "+" terminal of a 10-V voltage to terminal A and the "−" to terminal C. Find the voltage v_{AB}.

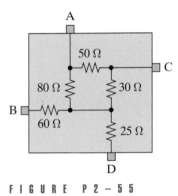

FIGURE P2 – 55

2–56 The resistance array circuit in Figure P2–55 has external terminals at pads A, B, C, and D. Connect the "+" terminal of a 10-V voltage to terminal A and the "−" to terminal B. Find the voltage v_{DB}.

2–57 The resistance array circuit in Figure P2–55 has external terminals at pads A, B, C, and D. Connect the "+" terminal of a 5-V voltage to terminal C and the "−" to ground. Connect terminal B to ground. Find the voltage v_{DB}.

2–58 Select the value of R_L in Figure P2–58 so that the power delivered to R_L is at least 50 mW.

2–59 Select the value of R_L in Figure P2–58 so that the voltage delivered to R_L is at least 2.5 V.

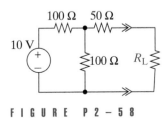

FIGURE P2 – 58

2–60 The box in the circuit in Figure P2–60 is a resistor whose value can be anywhere between 8 kΩ and 80 kΩ. Use circuit reduction to find the range of values of the outputs v_x and i_x.

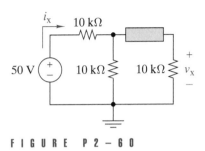

FIGURE P2 – 60

INTEGRATING PROBLEMS

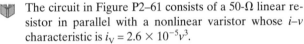

2–61 Device Modeling Ⓐ

The circuit in Figure P2–61 consists of a 50-Ω linear resistor in parallel with a nonlinear varistor whose i–v characteristic is $i_V = 2.6 \times 10^{-5} v^3$.

(a) Plot the i–v characteristic of the parallel combination.

(b) State whether the parallel combination is linear or nonlinear, active or passive, and bilateral or nonbilateral.

(c) Identify a range of voltages over which the parallel combination can be modeled within ±10% by a linear resistor.

(d) Identify a range of voltages over which the parallel combination can be safely operated if both devices are rated at 50 W. Which device limits this range?

(e) The parallel combination is connected in series with a 50-Ω resistor and a 5-V voltage source. In this circuit, how would you model the parallel combination and why?

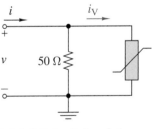

FIGURE P2 – 61

2–62 Wheatstone Bridge Ⓐ

The Wheatstone bridge circuit in Figure P2–62(a) is used in instrumentation systems. The resistance R_x is the equivalent resistance of a transducer (a device that converts energy from one form to another). The value of R_x varies in relation to an external physical phenomenon such as temperature, pressure, or light. The resistance R_M is the equivalent resistance of a measuring instrument such as a D'Arsonval meter. Prior to any measurements, one of the other resistors (usually R_3) is adjusted until the current i_M is zero. The resistance of the transducer changes when exposed to the physical phenomenon it is designed to measure. This change causes the bridge to become unbalanced, and the meter indicates the resulting current through R_M. The deflection of the meter is calibrated to indicate the value of the physical phenomenon measured by the transducer.

(a) **A** Derive the relationship between R_1, R_2, R_3, and R_x when $i_M = 0$ A.

(b) **A** Suppose the transducer resistance R_x varies with temperature, as shown in Figure P2–62(b). With $R_1 = R_2 = 2.2$ kΩ, find the value of R_3 that produces $i_M = 0$ at a temperature of 57.5° C.

(c) **A** A current $i_M = -1.5$ mA is observed when $R_1 = R_2 = 2.2$ kΩ and R_3 is set to the value found in part (b). Is the temperature higher or lower than 57.5° C?

(d) **A** For $R_1 = R_2 = 2.2$ kΩ and $R_3 = 2.4$ kΩ, find the temperature at which $i_M = 0$.

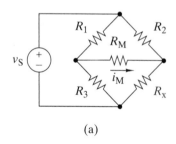

(a)

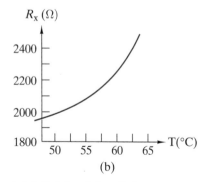

(b)

F I G U R E   P 2 – 6 2

2–63 Three-Terminal Equivalence **A**

Two circuits are said to be equivalent when they have the same i–v characteristics between specified terminal pairs. In this chapter we applied this definition to two-terminal circuits such as resistors in series or parallel and source transformations. The concept can be extended to the three-terminal circuits in Figure P2–63. These three-terminal circuits will be equivalent if the equivalent resistances seen between terminal pairs A and B, B and C, and C and A are the same.

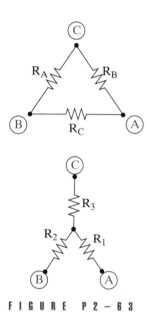

F I G U R E   P 2 – 6 3

(a) **A** Show that the two circuits are equivalent when

$$R_1 = \frac{R_B R_C}{R_A + R_B + R_C} \quad \text{and}$$

$$R_2 = \frac{R_C R_A}{R_A + R_B + R_C} \quad \text{and}$$

$$R_3 = \frac{R_A R_B}{R_A + R_B + R_C}$$

(b) **A** Equivalence is useful because replacing one circuit by an equivalent circuit does not change the response of the rest of the circuit and may simplify an analysis problem. For example, the equations in (a) tell us how to create an equivalent Y-connected subcircuit like Figure P2–63(b) to replace the Δ-connected subcircuit in Figure P2–63(a). Such a replacement is called a Δ-to-Y transformation. Show that a Y-to-Δ transformation is also possible provided

$$G_A = \frac{G_2 G_3}{G_1 + G_2 + G_3} \quad \text{and}$$

$$G_B = \frac{G_1 G_3}{G_1 + G_2 + G_3} \quad \text{and}$$

$$G_C = \frac{G_1 G_2}{G_1 + G_2 + G_3}$$

2–64 Digital-to-Analog Conversion **A**

Digital-to-analog (D-to-A) conversion provides a link between the digital and analog worlds. A D-to-A converter produces a single analog output v_O from a multibit digital input $\{b_0, b_1, b_2, \ldots b_{N-1}\}$, where the bits b_j $(j = 0, 1, \ldots N - 1)$ are either 0 or 1. One method is to produce an analog output that is proportional to a fixed reference voltage V_{REF} and related to the digital inputs by the following algorithm

$$v_O = K V_{REF} \sum_{j=0}^{N-1} b_j 2^j$$

Figure P2–64 shows a programmable voltage divider in which two digital inputs control complementary analog switches connecting a multitap voltage divider to the output terminal. Show that the programmable voltage divider implements this D-to-A algorithm with $K = 0.25$ and $N = 2$.

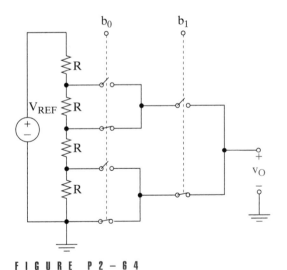

FIGURE P2-64

2–65 Analog Voltmeter Design **A D E**

A voltmeter can be made using a series resistor and a D'Arsonval meter. Figure P2–65(a) shows a voltmeter circuit consisting of a D'Arsonval meter, a two-position selector switch, and two series resistors. A current of 500 µA produces full-

scale deflection of the D'Arsonval meter, whose internal resistance is $R_M = 50\ \Omega$.

(a) **D** Select the series resistance R_1 and R_2 so a voltage $v_x = 50$ V produces full-scale deflection when the switch is in position A, and voltage $v_x = 10$ V produces full-scale deflection when the switch is in position B.

(b) **A** By voltage division the voltage across the 20-kΩ resistor in Figure P2–65(b) is 20 V when the voltmeter is disconnected. What voltage reading is obtained when the voltmeter designed in part (a) is connected across the 20-kΩ resistor? What is the percentage error in the voltmeter reading?

(c) **E** A D'Arsonval meter with an internal resistance of 200 Ω and a full-scale deflection current of 100 µA is available. If the voltmeter in part (a) is redesigned using this D'Arsonval meter, would the error found in part (b) be smaller or larger? Explain.

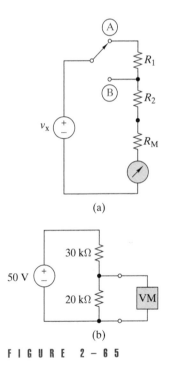

FIGURE 2-65

CHAPTER 3

CIRCUIT ANALYSIS
TECHNIQUES

Assuming any system of linear conductors connected in such a manner that to the extremities of each one of them there is connected at least one other, a system having electromotive forces $E_1, E_2 \ldots E_3$, no matter how distributed, we consider two points A and A′ belonging to the system and having potentials V and V′. If the points A and A′ are connected by a wire ABA′, which has a resistance r, with no electromotive forces, the potentials of points A and A′ assume different values from V and V′, but the current i flowing through this wire is given by $i = (V - V')/(r + R)$ in which R represents the resistance of the original wire, this resistance being measured between the points A and A′, which are considered to be electrodes.

Leon Charles Thévenin, 1883,
French Telegraph Engineer

3–1 Node-voltage Analysis

3–2 Mesh-current Analysis

3–3 Linearity Properties

3–4 Thévenin and Norton Equivalent Circuits

3–5 Maximum Signal Transfer

3–6 Interface Circuit Design

Summary

Problems

Integrating Problems

Thévenin's theorem is one of the important circuit analysis concepts developed in this chapter. Leon Charles Thévenin (1857–1926), a distinguished French telegraph engineer and teacher, was led to his theorem in 1883 following an extensive study of Kirchhoff's laws. Norton's theorem, which is the dual of Thévenin's theorem, was not proposed until 1926 by Edward L. Norton, an American electrical engineer working on long-distance telephony. Curiously, it turns out that the basic concept had been discovered in 1853 by Herman von Helmholtz. The earlier discovery by Helmholtz is not recognized in engineering terminology possibly because Thévenin and Norton both worked in areas of technology that offered immediate applications for their results, whereas Helmholtz was studying electricity in animal tissue at the time.

The analysis methods developed in the previous chapter offer insight into circuit analysis and design because we work directly with the circuit model to find responses. With practice and experience we learn which tools to use and in what order to avoid going down blind alleys. This ad hoc approach is practical as long as the circuits are fairly simple. As circuit complexity increases, however, a more systematic approach is needed.

Two basic methods of systematic circuit analysis—node-voltage analysis and mesh-current analysis—are developed in the first two sections of this chapter. With these methods the device and connection constraints are used to formulate a set of linear algebraic equations that characterize the circuit. Solving these equations simultaneously then yields the desired circuit responses. General methods of circuit analysis are necessarily more abstract because we manipulate sets of equations, rather than the circuit itself. In doing so we lose some contact with the more intuitive approach developed in the previous chapter. However, systematic methods make it possible for us to treat a wider range of applications and provide a framework for exploiting other properties of linear circuits.

Two key properties of linear circuits are developed in the third section. The superposition and proportionality properties lead to new circuit analysis techniques and are often used to derive other properties of linear circuits. For example, the superposition property is used in the fourth section to derive Thévenin's theorem. This theorem provides a viewpoint for dealing with circuit interfaces, and that leads directly to the maximum signal transfer properties developed in the fifth section. In the final section, circuit design is introduced by showing how the maximum signal transfer principles guide the design of interface circuit.

3-1   NODE-VOLTAGE ANALYSIS

Before describing node-voltage analysis, we first review the foundation for every method of circuit analysis. As noted in Sec. 2–3, circuit behavior is based on constraints of two types: (1) connection constraints (Kirchhoff's laws) and (2) device constraints (element *i–v* relationships). As a practical

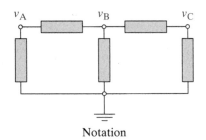

Notation

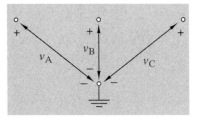

Interpretation

F I G U R E   3 – 1   *Node-voltage definition and notation.*

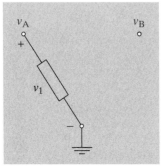

Case A

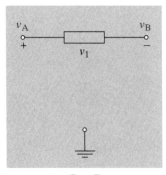

Case B

F I G U R E   3 – 2   *Two possible connections of a two-terminal element.*

matter, however, using element voltages and currents to express the circuit constraints produces a large number of equations that must be solved simultaneously to find the circuit responses. For example, a circuit with only six devices requires us to treat 12 equations with 12 unknowns. Although this is not an impossible task using software tools like Mathcad, it is highly desirable to reduce the number of equations that must be solved simultaneously.

You should not abandon the concept of element and connection constraints. This method is vital because it provides the foundation for all methods of circuit analysis. In subsequent chapters, we use element and connection constraints many times to develop important ideas in circuit analysis.

Using node voltages instead of element voltages as circuit variables can reduce the number of equations that must be treated simultaneously. To define a set of node voltages we first select a reference node. The **node voltages** are then defined as the voltages between the remaining nodes and the selected reference node. Figure 3–1 shows the notation used to define node-voltage variables. In this figure the reference node indicated by the ground symbol and the node voltages are identified by a voltage symbol next to all the other nodes. This notation means that the positive reference mark for the node voltage is located at the node in question while the negative mark is at the reference node. Obviously, any circuit with N nodes involves $N-1$ node voltages.

A fundamental property of node voltages needs to be covered at the outset. Suppose we are given a two-terminal element whose element voltage is labeled v_1. Suppose further that the terminal with the plus reference mark is connected to a node, say node A. The two cases shown in Figure 3–2 are the only two possible ways the other element terminal can be connected. In case A, the other terminal is connected to the reference node, in which case KVL requires $v_1 = v_A$. In case B, the other terminal is connected to a non-reference node, say node B, in which case KVL requires $v_1 = v_A - v_B$. This example illustrates the following fundamental property of node voltages:

> *If the Kth two-terminal element is connected between nodes X and Y, then the element voltage can be expressed in terms of the two node voltages as*

$$v_K = v_X - v_Y \tag{3–1}$$

> *where X is the node connected to the positive reference for element voltage v_K.*

Equation (3–1) is a KVL constraint at the element level. If node Y is the reference node, then by definition $v_Y = 0$ and Eq. (3–1) reduces to $v_K = v_X$. On the other hand, if node X is the reference node, then $v_X = 0$ and therefore $v_K = -v_Y$. The minus sign occurs here because the positive reference for the element is connected to the reference node. In any case, the important fact is that the voltage across any two-terminal element can be expressed as the difference of two node voltages, one of which may be zero.

Exercise 3–1

The reference node and node voltages in the bridge circuit of Figure 3–3 are $v_A = 5$ V, $v_B = 10$ V, and $v_C = -3$ V. Find the element voltages.

FIGURE 3 - 3

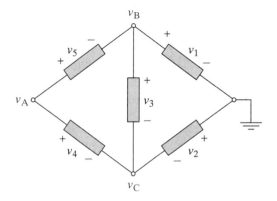

Answers:

$$v_1 = 10 \text{ V}, v_2 = 3 \text{ V}, v_3 = 13 \text{ V}, v_4 = 8 \text{ V, and } v_5 = -5 \text{ V}.$$

FORMULATING NODE-VOLTAGE EQUATIONS

To formulate a circuit description using node voltages, we use element and connection constraints, except that the KVL connection equations are not explicitly written. Instead we use the fundamental property of node analysis to express the element voltages in terms of the node voltages.

The circuit in Figure 3–4 will demonstrate the formulation of node-voltage equations. In Figure 3–4 we have identified a reference node (indicated by the ground symbol), four element currents (i_0, i_1, i_2, and i_3), and two node voltages (v_A and v_B).

The KCL constraints at the two nonreference nodes are

$$\text{Node A: } -i_0 - i_1 - i_2 = 0$$

$$\text{Node B: } i_2 - i_3 = 0 \tag{3–2}$$

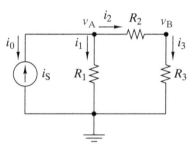

FIGURE 3 - 4 *Circuit for demonstrating node-voltage analysis.*

Using the fundamental property of node analysis, we use the element equations to relate the element currents to the node voltages.

$$\text{Resistor } R_1: i_1 = G_1 v_A$$

$$\text{Resistor } R_2: i_2 = G_2(v_A - v_B)$$

$$\text{Resistor } R_3: i_3 = G_3 v_B$$

$$\text{Current Source: } i_0 = -i_S \tag{3–3}$$

We have written six equations in six unknowns—four element currents and two node voltages. The right side of the element equations in Eq. (3–3) involves unknown node voltages and the input signal i_S. Substituting the element constraints in Eq. (3–3) into the KCL connection constraints in Eq. (3–2) yields

$$\text{Node A: } i_S - G_1 v_A - G_2(v_A - v_B) = 0$$

$$\text{Node B: } G_2(v_A - v_B) - G_3 v_B = 0$$

which can be arranged in the following standard form:

$$\text{Node A:} \quad (G_1 + G_2)v_A - G_2v_B = i_S$$

$$\text{Node B:} \quad -G_2v_A + (G_2 + G_3)v_B = 0 \tag{3-4}$$

In this standard form all of the unknown node voltages are grouped on one side and the independent sources on the other.

By systematically eliminating the element currents, we have reduced the circuit description to two linear equations in the two unknown node voltages. The coefficients in the equations on the left side ($G_1 + G_2$, G_2, $G_2 + G_3$) depend only on circuit parameters, while the right side contains the known input driving force i_S.

As noted previously, every method of circuit analysis must satisfy KVL, KCL, and the element i–v relationships. In developing the node-voltage equations in Eq. (3–4), it may appear that we have not used KVL. However, KVL is satisfied because the equations $v_1 = v_A$, $v_2 = v_A - v_B$, and $v_3 = v_B$ were used to write the right side of the element equations in Eq. (3–3). The KVL constraints do not appear explicitly in the formulation of node equations, but are implicitly included when the fundamental property of node analysis is used to write the element voltages in terms of the node voltages.

In summary, four steps are needed to develop node-voltage equations.

STEP 1 Select a reference node. Identify a node voltage at each of the remaining $N - 1$ nodes and a current with every element in the circuit.

STEP 2 Write KCL connection constraints in terms of the element currents at the $N - 1$ nonreference nodes.

STEP 3 Use the element i–v relationships and the fundamental property of node analysis to express the element currents in terms of the node voltages.

STEP 4 Substitute the element constraints from step 3 into the KCL connection constraints from step 2 and arrange the resulting $N - 1$ equations in a standard form.

Writing node-voltage equations leads to $N - 1$ equations that must be solved simultaneously. If we write the element and connection constraints in terms of element voltages and currents, we must solve $2E$ simultaneous equations. The reduction from $2E$ to $N - 1$ is particularly impressive in circuits with a large number of elements (large E) connected in parallel (small N).

EXAMPLE 3–1

Formulate node-voltage equations for the bridge circuit in Figure 3–5.

SOLUTION:

Step 1: The reference node, node voltages, and element currents are shown in Figure 3–5.

Step 2: The KCL constraints at the three nonreference nodes are:

$$\text{Node A: } i_0 - i_1 - i_2 = 0$$

$$\text{Node B: } i_1 - i_3 + i_5 = 0$$

$$\text{Node C: } i_2 - i_4 - i_5 = 0$$

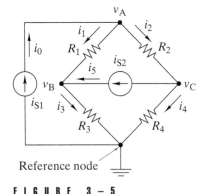

Reference node

FIGURE 3 – 5

Step 3: We write the element equations in terms of the node voltages and input signal sources.

$$i_0 = i_{S1} \qquad\qquad i_3 = G_3 v_B$$
$$i_1 = G_1(v_A - v_B) \quad i_4 = G_4 v_C$$
$$i_2 = G_2(v_A - v_C) \quad i_5 = i_{S2}$$

Step 4: Substituting the element equations into the KCL constraints and arranging the result in standard form yields three equations in the three unknown node voltages.

$$\text{Node A: } (G_1 + G_2)v_A - G_1 v_B - G_2 v_C = i_{S1}$$
$$\text{Node B: } \qquad -G_1 v_A + (G_1 + G_3)v_B = i_{S2}$$
$$\text{Node C: } \qquad -G_2 v_A + (G_2 + G_4)v_C = -i_{S2} \qquad\blacksquare$$

WRITING NODE-VOLTAGE EQUATIONS BY INSPECTION

The node-voltage equations derived in Example 3–1 have a symmetrical pattern. The coefficient of v_B in the node A equation and the coefficient of v_A in the node B equation are both the negative of the conductance connected between the nodes ($-G_1$). Likewise, the coefficients of v_A in the node C equation and v_C in the node A equation are both $-G_2$. Finally, coefficients of v_A in the node A equation, v_B in the node B equation, and v_C in the node C equation are the sum of the conductances connected to the node in question.

The symmetrical pattern always occurs in circuits containing only resistors and independent current sources. To understand why, consider any general two-terminal conductance G with one terminal connected to, say, node A. According to the fundamental property of node analysis, there are only two possibilities. Either the other terminal of G is connected to the reference node, in which case the current *leaving* node A via conductance G is

$$i = G(v_A - 0) = Gv_A$$

or else it is connected to another nonreference node, say, node B, in which case the current *leaving* node A via G is

$$i = G(v_A - v_B)$$

The pattern for node equations follows from these observations. The sum of the currents leaving any node A via conductances involves the following terms:

1. v_A times the sum of conductances connected to node A

2. Minus v_B times the sum of conductances connected between nodes A and B and similar terms for all other nodes connected to node A by conductances.

Because of KCL, the sum of currents leaving node A via conductances plus the sum of currents directed away from node A by independent current sources must equal zero.

The aforementioned process allows us to write node-voltage equations by inspection without going through the intermediate steps involving the

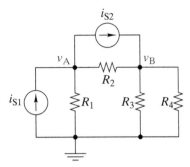

FIGURE 3-6 *Circuit for demonstrating writing node-voltage equations by inspection.*

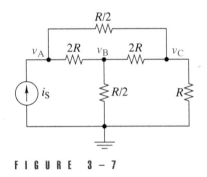

FIGURE 3-7

KCL constraints and the element equations. For example, the circuit in Figure 3–6 contains two independent current sources and four resistors. Starting with node A, the sum of conductances connected to node A is $G_1 + G_2$. The conductances between nodes A and B is G_2. The reference direction for the source current i_{S1} is into node A, while the reference direction for i_{S2} is directed away from node A. Pulling all of the observations together, we write the sum of currents directed out of node A as

$$\text{Node A:} \ (G_1 + G_2)v_A - G_2v_B - i_{S1} + i_{S2} = 0 \qquad (3\text{–}5)$$

Similarly, the sum of conductances connected to node B is $G_2 + G_3 + G_4$, the conductance connected between nodes B and A is again G_2, and the source current i_{S2} is directed away from node B. These observations yield the following node-voltage equation:

$$\text{Node B:} \ (G_2 + G_3 + G_4)v_B - G_2v_A - i_{S2} = 0 \qquad (3\text{–}6)$$

Rearranging Eqs. (3–5) and (3–6) in standard form yields

$$\text{Node A:} \qquad (G_1 + G_2)v_A - G_2v_B = i_{S1} - i_{S2}$$
$$\text{Node B:} \ - G_2v_A + (G_2 + G_3 + G_4)v_B = i_{S2} \qquad (3\text{–}7)$$

We have two symmetrical equations in the two unknown node voltages. The equations are symmetrical because the conductance G_2 connected between nodes A and B appears as the cross-coupling term in each equation.

EXAMPLE 3-2

Formulate node-voltage equations for the bridged-T circuit in Figure 3–7.

SOLUTION:
The total conductance connected to node A is $1/2R + 2/R = 2.5G$, to node B is $1/2R + 1/2R + 2/R = 3G$, and to node C is $1/R + 2/R + 1/2R = 3.5G$. The conductance connected between nodes A and B is $1/2R = 0.5G$, between nodes A and C is $2/R = 2G$, and between nodes B and C is $1/2R = 0.5G$. The independent current source is directed into node A. By inspection, the node-voltage equations are

$$\text{Node A:} \quad 2.5Gv_A - 0.5Gv_B - 2Gv_C = i_S$$
$$\text{Node B:} \ -0.5Gv_A + 3Gv_B - 0.5Gv_C = 0$$
$$\text{Node C:} \ -2Gv_A - 0.5Gv_B + 3.5Gv_C = 0$$

Written in matrix form,

$$\begin{bmatrix} 2.5G & -0.5G & -2G \\ -0.5G & 3G & -0.5G \\ -2G & -0.5G & 3.5G \end{bmatrix} \begin{bmatrix} v_A \\ v_B \\ v_C \end{bmatrix} = \begin{bmatrix} i_S \\ 0 \\ 0 \end{bmatrix}$$

This matrix equation is of the form $\mathbf{Ax = B}$, where $\mathbf{A}$ is a 3×3 square matrix describing the circuit, $\mathbf{x}$ is a 3×1 column matrix of unknown node voltages, and **B** is a 3×1 column matrix of known inputs. Note that the **A** matrix is symmetrical about its main diagonal.[1]   ■

Exercise 3–2

Formulate node-voltage equations for the circuit in Figure 3–8.

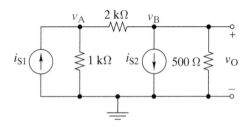

FIGURE 3 – 8

Answers:

$$(1.5 \times 10^{-3})v_A - (0.5 \times 10^{-3})v_B = i_{S1}$$

$$-(0.5 \times 10^{-3})v_A + (2.5 \times 10^{-3})v_B = -i_{S2}$$

SOLVING LINEAR ALGEBRAIC EQUATIONS

So far, we have only dealt with the problem of formulating node-voltage equations. To complete a circuit analysis problem, we must solve these linear equations for selected responses. Cramer's rule and Gaussian elimination are standard mathematical tools commonly used for hand solution of circuit equations. These tools are assumed to be part of the reader's mathematical background. Those needing a review of these matters are referred to Appendix B.

Cramer's rule and Gaussian elimination are suitable for hand calculations involving up to three or perhaps four simultaneous equations. Cramer's rule is useful when the circuit parameters are left in symbolic form, while the Gaussian method is more efficient for numerical examples involving four or more equations. However, any problem in which Gaussian methods offer significant advantages over Cramer's rule is probably best handled using computer tools. In other words, the ready availability of computer tools makes hand solution by Gaussian elimination an obsolete skill.

At about four or five simultaneous equations, numerical solutions are best obtained using computer-aided analysis. Many scientific hand-held calculators have a built-in capability to solve five or more linear equations. Virtually all PC-based mathematical software can solve systems of linear equations or, what is equivalent, perform matrix manipulations. In particu-

---

1 See Appendix B for a discussion of matrix algebra, including the definition of a symmetrical matrix.

lar, Appendix B illustrates that both MATLAB and Mathcad can solve systems of linear equations written in matrix form.

In this book we often use Cramer's rule to solve simultaneous equations. This does not mean Cramer's rule is the optimum method, but only that it can easily handle and compactly document the solution of the class of problems treated in this book. We will also use MATLAB and Mathcad to demonstrate the use of computer tools to solve circuit equations in the matrix form. The student solutions manual supplement has additional examples of solving matrix equations using computer tools. Readers needing a review of or an introduction to matrix methods are advised to study Appendix B.

Earlier in this section we formulated node-voltage equations for the circuit in Figure 3–4 [See Eq. (3–4).]

$$\text{Node A:} \quad (G_1 + G_2)v_A - G_2 v_B = i_S$$

$$\text{Node B:} \quad -G_2 v_A + (G_2 + G_3)v_B = 0$$

We use Cramer's rule to solve these equations because it easily handles the case in which the circuit parameters are left in symbolic form:

$$v_A = \frac{\Delta_A}{\Delta} = \frac{\begin{vmatrix} i_S & -G_2 \\ 0 & G_2 + G_3 \end{vmatrix}}{\begin{vmatrix} G_1 + G_2 & -G_2 \\ -G_2 & G_2 + G_3 \end{vmatrix}} = \left(\frac{G_2 + G_3}{G_1 G_2 + G_1 G_3 + G_2 G_3} \right) i_S \tag{3-8}$$

$$v_B = \frac{\Delta_B}{\Delta} = \frac{\begin{vmatrix} G_1 + G_2 & i_S \\ -G_2 & 0 \end{vmatrix}}{\Delta} = \left(\frac{G_2}{G_1 G_2 + G_1 G_3 + G_2 G_3} \right) i_S \tag{3-9}$$

These results express the two node voltages in terms of the circuit parameters and the input signal. Given the two node voltages v_A and v_B, we can now determine every element voltage and every current using Ohm's law and the fundamental property of node voltages.

$$v_1 = v_A \qquad v_2 = v_A - v_B \qquad v_3 = v_B$$

$$i_1 = G_1 v_A \qquad i_2 = G_2(v_A - v_B) \qquad i_3 = G_3 v_B$$

In solving the node equations, we left everything in symbolic form to emphasize that responses depend on the values of the circuit parameters (G_1, G_2, G_3) and the input signal (i_S). Even when numerical values are given, it is sometimes useful to leave some parameters in symbolic form to obtain input-output relationships or to reveal the effect of specific parameters on circuit response.

EXAMPLE 3–3

Given the circuit in Figure 3–9, find the input resistance R_{IN} seen by the current source and the output voltage v_O.

FIGURE 3-9

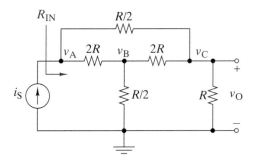

SOLUTION:
In Example 3–2 we formulated node-voltage equations for this circuit as
follows:

$$\text{Node A:} \quad 2.5Gv_\text{A} - 0.5Gv_\text{B} - 2Gv_\text{C} = i_\text{S}$$

$$\text{Node B:} \quad -0.5Gv_\text{A} + 3Gv_\text{B} - 0.5Gv_\text{C} = 0$$

$$\text{Node C:} \quad -2Gv_\text{A} - 0.5Gv_\text{B} + 3.5Gv_\text{C} = 0$$

The input resistance is the ratio v_A/i_S, so we first solve for v_A:

$$v_\text{A} = \frac{\Delta_\text{A}}{\Delta} = \frac{\begin{vmatrix} i_\text{S} & -0.5G & -2G \\ 0 & 3G & -0.5G \\ 0 & -0.5G & 3.5G \end{vmatrix}}{\begin{vmatrix} 2.5G & -0.5G & -2G \\ -0.5G & 3G & -0.5G \\ -2G & -0.5G & 3.5G \end{vmatrix}}$$

$$= \frac{i_\text{S} \times G^2 \begin{vmatrix} 3 & -0.5 \\ -0.5 & 3.5 \end{vmatrix}}{2.5G^3 \begin{vmatrix} 3 & -0.5 \\ -0.5 & 3.5 \end{vmatrix} + 0.5G^3 \begin{vmatrix} -0.5 & -2 \\ -0.5 & 3.5 \end{vmatrix} - 2G^3 \begin{vmatrix} -0.5 & -2 \\ 3 & -0.5 \end{vmatrix}}$$

$$= \frac{10.25i_\text{S}}{11.75G}$$

Hence the input resistance is

$$R_\text{IN} = \frac{v_\text{A}}{i_\text{S}} = \frac{10.25}{11.75G} = 0.872R$$

To find the output voltage, we solve for v_C:

$$v_\text{C} = \frac{\Delta_\text{C}}{\Delta} = \frac{\begin{vmatrix} 2.5G & -0.5G & i_\text{S} \\ -0.5G & 3G & 0 \\ -2G & -0.5G & 0 \end{vmatrix}}{\Delta} = \frac{i_\text{S} \times G^2 \begin{vmatrix} -0.5 & 3 \\ -2 & -0.5 \end{vmatrix}}{\Delta}$$

$$= \frac{6.25G^2 i_\text{S}}{11.75G^3} = 0.532i_\text{S}R$$ ∎

Exercise 3–3

Solve for the node-voltage equations in Exercise 3–2 for v_O in Figure 3–8.

Answer:

$$v_O = 1000(i_{S1} - 3i_{S2})/7$$

Exercise 3–4

Use node-voltage equations to solve for v_1, v_2, and i_3 in Figure 3–10.

F I G U R E 3 – 1 0

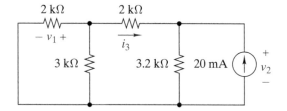

Answers:

$$v_1 = 12 \text{ V}, v_2 = 32 \text{ V}, \text{ and } i_3 = -10 \text{ mA}$$

NODE ANALYSIS WITH VOLTAGE SOURCES

Up to this point we have analyzed circuits containing only resistors and independent current sources. Applying KCL in such circuits is simplified because the sum of currents at a node only involves the output of current sources or resistor currents expressed in terms of the node voltages. Adding voltage sources to circuits modifies node analysis procedures because the current through a voltage source is not directly related to the voltage across it. While initially it may appear that voltage sources complicate the situation, they actually simplify node analysis by reducing the number of equations required.

Figure 3–11 shows three ways to deal with voltage sources in node analysis. Method 1 uses a source transformation to replace the voltage source and series resistance with an equivalent current source and parallel resistance. We can then formulate node equations at the remaining nonreference nodes in the usual way. The source transformation eliminates node C, so there are only $N - 2$ nonreference nodes left in the circuit. Obviously, method 1 only applies when there is a resistance in series with the voltage source.

Method 2 in Figure 3–11 can be used whether or not there is a resistance in series with the voltage source. When node B is selected as the reference node, then by definition $v_B = 0$ and the fundamental property of node voltages says that $v_A = v_S$. We do not need a node-voltage equation at node A because its voltage is known to be equal to the source voltage. We write the node equations at the remaining $N - 2$ nonreference nodes in the usual way. In the final step, we move all terms involving v_A to the right side, since it is a known input and not an unknown response. Method 2 reduces the number of node equations by 1 since no equation is needed at node A.

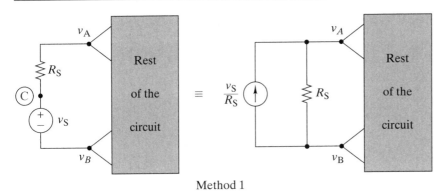

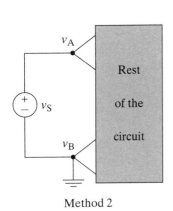

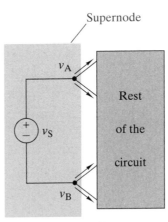

FIGURE 3–11 *Three methods of treating voltage sources in node analysis.*

The third method in Figure 3–11 is needed when neither node A nor node B can be selected as the reference and the source is not connected in series with a resistance. In this case we combine nodes A and B into a **supernode**, indicated by the boundary in Figure 3–11. We use the fact that KCL applies to the currents penetrating this boundary to write a node equation at the supernode. We then write node equations at the remaining $N-3$ nonreference nodes in the usual way. We now have $N-3$ node equations plus one supernode equation, leaving us one equation short of the $N-1$ required. Using the fundamental property of node voltages, we can write

$$v_A - v_B = v_S \qquad (3\text{–}10)$$

The voltage source inside the supernode constrains the difference between the node voltages at nodes A and B. The voltage source constraint provides the additional relationship needed to write $N-1$ independent equations in $N-1$ node voltages.

For reference purposes we will call these modified node equations, since we either modify the circuit (method 1), use voltage source constraints to define node voltage at some nodes (method 2), or combine nodes to produce a supernode (method 3). The three methods are not mutually exclusive. We frequently use a combination of methods, as illustrated in the following examples.

EXAMPLE 3-4

Use node-voltage analysis to find v_O in the circuit in Figure 3–12(a).

FIGURE 3 - 1 2

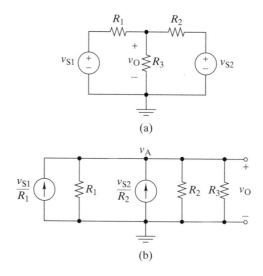

(a)

(b)

SOLUTION:

The given circuit in Figure 3–12(a) has four nodes, so we appear to need $N - 1 = 3$ node-voltage equations. However, applying source transformations to the two voltage sources (method 1) produces the two-node circuit in Figure 3–12(b). For the modified circuit, we need only one node equation.

$$(G_1 + G_2 + G_3)v_A = G_1 v_{S1} + G_2 v_{S2}$$

To find the output voltage, we solve for v_A:

$$v_O = v_A = \frac{G_1 v_{S1} + G_2 v_{S2}}{G_1 + G_2 + G_3}$$

Because of the two voltage sources, we need only one node equation in what appears to be a three-node circuit. The two voltage sources have a common node, so the number of unknown node voltages is reduced from three to one. The general principle illustrated is that the number of independent KCL constraints in a circuit containing N nodes and N_V voltage sources is $N - 1 - N_V$. ■

EXAMPLE 3-5

Find the input resistance of the circuit in Figure 3–13.

FIGURE 3 - 1 3

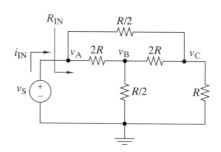

SOLUTION:

Method 1 of handling voltage sources will not work here because the source in Figure 3–13 is not connected in series with a resistor. Method 2 will work in this case because the voltage source is connected to the reference node. As a result, we only need node equations at nodes B and C since the node A voltage is $v_A = v_S$. By inspection, the two required node equations are

$$\text{Node B:} \quad -0.5Gv_A + 3Gv_B - 0.5Gv_C = 0$$

$$\text{Node C:} \quad -2Gv_A - 0.5Gv_B + 3.5Gv_C = 0$$

Since $v_A = v_S$, these equations can be written in standard form as follows:

$$\text{Node B:} \quad 3Gv_B - 0.5Gv_C = 0.5Gv_S$$

$$\text{Node C:} \quad -0.5Gv_B + 3.5Gv_C = 2Gv_S$$

Solving for the two unknown node voltages yields

$$v_B = \frac{\Delta_B}{\Delta} = \frac{\begin{vmatrix} 0.5Gv_S & -0.5G \\ 2Gv_S & 3.5G \end{vmatrix}}{\begin{vmatrix} 3G & -0.5G \\ -0.5G & 3.5G \end{vmatrix}} = \frac{2.75G^2v_S}{10.25G^2} = \frac{2.75v_S}{10.25}$$

$$v_C = \frac{\Delta_C}{\Delta} = \frac{\begin{vmatrix} 3G & 0.5Gv_S \\ -0.5G & 2Gv_S \end{vmatrix}}{\Delta} = \frac{6.25G^2v_S}{10.25G^2} = \frac{6.25v_S}{10.25}$$

Given the two node voltages, we can now solve for the input current.

$$i_{IN} = \frac{v_S - v_B}{2R} + \frac{v_S - v_C}{R/2} = \frac{3.75v_S}{10.25R} + \frac{8v_S}{10.25R} = \frac{11.75v_S}{10.25R}$$

Hence, the input resistance is

$$R_{IN} = \frac{v_S}{i_{IN}} = \frac{10.25R}{11.75} = 0.872R$$

This is the same answer as in Example 3–3, where the same circuit was driven by a current source rather than a voltage source. Input resistance is an intrinsic property of a circuit that does not depend on how the circuit is driven. ■

EXAMPLE 3–6

For the circuit in Figure 3–14,

(a) Formulate node-voltage equations.
(b) Solve for the output voltage v_O using $R_1 = R_4 = 2$ kΩ and $R_2 = R_3 = 4$ kΩ.

SOLUTION:

(a) The voltage sources in Figure 3–14 do not have a common node, and we cannot select a reference node that includes both sources. Selecting node D as the reference forces the condition $v_B = v_{S2}$ (method 2) but leaves the other source v_{S1} ungrounded. We surround the ungrounded source,

FIGURE 3–14

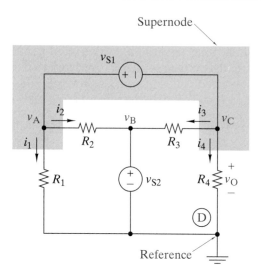

and all wires leading to it, by the supernode boundary shown in Figure 3–14 (method 3). KCL applies to the four element currents that penetrate the supernode boundary, and we can write

$$i_1 + i_2 + i_3 + i_4 = 0$$

These currents can easily be expressed in terms of the node voltages.

$$G_1 v_A + G_2(v_A - v_B) + G_3(v_C - v_B) + G_4 v_C = 0$$

Since $v_B = v_{S2}$, the standard form of this equation is

$$(G_1 + G_2)v_A + (G_3 + G_4)v_C = (G_2 + G_3)v_{S2}$$

We have one equation in the two unknown node voltages v_A and v_C. Applying the fundamental property of node voltages inside the supernode, we can write

$$v_A - v_C = v_{S1}$$

That is, the ungrounded voltage source constrains the difference between the two unknown node voltages inside the supernode. It thereby supplies the relationship needed to obtain two equations in two unknowns.

(b) Inserting the given numerical values yields

$$(7.5 \times 10^{-4})v_A + (7.5 \times 10^{-4})v_C = (5 \times 10^{-4})v_{S2}$$

$$v_A - v_C = v_{S1}$$

To find the output v_O, we need to solve these equations for v_C. The second equation yields $v_A = v_C + v_{S1}$, which, when substituted into the first equation, yields the required output:

$$v_O = v_C = \frac{v_{S2}}{3} - \frac{v_{S1}}{2} \qquad \blacksquare$$

> **Exercise 3–5**
>
> Find v_O in Figure 3–15 when the element E is
>
> (a) A 10-kΩ resistance,
> (b) A 4-mA independent current source with reference arrow pointing left.
>
> **Answers:**
>
> (a) 2.53 V
> (b) −17.3 V

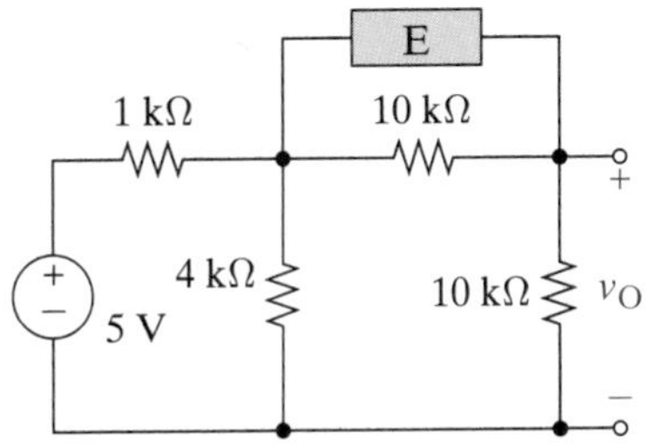

FIGURE 3 − 1 5

> **Exercise 3–6**
>
> Find v_O in Figure 3–15 when the element E is
>
> (a) An open circuit
> (b) A 10-V independent voltage source with the plus reference on the right.
>
> **Answers:**
>
> (a) 1.92 V
> (b) 12.96 V

SUMMARY OF NODE-VOLTAGE ANALYSIS

We have seen that node-voltage equations are very useful in the analysis of a variety of circuits. These equations can always be formulated using KCL, the element constraints, and the fundamental property of node voltages. When in doubt, always fall back on these principles to formulate node equations in new situations. With practice and experience, however, we eventually develop an analysis approach that allows us to recognize short-cuts in the formulation process. The following guidelines summarize our approach and may help you develop your own analysis style:

1. Simplify the circuit by combining elements in series and parallel wherever possible.

2. If not specified, select a reference node so that as many voltage sources as possible are directly connected to the reference.

3. Node equations are required at supernodes and all other nonref-erence nodes except those that are directly connected to the reference by voltage sources.

4. Use KCL to write node equations at the nodes identified in step 3. Express element currents in terms of node voltages or the cur-rents produced by independent current sources.

5. Write expressions relating the node voltages to the voltages pro-duced by independent voltage sources.

6. Substitute the expressions from step 5 into the node equations from step 4 and arrange the resulting equations in standard form.

7. Solve the equations from step 6 for the node voltages of interest. Cramer's rule is often useful when circuit parameters are left in symbolic form. Computer tools are useful when there are four or more equations and numerical values are given.

3–2   M E S H - C U R R E N T   A N A L Y S I S

Mesh currents are analysis variables that are useful in circuits containing many elements connected in series. To review terminology, a loop is a closed path formed by passing through an ordered sequence of nodes without passing through any node more than once. A mesh is a special type of loop that does not enclose any elements. For example, loops A and B in Figure 3–16 are meshes, while the loop X is not a mesh because it encloses an element.

Mesh-current analysis is restricted to planar circuits. A **planar circuit** can be drawn on a flat surface without crossovers in the "window pane" fashion shown in Figure 3–16. To define a set of variables, we associate a **mesh current** (i_A, i_B, i_C, etc.) with each window pane and assign a reference direction. The reference directions for all mesh currents are customarily taken in a clockwise sense. There is no momentous reason for this, except perhaps tradition.

We think of these mesh currents as circulating through the elements in their respective meshes, as suggested by the reference directions shown in Figure 3–16. We should emphasize that this viewpoint is not based on the physics of circuit behavior. There are not red and blue electrons running around that somehow get assigned to mesh currents i_A or i_B. Mesh currents are variables used in circuit analysis. They are only somewhat abstractly related to the physical operation of a circuit and may be impossible to measure directly. For example, there is no way to cut the circuit in Figure 3–16 to insert an ammeter that only measures i_E.

Mesh currents have a unique feature that is the dual of the fundamental property of node voltages. If we examine Figure 3–16, we see the elements around the perimeter are contained in only one mesh, while those in the interior are in two meshes. In a planar circuit any given element is contained in at most two meshes. When an element is in two meshes, the two mesh currents circulate through the element in opposite directions. In such cases KCL declares that the net current through the element is the difference of the two mesh currents.

These observations lead us to the fundamental property of mesh currents:

F I G U R E   3 – 1 6   *Meshes in a planar circuit.*

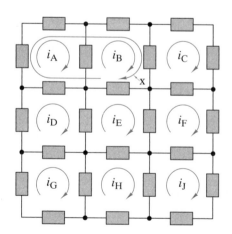

If the Kth two-terminal element is contained in meshes X and
Y, then the element current can be expressed in terms of the
two mesh currents as

$$i_K = i_X - i_Y \qquad (3–11)$$

where X is the mesh whose reference direction agrees with the
reference direction of i_K.

Equation (3–11) is a KCL constraint at the element level. If the element is
contained in only one mesh, then $i_K = i_X$ or $i_K = -i_Y$, depending on whether
the reference direction for the element current agrees or disagrees with the
reference direction of the mesh current. The key idea is that the current
through every two-terminal element in a planar circuit can be expressed in
terms of no more than two mesh currents.

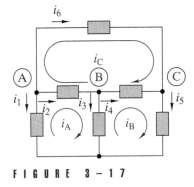

F I G U R E   3 – 1 7

Exercise 3–7

In Figure 3–17 the mesh currents are $i_A = 10$ A, $i_B = 5$ A, and $i_C = -3$ A. Find the
element currents i_1 through i_6 and show that KCL is satisfied at nodes A, B, and C.

Answers:

$i_1 = -10$ A, $i_2 = 13$ A, $i_3 = 5$ A, $i_4 = 8$ A, $i_5 = 5$ A, and $i_6 = -3$ A.

To use mesh currents to formulate circuit equations, we use elements
and connection constraints, except that the KCL constraints are not explicitly written. Instead, we use the fundamental property of mesh currents to
express the element voltages in terms of the mesh currents. By doing so we
avoid using the element currents and work only with the element voltages
and mesh currents.

For example, the planar circuit in Figure 3–18 can be analyzed using the
mesh-current method. In the figure we have defined two mesh currents

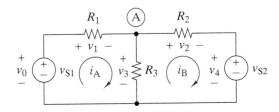

F I G U R E   3 – 1 8 *Circuit for demonstrat-*
ing mesh-current analysis.

and five element voltages. We write KVL constraints around each mesh
using the element voltages.

Mesh A: $-v_0 + v_1 + v_3 = 0$

Mesh B: $-v_3 + v_2 + v_4 = 0$ $\qquad (3–12)$

Using the fundamental property of mesh currents, we write the element
voltages in terms of the mesh currents and input voltages:

$$v_1 = R_1 i_A \qquad\qquad v_0 = v_{S1}$$
$$v_2 = R_2 i_B \qquad\qquad v_4 = v_{S2} \qquad (3–13)$$
$$v_3 = R_3(i_A - i_B)$$

We substitute these element equations into the KVL connection equations and arrange the result in standard form.

$$(R_1 + R_3)i_A - R_3 i_B = v_{S1}$$
$$-R_3 i_A + (R_2 + R_3)i_B = -v_{S2} \qquad (3-14)$$

We have completed the formulation process with two equations in two unknown mesh currents.

As we have previously noted, every method of circuit analysis must satisfy KCL, KVL, and the element $i-v$ relationships. When formulating mesh equations, it may appear that we have not used KCL. However, writing the element constraints in the form in Eq. (3–13) requires the KCL equations $i_1 = i_A$, $i_2 = i_B$, and $i_3 = i_A - i_B$. Mesh-current analysis implicitly satisfies KCL when the element constraints are expressed in terms of the mesh currents. In effect, the fundamental property of mesh currents ensures that the KCL constraints are satisfied.

We use Cramer's rule to solve for the mesh currents in Eq. (3–14):

$$i_A = \frac{\Delta_A}{\Delta} = \frac{\begin{vmatrix} v_{S1} & -R_3 \\ -v_{S2} & R_2 + R_3 \end{vmatrix}}{\begin{vmatrix} R_1 + R_3 & -R_3 \\ -R_3 & R_2 + R_3 \end{vmatrix}} = \frac{(R_2 + R_3)v_{S1} - R_3 v_{S2}}{R_1 R_2 + R_1 R_3 + R_2 R_3} \qquad (3-15)$$

and

$$i_B = \frac{\Delta_B}{\Delta} = \frac{\begin{vmatrix} R_1 + R_3 & v_{S1} \\ -R_3 & -v_{S2} \end{vmatrix}}{\begin{vmatrix} R_1 + R_3 & -R_3 \\ -R_3 & R_2 + R_3 \end{vmatrix}} = \frac{R_3 v_{S1} - (R_1 + R_3)v_{S2}}{R_1 R_2 + R_1 R_3 + R_2 R_3} \qquad (3-16)$$

Equations (3–15) and (3–16) can now be substituted into the element constraints in Eq. (3–13) to solve for every voltage in the circuit. For instance, the voltage across R_3 is

$$v_3 = R_3(i_A - i_B) = \frac{R_2 R_3 v_{S1} + R_1 R_3 v_{S2}}{R_1 R_2 + R_1 R_3 + R_2 R_3} \qquad (3-17)$$

You are invited to show that the result in Eq. (3–17) agrees with the node analysis result obtained in Example 3–4 for the same circuit.

The mesh-current analysis approach just illustrated can be summarized in four steps:

S T E P   1: Identify a mesh current with every mesh and a voltage across every circuit element.

S T E P   2: Write KVL connection constraints in terms of the element voltages around every mesh.

S T E P   3: Use KCL and the $i-v$ relationships of the elements to express the element voltages in terms of the mesh currents.

S T E P   4: Substitute the element constraints from step 3 into the connection constraints from step 2 and arrange the resulting equations in standard form.

The number of mesh-current equations derived in this way equals the number of KVL connection constraints in step 2. When discussing combined constraints in Chapter 2, we noted that there are $E - N + 1$ independent KVL constraints in any circuit. Using the window panes in a planar circuit generates $E - N + 1$ independent mesh currents. Mesh analysis works best when the circuit has many elements (E large) connected in series (N also large).

WRITING MESH-CURRENT EQUATIONS BY INSPECTION

The mesh equations in Eq. (3–14) have a symmetrical pattern that is similar to the coefficient symmetry observed in node equations. The coefficients of i_B in the first equation and i_A in the second equation are the negative of the resistance common to meshes A and B. The coefficients of i_A in the first equation and i_B in the second equation are the sum of the resistances in meshes A and B, respectively.

This pattern will always occur in planar circuits containing resistors and independent voltage sources when the mesh currents are defined in the window panes of a planar circuit, as shown in Figure 3–16. To see why, consider a general resistance R that is contained in, say, mesh A. There are only two possibilities. Either R is not contained in any other mesh, in which case the voltage across it is

$$v = R(i_A - 0) = Ri_A$$

or else it is also contained in only one adjacent mesh, say mesh B, in which case the voltage across it is

$$v = R(i_A - i_B)$$

These observations lead to the following conclusions. The voltage across resistance in mesh A involves the following terms:

1. i_A times the sum of the resistances in mesh A

2. $-i_B$ times the sum of resistances common to mesh A and mesh B, and similar terms for any other mesh adjacent to mesh A.

The sum of the voltages across resistors plus the sum of the independent voltage sources around mesh A must equal zero.

The aforementioned process makes it possible for us to write mesh-current equations by inspection without going through the intermediate steps involving the KVL connection constraints and the element constraints.

EXAMPLE 3-7

For the circuit of Figure 3–19,

(a) Formulate mesh-current equations.
(b) Find the output v_O using $R_1 = R_4 = 2$ kΩ and $R_2 = R_3 = 4$ kΩ.

FIGURE 3-19

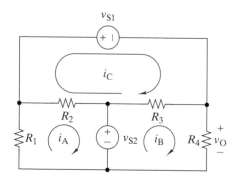

SOLUTION:

(a) To write mesh-current equations by inspection, we note that the total resistances in meshes A, B, and C are $R_1 + R_2$, $R_3 + R_4$, and $R_2 + R_3$, respectively. The resistance common to meshes A and C is R_2. The resistance common to meshes B and C is R_3. There is no resistance common to meshes A and B. Using these observations, we write the mesh equations as

$$\text{Mesh A: } (R_1 + R_2)i_A - 0\, i_B - R_2 i_C + v_{S2} = 0$$

$$\text{Mesh B: } (R_3 + R_4)i_B - 0\, i_A - R_3 i_C - v_{S2} = 0$$

$$\text{Mesh C: } (R_2 + R_3)i_C - R_2 i_A - R_3 i_B + v_{S1} = 0$$

The algebraic signs assigned to voltage source terms follow the passive convention for the mesh current in question. Arranged in standard form, these equations become

$$(R_1 + R_2)i_A - R_2 i_C = -v_{S2}$$

$$+ (R_3 + R_4)i_B - R_3 i_C = +v_{S2}$$

$$- R_2 i_A - R_3 i_B + (R_2 + R_3)i_C = -v_{S1}$$

Coefficient symmetry greatly simplifies the formulation of these equations compared with the more fundamental, but time-consuming, process of writing element and connection constraints. (b) Inserting the numerical values into these equations yields

$$6000\, i_A \qquad\qquad - 4000\, i_C = -v_{S2}$$

$$6000\, i_B - 4000\, i_C = v_{S2}$$

$$-4000\, i_A - 4000\, i_B + 8000\, i_C = -v_{S1}$$

Placing these three mesh equations in matrix form produces

$$\begin{bmatrix} 6000 & 0 & -4000 \\ 0 & 6000 & -4000 \\ -4000 & -4000 & +8000 \end{bmatrix} \begin{bmatrix} i_A \\ i_B \\ i_C \end{bmatrix} = \begin{bmatrix} -v_{S2} \\ v_{S2} \\ -v_{S1} \end{bmatrix}$$

This is a matrix equation of the form $\mathbf{AX} = \mathbf{B}$, where

$$\mathbf{A} = \begin{bmatrix} 6000 & 0 & -4000 \\ 0 & 6000 & -4000 \\ -4000 & -4000 & +8000 \end{bmatrix} \quad \mathbf{X} = \begin{bmatrix} i_A \\ i_B \\ i_C \end{bmatrix} \quad \mathbf{B} = \begin{bmatrix} -v_{S2} \\ v_{S2} \\ -v_{S1} \end{bmatrix}$$

The matrix **A** is a square matrix of the coefficients on the left side of the mesh equations, **X** is a column matrix of the unknown mesh currents, and **B** is a column matrix of the input voltages on the right side of the mesh equations. To solve this matrix equation we multiply by the inverse of the coefficient matrix ($\mathbf{A}^{-1}$). On the left side the result is $\mathbf{A}^{-1}\mathbf{AX} = \mathbf{X}$, and the right side becomes $\mathbf{A}^{-1}\mathbf{B}$. In other words, multiplying by $\mathbf{A}^{-1}$ yields the solution for the unknown mesh currents as

$$\mathbf{X} = \mathbf{A}^{-1}\mathbf{B}$$

In sum, to solve a system of linear equations by matrix methods we form the matrix product $\mathbf{A}^{-1}\mathbf{B}$. It is at this point that MATLAB comes into play.

Using MATLAB to solve for the mesh currents, we first enter the **A** matrix by the statement

```
A=[6000 0 -4000;0 6000 -4000;-4000 -4000 8000];
```

The elements in **B** matrix are the symbolic variables v_{S1} and v_{S2}. These quantities are not unknowns, but symbols that represent all possible values of the input voltages. Thus, we must first write the statement

```
syms VS1 VS2
```

which declares these identifiers to be symbolic rather than numerical quantities. We can now enter the **B** matrix as

```
B=[-VS2;VS2;-VS1];
```

and solve for the unknown mesh currents using MATLAB statement

```
X=inv(A)*B
```

which yields

```
X =

[-1/6000*VS2-1/4000*VS1]
[1/6000*VS2-1/4000*VS1]
[          -3/8000*VS1]
```

The elements of the column matrix **X** are the three unknown mesh currents expressed in terms of the input voltages. The output voltage in Figure 3–19 is written in terms of the mesh currents as $v_O = R_4 i_B$. In MATLAB notation the required mesh current is

```
iB = X(2)=1/6000*VS2-1/4000*VS1
```

Since $R_4 = 2000$ we conclude that

$$v_O = \frac{v_{S2}}{3} - \frac{v_{S1}}{2}$$

The mesh current analysis result obtained here is the same as the node voltage result obtained in Example 3–6. Either approach produces the same answer, but which method do you think is easier?   ■

MESH EQUATIONS WITH CURRENT SOURCES

In developing mesh analysis, we assumed that circuits contain only voltage sources and resistors. This assumption simplifies the formulation process because the sum of voltages around a mesh is determined by voltage sources and the mesh currents through resistors. A current source complicates the picture because the voltage across it is not directly related to its current. We need to adapt mesh analysis to accommodate current sources just as we revised node analysis to deal with voltage sources.

There are three ways to handle current sources in mesh analysis:

1. If the current source is connected in parallel with a resistor, then it can be converted to an equivalent voltage source by source transformation. Each source conversion eliminates a mesh and reduces the number of equations required by 1. This method is the dual of method 1 for node analysis.

2. If a current source is contained in only one mesh, then that mesh current is determined by the source current and is no longer an unknown. We write mesh equations around the remaining meshes in the usual way and move the known mesh current to the source side of the equations in the final step. The number of equations obtained is one less than the number of meshes. This method is the dual of method 2 for node analysis.

3. Neither of the first two methods will work when a current source is contained in two meshes or is not connected in parallel with a resistance. In this case we create a **supermesh** by excluding the current source and any elements connected in series with it, as shown in Figure 3–20. We write one mesh equation around the

F I G U R E   3 – 2 0   *Example of a supermesh.*

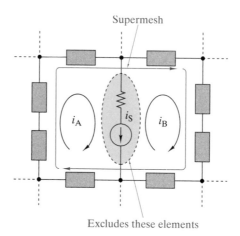

supermesh using the currents i_A and i_B. We then write mesh equations of the remaining meshes in the usual way. This leaves us one equation short because parts of meshes A and B are included in the supermesh. However, the fundamental property of mesh currents relates the currents i_S, i_A, and i_B as

$$i_A - i_B = i_S$$

This equation supplies the one additional relationship needed to get the requisite number of equations in the unknown mesh currents.

The aforementioned three methods are not mutually exclusive. We can use more than one method in a circuit, as the following examples illustrate.

EXAMPLE 3–8

Use mesh-current equations to find i_O in the circuit in Figure 3–21(a).

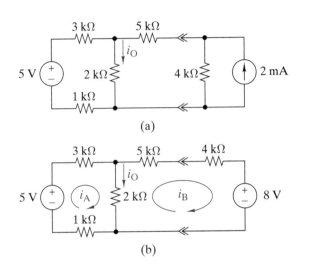

FIGURE 3–21

(a)

(b)

SOLUTION:

The current source in this circuit can be handled by a source transformation (method 1). The 2-mA source in parallel with the 4-kΩ resistor in Figure 3–21(a) can be replaced by an equivalent 8-V voltage source in series with the same resistor, as shown in Figure 3–21(b). In this circuit the total resistance in mesh A is 6 kΩ, the total resistance in mesh B is 11 kΩ, and the resistance contained in both meshes is 2 kΩ. By inspection, the mesh equations for this circuit are

$$(6000)i_A - (2000)i_B = 5$$

$$-(2000)i_A + (11000)i_B = -8$$

Solving for the two mesh currents yields $i_A = 0.6290$ mA and $i_B = -0.6129$ mA. By KCL the desired current is $i_O = i_A - i_B = 1.2419$ mA. The given circuit in Figure 3–21(a) has three meshes and one current source. The source transformation leading to Figure 3–21(b) produces a circuit with only two meshes. The general principle illustrated is that the number of independent mesh equations in a circuit containing E elements, N nodes, and N_I current sources is $E - N + 1 - N_I$.   ∎

EXAMPLE 3–9

Use mesh-current equations to find the v_O in Figure 3–22.

SOLUTION:

Source transformation (method 1) is not possible here since neither current source is connected in parallel with a resistor. The current source i_{S2} is in both mesh B and mesh C, so we exclude this element and create the

FIGURE 3 - 2 2

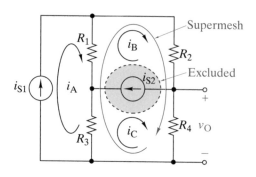

supermesh (method 3) shown in the figure. The sum of voltages around the supermesh is

$$R_1(i_B - i_A) + R_2(i_B) + R_4(i_C) + R_3(i_C - i_A) = 0$$

The supermesh voltage constraint yields one equation in the three unknown mesh currents. Applying KCL to each of the current sources yields

$$i_A = i_{S1}$$

$$i_B - i_C = i_{S2}$$

Because of KCL the two current sources force constraints that supply two more equations. Using these two KCL constraints to eliminate i_A and i_B from the supermesh KVL constraint yields

$$(R_1 + R_2 + R_3 + R_4)i_C = (R_1 + R_3)i_{S1} - (R_1 + R_2)i_{S2}$$

Hence, the required output voltage is

$$v_O = R_4 i_C = R_4 \times \left[\frac{(R_1 + R_3)i_{S1} - (R_1 + R_2)i_{S2}}{R_1 + R_2 + R_3 + R_4} \right] \qquad \blacksquare$$

Exercise 3–8

Use mesh analysis to find the current i_O in Figure 3–23 when the element E is

(a) A 5-V voltage source with the positive reference at the top
(b) A 10-kΩ resistor.

Answers:

(a) –0.136 mA
(b) –0.538 mA

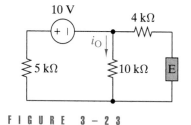

FIGURE 3 - 2 3

Exercise 3–9

Use mesh analysis to find the current i_O in Figure 3–23 when the element E is

(a) A 1-mA current source with the reference arrow directed down
(b) Two 20-kΩ resistors in parallel.

Answers:

(a) –1 mA
(b) –0.538 mA

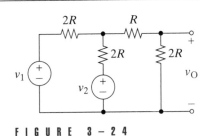

FIGURE   3 – 2 4

Exercise 3–10

Use mesh-current equations to find v_O in Figure 3–24.

Answer:

$$v_O = (v_1 + v_2)/4$$

SUMMARY OF MESH-CURRENT ANALYSIS

Mesh-current equations can always be formulated from KVL, the element constraints, and the fundamental property of mesh currents. When in doubt, always fall back on these principles to formulate mesh equations in new situations. The following guidelines summarize an approach to formulating mesh equations for resistance circuits:

1. Simplify the circuit by combining elements in series or parallel wherever possible.

2. Mesh equations are required for supermeshes and all other meshes except those where current sources are contained in only one mesh.

3. Use KVL to write mesh equations for the meshes identified in step 2. Express element voltages in terms of mesh currents or the voltage produced by independent voltage sources.

4. Write expressions relating the mesh currents to the currents produced by independent current sources.

5. Substitute the expressions from step 4 into the mesh equations from step 3 and place the result in standard form.

6. Solve the equations from step 5 for the mesh currents of interest. Cramer's rule is often useful when circuit parameters are left in symbolic form. Computer tools are useful when there are four or more equations and numerical values are given.

3–3 LINEARITY PROPERTIES

This book treats the analysis and design of **linear circuits**. A circuit is said to be linear if it can be adequately modeled using only linear elements and independent sources. The hallmark feature of a linear circuit is that outputs are linear functions of the inputs. Circuit **inputs** are the signals produced by external sources, and **outputs** are any other designated signals. Mathematically, a function is said to be linear if it possesses two properties—homogeneity and additivity. In linear circuits, **homogeneity** means that the output is proportional to the input. **Additivity** means that the output due to two or more inputs can be found by adding the outputs obtained when each input is applied separately. Mathematically, these properties are written as follows:

$$f(Kx) = Kf(x) \text{ (homogeneity)} \qquad (3\text{–}18)$$

and

$$f(x_1 + x_2) = f(x_1) + f(x_2) \text{ (additivity)} \qquad (3\text{–}19)$$

where K is a scalar constant. In circuit analysis the homogeneity property is called **proportionality**, while the additivity property is called **superposition**.

THE PROPORTIONALITY PROPERTY

The **proportionality property** applies to linear circuits with one input. For linear resistive circuits, proportionality states that every input-output relationship can be written as

$$y = Kx \qquad (3\text{--}20)$$

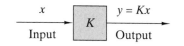

x — Input — K — $y = Kx$ — Output

FIGURE 3–25 *Block diagram representation of the proportionality property.*

where x is the input current or voltage, y is an output current or voltage, and K is a constant. The block diagram in Figure 3–25 describes this linear input–output relationship. In a block diagram the lines headed by arrows indicate the direction of signal flow. The arrow directed into the block indicates the input, while the output is indicated by the arrow directed out of the block. The variable names written next to these lines identify the input and output signals. The scalar constant K written inside the block indicates that the input signal x is multiplied by K to produce the output signal as $y = Kx$.

Caution: Proportionality only applies when the input and output are current or voltage. It does not apply to output power since power is equal to the product of current and voltage. In other words, output power is not linearly related to the input current or voltage.

We have already seen several examples of proportionality. For instance, using voltage division in Figure 3–26(a) produces

$$v_O = \left(\frac{R_2}{R_1 + R_2}\right)v_S$$

which means

$$x = v_S \qquad y = v_O$$

$$K = \frac{R_2}{R_1 + R_2}$$

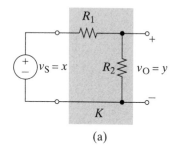

(a)

Similarly, applying current division in Figure 3–26(b) yields

$$i_O = \left(\frac{G_2}{G_1 + G_2}\right)i_S$$

so that

$$x = i_S \qquad y = i_O$$

$$K = \frac{G_2}{G_1 + G_2}$$

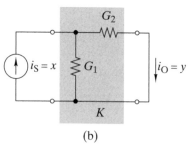

(b)

FIGURE 3–26 *Examples of circuit exhibiting of proportionality: (a) Voltage divider. (b) Current divider.*

In these two examples the proportionality constant K is dimensionless because the input and output have the same units. In other situations K could have the units of ohms or siemens when the input and output have different units.

The next example illustrates that the proportionality constant K can be positive, negative, or even zero.

EXAMPLE 3-10

Given the bridge circuit of Figure 3–27,

(a) Find the proportionality constant K in the input-output relationship
$v_O = Kv_S$
(b) Find the sign of K when $R_2R_3 > R_1R_4$, $R_2R_3 = R_1R_4$, and $R_2R_3 < R_1R_4$.

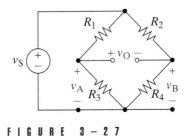

FIGURE 3-27

SOLUTION:
(a) We observe that the circuit consists of two voltage dividers. Applying the voltage division rule to each side of the bridge circuit yields

$$v_A = \frac{R_3}{R_1 + R_3}v_S \quad \text{and} \quad v_B = \frac{R_4}{R_2 + R_4}v_S$$

The fundamental property of node voltages allows us to write

$$v_O = v_A - v_B$$

Substituting the equations for v_A and v_B into this KVL equation yields

$$v_O = \left(\frac{R_3}{R_1 + R_3} - \frac{R_4}{R_2 + R_4}\right)v_S$$

$$= \left(\frac{R_2R_3 - R_1R_4}{(R_1 + R_3)(R_2 + R_4)}\right)v_S$$

$$= \qquad (K) \qquad v_S$$

(b) The proportionality constant K can be positive, negative, or zero. Specifically,

$$\text{If } R_2R_3 > R_1R_4 \text{ then } K > 0$$

$$\text{If } R_2R_3 = R_1R_4 \text{ then } K = 0$$

$$\text{If } R_2R_3 < R_1R_4 \text{ then } K < 0$$

When the products of the resistances in opposite legs of the bridge are equal, then $K = 0$ and the bridge is said to be balanced. ∎

UNIT OUTPUT METHOD

The **unit output method** is an analysis technique based on the proportionality property of linear circuits. The method involves finding the input-output proportionality constant K by assuming an output of one unit and determining the input required to produce that unit output. This technique is most useful when applied to ladder circuits, and it involves the following steps:

1. A unit output is assumed; that is, $v_O = 1$ V or $i_O = 1$ A.

2. The input required to produce the unit output is then found by successive application of KCL, KVL, and Ohm's law.

3. Because the circuit is linear, the proportionality constant relating input and output is

$$K = \frac{\text{Output}}{\text{Input}} = \frac{1}{\text{Input for unit output}}$$

Given the proportionality constant K, we can find the output for any input using Eq. (3–20).

In a way, the unit output method solves the circuit response problem backwards—that is, from output to input—as illustrated by the next example.

EXAMPLE 3–11

Use the unit output method to find v_O in the circuit shown in Figure 3–28(a).

FIGURE 3 – 28

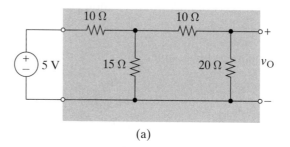

(a)

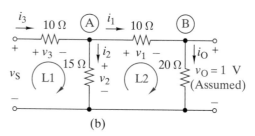

(b)

SOLUTION:
We start by assuming $v_O = 1$, as shown in Figure 3–28(b). Then, using Ohm's law, we find i_O.

$$i_O = \frac{v_O}{20} = 0.05 \text{ A}$$

Next, using KCL at node B, we find i_1.

$$i_1 = i_O = 0.05 \text{ A}$$

Again, using Ohm's law, we find v_1.

$$v_1 = 10i_1 = 0.5 \text{ V}$$

Then, writing a KVL equation around loop L2, we find v_2 as

$$v_2 = v_1 + v_O = 0.5 + 1.0 = 1.5 \text{ V}$$

Using Ohm's law once more produces

$$i_2 = \frac{v_2}{15} = \frac{1.5}{15} = 0.1 \text{ A}$$

Next, writing a KCL equation at node A yields

$$i_3 = i_1 + i_2 = 0.05 + 0.1 = 0.15 \text{ A}$$

Using Ohm's law one last time,

$$v_3 = 10i_3 = 1.5 \text{ V}$$

We can now find the source voltage v_S by applying KVL around loop L1:

$$v_S|_{\text{for } v_O = 1V} = v_3 + v_2 = 1.5 + 1.5 = 3 \text{ V}$$

A 3-V source voltage is required to produce a 1-V output. From this result, we calculate the proportionality constant K to be

$$K = \frac{v_O}{v_S} = \frac{1}{3}$$

Once K is known, the output for the specified 5-V input is $v_O = (\frac{1}{3})5 = 1.667$ V. ∎

Exercise 3–11

Find v_O in the circuit of Figure 3–28(a) when v_S is –5 V, 10 mV, and 3 kV.

Answers:

$$v_O = -1.667 \text{ V}, 3.333 \text{ mV}, \text{ and } 1 \text{ kV}$$

Exercise 3–12

Use the unit output method to find $K = i_O/i_{\text{IN}}$ for the circuit in Figure 3–29. Then use the proportionality constant K to find i_O for the input current shown in the figure.

FIGURE 3 - 29

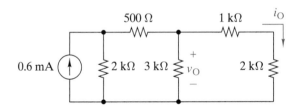

Answers:

$$K = \frac{1}{4}; i_O = 0.15 \text{ mA}$$

Exercise 3–13

Use the unit output method to find $K = v_O/i_{\text{IN}}$ for the circuit in Figure 3–29. Then use K to find v_O for the input current shown in the figure.

Answers:

$$K = 750 \text{ }\Omega; v_O = 450 \text{ mV}$$

Note: In this exercise K has the dimensions of ohms because the input is a current and the output a voltage.

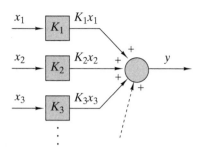

FIGURE 3-30 *Block diagram representation of the additivity property.*

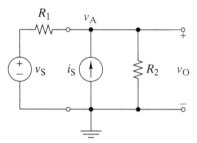

FIGURE 3-31 *Circuit used to demonstrate superposition.*

ADDITIVITY PROPERTY

The **additivity property** states that any output current or voltage of a linear resistive circuit with multiple inputs can be expressed as a linear combination of the several inputs:

$$y = K_1 x_1 + K_2 x_2 + K_3 x_3 + \dots \qquad (3-21)$$

where $x_1, x_2, x_3, \dots$ are current or voltage inputs, and $K_1, K_2, K_3 \dots$ are constants that depend on the circuit parameters. Figure 3–30 shows how we represent this relationship in block diagram form. Again the arrows indicate the direction of signal flow and the K's within the blocks are scalar multipliers. The circle in Figure 3–30 is a new block diagram element called a summing point that implements the operation $y = \sum K_i x_i$. Although the block diagram in Figure 3–30 is nothing more than a pictorial representation of Eq. (3–21), the diagram often helps us gain a clearer picture of how signals interact in different parts of a circuit.

To illustrate this property, we analyze the two-input circuit in Figure 3–31 using node-voltage analysis. Applying KCL at node A, we obtain

$$\frac{v_A - v_S}{R_1} - i_S + \frac{v_A}{R_2} = 0$$

Moving the inputs to the right side of this equation yields

$$\left[\frac{1}{R_1} + \frac{1}{R_2}\right] v_A = \frac{v_S}{R_1} + i_S$$

Since $v_O = v_A$, we obtain the input-output relationship in the form

$$v_O = \left[\frac{R_2}{R + R_2}\right] v_S + \left[\frac{R_1 R_2}{R_1 + R_2}\right] i_S$$

$$y = [K_1] x_1 + [K_2] x_2 \qquad (3-22)$$

This result shows that the output is a linear combination of the two inputs. Note that K_1 is dimensionless since its input and output are voltages, and that K_2 has the units of ohms since its input is a current and its output is a voltage.

SUPERPOSITION PRINCIPLE

Since the output in Eq. (3–21) is a linear combination, the contribution of each input source is independent of all other inputs. This means that the output can be found by finding the contribution from each source acting alone and then adding the individual responses to obtain the total response. This suggests that the output of a multiple-input linear circuit can be found by the following steps:

    STEP 1: "Turn off" all independent sources except one and find the output of the circuit due to that source acting alone.

    STEP 2: Repeat the process in step 1 until each independent source has been turned on and the output due to that source found.

    STEP 3: The total output with all independent sources turned on is the algebraic sum of the outputs caused by each source acting alone.

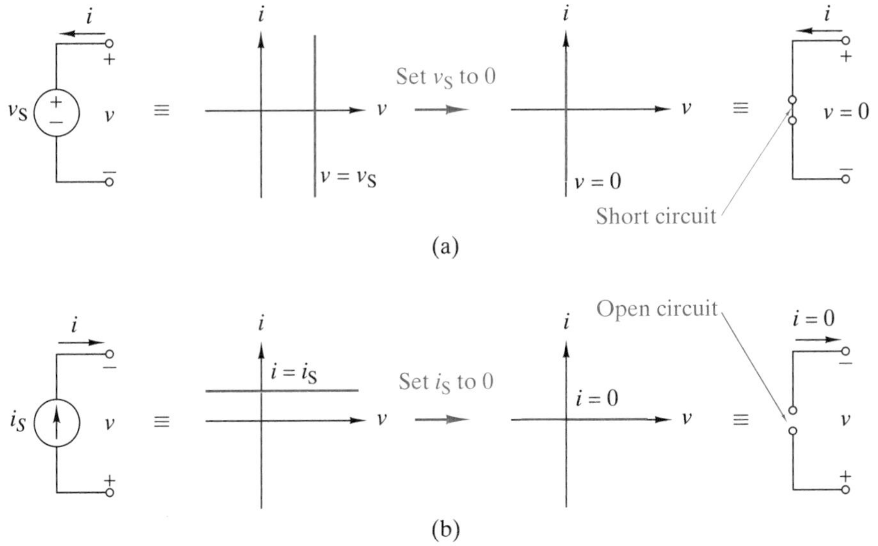

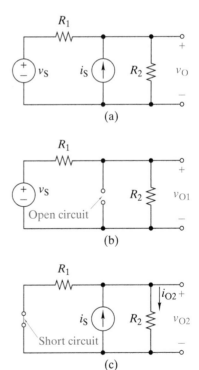

FIGURE 3 - 32 *Turning off an independent source: (a) Voltage source. (b) Current source.*

These steps describe a circuit analysis technique called the **superposition principle**. Before applying this method, we must discuss what happens when a voltage or current source is "turned off."

The i–v characteristics of voltage and current sources are shown in Figure 3–32. A voltage source is "turned off" by setting its voltage to zero ($v_S = 0$). This step translates the voltage source i–v characteristic to the i–axis, as shown in Figure 3–32(a). In Chapter 2 we found that a vertical line on the i–axis is the i–v characteristic of a short circuit. Similarly, "turning off" a current source ($i_S = 0$) in Figure 3–32(b) translates its i–v characteristic to the v-axis, which is the i–v characteristic of an open circuit. Therefore, when a voltage source is "turned off" we replace it by a short circuit, and when a current source is "turned off" we replace it by an open circuit.

The superposition principle is now applied to the circuit in Figure 3–31 to duplicate the response in Eq. (3–22), which was found by node analysis. Figure 3–33 shows the steps involved in applying superposition to the circuit in Figure 3–31. Figure 3–33(a) shows that the circuit has two input sources. We will first "turn off" i_S and replace it with an open circuit, as shown in Figure 3–33(b). The output of the circuit in Figure 3–33(b) is called v_{O1} and represents that part of the total output caused by the voltage source. Using voltage division in Figure 3–33(b) yields v_{O1} as

$$v_{O1} = \frac{R_2}{R_1 + R_2} v_S$$

Next we "turn off" the voltage source and "turn on" the current source, as shown in Figure 3–33(c). Using Ohm's law, we get $v_{O2} = i_{O2}R_2$. We use current division to express i_{O2} in terms of i_S to obtain v_{O2}:

$$v_{O2} = i_{O2}R_2 = \left[\frac{R_1}{R_1 + R_2} i_S \right] R_2 = \frac{R_1 R_2}{R_1 + R_2} i_S$$

FIGURE 3 - 33 *Circuit analysis using superposition: (a) Current source off. (b) Voltage source off.*

Applying the superposition principle, we find the response with both sources "turned on" by adding the two responses v_{O1} and v_{O2}.

$$v_O = v_{O1} + v_{O2}$$

$$= \left[\frac{R_2}{R_1 + R_2}\right]v_S + \left[\frac{R_1 R_2}{R_1 + R_2}\right]i_S$$

This superposition result is the same as the circuit reduction result given in Eq. (3–22).

EXAMPLE 3–12

Figure 3–34(a) shows a resistance circuit used to implement a signal-summing function. Use superposition to show that the output v_O is a weighted sum of the inputs v_{S1}, v_{S2}, and v_{S3}.

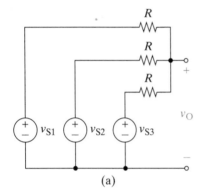

(a)

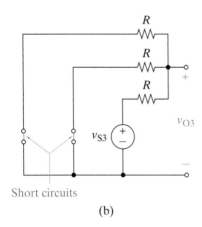

Short circuits

(b)

F I G U R E   3 – 3 4

SOLUTION:

To determine v_O using superposition, we first turn off sources 1 and 2 ($v_{S1} = 0$ and $v_{S2} = 0$) to obtain the circuit in Figure 3–34(b). This circuit is a voltage divider in which the output leg consists of two equal resistors in parallel. The equivalent resistance of the output leg is $R/2$, so the voltage division rule yields

$$v_{O3} = \frac{R/2}{R + R/2}v_{S3} = \frac{v_{S3}}{3}$$

Because of the symmetry of the circuit, it can be seen that the same technique applies to all three inputs; therefore

$$v_{O2} = \frac{v_{S2}}{3} \quad \text{and} \quad v_{O1} = \frac{v_{S1}}{3}$$

Applying the superposition principle, the output with all sources "turned on" is

$$v_O = v_{O1} + v_{O2} + v_{O3}$$

$$= \frac{1}{3}[v_{S1} + v_{S2} + v_{S3}]$$

That is, the output is proportional to the sum of the three input signals with $K_1 = K_2 = K_3 = \frac{1}{3}$. ∎

Exercise 3–14

The circuit of Figure 3–35 contains two of the R-$2R$ modules discussed in Problem 2–35. Use superposition to find v_O.

Answer:

$$v_O = \frac{1}{2}v_{S1} + \frac{1}{4}v_{S2}$$

FIGURE 3-35

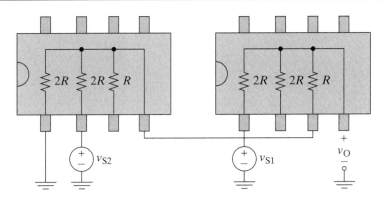

Exercise 3–15

Repeat Exercise 3–14 with the voltage source v_{S2} replaced by a current source i_{S2} with the current reference arrow directed toward ground.

Answer:

$$v_O = 3v_{S1}/5 - 4i_{S2}R/5$$

The preceding examples and exercises illustrate the applications of the superposition principle. You should not conclude that superposition is used primarily to solve for the response of circuits with multiple independent sources. In fact, superposition is not a particularly attractive method of analysis since a circuit with N sources requires N different circuit analyses to obtain the final result. Unless the circuit is relatively simple, superposition does not reduce the analysis effort compared with, say, node-voltage analysis. Superposition is still an important property of linear circuits because it is often used as a conceptual tool to develop other circuit analysis techniques. For example, superposition is used in the next section to prove Thévenin's theorem.

3–4 THÉVENIN AND NORTON EQUIVALENT CIRCUITS

An *interface* is a connection between circuits. Circuit interfaces occur frequently in electrical and electronic systems, so special analysis methods are used to handle them. For the two-terminal interface shown in Figure 3–36, we normally think of one circuit as the source S and the other as the load L. We think of signals as being produced by the source circuit and delivered to the load circuit. The source-load interaction at an interface is one of the central problems of circuit analysis and design.

The Thévenin and Norton equivalent circuits shown in Figure 3–37 are valuable tools for dealing with circuit interfaces. The conditions under which these equivalent circuits exist can be stated as a theorem:

> *If the source circuit in a two-terminal interface is linear, then the interface signals v and i do not change when the source circuit is replaced by its Thévenin or Norton equivalent circuit.*

The equivalence requires the source circuit to be linear, but places no restriction on the linearity of the load circuit. Later in this section we

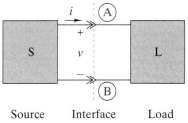

Source   Interface   Load

FIGURE 3-36 *A two-terminal interface.*

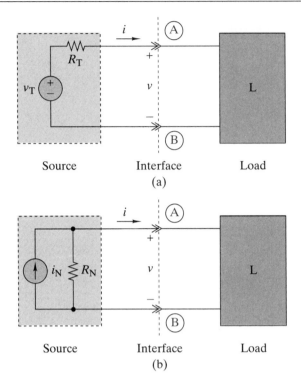

consider cases in which the load is nonlinear. In subsequent chapters we will study circuits in which the loads are energy storage elements called capacitors and inductors.

The Thévenin equivalent circuit consists of a voltage source (v_T) in series with a resistance (R_T). The Norton equivalent circuit is a current source (i_N) in parallel with a resistance (R_N). Note that the Thévenin and Norton equivalent circuits are practical sources in the sense discussed in Chapter 2.

The two circuits have the same $i-v$ characteristics, since replacing one by the other leaves the interface signals unchanged. To derive the equivalency conditions, we apply KVL and Ohm's law to the Thévenin equivalent in Figure 3–37(a) to obtain its $i-v$ relationship at the terminals A and B:

$$v = v_T - iR_T \tag{3-23}$$

Next, applying KCL and Ohm's law to the Norton equivalent in Figure 3–37(b) yields its $i-v$ relationship at terminals A–B:

$$i = i_N - \frac{v}{R_N} \tag{3-24}$$

Solving Eq. (3–24) for v yields

$$v = i_N R_N - iR_N \tag{3-25}$$

The Thévenin and Norton circuits have identical $i-v$ relationships. Comparing Eqs. (3–23) and (3–25), we conclude that

$$R_N = R_T$$

$$i_N R_N = v_T \tag{3-26}$$

In essence, the Thévenin and Norton equivalent circuits are related by the source transformation studied in Chapter 2. We do not need to find both equivalent circuits. Once one of them is found, the other can be determined by a source transformation. The Thévenin and Norton circuits involve four parameters (v_T, R_T, i_N, R_N) and Eq. (3–26) provides two relations between the four parameters. Therefore, only two parameters are needed to specify either equivalent circuit.

In circuit analysis problems it is convenient to use the short-circuit current and open-circuit voltage to specify Thévenin and Norton circuits. The circuits in Figure 3–38(a) show that when the load is an open circuit the interface voltage equals the Thévenin voltage; that is, $v_{OC} = v_T$, since there is no voltage across R_T when $i = 0$. Similarly, the circuits in Figure 3–38(b) show that when the load is a short circuit the interface current equals the Norton current; that is, $i_{SC} = i_N$, since all the source current i_N is diverted through the short-circuit load.

In summary, the parameters of the Thévenin and Norton equivalent circuits at a given interface can be found by determining the open-circuit voltage and the short-circuit current.

$$v_T = v_{OC}$$

$$i_N = i_{SC} \tag{3–27}$$

$$R_N = R_T = v_{OC}/i_{SC}$$

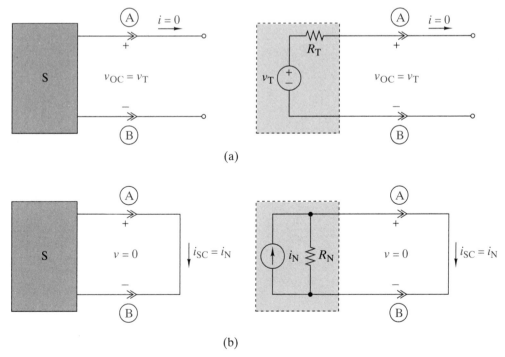

(a)

(b)

FIGURE 3 – 3 8 *Loads used to find Thévenin and Norton equivalent circuits: (a) Open circuit yields the Thévenin voltage. (b) Short-circuit yields the Norton current.*

APPLICATIONS OF THÉVENIN AND NORTON EQUIVALENT CIRCUITS

Replacing a complex circuit by its Thévenin or Norton equivalent can greatly simplify the analysis and design of interface circuits. For example, suppose we need to select a load resistance in Figure 3–39(a) so the source circuit to the left of the interface A–B delivers 4 volts to the load. This task is easily handled once we have the Thévenin or Norton equivalent for the source circuit.

To obtain the Thévenin and Norton equivalents, we need v_{OC} and i_{SC}. The open-circuit voltage v_{OC} is found by disconnecting the load at the terminals A and B, as shown in Figure 3–39(b). The voltage across the 15-Ω resistor is zero because the open circuit causes the current through the resistor to be zero. The open-circuit voltage at the interface is the same as the voltage across the 10-Ω resistor. Using voltage division, this voltage is

$$v_T = v_{OC} = \frac{10}{10 + 5} \times 15 = 10 \text{ V}$$

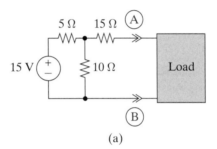

(a)

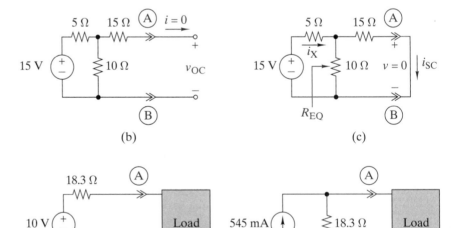

(b)   (c)

(d)   (e)

FIGURE 3 – 3 9   *Example of finding the Thévenin and Norton equivalent circuits: (a) The given circuit. (b) Open circuit yields the Thévenin voltage. (c) Short circuit yields the Norton current. (d) Thévenin equivalent circuit. (e) Norton equivalent circuit.*

Next we find the short-circuit current i_{SC} using the circuit in Figure 3–39(c). The total current i_X delivered by the 15-V voltage source is

$$i_X = 15/R_{EQ}$$

where R_{EQ} is the equivalent resistance seen by the voltage source with a short circuit at the interface.

$$R_{EQ} = 5 + \cfrac{1}{\cfrac{1}{10} + \cfrac{1}{15}} = 11 \ \Omega$$

We find $i_X = 15/11 = 1.36$ A. Given i_X, we now use current division to obtain the short-circuit current:

$$i_N = i_{SC} = \frac{10}{10 + 15} \times i_X = 0.545 \ A$$

Finally, we compute the Thévenin and Norton resistances:

$$R_T = R_N = \frac{v_{OC}}{i_{SC}} = 18.3 \ \Omega$$

The resulting Thévenin and Norton equivalent circuits are shown in Figures 3–39(d) and 3–39(e).

It now is an easy matter to select a load R_L so 4 V is supplied to the load. Using the Thévenin equivalent circuit, the problem reduces to a voltage divider:

$$\frac{R_L}{R_L + R_T} \times v_T = \frac{R_L}{R_L + 18.3} \times 10 = 4 \ V$$

Solving for R_L yields $R_L = 12.2 \ \Omega$.

The Thévenin or Norton equivalent can always be found from the open-circuit voltage and short-circuit current at the interface. The following examples illustrate other methods of determining these equivalent circuits.

EXAMPLE 3−13

(a) Find the Thévenin equivalent circuit of the source circuit to the left of the interface in Figure 3–40(a).
(b) Use the Thévenin equivalent to find the power delivered to two different loads. The first load is a 10-kΩ resistor and the second is a 5-V voltage source whose positive terminal is connected to the upper interface terminal.

SOLUTION:

(a) To find the Thévenin equivalent, we use the sequence of circuit reductions shown in Figure 3–40. In Figure 3–40(a) the 15-V voltage source in series with the 3-kΩ to the left of terminals A and B is replaced by a 3-kΩ resistor in parallel with an equivalent current source with $i_S = 15/3000 = 5$ mA. In Figure 3–40(b), looking to the left at terminals C and D, we see two resistors in parallel whose equivalent resistance is $(3 \ k\Omega)\|(6 \ k\Omega) = 2 \ k\Omega$. We also see two current sources in parallel whose equivalent is $i_S = 5$ mA $- 2$ mA $= 3$ mA. This equivalent current source is shown in Figure 3–40(c) to the left of terminals C and D. Figure 3–40(d) shows this current source converted to an equivalent voltage source $v_S = 3$ mA $\times 2$ k$\Omega = 6$ V in series with 2 kΩ. In Figure 3–40(d) the three resistors are connected in series and can be replaced by an equivalent resistance $R_{EQ} = 2$ k$\Omega + 3$ k$\Omega + 4$ k$\Omega = 9$ kΩ. This step produces the Thévenin equivalent shown in Figure 3–40(e).

Note: The steps leading from Figure 3–40(a) to 3–40(e) involve circuit reduction techniques studied in Chapter 2, so we know that this approach only works on ladder circuits like the one in Figure 3–40(a).

FIGURE 3 – 40

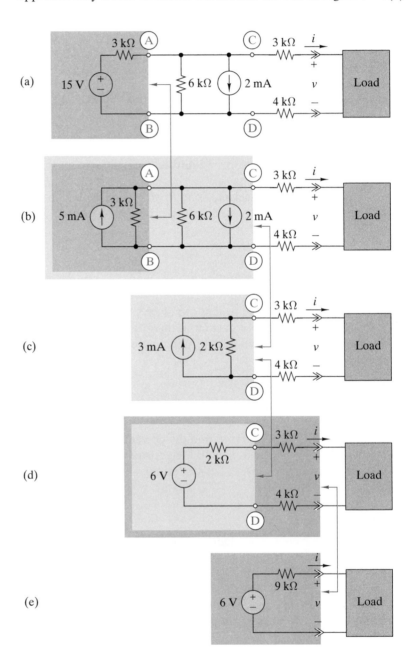

(b) Figure 3–41 shows the Thévenin equivalent found in (a) and the two loads. When the load is a 10-kΩ resistor, the interface current is $i = (6)/(9000 + 10,000) = 0.3158$ mA, and the power delivered to the load is $i^2 R_L = 0.9973$ mW. When the load is a 5-V source, the interface voltage and current are $v = 5$ V and $i = (6 - 5)/9000 = 0.1111$ mA, and the power to the load is $vi = 0.5555$ mW. Since $p > 0$ in the latter case, we see that the voltage source load is absorbing rather than delivering power. A practical example of this situation is a battery charger.

Caution: The Thévenin equivalent allows us to calculate the power delivered to a load, but it does not tell us what power is dissipated in the source circuit. For instance, if the load in Figure 3–40(e) is an open circuit, then no power is dissipated in the Thévenin equivalent since $i = 0$. This does not mean that the power dissipated in the source circuit is zero, as we can easily see by looking back at Figure 3–40(a). The Thévenin equivalent circuit has the same i–v characteristic at the interface, but it does not duplicate the internal characteristics of the source circuit. ∎

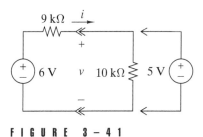

FIGURE 3 – 41

EXAMPLE 3–14

(a) Find the Norton equivalent of the source circuit to the left of the interface in Figure 3–42.
(b) Find the interface current i when the power delivered to the load is 5 W.

FIGURE 3 – 42

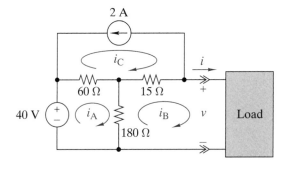

SOLUTION:
(a) The circuit reduction method will not work here since the source circuit is not a ladder. In this example we write mesh-current equations and solve directly for the source circuit i–v relationship. We only need to write equations for meshes A and B since the 2-A current source determines the mesh C current. The voltages sums around these meshes are

$$\text{Mesh A:} \quad -40 + 60(i_A - i_C) + 180(i_A - i_B) = 0$$

$$\text{Mesh B:} \quad -180(i_A - i_B) + 15(i_B - i_C) + v = 0$$

But since $i_B = i$ and the current source forces the condition $i_C = -2$, these equations have the form

$$240 i_A - 180 i = -80$$

$$-180 i_A + 195 i = -30 - v$$

Solving for i in terms of v yields

$$i = \frac{\begin{vmatrix} 240 & -80 \\ -180 & -30-v \end{vmatrix}}{\begin{vmatrix} 240 & -180 \\ -180 & 195 \end{vmatrix}} = \frac{-21600 - 240v}{14400}$$

$$= -1.5 - \frac{v}{60}$$

At the interface the i–v relationship of the source circuit is $i = -1.5 - v/60$. Equation (3–24) gives the i–v relationship of the Norton circuit as $i = i_N - v/R_N$. By direct comparison, we conclude that $i_N = -1.5$ A and $R_N = 60\ \Omega$. This equivalent circuit is shown in Figure 3–43.

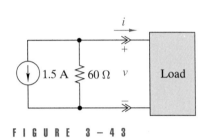

FIGURE 3–43

(b) When 5 W is delivered to the load, we have $vi = 5$ or $v = 5/i$. Substituting $v = 5/i$ into the source i–v relationship $i = -1.5 - v/60$ yields a quadratic equation

$$12i^2 + 18i + 1 = 0$$

whose roots are $i = -0.05778$ A and -1.442 A. Thus, there are two values of interface current that deliver 5 W to the load. ∎

DERIVATION OF THÉVENIN'S THEOREM

The derivation of Thévenin's theorem is based on the superposition principle. We begin with the circuit in Figure 3–44(a), where the source circuit S is linear. Our approach is to use superposition to show that the source circuit and the Thévenin circuit have the same i–v relationship at the interface. To find the source circuit i–v relationship, we first disconnect the load and apply a current source i_{TEST}, as shown in Figure 3–44(b). Using superposition to find v_{TEST}, we first turn i_{TEST} off and leave all the sources inside S on, as shown in Figure 3–44(c). Turning a current source off leaves an open circuit, so

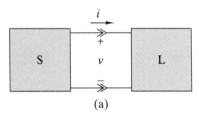

(a)

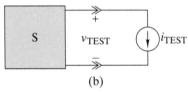

(b)

$$v_{TEST1} = v_{OC}$$

Next we turn i_{TEST} back on and turn off all of the sources inside S. Since the source circuit S is linear, it reduces to the equivalent resistance shown in Figure 3–44(d) when all internal sources are turned off. Using Ohm's law, we write

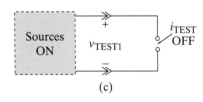

(c)

$$v_{TEST2} = (R_{EQ})(-i_{TEST})$$

The minus sign in this equation results from the reference directions originally assigned to i_{TEST} and v_{TEST} in Figure 3–44(b). Using the superposition principle, we find the i–v relationship of the source circuit at the interface to be

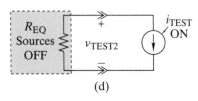

(d)

FIGURE 3–44 *Using superposition to prove Thévenin's theorem.*

$$v_{TEST} = v_{TEST1} + v_{TEST2}$$

$$= v_{OC} - R_{EQ}i_{TEST}$$

This equation has the same form as the i–v relationship of the Thévenin equivalent circuit in Eq. (3–23) when $v_{TEST} = v$, $i_{TEST} = i$, $v_{OC} = v_T$, and $R_T = R_{EQ}$.

The derivation points out another method of finding the Thévenin resistance. As indicated in Figure 3–44(d), when all the sources are turned off, the i–v relationship of the source circuit reduces to $v = -iR_{EQ}$. Similarly, the i–v relationship of a Thévenin equivalent circuit reduces to $v = -iR_T$ when $v_T = 0$. We conclude that

$$R_T = R_{EQ} \qquad \text{(3–28)}$$

We can find the value of R_T by determining the resistance seen looking back into the source circuit with all sources turned off. For this reason the Thévenin resistance R_T is sometimes called the **lookback resistance**.

The next example shows how lookback resistance contributes to finding a Thévenin equivalent circuit.

EXAMPLE 3–15

(a)　Find the Thévenin equivalent of the source circuit to the left of the interface in Figure 3–45.
(b)　Use the Thévenin equivalent to find the voltage delivered to the load.

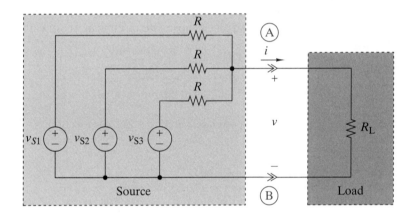

FIGURE 3–45

SOLUTION:
(a)　The source circuit in Figure 3–45 is treated in Example 3–12 by using superposition to calculate the open-circuit voltage between terminals A and B. Using the results from Example 3–12, we have

$$v_T = v_{OC} = \frac{1}{3}(v_{S1} + v_{S2} + v_{S3})$$

Turning all sources off in Figure 3–45 leads to the resistance circuit in Figure 3–46. Looking back into the source circuit in Figure 3–46, we see three equal resistances connected in parallel whose equivalent resistance is $R/3$. Hence, the Thévenin resistance is

$$R_T = R_{EQ} = \frac{R}{3}$$

(b)　Given the Thévenin circuit parameters v_T and R_T, we apply voltage division in Figure 3–45 to find the interface voltage.

FIGURE 3 – 46

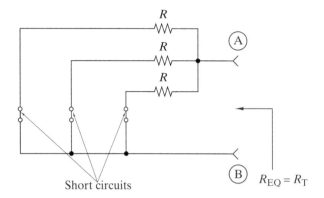

$$v = \frac{R_L}{R_L + R_T}v_T = \left(\frac{R_L}{R_L + R/3}\right)\left(\frac{v_{S1} + v_{S2} + v_{S3}}{3}\right)$$

$$= \left(\frac{R_L}{3R_L + R}\right)(v_{S1} + v_{S2} + v_{S3})$$

The interface voltage is proportional to the sum of the three source voltages. The proportionality constant $K = R_L/(3R_L + R)$ depends on both the source and the load since these two circuits are connected at the interface. ∎

Exercise 3–16

(a) Find the Thévenin and Norton equivalent circuits seen by the load in Figure 3–47.
(b) Find the voltage, current, and power delivered to a 50-Ω load resistor.

FIGURE 3 – 47

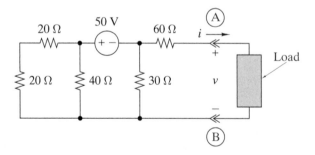

Answers:
(a) $v_T = -30$ V, $i_N = -417$ mA, $R_N = R_T = 72$ Ω
(b) $v = -12.3$ V, $i = -246$ mA, $p = 3.03$ W

Exercise 3–17

Find the current and power delivered to an unknown load in Figure 3–47 when $v = +6$ V.

Answers:

$$i = -\tfrac{1}{2}\text{A}, p = -3 \text{ W}$$

APPLICATION TO NONLINEAR LOADS

Thévenin and Norton equivalent circuits can be used to find the response of a two-terminal nonlinear element (NLE). The method of analysis is a straightforward application of device and interface i–v characteristics. An interface is defined at the terminals of the nonlinear element, and the linear part of the circuit is reduced to the Thévenin equivalent in Figure 3–48(a). The i–v relationship of the Thévenin equivalent can be written with interface current as the dependent variable:

$$i = \left(-\frac{1}{R_T}\right)v + \left(\frac{v_T}{R_T}\right) \qquad (3\text{--}29)$$

This is the equation of a straight line in the i–v plane shown in Figure 3–48(b). The line intersects the i-axis ($v = 0$) at $i = v_T/R_T = i_{SC}$ and inter-

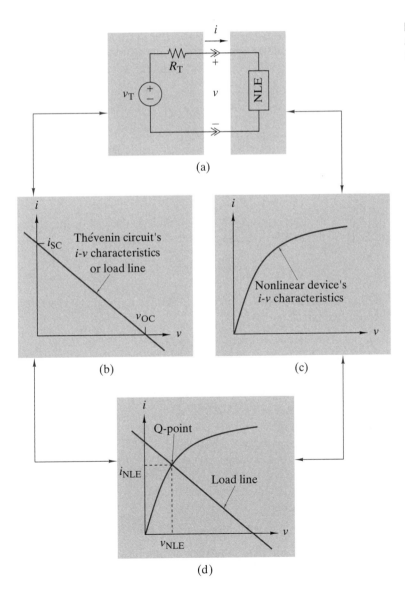

(a)

(b)

(c)

(d)

FIGURE 3 – 48 *Graphical analysis of a nonlinear circuit: (a) Given circuit. (b) Load line. (c) Nonlinear device i–v characteristics. (d) Q-point.*

sects the v–axis ($i = 0$) at $v = v_T = v_{OC}$. This line could logically be called the source line since it is determined by the Thévenin parameters of the source circuit. Logic notwithstanding, electrical engineers call this the **load line** for reasons that have blurred with the passage of time.

The nonlinear element has the i–v characteristic shown in Figure 3–48(c). Mathematically, this nonlinear characteristic has the form

$$i = f(v) \tag{3–30}$$

To find the circuit response, we must solve Eqs. (3–29) and (3–30) simultaneously. Computer software tools like Mathcad can easily solve this problem when a numerical expression for the function $f(v)$ is known explicitly. However, in practice an approximate graphical solution is often adequate, particularly when $f(v)$ is only given in graphical form.

In Figure 3–48(d) we superimpose the load line on the i–v characteristic curve of the nonlinear element. The two curves intersect at the point $i = i_{NLE}$ and $v = v_{NLE}$, which yields the values of interface variables that satisfy both the source constraints in Eq. (3–29) and the nonlinear element constraints in Eq. (3–30). In the terminology of electronics, the point of intersection is called the operating point or **Q-point**, where Q stands for quiescent.

EXAMPLE 3–16

Find the voltage, current, and power delivered to the diode in Figure 3–49(a).

SOLUTION:

We first find the Thévenin equivalent of the circuit to the left of the terminals A and B. By voltage division, the open-circuit voltage is

$$v_T = v_{OC} = \frac{100}{100 + 100} \times 5 = 2.5 \text{ V}$$

When the voltage source is turned off, the lookback equivalent resistance seen between terminals A and B is

$$R_T = 10 + 100\|100 = 60 \text{ }\Omega$$

The source circuit load line is given by

$$i = -\frac{1}{60}v + \frac{1}{60} \times 2.5$$

This line intersects the i-axis ($v = 0$) at $i = i_{SC} = 2.5/60 = 41.7$ mA and intersects the v–axis ($i = 0$) at $v = v_{OC} = 2.5$ V. Figure 3–49(b) superimposes the source circuit load line on the diode's i–v curve. The intersection (Q-point) is at $i = i_D = 15$ mA and $v = v_D = 1.6$ V. This is the point (i_D, v_D) at which both the source and diode device constraints are satisfied. Finally, the power delivered to the diode is

$$p_D = i_D v_D = (15 \times 10^{-3})(1.6) = 24 \text{ mW}$$

Because of the nonlinear element, the proportionality and superposition properties do not apply to this circuit. For instance, if the source voltage in

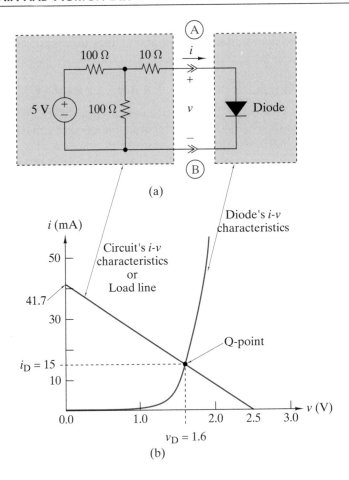

(a)

(b)

Figure 3–49 is decreased from 5 V to 2.5 V, the diode current and voltage do not decrease by one-half. Try it.   ■

Exercise 3–18

Find the voltage, current, and power delivered to the diode in Figure 3–49(a) when the 10-Ω resistor is replaced by a short circuit.

Answer:

$$v_D = 1.7 \text{ V}, i_D = 18 \text{ mA, and } p_D = 30.6 \text{ mW}$$

In summary, any two of the following parameters determine the Thévenin or Norton equivalent circuit at a specified interface:

- The open-circuit voltage at the interface
- The short-circuit current at the interface
- The source circuit lookback resistance.

Alternatively, for ladder circuits the Thévenin or Norton equivalent circuit can be found by a sequence of circuit reductions (see Example 3–13). For general circuits they can always be found by directly solving for the *i–v* relationship of the source circuit using node-voltage or mesh-current equations that include the interface current and voltage as unknowns (see Example 3–14).

3–5 MAXIMUM SIGNAL TRANSFER

An interface is a connection between two circuits at which the signal levels may be observed or specified. In this regard an important consideration is the maximum signal levels that can be transferred across a given interface. This section defines the maximum voltage, current, and power available at an interface between a *fixed source* and an *adjustable load*.

For simplicity we will treat the case in which both the source and load are linear resistance circuits. The source can be represented by a Thévenin equivalent and the load by an equivalent resistance R_L, as shown in Figure 3–50. For a fixed source, the parameters v_T and R_T are given and the interface signal levels are functions of the load resistance R_L.

By voltage division, the interface voltage is

$$v = \frac{R_L}{R_L + R_T} v_T \qquad (3\text{–}31)$$

For a fixed source and a variable load, the voltage will be a maximum if R_L is made very large compared with R_T. Ideally, R_L should be made infinite (an open circuit), in which case

$$v_{MAX} = v_T = v_{OC} \qquad (3\text{–}32)$$

Therefore, the maximum voltage available at the interface is the source open-circuit voltage v_{OC}.

The current delivered at the interface is

$$i = \frac{v_T}{R_L + R_T} \qquad (3\text{–}33)$$

For a fixed source and a variable load, the current will be a maximum if R_L is made very small compared with R_T. Ideally, R_L should be zero (a short circuit), in which case

$$i_{MAX} = \frac{v_T}{R_T} = i_N = i_{SC} \qquad (3\text{–}34)$$

FIGURE 3–50 *Two-terminal interface for deriving the maximum signal transfer conditions.*

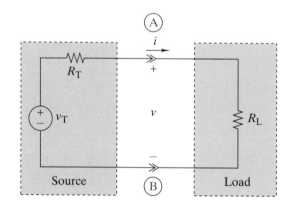

Therefore, the maximum current available at the interface is the source short-circuit current i_{SC}.

The power delivered at the interface is equal to the product $v \times i$. Using Eqs. (3–31) and (3–33), the power is

$$p = v \times i$$
$$= \frac{R_L v_T^2}{(R_L + R_T)^2} \tag{3–35}$$

For a given source, the parameters v_T and R_T are fixed and the delivered power is a function of a single variable R_L. The condition for maximum voltage ($R_L \to \infty$) and the condition for maximum current ($R_L = 0$) both produce zero power. The value of R_L that maximizes the power lies somewhere between these two extremes. To find this value, we differentiate Eq. (3–35) with respect to R_L and solve for the value of R_L for which $dp/dR_L = 0$.

$$\frac{dp}{dR_L} = \frac{[(R_L + R_T)^2 - 2R_L(R_L + R_T)]v_T^2}{(R_L + R_T)^4} = \frac{R_T - R_L}{(R_L + R_T)^3} v_T^2 = 0 \tag{3–36}$$

Clearly, the derivative is zero when $R_L = R_T$. Therefore, **maximum power transfer** occurs when the load resistance equals the Thévenin resistance of the source. When the condition $R_L = R_T$ exists, the source and load are said to be **matched**.

Substituting the condition $R_L = R_T$ back into Eq. (3–35) shows the maximum power to be

$$p_{MAX} = \frac{v_T^2}{4R_T} \tag{3–37}$$

Since $v_T = i_N R_T$, this result can also be written as

$$p_{MAX} = \frac{i_N^2 R_T}{4} \tag{3–38}$$

or

$$p_{MAX} = \frac{v_T i_N}{4} = \left[\frac{v_{OC}}{2}\right]\left[\frac{i_{SC}}{2}\right] \tag{3–39}$$

These equations are consequences of what is known as the **maximum power transfer theorem**:

> *A source with a fixed Thévenin resistance $\mathbf{R}_T$ delivers maximum power to an adjustable load $\mathbf{R}_L$ when $\mathbf{R}_L = \mathbf{R}_T$.*[1]

To summarize, at an interface with a fixed source,

1. The maximum available voltage is the open-circuit voltage.

2. The maximum available current is the short-circuit current.

3. The maximum available power is the product of one-half the open-circuit voltage times one-half the short-circuit current.

---

1 An ideal voltage source has zero internal resistance, hence $R_T = 0$. Equation (3–37) points out that $R_T = 0$ implies an infinite p_{MAX}. Infinite power is a physical impossibility, which reminds us that all ideal circuit models have some physical limitations.

FIGURE 3 – 51 *Normalized plots of current, voltage, and power versus* R_L/R_T.

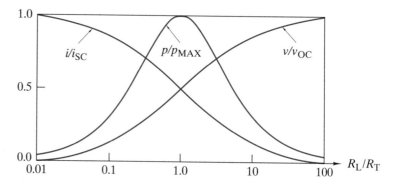

Figure 3–51 shows plots of the interface voltage, current, and power as functions of R_L/R_T. The plots of v/v_{OC}, i/i_{SC}, and p/p_{MAX} are normalized to the maximum available signal levels, so the ordinates in Figure 3–51 range from 0 to 1. The plot of the normalized power p/p_{MAX} in the neighborhood of the maximum is not a particularly strong function of R_L/R_T. Changing the ratio R_L/R_T by a factor of 2 in either direction from the maximum reduces p/p_{MAX} by less than 20%. The normalized voltage v/v_{OC} is within 20% of its maximum when $R_L/R_T = 4$. Similarly, the normalized current is within 20% of its maximum when $R_L/R_T = \frac{1}{4}$. In other words, for engineering purposes we can get close to the maximum signal levels with load resistances that only approximate the theoretical requirements.

EXAMPLE 3–17

A source circuit with $v_T = 2.5$ V and $R_T = 60\ \Omega$ drives a load with $R_L = 30\ \Omega$.

(a) Determine the maximum signal levels available from the source circuit.

(b) Determine the actual signal levels delivered to the load.

SOLUTION:

(a) The maximum available voltage and current are

$$v_{MAX} = v_{OC} = v_T = 2.5 \text{ V } (R_L \to \infty)$$

$$i_{MAX} = i_{SC} = \frac{v_T}{R_T} = 41.7 \text{ mA } (R_L = 0)$$

The maximum available power is found using Eq. (3–39).

$$p_{MAX} = \left[\frac{v_{OC}}{2}\right]\left[\frac{i_{SC}}{2}\right] = 26.0 \text{ mW } (R_L = R_T = 60\ \Omega)$$

(b) The actual signal levels delivered to the 30-Ω load are

$$v_L = \frac{30}{30 + 60} 2.5 = 0.833 \text{ V}$$

$$i_L = \frac{2.5}{30 + 60} = 27.8 \text{ mA}$$

$$p_L = v_L i_L = 23.1 \text{ mW}$$

Although these levels are less than the maximum available values, the power delivered to the 30-Ω load is nearly 90% of the maximum.   ■

Exercise 3–19

A source circuit delivers 4 V when a 50-Ω resistor is connected across its output and 5 V when a 75-Ω resistor is connected. Find the maximum voltage, current, and power available from the source.

Answers:
10 V, 133 mA, and 333 mW

Remember that the maximum signal levels just derived are for a fixed source resistance and an adjustable load resistance. This situation often occurs in communication systems where devices such as antennas, transmitters, and signal generators have fixed source resistances such as 50, 75, 300, or 600 ohms. In such cases the load resistance is selected to achieve the desired interface conditions, which often involves matching.

Matching source and load applies when the load resistance R_L in Figure 3–50 is adjustable and the Thévenin source resistance R_T is fixed. When R_L is fixed and R_T is adjustable, then Eqs. (3–31), (3–33), and (3–35) point out that the maximum voltage, current, and power are delivered when the Thévenin source resistance is zero. If the source circuit at an interface is adjustable, then ideally the Thévenin source resistance should be zero. In the next chapter we will see that OP AMP circuits approach this ideal.

3–6   INTERFACE CIRCUIT DESIGN

The maximum signal levels discussed in the previous section place bounds on what is achievable at an interface. However, those bounds are based on a fixed source and an adjustable load. In practice, there are circumstances in which the source or the load, or both, can be adjusted to produce prescribed interface signal levels. Sometimes it is necessary to insert a circuit between the source and the load to achieve the desired results. Figure 3–52 shows the general situations and some examples of resistive interface circuits. By its nature, the inserted circuit has two terminal pairs, or interfaces, at which voltage and current can be observed or specified. These terminal pairs are also called *ports*, and the interface circuit is referred to as a **two-port network**. The port connected to the source is called the input, and the port connected to the load is called the output. The purpose of this two-port network is to make certain that the source and load interact in a prescribed way.

BASIC CIRCUIT DESIGN CONCEPTS

Before we treat examples of different interface situations, you should recognize that we are now discussing a limited form of circuit design, as contrasted with circuit analysis. Although we use circuit analysis tools in design, there are important differences. A linear circuit analysis problem generally has a unique solution. A circuit design problem may have many solutions or even no solution. The maximum available signal levels found in the preceding section provide bounds that help us test for the existence of a solution. Generally there will be several ways to meet the interface

F I G U R E   3 – 5 2   *A general interface cir-cuit and some examples.*

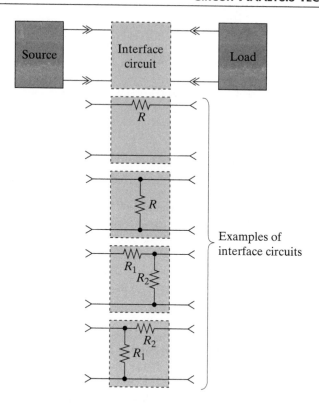

Examples of interface circuits

constraints, and it then becomes necessary to evaluate the alternatives using other factors, such as cost, power consumption, or reliability.

At this point in our study, resistors are the only elements we can use to design interface circuits. In subsequent chapters we will introduce other devices, such as OP AMPs (Chapter 4) and capacitors and inductors (Chapter 6). In a design situation the engineer must choose the resistance values in a proposed circuit. This decision is influenced by a host of practical considerations, such as standard values and tolerances, power ratings, temperature sensitivity, cost, and fabrication methods. We will occasionally introduce some of these considerations into our design examples. Gaining a full understanding of these practical matters is not one of our en route objectives. Rather, our goal is simply to illustrate how different constraints can influence the design process.

D   **D E S I G N   E X A M P L E   3 – 1 8**

Select the load resistance in Figure 3–53 so that the interface signals are in the range defined by $v \geq 4$ V and $i \geq 30$ mA.

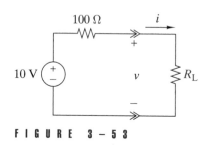

F I G U R E   3 – 5 3

SOLUTION:

In this design problem, the source circuit is given and we are free to select the load. For a fixed source the maximum signal levels available at the interface are

$$v_{MAX} = v_T = 10 \text{ V}$$

$$i_{MAX} = \frac{v_T}{R_T} = 100 \text{ mA}$$

The bounds given as design requirements are below the maximum available signal levels, so we should be able to find a suitable resistor. Using voltage division, the interface voltage constraint requires

$$\frac{R_L}{100 + R_L} \times 10 \geq 4$$

or

$$10R_L \geq 4R_L + 400$$

This condition yields $R_L \geq 400/6 = 66.7 \ \Omega$. The interface current constraint can be written as

$$\frac{10}{100 + R_L} \geq 0.03$$

or

$$10 \geq 3 + 0.03R_L$$

which requires $R_L \leq 7/0.03 = 233 \ \Omega$. In theory, any value of R_L between 66.7 Ω and 233 Ω will work. However, to allow for parameter variations we select $R_L = 150 \ \Omega$ because it lies at the arithmetic midpoint of the allowable range and is a standard value of resistance (see Table A–1, Appendix A). ■

D DESIGN EXAMPLE 3–19

Select the load resistor in Figure 3–53 so that the voltage delivered to the load is 3 V ±10%. Use only the standard resistance values given in Appendix A for ±10% tolerance.

SOLUTION:

The ±10% voltage tolerance means that the interface voltage must fall between 2.7 V and 3.3 V. The required range can be achieved since the open-circuit voltage of the source circuit is 10 V. Using voltage division, we can write the constraints on the value of R_L:

$$2.7 \leq \frac{R_L}{100 + R_L} \times 10 \leq 3.3$$

The left inequality requires

$$270 + 2.7R_L \leq 10R_L$$

or $R_L \geq 270/7.3 = 37.0 \ \Omega$. The right inequality requires

$$10R_L \leq 330 + 3.3R_L$$

or $R_L \leq 330/6.7 = 49.2 \ \Omega$. Thus, R_L must lie in the range from 37.0 Ω to 49.2 Ω, which has a midpoint at 43.1 Ω.

Commercial resistors are not available in infinitely many values. To limit inventory costs, the electronic industry has agreed on a finite set of standard resistance values. For resistors with ±10% tolerance the standard values are

$$10 \quad 12 \quad 15 \quad 18 \quad 22 \quad 27 \quad 33 \quad 39 \quad 47 \quad 56 \quad 68 \quad 82 \; \Omega$$

and multiples of 10 times these values. The standard values of 39 Ω and 47 Ω fall within the required 37.0 to 49.2 Ω range. However, these nominal values and the ±10% tolerance could carry the actual resistance value outside the range. Various series and parallel combinations of standard ±10% resistors produce values that fall in the desired range even with 10% tolerance. For example, two 22-Ω resistors connected in series produce a nominal equivalent resistance of 44 Ω with a ±10% range from 39.6 Ω to 48.4 Ω. Finally, it turns out that 43 Ω is a standard value for resistors with ±5% tolerance (see Appendix A, Table A–1). This option is attractive since 43 Ω falls almost exactly at the midpoint of the desired resistance range and the 5% tolerance easily meets the interface voltage range requirement.

In summary, either a standard 39-Ω resistor or a standard 47-Ω resistor produces nominal designs that meet the design requirements, but the 10% tolerance on the resistance could cause the voltage to fall outside the specified range. Series and parallel combinations of two 10% resistors meet all requirements but lead to a more complicated and costly design. A single 43-Ω 5% resistor meets all requirements, but its tighter tolerance could mean higher parts costs than two 10% resistors. The final design decision must take into account many factors, particularly the sensitivity of the system to changes in the specified interface voltage.   ■

D DESIGN EXAMPLE 3–20

The resistance array package shown in Figure 3–54 includes seven nearly identical resistors with nominal resistances of 1 kΩ. Use this resistance array to design a voltage divider that reduces the output of a 10-V dc power supply to 4.3 V.

FIGURE 3 - 5 4

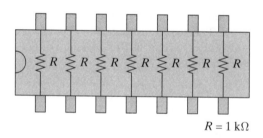

$$R = 1 \text{ k}\Omega$$

SOLUTION:
The required interface voltage is well within the capability of a 10-V source. Using voltage division, we can express the design requirement as

$$\frac{R_2}{R_1 + R_2} \times 10 = 4.3$$

where R_1 and R_2 are the series and shunt legs of the divider, respectively. This design constraint can be rearranged as

$$\frac{R_1}{R_2} = 1.33$$

In other words, the design requirement imposes a constraint on the ratio of R_1 and R_2.

Resistance array packages are particularly useful when the design constraints involve resistance ratios. The nominal resistance can change from one array to the next, but within a given array the resistances are all nearly equal, so their ratios are nearly unity. In effect, the resistance array in Figure 3–54 contains seven identical resistors, although we do not know the exact value of their resistance. Constructing the series leg using three resistors in parallel and the shunt leg using four resistors in parallel produces R_1 = R/3 and R_2 = R/4, where R is nominally 1 kΩ. But regardless of the actual value of R, the ratio R_1/R_2 = 1.33 as required.

The final design shown in Figure 3–55 may seem wasteful of resistors. However, in large-volume production an integrated circuit resistance array can cost less than a discrete resistor design. Resistance array packages are better suited to automated manufacturing than discrete resistors. ■

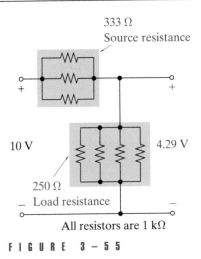

FIGURE 3 - 5 5

D DESIGN EXAMPLE 3–21

Design the two-port interface circuit in Figure 3–56 so the 10-A source delivers 100 V to the 50-Ω load.

FIGURE 3 - 5 6

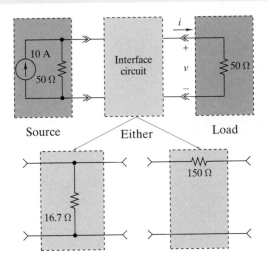

SOLUTION:
The problem requires that the current delivered to the load be i = 100/50 = 2 A, which is well within the maximum value of current available from the source. In fact, if the 10-A source is connected directly to the load, the source current divides equally between two 50-Ω resistances, producing 5 A through the load. Therefore, an interface circuit is needed to reduce the load current to the specified 2-A level. Two possible design solutions are shown in Figure 3–56. Applying current division to the parallel resistor case yields the following constraint:

$$i = \frac{1/50}{1/50 + 1/50 + 1/R_{PAR}} \times 10$$

After inserting the $i = 2$ A design requirement, this equation becomes

$$2 = \frac{10}{2 + 50/R_{PAR}}$$

Solving for R_{PAR} yields

$$R_{PAR} = \frac{50}{3} = 16.7 \ \Omega$$

Applying the two-path current division rule to the series resistor case yields the following constraint:

$$i = \frac{50}{50 + (50 + R_{SER})} \times 10 = 2 \text{ A}$$

Solving for R_{SER} yields

$$R_{SER} = 150 \ \Omega$$

We have two designs that meet the basic $i = 2$ A requirement. In practice, engineers evaluate alternative designs using additional criteria. One such consideration is the required power ratings of the resistors in each design. The voltage across the parallel resistor is $v = 100$ V, so the power loss is

$$P_{PAR} = \frac{100^2}{50/3} = 600 \text{ W}$$

The current through the series resistor interface is $i = 2$ A, so the power loss is

$$P_{SER} = 2^2 \times 150 = 600 \text{ W}$$

In either design the resistors must have a power rating of at least 600 W. Other factors besides power rating determine which design should be selected. ∎

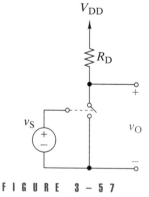

FIGURE 3–57

ⓓ DESIGN EXAMPLE 3–22

Figure 3–57 is a simplified circuit for a digital inverter. The basic performance requirement is that the output be low when the input is high and vice versa. In this example the switch closes when $v_S > 2.0$ V ($R_{ON} = 200 \ \Omega$) and opens when $v_S < 0.8$ V ($R_{OFF} = 10^{11} \ \Omega$). Stated in terms of voltages, the design requirements are (1) $v_O < 0.5$ V when $v_S > 2.0$ V and (2) $v_O > 3.5$ V when $v_S < 0.8$ V. In addition, the current through the switch must be less than 1 mA and the total power dissipated in the circuit less than 2.5 mW. For $V_{DD} = 5$ V, select the value of R_D to meet all of these requirements.

SOLUTION:
When $v_S > 2.0$ V the switch is closed and the output voltage is found by voltage division as

$$v_O = \frac{R_{ON}}{R_{ON} + R_D} \times V_{DD} = \frac{200}{200 + R_D} \times 5 < 0.5 \text{ V}$$

which yields the requirement $R_D > 1800 \ \Omega$. The switch current is a maximum when the switch is closed, hence

$$I_{SW} = \frac{V_{DD}}{R_{ON} + R_D} = \frac{5}{200 + R_D} < 1 \text{ mA}$$

which yields the requirement $R_D > 4800 \ \Omega$. Likewise, the maximum power is dissipated in the circuit when the switch is closed, hence

$$p_D = \frac{V_{DD}^2}{R_{ON} + R_D} = \frac{25}{200 + R_D} < 2.5 \text{ mW}$$

which yields the requirement $R_D > 9800 \ \Omega$. Clearly the power dissipation requirement places the greatest lower bound on R_D. When $v_S < 0.8$ V the switch is open and the output voltage is found by voltage division as

$$v_O = \frac{R_{OFF}}{R_{OFF} + R_D} \times V_{DD} = \frac{10^{11}}{10^{11} + R_D} \times 5 > 3.5 \text{ V}$$

which yields the requirement $R_D < 4.28 \times 10^{11} \ \Omega$. Thus, all requirements are met for

$$9800 < R_D < 4.28 \times 10^{11} \ \Omega$$

Any value of R_D in this very wide range would meet the design constraints given here. In later chapters we will see that the speed of response improves as we move toward the lower end of this range. ■

Exercise 3–20

Find the lower bound R_D in Example 3–22 when the allowable power dissipation is 5 mW.

Answer: 4800 Ω

D DESIGN EXAMPLE 3–23

Design the two-port interface circuit in Figure 3–58 so the load "sees" a Thévenin resistance of 50 Ω between terminals C and D, while simultaneously the source "sees" a load resistance of 300 Ω between A and B.

SOLUTION:
We first try a single resistor in the interface circuit. A 60-Ω parallel resistor in the interface circuit would make the load see 50 Ω Thévenin resistance, but then the source would see a load resistance much less than 300 Ω. A 250-Ω series resistor in the interface circuit would make the source see a 300-Ω load, but then the load would see a Thévenin resistance much greater than 50 Ω.

We next try an interface circuit containing two resistors. Since the load must see a smaller resistance than the source, it should "look" into a parallel resistor. On the other hand, since the source must see a larger resistance

FIGURE 3-58

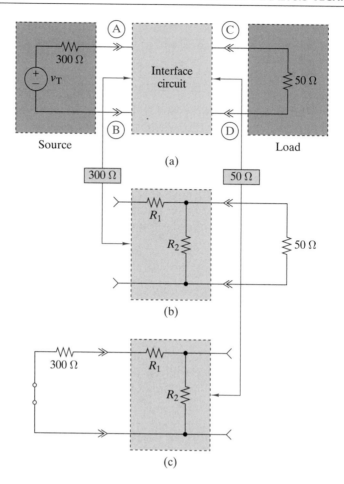

(a)

(b)

(c)

than the load, it should look into a series resistor. A configuration that meets these conditions is the L-circuit shown in Figures 3–58(b) and 3–58(c).

The preceding discussion can be summarized mathematically. Using the L-circuit, the design requirement at terminals C and D is

$$\frac{(R_1 + 300)\, R_2}{R_1 + 300 + R_2} = 50\ \Omega$$

At terminals A and B, the requirement is

$$R_1 + \frac{50\, R_2}{R_2 + 50} = 300\ \Omega$$

The design requirements yield two equations in two unknowns; what could be simpler? It turns out that solving these nonlinear equations by hand analysis is a bit of a chore. These equations can easily be solved using a math solver, as we will shortly demonstrate. However, at this point in your development we encourage you to think about the problem in physical terms.

Given the L-circuits in Figure 3–58(b), one approach goes as follows. If we let $R_2 = 50$ ohms, then the requirement at terminals C and D will be met, at least approximately. Similarly, if $R_1 + R_2 = 300\ \Omega$, the requirements at terminals A and B will be approximately satisfied. In other words, try

$R_1 = 250 \ \Omega$ and $R_2 = 50 \ \Omega$ as a first cut. These values yield equivalent resistances of $R_{CD} = 50 \| 550 = 45.8 \ \Omega$ and $R_{AB} = 250 + 50 \| 50 = 275 \ \Omega$. These equivalent resistances are not the exact values specified, but are within $\pm 10\%$. Since the tolerance on electrical components may be at least this high, a design using these values could be adequate.

Figure 3–59 shows how Mathcad solves the two design equations to find the exact values of R_1 and R_2. The approximate values derived previously are used as the first estimates required by a Mathcad solve block. Mathcad uses these estimates in an iterative process to solve the design equations listed between the Given and the Find statements. The final results at the bottom in Figure 3–59 show that the approximate values derived from physical reasoning are within 10% of the exact values derived using Mathcad.

In this example we used physical reasoning to find an approximation solution of a problem. Using physical reasoning to find practical solutions is what engineering is all about. ■

FIGURE 3 - 59

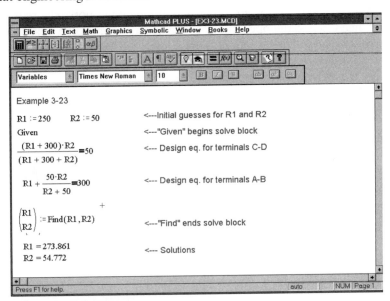

APPLICATION NOTE: EXAMPLE 3–24

An **attenuation pad** is a two-port resistance circuit that provides a non-adjustable reduction in signal level while also providing resistance matching at the input and output ports. Figure 3-60 shows an example of an attenuation pad. The manufacturer's data sheet for this pad specifies the following characteristics at the input and output ports.

Port	Characteristics	Condition	Value	Units
Output	Thévenin voltage	600-Ω source connected at the input port	$v_S/4$	V
Output	Thévenin resistance	600-Ω source connected at the input port	600	Ω
Input	Input resistance	600-Ω load connected at the output port	600	Ω

Use Orcad to verify these characteristics.

FIGURE 3-60

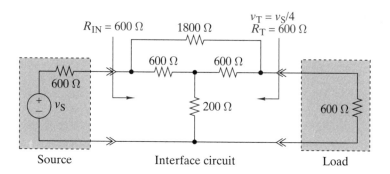

Source   Interface circuit   Load

SOLUTION:

Orcad Capture can calculate the two-port characteristics of the pad in Figure 3–60. A circuit diagram created using Orcad is shown in the upper right of Figure 3–61. In this schematic a 600-Ω source is connected at the pad's input port. A 600-Ω load is *not* connected at the output port. The reason is that we need to find the open-circuit (Thévenin) voltage at the output port. We have labeled the pad's output as node D. Orcad has labeled our node D as N03172. This Orcad assignment is not under our control and will change with every new simulation.[1]

FIGURE 3-61

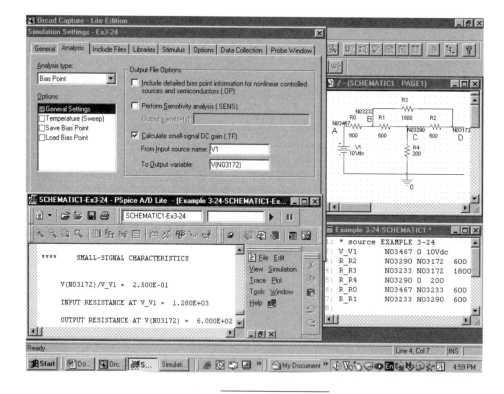

---

1 Every time a schematic is drawn (even the same schematic) Orcad generates a new random number for each node name. To display the Orcad assigned names, double-click on each node to open a new window that gives information about the node. Click on the node's name then select **Display**. This opens a second window. Select **Value Only**. Then click **OK**. Finally select **Apply** to place it on the drawing. Alternately, the Orcad assigned names can be viewed in the Net List file.

Selecting the **PSpice** and the **Edit Simulation Profile** commands causes the "Simulation Settings" dialog box shown in the upper left portion of Figure 3-61 to appear. We first check the "Calculate small-signal DC gain (.TF)" box as shown in the figure. To calculate a gain Orcad must know the input and output variables. We inform Orcad of these variables by supplying entries in the "From Input source name" and "to Output variable" boxes. In this example, voltage source V1 is the input source and the output variable is the node voltage V(N03172) (the voltage at our node D). Note that V(N03172) is the open-circuit voltage at the output port. With these entries Orcad PSpice will calculate three things: (1) the small-signal gain V(N03172)/V_V1, (2) the input resistance seen by V1, and (3) the output resistance.

Returning to the main menu, we select **PSpice/ Run**. Orcad PSpice performs a dc analysis of the circuit and writes the results to the output file. Included in the output file are the small signal characteristics shown at the bottom left in Figure 3-61. PSpice reports that the small-signal gain V(N03172)/V_V1 = 2.500E-01, the input resistance seen by V1 is 1.280E+03, and the output resistance is 6.000E+02. This means that the open-circuit (Thévenin) voltage at the output port is one fourth of the source voltage and the output (Thévenin) resistance is 600 Ω as stated in the data sheet. The input resistance (1280 Ω) given in the output file is the resistance seen by the source V1 when the output port is open-circuited. The manufacturer's data sheet specifies the input resistance when a 600-Ω load is connected at the output port. Thus, the simulation has not directly verified the input resistance characteristic.

However, a moment's reflection shows that the two-port pad circuit is symmetrical. The simulation results show that the output resistance is 600 Ω when a 600-Ω source is connected at the input. It follows from the circuit's symmetry that the input resistance is 600 Ω when a 600-Ω load is connected at the output. In summary, the simulation has verified all of the entries in the manufacturer's data sheet. ∎

D DESIGN EXERCISE: EXERCISE 3–21

Using only standard ±5% resistors, design the two-port interface circuit in Figure 3–62 to meet the following requirements:

(a) The power delivered to the 50-Ω load is 0.2 W ±5%.
(b) Repeat (a) for $p = 0.6$ W ±5%.

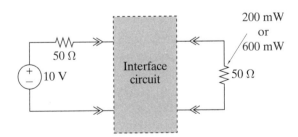

FIGURE 3 – 62

Answer: There are no unique answers to this problem. Some possibilities are discussed next.

DISCUSSION: *(a) A standard 56-Ω resistor in series (ideally 58.1 Ω) yields 200 mW ±5% into the 50-Ω load. (b) No resistive interface will work since the maximum power available from the source is only 0.5 W. Delivering more power than the signal source can provide requires an active device, such as an OP AMP or transistor (treated in the next chapter).*

SUMMARY

- Node-voltage analysis involves identifying a reference node and the node to datum voltages at the remaining $N - 1$ nodes. The KCL connection constraints at the $N - 1$ nonreference nodes combined with the element constraints written in terms of the node voltages produce $N - 1$ linear equations in the unknown node voltages.

- Mesh-current analysis involves identifying mesh currents that circulate around the perimeter of each mesh in a planar circuit. The KVL connection constraints around $E - N + 1$ meshes combined with the element constraints written in terms of the mesh currents produce $E - N + 1$ linear equations in the unknown mesh currents.

- Node and mesh analysis can be modified to handle both types of independent sources using a combination of three methods: (1) source transformations, (2) selecting circuit variables so independent sources specify the values of some of the unknowns, and (3) using supernodes or supermeshes.

- A circuit is linear if it contains only linear elements and independent sources. For single-input linear circuits, the proportionality property states that any output is proportional to the input. For multiple-input linear circuits, the superposition principle states that any output can be found by summing the output produced when each input acts alone.

- A Thévenin equivalent circuit consists of a voltage source in series with a resistance. A Norton equivalent circuit consists of a current source in parallel with a resistance. The Thévenin and Norton equivalent circuits are related by a source transformation.

- The parameters of the Thévenin and Norton equivalent circuits can be determined using any two of the following: (1) the open-circuit voltage at the interface, (2) the short-circuit current at the interface, and (3) the equivalent resistance of the source circuit with all sources turned off.

- The parameters of the Thévenin and Norton equivalent circuits can also be determined using circuit reduction methods or by directly solving for the source i–v relationship using node-voltage or mesh-current analysis.

- For a fixed source and an adjustable load, the maximum interface signal levels are $v_{MAX} = v_{OC}$ $(R_L = \infty)$, $i_{MAX} = i_{SC}$ $(R_L = 0)$, and $p_{MAX} = v_{OC}i_{SC}/4$ $(R_L = R_T)$. When $R_L = R_T$, the source and load are said to be matched.

- Interface signal transfer conditions are specified in terms of the voltage, current, or power delivered to the load. The design constraints depend on the signal conditions specified and the circuit parameters that are adjustable. Some design requirements may require a two-port interface circuit. An interface design problem may have one, many, or no solutions.

PROBLEMS

ERO 3–1 GENERAL CIRCUIT ANALYSIS (SECT. 3–1 TO 3–2)

Given a circuit consisting of linear resistors and independent sources,

(a) (Formulation) Write node-voltage or mesh-current equations for the circuit.

(b) (Solution) Solve the equations from (a) for selected signal variables or input-output relationships.

Node-voltage method:
See Examples 3–1, 3–2, 3–3, 3–4, 3–5, 3–6 and Exercises 3–2, 3–3, 3–4, 3–5, 3–6
Mesh-current method:
See Examples 3–7, 3–8, 3–9, and Exercises 3–8, 3–9, 3–10

3–1 (a) Formulate node-voltage equations for the circuit in Figure P3–1.

(b) Use these equations to find v_x and i_x.

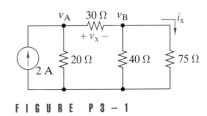

FIGURE P3–1

3–2 (a) Formulate node-voltage equations for the circuit in Figure P3–2.

(b) Use these equations to find v_x and i_x.

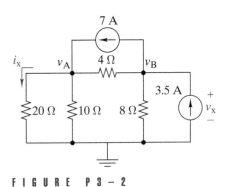

FIGURE P3–2

3–3 (a) Formulate node-voltage equations for the circuit in Figure P3–3.

(b) Use these equations to find v_x and i_x.

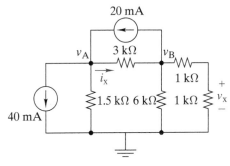

FIGURE P3–3

3–4 (a) Formulate a set of node-voltage equations for the circuit in Figure P3–4.

(b) Use these equations to find v_x and i_x.

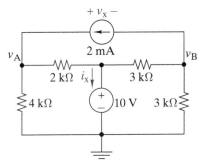

FIGURE P3–4

3–5 (a) Formulate node-voltage equations for the circuit in Figure P3–5.

(b) Solve for v_x and i_x when $v_S = 4$ V, $i_S = 2$ mA, $R_1 = R_2 = 10$ kΩ, and $R_3 = R_4 = 5$ kΩ.

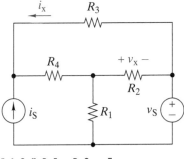

FIGURE P3–5

3–6 (a) Formulate node-voltage equations for the circuit in Figure P3–6.

(b) Solve for v_x and i_x using $R_1 = 10$ kΩ, $R_2 = 10$ kΩ, $R_3 = 40$ kΩ, $R_4 = 20$ kΩ, $v_1 = v_2 = 5$ V, and $v_3 = 15$ V.

(c) Find the power delivered to resistor R_1.

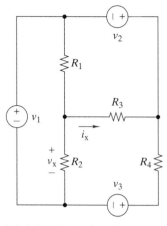

FIGURE P3-6

3-7 **(a)** Formulate node-voltage equations for the circuit in Figure P3-7.
(b) Solve for v_x and i_x for $R_1 = R_2 = R_3 = R_4 = 1\ k\Omega$, $R_5 = R_6 = 4\ k\Omega$, and $i_{S1} = i_{S2} = 50\ mA$.
(c) Find the total power delivered to the circuit by the two current sources.

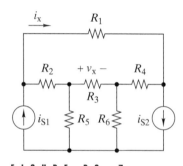

FIGURE P3-7

3-8 **(a)** Formulate node-voltage equations for the circuit in Figure P3-8.
(b) Solve for v_x and i_x using $R_1 = 10\ k\Omega$, $R_2 = 20\ k\Omega$, $R_3 = 60\ k\Omega$, $R_4 = 20\ k\Omega$, $R_x = 3\ k\Omega$, and $v_S = 15\ V$.
(c) Find the power absorbed by resistor R_2.

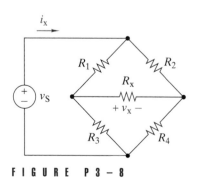

FIGURE P3-8

3-9 **(a)** Formulate mesh-current equations for the circuit in Figure P3-9.
(b) Use these equations to find v_x and i_x.

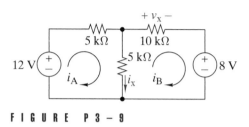

FIGURE P3-9

3-10 **(a)** Formulate mesh-current equations for the circuit in Figure P3-10.
(b) Use these equations to find v_x and i_x.

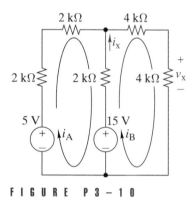

FIGURE P3-10

3-11 **(a)** Formulate mesh-current equations for the circuit in Figure P3-11.
(b) Use these equations to find v_x and i_x.

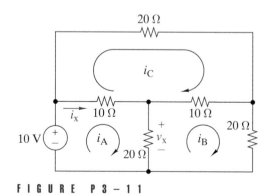

FIGURE P3-11

3–12 **(a)** Formulate mesh-current equations for the circuit in Figure P3–12.

(b) Solve for v_x and i_x using $R_1 = 200\ \Omega$, $R_2 = 500\ \Omega$, $R_3 = 60\ \Omega$, $R_4 = 240\ \Omega$, $R_5 = 200\ \Omega$, $i_S = 50$ mA, and $v_S = 15$ V.

(c) Find the total power dissipated in the circuit.

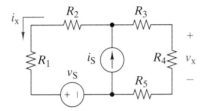

FIGURE P3–12

3–13 **(a)** Formulate mesh-current equations for the circuit in Figure P3–13.

(b) Solve for v_x and i_x using $R_1 = R_2 = 10$ kΩ, $R_3 = 2$ kΩ, $R_4 = 1$ kΩ, $i_S = 2.5$ mA, $v_{S1} = 12$ V, and $v_{S2} = 0.5$ V.

(c) Find the power supplied by v_{S1}.

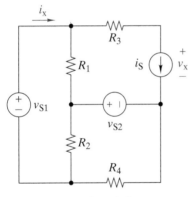

FIGURE P3–13

3–14 The circuit in Figure P3–14 seems to require two supermeshes since both current sources appear in two meshes. However, a circuit diagram can sometimes be rearranged to eliminate the need for supermesh equations.

(a) Show that supermeshes in Figure P3–14 can be avoided by connecting resistor R_6 between node A and node D via a different route.

(b) Formulate mesh-current equations for the modified circuit as redrawn in (a).

(c) Solve for v_x using $R_1 = R_2 = R_3 = R_4 = 2$ kΩ, $R_5 = R_6 = 1$ kΩ, $i_{S1} = 40$ mA, and $i_{S2} = 20$ mA.

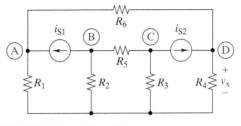

FIGURE P3–14

3–15 **(a)** Formulate mesh-current equations for the circuit in Figure P3–15.

(b) Use these equations to find v_x and i_x.

(c) Find the total power delivered to the resistors.

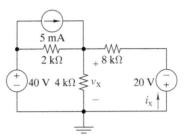

FIGURE P3–15

3–16 **(a)** Formulate mesh-current equations for the circuit in Figure P3–16.

(b) Use these equations to find the input resistance.

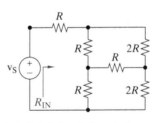

FIGURE P3–16

3–17 Use node-voltage or mesh-current analysis to find i_x in Figure P3–17.

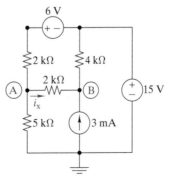

FIGURE P3–17

3–18 Use the node-voltage or mesh-current method in Figure P3–18 to find v_O in terms of v_1, v_2, and v_3.

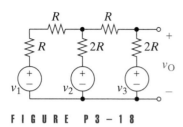

FIGURE P3 – 18

3–19 Find the node voltages v_A and v_B in Figure P3–19.

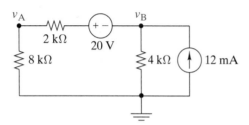

FIGURE P3 – 19

3–20 Find the mesh currents i_A, i_B, and i_C in Figure P3–20.

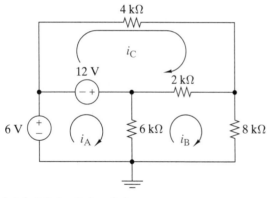

FIGURE P3 – 20

ERO 3–2 LINEARITY PROPERTIES (SECT. 3–3)

(a) Given a circuit containing linear resistors and one independent source, use the proportionality principle to find selected signal variables.
(b) Given a circuit containing linear resistors and two or more independent sources, use the superposition principle to find selected signal variables.

See Examples 3–10, 3–11, 3–12 and Exercises 3–11, 3–12, 3–13, 3–14, 3–15

3–21 Find the proportionality constant $K = v_O/i_S$ for the circuit in Figure P3–21.

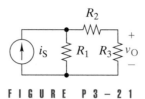

FIGURE P3 – 21

3–22 Find the proportionality constant $K = i_O/i_S$ for the circuit in Figure P3–22.

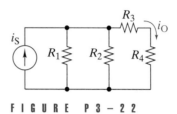

FIGURE P3 – 22

3–23 Find the proportionality constant $K = v_O/i_S$ for the circuit in Figure P3–23.

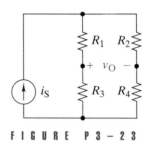

FIGURE P3 – 23

3–24 Use the unit output method to find v_O in the circuit in Figure P3–24.

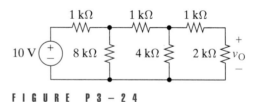

FIGURE P3 – 24

3–25 Use the unit output method to find i_O in the circuit in Figure P3–25.

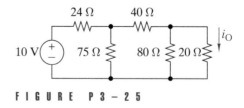

FIGURE P3 – 25

3–26 Use the unit output method to select R in the circuit in Figure P3–26 so that the proportionality constant $K = v_O/v_S = 1/4$.

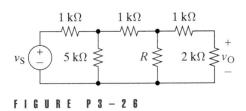

FIGURE P3-26

3–27 Find the proportionality constant $K = v_O/v_S$ for the circuit in Figure P3–27.

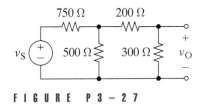

FIGURE P3-27

3–28 Use the superposition principle in the circuit of Figure P3–28 to find v_O.

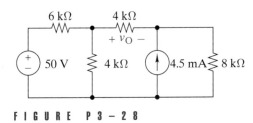

FIGURE P3-28

3–29 Use the superposition principle in the circuit of Figure P3–29 to find i_O.

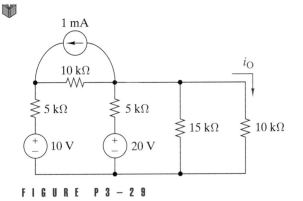

FIGURE P3-29

3–30 Use the superposition principle in the circuit of Figure P3–30 to find v_O in terms of v_1, v_2, and R.

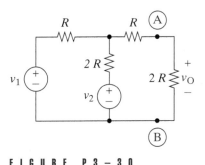

FIGURE P3-30

3–31 Use the superposition principle in the circuit of Figure P3–31 to find v_O in terms of i_1, i_2, and R.

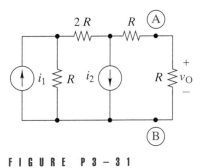

FIGURE P3-31

3–32 A linear circuit containing two sources drives a 100-Ω load resistor. Source number 1 delivers 250 mW to the load when source number 2 is off. Source number 2 delivers 4 W to the load when source number 1 is off. Find the power delivered to the load when both sources are on. *Hint*: The answer is not 4.25 W. Why?

3–33 A linear circuit is driven by an independent voltage source $v_S = 10$ V and an independent current source $i_S = 10$ mA. When the voltage source is on and the current source is off, the output voltage is $v_O = 2$ V. When both sources are on, the output is $v_O = 1$ V. Find the output when $v_S = 5$ V and $i_S = -10$ mA.

3–34 The following table lists test data of the output of a linear resistive circuit for different values of its three inputs. Find the input–output relationship for the circuit.

$v_{S1}(V)$	$v_{S2}(V)$	$v_{S3}(V)$	$v_O(V)$
0	4	−4	0
2	0	2	1.5
2	4	0	2

3–35 **D** This problem involves designing a resistive circuit with two inputs v_{S1} and v_{S2} and a single output voltage v_O. Design the circuit so that $v_O = K(v_{S1} + 3v_{S2})$ is delivered across a 500-Ω load. The value of K is not specified but should be greater than 1/20.

ERO 3–3 THÉVENIN AND NORTON EQUIVALENT CIRCUITS (SECT. 3–4)

Given a circuit containing linear resistors and independent sources,

(a) Find the Thévenin or Norton equivalent at a specified pair of terminals.
(b) Use the Thévenin or Norton equivalent to find the signals delivered to linear or nonlinear loads.

See Examples 3–13, 3–14, 3–15, 3–16 and Exercises 3–16, 3–17, 3–18

3–36 (a) Find the Thévenin or Norton equivalent circuit seen by R_L in Figure P3–36.
(b) Use the equivalent circuit found in (a) to find i_L in terms of i_S, R_1, R_2, and R_L.
(c) Check your answer in (b) using current division.

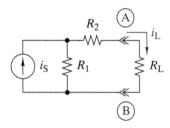

FIGURE P3–36

3–37 (a) Find the Thévenin or Norton equivalent circuit seen by R_L in Figure P3–37.
(b) Use the equivalent circuit found in (a) to find load power when $R_L = 50$ Ω, 100 Ω, and 500 Ω.

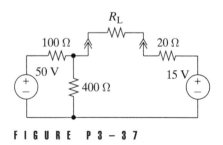

FIGURE P3–37

3–38 (a) Find the Thévenin or Norton equivalent seen by R_L in Figure P3–38.
(b) Use the equivalent circuit found in (a) to find load voltage when $R_L = 10$ kΩ, 25 kΩ, 50 kΩ, and 100 kΩ.

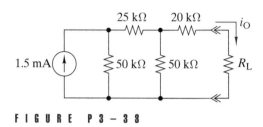

FIGURE P3–38

3–39 (a) Find the Thévenin or Norton equivalent seen by R_L in Figure P3–39.
(b) Use the equivalent circuit found in (a) to find i_O for $R_L = 6$ kΩ, $R_L = 12$ kΩ, 24 kΩ, and 48 kΩ.

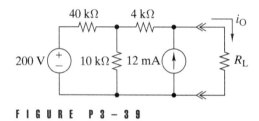

FIGURE P3–39

3–40 (a) Find the Thévenin or Norton equivalent at terminals A and B in Figure P3–40.
(b) Use the equivalent circuit to find interface power when a 10-Ω load is connected between terminals A and B.
(c) Repeat (b) when a 5-V source is connected between terminals A and B with the plus terminal at terminal A.

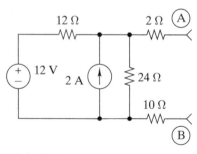

FIGURE P3–40

3–41 (a) Find the Thévenin or Norton equivalent seen to the left of terminals A and B in Figure P3–41. (*Hint*: Use source transformations and circuit reduction.)

(b) Use the equivalent found in (a) to find the output v_O in terms of v_1, v_2, and R.

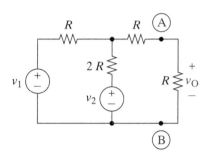

F I G U R E   P 3 – 4 1

3–42 Figure P3–42 shows a source circuit with two accessible terminals. When $i = 0$ the output voltage is $v = 10$ V. When a 2.4-kΩ resistor is connected between the terminals the output drops to 6 V.

(a) Find the Thévenin equivalent of the source.

(b) Use the equivalent circuit to find the power the source would deliver to resistive loads of 500 Ω, 1 kΩ, and 2 kΩ.

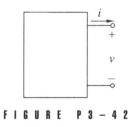

F I G U R E   P 3 – 4 2

3–43 Figure P3–42 shows a source circuit with two accessible terminals. When a 300-Ω resistor is connected across the accessible terminals, the output current is $i = 30$ mA. When a 500-Ω resistor is connected, the output current is $i = 20$ mA. How much current would this source deliver to a 10-V source?

3–44 Figure P3–42 shows a source circuit with two accessible terminals. Some voltage and current measurements at the accessible terminals are

v (V)	−10	−5	0	+5	+10	12	13	14
i (mA)	+5	+4	+3	+2	+1	0	−1	−2

(a) Plot the source i–v characteristic using these data.

(b) Develop a Thévenin equivalent circuit valid on the range $|v| < 10$ V.

(c) Use the equivalent circuit to predict the source v_{OC} and i_{SC}.

(d) Compare your results in (c) with the given measurements and explain any differences.

3–45 The Thévenin equivalent parameters of a two-terminal source are $v_T = 5$ V and $R_T = 150$ Ω. Find the minimum allowable load resistance if the delivered load voltage must exceed 3.5 V.

3–46 Use a sequence of source transformations to find the Thévenin equivalent at terminals A and B in Figure P3–46.

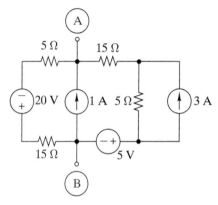

F I G U R E   P 3 – 4 6

3–47 (a) Find the Thévenin or Norton equivalent seen by R_L in Figure P3–47.

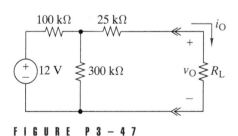

F I G U R E   P 3 – 4 7

(b) Use the equivalent circuit to find the value of R_L that produces $v_O = 6$ V.

3–48 The current delivered to R_L in Figure P3–47 is observed to be $i_O = 36$ μA. Find the value of R_L.

3–49 A nonlinear resistor is connected across a two-terminal source whose Thévenin equivalent is $v_T = 5$ V and $R_T = 500$ Ω.

(a) Plot the *i–v* characteristic of the source in the first quadrant ($i \geq 0$, $v \geq 0$).

(b) The *i–v* characteristic of the resistor is $i = 10^{-4}$ $(v + 2 v^{3.3})$. Plot this characteristic on the source plot obtained in (a) and graphically determine the voltage across and current through the nonlinear resistor.

3–50 A nonlinear resistor is connected across a two-terminal source whose Thévenin equivalent is $v_T = 10$ V and $R_T = 200$ Ω.

(a) Plot the *i–v* characteristic of the source in the first quadrant ($i \geq 0$, $v \geq 0$).

(b) The *i–v* characteristic of the resistor is $v = 4000 i^2$. Plot this characteristic on the source plot obtained in (a) and graphically determine the voltage across and current through the nonlinear resistor.

ERO 3–4 MAXIMUM SIGNAL TRANSFER (SECT. 3–5)

Given a circuit containing linear resistors and independent sources,

(a) Find the maximum voltage, current, and power available at a specified pair of terminals.

(b) Find the resistive loads required to obtain the maximum available signal levels.

See Example 3–17 and Exercise 3–19

3–51 (a) The load resistance in Figure P3–51 is adjusted until the maximum power is delivered. Find the power delivered and the value of R_L.

(b) The load resistance is adjusted until maximum voltage is delivered. Find the voltage delivered and the value of R_L.

(c) The load resistance is adjusted until maximum current is delivered. Find the current delivered and the value of R_L.

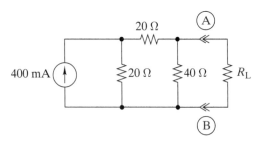

FIGURE P 3 – 5 1

3–52 Find the maximum power available to the load resistance in Figure P3–52. What value of R_L will extract maximum power?

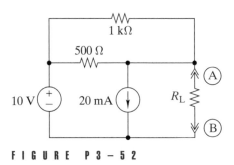

FIGURE P 3 – 5 2

3–53 Find the maximum power available to the load resistance in Figure P3–53. What value of R_L will extract maximum power?

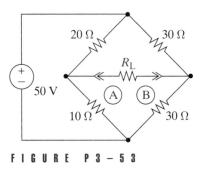

FIGURE P 3 – 5 3

3–54 The resistance R in Figure P3–54 is adjusted until maximum power is delivered across the interface to the load consisting of R and the 6-kΩ resistor in parallel. Find the voltage and power delivered to the load and the value of R.

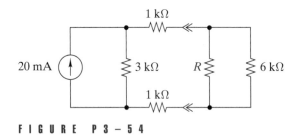

FIGURE P 3 – 5 4

3–55 When a 2-kΩ resistor is connected across a two-terminal source, a current of 0.75 mA is delivered to the load. When a second 2-kΩ resistor is connected in parallel with the first, a total current of 1 mA is delivered. Find the maximum power available from the source and specify the load resistance required to extract maximum power from the source.

3–56 **(a)** The potentiometer in Figure P3–56 is adjusted until maximum power is delivered to the 2-kΩ load. Find the wiper position x. *Caution*: The load is fixed.

(b) The potentiometer in Figure P3–56 is adjusted until maximum voltage is delivered to the 2-kΩ load. Find the wiper position x.

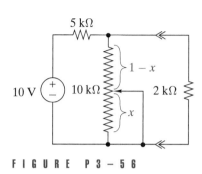

FIGURE P3–56

3–57 Find the value of R_L in Figure P3–57 such that $i_L = 3$ mA.

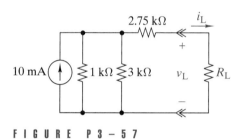

FIGURE P3–57

3–58 Find the value of R_L in Figure P3–57 such that $v_L = 5$ V.

3–59 A 15-V source with 100 Ω internal resistance is connected in series with a resistor R_S. Select the value of R_S so that outputs of the series combination are bounded by $i < 50$ mA and $p < 200$ mW for any load resistance.

3–60 A practical source delivers 50 mA to a 300-Ω load. The source delivers 12 V to a 120-Ω load. Find the maximum power available from the source.

ERO 3–5 INTERFACE CIRCUIT DESIGN (SECT. 3–6)

Given the signal transfer objectives at a source-load interface, adjust the circuit parameters or design one or more two-port interface circuits to achieve the specified objectives within stated constraints. See Examples 3–18, 3–19, 3–20, 3–21, 3–22 and Exercises 3–20, 3–21

3–61 **D** Figure P3–61 shows a two-port interface circuit connecting source and load circuits. In this problem $v_S = 10$ V, $R_S = 50$ Ω, and the load is a 50-Ω resistor. To

avoid damaging the source, its output current must be less than 100 mA. Design a resistive interface circuit so that the voltage delivered to the load is 4 V and the source current is less than 100 mA.

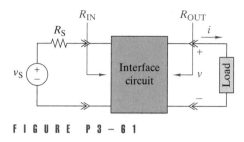

FIGURE P3–61

3–62 **D** Figure P3–61 shows a two-port interface circuit connecting source and load circuits. In this problem $v_S = 15$ V, $R_S = 1$ kΩ, and the load is a diode whose i–v characteristic is

$$i = 10^{-15}(e^{40v} - 1)$$

Design an interface circuit that dissipates less than 50 mW and produces an interface voltage of $v = 0.7$ V.

3–63 **D** Figure P3–61 shows a two-port interface circuit connecting source and load circuits. In this problem $v_S = 15$ V, $R_S = 100$ Ω, and the load is a 2-kΩ resistor. Design a resistive interface circuit so that the voltage delivered to the load is 10 V ±10% using one or more of the following standard resistors: 68 Ω, 100 Ω, 220 Ω, 330 Ω, 470 Ω, and 680 Ω. The resistors all have a tolerance of ±5%, which you must account for in your design.

3–64 **D** In Figure P3–61 the load is a 500-Ω resistor and $R_S = 75$ Ω. Design an interface circuit so that the input resistance of the two-port is 75 Ω ±10% and the output resistance seen by the load is 500 Ω ±10%.

3–65 **D** Use the resistor array in Figure P3–65 to design a two-port resistance circuit whose voltage gain $K = v_O/v_{IN}$ is at least 0.6 and whose input resistance is close to 100 Ω.

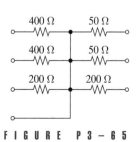

FIGURE P3–65

3–66 Ⓓ Figure P3–61 shows a two-port interface circuit connecting source and load circuits. In this problem $v_S = 12$ V, $R_S = 300$ Ω, and the load is a 50-Ω resistor. Design an interface circuit so that $v < 2$ V and $i > 40$ mA.

3–67 Ⓓ Figure P3–61 shows a two-port interface circuit connecting source and load circuits. In this problem $v_S = 10$ V, $R_S = 50$ Ω, and the load is a 50-Ω resistor. Design an interface so that 350 mW is delivered to the load.

3–68 Ⓓ Figure P3–61 shows a two-port interface circuit connecting a source and load. In this problem the source with $v_S = 5$ V and $R_S = 5$ Ω is to be used in production testing of two-terminal semiconductor devices. The devices are to be connected as the load in Figure P3–61 and have highly nonlinear and variable i–v characteristics. The normal operating range for acceptable devices is $\{i > 10$ mA or $v > 0.7$ V$\}$ and $\{p < 10$ mW$\}$. Design an interface circuit so that the operating point always lies within the specified normal range regardless of the test article's i–v characteristic.

3–69 Ⓓ Figure P3–69 shows an interface circuit connecting a source circuit and a load. Design an interface circuit so that $v < 4$ V regardless of the load resistance.

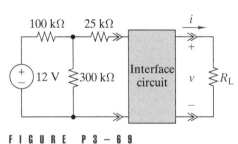

FIGURE P3-69

3–70 Ⓓ Design the interface circuit in Figure P3–69 so that the power delivered to the load never exceeds 1 mW regardless of the load resistance.

INTEGRATING PROBLEMS

3–71 Ⓐ COMPARISON OF ANALYSIS METHODS

The circuit in Figure P3–71 is called an R-$2R$ ladder for obvious reasons. The additive property of linear circuits states that the output of the form

$$v_O = k_1 v_1 + k_2 v_2 + k_3 v_3$$

We could use superposition to find the gains k_1, k_2, and k_3. But applying the input sources one at a time may not be the best way to find these gains. The purpose of this problem is to compare two other methods.

(a) Write node-voltage equations for the circuit in Figure P3–71 using node B as the reference node. Solve these equations to show that $v_O = v_1/8 + v_2/4 + v_3/2$.

(b) We can also use circuit reduction techniques since the circuit is a ladder. Use successive source transformations and series/parallel equivalence to reduce the circuit between nodes A and B to a single equivalent voltage source connected in series with an equivalent resistor. Use this equivalent circuit to show that $v_O = v_1/8 + v_2/4 + v_3/2$.

(c) Which method do you think is easier and why?

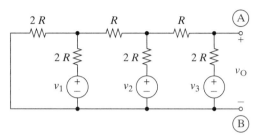

FIGURE P3-71

3–72 Ⓐ THREE-TERMINAL DEVICE MODELING

Figure P3–72 shows a three-terminal device with a voltage source v_S connected at the input and a 50-Ω load resistor connected at the output. The purpose of the problem is to develop an input–output relationship for the device using the following experimental data.

v_S(V)	−12	−9	−6	−3	0	+3	+6	+9	+12
v_O(V)	3	2.5	2	1	0	−1	−2	−2	−2

The device input resistance is $R_{IN} = 1$ kΩ when the 50-Ω load is connected at the output.

(a) Plot v_O versus v_S and state whether the graph is linear or nonlinear.

(b) Devise a linear model of the form $v_O = Kv_S$ that approximates the data for $|v_S| < 6$ V.

(c) Use your model from part (b) to predict the circuit output for $v_S = \pm 10$ V and explain why the predictions do not agree with the experimental results.

(d) For $v_S = 1$ V use your model from part (c) to find the circuit power gain = P_O/P_{IN}.

(e) In view of the result in (d), is the device active or passive?

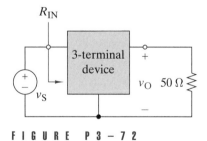

FIGURE P3-72

3–73 🅰 ADJUSTABLE THÉVENIN EQUIVALENT

Figure P3–73 shows an ideal voltage source in parallel with an adjustable potentiometer. This problem concerns the effect of adjusting the potentiometer on the interface Thévenin equivalent circuit and the signals available at the interface.

(a) Find the parameters v_T and R_T of the Thévenin equivalent circuit at the interface in terms of circuit parameters k, v_S, and R.

(b) Find the value of R_T for $k = 0$, $k = 0.5$, and $k = 1$. Explain your results physically in terms of the position of the movable arm of the potentiometer.

(c) What is the power available at the interface for $k = 0$, $k = 0.5$, and $k = 1$? Justify your answers physically in terms of the position of the movable arm of the potentiometer.

(d) A load resistor $R_L = R/4$ is connected across the interface. Where should the potentiometer be positioned ($k = ?$) to transfer the maximum power to this load?

(e) A load resistor $R_L = R/4$ is connected across the interface. Where should the potentiometer be positioned ($k = ?$) to deliver an interface voltage $v = v_S/4$ to this load?

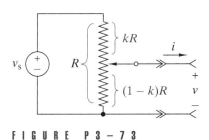

FIGURE P3–73

3–74 🅳 BATTERY DESIGN

A satellite requires a battery with an open-circuit voltage $v_{OC} = 36$ V and a Thévenin resistance $R_T \leq 10$ Ω. The battery is to be constructed using series and parallel combinations of one of two types of cells. The first type has $v_{OC} = 9$ V, $R_T = 4$ Ω, and a weight of 40 grams. The second type has $v_{OC} = 4$ V, $R_T = 0.5$ Ω, and a weight of 15 grams. Design a minimum weight battery to meet the open-circuit voltage and Thévenin resistance requirements.

3–75 🅔 TTL TO ECL CONVERTER

 It is claimed that the resistive circuit in Figure P3–75 converts transistor-transistor logic (TTL) input signals into output signals compatible with emitter coupled logic (ECL). Specifically, the claim is that for any output current in the range -0.025 mA $\leq i \leq 0.025$ mA, the circuit converts any input in the TTL low range ($0 \leq v_S \leq 0.4$ V) to an output in the ECL low range (-1.7 V $\leq v \leq -1.5$ V), and converts any input in the TTL high range ($3.0 \leq v_S \leq 3.8$ V) to an output in the ECL high range (-0.9 V $\leq v \leq -0.6$ V). The purpose of this problem is to verify this claim.

(a) The output voltage can be written in the form

$$v = k_1 v_S + k_2 i$$

Write a KCL equation at the output with the current i as an unknown and solve for the constants k_1 and k_2.

(b) Use the relationship found in (a) to verify that the output voltage falls in one of the allowed ECL ranges for every allowed combination of TTL inputs and load currents.

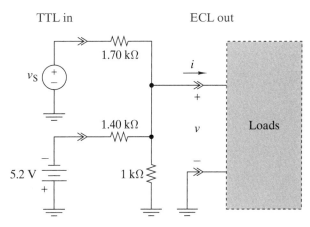

FIGURE P3–75

CHAPTER 4

ACTIVE CIRCUITS

Then came the morning of Tuesday, August 2, 1927, when the concept of the negative feedback amplifier came to me in a flash while I was crossing the Hudson River on the Lacakawana Ferry, on my way to work.

Harold S. Black, 1927,
American Electrical Engineer

4–1 Linear Dependent Sources

4–2 Analysis of Circuits with Dependent Sources

4–3 The Transistor

4–4 The Operational Amplifier

4–5 OP AMP Circuit Analysis

4–6 OP AMP Circuit Design

4–7 The Comparator

Summary

Problems

Integrating Problems

The integrated circuit operational amplifier (OP AMP) is the workhorse of present-day linear electronic circuits. However, to operate as a linear amplifier the OP AMP must be provided with "negative feedback." The negative feedback amplifier is one of the key inventions of all time. During the 1920s, Harold S. Black (1898–1983) worked for several years without much success on the problem of improving the performance of vacuum tube amplifiers in telephone systems. The feedback amplifier solution came to him suddenly on his way to work. He documented his invention by writing the key concepts of negative feedback on his morning copy of the *New York Times*. His invention paved the way for the development of worldwide communication systems and spawned whole new areas of technology, such as feedback control systems and robotics.

An **active circuit** contains one or more devices that require an external power supply to operate correctly. These active devices control the flow of electrical power from the external supply and can operate in both linear and nonlinear modes. In their linear range active devices can be modeled using the four dependent source elements introduced in the first section. The second section shows how dependent sources change the properties of resistive circuits and how we modify our methods of analysis to cover these new linear elements. Dependent sources and resistors are used to model many active devices, including the transistor and OP AMP devices discussed in the third and fourth sections. The fourth section also develops an ideal OP AMP model that is useful in circuit analysis and design. In the fifth section the ideal OP AMP model is used to analyze circuit realizations of basic analog signal-processing functions, such as amplifiers, summers, subtractors, and inverters. These basic building blocks are then used to design OP AMP circuits that realize specified signal-processing functions. The comparator section treats a case in which the OP AMP intentionally operates as a nonlinear element.

4–1   LINEAR DEPENDENT SOURCES

This chapter treats the analysis and design of circuits containing active devices, such as transistors or operational amplifiers (OP AMPs). An **active device** is a component that requires an external power supply to operate correctly. An **active circuit** is one that contains one or more active devices. An important property of active circuits is that they are capable of providing signal amplification, one of the most important signal-processing functions in electrical engineering. Linear active circuits are governed by the proportionality property, so their input-output relationships are of the form $y = Kx$. The term **signal amplification** means the proportionality factor K is greater than 1 when the input x and output y have the same dimensions. Thus, active circuits can deliver more signal voltage, current, and power at their output than they receive from the input signal. The passive resistance circuits studied thus far cannot produce voltage, current, or power gains greater than unity.

Active devices operating in a linear mode are modeled using resistors and one or more of the dependent sources shown in Figure 4–1. A **dependent source** is a voltage or current source whose output is controlled by a voltage or current in a different part of the circuit. As a result, there are

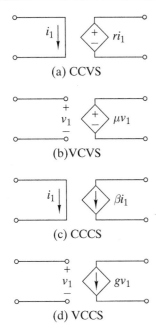

FIGURE 4–1 *Dependent source circuit symbols: (a) Current-controlled voltage source. (b) Voltage-controlled voltage source. (c) Current-controlled current source. (d) Voltage-controlled current source.*

four possible types of dependent sources: a current-controlled voltage source (CCVS), a voltage-controlled voltage source (VCVS), a current-controlled current source (CCCS), and a voltage-controlled current source (VCCS). The properties of these dependent sources are very different from those of the independent sources described in Chapter 2. The output voltage (current) of an independent voltage (current) source is a specified value that does not depend on the circuit to which it is connected. To distinguish between the two types of sources, the dependent sources are represented by the diamond symbol in Figure 4–1, in contrast to the circle symbol used for independent sources.

Caution: This book uses the diamond symbol to indicate a dependent source and the circle to show an independent source. However, some books and circuit analysis programs use the circle symbol for both dependent and independent sources.

A **linear dependent source** is one whose output is proportional to the controlling voltage or current. The defining relationship for dependent sources in Figure 4–1 are all of the form $y = Kx$, where x is the controlling variable, y is the source output variable, and K is the proportionality factor. Each type of dependent source is characterized by a proportionality factor, either μ, β, r, or g. These parameters are often called simply the **gain** of the controlled source. Strictly speaking, the parameters μ and β are dimensionless quantities called the **voltage gain** and **current gain**, respectively. The parameter r has the dimensions of ohms and is called the **transresistance**, a contraction of transfer resistance. The parameter g is then called **transconductance** and has the dimensions of siemens.

Although dependent sources are elements used in circuit analysis, they are conceptually different from the other circuit elements we have studied. The linear resistor and ideal switch are models of actual devices called resistors and switches. However, you will not find dependent sources listed in electronic part catalogs. For this reason, dependent sources are more abstract, since they are not models of identifiable physical devices. Dependent sources are used in combination with other resistive elements to create models of active devices.

In Chapter 3 we found that a voltage source acts as a short circuit when it is turned off. Likewise, a current source behaves as an open circuit when it is turned off. The same results apply to dependent sources, with one important difference. Dependent sources cannot be turned on and off individually because they depend on excitation supplied by independent sources.

Some consequences of this dependency are illustrated in Figure 4–2. When the independent current source is turned on, KCL requires that $i_1 = i_S$. Through controlled source action, the current controlled voltage source is on and its output is

$$v_O = r i_1 = r i_S$$

When the independent current source is off ($i_S = 0$), it acts as an open circuit and KCL requires that $i_1 = 0$. The dependent source is now off and its output is

$$v_O = r i_1 = 0$$

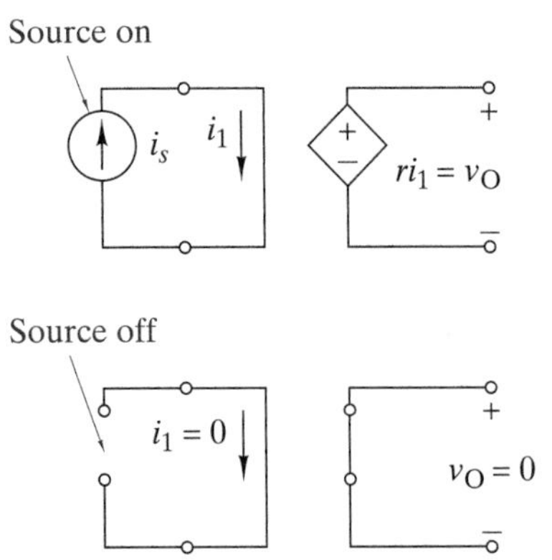

Source on

Source off

FIGURE 4–2 *Turning off the independent source affects the dependent source.*

When the independent voltage source is off, the dependent source acts as a short circuit.

In other words, turning the independent source on and off turns the dependent source on and off as well. We must be careful when applying the superposition principle and Thévenin's theorem to active circuits, since the state of a dependent source depends on the excitation supplied by independent sources. To account for this possibility, we modify the superposition principle to state that the response due to all *independent* sources acting simultaneously is equal to the sum of the responses due to each *independent* source acting one at a time.

EXAMPLE 4–1

Figure 4–3(a) shows the symbols used in Orcad Capture to represent the four dependent sources. These elements are found in the **Place Part/Analog** symbol library as part names starting with the letters:

E/Analog for a voltage-controlled voltage source   (VCVS)

F/Analog for a current-controlled current source   (CCCS)

G/Analog for a voltage-controlled current source   (VCCS)

H/Analog for a current-controlled voltage source   (CCVS)

The input ports are open circuits for the voltage-controlled sources and short-circuits for the current-controlled sources. The output ports are voltage sources or current sources depending on the controlled variable. Note that Orcad uses circles rather than diamonds to represent dependent sources. All four dependent sources are characterized by a single parameter called "GAIN." This parameter is found in the "Property Editor" which appears when you double click on the controlled source symbol. Once the desired gain is entered, click on **Apply** to update the value for subsequent simulations. The units of the gain depend on the units of the signals at the input and output ports.

Figure 4–3(b) shows the same symbols for Electronics Workbench. These elements are available in the **Sources** group in the **Parts Bin** menu bar. The Electronics Workbench screen symbols for dependent sources are diamonds.

4–2 ANALYSIS OF CIRCUITS WITH DEPENDENT SOURCES

With certain modifications the analysis tools developed for passive circuits apply to active circuits as well. Circuit reduction applies to active circuits, but in so doing we must not eliminate the control variable for a dependent source. As noted previously, when applying the superposition principle or Thévenin's theorem, we must remember that dependent sources cannot be turned on and off independently since their states depend on excitation supplied by one or more independent sources. Applying a source transformation to a dependent source is sometimes helpful, but again we must not lose the identity of a controlling signal for a dependent source. Methods like node and mesh analysis can be adapted to include dependent sources as well.

FIGURE 4-3 *Dependent source representations. (a) Orcad Capture. (b) Electronics Workbench.*

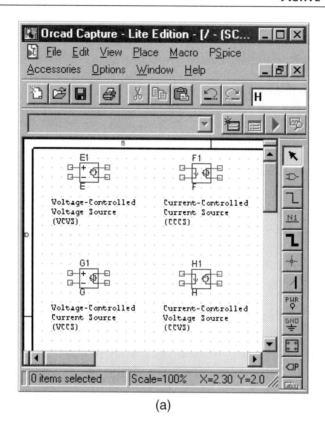

(a)

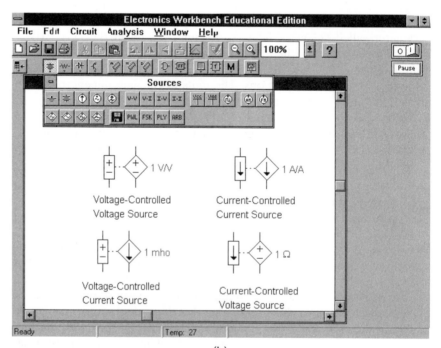

(b)

However, the main difference is that the properties of active circuits can be significantly different from those of the passive circuits treated in Chapters 2 and 3. Our analysis examples are chosen to highlight these differences.

In our first example the objective is to determine the current, voltage, and power delivered to the 500-Ω output load in Figure 4–4. The control current i_x is found using current division in the input circuit:

$$i_x = \left(\frac{50}{50 + 25}\right) i_S = \frac{2}{3} i_S \qquad (4\text{--}1)$$

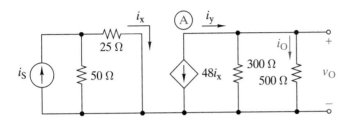

FIGURE 4 – 4 *A circuit with a dependent source.*

Similarly, the output current i_O is found using current division in the output circuit:

$$i_O = \left(\frac{300}{300 + 500}\right) i_y = \frac{3}{8} i_y \qquad (4\text{--}2)$$

At node A, KCL requires that $i_y = -48i_x$. Combining this result with Eqs. (4–1) and (4–2) yields the output current:

$$i_O = \left(\frac{3}{8}\right)(-48)i_x = (-18)\left(\frac{2}{3}i_S\right)$$

$$= -12\,i_S \qquad (4\text{--}3)$$

The output voltage v_O is found using Ohm's law:

$$v_O = i_O\,500 = -6000\,i_S \qquad (4\text{--}4)$$

The input-output relationships in Eqs. (4–3) and (4–4) are of the form $y = Kx$ with $K < 0$. The proportionality constants are negative because the reference direction for i_O in Figure 4–4 is the opposite of the orientation of the dependent source reference arrow. Active circuits often produce negative values of K, which means that the input and output signals have opposite algebraic signs. Circuits for which $K < 0$ are said to provide **signal inversion**. In the analysis and design of active circuits, it is important to keep track of signal inversions.

Using Eqs. (4–3) and (4–4), the power delivered to the 500-Ω load in Figure 4–4 is

$$p_O = v_O i_O = (-6000\,i_S)(-12\,i_S) = 72{,}000 i_S^2 \qquad (4\text{--}5)$$

The independent source at the input delivers its power to the parallel combination of 50 Ω and 25 Ω. Hence, the input power supplied by the independent source is

$$p_S = (50 \parallel 25)i_S^2 = \left(\frac{50}{3}\right)i_S^2$$

Given the input power and output power, we find the power gain in the circuit:

$$\text{Power gain} = \frac{p_O}{p_S} = \frac{72{,}000i_S^2}{(50/3)i_S^2} = 4320$$

A power gain greater than unity means that the circuit delivers more power at its output than it receives from the input source. At first glance this appears to be a violation of energy conservation, until we remember that dependent sources are models of active devices that require an external power supply to operate. Usually the external power supply is not shown in circuit diagrams. When using a dependent source to model an active circuit, we assume that the external supply and the active device itself can handle whatever power is required by the circuit. When designing the actual circuit, the engineer must make certain that the active device and its power supply operate within their power ratings.

Exercise 4–1

Find the output v_O in terms of the input v_S in the circuit in Figure 4–5.

F I G U R E   4 – 5

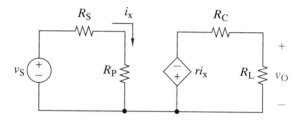

Answer:

$$v_O = \left[\frac{-R_L r}{(R_S + R_P)(R_C + R_L)}\right] v_S$$

NODE-VOLTAGE ANALYSIS WITH DEPENDENT SOURCES

Node analysis of active circuits is much the same as for passive circuits except that we must account for the additional constraints caused by the dependent sources. For example, the circuit in Figure 4–6 has five nodes. With node E as the reference, each independent voltage source has one terminal connected to ground. These connections force the conditions $v_A = v_{S1}$ and $v_B = v_{S2}$. Therefore, we only need to write node equations at nodes C and D because voltages at nodes A and B are already known.

Node analysis involves expressing element currents in terms of the node voltages and applying KCL at each unknown node. The sum of the currents leaving node C is

$$G_1(v_C - v_{S1}) + G_2(v_C - v_{S2}) + G_B v_C + G_P(v_C - v_D) = 0$$

Similarly, the sum of currents leaving node D is

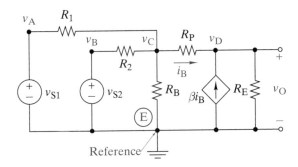

$$G_P(v_D - v_C) + G_E v_D - \beta i_B = 0$$

These two node equations can be rearranged into the form

Node C: $(G_1 + G_2 + G_B + G_P)v_C - G_P v_D = G_1 v_{S1} + G_2 v_{S2}$

Node D: $- G_P v_C + (G_P + G_E)v_D = \beta i_B$ (4–6)

We could write these two symmetrical node equations by inspection if the dependent current source βi_B had been an independent source. Since it is not independent, we must express its controlling variable i_B in terms of the unknown node voltages. Using the fundamental property of node voltages and Ohm's law, we express the current i_B in terms of the node voltages as

$$i_B = G_P(v_C - v_D)$$

Substituting this expression for i_B into Eq. (4–6) and putting the results in standard form yields

Node C: $(G_1 + G_2 + G_B + G_P)v_C - G_P v_D = G_1 v_{S1} + G_2 v_{S2}$

Node D: $-(\beta + 1)G_P v_C + [(\beta + 1)G_P + G_E]v_D = 0$ (4–7)

The result in Eq. (4–7) involves two equations in the two unknown node voltages and includes the effect of the dependent source. Note that including the dependent source constraint destroys the coefficient symmetry in Eq. (4–6).

   This example illustrates a general approach to writing node-voltage equations for circuits with dependent sources. We start out treating the dependent sources as if they are independent sources and write node equations for the resulting passive circuit using the inspection method developed in Chapter 3. This step produces a set of symmetrical node-voltage equations with the independent and dependent source terms on the right-hand side. Then we express the dependent source terms in terms of the unknown node voltages and move them to the left-hand side of the equations with the other terms involving the unknown node voltages. This step destroys the coefficient symmetry but leads to a set of node-voltage equations that describe the active circuit.

EXAMPLE 4−2

For the circuit in Figure 4–6, use the node-voltage equations in Eq. (4–7) to find the output voltage v_O when $R_1 = 1$ kΩ, $R_2 = 3$ kΩ, $R_B = 100$ kΩ, $R_P = 1.3$ kΩ, $R_E = 3.3$ kΩ, and $\beta = 50$.

SOLUTION:

Substituting the given numerical values into Eq. (4–7) yields

$$(2.11 \times 10^{-3})v_C - (7.69 \times 10^{-4})v_D = (10^{-3})v_{S1} + (3.33 \times 10^{-4})v_{S2}$$

$$-(3.92 \times 10^{-2})v_C + (3.95 \times 10^{-2})v_D = 0$$

We solve the second equation for $v_C = 1.008v_D$. When this equation is substituted into the first equation, we obtain

$$v_O = v_D = 0.736v_{S1} + 0.245v_{S2}$$

This circuit is a signal summer that does not involve a signal inversion. The fact that the output is a linear combination of the two inputs reminds us that the circuit is linear. ∎

EXAMPLE 4–3

The circuit in Figure 4–7(a) is a model of an inverting OP AMP circuit.

FIGURE 4 – 7

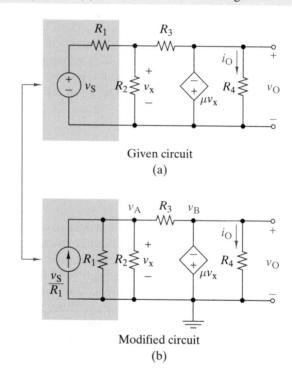

Given circuit
(a)

Modified circuit
(b)

(a) Use node-voltage analysis to find the output v_O in terms of the in-put v_S.
(b) Evaluate the input-output relationship found in (a) as the gain μ becomes very large.

SOLUTION:

(a) Applying a source transformation to the independent source leads to the modified three-node circuit shown in Figure 4–7(b). With the indicated reference node the dependent voltage source constrains the voltage at node B. The control voltage is $v_x = v_A$, and the controlled source forces the node B voltage to be

$$v_B = -\mu v_x = -\mu v_A$$

Thus, node A is the only independent node in the circuit. We can write the node A equation by inspection as

$$(G_1 + G_2 + G_3)v_A - G_3 v_B = G_1 v_S$$

Substituting in the control source constraint yields the standard form for this equation:

$$[G_1 + G_2 + G_3(1 + \mu)] v_A = G_1 v_S$$

We end up with only one node equation even though at first glance the given circuit appears to need three node equations. The reason is that there are two voltage sources in the original circuit in Figure 4–7(a). Since the two sources share the reference node, the number of unknown node voltages is reduced from three to one. The general principle illustrated is that the number of independent KCL constraints in a circuit containing N nodes and N_V voltage sources (dependent or independent) is $N - 1 - N_V$.

The one-node equation can easily be solved for the output voltage $v_O = v_B$.

$$v_O = v_B = -\mu v_A = \left(\frac{-\mu G_1}{G_1 + G_2 + G_3(1 + \mu)} \right) v_S$$

The minus sign in the numerator means that the circuit provides signal inversion. The output voltage does not depend on the value of the load resistor R_4, since the load is connected across an ideal (though dependent) voltage source.

(b) For large gains μ we have $(1 + \mu)G_3 \gg G_1 + G_2$ and the input-output relationship reduces to

$$v_O \approx \left[\frac{-\mu G_1}{(1 + \mu) G_3} \right] v_S \approx -\left[\frac{R_3}{R_1} \right] v_S$$

That is, when the active device gain is large, the voltage gain of the active circuit depends on the ratio of two resistances. We will see this situation again with OP AMP circuits. ∎

Exercise 4–2

(a) Formulate node-voltage equations for the circuit in Figure 4–8.

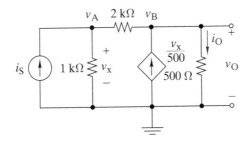

FIGURE 4 - 8

(b) Solve the node-voltage equations for v_O and i_O in terms of i_S.

Answers:

(a)

$$(1.5 \times 10^{-3}) v_A - (0.5 \times 10^{-3}) v_B = i_S$$

$$-(2.5 \times 10^{-3}) v_A + (2.5 \times 10^{-3}) v_B = 0$$

(b) $v_O = 1000 i_S$; $i_O = 2i_S$.

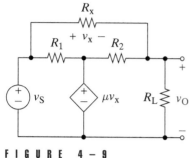

FIGURE 4–9

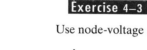

Use node-voltage analysis to find v_O in Figure 4–9.

Answer:

$$v_O = \frac{G_x + \mu G_2}{G_x + G_L + (\mu + 1)G_2} \, v_S$$

MESH-CURRENT ANALYSIS WITH DEPENDENT SOURCES

Mesh-current analysis of active circuits follows the same pattern noted for node-voltage analysis. We initially treat the dependent sources as independent sources and write the mesh equations of the resulting passive circuit using the inspection method from Chapter 3. We then account for the dependent sources by expressing their constraints in terms of unknown mesh currents. The following example illustrates the method.

EXAMPLE 4–4

(a) Formulate mesh-current equations for the circuit in Figure 4–10.
(b) Use the mesh equations to find v_O and R_{IN} when $R_1 = 50\ \Omega$, $R_2 = 1\ \text{k}\Omega$, $R_3 = 100\ \Omega$, $R_4 = 5\ \text{k}\Omega$, and $g = 100\ \text{mS}$.

FIGURE 4–10

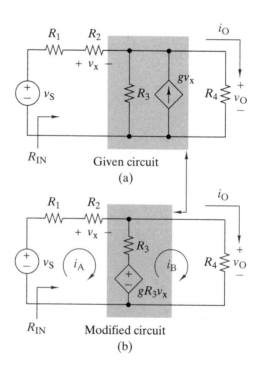

Given circuit
(a)

Modified circuit
(b)

SOLUTION:
(a) Applying source transformation to the parallel combination of R_3 and gv_x in Figure 4–10(a) produces the dependent voltage source $R_3gv_x = \mu v_x$ in Figure 4–10(b). In the modified circuit we have identified two mesh currents. Initially treating the dependent source $(gR_3)v_x$ as an independent source leads to two symmetrical mesh equations.

$$\text{Mesh A: } (R_1 + R_2 + R_3)i_A - R_3 i_B = v_S - (gR_3)v_x$$

$$\text{Mesh B: } \quad -R_3 i_A + (R_3 + R_4)i_B = (gR_3)v_x$$

The control voltage v_x can be written in terms of mesh currents.

$$v_x = R_2 i_A$$

Substituting this equation for v_x into the mesh equations and putting the equations in standard form yields

$$(R_1 + R_2 + R_3 + gR_2R_3)i_A - R_3 i_B = v_S$$

$$-(R_3 + gR_2R_3)i_A + (R_3 + R_4)i_B = 0$$

The resulting mesh equations are not symmetrical because of the controlled source.

(b) Substituting the numerical values into the mesh equations gives

$$(1.115 \times 10^4)i_A - (10^2)i_B = v_S$$

$$-(1.01 \times 10^4)i_A + (5.1 \times 10^3)i_B = 0$$

Solving for the two mesh currents yields

$$i_A = \frac{\Delta_A}{\Delta} = \frac{\begin{vmatrix} v_S & -10^2 \\ 0 & 5.1 \times 10^3 \end{vmatrix}}{\begin{vmatrix} 1.115 \times 10^4 & -10^2 \\ -1.01 \times 10^4 & 5.1 \times 10^3 \end{vmatrix}} = \frac{5.1 \times 10^3 v_S}{5.5855 \times 10^7}$$

$$= 0.9131 \times 10^{-4} v_S$$

$$i_B = \frac{\Delta_B}{\Delta} = \frac{\begin{vmatrix} 1.115 \times 10^4 & v_S \\ -1.01 \times 10^4 & 0 \end{vmatrix}}{5.5885 \times 10^7} = 1.808 \times 10^{-4} v_S$$

The output voltage and input resistance are found using Ohm's law.

$$v_O = R_4 i_B = 0.904 v_S$$

$$R_{IN} = \frac{v_S}{i_A} = 10.95 \text{ k}\Omega \qquad \blacksquare$$

EXAMPLE 4–5

The circuit in Figure 4–11 is a model of a bipolar junction transistor operating in the active mode. Use mesh analysis to find the transistor base current i_B.

FIGURE 4 – 1 1 *Junction transistor cir-cuit model.*

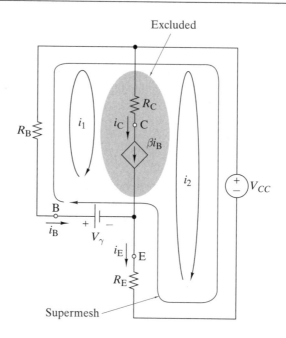

SOLUTION:

The two mesh currents in Figure 4–11 are labeled i_1 and i_2 to avoid possible confusion with the transistor base current i_B. As drawn, the circuit requires a supermesh since the dependent current source βi_B is included in both meshes and is not connected in parallel with a resistance. A supermesh is created by combining meshes 1 and 2 after excluding the series subcircuit consisting of βi_B and R_C. Beginning at the bottom of the circuit, we write a KVL mesh equation around the supermesh using unknowns i_1 and i_2:

$$i_2 R_E - V_\gamma + i_1 R_B + V_{CC} = 0$$

This KVL equation provides one equation in the two unknown mesh currents. Since the two mesh currents have opposite directions through the dependent current source βi_B, the currents i_1, i_2, and βi_B are related by KCL as

$$i_1 - i_2 = \beta i_B$$

This constraint supplies the additional relationship needed to obtain two equations in the two unknown mesh-current variables. Since $i_B = -i_1$, the preceding KCL constraint means that $i_2 = (\beta + 1)i_1$. Substituting $i_2 = (\beta + 1)i_1$ into the supermesh KVL equation and solving for i_B yields

$$i_B = -i_1 = \frac{V_{CC} - V_\gamma}{R_B + (\beta + 1)R_E} \qquad \blacksquare$$

Exercise 4–4

Use mesh analysis to find the current i_O in Figure 4–12 when the element E is a dependent current source $2i_x$ with the reference arrow directed down.

   Answer:
–0.857 mA

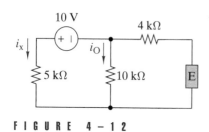

FIGURE 4 – 1 2

Exercise 4-5

Use mesh analysis to find the current i_O in Figure 4–12 when the element E is a dependent voltage source $2000i_X$ with the plus reference at the top.

Answer:
–0.222 mA

EXAMPLE 4-6

The circuit in Fig. 4–13 is a small signal model of a field effect transistor (FET) amplifier with two inputs v_{S1} and v_{S2}. Use Orcad Capture to find the input-output relationship of the circuit.

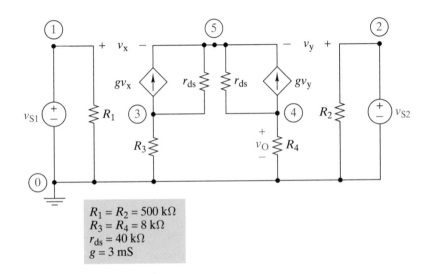

$R_1 = R_2 = 500 \text{ k}\Omega$
$R_3 = R_4 = 8 \text{ k}\Omega$
$r_{ds} = 40 \text{ k}\Omega$
$g = 3 \text{ mS}$

FIGURE 4-13 *Field effect transistor amplifier.*

SOLUTION:
Since the circuit is linear the input-output relationship is of the form

$$v_O = K_1 v_{S1} + K_2 v_{S2}$$

The gain K_1 can be found by setting $v_{S1} = 1$ and $v_{S2} = 0$ and solving for v_O. The gain K_2 is then found by setting $v_{S1} = 0$ and $v_{S2} = 1$ and again solving for v_O. We now turn to Orcad to perform these operations.

Figure 4–14 shows an Orcad Schematics circuit diagram for the case $v_{S1} = 1$ V and $v_{S2} = 0$. In Figure 4–14 the node voltage V(N00521) corresponds to the output v_O in Figure 4–13. Figure 4–14 also shows the resulting netlist and a portion of the output file that includes the dc solution for the five node voltages. In particular the output file reports that V(N00521) = 10, which means that $K_1 = $ V(N00521)$/v_{S1} = 10$.

To find K_2 we change the source attributes to $v_{S1} = 0$ and $v_{S2} = 1$ and again simulate the circuit using Orcad Capture. For this case the output file reports V(N00521) = –10.000, which means that $K_2 = $ V(N00521)$/v_{S2} = -10$. Taken together, the two Orcad simulations mean that the input-output relationship for the circuit is

FIGURE 4-14 *Orcad Capture schematic diagram of the circuit in Figure 4–13.*

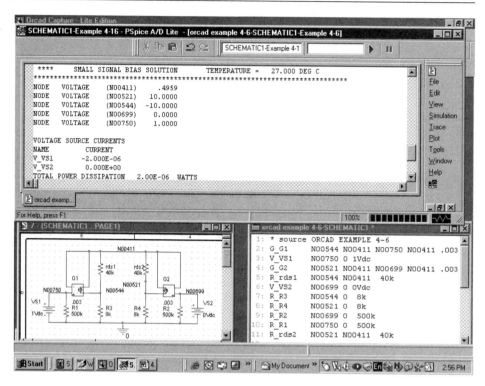

$$v_O = 10(v_{S1} - v_{S2})$$

The circuit is a differential amplifier of a type often used as the input stage of an OP AMP.   ■

THÉVENIN EQUIVALENT CIRCUITS WITH DEPENDENT SOURCES

To find the Thévenin equivalent of an active circuit, we must leave the independent sources on or else supply excitation from an external test source. This means that the Thévenin resistance cannot be found by the lookback method because that method requires that all independent sources be turned off. Turning off the independent sources deactivates the dependent sources as well and can result in a profound change in the input and output characteristics of an active circuit. Thus, there are two ways of finding active circuit Thévenin equivalents. We can either find the open-circuit voltage and short-circuit current at the interface or directly solve for the interface i–v relationship.

EXAMPLE 4-7

Find the input resistance of the circuit in Figure 4–15.

SOLUTION:

With the independent source turned off ($i_{IN} = i_S = 0$), the resistance seen at the input port is R_E since the dependent current source βi_{IN} is inactive and acts like an open circuit. Applying KCL at node A with the input source turned on yields

$$i_E = i_{IN} + \beta i_{IN} = (\beta + 1)i_{IN}$$

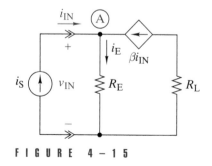

FIGURE 4-15

By Ohm's law, the input voltage is

$$v_{IN} = i_E R_E = (\beta + 1)i_{IN} R_E$$

Hence, the active input resistance is

$$R_{IN} = \frac{v_{IN}}{i_{IN}} = (\beta + 1)R_E$$

The circuit in Figure 4–15 is a model of a transistor circuit in which the gain parameter β typically lies between 50 and 100. The input resistance with external excitation is $(\beta + 1)R_E$, which is significantly different from the value of R_E without external excitation. ∎

EXAMPLE 4–8

Find the Thévenin equivalent at the output interface of the circuit in Figure 4–16.

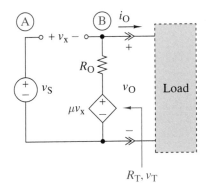

FIGURE 4–16

SOLUTION:
In this circuit the controlled voltage v_x appears across an open circuit between nodes A and B. By the fundamental property of node voltages, $v_x = v_S - v_O$. With the load disconnected and the input source turned off ($v_x = 0$), the dependent voltage source μv_x acts like a short circuit, and the Thévenin resistance looking back into the output port is R_O. With the load connected and the input source turned on, the sum of currents leaving node B is

$$\frac{v_O - \mu v_x}{R_O} + i_O = 0$$

Using the relationship $v_x = v_S - v_O$ to eliminate v_x and then solving for v_O produces the i–v characteristic at the output interface as

$$v_O = \frac{\mu v_S}{\mu + 1} - i_O \left[\frac{R_O}{\mu + 1} \right]$$

The i–v relationship of a Thévenin circuit is $v = v_T - iR_T$. By direct comparison, we find the Thévenin parameters of the active circuit to be

$$v_T = \frac{\mu v_S}{\mu + 1} \quad \text{and} \quad R_T = \frac{R_O}{\mu + 1}$$

The circuit in Figure 4–16 is a model of an OP AMP circuit called a voltage follower. The resistance R_O for a general-purpose OP AMP is around 100 Ω, while the gain μ is about 10^5. Thus, the active Thévenin resistance of the voltage follower is not 100 Ω, as the lookback method suggests, but is only a milliohm.

Exercise 4–6

Find the input resistance and output Thévenin equivalent circuit of the circuit in Figure 4–17.

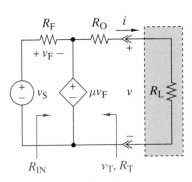

FIGURE 4–17

Answers:

$$R_{IN} = (1 + \mu)R_F$$

$$v_T = \frac{\mu}{\mu + 1} v_S$$

$$R_T = R_O$$

4–3 THE TRANSISTOR

We have defined four linear dependent sources and shown how to analyze circuits containing these active elements. In this and the next section we show how dependent sources are used to model semiconductor devices like transistors and OP AMPs. The transistor model used here describes the voltages and currents at its external terminals. The model does not describe the transistor's physical structure or internal charge flow. Those subjects are left to subsequent courses in semiconductor materials and devices.

The two basic transistor types are the *bipolar junction transistor* (BJT) and the *field effect transistor* (FET). Both types have several possible operating modes, each with a different set of *i–v* characteristics. This is something new in our study. Up to this point the characteristics of circuit elements have been fixed. With the transistor we encounter a device whose *i–v* characteristics can change. We concentrate on the BJT because its *i–v* characteristics are much easier to understand than the FET. Because it is easier to understand, the simpler BJT best serves as a prelude to our study of the OP AMP—an important semiconductor device that also has several possible operating modes.

The circuit symbol of the BJT is shown in Figure 4–18. The device has three terminals called the **emitter (E)**, the **base (B)**, and the **collector (C)**. The voltages v_{BE} and v_{CE} are called the *base-emitter* and *collector-emitter* voltage, respectively. The three currents i_E, i_B, and i_C are called the emitter, base, and collector currents.

Applying KCL to the BJT as a whole yields

$$i_E = i_B + i_C$$

which means that only two of the three currents can be independently specified. We normally work with i_B and i_C, and use KCL to find i_E when it is needed.

The BJT's large-signal model is defined in terms of input signals i_B and v_{BE}, and output signals i_C and v_{CE}. For the BJT shown in Figure 4–18 the model applies to a region in which these signals are never negative. Within this region there are three possible operating modes. The **active mode** is the dominant feature of a BJT. In this mode the collector current i_C is controlled by the base current i_B and v_{BE} is constant.

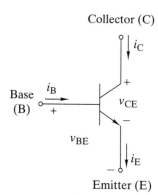

$$\text{Active Mode:} \quad i_C = \beta i_B \quad \text{and} \quad v_{BE} = V_\gamma \qquad (4\text{–}8)$$

The proportionality factor β is called the *forward current gain* and typically ranges from about 50 to several hundred. The constant V_γ is called the *threshold voltage,* which is normally less than a volt. Figure 4–19(a) shows the circuit elements that model the active mode *i–v* characteristics as defined in Eq. (4–8). In the active mode i_B and v_{CE} are determined by the interaction of these *i–v* characteristics with the rest of the circuit.

FIGURE 4 – 1 8 *Circuit symbol for the bipolar junction transistor.*

Two additional operating modes exist at the boundary of the BJT's operating region. When $i_B = 0$ and $i_C = 0$, the transistor is in the **cutoff mode**, and the device acts like an open circuit between the collector and emitter. When $v_{CE} = 0$ and $v_{BE} = V_\gamma$ the transistor is in the **saturation mode,** and the device acts like a short circuit between the collector and emitter. These two modes are summarized by writing

$$\text{Cutoff Mode:} \quad i_B = 0 \quad \text{and} \quad i_C = 0$$
$$\text{Saturation Mode:} \quad v_{CE} = 0 \quad \text{and} \quad v_{BE} = V_\gamma \tag{4-9}$$

Figure 4–19(b) and (c) show the circuit elements that model the i–v characteristics defined in Eq. (4–9).

The circuit in Figure 4–19(b) points out that in the cutoff mode v_{CE} must equal the open-circuit voltage available from the external circuit. The circuit in Figure 4–19(c) points out that in the saturation mode, i_C must equal the short-circuit current available from the external circuit. The net result is that the BJT's output variables must fall within the following bounds:

$$
\begin{array}{cc}
\text{Cutoff} & \text{Saturation} \\
\text{Bounds} & \text{Bounds} \\
\end{array} \tag{4-10}
$$

$$
\begin{aligned}
0 &\leq i_C \leq i_{SC} \\
v_{OC} &\geq v_{CE} \geq 0
\end{aligned}
$$

where v_{OC} and i_{SC} are the open-circuit voltage and short-circuit current available between the collector and emitter terminals. In the cutoff mode the transistor outputs i_C and v_{CE} are equal to their respective cutoff bounds. In saturation mode the outputs equal their saturation bounds. In the active mode the outputs fall between the cutoff and saturation bounds.

With this background we are prepared to analyze the transistor circuit in Figure 4–20.[1] Our analysis objective is to find the outputs i_C and v_{CE}. To do this we must know the transistor's operating mode. To find the operating mode we make use of two facts:

1. The lower bounds in Eq. (4–10) mean that i_C and v_{CE} cannot be negative.

2. The upper bounds in Eq. (4–10) depend on the rest of the circuit.

For the circuit in Figure 4–20 these upper bounds are $v_{OC} = V_{CC}$ and $i_{SC} = V_{CC}/R_C$.

Our analysis strategy *assumes* the device is in the active mode and uses the active mode device equations to find i_C. According to Eq. (4–8) the active mode element equations are $v_{BE} = V_\gamma$ and $i_C = \beta i_B$. Using these element constraints and applying KVL around the input loop in Figure 4–20 yields the collector current as

$$i_C = \beta i_B = \beta \left(\frac{v_S - V_\gamma}{R_B} \right) \tag{4-11}$$

This equation indicates that if $v_S > V_\gamma$ then $i_C > 0$. However, if $v_S < V_\gamma$ then $i_C < 0$, which would violate its cutoff bound. Thus, if the input voltage v_S is

---

1 This circuit is called the common-emitter configuration because the emitter terminal is common to the input and output loops.

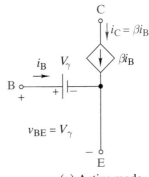

(a) Active mode

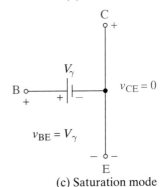

(b) Cutoff mode

(c) Saturation mode

FIGURE   4 – 1 9   *Circuit models for transistor operating modes.*

FIGURE 4–20

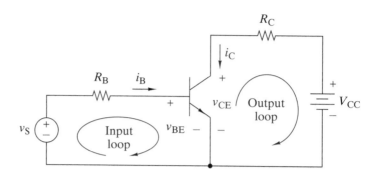

greater than the threshold voltage V_γ, then the BJT can be in the active mode. But, if $v_S < V_\gamma$ the BJT is in the cutoff mode and the outputs equal their cutoff bounds in Eq. (4–10), namely $i_C = 0$ and $v_{CE} = v_{OC} = V_{CC}$.

When $v_S > V_\gamma$, Eq. (4–11) predicts a positive collector current that increases linearly with v_S. To find the collector-emitter voltage we apply KVL around the output loop in Figure 4–20 to obtain

$$v_{CE} = V_{CC} - i_C R_C \qquad (4\text{--}12)$$

This equation predicts that $v_{CE} > 0$ as long as $i_C < V_{CC}/R_C$. But V_{CC}/R_C is the short-circuit current available from the external circuit. Thus, as long as $v_S > V_\gamma$ and $i_C < i_{SC}$, the BJT is in the active mode and Eqs. (4–11) and (4–12) correctly predict the outputs i_C and v_{CE}. However, if Eq. (4–11) predicts that $i_C > i_{SC}$, then Eq. (4–12) says that $v_{CE} < 0$. Both of these results violate the saturation bounds in Eq. (4–10). When this happens the BJT is actually in the saturation mode and the outputs equal their saturation bounds in Eq. (4–10), namely $i_C = i_{SC} = V_{CC}/R_C$ and $v_{CE} = 0$.

Figure 4–21 summarizes this discussion using graphs of the outputs v_{CE} and i_C versus the input voltage v_S. When $v_S < V_\gamma$, the BJT is in the cutoff mode and the outputs are $i_C = 0$ and $v_{CE} = v_{OC} = V_{CC}$. When $v_S > V_\gamma$, the BJT enters the active mode and the outputs i_C and v_{CE} are governed by Eqs. (4–11) and (4–12). Under these equations, i_C increases linearly as v_S increases, with the result that v_{CE} decreases linearly. The collector current continues to increase as v_S increases until it reaches its saturation bound at

FIGURE 4–21 *Transfer characteristics of the transistor circuit in Figure 4–20.*

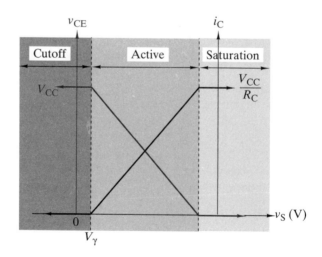

$i_C = i_{SC}$. At that point the transistor switches into the saturation mode and thereafter the outputs remain constant at $i_C = i_{SC} = V_{CC}/R_C$ and $v_{CE} = 0$.

Before working some examples it is worth noting that the graph of v_{CE} versus v_S is called the *transfer characteristic* of the transistor circuit in Figure 4–20. In digital applications the input voltage drives the transistor between the cutoff and saturation modes passing through the active mode as quickly as possible. In analog circuit applications the transistor remains in the active mode where the slope of the transfer characteristic provides voltage amplification. In the next section we find that the OP AMP has similar transfer characteristics.

EXAMPLE 4–9

Suppose that the circuit parameters in Figure 4–20 are $\beta = 100$, $V_\gamma = 0.7$ V, $R_B = 100$ kΩ, $R_C = 1$ kΩ, and $V_{CC} = 5$ V. Find i_C and v_{CE} when $v_S = 2$ V. Repeat when $v_S = 6$ V.

SOLUTION:
Since $v_S = 2$ is greater than $V_\gamma = 0.7$, the transistor is *not* in the cutoff mode. We assume that it is in the active mode and use Eq. (4–11) to calculate i_C.

$$i_C = \beta \left(\frac{v_S - V_\gamma}{R_B} \right) = 100 \left(\frac{2 - 0.7}{100 \times 10^3} \right) = 1.3 \text{ mA}$$

The available short-circuit current is $i_{SC} = V_{CC}/R_C = 5$ mA. Since the calculated i_C is less than i_{SC}, the transistor is in fact in the active mode and we use Eq. (4–12) to find v_{CE}.

$$v_{CE} = V_{CC} - i_C R_C = 5 - 1.3 \times 10^{-3} \times 1000 = 3.7 \text{ V}$$

For $v_S = 2$ V, the transistor is in the active mode and the outputs are $i_C = 1.3$ mA and $v_{CE} = 3.7$ V.

For $v_S = 6$ V, we again assume that the transistor is in the active mode and calculate the collector current from Eq. (4–11).

$$i_C = \beta \left(\frac{v_S - V_\gamma}{R_B} \right) = 100 \left(\frac{6 - 0.7}{100 \times 10^3} \right) = 5.3 \text{ mA}$$

The calculated i_C is greater than the available i_{SC}. For this input the transistor is in the saturation mode and the outputs equal their saturation bounds, namely $i_C = i_{SC} = 5$ mA and $v_{CE} = 0$. ∎

EXAMPLE 4–10

Suppose the circuit parameters in Figure 4–20 are $\beta = 50$, $V_\gamma = 0.7$ V, $R_B = 20$ kΩ, $R_C = 2.2$ kΩ, and $V_{CC} = 10$ V. Find the range of the input voltage v_S for which the transistor remains in the active mode.

SOLUTION:
To avoid the cutoff mode, the input voltage must exceed the transistor threshold voltage. Hence, the lower bound is $v_S > V_\gamma = 0.7$ V. To avoid saturation the collector current must be less than the available short-circuit

current of $i_{SC} = V_{CC}/R_C = 4.545$ mA. To ensure active mode operation, we use Eq. (4–11) to bound the collector current as

$$i_C = \beta \left(\frac{v_S - V_\gamma}{R_B} \right) = 50 \left(\frac{v_S - 0.7}{20 \times 10^3} \right) < 4.545 \times 10^{-3} \text{ A}$$

Solving the inequality for v_S yields

$$v_S < 0.7 + \frac{4.545 \times 20}{50} = 2.518 \text{ V}$$

The transistor operates in the active mode when the input falls in the range $0.7 < v_S < 2.518$ V. ∎

EXAMPLE 4-11

The known parameters in Figure 4–22 are $\beta = 100$, $V_\gamma = 0.7$ V, $R_C = 1$ kΩ, and $V_{CC} = 5$ V. The circuit is to function as a digital inverter that meets two conditions:

1. An input of $v_S = 0$ V must produce an output $v_{CE} = 5$ V.
2. An input of $v_S = 5$ V must produce an output $v_{CE} = 0$ V.

Select a value of R_B so that the circuit meets these conditions.

FIGURE 4-22

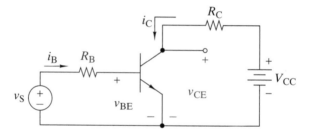

SOLUTION:
To meet the first condition, the transistor must be in the cutoff mode since the required output is $v_{CE} = 5$ V $= v_{OC} = V_{CC}$. For $v_S = 0$ the transistor will be in cutoff regardless of the value of R_B since the input is less than the threshold voltage $V_\gamma = 0.7$ V. To meet the second condition, the transistor must be in the saturation mode since the required output is $v_{CE} = 0$. In this case we know that the input is $v_S = 5$ V. We know that the outputs are $v_{CE} = 0$ and $i_C = i_{SC} = 5/1000 = 5$ mA. We also know that in the saturation mode $v_{BE} = V_\gamma = 0.7$ V. What we don't know is a value of R_B that will make all this happen.

To be in saturation the collector current *predicted* by the active mode i–v characteristics, namely, $i_C = \beta i_B$, must exceed the available short-circuit current. This does not mean that the collector current must exceed the short-circuit current. In fact, it can't because the transistor is in saturation with $i_C = i_{SC}$. What it does mean is that $\beta i_B > i_{SC} = 5$ mA. Using Eq. (5–11) to place a bound on the quantity βi_B yields

$$\beta i_B = \beta \left(\frac{v_S - V_\gamma}{R_B} \right) = 100 \left(\frac{5 - 0.7}{R_B} \right) > 5 \times 10^{-3} \text{ A}$$

Solving the inequality for R_B yields

$$R_B < \frac{100 \times 4.3}{5 \times 10^{-3}} = 86 \text{ k}\Omega$$

Any reasonable value less than 86 kΩ (say 56 kΩ, a standard value) will work. ∎

Exercise 4–7

The circuit parameters in Figure 4–22 are $\beta = 100$, $V_\gamma = 0.7$ V, $R_C = 1$ kΩ, $R_B = 100$ kΩ, and $V_{CC} = 5$ V. Find i_C and v_{CE} when $v_S = 5$ V.

Answers:

$i_C = 4.3$ mA; $v_{CE} = 0.7$ V.

Exercise 4–8

The circuit parameters in Figure 4–22 are $\beta = 100$, $V_\gamma = 0.7$ V, $R_C = 1$ kΩ, and $V_{CC} = 5$ V. Find the value of R_B so that $v_{CE} = 2.5$ V when $v_S = 5$ V.

Answer:

$R_B = 172$ kΩ.

4–4   THE OPERATIONAL AMPLIFIER

The integrated circuit operational amplifier is the premier linear active device in present-day analog circuit applications. The term *operational amplifier* was apparently first used in a 1947 paper by John R. Ragazzini and his colleagues, who reported on work carried out for the National Defense Research Council during World War II. The paper described high-gain dc amplifier circuits that perform mathematical operations (addition, subtraction, multiplication, division, integration, etc.), hence the name *operational amplifier.* For more than a decade the most important applications were general- and special-purpose analog computers using vacuum tube amplifiers. In the early 1960s general-purpose, discrete-transistor operational amplifiers became readily available, and by the mid-1960s the first commercial integrated circuit OP AMPs entered the market. The transition from vacuum tubes to integrated circuits decreased the size, power consumption, and cost of OP AMPs by nearly three orders of magnitude. By the early 1970s the integrated circuit version became the dominant active device in analog circuits.

The device itself is a complex array of transistors, resistors, diodes, and capacitors all fabricated and interconnected on a tiny silicon chip. Figure 4–23 shows examples of ways OP AMPs are packaged for use in circuits. In spite of its complexity, the device can be modeled by rather simple *i–v* characteristics. We do not need to concern ourselves with what is going on inside the package; rather, we treat the OP AMP using a behavioral model that constrains the voltages and currents at the external terminals of the device.

OP AMP NOTATION

Certain matters of notation and nomenclature must be discussed before developing a circuit model for the OP AMP. The OP AMP is a five-terminal device, as shown in Figure 4–24(a). The "+" and "–" symbols identify the

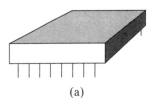

(a)

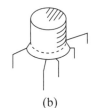

(b)

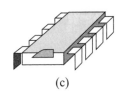

(c)

FIGURE   4 – 2 3   *Examples of integrated circuit OP AMP packages: (a) Encapsulated hybrid. (b) TO-5 can. (c) Dual in-line package.*

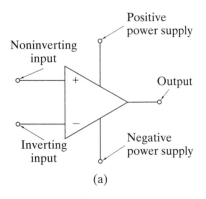

(a)

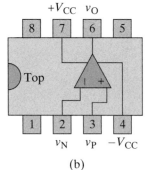

(b)

FIGURE 4-24 *The OP AMP: (a) Circuit symbol. (b) Pin out diagram for an eight-pin package.*

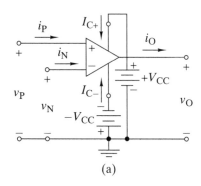

(a)

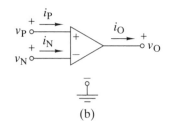

(b)

FIGURE 4-25 *OP AMP voltage and current definitions: (a) Complete set. (b) Shorthand set.*

input terminals and are a shorthand notation for the noninverting and inverting input terminals, respectively. These "+" and "−" symbols identify the two input terminals and have nothing to do with the polarity of the voltages applied. The other terminals are the output and the positive and negative supply voltages, usually labeled $+V_{CC}$ and $-V_{CC}$. While some OP AMPs have more than five terminals, these five are always present and are the only ones we will use in this text. Figure 4–24(b) shows how these terminals are arranged in a common eight-pin integrated circuit package.

The two power supply terminals in Figure 4–24 are not usually shown in circuit diagrams. Be assured that they are always there because the external power supplies are required for the OP AMP to operate as an active device. The power required for signal amplification comes through these terminals from an external power source. The $+V_{CC}$ and $-V_{CC}$ voltages applied to these terminals also determine the upper and lower limits on the OP AMP output voltage.

Figure 4–25(a) shows a complete set of voltage and current variables for the OP AMP, while Figure 4–25(b) shows the abbreviated set of signal variables we will use. All voltages are defined with respect to a common reference node, usually ground. Voltage variables v_P, v_N, and v_O are defined by writing a voltage symbol beside the corresponding terminals. This notation means the "+" reference mark is at the terminal in question and the "−" reference mark is at the reference or ground terminal. In this book the reference directions for the currents are directed in at input terminals and out at the output. At times the abbreviated set of current variables may appear to violate KCL. For example, a global KCL equation for the complete set of variables in Figure 4–25(a) is

$$i_O = I_{C+} + I_{C-} + i_P + i_N \quad \text{(correct equation)} \quad (4\text{–}13)$$

A similar equation using the shorthand set of current variables in Figure 4–25(b) reads

$$i_O = i_N + i_P \quad \text{(incorrect equation)} \quad (4\text{–}14)$$

This equation is *not* correct, since it does not include all the currents. What is more important, it implies that the output current comes from the inputs. In fact, this is wrong. The input currents are very small, ideally zero. The output current comes from the supply voltages, as Eq. (4–13) points out, even though these terminals are not shown on the abbreviated circuit diagram.

TRANSFER CHARACTERISTICS

The dominant feature of the OP AMP is the transfer characteristic shown in Figure 4–26. This characteristic provides the relationships between the **noninverting input** v_P, the **inverting input** v_N, and the **output voltage** v_O. The transfer characteristic is divided into three regions or modes called **+saturation**, **−saturation**, and **linear**. In the linear region the OP AMP is a **differential amplifier** because the output is proportional to the difference between the two inputs. The slope of the line in the linear range is called the voltage gain. In this linear region the input-output relation is

$$v_O = A(v_P - v_N) \quad (4\text{–}15)$$

The voltage gain of an OP AMP is very large, usually greater than 10^5. As long as the net input $(v_P - v_N)$ is very small, the output will be proportional to the input. However, when $A|v_P - v_N| > V_{CC}$, the OP AMP is saturated and the output voltage is limited by the supply voltages (less some small internal losses).

In the previous section, we found that the transistor has three operating modes. The input-output characteristic in Figure 4–26 points out that the OP AMP also has three operating modes:

1. +Saturation mode when $A(v_P - v_N) > V_{CC}$ and $v_O = +V_{CC}$.
2. –Saturation mode when $A(v_P - v_N) < -V_{CC}$ and $v_O = -V_{CC}$.
3. Linear mode when $A|v_P - v_N| < V_{CC}$ and $v_O = A(v_P - v_N)$.

Usually we analyze and design OP AMP circuits using the model for the linear mode. When the operating mode is not given, we use a self-consistent approach similar to the one used for the transistor. That is, we assume that the OP AMP is in the linear mode and then calculate the output voltage v_O. If it turns out that $-V_{CC} < v_O < +V_{CC}$, then the assumption is correct and the OP AMP is indeed in the linear mode. If $v_O < -V_{CC}$, then the assumption is wrong and the OP AMP is in the –saturation mode with $v_O = -V_{CC}$. If $v_O > +V_{CC}$, then the assumption is wrong and the OP AMP is in the +saturation mode with $v_O = +V_{CC}$.

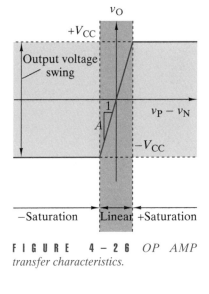

FIGURE 4–26 *OP AMP transfer characteristics.*

IDEAL OP AMP MODEL

A dependent-source model of an OP AMP operating in its linear range is shown in Figure 4–27. This model includes an input resistance (R_I), an output resistance (R_O), and a voltage-controlled voltage source whose gain is A. Numerical values of these OP AMP parameters typically fall in the following ranges:

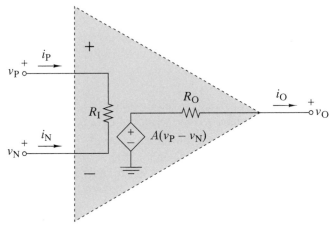

FIGURE 4–27 *Dependent source model of an OP AMP operating in the linear mode.*

$$10^6 < R_I < 10^{12} \ \Omega$$

$$10 < R_O < 100 \ \Omega$$

$$10^5 < A < 10^8$$

Clearly, high input resistance, low output resistances, and high voltage gain are the key attributes of an OP AMP.

The dependent-source model can be used to develop the i–v relationships of the ideal model. For the OP AMP to operate in its linear mode, the output voltage is bounded by

$$-V_{CC} \le v_O \le +V_{CC}$$

Using Eq. (4–15), we can write this bound as

$$-\frac{V_{CC}}{A} \le (v_P - v_N) \le +\frac{V_{CC}}{A}$$

The supply voltage V_{CC} is typically about 15 V, while A is a very large number, usually 10^5 or greater. Consequently, linear operation requires that $v_P \approx v_N$. In the ideal OP AMP model, the voltage gain is assumed to be infinite ($A \to \infty$), in which case linear operation requires $v_P = v_N$. The input resistance R_I of the ideal OP AMP is assumed to be infinite, so the currents entering both input terminals are both zero. In summary, the i–v relationships of the **ideal model** of the OP AMP are

$$v_P = v_N$$

$$i_P = i_N = 0 \tag{4–16}$$

The implications of these element equations are illustrated on the OP AMP circuit symbol in Figure 4–28.

At first glance the element constraints of the ideal OP AMP appear to be fairly useless. They look more like connection constraints and are totally silent about the output quantities (v_O and i_O), which are usually the signals of greatest interest. They seem to say that the OP AMP input terminals are simultaneously a short circuit ($v_P = v_N$) and an open circuit ($i_P = i_N = 0$). In practice, however, the ideal model of the OP AMP is very useful because in linear applications feedback is always present. That is, for the OP AMP to operate in a linear mode, it is necessary for there to be feedback paths from the output to one or both of the inputs. These feedback paths ensure that $v_P \approx v_N$ and make it possible for us to analyze OP AMP circuits using the ideal OP AMP element constraints in Eq. (4–16).

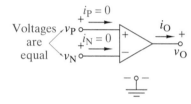

F I G U R E 4 – 2 8 *Ideal OP AMP characteristics.*

NONINVERTING OP AMP

To illustrate the effects of feedback, let us find the input-output characteristics of the circuit in Figure 4–29. In this circuit the voltage divider provided a feedback path from the output to the inverting input.[1] Since the ideal OP AMP draws no current at either input ($i_P = i_N = 0$), we can use voltage division to determine the voltage at the inverting input:

$$v_N = \frac{R_2}{R_1 + R_2} v_O \tag{4–17}$$

The input source connection at the noninverting input requires the condition

$$v_P = v_S \tag{4–18}$$

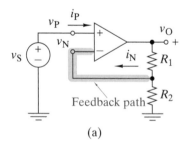

(a)

(b)

F I G U R E 4 – 2 9 *The noninverting amplifier circuit.*

---

1 The feedback must always be to the inverting input, otherwise the circuit will be unstable for reasons that cannot be explained by what we have learned up to now.

The ideal OP AMP element constraints demand that $v_P = v_N$, therefore, we can equate the right sides of Eqs. (4–17) and (4–18) to obtain the input–output relationship of the overall circuit.

$$v_O = \frac{R_1 + R_2}{R_2} v_S \qquad (4\text{--}19)$$

The preceding analysis illustrates a general strategy for analyzing OP AMP circuits. We use normal circuit analysis methods to express the OP AMP input voltages v_P and v_N in terms of circuit parameters. We then use the ideal OP AMP constraint $v_P = v_N$ to solve for the overall circuit input–output relationship.

The circuit in Figure 4–29(a) is called a **noninverting amplifier**. The input-output relationship is of the form $v_O = K v_S$, which reminds us that the circuit is linear. Figure 4–29(b) shows the functional building block for this circuit, where the proportionality constant K is

$$K = \frac{R_1 + R_2}{R_2} \qquad (4\text{--}20)$$

In an OP AMP circuit the proportionality constant K is sometimes called the **closed-loop** gain, because it defines the input-output relationship when the feedback loop is connected (closed).

When discussing OP AMP circuits, it is necessary to distinguish between two types of gains. The first is the large voltage gain provided by the OP AMP device itself. The second is the voltage gain of the OP AMP circuit with a negative feedback path. Note that Eq. (4–20) indicates that the circuit gain is determined by the resistors in the feedback path and not by the value of the OP AMP gain. The gain in Eq. (4–20) is really the voltage division rule upside down. Variation of the value of K depends on the tolerance on the resistors in the feedback path, and not the variation in the value of the OP AMP's gain. In effect, feedback converts the OP AMP's very large but variable gain into a much smaller but well-defined gain.

D DESIGN EXAMPLE 4–12

Design an amplifier with a gain of $K = 10$.

SOLUTION:
Using a noninverting OP AMP circuit, the design problem is to select the values of the resistors in the feedback path. From Eq. (4–20) the design constraint is

$$10 = \frac{(R_1 + R_2)}{R_2}$$

We have one constraint with two unknowns. Arbitrarily selecting $R_2 = 10 \text{ k}\Omega$, we find $R_1 = 90 \text{ k}\Omega$. These resistors would normally have low tolerances ($\pm 1\%$ or less) to produce a precisely controlled closed-loop gain.

Comment: The problem of choosing resistance values in OP AMP circuit design problems deserves some discussion. Although values of resistance from a few ohms to several hundred megohms are commercially available, we generally limit ourselves to the range from about 1 kΩ to perhaps 1 MΩ. The lower limit of 1 kΩ is imposed in part because of power dissipation in the resistors. Typically, we use resistors with ¼-W power ratings or less. The maximum voltage in OP AMP circuits is often around 15 V. The smallest ¼-W resistance we can use is $R_{MIN} \geq (15)^2/0.25 = 900\ \Omega$, or about 1 kΩ. The upper bound of 1 MΩ comes about because surface leakage makes it difficult to maintain the tolerance in a high-value resistance. High-value resistors are also noisy, which leads to problems when they are connected in the feedback path. The 1-kΩ to 1-MΩ range should be used as a guideline and not an inviolate design rule. Actual design choices are influenced by system-specific factors and changes in technology. ■

Exercise 4–9

The noninverting amplifier circuit in Figure 4–29(a) is operating with $R_1 = 2R_2$ and $V_{CC} = \pm12$ V. Over what range of input voltages v_S is the OP AMP in the linear mode?

Answer:
$-4\ V < v_S < +4\ V$

EFFECTS OF FINITE OP AMP GAIN

The ideal OP AMP model has an infinite gain. Actual OP AMP devices have very large, but finite voltage gains. We now address the effect of large but finite gain on the input-output relationships of OP AMP circuits.

The circuit in Figure 4–30 shows a finite gain OP AMP circuit model in which the input resistance R_I is infinite. The actual values of OP AMP input resistance range from 10^6 to 10^{12} Ω, so no important effect is left out by ignoring this resistance. Examining the circuit, we see that the noninverting input voltage is determined by the independent voltage source. The inverting input can be found by voltage division, since the current i_N is zero. In other words, Eqs. (4–17) and (4–18) apply to this circuit as well.

FIGURE 4–30 *The noninverting amplifier circuit with the dependent source model.*

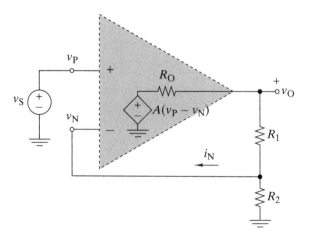

We next determine the output voltage in terms of the controlled source voltage using voltage division on the series connection of the three resistors R_O, R_1, and R_2:

$$v_O = \frac{R_1 + R_2}{R_O + R_1 + R_2} A(v_P - v_N)$$

Substituting v_P and v_N from Eqs. (4–17) and (4–18) yields

$$v_O = \left[\frac{R_1 + R_2}{R_O + R_1 + R_2}\right] A \left[v_S - \frac{R_2}{R_1 + R_2} v_O\right] \qquad (4\text{–}21)$$

The intermediate result in Eq. (4–21) shows that feedback is present since v_O appears on both sides of the equation. Solving for v_O yields

$$v_O = \frac{A(R_1 + R_2)}{R_O + R_1 + R_2(1 + A)} v_S \qquad (4\text{–}22)$$

In the limit, as $A \to \infty$, Eq. (4–22) reduces to

$$v_O = \frac{R_1 + R_2}{R_2} v_S = K v_S$$

where K is the closed-loop gain we previously found using the ideal OP AMP model.

To see the effect of a finite A, we ignore R_O in Eq. (4–22) since it is generally quite small compared with $R_1 + R_2$. With this approximation Eq. (4–22) can be written in the following form:

$$v_O = \frac{K}{1 + (K/A)} v_S \qquad (4\text{–}23)$$

When written in this form, we see that the closed-loop gain reduces to K as $A \to \infty$. Moreover, we see that the finite-gain model yields a good approximation to the ideal model results as long as $K \ll A$. In other words, the ideal model yields good results as long as the closed-loop gain is much less than the gain of the OP AMP device. One practical rule of thumb is to limit the closed-loop gain to less than 1% of the OP AMP gain (i.e., $K < A/100$).

The feedback path also affects the active output resistance. To see this, we construct a Thévenin equivalent circuit using the open-circuit voltage and the short-circuit current. Equation (4–23) is the open-circuit voltage, and we need only find the short-circuit current. Connecting a short-circuit at the output in Figure 4–30 forces $v_N = 0$ but leaves $v_P = v_S$. Therefore, the short-circuit current is

$$i_{SC} = A(v_S/R_O)$$

As a result, the Thévenin resistance is

$$R_T = \frac{v_{OC}}{i_{SC}} = \frac{K/A}{1 + K/A} R_O$$

When $K \ll A$, this expression reduces to

$$R_T = \frac{K}{A} R_O \approx 0 \ \Omega$$

The OP AMP circuit with feedback has an output Thévenin resistance that is much smaller than the output Thévenin resistance of the OP AMP device itself. In fact, the Thévenin resistance is very small since R_O is typically less than 100 Ω and A is greater than 10^5.

At this point we can summarize our discussion. We introduced the OP AMP as an active five-terminal device including two supply terminals not normally shown on the circuit diagram. We then developed an ideal model of this device that is used to analyze and design circuits that have feedback. Feedback must be present for the device to operate in the linear mode. The most dramatic feature of the ideal model is the assumption of infinite gain. Using a finite-gain model, we found that the ideal model predicts the circuit input-output relationship quite closely as long as the circuit gain K is much smaller than the OP AMP gain A. We also discovered that the Thévenin output resistance of an OP AMP with feedback is essentially zero.

In the rest of this book we use the ideal i–v constraints in Eq. (4–16) to analyze OP AMP circuits. The OP AMP circuits have essentially zero output resistance, which means that the output voltage does not change with different loads. Unless otherwise stated, from now on the term *OP AMP* refers to the ideal model.

4–5 OP AMP CIRCUIT ANALYSIS

This section introduces OP AMP circuit analysis using examples that are building blocks for analog signal-processing systems. We have already introduced one of these circuits—the noninverting amplifier discussed in the preceding section. The other basic circuits are the voltage follower, the inverting amplifier, the summer, and the subtractor. The key to using the building block approach is to recognize the feedback pattern and to isolate the basic circuit as a building block. The first example illustrates this process.

EXAMPLE 4–13

Find the input-output relationship of the circuit in Figure 4–31(a).

FIGURE 4–31

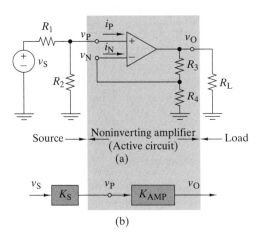

Source ⟶ Noninverting amplifier
(Active circuit)
(a) ⟵ Load

(b)

SOLUTION:
When the circuit is partitioned as shown in Figure 4–31(a), we recognize two building block gains: (1) K_S, the proportionality constant of the source circuit, and (2) K_{AMP}, the gain of the noninverting amplifier. The OP AMP circuit input current $i_P = 0$; hence, we can use voltage division to find K_S as

$$K_S = \frac{v_P}{v_S} = \frac{R_2}{R_1 + R_2}$$

Since the noninverting amplifier has zero output resistance, the load R_L has no effect on the output voltage v_O. Using Eq. (4–19), the gain of the noninverting amplifier circuit is

$$K_{AMP} = \frac{v_O}{v_P} = \frac{R_3 + R_4}{R_4}$$

The overall circuit gain is found as

$$K_{CIRCUIT} = \frac{v_O}{v_S} = \left[\frac{v_P}{v_S}\right]\left[\frac{v_O}{v_P}\right]$$

$$= K_S \times K_{AMP}$$

$$= \left[\frac{R_2}{R_1 + R_2}\right]\left[\frac{R_3 + R_4}{R_4}\right]$$

The gain $K_{CIRCUIT}$ is the product of K_S times K_{AMP} because the amplifier circuit does not load the source circuit since $i_P = 0$. ■

VOLTAGE FOLLOWER

The OP AMP in Figure 4–32(a) is connected as a **voltage follower** or **buffer**. In this case, the feedback path is a direct connection from the output to the inverting input. The feedback connection forces the condition $v_N = v_O$. The input current $i_P = 0$, so there is no voltage across the source resistance R_S. Applying KVL, we have the input condition $v_P = v_S$. The ideal OP AMP model requires $v_P = v_N$, so we conclude that $v_O = v_S$. By inspection, the closed-loop gain is $K = 1$. Since the output exactly equals the input, we say that the output follows the input (hence the name *voltage follower*).

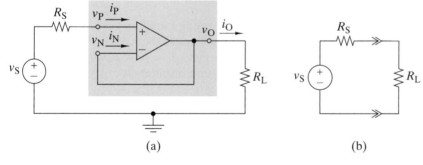

(a)   (b)

FIGURE 4–32 (a) Source-load interface with a voltage follower. (b) Interface without the voltage follower.

The voltage follower is used in interface circuits because it isolates the source from the load. Note that the input-output relationship $v_O = v_S$ does not depend on the source or load resistance. When the source is connected directly to the load, as in Figure 4–32(b), the voltage delivered to the load depends on R_S and R_L. The source and load interaction limits the signals that can be transferred across the interface, as discussed in Chapter 3. When the voltage follower is inserted between the source and load, the signal levels are limited by the capability of the OP AMP.

By Ohm's law, the current delivered to the load is $i_O = v_O/R_L$. But since $v_O = v_S$, the output current can be written in the form

$$i_O = v_S/R_L$$

Applying KCL at the reference node, we discover an apparent dilemma:

$$i_P = i_O$$

For the ideal model $i_P = 0$, but the preceding equations say that i_O cannot be zero unless v_S is zero. It appears that KCL is violated.

The dilemma is resolved by noting that the circuit diagram does not include the supply terminals. The output current comes from the power supply and not from the input. This dilemma arises only at the reference node (the ground terminal). In OP AMP circuits, as in all circuits, KCL must be satisfied. However, we must be alert to the fact that a KCL equation at the reference node could yield misleading results because the power supply terminals are not usually included in circuit diagrams.

Exercise 4–10

The circuits in Figure 4–32 have $v_S = 1.5$ V, $R_S = 2$ kΩ, and $R_L = 1$ kΩ. Compute the maximum power available from the source. Compute the power absorbed by the load resistor in the direct connection in Figure 4–32(b) and in the voltage follower circuit in Figure 4–32(a). Discuss any differences.

> **Answers:**
> 281 μW; 250 μW; 2250 μW.

DISCUSSION: With the direct connection, the power delivered to the load is less than the maximum power available. With the voltage follower circuit, the power delivered to the load is greater than the maximum value specified by the maximum power transfer theorem. However, the maximum power transfer theorem does not apply to the voltage follower circuit since the load power comes from the OP AMP power supply rather than the signal source.

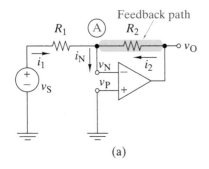

(a)

(b)

FIGURE 4–33 *The inverting amplifier circuit.*

THE INVERTING AMPLIFIER

The circuit in Figure 4–33 is called an **inverting amplifier**. The key feature of this circuit is that the input signal and the feedback are both applied at the inverting input. Since the noninverting input is grounded, we have $v_P = 0$, an observation we will use shortly. The sum of currents entering node A can be written as

$$\frac{v_S - v_N}{R_1} + \frac{v_O - v_N}{R_2} - i_N = 0 \tag{4–24}$$

The element constraints for the OP AMP are $v_P = v_N$ and $i_P = i_N = 0$. Since $v_P = 0$, it follows that $v_N = 0$. Substituting the OP AMP constraints into Eq. (4–24) and solving for the input-output relationship yields

$$v_O = -\left(\frac{R_2}{R_1}\right) v_S \tag{4–25}$$

This result is of the form $v_O = Kv_S$, where K is the closed-loop gain. However, in this case the voltage gain $K = -R_2/R_1$ is negative, indicating a signal inversion (hence the name *inverting amplifier*). We use the block diagram symbol in Figure 4–33(b) to indicate either the inverting or the noninverting OP AMP configuration.

The OP AMP constraints mean that the input current i_1 in Figure 4–33(a) is

$$i_1 = \frac{v_S - v_N}{R_1} = \frac{v_S}{R_1}$$

This, in turn, shows that the input resistance seen by the source v_S is

$$R_{IN} = \frac{v_S}{i_1} = R_1 \qquad (4\text{--}26)$$

In other words, the inverting amplifier has a finite input resistance determined by the external resistor R_1.

The next example shows that the finite input resistance must be taken into account when analyzing circuits with OP AMPs in the inverting amplifier configuration.

EXAMPLE 4−14

Find the input-output relationship of the circuit in Figure 4–34(a).

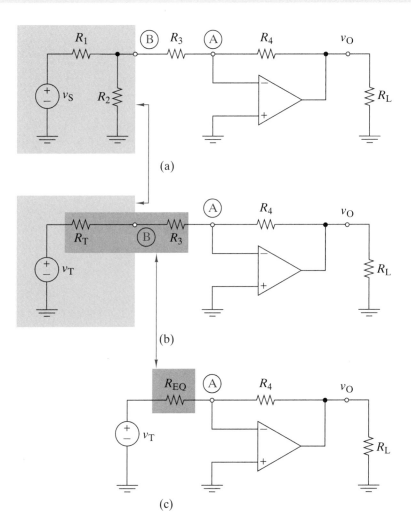

(a)

(b)

(c)

FIGURE 4 − 3 4

SOLUTION:

The circuit to the right of node B is an inverting amplifier. The load resistance R_L has no effect on the circuit transfer characteristics since the OP AMP has zero output resistance. However, the source circuit to the left of node B is influenced by the input resistance of the inverting amplifier circuit. The effect can be seen by constructing a Thévenin equivalent of the circuit to the left of node B, as shown in Figure 4–34(b). By inspection of Figure 4–34(a),

$$v_T = \frac{R_2}{R_1 + R_2} v_S$$

$$R_T = \frac{R_1 R_2}{R_1 + R_2}$$

In Figure 4–34(b) this Thévenin resistance is connected in series with the input resistor R_3, yielding the equivalence resistance $R_{EQ} = R_T + R_3$ shown in Figure 4–34(c). This reduced circuit is in the form of an inverting amplifier, so we can write the input-output relationship relating v_O and v_T as

$$K_1 = \frac{v_O}{v_T} = -\frac{R_4}{R_{EQ}} = -\frac{R_4(R_1 + R_2)}{R_1 R_2 + R_1 R_3 + R_2 R_3}$$

The overall input-output relationship from the input source v_S to the OP AMP output v_O is obtained by writing

$$
\begin{aligned}
K_{CIRCUIT} = \frac{v_O}{v_S} &= \left[\frac{v_O}{v_T}\right]\left[\frac{v_T}{v_S}\right] \\
&= -\left[\frac{R_4(R_1 + R_2)}{R_1 R_2 + R_1 R_3 + R_2 R_3}\right]\left[\frac{R_2}{R_1 + R_2}\right] \\
&= -\left[\frac{R_2 R_4}{R_1 R_2 + R_1 R_3 + R_2 R_3}\right]
\end{aligned}
$$

It is important to note that the overall gain is *not* the product of the source circuit voltage gain $R_2/(R_1 + R_2)$ and the inverting amplifier gain $-R_4/R_3$. In this circuit the two building blocks interact because the input resistance of the inverting amplifier circuit loads the source circuit. ∎

Exercise 4–11

Sketch the transfer characteristic of the OP AMP circuit in Figure 4–35(a) for $-10\text{ V} < v_S < 10\text{ V}$.

Answer:
The solution is shown in Figure 4–35(b).

THE SUMMING AMPLIFIER

The **summing amplifier** or **adder** circuit is shown in Figure 4–36(a). This circuit has two inputs connected at node A, which is called the **summing point**. Since the noninverting input is grounded, we have the condition $v_P = 0$. This configuration is similar to the inverting amplifier, so we start by applying KCL to write the sum of currents entering the node A summing point.

$$\frac{v_1 - v_N}{R_1} + \frac{v_2 - v_N}{R_2} + \frac{v_O - v_N}{R_F} - i_N = 0 \tag{4–27}$$

With the noninverting input grounded, the OP AMP element constraints are $v_N = v_P = 0$ and $i_N = 0$. Substituting these OP AMP constraints into Eq. (4–27), we can solve for the circuit input-output relationship.

FIGURE 4–35

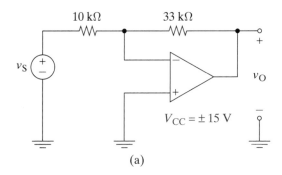

(a)

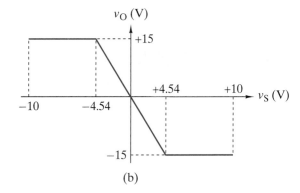

(b)

$$v_O = \left(-\frac{R_F}{R_1}\right) v_1 + \left(-\frac{R_F}{R_2}\right) v_2$$

$$= (K_1) v_1 + (K_2) v_2$$

(4–28)

The output is a weighted sum of the two inputs. The scale factors (or gains, as they are called) are determined by the ratio of the feedback resistor R_F to the input resistor for each input: that is, $K_1 = -R_F/R_1$ and $K_2 = -R_F/R_2$. In the special case $R_1 = R_2 = R$, Eq. (4–28) reduces to

$$v_O = -\frac{R_F}{R}(v_1 + v_2)$$

In this special case the output is proportional to the sum of the two inputs (hence the name *summing amplifier* or, more precisely, *inverting summer*). A block diagram representation of this circuit is shown in Figure 4–36(b).

FIGURE 4–36 *The inverting summer.*

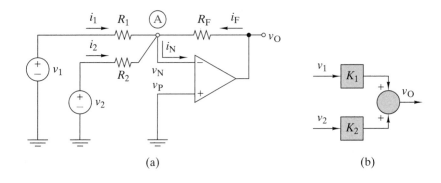

(a)   (b)

The summing amplifier in Figure 4–36 has two inputs, so there are two gains to contend with, one for each input. The input-output relationship in Eq. (4–28) is easily generalized to the case of n inputs as

$$v_O = \left(-\frac{R_F}{R_1}\right) v_1 + \left(-\frac{R_F}{R_2}\right) v_2 + \cdots + \left(-\frac{R_F}{R_n}\right) v_n$$

$$= K_1 v_1 + K_2 v_2 + \cdots + K_n v_n$$

(4–29)

where R_F is the feedback resistor and $R_1, R_2, \ldots R_n$ are the input resistors for the n input voltages $v_1, v_2, \ldots v_n$. You can easily verify this result by expanding the KCL sum in Eq. (4–27) to include n inputs, invoking the OP AMP constraints, and then solving for v_O.

D ⬤ DESIGN EXAMPLE 4–15

Design an inverting summer that implements the input-output relationship

$$v_O = -(5\,v_1 + 13\,v_2)$$

SOLUTION:

The design problem involves selecting the input and feedback resistors so that

$$\frac{R_F}{R_1} = 5 \quad \text{and} \quad \frac{R_F}{R_2} = 13$$

One solution is arbitrarily to select $R_F = 65$ kΩ, which yields $R_1 = 13$ kΩ and $R_2 = 5$ kΩ. The resulting circuit is shown in Figure 4–37(a). The design can be modified to use standard resistance values for resistors with ±5% tolerance (see Appendix A, Table A-1). Selecting the standard value $R_F = 56$ kΩ requires $R_1 = 11.2$ kΩ and $R_2 = 4.31$ kΩ. The nearest standard values are 11 kΩ and 4.3 kΩ. The resulting circuit shown in Figure 4–37(b) incorporates standard value resistors and produces gains of $K_1 = 56/11 = 5.09$ and $K_2 = 56/4.3 = 13.02$. These nominal gains are within 2% of the values in the specified input-output relationship. ∎

FIGURE 4 – 37

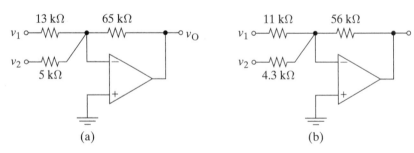

(a)   (b)

Exercise 4–12

(a) Find v_O in Figure 4–37(a) when $v_1 = 2$ V and $v_2 = -0.5$ V.
(b) If $v_1 = 400$ mV and $V_{CC} = \pm15$ V, what is the maximum value of v_2 for linear mode operation?

(c) If $v_1 = 500$ mV and $V_{CC} = \pm 15$ V, what is the minimum value of v_2 for linear mode operation?

Answers:
(a) –3.5 V; (b) 1 V; (c) –1.346 V

THE DIFFERENTIAL AMPLIFIER

The circuit in Figure 4–38(a) is called a **differential amplifier** or **subtractor**. Like the summer, this circuit has two inputs, one applied at the inverting input and one at the noninverting input of the OP AMP. The input-output relationship can be obtained using the superposition principle.

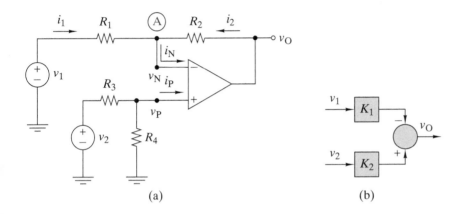

FIGURE 4-38 *The differential amplifier.*

(a)

(b)

First, we turn off source v_2, in which case there is no excitation at the noninverting input and $v_P = 0$. In effect, the noninverting input is grounded and the circuit acts like an inverting amplifier with the result that

$$v_{O1} = -\frac{R_2}{R_1} v_1 \tag{4–30}$$

Next, turning v_2 back on and turning v_1 off, we see that the circuit looks like a noninverting amplifier with a voltage divider connected at its input. This case was treated in Example 4–13, so we can write

$$v_{O2} = \left[\frac{R_4}{R_3 + R_4}\right]\left[\frac{R_1 + R_2}{R_1}\right] v_2 \tag{4–31}$$

Using superposition, we add outputs in Eqs. (4–30) and (4–31) to obtain the output with both sources on:

$$
\begin{aligned}
v_O &= v_{O1} + v_{O2} \\
&= -\left[\frac{R_2}{R_1}\right] v_1 + \left[\frac{R_4}{R_3 + R_4}\right]\left[\frac{R_1 + R_2}{R_1}\right] v_2 \tag{4–32} \\
&= -[K_1] v_1 + [K_2] v_2
\end{aligned}
$$

where K_1 and K_2 are the inverting and noninverting gains. Figure 4–38(b) shows how the differential amplifier is represented in a block diagram.

For the special case of $R_3/R_1 = R_4/R_2$, Eq. (4–32) reduces to

$$v_O = \frac{R_2}{R_1}(v_2 - v_1) \qquad (4\text{–}33)$$

In this case the output is proportional to the difference between the two inputs (hence the name *differential amplifier* or *subtractor*).

Exercise 4–13

(a) Find v_O in Figure 4–39 when $v_S = 10$ V.
(b) When $V_{CC} = \pm 15$ V, find the maximum value of v_S for linear mode operation.
(c) When $V_{CC} = \pm 15$ V, find the minimum value of v_S for linear mode operation.

Answers:
(a) –5 V; (b) 30 V; (c) –30 V

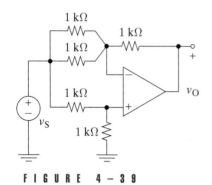

F I G U R E   4 – 3 9

BASIC OP AMP BUILDING BLOCKS

The block diagram representations of the basic OP AMP circuit configurations are shown in Figure 4–40. The noninverting and inverting amplifiers are represented as gain blocks. The summing amplifier and differential amplifier require both gain blocks and the summing point symbol. Considerable care must be used when translating from a block diagram to a circuit, or vice versa, since some gain blocks involve negative gains. For example, the gains of the inverting summer are negative. The required minus sign is sometimes moved to the summing point and the value of K within the gain block changed to a positive number. Since there is no standard convention for doing this, it is important to keep track of the signs associated with gain blocks and summing points.

EXAMPLE 4–16

Find the input-output relationship of the circuit in Figure 4–41(a).

SOLUTION:
This circuit is a cascade connection of three OP AMP circuits. The first circuit is an inverting amplifier with $K_1 = -0.33$, the second is a unity-gain inverting summer, and the final circuit is another inverting amplifier with $K_3 = -5/9$. Given these observations, we construct the block diagram shown in Figure 4–41(b). To trace the signal from input to output, we note that the input to the first circuit is 9.7 V and its output is $K_1 \times 9.7 = (-0.33) \times (9.7) = -3.2$ V. This voltage is added to the variable input v_F in the second circuit to produce $(-1) \times (-3.2) + (-1) \times (v_F) = 3.2 - v_F$. This result is then multiplied by the gain of the final circuit ($K_3 = -5/9$) to obtain the overall circuit output.

$$v_C = -\frac{5}{9}(3.2 - v_F) = \frac{5}{9}(v_F - 3.2)$$

This circuit implements a conversion of temperature from degrees Fahrenheit to degrees Celsius. Specifically, when the input voltage

Circuit	Block diagram	Gains
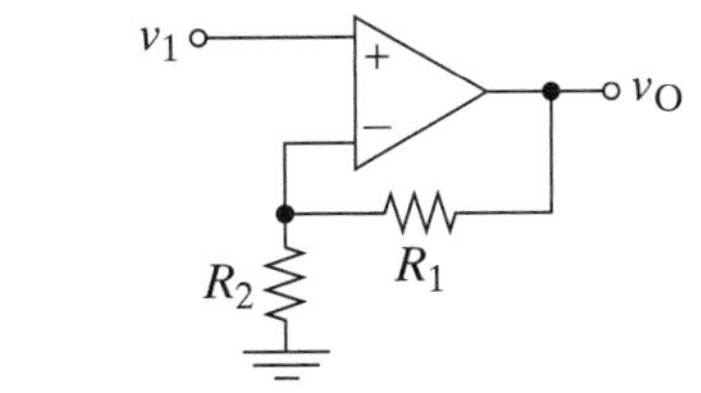 v_1 ... v_O ; R_2, R_1	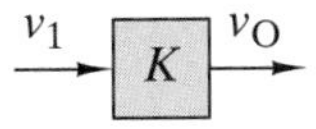 $v_1 \rightarrow K \rightarrow v_O$ NONINVERTING	$K = \dfrac{R_1 + R_2}{R_2}$
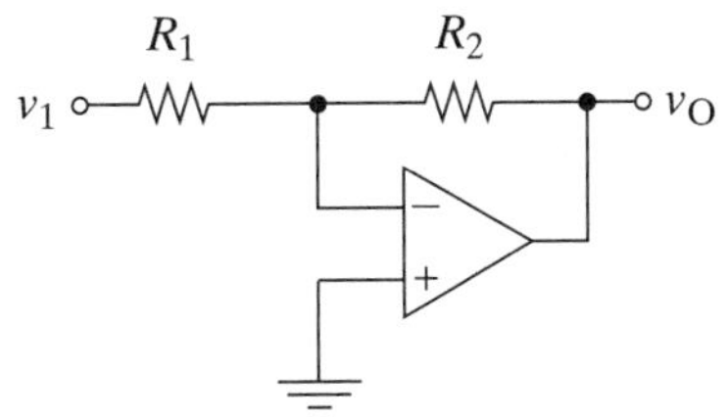 R_1, R_2 ; v_1 ... v_O	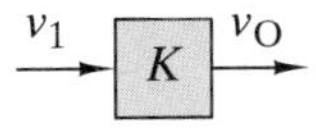 $v_1 \rightarrow K \rightarrow v_O$ INVERTING	$K = -\dfrac{R_2}{R_1}$
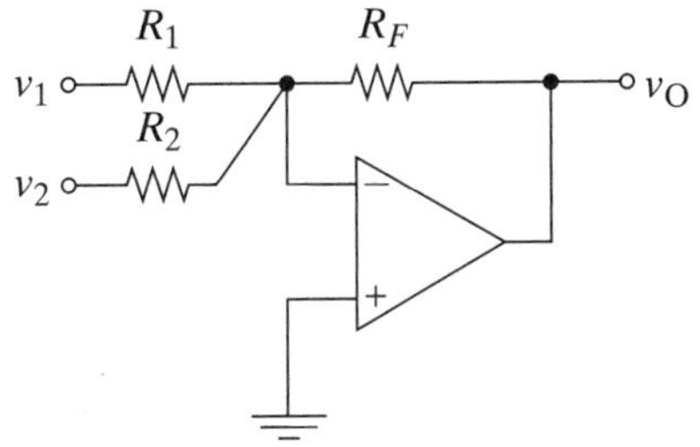 R_1, R_F ; v_1 ; R_2 ; v_2 ... v_O	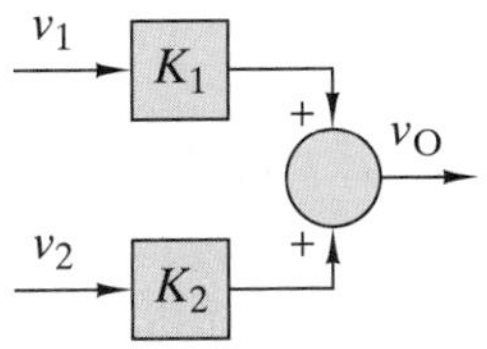 $v_1 \rightarrow K_1$, $v_2 \rightarrow K_2 \rightarrow v_O$ SUMMER	$K_1 = -\dfrac{R_F}{R_1}$ $K_2 = -\dfrac{R_F}{R_2}$
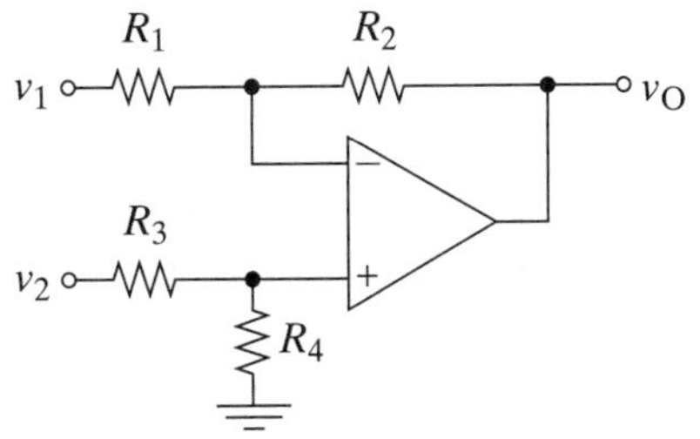 R_1, R_2 ; v_1 ; R_3 ; v_2 ; R_4 ... v_O	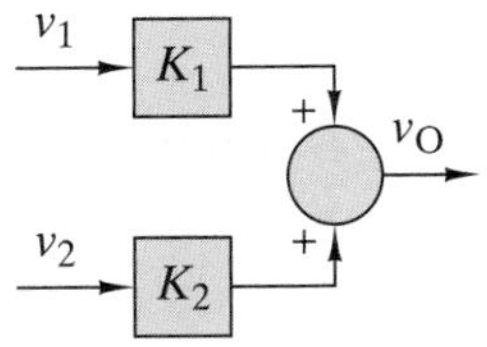 $v_1 \rightarrow K_1$, $v_2 \rightarrow K_2 \rightarrow v_O$ SUBTRACTOR	$K_1 = -\dfrac{R_2}{R_1}$ $K_2 = \left(\dfrac{R_1 + R_2}{R_1}\right)\left(\dfrac{R_4}{R_3 + R_4}\right)$

FIGURE   4 – 40   *Summary of basic OP AMP signal processing circuits.*

is $v_F = \theta_F/10$, where θ_F is a temperature in degrees Fahrenheit, then the output voltage is $v_C = \theta_C/10$, where θ_C is the equivalent temperature in degrees Celsius.

Each circuit in Figure 4–41 is an inverting amplifier with input resistances of about 10 kΩ. Since $V_{CC} = 15$ V, this means that the maximum current drawn at the input of any stage is about $15 \div 10^4 = 1.5$ mA, which is well within the power capabilities of general-purpose OP AMP devices.

Show that all OP AMPs are in the linear mode when the input temperature is in the range from –128°F to +182°F.

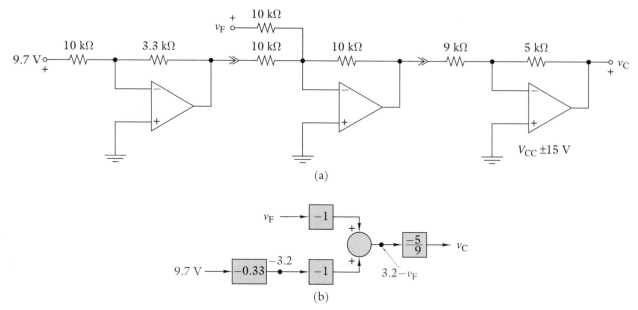

(a)

(b)

FIGURE 4-41

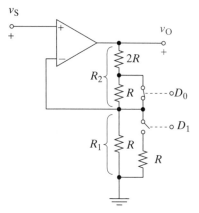

FIGURE 4-42

EXAMPLE 4-17

Figure 4–42 shows a noninverting amplifier in which the digital signals D_1 and D_0 adjust the effective feedback resistance R_1 and R_2 by opening and closing the analog switches shown. Determine the amplifier gain $K = v_O/v_S$ for $(D_1, D_0) = (0, 0)$, $(0, 1)$, $(1, 0)$, and $(1, 1)$. Assume an ideal analog switch with $R_{ON} = 0$ and $R_{OFF} = \infty$.

SOLUTION:

Note that D_1 controls a normally open switch, whereas D_0 controls a normally closed switch. The following truth table describes the switch states for each of the four digital input combinations.

DIGITAL INPUTS		SWITCH STATE	
D_1	D_0	LOWER	UPPER
0	0	open	closed
0	1	open	open
1	0	closed	closed
1	1	closed	open

When $(D_1, D_0) = (0, 0)$ the lower switch is open, the upper switch is closed, the feedback resistances are $R_1 = R$ and $R_2 = 2R$, and the amplifier gain is

$$K = \frac{R_1 + R_2}{R_1} = \frac{3R}{R} = 3$$

When $(D_1, D_0) = (0, 1)$ both switches are open, the feedback resistances are $R_1 = R$ and $R_2 = 3R$, and the amplifier gain is

$$K = \frac{R_1 + R_2}{R_1} = \frac{4R}{R} = 4$$

When $(D_1, D_0) = (1, 0)$, both switches are closed, the feedback resistances are $R_1 = R/2$ and $R_2 = 2R$, and the amplifier gain is

$$K = \frac{R_1 + R_2}{R_1} = \frac{2.5R}{0.5R} = 5$$

Finally, when $(D_1, D_0) = (1, 1)$, the lower switch is closed, the upper switch is open, the feedback resistances are $R_1 = R/2$ and $R_2 = 3R$, and the amplifier gain is

$$K = \frac{R_1 + R_2}{R_1} = \frac{3.5R}{0.5R} = 7$$

This example illustrates how digital signals can control an analog signal processor. ∎

NODE-VOLTAGE ANALYSIS WITH OP AMPs

In many applications we encounter OP AMP circuits that are more complicated than the four basic configurations in Figure 4–40. In such cases we use a modified form of node-voltage analysis that is based on the OP AMP connections in Figure 4–43. The overall circuit contains N nodes, including the three associated with the OP AMP. Normally the objective is to find the OP AMP output voltage (v_O) relative to the reference node (ground). We assign node voltage variables to the $N - 1$ nonreference nodes, including a variable at the OP AMP output. However, an ideal OP AMP acts like a dependent voltage source connected between the output terminal and ground. As a result, the OP AMP output voltage is determined by the other node voltages, so we do not need to write a node equation at the OP AMP output node.

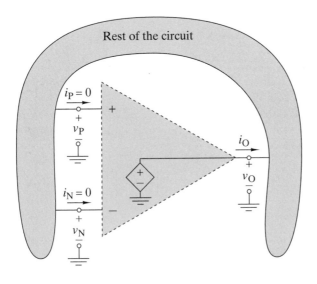

FIGURE 4–43 *General OP AMP circuit analysis.*

We formulate node equations at the other $N - 2$ nonreference nodes in the usual way. Since there are $N - 1$ node voltages, we seem to have more unknowns than equations. However, the OP AMP forces the condition $v_P = v_N$ in Figure 4–43. This eliminates one unknown node voltage since these two nodes are forced to have identical voltages. Finally, remember that the ideal OP AMP draws no current at its inputs ($i_P = i_N = 0$) in Figure 4–43, so these currents can be ignored when formulating node equations.

The following steps outline an approach to the formulation of node equations for OP AMP circuits:

STEP 1: Identify a node voltage at all nonreference nodes, including OP AMP outputs, but do *not* formulate node equations at the OP AMP output nodes.

STEP 2: Formulate node equations at the remaining nonreference nodes and then use the ideal OP AMP voltage constraint $v_P = v_N$ to reduce the number of unknowns.

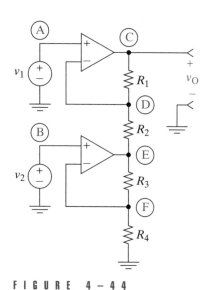

FIGURE 4-44

EXAMPLE 4–18

Find the input-output relationship of the circuit in Figure 4–44.

SOLUTION:

The circuit contains two OP AMPs and a total of six nonreference nodes. Nodes C and E are connected to OP AMP outputs, and nodes A and B are connected to the grounded independent voltage sources. As a result, we only require node equations at nodes D and F. By inspection,

$$\text{Node D: } (G_1 + G_2)v_D - G_1v_C - G_2v_E = 0$$

$$\text{Node F: } \qquad (G_3 + G_4)v_F - G_3v_E = 0$$

This formulation yields two equations in four unknowns. However, the noninverting inputs are connected to independent voltage sources, so $v_A = v_1$ and $v_B = v_2$. The OP AMP voltage constraint ($v_P = v_N$) means $v_D = v_A = v_1$ and $v_F = v_B = v_2$. Substituting these constraints reduces the two node equations to

$$G_1v_C + G_2v_E = (G_1 + G_2)\, v_1$$

$$G_3\, v_E = (G_3 + G_4)v_2$$

Node analysis quickly reduces this rather formidable appearing OP AMP circuit to two equations in two unknowns.

The circuit output voltage is $v_O = v_C$, so we use the second equation to eliminate v_E from the first equation and then solve for the input-output relationship as

$$v_O = v_C = \left[\frac{G_1 + G_2}{G_1}\right]v_1 - \frac{G_2}{G_1}\left[\frac{G_3 + G_4}{G_3}\right]v_2$$

For the special case $R_1 = R_4$ and $R_2 = R_3$, this equation reduces to

$$v_O = \frac{R_1 + R_2}{R_2}[v_1 - v_2]$$

The configuration in Figure 4–44 is a two-OP-AMP subtractor. The advantage of this circuit over the one-OP-AMP configuration in Figure 4–40 is that both inputs are connected to noninverting inputs, so the two-OP-AMP subtractor does not load the input sources. ■

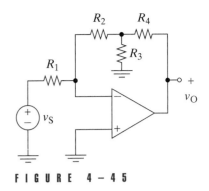

FIGURE 4-45

Exercise 4–14

Find the input-output relationship for the circuit in Figure 4–45.

Answer:

$$v_O = -\left[\frac{R_2R_3 + R_2R_4 + R_3R_4}{R_1R_3}\right]v_S$$

EXAMPLE 4–19

Circuit simulation programs like PSpice or Electronics Workbench do not support the theoretical characteristics of the ideal OP AMP model we use in circuit analysis and design. The default OP AMP in the Electronics Workbench library lists its model as "ideal." The purpose of this example is to evaluate this default model as a stand-in for the theoretical ideal OP AMP.

(a) Determine the parameters of the default OP AMP model in Electronics Workbench.
(b) Compare the theoretical voltage gain, input resistance, and output resistance of an inverting OP AMP circuit with $R_1 = 10$ kΩ and $R_2 = 50$ kΩ with the simulation results predicted by Electronics Workbench using its default model.

SOLUTION:
(a) Figure 4–46 shows the circuit symbol for the three-terminal OP AMP component available in the Analog IC's group of the Parts Bin toolbar. Double-clicking on the circuit symbol brings up a menu of OP AMPs including the default OP AMP whose model is listed as "ideal." Clicking on edit brings up the parameter list shown in Figure 4–46. The voltage gain, input resistance, and output resistance of the "ideal" model are listed as $A = 10^6$, $R_I = 10^{10}$, and $R_O = 1$. The corresponding theoretical values for an ideal OP AMP are $A = \infty$, $R_I = \infty$, and $R_O = 0$. Thus, the default OP AMP approximates, but does not duplicate, the theoretical ideal. The question is, how much difference does this make?
(b) Figure 4–47 shows different stages of constructing the circuit diagram for an inverting amplifier with $R_1 = 10$ kΩ and $R_2 = 50$ kΩ. The OP AMP is the default model whose parameters were found in (a). The left side of the figure shows the component placements prior to interconnection. The resistors are available in the Basic Group of the Parts Bin toolbar. The input voltage source and ground are found in the Sources Group. To wire components together, we point at a component terminal, click and drag a wire to the terminal of another component, and release the mouse button. The program then automatically routes the wire at right angles,

FIGURE 4-46

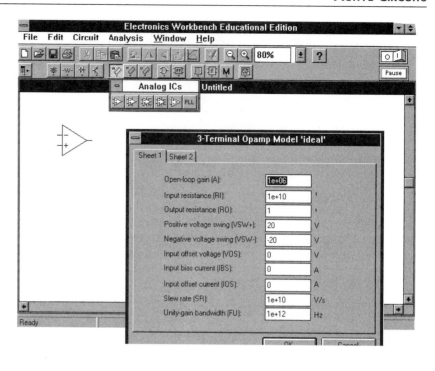

without overlapping any of the components. The circuit with all components interconnected is shown in the right side of Figure 4–47.

From the Analysis Menu we select the Transfer Function option, which brings up the window shown at the top of Figure 4–48. In this window we identify the output at the voltage between node 3 and node 0 (ground), and the input as the voltage source VS. Clicking on the simulate button causes the program to perform a dc analysis of the

FIGURE 4-47

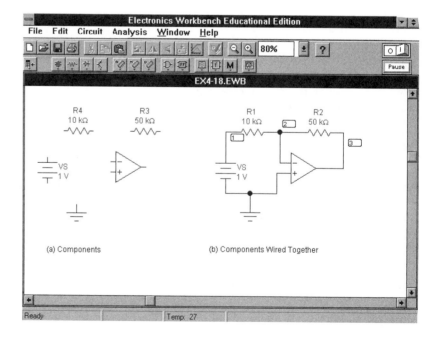

FIGURE 4-48

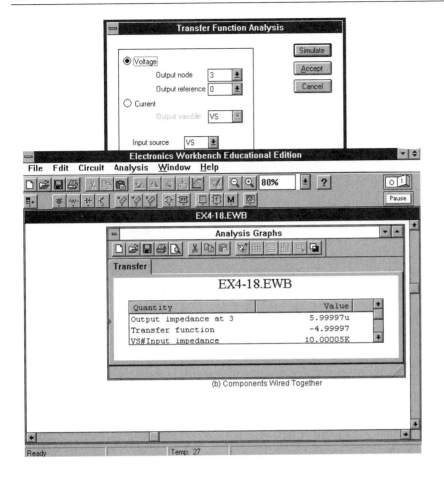

(b) Components Wired Together

circuit and report the results shown at the bottom of Figure 4–48. Using the default model the program predicts an output resistance of 5.999997 $\mu\Omega$, a voltage gain of –4.99997, and an input resistance of 10.00005 kΩ. An ideal OP AMP in the same inverting configuration would have an output resistance of zero, a voltage gain of $-R_2/R_1 = -5$, and an input resistance of $R_{IN} = R_1 = 10$ kΩ. The results using the default model agree with those of the ideal model to within four significant figures for voltage gain and input resistance. The output resistance is not exactly zero, but the value 6 $\mu\Omega$ is negligible compared with other circuit resistances. In summary, the default model and the ideal model predict essentially identical results in typical OP AMP circuit applications. From this point on we will use the default model in circuit simulations as a stand-in for the ideal OP AMP. ■

4–6 OP AMP CIRCUIT DESIGN

With OP AMP circuit analysis, we are asked to find the input-output relationship for a given circuit configuration. An OP AMP circuit analysis problem has a unique answer. In OP AMP circuit design, we are given an equation or block diagram representation of a signal-processing function and asked to devise a circuit configuration that implements the desired function. Circuit design can be accomplished by interconnecting the amplifier, summer, and subtractor building blocks shown in Figure 4–40. The design process is greatly simplified by the nearly one-to-one correspondence between the OP AMP circuits and the elements in a block diagram. However, a design problem may not have a unique answer since often there are several OP AMP circuits that meet the design objective. The following example illustrates the design process.

D **DESIGN EXAMPLE 4–20**

Design an OP AMP circuit that implements the block diagram in Figure 4–49.

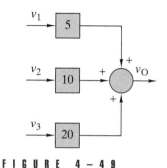

FIGURE 4 – 49

SOLUTION:
The input-output relationship represented by the block diagram is

$$v_O = 5v_1 + 10v_2 + 20v_3$$

The inverting summer can be used to produce the summation required in this relationship. A three-input adder implements the relationship

$$v_O = -\left[\frac{R_F}{R_1}v_1 + \frac{R_F}{R_2}v_2 + \frac{R_F}{R_3}v_3\right]$$

The required scale factors are obtained by first selecting $R_F = 100$ kΩ, and then choosing $R_1 = 20$ kΩ, $R_2 = 10$ kΩ, and $R_3 = 5$ kΩ. However, the circuit involves a signal inversion. To implement the block diagram, we must add an inverting amplifier ($K = -R_2/R_1$) with $R_1 = R_2 = 100$ kΩ. The final implementation is shown in Figure 4–50. ■

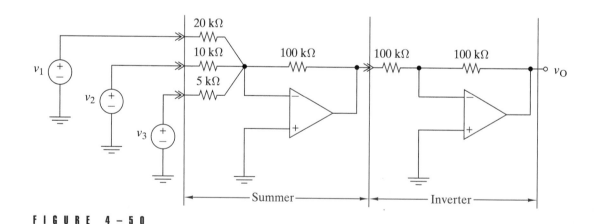

FIGURE 4 – 50

EVALUATION EXAMPLE 4–21

It is claimed that the OP AMP circuit in Figure 4–51 implements the same input-output function as the circuit in Figure 4–50. Verify this claim and compare the two circuits in terms of element count.

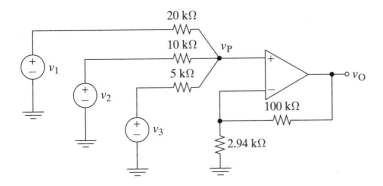

FIGURE 4-51

SOLUTION:

The circuit in Figure 4–51 is a modification of the OP AMP subtractor in Figure 4–40. In the modified circuit, several inputs are connected to the noninverting input and the negative input is grounded. The output voltage is related to the signal v_P by a noninverting amplifier gain; that is,

$$v_O = Kv_P = \frac{100 \times 10^3 + 2.94 \times 10^3}{2.94 \times 10^3} v_P$$

$$= 35v_P$$

Applying KCL, the sum of currents entering the noninverting input is

$$\frac{v_1 - v_P}{2 \times 10^4} + \frac{v_2 - v_P}{10^4} + \frac{v_3 - v_P}{0.5 \times 10^4} = 0$$

Multiplying this equation by 10^4 and solving for v_P yields

$$v_P = \frac{0.5}{3.5} v_1 + \frac{1}{3.5} v_2 + \frac{2}{3.5} v_3$$

Thus the overall input-output relationship is

$$v_O = 35v_P = 5v_1 + 10v_2 + 20v_3$$

and the claim is verified. The circuit in Figure 4–51 has the same input–output relationship as the circuit in Figure 4–50. The circuit in Figure 4–51 requires only one OP AMP since the inverting stage in Figure 4–50 is not required.

The circuit in Figure 4–51 is an example of a **noninverting summer**. The input-output relationship for a general noninverting summer in Figure 4–52 is

FIGURE 4-52

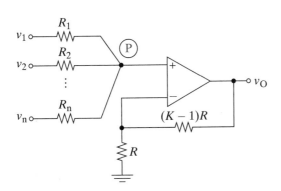

$$v_O = K\left[\left(\frac{R_{EQ}}{R_1}\right)v_1 + \left(\frac{R_{EQ}}{R_2}\right)v_2 + \cdots + \left(\frac{R_{EQ}}{R_n}\right)v_n\right] \quad (4\text{--}34)$$

where R_{EQ} is the Thévenin resistance looking to the left at point P with all sources turned off (i.e., $R_{EQ} = R_1 \| R_2 \| R_3 \ldots \| R_n$) and K is the gain of the non-inverting amplifier circuit to the right of point P. There are several similarities between this equation and the general inverting summer result in Eq. (4–29). In both cases the gain factor assigned to each input voltage is inversely proportional to the input resistance to which it is connected. The gain factor is directly proportional to the feedback resistor R_F in the inverting summer and is directly proportional to R_{EQ} in the noninverting summer. ■

DIGITAL-TO-ANALOG CONVERTERS

A **digital-to-analog converter** (DAC) is a signal processor whose input is an n-bit digital word and whose output is an analog signal proportional to the binary value of the digital input. For example, the parallel four-bit digital signal in Figure 4–53 represents the value of a signal. Each bit can only have two values: (1) a high or "1" (e.g., +5 V) and (2) a low or "0" (e.g., 0 V). The bits have binary weights, so v_1 is worth $(2)^3 = 8$ times as much v_4, v_2 is worth $(2)^2 = 4$ times as much as v_4, and v_3 is worth $(2)^1$ times as much v_4. We call v_4 the **least significant bit** (LSB) and call v_1 the **most significant bit** (MSB). To convert the digital representation of the signal to analog form, we must weight the bits so that the analog output v_O is

$$v_O = \pm K(8v_1 + 4v_2 + 2v_3 + v_4) \quad (4\text{--}35)$$

where K is an overall scale factor. This result is the input-output relationship of a 4-bit DAC.

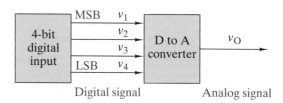

FIGURE 4–53 *A digital-to-analog converter (DAC).*

One way to implement Eq. (4–35) is to use an inverting summer with binary-weighted input resistors. Figure 4–54 shows the OP AMP circuit and a block diagram of the circuit input-output relationship. In either form, the output is seen to be a binary-weighted sum of the digital input scaled by $-R_F/R$. That is, the output voltage is

$$v_O = \frac{-R_F}{R}(8v_1 + 4v_2 + 2v_3 + v_4) \quad (4\text{--}36)$$

The *R*-2*R* ladder circuit in Figure 4–55(a) also produces a 4-bit DAC. The resistance seen looking back into the *R*-2*R* ladder at point A, with all

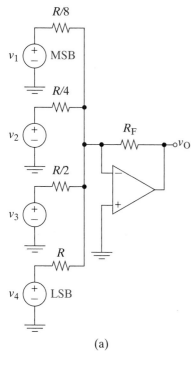

(a)

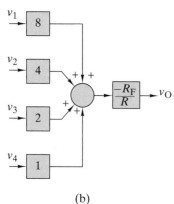

(b)

FIGURE 4–54 *A binary-weighted DAC: (a) Circuit diagram. (b) Block diagram.*

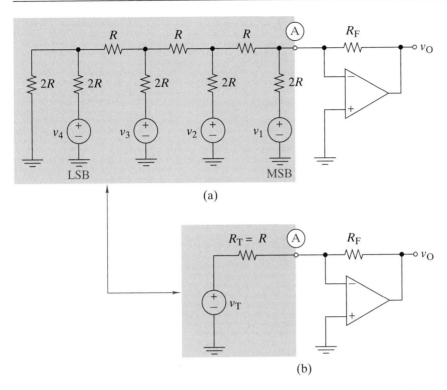

sources turned off, is seen to be $R_T = R$. The Thévenin equivalent circuit of the R-$2R$ circuit is shown in Figure 4–55(b), where

$$v_T = \frac{v_1}{2} + \frac{v_2}{4} + \frac{v_3}{8} + \frac{v_4}{16}$$

The output voltage is found using the inverting amplifier gain relationship:

$$v_O = \frac{-R_F}{R} v_T = \frac{-R_F}{R}\left(\frac{v_1}{2} + \frac{v_2}{4} + \frac{v_3}{8} + \frac{v_4}{16}\right) \tag{4-37}$$

Using $R_F = R$ yields

$$v_O = -\frac{1}{16}(8v_1 + 4v_2 + 2v_3 + v_4) \tag{4-38}$$

This equation shows that the ladder assigns the correct binary weights to the digital inputs—namely, 8, 4, 2, and 1.

In theory, the circuits in Figures 4–54 and 4–55 perform the same signal-processing function—namely, a 4-bit DAC. However, there are important practical differences between the two circuits. The inverting summer in Figure 4–54 requires precision resistors with four different values spanning an 8:1 range. An 8-bit converter would require eight precision resistors spanning a 256:1 range. Moreover, the digital voltage sources in Figure 4–54 see input resistances that span an 8:1 range; therefore, the source-load interface is not the same for each bit. On the other hand, the resistances in the R-$2R$ ladder converter in Figure 4–55 only span a 2:1 range regardless of the number of digital bits. Another desirable feature of the R-$2R$ ladder is that it presents the same input resistance to each binary input.

Exercise 4–15

Find the output voltage when the following inputs are applied to the R-$2R$ ladder D/A converter in Figure 4–55(a) with $R_F = 16R/5$.
(a) $v_1 = 0$ V; $v_2 = 0$ V; $v_3 = 0$ V; $v_4 = 5$ V.
(b) $v_1 = 0$ V; $v_2 = 0$ V; $v_3 = 5$ V; $v_4 = 5$ V.
(c) $v_1 = 0$ V; $v_2 = 5$ V; $v_3 = 0$ V; $v_4 = 5$ V.
(d) $v_1 = 5$ V; $v_2 = 0$ V; $v_3 = 0$ V; $v_4 = 5$ V.

Answers:
(a) $v_O = -1$ V, (b) $v_O = -3$ V, (c) $v_O = -5$ V, (d) $v_O = -9$ V.

D DESIGN EXAMPLE 4–22

Design a noninverting summer that implements a 3-bit D/A converter defined by

$$v_O = \frac{1}{5}[4v_1 + 2v_2 + v_3]$$

SOLUTION:
From Eq. (4–34) the input-output relationship for a three-input noninverting summer is

$$v_O = K\left[\left(\frac{R_{EQ}}{R_1}\right)v_1 + \left(\frac{R_{EQ}}{R_2}\right)v_2 + \left(\frac{R_{EQ}}{R_3}\right)v_3\right]$$

To implement the given relationship, we first select the input resistors so that the correct weight is assigned to each input. One way to accomplish this is to assign $R_1 = 10$ kΩ, $R_2 = 20$ kΩ, and $R_3 = 40$ kΩ. For this assignment $R_{EQ} = R_1\|R_2\|R_3 = 1/0.175$ kΩ, and the input-output relationship is

$$v_O = K\left[\frac{1}{1.75}v_1 + \frac{1}{3.5}v_2 + \frac{1}{7}v_3\right]$$

$$= \frac{K}{7}[4v_1 + 2v_2 + v_3]$$

To achieve the specified scale factor, we need $K = 7/5 = 1.4$. Figure 4–56 shows the circuit diagram for the noninverting summer design. ∎

FIGURE 4–56

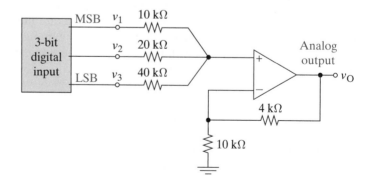

INSTRUMENTATION SYSTEMS

One of the most interesting and useful applications of OP AMP circuits is in instrumentation systems that collect and process data about physical phenomena. Such systems normally involve the signal processing elements shown in Figure 4–57. For our purposes we define a **transducer** as a device that converts a change in some physical quantity, such as temperature or pressure, into a change in an electrical signal. The transducer's electrical output signal is processed by OP AMP circuits that perform signal conditioning and amplification. **Signal conditioning** compensates for transducer shortcomings and produces a signal suitable for further electrical processing. Signal **amplification** is required because the transducer output voltage is quite small, often in the millivolt range or less. In simple systems the amplified analog output goes to a **display**, such as a digital voltmeter. In data acquisition systems with many transducers, the analog signals are converted to digital form and sent to a **computer** for further processing and storage. In some systems the processed data may be used in a feedback system to control the physical process monitored by the transducers.

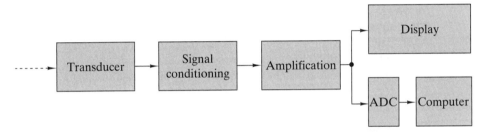

FIGURE 4 – 5 7 *A block diagram of a basic instrumentation system.*

Our interest here is the design of the OP AMP circuits that perform the signal conditioning and amplification functions in an instrumentation system. The overall objective of these circuits is to deliver an analog output signal that falls within a prescribed range and is directly proportional to the value of the physical variable measured by the input transducer. The input transducer converts a physical variable x into an electrical voltage v_{TR}. For many transducers there is a linear range in which this voltage is of the form

$$v_{TR} = mx + b$$

where m is a calibration constant and b is a constant offset or bias. The transducer voltage is usually quite small and must be amplified by a gain K. Direct amplification of the transducer output would produce both the desired signal component $K(mx)$ and an unwanted bias component $K(b)$. The signal conditioning circuit removes the bias component so that the analog output $K(mx)$ is directly proportional to the physical variable measured.

Let's look at a specific example. In a laboratory experiment the amount (5 to 20 lumens) of incident light is to be measured using a photocell as the input transducer. The system output is to be sent to an analog-to-digital converter whose input range is 0 to 5 V. The transducer characteristics of the photocell are shown in Figure 4–58. The overall design requirements are that 5 lumens produce an analog output of 0 V, and

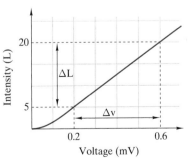

FIGURE 4 – 5 8 *Photocell transducer characteristics.*

that 20 lumens produce 5 V. From the transducer characteristics we see that a change in light intensity $\Delta\Phi = (20 - 5) = 15$ lumens will produce a change in transducer output of $\Delta v = (0.6 - 0.2) = 0.4$ mV. This 0.4-mV change must be translated into a 5-V change at the output. To accomplish this, the transducer voltage must be amplified by an overall gain of

$$K = \frac{\text{specified output range}}{\text{available input range}} = \frac{(5 - 0)}{(0.6 - 0.2) \times 10^{-3}} = 1.25 \times 10^4$$

If the transducer's output voltage range (0.2 to 0.6 mV) is multiplied by the gain $K = 1.25 \times 10^4$, the resulting output voltage range is 2.5 to 7.5 V, rather than the required range of 0 to 5 V. The bias in the output range is removed by subtracting a constant voltage at some point in the signal conditioning circuit.

Figure 4–59(a) shows a block diagram of one possible way to distribute the required signal-processing functions. The required gain is divided between two stages because the overall voltage gain of $K = 1.25 \times 10^4$ is too large to achieve with a single-stage OP AMP circuit. The transducer bias is removed by subtracting a constant voltage following the first-stage amplification ($K_1 = 100$). The second-stage amplification ($K_2 = 125$) then produces the required output voltage range. Figure 4–59(b) shows one possible OP AMP circuit that implements this block diagram. The first stage is a subtractor circuit, and the second stage is an inverting amplifier. This arrangement has two signal inversions so that the overall gain in the signal path is

$$K = (-100)(-125) = 1.25 \times 10^4$$

which is positive as required. As always, many other alternative designs are possible.

FIGURE 4 – 59 *Transducer signal conditioning design. (a) Block diagram. (b) Circuit diagram.*

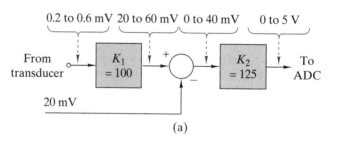

(a)

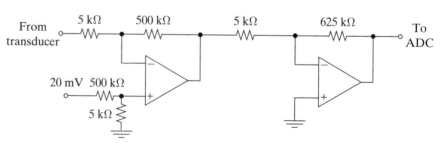

(b)

D (**DESIGN EXAMPLE 4–23**)

A strain gauge is a resistive device that measures the elongation (strain) of a solid material caused by applied forces (stress). A typical strain gauge consists of a thin film of conducting material deposited on an insulating substrate. When bonded to a member under stress, the resistance of the gauge changes by an amount

$$\Delta R = 2R_G \frac{\Delta L}{L}$$

where R_G is the resistance of the gauge with no applied stress and $\Delta L/L$ is the elongation of the material expressed as a fraction of the unstressed length L. The change in resistance ΔR is only a few tenths of a milliohm, far too small a value to be measured with a standard ohmmeter. To detect such a small change, the strain gauge is placed in a Wheatstone bridge circuit like the one shown in Figure 4–60. The bridge contains fixed resistors R_A and R_B, matched strain gauges R_{G1} and R_{G2}, and a precisely controlled reference voltage V_{REF}. The values of R_A and R_B are chosen so that the bridge is balanced ($v_1 = v_2$) when no stress is applied. When stress is applied, the resistance of the stressed gauge changes to $R_{G2} + \Delta R$, and the bridge is unbalanced ($v_1 \neq v_2$). The differential signal ($v_2 - v_1$) indicates the strain resulting from the applied stress.

Design an OP AMP circuit to translate strains over the range $0 < \Delta L/L < 0.02\%$ into an output voltage on the range $0 < v_O < 4$ V, for $R_G = 120$ Ω and $V_{REF} = 25$ V.

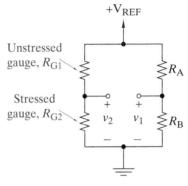

FIGURE   4–60 *Strain gauges in a Wheatstone bridge.*

SOLUTION:

When an external stress is applied, the resistance R_{G2} changes to $R_{G2} + \Delta R$. Applying voltage division to each leg of the bridge yields

$$v_2 = \frac{R_{G2} + \Delta R}{R_{G1} + R_{G2}} V_{REF}$$

$$v_1 = \frac{R_B}{R_A + R_B} V_{REF}$$

The differential voltage ($v_2 - v_1$) can be written as

$$v_2 - v_1 = V_{REF} \left[\frac{R_{G2} + \Delta R}{R_{G1} + R_{G2}} - \frac{R_A}{R_A + R_B} \right]$$

To achieve bridge balance in the unstressed state, we select $R_{G1} = R_{G2} = R_A = R_B = R_G$, in which case the differential voltage reduces to

$$v_2 - v_1 = V_{REF} \left[\frac{\Delta R}{2R_G} \right] = V_{REF} \left[\frac{\Delta L}{L} \right]$$

Thus, the differential voltage is directly proportional to the strain $\Delta L/L$. However, for $V_{REF} = 25$ V and $\Delta L/L = 0.02\%$, the differential voltage is only $(V_{REF})(\Delta L/L) = 25 \times 0.0002 = 5$ mV. To obtain the required 4-V output, we need a voltage gain of $K = 4/0.005 = 800$.

The OP AMP subtractor in Figure 4–40 is ideally suited to the task of amplifying differential signals. Selecting $R_1 = R_3 = 10$ kΩ and $R_2 = R_4 = 8$ MΩ produces an input-output relationship for the subtractor circuit of

$$v_O = 800(v_2 - v_1)$$

Figure 4–61 shows the basic design.

F I G U R E   4 – 6 1

The input resistance of the subtractor circuit must be large to avoid loading the bridge circuit. The Thévenin resistance looking back into the bridge circuit is

$$R_T = R_{G1}\|R_{G2} + R_A\|R_B$$
$$= R_G\|R_G + R_G\|R_G$$
$$= R_G = 120 \ \Omega$$

This value is small compared with the 10-kΩ input resistance of the subtractor's inverting input.

Comment: The transducer in this example is the resistor R_{G2}. In the unstressed state the voltage across this resistor is $v_2 = 12.5$ V. In the stressed state the voltage is $v_2 = 12.5$ V plus a 5-mV signal. In other words, the transducer's 5-mV signal component is accompanied by a much larger bias component. We cannot afford to amplify the 12.5-V bias component by $K = 800$ before subtracting it out. The bias is eliminated at the input by using a bridge circuit in which $v_1 = 12.5$ V, and then processing the differential signal $v_2 - v_1$. The situation illustrated in this example is quite common. Consequently, the first stage in most instrumentation systems is a differential amplifier that removes the transducer bias. ∎

D **D E S I G N   E X E R C I S E :   4–16**

Design an OP AMP circuit to translate the temperature range from –20°C to +120°C onto a 0- to 1-V output signal using the transducer characteristics shown in Figure 4–62(a). The transducer has a Thévenin resistance of 880 Ω.

Answer: One possible design is shown in Figure 4–62(b).

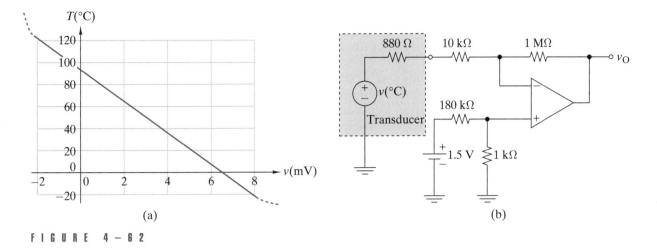

FIGURE 4–62

DISCUSSION: *It is important for the input resistor of the inverting amplifier to be at least 10 times as large as the Thévenin resistance of the transducer. Our design uses an input resistance of 10 kΩ compared with a transducer Thévenin resistance of 880 Ω.*

4–7 THE COMPARATOR

Up to this point we have treated circuits in which negative feedback keeps the OP AMP in the linear mode with $v_P \approx v_N$. When v_P and v_N differ by more than a few millivolts, the OP AMP is driven into one of its two saturation modes:

1. +Saturation with $v_O = +V_{CC}$ when $(v_P - v_N) > 0$
2. –Saturation with $v_O = -V_{CC}$ when $(v_P - v_N) < 0$.

With no feedback the OP AMP operates in one of these two saturation modes whenever the inputs are not equal. Under saturation conditions the output voltage indicates whether $v_P > v_N$ or $v_P < v_N$. A device that discriminates between two unequal voltages is called a **comparator**.

Figure 4–63 shows an example of an OP AMP operating as a comparator. First note that there is no feedback and $v_N = 0$, since the inverting input is grounded. The voltage source connected to the noninverting input means that when $v_P = v_S > 0$, the OP AMP is driven into +saturation with $v_O = +V_{CC}$. Conversely, when $v_P = v_S < 0$, the OP AMP is in –saturation with $v_O = -V_{CC}$. Figure 4–64 shows an example of a plot of $v_S(t)$ versus time plus the resulting comparator output. The circuit in Figure 4–63 is called a **zero crossing detector** because the comparator output changes its saturation state whenever the input v_S crosses through zero. Notice that the comparator output is not proportional to the input, which shows that the comparator is a nonlinear device.

A modified version of the zero-crossing detector is shown in Figure 4–65. In this circuit a constant 2-V source is applied at the inverting input

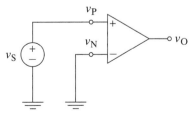

FIGURE 4–63 *An OP AMP comparator.*

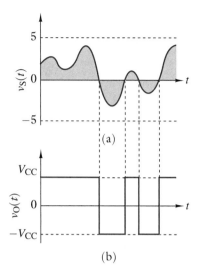

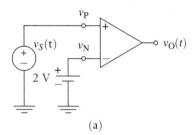

FIGURE 4-64 *Input and output signals of an OP AMP comparator.*

and the input signal v_S from Figure 4–64 is applied to the noninverting input. The input signal is now compared with 2 V rather than with zero. When $v_P = v_S > 2$ V, the OP AMP is in +saturation with $v_O = +V_{CC}$, and when $v_P = v_S < 2$ V, the OP AMP is in –saturation with $v_O = -V_{CC}$. A plot of the resulting output voltage is shown in Figure 4–65(c). The value of the fixed source determines the input signal level at which the comparator switches from one saturation state to the other. For example, connecting a 10-V fixed source to the inverting input causes the comparator output to remain at $-V_{CC}$, since v_S never exceeds 10 V.

Although ordinary OP AMPs can be used as comparators, there are integrated circuit devices specifically designed to operate in saturation. These comparator devices are designed to switch rapidly from one saturation state to the other and to have output saturation levels that are compatible with digital circuits.

Using digital circuit terminology, the input-output characteristics of an **ideal comparator** can be written as

$$\text{If } v_P > v_N, \text{ then } v_O = V_{OH}, \text{ else}$$

$$\text{If } v_P < v_N, \text{ then } v_O = V_{OL} \tag{4-38}$$

where V_{OH} and V_{OL} are the high and low saturation levels of the element. For example, a comparator with $V_{OH} = +5$ V and $V_{OL} = 0$ V is compatible with commonly used TTL (transistor–transistor logic) digital circuits. Note that the equal sign is omitted in the conditional part of the statement because we cannot be certain what the output will be if $v_P = v_N$.

EXAMPLE 4-24

Figure 4–66 shows an OP AMP whose output is connected to two 15-V lamps. Analyze the circuit and find the input signal condition required to turn each of the lamps on.

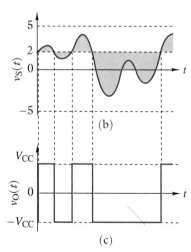

FIGURE 4-65 *Comparator with a 2-V offset: (a) Circuit diagram. (b) Input voltage. (c) Output voltage.*

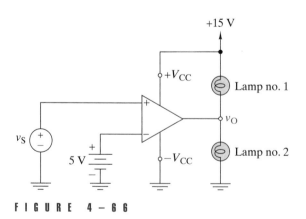

FIGURE 4-66

SOLUTION:

The circuit has no feedback so the OP AMP operates as a comparator. Figure 4–66 shows the power supply inputs to the OP AMP. The positive supply terminal is connected to a +15-V source and the negative supply terminal is grounded. Therefore, the two output saturation levels are $V_{OH} = +15$ V and $V_{OL} = 0$ V.

Because the 5-V source is connected to the inverting input, the comparator changes its state whenever the input voltage v_S passes through 5 V. When $v_S < 5$ V, the comparator is in negative saturation with $v_O = V_{OL} = 0$ V. In this case, the voltages across lamps 1 and 2 are 15 V and 0 V, respectively. Conversely, when $v_S > 5$ V, the comparator is in positive saturation with $v_O = V_{OH} = +15$ V, and hence the voltages across lamps 1 and 2 are 0 V and 15 V, respectively. These observations can be summarized as follows:

INPUT RANGE	COMPARATOR STATE	OUTPUT VOLTAGE	LAMP 1	LAMP 2
$v_S < 5$ V	−Saturation	$v_O = 0$ V	On	Off
$v_S > 5$ V	+Saturation	$v_O = 15$ V	Off	On

APPLICATION NOTE: EXAMPLE 4−25

The circuit in Figure 4–67 is an analog-to-digital converter (ADC) that changes the analog voltage v_S into a 3-bit digital output v_{O1}, v_{O2}, and v_{O3}. The inverting inputs to the comparators are connected to a voltage divider. Using voltage division, we see that $v_{N1} = 1$ V, $v_{N2} = 3$ V, and $v_{N3} = 5$ V. The analog input v_S is applied simultaneously to the noninverting inputs of all three comparators. When $v_S < 1$ V, all three comparators are in −saturation with $v_{O1} = v_{O2} = v_{O3} = V_{OL} = 0$. When the analog input is in the range $3 > v_S > 1$, comparator 1 switches to +saturation and the three outputs are $v_{O1} = V_{OH} = 5$ V and $v_{O2} = v_{O3} = V_{OL} = 0$ V. When $5 > v_S > 3$, comparators 1 and 2 are in +saturation so the three outputs are $v_{O1} = v_{O2} = V_{OH} = 5$ V and $v_{O3} = V_{OL} = 0$ V. Finally, when $v_S > 5$ V, all three comparators are in +saturation with $v_{O1} = v_{O2} = v_{O3} = V_{OH} = 5$ V. These observations can be summarized as follows:

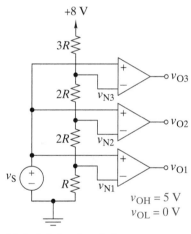

FIGURE 4−67

INPUT RANGE	COMPARATOR 1 V_{O1}	COMPARATOR 2 V_{O2}	COMPARATOR 3 V_{O3}
$1 > v_S$	0	0	0
$3 > v_S > 1$	5	0	0
$5 > v_S > 3$	5	5	0
$v_S > 5$	5	5	5

Note that each analog voltage range is converted into a unique 3-bit digital word. This 3-bit code can be converted to a 2-bit binary code using standard combinatorial logic design methods. The circuit is called a flash converter because the comparators operate in parallel and the analog-to-digital conversion takes place almost instantaneously.

SUMMARY

- The output of a dependent source is controlled by a signal in a different part of the circuit. Linear dependent sources are circuit elements used to model active devices and are represented in this text by a diamond-shaped source symbol. Each type of controlled source is characterized by a single-gain parameter μ, β, r, or g.

- The Thévenin resistance of a circuit containing dependent sources can be found using the open-circuit voltage and the short-circuit current, or by directly solving for the interface i–v characteristic. The active Thévenin resistance may be significantly different from the passive lookback resistance.

- The large-signal model describes the BJT in terms of cutoff, active, and saturation modes. Each mode has a unique set of i–v characteristics. The operating mode must be determined to correctly analyze a transistor circuit. The operating mode can be determined by calculating the circuit responses assuming that the device is in the active mode and then comparing the calculated responses with known bounds.

- The OP AMP is an active device with at least five terminals: the inverting input, the noninverting input, the output, and two power supply terminals. The power supply terminals are not usually shown in circuit diagrams. The integrated circuit OP AMP is a differential amplifier with a very high voltage gain.

- The OP AMP can operate in a linear mode when there is a feedback path from the output to the inverting input. To remain in the linear mode, the output voltage is limited to the range $-V_{CC} \le v_O \le +V_{CC}$, where $\pm V_{CC}$ are the supply voltages.

- The i–v characteristics of the ideal model of an OP AMP are $i_P = i_N = 0$ and $v_P = v_N$. The ideal OP AMP has an infinite voltage gain, an infinite input resistance, and zero output resistance. The ideal model is a good approximation to real devices as long as the circuit gain is much smaller than the OP AMP gain.

- Four basic signal-processing functions performed by OP AMP circuits are the inverting amplifier, noninverting amplifier, inverting summer, and differential amplifier. These arithmetic operations can also be represented in block diagram form.

- OP AMP circuits can be connected in cascade to obtain more complicated signal-processing functions. The analysis and design of the individual stages in the cascade can be treated separately, provided the input resistance of the following stage is kept sufficiently high.

- OP AMP circuits are easily treated using node analysis. A node voltage is identified at each OP AMP output, but a node equation is not written at these nodes. Node equations are then written at the remaining nodes, and the ideal OP AMP input voltage constraint ($v_N = v_P$) is used to reduce the number of unknowns.

- The comparator is a nonlinear signal-processing device obtained by operating an OP AMP device without feedback. The comparator has two analog inputs and a two-state digital output.

PROBLEMS

ERO 4–1 LINEAR ACTIVE CIRCUITS (SECTS. 4–1, 4–2)

Given a circuit containing linear resistors, dependent sources, and independent sources, find selected output signal variables, input-output relationships, or input-output resistances.
See Examples 4–2, 4–3, 4–4, 4–5, 4–6, 4–7, 4–8 and Exercises 4–1, 4–2, 4–3, 4–5, 4–6

4–1 Find the voltage gain v_O/v_S and current gain i_O/i_x in Figure P4–1.

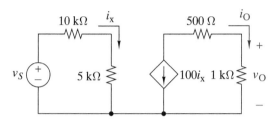

FIGURE   P4–1

4–2 Find the voltage gain v_O/v_1 and the current gain i_O/i_S in Figure P4–2. For $i_S = 2$ mA, find the power supplied by the input source i_S and the power delivered to the 2-kΩ load resistor.

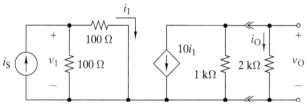

FIGURE   P4–2

4–3 The circuit in Figure P4–3 is a dependent-source model of a two-stage amplifier.
Find the output voltage v_3 and the current gain i_3/i_1 when $v_S = 1$ mV.

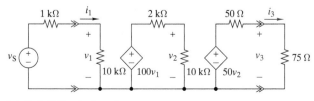

FIGURE   P4–3

4–4 The circuit in Figure P4–4 is an ideal voltage amplifier with negative feedback provided via the resistor R_F.
  (a) Find the output voltage v_2 and the current gain i_2/i_1 when $v_S = 10$ mV and $R_F = 10$ kΩ.
  (b) Find the input resistance $R_{IN} = v_1/i_1$.

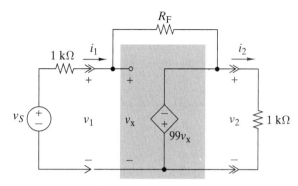

FIGURE   P4–4

4–5 The circuit in Figure P4–5 is an ideal current amplifier with negative feedback via the resistor R_E.
  (a) Find the output current i_2 and the voltage gain v_2/v_1 when $i_S = 25$ μA and $R_E = 450$ Ω.
  (b) Find the input resistance $R_{IN} = v_1/i_1$.

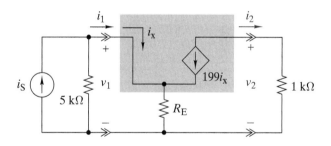

FIGURE   P4–5

4–6 Find the voltage gain v_O/v_S and the current gain i_O/i_S in Figure P4–6.

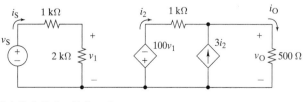

FIGURE   P4–6

4–7 Find the voltage gain v_O/v_S in Figure P4–7.

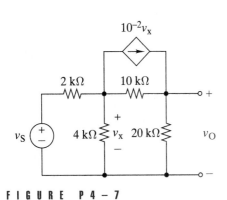

FIGURE P4–7

4–8 The circuit in Figure P4–8 is a model of a feedback amplifier using two identical transistors. Formulate either node-voltage or mesh-current equations for this circuit. Use these equations to solve for the input-output relationship $v_O = Kv_S$ using $r_\pi = 1$ kΩ, $R_E = 200$ Ω, $R_C = 10$ kΩ, $R_L = 5$ kΩ, $R_F = 5$ kΩ, and $\beta = 100$.

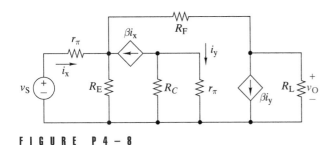

FIGURE P4–8

4–9 Find an expression for the voltage gain v_O/v_S in Figure P4–9.

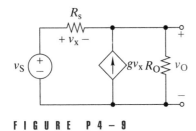

FIGURE P4–9

4–10 Find an expression for the current gain i_O/i_S in Figure P4–10.

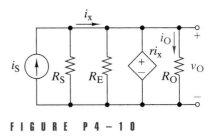

FIGURE P4–10

4–11 The circuit in Figure P4–11 is a model of a bipolar junction transistor operating in the active mode. The input signal is the current source i_S, and the voltage source V_{CC} supplies power. Find expressions for the base current i_B and the output voltage v_O in terms of V_{CC} and i_S.

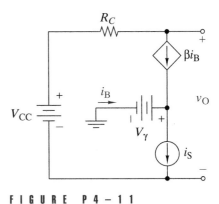

FIGURE P4–11

4–12 Find the Thévenin equivalent circuit seen by the resistor R_L in Figure P4–12.

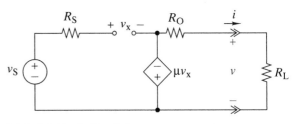

FIGURE P4–12

ERO 4–2 TRANSISTOR CIRCUITS (SECT. 4–3)

Given a linear resistive circuit with one transistor,

(a) Find the transistor operating mode and circuit responses.
(b) Select circuit parameters to obtain a specified operating mode.

See Examples 4–9, 4–10, 4–11 and Exercises 4–7, 4–8

4–13 In Figure P4–13 the circuit parameters are $R_B = 50$ kΩ, $R_C = 2$ kΩ, β = 100, $V_\gamma = 0.7$ V, and $V_{CC} = 10$ V. Find i_C and v_{CE} for $v_S = 2$ V. Repeat for $v_S = 5$ V.

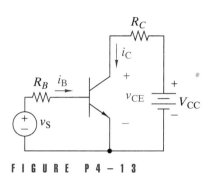

F I G U R E   P 4 – 1 3

4–14 In Figure P4–13 the circuit parameters are $R_B = 50$ kΩ, $R_C = 5$ kΩ, β = 50, $V_\gamma = 0.6$ V, and $V_{CC} = 15$ V. Find the range of v_S for which the transistor operates in the active mode.

4–15 In Figure P4–13 the circuit parameters are $R_B = 20$ kΩ, $R_C = 470$ Ω, β = 150, $V_\gamma = 0.7$ V, and $V_{CC} = 15$ V. Find the range of v_S for which the transistor operates in the saturation mode.

4–16 In Figure P4–13 the circuit parameters are $R_C = 1$ kΩ, β = 75, $V_\gamma = 0.7$ V, $V_{CC} = 20$ V, and $v_S = 2.5$ V. Select a value of R_B so that the transistor is in the active mode with $v_{CE} = V_{CC}/2$.

4–17 In Figure P4–17 the transistor parameters are β = 150 and $V_\gamma = 0.7$ V. Find i_C and v_{CE} for $v_S = 0.5$ V. Repeat for $v_S = 1$ V.

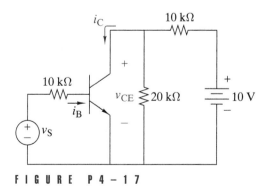

F I G U R E   P 4 – 1 7

4–18 In Figure P4–18 the transistor parameters are β = 80 and $V_\gamma = 0.7$ V. Find i_C and v_{CE} for $v_S = 1$ V. Repeat for $v_S = 4$ V.

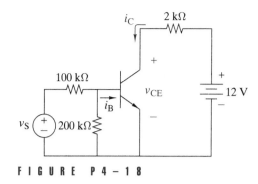

F I G U R E   P 4 – 1 8

4–19 In Figure P4–19 the transistor parameters are β = 100 and $V_\gamma = 0.7$ V. Find i_C and v_{CE}.

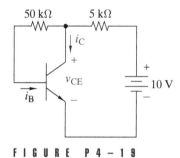

F I G U R E   P 4 – 1 9

4–20 The input in Figure P4–20 is a series connection of a dc source V_{BB} and a signal source v_S. The circuit parameters are $R_B = 500$ kΩ, $R_C = 5$ kΩ, β = 100, $V_\gamma = 0.7$ V, and $V_{CC} = 15$ V.

(a) With $v_S = 0$, select the value of V_{BB} so that the transistor is in the active mode with $v_{CE} = V_{CC}/2$.

(b) Using the value of V_{BB} found in (a), find the range of values of the signal voltage v_S for which the transistor remains in the active mode.

(c) Plot the transfer characteristic v_{CE} versus v_S as the signal voltage sweeps across the range from −10 V to +10 V.

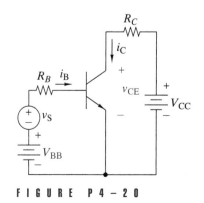

F I G U R E   P 4 – 2 0

ERO 4–3 OP AMP CIRCUIT ANALYSIS (SECTS. 4–4, 4–5)

Given a circuit consisting of linear resistors, OP AMPs, and independent sources, find selected output signals or input-output relationships in equation or block diagram form.
See Examples 4–13, 4–14, 4–16, 4–17, 4–18, 4–19 and Exercises 4–10, 4–11, 4–12, 4–13, 4–15, 4–16

4–21   Find v_O in terms of v_S in Figure P4–21.

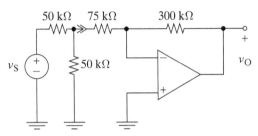

FIGURE   P4–21

4–22   Find v_O in terms of v_S in Figure P4–22.

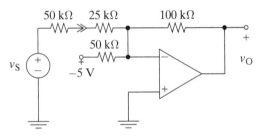

FIGURE   P4–22

4–23   (a) Find v_O in terms v_S in Figure P4–23.
(b) Find i_O for $v_S = 1.5$ V.

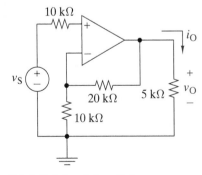

FIGURE   P4–23

4–24   (a) Find v_O in terms of v_{S1} and v_{S2} in Figure P4–24.
(b) For $V_{CC} = \pm15$ V and $v_{S2} = 10$ V, find the allowable range of v_{S1} for linear operation.

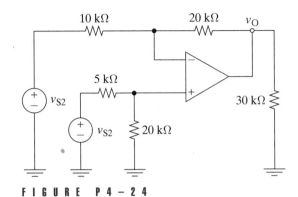

FIGURE   P4–24

4–25   The input-output relationship for a three-input inverting summer is

$$v_O = -[v_1 + 3v_2 + 5v_3]$$

The resistance of the feedback resistor is 75 kΩ, and the supply voltages are $V_{CC} = \pm15$ V.
(a) Find the values of the input resistors R_1, R_2, and R_3.
(b) For $v_2 = 0.5$ V and $v_3 = -1$ V, find the allowable range of v_1 for linear operation.

4–26   (a) Find v_O in terms v_S and V_{BB} in Figure P4–26.
(b) For $V_{CC} = \pm15$ V, $V_{BB} = 5$ V, and $R_1 = R_2$, sketch the v_O versus v_S for v_S on the range from –15 V to +15 V.

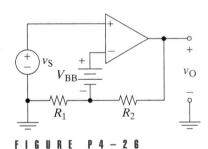

FIGURE   P4–26

4–27   Find v_O in terms of v_{S1} and v_{S2} in Figure P4–27

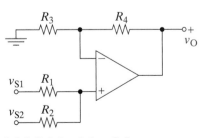

FIGURE   P4–27

4–28 **E** It is claimed that $v_O = v_S$ when the switch is closed in Figure P4–28 and that $v_O = -v_S$ when the switch is open. Prove or disprove this claim.

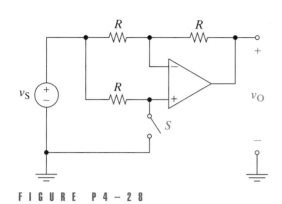

FIGURE P4–28

4–29 What range of gain is available from the circuit in Figure P4–29?

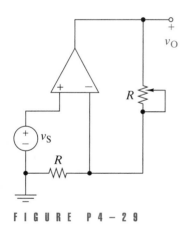

FIGURE P4–29

4–30 Find v_O in terms of v_{S1} and v_{S2} in Figure P4–30.

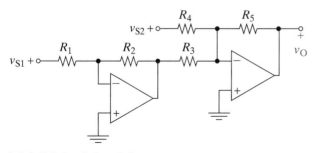

FIGURE P4–30

4–31 Find the output v_2 in terms of the input v_1 in Figure P4–31.

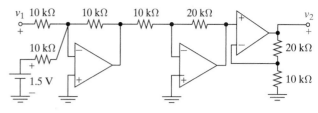

FIGURE P4–31

4–32 Find the output v_2 in terms of the input v_1 in Figure P4–32.

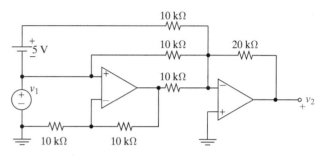

FIGURE P4–32

4–33 Find the output v_O in terms of the input v_S in Figure P4–33.

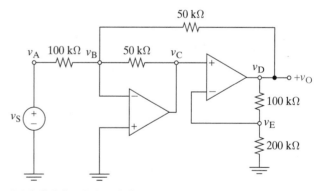

FIGURE P4–33

4–34 Find the output v_O in terms of v_1 and v_2 in Figure P4–34.

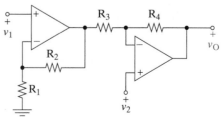

FIGURE P4–34

4–35 Find the output v_O in terms of v_1 and v_2 in Figure P4–35.

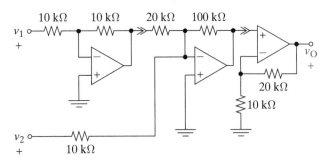

FIGURE P4–35

ERO 4–4 OP AMP CIRCUIT DESIGN (SECT. 4–6)

Given an input-output relationship, use resistors and OP AMPs to design one or more circuits that implement the relationship within stated constraints.
See Examples 4–15, 4–20, 4–21, 4–22, 4–23 and Exercise 4–16

4–36 ⒹShow how to interconnect a single OP AMP and the R-$2R$ resistor array shown in Figure P4–36 to obtain voltage gains of +1/4, +4, −1/3, and −3.

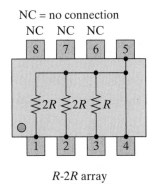

R-2R array

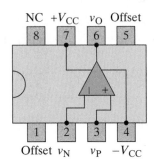

OP AMP

FIGURE P4–36

4–37 ⒹDesign circuits using resistors and OP AMPs to implement each of the following input-output relationships:
  (a) $v_O = 3v_1 - 2v_2$
  (b) $v_O = 2v_1 + v_2$

4–38 ⒹDesign a signal conditioning circuit for the temperature transducer whose characteristics are shown in Figure P4–38. The conditioning circuit must convert the transducer output for temperatures between −300 °C and −100 °C to a range of 0 to 5 V. The voltage gain for all stages must be less than 1000.

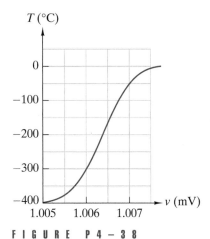

FIGURE P4–38

4–39 ⒹThe resistance of a pressure transducer varies from 5 kΩ to 15 kΩ when the pressure varies over its specified operating range. Design a signal conditioning circuit to convert the sensor resistance variation to a voltage signal on the range from 0 to 5 V.

4–40 ⒹDesign an OP AMP circuit that implements the block diagram in Figure P4–40 using only standard resistance values for ±5% tolerance (see Appendix A Table A–1).

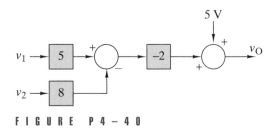

FIGURE P4–40

4–41 ⒹDesign an OP AMP circuit summer that implements a 3-bit D/A converter defined by

$$v_O = K(4v_1 + 2v_2 + v_3)$$

The digital inputs (v_1, v_2, v_3) are either 0 V or 5 V. The analog output (v_O) must fall in the range of 0 V to 10 V for all possible inputs.

4–42 **D** An instrumentation system combines the outputs of three transducers into a single output using gains of -20, 3, and 18. Each transducer has a Thévenin resistance of 600 Ω and a Thévenin voltage in the range ±120 mV. Design an OP AMP circuit to meet these requirements.

4–43 **D** A requirement exists for an amplifier with a gain of –12,000 and an input resistance of at least 300 kΩ. Design an OP AMP circuit that meets the requirements using general-purpose OP AMPs with voltage gains of $A = 2 \times 10^5$, input resistances of $R_I = 4 \times 10^8$ Ω, and output resistances $R_O = 20$ Ω.

4–44 **D** The potentiometer in Figure P4–44 is used as a position sensor. The range of mechanical input moves the wiper between the bottom and the top of the potentiometer. An interface circuit is required to convert the v_P range (–15 to +15 V) to a v_O range (0 to 5 V) suitable for input to the analog-to-digital converter. Design a suitable interface circuit.

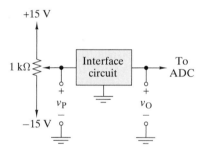

FIGURE P4 – 44

4–45 **D** Select the resistances in Figure P4–45 to produce the following gains.

S_1 open	S_1 open	K
open	open	10
open	closed	5
closed	x	2

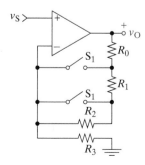

FIGURE P4 – 45

ERO 4–5 THE COMPARATOR (SECT. 4–7)

Given a circuit with one or more comparators, find the circuit input-output relationship.
See Examples 4–25, 4–26

4–46 The circuit in Figure P4–46 has $V_{OH} = 15$ V and $V_{OL} = 0$ V.
 (a) Determine the input voltage ranges for which $v_O = V_{OH}$ and $v_O = V_{OL}$.
 (b) Sketch the circuit transfer characteristics for v_S on the range from –15 V to +15 V.

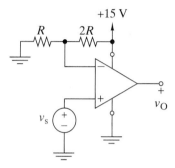

FIGURE P4 – 46

4–47 The circuit in Figure P4–47 has $V_{OH} = 10$ V and $V_{OL} = 0$ V.
 (a) Determine the input voltage ranges for which $v_O = V_{OH}$ and $v_O = V_{OL}$.
 (b) Sketch the circuit transfer characteristics for v_S on the range from –15 V to +15 V.

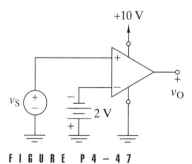

FIGURE P4 – 47

4–48 The circuit in Figure P4–48 is called a window detector. In this circuit the OP AMP saturation levels are $\pm V_{CC}$. For $V_{CC} = 15$ V, $V_{AA} = 10$ V, and $V_{BB} = 5$ V, show that output is determined by the following statement:

If $(-V_{BB} < v_S < V_{AA})$ then $v_O = 15$ V else $v_O = 0$.

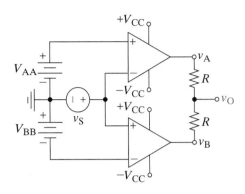

FIGURE P 4 - 4 8

4–49 The circuit in Figure P4–49 has $V_{CC} = 15$ V and $V_{BB} = -5$ V.

(a) Determine the input voltage ranges for which $v_O = V_{OH}$ and $v_O = V_{OL}$.

(b) Sketch the circuit transfer characteristics for v_S on the range from -10 V to $+10$ V.

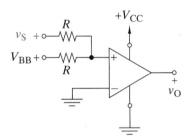

FIGURE P 4 - 4 9

4–50 Repeat Problem 4–49 with $V_{BB} = 5$ V.

4–51 The circuit in Figure P4–51 has $V_{CC} = 15$ V and $V_{BB} = 5$ V. Sketch the output voltage v_O on the range $0 \leq t \leq 2$ s for $v_S = 10 \sin(2\pi t)$ V.

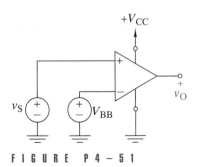

FIGURE P 4 - 5 1

4–52 Repeat Problem 4–51 with $V_{BB}(t) = 10t$ V.

INTEGRATING PROBLEMS

4–53 **A** **D** **E** BRIDGED-T INVERTING AMPLIFIER

Using the basic inverting OP AMP configuration to obtain a large voltage gain requires a small input resistor, a large feedback resistor, or both. Small input resistors load the input source, and large ($R > 1$ MΩ) feedback resistors have more noise and exhibit greater life-cycle variations. The circuit in Figure P4–53 circumvents these problems by using a bridged-T circuit in the feedback path. Note that R_3 occurs twice in this diagram.

(a) **A** Show that the gain of the circuit in Figure P4–53 can be written as $K = -R_{FDBK}/R_1$, where R_{FDBK} is the effective feedback resistance defined as:

$$R_{FDBK} = R_3 \left(2 + \frac{R_3}{R_2} \right)$$

(b) **D** Design a basic inverting amplifier to achieve $K = -400$ and $R_{IN} \geq 20$ kΩ.

(c) **D** Design a bridged-T inverting amplifier to achieve $K = -400$ and $R_{IN} \geq 20$ kΩ.

(d) **E** Evaluate the two designs by comparing their element counts, element spreads (ratio of largest over smallest resistance), and total resistances (sum of all resistances).

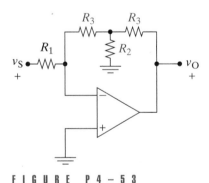

FIGURE P 4 - 5 3

4–54 **A** HYBRID CIRCUIT ANALYSIS

The two-port interface circuit in Figure P4–54 is a small-signal model of a bipolar junction transistor using what are called hybrid parameters. The analysis of this circuit illustrates that it can be useful to use a mixture (or hybrid) of mesh-current and node-voltage equations.

(a) Using the symbolic notation in the figure, write a mesh-current equation for the input circuit and a node-voltage equation for the output circuit.

(b) Using $R_S = 2$ kΩ, $R_1 = 5$ kΩ, $R_2 = 50$ kΩ, $R_L = 150$ kΩ, $\mu = 10^{-3}$, and $\beta = 50$, solve the equations from (a) for the input current i_1 and the output voltage v_2 in terms of the input v_S.

(c) Using the results from (b), solve for the input resistance and the voltage gain of the circuit.

(d) Calculate the power gain defined as the ratio of the power delivered to R_L divided by the power supplied by v_S.

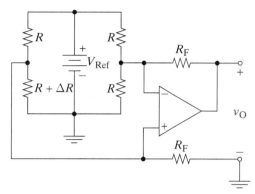

F I G U R E   P 4 – 5 5

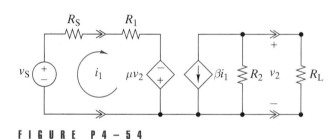

F I G U R E   P 4 – 5 4

4–55   A D WHEATSTONE BRIDGE AMPLIFIER

The circuit in Figure P4–55 shows a Wheatstone bridge consisting of three equal resistors and a fourth resistor that is a transducer whose resistance is $R + \Delta R$, where $\Delta R << R$. The bridge is excited by a constant reference voltage source V_{REF} and is connected to the inputs of an OP AMP. For $\Delta R << R$ the OP AMP output voltage can be expressed in the form

$$v_O = K \left[\frac{\Delta R}{R} \right] V_{REF}$$

(a) A Verify that this expression is correct and express K in terms of circuit parameters.

(b) D For $V_{REF} = 15$ V, $R = 100$ Ω and $\Delta R/R$ in the range $\pm 0.04\%$, select value R_F so that the output voltage v_O falls in the range ± 3 V.

4–56   D TEMPERATURE SENSOR DESIGN

Figure P4–56 shows a circuit with a semiconductor temperature sensor modeled as a temperature-controlled current source. The device senses absolute temperature T_A (°K) and delivers a current kT_A, where $k = 1$ μA/°K. The purpose of the OP AMP circuit is to make the output voltage proportional to °C. For $V_{CC} = 10$ V, select values for R_1 and R_2 so that output voltage sensitivity is 100 mV/°C.

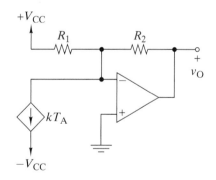

F I G U R E   P 4 – 5 6

4-57   A D E SUBTRACTOR CIRCUITS

The input-output relationship for both circuits in Figure P4–57 are of the form $v_O = K_2 v_2 + K_1 v_1$.

(a) A For circuits C1 and C2, determine the constants K_1 and K_2 in terms of circuit parameters.

(b) D In circuit C1 with $R_1 = R_2 = 1$ kΩ and $R_3 = R_5 = 10$ kΩ, select the values R_4 and R_6 that produce $v_O = 5(v_2 - v_1)$.

(c) D Repeat part (b) for circuit C2.

(d) **E** Evaluate the two designs by comparing the number of devices required and the load they impose on the input signal sources.

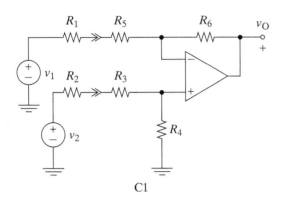

C1

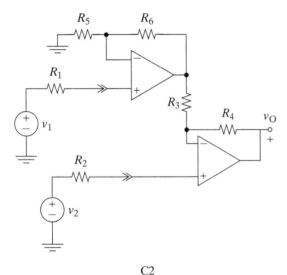

C2

FIGURE P4–57

4-58   A D E PHOTORESISTOR INSTRUMENTATION

Both circuits in Figure P4–58 contain a photoresistor R_X whose resistance varies inversely with the intensity of the incident light. In complete darkness its resistance is 10 kΩ. In bright sunlight its resistance is 2 kΩ. At any given light level the circuit is linear, so its input-output relationship is of the form $v_O = Kv_1$.

(a) **A** For circuits C1 and C2, determine the constant K in terms of circuit resistances.

(b) **D** For circuit C1 with $v_1 = +15$ V, select the values of R and R_F so that $v_O = -10$ V in bright sunlight and +10 V in complete darkness.

(c) **D** Repeat part (b) for circuit C2.

(d) **E** Evaluate the two designs by comparing the number of devices required and the total power dissipated.

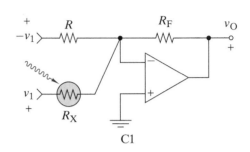

C1

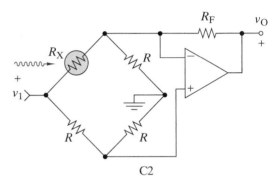

C2

FIGURE P4–58

4-59   A E DESIGN EVALUATION

As chief engineer of a small electronics company, you find yourself with a dilemma. Your two engineering interns have worked independently and have produced different solutions for the design problem you gave them. Their proposed solutions are shown in Figure 4–59.

You asked them to design a circuit with three inputs v_1, v_2, v_3, and an output v_O. The output is to be proportional to $v_1/16 + v_2/8 + v_3/4$. The output can be either positive or negative, but must be able to drive a 500-Ω load. A ±15 V dc power supply is available within the system without added cost. You need to manufacture 2 million of these circuits, so part costs are a concern, but performance is the top priority. Electrical characteristics and cost data for available parts are as follows.

PART TYPE	PART CHARACTERISTICS	RELATIVE COST
Standard resistors	1.0, 1.5, 2.2, 3.3, 4.7, 6.8 Ω and multiples of 10 thereof	$4000 per 100,000
Custom resistors	Values to three significant figures from 100 Ω to 200 kΩ	$4500 per 100,000
OP AMPs	Minimum voltage gain 100,000	$11,500 per 100,000
	Output voltage range ±15 V	
	Maximum output current 10 mA	

(a) **A** Verify whether both circuits comply with the performance requirements. If not, explain how one or both can be modified to comply.

(b) **E** Which of the two circuits would you select for production and why?

4–60 **D** COMPUTER-AIDED CIRCUIT DESIGN

Use computer-aided circuit analysis to find the value of R_F in Figure P4–60 that causes the input resistance seen by i_S to be 50 Ω. Find the current gain i_O/i_S for this value of R_F. Use $\beta = 100$, $r_\pi = 1.1$ kΩ, $R_C = 10$ kΩ, $R_E = 100$ Ω, and $R_L = 100$ Ω.

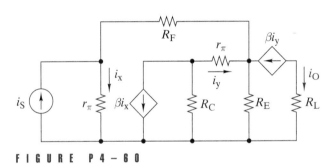

FIGURE P4–60

FIGURE P4–59

CHAPTER 5

SIGNAL WAVEFORMS

Under the sea, under the sea
mark how the telegraph motions to me.
Under the sea, under the sea
signals are coming along.

James Clerk Maxwell, 1873,
Scottish Physicist and
Occasional Humorous Poet

5–1 Introduction
5–2 The Step Waveform
5–3 The Exponential Waveform
5–4 The Sinusoidal Waveform
5–5 Composite Waveforms
5–6 Waveform Partial Descriptors
Summary
Problems
Integrating Problems

James Clerk Maxwell (1831–1879) is considered the unifying founder of the mathematical theory of electromagnetics. This genial Scotsman often communicated his thoughts to friends and colleagues via whimsical poetry. In the preceding short excerpt, Maxwell reminds us that the purpose of a communication system (the submarine cable telegraph in this case) is to transmit signals and that those signals must be changing, or *in motion*, as he put it.

This chapter marks the beginning of a new phase of our study of circuits. Up to this point we have dealt with resistive circuits, in which voltages and currents are constant (for example, +15 V or –3 mA). From this point forward we will be dealing with dynamic circuits, in which voltages and currents vary as functions of time. To analyze dynamic circuits, we need models for the time-varying signals and models for devices that describe the effects of time-varying signals in circuits. Signal models are introduced in this chapter. The dynamic circuit elements called capacitance and inductance are introduced in the next chapter. Chapters 7, 8, 9, and 10 then show how we combine signal models and device models to analyze dynamic circuits.

This chapter introduces the basic signal models used in the remainder of the book. We devote a chapter to this so you can master these models prior to launching into the complexities of dynamic circuits. We first introduce three key waveforms: the step, exponential, and sinusoid functions. By combining these three models we can build composite waveforms for all signals encountered in this book. Descriptors used to classify and describe waveforms are introduced because they highlight important signal attributes.

5–1 INTRODUCTION

We normally think of a signal as an electrical current $i(t)$ or voltage $v(t)$. The time variation of the signal is called a waveform. More formally,

> **A waveform** *is an equation or graph that defines the signal as a function of time.*

Up to this point our study has been limited to the type of waveform shown in Figure 5–1. Waveforms that are constant for all time are called **dc signals**. The abbreviation *dc* stands for direct current, but it applies to either voltage or current. Mathematical expressions for a dc voltage $v(t)$ or current $i(t)$ take the form

$$\left.\begin{array}{r} v(t) = V_0 \\ i(t) = I_0 \end{array}\right\} \quad \text{for } -\infty < t < \infty \qquad (5\text{--}1)$$

This equation is only a model. No physical signal can remain constant forever. It is a useful model, however, because it approximates the signals produced by physical devices such as batteries.

There are two matters of notation and convention that must be discussed before continuing. First, quantities that are constant (non-time-varying)

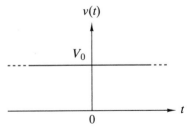

FIGURE 5–1 *A constant or dc waveform.*

are usually represented by uppercase letters (V_A, I, T_O) or lowercase letters in the early part of the alphabet (a, b_7, f_0). Time-varying electrical quantities are represented by the lowercase letters i, v, p, q, and w. The time variation is expressly indicated when we write these quantities as $v_1(t)$, $i_A(t)$, or $w_C(t)$. Time variation is implicit when they are written as v_1, i_A, or w_C.

Second, in a circuit diagram signal variables are normally accompanied by the reference marks (+, −) for voltage and (→) for current. It is important to remember that these reference marks *do not* indicate the polarity of a voltage or the direction of current. The marks provide a baseline for determining the sign of the numerical value of the actual waveform. When the actual voltage polarity or current direction coincides with the reference directions, the signal has a positive value. When the opposite occurs, the value is negative. Figure 5–2 shows examples of voltage waveforms, including some that assume both positive and negative values. The bipolar waveforms indicate that the actual voltage polarity is changing as a function of time.

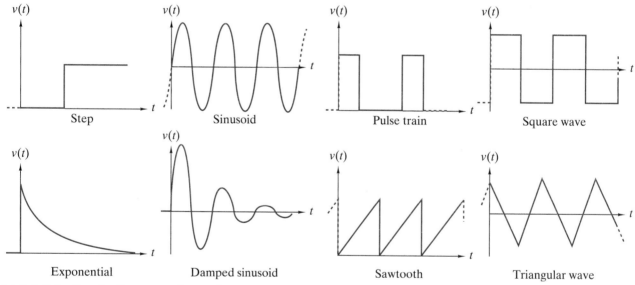

F I G U R E   5 · − 2   *Some example waveforms.*

The waveforms in Figure 5–2 are examples of signals used in electrical engineering. Since there are many such signals, it may seem that the study of signals involves the uninviting task of compiling a lengthy catalog of waveforms. However, it turns out that a long list is not needed. In fact, we can derive most of the waveforms of interest using just three basic signal models: the step, exponential, and sinusoidal functions. The small number of basic signals illustrates why models are so useful to engineers. In reality, waveforms are very complex, but their time variation can be approximated adequately using only a few basic building blocks.

Finally, in this chapter we will generally use voltage $v(t)$ to represent a signal waveform. Remember, however, that a signal can be either a voltage $v(t)$ or current $i(t)$.

5−2   THE STEP WAVEFORM

The first basic signal in our catalog is the step waveform. The general step function is based on the **unit step function** defined as

$$u(t) = \begin{cases} 0 & \text{for } t < 0 \\ 1 & \text{for } t > 0 \end{cases} \qquad (5\text{--}2)$$

The step function waveform is equal to zero when its argument t is negative, and is equal to unity when its argument is positive. Mathematically, the function $u(t)$ has a jump discontinuity at $t = 0$.

Strictly speaking, it is impossible to generate a true step function since signal variables like current and voltage cannot jump from one value to another in zero time. Practically speaking, we can generate very good approximations to the step function. What is required is that the transition time be short compared with other response times in the circuit. Actually, the generation of approximate step functions is an everyday occurrence since people frequently turn things like TVs, stereos, and lights on and off.

On the surface, it may appear that the step function is not a very exciting waveform or, at best, only a source of temporary excitement. However, the step waveform is a versatile signal used to construct a wide range of useful waveforms. Multiplying $u(t)$ by a constant V_A produces the waveform

$$V_A u(t) = \begin{cases} 0 & \text{for } t < 0 \\ V_A & \text{for } t \geqslant 0 \end{cases} \qquad (5\text{--}3)$$

Replacing t by $(t - T_S)$ produces a waveform $V_A u(t - T_S)$, which takes on the values

$$V_A u(t - T_S) = \begin{cases} 0 & \text{for } t < T_S \\ V_A & \text{for } t \geqslant T_S \end{cases} \qquad (5\text{--}4)$$

The **amplitude** V_A scales the size of the step discontinuity, and the **time-shift** parameter T_S advances or delays the time at which the step occurs, as shown in Figure 5–3.

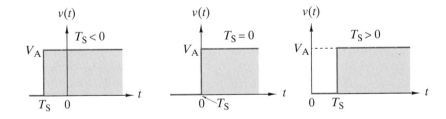

FIGURE   5 − 3 *Effect of changing time shifting on step function waveform.*

Amplitude and time-shift parameters are required to define the general step function. The amplitude V_A carries the units of volts. The amplitude of step function in electric current is I_A and carries the units of amperes. The constant T_S carries the units of time, usually seconds. The parameters V_A (or I_A) and T_S can be positive, negative, or zero. By combining several step functions, we can represent a number of important waveforms. One possibility is illustrated in the following example:

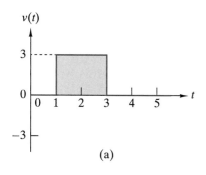

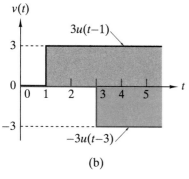

FIGURE 5–4

EXAMPLE 5–1

Express the waveform in Figure 5–4(a) in terms of step functions.

SOLUTION:

The amplitude of the pulse jumps to a value of 3 V at $t = 1$ s; therefore, 3 $u(t-1)$ is part of the equation for the waveform. The pulse returns to zero at $t = 3$ s, so an equal and opposite step must occur at $t = 3$ s. Putting these observations together, we express the rectangular pulse as

$$v(t) = 3u(t-1) - 3u(t-3)$$

Figure 5–4(b) shows how the two step functions combine to produce the given rectangular pulse. ∎

THE IMPULSE FUNCTION

The generalization of Example 5–1 is the waveform

$$v(t) = V_A[u(t-T_1) - u(t-T_2)]$$

This waveform is a rectangular pulse of amplitude V_A that turns on at $t = T_1$ and off at $t = T_2$. The pulse train and square wave signals in Figure 5–2 can be generated by a series of these pulses. Pulses that turn on at some time T_1 and off at some later time T_2 are sometimes called **gating functions** because they are used in conjunction with electronic switches to enable or inhibit the passage of another signal.

A unit-area pulse centered on $t = 0$ is written in terms of step functions as

$$v(t) = \frac{1}{T}\left[u\left(t + \frac{T}{2}\right) - u\left(t - \frac{T}{2}\right)\right] \qquad (5\text{–}5)$$

The pulse in Eq. (5–5) is zero everywhere except in the range $-T/2 \leq t \leq T/2$, where its value is $1/T$. The area under the pulse is 1 because its scale factor is inversely proportional to its duration. As shown in Figure 5–5(a), the pulse becomes narrower and higher as T decreases but maintains its unit area. In the limit as $T \to 0$ the scale factor approaches infinity but the area remains 1. The function obtained in the limit is called a **unit impulse**, symbolized as $\delta(t)$. The graphical representation of $\delta(t)$ is shown in Figure 5–5(b). The impulse is an idealized model of a large-amplitude, short-duration pulse.

A formal definition of the unit impulse is

$$\delta(t) = 0 \text{ for } t \neq 0 \quad \text{and} \quad \int_{-\infty}^{t} \delta(x)dx = u(t) \qquad (5\text{–}6)$$

The first condition says the impulse is zero everywhere except at $t = 0$. The second condition suggests that the unit impulse is the derivative of a unit step function:

$$\delta(t) = \frac{du(t)}{dt} \qquad (5\text{–}7)$$

The conclusion in Eq. (5–7) cannot be justified using elementary mathematics since the function $u(t)$ has a discontinuity at $t = 0$ and its derivative at that point does not exist in the usual sense. However, the concept can be justified using limiting conditions on continuous functions, as discussed in texts on signals and systems.[1] Accordingly, we defer the question of mathematical rigor to later courses and think of the unit impulse as the derivative of a unit step function. Note that this means that the unit impulse $\delta(t)$ has units of reciprocal time, or s^{-1}.

An impulse of strength K is denoted $v(t) = K\delta(t)$. Consequently, the scale factor K has the units of V-s and is the area under the impulse $K\delta(t)$. In the graphical representation of the impulse the value of K is written in parentheses beside the arrow, as shown in Figure 5–5(b).

EXAMPLE 5–2

Calculate and sketch the derivative of the pulse in Figure 5–6(a).

SOLUTION:
In Example 5–1 the pulse waveform was written as

$$v(t) = 3u(t - 1) - 3u(t - 3) \text{ V}$$

Using the derivative property of the step function, we write

$$\frac{dv(t)}{dt} = 3\delta(t - 1) - 3\delta(t - 3)$$

The derivative waveform consists of a positive-going impulse at $t = 1$ s and a negative-going impulse at $t = 3$ s. Figure 5–6(b) shows how the impulse train is represented graphically. The waveform $v(t)$ has the units of volts (V), so its derivative $dv(t)/dt$ has the units of V/s. ∎

THE RAMP FUNCTION

The **unit ramp** is defined as the integral of a step function:

$$r(t) = \int_{-\infty}^{t} u(x)dx = tu(t) \qquad (5\text{–}8)$$

The unit ramp waveform $r(t)$ in Figure 5–7(a) is zero for $t < 0$ and is equal to t for $t > 0$. Notice that the slope of $r(t)$ is 1 and has the units of time, or s. A ramp of strength K is denoted $v(t) = Kr(t)$, where the scale factor K has the units of V/s and is the slope of the ramp. The general ramp waveform shown in Figure 5–7(b) written as $v(t) = Kr(t - T_S)$ is zero for $t < T_S$ and equal to $K(t - T_S)$ for $t \geq T_S$. By adding a sequence of ramps, we can create the triangular and sawtooth waveforms shown in Figure 5–2.

SINGULARITY FUNCTIONS

The unit impulse, unit step, and unit ramp form a triad of related signals that are referred to as **singularity functions**. They are related by integration as

1 For example, see Alan V. Oppenheim and Allan S. Willsky, *Signals and Systems Analysis*, (Englewood Cliffs, N.J.: Prentice Hall, 1983), pp. 22–23.

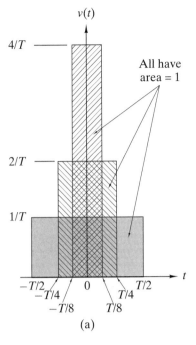

(a)

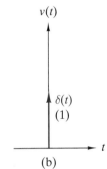

(b)

FIGURE   5 – 5 *Rectangular pulse waveforms and the impulse.*

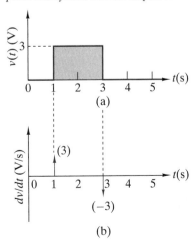

FIGURE   5 – 6

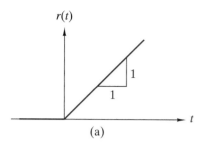

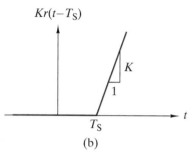

FIGURE 5–7 *(a) Unit ramp waveform. (b) General ramp waveform.*

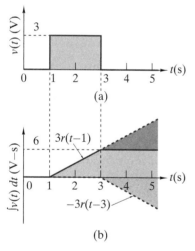

FIGURE 5–8

$$u(t) = \int_{-\infty}^{t} \delta(x)dx$$

$$r(t) = \int_{-\infty}^{t} u(x)dx \tag{5–9}$$

or by differentiation as

$$\delta(t) = \frac{du(t)}{dt}$$

$$u(t) = \frac{dr(t)}{dt} \tag{5–10}$$

These signals are used to generate other waveforms and as test inputs to linear systems to characterize their responses. When applying the singularity functions in circuit analysis, it is important to remember that $u(t)$ is a dimensionless function. But Eqs. (5–9) and (5–10) point out that $\delta(t)$ carries the units of s^{-1} and $r(t)$ carries units of seconds.

EXAMPLE 5–3

Derive an expression for the waveform for the integral of the pulse in Figure 5–8(a).

SOLUTION:
In Example 5–1 the pulse waveform was written as

$$v(t) = 3u(t - 1) - 3u(t - 3) \text{ V}$$

Using the integration property of the step function, we write

$$\int_{-\infty}^{t} v(x)dx = 3r(t - 1) - 3r(t - 3)$$

The integral is zero for $t < 1$ s. For $1 < t < 3$ the waveform is $3(t - 1)$. For $t > 3$ it is $3(t - 1) - 3(t - 3) = 6$. These two ramps produce the pulse integral shown in Figure 5–8(b). The waveform $v(t)$ has the units of volts (V), so the units of its integral are V-s. ■

EXAMPLE 5–4

Figure 5–9(a) shows an ideal electronic switch whose input is a ramp $2r(t)$, where the scale factor $K = 2$ carries the units of V/s. Find the switch output $v_O(t)$ when the gate function in Example 5–1 is applied to the control terminal (G) of the switch.

SOLUTION:
In Example 5–1 the gate function was written as

$$v_G(t) = 3u(t - 1) - 3u(t - 3) \text{ V}$$

The gate function turns the switch on at $t = 1$ s and off at $t = 3$ s. The output voltage of the switch is

$$v_O(t) = \begin{cases} 0 & t < 1 \\ 2t & 1 < t < 3 \\ 0 & 3 < t \end{cases}$$

Only the portion of the input waveform within the gate interval appears at the output. Figures 5–9(b), 5–9(c), and 5–9(d) show how the gate function $v_G(t)$ controls the passage of the input signal through the electronic switch. ∎

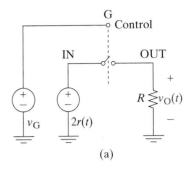

(a)

Exercise 5-1

Express the following signals in terms of singularity functions:

(a) $v_1(t) = \begin{cases} 0 & t < 2 \\ 4 & 2 < t < 4 \\ -4 & 4 < t \end{cases}$ (b) $v_2(t) = \begin{cases} 0 & t < 2 \\ 4 & 2 < t < 4 \\ -2t + 12 & 4 < t \end{cases}$

(c) $v_3(t) = \int_{-\infty}^{t} v_1(x)dx$ (d) $v_4(t) = \dfrac{dv_2(t)}{dt}$

Answers:

(a) $v_1(t) = 4\,u(t-2) - 8\,u(t-4)$ (b) $v_2(t) = 4\,u(t-2) - 2\,r(t-4) + 8\,u(t-4)$
(c) $v_3(t) = 4\,r(t-2) - 8\,r(t-4)$ (d) $v_4(t) = 4\,\delta(t-2) - 2\,u(t-4)$

Exercise 5-2

(a) Write an expression for a rectangular pulse with an amplitude of 15 V that begins at $t = -5$ s and ends at $t = 10$ s.
(b) Write an expression for the derivative of the pulse defined in (a).
(c) Write an expression for the integral of the pulse in (a).

Answers:

(a) $15[u(t+5) - u(t-10)]$
(b) $15[\delta(t+5) - \delta(t-10)]$
(c) $15(t+5)u(t+5) - 15(t-10)u(t-10) = 15[r(t+5) - r(t-10)]$

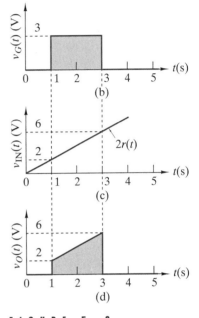

FIGURE 5-9

5-3 THE EXPONENTIAL WAVEFORM

The **exponential waveform** is a step function whose amplitude factor gradually decays to zero. The equation for this waveform is

$$v(t) = \left[V_A e^{-t/T_C}\right] u(t) \qquad (5-11)$$

A graph of $v(t)$ versus t/T_C is shown in Figure 5–10. The exponential starts out like a step function. It is zero for $t < 0$ and jumps to a maximum amplitude of V_A at $t = 0$. Thereafter it monotonically decays toward zero as time marches on. The two parameters that define the waveform are the **amplitude** V_A (in volts) and the **time constant** T_C (in seconds). The amplitude of a current exponential would be written I_A and carry the units of amperes.

The time constant is of special interest, since it determines the rate at which the waveform decays to zero. An exponential decays to about 37%

FIGURE 5 – 1 0 *The exponential wave-form.*

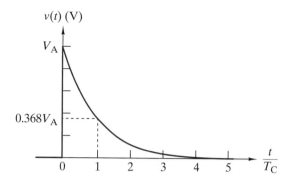

of its initial amplitude $v(0) = V_A$ in one time constant, because at $t = T_C$, $v(T_C) = V_A e^{-1}$, or approximately $0.368 \times V_A$. At $t = 5T_C$, the value of the waveform is $V_A e^{-5}$, or approximately $0.00674\ V_A$. An exponential signal decays to less than 1% of its initial amplitude in a time span of five time constants. In theory, an exponential endures forever, but practically speaking after about $5T_C$ the waveform amplitude becomes negligibly small. We define the **duration** of a waveform to be the interval of time outside of which the waveform is everywhere less than a stated value. Using this concept, we say the duration of an exponential waveform is $5T_C$.

EXAMPLE 5–5

Plot the waveform $v(t) = [-17e^{-100t}]u(t)$ V.

SOLUTION:

From the form of $v(t)$, we recognize that $V_A = -17$ V and $T_C = 1/100$ s or 10 ms. The minimum value of $v(t)$ is $v(0) = -17$ V, and the maximum value is approximately 0 V as t approaches $5T_C = 50$ ms. These observations define appropriate scales for plotting the waveform. Spreadsheet programs are especially useful for the repetitive calculations and graphical functions involved in waveform plotting. Figure 5–11 shows how this example can be handled using Excel. We begin by filling in the "A" column with time values ranging from $t = 0$ to $t = 50$ ms. The data in the "B" column are obtained by calculating the value of $-17e^{-1001t}$ for each of the values of t (ms) in the "A" column. Then using the graphing tool, we create a plot showing that the waveform starts out at $v(t) = -17$ V at $t = 0$ and then increases toward $v(t) = 0$ as $t \rightarrow 50$ ms. ∎

PROPERTIES OF EXPONENTIAL WAVEFORMS

The **decrement property** describes the decay rate of an exponential signal. For $t > 0$ the exponential waveform is given by

$$v(t) = V_A e^{-t/T_c} \tag{5–12}$$

The step function can be omitted since it is unity for $t > 0$. At time $t + \Delta t$ the amplitude is

$$v(t + \Delta t) = V_A e^{-(t + \Delta t)/T_c} = V_A e^{-t/T_c} e^{-\Delta t/T_c} \tag{5–13}$$

FIGURE   5 – 1 1

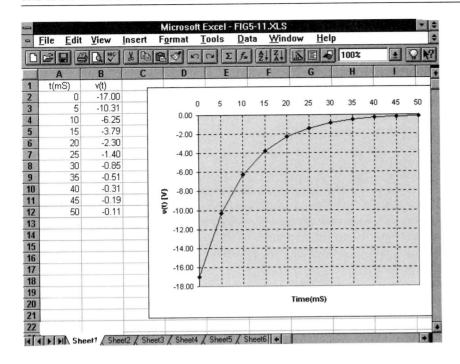

The ratio of these two amplitudes is

$$\frac{v(t + \Delta t)}{v(t)} = \frac{V_A e^{-t/T_C} e^{-\Delta t/T_C}}{V_A e^{-t/T_C}} = e^{-\Delta t/T_C} \tag{5–14}$$

The decrement ratio is independent of amplitude and time. In any fixed time period Δt, the fractional decrease depends only on the time constant. The decrement property states that the same percentage decay occurs in equal time intervals.

The slope of the exponential waveform (for $t > 0$) is found by differentiating Eq. (5–12) with respect to time:

$$\frac{dv(t)}{dt} = -\frac{V_A}{T_C} e^{-t/T_C} = -\frac{v(t)}{T_C} \tag{5–15}$$

The **slope property** states that the time rate of change of the exponential waveform is inversely proportional to the time constant. Small time constants lead to large slopes or rapid decays, while large time constants produce shallow slopes and long decay times.

Equation (5–15) can be rearranged as

$$\frac{dv(t)}{dt} + \frac{v(t)}{T_C} = 0 \tag{5–16}$$

When $v(t)$ is an exponential of the form in Eq. (5–12), then $dv/dt + v/T_C = 0$. That is, the exponential waveform is a solution of the first-order linear differential equation in Eq. (5–16). We will make use of this fact in Chapter 7.

The time-shifted exponential waveform is obtained by replacing t in Eq. (5–11) by $t - T_S$. The general exponential waveform is written as

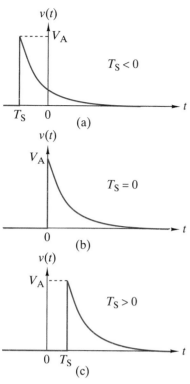

FIGURE 5–12 *Effect of time shifting on the exponential waveform.*

$$v(t) = \left[V_A e^{-(t-T_S)/T_C}\right] u(t - T_S) \qquad (5\text{–}17)$$

where T_S is the time-shift parameter for the waveform. Figure 5–12 shows exponential waveforms with the same amplitude and time constant but different values of T_S. Time shifting translates the waveform to the left or right depending on whether T_S is negative or positive. *Caution:* The factor $t - T_S$ must appear in both the argument of the step function and the exponential, as shown in Eq. (5–17).

EXAMPLE 5–6

An oscilloscope is a laboratory instrument that displays the instantaneous value of waveform versus time. Figure 5–13 shows an oscilloscope display of a portion of an exponential waveform. In the figure the vertical (amplitude) axis is calibrated at 2 V per division, and the horizontal (time) axis is calibrated at 1 ms per division. Find the time constant of the exponential.

SOLUTION:

For $t > 0$ the general expression for an exponential in Eq. (5–17) becomes

$$v(t) = V_A e^{-(t-T_S)/T_C}$$

We have only a portion of the waveform, so we do not know the location of the $t = 0$ time origin; hence, we cannot find the amplitude V_A or the time shift T_S from the display. But, according to the decrement property, we should be able to determine the time constant since the decrement ratio is independent of amplitude and time. Specifically, Eq. (5–14) points out that

$$\frac{v(t + \Delta t)}{v(t)} = e^{-\Delta t/T_C}$$

Solving for the time constant T_C yields

$$T_C = \frac{\Delta t}{\ln\left[\dfrac{v(t)}{v(t + \Delta t)}\right]}$$

FIGURE 5–13

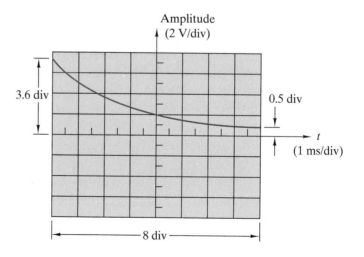

Taking the starting point at the left edge of the oscilloscope display yields

$$v(t) = (3.6 \text{ div})(2 \text{ V/div}) = 7.2 \text{ V}$$

Next, defining Δt to be the full width of the display produces

$$\Delta t = (8 \text{ div})(1 \text{ ms/div}) = 8 \text{ ms}$$

and

$$v(t + \Delta t) = (0.5 \text{ div})(2 \text{ V/div}) = 1 \text{ V}$$

As a result, the time constant of the waveform is found to be

$$T_C = \frac{\Delta t}{\ln\left[\dfrac{v(t)}{v(t + \Delta t)}\right]} = \frac{8 \times 10^{-3}}{\ln(7.2/1)} = 4.05 \text{ ms} \qquad \blacksquare$$

Exercise 5–3

(a) An exponential waveform has $v(0) = 1.2$ V and $v(3) = 0.5$ V. What are V_A and T_C for this waveform?

(b) An exponential waveform has $v(0) = 5$ V and $v(2) = 1.25$ V. What are values of $v(t)$ at $t = 1$ and $t = 4$?

(c) An exponential waveform has $v(0) = 5$ and an initial ($t = 0$) slope of -25 V/s. What are V_A and T_C for this waveform?

(d) An exponential waveform decays to 10% of its initial value in 3 ms. What is T_C for this waveform?

(e) A waveform has $v(2) = 4$ V, $v(6) = 1$ V, and $v(10) = 0.5$ V. Is it an exponential waveform?

Answers:

(a) $V_A = 1.2$ V, $T_C = 3.43$ s
(b) $v(1) = 2.5$ V, $v(4) = 0.3125$ V
(c) $V_A = 5$ V, $T_C = 200$ ms
(d) $T_C = 1.303$ ms
(e) No, it violates the decrement property

Exercise 5–4

Find the amplitude and time constant of the following exponential signals:

(a) $v_1(t) = [-15e^{-1000t}]u(t)$ V
(b) $v_2(t) = [+12e^{-t/10}]u(t)$ mV
(c) $i_3(t) = [15e^{-500t}]u(-t)$ mA
(d) $i_4(t) = [4e^{-200(t-100)}]u(t-100)$ A

Answers:

(a) $V_A = -15$ V, $T_C = 1$ ms   (b) $V_A = 12$ mV, $T_C = 10$ s

(c) $I_A = 15$ mA, $T_C = 2$ ms   (d) $I_A = 4$ A, $T_C = 5$ ms

5–4   THE SINUSOIDAL WAVEFORM

The cosine and sine functions are important in all branches of science and engineering. The corresponding time-varying waveform in Figure 5–14 plays an especially prominent role in electrical engineering.

FIGURE 5-14 *The eternal sinusoid.*

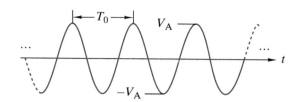

In contrast with the step and exponential waveforms studied earlier, the sinusoid, like the dc waveform in Figure 5–1, extends indefinitely in time in both the positive and negative directions. The sinusoid has neither a beginning nor an end. Of course, real signals have finite durations. They were turned on at some finite time in the past and will be turned off at some time in the future. While it may seem unrealistic to have a signal model that lasts forever, it turns out that the eternal sinewave is a very good approximation in many practical applications.

The sinusoid in Figure 5–14 is an endless repetition of identical oscillations between positive and negative peaks. The **amplitude** V_A (in volts) defines the maximum and minimum values of the oscillations. The **period** T_0 (usually seconds) is the time required to complete one cycle of the oscillation. The sinusoid can be expressed mathematically using either the sine or the cosine function. The choice between the two depends on where we choose to define $t = 0$. If we choose $t = 0$ at a point where the sinusoid is zero, then it can be written as

$$v(t) = V_A \sin(2\pi t/T_0) \qquad (5\text{--}18a)$$

On the other hand, if we choose $t = 0$ at a point where the sinusoid is at a positive peak, we can write an equation for it in terms of a cosine function:

$$v(t) = V_A \cos(2\pi t/T_0) \qquad (5\text{--}18b)$$

Although either choice will work, it is common practice to choose $t = 0$ at a positive peak; hence Eq. (5–18b) applies. Thus, we will continue to call the waveform a sinusoid even though we use a cosine function to describe it.

As in the case of the step and exponential functions, the general sinusoid is obtained by replacing t by $(t - T_S)$. Inserting this change in Eq. (5–18) yields a general expression for the sinusoid as

$$v(t) = V_A \cos[2\pi(t - T_S)/T_0] \qquad (5\text{--}19)$$

where the constant T_S is the time-shift parameter. Figure 5–15 shows that the sinusoid shifts to the right when $T_S > 0$ and to the left when $T_S < 0$. In effect, time shifting causes the positive peak nearest the origin to occur at $t = T_S$.

The time-shifting parameter can also be represented by an angle:

$$v(t) = V_A \cos[2\pi t/T_0 + \phi] \qquad (5\text{--}20)$$

The parameter ϕ is called the **phase angle**. The term *phase angle* is based on the circular interpretation of the cosine function. We think of the period as being divided into 2π radians. In this sense the phase angle is the angle between $t = 0$ and the nearest positive peak. Comparing Eqs. (5–19) and (5–20), we find the relation between T_S and ϕ to be

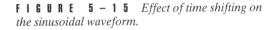

FIGURE 5–15 *Effect of time shifting on the sinusoidal waveform.*

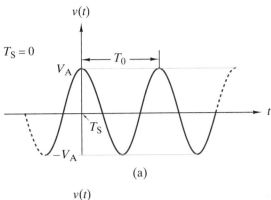

(a)

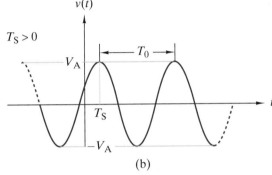

(b)

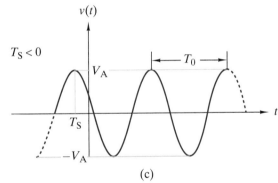

(c)

$$\phi = -2\pi \frac{T_S}{T_0} \qquad (5\text{--}21)$$

Changing the phase angle moves the waveform to the left or right, revealing different phases of the oscillating waveform (hence the name *phase angle*).

The phase angle should be expressed in radians, but is often reported in degrees. Care must be used when numerically evaluating the argument of the cosine $(2\pi t/T_0 + \phi)$ to ensure that both terms have the same units. The term $2\pi t/T_0$ has the units of radians, so it is necessary to convert ϕ to radians when it is given in degrees.

An alternative form of the general sinusoid is obtained by expanding Eq. (5–20) using the identity $\cos(x + y) = \cos(x)\cos(y) - \sin(x)\sin(y)$,

$$v(t) = [V_A \cos \phi] \cos (2\pi t/T_0) + [-V_A \sin \phi] \sin (2\pi t/T_0)$$

The quantities inside the brackets in this equation are constants; therefore, we can write the general sinusoid in the form

$$v(t) = a \cos(2\pi t/T_0) + b \sin(2\pi t/T_0) \qquad (5\text{–}22)$$

The two amplitudelike parameters a and b have the same units as the waveform (volts in this case) and are called Fourier coefficients. By definition, the Fourier coefficients are related to the amplitude and phase parameters by the equations

$$a = V_A \cos \phi$$
$$b = -V_A \sin \phi \qquad (5\text{–}23)$$

The inverse relationships are obtained by squaring and adding the expressions in Eq. (5–23):

$$V_A = \sqrt{a^2 + b^2} \qquad (5\text{–}24)$$

and by dividing the second expression in Eq. (5–23) by the first:

$$\phi = \tan^{-1} \frac{-b}{a} \qquad (5\text{–}25)$$

Caution: The inverse tangent function on a calculator has a ±180° ambiguity that can be resolved by considering the signs of the Fourier coefficients a and b.

It is customary to describe the time variation of the sinusoid in terms of a frequency parameter. **Cyclic frequency** f_0 is defined as the number of periods per unit time. By definition, the period T_0 is the number of seconds per cycle; consequently, the number of cycles per second is

$$f_0 = \frac{1}{T_0} \qquad (5\text{–}26)$$

where f_0 is the cyclic frequency or simply the frequency. The unit of frequency (cycles per second) is the **hertz** (Hz). The **angular frequency** ω_0 in radians per second is related to the cyclic frequency by the relationship

$$\omega_0 = 2\pi f_0 = \frac{2\pi}{T_0} \qquad (5\text{–}27)$$

because there are 2π radians per cycle.

There are two ways to express the concept of sinusoidal frequency: cyclic frequency (f_0, hertz) and angular frequency (ω_0, radians per second). When working with signals, we tend to use the former. For example, radio stations transmit carrier signals at frequencies specified as 690 kHz (AM band) or 101 MHz (FM band). Radian frequency is more convenient when describing the characteristics of circuits driven by sinusoidal inputs.

In summary, there are several equivalent ways to describe the general sinusoid:

$$v(t) = V_A \cos\left[\frac{2\pi(t - T_S)}{T_0}\right] = V_A \cos\left(\frac{2\pi t}{T_0} + \phi\right) = a \cos\left(\frac{2\pi t}{T_0}\right) + b \sin\left(\frac{2\pi t}{T_0}\right)$$

$$= V_A \cos[2\pi f_0(t - T_S)] = V_A \cos(2\pi f_0 t + \phi) = a \cos(2\pi f_0 t) + b \sin(2\pi f_0 t)$$

$$= V_A \cos[\omega_0(t - T_S)] = V_A \cos(\omega_0 t + \phi) = a \cos(\omega_0 t) + b \sin(\omega_0 t)$$

To use any one of these expressions, we need three types of parameters:

1. *Amplitude:* either V_A or the Fourier coefficients a and b
2. *Time shift:* either T_S or the phase angle ϕ
3. *Time/frequency:* either T_0, f_0, or ω_0.

In different parts of this book we use different forms to represent a sinusoid. Therefore, it is important for you to understand thoroughly the relationships among the various parameters in Eqs. (5–21) through (5–27).

EXAMPLE 5–7

Figure 5–16 shows an oscilloscope display of a sinusoid. The vertical axis (amplitude) is calibrated at 5 V per division, and the horizontal axis (time) is calibrated at 0.1 ms per division. Derive an expression for the sinusoid displayed in Figure 5–16.

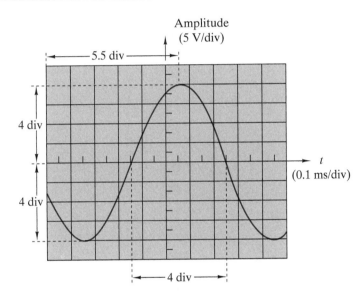

FIGURE 5–16

SOLUTION:

The maximum amplitude of the waveform is seen to be four vertical divisions; therefore,

$$V_A = (4\ \text{div})(5\ \text{V/div}) = 20\ \text{V}$$

There are four horizontal divisions between successive zero crossings, which means there are a total of eight divisions in one cycle. The period of the waveform is

$$T_0 = (8\ \text{div})(0.1\ \text{ms/div}) = 0.8\ \text{ms}$$

The two frequency parameters are $f_0 = 1/T_0 = 1.25$ kHz and $\omega_0 = 2\pi f_0 = 7854$ rad/s. The parameters V_A, T_0, f_0, and ω_0 do not depend on the location of the $t = 0$ axis.

To determine the time shift T_S, we need to define a time origin. The $t = 0$ axis is arbitrarily taken at the left edge of the display in Figure 5–16. The positive peak shown in the display is 5.5 divisions to the right of $t = 0$, which is more than half a cycle (four divisions). The positive peak closest to $t = 0$ is not shown in Figure 5–16 because it must lie beyond the left edge of the display. However, the positive peak shown in the display is located at $t = T_S + T_0$ since it is one cycle after $t = T_S$. We can write

$$T_S + T_0 = (5.5 \text{ div})(0.1 \text{ ms/div}) = 0.55 \text{ ms}$$

which yields $T_S = 0.55 - T_0 = -0.25$ ms. As expected, T_S is negative because the nearest positive peak is to the left of $t = 0$.

Given T_S, we can calculate the remaining parameters of the sinusoid as follows:

$$\phi = -\frac{2\pi T_S}{T_0} = 1.96 \text{ rad or } 112.5°$$

$$a = V_A \cos \phi = -7.65 \text{ V}$$

$$b = -V_A \sin \phi = -18.5 \text{ V}$$

Finally, the three alternative expressions for the displayed sinusoid are

$$v(t) = 20 \cos[(7854 (t + 0.25 \times 10^{-3})]$$

$$= 20 \cos(7854 t + 112.5°)$$

$$= -7.65 \cos 7854t - 18.5 \sin 7854t \qquad \blacksquare$$

Exercise 5–5

Derive an expression for the sinusoid displayed in Figure 5–16 when $t = 0$ is placed in the middle of the display.

Answer: $v(t) = 20 \cos(7854t - 22.5°)$

PROPERTIES OF SINUSOIDS

In general, a waveform is said to be **periodic** if

$$v(t + T_0) = v(t)$$

for all values of t. The constant T_0 is called the period of the waveform if it is the smallest nonzero interval for which $v(t + T_0) = v(t)$. Since this equality must be valid for all values of t, it follows that periodic signals must have eternal waveforms that extend indefinitely in time in both directions. Signals that are not periodic are called **aperiodic**.

The sinusoid is a periodic signal since

$$v(t + T_0) = V_A \cos[2\pi (t + T_0)/T_0 + \phi]$$

$$= V_A \cos[2\pi (t)/T_0 + \phi + 2\pi]$$

But $\cos(x + 2\pi) = \cos(x)$. Consequently,

$$v(t + T_0) = V_A \cos(2\pi t/T_0 + \phi) = v(t)$$

for all t.

The **additive property** of sinusoids states that summing two or more sinusoids with the same frequency yields a sinusoid with different amplitude and phase parameters but the same frequency. To illustrate, consider two sinusoids

$$v_1(t) = a_1 \cos(2\pi f_0 t) + b_1 \sin(2\pi f_0 t)$$
$$v_2(t) = a_2 \cos(2\pi f_0 t) + b_2 \sin(2\pi f_0 t)$$

The waveform $v_3(t) = v_1(t) + v_2(t)$ can be written as

$$v_3(t) = (a_1 + a_2) \cos(2\pi f_0 t) + (b_1 + b_2) \sin(2\pi f_0 t)$$

because cosine and sine are linearly independent functions. We obtain the Fourier coefficients of the sum of two sinusoids by adding their Fourier coefficients, provided the two have the same frequency. *Caution:* The summation must take place with the sinusoids in Fourier coefficient form. Sums of sinusoids *cannot* be found by adding amplitudes and phase angles.

The **derivative** and **integral** properties state that when we differentiate or integrate a sinusoid, the result is another sinusoid with the same frequency:

$$\frac{d(V_A \cos \omega t)}{dt} = -\omega V_A \sin \omega t = \omega V_A \cos(\omega t + \pi/2)$$

$$\int V_A \cos(\omega t)\, dt = \frac{V_A}{\omega} \sin \omega t = \frac{V_A}{\omega} \cos(\omega t - \pi/2)$$

These operations change the amplitude and phase angle but do not change the frequency. The fact that differentiation and integration preserve the underlying waveform is a key property of the sinusoid. No other periodic waveform has this shape-preserving property.

EXAMPLE 5–8

(a) Find the periods cyclic and radian frequencies of the sinusoids
$v_1(t) = 17 \cos(2000t - 30°)$
$v_2(t) = 12 \cos(2000t + 30°)$.
(b) Find the waveform of $v_3(t) = v_1(t) + v_2(t)$.

SOLUTION:
(a) The two sinusoids have the same frequency $\omega_0 = 2000$ rad/s since a term $2000t$ appears in the arguments of $v_1(t)$ and $v_2(t)$. Therefore, $f_0 = \omega_0/2\pi = 318.3$ Hz and $T_0 = 1/f_0 = 3.14$ ms.
(b) We use the additive property, since the two sinusoids have the same frequency. Beyond this checkpoint, the frequency plays no further role in the calculation. The two sinusoids must be converted to the Fourier coefficient form using Eq. (5–23).

$$a_1 = 17 \cos(-30°) = +14.7 \text{ V}$$
$$b_1 = -17 \sin(-30°) = +8.50 \text{ V}$$
$$a_2 = 12 \cos(30°) = +10.4 \text{ V}$$
$$b_2 = -12 \sin(30°) = -6.00 \text{ V}$$

The Fourier coefficients of the signal $v_3 = v_1 + v_2$ are found as

$$a_3 = a_1 + a_2 = 25.1 \text{ V}$$

$$b_3 = b_1 + b_2 = 2.50 \text{ V}$$

The amplitude and phase angle of $v_3(t)$ are found using Eqs. (5–24) and (5–25):

$$V_A = \sqrt{a_3^2 + b_3^2} = 25.2 \text{ V}$$

$$\phi = \tan^{-1}(-2.5/25.1) = -5.69°$$

Two equivalent representations of $v_3(t)$ are

$$v_3(t) = 25.1 \cos(2000t) + 2.5 \sin(2000t) \text{ V}$$

and

$$v_3(t) = 25.2 \cos(2000t - 5.69°) \text{ V}$$

EXAMPLE 5–9

The balanced three-phase voltages used in electrical power systems can be written as

$$v_A(t) = V_m \cos(2\pi f_0 t)$$

$$v_B(t) = V_m \cos(2\pi f_0 t + 120°)$$

$$v_C(t) = V_m \cos(2\pi f_0 t + 240°)$$

Show that the sum of these voltages is zero.

SOLUTION:
The three voltages are given in amplitude/phase angle form. They can be converted to the Fourier coefficient form using Eq. (5–23):

$$v_A(t) = + V_m \cos(2\pi f_0 t)$$

$$v_B(t) = - \frac{V_m}{2} \cos(2\pi f_0 t) + \frac{\sqrt{3}V_m}{2} \sin(2\pi f_0 t)$$

$$v_C(t) = - \frac{V_m}{2} \cos(2\pi f_0 t) - \frac{\sqrt{3}V_m}{2} \sin(2\pi f_0 t)$$

The sum of the cosine and sine terms is zero; consequently, $v_A(t) + v_B(t) + v_C(t) = 0$. The zero sum occurs because the three sinusoids have equal amplitudes and are disposed at 120° intervals. If the amplitudes are not equal or the phase angles do not differ by 120°, then the voltages are said to be unbalanced. ■

Exercise 5–6

Write an equation for the waveform obtained by integrating and differentiating the following signals:

(a) $v_1(t) = 30 \cos(10t - 60°)$

(b) $v_2(t) = 3 \cos(4000\pi t) - 4 \sin(4000\pi t)$

Answers:

$$(a) \frac{dv_1}{dt} = 300 \cos(10t + 30°)$$

$$\int v_1(t)\, dt = 3 \cos(10t - 150°)$$

$$(b) \frac{dv_2}{dt} = 2\pi \times 10^4 \cos(4000\pi t + 143.1°)$$

$$\int v_2\, dt = \frac{1}{800\pi} \cos(4000\pi t - 36.87°)$$

Exercise 5–7

A sinusoid has a period of 5 μs. At $t = 0$ the amplitude is 12 V. The waveform reaches its first positive peak after $t = 0$ at $t = 4$ μs. Find its amplitude, frequency, and phase angle.

Answers: $V_A = 38.8$ V, $f_0 = 200$ kHz, $\phi = +72°$

5–5 COMPOSITE WAVEFORMS

In the previous sections we introduced the step, exponential, and sinusoidal waveforms. These waveforms are basic signals because they can be combined to synthesize all other signals used in this book. Signals generated by combining the three basic waveforms are called **composite signals**. This section provides examples of composite waveforms.

EXAMPLE 5–10

Characterize the composite waveform generated by

$$v(t) = V_A u(t) - V_A u(-t)$$

SOLUTION:

The first term in this waveform is simply a step function of amplitude V_A that occurs at $t = 0$. The second term involves the function $u(-t)$, whose waveform requires some discussion. Strictly speaking, the general step function $u(x)$ is unity when $x > 0$ and zero when $x < 0$. That is, $u(x)$ is unity when its argument is positive and zero when it is negative. Under this rule the function $u(-t)$ is unity when $-t > 0$ and zero when $-t < 0$, that is,

$$u(-t) = \begin{cases} 1 & \text{for } t < 0 \\ 0 & \text{for } t > 0 \end{cases}$$

which is the reverse of the step function $u(t)$. Figure 5–17 shows how the two components combine to produce a composite waveform that extends indefinitely in both directions and has a jump discontinuity of $2V_A$ at $t = 0$. This composite waveform is called a **signum** function.

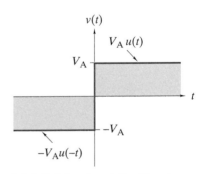

FIGURE 5 – 1 7 *The signum waveform.*

EXAMPLE 5–11

Characterize the composite waveform generated by subtracting an exponential from a step function with the same amplitude.

SOLUTION:
The equation for this composite waveform is

$$v(t) = V_A u(t) - [V_A e^{-t/T_C}] u(t)$$
$$= V_A [1 - e^{-t/T_C}] u(t)$$

For $t < 0$ the waveform is zero because of the step function. At $t = 0$ the waveform is still zero since the step and exponential cancel:

$$v(0) = V_A [1 - e^0](1) = 0$$

For $t \gg T_C$ the waveform approaches a constant value V_A because the exponential term decays to zero. For practical purposes $v(t)$ is within less than 1% of its final value V_A when $t = 5T_C$. At $t = T_C$, $v(T_C) = V_A(1 - e^{-1}) = 0.632 V_A$. The waveform rises to about 63% of its final value in one time constant. All of the observations are summarized in the plot shown in Figure 5–18. This waveform is called an **exponential rise**. It is also sometimes referred to as a "charging exponential," since it represents the behavior of signals that occur during the buildup of voltage in resistor-capacitor circuits studied in Chapter 7. ■

FIGURE 5 – 1 8 *The exponential rise waveform.*

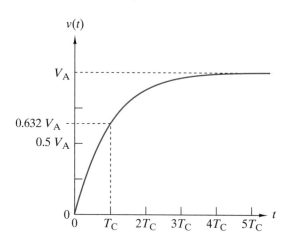

EXAMPLE 5–12

Characterize the composite waveform obtained by multiplying the ramp $r(t)/T_C$ times an exponential.

SOLUTION:
The equation for this composite waveform is

$$v(t) = \frac{r(t)}{T_C} [V_A e^{-t/T_C}] u(t)$$
$$= V_A [(t/T_C) e^{-t/T_C}] u(t)$$

For $t < 0$ the waveform is zero because of the step function. At $t = 0$ the waveform is zero because $r(0) = 0$. For $t > 0$ there is a competition between two effects—the ramp increases linearly with time while the exponential decays to zero. Since the composite waveform is the product of these terms, it is important to determine which effect dominates. In the limit, as $t \to \infty$, the product of the ramp and exponential takes on the indeterminate form of infinity times zero. A single application of *l'Hôpital's* rule, then, shows that the exponential dominates, forcing the $v(t)$ to zero as t becomes large. That is, the exponential decay overpowers the linearly increasing ramp, as shown by the graph in Figure 5–19. The waveform obtained by multiplying a ramp by a decaying exponential is called a **damped ramp**. ■

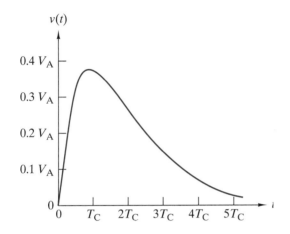

FIGURE   5 – 1 9   *The damped ramp waveform.*

EXAMPLE 5–13

Characterize the composite waveform obtained by multiplying $\sin \omega_0 t$ by an exponential.

SOLUTION:
In this case the composite waveform is expressed as

$$v(t) = \sin \omega_0 t [V_A e^{-t/T_C}] u(t)$$
$$= V_A [e^{-t/T_C} \sin \omega_0 t] u(t)$$

Figure 5–20 shows a graph of this waveform for $T_0 = 2T_C$. For $t < 0$ the step function forces the waveform to be zero. At $t = 0$, and periodically thereafter, the waveform passes through zero because $\sin(n\pi) = 0$. The waveform is not periodic, however, because the decaying exponential gradually reduces the amplitude of the oscillation. For all practical purposes the oscillations become negligibly small for $t > 5T_C$. The waveform obtained by multiplying a sinusoid by a decaying exponential is called a **damped sine**. ■

EXAMPLE 5–14

Characterize the composite waveform obtained as the difference of two exponentials with the same amplitude.

FIGURE 5 – 20 *The damped sine waveform.*

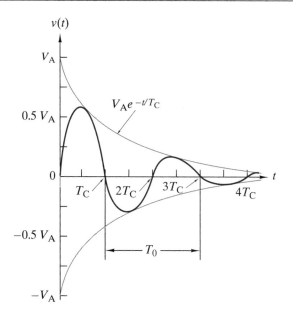

SOLUTION:

The equation for this composite waveform is

$$v(t) = [V_A\, e^{-t/T_1}]\, u(t) - [V_A\, e^{-t/T_2}]\, u(t)$$
$$= V_A\, (e^{-t/T_1} - e^{-t/T_2})\, u(t)$$

For $T_1 > T_2$ the resulting waveform is illustrated in Figure 5–21 (plotted for $T_1 = 2T_2$). For $t < 0$ the waveform is zero. At $t = 0$ the waveform is still zero, since

$$v(0) = V_A\, (e^{-0} - e^{-0})$$
$$= V_A\, (1 - 1) = 0$$

FIGURE 5 – 21 *The double exponential waveform.*

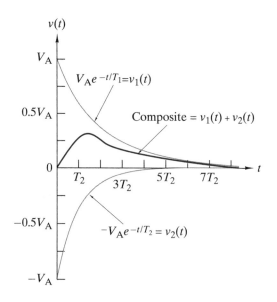

For $t \gg T_1$ the waveform returns to zero because both exponentials decay to zero. For $5T_1 > t > 5T_2$ the second exponential is negligible and the waveform essentially reduces to the first exponential. Conversely, for $t \ll T_1$ the first exponential is essentially constant, so the second exponential determines the early time variation of the waveform. The waveform is called a **double exponential**, since both exponential components make important contributions to the waveform. ■

EXAMPLE 5–15

Characterize the composite waveform defined by

$$v(t) = 5 - \frac{10}{\pi} \sin (2 \pi 500\, t) - \frac{10}{2\pi} \sin (2 \pi 1000\, t) - \frac{10}{3\pi} \sin (2 \pi 1500\, t)$$

SOLUTION:
The waveform is the sum of a constant (dc) term and three sinusoids at different frequencies. The first sinusoidal component is called the **fundamental** because it has the lowest frequency. As a result the frequency $f_0 = 500$ Hz is called the **fundamental frequency**. The other sinusoidal terms are said to be harmonics because their frequencies are integer multiples of f_0. Specifically, the second sinusoidal term is called the **second harmonic** ($2f_0 = 1000$ Hz) while the third term is the **third harmonic** ($3f_0 = 1500$ Hz). Figure 5–22 shows a plot of this waveform. Note that the waveform is periodic with a period equal to that of the fundamental component, namely, $T_0 = 1/f_0 = 2$ ms. The decomposition of a periodic waveform into a sum of harmonic sinusoids is called a **Fourier series**, a topic we will study in detail in Chapter 13. In fact, the waveform in this example is the first four terms in the Fourier series for a 10-V sawtooth wave of the type shown in Figure 5–1. ■

FIGURE   5 – 2 2

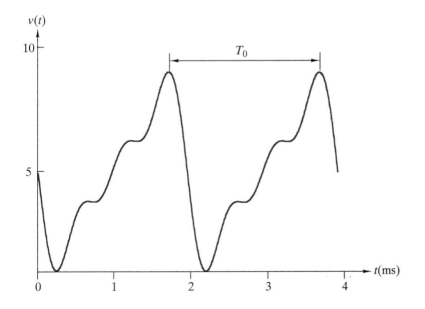

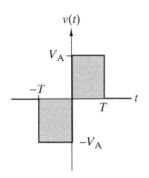

FIGURE 5 – 2 3

Write an expression for the pulse in Figure 5–23 using step functions.

> **Answer:**
>
> $$v(t) = V_A[u(t) - u(t - T) - u(-t) + u(-t - T)]$$
>
> or
>
> $$v(t) = V_A[-u(t + T) + 2u(t) - u(t - T)]$$

Find the maximum amplitude and the approximate duration of the following composite waveforms:

(a) $v_1(t) = [25 \sin 1000t][u(t) - u(t - 10)]$ V
(b) $v_2(t) = [50 \cos 1000t][e^{-200t}]u(t)$ V
(c) $i_3(t) = [3000te^{-1000t}]u(t)$ mA
(d) $i_4(t) = [10e^{5000t}]u(-t) + [10e^{-5000t}]u(t)$ A

> **Answers:**
>
> (a) 25 V, 10 s
> (b) 50 V, 25 ms
> (c) 1.10 mA, 5 ms
> (d) 10 A, 2 ms

5–6 WAVEFORM PARTIAL DESCRIPTORS

An equation or graph defines a waveform for all time. The value of a waveform $v(t)$ or $i(t)$ at time t is called the **instantaneous value** of the waveform. We often use parameters called **partial descriptors** that characterize important features of a waveform but do not give a complete description. These partial descriptors fall into two categories: (1) those that describe temporal features and (2) those that describe amplitude features.

TEMPORAL DESCRIPTORS

Temporal descriptors identify waveform attributes relative to the time axis. For example, waveforms that repeat themselves at fixed time intervals are said to be **periodic**. Stated formally,

> *A signal v(t) is periodic if v(t + T₀) = v(t) for all t, where the period T₀ is the smallest value that meets this condition. Signals that are not periodic are called aperiodic.*

The fact that a waveform is periodic provides important information about the signal, but does not specify all of its characteristics. Thus, the fact that a signal is periodic is itself a partial description, as is the value of the period. The eternal sinewave is the premier example of a periodic signal. The square wave and triangular wave in Figure 5–2 are also periodic. Examples of aperiodic waveforms are the step function, exponential, and damped sine.

Waveforms that are identically zero prior to some specified time are said to be **causal**. Stated formally,

*A signal $v(t)$ is causal if there exists a value of **T** such that $v(t)$*
*$\equiv 0$ for all **t** < **T**; otherwise it is noncausal.*

It is usually assumed that a causal signal is zero for $t < 0$, since we can always use time shifting to make the starting point of a waveform at $t = 0$. Examples of causal waveforms are the step function, exponential, and damped sine. The eternal sinewave is, of course, noncausal.

Causal waveforms play a central role in circuit analysis. When the input driving force $x(t)$ is causal, the circuit response $y(t)$ must also be causal. That is, a physically realizable circuit cannot anticipate and respond to an input before it is applied. Causality is an important temporal feature, but only a partial description of the waveform.

AMPLITUDE DESCRIPTORS

Amplitude descriptors are positive scalars that describe signal strength. Generally, a waveform varies between two extreme values denoted as V_{MAX} and V_{MIN}. The **peak-to-peak value** (V_{pp}) describes the total excursion of $v(t)$ and is defined as

$$V_{pp} = V_{MAX} - V_{MIN} \qquad (5\text{–}28)$$

Under this definition V_{pp} is always positive even if V_{MAX} and V_{MIN} are both negative. The **peak value** (V_p) is the maximum of the absolute value of the waveform. That is,

$$V_p = \text{Max}\{|V_{MAX}|, |V_{MIN}|\} \qquad (5\text{–}29)$$

The peak value is a positive number that indicates the maximum absolute excursion of the waveform from zero. Figure 5–24 shows examples of these two amplitude descriptors.

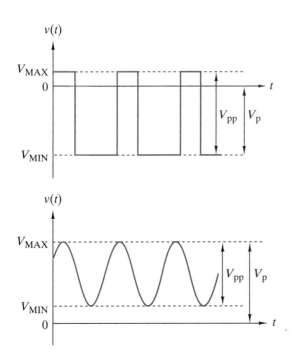

FIGURE 5 – 24 *Peak value (V_p) and peak-to-peak value (V_{pp}).*

The peak and peak-to-peak values describe waveform variation using the extreme values. The average value smooths things out to reveal the underlying waveform baseline. Average value is the area under the waveform over some period of time T, divided by that time period. Mathematically, we define **average value** (V_{avg}) over the time interval T as

$$V_{avg} = \frac{1}{T}\int_{t}^{t+T} v(x)\,dx \tag{5–30}$$

For periodic signals the period T_0 is used as the averaging interval T.

For some periodic waveforms the integral in Eq. (5–30) can be estimated graphically. The net area under the waveform is the area above the time axis minus the area below the time axis. For example, the two waveforms in Figure 5–24 obviously have nonzero average values. The top waveform has a negative average value because the negative area below the time axis more than cancels the area above the axis. Similarly, the bottom waveform clearly has a positive average value.

The average value indicates whether the waveform contains a constant, non-time-varying component. The average value is also called the **dc component** because dc signals are constant for all t. On the other hand, the **ac components** have zero average value and are periodic. For example, the waveform in Example 5–15

$$v(t) = 5 - \frac{10}{\pi}\sin(2\pi 500\,t) - \frac{10}{2\pi}\sin(2\pi 1000\,t) - \frac{10}{3\pi}\sin(2\pi 1500\,t)$$

has a 5-V average value due to its dc component. The three sinusoids are ac components because they are periodic and have zero average value. Sinusoids have zero average value because over any given cycle the positive area above the time axis is exactly canceled by the negative area below.

EXAMPLE 5–16

Find the peak, peak-to-peak, and average values of the periodic input and output waveforms in Figure 5–25.

SOLUTION:
The input waveform is a sinusoid whose amplitude descriptors are

$$V_{pp} = 2V_A \quad V_p = V_A \quad V_{avg} = 0$$

The output waveform is obtained by clipping off the negative half-cycle of the input sinusoid. The amplitude descriptors of the output waveform are

$$V_{pp} = V_p = V_A$$

The output has a nonzero average value, since there is a net positive area under the waveform. The upper limit in Eq. (5–30) can be taken as $T_0/2$, since the waveform is zero from $T_0/2$ to T_0.

$$V_{avg} = \frac{1}{T_0}\int_{0}^{T_0/2} V_A \sin(2\pi t/T_0)\,dt = \left. -\frac{V_A}{2\pi}\cos(2\pi\,t/T_0)\right|_{0}^{T_0/2}$$

$$= \frac{V_A}{\pi}$$

The signal processor produces an output with a dc value from an input with no dc component. Rectifying circuits described in electronics courses produce waveforms like the output in Figure 5–25. ■

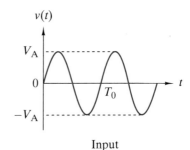

Input

ROOT-MEAN-SQUARE VALUE

The **root-mean-square value** (V_{rms}) is a measure of the average power carried by the signal. The instantaneous power delivered to a resistor R by a voltage $v(t)$ is

$$p(t) = \frac{1}{R}[v(t)]^2 \qquad (5\text{–}31)$$

The average power delivered to the resistor in time span T is defined as

$$P_{avg} = \frac{1}{T}\int_{t}^{t+T} p(t)dt \qquad (5\text{–}32)$$

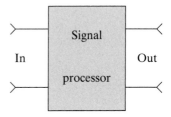

Combining Eqs. (5–31) and (5–32) yields

$$P_{avg} = \frac{1}{R}\left[\frac{1}{T}\int_{t}^{t+T}[v(t)]^2\,dt\right] \qquad (5\text{–}33)$$

The quantity inside the large brackets in Eq. (5–33) is the average value of the square of the waveform. The units of the bracketed term are volts squared. The square root of this term defines the amplitude partial descriptor V_{rms}:

$$V_{rms} = \sqrt{\frac{1}{T}\int_{t}^{t+T}[v(t)]^2\,dt} \qquad (5\text{–}34)$$

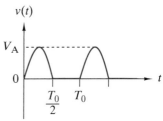

Output

FIGURE 5–25

The amplitude descriptor V_{rms} is called the root-mean-square (rms) value because it is obtained by taking the square root of the average (mean) of the square of the waveform. For periodic signals the averaging interval is one cycle since such a waveform repeats itself every T_0 seconds.

We can express the average power delivered to a resistor in terms of V_{rms} as

$$P_{avg} = \frac{1}{R}V_{rms}^2 \qquad (5\text{–}35)$$

The equation for average power in terms of V_{rms} has the same form as the power delivered by a dc signal. For this reason the rms value was originally called the **effective value**, although this term is no longer common. If the waveform amplitude is doubled, its rms value is doubled, and the average power is quadrupled. Commercial electrical power systems use transmission voltages in the range of several hundred kilovolts (rms).

EXAMPLE 5–17

Find the average and rms values of the sinusoid and sawtooth in Figure 5–26.

FIGURE 5–26

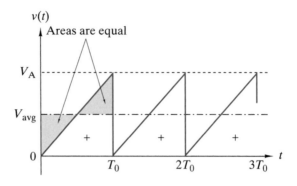

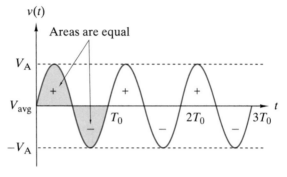

SOLUTION:

As noted previously, the sinusoid has an average value of zero. The sawtooth clearly has a positive average value. By geometry, the net area under one cycle of the sawtooth waveform is $V_A T_0/2$, so its average value is $(1/T_0)(V_A T_0/2) = V_A/2$. To obtain the rms value of the sinusoid we apply Eq. (5–34) as

$$V_{rms} = \sqrt{\frac{(V_A)^2}{T_0} \int_0^{T_0} \sin^2(2\pi t/T_0)dt}$$

$$= \sqrt{\frac{(V_A)^2}{T_0} \left[\frac{t}{2} - \frac{\sin(4\pi t/T_0)}{8\pi/T_0} \right]_0^{T_0}} = \frac{V_A}{\sqrt{2}}$$

For the sawtooth the rms value is found as:

$$V_{rms} = \sqrt{\frac{1}{T_0} \int_0^{T_0} (V_A t/T_0)^2 \, dt} = \sqrt{\frac{(V_A)^2}{T_0^3} \left[\frac{t^3}{3} \right]_0^{T_0}} = \frac{V_A}{\sqrt{3}}$$ ∎

Exercise 5–10

Find the peak, peak-to-peak, average, and rms values of the periodic waveform in Figure 5–27.

Answers:

$$V_p = 2V_A, \quad V_{pp} = 3V_A, \quad V_{avg} = \frac{V_A}{4}, \quad V_{rms} = \frac{\sqrt{5}}{2}V_A$$

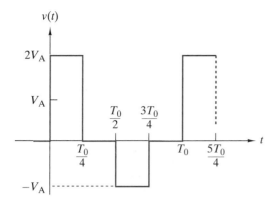

Exercise 5–11

Classify each of the following signals as periodic or aperiodic and causal or noncausal. Then calculate the average and rms values of the periodic waveforms, and the peak and peak-to-peak values of the other waveforms.

(a) $v_1(t) = 99 \cos 3000t − 132 \sin 3000t$ V
(b) $v_2(t) = 34 \, [\sin 800\pi t][u(t) − u(t − 0.03)]$ V
(c) $i_3(t) = 120[u(t + 5) − u(t − 5)]$ mA
(d) $t_4(t) = 50$ A

Answers:
(a) Periodic, noncausal, $V_{avg} = 0$ and $V_{rms} = 117$ V
(b) Aperiodic, causal, $V_p = 34$ V and $V_{pp} = 68$ V
(c) Aperiodic, causal, $V_p = V_{pp} = 120$ mA
(d) Aperiodic, noncausal, $V_p = 50$ A and $V_{pp} = 0$

Exercise 5–12

Construct waveforms that have the following characteristics:

(a) Aperiodic and causal with $V_p = 8$ V and $V_{pp} = 15$ V
(b) Periodic and noncausal with $V_{avg} = 10$ V and $V_{pp} = 50$ V
(c) Periodic and noncausal with $V_{avg} = V_p/2$
(d) Aperiodic and causal with $V_p = V_{pp} = 10$ V

Answers: There are many possible correct answers since the given parameters are only partial descriptions of the required waveforms. Some examples that meet the requirements are as follows:

(a) $v(t) = 8 \, u(t) − 15 \, u(t − 1) + 7 \, u(t − 2)$
(b) $v(t) = 10 + 25 \sin 1000t$
(c) A sawtooth wave
(d) $v(t) = 10[e^{−100t}]u(t)$

APPLICATION NOTE: EXAMPLE 5−18

The operation of a digital system is coordinated and controlled by a periodic waveform called a clock. The *clock waveform* provides a standard timing reference to maintain synchronization between signal processing results that become valid at different times during the clock cycle. Because of differences in digital circuit delays, there must be agreed-upon instants of time when circuit outputs can be treated as valid. The clock defers further signal processing until slower and faster outputs settle down when the clock signals the start of the next signal processing cycle.

Figure 5–28 shows an idealized clock waveform as a periodic sequence of rectangular pulses. While we could easily write an exact expression for the clock waveform, we are interested here in discussing its partial descriptors. The first descriptor is the period T_0 or equivalently the **clock frequency** $f_0 = 1/T_0$. Clock frequency is a common measure of signal processing speed and can range up into the hundreds of MHz. The pulse duration T is the time interval in each cycle when the pulse amplitude is high (not zero). In waveform terminology the ratio of the time in the high state to the period, that is, T/T_0, is called the **duty cycle**, usually expressed as a percentage. The **pulse edges** are the transition points at which the pulse changes states. There is a **rising edge** at the low-to-high transition and a **falling edge** at the high-to-low transition.

The pulse edges define the agreed-upon time instants at which the circuit outputs can be treated as valid inputs to other circuits. This means that circuit outputs must settle down during the time period between successive edges. Some synchronous operations are triggered by the rising edge and others by the falling edge. To provide equal settling times for both cases requires equal time between edges. In other words, it is desirable for the clock duty cycle to be 50%. As a result, the clock waveform is essentially a raised square wave whose dc offset equals one half of the peak-to-peak value.

FIGURE 5 – 28

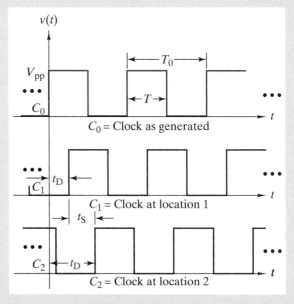

C_0 = Clock as generated

C_1 = Clock at location 1

C_2 = Clock at location 2

The system clock C_0 in Figure 5–28 is generated at some point in a circuit and then distributed to other locations. The clock distribution network almost invariably introduces delays, as illustrated by C_1 and C_2 in Figure 5-28. **Clock delay (t_D)** is defined as the time difference between a clock edge at a given location and the corresponding edge in the system clock at the point where it was generated. Delay is not necessarily a bad thing unless unequal delays cause the edges to be skewed, as indicated by the offset between C_1 and C_2 in Figure 5–28. When delays are significantly different, there is uncertainty as to instants of time at which further signal processing can safely proceed. This delay dispersion is called **clock skew (t_S)**, defined as the time difference between a clock edge at a given location and the corresponding edge at another location. Controlling clock skew is an important consideration in the design of the clock distribution network in high-speed VLSI circuits.

Thus, partial descriptors of clock waveforms include *frequency*, *duty cycle*, *edges*, *delay*, and *skew*. The coming chapters treat dynamic circuits that modify input waveforms to produce outputs with different partial descriptors. In particular, dynamic circuit elements cause changes in a clock waveform, especially the partial descriptors of edges, delay, and skew.

APPLICATION NOTE: **EXAMPLE 5–19**

An electrocardiogram (ECG) is a valuable diagnostic tool used in cardiovascular medicine. The ECG is based on the fact that the heart emits measurable bioelectric signals that can be recorded to evaluate the functioning of the heart as a mechanical pump. These signals were first observed in the late 19th century and subsequent signal processing developments have led to the advanced technology of present day ECG equipment.

The bioelectric signals of the heart muscle are measured and recorded through the placement of skin electrodes at various sites on the surface of the body. The site selection as well as discussion of the functions of the cardiac muscle are beyond the scope of this example. Rather, our purpose is to introduce some of the useful partial descriptors of ECG waveforms.

In bioelectric terminology the normal ECG waveform in Figure 5–29 is composed of a P wave, a QRS complex, and a T wave. This sequence of pulses depicts the electrical activity that stimulates the correct functioning of the cardiac muscle. The flat baseline between successive events is called isoelectric, which means there is no bioelectric activity and the heart muscle returns to a resting state. The body's natural pacemaker produces a nominally periodic waveform under the resting conditions used with ECG tests.

Partial waveform descriptors used to analyze ECG waveforms include:

1. The **heart rate** ($1/T_0$), which is normally between 60 and 100 beats per minute.
2. The **PR interval** (normally 0.12–0.20 seconds), which is the time between the start of the P wave and the start of the QRS complex.
3. The **QRS interval** (normally 0.06–0.10 seconds), which is the time between onset and end of the QRS complex.
4. The **ST segment** is the signal level between the end of the QRS complex and the start of the T wave. This level should be the same as the isoelectric baseline between successive pulses.

Departures from these normal conditions serve as diagnostic tools in cardiovascular medicine. Some of the abnormal waveform features of concern include an irregular heart rate, a missing P wave, a prolonged QRS interval, or an elevated ST segment. Departures from nominal conditions allow the trained clinician to diagnose the situation, especially when abnormal features occur in certain combinations. However, it is not our purpose to discuss the medical interpretation of ECG waveform abnormalities. Rather, this examples illustrates that bioelectric signals carry information and that the information is decoded by analyzing the signal's partial waveform descriptors.

FIGURE 5-28

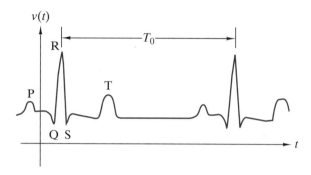

SUMMARY

- A waveform is an equation or graph that describes a voltage or current as a function of time. Most signals of interest in electrical engineering can be derived using three basic waveforms: the step, exponential, and sinusoid.

- The step function is defined by its amplitude and time-shift parameters. The impulse, step, and ramp are called singularity functions and are often used as test inputs for circuit analysis purposes.

- The exponential waveform is defined by its amplitude, time constant, and time-shift parameter. For practical purposes, the duration of the exponential waveform is five time constants.

- A sinusoid can be defined in terms of three types of parameters: *amplitude* (either V_A or the Fourier coefficients a and b), *time shift* (either T_S or the phase angle ϕ), and *time/frequency* (either T_0 or f_0 or ω_0).

- Many composite waveforms can be derived using the three basic waveforms. Some examples are the impulse, ramp, damped ramp, damped sinusoid, exponential rise, and double exponential.

- Partial descriptors are used to classify or describe important signal attributes. Two important temporal attributes are periodicity and causality. Periodic waveforms repeat themselves every T_0 seconds. Causal signals are zero for $t < 0$. Some important amplitude descriptors are peak value V_p, peak-to-peak value V_{pp}, average value V_{avg}, and root-mean-square value V_{rms}.

- A spectrum is an equation or graph that defines the amplitudes and phase angles of sinusoidal components contained in a signal. The signal bandwidth (B) is a partial descriptor that defines the range of frequencies outside of which the component amplitudes are less than a specified value. Periodic signals can be resolved into a dc component and a sum of ac components at harmonic frequencies.

PROBLEMS

ERO 5–1 BASIC WAVEFORMS (SECTS. 5–2, 5–3, 5–4)

Given an equation, graph, or word description of a step, exponential, or sinusoid waveform,

(a) Construct an alternative description of the waveform.
(b) Find the parameters or properties of the waveform.
(c) Find new waveforms by summing, integrating, or differentiating the given waveform.

See Examples 5–1, 5–2, 5–3, 5–5, 5–6, 5–7, 5–8, 5–9 and Exercises 5–1, 5–2, 5–3, 5–4, 5–5, 5–6, 5–7

5–1 Graph the following step function waveforms:
   (a) $v_1(t) = 5u(t)$ V
   (b) $v_2(t) = -5u(t-2)$ V
   (c) $v_3(t) = v_1(t) + v_2(t)$ V
   (d) $v_4(t) = v_1(t) - v_2(t)$ V

5–2 Graph the waveforms obtained by integrating the waveforms in Problem 5-1.

5–3 Use step functions to write an expression for a rectangular pulse with amplitude V_A, duration T, and centered at $t = 0$.

5–4 A waveform $v(t)$ is –5 V for t in the range 5 ms ≤ t ≤ 3 ms and +5 V elsewhere. Write an equation for the waveform using step functions.

5–5 A waveform $v(t)$ is zero for $t = 0$, rises linearly to 5 V at $t = 2$ ms, and abruptly drops to zero thereafter. Write an equation for the waveform using singularity functions.

5–6 Find the amplitude and time constant of each of the following exponential waveforms. Graph each of the waveforms.
   (a) $v_1(t) = [10\,e^{-2t}]u(t)$ V
   (b) $v_2(t) = [10\,e^{-t/2}]u(t)$ V
   (c) $v_3(t) = [-10\,e^{-20t}]u(t)$ V
   (d) $v_4(t) = [-10\,e^{-t/20}]u(t)$ V

5–7 Graph the waveforms obtained by differentiating the first two waveforms in Problem 5-6.

5–8 An exponential waveform starts at $t = 0$ and decays to 4 V at $t = 5$ ms and 2 V at $t = 6$ ms. Find the amplitude and time constant of the waveform.

5–9 The value of an exponential waveform is 5 V at $t = 5$ ms and 3.5 V at $t = 7$ ms. What is its value at $t = 2$ ms?

5–10 By direct substitution show that the exponential function $v(t) = V_A e^{-\alpha t}$ satisfies the following first-order differential equation:

$$\frac{dv(t)}{dt} + \alpha v(t) = 0$$

5–11 Find the period, frequency, amplitude, time shift, and phase angle of the following sinusoids:
   (a) $v_1(t) = 10 \cos(2000\pi t) + 10 \sin(2000\pi t)$ V
   (b) $v_2(t) = -30 \cos(2000\pi t) - 20 \sin(2000\pi t)$ V

   (c) $v_3(t) = 10 \cos(2\pi t/10) - 10 \sin(2\pi t/10)$ V
   (d) $v_4(t) = -20 \cos(800\pi t) + 30 \sin(800\pi t)$ V

5–12 Find the period, frequency, amplitude, time shift, and phase angle of the sum of the first two sinusoids in Problem 5–11.

5–13 A sinusoid has an amplitude of 15 V. At $t = 0$ the value of a sinusoid is 10 V and its slope is positive. Find the Fourier coefficients of the waveform.

5–14 A sinusoid has a frequency of 5 MHz, a value of –10 V at $t = 0$, and reaches its first positive peak at $t = 125$ ns. Find its amplitude, phase angle, and Fourier coefficients.

5–15 Find the frequency, period, and Fourier coefficients of the following sinusoids:
   (a) $v_1(t) = 20 \cos(4000\pi t - 180°)$ V
   (b) $v_2(t) = 20 \cos(4000\pi t - 90°)$ V
   (c) $v_3(t) = 30 \cos(2\pi t/400 - 45°)$ V
   (d) $v_4(t) = 60 \sin(2000\pi t + 45°)$ V

5–16 Find the frequency, period, phase angle, and amplitude of the sum of the first two sinusoids in Problem 5-15.

5–17 A 100 kHz sinusoid has an amplitude of 75 V and passes through 0 V with a positive slope at $t = 5$ μs. Find the Fourier coefficients, phase angle, and time shift of the waveform.

ERO 5–2 COMPOSITE WAVEFORMS (SECT. 5–5)

Given an equation, graph, or word description of a composite waveform,

(a) Construct an alternative description of the waveform.
(b) Find the parameters or properties of the waveform.
(c) Find new waveforms by integrating, or differentiating the given waveform.

See Examples 5–10, 5–11, 5–12, 5–13, 5–14, 5–15 and Exercises 5–8, 5–9

5–18 Graph the following waveforms:
   (a) $v_1(t) = 10\,[1 - 2e^{-200t} \sin(1000\pi t)]u(t)$ V
   (b) $v_2(t) = [20 - 10e^{-1000t}]u(t)$ V
   (c) $v_3(t) = 10\,[2 - \sin(1000\pi t)]u(t)$ V
   (d) $v_4(t) = 10\,[4 - 2e^{-1000t}]u(t)$ V

5–19 The damped ramp waveform can be written as $v(t) = V_A \alpha t e^{-\alpha t}u(t)$. Find the maximum value of the waveform and the time at which the maximum occurs.

5–20 A waveform is known to be of the form $v(t) = V_A - V_B e^{-\alpha t}$. At $t = 0$ the value of the waveform is 8 V, at $t = 5$ μs its value is 9 V, and it approaches 10 V for large values of t. Find the values of V_A, V_B, and α.

5–21 A waveform is known to be of the form $v(t) = V_A e^{-\alpha t} \sin \beta t$. The waveform periodically passes through zero every 2.5 ms. At $t = 1$ ms its value is 3.5 V and at $t = 2$ ms it is 0.8 V. Find the values of V_A, α, and β, and then graph the waveform.

5–22 Write an expression for the first cycle of the periodic waveform in Figure P5–22.

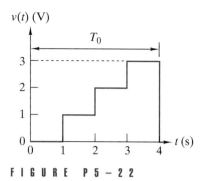

FIGURE P5-22

5-23 Write an expression for the first cycle of the periodic waveform in Figure P5-23.

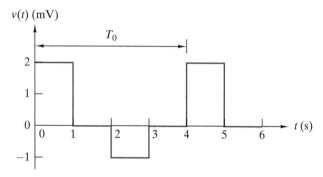

FIGURE P5-23

5-24 A waveform is known to be of the form $v(t) = V_A - V_B \sin \beta t$. At $t = 0$ the value of the waveform is 5 V, and it periodically reaches a minimum value of -2 V every 5 μs. Find the values of V_A, V_B, and β, and then graph the waveform.

5-25 Write an equation for the periodic waveform in Figure P5-25.

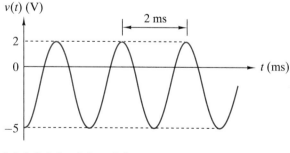

FIGURE P5-25

5-26 Write an expression for the derivative of the periodic waveform in Figure P5-25.

5-27 Write an equation for the first cycle of the periodic waveform in Figure P5-27 and then graph the waveform obtained by differentiating the signal.

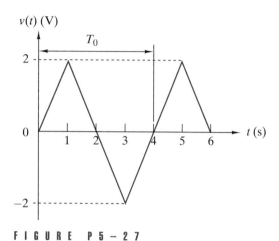

FIGURE P5-27

5-28 The exponential rise waveform $v(t) = V_A(1 - e^{-\alpha t})u(t)$ is zero at $t = 0$ and monotonically approaches a final value as $t \to \infty$. What is the final value and for what value of t does the waveform reach 50% of this value?

5-29 Figure P5-29 shows a plot of the waveform $v(t) = V_A[e^{-\alpha t} \sin \beta t]u(t)$. Find the values of V_A, α, and β.

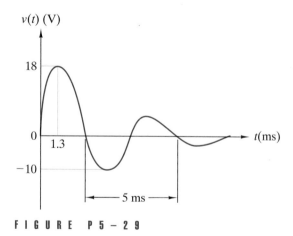

FIGURE P5-29

5-30 The basic waveform $u(x)$ is zero for all $x < 0$. The function $f(t) = u(\cos t)$ defines a periodic waveform. Sketch the waveform of cos *t* and then sketch *f*(*t*). Is the *f*(*t*) zero for all *t* < 0? If not, why not? How would you describe *f*(*t*)? Is it periodic? If so, what is the period of *f*(*t*)? How would you modify the function to change its period and peak amplitude?

ERO 5–3 WAVEFORM PARTIAL DESCRIPTORS (SECT. 5–6)

Given a complete description of a basic or composite waveform,

(a) Classify the waveform as periodic or aperiodic and causal or noncausal.
(b) Find the applicable partial waveform descriptors.
(c) Find the parameters of a waveform given its partial descriptors.

See Examples 5–16, 5–17 and Exercises 5–10, 5–11, and 5–12

5–31 Classify the following waveforms as periodic or aperiodic and causal or noncausal. Find V_p and V_{pp} for all of the waveforms. Find V_{rms} and V_{avg} for those waveforms that are periodic.
  (a) $v_1(t) = [150 - 80 \sin(2000\pi t)]u(t)$ V
  (b) $v_2(t) = 40[\sin(2000\pi t)][u(t) - u(t - 1)]$ V
  (c) $v_3(t) = 15 \cos(2000\pi t) + 10 \sin(2000\pi t)$ V
  (d) $v_4(t) = [10 - 5e^{-400t}]u(t)$ V

5–32 Find V_p, V_{pp}, V_{rms}, and V_{avg} for each of the sinusoids in Problem 5–11.

5–33 Find V_p, V_{pp}, V_{rms}, and V_{avg} for each of the sinusoids in Problem 5–15.

5–34 Find V_p, V_{pp}, V_{rms}, and V_{avg} for the periodic waveform Figure P5–22.

5–35 Find V_p, V_{pp}, V_{rms}, and V_{avg} for the periodic waveform Figure P5–23.

5–36 In digital data communications the *time constant of fall* is defined as the time required for a pulse to fall from 70.7% to 26.0% of its maximum value. Assuming that the pulse decay is exponential, find the relationship between the time constant of fall and the time constant of the exponential decay.

5–37 The waveform $v(t) = V_0 + 10 \cos(200\pi t)$ is applied at the input of an OP AMP voltage follower with $V_{CC} = \pm 15$ V. What range of values of the dc component V_0 ensures that the OP AMP does not saturate?

5–38 The first cycle ($t > 0$) of a periodic waveform with $T_0 = 70$ ms can be expressed as

$$v(t) = 2u(t) - 3u(t - 0.01) + 5u(t - 0.06)   \text{V}$$

Sketch the waveform and find V_{max}, V_{min}, V_p, V_{pp}, and V_{avg}.

5–39 A periodic waveform can be expressed as

$$v(t) = 100 - 200 \cos 2000 \pi t - 75 \sin 40000 \pi t$$
$$+ 35 \cos 80000 \pi t   \text{mV}$$

What is the period of the waveform? What is the average value of the waveform? What is the amplitude of the fundamental component? What is the highest frequency in the waveform?

5–40 The first cycle ($0 \le t < T_0$) of a periodic clock pulse train is

$$v(t) = V_A[u(t) - u(t - T)]   \text{V}$$

The duty cycle of the clock pulse is defined as the fraction of the period during which the pulse is not zero. Derive an expression for the average value of the clock pulse train in terms of V_A and the duty cycle $D = T/T_0$.

INTEGRATING PROBLEMS

5–41 ⓐ THE SPICE EXPONENTIAL SIGNAL

The SPICE circuit simulation language defines a composite exponential waveform as

$$v(t) = \begin{vmatrix} V_0 + (V_1 - V_0)(1 - e^{-t/T_{C1}}) & 0 \le t \le T_D \\ V_0 + (V_1 - V_0)(1 - e^{-T_D/T_{C1}})e^{-(t-T_D)/T_{C2}} & T_D < t \end{vmatrix}$$

What is the value of the waveform at $t = 0$? At $t = T_D$? What is the value of the waveform as t approaches infinity? What is the maximum amplitude of the waveform? Sketch the waveform on the range from $t = 0$ to $t = T_D + 2T_{C2}$. What is the practical duration of the waveform?

5–42 ⓐ ANALOG-TO-DIGITAL CONVERSION

Figure P5–42 shows a circuit diagram of an analog-to-digital converter based on a voltage divider and OP AMPs operating without feedback. Each OP AMP operates as a comparator whose output is $V_{OH} = 10$ V when $v_P > v_N$ and $V_{OL} = 0$ V when $v_P < v_N$. The input signal $v_S(t)$ is applied to all of the noninverting inputs simultaneously. The voltages applied to the inverting inputs come from the four-resistor voltage divider shown. For an input voltage of $v_S(t) = 25e^{-2000t}$ V, express the output voltages $v_A(t)$, $v_B(t)$, $v_C(t)$, and $v_D(t)$ in terms of step functions.

5–43 ⓐ DIGITAL-TO-ANALOG CONVERSION

Figure P5–43 shows a 3-bit digital-to-analog converter of the type discussed in Sect. 4–6. The circuit consists of an *R-2R* ladder connected to the inverting input to the OP AMP. The ladder circuit is driven by three digital input signals $v_1(t)$, $v_2(t)$, and $v_3(t)$.
  (a) Show that the input-output relationship for the OP AMP circuit is

$$v_O(t) = -\frac{R_F}{8R}[4v_1 + 2v_2 + v_3]$$

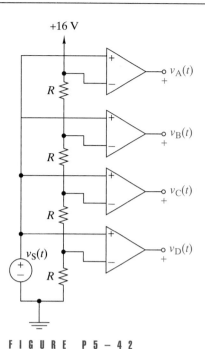

FIGURE P 5 – 4 2

(b) The digital input signals can be expressed in terms of step functions as follows:

$$v_1(t) = 5u(t - 1) - 5u(t - 5) \text{ V}$$

$$v_2(t) = 5u(t - 3) - 5u(t - 4) + 5u(t - 6) - 5u(t - 10) \text{ V}$$

$$v_3(t) = 5u(t) - 5u(t - 4) + 5u(t - 8) \text{ V}$$

Plot each digital waveform separately and stack the plots on top of each other with $v_1(t)$ at the bottom and $v_3(t)$ at the top. A stacked plot of digital signals is called a *timing diagram*.

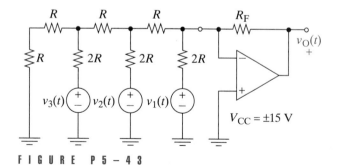

FIGURE P 5 – 4 3

(c) For $R_F = R$, use the input-output relationship in (a) and the input timing diagram in (b) to obtain a plot of the output voltage in the range $0 < t < 10$ s.

5–44 ⬥ TIMING DIAGRAMS

Timing diagrams are separate plots of digital waveforms that are stacked on top of each other to show how signals change over time and to reveal temporal relationships between signals. Timing diagrams are widely used to analyze, design, and evaluate digital circuits and systems. Timing diagrams waveforms can take on only two states: 0 (low) and 1 (high). The temporal points at which waveforms transition from one state to another are called **edges**. Thus, waveforms are described by the temporal location of their edges and the state of the signals following each edge.

A 3-bit binary counter has outputs $(D_2 \, D_1 \, D_0)$. At $t = 0$ the outputs are in the zero state, that is, $(D_2 \, D_1 \, D_0) = (0 \ 0 \ 0)$. Thereafter, the outputs undergo the following state transitions.

$$000 \text{---}{>}001 \text{---}{>}010 \text{---}{>}011 \text{---}{>}100 \text{---}{>}$$
$$101 \text{---}{>}110 \text{---}{>}111 \text{---}{>}000$$

The first transition occurs at $t = 50$ ns and subsequent transitions occur every 50 ns.
(a) Sketch a timing diagram for these signals with D_2 at the bottom and D_0 at the top.
(b) A digital signal D_4 is 1 (high) when D_2 AND D_0 are high and is 0 (low) otherwise. Add D_4 at the bottom of the timing diagram in (a).
(c) The outputs feed a digital-to-analog converter whose output is

$$v_O(t) = D_0 + 2D_1 + 4D_2$$

Add the analog waveform $v_O(t)$ at the bottom of the timing diagram in (b).

5–45 ⬥ FOURIER SERIES

Reasonably well behaved periodic waveforms can be expressed as a sum of sinusoidal components called a Fourier series. For example, the sum

$$v(t) = \frac{V_A}{2} - \sum_{n=1}^{n=\infty} \frac{V_A}{n\pi} \sin\left(2\pi \, n f_0 t\right)$$

is the Fourier series of one of the periodic waveforms in Figure 5–2. Use a spreadsheet or math solver to plot the sum of the first 10 terms of this series for $V_A = 10$ V and $f_0 = 50$ kHz. Use a time interval $0 \leq t \leq 2T_0$. Can you identify this waveform in Figure 5–2?

5–46 A VOLTMETER CALIBRATION

Most dc voltmeters measure the average value of the applied signal. A dc meter that measures the average value can be adapted to indicate the rms value of an ac signal. The input is passed through a rectifier circuit. The rectifier output is the absolute value of the input and is applied to a dc meter whose deflection is proportional to the average value of the rectified signal. The meter scale is calibrated to indicate the rms value of the input signal. A calibration factor is needed to convert the average absolute value into the rms value of the ac signal. What is the required calibration factor for a sinusoid? Would the same calibration factor apply to a square wave?

CHAPTER 6

CAPACITANCE AND INDUCTANCE

From the foregoing facts, it appears that a current of electricity is produced, for an instant, in a helix of copper wire surrounding a piece of soft iron whenever magnetism is induced in the iron; also that an instantaneous current in one or the other direction accompanies every change in the magnetic intensity of the iron.

Joseph Henry, 1831,
American Physicist

6–1 The Capacitor

6–2 The Inductor

6–3 Dynamic OP AMP Circuits

6–4 Equivalent Capacitance and Inductance

Summary

Problems

Integrating Problems

Joseph Henry (1797–1878) and the British physicist Michael Faraday (1791–1867) independently discovered magnetic induction almost simultaneously. The foregoing quotation is Henry's summary of the experiments leading to his discovery of magnetic induction. Although Henry and Faraday used similar apparatus and observed almost the same results, Henry was the first to recognize the importance of the discovery. The unit of circuit inductance (henry) honors Henry, while the mathematical generalization of magnetic induction is called Faraday's law. Michael Faraday had wide-ranging interests and performed many fundamental experiments in chemistry and electricity. His electrical experiments often used capacitors, or Leyden jars as they were called in those days. Faraday was a meticulous experimenter, and his careful characterization of these devices may be the reason that the unit of capacitance (the farad) honors Faraday.

The dynamic circuit responses involve memory effects that cannot be explained by the circuit elements we included in resistance circuits. The two new circuit elements required to explain memory effects are the capacitor and the inductor. The purpose of this chapter is to introduce the *i–v* characteristics of these dynamic elements and to explore circuit applications that illustrate dynamic behavior.

The first two sections of this chapter develop the device constraints for the linear capacitor and inductor. The third section is a circuit application in which resistors, capacitors, and operational amplifiers (OP AMPs) perform waveform integration and differentiation. The fourth section shows that capacitors or inductors connected in series or parallel can be replaced by a single equivalent capacitance or inductance.

6-1 THE CAPACITOR

A capacitor is a dynamic element involving the time variation of an electric field produced by a voltage. Figure 6–1(a) shows the parallel plate capacitor, which is the simplest physical form of a capacitive device. Figure 6–1 also shows two alternative circuit symbols and sketches of actual devices. Some of the physical features of commercially available devices are given in Appendix A.

Electrostatics shows that a uniform electric field $\mathcal{E}(t)$ exists between the metal plates in Figure 6–1(a) when a voltage exists across the capacitor.[1] The electric field produces charge separation with equal and opposite charges appearing on the capacitor plates. When the separation *d* is small compared with the dimension of the plates, the electric field between the plates is

$$\mathcal{E}(t) = \frac{q(t)}{\varepsilon A} \tag{6-1}$$

where ε is the permittivity of the dielectric, *A* is the area of the plates, and $q(t)$ is the magnitude of the electric charge on each plate. The relationship between the electric field and the voltage across the capacitor $v_C(t)$ is given by

$$\mathcal{E}(t) = \frac{v_C(t)}{d} \tag{6-2}$$

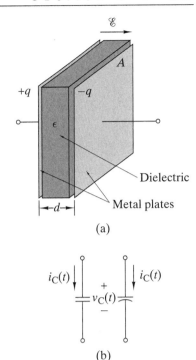

(a)

(b)

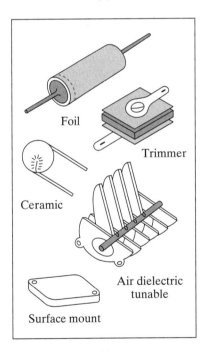

(c)

FIGURE 6-1 *The capacitor: (a) Parallel plate device. (b) Circuit symbol. (c) Example devices.*

---

1 An electric field is a vector quantity. In Figure 6–1(a) the field is confined to the space between the two plates and is perpendicular to the plates.

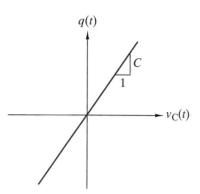

F I G U R E 6 – 2 *Graph of the defining relationship of a linear capacitor.*

Substituting Eq. (6–2) into Eq. (6–1) and solving for the charge $q(t)$ yields

$$q(t) = \left[\frac{\varepsilon A}{d}\right]v_C(t) \tag{6–3}$$

The proportionality constant inside the brackets in this equation is the **capacitance** C of the capacitor. That is, by definition,

$$C = \frac{\varepsilon A}{d} \tag{6–4}$$

The unit of capacitance is the **farad** (F), a term that honors the British physicist Michael Faraday. Values of capacitance range from a few pF (10^{-12} F) in semiconductor devices to tens of mF (10^{-3} F) in industrial capacitor banks. Using Eq. (6–4), the defining relationship for the capacitor becomes

$$q(t) = Cv_C(t) \tag{6–5}$$

Figure 6–2 graphically displays the element constraint in Eq. (6–5). The graph points out that the capacitor is a linear element since the defining relationship between voltage and charge is a straight line through the origin.

I–V RELATIONSHIP

To express the element constraint in terms of voltage and current, we differentiate Eq. (6–5) with respect to time t:

$$\frac{dq(t)}{dt} = \frac{d[Cv_C(t)]}{dt}$$

Since C is constant and $i_C(t)$ is the time derivative of $q(t)$, we obtain a capacitor i–v relationship in the form

$$i_C(t) = C\frac{dv_C(t)}{dt} \tag{6–6}$$

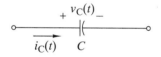

F I G U R E 6 – 3 *Capacitor current and voltage.*

The relationship assumes that the reference marks for the current and voltage follow the passive sign convention shown in Figure 6–3.

The time derivative in Eq. (6–6) means the current is zero when the voltage across the capacitor is constant, and vice versa. In other words, the capacitor acts like an open circuit ($i_C = 0$) when dc excitations are applied. The capacitor is a dynamic element because the current is zero unless the voltage is changing. However, a discontinuous change in voltage would require an infinite current, which is physically impossible. Therefore, the capacitor voltage must be a continuous function of time.

Equation (6–6) relates the capacitor current to the rate of change of the capacitor voltage. To express the voltage in terms of the current, we multiply both sides of Eq. (6–6) by dt, solve for the differential dv_C, and integrate:

$$\int dv_C = \frac{1}{C}\int i_C(t)dt$$

Selecting the integration limits requires some discussion. We assume that at some time t_0 the voltage across the capacitor $v_C(t_0)$ is known and we want to determine the voltage at some later time $t > t_0$. Therefore, the integration limits are

$$\int_{v_C(t_0)}^{v_C(t)} dv_C = \int_{t_0}^{t} i_C(x)dx$$

where x is a dummy integration variable. Integrating the left side of this equation yields

$$v_C(t) = v_C(t_0) + \frac{1}{C}\int_{t_0}^{t} i_C(x)dx \qquad (6-7)$$

In practice, the time t_0 is established by a physical event such as closing a switch or the start of a particular clock pulse. Nothing is lost in the integration in Eq. (6–7) if we arbitrarily define t_0 to be zero. Using $t_0 = 0$ in Eq. (6–7) yields

$$v_C(t) = v_C(0) + \frac{1}{C}\int_{0}^{t} i_C(x)dx \qquad (6-8)$$

Equation (6–8) is the integral form of the capacitor i–v constraint. Both the integral form and the derivative form in Eq. (6–6) assume that the reference marks for current and voltage follow the passive sign convention in Figure 6–3.

POWER AND ENERGY

With the passive sign convention the capacitor power is

$$p_C(t) = i_C(t)v_C(t) \qquad (6-9)$$

Using Eq. (6–6) to eliminate $i_C(t)$ from Eq. (6–9) yields the capacitor power in the form

$$p_C(t) = Cv_C(t)\frac{dv_C(t)}{dt} = \frac{d}{dt}[^1/_2 Cv_C^2(t)] \qquad (6-10)$$

This equation shows that the power can be either positive or negative because the capacitor voltage and its time rate of change can have opposite signs. With the passive sign convention, a positive sign means the element absorbs power, while a negative sign means the element delivers power. The ability to deliver power implies that the capacitor can store energy.

To determine the stored energy, we note that the expression for power in Eq. (6–10) is a perfect derivative. Since power is the time rate of change of energy, the quantity inside the brackets must be the energy stored in the capacitor. Mathematically, we can infer from Eq. (6–10) that the energy at time t is

$$w_C(t) = {}^1/_2 Cv_C^2(t) + \text{constant}$$

The constant in this equation is the value of stored energy at some instant t when $v_C(t) = 0$. At such an instant the electric field is zero; hence the stored energy is also zero. As a result, the constant is zero and we write the capacitor energy as

$$w_C(t) = \frac{1}{2}Cv_C^2(t) \qquad (6-11)$$

The stored energy is never negative, since it is proportional to the square of the voltage. The capacitor absorbs power from the circuit when storing energy and returns previously stored energy when delivering power to the circuit.

The relationship in Eq. (6–11) also implies that voltage is a continuous function of time, since an abrupt change in the voltage implies a discontinuous change in energy. Since power is the time derivative of energy, a discontinuous change in energy implies infinite power, which is physically impossible. The capacitor voltage is called a **state variable** because it determines the energy state of the element.

To summarize, the capacitor is a dynamic circuit element with the following properties:

1. *The current through the capacitor is zero unless the voltage is changing. The capacitor acts like an open circuit to dc excitations.*

2. *The voltage across the capacitor is a continuous function of time. A discontinuous change in capacitor voltage would require infinite current and power, which is physically impossible.*

3. *The capacitor absorbs power from the circuit when storing energy and returns previously stored energy when delivering power. The net energy transfer is nonnegative, indicating that the capacitor is a passive element.*

The following examples illustrate these properties.

EXAMPLE 6–1

The voltage in Figure 6–4(a) appears across a ½-µF capacitor. Find the current through the capacitor.

SOLUTION:

The capacitor current is proportional to the time rate of change of the voltage. For $0 < t < 2$ ms the slope of the voltage waveform has a constant value

$$\frac{dv_C}{dt} = \frac{10}{2 \times 10^{-3}} = 5000 \text{ V/s}$$

The capacitor current during this interval is

$$i_C(t) = C\frac{dv_C}{dt} = (0.5 \times 10^{-6}) \times (5 \times 10^3) = 2.5 \text{ mA}$$

For $2 < t < 3$ ms the rate of change of the voltage is −5000 V/s. Since the rate of change of voltage is negative, the current changes direction and takes on the value $i_C(t) = -2.5$ mA. For $t > 3$ ms, the voltage is constant, so its slope is zero; hence the current is zero. The resulting current waveform is shown in Figure 6–4(b). Note that the voltage across the capacitor (the state variable) is continuous, but the capacitor current can be, and in this case is, discontinuous. ∎

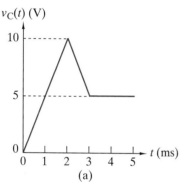

(a)

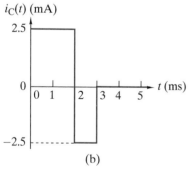

(b)

FIGURE 6–4

EXAMPLE 6–2

The $i_C(t)$ in Figure 6–5(a) is given by

$$i_C(t) = I_0(e^{-t/T_C})u(t)$$

Find the voltage across the capacitor if $v_C(0) = 0$ V.

SOLUTION:

Using the capacitor i–v relationship in integral form,

$$v_C(t) = v_C(0) + \frac{1}{C}\int_0^t i_C(x)dx$$

$$= 0 + \frac{1}{C}\int_0^t I_0 e^{-x/T_C}\,dx = \frac{I_0 T_C}{C}\left(-e^{-x/T_C}\right)\Big|_0^t$$

$$= \frac{I_0 T_C}{C}\left(1 - e^{-t/T_C}\right)$$

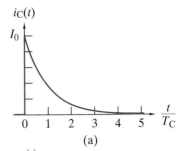

(a)

The graphs in Figure 6–5(b) show that the voltage is continuous while the current is discontinuous. ∎

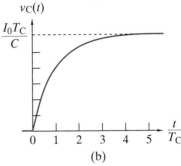

(b)

FIGURE 6–5

Exercise 6–1

(a) The voltage across a 10-µF capacitor is $25[\sin 2000t]u(t)$. Derive an expression for the current through the capacitor.
(b) At $t = 0$ the voltage across a 100-pF capacitor is –5 V. The current through the capacitor is $10[u(t) - u(t - 10^{-4})]$ µA. What is the voltage across the capacitor for $t > 0$?

Answers:

(a) $i_C(t) = 0.5\,[\cos 2000t]u(t)$ A.
(b) $v_C(t) = -5 + 10^5 t$ V for $0 < t < 0.1$ ms and $v_C(t) = 5$ V for $t > 0.1$ ms

Exercise 6–2

For $t \geq 0$ the voltage across a 200-pF capacitor is $5e^{-4000t}$ V.

(a) What is the charge on the capacitor at $t = 0$ and $t = +\infty$?
(b) Derive an expression for the current through the capacitor for $t \geq 0$.
(c) For $t > 0$ is the device absorbing or delivering power?

Answers:

(a) 1 nC and 0 C
(b) $i_C(t) = -4e^{-4000t}$ µA
(c) Delivering

EXAMPLE 6–3

Figure 6–6(a) shows the voltage across 0.5-µF capacitor. Find the capacitor's energy and power.

SOLUTION:

The current through the capacitor was found in Example 6–1. The power waveform is the point-by-point product of the voltage and current waveforms. The energy is found by either integrating the power waveform or by calculating $\frac{1}{2}C[v_C(t)]^2$ point by point. The current, power, and energy are shown in Figures 6–6(b), 6–6(c), and 6–6(d). Note that the capacitor energy increases when it is absorbing power $[p_C(t) > 0]$ and decreases when delivering power $[p_C(t) < 0]$. ∎

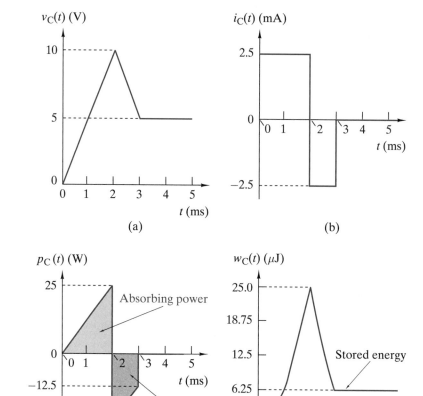

FIGURE 6 – 6

EXAMPLE 6–4

The current through a capacitor is given by

$$i_C(t) = I_0[e^{-t/T_C}]u(t)$$

Find the capacitor's energy and power.

SOLUTION:

The current and voltage were found in Example 6–2 and are shown in Figures 6–7(a) and 6–7(b). The power waveform is found as the product of current and voltage:

$$p_C(t) = i_C(t)v_C(t)$$

$$= [I_0 e^{-t/T_C}]\left[\frac{I_0 T_C}{C}(1 - e^{-t/T_C})\right]$$

$$= \frac{I_0^2 T_C}{C}(e^{-t/T_C} - e^{-2t/T_C})$$

The waveform of the power is shown in Figure 6–7(c). The energy is

$$w_C(t) = \frac{1}{2} C v_C^2(t) = \frac{(I_0 T_C)^2}{2C} (1 - e^{-t/T_C})^2$$

The time history of the energy is shown in Figure 6–7(d). In this example both power and energy are always positive. ∎

Exercise 6–3

Find the power and energy for the capacitors in Exercise 6–1.

Answers:

(a) $p_C(t) = 6.25[\sin 4000t]u(t)$ W
   $w_C(t) = 3.125 \sin^2 2000t$ mJ
(b) $p_C(t) = -0.05 + 10^3 t$ mW for $0 < t < 0.1$ ms
   $p_C(t) = 0$ for $t > 0.1$ ms
   $w_C(t) = 1.25 - 5 \times 10^4 t + 5 \times 10^8 t^2$ nJ for $0 < t < 0.1$ ms
   $w_C(t) = 1.25$ nJ for $t > 0.1$ ms

Exercise 6–4

Find the power and energy for the capacitor in Exercise 6–2.

Answers:

$$p_C(t) = -20 e^{-8000t} \; \mu W$$

$$w_C(t) = 2.5 e^{-8000t} \; nJ$$

APPLICATION NOTE: EXAMPLE 6–5

A **sample-and-hold circuit** is usually found at the input to an analog-to-digital converter (ADC). The purpose of the circuit is to sample a time-varying input waveform at a specified instant and then hold that value constant until conversion to digital form is complete. This example discusses the role of a capacitor in such a circuit.

The basic sample-and-hold circuit in Figure 6–8(a) includes an input buffer, a digitally controlled electronic switch, a holding capacitor, and an output buffer. The input buffer is a voltage follower whose output replicates the analog input $v_S(t)$ and supplies charging current to the capacitor. The output buffer is also a voltage follower whose output replicates the capacitor voltage.

To see how the circuit operates, we describe one cycle of the sample-hold process. At time t_1 shown in Figure 6–8(b), the digital control $v_G(t)$ goes high, which causes the switch to close. Thereafter, the input buffer supplies a charging current $i_C(t)$ to drive the capacitor voltage to the level of the analog input. At time t_2 shown in Figure 6–8(b) the digital control goes low, the switch opens, and thereafter the capacitor current $i_C(t) = 0$. Zero current means that capacitor voltage is constant since dv_C/dt is zero. In sum, closing the switch causes the capacitor voltage to track the input and opening the switch causes the capacitor voltage to hold a sample of the input.

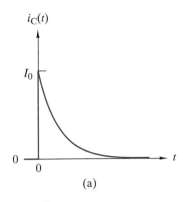

(a)

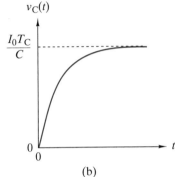

(b)

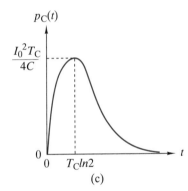

(c)

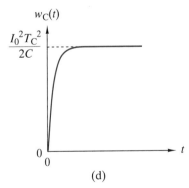

(d)

FIGURE 6-7

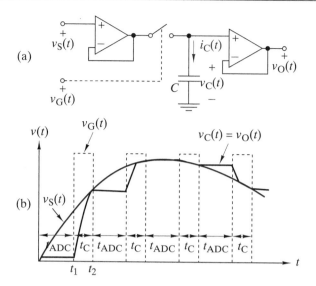

(a)

(b)

Figure 6–8(b) shows several more cycles of the sample-and-hold process. Samples of the input waveform are acquired during the time intervals labelled t_C. During these intervals the control signal is high, the switch is closed, and the capacitor charges or discharges in order to track the analog input voltage. Analog-to-digital conversion of the circuit output voltage takes place during the time intervals t_{ADC}. During these intervals the control signal is low, the switch is open, and the capacitor holds the output voltage constant.

Sample-and-hold circuits are available as monolithic integrated circuits that include the two buffers, the electronic switch, but not the holding capacitor. The capacitor is supplied externally, and its selection involves a trade-off. In an ideal sample-and-hold circuit, the capacitor voltage tracks the input when the switch is closed (sample mode) and holds the value indefinitely when the switch is open (hold mode). In real circuits the input buffer has a maximum output current, which means that some time is needed to charge the capacitor in the sample mode. Minimizing this sample acquisition time argues for a small capacitor. On the other hand, in the hold mode the output buffer draws a small current which gradually discharges the capacitor causing the output voltage to slowly decrease. Minimizing this output droop calls for a large capacitor. Thus, selecting the capacitance of the holding capacitor involves a compromise between the sample acquisition time and the output voltage droop in the hold mode.

6–2 THE INDUCTOR

The inductor is a dynamic circuit element involving the time variation of the magnetic field produced by a current. Magnetostatics shows that a magnetic flux ϕ surrounds a wire carrying an electric current. When the wire is wound into a coil, the lines of flux concentrate along the axis of the coil, as shown in Figure 6–9. In a linear magnetic medium, the flux is proportional to both the current and the number of turns in the coil. Therefore, the total flux is

$$\phi(t) = k_1 N i_L(t) \tag{6–12}$$

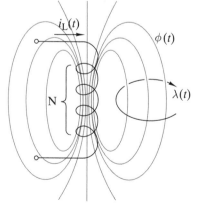

F I G U R E   6 – 9   *Magnetic flux surrounding a current-carrying coil.*

where k_1 is a constant of proportionality.

The magnetic flux intercepts or links the turns of the coil. The flux linkage in a coil is represented by the symbol λ, with units of webers (Wb), named after the German scientist Wilhelm Weber (1804–1891). The flux linkage is proportional to the number of turns in the coil and to the total magnetic flux, so λ(t) is

$$\lambda(t) = N\phi(t) \tag{6–13}$$

Substituting Eq. (6–12) into Eq. (6–13) gives

$$\lambda(t) = [k_1 N^2]i_L(t) \tag{6–14}$$

The proportionality constant inside the brackets in this equation is the **inductance** L of the coil. That is, by definition

$$L = k_1 N^2 \tag{6–15}$$

The unit of inductance is the henry (H) (plural henrys), a name that honors American scientist Joseph Henry. Figure 6–10 shows the circuit symbol for an inductor and some examples of actual devices.

Using Eq. (6–15), the defining relationship for the inductor becomes

$$\lambda(t) = Li_L(t) \tag{6–16}$$

Figure 6–11 graphically displays the inductor's element constraint in Eq. (6–16). The graph points out that the inductor is a linear element since the defining relationship is a straight line through the origin.

I−V RELATIONSHIP

Equation (6–16) is the inductor element constraint in terms of current and flux linkage. To obtain the element characteristic in terms of voltage and current, we differentiate Eq. (6–16) with respect to time:

$$\frac{d[\lambda(t)]}{dt} = \frac{d[Li_L(t)]}{dt} \tag{6–17}$$

The inductance L is a constant. According to Faraday's law, the voltage across the inductor is equal to the time rate of change of flux linkage. Therefore, we obtain an inductor *i–v* relationship in the form

$$v_L(t) = L\frac{di_L(t)}{dt} \tag{6–18}$$

The time derivative in Eq. (6–18) means that the voltage across the inductor is zero unless the current is time varying. Under dc excitation the current is constant and $v_L = 0$, so the inductor acts like a short circuit. The inductor is a dynamic element because only a changing current produces a nonzero voltage. However, a discontinuous change in current would produce an infinite voltage, which is physically impossible. Therefore, the current $i_L(t)$ must be a continuous function of time t.

Equation (6–18) relates the inductor voltage to the rate of change of the inductor current. To express the inductor current in terms of the voltage, we multiply both sides of Eq. (6–18) by dt, solve for the differential di_L, and integrate:

$$\int di_L = \frac{1}{L} \int v_L(t)dt \tag{6–19}$$

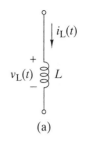

(a)

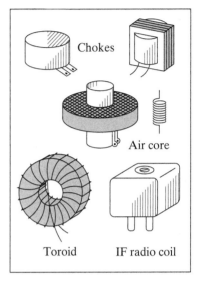

Chokes

Air core

Toroid   IF radio coil

(b)

FIGURE 6 − 1 0 *The inductor: (a) Circuit symbol. (b) Example devices.*

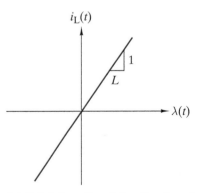

FIGURE 6 − 1 1 *Graph of the defining relationship of a linear inductor.*

To set the limits of integration, we assume that the inductor current $i_L(t_0)$ is known at some time t_0. Under this assumption the integration limits are

$$\int_{i_L(t_0)}^{i_L(t)} di_L = \frac{1}{L} \int_{t_0}^{t} v_L(x)dx \tag{6–20}$$

where x is a dummy integration variable. The left side of Eq. (6–20) integrates to produce

$$i_L(t) = i_L(t_0) + \frac{1}{L} \int_{t_0}^{t} v_L(x)dx \tag{6–21}$$

The reference time t_0 is established by some physical event, such as closing or opening a switch. Without losing any generality, we can assume $t_0 = 0$ and write Eq. (6–21) in the form

$$i_L(t) = i_L(0) + \frac{1}{L} \int_{0}^{t} v_L(x)dx \tag{6–22}$$

Equation (6–22) is the integral form of the inductor i–v characteristic. Both the integral form and the derivative form in Eq. (6–18) assume that the reference marks for the inductor voltage and current follow the passive sign convention shown in Figure 6–10.

POWER AND ENERGY

With the passive sign convention the inductor power is

$$p_L(t) = i_L(t)v_L(t) \tag{6–23}$$

Using Eq. (6–18) to eliminate $v_L(t)$ from this equation puts the inductor power in the form

$$p_L(t) = [i_L(t)]\left[L\frac{di_L(t)}{dt}\right] = \frac{d}{dt}[{}^1/_2 Li_L^2(t)] \tag{6–24}$$

This expression shows that power can be positive or negative because the inductor current and its time derivative can have opposite signs. With the passive sign convention a positive sign means the element absorbs power, while a negative sign means the element delivers power. The ability to deliver power indicates that the inductor can store energy.

To find the stored energy, we note that the power relation in Eq. (6–24) is a perfect derivative. Since power is the time rate of change of energy, the quantity inside the brackets must represent the energy stored in the magnetic field of the inductor. From Eq. (6–24), we infer that the energy at time t is

$$w_L(t) = \frac{1}{2}Li_L^2(t) + \text{constant}$$

As is the case with capacitor energy, the constant in this expression is zero since it is the energy stored at an instant t at which $i_L(t) = 0$. As a result, the energy stored in the inductor is

$$w_L(t) = \frac{1}{2}Li_L^2(t) \tag{6–25}$$

The energy stored in an inductor is never negative because it is proportional to the square of the current. The inductor stores energy when absorbing power and returns previously stored energy when delivering power, so that the net energy transfer is never negative.

Equation (6–25) implies that inductor current is a continuous function of time because an abrupt change in current causes a discontinuity in the energy. Since power is the time derivative of energy, an energy discontinuity implies infinite power, which is physically impossible. Current is called the **state variable** of the inductor because it determines the energy state of the element.

In summary, the inductor is a dynamic circuit element with the following properties:

1. *The voltage across the inductor is zero unless the current through the inductor is changing. The inductor acts like a short circuit for dc excitations.*

2. *The current through the inductor is a continuous function of time. A discontinuous change in inductor current would require infinite voltage and power, which is physically impossible.*

3. *The inductor absorbs power from the circuit when storing energy and delivers power to the circuit when returning previously stored energy. The net energy is nonnegative, indicating that the inductor is a passive element.*

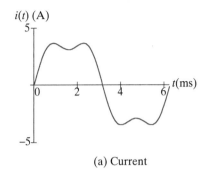

(a) Current

EXAMPLE 6–6

The current through a 2-mH inductor is $i_L(t) = 4 \sin 1000t + 1 \sin 3000t$ A. Find the resulting inductor voltage.

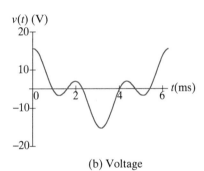

(b) Voltage

FIGURE 6–12

SOLUTION:
The voltage is found from the derivative form of the i–v relationship:

$$v_L(t) = L\frac{di_L(t)}{dt} = 0.002\,[4 \times 1000 \cos 1000\,t + 1 \times 3000 \cos 3000\,t]$$

$$= 8 \cos 1000\,t + 6 \cos 3000\,t \text{ V}$$

The current and voltage waveforms are shown in Figure 6–12. Note that the current and voltage each contain two sinusoids at different frequencies. However, the relative amplitudes of the two sinusoids are different. In $i_L(t)$ the ratio of the amplitude of the ac component at $\omega = 3$ krad/s to the ac component at $\omega = 1$ krad/s is 1-to-3, whereas in $v_L(t)$ this ratio is 3-to-4. The fact that the ac responses of energy storage elements depend on frequency allows us to create frequency selective signal processors called filters. ∎

EXAMPLE 6–7

Figure 6–13 shows the current through and voltage across an unknown energy storage element.
(a) What is the element and what is its numerical value?
(b) If the energy stored in the element at $t = 0$ is zero, how much energy is stored in the element at $t = 1$ s?

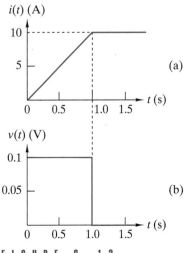

FIGURE 6–13

SOLUTION:

(a) By inspection, the voltage across the device is proportional to the derivative of the current, so the element is a linear inductor. During the interval $0 < t < 1$ s, the slope of the current waveform is 10 A/s. During the same interval the voltage is a constant 100 mV. Therefore, the inductance is

$$L = \frac{v}{di/dt} = \frac{0.1 \text{ V}}{10 \text{ A/s}} = 10 \text{ mH}$$

(b) The energy stored at $t = 1$ s is

$$w_L(1) = \frac{1}{2}Li_L^2(1) = 0.5(0.01)(10)^2 = 0.5 \text{ J} \qquad ■$$

EXAMPLE 6−8

The current through a 2.5-mH inductor is a damped sine $i(t) = 10e^{-500t}$ sin 2000t. Plot the waveforms of the element current, voltage, power, and energy.

SOLUTION:

Figure 6–14 shows a Mathcad file that produces the required graphs. The first two lines in the file define the inductance and the current. The inductor voltage is found by differentiation, the power as the product of voltage and current, and the energy by integrating the power. In the plots shown in Figure 6–14, note that the current, voltage, and power alternate signs, whereas the total energy is always positive.

In Mathcad the time scale statement shown in the figure uses the following format:

$$t = T_{\text{Start}}, \ T_{\text{Start}} + \Delta t \ldots T_{\text{Stop}}$$

Numerical values for T_{Start}, Δt, and T_{Stop} are chosen by considering the damped sine waveform. The period of the sinusoid is

$$T_0 = \frac{2 \times \pi}{2000} = 0.00314 \text{ s}$$

The plots start at $T_{\text{Start}} = 0$. To include at least one cycle requires an end time of $T_{\text{End}} > T_0 = 0.00314$ s; hence we select $T_{\text{Stop}} = 0.004$ s. To include at least 20 points per cycle requires a time increment ($t < T_0/20 = 0.000158$; hence we select $\Delta t = 0.0001$ s. ■

Exercise 6–5

For $t > 0$, the voltage across a 4-mH inductor is $v_L(t) = 20e^{-2000t}$ V. The initial current is $i_L(0) = 0$.

(a) What is the current through the inductor for $t > 0$?
(b) What is the power and (c) what is the energy for $t > 0$?

FIGURE   6 – 1 4

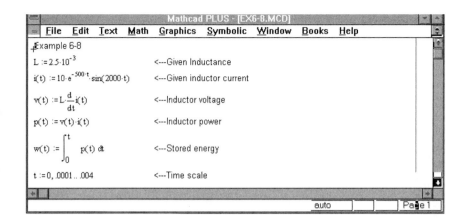

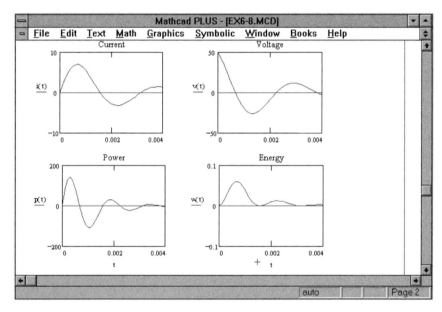

Answers:

(a) $i_L(t) = 2.5(1 - e^{-2000t})$ A
(b) $p_L(t) = 50(e^{-2000t} - e^{-4000t})$ W
(c) $w_L(t) = 12.5(1 - 2e^{-2000t} + e^{-4000t})$ mJ

Exercise 6–6

For $t < 0$, the current through a 100-mH inductor is zero. For $t \geq 0$, the current is
$i_L(t) = 20e^{-2000t} - 20e^{-4000t}$ mA.

(a) Derive an expression for the voltage across the inductor for $t > 0$.
(b) Find the time $t > 0$ at which the inductor voltage passes through zero.
(c) Derive an expression for the inductor power for $t > 0$.
(d) Find the time interval over which the inductor absorbs power and the interval over which it delivers power.

Answers:

(a) $v_L(t) = -4e^{-2000t} + 8e^{-4000t}$ V

(b) $t = 0.347$ ms

(c) $p_L(t) = -80e^{-4000t} + 240e^{-6000t} - 160e^{-8000t}$ mW

(d) Absorbing for $0 < t < 0.347$ ms, delivering for $t > 0.347$ ms

MORE ABOUT DUALITY

The capacitor and inductor characteristics are quite similar. Interchanging C and L, and i and v converts the capacitor equations into the inductor equations, and vice versa. This interchangeability illustrates the principle of duality. The dual concepts seen so far are as follows:

KVL	$\leftrightarrow$	KCL
Loop	$\leftrightarrow$	Node
Resistance	$\leftrightarrow$	Conductance
Voltage source	$\leftrightarrow$	Current source
Thévenin	$\leftrightarrow$	Norton
Short circuit	$\leftrightarrow$	Open circuit
Series	$\leftrightarrow$	Parallel
Capacitance	$\leftrightarrow$	Inductance
Flux linkage	$\leftrightarrow$	Charge

The term in one column is the dual of the term in the other column. The **principle of duality** states that

> *If every electrical term in a correct statement about circuit behavior is replaced by its dual, then the result is another correct statement.*

This principle may help beginners gain confidence in their understanding of circuit analysis. When the concept in one column is understood, the dual concept in the other column becomes easier to remember and apply.

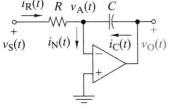

F I G U R E   6 – 1 5 *The inverting OP AMP integrator.*

6–3 DYNAMIC OP AMP CIRCUITS

The dynamic characteristics of capacitors and inductors produce signal processing functions that cannot be obtained using resistors. The OP AMP circuit in Figure 6–15 is similar to the inverting amplifier circuit except for the capacitor in the feedback path. To determine the signal-processing function of the circuit, we need to find its input-output relationship.

We begin by writing a KCL equation at node A.

$$i_R(t) + i_C(t) = i_N(t)$$

The resistor and capacitor device equations are written using their i–v relationships and the fundamental property of node voltages:

$$i_C(t) = C\frac{d[v_O(t) - v_A(t)]}{dt}$$

$$i_R(t) = \frac{1}{R}[v_S(t) - v_A(t)]$$

The ideal OP AMP device equations are $i_N(t) = 0$ and $v_A(t) = 0$. Substituting all of the element constraints into the KCL connection constraint produces

$$\frac{v_S(t)}{R} + C\frac{dv_O(t)}{dt} = 0$$

To solve for the output $v_O(t)$, we multiply this equation by dt, solve for the differential dv_O, and integrate:

$$\int dv_0 = -\frac{1}{RC}\int v_S(t)dt$$

Assuming the output voltage is known at time $t_0 = 0$, the integration limits are

$$\int_{v_0(0)}^{v_0(t)} dv_O = -\frac{1}{RC}\int_0^t v_S(x)dx$$

which yields

$$v_O(t) = v_O(0) - \frac{1}{RC}\int_0^t v_S(x)dx$$

The initial condition $v_O(0)$ is actually the voltage on the capacitor at $t = 0$, since by KVL, we have $v_C(t) = v_O(t) - v_A(t)$. But $v_A = 0$ for the OP AMP, so in general $v_O(t) = v_C(t)$. When the voltage on the capacitor is zero at $t = 0$, the circuit input-output relationship reduces to

$$v_O(t) = -\frac{1}{RC}\int_0^t v_S(x)dx \qquad (6\text{--}26)$$

The output voltage is proportional to the integral of the input voltage when the initial capacitor voltage is zero. The circuit in Figure 6–15 is an **inverting integrator** since the proportionality constant is negative. The constant $1/RC$ has the units of reciprocal seconds (s^{-1}) so that both sides of Eq. (6–26) have the units of volts.

Interchanging the resistor and capacitor in Figure 6–15 produces the OP AMP differentiator in Figure 6–16. To find the input-output relationship of this circuit, we start by writing the element and connection equations. The KCL connection constraint at node A is

$$i_R(t) + i_C(t) = i_N(t)$$

The device equations for the input capacitor and feedback resistor are

$$i_C(t) = C\frac{d[v_S(t) - v_A(t)]}{dt}$$

$$i_R(t) = \frac{1}{R}[v_O(t) - v_A(t)]$$

The device equations for the OP AMP are $i_N(t) = 0$ and $v_A(t) = 0$. Substituting all of these element constraints into the KCL connection constraint produces

$$\frac{v_O(t)}{R} + C\frac{dv_S(t)}{dt} = 0$$

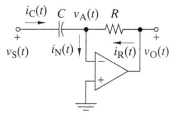

FIGURE 6–16 *The inverting OP AMP differentiator.*

Circuit	Block diagram	Gains
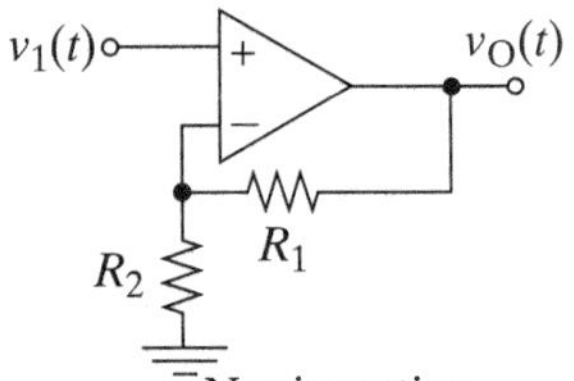 Noninverting	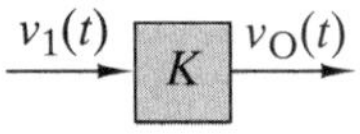 $v_1(t) \rightarrow \boxed{K} \rightarrow v_O(t)$	$K = \dfrac{R_1 + R_2}{R_2}$
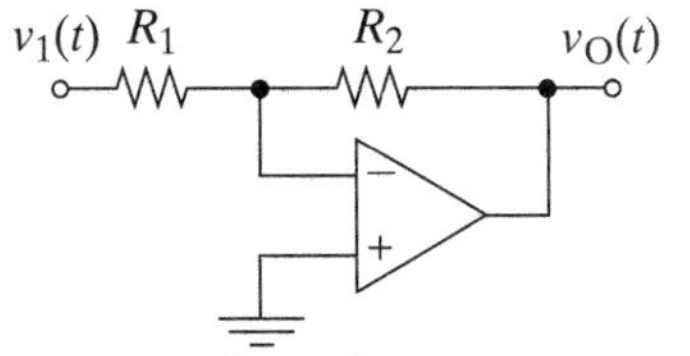 Inverting	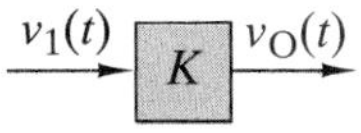 $v_1(t) \rightarrow \boxed{K} \rightarrow v_O(t)$	$K = -\dfrac{R_2}{R_1}$
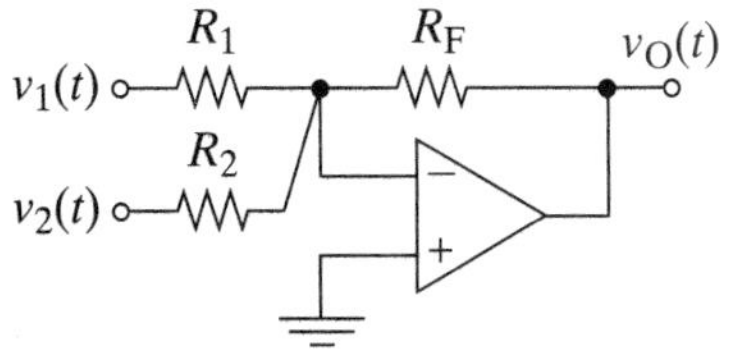 Summer	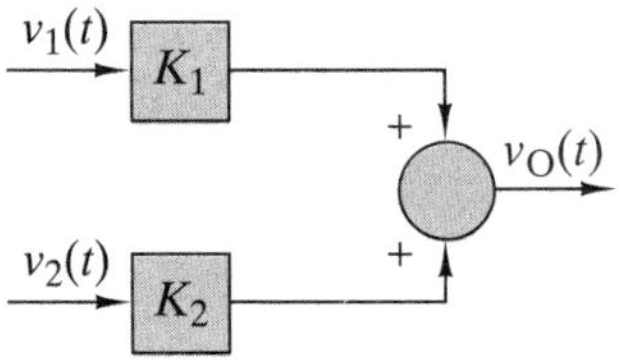 $v_1(t) \rightarrow \boxed{K_1}$, $v_2(t) \rightarrow \boxed{K_2}$ sum $\rightarrow v_O(t)$	$K_1 = -\dfrac{R_F}{R_1}$ $K_2 = -\dfrac{R_F}{R_2}$
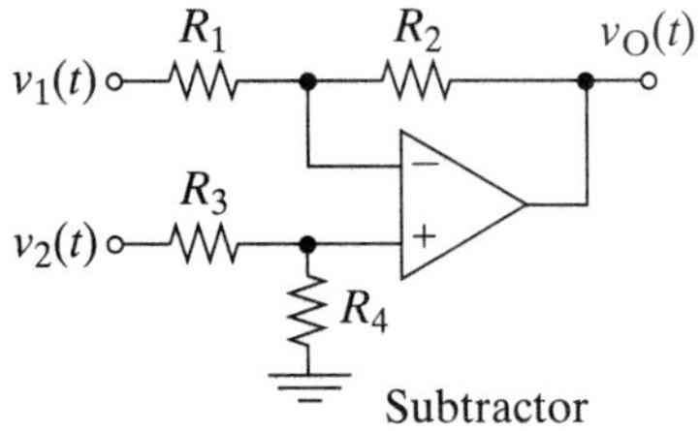 Subtractor	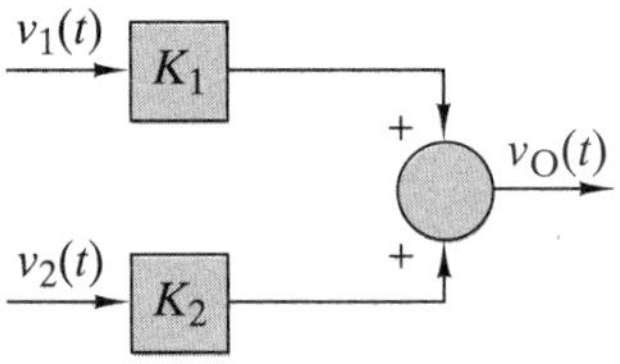 $v_1(t) \rightarrow \boxed{K_1}$, $v_2(t) \rightarrow \boxed{K_2}$ sum $\rightarrow v_O(t)$	$K_1 = -\dfrac{R_2}{R_1}$ $K_2 = \left(\dfrac{R_1 + R_2}{R_1}\right)\left(\dfrac{R_4}{R_3 + R_4}\right)$
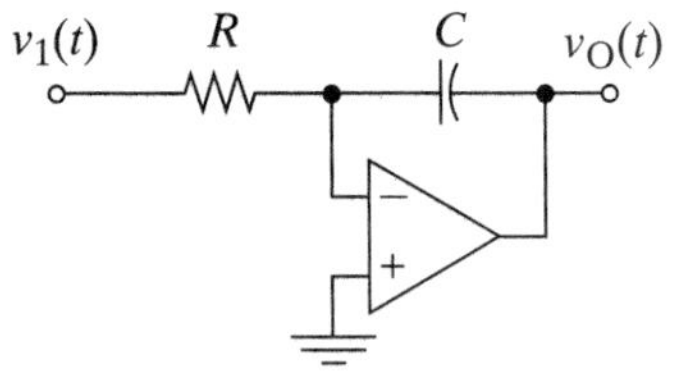 Integrator	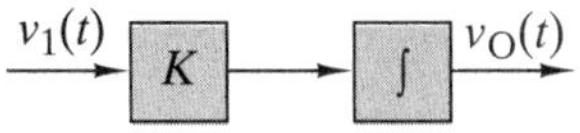 $v_1(t) \rightarrow \boxed{K} \rightarrow \boxed{\int} \rightarrow v_O(t)$	$K = -\dfrac{1}{RC}$
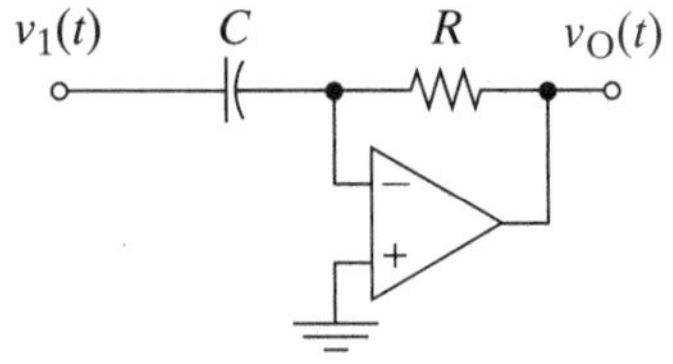 Differentiator	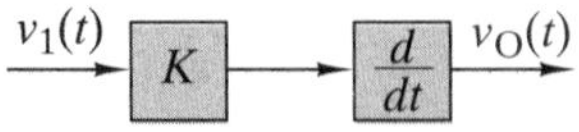 $v_1(t) \rightarrow \boxed{K} \rightarrow \boxed{\dfrac{d}{dt}} \rightarrow v_O(t)$	$K = -RC$

FIGURE 6–17 *Summary of basic OP AMP signal processing circuits.*

Solving this equation for $v_O(t)$ produces the circuit input-output relationship:

$$v_O(t) = -RC\frac{dv_S(t)}{dt} \qquad (6-27)$$

The output voltage is proportional to the derivative of the input voltage. The circuit in Figure 6-16 is an **inverting differentiator** since the proportionality constant $(-RC)$ is negative. The units of the constant RC are seconds so that both sides of Eq. (6-27) have the units of volts.

There are OP AMP inductor circuits that produce the inverting integrator and differentiator functions; however, they are of little practical interest because of the physical size and resistive losses in real inductor devices.

Figure 6-17 shows OP AMP circuits and block diagrams for the inverting integrator and differentiator, together with signal-processing functions studied in Chapter 4. The term *operational amplifier* results from the various mathematical operations implemented by these circuits. The following examples illustrate using the collection of circuits in Figure 6-17 in the analysis and design of signal-processing functions.

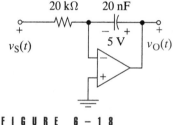

FIGURE 6-18

EXAMPLE 6-9

The input to the circuit in Figure 6-18 is $v_S(t) = 10u(t)$. Derive an expression for the output voltage. The OP AMP saturates when $v_O(t) = \pm15$ V.

SOLUTION:

The circuit is the inverting integrator with an initial voltage of 5 V across the capacitor. For the reference marks shown in Figure 6-18, this means that $v_O(0) = +5$ V. Assuming the OP AMP is operating in the linear mode, the output voltage is

$$v_O(t) = v_O(0) - \frac{1}{RC}\int_0^t v_S \, dt$$

$$= 5 - 2500 \int_0^t 10 \, dt$$

$$= 5 - 25,000t \quad t > 0$$

The output contains a negative going ramp because the circuit is an inverting integrator. The ramp output response is valid only as long as the OP AMP remains in its linear range. Negative saturation will occur when $5 - 25,000t = -15$, or at $t = 0.8$ ms. For $t > 0.8$ ms, the OP AMP is in the negative saturation mode with $v_O = -15$ V.

This example illustrates that dynamic circuits with bounded inputs may have unbounded responses. The circuit input here is a 10-V step function that has a bounded amplitude. The circuit output is a ramp whose output would be unbounded except that the OP AMP saturates. ∎

Exercise 6-7

The input to the circuit in Figure 6-18 is $v_S(t) = 10 \, [e^{-5000t}]u(t)$ V.

(a) For $v_C(0) = 0$, derive an expression for the output voltage, assuming the OP AMP is in its linear range.

(b) Does the OP AMP saturate with the given input?

Answers:

(a) $v_O(t) = 5(e^{-5000t} - 1)u(t)$

(b) Does not saturate

EXAMPLE 6–10

The input to the circuit in Figure 6–19(a) is a trapezoidal waveform shown in (b). Find the output waveform. The OP AMP saturates when $v_O(t) = \pm 15$ V.

FIGURE 6–19

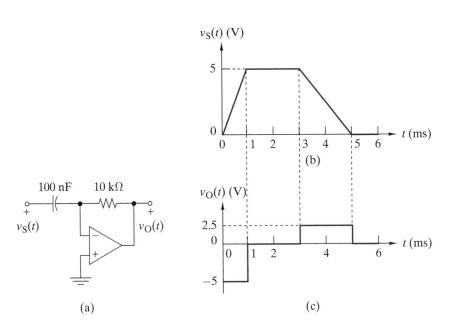

(b)

(a)

(c)

SOLUTION:

The circuit is the inverting differentiator with the following input-output relationship:

$$v_O(t) = -RC\frac{dv_S(t)}{dt} = -\frac{1}{1000}\frac{dv_S(t)}{dt}$$

The output voltage is constant over each of the following three time intervals:

1. For $0 < t < 1$ ms, the input slope is 5000 V/s and the output is $v_O = -5$ V.
2. For $1 < t < 3$ ms, the input slope is zero, so the output is zero as well.
3. For $3 < t < 5$ ms, the input slope is -2500 V/s and the output is $+2.5$ V.

The resulting output waveform is shown in Figure 6–19(c).

The output voltage remains within ± 15-V limits, so the OP AMP operates in the linear mode. ∎

Exercise 6-8

The input to the circuit in Figure 6–19 is $v_S(t) = V_A \cos 2000t$. The OP AMP saturates when $v_O = \pm 15$ V.

(a) Derive an expression for the output, assuming that the OP AMP is in the linear mode.
(b) What is the maximum value of V_A for linear operation?

Answers:
(a) $v_O(t) = 2V_A \sin 2000t$
(b) $|V_A| \le 7.5$ V

EXAMPLE 6–11

Determine the input-output relationship of the circuit in Figure 6–20(a).

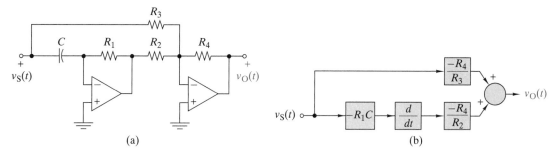

(a)   (b)

FIGURE 6–20

SOLUTION:
The circuit contains an inverting differentiator and an inverting summer. To find the input-output relationship, it is helpful to develop a block diagram for the circuit. Figure 6–20(b) shows a block diagram using the functional blocks in Figure 6–17. The product of the gains along the lower path yields its contribution to the output as

$$(-R_1 C)\left[\frac{d}{dt}\left(-\frac{R_4}{R_2}\right)v_S(t)\right]$$

The upper path contributes $(-R_4/R_3)v_S(t)$ to the output. The total output is the sum of the contributions from each path:

$$v_O(t) = \left(\frac{R_1 C R_4}{R_2}\right)\frac{dv_S(t)}{dt} - \left(\frac{R_4}{R_3}\right)v_S(t)$$

This equation assumes that both OP AMPs remain within their linear range. ∎

Exercise 6-9

Find the input-output relationship of the circuit in Figure 6–21.

Answer:

$$v_O(t) = v_O(0) + \frac{1}{RC}\int_0^t (v_{S1} - v_{S2})dt$$

FIGURE 6-21

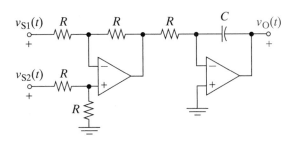

DESIGN EXAMPLE 6-12

Use the functional blocks in Figure 6–17 to design an OP AMP circuit to implement the following input-output relationship:

$$v_O(t) = 10v_S(t) + \int_0^t v_S \, dt$$

FIGURE 6-22

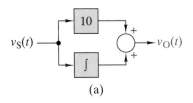

(a)

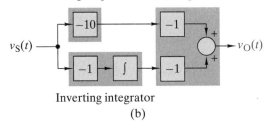

(b)

SOLUTION:
There is no unique solution to this design problem. We begin by drawing the block diagram in Figure 6–22(a), which shows that we need a gain block, an integrator, and a summer. However, the integrator and summer in Figure 6–17 are inverting circuits. Figure 6–22(b) shows how to overcome this problem by including an even number of signal inversions in each path. The inverting building blocks are realizable using OP AMP circuits in Figure 6–17, and the overall transfer characteristic is noninverting as required. One (of many) possible circuit realization of these processors is shown in Figure 6–23. The element parameter constraint on this circuit is $RC = 1$. Selecting the OP AMPs and the values of R and C depends on many additional factors, such as accuracy, internal resistance of the input source, and output load. ∎

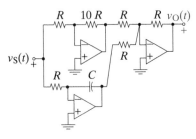

FIGURE 6-23

APPLICATION NOTE: EXAMPLE 6–13

An **integrating type analog-to-digital converter** (ADC) uses an integrator, a comparator, and a binary counter to produce an n-bit digital word representing the amplitude of an analog input. Figure 6–24 shows a block diagram of an integrating ADC. At $t = 0$, the start command resets the integrator output to zero and causes the sample-and-hold (S/H) circuit to acquire and hold the analog input $V_{IN} = v_S(0)$. At the same time, the integrator begins to produce a ramp

$$\frac{1}{RC} \int_0^t V_{REF}\, dx = V_{REF}\, \frac{t}{RC}$$

where V_{REF} is constant reference voltage. This ramp is fed to the negative input of a comparator (COMP). The comparator's output is high when the signal at its plus input (the analog sample) is greater than the signal at its negative input (the integrator ramp). At $t = 0$, the analog sample exceeds the ramp so the comparator goes high. As time marches on, the ramp increases and eventually exceeds the analog input, causing the output of the comparator to go low. This occurs at time $t = T$, where

$$T = RC\, \frac{V_{IN}}{V_{REF}}$$

The comparator output is a pulse whose duration T is proportional to the amplitude of the analog input. This pulse enables an n-bit binary counter whose initial count has been set to all zeros by the start command. As long as the enabling pulse is high, the count advances by one binary digit for each clock pulse. When the enabling pulse goes low, the accumulated count is an n-bit digital word whose binary weight is proportional to the duration of the enabling pulse—hence proportional to the amplitude of the analog input.

The block diagram in Figure 6–24 is called a single-slope ADC. Most commercially available integrating ADCs are the more complicated dual-slope type. In either case the basic principle of operation remains the same. Namely, an integrator and comparator convert the amplitude of the analog input into the duration of a pulse that enables a binary counter.

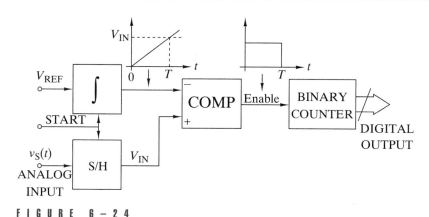

FIGURE 6–24

6–4 EQUIVALENT CAPACITANCE AND INDUCTANCE

In Chapter 2 we found that resistors connected in series or parallel can be replaced by equivalent resistances. The same principle applies to connections of capacitors and inductors—for example, to the parallel connection of capacitors in Figure 6–25(a). Applying KCL at node A yields

$$i(t) = i_1(t) + i_2(t) + \ldots + i_N(t)$$

Since the elements are connected in parallel, KVL requires

$$v_1(t) = v_2(t) = \ldots = v_N(t) = v(t)$$

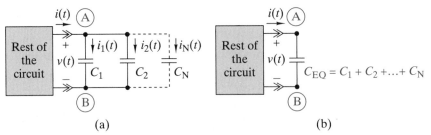

(a)   (b)

FIGURE 6–25 *Capacitors connected in parallel. (a) Given circuit. (b) Equivalent circuit.*

Because the capacitors all have the same voltage, their *i–v* relationships are all of the form $i_k(t) = C_k dv(t)/dt$. Substituting the *i–v* relationships into the KCL equation yields

$$i(t) = C_1 \frac{dv(t)}{dt} + C_2 \frac{dv(t)}{dt} + \ldots + C_N \frac{dv(t)}{dt}$$

Factoring the derivative out of each term produces

$$i(t) = (C_1 + C_2 + \ldots + C_N) \frac{dv(t)}{dt}$$

This equation states that the responses $v(t)$ and $i(t)$ in Figure 6–25(a) do not change when the N parallel capacitors are replaced by an equivalent capacitance:

$$C_{EQ} = C_1 + C_2 + \ldots + C_N \quad \text{(parallel connection)} \quad \text{(6–28)}$$

The equivalent capacitance simplification is shown in Figure 6–25(b). The initial voltage, if any, on the equivalent capacitance is $v(0)$, the common voltage across all of the original N capacitors at $t = 0$.

Next consider the series connection of N capacitors in Figure 6–26(a). Applying KVL around loop 1 in Figure 6–26(a) yields the equation

$$v(t) = v_1(t) + v_2(t) + \ldots + v_N(t)$$

Since the elements are connected in series, KCL requires

$$i_1(t) = i_2(t) = \ldots = i_N(t) = i(t)$$

Since the same current exists in all capacitors, their *i–v* relationships are all of the form

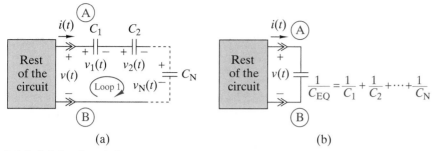

F I G U R E   6 – 2 6   *Capacitors connected in series. (a) Given circuit. (b) Equivalent circuit.*

$$v_k(t) = v_k(0) + \frac{1}{C_k} \int_0^t i(x)dx$$

Substituting these i–v relationships into the loop 1 KVL equation yields

$$v(t) = v_1(0) + \frac{1}{C_1} \int_0^t i(x)dx + v_2(0) + \frac{1}{C_2} \int_0^t i(x)dx$$

$$+ \cdots + v_N(0) + \frac{1}{C_N} \int_0^t i(x)dx$$

We can factor the integral out of each term to obtain

$$v(t) = [v_1(0) + v_2(0) + \ldots + v_N(0)] + \left(\frac{1}{C_1} + \frac{1}{C_2} + \ldots + \frac{1}{C_N} \right) \int_0^t i(x)dx$$

This equation indicates that the responses $v(t)$ and $i(t)$ in Figure 6–26(a) do not change when the N series capacitors are replaced by an equivalent capacitance:

$$\frac{1}{C_{EQ}} = \frac{1}{C_1} + \frac{1}{C_2} + \ldots + \frac{1}{C_N} \qquad \text{(series connection)} \qquad (6\text{–}29)$$

The equivalent capacitance is shown in Figure 6–26(b). The initial voltage on the equivalent capacitance is the sum of the initial voltages on each of the original N capacitors.

The equivalent capacitance of a parallel connection is the sum of the individual capacitances. The reciprocal of the equivalent capacitance of a series connection is the sum of the reciprocals of the individual capacitances. Since the capacitor and inductor are dual elements, the corresponding results for inductors are found by interchanging the series and parallel equivalence rules for the capacitor. That is, in a series connection the equivalent inductance is the sum of the individual inductances:

$$L_{EQ} = L_1 + L_2 + \ldots + L_N \qquad \text{(series connection)} \qquad (6\text{–}30)$$

For the parallel connection, the reciprocals add to produce the reciprocal of the equivalent inductance:

$$\frac{1}{L_{EQ}} = \frac{1}{L_1} + \frac{1}{L_2} + \ldots + \frac{1}{L_N} \qquad \text{(parallel connection)} \qquad (6\text{–}31)$$

Derivation of Eqs. (6–30) and (6–31) uses the approach given previously for the capacitor except that the roles of voltage and current are interchanged. Completion of the derivation is left as a problem for the reader.

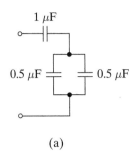

(a)

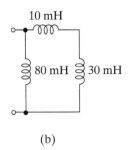

(b)

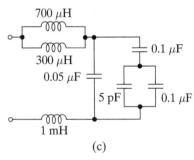

(c)

FIGURE 6 – 27

EXAMPLE 6–14

Find the equivalent capacitance and inductance of the circuits in Figure 6–27.

SOLUTION:

(a) For the circuit in Figure 6–27(a), the two 0.5-μF capacitors in parallel combine to yield an equivalent $0.5 + 0.5 = 1$-μF capacitance. This 1-μF equivalent capacitance is in series with a 1-μF capacitor, yielding an overall equivalent of $C_{EQ} = 1/(1/1 + 1/1) = 0.5$ μF.

(b) For the circuit of Figure 6–27(b), the 10-mH and the 30-mH inductors are in series and add to produce an equivalent inductance of 40 mH. This 40-mH equivalent inductance is in parallel with the 80-mH inductor. The equivalent inductance of the parallel combination is $L_{EQ} = 1/(1/40 + 1/80) = 26.67$ mH.

(c) The circuit of Figure 6–27(c) contains both inductors and capacitors. In later chapters, we will learn how to combine all of these into a single equivalent element. For now, we combine the inductors and the capacitors separately. The 5-pF capacitor in parallel with the 0.1-μF capacitor yields an equivalent capacitance of 0.100005 μF. For all practical purposes, the 5-pF capacitor can be ignored, leaving two 0.1-μF capacitors in series with equivalent capacitance of 0.05 μF. Combining this equivalent capacitance in parallel with the remaining 0.05-μF capacitor yields an overall equivalent capacitance of 0.1 μF. The parallel 700-μH and 300-μH inductors yield an equivalent inductance of $1/(1/700 + 1/300) = 210$ μH. This equivalent inductance is effectively in series with the 1-mH inductor at the bottom, yielding $1000 + 210 = 1210$ μH as the overall equivalent inductance.

Figure 6–28 shows the simplified equivalent circuits for each of the circuits of Figure 6–27. ∎

Exercise 6–10

The current through a series connection of two 1-μF capacitors is a rectangular pulse with an amplitude of 2 mA and a duration of 10 ms. At $t = 0$ the voltage across the first capacitor is +10 V and across the second is zero.

(a) What is the voltage across the series combination at $t = 10$ ms?
(b) What is the maximum instantaneous power delivered to the series combination?
(c) What is the energy stored on the first capacitor at $t = 0$ and $t = 10$ ms?

Answers:
(a) 50 V
(b) 100 mW at $t = 10$ ms
(c) 50 μJ and 450 μJ

DC EQUIVALENT CIRCUITS

Sometimes we need to find the dc response of circuits containing capacitors and inductors. In the first two sections of this chapter, we found that under dc conditions a capacitor acts like an open circuit and an inductor acts like a short circuit. In other words, under dc conditions, an equivalent circuit for a capacitor is an open circuit and an equivalent circuit of an inductor is a short circuit.

To determine dc responses, we replace capacitors by open circuits and inductors by short circuits and analyze the resulting resistance circuit using any of the methods in Chapters 2 through 4. The circuit analysis involves only resistance circuits and yields capacitor voltages and inductor currents along with any other variables of interest. Computer programs like SPICE use this type of dc analysis to find the initial operating point of a circuit to be analyzed. The dc capacitor voltages and inductor currents become initial conditions for a transient response that begins at $t = 0$ when something in the circuit changes, such as the position of a switch.

EXAMPLE 6–15

Determine the voltage across the capacitors and current through the inductors in Figure 6–29(a).

SOLUTION:

The circuit is driven by a 5-V dc source. Figure 6–29(b) shows the equivalent circuit under dc conditions. The current in the resulting series circuit is $5/(50 + 50) = 50$ mA. This dc current exists in both inductors, so $i_{L1} = i_{L2} = 50$ mA. By voltage division the voltage across the 50-Ω output resistor is $v = 5 \times 50/(50 + 50) = 2.5$ V; therefore, $v_{C1}(0) = 2.5$ V. The voltage across C_2 is zero because of the short circuits produced by the two inductors. ∎

Exercise 6–11

Find the OP AMP output voltage in Figure 6–30.

Answer:

$$v_O = \frac{R_2 + R_1}{R_1} V_{dc}$$

SUMMARY

- The linear capacitor and inductor are dynamic circuit elements that can store energy. The instantaneous element power is positive when they are storing energy and negative when they are delivering previously stored energy. The net energy transfer is never negative because inductors and capacitors are passive elements.

- The current through a capacitor is zero unless the voltage is changing. A capacitor acts like an open circuit to dc excitations.

- The voltage across an inductor is zero unless the current is changing. An inductor acts like a short circuit to dc excitations.

- Capacitor voltage and inductor current are called state variables because they define the energy state of a circuit. Circuit state variables are continuous functions of time as long as the circuit driving forces are finite.

- OP AMP capacitor circuits perform signal integration or differentiation. These operations, together with the summer and gain functions, provide the building blocks for designing dynamic input-output characteristics.

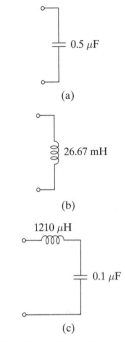

(a)

(b)

(c)

FIGURE 6–28

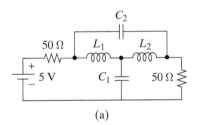

(a)

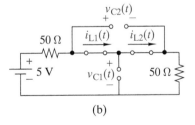

(b)

FIGURE 6–29

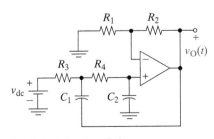

FIGURE 6–30

• Capacitors or inductors in series or parallel can be replaced with an equivalent element found by adding the individual capacitances or inductances or their reciprocals. The dc response of a dynamic circuit can be found by replacing all capacitors with open circuits and all inductors with short circuits.

PROBLEMS

ERO 6–1 CAPACITOR AND INDUCTOR RESPONSES (SECTS. 6–1, 6–2)

(a) Given the current through a capacitor or an inductor, find the voltage across the element.
(b) Given the voltage across a capacitor or an inductor, find the current through the element.
(c) Find the power and energy associated with a capacitor or inductor.

See Examples 6–1, 6–2, 6–3, 6–4, 6–6, 6–7, 6–8 and Exercises 6–1, 6–2, 6–3, 6–4, 6–5, 6–6

6–1 For $t \geq 0$ the voltage across a 2-μF capacitor is $v_C(t) = 3\,e^{-4000t}$ V. Derive expressions for $i_C(t)$, $p_C(t)$, and $w_C(t)$. Is the capacitor absorbing power, delivering power, or both?

6–2 A voltage of $15\cos(2\pi 1000t)$ appears across a 3.3-μF capacitor. Find the energy stored on the capacitor at $t = 0.5, 0.75,$ and 1 ms.

6–3 For $t \geq 0$ the current through a 100-mH inductor is $i_L(t) = 30\,e^{-4000t}$ mA. Derive expressions for $v_L(t)$, $p_L(t)$, and $w_L(t)$. Is the inductor absorbing power, delivering power, or both?

6–4 For $t \geq 0$ the current through a 25-μF capacitor is $i_C(t) = 50[u(t) - u(t - 5\times10^{-3})]$ mA. Prepare sketches of $v_C(t)$ and $p_C(t)$. Is the capacitor absorbing power, delivering power, or both?

6–5 At $t = 0$ the voltage across a 200-nF capacitor is $v_C(0) = 30$ V. For $t > 0$, the current through the capacitor is $i_C(t) = 0.4\cos 10^5 t$ A. Derive an expression for $v_C(t)$.

6–6 The voltage across a 0.5-μF capacitor is shown in Figure P6–6. Prepare sketches of $i_C(t)$, $p_C(t)$, and $w_C(t)$. Is the capacitor absorbing power, delivering power, or both?

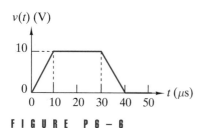

$v(t)$ (V)

FIGURE P6–6

6–7 The current through a 10-nF capacitor is shown in Figure P6–7. Given that $v_C(0) = -5$ V, find the value of $v_C(t)$ at $t = 5, 10,$ and 20 μs.

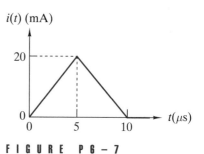

$i(t)$ (mA)

FIGURE P6–7

6–8 At $t = 0$ the current through a 10-mH inductor is $i_L(0) = 1$ mA. For $t \geq 0$ the voltage across the inductor is $v_L(t) = 50e^{-5000t}$ mV. Derive an expression for $i_L(t)$. Is the inductor absorbing power, delivering power, or both?

6–9 For $t > 0$ the current through a 500-mH inductor is $i_L(t) = 10\,e^{-2000t}\sin 1000t$ mA. Derive expressions for $v_L(t)$, $p_L(t)$, and $w_L(t)$. Is the inductor absorbing power, delivering power, or both?

6–10 A voltage $v_L(t) = 5\cos(1000t) - 2\sin(3000t)$ V appears across a 50-mH inductor. Derive an expression for $i_L(t)$. Assume $i_L(0) = 0$. Discuss the effect of frequency on the relative amplitudes of the sinusoidal components in $v_L(t)$ and $i_L(t)$.

6–11 For $t > 0$ the voltage across a 50-nF capacitor is $v_C(t) = -100e^{-1000t}$ V. Derive an expression for $i_C(t)$. Is the capacitor absorbing power, delivering power, or both?

6–12 The current through a 25-mH inductor is shown in Figure P6–7. Prepare sketches of $v_L(t)$, $p_L(t)$, and $w_L(t)$.

6–13 The voltage across a 100-μH inductor is shown in Figure P6–6. The inductor current is observed to be zero at $t = 5$ μs. What is the value of $i_L(0)$?

6–14 The capacitor in Figure P6–14 carries an initial voltage $v_C(0) = 25$ V. At $t = 0$, the switch is closed, and thereafter the voltage across the capacitor is $v_C(t) = 50 - 25\,e^{-5000t}$ V. Derive expressions for $i_C(t)$ and $p_C(t)$ for $t > 0$. Is the capacitor absorbing power, delivering power, or both?

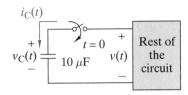

FIGURE P6-14

6-15 The capacitor in Figure P6–14 carries an initial voltage $v_C(0) = 0$ V. At $t = 0$, the switch is closed, and thereafter the voltage across the capacitor is $v_C(t) = 10 (1 - e^{-1000t})$ V. Derive expressions for $i_C(t)$ and $p_C(t)$ for $t > 0$. Is the capacitor absorbing or delivering power?

6-16 The inductor in Figure P6–16 carries an initial current of $i_L(0) = 20$ mA. At $t = 0$, the switch opens, and thereafter the current into the rest of the circuit is $i(t) = -20 e^{-500t}$ mA. Derive expressions for $v_L(t)$ and $p_L(t)$ for $t > 0$. Is the inductor absorbing or delivering power?

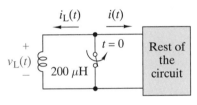

FIGURE P6-16

6-17 The inductor in Figure P6–16 carries an initial current of $i_L(0) = -20$ mA. At $t = 0$, the switch opens, and thereafter the current into the rest of the circuit is $i(t) = 40 - 20 e^{-500t}$ mA. Derive expressions for $v_L(t)$ and $p_L(t)$ for $t > 0$. Is the inductor absorbing or delivering power?

6-18 A 2.2-μF capacitor is connected in series with a 200-Ω resistor. The voltage across the capacitor is $v_C(t) = 10 \cos(2000t)$ V. What is the voltage across the resistor?

6-19 A 300-mH inductor is connected in parallel with a 10-kΩ resistor. The current through the inductor is $i_L(t) = 10 e^{-1000t}$ mA. What is the current through the resistor?

6-20 For $t > 0$ the current through a 2-mH inductor is $i_L(t) = 100te^{-1000t}$ A. Derive an expression for $v_L(t)$. Is the inductor absorbing power or delivering power at $t = 0.5$, 1, and 2 ms?

ERO 6-2 DYNAMIC OP AMP CIRCUITS (SECT. 6-3)

(a) Given a circuit consisting of resistors, capacitors, and OP AMPs, determine its input–output relationship and use the relationship to find the output for specified inputs.

(b) Design an OP AMP circuit to implement a given input–output relationship or a block diagram.

See Examples 6–9, 6–10, 6–11, 6–12 and Exercises 6–7, 6–8, 6–9

6-21 Find the input–output relationship of the OP AMP circuit in Figure P6–21. *Hint*: the voltage at the inverting input is $v_N(t) = v_S(t)$.

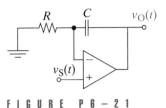

FIGURE P6-21

6-22 Show that the OP AMP capacitor circuit in Figure P6–22 is a noninverting integrator whose input–output relationship is

$$v_O(t) = \frac{1}{RC} \int_0^t v_S(x)dx$$

Hint: By voltage division the voltage at the inverting input is $v_N(t) = v_O(t)/2$.

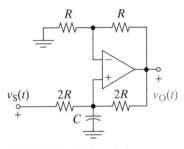

FIGURE P6-22

6-23 Find the input–output relationship of the OP AMP circuit in Figure P6–23. *Hint*: the circuit has two inputs.

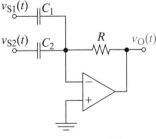

FIGURE P6-23

6-24 In Figure P6–24 the voltage across the capacitor at $t = 0$ is such that $v_O(0) = -10$ V. The input signal is $v_S(t) = 5u(t)$ V. Derive an equation for the output voltage for the OP AMP in its linear range. If the OP AMP saturates at ±15 V, find the time when the OP AMP saturates.

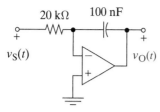

FIGURE P6-24

6–25 In Figure P6–24 the voltage across the capacitor at $t = 0$ is 0 V. The input signal is a rectangular pulse $v_S(t) = 5[u(t) - u(t - T)]$ V. The OP AMP saturates at ±15 V. What is the maximum pulse duration for linear operation?

6–26 At $t = 0$ the voltage across the capacitor in Figure P6–24 is zero. The OP AMP saturates at ±15 V. For $v_S(t) = 5[\sin \omega t]u(t)$ V, derive an expression for the output voltage for the OP AMP in its linear range. What is the minimum value of ω for linear operation?

6–27 The input to the circuit in Figure P6–27 is $v_S(t) = V_A[\sin 10^6 t]u(t)$ V. Derive an expression for the output voltage for the OP AMP in its linear range. The OP AMP saturates at ±15 V. What is the maximum value of V_A for linear operation?

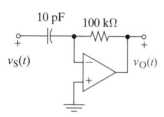

FIGURE P6-27

6–28 The input to the circuit in Figure P6–27 is $v_S(t) = 5[\sin \omega t]u(t)$ V. Derive an expression for the output voltage for the OP AMP in its linear range. The OP AMP saturates at ±15 V. What is the maximum value of ω for linear operation?

6–29 The input to the circuit in Figure P6–27 is $v_S(t) = 5[e^{-\alpha t}]u(t)$ V. Derive an expression for the output voltage for the OP AMP in its linear range. If the OP AMP saturates at ±15 V, what is the maximum value of α for linear operation?

6–30 Construct a block diagram for the following input–output relationship:

$$v_O(t) = -5v_S(t) - 20 \int_0^t v_S(x)dx$$

6–31 Construct a block diagram for the following input–output relationship:

$$v_O(t) = 5v_S(t) + \frac{1}{20} \frac{dv_S(t)}{dt}$$

6–32 What is the input–output relationship of the block diagram in Figure P6–32.

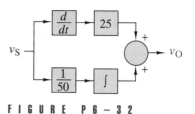

FIGURE P6-32

6–33 **D** Design an OP AMP circuit to implement the input–output relationship in Problem 6–30.

6–34 **D** Design an OP AMP circuit to implement the input–output relationship in Problem 6–31.

6–35 **D** Design OP AMP *RC* circuits to implement the input–output relationship of the block diagram in Figure P6–32.

ERO 6–3 EQUIVALENT INDUCTANCE AND CAPACITANCE (SECT. 6–4)

(a) Derive equivalence properties of inductors and capacitors or use equivalence properties to simplify *LC* circuits.
(b) Solve for currents and voltages in *RLC* circuits with dc input signals.

See Examples 6–14, 6–15, and Exercise 6–10, 6–11

6–36 Find a single equivalent element for each circuit in Figure P6–36.

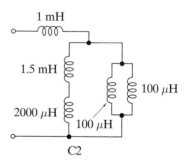

FIGURE P6-36

6–37 Verify Eqs. (6–30) and (6–31).

6–38 A 3-mH inductor is connected in series with a 100-μH inductor and the combination connected in parallel with a 12-mH inductor. Find the equivalent inductance of the connection.

6–39 A series connection of a 3.3-μF capacitor and a 4.7-μF capacitor is connected in parallel with a series connection of two 6.8-μF capacitors. Find the equivalent capacitance of the connection.

6–40 Find the equivalent capacitance between terminals A and B in Figure P6–40.

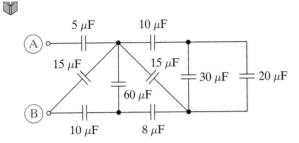

F I G U R E   P 6 – 4 0

6–41 Figure P6–41 is the equivalent circuit of a two-wire feed through capacitor.
(a) What is the capacitance between terminal 1 and ground?
(b) What is the capacitance between terminal 1 and ground when terminal 2 is grounded?

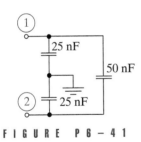

F I G U R E   P 6 – 4 1

6–42 🖻 A capacitor bank is required that can be charged to 5 kV and store at least 250 J of energy. Design a series/parallel combination that meets the voltage and energy requirements using 20-μF capacitors each rated at 1.5 kV max.

6–43 Figure P6–43 shows a power inductor package containing two identical inductors. When terminal 2 is connected to terminal 4 the inductance between terminals 1 and 3 is 260-μH.
(a) What is the inductance between terminals 1 and 3 when terminal 1 is connected to terminal 4?
(b) What is the inductance between terminals 1 and 3 when terminal 1 is connected to terminal 4 and terminal 2 is connected to terminal 3?

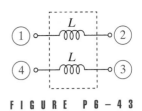

F I G U R E   P 6 – 4 3

6–44 The circuits in Figure P6–44 are driven by dc sources. Find the voltage across capacitors and the current through inductors.

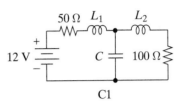

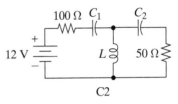

F I G U R E   P 6 – 4 4

6–45 The OP AMP circuits in Figure P6–45 are driven by dc sources. Find the output voltage v_O.

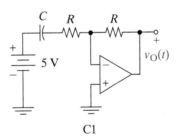

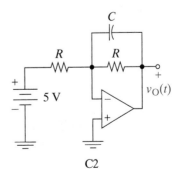

F I G U R E   P 6 – 4 5

INTEGRATING PROBLEMS

6–46 ▲ CAPACITIVE DISCHARGE PULSER

A capacitor bank for a large pulse generator consists of 11 capacitor strings connected in parallel. Each string consists of 16 1.5-mF capacitors connected in series. The purpose of this problem is to calculate important characteristics of the pulser.

(a) What is the total equivalent capacitance of the bank?

(b) If each capacitor in a series string is charged to 300 V, what is the total energy stored in the bank?

(c) In the discharge mode the voltage across the capacitor bank is $v(t) = 4.8[e^{-500t}]u(t)$ kV. What is the peak power delivered by the capacitor bank?

(d) For practical purposes the capacitor bank is completely discharged after about five time constants. What is the average power delivered during that interval?

6–47 ▲ LC CIRCUIT RESPONSE

In the circuit shown in Figure P6–47 the initial capacitor voltage is $v_C(0) = 30$ V. At $t = 0$ the switch is closed, and thereafter the current into the rest of the circuit is

$$i(t) = 2(e^{-2000t} - e^{-8000t}) \quad \text{A}$$

The purpose of this problem is to find the voltage $v(t)$ and the equivalent resistance looking into the rest of the circuit.

(a) Use the inductor's i–v characteristic to find $v_L(t)$ for $t \geq 0$. Find the value of $v_L(t)$ at $t = 0$.

(b) Use the capacitor's i–v characteristic to find $v_C(t)$ for $t \geq 0$. Find the value of $v_C(t)$ at $t = 0$. Does this value agree with the initial condition given in the problem statement? If not, you need to review your work to find the error.

(c) Use KVL and the results from (a) and (b) to find the voltage $v(t)$ delivered to the rest of the circuit. What is the value of $v(t)$ at $t = 0$?

(d) The $v(t)$ found in (c) should be proportional to the $i(t)$ given in the problem statement. If so, what is the equivalent resistance looking into the rest of the circuit?

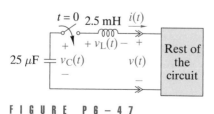

FIGURE P 6 – 4 7

6–48 ▲ CAPACITOR MULTIPLIER

It is claimed that the i–v characteristic at the input interface in Figure P6–48 is

$$i(t) = C_{EQ} \frac{dv(t)}{dt}$$

where $C_{EQ} = (1 + R_2/R_1)C$. Since $C_{EQ} > C$ the circuit is called a capacitor multiplier.

(a) Prove or disprove this claim.

(b) If the initial capacitor voltage $v_C(0) = V_0$, what is the initial value of the input voltage $v(0)$?

(c) If the initial energy stored in the capacitor is $w_C(0) = W_0$, what is the initial energy stored in equivalent capacitor C_{EQ}?

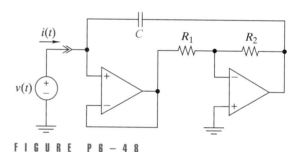

FIGURE P 6 – 4 8

6–49 ▲ INDUCTOR SIMULATION CIRCUIT

Circuits that simulate an inductor can be produced using resistors, OP AMPs, and a capacitor. It is claimed that the i–v characteristic at the input interface in Figure P6–49 is

$$i(t) = \frac{1}{L_{EQ}} \int_0^t v(x)\, dx$$

where $L_{EQ} = R_1 R_2 C$.

(a) Prove or disprove this claim.

(b) If the initial capacitor voltage $v_C(0) = V_0$, what is the initial value of the input current $i(0)$?

(c) If the initial energy stored in the capacitor is $w_C(0) = W_0$, what is the initial energy stored in equivalent inductor L_{EQ}?

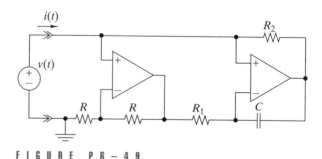

FIGURE P 6 – 4 9

6–50 Ⓐ Ⓓ Ⓔ RC OP AMP Circuit Design

An upgrade to one of your company's robotics products requires a proportional plus integral compensator that implements the input–output relationship

$$v_O(t) = v_S(t) + 50 \int_0^t v_S(x)\, dx$$

The input voltage $v_S(t)$ comes from an OP AMP, and the output voltage $v_O(t)$ drives a 10-kΩ resistive load. As the junior engineer in the company, you have been given the responsibility of developing a preliminary design.

(a) Ⓓ Design a circuit that implements the relationship using the standard OP AMP building blocks in Figure 6–17. Minimize the parts count in your design.

(b) Ⓐ The RonAl Corporation (founded by two well-known authors) has given you an unsolicited proposal claiming that their standard DIFF AMP-10 product can realize the required relationship. Their proposal is shown in Figure P6-50. Verify their claim.

(c) Ⓔ Compare your design in part (a) with the RonAl proposal in part (b) in terms of the total part costs. Part costs in your company are as follows: resistors, $400 per 10,000, capacitors, $1,500 per 10,000; and OP AMPs, $2500 per 10,000. RonAl's proposal offers to provide the company's standard DIFF AMP product for $5000 per 10,000. Note that for the RonAl product you must supply the capacitor shown in Figure P6–50.

RonAl Corporation

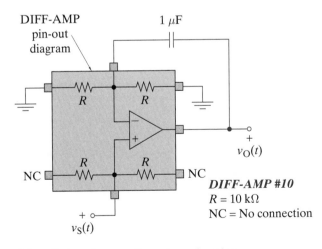

DIFF-AMP pin-out diagram

1 μF

$v_O(t)$

$v_S(t)$

DIFF-AMP #10
$R = 10\ \text{k}\Omega$
NC = No connection

Note: External connections shown in color

FIGURE P6–50

6–51 Ⓐ Super Capacitor

Super capacitors have very large capacitance (typically from 0.1 to 50 F), very long charge holding times, and small sizes making them useful in nonbattery backup power applications. The circuit in Figure P6–51 is a standard method of measuring the capacitance of such a device. Initially the switch is held in Position A until the unknown capacitance is charged to a specified voltage $v_C = V_0$. At $t = 0$ the switch is moved to Position B and thereafter the load resistor R_L is continuously adjusted to maintain a constant discharge current $i_D = I_0 > 0$. The capacitor voltage is monitored until at time $t = T_1$ it decreases to a prescribed level $v_C = V_1$. For $V_0 = 5.5$ V, $V_1 = 3$ V, $I_0 = 1$ mA, and $T_1 = 3000$ s,

(a) Find the time derivative of the capacitor voltage for $0 < t < T_1$.

(b) Use the result in (a) to find the capacitance.

(c) Using the result in (b) calculate the amount of energy dissipated in R_L between $t = 0$ and $t = 3000$ s. *Caution:* R_L is not constant.

(d) Suppose the 1 mA constant current discharge continues after $t = 3000$ s. At $t = 4200$ s the capacitor voltage decreases to $v_C = 2$ V. Are these results consistent with the capacitance found in part (b)?

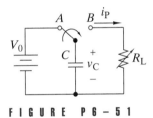

FIGURE P6–51

6–52 Ⓓ Differentiator Design

The input to the OP AMP differentiator in Figure 6–17 is a sinusoid with a peak-to-peak amplitude of 10 V. Select the values of R and C so that the OP AMP operates in its linear mode for all input frequencies less than 1 kHz. Assume the OP AMP saturates at ±15 V.

CHAPTER 7

FIRST-ORDER CIRCUITS

"The most important topic to be taught is that of the time constant, and so we have stressed first-order circuits."[1]

M. E. Van Valkenburg
American Engineer

7-1 *RC* and *RL* Circuits

7-2 First-order Circuit Step Response

7-3 Initial and Final Conditions

7-4 First-order Circuit Sinusoidal Response

Summary

Problems

Integrating Problems

Mac Elwyn Van Valkenburg (1921–1997) was a world-renowned engineering educator, scholar and curricular innovator. His most lasting legacy is the number of students influenced by his seven textbooks, most notably his classic *Network Analysis* first published in 1955. Over the next twenty years this book appeared in three editions and profoundly influenced the way undergraduate circuit analysis is taught around the world.

In this chapter we begin the analysis of circuits containing energy stor-

---

1 Quotation from M. E. Van Valkenburg, *Linear Circuits,* Englewood Cliffs, NJ: Prentice Hall, 1982, p. xii.

age elements. In an analysis problem the circuit is initially in equilibrium in a known state. At an instant of time designated $t = 0$, this equilibrium is disturbed by an event such as the opening or closing of a switch. The objective of the analysis is to find mathematical expressions for the circuit voltages and currents that are valid for $t \geq 0$. Finding these expressions involves solving a differential equation that governs the circuit response. Once these responses are understood, they become the design tools used to create dynamic circuits with prescribed responses.

In this chapter we restrict ourselves to circuits containing the equivalent of one capacitor or one inductor. These *RC* (resistor-capacitor) and *RL* (resistor-inductor) circuits are described by first-order differential equations, so they are also called *first-order circuits*. We begin by showing how to derive the first-order differential equation for an *RC* or *RL* circuit. We then solve this equation to develop the concepts of natural response, forced response, and complete response. The key response parameter is the circuit time constant which provides a measure of the time scale for the natural response. We conclude by finding the first-order circuit response for a sinusoidal input.

7–1   *R C* A N D   *R L*   C I R C U I T S

The flow diagram in Figure 7–1 shows the two major steps in the analysis of a dynamic circuit. In the first step we use device and connection equations to formulate a differential equation describing the circuit. In the second step we solve the differential equation to find the circuit response. In this chapter we examine basic methods of formulating circuit differential equations and the time-honored, classical methods of solving for responses. Solving for the responses of simple dynamic circuits gives us insight into the physical behavior of the basic modules of the complex networks in subsequent chapters. This insight will help us correlate circuit behavior with the results obtained by other methods of dynamic circuit analysis.

FORMULATING *RC* AND *RL* CIRCUIT EQUATIONS

RC and *RL* circuits contain linear resistors and a single capacitor or a single inductor. Figure 7–2 shows how we can divide *RC* and *RL* circuits into two parts: (1) the dynamic element and (2) the rest of the circuit, containing only linear resistors and sources. To formulate the equation governing either of these circuits, we replace the resistors and sources by their Thévenin and Norton equivalents shown in Figure 7–2.

Dealing first with the *RC* circuit in Figure 7–2(a), we note that the Thévenin equivalent source is governed by the constraint

$$R_T i(t) + v(t) = v_T(t) \tag{7–1}$$

The capacitor *i–v* constraint is

$$i(t) = C \frac{dv(t)}{dt} \tag{7–2}$$

Substituting the *i–v* constraint into the source constraint yields

$$R_T C \frac{dv(t)}{dt} + v(t) = v_T(t) \tag{7–3}$$

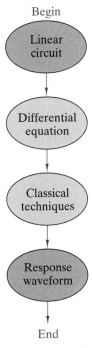

Begin

Linear circuit

Differential equation

Classical techniques

Response waveform

End

FIGURE   7 – 1   *Flow diagram for dynamic circuit analysis.*

FIGURE 7-2 *First-order circuits:*
(a) RC *circuit. (b)* RL *circuit.*

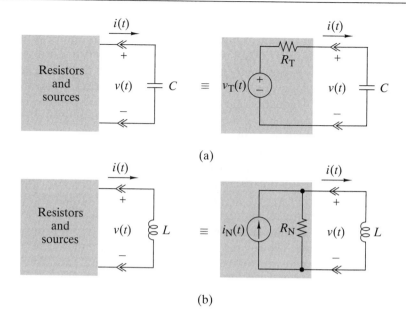

(a)

(b)

Combining the source and element constraints produces the equation governing the *RC* series circuit. The unknown in Eq. (7–3) is the capacitor voltage $v(t)$ that determines the amount of energy stored in the *RC* circuit and is referred to as the **state variable**.

Mathematically, Eq. (7–3) is a first-order linear differential equation with constant coefficients. The equation is first order because the first derivative of the dependent variable is the highest order derivative in the equation. The product R_TC is a constant coefficient because it depends on fixed circuit parameters. The signal $v_T(t)$ is the Thévenin equivalent of the independent sources driving the circuit. The voltage $v_T(t)$ is the input, and the capacitor voltage $v(t)$ is the circuit response.

The Norton equivalent source in the *RL* circuit in Figure 7–2(b) is governed by the constraint

$$G_Nv(t) + i(t) = i_N(t) \tag{7–4}$$

The element constraint for the inductor can be written

$$v(t) = L\frac{di(t)}{dt} \tag{7–5}$$

Combining the element and source constraints produces the differential equation for the *RL* circuit:

$$G_NL\frac{di(t)}{dt} + i(t) = i_N(t) \tag{7–6}$$

The response of the *RL* circuit is also governed by a first-order linear differential equation with constant coefficients. The dependent variable in Eq. (7–6) is the inductor current. The circuit parameters enter as the constant product G_NL, and the driving forces are represented by a Norton equivalent current $i_N(t)$. The unknown in Eq. (7–6) is the inductor current

$i(t)$. This current determines the amount of energy stored in the *RL* circuit and is referred to as the **state variable**.

The state variables in first-order circuits are the capacitor voltage in the *RC* circuit and the inductor current in the *RL* circuit. As we will see, these state variables contain sufficient information about the past to determine future circuit responses.

We observe that Eqs. (7–3) and (7–6) have the same form. In fact, interchanging the quantities

$$R_T \leftrightarrow G_N \quad C \leftrightarrow L \quad v \leftrightarrow i \quad v_T \leftrightarrow i_N$$

converts one equation into the other. This interchange is another example of the principle of duality. Because of duality we do not need to study the *RC* and *RL* circuits as independent problems. Everything we learn by solving the *RC* circuit can be applied to the *RL* circuit as well.

We refer to the *RC* and *RL* circuits as **first-order circuits** because they are described by a first-order differential equation. The first-order differential equations in Eqs. (7–3) and (7–6) describe general *RC* and *RL* circuits shown in Figure 7–2. Any circuit containing a single capacitor or inductor and linear resistors and sources is a first-order circuit.

ZERO-INPUT RESPONSE OF FIRST-ORDER CIRCUITS

The response of a first-order circuit is found by solving the circuit differential equation. For the *RC* circuit the response $v(t)$ must satisfy the differential equation in Eq. (7–3) and the initial condition $v(0)$. By examining Eq. (7–3) we see that the response depends on three factors:

1. The inputs driving the circuit $v_T(t)$
2. The values of the circuit parameters R_T and C
3. The value of $v(t)$ at $t = 0$ (i.e., the initial condition).

The first two factors apply to any linear circuit, including resistance circuits. The third factor relates to the initial energy stored in the circuit. The initial energy can cause the circuit to have a nonzero response even when the input $v_T(t) = 0$ for $t \geq 0$. The existence of a response with no input is something new in our study of linear circuits.

To explore this discovery we find the **zero-input response**. Setting all independent sources in Figure 7–2 to zero makes $v_T = 0$ in Eq. (7–3):

$$R_T C \frac{dv}{dt} + v = 0 \tag{7–7}$$

Mathematically, Eq. (7–7) is a **homogeneous equation** because the right side is zero. The classical approach to solving a linear homogeneous differential equation is to try a solution in the form of an exponential

$$v(t) = K e^{st} \tag{7–8}$$

where K and s are constants to be determined.

The form of the homogenous equation suggests an exponential solution for the following reasons. Equation (7–7) requires that $v(t)$ plus $R_T C$ times its derivative must add to zero for all time $t \geq 0$. This can only occur if $v(t)$ and its derivative have the same form. In Chapter 5 we saw that an exponential signal and its derivative are both of the form e^{-t/T_C}. Therefore, the exponential is a logical starting place.

If Eq. (7–8) is indeed a solution, then it must satisfy the differential equation in Eq. (7–7). Substituting the trial solution into Eq. (7–7) yields

$$R_T C K s e^{st} + K e^{st} = 0$$

or

$$K e^{st}(R_T C s + 1) = 0$$

The exponential function e^{st} cannot be zero for all t. The condition $K = 0$ is a trivial solution because it implies that $v(t)$ is zero for all time t. The only nontrivial way to satisfy the equation involves the condition

$$R_T C s + 1 = 0 \qquad (7–9)$$

Equation (7–9) is the circuit **characteristic equation** because its root determines the attributes of $v(t)$. The characteristic equation has a single root at $s = -1/R_T C$ so the zero-input response of the RC circuit has the form

$$v(t) = K e^{-t/R_T C} \qquad t \geqslant 0$$

The constant K can be evaluated using the value of $v(t)$ at $t = 0$. Using the notation $v(0) = V_0$ yields

$$v(0) = K e^0 = K = V_0$$

The final form of the zero-input response is

$$v(t) = V_0 e^{-t/R_T C} \qquad t \geqslant 0 \qquad (7–10)$$

The zero-input response of the RC circuit is the familiar exponential waveform shown in Figure 7–3. At $t = 0$ the exponential response starts out at $v(0) = V_0$ and then decays to zero at $t \to \infty$. The time constant $T_C = R_T C$ depends only on fixed circuit parameters. From our study of the exponential signals in Chapter 5, we know that the $v(t)$ decays to about 37% of its initial amplitude in one time constant and to essentially zero after about five time constants. The zero-input response of the RC circuit is determined by two quantities: (1) the circuit time constant and (2) the value of the capacitor voltage at $t = 0$.

FIGURE 7–3 *First-order RC circuit zero-input response.*

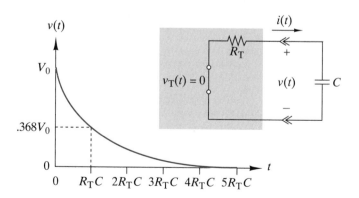

The zero-input response of the RL circuit in Figure 7–2(b) is found by setting the Norton current $i_N(t) = 0$ in Eq. (7–6).

$$G_N L \frac{di}{dt} + i = 0 \qquad (7\text{--}11)$$

The unknown in this homogeneous differential equation is the inductor current $i(t)$. Equation (7–11) has the same form as the homogeneous equation for the RC circuit, which suggests a trial solution of the form

$$i(t) = Ke^{st}$$

where K and s are constants to be determined. Substituting the trial solution into Eq. (7–11) yields the RL circuit characteristic equation.

$$G_N Ls + 1 = 0 \qquad (7\text{--}12)$$

The root of this equation is $s = -1/G_N L$. Denoting the initial value of the inductor current by I_0, we evaluate the constant K:

$$i(0) = I_0 = Ke^0 = K$$

The final form of the zero-input response of the RL circuit is

$$i(t) = I_0 e^{-t/G_N L} \qquad t \geqslant 0 \qquad (7\text{--}13)$$

For the RL circuit the zero-input response of the state variable $i(t)$ is an exponential function with a time constant of $T_C = G_N L = L/R_T$. This response connects the initial state $i(0) = I_0$ with the final state $i(\infty) = 0$.

The zero-input responses in Eqs. (7–10) and (7–13) show the duality between first-order RC and RL circuits. These results point out that the zero-input response in a first-order circuit depends on two quantities: (1) the circuit time constant and (2) the value of the state variable at $t = 0$. Capacitor voltage and inductor current are called state variables because they determine the amount of energy stored in the circuit at any time t. The following examples show that the zero-input response of the state variable provides enough information to determine the zero-input response of every other voltage and current in the circuit.

EXAMPLE 7-1

The switch in Figure 7–4 is closed at $t = 0$, connecting a capacitor with an initial voltage of 30 V to the resistances shown. Find the responses $v_C(t)$, $i(t)$, $i_1(t)$, and $i_2(t)$ for $t \geq 0$.

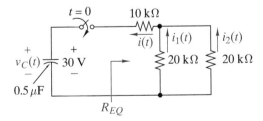

FIGURE 7 - 4

SOLUTION:
This problem involves the zero-input response of an RC circuit since there is no independent source in the circuit. To find the required responses, we first determine the circuit time constant with the switch closed ($t \geq 0$). The equivalent resistance seen by the capacitor is

$$R_{EQ} = 10 + (20\|20) = 20 \text{ k}\Omega$$

For $t \geq 0$ the circuit time constant is

$$T_C = R_T C = 20 \times 10^3 \times 0.5 \times 10^{-6} = 10 \text{ ms}$$

The initial capacitor voltage is given by $V_0 = 30$ V. Using Eq. (7–10), the zero-input response of the capacitor voltage is

$$v_C(t) = 30e^{-100t} \text{ V} \quad t \geq 0$$

The capacitor voltage provides the information needed to solve for all other zero-input responses. The current $i(t)$ through the capacitor is

$$i(t) = C\frac{dv_C}{dt} = (0.5 \times 10^{-6})(30)(-100) \, e^{-100t}$$

$$= -1.5 \times 10^{-3} \, e^{-100t} \text{ A} \quad t \geq 0$$

The minus sign means the actual current direction is opposite of the reference direction shown in Figure 7–4. The minus sign makes physical sense because the initial voltage on the capacitor is positive, which forces current into the resistances to the right of the switch. The other current responses are found by current division.

$$i_1(t) = i_2(t) = \frac{20}{20 + 20}i(t) = -0.75 \times 10^{-3}e^{-100t} \text{ A} \quad t \geq 0$$

Notice the analysis pattern. We first determine the zero-input response of the capacitor voltage. The state variable response together with resistance circuit analysis techniques were then used to find other voltages and currents. The circuit time constant and the value of the state variable at $t = 0$ provide enough information to determine the zero-input response of every voltage or current in the circuit. ∎

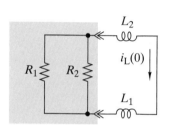

Given circuit

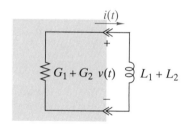

Equivalent circuit

FIGURE 7–5

EXAMPLE 7–2

Find the response of the state variable of the RL circuit in Figure 7–5 using $L_1 = 10$ mH, $L_2 = 30$ mH, $R_1 = 2$ kΩ, $R_2 = 6$ kΩ, and $i_L(0) = 100$ mA.

SOLUTION:
The inductors are connected in series and can be replaced by an equivalent inductor

$$L_{EQ} = L_1 + L_2 = 10 + 30 = 40 \text{ mH}$$

Likewise, the resistors are connected in parallel and the conductance seen by L_{EQ} is

$$G_{EQ} = G_1 + G_2 = 10^{-3}/2 + 10^{-3}/6 = 2 \times 10^{-3}/3 \text{ S}$$

Figure 7–5 shows the resulting equivalent circuit. The interface signals $v(t)$ and $i(t)$ are the voltage across and current through $L_{EQ} = L_1 + L_2$. The time constant of the equivalent RL circuit is

$$T_C = G_{EQ}L_{EQ} = 8 \times 10^{-5}/3 \text{ s} = 1/37500 \text{ s}$$

The initial current through L_{EQ} is $i_{\text{L}}(0) = 0.1$ A. Using Eq. (7–13) with $I_0 = 0.1$ yields the zero-state response of the inductor current.

$$i(t) = 0.1e^{-37500t} \text{ A} \qquad t \geqslant 0$$

Given the state variable response, we can find every other response in the original circuit. For example, by KCL and current division the currents through R_1 and R_2 are

$$i_{\text{R}_1}(t) = \frac{R_2}{R_1 + R_2}i(t) = 0.075 \, e^{-37500t} \text{ A} \qquad t \geqslant 0$$

$$i_{\text{R}_2}(t) = \frac{R_1}{R_1 + R_2}i(t) = 0.025 \, e^{-37500t} \text{ A} \qquad t \geqslant 0 \qquad \blacksquare$$

Example 7–2 illustrates an important point. The RL circuit in Figure 7–5 is a first-order circuit even though it contains two inductors. The two inductors are connected in series and can be replaced by a single equivalent inductor. In general, capacitors or inductors in series and parallel can be replaced by a single equivalent element. Thus, any circuit containing the *equivalent* of a single inductor or a single capacitor is a first-order circuit.

Sometimes it may be difficult to determine the Thévenin or Norton equivalent seen by the dynamic element in a first-order circuit. In such cases we use other circuit analysis techniques to derive the differential equation in terms of a more convenient signal variable. For example, the OP AMP RC circuit in Figure 7–6 is a first-order circuit because it contains a single capacitor.

From previous experience we know that the key to analyzing an inverting OP AMP circuit is to write a KCL equation at the inverting input. The sum of currents entering the inverting input is

$$\underbrace{G_1(v_{\text{S}} - v_{\text{N}})}_{i_1(t)} + \underbrace{G_2(v_{\text{O}} - v_{\text{N}})}_{i_2(t)} + \underbrace{C\frac{d(v_{\text{O}} - v_{\text{N}})}{dt}}_{i_{\text{C}}(t)} - i_{\text{N}}(t) = 0$$

The element equations for the OP AMP are $i_{\text{N}}(t) = 0$ and $v_{\text{N}}(t) = v_{\text{P}}(t)$. However, the noninverting input is grounded; hence $v_{\text{N}}(t) = v_{\text{P}}(t) = 0$. Substituting the OP AMP element constraints into the KCL constraint yields

$$G_1 v_{\text{S}} + G_2 v_{\text{O}} + C\frac{dv_{\text{O}}}{dt} = 0$$

which can be rearranged in standard form as

$$R_2 C\frac{dv_{\text{O}}}{dt} + v_{\text{O}} = -\frac{R_2}{R_1}v_{\text{S}}(t) \qquad (7\text{–}14)$$

The unknown Eq. (7–14) is the OP AMP output voltage rather than the capacitor voltage. The form of the differential equation indicates that the circuit time constant is $T_{\text{C}} = R_2 C$.

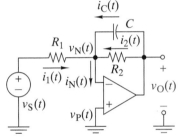

FIGURE   7 – 6   *First-order OP AMP RC circuit.*

EXAMPLE 7–3

Use Orcad Capture to calculate the response $v_C(t)$ in the circuit in Fig. 7–4 for $t \geq 0$.

SOLUTION:

An analytical solution for this problem was given in Example 7–1. The problem is repeated here to introduce computer simulation of dynamic circuits. The simulation option used is called *Time Domain (Transient)* analysis which predicts the variation of currents, voltages, and powers as a function of time. Figure 7–7 shows the circuit diagram as drawn in Orcad Capture *Schematics*. This diagram does not include the switch in Figure 7–4 since for $t \geq 0$ the switch is closed. The netlist in Figure 7–7 shows that the 0.5 μF is connected between Nodes N00102 and Node 0 (ground) and has an initial condition of IC = 30 V. The **Property Editor** dialog box in Figure 7–7 shows how the capacitor's initial condition attribute is given a numerical value. The **Property Editor** is accessed by double-clicking on the desired part—C1 in this case (highlighted by a "dashed square" in the figure). The 30-V initial condition is entered in the **IC** box, and then the "Apply" button is pressed. With capacitors and inductors simulation programs require the user to specify both the element value and the initial condition attributes.

FIGURE 7–7 *Schematic, net list, and part dialog box.*

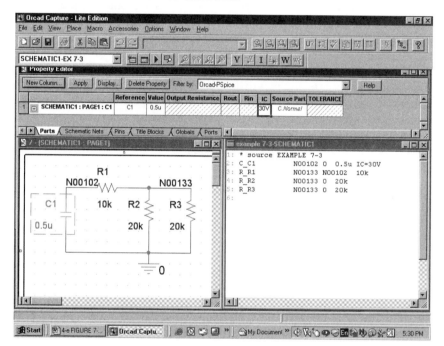

Once the circuit diagram is complete, we must set up a transient analysis run. Selecting **PSpice/New Simulation/Time Domain (Transient)** from the schematics menu bar brings up the dialog box in Figure 7–8. The key entry in this box is the **Run to time**, which specifies the time duration of simulation. A PSpice transient analysis always starts at $t = 0$ and runs to $t =$ **(TSTOP)**. In the present case we know that the circuit time constant is 10 ms, so we can safely specify **Run to time** $= 5T_C = 50$ ms.[2]

---

2 Specifying a transient analysis **Run to time** presents a dilemma. We must know something about the response to specify an appropriate simulation time. If we know everything about the response, there is no need to run the simulation. If we know absolutely nothing, we may specify a simulation time that is too short or too long to observe the response. Using circuit simulation tools effectively requires that we operate in a wide grey area between no knowledge and complete knowledge.

The **Start saving data after [ ] seconds** and the **Maximum step size: [ ] seconds** parameters in Figure 7–8 are not important to us here since we observe the calculated response using Orcad **Probe**.

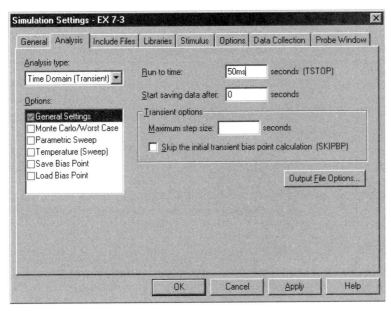

FIGURE 7 – 8 *Analysis setup.*

With the circuit defined and the transient analysis run set up, we can start Orcad Capture PSpice by selecting **PSpice/Run** from the schematics menu bar. Orcad Capture **Probe** starts automatically once PSpice successfully completes the transient simulation. The results shown in Figure 7–9 were created using the probe graphical interface. Probe first displays a blank plot with only the time axis labeled. To add circuit responses to the plot, select **Trace/Add Trace**. In this example this brings up a menu of 28 possible responses to choose from. Selecting node voltages V(N00102) and V(N00123) produce the two exponential voltage traces in Figure 7–9. We can also display current responses on the same plot. To display voltages and currents together, we first select **Plot** and **Add Y Axis**; this adds a second axis that is automatically scaled to the values of the currents. Adding an axis for currents must be done after selecting the voltage responses but before selecting the current responses. The legend at the bottom of Figure 7–9 indicates we have chosen a total of five responses—the two node voltage traces referenced to the first Y-axis and the three current traces referenced to the second Y-axis.

The node voltage V(N00102) is the voltage across the capacitor. This trace begins at 30 V and decays to zero in about 50 ms. This agrees with our analytical solution $v_C(t) = 30\,e^{-100t}$ V, $t \geq 0$. The node voltage V(N00133) is the voltage across the two 20-kΩ resistors in parallel. This exponentially decaying trace begins at 15 V and is exactly half of the capacitor voltage because of the voltage divider formed by the three resistors. The capacitor current trace I(C1) is negative, indicating that the capacitor is delivering power to the circuit. The two resistor currents I(R2) and I(R3) are positive, indicating that they are absorbing power. The traces for I(R2) and I(R3) fall on top of each other since the two

F I G U R E 7 – 9 *Response plots.*

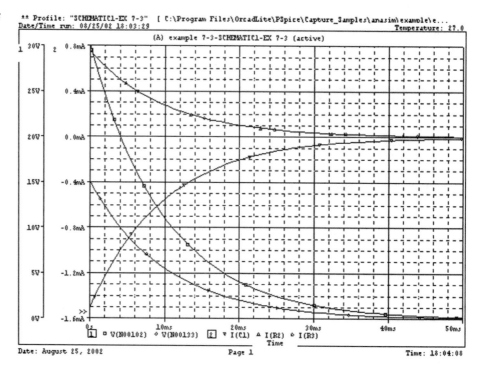

F I G U R E 7 – 9 *Response plots.*

resistors have the same value. Although not shown in the figure, the waveform of the power provided or dissipated by each element can also be displayed.

Exercise 7–1

Find the time constants of the circuits in Figure 7–10.

Answers:

C1: $\dfrac{2L}{3R}$ C2: $\dfrac{RC}{4}$ C3: $\dfrac{L}{4R}$

Exercise 7–2

The switch in Figure 7–11 closes at $t = 0$. For $t \geq 0$ the current through the resistor is $i_R(t) = e^{-100t}$ mA.

(a) What is the capacitor voltage at $t = 0$?
(b) Write an equation for $v(t)$ for $t \geq 0$.
(c) Write an equation for the power absorbed by the resistor for $t \geq 0$.
(d) How much energy does the resistor dissipate for $t \geq 0$?
(e) How much energy is stored in the capacitor at $t = 0$?

Answers:
(a) 10 V
(b) $v(t) = 10e^{-100t}$ V
(c) $p_R(t) = 10e^{-200t}$ mW
(d) 50 μJ
(e) 50 μJ

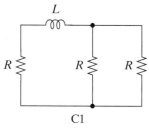

C1

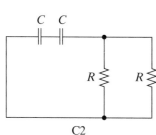

C2

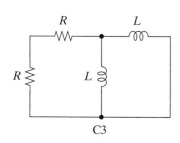

C3

F I G U R E 7 – 1 0

For $t > 0$ the current through the 40-mH inductor in a first-order circuit is $20e^{-500t}$ mA.

(a) What is the circuit time constant?
(b) How much energy is stored in the inductor at $t = 0$, $t = T_C$, and $t = 5T_C$?
(c) Write an equation for the voltage across the inductor.
(d) What is the equivalent resistance seen by the inductor?

Answers:
(a) 2 ms
(b) 8, 1.08, 0.000363 μJ
(c) $-400e^{-500t}$ mV
(d) 20 Ω

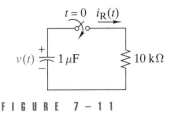

F I G U R E   7 – 1 1

7–2 FIRST-ORDER CIRCUIT STEP RESPONSE

Linear circuits are often characterized by applying step function and sinusoid inputs. This section introduces the step response of first-order circuits. Later in this chapter we treat the sinusoidal response of first-order circuits and step response of second-order circuits. The step response analysis introduces the concepts of forced, natural, and zero-state responses that appear extensively in later chapters.

Our development of first-order step response treats the RC circuit in detail and then summarizes the corresponding results for its dual, the RL circuit. When the input to the RC circuit in Figure 7–2 is a step function, we can write the Thévenin source as $v_T(t) = V_A u(t)$. The circuit differential equation in Eq. (7–3) becomes

$$R_T C \frac{dv}{dt} + v = V_A u(t) \qquad (7\text{–}15)$$

The step response is a function $v(t)$ that satisfies this differential equation for $t \geq 0$ and meets the initial condition $v(0)$. Since $u(t) = 1$ for $t \geq 0$ we can write Eq. (7–15) as

$$R_T C \frac{dv(t)}{dt} + v(t) = V_A \qquad \text{for } t \geq 0 \qquad (7\text{–}16)$$

Mathematics provides a number of approaches to solving this equation, including separation of variables and integrating factors. However, because the circuit is linear we chose a method that uses superposition to divide solution $v(t)$ into two components:

$$v(t) = v_N(t) + v_F(t) \qquad (7\text{–}17)$$

The first component $v_N(t)$ is the **natural response** and is the general solution of Eq. (7–16) when the input is set to zero. The natural response has its origin in the physical characteristic of the circuit and does not depend on the form of the input. The component $v_F(t)$ is the **forced response** and is a particular solution of Eq. (7–16) when the input is the step function. We call this the forced response because it represents what the circuit is compelled to do by the form of the input.

Finding the natural response requires the general solution of Eq. (7–16) with the input set to zero:

$$R_T C \frac{dv_N(t)}{dt} + v_N(t) = 0 \qquad t \geq 0$$

But this is the homogeneous equation that produces the zero-input response in Eq. (7–8). Therefore, we know that the natural response takes the form

$$v_N(t) = K e^{-t/R_T C} \qquad t \geq 0 \qquad (7\text{–}18)$$

This is a general solution of the homogeneous equation because it contains an arbitrary constant K. At this point we cannot evaluate K from the initial condition, as we did for the zero-input response. The initial condition applies to the total response (natural plus forced), and we have yet to find the forced response.

Turning now to the forced response, we seek a particular solution of the equation

$$R_T C \frac{dv_F(t)}{dt} + v_F(t) = V_A \qquad t \geq 0 \qquad (7\text{–}19)$$

The equation requires that a linear combination of $v_F(t)$ and its derivative equal a constant V_A for $t \geq 0$. Setting $v_F(t) = V_A$ meets this condition since $dv_F/dt = dV_A/dt = 0$. Substituting $v_F = V_A$ into Eq. (7–19) reduces it to the identity $V_A = V_A$.

Now combining the forced and natural responses, we obtain

$$v(t) = v_N(t) + v_F(t)$$
$$= K e^{-t/R_T C} + V_A \qquad t \geq 0$$

This equation is the general solution for the step response because it satisfies Eq. (7–16) and contains an arbitrary constant K. This constant can now be evaluated using the initial condition:

$$v(0) = V_0 = K e^0 + V_A = K + V_A$$

The initial condition requires that $K = (V_0 - V_A)$. Substituting this conclusion into the general solution yields the step response of the RC circuit.

$$v(t) = (V_0 - V_A) e^{-t/R_T C} + V_A \qquad t \geq 0 \qquad (7\text{–}20)$$

A typical plot of $v(t)$ is shown in Figure 7–12.

F I G U R E 7 – 1 2 *Step response of a first-order* RC *circuit.*

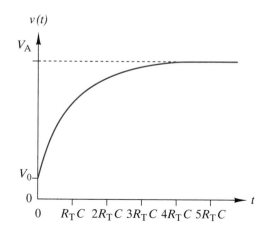

The RC circuit step response in Eq. (7–20) starts out at the initial condition V_0 and is driven to a final condition V_A, which is determined by the amplitude of the step function input. That is, the initial and final values of the response are

$$\lim_{t \to 0+} v(t) = (V_0 - V_A)e^{-0} + V_A = V_0$$
$$\lim_{t \to \infty} v(t) = (V_0 - V_A)e^{-\infty} + V_A = V_A$$

The path between the two end points is an exponential waveform whose time constant is the circuit time constant. We know from our study of exponential signals that the step response will reach its final value after about five time constants. In other words, after about five time constants the natural response decays to zero and we are left with a constant forced response caused by the step function input.

The RL circuit in Figure 7–2 is the dual of the RC circuit, so the development of its step responses follows the same pattern discussed previously. Briefly sketching the main steps, the Norton equivalent input is a step function $I_A u(t)$, and for $t \geq 0$ the RL circuit differential equation Eq. (7–6) becomes

$$G_N L \frac{di(t)}{dt} + i(t) = I_A \qquad t \geq 0 \qquad (7\text{–}21)$$

The solution of this equation is found by superimposing the natural and forced components. The natural response is the solution of the homogeneous equation [right side of Eq. (7–21) set to zero] and takes the same form as the zero-input response found in the previous section.

$$i_N(t) = K e^{-t/G_N L} \qquad t \geq 0$$

where K is a constant to be evaluated from the initial condition once the complete response is known. The forced response is a particular solution of the equation

$$G_N L \frac{di_F(t)}{dt} + i_F(t) = I_A \qquad t \geq 0$$

Setting $i_F = I_A$ satisfies this equation since $dI_A/dt = 0$.

Combining the forced and natural responses, we obtain the general solution of Eq. (7–21) in the form

$$i(t) = i_N(t) + i_F(t)$$
$$= K e^{-t/G_N L} + I_A \qquad t \geq 0$$

The constant K is now evaluated from the initial condition:

$$i(0) = I_0 = K e^{-0} + I_A = K + I_A$$

The initial condition requires that $K = I_0 - I_A$, so the step response of the RL circuit is

$$i(t) = (I_0 - I_A)e^{-t/G_N L} + I_A \qquad t \geq 0 \qquad (7\text{–}22)$$

The RL circuit step response has the same form as the RC circuit step response in Eq. (7–20). At $t = 0$ the starting value of the response is $i(0) = I_0$, as required by the initial condition. The final value is the forced response $i(\infty) = i_F = I_A$, since the natural response decays to zero as time increases.

A step function input to the RC or RL circuit drives the state variable from an initial value determined by what happened prior to $t = 0$ to a final value determined by amplitude of the step function applied at $t = 0$. The time needed to transition from the initial to the final value is about $5T_C$, where T_C is the circuit time constant. We conclude that the step response of a first-order circuit depends on three quantities:

1. The amplitude of the step input (V_A or I_A)
2. The circuit time constant ($R_T C$ or $G_N L$)
3. The value of the state variable at $t = 0$ (V_0 or I_0).

EXAMPLE 7–4

Find the response of the RC circuit in Figure 7–13.

SOLUTION:

The circuit is first order, since the two capacitors in series can be replaced by a single equivalent capacitor

$$C_{EQ} = \frac{1}{\dfrac{1}{C_1} + \dfrac{1}{C_2}} = 0.0833 \ \mu F$$

The initial voltage on C_{EQ} is the sum of the initial voltages on the original capacitors.

$$V_0 = V_{01} + V_{02} = 5 + 10 = 15 \text{ V}$$

To find the Thévenin equivalent seen by C_{EQ}, we first find the open-circuit voltage. Disconnecting the capacitors in Figure 7–13 and using voltage division at the interface yields

$$v_T = v_{OC} = \frac{R_2}{R_1 + R_2} V_A u(t) = \frac{10}{40}100u(t) = 25u(t) \text{ V}$$

Replacing the voltage source by a short circuit and looking to the left at the interface, we see R_1 in parallel with R_2. The Thévenin resistance of this combination is

$$R_T = \frac{1}{\dfrac{1}{R_1} + \dfrac{1}{R_2}} = 7.5 \text{ k}\Omega$$

The circuit time constant is

$$T_C = R_T C_{EQ} = (7.5 \times 10^3)(8.33 \times 10^{-8}) = \frac{1}{1600} \text{ s}$$

For the Thévenin equivalent circuit, the initial capacitor voltage is $V_0 = 15$ V, the step input is $25u(t)$, and the time constant is $1/1600$ s. Using the RC circuit step response in Eq. (7–20) yields

$$v(t) = (15 - 25) e^{-1600t} + 25$$

$$= 25 - 10 e^{-1600t} \text{ V} \qquad t \geq 0$$

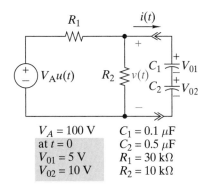

$V_A = 100$ V	$C_1 = 0.1 \ \mu F$
at $t = 0$	$C_2 = 0.5 \ \mu F$
$V_{01} = 5$ V	$R_1 = 30$ kΩ
$V_{02} = 10$ V	$R_2 = 10$ kΩ

FIGURE 7–13

The initial ($t = 0$) value of $v(t)$ is $25 - 10 = 15$ V, as required. The equivalent capacitor voltage is driven to a final value of 25 V by the step input in the Thévenin equivalent circuit. For practical purposes, $v(t)$ reaches 25 V after about $5T_C = 3.125$ ms. ■

EXAMPLE 7–5

Find the step response of the RL circuit in Figure 7–14(a). The initial condition is $i(0) = I_0$.

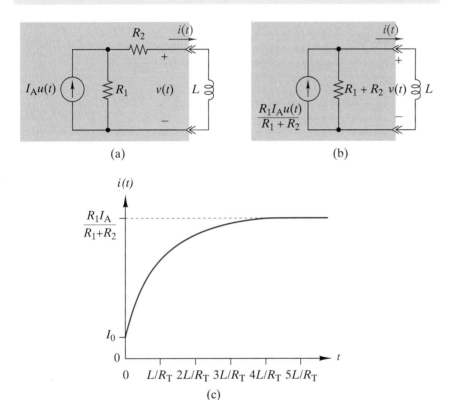

FIGURE 7 – 14

(a)

(b)

(c)

SOLUTION:

We first find the Norton equivalent to the left of the interface. By current division, the short-circuit current at the interface is

$$i_{SC}(t) = \frac{R_1}{R_1 + R_2} I_A u(t)$$

Looking to the left at the interface with the current source off (replaced by an open circuit), we see R_1 and R_2 in series producing a Thévenin resistance

$$R_T = \frac{1}{G_N} = R_1 + R_2$$

The time constant of the Norton equivalent circuit in Figure 7–14(b) is

$$T_C = G_N L = \frac{L}{(R_1 + R_2)}$$

The natural response of the Norton equivalent circuit is

$$i_N(t) = Ke^{-(R_1+R_2)t/L} \qquad t \geq 0$$

The short-circuit current $i_{SC}(t)$ is the step function input in the Norton circuit. Therefore, the forced response is

$$i_F(t) = i_{SC}(t) = \frac{R_1}{(R_1+R_2)}I_A u(t)$$

Superimposing the natural and forced responses yields

$$i(t) = Ke^{-(R_1+R_2)t/L} + \frac{R_1 I_A}{R_1+R_2} \qquad t \geq 0$$

The constant K can be evaluated from the initial condition:

$$i(0) = I_0 = K + \frac{R_1 I_A}{R_1+R_2}$$

which requires that

$$K = I_0 - \frac{R_1 I_A}{R_1+R_2}$$

So circuit step response is

$$i(t) = \left[I_0 - \frac{R_1 I_A}{R_1+R_2}\right]e^{-(R_1+R_2)t/L} + \frac{R_1 I_A}{R_1+R_2} \qquad t \geq 0$$

An example of this response is shown in Figure 7–14(c). ■

EXAMPLE 7–6

The state variable response of a first-order RC circuit for a step function input is

$$v_C(t) = 20e^{-200t} - 10 \text{ V} \qquad t \geq 0$$

(a) What is the circuit time constant?
(b) What is the initial voltage across the capacitor?
(c) What is the amplitude of the forced response?
(d) At what time is $v_C(t) = 0$?

SOLUTION:
(a) The natural response of a first-order circuit is of the form Ke^{-t/T_C}. Therefore, the time constant of the given responses is $T_C = 1/200 = 5$ ms.
(b) The initial ($t = 0$) voltage across the capacitor is

$$v_C(0) = 20e^{-0} - 10 = 20 - 10 = 10 \text{ V}$$

(c) The natural response decays to zero, so the forced response is the final value $v_C(t)$.

$$v_C(\infty) = 20e^{-\infty} - 10 = 0 - 10 = -10 \text{ V}$$

(d) The capacitor voltage must pass through zero at some intermediate time, since the initial value is positive and the final value negative. This time is found by setting the step response equal to zero:

$$20e^{-200t} - 10 = 0$$

which yields the condition $e^{200t} = 2$ or $t = \ln 2/200 = 3.47$ ms.   ∎

Exercise 7–4

Given the following first-order circuit step response

$$v_C(t) = 20 - 20e^{-1000t} \text{ V} \quad t \geq 0$$

(a) What is the amplitude of the step input?
(b) What is the circuit time constant?
(c) What is the initial value of the state variable?
(d) What is the circuit differential equation?

Answers:
(a) 20 V
(b) 1 ms
(c) 0 V
(d) $10^{-3} \, dv_C/dt + v_C = 20u(t)$

Exercise 7–5

Find the solution of the following first-order differential equations:

(a) $10^{-4}\dfrac{dv_C}{dt} + v_C = -5u(t)$ $\qquad v_C(0) = 5$ V

(b) $5 \times 10^{-2}\dfrac{di_L}{dt} + 2000 \, i_L = 10u(t)$ $\qquad i_L(0) = -5$ mA

Answers:
(a) $v_C(t) = -5 + 10e^{-10000t}$ V $\qquad t \geq 0$
(b) $i_L(t) = 5 - 10e^{-40000t}$ mA $\qquad t \geq 0$

ZERO-STATE RESPONSE

Additional properties of dynamic circuit responses are revealed by rearranging the RC and RL circuit step responses in Eqs. (7–20) and (7–22) in the following way:

$$RC \text{ circuit: } v(t) = \underbrace{V_0 e^{-t/R_TC}}_{\substack{\text{Zero-input} \\ \text{response}}} + \underbrace{V_A(1 - e^{-t/R_TC})}_{\substack{\text{Zero-state} \\ \text{response}}} \qquad t \geq 0$$

$$RL \text{ circuit: } i(t) = \overbrace{I_0 e^{-t/G_NL}} + \overbrace{I_A(1 - e^{-t/G_NL})} \qquad t \geq 0$$

We recognize the first term on the right in each equation as the zero-input response discussed in Sect. 7–1. By definition, the **zero-input response** occurs when the input is zero ($V_A = 0$ or $I_A = 0$). The second term on the right in each equation is called the **zero-state response** because this part occurs when the initial state of the circuit is zero ($V_0 = 0$ or $I_0 = 0$).

The zero-state response is proportional to the amplitude of the input step function (V_A or I_A). However, the total response (zero input plus zero state) is not directly proportional to the input amplitude. When the initial state is not zero, the circuit appears to violate the proportionality property of linear circuits. However, bear in mind that the proportionality property applies to linear circuits with only one input.

The RC and RL circuits can store energy and have memory. In effect, they have two inputs: (1) the input that occurred before $t = 0$, and (2) the step function applied at $t = 0$. The first input produces the initial energy state of the circuit at $t = 0$, and the second causes the zero-state response for $t \geq 0$. In general, for $t \geq 0$, the total response of a dynamic circuit is the sum of two responses: (1) the zero-input response caused by the initial conditions produced by inputs applied before $t = 0$, and (2) the zero-state response caused by inputs applied after $t = 0$.

APPLICATION NOTE: **EXAMPLE 7–7**

The operation of a digital system is controlled by a clock waveform that provides a standard timing reference. At its source a clock waveform can be described by a rectangular pulse of the form

$$v_S(t) = V_A \left[u(t) - u(t - T) \right]$$

In this example the pulse amplitude is $V_A = 5$ V and the pulse duration is $T = 10$ ns. This clock pulse drives a digital device that can be modeled by the circuit in Figure 7–15(a). In this model $v_S(t)$ is the rectangular clock pulse defined above and $v(t)$ is the clock waveform as received at the input to the digital device. The presence of a clock pulse at the device input will be detected only if $v(t)$ exceeds a specified logic "1" threshold level.

Find the zero-state response of the voltage $v(t)$ when $RC = 10$ ns. Will the clock pulse be detected if the logic "1" threshold level is 3.7 V?

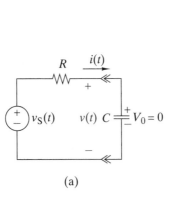

(a)

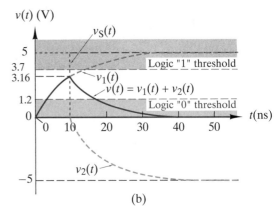

(b)

FIGURE 7–15

SOLUTION:

The rectangular pulse input $v_S(t)$ is indicated by dashed lines in Figure 7–15(b). The initial capacitor voltage is zero since we seek the zero-state response. The total response can be found as the sum of the zero-state responses cause by two inputs:

1. A positive 5-V step function applied at $t = 0$.

2. A negative 5-V step function applied at $t = 10$ ns.

The first input causes a zero-state response of

$$v_1(t) = V_A(1 - e^{-t/RC})u(t)$$
$$= 5(1 - e^{-10^8 t})u(t)$$

The second input causes a zero-state response of

$$v_2(t) = -V_A(1 - e^{-(t-T)/RC})u(t - T)$$
$$= -5[1 - e^{-10^8(t-10^{-8})}]u(t - 10^{-8})$$

Notice that $v_2(t) = -v_1(t - T)$, that is, $v_2(t)$ is obtained by inverting and delaying $v_1(t)$ by $T = 10$ ns. The total response is the superposition of these two responses.

$$v(t) = v_1(t) + v_2(t)$$

Figure 7–15(b) shows how the two responses combine to produce the overall pulse response of the circuit. The response $v_1(t)$ begins at zero and eventually reaches a final value of +5 V. At $t = T = 10$ ns the first response reaches $v_1(T) = 5(1 - e^{-1}) = 3.16$ V. The second response $v_2(t)$ begins at $t = T = 10$ ns, and thereafter is equal and opposite to $v_1(t)$ except that it is delayed by $T = 10$ ns. The net result is that the total response reaches a maximum of $v(T) = 3.16$ V. In this example the clock pulse will not be detected since the logic "1" threshold level is 3.7 V. Clock pulse detection would be made possible by increasing the pulse duration so that $v(T) > 3.7$ V. This requires that

$$5(1 - e^{-10^8 T}) > 3.7 \quad V$$

or $1.3 > 5e^{-10^8 T}$ which yields $T > 1.347 \times 10^{-8}$. For the digital device in this example, the minimum detectable clock pulse duration is about 13.5 ns. ∎

Exercise 7–6

The switch in Figure 7–16 closes at $t = 0$. Find the zero-state response of the capacitor voltage for $t \geq 0$

Answer:

$$v_C(t) = 2.5(1 - e^{-200t}) \quad V$$

Exercise 7–7

The switch in Figure 7–17 opens at $t = 0$. Find the zero-state response of the inductor current for $t \geq 0$.

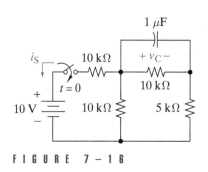

FIGURE 7–16

FIGURE 7-17

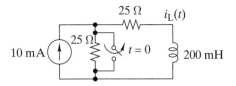

Answer:

$i_L(t) = 5(1 - e^{-10t})$ mA

7-3 INITIAL AND FINAL CONDITIONS

Reviewing the first-order step responses of the last section shows that for $t \geq 0$ the state variable responses can be written in the form

$$RC \text{ circuit: } v_C(t) = [v_C(0) - v_C(\infty)]e^{-t/T_C} + v_C(\infty) \qquad t \geq 0$$

$$RL \text{ circuit: } i_L(t) = [i_L(0) - i_L(\infty)]e^{-t/T_C} + i_L(\infty) \qquad t \geq 0$$

(7–23)

In both circuits the step response is of the general form

$$\begin{bmatrix} \text{The state} \\ \text{variable} \\ \text{response} \end{bmatrix} = \begin{bmatrix} \text{The initial} & \text{The final} \\ \text{value of the} & - & \text{value of the} \\ \text{state variable} & & \text{state variable} \end{bmatrix} \times e^{-t/T_C} + \begin{matrix} \text{The final} \\ \text{value of the} \\ \text{state variable} \end{matrix}$$

To determine the step response of a first-order circuit, we need three quantities: the initial value of the state variable, the final value of the state variable, and the time constant. Since we know how to get the time constant directly from the circuit, it would be useful to have a direct way to determine the initial and final values by inspecting the circuit itself.

The final value can be calculated directly from the circuit by observing that for $t > 5T_C$ the step responses approach a constant value or dc value. Under dc conditions a capacitor acts like an open circuit and an inductor acts like a short circuit. As a result, the final value of the state variable is found by applying dc analysis methods to the circuit configuration for $t > 0$, with capacitors replaced by open circuits and inductors replaced by short circuits.

We can also use dc analysis to determine the initial value in many practical situations. A common situation is a circuit containing dc sources and a switch that is in one position for a period of time much greater than the circuit time constant, and then is moved to a new position at $t = 0$. For example, if the switch is closed for a long period of time, then the dc sources drive the state variable to a final value. If the switch is now opened at $t = 0$, then the dc sources drive the state variable to a new final condition appropriate to the new circuit configuration for $t > 0$.

Note: The initial condition at $t = 0$ is the dc value of the state variable for the circuit configuration that existed before the switch changed positions at $t = 0$. The switching action cannot cause an instantaneous change in the initial condition because capacitor voltage and inductor current are continuous functions of time. In other words, opening a switch at $t = 0$ marks the boundary between two eras. The final condition of the state variable for the $t < 0$ era is the initial condition for the $t > 0$ era that follows.

The usual way to state a switched circuit problem is to say that a switch has been closed (open) for a long time and then is opened (closed) at $t = 0$. In this context, a long time means at least five time constants. Time constants rarely exceeds a few hundred milliseconds in electrical circuits, so a long time passes rather quickly.

The state variable response in switched dynamic circuits is found using the following steps:

STEP 1: Find the initial value by applying dc analysis to the circuit configuration for $t < 0$.

STEP 2: Find the final value by applying dc analysis to the circuit configuration for $t > 0$.

STEP 3: Find the time constant T_C of the circuit in the configuration for $t > 0$.

STEP 4: Write the step response directly using Eq. (7–23) without formulating and solving the circuit differential equation.

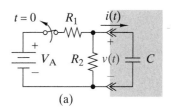

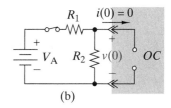

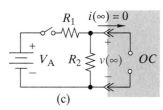

FIGURE   7–18 *Solving a switched dynamic circuit using the initial and final conditions.*

For example, the switch in Figure 7–18(a) has been closed for a long time and is opened at $t = 0$. We want to find the capacitor voltage $v(t)$ for $t \geq 0$.

STEP 1: The initial condition is found by dc analysis of the circuit configuration in Figure 7–18(b), where the switch is closed. Using voltage division, the initial capacitor voltage in found to be

$$v(0) = \frac{R_2 V_A}{R_1 + R_2}$$

STEP 2: The final condition is found by dc analysis of the circuit configuration in Figure 7–18(c), where the switch is open. When the switch is open the circuit has no dc excitation, so the final value of the capacitor voltage is zero.

STEP 3: The circuit in Figure 7–18(c) also gives us the time constant. Looking back at the interface, we see an equivalent resistance of R_2, since R_1 is connected in series with an open switch. For $t \geq 0$ the time constant is $R_2 C$. Using Eq. (7–23), the capacitor voltage for $t \geq 0$ is

$$v(t) = [v(0) - v(\infty)]e^{-t/T_C} + v_C(\infty)$$

$$= \frac{R_2 V_A}{R_1 + R_2}e^{-t/R_2 C} \qquad t \geq 0$$

The result is a zero-input response, since there is no excitation for $t \geq 0$. But now we see how the initial condition for the zero-input response could be produced physically by opening a switch that has been closed for a long time.

To continue the analysis, we find the capacitor current using its element constraint:

$$i(t) = C\frac{dv}{dt} = -\frac{V_A}{R_1 + R_2}e^{-t/R_2 C} \qquad t \geq 0$$

FIGURE 7 – 1 9 *Two responses in the*
FIGURE 7 – 1 9 *Two responses in the RC circuit of Figure 7–18.*

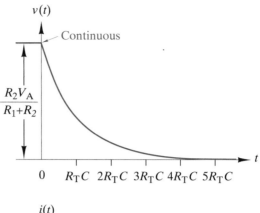

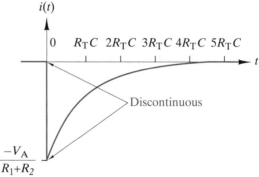

This is the capacitor current for $t \geq 0$. For $t < 0$ the circuit in Figure 7–18(b) points out that the capacitor current is zero since the capacitor acts like an open circuit.

The capacitor voltage and current responses are plotted in Figure 7–19. The capacitor voltage is continuous at $t = 0$, but the capacitor current has a jump discontinuity at $t = 0$. In other words, state variables are continuous, but nonstate variables can have discontinuities at $t = 0$. Since the state variable is continuous, we first find the circuit state variable and then solve for other circuit variables using the element and connection constraints.

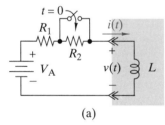

(a)

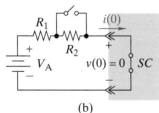

(b)

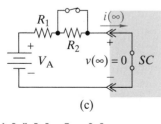

(c)

FIGURE 7 – 2 0

EXAMPLE 7–8

The switch in Figure 7–20(a) has been open for a long time and is closed at $t = 0$. Find the inductor current for $t > 0$.

SOLUTION:

We first find the initial condition using the circuit in Figure 7–20(b). By series equivalence the initial current is

$$i(0) = \frac{V_A}{R_1 + R_2}$$

The final condition and the time constant are determined from the circuit in Figure 7–17(c). Closing the switch shorts out R_2, and the final condition and time constant for $t > 0$ are

$$i(\infty) = \frac{V_A}{R_1} \quad \text{and} \quad T_C = G_N L = \frac{L}{R_1}$$

Using Eq. (7–23), the inductor current for $t \geq 0$ is

$$i(t) = [i(0) - i(\infty)]e^{-t/T_C} + i(\infty)$$

$$= \left[\frac{V_A}{R_1 + R_2} - \frac{V_A}{R_1}\right]e^{-R_1 t/L} + \frac{V_A}{R_1} \qquad t \geq 0 \qquad \blacksquare$$

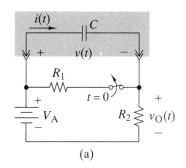

(a)

EXAMPLE 7-9

The switch in Figure 7–21(a) has been closed for a long time and is opened at $t = 0$. Find the voltage $v_O(t)$.

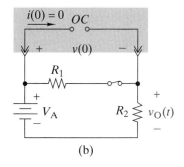

(b)

SOLUTION:

The problem asks for voltage $v_O(t)$, which is not the circuit state variable. Our approach is first to find the state variable response and then use this response to solve for the required nonstate variable.

For $t < 0$ the circuit in Figure 7–21(b) applies. By voltage division, the initial capacitor voltage is

$$v(0) = \frac{R_1 V_A}{R_1 + R_2}$$

The final value and time constant are found from the circuit in Figure 7–18(c).

$$v(\infty) = V_A \quad \text{and} \quad T_C = R_T C = R_2 C$$

Using Eq. (7–23), the capacitor voltage for $t \geq 0$ is

$$v(t) = [v(0) - v(\infty)]e^{-t/T_C} + v(\infty)$$

$$= \left[\frac{R_1 V_A}{R_1 + R_2} - V_A\right]e^{-t/R_2 C} + V_A$$

$$= V_A + \frac{R_2 V_A}{R_1 + R_2}e^{-t/R_2 C} \qquad t \geq 0$$

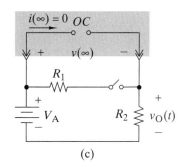

(c)

FIGURE 7-21

Given the state variable, we can find the voltage $v_O(t)$ by writing a KVL equation around the perimeter of the circuit in Figure 7–21(a):

$$-V_A + v(t) + v_O(t) = 0$$

or

$$v_O(t) = V_A - v(t) = -\frac{R_2 V_A}{R_1 + R_2}e^{-t/R_2 C} \qquad t \geq 0$$

The output voltage response looks like a zero-input response even though the circuit input is not zero for $t \geq 0$. However, $v_O(t)$ is not the state variable but the voltage across the resistor R_2. The voltage across R_2 is proportional to the capacitor current, which eventually decays to zero in the final circuit configuration in Figure 7–21(c) because the capacitor acts like an open circuit.   ■

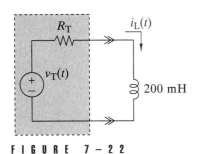

FIGURE 7–22

EXAMPLE 7–10

For $t \geq 0$ the state-variable response of the *RL* circuit in Figure 7–22 is observed to be

$$i_L(t) = 50 + 100 \, e^{-5000t} \quad \text{mA}$$

(a) Identify the forced and natural components of the response.
(b) Find the circuit time constant.
(c) Find the Thévenin equivalent circuit seen by the inductor.

SOLUTION:

(a) The natural component is the exponential term $100e^{-5000\,t}$ mA. The forced component is what remains after the natural component dies out as $t \to \infty$, namely $i_L(\infty) = 50$ mA. The forced response is a constant 50 mA, which means that the Thévenin equivalent is a dc source.
(b) The time constant is the reciprocal of the coefficient of t in $e^{-5000\,t}$, i.e., $T_C = 5000^{-1} = 0.2$ ms.
(c) Expressed in terms of circuit parameters the time constant is $T_C = L/R_T$, which yields the Thévenin resistance as $R_T = L/T_C = 100\ \Omega$. For dc excitation the inductor acts like a short circuit at $t = \infty$. Hence, $i_L(\infty) = v_T/R_T$ and the Thévenin voltage is

$$v_T = R_T i_L(\infty) = 100 \times 0.05 = 5 \quad \text{V}$$

APPLICATION NOTE: **EXAMPLE 7–11**

Figure 7–23 shows a circuit model of an interface between CMOS digital circuits. The circuit to the left of the interface represents the output of a CMOS driver. The capacitor represents the input capacitance of a CMOS receiver. Some representative values of the model parameters are $V_{DD} = 5$ V, $R_A = 150\ \Omega$, $R_B = 100\ \Omega$, and $C = 10$ pF.

When the driver switch is in Position B, the capacitor voltage decays toward zero through the resistance R_B. When the driver is in Position A, the capacitor voltage is driven toward V_{DD} via the resistance R_A. Thus, the capacitor voltage (input to the receiver) varies between 0 V and $V_{DD} = 5$ V. In CMOS digital circuits a voltage is interpreted as a logic high when it is greater than $V_{OH} < V_{DD}$ and as a logic zero when it is less than $V_{OL} > 0$. When the switch is in Position A, the receiver input is pulled up into the logic high range. Conversely, when it is in Position B the receiver input is pulled down into the low range. However, switching logic states takes time because the capacitor voltage can not change in zero time. The purpose of this note is to calculate the time needed to transition from low-to-high and high-to-low.

If the switch has been in Position B for a long time and is moved to Position A at $t = 0$, then the capacitor voltage is driven from 0 V to 5 V with a time constant of $T_C = R_A C = 1.5 \times 10^{-9}$, and we have

$$v_O(t) = 5\left(1 - e^{-\frac{10^9 t}{1.5}}\right)$$

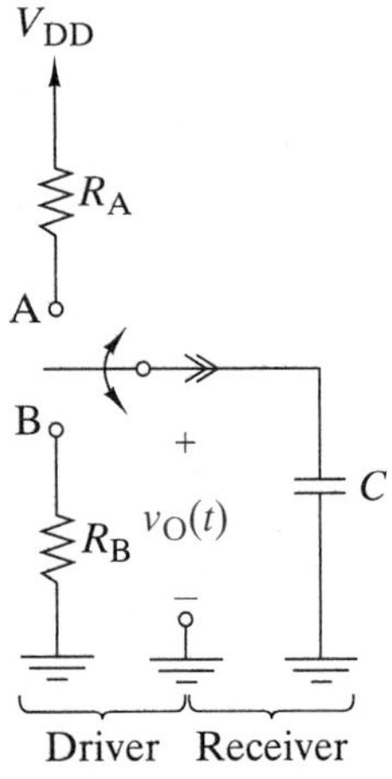

FIGURE 7–23

Driver Receiver

If V_{OH} is 3.7 V, then the low-to-high transition time t_{01} is found from

$$5\left(1 - e^{-\frac{10^9 t_{01}}{1.5}}\right) = 3.7$$

or

$$5 e^{-\frac{10^9 t_{01}}{1.5}} = 1.3$$

which yields $t_{01} = 1.5 \times 10^{-9} \ln(5/1.3) = 2.02 \times 10^{-9}$ s. Thus, it takes about 2 ns to transition from zero volts into the logic high range.

Conversely, if the switch has been in Position A for a long time and is moved to Position B at $t = 0$, then the capacitor voltage decreases from 5 V to 0 V with a time constant of $T_C = R_B C = 10^{-9}$, and we have

$$v_O(t) = 5 e^{-10^9 t}$$

If V_{OL} is 1 V, then the high-to-low transition time t_{10} is found from

$$5 e^{-10^9 t_{01}} = 1$$

or $t_{10} = 10^{-9} \ln(5) = 1.61 \times 10^{-9}$ s. The low-to-high transition time ($t_{01} = 2.02$ ns) is greater than the high-to-low transition time ($t_{10} = 1.61$ ns). The switching rate in a digital circuit is limited by the worst case transition time, which is t_{01} in this example.

Exercise 7–8

In each circuit shown in Figure 7–24 the switch has been in position A for a long time and is moved to position B at $t = 0$. Find the circuit state variable for each circuit for $t \geq 0$.

Answers:

$$\text{(a) } v_C(t) = V_A e^{-t/(R_1 + R_2)C}$$

$$\text{(b) } i_L(t) = \frac{V_A}{R_2} e^{-(R_1 + R_2)t/L}$$

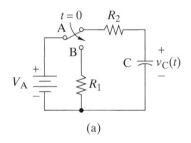

(a)

Exercise 7–9

In each circuit shown in Figure 7–24 the switch has been in position B for a long time and is moved to position A at $t = 0$. Find the circuit state variable for each circuit for $t \geq 0$.

Answers:

$$\text{(a) } v_C(t) = V_A (1 - e^{-t/R_2 C})$$

$$\text{(b) } i_L(t) = \frac{V_A}{R_2} (1 - e^{-R_2 t/L})$$

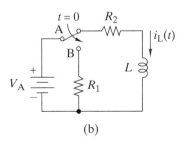

(b)

FIGURE 7 – 24

Exercise 7–10

In the circuit in Figure 7–25 the switch has been in position A for a long time and is moved to position B at $t = 0$. For $t \geq 0$ find the output voltage $v_O(t)$.

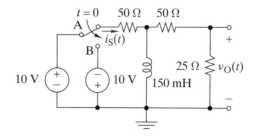

Answer:
$$v_O(t) = -4e^{-200t} \text{ V}$$

7–4 FIRST-ORDER CIRCUIT SINUSOIDAL RESPONSE

The response of linear circuits to sinusoidal inputs is one of the central themes of electrical engineering. In this introduction to the concept we treat the sinusoidal response of first-order circuits using differential equations. In later chapters we see that sinusoidal response can be found using other techniques. But for the moment, we concentrate on the classical method of finding the forced response from the circuit differential equation.

If the input to the *RC* circuit in Figure 7–2 is a casual sinusoid, then the circuit differential equation in Eq. (7–3) is written as

$$R_T C \frac{dv(t)}{dt} + v(t) = V_A[\cos \omega t]u(t) \tag{7–24}$$

The input on the right side of Eq. (7–24) is *not* an eternal sinewave but a casual sinusoid that starts at $t = 0$, through some action such as closing a switch. We seek a solution function $v(t)$ that satisfies Eq. (7–24) for $t \geq 0$ and that meets the prescribed initial condition $v(0) = V_0$.

As with the step response, we find the solution in two parts: natural response and forced response. The natural response is of the form

$$v_N(t) = K e^{-t/R_T C} \qquad t \geq 0$$

The natural response of a first-order circuit always has this form because it is a general solution of the homogeneous equation with input set to zero. The form of the natural response depends on the physical characteristics of the circuit and is independent of the input.

The forced response depends on both the circuit and the nature of the forcing function. The forced response is a particular solution of the equation

$$R_T C \frac{dv_F(t)}{dt} + v_F(t) = V_A \cos \omega t \qquad t \geq 0$$

This equation requires that $v_F(t)$ plus $R_T C$ times its first derivative add to produce a cosine function for $t \geq 0$. The only way this can happen is for $v_F(t)$ and its derivative to be sinusoids of the same frequency. This requirement brings to mind the derivative property of the sinusoid. So we try a solution in the form of a general sinusoid. As noted in Chapter 5, a general sinusoid can be written in amplitude and phase angle form as

$$v_F(t) = V_F \cos(\omega t + \phi) \qquad (7\text{–}25a)$$

or in terms of Fourier coefficients as

$$v_F(t) = a \cos \omega t + b \sin \omega t \qquad (7\text{–}25b)$$

While either form will work, it is somewhat easier to work with the Fourier coefficient format.

The approach we are using is called the method of undetermined coefficients, where the unknown coefficients are the Fourier coefficients a and b in Eq. (7–25b). To find these unknowns we insert the proposed forced response in Eq. (7–25b) into the differential equation to obtain

$$R_T C \frac{d}{dt}(a \cos \omega t + b \sin \omega t) + (a \cos \omega t + b \sin \omega t) = V_A \cos \omega t \quad t \geq 0$$

Performing the differentiation gives

$$R_T C(-\omega a \sin \omega t + \omega b \cos \omega t) + (a \cos \omega t + b \sin \omega t) = V_A \cos \omega t$$

We next gather all sine and cosine terms on one side of the equation.

$$[R_T C \omega b + a - V_A]\cos \omega t + [-R_T C \omega a + b]\sin \omega t = 0$$

The left side of this equation is zero for all $t \geq 0$ only when the coefficients of the cosine and sine terms are identically zero. This requirement yields two linear equations in the unknown coefficients a and b:

$$a + (R_T C \omega)b = V_A$$

$$-(R_T C \omega)a + b = 0$$

The solutions of these linear equations are

$$a = \frac{V_A}{1 + (\omega R_T C)^2} \qquad b = \frac{\omega R_T C V_A}{1 + (\omega R_T C)^2}$$

These equations express the unknowns a and b in terms of known circuit parameters $(R_T C)$ and known input signal parameters (ω and V_A).

We combine the forced and natural responses as

$$v(t) = Ke^{-t/R_T C} + \frac{V_A}{1 + (\omega R_T C)^2}(\cos \omega t + \omega R_T C \sin \omega t) \qquad t \geq 0$$

$$(7\text{–}26)$$

The initial condition requires

$$v(0) = V_0 = K + \frac{V_A}{1 + (\omega R_T C)^2}$$

which means K is

$$K = V_0 - \frac{V_A}{1 + (\omega R_T C)^2}$$

We substitute this value of K into Eq. (7–26) to obtain the function $v(t)$ that satisfies the differential equation and the initial conditions.

$$v(t) = \underbrace{\left[V_0 - \frac{V_A}{1 + (\omega R_T C)^2}\right] e^{-t/R_T C}}_{\text{Natural response}} +$$

$$\underbrace{\frac{V_A}{1 + (\omega R_T C)^2}(\cos \omega t + \omega R_T C \sin \omega t)}_{\text{Forced response}} \qquad t \geq 0$$

This expression seems somewhat less formidable when we convert the forced response to an amplitude and phase angle format

$$v(t) = \underbrace{\left[V_0 - \frac{V_A}{1 + (\omega R_T C)^2}\right] e^{-t/R_T C}}_{\text{Natural Response}} + \underbrace{\frac{V_A}{\sqrt{1 + (\omega R_T C)^2}}\cos(\omega t + \theta)}_{\text{Forced Response}} \quad t \geq 0$$

(7–27)

where

$$\theta = \tan^{-1}(-b/a) = \tan^{-1}(-\omega R_T C)$$

Equation (7–27) is the complete response of the *RC* circuit for an initial condition V_0 and a sinusoidal input $[V_A \cos \omega t]u(t)$. Several aspects of the response deserve comment:

1. After roughly five time constants the natural response decays to zero but the sinusoidal forced response persists.

2. The forced response is a sinusoid with the same frequency (ω) as the input but with a different amplitude and phase angle.

3. The forced response is proportional to V_A. This means that the amplitude of the forced component has the proportionality property because the circuit is linear.

In the terminology of electrical engineering, the forced component is called the **sinusoidal steady-state response**. The words *steady state* may be misleading since together they seem to imply a constant or "steady" value, whereas the forced response is a sustained oscillation. To electrical engineers *steady state* means the conditions reached after the natural response has died out. The sinusoidal steady-state response is also called the **ac steady-state response**. Often the words *steady state* are dropped and it is called simply the **ac response**. Hereafter, ac response, sinusoidal steady-state response, and the forced response for a sinusoidal input will be used interchangeably.

Finally, the forced response due to a step function input is called the **zero-frequency** or **dc steady-state response**. The zero-frequency terminology means that we think of a step function as a cosine $V_A[\cos \omega t]u(t)$ with $\omega = 0$. The reader can easily show that inserting $\omega = 0$ reduces Eq. (7–27) to the *RC* circuit step response in Eq. (7–20).

EXAMPLE 7–12

The switch in Figure 7–26 has been open for a long time and is closed at $t = 0$. Find the voltage $v(t)$ for $t \geq 0$ when $v_S(t) = [20 \sin 1000t]u(t)$ V.

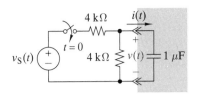

FIGURE 7–26

SOLUTION:
We first derive the circuit differential equation. By voltage division, the Thévenin voltage seen by the capacitor is

$$v_T(t) = \frac{4}{4 + 4}v_S(t) = 10 \sin 1000t \quad V$$

The Thévenin resistance (switch closed and source off) looking back into the interface is two 4-kΩ resistors in parallel, so $R_T = 2$ kΩ. The circuit time constant is

$$T_C = R_T C = (2 \times 10^3)(1 \times 10^{-6}) = 2 \times 10^{-3} = 1/500 \text{ s}$$

Given the Thévenin equivalent seen by the capacitor and the circuit time constant, the circuit differential equation is

$$2 \times 10^{-3}\frac{dv(t)}{dt} + v(t) = 10 \sin 1000t \qquad t \geq 0$$

Note that the right side of the circuit differential equation is the Thévenin voltage $v_T(t)$ and not the original source input $v_S(t)$. The natural response is of the form

$$v_N(t) = Ke^{-500t} \qquad t \geq 0$$

The forced response with undetermined Fourier coefficients is

$$v_F(t) = a \cos 1000t + b \sin 1000t$$

Substituting the forced response into the differential equation produces

$$2 \times 10^{-3}(-1000\, a \sin 1000t + 1000b \cos 1000t) +$$

$$a \cos 1000t + b \sin 1000t = 10 \sin 1000t$$

Collecting all sine and cosine terms on one side of this equation yields

$$(a + 2b) \cos 1000t + (-2a + b - 10)\sin 1000t = 0$$

The left side of this equation is zero for all $t \geq 0$ only when the coefficient of the sine and cosine terms vanish:

$$a + 2b = 0$$

$$-2a + b = 10$$

The solutions of these two linear equations are $a = -4$ and $b = 2$. We combine the forced and natural responses

$$v(t) = Ke^{-500t} - 4 \cos 1000t + 2 \sin 1000t \qquad t \geq 0$$

The constant K is found from the initial conditions

$$v(0) = V_0 = K - 4$$

The initial condition is $V_0 = 0$ because with the switch open the capacitor had no input for a long time prior to $t = 0$. The initial condition $v(0) = 0$ requires $K = 4$, so we can now write the complete response in the form

$$v(t) = 4e^{-500t} - 4\cos 1000t + 2\sin 1000t \text{ V} \qquad t \geq 0$$

or, in an amplitude, phase angle format as

$$v(t) = 4e^{-500t} + 4.47\cos(1000t + 153°) \text{ V} \qquad t \geq 0$$

Figure 7–27 shows an Excel worksheet that generates plots of the natural response, forced response, and total response. Column A is the time at 0.25-ms intervals. Columns B and C calculate the natural response ($4e^{-500t}$) and the forced response ($-4\cos 1000t + 2\sin 1000t$) at each of the times given in column A. The total response in column D is the sum of the entries in columns B and C. The plots show that the total response merges into the sinusoidal forced response since the natural response decays to zero after about $5T_C = 10$ ms. That is, after about 10 ms or so the circuit settles down to an ac steady-state condition. ∎

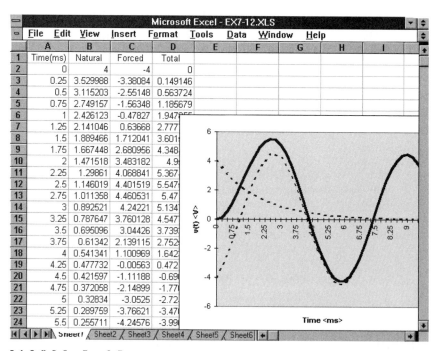

FIGURE 7–27

FIGURE 7–28

EXAMPLE 7–13

Find the sinusoidal steady-state response of the output voltage $v_O(t)$ in Figure 7–28 when the input current is $i_S(t) = [I_A \cos \omega t]u(t)$.

SOLUTION:

In keeping with our general analysis approach, we first find the steady-state response of the state variable $i(t)$ and use it to determine the required output voltage. The differential equation of the circuit in terms of the inductor current is

$$GL\frac{di}{dt} + i = I_A \cos \omega t \qquad t \geq 0$$

To find the steady-state response, we need to find the unknown Fourier coefficients a and b in the forced component:

$$i_F(t) = a \cos \omega t + b \sin \omega t \qquad t \geq 0$$

Substituting this expression into the differential equation yields

$$GL(-a\omega \sin \omega t + b\omega \cos \omega t) +$$

$$a \cos \omega t + b \sin \omega t = I_A \cos \omega t$$

Collecting sine and cosine terms produces

$$(a + GLb\omega - I_A)\cos \omega t + (-GLa\omega + b)\sin \omega t = 0$$

The left side of this equation is zero for all $t \geq 0$ only if

$$a + (GL\omega)b = I_A$$

$$-(GL\omega)a + b = 0$$

The solutions of these linear equations are

$$a = \frac{I_A}{1 + (GL\omega)^2} \qquad b = \frac{\omega GLI_A}{1 + (GL\omega)^2}$$

Therefore, the forced component of the inductor current is

$$i_F(t) = \frac{I_A}{1 + (\omega GL)^2}(\cos \omega t + \omega GL \sin \omega t) \qquad t \geq 0$$

The prescribed output is the voltage across the inductor. The steady-state output voltage is found using the inductor element equation:

$$v_O = L\frac{di_F}{dt} = \left[\frac{I_A L}{1 + (\omega GL)^2}\right]\frac{d}{dt}[\cos \omega t + \omega GL \sin \omega t]$$

$$= \left[\frac{I_A L}{1 + (\omega GL)^2}\right][-\omega \sin \omega t + \omega^2 GL \cos \omega t]$$

$$= \frac{I_A \omega L}{\sqrt{1 + (\omega GL)^2}} \cos (\omega t + \theta) \qquad t \geq 0$$

where $\theta = \tan^{-1}(1/\omega GL)$. The output voltage is a sinusoid with the same frequency as the input signal, but with a different amplitude and phase angle. In fact, in the sinusoidal steady state every voltage and current in a linear circuit is sinusoidal with the same frequency.

Notice that the amplitude of the steady-state output takes the form

$$\frac{I_A \omega L}{\sqrt{1 + (\omega GL)^2}}$$

Thus, the amplitude changes with the frequency of the sinusoidal current input. At $\omega = 0$ the input $i_S(t) = I_A \cos(0) = I_A$ is a constant dc waveform and the steady-state output is $v_O = 0$. This makes sense because at dc the inductor acts like a short circuit that forces the steady-state output in Figure 7–28 to be zero. At very high frequencies ($\omega GL \gg 1$) the steady-state output approaches $v_O = I_A R$. This also makes sense because at very high frequency the inductor acts like an open circuit that forces all of the input current to pass through the resistor in Figure 7–28. In between these two extremes the input current divides between the two paths in a manner that depends on the frequency. We will study frequency dependent responses in detail in later chapters. ■

Exercise 7–11

Find the forced component solution of the differential equation

$$10^{-3}\frac{dv}{dt} + v = 10 \cos \omega t$$

for the following frequencies:

(a) $\omega = 500$ rad/s
(b) $\omega = 1000$ rad/s
(c) $\omega = 2000$ rad/s

Answers:

(a) $v_F(t) = 8 \cos 500t + 4 \sin 500t$ $t \geq 0$
(b) $v_F(t) = 5 \cos 1000t + 5 \sin 1000t$ $t \geq 0$
(c) $v_F(t) = 2 \cos 2000t + 4 \sin 2000t$ $t \geq 0$

DISCUSSION: Converting these answers to an amplitude and phase angle as

(a) $v_F(t) = 8.94 \cos (500t - 26.6°)$ $t \geq 0$
(b) $v_F(t) = 7.07 \cos (1000t - 45°)$ $t \geq 0$
(c) $v_F(t) = 4.47 \cos (2000t - 63.4°)$ $t \geq 0$

we see that increasing the frequency of the input sinusoid decreases the amplitude and phase angle of the sinusoidal steady-state output of the circuit.

Exercise 7–12

The circuit in Figure 7–29 is operating in the sinusoidal steady state with

$$v_O(t) = 10 \cos(100t - 45°) \ \text{V}$$

Find the source voltage $v_S(t)$.

Answer:
$$v_S(t) = 10\sqrt{2} \cos 100t \ \text{V}$$

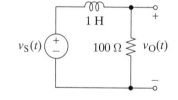

$v_S(t)$ 1 H   100 Ω $v_O(t)$

FIGURE 7 – 29

SUMMARY

- First-order circuits contain one capacitor or one inductor and are described by first-order differential equations. The dependent variable in the differential equation is the capacitor voltage or inductor current.

- The zero-input response of a first-order circuit is an exponential whose time constant depends on circuit parameters. The amplitude of the exponential is equal to the initial value of the capacitor voltage or inductor current.

- The natural response is the general solution of the homogeneous differential equation obtained by setting the input to zero. The forced response is a particular solution of the differential equation for the given input. For linear circuits the total response is the sum of the forced and natural responses.

- For linear circuits the total response is also the sum of a zero-input and zero-state response. The zero-input response results when no input is applied at $t = 0$ and there is initial energy stored in the capacitor or inductor. The zero-state response results when an input applied at $t = 0$ and there is no initial energy stored in the capacitor or inductor.

- The state-variable response of circuits containing constant sources and one or more switches involves three quantities: the initial value of the state variable, the final value of the state variable, and the circuit time constant. These quantities can be found directly from the circuit without formulating the circuit differential equation.

- For a sinusoidal input the forced response is called the sinusoidal steady-state response or the ac response. The ac response is a sinusoid with the same frequency as the input but with a different amplitude and phase angle. The ac response can be found from the circuit differential equation using the method of undetermined coefficients.

PROBLEMS

ERO 7–1 FIRST-ORDER CIRCUIT ANALYSIS (SECTS. 7–1, 7–2, 7–3, 7–4)

Given a first-order RC or RL circuit

(a) Find the circuit differential equation, the circuit characteristic equation, the circuit time constant, and the initial conditions (if not given).
(b) Find the zero-input response.
(c) Find the complete response for step function and sinusoidal inputs.

See Examples 7–1, 7–2, 7–3, 7–4, 7–5, 7–7, 7–8, 7–9, 7–12, 7–13 and Exercises 7–1, 7–2, 7–5, 7–6

7–1 Find the function that satisfies the following differential equation and the initial condition.

$$\frac{dv(t)}{dt} + 1500\, v(t) = 0, \quad v(0) = -15 \quad \text{V}$$

7–2 Find the function that satisfies the following differential equation and initial condition.

$$10^{-4}\frac{di(t)}{dt} + 10^{-1} i(t) = 0 \quad i(0) = -20 \quad \text{mA}$$

7–3 Find the time constants of the circuits in Figure P7–3.

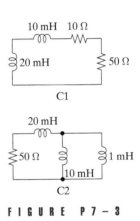

FIGURE P7-3

7–4 Find the time constants of the circuits in Figure P7–4.

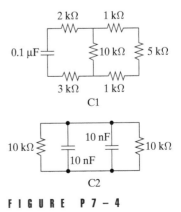

FIGURE P7-4

7–5 The switch in Figure P7–5 is at $t = 0$. The initial voltage on the capacitor is $v_C(0) = 15$ V. Find $v_C(t)$ and $i_O(t)$ for $t \geq 0$.

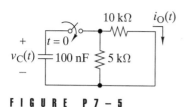

F I G U R E   P 7 – 5

7–6 In Figure P7–6 the initial current through the inductor is $i_L(0) = 25$ mA. Find $i_L(t)$ and $v_O(t)$ for $t \geq 0$.

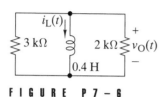

F I G U R E   P 7 – 6

7–7 The switch in each circuit in Figure P7–7 has been in position A for a long time and is moved to position B at $t = 0$. For each circuit develop an expression for the state variable for $t \geq 0$.

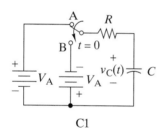

F I G U R E   P 7 – 7

7–8 Repeat Problem 7–7 when the switch in each circuit has been in position B for a long time and is moved to position A at $t = 0$.

7–9 The circuit in Figure P7–9 is in the zero state when the input $i_S(t) = I_A u(t)$ is applied. Find the voltage $v_O(t)$ for $t \geq 0$. Identify the forced and natural components.

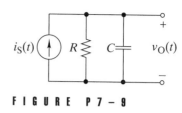

F I G U R E   P 7 – 9

7–10 The circuit in Figure P7–10 is in the zero state when the input $v_S(t) = V_A u(t)$ is applied. Find $v_O(t)$ for $t \geq 0$. Identify the forced and natural components.

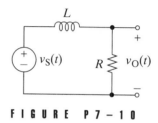

F I G U R E   P 7 – 1 0

7–11 The switch in Figure P7–11 has been in position A for a long time and is moved to position B at $t = 0$. Find $v_C(t)$ for $t \geq 0$. Identify the forced and natural components.

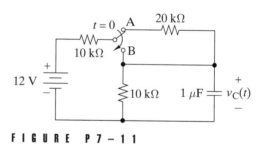

F I G U R E   P 7 – 1 1

7–12 Repeat Problem 7–11 when the switch has been in position B for a long time and is moved to position A at $t = 0$.

7–13 The input in Figure P7–13 is $v_S(t) = 15$ V. The switch has been open for a long time and is closed at $t = 0$. Find $v_C(t)$ for $t \geq 0$. Identify the forced and natural components, and sketch their waveforms.

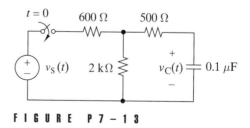

F I G U R E   P 7 – 1 3

7–14 The input in Figure P7–13 is $v_S(t) = 15$ V. The switch has been open for a long time and is closed at $t = 0$. The switch is reopened at $t = 200$ μs. Find $v_C(t)$ for $t \geq 0$ and sketch its waveform.

7–15 Find the function that satisfies the following differential equation and the initial condition for an input $v_S(t) = 25 \cos(100t)$ V.

$$\frac{dv(t)}{dt} + 200\, v(t) = v_S(t) \qquad v(0) = 0 \quad \text{V}$$

7–16 Repeat Problem 7–15 for $v_S(t) = 25 \sin(100t)$ V.

7–17 The input in Figure P7–17 is $v_S(t) = 20 \cos 5t$ V. The switch has been open for a long time and is closed at $t = 0$. Find $i_L(t)$ for $t \geq 0$.

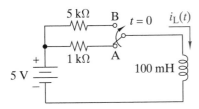

FIGURE P7–17

7–18 The switch in Figure P7–18 has been in position A for a long time and is moved to position B at $t = 0$. Find $i_L(t)$ for $t \geq 0$ and sketch its waveform.

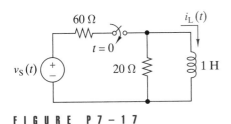

FIGURE P7–18

7–19 The switch in Figure P7–19 has been in position A for a long time and is moved to position B at $t = 0$. Find $v_C(t)$ for $t \geq 0$ and sketch its waveform.

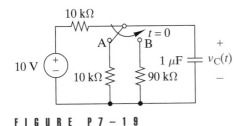

FIGURE P7–19

7–20 Switches 1 and 2 in Figure P7–20 have both been in position A for a long time. Switch 1 is moved to posi-

tion B at $t = 0$ and Switch 2 is moved to position B at $t = 20$ ms. Find the voltage across the 0.1-μF capacitor for $t > 0$ and sketch its waveform.

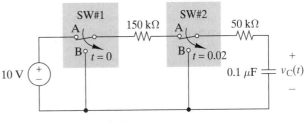

FIGURE P7–20

7–21 The switch in Figure P7–21 has been open for a long time and is closed at $t = 0$. The switch is reopened at $t = 2$ ms. Find $v_C(t)$ for $t \geq 2$ ms.

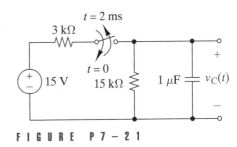

FIGURE P7–21

7–22 The input in Figure P7–22 is a rectangular pulse of the form

$$v_S(t) = V_A[u(t) - u(t - T)]$$

where $V_A = 5$ V and T is the pulse duration. A digital device with $RC = 20$ ns can detect a pulse input only if $v_C(t)$ exceeds 4 V. Find the minimum detectable pulse duration. Assume $v_C(0) = 0$.

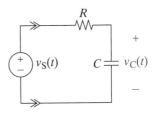

FIGURE P7–22

ERO 7–2 First-order Circuit Responses (Sects. 7–1, 7–2, 7–3)

Given responses in a first-order RC or RL circuit,

(a) Find the circuit parameters or other responses.
(b) Select element values to produce a given response.

See Examples 7–6, 7–10 and Exercises 7–2, 7–3, 7–4, 7–12

7–23   For $t \geq 0$ the zero-input responses of the circuit in Figure P7–23 are

$$v_C(t) = 10e^{-1000t} \text{ V and } i_C(t) = -20e^{-1000t} \text{ mA}$$

(a) Find the circuit time constant.
(b) Find the initial value of the state variable.
(c) Find R and C.
(d) Find the energy stored in the inductor at $t = 1$ ms.

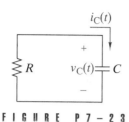

F I G U R E   P 7 – 2 3

7–24   For $t \geq 0$ the zero-input responses of the circuit in Figure P7–24 are

$$i_L(t) = 5e^{-5000t} \text{ mA and } v_L(t) = -10e^{-5000t} \text{ V} \qquad t \geq 0$$

(a) Find the circuit time constant.
(b) Find the initial value of the state variable.
(c) Find R and L.
(d) Find the energy stored in the inductor at $t = 0$.

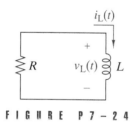

F I G U R E   P 7 – 2 4

7–25   For $t \geq 0$ the voltage across and current though the capacitor in Figure P7–25 are

$$v_C(t) = 15 - 10e^{-2000t} \text{ V and } i_C(t) = 10e^{-2000t} \text{ mA}$$

(a) Find the circuit time constant.
(b) Find the initial value of the state variable.
(c) Find v_S, R, and C.
(d) Find the energy stored in the capacitor at $t = 1$ ms.

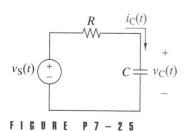

F I G U R E   P 7 – 2 5

7–26   For $t \geq 0$ the voltage across the 100 nF capacitor in Figure P7–25 is:

$$v_C(t) = 10 - 10e^{-500t} + [15e^{-500(t - 0.005)} - 15]u(t - 0.005) \text{ V}$$

(a) Find the circuit time constant.
(b) Find the initial and final value of the state variable.
(c) Find v_S and R.

7–27   For $t \geq 0$ the current through and voltage across the inductor in Figure P7–27 is

$$i_L(t) = 5 - 10e^{-1000t} \text{ mA and } v_L(t) = e^{-1000t} \text{ V}$$

(a) Find the circuit time constant.
(b) Find the initial and final value of the state variable.
(c) Find v_S, R, and L.
(d) Find the energy stored in the inductor at $t = 0$ and $t = \infty$.

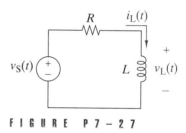

F I G U R E   P 7 – 2 7

7–28   For $t \geq 0$ the zero-input voltage across the inductor in Figure P7–27 is

$$v_L(t) = -10e^{-500t} \text{ V}$$

(a) Find the circuit time constant.
(b) The initial value of the state variable is $i_L(0) = 10$ mA. Find R and L.

7–29   Select values for v_S, R, and C in Figure P7–25 to produce the following zero-state response.

$$v_C(t) = 10 - 10e^{-100t} \text{ V} \quad t \geq 0$$

7–30   The RC circuit in Figure P7–25 is in the zero state and $v_S(t) = 5u(t)$. Select values of R and C such that $v_C(t) = 2.5$ V at $t = 5$ ms.

INTEGRATING PROBLEMS

7–31   ⬡ RC CIRCUIT DESIGN

Design an RC circuit whose step response fits neatly in the unshaded region in Figure P7–31. The source you must use is a 1.5-V battery. Your design must include a method of generating the required step function and can use resistors in the range from 1 kΩ to 100 kΩ. Test your design by finding its step response and calculating its value at $t = 1$ ms, 2 ms, and 10 ms.

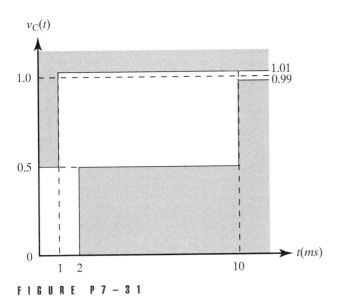

$v_C(t)$

1.01
0.99

FIGURE P7–31

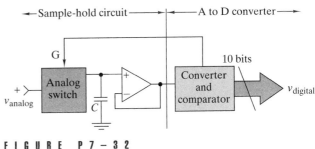

←— Sample-hold circuit —→ ←— A to D converter —→

FIGURE P7–32

7–32 ▲ SAMPLE HOLD CIRCUIT

Figure P7–32 shows a sample-hold circuit at the input of an analog-to-digital converter. In the sample mode the converter commands the analog switch to close (ON), and the capacitor charges up to the value of the input signal. In the hold mode the converter commands the analog switch to open (OFF), and the capacitor holds and feeds the input signal value to the converter via the OP AMP voltage follower. When conversion is completed the analog switch is turned ON again and the sample-hold cycle repeats.

(a) The series resistances of the analog switch are $R_{ON} = 50\ \Omega$ and $R_{OFF} = 100\ M\Omega$. When $C = 20$ pF, what is the time constant in the sample mode and the time constant in the hold mode?

(b) The number of sample-hold cycles per second must be at least twice the highest frequency in the analog input signal. What is the minimum number of sample-hold cycles per second for an input $v_S(t) = 5 + 5 \sin 2\pi 1000t$?

(c) Sampling at 10 times the minimum number of sample-hold cycles per second, what is the duration of the sample mode if the hold mode lasts nine times as long as the sample mode?

(d) For the input in (b), will the capacitor voltage reach a steady-state condition during the sample mode?

(e) What fraction of the capacitor voltage will be lost during the hold mode?

7–33 ▣ SUPER-CAPACITOR SPECIFICATION

Super-capacitors have very large capacitance (0.1 to 50 F), very long charge holding times (several days), and are several thousand times smaller than conventional capacitors with the same energy storage capacity. These attributes make them useful in nonbattery backup power applications such as VCRs, cash registers, computer terminals, and programmable consumer products. Because of their unique capabilities, special parameters are used to characterize these devices. One of the standard parameters is the so-called 30-minute current. The circuit in Figure P7–33 is the industry standard method of measuring this parameter. With the capacitor completely discharged (shorted for at least 24 hours), the switch is closed and the capacitor current measured by observing the voltage across the resistor. As the name suggests, the 30-minute current is the charging current observed 30 minutes after the switch is closed.

The following data are specification values for a certain super capacitor:

$C = 1.4$ F $+80\%, -20\%$
Max rated voltage $= 11$ V
Max current at 30 minutes $= 1.5$ mA
Test circuit conditions $V_{test} = 10$ V (dc) and $R_{test} = 200\ \Omega$.

Are these values internally consistent?

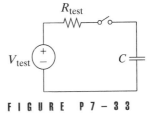

FIGURE P7–33

CHAPTER 8

SECOND-ORDER CIRCUITS

When a mathematician engaged in investigating physical actions and results has arrived at his own conclusions, may they not be expressed in common language as fully, clearly and definitely as in mathematical formula? If so, would it not be a great boon to such as we to express them so—translating them out of their hieroglyphics that we also might work upon them by experiment.

Michael Faraday, 1857,
British Physicist

8–1 The Series *RLC* Circuit

8–2 The Parallel *RLC* Circuit

8–3 Second-order Circuit Step Response

8–4 Second-order OP AMP Circuits

Summary

Problems

Integrating Problems

Michael Faraday (1791–1867) was appointed a Fellow in the Royal Society at age 32 and was a lecturer at the Royal Institution in London for more than 50 years. During this time he published over 150 papers on chemistry and electricity. The most important of these papers was the series *Experimental Researches in Electricity,* which included a description of his discovery of magnetic induction. A gifted experimentalist, Faraday had no formal education in mathematics and apparently felt that mathematics obscured the physical truths he discovered through experimentation. The foregoing quotation is taken from a letter written by Faraday to James Clerk Maxwell in 1857. Faraday's comments are perhaps ironic because Maxwell is best known today for his formulation of the mathematical theory of electromagnetics, which he published in 1873.

The dynamic circuits treated in Chapter 7 were limited to those governed by first-order differential equations. In this chapter we continue the study of differential equations by treating circuits described by second-order differential equations. The mathematical procedures in these two chapters come under the heading of classical methods. Classical methods offer insight into the physical information needed to analyze dynamic circuits and the types of possible responses.

Second-order circuits contain two energy storage elements that cannot be combined into a single equivalent element. Although there is an endless number of such circuits, in this chapter we focus on two time-honored cases: the series *RLC* circuit and the parallel *RLC* circuit. As with first-order circuits, we begin by showing how to derive the second-order differential equations for these two circuits. Solving these differential equations using classical methods introduces the basic techniques and terminology of second-order circuit analysis. We conclude by showing that concepts developed for the series and parallel *RLC* circuits apply to other second-order circuits as well.

8–1 THE SERIES *RLC* CIRCUIT

Our treatment of second-order circuits will parallel that used for first-order circuits. We first *formulate* a differential equation describing the circuit using basic physical concepts and relationships. Proper problem formulation is critical because mathematics cannot provide what is not provided by the physical constraints in the problem statement. We then employ classical methods to *solve* the differential equation for two situations: (1) the zero-input response and (2) the zero-state response for a step function input. Classical methods also work for other inputs such as ramps, exponential, and sinusoid. However, these situations are more easily handled using Laplace transform methods presented in the next chapter.

FORMULATING SERIES *RLC* CIRCUIT EQUATIONS

We begin with the circuit in Figure 8–1(a), where the inductor and capacitor are connected in series. The source-resistor circuit can be reduced to the Thévenin equivalent shown in Figure 8–1(b). The result is a circuit in which a voltage source, resistor, inductor, and capacitor are connected in series (hence the name **series *RLC* circuit**).

The first task is to develop the equations that describe the series *RLC* circuit. The Thévenin equivalent to the left of the interface in Figure 8–1(b) produces the KVL constraint

$$v + R_\text{T} i = v_\text{T} \qquad (8\text{–}1)$$

Applying KVL around the loop on the right side of the interface yields

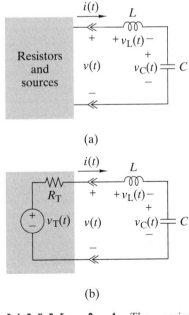

(a)

(b)

FIGURE 8 – 1 *The series RLC circuit.*

$$v = v_L + v_C \tag{8-2}$$

Finally, the i–v characteristics of the inductor and capacitor are

$$v_L = L\frac{di}{dt} \tag{8-3}$$

$$i = C\frac{dv_C}{dt} \tag{8-4}$$

Equations (8–1) through (8–4) are four independent equations in four unknowns (i, v, v_L, v_C). Collectively, this set of equations provides a complete description of the dynamics of the series RLC circuit. To find the circuit response using classical methods, we must derive a circuit equation containing only one of these unknowns.

We use circuit state variables as solution variables because they are continuous functions of time. In the series RLC circuit in Figure 8–1(b), there are two state variables: (1) the capacitor voltage $v_C(t)$ and (2) the inductor current $i(t)$. We first show how to describe the circuit using the capacitor voltage as the solution variable.

To derive a single equation in $v_C(t)$, we substitute Eqs. (8–2) and (8–4) into Eq. (8–1).

$$v_L + v_C + R_T C\frac{dv_C}{dt} = v_T \tag{8-5}$$

These substitutions eliminate the unknowns except v_C and v_L. To eliminate the inductor voltage, we substitute Eq. (8–4) into Eq. (8–3) to obtain

$$v_L = LC\frac{d^2v_C}{dt^2}$$

Substituting this result into Eq. (8–5) produces

$$LC\frac{d^2v_C}{dt^2} + R_T C\frac{dv_C}{dt} + v_C = v_T$$

$$v_L \quad + \quad\quad v_R \quad + v_C = v_T \tag{8-6}$$

In effect, this is a KVL equation around the loop in Figure 8–1(b), where the inductor and resistor voltages have been expressed in terms of the capacitor voltage.

Equation (8–6) is a second-order linear differential equation with constant coefficients. It is a second-order equation because the highest order derivative is the second derivative of the dependent variable $v_C(t)$. The coefficients are constant because the circuit parameters L, C, and R_T do not change. The Thévenin voltage $v_T(t)$ is a known driving force. The initial conditions

$$v_C(0) = V_0 \text{ and } \frac{dv_C}{dt}(0) = \frac{1}{C}i(0) = \frac{I_0}{C} \tag{8-7}$$

are determined by the values of the capacitor voltage and inductor current at $t = 0$, V_0 and I_0.

In summary, the second-order differential equation in Eq. (8–6) characterizes the response of the series RLC circuit in terms of the capacitor voltage $v_C(t)$. Once the solution $v_C(t)$ is found, we can solve for every other voltage or current, including the inductor current, using the element and connection constraints in Eqs. (8–1) to (8–4).

Alternatively, we can characterize the series *RLC* circuit using the inductor current. We first write the capacitor *i–v* characteristics in integral form:

$$v_C(t) = \frac{1}{C}\int_0^t i(x)dx + v_C(0) \qquad (8\text{--}8)$$

Equations (8–2), (8–3), and (8–8) are inserted into the interface constraint of Eq. (8–1) to obtain a single equation in the inductor current $i(t)$:

$$L\frac{di}{dt} + \frac{1}{C}\int_0^t i(x)dx + v_C(0) + R_T i = v_T$$

$$v_L + \qquad\quad v_C \qquad\quad + v_R \; = v_T \qquad (8\text{--}9)$$

In effect, this a KVL equation around the loop in Figure 8–1(b), where the capacitor and resistor voltages have been expressed in terms of the inductor current.

Equation (8–9) is a second-order linear integro-differential equation with constant coefficients. It is second order because it involves the first derivative and the first integral of the dependent variable $i(t)$. The coefficients are constant because the circuit parameters L, C, and R_T do not change. The Thévenin equivalent voltage $v_T(t)$ is a known driving force, and the initial conditions are $v_C(0) = V_0$ and $i(0) = I_0$.

Equations (8–6) and (8–9) involve the same basic ingredients: (1) an unknown state variable, (2) three circuit parameters (R_T, L, C), (3) a known input $v_T(t)$, and (4) two initial conditions (V_0 and I_0). The only difference is that one expresses the sum of voltages around the loop in terms of the capacitor voltage, while the other uses the inductor current. Either equation characterizes the dynamics of the series *RLC* circuit because once a state variable is found, every other voltage or current can be found using the element and connection constraints.

ZERO-INPUT RESPONSE OF THE SERIES *RLC* CIRCUIT

The circuit dynamic response for $t \geq 0$ can be divided into two components: (1) the zero-input response caused by the initial conditions and (2) the zero-state response caused by driving forces applied after $t = 0$. Because the circuit is linear, we can solve for these responses separately and superimpose them to get the total response. We first deal with the zero-input response.

With $v_T = 0$ (zero-input) Eq. (8–6) becomes

$$LC\frac{d^2 v_C}{dt^2} + R_T C\frac{dv_C}{dt} + v_C = 0 \qquad (8\text{--}10)$$

This result is a second-order homogeneous differential equation in the capacitor voltage. Alternatively, we set $v_T = 0$ in Eq. (8–9) and differentiate once to obtain the following homogeneous differential equation in the inductor current:

$$LC\frac{d^2 i}{dt^2} + R_T C\frac{di}{dt} + i = 0 \qquad (8\text{--}11)$$

We observe that Eqs. (8–10) and (8–11) have exactly the same form except that the dependent variables are different. The zero-input response of the capacitor voltage and inductor current have the same general form. We do not need to study both to understand the dynamics of the series RLC circuit. In other words, in the series RLC circuit we can use either state variable to describe the zero-input response.

In the following discussion we will concentrate on the capacitor voltage response. Equation (8–10) requires the capacitor voltage, plus RC times its first derivative, plus LC times its second derivative to add to zero for all $t \geq 0$. The only way this can happen is for $v_C(t)$, its first derivative, and its second derivative to have the same waveform. No matter how many times we differentiate an exponential of the form e^{st}, we are left with a signal with the same waveform. This observation, plus our experience with first-order circuits, suggests that we try a solution of the form

$$v_C(t) = Ke^{st}$$

where the parameters K and s are to be evaluated. When the trial solution is inserted in Eq. (8–10), we obtain the condition

$$Ke^{st}(LCs^2 + R_TCs + 1) = 0$$

The function e^{st} cannot be zero for all $t \geq 0$. The condition $K = 0$ is not allowed because it is a trivial solution declaring that $v_C(t)$ is zero for all t. The only useful way to meet the condition is to require

$$LCs^2 + R_TCs + 1 = 0 \qquad (8\text{–}12)$$

Equation (8–12) is the **characteristic equation** of the series RLC circuit. The characteristic equation is a quadratic because the circuit contains two energy storage elements. Inserting Ke^{st} into the homogeneous equation of the inductor current in Eq. (8–11) produces the same characteristic equation. Thus, Eq. (8–12) relates the zero-input response to circuit parameters for both state variables (hence the name *characteristic equation*).

In general, a quadratic characteristic equation has two roots:

$$s_1,\ s_2 = \frac{-R_TC \pm \sqrt{(R_TC)^2 - 4LC}}{2LC} \qquad (8\text{–}13)$$

From the form of the expression under the radical in Eq. (8–13), we see that there are three distinct possibilities:

Case A: If $(R_TC)^2 - 4LC > 0$, there are two real, unequal roots ($s_1 = -\alpha_1 \neq s_2 = -\alpha_2$).

Case B: If $(R_TC)^2 - 4LC = 0$, there are two real, equal roots ($s_1 = s_2 = -\alpha$).

Case C: If $(R_TC)^2 - 4LC < 0$, there are two complex conjugate roots ($s_1 = -\alpha - j\beta$ and $s_2 = -\alpha + j\beta$).

Where the symbol j represents the imaginary number $\sqrt{-1}$.[1] Before dealing with the form of the zero-input response for each case, we consider an example.

---

1 Mathematicians use the letter i to represent $\sqrt{-1}$. Electrical engineers use j, since the letter i represents electric current.

EXAMPLE 8–1

A series *RLC* circuit has $C = 0.25$ μF and $L = 1$ H. Find the roots of the characteristic equation for $R_T = 8.5$ kΩ, 4 kΩ, and 1 kΩ.

SOLUTION:

For $R_T = 8.5$ kΩ, the characteristic equation is

$$0.25 \times 10^{-6}s^2 + 2.125 \times 10^{-3}s + 1 = 0$$

whose roots are

$$s_1, s_2 = -4250 \pm \sqrt{(3750)^2} = -500, \quad -8000$$

These roots illustrate case A. The quantity under the radical is positive, and there are two real, unequal roots at $s_1 = -500$ and $s_2 = -8000$.

For $R_T = 4$ kΩ, the characteristic equation is

$$0.25 \times 10^{-6}s^2 + 10^{-3}s + 1 = 0$$

whose roots are

$$s_1, s_2 = -2000 \pm \sqrt{4 \times 10^6 - 4 \times 10^6} = -2000$$

This is an example of case B. The quantity under the radical is zero, and there are two real, equal roots at $s_1 = s_2 = -2000$.

For $R_T = 1$ kΩ the characteristic equation is

$$0.25 \times 10^{-6}s^2 + 0.25 \times 10^{-3}\,s + 1 = 0$$

whose roots are

$$s_1, s_2 = -500 \pm 500\sqrt{-15}$$

The quantity under the radical is negative, illustrating case C.

$$s_1, s_2 = -500 \pm j500\sqrt{15}$$

In case C the two roots are complex conjugates.   ∎

Exercise 8–1

For a series *RLC* circuit:

(a) Find the roots of the characteristic equation when $R_T = 2$ kΩ, $L = 100$ mH, and $C = 0.4$ μF.
(b) For $L = 100$ mH, select the values of R_T and C so the roots of the characteristic equation are $s_1, s_2 = -1000 \pm j2000$.
(c) Select the values of R_T, L, and C so $s_1 = s_2 = -10^4$.

Answers:

(a) $s_1 = -1340$, $s_2 = -18{,}660$
(b) $R_T = 200$ Ω, $C = 2$ μF
(c) There is no unique answer to part (c) since the requirement

$$(10^{-4}s + 1)^2 = LCs^2 + R_TCs + 1$$
$$= 10^{-8}s^2 + 10^{-4}s + 1$$

gives two equations $R_TC = 10^{-4}$ and $LC = 10^{-8}$ in three unknowns. One solution is to select $C = 1$ μF, which yields $L = 10$ mH and $R_T = 200$ Ω.

We have not introduced complex numbers simply to make things complex. Complex numbers arise quite naturally in practical physical situations involving nothing more than factoring a quadratic equation. The ability to deal with complex numbers is essential to our study. For those who need a review of such matters, there is a concise discussion in Appendix C.

FORM OF THE ZERO-INPUT RESPONSE

Since the characteristic equation has two roots, there are two solutions to the homogeneous differential equation:

$$v_{C1}(t) = K_1 e^{s_1 t}$$

$$v_{C2}(t) = K_2 e^{s_2 t}$$

That is,

$$LC \frac{d^2}{dt^2}(K_1 e^{s_1 t}) + R_T C \frac{d}{dt}(K_1 e^{s_1 t}) + K_1 e^{s_1 t} = 0$$

and

$$LC \frac{d^2}{dt^2}(K_2 e^{s_2 t}) + R_T C \frac{d}{dt}(K_2 e^{s_2 t}) + K_2 e^{s_2 t} = 0$$

The sum of these two solutions is also a solution since

$$LC \frac{d^2}{dt^2}(K_1 e^{s_1 t} + K_2 e^{s_2 t}) + R_T C \frac{d}{dt}(K_1 e^{s_1 t} + K_2 e^{s_2 t}) + K_1 e^{s_1 t} + K_2 e^{s_2 t} = 0$$

Therefore, the general solution for the zero-input response is of the form

$$v_C(t) = K_1 e^{s_1 t} + K_2 e^{s_2 t} \qquad (8\text{--}14)$$

The constants K_1 and K_2 can be found using the initial conditions given in Eq. (8–7). At $t = 0$ the condition on the capacitor voltage yields

$$v_C(0) = V_0 = K_1 + K_2 \qquad (8\text{--}15)$$

To use the initial condition on the inductor current, we differentiate Eq. (8–14).

$$\frac{dv_C}{dt} = K_1 s_1 e^{s_1 t} + K_2 s_2 e^{s_2 t}$$

Using Eq. (8–7) to relate the initial value of the derivative of the capacitor voltage to the initial inductor current $i(0)$ yields

$$\frac{dv_C(0)}{dt} = \frac{I_0}{C} = K_1 s_1 + K_2 s_2 \qquad (8\text{--}16)$$

Equations (8–15) and (8–16) provide two equations in the two unknown constants K_1 and K_2:

$$K_1 + K_2 = V_0$$

$$s_1 K_1 + s_2 K_2 = I_0/C$$

The solutions of these equations are

$$K_1 = \frac{s_2 V_0 - I_0/C}{s_2 - s_1} \quad \text{and} \quad K_2 = \frac{-s_1 V_0 + I_0/C}{s_2 - s_1}$$

Inserting these solutions back into Eq. (8–14) yields

$$v_C(t) = \frac{s_2 V_0 - I_0/C}{s_2 - s_1} e^{s_1 t} + \frac{-s_1 V_0 + I_0/C}{s_2 - s_1} e^{s_2 t} \qquad t \geq 0 \qquad (8\text{–}17)$$

Equation (8–17) is the general zero-input response of the series *RLC* circuit. The response depends on two initial conditions V_0 and I_0, and the circuit parameters R_T, L, and C since s_1 and s_2 are the roots of the characteristic equation $LCs^2 + R_T Cs + 1 = 0$. The response takes on different forms depending on whether the roots s_1 and s_2 fall under case A, B, or C.

For case A the two roots are real and distinct. Using the notation $s_1 = -\alpha_1$ and $s_2 = -\alpha_2$, the form zero-input response for $t \geq 0$ is

$$v_C(t) = \left[\frac{\alpha_2 V_0 + I_0/C}{\alpha_2 - \alpha_1} \right] e^{-\alpha_1 t} + \left[\frac{-\alpha_1 V_0 - I_0/C}{\alpha_2 - \alpha_1} \right] e^{-\alpha_2 t} \qquad (8\text{–}18)$$

For case A the response is the sum of two exponential functions similar to the double exponential signal treated in Example 5–14. The function has two time constants $1/\alpha_1$ and $1/\alpha_2$. The time constants can be greatly different, or nearly equal, but they cannot be equal because we would have case B.

With case B the roots are real and equal. Using the notation $s_1 = s_2 = -\alpha$, the general form in Eq. (8–17) becomes

$$v_C(t) = \frac{(\alpha V_0 + I_0/C)e^{-\alpha t} + (-\alpha V_0 - I_0/C)e^{-\alpha t}}{\alpha - \alpha}$$

We immediately see a problem here because the denominator vanishes. However, a closer examination reveals that the numerator vanishes as well, so the solution reduces to the indeterminate form 0/0. To investigate the indeterminacy, we let $s_1 = -\alpha$ and $s_2 = -\alpha + x$, and we explore the situation as x approaches zero. Inserting s_1 and s_2 in this notation in Eq. (8–17) produces

$$v_C(t) = V_0 e^{-\alpha t} + \left[\frac{-\alpha V_0 - I_0/C}{x} \right] e^{-\alpha t} + \left[\frac{\alpha V_0 + I_0/C}{x} \right] e^{-\alpha t} e^{xt}$$

which can be arranged in the form

$$v_C(t) = e^{-\alpha t}\left[V_0 - (\alpha V_0 + I_0/C) \frac{1 - e^{xt}}{x} \right]$$

We see that the indeterminacy comes from the term $(1 - e^{xt})/x$, which reduces to 0/0 as x approaches zero. Application of l'Hôpital's rule reveals

$$\lim_{x \to 0} \frac{1 - e^{xt}}{x} = \lim_{x \to 0} \frac{-t e^{xt}}{1} = -t$$

This result removes the indeterminacy, and as x approaches zero the zero-input response reduces to

$$v_C(t) = V_0 e^{-\alpha t} + (\alpha V_0 + I_0/C)\, t e^{-\alpha t} \qquad t \geq 0 \qquad (8\text{–}19)$$

For case B the response includes an exponential and the damped ramp studied in Example 5–12. The damped ramp is required, rather than two exponentials, because in case B the two equal roots produce the same exponential function.

Case C produces complex conjugate roots of the form

$$s_1 = -\alpha - j\beta \quad \text{and} \quad s_2 = -\alpha + j\beta$$

Inserting these roots into Eq. (8–17) yields

$$v_C(t) = \left[\frac{(-\alpha + j\beta)\,V_0 - I_0/C}{j2\beta}\right]e^{-\alpha t}\,e^{-j\beta t} + \left[\frac{(\alpha + j\beta)\,V_0 + I_0/C}{j2\beta}\right]e^{-\alpha t}\,e^{j\beta t}$$

which can be arranged in the form

$$v_C(t) = V_0 e^{-\alpha t}\left[\frac{e^{j\beta t} + e^{-j\beta t}}{2}\right] + \frac{\alpha V_0 + I_0/C}{\beta}e^{-\alpha t}\left[\frac{e^{j\beta t} - e^{-j\beta t}}{j2}\right] \qquad (8\text{--}20)$$

The expressions within the brackets have been arranged in a special way for the following reasons. Euler's relationships for an imaginary exponential are written as

$$e^{j\theta} = \cos\theta + j\sin\theta$$

and

$$e^{-j\theta} = \cos\theta - j\sin\theta$$

When we add and subtract these equations, we obtain

$$\cos\theta = \frac{e^{j\theta} + e^{-j\theta}}{2} \quad \text{and} \quad \sin\theta = \frac{e^{j\theta} - e^{-j\theta}}{2j}$$

Comparing these expressions for $\sin\theta$ and $\cos\theta$ with the complex terms in Eq. (8–20) reveals that we can write $v_C(t)$ in the form

$$v_C(t) = V_0\,e^{-\alpha t}\cos\beta t + \frac{\alpha V_0 + I_0/C}{\beta}e^{-\alpha t}\sin\beta t \qquad t \geq 0$$

For case C the response contains the damped sinusoid studied in Example 5–13. The real part of the roots (α) provides the exponent coefficient in the exponential function, while the imaginary part (β) defines the frequency of the sinusoidal oscillation.

In summary, the roots of the characteristic equation affect the form of the zero-input response in the following ways. In case A the two roots are real and unequal ($s_1 = -\alpha_1 \neq s_2 = -\alpha_2$) and the zero-input response is the sum of two exponentials of the form

$$v_C(t) = K_1 e^{-\alpha_1 t} + K_2 e^{-\alpha_2 t} \qquad (8\text{--}21a)$$

In case B the two roots are real and equal ($s_1 = s_2 = -\alpha$) and the zero-input response is the sum of an exponential and a damped ramp.

$$v_C(t) = K_1 e^{-\alpha t} + K_2 t e^{-\alpha t} \qquad (8\text{--}21b)$$

In case C the two roots are complex conjugates ($s_1 = -\alpha - j\beta,\, s_2 = -\alpha + j\beta$) and the zero-input response is the sum of a damped cosine and a damped sine.

$$v_C(t) = K_1 e^{-\alpha t}\cos\beta t + K_2 e^{-\alpha t}\sin\beta t \qquad (8\text{--}21c)$$

In determining the zero-input response we use the parameters s, α, and β. At various points in the development these parameters appear in expressions such as e^{st}, $e^{-\alpha t}$, and $e^{j\beta t}$. Since the exponent of e must be dimensionless, the parameters s, α, and β all have the dimensions of the reciprocal of time, or equivalently, frequency. Collectively, we say that s, α, and β define the **natural frequencies** of the circuit. When it is necessary to distinguish between these three parameters we say that s is the **complex frequency**, α is the **neper frequency**, and β is the **radian frequency**. The importance of this notation will become clear as we proceed through subsequent chapters of this book. To be consistent with expressions such as $s = -\alpha + j\beta$, we specify numerical values of s, α, and β in units of radians per second (rad/s).[2]

The constants K_1 and K_2 in Eqs. (8–21a), (8–21b), and (8–21c) are determined by the initial conditions on two state variables, as illustrated in the following example.

EXAMPLE 8−2

The circuit of Figure 8–2 has $C = 0.25\ \mu\text{F}$ and $L = 1\ \text{H}$. The switch has been open for a long time and is closed at $t = 0$. Find the capacitor voltage for $t \geq 0$ for (a) $R = 8.5\ \text{k}\Omega$, (b) $R = 4\ \text{k}\Omega$, and (c) $R = 1\ \text{k}\Omega$. The initial conditions are $I_0 = 0$ and $V_0 = 15\ \text{V}$.

FIGURE 8−2

SOLUTION:
The roots of the characteristic equation for these three values of resistance are found in Example 8–1. We are now in a position to use those results to find the corresponding zero-input responses.

(a) In Example 8–1 the value $R = 8.5\ \text{k}\Omega$ yields case A with roots $s_1 = -500$ and $s_2 = -8000$. The corresponding zero-input solution takes the form in Eq. (8–21a).

$$v_C(t) = K_1 e^{-500t} + K_2 e^{-8000t}$$

The initial conditions yield two equations in the constants K_1 and K_2:

$$v_C(0) = V_0 = 15 = K_1 + K_2$$

$$\frac{dv_C(0)}{dt} = \frac{I_0}{C} = 0 = -500K_1 - 8000K_2$$

Solving these equations yields $K_1 = 16$ and $K_2 = -1$, so that the zero-input response is

$$v_C(t) = 16e^{-500t} - e^{-8000t}\ \text{V} \qquad t \geq 0$$

---

2 The term *neper frequency* honors the sixteenth-century mathematician John Napier, who invented the base e or natural logarithms. The term *complex frequency* was apparently first used about 1900 by the British engineer Oliver Heaviside.

(b) In Example 8–1 the value $R = 4 \text{ k}\Omega$ yields case B with roots $s_1 = s_2 = -2000$. The zero-input response takes the form in Eq. (8–21b):

$$v_C(t) = K_1 e^{-2000t} + K_2 t e^{-2000t}$$

The initial conditions yield two equations in the constants K_1 and K_2:

$$v_C(0) = V_0 = 15 = K_1$$

$$\frac{dv_C(0)}{dt} = \frac{I_0}{C} = 0 = -2000 K_1 + K_2$$

Solving these equations yields $K_1 = 15$ and $K_2 = 2000 \times 15$, so the zero-input response is

$$v_C(t) = 15 e^{-2000t} + 15(2000t)e^{-2000t} \text{ V} \qquad t \geqslant 0$$

(c) In Example 8–1 the value $R = 1 \text{ k}\Omega$ yields case C with roots $s_1, s_2 = -500 \pm j500\sqrt{15}$. The zero-input response takes the form in Eq. (8–21c):

$$v_C(t) = K_1 e^{-500t} \cos(500\sqrt{15})t + K_2 e^{-500t} \sin(500\sqrt{15})t$$

The initial conditions yield two equations in the constants K_1 and K_2:

$$v_C(0) = V_0 = 15 = K_1$$

$$\frac{dv_C(0)}{dt} = \frac{I_0}{C} = 0 = -500K_1 + 500\sqrt{15}K_2$$

which yield $K_1 = 15$ and $K_2 = \sqrt{15}$, so the zero-input response is

$$v_C(t) = 15 e^{-500t} \cos(500\sqrt{15})t + \sqrt{15} \, e^{-500t} \sin(500\sqrt{15})t \text{ V} \qquad t \geqslant 0$$

Figure 8–3 shows plots of these responses. All three responses start out at 15 V (the initial condition) and all eventually decay to zero. The temporal decay of the responses is caused by energy loss in the circuit and is called **damping**. The case A response does not change sign and is called the **overdamped** response. The case C response undershoots and then oscillates about the final value. This response is said to be **underdamped** because there is not enough damping to prevent these oscillations. The case B response is said to be **critically damped** since it is a special case at the boundary between overdamping and underdamping. ■

FIGURE 8 – 3

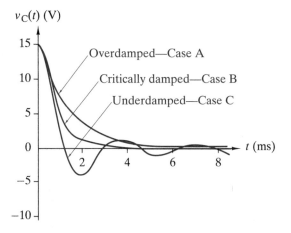

$v_C(t)$ (V)

Overdamped—Case A

Critically damped—Case B

Underdamped—Case C

EXAMPLE 8-3

In a series RLC circuit the zero-input voltage across the 1 μF capacitor is

$$v_C(t) = 10e^{-1000t} \sin 2000t \quad \text{V} \quad t \geq 0$$

(a) Find the circuit characteristic equation.
(b) Find the R and L.
(c) Find $i_L(t)$ for $t \geq 0$.
(d) Find the initial values of the state variables.

SOLUTION:

(a) The circuit is underdamped since the zero-input response is a damped sine with $\alpha = 1000$ and $\beta = 2000$ rad/s. The characteristic equation is

$$(s + 1000 - j2000)(s + 1000 + j2000) = s^2 + 2000s + 5 \times 10^6 = 0$$

(b) The characteristic equation of a series RLC circuit [Eq. (8-12)] can be written as

$$s^2 + \frac{R}{L}s + \frac{1}{LC} = 0$$

Comparing this term by term to the result in (a) yields the constraints

$$\frac{R}{L} = 2000 \quad \text{and} \quad \frac{1}{LC} = 5 \times 10^6$$

Since $C = 1$ μF we find that $L = 0.2$ H and $R = 400$ Ω.

(c) In a series circuit KCL requires $i_L(t) = i_C(t)$. Hence, the inductor current is

$$i_L(t) = C\frac{dv_C(t)}{dt} = 10^{-6}\frac{d}{dt}[10e^{-1000t}\sin 2000t]$$

$$= -10e^{-1000t}\sin 2000t + 20e^{-1000t}\cos 2000t \quad \text{mA} \qquad t \geq 0$$

(d) By inspection the initial values of the state variables are $v_C(0) = 0$ and $i_L(0) = 20$ mA.   ∎

Exercise 8-2

In a series RLC circuit, $R = 250$ Ω, $L = 10$ mH, $C = 1$μF, $V_0 = 0$, and $I_0 = 30$ mA. Find the capacitor voltage and inductor current for $t \geq 0$.

Answers:

$$v_C(t) = 2e^{-5000t} - 2e^{-20000t} \quad \text{V}$$

$$i_L(t) = -10e^{-5000t} + 40e^{-20000t} \quad \text{mA}$$

Exercise 8-3

In a series RLC circuit the zero-input responses are:

$$v_C(t) = 2000te^{-500t} \quad \text{V}$$

$$i_L(t) = 3.2e^{-500t} - 1600te^{-500t} \quad \text{mA}$$

(a) Find the circuit characteristic equation.
(b) Find the initial values of the state variables.
(c) Find R, L, and C.

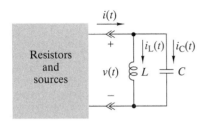

(a)

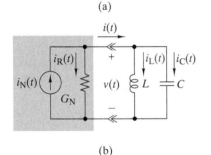

(b)

F I G U R E   8 – 4 *The parallel RLC circuit.*

8–2 THE PARALLEL *RLC* CIRCUIT

The inductor and capacitor in Figure 8–4(a) are connected in parallel. The source-resistor circuit can be reduced to the Norton equivalent shown in Figure 8–4(b). The result is a **parallel *RLC* circuit** consisting of a current source, resistor, inductor, and capacitor. Our first task is to develop a differential equation for this circuit. We expect to find a second-order differential equation because there are two energy storage elements.

The Norton equivalent to the left of the interface introduces the constraint

$$i + G_N v = i_N \qquad (8\text{–}22)$$

Writing a KCL equation at the interface yields

$$i = i_L + i_C \qquad (8\text{–}23)$$

The *i–v* characteristics of the inductor and capacitor are

$$i_C = C\frac{dv}{dt} \qquad (8\text{–}24)$$

$$v = L\frac{di_L}{dt} \qquad (8\text{–}25)$$

Equations (8–22) through (8–25) provide four independent equations in four unknowns (i, v, i_L, i_C). Collectively these equations describe the dynamics of the parallel *RLC* circuit. To solve for the circuit response using classical methods, we must derive a circuit equation containing only one of these four variables.

We prefer using state variables because they are continuous. To obtain a single equation in the inductor current, we substitute Eqs. (8–23) and (8–25) into Eq. (8–22):

$$i_L + i_C + G_N L\frac{di_L}{dt} = i_N \qquad (8\text{–}26)$$

The capacitor current can be eliminated from this result by substituting Eq. (8–25) into Eq. (8–24) to obtain

$$i_C = LC\frac{d^2 i_L}{dt^2} \qquad (8\text{–}27)$$

Inserting this equation into Eq. (8–26) produces

$$LC\frac{d^2 i_L}{dt^2} + G_N L\frac{di_L}{dt} + i_L = i_N \qquad (8\text{–}28)$$

$$i_C \quad + \quad i_R \quad + i_L = i_N$$

This result is a KCL equation in which the resistor and capacitor currents are expressed in terms of the inductor current.

Equation (8–28) is a second-order linear differential equation of the same form as the series *RLC* circuit equation in Eq. (8–6). In fact, if we interchange the following quantities:

$$v_C \leftrightarrow i_L \qquad L \leftrightarrow C \qquad R_T \leftrightarrow G_N \qquad v_T \leftrightarrow i_N$$

we change one equation into the other. The two circuits are duals, which means that the results developed for the series case apply to the parallel circuit with the preceding duality interchanges.

However, it is still helpful to outline the major features of the analysis of the parallel *RLC* circuit. The initial conditions in the parallel circuit are the initial inductor current I_0 and capacitor voltage V_0. The initial inductor current provides the condition $i_L(0) = I_0$ for the differential equation in Eq. (8–28). By using Eq. (8–25), the initial capacitor voltage specifies the initial rate of change of the inductor current as

$$\frac{di_L(0)}{dt} = \frac{1}{L}v_C(0) = \frac{1}{L}V_0$$

These initial conditions are the dual of those obtained for the series *RLC* circuit in Eq. (8–7).

To solve for the zero-input response, we set $i_N = 0$ in Eq. (8–28) and obtain a homogeneous equation in the inductor current:

$$LC\frac{d^2i_L}{dt^2} + G_NL\frac{di_L}{dt} + i_L = 0$$

A trial solution of the form $i_L = Ke^{st}$ leads to the characteristic equation

$$LCs^2 + G_NLs + 1 = 0 \qquad\qquad (8\text{–}29)$$

The characteristic equation is quadratic because there are two energy storage elements in the parallel *RLC* circuit. The characteristic equation has two roots:

$$s_1, s_2 = \frac{-G_NL \pm \sqrt{(G_NL)^2 - 4LC}}{2LC}$$

and, as in the series case, there are three distinct cases:

Case A: If $(G_NL)^2 - 4LC > 0$, there are two unequal real roots $s_1 = -\alpha_1$ and $s_2 = -\alpha_2$ and the zero-input response is the overdamped form

$$i_L(t) = K_1e^{-\alpha_1 t} + K_2e^{-\alpha_2 t} \qquad t \geq 0 \qquad (8\text{–}30)$$

Case B: If $(G_NL)^2 - 4LC = 0$, there are two real equal roots $s_1 = s_2 = -\alpha$ and the zero-input response is the critically damped form

$$i_L(t) = K_1e^{-\alpha t} + K_2te^{-\alpha t} \qquad t \geq 0 \qquad (8\text{–}31)$$

Case C: If $(G_NL)^2 - 4LC < 0$, there are two complex, conjugate roots $s_1, s_2 = -\alpha \pm j\beta$ and the zero-input response is the underdamped form

$$i_L(t) = K_1e^{-\alpha t}\cos\beta t + K_2e^{-\alpha t}\sin\beta t \qquad t \geq 0 \qquad (8\text{–}32)$$

The analysis results for the series *RLC* circuit apply to the parallel *RLC* case with the appropriate duality replacements. In particular, the form of overdamped, critically damped, and underdamped response applies to both circuits.

The forms of the responses in Eqs. (8–30), (8–31), and (8–32) have been written with two arbitrary constants K_1 and K_2. The following example shows how to evaluate these constants using the initial conditions for the two state variables.

EXAMPLE 8–4

In a parallel RLC circuit $R_T = 1/G_N = 500\ \Omega$, $C = 1\ \mu F$, $L = 0.2\ H$. The initial conditions are $I_0 = 50\ mA$ and $V_0 = 0$. Find the zero-input response of inductor current, resistor current, and capacitor voltage.

SOLUTION:
From Eq. (8–29) the circuit characteristic equation is

$$LCs^2 + G_NLs + 1 = 2 \times 10^{-7}s^2 + 4 \times 10^{-4}s + 1 = 0$$

The roots of the characteristic equation are

$$s_1,\ s_2 = \frac{-4 \times 10^{-4} \pm \sqrt{16 \times 10^{-8} - 8 \times 10^{-7}}}{4 \times 10^{-7}} = -1000 \pm j2000$$

Since roots are complex conjugates, we have the underdamped case. The zero-input response of the inductor current takes the form of Eq. (8–32).

$$i_L(t) = K_1e^{-1000t}\cos 2000t + K_2e^{-1000t}\sin 2000t \qquad t \geqslant 0$$

The constants K_1 and K_2 are evaluated from the initial conditions. At $t = 0$ the inductor current reduces to

$$i_L(0) = I_0 = K_1e^0\cos 0 + K_2e^0\sin 0 = K_1$$

We conclude that $K_1 = I_0 = 50\ mA$. To find K_2 we use the initial capacitor voltage. In a parallel RLC circuit the capacitor and inductor voltages are equal.

$$L\frac{di_L}{dt} = v_C(t)$$

In this example the initial capacitor voltage is zero, so the initial rate of change of inductor current is zero at $t = 0$. Differentiating the zero-input response produces

$$\frac{di_L}{dt} = -2000K_1e^{-1000t}\sin 2000t - 1000K_1e^{-1000t}\cos 2000t$$
$$- 1000K_2e^{-1000t}\sin 2000t + 2000K_2e^{-1000t}\cos 2000t$$

Evaluating this expression at $t = 0$ yields

$$\frac{di_L}{dt}(0) = -2000K_1e^0\sin 0 - 1000K_1e^0\cos 0$$
$$- 1000K_2e^0\sin 0 + 2000K_2e^0\cos 0$$
$$= -1000K_1 + 2000K_2 = 0$$

The derivative initial condition gives condition $K_2 = K_1/2 = 25\ mA$. Given the values of K_1 and K_2, the zero-input response of the inductor current is

$$i_L(t) = 50e^{-1000t}\cos 2000t + 25e^{-1000t}\sin 2000t \text{ mA} \qquad t \geq 0$$

The zero-input response of the inductor current allows us to solve for every voltage and current in the parallel *RLC* circuit. For example, using the *i–v* characteristic of the inductor, we obtain the inductor voltage:

$$v_L(t) = L\frac{di_L}{dt} = -25e^{-1000t}\sin 2000t \text{ V} \qquad t \geq 0$$

Since the elements are connected in parallel, we obtain the capacitor voltage and resistor current as

$$v_C(t) = v_L(t) = L\frac{di_L}{dt} = -25e^{-1000t}\sin 2000t \text{ V} \qquad t \geq 0$$

$$i_R(t) = \frac{v_L(t)}{R} = -50e^{-1000t}\sin 2000t \text{ mA} \qquad t \geq 0 \qquad\blacksquare$$

EXAMPLE 8–5

The switch in Figure 8–5 has been open for a long time and is closed at $t = 0$.

(a) Find the initial conditions at $t = 0$.
(b) Find the inductor current for $t \geq 0$.
(c) Find the capacitor voltage and current through the switch for $t \geq 0$.

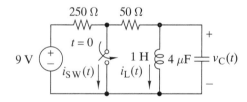

FIGURE 8 – 5

SOLUTION:
(a) For $t < 0$ the circuit is in the dc steady state, so the inductor acts like a short circuit and the capacitor like an open circuit. Since the inductor shorts out the capacitor, the initial conditions just prior to closing the switch at $t = 0$ are

$$v_C(0) = 0 \qquad i_L(0) = \frac{9}{250 + 50} = 30 \text{ mA}$$

(b) For $t \geq 0$ the circuit is a zero-input parallel *RLC* circuit with initial conditions found in (a). The circuit characteristic equation is

$$LCs^2 + GLs + 1 = 4 \times 10^{-6}s^2 + 2 \times 10^{-2}s + 1 = 0$$

The roots of this equation are

$$s_1 = -50.51 \quad \text{and} \quad s_2 = -4950$$

The circuit is overdamped (case A), since the roots are real and unequal. The general form of the inductor current zero-input response is

$$i_L(t) = K_1 e^{-50.51t} + K_2 e^{-4950t} \qquad t \geq 0$$

The constants K_1 and K_2 are found using the initial conditions. At $t = 0$ the zero-input response is

$$i_L(0) = K_1 e^0 + K_2 e^0 = K_1 + K_2 = 30 \times 10^{-3}$$

The initial capacitor voltage establishes an initial condition on the derivative of the inductor current since

$$L\frac{di_L}{dt}(0) = v_C(0) = 0$$

The derivative of the inductor response at $t = 0$ is

$$\frac{di_L}{dt}(0) = (-50.51 K_1 e^{-50.51t} - 4950 K_2 e^{-4950t})\big|_{t=0}$$
$$= -50.51 K_1 - 4950 K_2 = 0$$

The initial conditions on inductor current and capacitor voltage produce two equations in the unknown constants K_1 and K_2:

$$K_1 + K_2 = 30 \times 10^{-3}$$
$$-50.51 K_1 - 4950 K_2 = 0$$

Solving these equations yields $K_1 = 30.3$ mA and $K_2 = -0.309$ mA. The zero-input response of the inductor current is

$$i_L(t) = 30.3 e^{-50.51t} - 0.309 e^{-4950t} \text{ mA} \qquad t \geq 0$$

(c) Given the inductor current in (b), the capacitor voltage is

$$v_C(t) = L\frac{di_L}{dt} = -1.53 e^{-50.51t} + 1.53 e^{-4950t} \text{ V} \qquad t \geq 0$$

For $t \geq 0$ the current $i_{SW}(t)$ is the current through the 50-Ω resistor plus the current through the 250-Ω resistor.

$$i_{SW}(t) = i_{250} + i_{50} = \frac{9}{250} + \frac{v_C(t)}{50}$$
$$= 36 - 30.6 e^{-50.51t} + 30.6 e^{-4950t} \text{ mA} \qquad t \geq 0 \qquad \blacksquare$$

Exercise 8–4

The zero-input responses of a parallel RLC circuit are observed to be

$$i_L(t) = 10t e^{-2000t} \text{ A}$$
$$v_C(t) = 10 e^{-2000t} - 20000 t e^{-2000t} \text{ V} \qquad t \geq 0$$

(a) What is the circuit characteristic equation?
(b) What are the initial values of the state variables?
(c) What are the values of R, L, and C?
(d) Write an expression for the current through the resistor.

Answers:
(a) $s^2 + 4000s + 4 \times 10^6 = 0$
(b) $i_L(0) = 0$, $v_C(0) = 10$ V
(c) $L = 1$ H, $C = 0.25$ μF, $R = 1$ kΩ
(d) $i_R(t) = 10e^{-2000t} - 20000te^{-2000t}$ mA $t \geq 0$

8–3 SECOND-ORDER CIRCUIT STEP RESPONSE

The step response provides important insights into the response of dynamic circuits in general. So it is natural that we investigate the step response of second-order circuits. In Chapter 11 we will develop general techniques for determining the step response of any linear circuit. However, in this introduction we use classical methods of solving differential equations to find the step response of second-order circuits.

The general second-order linear differential equation with a step function input has the form

$$a_2 \frac{d^2y(t)}{dt^2} + a_1 \frac{dy(t)}{dt} + a_0 y(t) = Au(t) \tag{8–33}$$

where $y(t)$ is a voltage or current response, $Au(t)$ is the step function input, and a_2, a_1, and a_0 are constant coefficients. The step response is the general solution of this differential equation for $t \geq 0$. The step response can be found by partitioning $y(t)$ into forced and natural components:

$$y(t) = y_N(t) + y_F(t) \tag{8–34}$$

The natural response $y_N(t)$ is the general solution of the homogeneous equation (input set to zero), while the forced response $y_F(t)$ is a particular solution of the equation

$$a_2 \frac{d^2y_F}{dt^2} + a_1 \frac{dy_F}{dt} + a_0 y_F = A \qquad t \geq 0$$

Since A is a constant, it follows that dA/dt and d^2A/dt^2 are both zero, so it is readily apparent that $y_F = A/a_0$ is a particular solution of this differential equation. So much for the forced response.

Turning now to the natural response, we seek a general solution of the homogeneous equation. The natural response has the same form as the zero-state response studied in the previous section. In a second-order circuit the zero-state and natural responses take one of the three possible forms: overdamped, critically damped, or underdamped. To describe the three possible forms, we introduce two new parameters: ω_0 (omega zero) and ζ (zeta). These parameters are defined in terms of the coefficients of the general second-order equation in Eq. (8–33):

$$\omega_0^2 = \frac{a_0}{a_2} \quad \text{and} \quad 2\zeta\omega_0 = \frac{a_1}{a_2} \tag{8–35}$$

The parameter ω_0 is called the **undamped natural frequency** and ζ is called the **damping ratio**. Using these two parameters, the general homogeneous equation is written in the form

$$\frac{d^2y_N(t)}{dt^2} + 2\zeta\omega_0 \frac{dy_N(t)}{dt} + \omega_0^2 y_N(t) = 0 \tag{8-36}$$

The left side of Eq. (8–36) is called the **standard form** of the second-order linear differential equation. When a second-order equation is arranged in this format, we can determine its damping ratio and undamped natural frequency by equating its coefficients with those in the standard form. For example, in standard form the homogeneous equation for the series RLC circuit in Eq. (8–10) is

$$\frac{d^2v_C}{dt^2} + \frac{R_T}{L}\frac{dv_C}{dt} + \frac{1}{LC}v_C = 0$$

Equating like terms yields

$$\omega_0^2 = \frac{1}{LC} \quad \text{and} \quad 2\zeta\omega_0 = \frac{R_T}{L}$$

for the series RLC circuit. In an analysis situation the circuit element values determine the value of the parameters ω_0 and ζ. In a design situation we select L and C to obtain a specified ω_0 and then select R_T to obtain a specified ζ.

To determine the form of the natural response using ω_0 and ζ, we insert a trial solution $y_N(t) = Ke^{st}$ into the standard form in Eq. (8–36). The trial function Ke^{st} is a solution provided that

$$Ke^{st}[s^2 + 2\zeta\omega_0 s + \omega_0^2] = 0$$

Since $K = 0$ is the trivial solution and $e^{st} \neq 0$ for all $t \geq 0$, the only useful way for the right side of this equation to be zero for all t is for the quadratic expression within the brackets to vanish. The quadratic expression is the characteristic equation for the general second-order differential equation:

$$s^2 + 2\zeta\omega_0 s + \omega_0^2 = 0$$

The roots of the characteristic equation are

$$s_1, s_2 = \omega_0(-\zeta \pm \sqrt{\zeta^2 - 1})$$

We begin to see the advantage of using the parameters ω_0 and ζ. The constant ω_0 is a scale factor that designates the size of the roots. The expression under the radical defines the form of the roots and depends only on the damping ratio ζ. As a result, we can express the three possible forms of the natural response in terms of the damping ratio.

Case A: For $\zeta > 1$ the discriminant is positive, there are two unequal, real roots

$$s_1, s_2 = -\alpha_1, -\alpha_2 = \omega_0(-\zeta \pm \sqrt{\zeta^2 - 1}) \tag{8-37a}$$

and the natural response has the overdamped form

$$y_N(t) = K_1 e^{-\alpha_1 t} + K_2 e^{-\alpha_2 t} \quad t \geq 0 \tag{8-37b}$$

Case B: For $\zeta = 1$ the discriminant vanishes, there are two real, equal roots

$$s_1 = s_2 = -\alpha = -\zeta\omega_0 \tag{8-38a}$$

and the natural response has the critically damped form

$$y_N(t) = K_1 e^{-\alpha t} + K_2 t e^{-\alpha t} \quad t \geq 0 \qquad (8\text{–}38b)$$

Case C: For $\zeta < 1$ the discriminant is negative, leading to two complex, conjugate roots $s_1, s_2 = -\alpha \pm j\beta$, where

$$\alpha = \zeta\omega_0 \quad \text{and} \quad \beta = \omega_0\sqrt{1 - \zeta^2} \qquad (8\text{–}39a)$$

and the natural response has the underdamped form

$$y_N(t) = K_1 e^{-\alpha t}\cos\beta t + K_2 e^{-\alpha t}\sin\beta t \quad t \geq 0 \qquad (8\text{–}39b)$$

Equations (8–37a), (8–38a), and (8–39a) provide relationships between the natural frequency parameters α and β and the new parameters ζ and ω_0. The reasons for using two equivalent sets of parameters to describe the natural frequencies of a second-order circuit will become clear as we continue our study of dynamic circuits. Since the units of complex frequency s are radians per second, the standard form of the characteristic equation $s^2 + 2\zeta\omega_0 s + \omega_0^2$ shows that ω_0 is specified in radians per second and ζ is dimensionless.

Combining the forced and natural responses yields the step response of the general second-order differential equation in the form

$$y(t) = y_N(t) + A/a_0 \quad t \geq 0 \qquad (8\text{–}40)$$

The factor A/a_0 is the forced response. The natural response $y_N(t)$ takes one of the forms in Eqs. (8–37b), (8–38b), or (8–39b), depending on the value of the damping ratio. The constants K_1 and K_2 in natural response can be evaluated from the initial conditions.

In summary, the step response of a second-order circuit is determined by

1. The amplitude of the step function input $Au(t)$
2. The damping ratio ζ and natural frequency ω_0
3. The initial conditions $y(0)$ and $dy/dt\,(0)$.

In this regard the damping ratio and natural frequency play the same role for second-order circuits that the time constant plays for first-order circuits. That is, these circuit parameters determine the basic form of the natural response, just as the time constant defines the form of the natural response in a first-order circuit. It is not surprising that a second-order circuit takes two parameters, since it contains two energy storage elements.

EXAMPLE 8–6

The series RLC circuit in Figure 8–6 is driven by a step function and is in the zero state at $t = 0$. Find the capacitor voltage for $t \geq 0$.

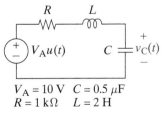

$V_A = 10\text{ V} \quad C = 0.5\ \mu\text{F}$
$R = 1\text{ k}\Omega \quad L = 2\text{ H}$

FIGURE 8 – 6

SOLUTION:

This is a series RLC circuit, so the differential equation for the capacitor voltage is

$$10^{-6}\frac{d^2 v_C}{dt^2} + 0.5 \times 10^{-3}\frac{dv_C}{dt} + v_C = 10 \quad t \geq 0$$

By inspection, the forced response is $v_{CF} = 10$ V. In standard format the homogeneous equation is

$$\frac{d^2v_{CN}}{dt^2} + 500\frac{dv_{CN}}{dt} + 10^6\,v_{CN} = 0 \qquad t \geq 0$$

Comparing this format to the standard form in Eq. (8–36) yields

$$\omega_0^2 = 10^6 \quad \text{and} \quad 2\zeta\omega_0 = 500$$

so $\omega_0 = 1000$ and $\zeta = 0.25$. Since $\zeta < 1$, the natural response is underdamped (case C). Using Eqs. (8–39a) and (8–39b), we have

$$\alpha = \zeta\omega_0 = 250$$

$$\beta = \omega_0\sqrt{1 - \zeta^2} = 968$$

$$v_{CN}(t) = K_1 e^{-250t}\cos 968t + K_2 e^{-250t}\sin 968t$$

The general solution of the circuit differential equation is the sum of the forced and natural responses:

$$v_C(t) = 10 + K_1 e^{-250t}\cos 968t + K_2 e^{-250t}\sin 968t \qquad t \geq 0$$

The constants K_1 and K_2 are determined by the initial conditions. The circuit is in the zero state at $t = 0$, so the initial conditions are $v_C(0) = 0$ and $i_L(0) = 0$. Applying the initial condition constraints to the general solution yields two equations in the constants K_1 and K_2:

$$v_C(0) = 10 + K_1 = 0$$

$$\frac{dv_C}{dt}(0) = -250\,K_1 + 968\,K_2 = 0$$

These equations yield $K_1 = -10$ and $K_2 = -2.58$. The step response of the capacitor voltage step response is

$$v_C(t) = 10 - 10e^{-250t}\cos 968t - 2.58e^{-250t}\sin 968t \text{ V} \qquad t \geq 0$$

A plot of $v_C(t)$ versus time is shown in Figure 8–7. The response and its first derivative at $t = 0$ satisfy the initial conditions. The natural response decays to zero, so the forced response determines the final value of $v_C(\infty)$ = 10 V. Beginning at $t = 0$ the response climbs rapidly but overshoots the mark several times before eventually settling down to the final value. The damped sinusoidal behavior results from the fact that $\zeta < 1$, producing an underdamped natural response. ■

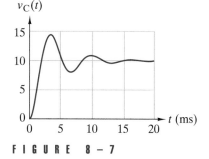

FIGURE 8–7

EXAMPLE 8–7

The step response of the series RLC circuit in Figure 8–6 is found analytically in Example 8–6. In this example we explore the effect of varying the resistance on the step response of this series RLC circuit. Use Electronics Workbench to calculate the zero-state step response of the capacitor voltage as R sweeps across the range from 1 kΩ to 8 kΩ with $L = 2$ H, $C = 0.5$ μF, and $V_A = 10$ V.

SOLUTION:

The Electronics Workbench circuit diagram in Figure 8–8 includes numerical values for the inductor, capacitor, voltage source, and the starting value of the resistance. The 10-V dc source simulates the step function input for $t \geq 0$. To get the zero-state response the initial inductor current and capaci-

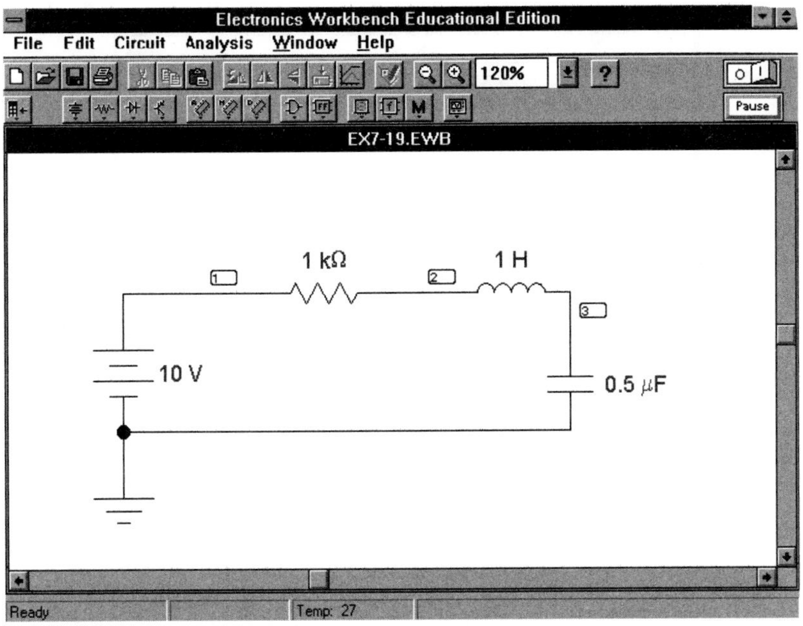

FIGURE 8-8

tor voltage must be set to zero. Note that the node voltage at node 3 is the voltage across the capacitor.

To obtain step responses as the resistance varies, we select the **Parameter Sweep** option from the **Analysis** menu. This brings up the parameter sweep dialog box shown at the top of Figure 8–9. Here we select resistance **R1** as the component to be varied, specify an analysis **Start value** of 1 kΩ and an **End value** of 8 kΩ. Specifying an octave **Sweep type** means that the parameter sweep values are obtained by doubling the start value until the end value is reached. That is, separate analyses will be run for R1 = 1, 2, 4, and 8 kΩ. Finally, the **Output node** to be monitored is set to node 3, which is the voltage across the capacitor in the series *RLC* circuit in Figure 8–8.

With the parameter sweep conditions set, we next press the **Set transient analysis** button, which brings up the transient analysis dialog box shown at the bottom of Figure 8–9. In this box the **Initial conditions** are set to zero (to get the zero-state response), the **Start time** is set to zero, the **End time** set to 0.02 s = 20 ms, and the **Minimum number of time points** set to 100. Pressing the **Accept** button returns us to the parameter sweep dialog box, where pressing the **Simulate** button launches a sequence of transient analysis runs as R1 sweeps across the specified range.

Figure 8–10 shows the family of step responses for R1 = 1, 2, 4, and 8 kΩ. As should be expected, increasing the resistance increases the damping so that the step response goes from the underdamped case at R1 = 1 kΩ to being overdamped at R1 = 8 kΩ.

We can qualitatively check these results by calculating ω_0 and ζ.

$$\omega_0 = \frac{1}{\sqrt{LC}} = \frac{1}{\sqrt{2 \times 0.5 \times 10^{-6}}} = 1000 \text{ rad/s}$$

and

FIGURE 8–9

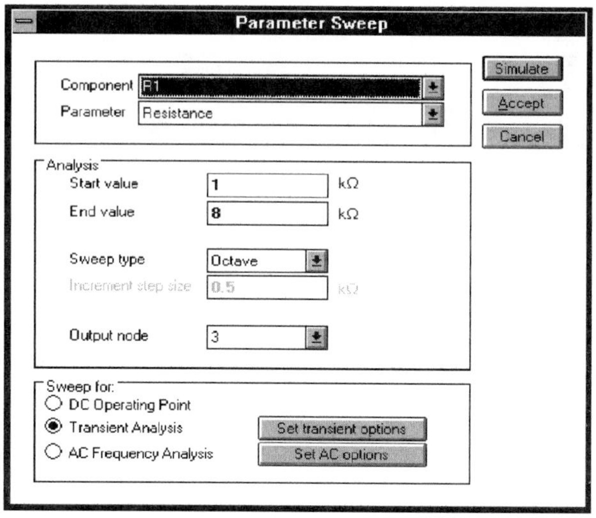

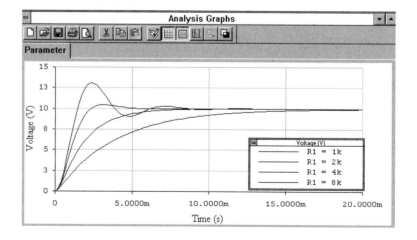

(a)

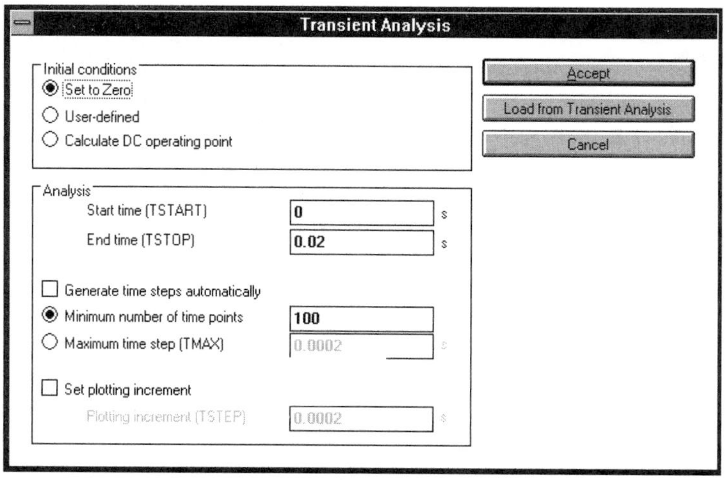

(b)

FIGURE 8–10

$$\zeta = \frac{R_1}{2\omega_0 L} = \frac{R_1}{4000}$$

Hence, $R_1 = 1$ kΩ yields $\zeta = 0.25$, $R_1 = 2$ kΩ yields $\zeta = 0.5$, $R_1 = 4$ kΩ yields $\zeta = 1$, and $R_1 = 8$ kΩ yields $\zeta = 2$. These calculations confirm that sweeping the resistance from 1 kΩ to 8 kΩ takes the step response from an underdamped case to an overdamped case, passing through the critically damped case ($\zeta = 1$) at $R_1 = 4$ kΩ. ∎

Exercise 8–5

Find the zero-state solution of the following differential equations.

(a) $10^{-4}\dfrac{d^2v}{dt^2} + 2 \times 10^{-2}\dfrac{dv}{dt} + v = 100u(t)$

(b) $\dfrac{d^2i}{dt^2} + 2400\dfrac{di}{dt} + 4 \times 10^6 i = 8 \times 10^4 u(t)$

Answers:

(a) $v(t) = 100 - 100e^{-100t} - 10^4 te^{-100t}$ V $t \geq 0$

(b) $i(t) = 20 - 20e^{-1200t}\cos 1600\,t - 15e^{-1200t}\sin 1600t$ mA $t \geq 0$

Exercise 8–6

The step response of a series RLC circuit is observed to be

$$v_C(t) = 15 - 15e^{-1000t}\cos 1000t \text{ V} \qquad t \geq 0$$
$$i_L(t) = 45e^{-1000t}\cos 1000t + 45e^{-1000t}\sin 1000t \text{ mA} \qquad t \geq 0$$

(a) What is the circuit characteristic equation?
(b) What are the initial values of the state variables?
(c) What is the amplitude of the step input?
(d) What are the values of R, L, and C?
(e) What is the voltage across the resistor?

Answers:
(a) $s^2 + 2000s + 2 \times 10^6 = 0$
(b) $v_C(0) = 0$, $i_L(0) = 45$ mA
(c) $V_A = 15$ V
(d) $R = 333\ \Omega$, $L = 167$ mH, $C = 3\ \mu$F
(e) $v_R(t) = 15e^{-1000t}\cos 1000t + 15e^{-1000t}\sin 1000t$ V

8–4 SECOND-ORDER OP AMP CIRCUITS

Up to this point all of the second-order circuits considered have been series or parallel RLC circuits. These two time-honored cases introduce the role of differential equations in the analysis of dynamic circuits. This introduction is useful even though in subsequent chapters we use the more powerful Laplace transforms to treat dynamic circuits in general. To see why the study of RLC circuits is useful, we conclude this chapter by studying two second-order OP AMP circuits. Our purpose is to show that other second-order circuits can be described and understood in terms of concepts learned in the analysis of *RLC* circuits. We use OP AMP circuits for this purpose because their modular nature makes them easy to analyze.

The circuit in Figure 8–11 consists of two resistors, two capacitors, and an OP AMP connected as voltage follower. This is a second-order circuit because the two capacitors can not be combined into a single equivalent element. Our objective is to derive the differential equation relating the circuit input $v_S(t)$ and output $v_2(t)$.

Note that the node voltage $v_1(t)$ appears at both the input and output of the OP AMP voltage follower. Using the ideal OP AMP model (no input current) and the current references shown in the figure, the KCL equations for this circuit are

$$\text{Node 1 } C_1 \frac{dv_1(t)}{dt} = \frac{v_S(t) - v_1(t)}{R_1}$$

$$\text{Node 2 } C_2 \frac{dv_2(t)}{dt} = \frac{v_1(t) - v_2(t)}{R_2}$$

(8–41)

These are two coupled first-order differential equations. To get a single second-order equation, we first solve the node 2 equation for $v_1(t)$,

$$v_1(t) = R_2 C_2 \frac{dv_2(t)}{dt} + v_2(t)$$

and use this result to eliminate $v_1(t)$ from the node 1 equation:

$$C_1 \frac{d}{dt}\left[R_2 C_2 \frac{dv_2(t)}{dt} + v_2(t)\right] = \frac{v_S(t) - \left[\dfrac{R_2 C_2 dv_2(t)}{dt} + v_2(t)\right]}{R_1}$$

This result reduces to the following equation:

$$R_1 C_1 R_2 C_2 \frac{d^2 v_2(t)}{dt^2} + (R_1 C_1 + R_2 C_2)\frac{dv_2(t)}{dt} + v_2(t) = v_S(t)$$

This is a second-order linear differential equation relating the circuit input $v_S(t)$ and output $v_2(t)$.

The form of the natural response is determined by the characteristic equation, which is easily obtained from the differential equation. In the present case the equation is

$$R_1 C_1 R_2 C_2 s^2 + (R_1 C_1 + R_2 C_2)s + 1 = 0$$

which can be factored as

$$(R_1 C_1 s + 1)(R_2 C_2 s + 1) = 0$$

The characteristic equation has two real roots located at

$$s = \frac{-1}{R_1 C_1} \quad \text{and} \quad s = \frac{-1}{R_2 C_2}$$

From the location of these roots, we see that the natural response has two possible forms:

Case A: If $R_1 C_1 \neq R_2 C_2$, the two real roots are distinct and the natural response is overdamped.

Case B: If $R_1 C_1 = R_2 C_2$, the two real roots are equal and the natural response is critically damped.

Notice how the RLC circuit concepts of overdamped and critically damped apply in this case. However, this particular OP AMP circuit can not produce the underdamped Case C because the two roots are always real, never complex. In fact, the two roots are the natural frequencies of the two first-order RC subcircuits that make up this second-order OP AMP circuit.

Exercise 8–7

(a) What is the general form of the natural response of the circuit in Figure 8–11 for $R_1 = R_2 = 10 \text{ k}\Omega$, $C_1 = 2 \text{ μF}$, and $C_2 = 10 \text{ μF}$?
(b) What value of R_1 is required to produce Case B?

Answers:
(a) $v_{2N}(t) = K_1 e^{-50 t} + K_2 e^{-10 t}$
(b) $R_1 = 50 \text{ k}\Omega$

Our next example is the OP AMP circuit in Figure 8–12. Like the previous example, this circuit is second order because the two capacitors can not be combined into a single equivalent element. However, in this case the OP AMP is at the circuit output and there is a feedback back path from the output to node 1 via capacitor C_1. These differences have a profound effect on the circuit responses.

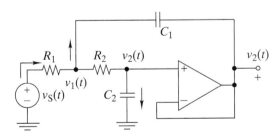

FIGURE 8 – 1 2

Note that the node voltage $v_2(t)$ appears at both the input and output of the OP AMP voltage follower. Again using the ideal OP AMP model and the current references indicated in the figure, the KCL equations for the circuit are

$$\text{Node 1} \quad C_1 \frac{d}{dt}[v_1(t) - v_2(t)] + \frac{v_1(t) - v_2(t)}{R_2} = \frac{v_S(t) - v_1(t)}{R_1}$$

$$\text{Node 2} \quad C_2 \frac{dv_2(t)}{dt} = \frac{v_1(t) - v_2(t)}{R_2} \tag{8–42}$$

Again we get two first-order equations. To get a single second-order equation we first solve the node 2 equation for $v_1(t)$.

$$v_1(t) = R_2 C_2 \frac{dv_2(t)}{dt} + v_2(t)$$

and use this result to eliminate $v_1(t)$ from the node 1 equation.

$$C_1 \frac{d}{dt}\left[R_2 C_2 \frac{dv_2(t)}{dt} + v_2(t) - v_2(t) \right] + \frac{R_2 C_2 \dfrac{dv_2(t)}{dt} + v_2(t) - v_2(t)}{R_2}$$

$$= \frac{v_S(t) - R_2 C_2 \dfrac{dv_2(t)}{dt} - v_2(t)}{R_1}$$

This rather formidable appearing result is easily seen to reduce to

$$R_1 C_1 R_2 C_2 \frac{d^2 v_2(t)}{dt^2} + (R_1 C_1 + R_2 C_2) \frac{dv_2(t)}{dt} + v_2(t) = v_S(t)$$

This is the second-order linear differential equation relating the output $v_2(t)$ and the input $v_S(t)$.

The characteristic equation for the differential equation is

$$R_1 C_1 R_2 C_2 s^2 + (R_1 C_2 + R_2 C_2)s + 1 = 0$$

which can be put in the form

$$s^2 + \left(\frac{1}{R_2 C_1} + \frac{1}{R_1 C_1} \right)s + \frac{1}{R_1 C_1 R_2 C_2} = 0$$

Comparing this to the standard second-order form $s^2 + 2\zeta \omega_0 s + \omega_0^2$ leads to the following:

$$\omega_0^2 = \frac{1}{R_1 C_1 R_2 C_2} \quad \text{and} \quad 2\zeta\omega_0 = \frac{1}{R_2 C_1} + \frac{1}{R_1 C_1}$$

or

$$\omega_0 = \frac{1}{\sqrt{R_1 C_1 R_2 C_2}} \quad \text{and} \quad 2\zeta = \sqrt{\frac{R_1 C_2}{R_2 C_1}} + \sqrt{\frac{R_2 C_2}{R_1 C_1}}$$

The undamped natural frequency ω_0 depends on the RC products and the damping ratio ζ depends on the ratio of the RC products. In practical applications of this circuit it is common to have $R_1 = R_2$, in which case the damping ratio reduces to $\zeta = \sqrt{C_2/C_1}$. For this useful special case the natural response has three possible forms.

Case A: If $C_2 > C_1$, then $\zeta > 1$ and the response is overdamped.

Case B: If $C_2 = C_1$, then $\zeta = 1$ and the response is critically damped.

Case C: If $C_2 < C_1$, then $\zeta < 1$ and the response is underdamped.

The natural response of this second-order OP AMP circuit has the same three forms as series or parallel RLC circuits.

These two OP AMP examples drive home three points:

1. Circuits with two independent energy storage elements are second order.

2. The form of the natural response is determined by the roots of the characteristic equation.

3. The roots can be real and distinct (ovedamped), real and equal (critically damped), or complex conjugates (underdamped).

We can also infer that the complete response also depends on the initial energy storage and the nature of the input driving force. All of this follows logically from our study of *RLC* circuits. We will revisit these concepts in our study of Laplace transforms.

Exercise 8–8

(a) What is the general form of the natural response of the circuit in Figure 8–12 for $R_1 = R_2 = 10\,\text{k}\Omega$, $C_1 = 10\,\mu\text{F}$, and $C_2 = 5\,\mu\text{F}$?

(b) What value of C_1 is needed to produce $\pi = 0.5$?

Answers:

(a) $v_{2N}(t) = K_1 e^{-10t}\cos 10t + K_2 e^{-10t}\sin 10t$

(b) $C_1 = 20\,\mu\text{F}$

SUMMARY

- Second-order circuits contain the equivalent of two energy storage elements and are described by second-order differential equations. The series *RLC* circuit and the parallel *RLC* circuit are two preeminent examples of second-order circuits.

- In the series *RLC* circuit the capacitor voltage is the dependent variable in the second-order differential equation. In the parallel *RLC* circuit the inductor current is the dependent variable in the second-order differential equation. In both circuits the initial conditions are the capacitor voltage and inductor current at $t = 0$.

- The two roots of the characteristic equation determine the form of the natural response of a second-order circuit. Unequal real roots yield an overdamped response, equal real roots yield a critically damped response, and complex conjugate roots yield an underdamped response. The initial conditions are used to find the two unknown coefficients in the natural response.

- The step response of a second-order circuit is the sum of the forced and natural responses. The forced response is a constant determined by the amplitude of the input step function. The form of the natural response is determined by the damping ratio ζ and undamped natural frequency ω_0. The natural response is overdamped if $\zeta > 1$, critically damped if $\zeta = 1$, and underdamped if $\zeta < 1$.

- OP AMP circuits with two capacitors are described by second-order differential equations. The natural response of these circuits are described by the same concepts used in the analysis of *RLC* circuits.

PROBLEMS

ERO 8–1 SECOND-ORDER CIRCUIT ANALYSIS (SECTS. 8–1, 8–2, 8–3, 8–4)

Given a second-order circuit,

(a) Find the circuit differential equation, the circuit characteristic equation, and the initial conditions (if not given).
(b) Find the zero-input response.
(c) Find the complete response for a step function input.

See Examples 8–1, 8–2, 8–3, 8–4, 8–5, 8–6 and Exercises 8–1, 8–2, 8–3

8–1 Find the response $v(t)$ that satisfies the following differential equation and meets the initial conditions.

$$\frac{d^2v}{dt^2} + 16\frac{dv}{dt} + 64v = 0, \quad v(0) = 0 \text{ V}, \quad \frac{dv}{dt}(0) = 12 \text{ V/s}$$

8–2 Find the function that satisfies the following differential equation and the initial conditions.

$$\frac{d^2v}{dt^2} + 20\frac{dv}{dt} + 500v = 0, \quad v(0) = 5 \text{ V}, \quad \frac{dv}{dt}(0) = 30 \text{ V/s}$$

8–3 Find the roots of the characteristic equation for each circuit in Figure P8–3. Specify whether the circuits are overdamped, underdamped, or critically damped.

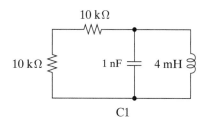

C1

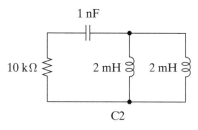

C2

FIGURE   P 8 – 3

8–4 The switch in Figure P8–4 has been open for a long time and is closed at $t = 0$. The circuit parameters are $L = 1$ H, $C = 0.5$ μF, $R_1 = 2$ kΩ, $R_2 = 3$ kΩ, and $V_A = 10$ V. Find $v_C(t)$ and $i_L(t)$ for $t \geq 0$. Is the circuit overdamped or underdamped?

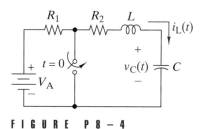

FIGURE   P 8 – 4

8–5 The switch in Figure P8–4 has been open for a long time and is closed at $t = 0$. The circuit parameters are $L = 1$ H, $C = 0.2$ μF, $R_1 = 3$ kΩ, $R_2 = 2$ kΩ, and $V_A = 10$ V. Find $v_C(t)$ and $i_L(t)$ for $t \geq 0$. Is the circuit overdamped or underdamped?

8–6 The switch in Figure P8–6 has been open for a long time and is closed at $t = 0$. The circuit parameters are $L = 0.5$ H, $C = 25$ nF, $R_1 = 5$ kΩ, $R_2 = 12$ kΩ, and $V_A = 10$ V.

(a) Find the initial values of v_C and i_L at $t = 0$ and the final values of v_C and i_L as $t \to \infty$.
(b) Find the differential equation for $v_C(t)$ and the circuit characteristic equation.
(c) Find $v_C(t)$ and $i_L(t)$ for $t \geq 0$. Is the circuit overdamped or underdamped?

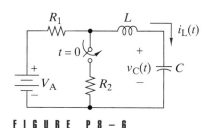

FIGURE   P 8 – 6

8–7 The switch in Figure P8–6 has been closed for a long time and is opened at $t = 0$. The circuit parameters are $L = 0.5$ H, $C = 0.4$ μF, $R_1 = 1$ kΩ, $R_2 = 2$ kΩ, and $V_A = 15$ V.

(a) Find the initial values of v_C and i_L at $t = 0$ and the final values of v_C and i_L as $t \to \infty$.
(b) Find the differential equation for $v_C(t)$ and the circuit characteristic equation.
(c) Find $v_C(t)$ and $i_L(t)$ for $t \geq 0$. Identify the forced and natural components of the responses and plot $v_C(t)$ and $i_L(t)$. Is the circuit overdamped or underdamped?

8–8 The switch in Figure P8–8 has been open for a long time and is closed at $t = 0$. The circuit parameters are $L = 0.4$ H, $C = 0.25$ μF, $R_1 = 3$ kΩ, $R_2 = 2$ kΩ, and $V_A = 15$ V.

Find $v_C(t)$ and $i_L(t)$ for $t \geq 0$. Is the circuit overdamped or underdamped?

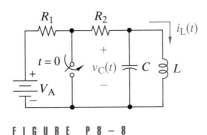

F I G U R E   P 8 – 8

8–9 Repeat Problem 8–8 with $L = 80$ mH.

8–10 The switch in Figure P8–10 has been open for a long time and is closed at $t = 0$. The circuit parameters are $L = 0.8$ H, $C = 50$ nF, $R_1 = 4$ kΩ, $R_2 = 4$ kΩ, and $V_A = 20$ V.

Find $v_C(t)$ and $i_L(t)$ for $t \geq 0$. Is the circuit over-damped or underdamped?

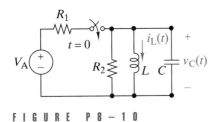

F I G U R E   P 8 – 1 0

8–11 Repeat Problem 8–10 with $L = 1.25$ H.

8–12 The switch in Figure P8–12 has been in position A for a long time. At $t = 0$ it is moved to position B. The circuit parameters are $R_1 = 1$ kΩ, $R_2 = 4$ kΩ, $L = 0.625$ H, $C = 6.25$ nF, and $V_A = 15$ V. Find $v_C(t)$ and $i_L(t)$ for $t > 0$. Identify the forced and natural components of the responses and plot $v_C(t)$ and $i_L(t)$. Is the circuit over-damped or underdamped?

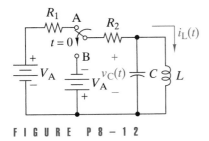

F I G U R E   P 8 – 1 2

8–13 The switch in Figure P8–12 has been in position B for a long time. At $t = 0$ it is moved to position A. The circuit parameters are $R_1 = 10$ Ω, $R_2 = 40$ Ω, $L = 1$ H, $C = 100$ μF, and $V_A = 12$ V. Find $v_C(t)$ and $i_L(t)$ for $t > 0$. Is the circuit overdamped or underdamped?

8–14 The switch in Figure P8–14 has been in position A for a long time and is moved to position B at $t = 0$. The circuit parameters are $R_1 = 500$ Ω, $R_2 = 500$ Ω, $L = 250$ mH, $C = 3.2$ μF, and $V_A = 5$ V. Find $v_C(t)$ and $i_L(t)$ for $t > 0$. Is the circuit overdamped or underdamped?

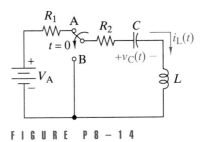

F I G U R E   P 8 – 1 4

8–15 The switch in Figure P8–14 has been in position B for a long time and is moved to position A at $t = 0$. The circuit parameters are $R_1 = 500$ Ω, $R_2 = 500$ Ω, $L = 250$ mH, $C = 1$ μF, and $V_A = 5$ V. Find $v_C(t)$ and $i_L(t)$ for $t > 0$. Is the circuit overdamped or underdamped?

8–16 The circuit in Figure P8–16 is in the zero state when the step function input is applied. The element values are $L = 2.5$ H, $C = 2$ μF, $R = 3$ kΩ, and $V_A = 100$ V. Find the output $v_O(t)$.

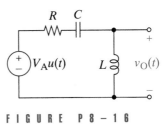

F I G U R E   P 8 – 1 6

8–17 The circuit in Figure P8–17 is in the zero state when the step function input is applied. The element values are $R = 4$ kΩ, $L = 160$ mH, $C = 25$ nF, and $V_A = 100$ V. Find the output $v_O(t)$.

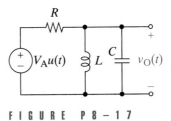

F I G U R E   P 8 – 1 7

8–18 The circuit in Figure P8–18 is in the zero state when the step function input is applied. The element values are $R = 4$ kΩ, $L = 400$ mH, $C = 12.5$ nF, and $V_A = 100$ V. Find the output $v_O(t)$.

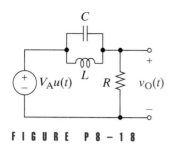

FIGURE P8-18

8-19 Find the differential equation relating v_O and v_S in Figure P8-19. Find the damping ratio and undamped natural frequency of the circuit.

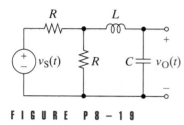

FIGURE P8-19

8-20 Find the differential equation relating v_O and v_S in Figure P8-20. Find the damping ratio and undamped natural frequency of the circuit.

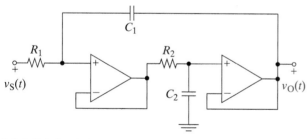

FIGURE P8-20

ERO 8-2 SECOND-ORDER CIRCUIT RESPONSES (SECTS. 8-1, 8-2, 8-3, 8-4)

Given the response of a second-order RLC circuit,

(a) Find the circuit parameters or other responses.
(b) Select element values to produce a given response.

See Example 8-3 and Exercises 8-1, 8-3, 8-4, 8-6

8-21 In a series RLC circuit the zero-state step response across the 1-μF capacitor is

$$v_C(t) = 10 - 50e^{-4000t} + 40e^{-5000t} \text{ V } \quad t \geq 0$$

(a) Find v_T, R, L.
(b) Find $i_L(t)$ for $t \geq 0$.

8-22 In a series RLC circuit the zero-input response in the 100-mH inductor is

$$i_L(t) = 5e^{-2000t}(\sin 1000t - \cos 1000t) \text{ mA } \quad t \geq 0$$

(a) Find R, C.
(b) Find $v_C(t)$ for $t \geq 0$.

8-23 In a parallel RLC circuit the step responses are

$$v_C(t) = e^{-100t}(5 \cos 500t + 25 \sin 500t) \text{ V } \quad t \geq 0$$
$$i_L(t) = 20 - 25e^{-100t} \cos 500t \text{ mA } \quad t \geq 0$$

Find R, L, and C.

8-24 The zero-input responses of an RLC circuit are

$$v_C(t) = 2e^{-2000t} \cos 1000t - 4e^{-2000t} \sin 1000t \text{ V } t \geq 0$$
$$i_L(t) = -80e^{-2000t} \cos 1000t + 60e^{-2000t} \sin 1000t \text{ mA } t \geq 0.$$

(a) Is this a series or a parallel RLC circuit?
(b) Find R, L, and C.

8-25 In a parallel RLC circuit with $v_C(0) = -10$ V, the inductor current is observed to be

$$i_L(t) = 10e^{-10t} \cos(20t) \text{ mA } \quad t \geq 0$$

Find $v_C(t)$.

8-26 A series RLC circuit has $L = 400$ mH. What value of R and C will produce a zero-input response of the form

$$v_L(t) = K_1 e^{-5000t} + K_2 e^{-10000t} \text{ V } \quad t > 0$$

8-27 Select values for R, L, and C in a series circuit so that $\zeta = 0.5$ and $\omega_0 = 4$ Mrad/s.

8-28 Select values for R, L, and C in a series circuit so that its step response has the form

$$v_C(t) = V_A - 2V_A e^{-200t} + V_A e^{-800t} \text{ V } \quad t > 0$$

where V_A is the amplitude of the input.

8-29 Select values for R, L, and C in a parallel RLC circuit so that its zero-input response has the form

$$i_L(t) = K_1 e^{-500t} + K_2 t e^{-500t} \text{ V } \quad t > 0$$

8-30 What range of damping ratios is available in the circuit in Figure P8-30.

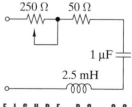

FIGURE P8-30

INTEGRATING PROBLEMS

8–31 ◗ LIGHTNING PULSER DESIGN

The circuit in Figure P8–31 is a simplified diagram of a pulser that delivers simulated lightning transients to the test article at the output interface. The pulser performance specification states that when the switch closes, the short-circuit current delivered by the pulser must be of the form $i_{SC}(t) = I_A e^{-\alpha t} \cos \beta t$, with $\alpha = 200$ krad/s, $\beta = 10$ Mrad/s, and $I_A = 2$ kA. Select the values of L, C, and the initial charge on the capacitor.

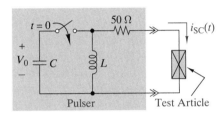

F I G U R E   P 8 – 3 1

8–32 ◢ EXPERIMENTAL SECOND-ORDER RESPONSE

Figure P8–32 shows an oscilloscope display of the voltage across the resistor in a series RLC circuit.

(a) Estimate the values of α and β of the damped sine signal.

(b) Use the values of α and β from (a) to write the characteristic equation of the circuit.

(c) The resistor is known to be 2.2 kΩ. Use the characteristic equation from (b) to determine the values of L and C.

(d) If the display shows the zero-input response of the circuit, what are the values of the initial conditions $v_C(0)$ and $i_L(0)$?

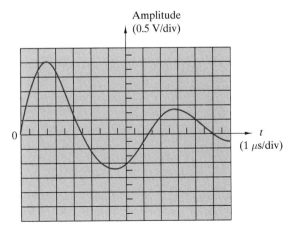

F I G U R E   P 8 – 3 2

8–33 ◢ SECOND-ORDER OP AMP CIRCUIT

The circuit in Figure P8–33 is a second-order circuit consisting of a cascade connection of a first-order RC circuit and a first-order RL circuit.

(a) Find the second-order differential equation relating the output voltage $v_2(t)$ to the input $v_S(t)$.

(b) Show that the roots of the characteristic equation are $-1/R_1C_1$ and $-R_2/L$. That is, show that the natural frequencies of the second-order circuit are the natural frequencies of the two first-order circuits.

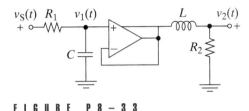

F I G U R E   P 8 – 3 3

CHAPTER 9

LAPLACE TRANSFORMS

My method starts with a complex integral; I fear this sounds rather formidable; but it is really quite simple . . . I am afraid that no physical people will ever try to make out my method: but I am hoping that it may give them confidence to try your methods.

Thomas John Bromwich, 1915,
British Mathematician

9–1 Signal Waveforms and Transforms

9–2 Basic Properties and Pairs

9–3 Pole-zero Diagrams

9–4 Inverse Laplace Transforms

9–5 Some Special Cases

9–6 Circuit Response Using Laplace Transforms

9–7 Initial Value and Final Value Properties

Summary

Problems

Integrating Problems

Laplace transforms have their roots in the pioneering work of the eccentric British engineer Oliver Heaviside (1850–1925). His operational calculus was essentially a collection of intuitive rules that allowed him to formulate and solve a number of the important technical problems of his day. Heaviside was a practical man with no interest in mathematical elegance. His intuitive approach drew bitter criticism from the mathematicians of his day. However, mathematicians like John Bromwich and others eventually recognized the importance of Heaviside's methods and began to supply the necessary mathematical foundations. The foregoing quotation is taken from a 1915 letter from Bromwich to Heaviside in which he described what we now call the Laplace transformation. The transformation is named for Laplace because a complete mathematical development of Heaviside's methods was eventually found in the 1780 writings of the French mathematician Pierre Simon Laplace.

The Laplace transformation provides a new and important method of representing circuits and signals. The transform approach offers a viewpoint and terminology that pervades electrical engineering, particularly in linear circuit analysis and design. The first two sections of this chapter present the basic properties of the Laplace transformation and the concept of converting signals from the time domain to the frequency domain. The third section introduces frequency domain signal description via the pole-zero diagram. The fourth and fifth sections treat the inverse procedure for transforming signals from the frequency domain back into the time domain. In the sixth section, transform methods are used to solve differential equations that describe the response of linear dynamic circuits. The last section treats additional properties of the Laplace transformation that provide important insights into the relationship between the time and frequency domains.

9-1 SIGNAL WAVEFORMS AND TRANSFORMS

A mathematical transformation employs rules to change the form of data without altering its meaning. An example of a transformation is the conversion of numerical data from decimal to binary form. In engineering circuit analysis, transformations are used to obtain alternative representations of circuits and signals. These alternate forms provide a different perspective that can be quite useful or even essential. Examples of the transformations used in circuit analysis are the Fourier transformation, the Z-transformation, and the Laplace transformation. These methods all involve specific transformation rules, make certain analysis techniques more manageable, and provide a useful viewpoint for circuit and system design.

This chapter deals with the Laplace transformation. The discussion of the Laplace transformation follows the path shown in Figure 9–1 by the solid arrow. The process begins with a linear circuit. We derive a differential equation describing the circuit response and then transform this equation into the frequency domain, where it becomes an algebraic equation. Algebraic techniques are then used to solve the transformed equation for the circuit response. The inverse Laplace transformation then changes the frequency domain response into the response waveform in the time domain. The dashed arrow in Figure 9–1 shows that there is another route to the time domain response using the classical techniques discussed in Chapter 7. The classical approach appears to be more direct, but the advantage of the Laplace transformation is that solving a differential equation becomes an algebraic process.

F I G U R E 9 – 1 *Flow diagram dynamic circuit analysis with Laplace transforms.*

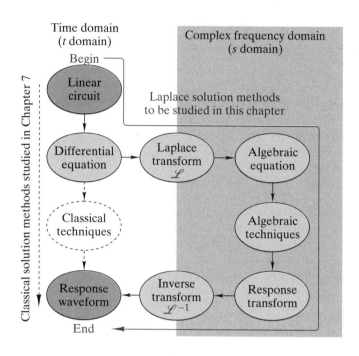

Symbolically, we represent the Laplace transformation as

$$\mathcal{L}\{f(t)\} = F(s) \qquad (9\text{–}1)$$

This expression states that $F(s)$ is the Laplace transform of the waveform $f(t)$. The transformation operation involves two domains: (1) the time domain, in which the signal is characterized by its **waveform** $f(t)$, and (2) the complex frequency domain, in which the signal is represented by its **transform** $F(s)$.

The symbol s stands for the complex frequency variable, a notation we first introduced in Chapter 7 in connection with the zero-state response of linear circuits. The variable s has the dimensions of reciprocal time, or frequency, and is expressed in units of radians per second. In this chapter the complex frequency variable is written as $s = \sigma + j\omega$, where $\sigma = \text{Re}\{s\}$ is the real part and $\omega = \text{Im}\{s\}$ is the imaginary part. This variable is the independent variable in the s domain, just as t is the independent variable in the time domain. Although we cannot physically measure complex frequency in the same sense that we measure time, it is an extremely useful concept that pervades the analysis and design of linear systems.

A signal can be expressed as a waveform or a transform. Collectively, $f(t)$ and $F(s)$ are called a **transform pair**, where the pair involves two representations of the signal. To distinguish between the two forms, a lowercase letter denotes a waveform and an uppercase a transform. For electrical waveforms such as current $i(t)$ or voltage $v(t)$, the corresponding transforms are denoted $I(s)$ and $V(s)$. In this chapter we will use $f(t)$ and $F(s)$ to stand for signal waveforms and transforms in general.

The **Laplace transformation** is defined by the integral

$$F(s) = \int_{0-}^{\infty} f(t)e^{-st}dt \qquad (9\text{–}2)$$

Since the definition involves an improper integral (the upper limit is infinite), we must discuss the conditions under which the integral exists (converges). The integral exists if the waveform $f(t)$ is piecewise continuous and of exponential order. **Piecewise continuous** means that $f(t)$ has a finite number of steplike discontinuities in any finite interval. **Exponential order** means that constants K and b exist such that $|f(t)| < Ke^{bt}$ for all $t > 0$. As a practical matter, the signals encountered in engineering applications meet these conditions.

From the integral definition of the Laplace transformation, we see that when a voltage waveform $v(t)$ has units of volts (V) the corresponding voltage transform $V(s)$ has units of volt-seconds (V-s). Similarly, when a current waveform $i(t)$ has units of amperes (A), the corresponding current transform $I(s)$ has units of ampere-seconds (A-s). Thus, waveforms and transforms do not have the same units. Even so, we often refer to both $V(s)$ and $v(t)$ as voltages and both $I(s)$ and $i(t)$ as currents despite the fact that they have different units. The reason is simply that it is awkward to keep adding the words *waveform* and *transform* to statements when the distinction is clear from the context.

Equation (9–2) uses a lower limit denoted $t = 0-$ to indicate a time just a whisker before $t = 0$. We use $t = 0-$ because in circuit analysis $t = 0$ is defined by a discrete event, such as closing a switch. Such an event may cause a discontinuity in $f(t)$ at $t = 0$. To capture this discontinuity, we set the lower limit at $t = 0-$, just prior to the event. Fortunately, in many situations there is no discontinuity so we will not distinguish between $t = 0-$ and $t = 0$ unless it is crucial.

Equally fortunate is the fact that the number of different waveforms encountered in linear circuits is relatively small. The list includes the three basic waveforms from Chapter 5 (the step, exponential, and sinusoid), as well as composite waveforms such as the impulse, ramp, damped ramp, and damped sinusoid. Since the number of waveforms of interest is relatively small, we do not often use the integral definition in Eq. (9–2) to find Laplace transforms. Once a transform pair has been found, it can be cataloged in a table for future reference and use. Tables 9–2 and 9–3 in this chapter are sufficient for our purposes.

EXAMPLE 9–1

Show that the Laplace transform of the unit step function $f(t) = u(t)$ is $F(s) = 1/s$.

SOLUTION:
Applying Eq. (9–2) yields

$$F(s) = \int_{0}^{\infty} u(t)e^{-st}dt$$

Since $u(t) = 1$ throughout the range of integration this integral becomes

$$F(s) = \int_0^\infty e^{-st}dt = \left. -\frac{e^{-st}}{s} \right|_0^\infty = \left. -\frac{e^{-(\sigma+j\omega)t}}{\sigma+j\omega} \right|_0^\infty$$

The last expression on the right side vanishes at the upper limit since e^{-st} goes to zero as t approaches infinity provided that $\sigma > 0$. At the lower limit the expression reduces to $1/s$. The integral used to calculate $F(s)$ is only valid in the region for which $\sigma > 0$. However, once evaluated, the result $F(s) = 1/s$ can be extended to neighboring regions provided that we avoid the point at $s = 0$ where the function blows up.   ■

EXAMPLE 9–2

Show that the Laplace transform $f(t) = [e^{-\alpha t}]u(t)$ is $F(s) = 1/(s + \alpha)$.

SOLUTION:
Applying Eq. (9–2) yields

$$F(s) = \int_0^\infty e^{-\alpha t}e^{-st}\,dt = \int_0^\infty e^{-(s+\alpha)t}dt = \left. \frac{e^{-(s+\alpha)t}}{-(s+\alpha)} \right|_0^\infty$$

The last term on the right side vanishes at the upper limit since $e^{-(s+\alpha)t}$ vanishes as t approaches infinity provided that $\sigma > \alpha$. At the lower limit the last term reduces to $1/(s + \alpha)$. Again, the integral is only valid for a limited region, but the result $F(s) = 1/(s + \alpha)$ can be extended outside this region if we avoid the point at $s = -\alpha$.   ■

EXAMPLE 9–3

Show that the Laplace transform of the impulse function $f(t) = \delta(t)$ is $F(s) = 1$.

SOLUTION:
Applying Eq. (9–2) yields

$$F(s) = \int_{0-}^\infty \delta(t)e^{-st}\,dt = \int_{0-}^{0+} \delta(t)e^{-st}\,dt = \int_{0-}^{0+} \delta(t)dt = 1$$

The difference between $t = 0-$ and $t = 0$ is important since the impulse is zero everywhere except at $t = 0$. To capture the impulse in the integration, we take a lower limit at $t = 0-$ and an upper limit at $t = 0+$. Since $e^{-st} = 1$ and $\int \delta(t)dt = 1$ on this integration interval, we find that $F(s) = 1$.   ■

INVERSE TRANSFORMATION

So far we have used the direct transformation to convert waveforms into transforms. But Figure 9–1 points out the need to perform the inverse transformation to convert transforms into waveforms. Symbolically, we represent the inverse process as

$$\mathcal{L}^{-1}\{F(s)\} = f(t) \qquad (9\text{–}3)$$

This equation states that $f(t)$ is the inverse Laplace transform of $F(s)$. The **inverse Laplace transformation** is defined by the complex inversion integral

$$f(t) = \frac{1}{2\pi j} \int_{\alpha - j\infty}^{\alpha + j\infty} F(s)e^{st}\, ds \qquad (9\text{–}4)$$

The Laplace transformation is an integral transformation since both the direct process in Eq. (9–2) and the inverse process in Eq. (9–4) involve integrations.

Happily, formal evaluation of the complex inversion integral is not necessary because of the uniqueness property of the Laplace transformation. A symbolical statement of the **uniqueness property** is

$$\text{IF } \mathcal{L}\{f(t)\} = F(s) \text{ THEN } \mathcal{L}^{-1}\{F(s)\}(=)u(t)f(t)$$

The mathematical justification for this statement is beyond the scope of our treatment.[1] However, the notation $(=)$ means "equal almost everywhere." The only points where equality may not hold is at the discontinuities of $f(t)$.

If we just look at the definition of the direct transformation in Eq. (9–2), we could conclude that $F(s)$ is not affected by the values of $f(t)$ for $t < 0$. However, when we use Eq. (9–2) we are not just looking for the Laplace transform of $f(t)$, but a Laplace transform pair such that $\mathcal{L}\{f(t)\} = F(s)$ and $\mathcal{L}^{-1}\{F(s)\} = f(t)$. The inverse Laplace transformation in Eq. (9–4) always produces a causal waveform, one that is zero for $t < 0$. Hence a transform pair $[f(t) \leftrightarrow F(s)]$ is unique if and only if $f(t)$ is causal. For instance, in Example 9–1 we show that $\mathcal{L}\{u(t)\} = 1/s$; hence by the uniqueness property we know that $\mathcal{L}^{-1}\{1/s\}\ (=) u(t)$.

For this reason Laplace transform–related waveforms are written as $[f(t)]u(t)$ to make their causality visible. For example, in the next section we find the Laplace transform of the sinusoid waveform $\cos \beta t$. In the context of Laplace transforms this signal is not an eternal sinusoid but a causal waveform $f(t) = [\cos \beta t]u(t)$. It is important to remember that causality and Laplace transforms go hand in hand when interpreting the results of circuit analysis.

9–2   Basic Properties and Pairs

The previous section gave the definition of the Laplace transformation and showed that the transforms of some basic signals can be found using the integral definition. In this section we develop the basic properties of the Laplace transformation and show how these properties can be used to obtain additional transform pairs.

The **linearity** property of the Laplace transformation states that

$$\mathcal{L}\{Af_1(t) + Bf_2(t)\} = AF_1(s) + BF_2(s) \qquad (9\text{–}5)$$

---

[1] See Wilber R. LePage, *Complex Variables and Laplace Transform for Engineering*, Dover Publishing Co., New York, 1980, p. 318.

where A and B are constants. This property is easily established using the integral definition in Eq. (9–2):

$$\mathcal{L}\{Af_1(t) + Bf_2(t)\} = \int_0^\infty [Af_1(t) + Bf_2(t)]e^{-st}dt$$

$$= A\int_0^\infty f_1(t)e^{-st}dt + B\int_0^\infty f_2(t)e^{-st}dt$$

$$= AF_1(s) + BF_2(s)$$

The integral definition of the inverse transformation in Eq. (9–4) is also a linear operation, so it follows that

$$\mathcal{L}^{-1}\{AF_1(s) + BF_2(s)\} = Af_1(t) + Bf_2(t) \qquad (9\text{--}6)$$

An important consequence of linearity is that for any constant K

$$\mathcal{L}\{Kf(t)\} = KF(s) \text{ and } \mathcal{L}^{-1}\{KF(s)\} = Kf(t) \qquad (9\text{--}7)$$

The linearity property is an extremely important feature that we will use many times in this and subsequent chapters. The next two examples show how this property can be used to obtain the transform of the exponential rise waveform and a sinusoidal waveform.

EXAMPLE 9–4

Show that the Laplace transform of $f(t) = A(1 - e^{-\alpha t})u(t)$ is

$$F(s) = \frac{A\alpha}{s(s + \alpha)}$$

SOLUTION:
This waveform is the difference between a step function and an exponential. We can use the linearity property of Laplace transforms to write

$$\mathcal{L}\{A(1 - e^{-\alpha t})u(t)\} = A\mathcal{L}\{u(t)\} - A\mathcal{L}\{e^{-\alpha t}u(t)\}$$

The transforms of the step and exponential functions were found in Examples 9–1 and 9–2. Using linearity, we find that the transform of the exponential rise is

$$F(s) = \frac{A}{s} - \frac{A}{s + \alpha} = \frac{A\alpha}{s(s + \alpha)} \qquad\blacksquare$$

EXAMPLE 9–5

Show that the Laplace transform of the sinusoid $f(t) = A[\sin(\beta t)]u(t)$ is $F(s) = A\beta/(s^2 + \beta^2)$.

SOLUTION:
Using Euler's relationship, we can express the sinusoid as a sum of exponentials.

$$e^{+j\beta t} = \cos\beta t + j\sin\beta t$$

$$e^{-j\beta t} = \cos\beta t - j\sin\beta t$$

Subtracting the second equation from the first yields

$$f(t) = A \sin \beta t = \frac{A(e^{j\beta t} - e^{-j\beta t})}{2j} = \frac{A}{2j}e^{j\beta t} - \frac{A}{2j}e^{-j\beta t}$$

The transform pair $\mathcal{L}\{e^{-at}\} = 1/(s + \alpha)$ in Example 9–2 is valid even if the exponent α is complex. Using this fact and the linearity property, we obtain the transform of the sinusoid as

$$\mathcal{L}\{A \sin \beta t\} = \frac{A}{2j}\mathcal{L}\{e^{j\beta t}\} - \frac{A}{2j}\mathcal{L}\{e^{-j\beta t}\}$$

$$= \frac{A}{2j}\left[\frac{1}{s - j\beta} - \frac{1}{s + j\beta}\right]$$

$$= \frac{A\beta}{s^2 + \beta^2}$$ ■

INTEGRATION PROPERTY

In the time domain the i–v relationships for capacitors and inductors involve integration and differentiation. Since we will be working in the s domain, it is important to establish the s-domain equivalents of these mathematical operations. Applying the integral definition of the Laplace transformation to a time-domain integration yields

$$\mathcal{L}\left[\int_0^t f(\tau)d\tau\right] = \int_0^\infty \left[\int_0^t f(\tau)d\tau\right]e^{-st}\,dt \tag{9–8}$$

The right side of this expression can be integrated by parts using

$$y = \int_0^t f(\tau)d\tau \quad \text{and} \quad dx = e^{-st}\,dt$$

These definitions result in

$$dy = f(t)dt \quad \text{and} \quad x = \frac{-e^{-st}}{s}$$

Using these factors reduces the right side of Eq. (9–8) to

$$\mathcal{L}\left[\int_0^t f(\tau)d\tau\right] = \left[\frac{-e^{-st}}{s}\int_0^t f(\tau)d\tau\right]_0^\infty + \frac{1}{s}\int_0^\infty f(t)e^{-st}\,dt \tag{9–9}$$

The first term on the right in Eq. (9–9) vanishes at the lower limit because the integral over a zero-length interval is zero provided that $f(t)$ is finite at $t = 0$. It vanishes at the upper limit because e^{-st} approaches zero as t goes to infinity for $\sigma > 0$. By the definition of the Laplace transformation, the second term on the right is $F(s)/s$. We conclude that

$$\mathcal{L}\left[\int_0^t f(\tau)d\tau\right] = \frac{F(s)}{s} \tag{9–10}$$

The **integration property** states that time-domain integration of a waveform $f(t)$ can be accomplished in the s domain by the algebraic process of dividing its transform $F(s)$ by s. The next example applies the integration property to obtain the transform of the ramp function.

EXAMPLE 9–6

Show that the Laplace transform of the ramp function $r(t) = tu(t)$ is $1/s^2$.

SOLUTION:

From our study of signals, we know that the ramp waveform can be obtained from $u(t)$ by integration.

$$r(t) = \int_0^t u(\tau)\, d\tau$$

In Example 9–1 we found $\mathcal{L}\{u(t)\} = 1/s$. Using these facts and the integration property of Laplace transforms, we obtain

$$\mathcal{L}\{r(t)\} = \mathcal{L}\left[\int_0^t u(\tau)\, d\tau\right] = \frac{1}{s}\mathcal{L}\{u(t)\} = \frac{1}{s^2} \qquad \blacksquare$$

DIFFERENTIATION PROPERTY

The time-domain differentiation operation transforms into the s domain as follows:

$$\mathcal{L}\left[\frac{df(t)}{dt}\right] = \int_0^\infty \left[\frac{df(t)}{dt}\right] e^{-st}\, dt \qquad (9\text{–}11)$$

The right side of this equation can be integrated by parts using

$$y = e^{-st} \text{ and } dx = \frac{df(t)}{dt}\, dt$$

These definitions result in

$$dy = -se^{-st}\, dt \quad \text{and} \quad x = f(t)$$

Inserting these factors reduces the right side of Eq. (9–11) to

$$\mathcal{L}\left[\frac{df(t)}{dt}\right] = f(t)e^{-st}\Big|_{0-}^{\infty} + s\int_{0-}^{\infty} f(t)e^{-st}\, dt \qquad (9\text{–}12)$$

For $\sigma > 0$ the first term on the right side of Eq. (9–12) is zero at the upper limit because e^{-st} approaches zero as t goes to infinity. At the lower limit it reduces to $-f(0-)$. By the definition of the Laplace transform, the second term on the right side is $sF(s)$. We conclude that

$$\mathcal{L}\left[\frac{df(t)}{dt}\right] = sF(s) - f(0-) \qquad (9\text{–}13)$$

The **differentiation property** states that time-domain differentiation of a waveform $f(t)$ is accomplished in the s domain by the algebraic process of multiplying the transform $F(s)$ by s and subtracting the constant $f(0-)$. Note that the constant $f(0-)$ is the value of $f(t)$ at $t = 0-$ just prior to $t = 0$.

The s-domain equivalent of a second derivative is obtained by repeated application of Eq. (9–13). We first define a waveform $g(t)$ as

$$g(t) = \frac{df(t)}{dt} \quad \text{hence} \quad \frac{d^2f(t)}{dt^2} = \frac{dg(t)}{dt}$$

Applying the differentiation rule to these two equations yields

$$G(s) = sF(s) - f(0-) \text{ and } \mathcal{L}\left[\frac{d^2f(t)}{dt^2}\right] = sG(s) - g(0-)$$

Substituting the first of these equations into the second results in

$$\mathcal{L}\left[\frac{d^2f(t)}{dt^2}\right] = s^2F(s) - sf(0-) - f'(0-)$$

where

$$f'(0-) = \frac{df}{dt}\bigg|_{t=0-}$$

Repeated application of this procedure produces the nth derivative:

$$\mathcal{L}\left[\frac{d^nf(t)}{dt^n}\right] = s^nF(s) - s^{n-1}f(0-) - s^{n-2}f'(0-)\ldots - f^{(n)}(0-)$$

$$(9-14)$$

where $f^{(n)}(0-)$ is the nth derivative of $f(t)$ evaluated at $t = 0-$.

A hallmark feature of the Laplace transformation is the fact that time integration and differentiation change into algebraic operations in the s domain. This observation gives us our first hint as to why it is often easier to work with circuits and signals in the s domain. The next example shows how the differentiation rule can be used to obtain additional transform pairs.

EXAMPLE 9-7

Show that the Laplace transform of $f(t) = [\cos \beta t]u(t)$ is $F(s) = s/(s^2 + \beta^2)$.

SOLUTION:
We can express $\cos \beta t$ in terms of the derivative of $\sin \beta t$ as

$$\cos \beta t = \frac{1}{\beta}\frac{d}{dt}\sin \beta t$$

In Example 9-5 we found $\mathcal{L}\{\sin \beta t\} = \beta/(s^2 + \beta^2)$. Using these facts and the differentiation rule, we can find the Laplace transform of $\cos \beta t$ as follows:

$$\mathcal{L}\{\cos \beta t\} = \frac{1}{\beta}\mathcal{L}\left\{\frac{d}{dt}\sin \beta t\right\} = \frac{1}{\beta}\left[s\left(\frac{\beta}{s^2 + \beta^2}\right) - \sin (0-)\right]$$

$$= \frac{s}{s^2 + \beta^2} \qquad \blacksquare$$

TRANSLATION PROPERTIES

The **s-domain translation property** of the Laplace transformation is

$$\text{IF } \mathcal{L}\{f(t)\} = F(s) \text{ THEN } \mathcal{L}\{e^{-\alpha t}f(t)\} = F(s + \alpha)$$

This theorem states that multiplying $f(t)$ by $e^{-\alpha t}$ is equivalent to replacing s by $s + \alpha$ (that is, translating the origin in the s plane by an amount α). In engineering applications the parameter α is always a real number, but it can be either positive or negative so the origin in the s domain can be translated to the left or right. Proof of the theorem follows almost immediately from the definition of the Laplace transformation.

$$\mathcal{L}\{e^{-\alpha t}f(t)\} = \int_0^\infty e^{-\alpha t}f(t)e^{-st}dt$$

$$= \int_0^\infty f(t)e^{-(s+\alpha)t}\, dt$$

$$= F(s + \alpha)$$

The s-domain translation property can be used to derive transforms of damped waveforms from undamped prototypes. For instance, the Laplace transform of the ramp, cosine, and sine functions are

$$\mathcal{L}\{tu(t)\} = \frac{1}{s^2}$$

$$\mathcal{L}\{[\cos \beta t]u(t)\} = \frac{s}{s^2 + \beta^2}$$

$$\mathcal{L}\{[\sin \beta t]u(t)\} = \frac{\beta}{s^2 + \beta^2}$$

To obtain the damped ramp, damped cosine, and damped sine functions, we multiply each undamped waveform by $e^{-\alpha t}$. Using the s-domain translation property, we replace s by $s + \alpha$ to obtain transforms of the corresponding damped waveforms.

$$\mathcal{L}\{te^{-\alpha t}u(t)\} = \frac{1}{(s + \alpha)^2}$$

$$\mathcal{L}\{[e^{-\alpha t}\cos \beta t]u(t)\} = \frac{s + \alpha}{(s + \alpha)^2 + \beta^2}$$

$$\mathcal{L}\{[e^{-\alpha t}\sin \beta t]u(t)\} = \frac{\beta}{(s + \alpha)^2 + \beta^2}$$

This completes the derivation of a basic set of transform pairs.

The **time-domain translation property** of the Laplace transformation is

IF $\mathcal{L}\{f(t)\} = F(s)$ THEN for $a > 0$ $\mathcal{L}\{f(t - a)u(t - a)\} = e^{-as}F(s)$

The theorem states that multiplying $F(s)$ by e^{-as} is equivalent to shifting $f(t)$ to the right in the time domain by an amount $a > 0$. In other words, it is equivalent to delaying $f(t)$ in time by an amount $a > 0$. Proof of this property follows from the definition of the Laplace transformation.

$$\mathcal{L}\{f(t - a)u(t - a)\} = \int_{0-}^\infty f(t - a)u(t - a)e^{-st}dt = \int_a^\infty f(t - a)e^{-st}dt$$

In this equation we have used the fact that $u(t - a)$ is zero for $t < a$ and is unity for $t \geq a$. We now change the integration variable from t to $\tau = t - a$.

With this change of variable the last integral in this equation takes the form

$$\mathcal{L}\{f(t-a)u(t-a)\} = \int_0^\infty f(\tau)e^{-s\tau}e^{-as}d\tau$$

$$= e^{-as}\int_0^\infty f(\tau)e^{-s\tau}\,d\tau$$

$$= e^{-as}F(s)$$

which confirms the statement of the time-domain translation property. A simple application of this property is finding the Laplace transform of the delayed step function.

$$\mathcal{L}\{u(t-T)\} = e^{-sT}\mathcal{L}\{u(t)\} = \frac{e^{-sT}}{s}$$

In this section we derived the basic transform properties listed in Table 9–1.

The Laplace transformation has other properties that are useful in signal-processing applications. We treat two of these properties in the last section of this chapter. However, the basic properties in Table 9–1 are used frequently in circuit analysis and are sufficient for nearly all of the applications in this book.

Similarly, Table 9–2 lists a basic set of Laplace transform pairs that is sufficient for most of the applications in this book. All of these pairs were derived in the preceding two sections.

T A B L E 9–1   BASIC LAPLACE TRANSFORMATION PROPERTIES

PROPERTIES	TIME DOMAIN	FREQUENCY DOMAIN
Independent variable	t	s
Signal representation	$f(t)$	$F(s)$
Uniqueness	$\mathcal{L}^{-1}\{F(s)\}\ (=)\ [f(t)]u(t)$	$\mathcal{L}\{f(t)\} = F(s)$
Linearity	$Af_1(t) + Bf_2(t)$	$AF_1(s) + BF_2(s)$
Integration	$\displaystyle\int_0^t f(\tau)\,d\tau$	$\dfrac{F(s)}{s}$
Differentiation	$\dfrac{df(t)}{dt}$	$sF(s) - f(0-)$
	$\dfrac{d^2f(t)}{dt^2}$	$s^2F(s) - sf(0-) - f'(0-)$
	$\dfrac{d^3f(t)}{dt^3}$	$s^3F(s) - s^2f(0-) - sf'(0-) - f''(0-)$
s-Domain translation	$e^{-\alpha t}f(t)$	$F(s + \alpha)$
t-Domain translation	$f(t-a)u(t-a)$	$e^{-as}F(s)$

TABLE 9–2   BASIC LAPLACE TRANSFORM PAIRS

SIGNAL	WAVEFORM f(t)	TRANSFORM F(s)
Impulse	$\delta(t)$	1
Step function	$u(t)$	$\dfrac{1}{s}$
Ramp	$tu(t)$	$\dfrac{1}{s^2}$
Exponential	$[e^{-\alpha t}]u(t)$	$\dfrac{1}{s + \alpha}$
Damped ramp	$[te^{-\alpha t}]u(t)$	$\dfrac{1}{(s + \alpha)^2}$
Sine	$[\sin \beta t]u(t)$	$\dfrac{\beta}{s^2 + \beta^2}$
Cosine	$[\cos \beta t]u(t)$	$\dfrac{s}{s^2 + \beta^2}$
Damped sine	$[e^{-\alpha t}\sin \beta t]u(t)$	$\dfrac{\beta}{(s + \alpha)^2 + \beta^2}$
Damped cosine	$[e^{-\alpha t}\cos \beta t]u(t)$	$\dfrac{(s + \alpha)}{(s + \alpha)^2 + \beta^2}$

All of the waveforms in Table 9–2 are causal. As a result, the Laplace transform pairs are unique and we can use the table in either direction. That is, given an $f(t)$ in the waveform column we find its Laplace transform in the right column, or given an $F(s)$ in the right column we find its inverse transform in the waveform column.

The last example in this section shows how to use the properties and pairs in Tables 9–1 and 9–2 to obtain the transform of a waveform not listed in the tables.

EXAMPLE 9–8

Find the Laplace transform of the waveform

$$f(t) = 2u(t) - 5[e^{-2t}]u(t) + 3[\cos 2t]u(t) + 3[\sin 2t]u(t)$$

SOLUTION:
Using the linearity property, we write the transform of $f(t)$ in the form

$$\mathcal{L}\{f(t)\} = 2\mathcal{L}\{u(t)\} - 5\mathcal{L}\{e^{-2t}u(t)\} + 3\mathcal{L}\{[\cos 2t]u(t)\} + 3\mathcal{L}\{[\sin 2t]u(t)\}$$

The transforms of each term in this sum are listed in Table 9–2:

$$F(s) = \frac{2}{s} - \frac{5}{s + 2} + \frac{3s}{s^2 + 4} + \frac{6}{s^2 + 4}$$

Normally, a Laplace transform is written as a quotient of polynom
rather than as a sum of terms. Rationalizing the preceding sum yields

$$F(s) = \frac{16(s^2 + 1)}{s(s + 2)(s^2 + 4)}$$

Exercise 9–1

Find the Laplace transforms of the following waveforms:

(a) $f(t) = [e^{-2t}]u(t) + 4tu(t) - u(t)$

(b) $f(t) = [2 + 2 \sin 2t - 2 \cos 2t]u(t)$

Answers:

(a) $F(s) = \dfrac{2(s + 4)}{s^2(s + 2)}$

(b) $F(s) = \dfrac{4(s + 2)}{s(s^2 + 4)}$

Exercise 9–2

Find the Laplace transforms of the following waveforms:

(a) $f(t) = [e^{-4t}]u(t) + 5 \displaystyle\int_0^t \sin 4x \, dx$

(b) $f(t) = 5[e^{-40t}]u(t) + \dfrac{d[5te^{-40t}]u(t)}{dt}$

Answers:

(a) $F(s) = \dfrac{s^3 + 36s + 80}{s(s + 4)(s^2 + 16)}$

(b) $F(s) = \dfrac{10s + 200}{(s + 40)^2}$

Exercise 9–3

Find the Laplace transforms of the following waveforms:

(a) $f(t) = A[\cos(\beta t - \phi)]u(t)$

(b) $f(t) = A[e^{-\alpha t} \cos (\beta t - \phi)]u(t)$

Answers:

(a) $F(s) = A \cos \phi \left[\dfrac{s + \beta \tan \phi}{s^2 + \beta^2} \right]$

(b) $F(s) = A \cos \phi \left[\dfrac{s + \alpha + \beta \tan \phi}{(s + \alpha)^2 + \beta^2} \right]$

Find the Laplace transforms of the following waveforms:

$$\text{(a) } f(t) = Au(t) - 2Au(t - T) + Au(t - 2T)$$

$$\text{(b) } f(t) = Ae^{-\alpha(t-T)}u(t - T)$$

Answers:

$$\text{(a) } F(s) = \frac{A(1 - e^{-Ts})^2}{s}$$

$$\text{(b) } F(s) = \frac{Ae^{-Ts}}{s + \alpha}$$

9–3 POLE-ZERO DIAGRAMS

The transforms for signals in Table 9–2 are ratios of polynomials in the complex frequency variable s. Likewise, the transform found in Example 9–8 takes the form of a ratio of two polynomials in s. These results illustrate that the signal transforms of greatest interest to us usually have the form

$$F(s) = \frac{b_m s^m + b_{m-1}s^{m-1} + \ldots + b_1 s + b_0}{a_n s^n + a_{n-1}s^{n-1} + \ldots + a_1 s + a_0} \tag{9–15}$$

If numerator and denominator polynomials are expressed in factored form, then $F(s)$ is written as

$$F(s) = K\frac{(s - z_1)(s - z_2)\ldots(s - z_m)}{(s - p_1)(s - p_2)\ldots(s - p_n)} \tag{9–16}$$

where the constant $K = b_m/a_n$ is called the **scale factor**.

The roots of the numerator and denominator polynomials, together with the scale factor K, uniquely define a transform $F(s)$. The denominator roots are called **poles** because for $s = p_i$ ($i = 1, 2, \ldots n$) the denominator vanishes and $F(s)$ becomes infinite. The roots of the numerator polynomial are called **zeros** because the transform $F(s)$ vanishes for $s = z_i$ ($i = 1, 2, \ldots m$). Collectively the poles and zeros are called **critical frequencies** because they are values of s at which $F(s)$ does dramatic things, like vanish or blow up.

In the s domain we can specify a signal transform by listing the location of its critical frequencies together with the scale factor K. That is, in the frequency domain we describe signals in terms of poles and zeros. The description takes the form of a **pole-zero diagram**, which shows the location of poles and zeros in the complex s plane. The pole locations in such plots are indicated by an $\times$ and the zeros by an $\bigcirc$. The independent variable in the frequency domain is the complex frequency variable s, so the poles or zeros can be complex as well. In the s plane we use a horizontal axis to plot the value of the real part of s and a vertical j-axis to plot the imaginary part. The j-axis is an important boundary in the frequency domain because it divides the s plane into two distinct half planes. The real part of s is negative in the left half plane and positive in the right half plane. As we will soon see, the sign of the real part of a pole has a profound effect on the form of the corresponding waveform.

For example, Table 9–2 shows that the transform of the exponential waveform $f(t) = e^{-\alpha t}u(t)$ is $F(s) = 1/(s + \alpha)$. The exponential signal has a sin-

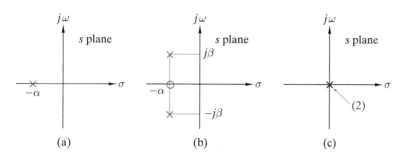

FIGURE 9–2 *Pole-zero diagrams in the s plane.*

gle pole at $s = -\alpha$ and no finite zeros. The pole-zero diagram in Figure 9–2(a) is the s-domain portrayal of the exponential signal. In this diagram the × identifies the pole located at $s = -\alpha + j0$, a point on the negative real axis in the left half plane.

The damped sinusoid $f(t) = [A\, e^{-\alpha t} \cos \beta t]u(t)$ is an example of a signal with complex poles. From Table 9–2 the corresponding transform is

$$F(s) = \frac{A(s + \alpha)}{(s + \alpha)^2 + \beta^2}$$

The transform $F(s)$ has a finite zero on the real axis at $s = -\alpha$. The roots of the denominator polynomial are $s = -\alpha \pm j\beta$. The resulting pole-zero diagram is shown in Figure 9–2(b). The poles of the damped cosine do not lie on either axis in the s plane because neither the real nor imaginary parts are zero.

Finally, the transform of a unit ramp $f(t) = tu(t)$ is $F(s) = 1/s^2$. This transform has no finite zeros and two poles at the origin ($s = 0 + j0$) in the s plane as shown in Figure 9–2(c). The poles in all of the diagrams of Figure 9–2 lie in the left half plane or on the j-axis boundary.

The diagrams in Figure 9–2 show the poles and zeros in the finite part of the s plane. Signal transforms may have poles or zeros at infinity as well. For example, the step function has a zero at infinity since $F(s) = 1/s$ approaches zero as $s \to \infty$. In general, a transform $F(s)$ given by Eq. 9–16 has a zero of order $n - m$ at infinity if $n > m$ and a pole of order $m - n$ at infinity if $n < m$. Thus, the number of zeros equals the number of poles if we include those at infinity.

The pole-zero diagram is the s-domain portrayal of the signal, just as a plot of the waveform versus time depicts the signal in the t domain. The utility of a pole-zero diagram as a description of circuits and signals will become clearer as we develop additional s-domain analysis and design concepts.

EXAMPLE 9–9

Find the poles and zeros of the waveform

$$f(t) = [e^{-2t} + \cos 2t - \sin 2t]u(t)$$

SOLUTION:
Using the linearity property and the basic pairs in Table 9–2, we write the transform in the form

$$F(s) = \frac{1}{s + 2} + \frac{s}{s^2 + 4} - \frac{2}{s^2 + 4}$$

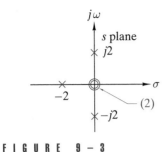

FIGURE 9–3

Rationalizing this expression yields $F(s)$.

$$F(s) = \frac{2s^2}{(s + 2)(s^2 + 4)} = \frac{2s^2}{(s + 2)(s + j2)(s - j2)}$$

This transform has three zeros and three poles. There are two zeros at $s = 0$ and one at $s = \infty$. There is a pole on the negative real axis at $s = -2 + j0$, and two poles on the imaginary axis at $s = \pm j2$. The resulting pole-zero diagram is shown in Figure 9–3. Reviewing the analysis, we can trace the poles to the components of $f(t)$. The pole on the real axis at $s = -2$ came from the exponential e^{-2t}, while the complex conjugate poles on the j-axis came from the sinusoid $\cos 2t - \sin 2t$. The zeros, however, are not traceable to specific components. Their locations depend on all three components. ■

EXAMPLE 9–10

Use a math analysis program to find the Laplace transform of the waveform

$$f(t) = [200te^{-25t} + 10e^{-50t} \sin (40t)]u(t)$$

and construct a pole-zero map of $F(s)$.

SOLUTION:
In this example we will illustrate the Laplace transform capabilities of MATLAB. Operating in the MATLAB command window, we first write the statement

```
syms t s
```

This defines the symbols t (for time) and s (for complex frequency) as symbolic variables rather than numerical quantities. We can then write the waveform $f(t)$ as

```
f=200*t*exp(-25*t)+10*exp(-50*t)*sin(40*t);
```

The Laplace transformation is performed by the statement

```
F=laplace (f, t, s);
```

This statement causes MATLAB to transform the previously defined waveform f from a domain where the independent variable is t to a domain in which the independent variable is s. The result of the transformation yields

```
F=
200/(s+25)^2+400/((s+50)^2+1600)
```

This result is clearly a term-by-term transformation of the waveform $f(t)$. The first term comes from the damped ramp and has a double pole at $s = -25$. The second term comes from the damped sine and has a pair of conjugate complex poles at $s = -50 \pm j40$. This set of poles can be predicted in advance from the form of the terms in the input waveform. To find the zeros of $F(s)$ we must express this result as a quotient of polynomials. To accomplish this we use a powerful MATLAB function called `simplify` that attempts to simplify an expression using many different mathematical tools. In the present case the result is

```
simplify(F)
ans=
200*(3*s^2+200*s+5350)/(s+25)^2/(s^2+100*s+4100)
```

Interpreting this result requires careful application of the MATLAB rules of precedence. Fortunately, MATLAB offers another way out. The function `pretty` attempts to place an expression in a form that more closely resembles type set mathematics. In the present case the result is

```
pretty(ans)
                  2
          3 s   + 200 s + 5350
    200  --------------------------
               2   2
          (s + 25) (s + 100 s + 4100)
```

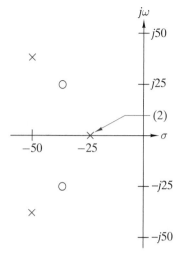

The zeros of $F(s)$ are roots of the numerator polynomial in this expression and are located at $s = -33.33 \pm j25.93$. Figure 9–4 shows the resulting pole-zero map. ∎

FIGURE 9 - 4

Exercise 9–5

Find the poles and zeros of transforms of each of the following waveforms:

(a) $f(t) = [-2e^{-t} - t + 2]u(t)$

(b) $f(t) = [4 - 3\cos \beta t]u(t)$

(c) $f(t) = [e^{-\alpha t}\cos \beta t + (\alpha/\beta)\, e^{-\alpha t}\sin \beta t]u(t)$

Answers:
(a) Zeros: $s = 1, s = \infty$ (2); poles: $s = 0$ (2), $s = -1$
(b) Zeros: $s = \pm j2\beta, s = \infty$; poles: $s = 0, s = \pm j\beta$
(c) Zeros: $s = -2\alpha, s = \infty$; poles: $s = -\alpha \pm j\beta$

Exercise 9–6

A transform has poles at $s = -3 \pm j6$ and $s = -2$ and finite zeros at $s = 0$ and $s = -1$. Write $F(s)$ as a quotient of polynomials in s.

Answer:

$$F(s) = K\frac{s^2 + s}{s^3 + 8s^2 + 57s + 90}$$

9-4 INVERSE LAPLACE TRANSFORMS

The inverse transformation converts a transform $F(s)$ into the corresponding waveform $f(t)$. Applying the inverse transformation in Eq. (9–4) requires knowledge of a branch of mathematics called the complex variable. Fortunately, we do not need Eq. (9–4) because the uniqueness of the Laplace transform pairs in Table 9–2 allows us to go from a transform to a waveform. This may not seem like much help since it does not take a very complicated circuit or signal before we exceed the listing in Table 9–2, or even the more extensive tables that are available. However, there is a general method of expanding $F(s)$ into a sum of terms that are listed in Table 9–2.

For linear circuits the transforms of interest are ratios of polynomials in s. In mathematics such functions are called **rational functions**. To perform the inverse transformation, we must find the waveform corresponding to rational functions of the form

$$F(s) = K\frac{(s - z_1)(s - z_2)\ldots(s - z_m)}{(s - p_1)(s - p_2)\ldots(s - p_n)} \tag{9–17}$$

where the K is the scale factor, z_i ($i = 1, 2, \ldots m$) are the zeros, and p_i ($i = 1, 2, \ldots n$) are the poles of $F(s)$.

If there are more finite poles than finite zeros ($n > m$), then $F(s)$ is called a **proper rational function**. If the denominator in Eq. (9–17) has no repeated roots ($p_i \neq p_j$ for $i \neq j$), then $F(s)$ is said to have **simple poles**. In this section we treat the problem of finding the inverse transform of proper rational functions with simple poles. The problem of improper rational functions and multiple poles is covered in the next section.

If a proper rational function has only simple poles, then it can be decomposed into a partial fraction expansion of the form

$$F(s) = \frac{k_1}{s - p_1} + \frac{k_2}{s - p_2} + \ldots + \frac{k_n}{s - p_n} \tag{9–18}$$

In this case, $F(s)$ can be expressed as a linear combination of terms with one term for each of its n simple poles. The k's associated with each term are called **residues**.

Each term in the partial fraction decomposition has the form of the transform of an exponential signal. That is, we recognize that $\mathscr{L}^{-1}\{k/(s + \alpha)\} = [ke^{-\alpha t}]u(t)$. We can now write the corresponding waveform using the linearity property

$$f(t) = [k_1 e^{p_1 t} + k_2 e^{p_2 t} + \ldots + k_n e^{p_n t}]u(t) \tag{9–19}$$

In the time domain, the s-domain poles appear in the exponents of exponential waveforms and the residues at the poles become the amplitudes.

Given the poles of $F(s)$, finding the inverse transform $f(t)$ reduces to finding the residues. To illustrate the procedure, consider a case in which $F(s)$ has three simple poles and one finite zero.

$$F(s) = K\frac{(s - z_1)}{(s - p_1)(s - p_2)(s - p_3)} = \frac{k_1}{s - p_1} + \frac{k_2}{s - p_2} + \frac{k_3}{s - p_3}$$

We find the residue k_1 by first multiplying this equation through by the factor $(s - p_1)$:

$$(s - p_1)F(s) = K\frac{(s - z_1)}{(s - p_2)(s - p_3)} = k_1 + \frac{k_2(s - p_1)}{s - p_2} + \frac{k_3(s - p_1)}{s - p_3}$$

If we now set $s = p_1$, the last two terms on the right vanish, leaving

$$k_1 = (s - p_1)F(s)\big|_{s = p_1} = K\frac{(s - z_1)}{(s - p_2)(s - p_3)}\bigg|_{s = p_1}$$

Using the same approach for k_2 yields

$$k_2 = (s - p_2)F(s)\big|_{s=p_2} = K\frac{(s - z_1)}{(s - p_1)(s - p_3)}\bigg|_{s=p_2}$$

The technique generalizes so that the residue at any simple pole p_i is

$$k_i = (s - p_i)F(s)\big|_{s=p_i} \qquad (9\text{--}20)$$

The process of determining the residue at any simple pole is sometimes called the **cover-up algorithm** because we temporarily remove (cover up) the factor $(s - p_i)$ in $F(s)$ and then evaluate the remainder at $s = p_i$.

EXAMPLE 9-11

Find the waveform corresponding to the transform

$$F(s) = 2\frac{(s + 3)}{s(s + 1)(s + 2)}$$

SOLUTION:
$F(s)$ is a proper rational function and has simple poles at $s = 0$, $s = -1$, $s = -2$. Its partial fraction expansion is

$$F(s) = \frac{k_1}{s} + \frac{k_2}{s + 1} + \frac{k_3}{s + 2}$$

The cover-up algorithm yields the residues as

$$k_1 = sF(s)\big|_{s=0} = \frac{2(s + 3)}{(s + 1)(s + 2)}\bigg|_{s=0} = 3$$

$$k_2 = (s + 1)F(s)\big|_{s=-1} = \frac{2(s + 3)}{s(s + 2)}\bigg|_{s=-1} = -4$$

$$k_3 = (s + 2)F(s)\big|_{s=-2} = \frac{2(s + 3)}{s(s + 1)}\bigg|_{s=-2} = 1$$

The inverse transform $f(t)$ is

$$f(t) = [3 - 4e^{-t} + e^{-2t}]u(t) \qquad \blacksquare$$

Exercise 9-7

Find the waveforms corresponding to the following transforms:

(a) $F_1(s) = \dfrac{4}{(s + 1)(s + 3)}$

(b) $F_2(s) = e^{-5s}\left[\dfrac{2s}{(s + 1)(s + 3)}\right]$

(c) $F_3(s) = \dfrac{4(s + 2)}{(s + 1)(s + 3)}$

Answers:

$$(a) f_1(t) = [2e^{-t} - 2e^{-3t}]u(t)$$

$$(b) f_2(t) = [-e^{-(t-5)} + 3e^{-3(t-5)}]u(t-5)$$

$$(c) f_3(t) = [2e^{-t} + 2e^{-3t}]u(t)$$

Comment: Note that

$$F_2(s) = e^{-5s}\left[\frac{2s}{(s+1)(s+3)}\right] = e^{-5s}\left[\frac{-1}{s+1} + \frac{3}{s+3}\right]$$

so that the delay factor e^{-5s} is not involved in the partial fraction expansion but simply flags the amount by which the resulting waveform $[-e^{-t} + 3e^{-3t}]u(t)$ is delayed to produce $f_2(t)$.

Exercise 9–8

Find the waveforms corresponding to the following transforms:

$$(a) F(s) = \frac{6(s+2)}{s(s+1)(s+4)}$$

$$(b) F(s) = \frac{4(s+1)}{s(s+1)(s+4)}$$

Answers:

$$(a) f(t) = [3 - 2e^{-t} - e^{-4t}]u(t)$$

$$(b) f(t) = [1 - e^{-4t}]u(t)$$

COMPLEX POLES

Special treatment is necessary when $F(s)$ has a complex pole. In physical situations the function $F(s)$ is a ratio of polynomials with real coefficients. If $F(s)$ has a complex pole $p = -\alpha + j\beta$, then it must also have a pole $p^* = -\alpha - j\beta$; otherwise the coefficients of the denominator polynomial would not be real. In other words, for physical signals the complex poles of $F(s)$ must occur in conjugate pairs. As a consequence, the partial fraction decomposition of $F(s)$ will contain two terms of the form

$$F(s) = \ldots + \frac{k}{s + \alpha - j\beta} + \frac{k^*}{s + \alpha + j\beta} + \ldots \qquad (9\text{–}21)$$

The residues k and k^* at the conjugate poles are themselves conjugates because $F(s)$ is a rational function with real coefficients. These residues can be calculated using the cover-up algorithm and, in general, they turn out to be complex numbers. If the complex residues are written in polar form as

$$k = |k|e^{j\theta} \text{ and } k^* = |k|e^{-j\theta}$$

then the waveform corresponding to the two terms in Eq. (9–21) is

$$f(t) = [\ldots + |k|e^{j\theta}e^{(-\alpha + j\beta)t} + |k|e^{-j\theta}e^{(-\alpha - j\beta)t} + \ldots]u(t)$$

This equation can be rearranged in the form

$$f(t) = \left[\ldots + 2|k|e^{-\alpha t}\left\{ \frac{e^{+j(\beta t+\theta)} + e^{-j(\beta t+\theta)}}{2} \right\} + \ldots \right]u(t) \qquad (9\text{--}22)$$

The expression inside the brackets is of the form

$$\cos x = \left\{ \frac{e^{+jx} + e^{-jx}}{2} \right\}$$

Consequently, we combine terms inside the braces as a cosine function with a phase angle:

$$f(t) = [\ldots + 2|k|e^{-\alpha t}\cos(\beta t + \theta) + \ldots]u(t) \qquad (9\text{--}23)$$

In summary, if $F(s)$ has a complex pole, then in physical applications there must be an accompanying conjugate complex pole. The inverse transformation combines the two poles to produce a damped cosine waveform. We only need to compute the residue at one of these poles because the residues at conjugate poles must be conjugates. Normally, we calculate the residue for the pole at $s = -\alpha + j\beta$ because its angle equals the phase angle of the damped cosine. Note that the imaginary part of this pole is positive, which means that the pole lies in the upper half of the s plane.

The inverse transform of a proper rational function with simple poles can be found by the partial fraction expansion method. The residues k at the simple poles can be found using the cover-up algorithm. The resulting waveform is a sum of terms of the form $[ke^{-\alpha t}]u(t)$ for real poles and $[2|k|e^{-\alpha t}\cos(\beta t + \theta)]u(t)$ for a pair of complex conjugate poles. The partial fraction expansion of the transform contains all of the data needed to construct the corresponding waveform.

EXAMPLE 9-12

Find the inverse transform of

$$F(s) = \frac{20(s + 3)}{(s + 1)(s^2 + 2s + 5)}$$

SOLUTION:
$F(s)$ has a simple pole at $s = -1$ and a pair of conjugate complex poles located at the roots of the quadratic factor

$$(s^2 + 2s + 5) = (s + 1 - j2)(s + 1 + j2)$$

The partial fraction expansion of $F(s)$ is

$$F(s) = \frac{k_1}{s + 1} + \frac{k_2}{s + 1 - j2} + \frac{k_2^*}{s + 1 + j2}$$

The residues at the poles are found from the cover-up algorithm.

$$k_1 = \left. \frac{20(s + 3)}{s^2 + 2s + 5} \right|_{s = -1} = 10$$

$$k_2 = \left. \frac{20(s + 3)}{(s + 1)(s + 1 + j2)} \right|_{s = -1 + j2} = -5 - j5 = 5\sqrt{2}e^{+j5\pi/4}$$

We now have all of the data needed to construct the inverse transform.

$$f(t) = [10e^{-t} + 10\sqrt{2}e^{-t}\cos(2t + 5\pi/4)]u(t)$$

In this example we used k_2 to obtain the amplitude and phase angle of the damped cosine term. The residue k_2^* is not needed, but to illustrate a point we note that its value is

$$k_2^* = (-5 - j5)^* = -5 + j5 = 5\sqrt{2}e^{-j5\pi/4}$$

If k_2^* is used instead, we get the same amplitude for the damped sine but the wrong phase angle. *Caution:* Remember that Eq. (9–23) uses the residue at the complex pole with a positive imaginary part. In this example this is the pole at $s = -1 + j2$, not the pole at $s = -1 - j2$.

Exercise 9–9

Find the inverse transforms of the following rational functions:

$$\text{(a) } F(s) = \frac{16}{(s + 2)(s^2 + 4)}$$

$$\text{(b) } F(s) = \frac{2(s + 2)}{s(s^2 + 4)}$$

Answers:

$$\text{(a) } f(t) = [2e^{-2t} + 2\sqrt{2}\cos(2t - 3\pi/4)]u(t)$$

$$\text{(b) } f(t) = [1 + \sqrt{2}\cos(2t - 3\pi/4)]u(t)$$

Exercise 9–10

Find the inverse transforms of the following rational functions:

$$\text{(a) } F(s) = \frac{8}{s(s^2 + 4s + 8)}$$

$$\text{(b) } F(s) = \frac{4s}{s^2 + 4s + 8}$$

Answers:

$$\text{(a) } f(t) = [1 + \sqrt{2}e^{-2t}\cos(2t + 3\pi/4)]u(t)$$

$$\text{(b) } f(t) = [4\sqrt{2}e^{-2t}\cos(2t + \pi/4)]u(t)$$

SUMS OF RESIDUES

The sum of the residues of a proper rational function are subject to certain conditions that are useful for checking the calculations in a partial fraction expansion. To derive these conditions, we multiply Eqs. (9–17) and (9–18) by s and take the limit as $s \to \infty$. These operations yield

$$\lim_{s \to \infty} sF(s) = \lim_{s \to \infty} \frac{Ks^{m+1}}{s^n} = \lim_{s \to \infty} \left(\frac{k_1 s}{s + p_1} + \ldots + \frac{k_n s}{s + p_n} \right)$$

In the limit this equation reduces to

$$K \left[\lim_{s \to \infty} \frac{s^{m+1}}{s^n} \right] = k_1 + k_2 + \ldots + k_n$$

Since $F(s)$ is a proper rational function with $n > m$, the limit process in this equation yields to the following conditions:

$$k_1 + k_2 + \ldots + k_n = \begin{cases} 0 \text{ if } n > m + 1 \\ K \text{ if } n = m + 1 \end{cases} \qquad (9\text{–}24)$$

For a proper rational function with simple poles, the sum of residues is either zero or else equal to the transform scale factor K.

Exercise 9–11

Use the sum of residues to find the unknown residue in the following expansions:

(a) $\dfrac{21 (s + 5)}{(s + 3)(s + 10)} = \dfrac{6}{s + 3} + \dfrac{k}{s + 10}$

(b) $\dfrac{58s}{(s + 2)(s^2 + 25)} = \dfrac{k}{s + 2} + \dfrac{2 + j5}{s + j5} + \dfrac{2 - j5}{s - j5}$

Answers:
(a) $k = 15$
(b) $k = -4$

9–5 SOME SPECIAL CASES

Most of the transforms encountered in physical applications are proper rational functions with simple poles. The inverse transforms of such functions can be handled by the partial fraction expansion method developed in the previous section. This section covers the problem of finding the inverse transform when $F(s)$ is an improper rational function or has multiple poles. These matters are treated as special cases because they only occur for certain discrete values of circuit or signal parameters. However, some of these special cases are important, so we need to learn how to handle improper rational functions and multiple poles.

$F(s)$ is an **improper rational function** when the order of the numerator polynomial equals or exceeds the order of the denominator ($m \geq n$). For example, the transform

$$F(s) = \frac{s^3 + 6s^2 + 12s + 11}{s^2 + 4s + 3} \qquad (9\text{–}25)$$

is improper because $m = 3$ and $n = 2$. Using long division this improper rational function can be changed into the sum of a quotient plus a remainder which is a proper rational function. We proceed as follows:

$$\begin{array}{r} s + 2 \\ s^2 + 4s + 3 \;\overline{)\; s^3 + 6s^2 + 12s + 11} \\ \underline{s^3 + 4s^2 + 3s} \\ 2s^2 + 9s + 8 \\ \underline{2s^2 + 8s + 6} \\ s + 2 \end{array}$$

which yields

$$F(s) = s + 2 + \frac{s + 5}{s^2 + 4s + 3}$$

$$= \text{Quotient} + \text{Remainder}$$

The remainder is a proper rational function, which can be expanded by partial fractions to produce

$$F(s) = s + 2 + \frac{2}{s + 1} - \frac{1}{s + 3}$$

All of the terms in this expansion are listed in Table 9–2 except the first term. The inverse transform of the first term is found using the transform of an impulse and the differentiation property. The Laplace transform of the derivative of an impulse is

$$\mathcal{L}\left[\frac{d\delta(t)}{dt}\right] = s\mathcal{L}[\delta(t)] - \delta(0-) = s$$

since $\mathcal{L}\{\delta(t)\} = 1$ and $\delta(0-) = 0$. By the uniqueness property of the Laplace transformation, we have $\mathcal{L}^{-1}\{s\} = d\delta(t)/dt$. The first derivative of an impulse is called a doublet. The inverse transform of the improper rational function in Eq. (9–25) is

$$f(t) = \frac{d\delta(t)}{dt} + 2\delta(t) + [2e^{-t} - 1e^{-3t}]u(t)$$

The method illustrated by this example generalizes in the following way. When $m = n$, long division produces a quotient K plus a proper rational function remainder. The constant K corresponds to an impulse $K\delta(t)$, and the remainder can be expanded by partial fractions to find the corresponding waveform. If $m > n$, then long division yields a quotient with terms like $s, s^2, \ldots s^{m-n}$ before a proper remainder function is obtained. These higher powers of s correspond to derivatives of the impulse. These pathological waveforms are theoretically interesting, but they do not actually occur in real circuits.

Improper rational functions can arise during mathematical manipulation of signal transforms. When $F(s)$ is improper, it is essential to reduce it by long division prior to expansion; otherwise the resulting partial fraction expansion will be incomplete.

Exercise 9–12

Find the inverse transforms of the following functions:

(a) $F(s) = \dfrac{s^2 + 4s + 5}{s^2 + 4s + 3}$

(b) $F(s) = \dfrac{s^2 - 4}{s^2 + 4}$

Answers:

$$(a) f(t) = \delta(t) + [e^{-t} - e^{-3t}]u(t)$$

$$(b) f(t) = \delta(t) - [4\sin(2t)]u(t)$$

Exercise 9–13

Find the inverse transforms of the following functions:

$$(a)\ F(s) = \frac{2s^2 + 3s + 5}{s}$$

$$(b)\ F(s) = \frac{s^3 + 2s^2 + s + 3}{s + 2}$$

Answers:

$$(a)\ f(t) = 2\frac{d\delta(t)}{dt} + 3\delta(t) + 5u(t)$$

$$(b)\ f(t) = \frac{d^2\delta(t)}{dt^2} + \delta(t) + [e^{-2t}]u(t)$$

MULTIPLE POLES

Under certain special conditions, transforms can have multiple poles. For example, the transform

$$F(s) = \frac{K(s - z_1)}{(s - p_1)(s - p_2)^2} \tag{9–26}$$

has a simple pole at $s = p_1$ and a pole of order 2 at $s = p_2$. Finding the inverse transform of this function requires special treatment of the multiple pole. We first factor out one of the two multiple poles.

$$F(s) = \frac{1}{s - p_2}\left[\frac{K(s - z_1)}{(s - p_1)(s - p_2)}\right] \tag{9–27}$$

The quantity inside the brackets is a proper rational function with only simple poles and can be expanded by partial fractions using the method of the previous section.

$$F(s) = \frac{1}{s - p_2}\left[\frac{c_1}{s - p_1} + \frac{k_{22}}{s - p_2}\right]$$

We now multiply through by the pole factored out in the first step to obtain

$$F(s) = \frac{c_1}{(s - p_1)(s - p_2)} + \frac{k_{22}}{(s - p_2)^2}$$

The first term on the right is a proper rational function with only simple poles, so it too can be expanded by partial fractions as

$$F(s) = \frac{k_1}{s - p_1} + \frac{k_{21}}{s - p_2} + \frac{k_{22}}{(s - p_2)^2}$$

After two partial fraction expansions, we have an expression in which every term is available in Table 9–2. The first two terms are simple poles that lead to exponential waveforms. The third term is of the form $k/(s + \alpha)^2$, which is the transform of a damped ramp waveform $[kt\ e^{-\alpha t}]u(t)$. Therefore, the inverse transform of $F(s)$ in Eq. (9–26) is

$$f(t) = [k_1 e^{p_1 t} + k_{12} e^{p_2 t} + k_{22} t e^{p_2 t}]u(t) \tag{9–28}$$

Caution: If $F(s)$ in Eq. (9–26) had another finite zero, then the term in the brackets in Eq. (9–27) would be an improper rational function. When this occurs, long division must be used to reduce the improper rational function before proceeding to the partial-fraction expansion in the next step.

As with simple poles, the *s*-domain location of multiple poles determines the exponents of the exponential waveforms. The residues at the poles are the amplitudes of the waveforms. The only difference here is that the double pole leads to two terms rather than a single waveform. The first term is an exponential of the form e^{pt}, and the second term is a damped ramp of the form te^{pt}.

EXAMPLE 9–13

Find the inverse transform of

$$F(s) = \frac{4(s + 3)}{s(s + 2)^2}$$

SOLUTION:
The given transform has a simple pole at $s = 0$ and a double pole at $s = -2$. Factoring out one of the multiple poles and expanding the remainder by partial fractions yields

$$F(s) = \frac{1}{s + 2}\left[\frac{4(s + 3)}{s(s + 2)}\right] = \frac{1}{s + 2}\left[\frac{6}{s} - \frac{2}{s + 2}\right]$$

Multiplying through by the removed factor and expanding again by partial fractions produces

$$F(s) = \frac{6}{s(s + 2)} - \frac{2}{(s + 2)^2} = \frac{3}{s} - \frac{3}{s + 2} - \frac{2}{(s + 2)^2}$$

The last expansion on the right yields the inverse transform as

$$f(t) = [3 - 3e^{-2t} - 2te^{-2t}]u(t) \qquad\blacksquare$$

In principle, the procedure illustrated in this example can be applied to higher-order poles, although the process rapidly becomes quite tedious. For example, an *n*th-order pole would require *n* partial-fraction expansions, which is not an idea with irresistible appeal. Mathematics offers other methods of determining multiple pole residues that reduce the computational burden somewhat. However, these advanced mathematical tools are probably not worth learning because computer tools like Mathcad and MATLAB readily handle multiple pole situations (see Example 9–14). Although we do encounter functions with high-order multiple poles in later chapters, we rarely need to find their inverse transforms.

Thus, for practical reasons our interest in multiple pole transforms is limited to two possibilities. First, a double pole on the negative real axis leads to the damped ramp:

$$\mathscr{L}^{-1}\left[\frac{k}{(s+\alpha)^2}\right] = [kte^{-\alpha t}]u(t) \qquad (9\text{-}29)$$

Second, a pair of double, complex poles leads to the damped cosine ramp:

$$\mathscr{L}^{-1}\left[\frac{k}{(s+\alpha-j\beta)^2} + \frac{k^*}{(s+\alpha+j\beta)^2}\right] = [2\,|k|\,te^{-\alpha t}\cos(\beta t + \angle k)]u(t)$$

$$(9\text{-}30)$$

These two cases illustrate a general principle: When a simple pole leads to a waveform $f(t)$, then a double pole at the same location leads to a waveform $tf(t)$. Multiplying a waveform by t tends to cause the waveform to increase without bound unless exponential damping is present. The following example illustrates a case in which the damping is not present.

EXAMPLE 9-14

Use a math analysis program to find the waveform corresponding to a transform with a zero at $s = -400$, a simple pole at $s = -1000$, a double pole at $s = j400$, a double pole at $s = -j400$, and a value at $s = 0$ of $F(0) = 2 \times 10^{-4}$. Plot the resulting waveform $f(t)$.

SOLUTION:

The symbolic analysis capability of Mathcad includes inverse Laplace transforms. Figure 9-5 is a worksheet demonstrating this capability. The specified critical frequencies define the $F(s)$ shown in the first line. The specified value at $s = 0$ allows us to evaluate the unknown scale factor K as shown in the second line. This leads directly to the transform in the third

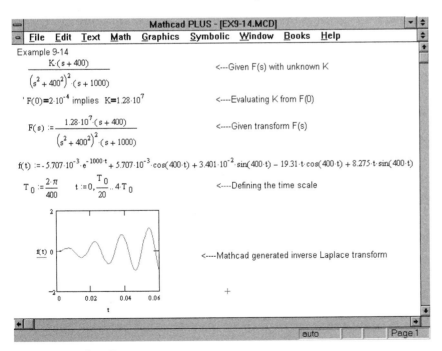

FIGURE 9-5

line. After highlighting (clicking on) an "s" in $F(s)$ to identify the independent variable, we select **Symbolic, Transforms, Inverse Laplace Transform** from the menu bar. Mathcad responds by producing the $f(t)$ shown in the figure. This waveform contains five terms: The exponential e^{-1000t} came from the simple pole at $s = -1000$, and the four sinusoidal terms $\cos(400t)$, $\sin(400t)$, $t \times \cos(400t)$, and $t \times \sin(400t)$ all came from the double poles at $s = \pm j400$. As time increases, the terms $t \times \cos(400t)$ and $t \times \sin(400t)$ increase without bound, leading to the linearly increasing amplitude shown in the waveform plot. A key point to remember is that poles on the j-axis lead to sustained waveforms, like the step function and sinusoid, that do not decay. Consequently, double poles on the j-axis lead to $tf(t)$ waveforms whose amplitudes increase without bound, producing an unstable response.

Exercise 9–14

Find the inverse transforms of the following functions:

$$\text{(a) } F(s) = \frac{s}{(s+1)(s+2)^2}$$

$$\text{(b) } F(s) = \frac{16}{s^2(s+4)}$$

$$\text{(c) } F(s) = \frac{800\,s(s+1)}{(s+2)(s+10)^2}$$

Answers:

$$\text{(a) } f(t) = [-e^{-t} + e^{-2t} + 2te^{-2t}]u(t)$$

$$\text{(b) } f(t) = [4t - 1 + e^{-4t}]u(t)$$

$$\text{(c) } f(t) = [25e^{-2t} + 775e^{-10t} - 9000te^{-10t}]u(t)$$

9–6 CIRCUIT RESPONSE USING LAPLACE TRANSFORMS

The payoff for learning about the Laplace transformation comes when we use it to find the response of dynamic circuits. The pattern for circuit analysis is shown by the solid line in Figure 9–1. The basic analysis steps are as follows:

STEP 1: Develop the circuit differential equation in the time domain.

STEP 2: Transform this equation into the s domain and algebraically solve for the response transform.

STEP 3: Apply the inverse transformation to this transform to produce the response waveform.

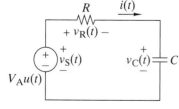

FIGURE 9 – 6 *First-order RC circuit.*

The first-order RC circuit in Figure 9–6 will be used to illustrate these steps.

STEP 1

The KVL equation around the loop and the element i–v relationship or element equations are

$$\text{KVL: } -v_S(t) + v_R(t) + v_C(t) = 0$$

$$\text{Source: } v_S(t) = V_A u(t)$$

$$\text{Resistor: } v_R(t) = i(t)R$$

$$\text{Capacitor: } i(t) = C\frac{dv_C(t)}{dt}$$

Substituting the i–v relationships into the KVL equation and rearranging terms produces a first-order differential equation

$$RC\frac{dv_C(t)}{dt} + v_C(t) = V_A u(t) \qquad (9\text{--}31)$$

with an initial condition $v_C(0-) = V_0$ V.

STEP 2

The analysis objective is to use Laplace transforms to find the waveform $v_C(t)$ that satisfies this differential equation and the initial condition. We first apply the Laplace transformation to both sides of Eq. (9–31):

$$\mathcal{L}\left[RC\frac{dv_C(t)}{dt} + v_C(t)\right] = \mathcal{L}[V_A u(t)]$$

Using the linearity property leads to

$$RC\mathcal{L}\left[\frac{dv_C(t)}{dt}\right] + \mathcal{L}[v_C(t)] = V_A\mathcal{L}[u(t)]$$

Using the differentiation property and the transform of a unit step function produces

$$RC[sV_C(s) - V_0] + V_C(s) = V_A\frac{1}{s} \qquad (9\text{--}32)$$

This result is an algebraic equation in $V_C(s)$, which is the transform of the response we seek. We rearrange Eq. (9–32) in the form

$$(s + 1/RC)V_C(s) = \frac{V_A/RC}{s} + V_0$$

and algebraically solve for $V_C(s)$.

$$V_C(s) = \frac{V_A/RC}{s(s + 1/RC)} + \frac{V_0}{s + 1/RC} \text{ V-s} \qquad (9\text{--}33)$$

The function $V_C(s)$ is the transform of the waveform $v_C(t)$ that satisfies the differential equation and the initial condition. The initial condition appears explicitly in this equation as a result of applying the differentiation rule to obtain Eq. (9–32).

STEP 3

To obtain the waveform $v_C(t)$, we find the inverse transform of the right side of Eq. (9–33). The first term on the right is a proper rational function with two simple poles on the real axis in the s plane. The pole at the origin was introduced by the step function input. The pole at $s = -1/RC$ came from the circuit. The partial fraction expansion of the first term in Eq. (9–33) is

$$\frac{V_A/RC}{s(s + 1/RC)} = \frac{k_1}{s} + \frac{k_2}{s + 1/RC}$$

The residues k_1 and k_2 are found using the cover-up algorithm.

$$k_1 = \left.\frac{V_A/RC}{s + 1/RC}\right|_{s=0} = V_A \quad \text{and} \quad k_2 = \left.\frac{V_A/RC}{s}\right|_{s=-1/RC} = -V_A$$

Using these residues, we expand Eq. (9–33) by partial fractions as

$$V_C(s) = \frac{V_A}{s} - \frac{V_A}{s + 1/RC} + \frac{V_0}{s + 1/RC} \tag{9–34}$$

Each term in this expansion is recognizable: The first is a step function and the next two are exponentials. Taking the inverse transform of Eq. (9–34) gives

$$\begin{aligned} v_C(t) &= [V_A - V_A e^{-t/RC} + V_0 e^{-t/RC}]u(t) \\ &= [V_A + (V_0 - V_A)e^{-t/RC}]u(t) \quad \text{V} \end{aligned} \tag{9–35}$$

The waveform $v_C(t)$ satisfies the differential equation in Eq. (9–31) and the initial condition $v_C(0-) = V_0$. The term $V_A u(t)$ is the forced response due to the step function input, and the term $[(V_0 - V_A)e^{t/RC}]u(t)$ is the natural response. The complete response depends on three parameters: the input amplitude V_A, the circuit time constant RC, and the initial condition V_0.

These results are identical to those found using the classical methods in Chapter 7. The outcome is the same, but the method is quite different. The Laplace transformation yields the complete response (forced and natural) by an algebraic process that inherently accounts for the initial conditions. The solid arrow in Figure 9–1 shows the overall procedure. Begin with Eq. (9–31) and relate each step leading to Eq. (9–35) to steps in Figure 9–1.

EXAMPLE 9–15

The switch in Figure 9–7 has been in position A for a long time. At $t = 0$ it is moved to position B. Find $i_L(t)$ for $t \geq 0$.

SOLUTION:

Step 1: The circuit differential equation is found by combining the KVL equation and element equations with the switch in position B.

$$\text{KVL: } v_R(t) + v_L(t) = 0$$

$$\text{Resistor: } v_R(t) = i_L(t)R$$

$$\text{Inductor: } v_L(t) = L\frac{di_L(t)}{dt}$$

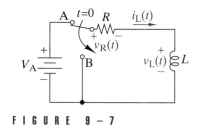

FIGURE 9–7

Substituting the element equations into the KVL equation yields

$$L\frac{di_L(t)}{dt} + Ri_L(t) = 0$$

Prior to $t = 0$, the circuit was in a dc steady-state condition with the switch in position A. Under dc conditions the inductor acts like a short circuit, and the inductor current just prior to moving the switch is $i_L(0-) = I_0 = V_A/R$.

Step 2: Using the linearity and differentiation properties, we transform the circuit differential equation into the s domain as

$$L[sI_L(s) - I_0] + RI_L(s) = 0$$

Solving algebraically for $I_L(s)$ yields

$$I_L(s) = \frac{I_0}{s + R/L} \quad \text{A-s}$$

Step 3: The inverse transform of $I_L(s)$ is an exponential waveform:

$$i_L(t) = [I_0 e^{-Rt/L}]u(t) \quad \text{A}$$

where $I_0 = V_A/R$. Substituting $i_L(t)$ back into the differential equation yields

$$L\frac{di_L(t)}{dt} + Ri_L(t) = -RI_0 e^{-Rt/L} + RI_0 e^{-Rt/L} = 0$$

The waveform found using Laplace transforms does indeed satisfy the circuit differential equation and the initial condition. ∎

EXAMPLE 9-16

The switch in Figure 9-8 has been open for a long time. At $t = 0$ the switch is closed. Find $i(t)$ for $t \geq 0$.

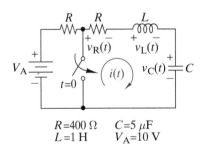

$R = 400 \ \Omega \qquad C = 5 \ \mu\text{F}$
$L = 1 \ \text{H} \qquad V_A = 10 \ \text{V}$

FIGURE 9-8

SOLUTION:
The governing equation for the second-order circuit in Figure 9-8 is found by combining the element equations and a KVL equation around the loop with the switch closed:

$$\text{KVL: } v_R(t) + v_L(t) + v_C(t) = 0$$

$$\text{Resistor: } v_R(t) = Ri(t)$$

$$\text{Inductor: } v_L(t) = L\frac{di(t)}{dt}$$

$$\text{Capacitor: } v_C(t) = \frac{1}{C}\int_0^t i(\tau)d\tau + v_C(0)$$

Substituting the element equations into the KVL equation yields

$$L\frac{di(t)}{dt} + Ri(t) + \frac{1}{C}\int_0^t i(\tau)d\tau + v_C(0) = 0$$

Using the linearity property, the differentiation property, and the integration property, we transform this second-order integrodifferential equation into the s domain as

$$L[sI(s) - i_L(0)] + RI(s) + \frac{1}{C}\frac{I(s)}{s} + v_C(0)\frac{1}{s} = 0$$

Solving for $I(s)$ results in

$$I(s) = \frac{si_L(0) - v_C(0)/L}{s^2 + \dfrac{R}{L}s + \dfrac{1}{LC}} \quad \text{A-s}$$

Prior to $t = 0$, the circuit was in a dc steady-state condition with the switch open. In dc steady state, the inductor acts like a short circuit and the capacitor like an open circuit, so the initial conditions are $i_L(0-) = 0$ A and $v_C(0-) = V_A = 10$ V. Inserting the initial conditions and the numerical values of the circuit parameters into the equation for $I(s)$ gives

$$I(s) = -\frac{10}{s^2 + 400s + 2 \times 10^5} \quad \text{A-s}$$

The denominator quadratic can be factored as $(s + 200)^2 + 400^2$ and $I(s)$ written in the following form:

$$I(s) = -\frac{10}{400}\left[\frac{400}{(s + 200)^2 + (400)^2}\right] \quad \text{A-s}$$

Comparing the quantity inside the brackets with the entries in the $F(s)$ column of Table 9–2, we find that $I(s)$ is a damped sine with $\alpha = 200$ and $\beta = 400$. By linearity, the quantity outside the brackets is the amplitude of the damped sine. The inverse transform is

$$i(t) = [-0.025\, e^{-200t}\sin 400t]u(t) \quad \text{A}$$

Substituting this result back into the circuit integrodifferential equation yields the following term-by-term tabulation:

$$L\frac{di(t)}{dt} = +5e^{-200t}\sin 400t - 10e^{-200t}\cos 400t$$

$$Ri(t) = -10e^{-200t}\sin 400t$$

$$\frac{1}{C}\int_0^t i(\tau)d\tau = +5e^{-200t}\sin 400t + 10e^{-200t}\cos 400t - 10$$

$$v_C(0) = +10$$

The sum of the right-hand sides of these equations is zero. This result shows that the waveform $i(t)$ found using Laplace transforms does indeed satisfy the circuit integrodifferential equation and the initial conditions. ■

Exercise 9–15

Find the transform $V(s)$ that satisfies the following differential equations and the initial conditions:

(a) $\dfrac{dv(t)}{dt} + 6v(t) = 4u(t)$ $v(0-) = -3$

(b) $4\dfrac{dv(t)}{dt} + 12v(t) = 16 \cos 3t,$ $v(0-) = 2$

Answers:

(a) $V(s) = \dfrac{4}{s(s + 6)} - \dfrac{3}{s + 6}$ V-s

(b) $V(s) = \dfrac{4s}{(s^2 + 9)(s + 3)} + \dfrac{2}{s + 3}$ V-s

Exercise 9–16

Find the $V(s)$ that satisfies the following equations:

(a) $\displaystyle\int_0^t v(\tau)d\tau + 10v(t) = 10u(t)$

(b) $\dfrac{d^2v(t)}{dt^2} + 4\dfrac{dv(t)}{dt} + 3v(t) = 5e^{-2t}$ $v'(0-) = 2$ $v(0-) = -2$

Answers:

(a) $V(s) = \dfrac{1}{s + 0.1}$ V-s

(b) $V(s) = \dfrac{5}{(s + 1)(s + 2)(s + 3)} - \dfrac{2}{s + 1}$ V-s

CIRCUIT RESPONSE WITH TIME-VARYING INPUTS

It is encouraging to find that the Laplace transformation yields results that agree with those obtained by classical methods. The transform method reduces solving circuit differential equations to an algebraic process that includes the initial conditions. However, before being overcome with euphoria we must remember that the Laplace transform method begins with the circuit differential equation and the initial conditions. It does not provide these quantities to us. The transform method simplifies the solution process, but it does not substitute for understanding how to formulate circuit equations.

The Laplace transform method is especially usefully when the circuit is driven by time-varying inputs. To illustrate, we return to the RC circuit in Figure 9–6 and replace the step function input by a general input signal denoted $v_S(t)$. The right side of the circuit differential equation in Eq. (9–31) changes to accommodate the new input by taking the form

$$RC\,\frac{dv_C(t)}{dt} + v_C(t) = v_S(t) \qquad (9–36)$$

with an initial condition $v_C(0) = V_0$ V.

The only change here is that the driving force on the right side of the differential equation is a general time-varying waveform $v_S(t)$. The objective is to find the capacitor voltage $v_C(t)$ that satisfies the differential equation and the initial conditions. The classical methods of solving for the forced response depend on the form of $v_S(t)$. However, with the Laplace transform method we can proceed without actually specifying the form of the input signal.

We first transform Eq. (9–36) into the s domain:

$$RC[sV_C(s) - V_0] + V_C(s) = V_S(s)$$

The only assumption here is that the input waveform is Laplace transformable, a condition met by all causal signals of engineering interest. We now algebraically solve for the response $V_C(s)$:

$$V_C(s) = \frac{V_S(s)/RC}{s + 1/RC} + \frac{V_0}{s + 1/RC} \quad \text{V-s} \qquad (9\text{–}37)$$

The function $V_C(s)$ is the transform of the response of the RC circuit in Figure 9–6 due to a general input signal $v_S(t)$. We have gotten this far without specifying the form of the input signal. In a sense, we have found the general solution in the s domain of the differential equation in Eq. (9–36) for any casual input signal.

All of the necessary ingredients are present in Eq. (9–37):

1. The transform $V_S(s)$ represents the applied input signal.

2. The pole at $s = -1/RC$ defines the circuit time constant.

3. The initial value $v_C(0-) = V_0$ summarizes all events prior to $t = 0$.

However, we must have a particular input in mind to solve for the waveform $v_C(t)$. The following examples illustrate the procedure for different input driving forces.

EXAMPLE 9–17

Find $v_C(t)$ in the RC circuit in Figure 9–6 when the input is the waveform $v_S(t) = [V_A e^{-\alpha t}]u(t)$.

SOLUTION:
The transform of the input is $V_S(s) = V_A/(s + \alpha)$. For the exponential input the response transform in Eq. (9–37) becomes

$$V_C(s) = \frac{V_A/RC}{(s + \alpha)(s + 1/RC)} + \frac{V_0}{s + 1/RC} \quad \text{V-s} \qquad (9\text{–}38)$$

If $\alpha \neq 1/RC$ then the first term on the right is a proper rational function with two simple poles. The pole at $s = -\alpha$ came from the input and the pole at $s = -1/RC$ from the circuit. A partial fraction expansion of the first term has the form

$$\frac{V_A/RC}{(s + \alpha)(s + 1/RC)} = \frac{k_1}{s + \alpha} + \frac{k_2}{s + 1/RC}$$

The residues in this expansion are

$$k_1 = \left. \frac{V_A/RC}{s + 1/RC} \right|_{s=-\alpha} = \frac{V_A}{1 - \alpha RC}$$

$$k_2 = \left. \frac{V_A/RC}{s + \alpha} \right|_{s=-1/RC} = \frac{V_A}{\alpha RC - 1}$$

The expansion of the response transform $V_C(s)$ is

$$V_C(s) = \frac{V_A/(1 - \alpha RC)}{s + \alpha} + \frac{V_A/(\alpha RC - 1)}{s + 1/RC} + \frac{V_0}{s + 1/RC} \quad \text{V-s}$$

The inverse transform of $V_C(s)$ is

$$v_C(t) = \left[\frac{V_A}{1 - \alpha RC} e^{-\alpha t} + \frac{V_A}{\alpha RC - 1} e^{-t/RC} + V_0 e^{-t/RC} \right] u(t) \quad \text{V}$$

The first term is the forced response, and the last two terms are the natural response. The forced response is an exponential because the input introduced a pole at $s = -\alpha$. The natural response is also an exponential, but its time constant depends on the circuit's pole at $s = -1/RC$. In this case the forced and natural responses are both exponential signals with poles on the real axis. However, the forced response comes from the pole introduced by the input, while the natural response depends on the circuit's pole.

If $\alpha = 1/RC$, then the response just given is no longer valid (k_1 and k_2 become infinite). To find the response for this condition, we return to Eq. (9–38) and replace α by $1/RC$:

$$V_C(s) = \frac{V_A/RC}{(s + 1/RC)^2} + \frac{V_0}{s + 1/RC} \quad \text{V-s}$$

We now have a double pole at $s = -1/RC = -\alpha$. The double pole term is the transform of a damped ramp, so the inverse transform is

$$v_C(t) = \left[V_A \frac{t}{RC} e^{-t/RC} + V_0 e^{-t/RC} \right] u(t) \quad \text{V}$$

When $\alpha = 1/RC$ the s-domain poles of the input and the circuit coincide and the zero-state ($V_0 = 0$) response has the form $\alpha t e^{-\alpha t}$. We cannot separate this response into forced and natural components since the input and circuit poles coincide. ∎

EXAMPLE 9-18

Find $v_C(t)$ when the input to the RC circuit in Figure 9–6 is $v_S(t) = [V_A \cos \beta t] u(t)$.

SOLUTION:
The transform of the input is $V_S(s) = V_A s/(s^2 + \beta^2)$. For a cosine input the response transform in Eq. (9–37) becomes

$$V_C(s) = \frac{sV_A/RC}{(s^2 + \beta^2)(s + 1/RC)} + \frac{V_0}{s + 1/RC} \quad \text{V-s}$$

The sinusoidal input introduces a pair of poles located at $s = \pm j\beta$. The first term on the right is a proper rational function with three simple poles. The partial fraction expansion of the first term is

$$\frac{sV_A/RC}{(s - j\beta)(s + j\beta)(s + 1/RC)} = \frac{k_1}{s - j\beta} + \frac{k_1^*}{s + j\beta} + \frac{k_2}{s + 1/RC}$$

To find the response, we need to find the residues k_1 and k_2:

$$k_1 = \left.\frac{sV_A/RC}{(s + j\beta)(s + 1/RC)}\right|_{s = j\beta} = \frac{V_A/2}{1 + j\beta RC} = |k_1|e^{j\theta}$$

where

$$|k_1| = \frac{V_A/2}{\sqrt{1 + (\beta RC)^2}} \quad \text{and} \quad \theta = -\tan^{-1}(\beta RC)$$

The residue k_2 at the circuit pole is

$$k_2 = \left.\frac{sV_A/RC}{s^2 + \beta^2}\right|_{s = -1/RC} = -\frac{V_A}{1 + (\beta RC)^2}$$

We now perform the inverse transform to obtain the response waveform:

$$v_C(t) = [2|k_1|\cos(\beta t + \theta) + k_2 e^{-t/RC} + V_0 e^{-t/RC}]u(t)$$

$$= \left[\frac{V_A}{\sqrt{1 + (\beta RC)^2}}\cos(\beta t + \theta) - \frac{V_A}{1 + (\beta RC)^2}e^{-t/RC} + V_0 e^{-t/RC}\right]u(t) \quad \text{V}$$

The first term is the forced response, and the remaining two are the natural response. The forced response is sinusoidal because the input introduces poles at $s = \pm j\beta$. The natural response is an exponential with a time constant determined by the location of the circuit's pole at $s = -1/RC$. ∎

9–7 I N I T I A L   V A L U E   A N D   F I N A L   V A L U E   P R O P E R T I E S

The **initial value** and **final value properties** can be stated as follows:

$$\text{Initial value:} \lim_{t \to 0+} f(t) = \lim_{s \to \infty} sF(s) \qquad (9\text{--}39)$$

$$\text{Final value:} \lim_{t \to \infty} f(t) = \lim_{s \to 0} sF(s)$$

These properties display the relationship between the origin and infinity in the time and frequency domains. The value of $f(t)$ at $t = 0+$ in the time domain (initial value) is the same as the value of $sF(s)$ at infinity in the s plane. Conversely, the value of $f(t)$ as $t \to \infty$ (final value) is the same as the value of $sF(s)$ at the origin in the s plane.

Proof of both the initial value and final value properties starts with the differentiation property:

$$sF(s) - f(0-) = \int_{0-}^{\infty} \frac{df}{dt}e^{-st}\,dt \qquad (9\text{--}40)$$

To establish the initial value property, we rewrite the integral on the right side of this equation and take the limit of both sides as $s \to \infty$.

$$\lim_{s\to\infty}[sF(s) - f(0-)] = \lim_{s\to\infty}\int_{0-}^{0+}\frac{df}{dt}e^{-st}dt + \lim_{s\to\infty}\int_{0+}^{\infty}\frac{df}{dt}e^{-st}\,dt \qquad (9\text{–}41)$$

The first integral on the right side reduces to $f(0+) - f(0-)$ since e^{-st} is unity on the interval from $t = 0-$ to $t = 0+$. The second integral vanishes because e^{-st} goes to zero as $s \to \infty$. In addition, on the left side of Eq. (9–41) the $f(0-)$ is independent of s and can be taken outside the limiting process. Inserting all of these considerations reduces Eq. (9–41) to

$$\lim_{s\to\infty} sF(s) = \lim_{t\to 0+} f(t) \qquad (9\text{–}42)$$

which completes the proof of the initial-value property.

Proof of the final value theorem begins by taking the limit of both sides of Eq. (9–40) as $s \to 0$:

$$\lim_{s\to 0}[sF(s) - f(0-)] = \lim_{s\to 0}\int_{0-}^{\infty}\frac{df}{dt}e^{-st}\,dt \qquad (9\text{–}43)$$

The integral on the right side of this equation reduces to $f(\infty) - f(0-)$ because e^{-st} becomes unity as $s \to 0$. Again, the $f(0-)$ on the left side is independent of s and can be taken outside of the limiting process. Inserting all of these considerations reduces Eq. (9–43) to

$$\lim_{s\to 0} sF(s) = \lim_{t\to\infty} f(t) \qquad (9\text{–}44)$$

which completes the proof of the final value property.

A damped cosine waveform provides an illustration of the application of these properties. The transform of the damped cosine is

$$\mathcal{L}\{[Ae^{-\alpha t}\cos\beta t]u(t)\} = \frac{A(s + \alpha)}{(s + \alpha)^2 + \beta^2}$$

Applying the initial and final value limits, we obtain

$$\text{Initial value: } \lim_{t\to 0} f(t) = \lim_{t\to 0} Ae^{-\alpha t}\cos\beta t = A$$

$$\lim_{s\to\infty} sF(s) = \lim_{s\to\infty}\frac{sA(s + \alpha)}{(s + \alpha)^2 + \beta^2} = A$$

$$\text{Final value: } \lim_{t\to\infty} f(t) = \lim_{t\to\infty} Ae^{-\alpha t}\cos\beta t = 0$$

$$\lim_{s\to 0} sF(s) = \lim_{s\to 0}\frac{sA(s + \alpha)}{(s + \alpha)^2 + \beta^2} = 0$$

Note the agreement between the t-domain and s-domain limits in both cases.

There are restrictions on the initial and final value properties. The initial value property is valid when $F(s)$ is a proper rational function or, equivalently, when $f(t)$ does not have an impulse at $t = 0$. The final value property is valid when the poles of $sF(s)$ are in the left half plane or, equivalently, when $f(t)$ is a waveform that approaches a final value at $t \to \infty$. Note that the final value restriction allows $F(s)$ to have a simple pole at the origin since the limitation is on the poles of $sF(s)$.

Caution: The initial and final value properties will appear to work when the aforementioned restrictions are not met. In other words, these properties do not tell you they are giving nonsense answers when you violate their limitations. You must always check the restrictions on $F(s)$ before applying either of these properties.

For example, applying the final value property to a cosine waveform yields

$$\lim_{t \to \infty} \cos \beta t = \lim_{s \to 0} s \left[\frac{s}{s^2 + \beta^2} \right] = 0$$

The final value property appears to say that $\cos \beta t$ approaches zero as $t \to \infty$. This conclusion is incorrect since the waveform oscillates between ± 1. The problem is that the final value property does not apply to cosine waveform because $sF(s)$ has poles on the j-axis at $s = \pm j\beta$.

EXAMPLE 9–19

Use the initial and final value properties to find the initial and final values of the waveform whose transform is

$$F(s) = 2 \frac{(s + 3)}{s(s + 1)(s + 2)}$$

SOLUTION:
The given $F(s)$ is a proper rational function so the initial value property can be applied as

$$f(0) = \lim_{s \to \infty} sF(s) = \lim_{s \to \infty} \left[2 \frac{(s + 3)}{(s + 1)(s + 2)} \right] = 0$$

The poles of $sF(s)$ are located in the left-half plane at $s = -1$ and $s = -2$; hence, the final value property can be applied as

$$f(\infty) = \lim_{s \to 0} sF(s) = \lim_{s \to 0} \left[2 \frac{(s + 3)}{(s + 1)(s + 2)} \right] = 3$$

In Example 9-11 the waveform corresponding to this transform was found to be

$$f(t) = [3 - 4e^{-t} + e^{-2t}]u(t)$$

from which we find

$$f(0) = 3 - 4e^{-0} + e^{-0} = 0$$

$$f(\infty) = 3 - 4e^{-\infty} + e^{-\infty} = 3$$

which confirms the results found directly from $F(s)$.   ■

Exercise 9–17

Find the initial and final values of the waveforms corresponding to the following transforms:

(a) $F_1(s) = 100 \dfrac{s + 3}{s(s + 5)(s + 20)}$

(b) $F_2(s) = 80 \dfrac{s(s + 5)}{(s + 4)(s + 20)}$

Answers:

(a) Initial value = 0, final value = 3.

(b) $F_2(s)$ is not a proper rational function, final value = 0.

APPLICATION NOTE: EXAMPLE 9-20

Biological processes such as cellular osmosis or glucose uptake involve the diffusion of a liquid (the solvent) containing a dissolved substance (the solute) from one volume to another. Figure 9–9 shows a schematic representation of two such volumes separated by a semipermeable membrane (γ). In this context V_1 and V_2 are the solvent volumes while m_1 and m_2 are the solute masses in each chamber.

The primary variables of interest are called the concentrations defined as

$$c_1 = \frac{m_1}{V_1} \qquad \text{and} \qquad c_2 = \frac{m_2}{V_2}$$

Units used to quantify concentration include such things as milligrams per deciliter. The diffusion through the membrane is governed by two coupled first-order differential equations.

$$\frac{dc_1(t)}{dt} = \alpha_1[c_2(t) - c_1(t)]$$

$$\frac{dc_2(t)}{dt} = \alpha_2[c_1(t) - c_2(t)]$$

where $\alpha_1 = \gamma/V_1$, $\alpha_2 = \gamma/V_2$, and is the permeability coefficient of the membrane. These equations point out that when the two concentrations are equal ($c_1 = c_2$) the diffusion process is in equilibrium since both time derivatives are zero. Note that

$$\text{If } c_1 > c_2, \text{ then } \frac{dv_1}{dt} < 0 \text{ and } \frac{dv_2}{dt} > 0$$

$$\text{and conversely}$$

$$\text{If } c_1 < c_2, \text{ then } \frac{dv_1}{dt} > 0 \text{ and } \frac{dv_2}{dt} < 0$$

That is, when the two concentrations are unequal the two derivatives have opposite signs and diffusion proceeds in such a direction as to bring the process back into equilibrium.

FIGURE 9-9

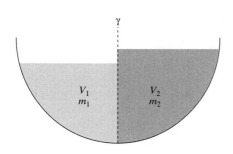

V_1
m_1

V_2
m_2

To describe the diffusion process we must solve these two differential equations simultaneously. One way to do this is to transform them into the s-domain where they become algebraic equations. Using the differentiation and linearity properties of the Laplace transformation, the two differential equations become

$$sC_1(s) - c_1(0) = \alpha_1[C_1(s) - C_2(s)]$$

$$sC_2(s) - c_2(0) = \alpha_2[C_2(s) - C_1(s)]$$

where $c_1(0)$ and $c_2(0)$ are the concentrations at $t = 0$. To solve these s-domain equations we rearrange them as two linear algebraic equations in the transforms $C_1(s)$ and $C_2(s)$.

$$(s + \alpha_1)C_1(s) \qquad - \alpha_1 C_2(s) = c_1(0)$$

$$- \alpha_2 C_1(s) + (s + \alpha_2)C_2(s) = c_2(0)$$

Applying Cramer's rule yields

$$C_1(s) = \frac{\Delta_1}{\Delta} = \frac{\begin{vmatrix} c_1(0) & - \alpha_1 \\ c_2(0) & s + \alpha_2 \end{vmatrix}}{\begin{vmatrix} s + \alpha_1 & - \alpha_1 \\ - \alpha_2 & s + \alpha_2 \end{vmatrix}} = \frac{(s + \alpha_2)c_1(0) + \alpha_1 c_2(0)}{s(s + \alpha_1 + \alpha_2)}$$

$$C_2(s) = \frac{\Delta_2}{\Delta} = \frac{\begin{vmatrix} s + \alpha_1 & c_1(0) \\ - \alpha_2 & c_2(0) \end{vmatrix}}{s(s + \alpha_1 + \alpha_2)} = \frac{(s + \alpha_1)c_2(0) + \alpha_2 c_1(0)}{s(s + \alpha_1 + \alpha_2)}$$

The corresponding time-domain waveforms can be found using the inverse Laplace transformation. However, let us see what we can infer by examining the response transforms. First, the two transforms have the same poles located at $s = 0$ and $s = -(\alpha_1 + \alpha_2)$. The first pole leads to a step function, and the second pole to an exponential of the form $\exp[-(\alpha_1 + \alpha_2)t]$. This tells us that in the time domain the final state $(t \to \infty)$ is a constant (the step function). The decaying exponential waveform connecting the initial $(t = 0)$ to the final state has a time constant of $1/(\alpha_1 + \alpha_2)$. Thus, the time needed to transition from the initial to the final state is about $5/(\alpha_1 + \alpha_2)$.

Since the pole at $s = -(\alpha_1 + \alpha_2)$ is in the left-half plane, we can apply the final value property to $sC_1(s)$ and $sC_2(s)$ to find the final state.

$$c_1(\infty) = \lim_{s \to 0} sC_1(s) = \lim_{s \to 0} \frac{(s + \alpha_2)c_1(0) + \alpha_1 c_2(0)}{(s + \alpha_1 + \alpha_2)} = \frac{\alpha_2 c_1(0) + \alpha_1 c_2(0)}{\alpha_1 + \alpha_2}$$

$$c_2(\infty) = \lim_{s \to 0} sC_2(s) = \lim_{s \to 0} \frac{(s + \alpha_1)c_2(0) + \alpha_2 c_1(0)}{(s + \alpha_1 + \alpha_2)} = \frac{\alpha_1 c_2(0) + \alpha_2 c_1(0)}{\alpha_1 + \alpha_2}$$

The two final values are equal, which means that the diffusion process proceeds from an initial state of $c_1(0) \neq c_2(0)$ to a final state with $c_1(\infty) = c_2(\infty)$, that is, from an unbalanced condition to an equilibrium condition.

This example illustrates that Laplace transforms have applications outside of linear circuit analysis. The example further illustrates that much can be learned about system responses by examining the response transforms themselves without formally performing the inverse transformation. This should not be surprising because transforms contain all of the information needed to determine the corresponding time-domain waveforms.

SUMMARY

- The Laplace transformation converts waveforms in the time domain to transforms in the s domain. The inverse transformation converts transforms into causal waveforms. A transform pair is unique if and only if $f(t)$ is casual.

- The Laplace transforms of basic signals like the step function, exponential, and sinusoid are easily derived from the integral definition. Other transform pairs can be derived using basic signal transforms and the uniqueness, linearity, time integration, time differentiation, and translation properties of the Laplace transformation.

- Proper rational functions with simple poles can be expanded by partial fraction to obtain inverse Laplace transforms. Simple real poles lead to exponential waveforms and simple complex poles to damped sinusoids. Partial-fraction expansions of improper rational functions and functions with multiple poles require special treatment.

- Using Laplace transforms to find the response of a linear circuit involves transforming the circuit differential equation into the s domain, algebraically solving for the response transform, and performing the inverse transformation to obtain the response waveform.

- The initial and final value properties determine the initial and final values of a waveform $f(t)$ from the value of $sF(s)$ at $s \to \infty$ and $s = 0$, respectively. The initial value property applies if $F(s)$ is a proper rational function. The final value property applies if all of the poles of $sF(s)$ are in the left-half plane.

PROBLEMS

ERO 9–1 LAPLACE TRANSFORM (SECTS. 9–1, 9–2, 9–3)

Find the Laplace transform of a given a signal waveform using transform properties and pairs, or using the integral definition of the Laplace transformation. Locate the poles and zeros of the transform and construct a pole-zero diagram.
See Examples 9–1, 9–2, 9–3, 9–4, 9–5, 9–6, 9–7, 9–8, 9–9, 9–10

9–1 Find the Laplace transform of $f(t) = A[e^{-\alpha t} - 2]u(t)$.

9–2 Find the Laplace transform of $f(t) = A[(1 + \alpha t)e^{-\alpha t}] u(t)$.

9–3 Find the Laplace transform of $f(t) = A[2 \sin(\beta t) + 1] u(t)$.

9–4 Find the Laplace transform of $f(t) = A[\cos(\beta t) - \sin(\beta t)]u(t)$. Locate the poles and zeros of $F(s)$.

9–5 Find the Laplace transform of $f(t) = A\delta(t) - A\beta e^{-\alpha t} \sin(\beta t)u(t)$. Locate the poles and zeros of $F(s)$.

9–6 Find the Laplace transform of $f(t) = A[2\alpha t - 1 + e^{-\alpha t}] u(t)$ for $\alpha > 0$. Locate the poles and zeros of $F(s)$.

9–7 Find the Laplace transforms of the following functions and plot their pole-zero diagrams:
(a) $f_1(t) = [5e^{-5t} - 10e^{-10t}]u(t)$
(b) $f_2(t) = [10 \cos(20t) - 16 \cos(10t)]u(t)$

9–8 Find the Laplace transforms of the following functions and plot their pole-zero diagrams:
(a) $f_1(t) = \delta(t) + [5e^{-50t} - 5e^{-20t}]u(t)$
(b) $f_2(t) = [25 - 25 \cos(500t)]u(t)$

9–9 Find the Laplace transform of the following function. Locate the poles and zeros of $F(s)$.

$$f(t) = \delta(t) + [25e^{-50t}(1 - 50t)]u(t)$$

9–10 Find the Laplace transform of the following function. Locate the poles and zeros of $F(s)$.

$$f(t) = [5 - 2e^{-5t} - 3 \cos(5t) - 2 \sin(5t)]u(t)$$

9–11 Find the Laplace transform of

$$f(t) = A\{e^{-\alpha(t-T)} \sin[\beta(t-T)]\} u(t-T)$$

9–12 The Laplace transform of $f(t) = 5[e^{-t} \sin 4t]u(t)$ is $F(s) = 20/(s^2 + 2s + 17)$.

(a) Use the transform $F(s)$ and the differentiation property of Laplace transforms to obtain the transform $G(s) = \mathcal{L}\{df(t)/dt\}$.

(b) Find the time-domain derivative of the waveform $f(t)$. Transform this waveform into the s domain and check that the result is $G(s)$ found in (a).

(c) Are there any poles or zeros in $G(s)$ that are not in $F(s)$? Explain.

9–13 (a) Write an expression for the waveform $f(t)$ in Figure P9–13 using only step functions and ramps.

(b) Use the time-domain translation property to find the Laplace transform of the waveform found in part (a).

(c) Verify your answer in (b) by applying the integral definition of the Laplace transformation to the $f(t)$ found in (a).

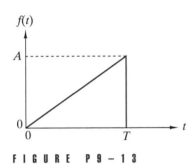

FIGURE P9-13

9–14 (a) Write an expression for the waveform $f(t)$ in Figure P9–14 using only step functions and ramps.

(b) Use the time-domain translation property to find the Laplace transform of the waveform found in part (a).

(c) Verify your answer in (b) by applying the integral definition of the Laplace transformation to the $f(t)$ found in (a).

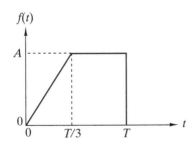

FIGURE P9-14

9–15 Find the Laplace transform of the following function. Locate the poles and zeros of $F(s)$.

$$f(t) = [1600\,t + 8 + 75\,e^{-10t} + 17e^{-50t}]u(t)$$

ERO 9–2 Inverse Transforms (Sects. 9–4, 9–5)

Find the inverse transform of a given Laplace transform using partial fraction expansion or basic transform properties and pairs.
See Examples 9–11, 9–12, 9–13, 9–14, and Exercises 9–7, 9–8, 9–9, 9–10, 9–11, 9–12, 9–13, 9–14

9–16 Find the inverse Laplace transforms of the following functions:

(a) $F_1(s) = \dfrac{s - 20}{s(s + 10)}$

(b) $F_2(s) = \dfrac{s^2 + 100}{s(s + 10)}$

9–17 Find the inverse Laplace transforms of the following functions:

(a) $F_1(s) = \dfrac{s + 10}{s(s + 5)}$

(b) $F_2(s) = \dfrac{s}{(s + 5)(s + 10)}$

9–18 Find the inverse Laplace transforms of the following functions:

(a) $F_1(s) = \dfrac{20\,(s + 20)}{(s + 10)^2 + 400}$

(b) $F_2(s) = \dfrac{20\,(s - 20)}{(s + 10)^2 + 400}$

9–19 A Laplace transform $F(s)$ has poles at $s = \pm j\beta$ and zeros at $s = 0$ and $s = \beta$. The scale factor is $K = 1$. Find the inverse transform $f(t)$.

9–20 A Laplace transform $F(s)$ has a simple pole at $s = -\alpha$ and a double pole at $s = 0$. The scale factor is $K = \alpha^2$. Find the inverse transform $f(t)$.

9–21 Plot the pole-zero diagrams of the following transforms, find the corresponding inverse transforms, and sketch their waveforms:

(a) $F_1(s) = \dfrac{900}{s^2 + 65s + 900}$

(b) $F_2(s) = \dfrac{900}{s^2 + 60s + 900}$

(c) $F_3(s) = \dfrac{900}{s^2 + 36s + 900}$

9–22 Find the inverse transforms of the following functions:

(a) $F_1(s) = \dfrac{16}{(s + 2)(s^2 + 10\,s + 24)}$

(b) $F_2(s) = \dfrac{(s^2 + 4\,s + 16)}{s(s^2 + 16)}$

9–23 Find the inverse transforms of the following functions:

(a) $F_1(s) = \dfrac{(s + 2)(s + 6)}{s(s + 4)(s + 8)}$

(b) $F_2(s) = \dfrac{(s^2 + 45)(s^2 + 80)}{s(s^2 + 20)(s^2 + 60)}$

9–24 Find the inverse transforms of the following functions:

(a) $F_1(s) = \dfrac{4(s+4)}{s(s^2+4s+8)}$

(b) $F_2(s) = \dfrac{(s-1)^2}{(s+4)(s^2+2s+17)}$

9–25 Find the inverse transforms of the following functions:

(a) $F_1(s) = \dfrac{2(s^2+4s+16)}{s(s^2+8s+32)}$

(b) $F_2(s) = \dfrac{s^2+20s+400}{s(s^2+50s+400)}$

9–26 Find the inverse transforms of the following functions:

(a) $F_1(s) = \dfrac{(s+50)^2}{(s+10)(s+100)}$

(b) $F_2(s) = \dfrac{s+1}{s^2-2s+1}$

9–27 A certain transform has a simple pole at $s = -20$, a simple zero at $s = -\gamma$, and a scale factor of $K = 1$. Select values for γ so the inverse transform is

(a) $f(t) = \delta(t) - 5e^{-20t}$ (b) $f(t) = \delta(t)$
(c) $f(t) = \delta(t) + 5e^{-20t}$

9–28 A transform has the form

$$F(s) = K\,\frac{(s^2+b_1 s+b_0)}{(s^2+400)(s+5)^2}$$

where the parameters K, b_1, and b_0 are real numbers. Select the values for K, b_1, and b_0 so that the inverse transform of $F(s)$ is

(a) $f(t) = [20te^{-5t}]u(t)$
(b) $f(t) = [5\sin(20t)]u(t)$
(c) $f(t) = [4\sin(20t) + \cos(20t) - e^{-5t}]u(t)$

9–29 Use Mathcad or MATLAB to find the inverse transform of the following function:

$$F(s) = \frac{s(s^2+3s+4)}{(s+2)(s^3+6s^2+16s+16)}$$

9–30 Use Mathcad or MATLAB to find the inverse transform of the following function:

$$F(s) = \frac{40(s^3+2s^2+s+2)}{s(s^3+4s^2+4s+16)}$$

ERO 9–3 CIRCUIT RESPONSE USING LAPLACE TRANSFORMS (SECT. 9–6)

Given a first- or second-order circuit,

(a) Determine the circuit differential equation and the initial conditions (if not given).
(b) Transform the differential equation into the s domain and solve for the response transform.
(c) Use the inverse transformation to find the response waveform.

See Examples 9–15, 9–16, 9–17, 9–18, 9–19

9–31 Use the Laplace transformation to find the $y(t)$ that satisfies the following first-order differential equations:

(a) $\dfrac{dy}{dt} + 20\,y = 0$ with $y(0-) = 5$

(b) $10^{-2}\dfrac{dy}{dt} + y = 10\,u(t)$ with $y(0-) = -10$

9–32 Use the Laplace transformation to find the $y(t)$ that satisfies the following first-order differential equation:

$$\frac{dy}{dt} + 500\,y = 2500\,[e^{-1000t}]u(t) \text{ with } y(0-) = 0$$

9–33 The switch in Figure P9–33 has been open for a long time and is closed at $t = 0$. For $R = 100\ \Omega$, $L = 100$ mH, and $V_A = 10$ V,

(a) Find the differential equation and initial condition for the inductor current $i_L(t)$.
(b) Solve for $i_L(t)$ using the Laplace transformation.

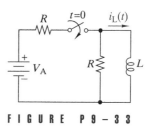

FIGURE P9–33

9–34 The switch in Figure P9–33 has been closed for a long time and is opened at $t = 0$. For $R = 50\ \Omega$, $L = 50$ mH, and $V_A = 100$ V,

(a) Find the differential equation and initial condition for the inductor current $i_L(t)$.
(b) Solve for $i_L(t)$ using the Laplace transformation.

9–35 The switch in Figure P9–35 has been open for a long time. At $t = 0$ the switch is closed.

(a) Find the differential equation for the circuit and initial condition.
(b) Find $v_O(t)$ using the Laplace transformation for $v_S(t) = 15[e^{-2500t}]u(t)$.

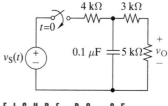

FIGURE P9–35

9–36 Repeat Problem 9–35 for the input waveform $v_S(t) = 10[\sin 1000t]u(t)$.

9–37 Use the Laplace transformation to find the $y(t)$ that satisfies the following second-order differential equation:

$$\frac{d^2y}{dt^2} + 20\frac{dy}{dt} + 75y = 0$$

with $y(0-) = 10$ and $y'(0-) = 0$.

9–38 Use the Laplace transformation to find the $y(t)$ that satisfies the following second-order differential equation:

$$\frac{d^2y}{dt^2} + 7\frac{dy}{dt} + 10\,y = 30\,u(t)$$

with $y(0-) = 0$ and $y'(0-) = 0$.

9–39 The switch in Figure P9–39 has been open for a long time and is closed at $t = 0$. The circuit parameters are $R = 1\,\text{k}\Omega$, $L = 400\,\text{mH}$, $C = 0.5\,\mu\text{F}$, and $V_A = 10\,\text{V}$.
  (a) Find the differential equation for the circuit and the initial conditions.
  (b) Use Laplace transforms to solve for the $i_L(t)$ for $t \geq 0$.

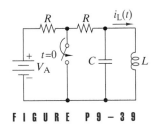

FIGURE P9–39

9–40 The switch in Figure P9–39 has been closed for a long time and is opened at $t = 0$. The circuit parameters are $R = 5\,\text{k}\Omega$, $L = 10\,\text{mH}$, $C = 16\,\text{pF}$, and $V_A = 10\,\text{V}$.
  (a) Find the differential equation for the circuit and the initial conditions.
  (b) Use Laplace transforms to solve for the $i_L(t)$ for $t \geq 0$.

9–41 There is no energy stored in the circuit in Figure P9–41 when the switch opens at $t = 0$. For $t > 0$ the sum of voltages around the loop yields the following integrodifferential equation.

$$\overbrace{L\frac{di_L(t)}{dt}}^{v_L(t)} + \overbrace{(R_1 + R_2)i_L(t)}^{v_R(t)} + \overbrace{\frac{1}{C}\int_0^t i_L(x)\,dx + v_C(0)}^{v_C(t)} - v_S(t) = 0$$

For $L = 0.5\,\text{H}$, $C = 250\,\text{nF}$, $R_1 = 1\,\text{k}\Omega$, $R_2 = 2\,\text{k}\Omega$, and $v_S(t) = [10e^{-1000t}]u(t)\,\text{V}$,
  (a) Transform the integro-differential equation into the s domain and solve for $I_L(s)$.
  (b) Construct a pole-zero map of $I_L(s)$ and identify the poles of the forced and natural responses.

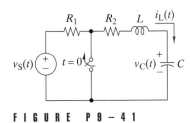

FIGURE P9–41

9–42 There is no energy stored in the circuit in Figure P9–41 when the switch opens at $t = 0$. For $t \geq 0$ the circuit differential equation in terms of the capacitor voltage is

$$LC\frac{d^2v_C(t)}{dt^2} + (R_1 + R_2)C\frac{dv_C(t)}{dt} + v_C(t) = v_S(t)$$

For $R_1 = 100\,\Omega$, $R_2 = 100\,\Omega$, $L = 100\,\text{mH}$, $C = 2\,\mu\text{F}$, and $v_S(t) = [10\sin(2000t)]u(t)\,\text{V}$,
  (a) Transform the circuit differential equation into the s-domain and solve for $V_C(s)$.
  (b) Construct a pole-zero map of $V_C(s)$ and identify the poles of the forced and natural responses.

9–43 The differential equation for the RL circuit in Figure P9–43 in terms of the current $i(t)$ is

$$L\frac{di(t)}{dt} + R\,i(t) = v_S(t)$$

  (a) Assume that the initial conditions are zero. Transform the circuit differential equation into the s-domain and solve for $I(s)$ in terms of the input $V_S(s)$.
  (b) Find the poles of $I(s)$ for $R = 1\,\text{k}\Omega$, $L = 0.5\,\text{H}$, and $v_S(t) = \{10e^{-1000t}\}u(t)\,\text{V}$.
  (c) Select a value for R and L, and define an input $v_S(t)$ so that $I(s)$ has three poles located at $s = -1000$ and $s = \pm j1000$.

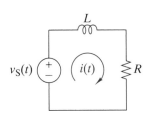

FIGURE P9–43

9–44 The integrodifferential equation for an undriven parallel RLC circuit takes the form

$$\overbrace{C\frac{dv_C(t)}{dt}}^{i_C(t)} + \overbrace{\frac{v_C(t)}{R}}^{+\ i_R(t)} + \overbrace{\frac{1}{L}\int_0^t v_C(x)dx + i_L(0)}^{+\ i_L(t)} = 0$$

  (a) With $v_C(0) = 0$ and $i_L(0) = I_0$, transform the equation into the s domain and solve for $V_C(s)$.
  (b) Find $v_C(t)$ for $R = 5\,\text{k}\Omega$, $L = 100\,\text{mH}$, $C = 10\,\text{nF}$, and $I_0 = 10\,\text{mA}$.

ERO 9–4 INITIAL AND FINAL VALUE (SECT. 9–7)

Find the initial and final values of waveforms using the initial and final value properties of Laplace transforms.

9–45 Use the initial and final value properties to find the initial and final values of the waveform corresponding to the transforms in Problem 9–22. If either property is not applicable, explain why.

9–46 Use the initial and final value properties to find the initial and final values of the waveform corresponding to the transforms in Problem 9–23. If either property is not applicable, explain why.

9–47 Use the initial and final value properties to find the initial and final values of the waveform corresponding to the following transforms. If either property is not applicable, explain why.

(a) $F_1(s) = \dfrac{2(s^2 + 5s + 6)}{(s + 2)(s + 6)(s + 12)}$

(b) $F_2(s) = \dfrac{10(s^2 + 10s + 40)}{s(s^2 - 100)}$

9–48 Use the initial and final value properties to find the initial and final values of the waveform corresponding to the following transforms. If either property is not applicable, explain why.

(a) $F_1(s) = \dfrac{s(s + 5)}{s^2 + 6s + 9}$

(b) $F_2(s) = \dfrac{10(s^2 + 10s - 20)}{s(s^2 + 100)}$

9–49 Use the initial and final value properties to find the initial and final values of the waveform corresponding to the transforms in Problem 9–24. If either property is not applicable, explain why.

9–50 Use the initial and final value properties to find the initial and final values of the waveform corresponding to the transforms in Problem 9–26. If either property is not applicable, explain why.

INTEGRATING PROBLEMS

9–51 ⚙ THE DOMINANT POLE APPROXIMATION

When a transform $F(s)$ has widely separated poles, then those closest to the j-axis tend to dominate the response because they have less damping. An approximation to the waveform can be obtained by ignoring the contributions of all except the dominant poles. We can ignore the nondominant poles by simply discarding their terms in the partial fraction expansion of $F(s)$. The purpose of this example is to examine a dominant pole approximation. Consider the transform

$$F(s) = 10^6 \frac{s + 1000}{(s + 4000)[(s + 25)^2 + 100^2]}$$

(a) Construct a partial-fraction expansion of $F(s)$ and find $f(t)$.
(b) Construct a pole-zero diagram of $F(s)$ and identify the dominant poles.
(c) Construct a dominant pole approximation $g(t)$ by discarding the nondominant poles in the partial fraction expansion in (a).
(d) Plot $f(t)$ and $g(t)$ and comment on the accuracy of the approximation.

9–52 ⚙ FIRST-ORDER CIRCUIT STEP RESPONSE

In Chapter 7 we found that the step response of a first-order circuit can be written as

$$f(t) = f(\infty) + [f(0) - f(\infty)]e^{-t/T_c}$$

where $f(0)$ is the initial value, $f(\infty)$ is the final value, and T_C is the time constant. Show that the corresponding transform has the form

$$F(s) = K\left[\frac{s + \gamma}{s(s + \alpha)}\right]$$

and relate the time-domain parameters $f(0)$, $f(\infty)$, and T_C to the s-domain parameters K, γ, and α.

9–53 ⚙ INVERSE TRANSFORM FOR COMPLEX POLES

In Section 9–4 we learned that complex poles occur in conjugate pairs and that for simple poles the partial fraction expansion of $F(s)$ will contain two terms of the form

$$F(s) = \dots \frac{k}{s + \alpha - j\beta} + \frac{k^*}{s + \alpha + j\beta} + \dots$$

Show that when the complex conjugate residues are written in rectangular form as

$$k = a + jb \quad \text{and} \quad k^* = a - jb$$

the corresponding term in the waveform $f(t)$ is

$$f(t) = \dots + 2e^{-\alpha t}[a\cos(\beta t) - b\sin(\beta t)] + \dots$$

9–54 ⚙ SOLVING STATE VARIABLE EQUATIONS

An undriven series RLC circuit can be described by the following coupled first-order equations in the inductor current $i_L(t)$ and capacitor voltage $v_C(t)$:

$$\frac{dv_C(t)}{dt} = \frac{1}{C} i_L(t)$$

$$\frac{di_L(t)}{dt} = -\frac{1}{L} v_C(t) - \frac{R}{L} i_L(t)$$

(a) Transform these equations into the s domain and solve for the transforms $I_L(s)$ and $V_C(s)$ in terms of the initial conditions $i_L(0) = I_0$ and $v_C(0) = V_0$.
(b) Find $i_L(t)$ and $v_C(t)$ for $R = 3$ kΩ, $L = 0.1$ H, $C = 50$ nF, $I_0 = 5$ mA, and $V_0 = -5$ V.

9–55 ⚙ COMPLEX DIFFERENTIATION PROPERTY

The complex differentiation property of the Laplace transformation states that

$$\text{If } \mathscr{L}\{f(t)\} = F(s) \text{ then } \mathscr{L}\{tf(t)\} = -\frac{d}{ds}F(s).$$

Use this property to find the Laplace transforms of $f(t) = \{tg(t)\}u(t)$ where $g(t) = e^{-\alpha t}$, $\sin \beta t$, and $\cos \beta t$.

CHAPTER 10

S-DOMAIN CIRCUIT ANALYSIS

The resistance operator Z is a function of the electrical constants of the circuit components and of d/dt, *the operator of time-differentiation, which will in the following be denoted by* p *simply.*

Oliver Heaviside, 1887,
British Engineer

10–1 Transformed Circuits

10–2 Basic Circuit Analysis in the *s* Domain

10–3 Circuit Theorems in the *s* Domain

10–4 Node-voltage Analysis in the *s* Domain

10–5 Mesh-current Analysis in the *s* Domain

10–6 Summary of *s*-Domain Circuit Analysis

Summary

Problems

Integrating Problems

The Laplace transform techniques in this chapter have their roots in the works of Oliver Heaviside (1850–1925). The preceding quotation was taken from his book *Electrical Papers* originally published in 1887. His resistance operator *Z*, which he later called impedance, is a central theme for much of electrical engineering. Heaviside does not often receive the recognition he deserves because his intuitive approach to mathematics was not accepted by most Victorian scientists of his day. Mathematical justification for his methods was eventually supplied by John Bromwich and others. However, no important errors were found in Heaviside's results.

In this chapter we transform circuits directly from the time domain into the *s* domain without first developing a circuit differential equation. The process begins in the first section, where we find that the transformed circuit obeys Kirchhoff's laws and that the passive element *i–v* characteristics become linear algebraic equations in transform variables. The algebraic form of the element constraints leads to the concept of *s*-domain impedance as a generalization of resistance. In the *s* domain the underlying element and connection constraints are similar to those for resistance circuits. As a result, *s*-domain circuit analysis uses techniques like equivalence, superposition, and Thévenin's theorem, which parallel those developed for resistance circuits. Most of this chapter is devoted to extending resistance circuit analysis methods to cover dynamic circuits in the *s* domain. The basic analysis tools and circuit theorems are discussed in the second and third sections. The fourth and fifth sections develop general methods of circuit analysis using node voltages and mesh currents.

10–1   TRANSFORMED CIRCUITS

So far we have used the Laplace transformation to change waveforms into transforms and convert circuit differential equations into algebraic equations. These operations provide a useful introduction to the *s* domain. However, the real power of the Laplace transformation emerges when we transform the circuit itself and study its behavior directly in the *s* domain.

The solid arrow in Figure 10–1 indicates the analysis path we will be following in this chapter. The process begins with a linear circuit in the time domain. We transform the circuit into the *s* domain, write the circuit equations directly in that domain, and then solve these algebraic equations for response transform. The inverse Laplace transformation then produces the response waveform. However, the *s*-domain approach is not just another way to derive response waveforms. This approach allows us to work directly with the circuit model using analysis tools such as voltage division and equivalence. By working directly with the circuit model, we gain insights into the interaction between circuits and signals that cannot be obtained using the classical approach indicated by the dotted path in Figure 10–1.

How are we to transform a circuit? We have seen several times that circuit analysis is based on device and connection constraints. The connection constraints are derived from Kirchhoff's laws and the device constraints from the *i–v* relationships used to model the physical devices in the circuit. To transform circuits, we must see how these two types of constraints are altered by the Laplace transformation.

FIGURE   10–1 *Flow   diagram   for* *s-domain circuit analysis.*

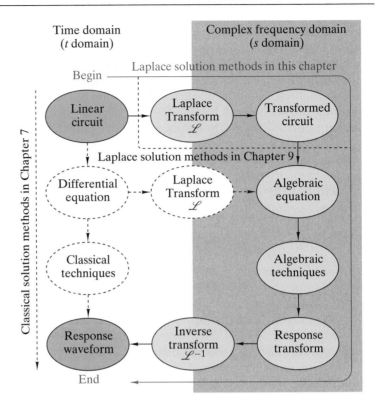

FIGURE   10–1 *Flow   diagram   for* *s-domain circuit analysis.*

CONNECTION CONSTRAINTS IN THE *S* DOMAIN

A typical KCL connection constraint could be written as

$$i_1(t) + i_2(t) - i_3(t) + i_4(t) = 0$$

This connection constraint requires that the sum of the current waveforms at a node be zero for all times t. Using the linearity property, the Laplace transformation of this equation is

$$I_1(s) + I_2(s) - I_3(s) + I_4(s) = 0$$

In the s domain the KCL connection constraint requires that the sum of the current transforms be zero for all values of s. This idea generalizes to any number of currents at a node and any number of nodes. In addition, this idea obviously applies to Kirchhoff's voltage law as well. The form of the connection constraints do not change because they are linear equations and the Laplace transformation is a linear operation. In summary, KCL and KVL apply to waveforms in the t domain and to transforms in the s domain.

ELEMENT CONSTRAINTS IN THE *S* DOMAIN

Turning now to the element constraints, we first deal with the independent signal sources shown in Figure 10–2. The i–v relationships for these elements are

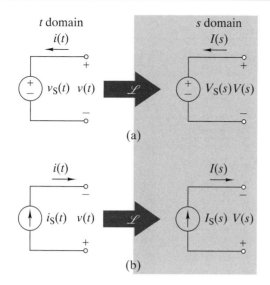

FIGURE 10 – 2 s-*Domain models of independent sources.*

Voltage source: $v(t) = v_S(t)$

$i(t)$ = Depends on Circuit   (10–1)

Current source: $i(t) = i_S(t)$

$v(t)$ = Depends on Circuit

Independent sources are two-terminal elements. In the t domain they constrain the waveform of one signal variable and adjust the unconstrained variable to meet the demands of the external circuit. We think of an independent source as a generator of a specified voltage or current waveform. The Laplace transformation of the expressions in Eq. (10–1) yields

Voltage source: $V(s) = V_S(s)$

$I(s)$ = Depends on Circuit   (10–2)

Current source: $I(s) = I_S(s)$

$V(s)$ = Depends on Circuit

In the s domain independent sources function the same way as in the t domain, except that we think of them as generating voltage or current transforms rather than waveforms.

Next we consider the active elements in Figure 10–3. In the time domain the element constraints for linear dependent sources are linear algebraic equations. Because of the linearity property of the Laplace transformation, the forms of these constraints are unchanged when they are transformed into the s domain:

	t domain	s domain
Voltage-controlled voltage source	$v_2(t) = \mu v_1(t)$	$V_2(s) = \mu V_1(s)$
Current-controlled current source	$i_2(t) = \beta i_1(t)$	$I_2(s) = \beta I_1(s)$

FIGURE 10–3 *s-Domain models of dependent sources and OP AMPs.*

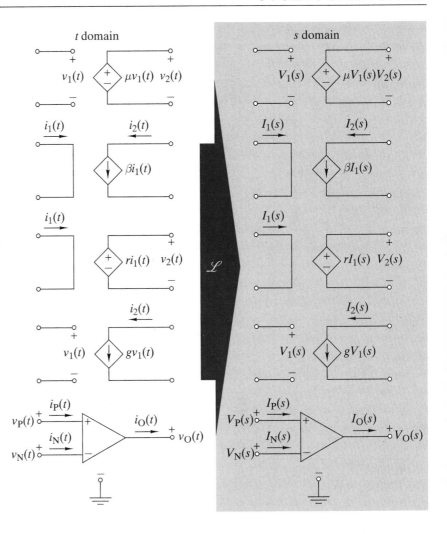

| Current-controlled voltage source | $v_2(t) = ri_1(t)$ | $V_2(s) = rI_1(s)$ | |
| Voltage-controlled current source | $i_2(t) = gv_1(t)$ | $I_2(s) = gV_1(s)$ | (10–3) |

Similarly, the element constraints of the ideal OP AMP are linear algebraic equations that are unchanged in form by the Laplace transformation:

t domain	*s* domain	
$v_P(t) = v_N(t)$	$V_P(s) = V_N(s)$	
$i_N(t) = 0$	$I_N(s) = 0$	(10–4)
$i_P(t) = 0$	$I_P(s) = 0$	

Thus, for linear active devices the only difference is that in the *s* domain the ideal element constraints apply to transforms rather than waveforms.

Finally, we consider the two-terminal passive circuit elements shown in Figure 10–4. In the time domain their *i–v* relationships are

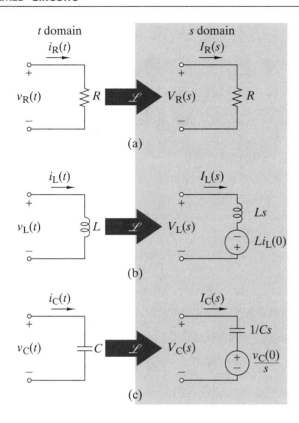

Resistor: $v_R(t) = R\, i_R(t)$

Inductor: $v_L(t) = L\dfrac{di_L(t)}{dt}$ (10–5)

Capacitor: $v_C(t) = \dfrac{1}{C}\displaystyle\int_0^t i_C(\tau)d\tau + v_C(0)$

These element constraints are transformed into the s domain by taking the Laplace transform of both sides of each equation using the linearity, differentiation, and integration properties.

Resistor: $V_R(s) = R\, I_R(s)$

Inductor: $V_L(s) = LsI_L(s) - Li_L(0)$ (10–6)

Capacitor: $V_C(s) = \dfrac{1}{Cs} I_C(s) + \dfrac{v_C(0)}{s}$

As expected, the element relationships are algebraic equations in the s domain. For the linear resistor the s domain version of Ohm's law says that the voltage transform $V_R(s)$ is proportional to the current transform $I_R(s)$. The element constraints for the inductor and capacitor also involve a proportionality between voltage and current, but include a term for the initial conditions as well.

The element constraints in Eq. (10–6) lead to the s-domain circuit models shown on the right side of Figure 10–4. The t-domain parameters L and

C are replaced by proportionality factors *Ls* and 1/*Cs* in the *s* domain. The initial conditions associated with the inductor and capacitor are modeled as voltage sources in series with these elements. The polarities of these sources are determined by the sign of the corresponding initial condition terms in Eq. (10–6). These initial condition voltage sources must be included when using these models to calculate the voltage transforms $V_L(s)$ or $V_C(s)$.

IMPEDANCE AND ADMITTANCE

The concept of impedance is a basic feature of *s*-domain circuit analysis. For zero initial conditions the element constraints in Eq. (10–6) reduce to

$$\text{Resistor: } V_R(s) = (R)\, I_R(s)$$

$$\text{Inductor: } V_L(s) = (Ls)I_L(s) \tag{10–7}$$

$$\text{Capacitor: } V_C(s) = (1/Cs)\, I_C(s)$$

In each case the element constraints are all of the form $V(s) = Z(s)I(s)$, which means that in the *s* domain the voltage across the element is proportional to the current through it. The proportionality factor is called the element **impedance** $Z(s)$. Stated formally,

> **Impedance** *is the proportionality factor relating the transform of the voltage across a two-terminal element to the transform of the current through the element with all initial conditions set to zero.*

The impedances of the three passive elements are

$$\text{Resistor: } Z_R(s) = R$$

$$\text{Inductor: } Z_L(s) = (Ls) \qquad \text{with } i_L(0) = 0 \tag{10–8}$$

$$\text{Capacitor: } Z_C(s) = (1/Cs) \qquad \text{with } v_C(0) = 0$$

It is important to remember that part of the definition of *s*-domain impedance is that the initial conditions are zero.

The *s*-domain impedance is a generalization of the *t*-domain concept of resistance. The impedance of a resistor is its resistance *R*. The impedance of the inductor and capacitor depend on the inductance *L* and capacitance *C* and the complex frequency variable *s*. Since a voltage transform has units of V-s and current transform has units of A-s, it follows that impedance has units of ohms since (V-s)/(A-s) = V/A = Ω.

Algebraically solving Eqs. (10–6) for the element currents in terms of the voltages produces alternative *s*-domain models.

$$\text{Resistor: } I_R(s) = \frac{1}{R}\, V_R(s)$$

$$\text{Inductor: } I_L(s) = \frac{1}{Ls}\, V_L(s) + \frac{i_L(0)}{s} \tag{10–9}$$

$$\text{Capacitor: } I_C(s) = Cs\, V_C(s) - C\, v_C(0)$$

In this form, the *i–v* relations lead to the *s*-domain models shown in the Figure 10–5. The reference directions for the initial condition current sources

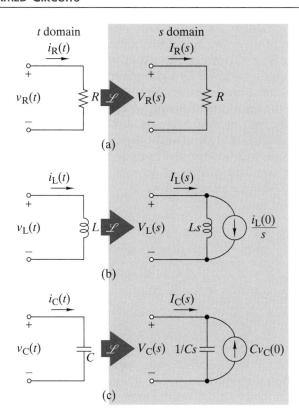

FIGURE 10–5 s-*Domain models of passive elements using current sources for initial conditions.*

are determined by the sign of the corresponding terms in Eqs. (10–9). The initial condition sources are in parallel with the element impedance.

Admittance $Y(s)$ is the s-domain generalization of the t-domain concept of conductance and can be defined as the reciprocal of impedance.

$$Y(s) = \frac{1}{Z(s)} \tag{10–10}$$

Using this definition, the admittances of the three passive elements are

$$\text{Resistor: } Y_R(s) = \frac{1}{Z_R(s)} = \frac{1}{R} = G$$

$$\text{Inductor: } Y_L(s) = \frac{1}{Z_L(s)} = \frac{1}{Ls} \quad \text{with } i_L(0) = 0 \tag{10–11}$$

$$\text{Capacitor: } Y_C(s) = \frac{1}{Z_C(s)} = Cs \quad \text{with } v_C(0) = 0$$

Since $Y(s)$ is the reciprocal of impedance, its units are siemens since $\Omega^{-1} = A/V = S$.

In summary, to transform a circuit into the s domain we replace each element by an s-domain model. For independent sources, dependent sources, OP AMPs, and resistors, the only change is that these elements now constrain transforms rather than waveforms. For inductors and capacitors, we can use either the model with a series initial condition voltage source (Figure 10–4), or the model with a parallel initial condition current source (Figure 10–5). However, to avoid possible confusion we always write the inductor impedance Ls and capacitor impedance $1/Cs$ beside the transformed element regardless of which initial condition source is used.

To analyze the transformed circuit, we can use the tools developed for resistance circuits in Chapters 2 through 4. These tools are applicable because KVL and KCL apply to transforms, and the *s*-domain element constraints are linear equations similar to those for resistance circuits. These features make *s*-domain analysis of dynamic circuits an algebraic process that is akin to resistance circuit analysis.

EXAMPLE 10–1

The switch in Figure 10–6 has been in position 1 for a long time and is moved to position 2 at $t = 0$. For $t > 0$ transform the circuit into the *s* domain and use Laplace transforms to solve for the voltage $v_C(t)$.

SOLUTION:

We have solved this type of problem before using various methods. In this example, we find the response by first transforming the circuit itself. For $t > 0$ the transformed circuit takes the form in Figure 10–7, where we have used a parallel current source $Cv_C(0)$ to account for the initial condition.

Applying KCL to the current transforms at node A produces

$$-I_1(s) - I_2(s) + Cv_C(0) = 0 \qquad \text{A-s}$$

For $t < 0$ the switch in Figure 10–6 is in position 1 and the circuit is in a dc steady-state condition. As a result, we have $v_C(0) = V_A$. In the *s*-domain circuit in Figure 10–7, the two branch current transforms can be written in terms of the capacitor voltage and element impedances as

$$\text{Resistor:} \quad I_1(s) = \frac{V_C(s)}{R}$$

$$\text{Capacitor:} \quad I_2(s) = \frac{V_C(s)}{1/Cs} = Cs\,V_C(s)$$

Substituting these observations into the KCL equation and solving for $V_C(s)$ yields

FIGURE 10 – 6

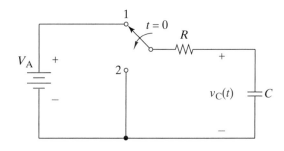

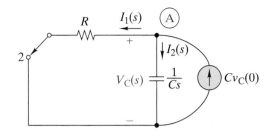

$$V_C(s) = \frac{CV_A}{Cs + \dfrac{1}{R}} = \frac{V_A}{s + \dfrac{1}{RC}} \quad \text{V-s}$$

Performing the inverse Laplace transformation leads to

$$v_C(t) = [V_A e^{-t/RC}]u(t) \quad \text{V}$$

The form of the response should be no great surprise. We could easily predict this response using the classical differential equation methods studied
in Chapter 7. What is important in this example is that we obtained the response using only basic circuit concepts applied in the s domain. ∎

EXAMPLE 10−2

(a) Transform the circuit in Figure 10–8(a) into the s domain.
(b) Solve for the current transform $I(s)$.
(c) Perform the inverse transformation to the waveform $i(t)$.

SOLUTION:

(a) Figure 10–8(b) shows the transformed circuit using a series voltage
source $Li_L(0)$ to represent the inductor initial condition. The impedances of the two passive elements are R and Ls. The independent
source voltage $V_A u(t)$ transforms as V_A/s.

(b) By KVL, the sum of voltage transforms around the loop is

$$-\frac{V_A}{s} + V_R(s) + V_L(s) = 0$$

Using the impedance models, the s-domain element constraints are

Resistor: $V_R(s) = RI(s)$
Inductor: $V_L(s) = LsI(s) - Li_L(0)$

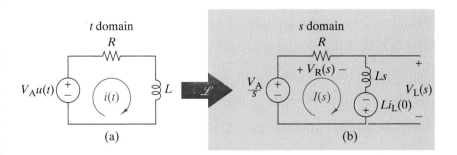

(a)

(b)

Substituting the element constraints into the KVL constraint and collecting terms yields

$$-\frac{V_A}{s} + (R + Ls)I(s) - Li_L(0) = 0$$

Solving for $I(s)$ produces

$$I(s) = \frac{V_A/L}{s(s + R/L)} + \frac{i_L(0)}{s + R/L} \quad \text{A-s}$$

The current $I(s)$ is the transform of the circuit response for a step function input. $I(s)$ is a rational function with simple poles at $s = 0$ and $s = -R/L$.

(c) To perform the inverse transformation, we expand $I(s)$ by partial fractions:

$$I(s) = \overbrace{\frac{V_A/R}{s}}^{\text{Forced}} \overbrace{- \frac{V_A/R}{s + R/L} + \frac{i_L(0)}{s + R/L}}^{\text{Natural}} \quad \text{A-s}$$

Taking the inverse transform of each term in this expansion gives

$$i(t) = \left[\overbrace{\frac{V_A}{R}}^{\text{Forced}} \overbrace{- \frac{V_A}{R} e^{-Rt/L} + i_L(0) e^{-Rt/L}}^{\text{Natural}} \right] u(t) \quad \text{A}$$

The forced response is caused by the step function input. The exponential terms in the natural response depend on the circuit time constant L/R. The step function and exponential components in $i(t)$ are directly related to the terms in the partial-fraction expansion of $I(s)$. The pole at the origin came from the step function input and leads to the forced response. The pole at $s = -R/L$ came from the circuit and leads to the natural response. Thus, in the *s* domain the forced response is that part of the total response that has the same poles as the input excitation. The natural response is that part of the total response whose poles came from the circuit. We say that the circuit contributes the natural poles because their locations depend on circuit parameters and not on the input. In other words, poles in the response do not occur by accident. They are present because the physical response depends on two things—(1) the input and (2) the circuit. ∎

10–2 BASIC CIRCUIT ANALYSIS IN THE *S* DOMAIN

In this section we develop the *s*-domain versions of series and parallel equivalence, and voltage and current division. These analysis techniques are the basic tools in *s*-domain circuit analysis, just as they are for resistance circuit analysis. These methods apply to circuits with elements connected in series or parallel. General analysis methods using node-voltage or mesh-current equations are covered later in Sects. 10–4 and 10–5.

SERIES EQUIVALENCE AND VOLTAGE DIVISION

The concept of a series connection applies in the *s* domain because Kirchhoff's laws do not change under the Laplace transformation. In Figure 10–9 the two-terminal elements are connected in series; hence by KCL the same current $I(s)$ exists in impedances $Z_1(s)$, $Z_2(s)$, ... $Z_N(s)$. Using KVL and the element constraints, the voltage across the series connection can be written as

$$V(s) = V_1(s) + V_2(s) + \cdots + V_N(s)$$
$$= Z_1(s)I(s) + Z_2(s)I(s) + \cdots + Z_N(s)I(s) \qquad (10\text{–}12)$$
$$= [Z_1(s) + Z_2(s) + \cdots + Z_N(s)]\,I(s)$$

The last line in this equation points out that the responses $V(s)$ and $I(s)$ do not change when the series-connected elements are replaced by an **equivalent impedance**:

$$Z_{EQ}(s) = Z_1(s) + Z_2(s) + \cdots + Z_N(s) \qquad (10\text{–}13)$$

In general, the equivalent impedance $Z_{EQ}(s)$ is a quotient of polynomials in the complex frequency variable of the form

$$Z_{EQ}(s) = \frac{b_m s^m + b_{m-1} s^{m-1} + \cdots + b_1 s + b_0}{a_n s^n + a_{n-1} s^{n-1} + \cdots + a_1 s + a_0} \qquad (10\text{–}14)$$

The roots of the numerator polynomial are the zeros of $Z_{EQ}(s)$, while the roots of the denominator are the poles.

Combining Eqs. (10–12) and (10–13), we can write the element voltages in the form

$$V_1(s) = \frac{Z_1(s)}{Z_{EQ}(s)}V(s) \quad V_2(s) = \frac{Z_2(s)}{Z_{EQ}(s)}V(s) \cdots V_N(s) = \frac{Z_N(s)}{Z_{EQ}(s)}V(s)$$

$$(10\text{–}15)$$

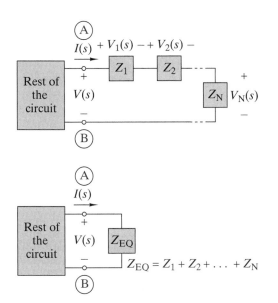

FIGURE 10–9 *Series equivalence in the s domain.*

These equations are the *s*-domain **voltage division principle**:

> *Every element voltage in a series connection is equal to its impedance divided by the equivalent impedance of the connection times the voltage across the series circuit.*

This statement parallels the corresponding rule for resistance circuits given in Chapter 4.

PARALLEL EQUIVALENCE AND CURRENT DIVISION

The parallel circuit in Figure 10–10 is the dual of the series circuit discussed previously. In this circuit the two-terminal elements are connected in parallel; hence by KVL the same voltage $V(s)$ appears across admittances $Y_1(s), Y_2(s), \ldots Y_N(s)$. Using KCL and the element constraints, the current into the parallel connection can be written as

$$
\begin{aligned}
I(s) &= I_1(s) + I_2(s) + \cdots + I_N(s) \\
&= Y_1(s)V(s) + Y_2(s)V(s) + \cdots + Y_N(s)V(s) \quad\quad (10\text{–}16) \\
&= [Y_1(s) + Y_2(s) + \cdots + Y_N(s)]V(s)
\end{aligned}
$$

The last line in this equation points out that the responses $V(s)$ and $I(s)$ do not change when the parallel connected elements are replaced by an **equivalent admittance**:

$$
Y_{EQ}(s) = Y_1(s) + Y_2(s) + \cdots + Y_N(s) \quad\quad (10\text{–}17)
$$

In general, the equivalent admittance $Y_{EQ}(s)$ is a quotient of polynomials in the complex frequency variable s. Since impedance and admittance are reciprocals, it turns out that if $Y_{EQ}(s) = p(s)/q(s)$, then the equivalent impedance at the same pair of terminals has the form $Z_{EQ}(s) = 1/Y_{EQ}(s) = q(s)/p(s)$. That is, at a given pair of terminals the poles of $Z_{EQ}(s)$ are zeros of $Y_{EQ}(s)$, and vice versa.

F I G U R E   1 0 – 1 0 *Parallel equivalence in the s domain.*

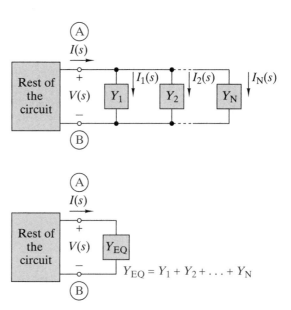

Combining Eqs. (10–16) and (10–17), we can write the element currents in the form

$$I_1(s) = \frac{Y_1(s)}{Y_{EQ}(s)}I(s) \quad I_2(s) = \frac{Y_2(s)}{Y_{EQ}(s)}I(s) \cdots I_N(s) = \frac{Y_N(s)}{Y_{EQ}(s)}I(s) \qquad \text{(10–18)}$$

These equations are the *s*-domain **current division principle**:

> *Every element current in a parallel connection is equal to its admittance divided by the equivalent admittance of the connection times the current into the parallel circuit.*

This statement is the dual of the results for a series circuit and parallels the current division rule for resistance circuits.

We begin to see that *s*-domain circuit analysis involves basic concepts that parallel the analysis of resistance circuits in the *t* domain. Repeated application of series/parallel equivalence and voltage/current division leads to an analysis approach called circuit reduction, discussed in Chapter 2. The major difference here is that we use impedance and admittances rather than resistance and conductance, and the analysis yields voltage and current transforms rather than waveforms.

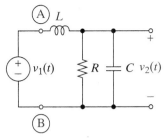

FIGURE 10–11

EXAMPLE 10–3

The inductor current and capacitor voltage in Figure 10–11 are zero at $t = 0$.

(a) Transform the circuit into the *s* domain and find the equivalent impedance between terminals A and B.

(b) Use voltage division to solve for the output voltage transform $V_2(s)$.

SOLUTION:

(a) Figure 10–12(a) shows the circuit in Figure 10–11 transformed into the *s* domain. As a first step we use parallel equivalence to find the equivalent impedance of the parallel resistor and capacitor.

$$Z_{EQ1}(s) = \frac{1}{Y_{EQ1}(s)} = \frac{1}{\dfrac{1}{R} + Cs} = \frac{R}{RCs + 1}$$

Figure 10–12(b) shows that the equivalent impedance $Z_{EQ1}(s)$ is connected in series with the inductor. This series combination can be replaced by an equivalent impedance

$$Z_{EQ}(s) = Ls + Z_{EQ1}(s) = Ls + \frac{R}{RCs + 1}$$

$$= \frac{RLCs^2 + Ls + R}{RCs + 1} \quad \Omega$$

as shown in Figure 10–12(c). The rational function $Z_{EQ}(s)$ is the impedance seen between terminals A and B in Figure 10–12(a).

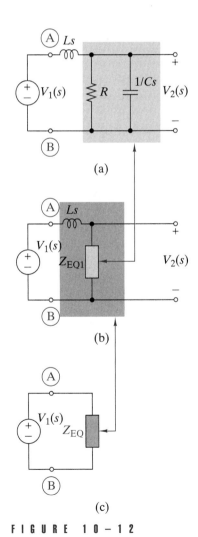

FIGURE 10–12

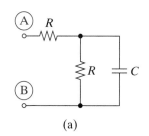

(a)

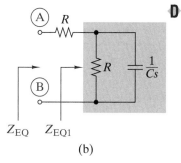

(b)

FIGURE 10-13

(b) Using voltage division in Figure 10–12(b), we find $V_2(s)$ as

$$V_2(s) = \left[\frac{Z_{EQ1}(s)}{Z_{EQ}(s)}\right]V_1(s) = \left[\frac{R}{RLCs^2 + Ls + R}\right]V_1(s)$$

Note that $Z_{EQ}(s)$ and $V_2(s)$ are rational functions of the complex frequency variable s. ■

EXAMPLE 10–4

In a circuit analysis problem we are required to find the poles and zeros of a circuit. In circuit design we are required to adjust circuit parameters to place the poles and zeros at specified s-plane locations. This example is a simple pole-placement design problem.

(a) Transform the circuit in Figure 10–13(a) into the s domain and find the equivalent impedance between terminals A and B.
(b) Select the values of R and C such that $Z_{EQ}(s)$ has a zero at $s = -5000$ rad/s.

SOLUTION:
(a) Figure 10–13(b) shows the circuit transformed to the s domain. The equivalent impedance $Z_{EQ1}(s)$ is

$$Z_{EQ1}(s) = \frac{1}{Y_R + Y_C} = \frac{1}{\dfrac{1}{R} + Cs}$$

$$= \frac{R}{RCs + 1}$$

Hence the equivalent impedance between terminals A and B is

$$Z_{EQ}(s) = R + Z_{EQ1}(s) = R + \frac{R}{RCs + 1}$$

$$= R\frac{RCs + 2}{RCs + 1} \quad \Omega$$

(b) For $Z_{EQ}(s)$ to have a zero at $s = -5000$ requires $2/RC = 5000$ or $RC = 4 \times 10^{-4}$. Selecting a standard value for the resistor $R = 10$ kΩ in turn requires $C = 40$ nF. ■

Exercise 10–1

The inductor current and capacitor voltage in Figure 10–14 are zero at $t = 0$.

(a) Transform the circuit into the s domain and find the equivalent admittance between terminals A and B.
(b) Solve for the output current transform $I_2(s)$ in terms of the input current $I_1(s)$.

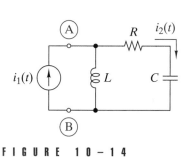

FIGURE 10-14

Answers:

$$\text{(a) } Y_{EQ}(s) = \frac{LCs^2 + RCs + 1}{Ls(RCs + 1)}$$

$$\text{(b) } I_2(s) = \left[\frac{LCs^2}{LCs^2 + RCs + 1}\right] I_1(s)$$

Exercise 10–2

The inductor current and capacitor voltage in Figure 10–15 are zero at $t = 0$.

(a) Transform the circuit into the s domain and find the equivalent impedance between terminals A and B.
(b) Solve for the output voltage transform $V_2(s)$ in terms of the input voltage $V_1(s)$.

Answers:

$$\text{(a) } Z_{EQ}(s) = \frac{(R_1Cs + 1)(Ls + R_2)}{LCs^2 + (R_1 + R_2)Cs + 1}$$

$$\text{(b) } V_2(s) = \left[\frac{Ls}{Ls + R_2}\right] V_1(s)$$

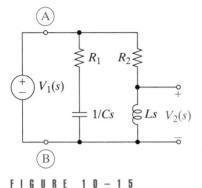

FIGURE 10 – 15

10–3 CIRCUIT THEOREMS IN THE s DOMAIN

In this section we study the s-domain versions of proportionality, superposition, and Thévenin/Norton equivalent circuits. These theorems define fundamental properties that provide conceptual tools for the analysis and design of linear circuits. With some modifications, all of the theorems studied in Chapter 3 apply to linear dynamic circuits in the s domain.

PROPORTIONALITY

For linear resistance circuits the **proportionality theorem** states that any output y is proportional to the input x:

$$y = Kx \qquad (10-19)$$

The same concept applies to linear dynamic circuits in the s domain except that the proportionality factor is a rational function of s rather than a constant. For instance, in Example 10–3 we found the output voltage $V_2(s)$ to be

$$V_2(s) = \left[\frac{R}{RLCs^2 + Ls + R}\right] V_1(s) \qquad (10-20)$$

where $V_1(s)$ is the transform of the input voltage. The quantity inside the brackets is a rational function that serves as the proportionality factor between the input and output transforms.

In the s-domain rational functions that relate inputs and outputs are called **network functions**. We begin the formal study network functions in Chapter 11. In this chapter we will simply illustrate network functions by an example.

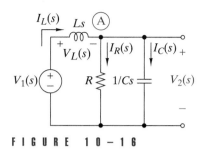

FIGURE 10-16

EXAMPLE 10-5

There is no initial energy stored in the circuit in Figure 10–16. Find the network functions relating $I_R(s)$ to $V_1(s)$ and $I_C(s)$ to $V_1(s)$.

SOLUTION:

The equivalent impedance seen by the voltage source is

$$Z_{EQ} = Ls + \frac{1}{\frac{1}{R} + Cs} = \frac{RLCs^2 + Ls + R}{RCs + 1}$$

Hence we can relate the $I_L(s)$ and $V_1(s)$ as

$$I_L(s) = \frac{V_1(s)}{Z_{EQ}(s)} = \left[\frac{RCs + 1}{RLCs^2 + Ls + R}\right]V_1(s)$$

Using s-domain current division, we can relate $I_R(s)$ and $I_C(s)$ to $I_L(s)$ as

$$I_R(s) = \frac{\frac{1}{R}}{\frac{1}{R} + Cs}I_L(s) = \left[\frac{1}{RCs + 1}\right]I_L(s)$$

$$I_C(s) = \frac{Cs}{\frac{1}{R} + Cs}I_L(s) = \left[\frac{RCs}{RCs + 1}\right]I_L(s)$$

Finally, using these relationships plus the relationship between $I_L(s)$ and $V_1(s)$ derived previously, we obtain the required network functions.

$$I_R(s) = \left[\frac{1}{RLCs^2 + Ls + R}\right]V_1(s)$$

$$I_C(s) = \left[\frac{RCs}{RLCs^2 + Ls + R}\right]V_1(s)$$ ∎

Exercise 10–3

In Figure 10–17 find the network function relating the output $V_2(s)$ to the input $I_1(s)$.

Answer:

$$V_2(s) = \left[\frac{R}{LCs^2 + RCs + 1}\right]I_1(s)$$

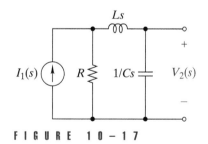

FIGURE 10-17

SUPERPOSITION

For linear resistance circuits the **superposition theorem** states that any output y of a linear circuit can be written as

$$y = K_1x_1 + K_2x_2 + K_3x_3 + \cdots \qquad (10\text{--}21)$$

where $x_1, x_2, x_3, \ldots$ are circuit inputs and $K_1, K_2. K_3, \ldots$ are weighting factors that depend on the circuit. The same concept applies to linear dynamic circuits in the *s* domain except that the weighting factors are rational functions of *s* rather than constants.

Superposition is usually thought of as a way to find the circuit response by adding the individual responses caused by each input acting alone. However, the principle applies to groups of sources as well. In particular, in the *s* domain there are two types of independent sources: (1) voltage and current sources representing the external driving forces for $t \geq 0$ and (2) initial condition voltage and current sources representing the energy stored at $t = 0$. As a result, the superposition principle states that the *s*-domain response can be found as the sum of two components: (1) the **zero-input response** caused by the initial condition sources with the external inputs turned off; or (2) the **zero-state response** caused by the external inputs with the initial condition sources turned off. Turning a source off means replacing voltage sources by short circuits $[V_S(s) = 0]$ and current sources by open circuits $[I_S(s) = 0]$.

The zero-input response is the response of a circuit to its initial conditions when the input excitations are set to zero. The zero-state response is the response of a circuit to its input excitations when all of the initial conditions are set to zero. The term *zero input* is self-explanatory. The term *zero state* is used because there is no energy stored in the circuit at $t = 0$.

The result is that voltage and current transform in a linear circuit can be found as the sum of two components of the form

$$V(s) = V_{zs}(s) + V_{zi}(s) \qquad I(s) = I_{zs}(s) + I_{zi}(s) \qquad \text{(10–22)}$$

where the subscript zs stands for zero state and zi for zero input. An important corollary is that the time-domain response can also be partitioned into zero-state and zero-input components because the inverse Laplace transformation is a linear operation.

We analyze the circuit treated in Example 10–2 to illustrate the superposition of zero-state and zero-input responses. The transformed circuit in Figure 10–18 has two independent voltage sources: (1) an input voltage source and (2) a voltage source representing the initial inductor current. The resistor and inductor are in series, so these two elements can be replaced by an impedance $Z_{EQ}(s) = Ls + R$.

First we turn off the initial condition source and replace its voltage source by a short circuit. Using the resulting zero-state circuit shown in Figure 10–18, we obtain the zero-state response:

$$I_{zs}(s) = \frac{V_A/s}{Z_{EQ}(s)} = \frac{V_A/L}{s(s + R/L)} \qquad \text{(10–23)}$$

The pole at $s = 0$ comes from the input source and the pole at $s = -R/L$ comes from the circuit. Next, we turn off the input source and use the zero-input circuit shown in Figure 10–18 to obtain the zero-input response:

$$I_{zi}(s) = \frac{L i_L(0)}{Z_{EQ}(s)} = \frac{i_L(0)}{s + R/L} \qquad \text{(10–24)}$$

The pole at $s = -R/L$ comes from the circuit. The zero-input response does not have a pole at $s = 0$ because the step function input is turned off.

Superposition states that the total response is the sum of the zero-state component in Eqs. (10–23) and the zero-input component in Eq. (10–24).

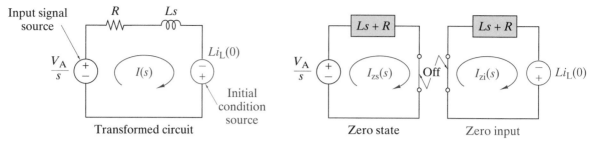

FIGURE 10–18 *Using superposition to find the zero-state and zero-input responses.*

$$I(s) = I_{zs}(s) + I_{zi}(s) = \frac{V_A/L}{s(s + R/L)} + \frac{i_L(0)}{s + R/L} \qquad (10\text{–}25)$$

The transform $I(s)$ in this equation is the same as found in Example 10–1. To derive the time-domain response, we expand $I(s)$ by partial fractions:

$$I(s) = \underbrace{\frac{V_A/R}{s} - \frac{V_A/R}{s + R/L}}_{\text{Zero State}} + \underbrace{\frac{i_L(0)}{s + R/L}}_{\text{Zero Input}} \qquad (10\text{–}26)$$

Performing the inverse transformation on each term yields

$$i(t) = \left[\underbrace{\frac{V_A}{R} - \frac{V_A}{R} e^{-Rt/L}}_{\text{Zero State}} + \underbrace{i_L(0) e^{-Rt/L}}_{\text{Zero Input}} \right] u(t) \quad \text{A} \qquad (10\text{–}27)$$

where the brace over the first two terms reads "Forced" and over the exponential terms reads "Natural".

Using superposition to partition the waveform into zero-state and zero-input components produces the same result as Example 10–1. The zero-state component contains the forced response. The zero-state and zero-input components both contain an exponential term due to the natural pole at $s = -R/L$ because both the external driving force and the initial condition source excite the circuit's natural response.

The superposition theorem helps us understand the response of circuits with multiple inputs, including initial conditions. It is a conceptual tool that helps us organize our thinking about s-domain circuits in general. It is not necessarily the most efficient analysis tool for finding the response of a specific multiple-input circuit.

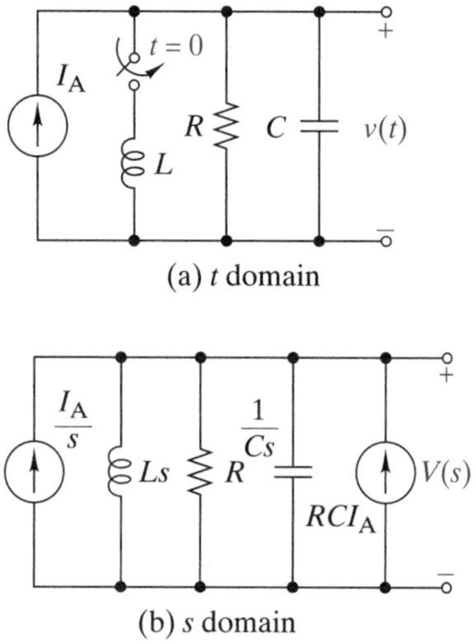

(a) t domain

(b) s domain

FIGURE 10–19

EXAMPLE 10–6

The switch in Figure 10–19(a) has been open for a long time and is closed at $t = 0$.

(a) Transform the circuit into the s domain.
(b) Find the zero-state and zero-input components of $V(s)$.
(c) Find $v(t)$ for $I_A = 1$ mA, $L = 2$ H, $R = 1.5$ kΩ, and $C = 1/6$ μF.

SOLUTION:

(a) To transform the circuit into the s domain, we must find the initial inductor current and capacitor voltage. For $t < 0$ the circuit is in a dc steady-state condition with the switch open. The inductor acts like a short circuit, and the capacitor acts like an open circuit. By inspection, the initial conditions at $t = 0-$ are $i_L(0) = 0$ and $v_C(0) = I_A R$. Figure 10–19(b) shows the s-domain circuit for these initial conditions. The current source version for the capacitor's initial condition is used here because the circuit elements are connected in parallel. The switch and constant current source combine to produce a step function $I_A u(t)$ whose transform is I_A/s.

(b) The resistor, capacitor, and inductor can be replaced by an equivalent impedance

$$Z_{EQ} = \frac{1}{Y_{EQ}} = \frac{1}{\dfrac{1}{Ls} + \dfrac{1}{R} + Cs}$$

$$= \frac{RLs}{RLCs^2 + Ls + R}$$

The zero-state response is found with the capacitor initial condition source replaced by an open circuit and the step function input source on:

$$V_{zs}(s) = Z_{EQ}(s)\frac{I_A}{s} = \left[\frac{RLs}{RLCs^2 + Ls + R}\right]\frac{I_A}{s} = \frac{I_A/C}{s^2 + \dfrac{s}{RC} + \dfrac{1}{LC}}$$

The pole in the input at $s = 0$ is canceled by the zero at the origin in $Z_{EQ}(s)$. As a result, the zero-state response does not have a forced pole at $s = 0$. The zero-input response is found by replacing the input source by an open circuit and turning the capacitor initial condition source on:

$$V_{zi}(s) = [Z_{EQ}(s)][CRI_A] = \frac{RI_A s}{s^2 + \dfrac{s}{RC} + \dfrac{1}{LC}}$$

(c) Inserting the given numerical values of the circuit parameters and expanding the zero-state and zero-input response transforms by partial fractions yields

$$V_{zs}(s) = \frac{6000}{(s + 1000)(s + 3000)} = \frac{3}{s + 1000} + \frac{-3}{s + 3000} \quad \text{V-s}$$

$$V_{zi}(s) = \frac{1.5\,s}{(s + 1000)(s + 3000)} = \frac{-0.75}{s + 1000} + \frac{2.25}{s + 3000} \quad \text{V-s}$$

The inverse transforms of these expansions are

$$v_{zs}(t) = [3e^{-1000t} - 3e^{-3000t}]\,u(t) \quad \text{V}$$

$$v_{zi}(t) = [-0.75e^{-1000t} + 2.25e^{-3000t}]\,u(t) \quad \text{V}$$

Note that the circuit responses contain only transient terms that decay to zero. There is no forced response because in the dc steady state the induc-

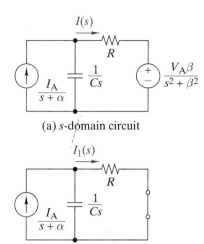

(a) *s*-domain circuit

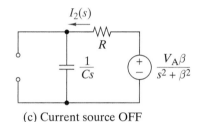

(b) Voltage source OFF

$I_2(s)$

(c) Current source OFF

F I G U R E   1 0 – 2 0

tor acts like a short circuit, forcing $v(t)$ to zero for $t \to \infty$. From an *s* domain viewpoint there is no forced response because the forced pole at $s = 0$ is canceled by a zero in the network function. ∎

EXAMPLE 10–7

Use superposition to find the zero-state component of $I(s)$ in the *s*-domain circuit shown in Figure 10–20(a).

SOLUTION:
Turning the voltage source off produces the circuit in Figure 10–20(b). In this circuit the resistor and capacitor are connected in parallel, so current division yields $I_1(s)$ in the form

$$I_1(s) = \frac{Y_R}{Y_C + Y_R} \frac{I_A}{(s + \alpha)} = \frac{I_A}{(RCs + 1)(s + \alpha)}$$

Turning the voltage source on and the current source off produces the circuit in Figure 10–20(c). In this case the resistor and capacitor are connected in series, and series equivalence gives the current $I_2(s)$ as

$$I_2(s) = \frac{1}{Z_R + Z_C} \frac{V_A \beta}{s^2 + \beta^2} = \frac{CsV_A\beta}{(RCs + 1)(s^2 + \beta^2)}$$

Using superposition, the total zero-state response is

$$I_{zs}(s) = I_1(s) - I_2(s)$$

$$= \frac{I_A}{(RCs + 1)(s + \alpha)} - \frac{CsV_A\beta}{(RCs + 1)(s^2 + \beta^2)}$$

There is a minus in this equation because $I_1(s)$ and $I_2(s)$ were assigned opposite reference directions in Figures 10–20(b) and 10–20(c). The total zero-state response has four poles. The natural pole at $s = -1/RC$ came from the circuit. The forced pole at $s = -\alpha$ came from the current source, and the two forced poles at $s = \pm j\beta$ came from the voltage source.

In this example the time-domain response would have a transient component $Ke^{-t/RC}$ due to the natural pole, a forced component $Ke^{-\alpha t}$ due to the current source, and a forced component of the form $K_A \cos \beta t + K_B \sin \beta t$ due to the voltage source. We can infer these general conclusions regarding the time-domain response by simply examining the poles of the *s*-domain response. ∎

Exercise 10–4

The initial conditions for the circuit in Figure 10–21 are $v_C(0-) = 0$ and $i_L(0-) = I_0$. Transform the circuit into the *s* domain and find the zero-state and zero-input components of $V(s)$.

Answers:

$$V_{zs}(s) = \left[\frac{1}{LCs^2 + RCs + 1} \right] \frac{V_A}{s}$$

$$V_{zi}(s) = \frac{LI_0}{LCs^2 + RCs + 1}$$

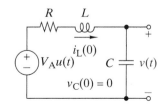

F I G U R E   1 0 – 2 1

The initial conditions for the circuit in Figure 10–22 are $v_C(0-) = 0$ and $i_L(0-) = I_0$. Transform the circuit into the *s* domain and find the zero-state and zero-input components of $I(s)$.

Answers:

$$I_{zs}(s) = \left[\frac{LCs^2}{LCs^2 + RCs + 1} \right] \frac{I_A}{s}$$

$$I_{zi}(s) = \left[\frac{LCs^2}{LCs^2 + RCs + 1} \right] \frac{I_0}{s}$$

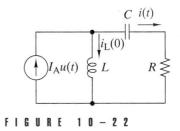

FIGURE   10–22

THÉVENIN AND NORTON EQUIVALENT CIRCUITS
AND SOURCE TRANSFORMATIONS

When the source circuit in Figure 10–23(a) contains only independent sources and linear elements, it can be replaced by the Thévenin equivalent circuit in Figure 10–23(b) or the Norton equivalent circuit in Figure 10–23(c). In the *s* domain the methods of finding and using Thévenin or Norton equivalent circuits are similar to those for resistive circuits. The important differences are that the signals are transforms rather than waveforms and the circuit elements are impedances rather than resistances.

The open-circuit voltage and short-circuit current at the interface provided enough information to define the Thévenin and Norton equivalent circuits. With an open-circuit load connected in Figures 10–23(b) and 10–23(c), we see that the interface voltage is

$$V(s) = V_{OC}(s) = I_N(s)Z_N(s) = V_T(s) \tag{10–28}$$

With a short-circuit load connected, the interface current is

$$I(s) = I_{SC}(s) = V_T(s)/Z_T(s) = I_N(s) \tag{10–29}$$

Taken together, Eqs. (10–28) and (10–29) yield the conditions

$$V_T(s) = V_{OC}(s) \quad I_N(s) = I_{SC}(s) \qquad Z_T(s) = Z_N(s) = \frac{V_{OC}(s)}{I_{SC}(s)} \tag{10–30}$$

Algebraically, the results in Eq. (10–30) are identical to the corresponding equations for resistance circuits. The important difference, as noted previously, is that these equations involve transforms and impedances rather than waveforms and resistances. In any case, the equations point out that the open-circuit voltage and short-circuit current are sufficient to define either equivalent circuit.

The Thévenin and Norton equivalent circuits are related by an *s*-domain source transformation. The *s*-domain **source transformation** relations in Eq. (10–30) allow us to transform a voltage source in series with an impedance into a current source in parallel with the same impedance, and vice versa. We use these source transformations when formulating node-voltage and mesh-current equations in the *s* domain.

FIGURE 10-23 *Thévenin and Norton equivalent circuits in the s domain.*

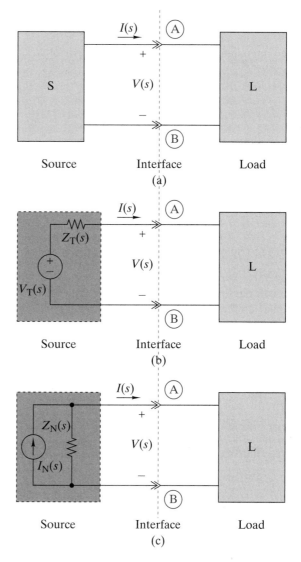

FIGURE 10-23 *Thévenin and Norton equivalent circuits in the s domain.*

Consider the interface shown in Figure 10–24. The source circuit is given in the *s* domain with zero initial voltage on the capacitor. The open-circuit voltage at the interface is found by voltage division:

$$V_{OC}(s) = \frac{1/Cs}{R + 1/Cs} \frac{V_A}{s} = \frac{V_A}{s(RCs + 1)} \qquad (10–31)$$

The pole in $V_{OC}(s)$ at $s = 0$ comes from the step function voltage source and the pole at $s = -1/RC$ from the RC circuit. Connecting a short circuit at the interface effectively removes the capacitor from the circuit. The short-circuit current is

$$I_{SC}(s) = \frac{1}{R} \frac{V_A}{s} \qquad (10–32)$$

This current does not have a pole at $s = -1/RC$ because the capacitor is shorted out. Taking the ratio of Eqs. (10–31) and (10–32) yields the Thévenin impedance:

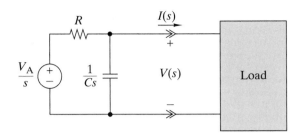

FIGURE 10–24 *A source-load interface in the s domain.*

$$Z_T(s) = \frac{V_{OC}(s)}{I_{SC}(s)} = \frac{R}{RCs + 1} \qquad (10\text{–}33)$$

When there are no dependent sources present, the Thévenin impedance can be found using the lookback method from Chapter 3. That is, the Thévenin impedance is the equivalent impedance seen looking back into the source circuit with the independent sources (external inputs and initial conditions) turned off. Replacing the voltage source in Figure 10–24 by a short circuit reduces the source circuit to a parallel combination of a resistor and capacitor. The equivalent lookback impedance of the combination is

$$Z_T(s) = \frac{1}{Y_R + Y_C} = \frac{1}{\dfrac{1}{R} + Cs} = \frac{R}{RCs + 1} \qquad (10\text{–}34)$$

which is the same result as Eq. (10–33).

This example illustrates that in the s domain, Thévenin or Norton theorems involve concepts similar to those discussed in Chapter 3. At a given interface these equivalent circuits are defined by any two of the following: (1) the open-circuit voltage, (2) the short-circuit current, or (3) the lookback impedance (when no dependent sources are involved). Thévenin and Norton equivalent circuits are useful tools that help us understand the operation of a circuit, especially when we must evaluate the circuit performance with different loads or parameter variations at an interface. They are not, in general, a way to reduce the computational burden of finding the response of a fixed circuit.

EXAMPLE 10–8

Find the interface voltage $v(t)$ in Figure 10–24 when the load is (a) a resistance R and (b) a capacitance C.

SOLUTION:

Figure 10–25 shows the s-domain circuit model of the source-load interface. The parameters of the Thévenin equivalent are given in Eqs. (10–31) and (10–33) as

$$V_T(s) = \frac{V_A}{s(RCs + 1)} \quad \text{and} \quad Z_T(s) = \frac{R}{RCs + 1}$$

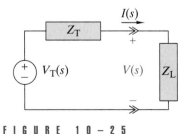

FIGURE 10–25

(a) For a resistance load, $Z_L(s) = R$ and the interface voltage is found by voltage division:

$$V(s) = \frac{Z_L(s)}{Z_T(s) + Z_L(s)} V_T(s)$$

$$= \left(\frac{R}{\frac{R}{RCs + 1} + R} \right) \frac{V_A}{s(RCs + 1)}$$

$$= \frac{V_A/RC}{s(s + 2/RC)} \quad \text{V-s}$$

The partial fraction expansion of $V(s)$ is

$$V(s) = \frac{V_A/2}{s} - \frac{V_A/2}{s + 2/RC}$$

Applying the inverse transform yields the interface voltage waveform as

$$v(t) = \frac{V_A}{2}[1 - e^{-2t/RC}] u(t) \quad \text{V}$$

(b) For a capacitance load, $Z_L(s) = 1/Cs$ and the interface voltage transform is

$$V(s) = \frac{Z_L(s)}{Z_T(s) + Z_L(s)} V_T(s) = \left(\frac{\frac{1}{Cs}}{\frac{R}{RCs + 1} + \frac{1}{Cs}} \right) \frac{V_A}{s(RCs + 1)}$$

$$= \frac{V_A/2RC}{s(s + 1/2 RC)} \quad \text{V-s}$$

The partial fraction expansion of $V(s)$ is

$$V(s) = \frac{V_A}{s} - \frac{V_A}{s + 1/2RC}$$

Applying the inverse transform to each term yields the interface voltage waveform for the capacitance load as

$$v(t) = V_A[1 - e^{-t/2RC}] u(t) \quad \text{V}$$

This example illustrates an important difference between resistive circuits and dynamic circuits. In a resistive circuit the load affects the amplitude but not the waveshape of the interface signals. In a dynamic circuit the load can change both amplitude and the waveshape.   ■

D DESIGN EXAMPLE 10–9

The shaded portion of the circuit in Figure 10–26(a) is a low-frequency model of a transistor. Select the value of C_E so that the circuit has a pole at $s = -300$ rad/s.

FIGURE   10 − 26

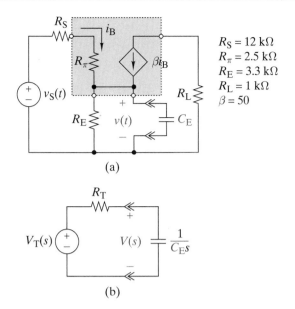

$R_S = 12 \text{ k}\Omega$
$R_\pi = 2.5 \text{ k}\Omega$
$R_E = 3.3 \text{ k}\Omega$
$R_L = 1 \text{ k}\Omega$
$\beta = 50$

(a)

(b)

SOLUTION:

Figure 10–26(b) shows the Thévenin equivalent seen by the capacitor C_E. In this circuit the voltage across the capacitor is

$$V_C(s) = \frac{1/C_E s}{R_T + 1/C_E s} V_T(s) = \frac{1}{R_T C_E s + 1} V_T(s)$$

The circuit has a simple pole at $s = -1/R_T C_E$, where R_T is the Thévenin equivalent resistance seen by the capacitor. To place the circuit pole at $s = -300$, we must select $C_E = 1/300 R_T$. Hence the problem reduces to finding the Thévenin equivalent resistance seen by the capacitor. With the capacitor in Figure 10–26(a) replaced by a short circuit, we find

$$i_{SC}(t) = \beta i_B(t) + i_B(t) = (\beta + 1)\left[\frac{v_S(t)}{R_S + R_\pi}\right]$$

$$= 3.52 v_S(t) \quad \text{mA}$$

Next, replacing the capacitor by an open circuit yields

$$v_{OC}(t) = i_B(t)R_E + \beta i_B(t)R_E$$

$$= (\beta + 1)R_E\left[\frac{v_S(t) - v_{OC}(t)}{R_S + R_\pi}\right]$$

Solving this expression for $v_{OC}(t)$, we obtain

$$v_{OC}(t) = \frac{(\beta + 1)R_E}{R_S + R_\pi + (\beta + 1)R_E} v_S(t)$$

$$= 0.921 v_S(t) \quad \text{V}$$

The Thévenin equivalent resistance seen by the capacitor is $R_T = v_{OC}/i_{SC} = 262 \ \Omega$, and the required value of capacitance is $C_E = 12.7 \ \mu\text{F}$. This example illustrates the general principle that we can select element values to position a circuit's poles at specified locations.   ■

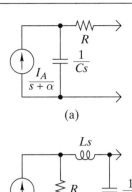

(a)

(b)

FIGURE   10 – 27

Find the Norton equivalent of the *s*-domain circuits in Figure 10–27.

Answers:

$$(a)\ I_N(s) = \frac{I_A}{(RCs + 1)(s + \alpha)} \qquad Z_N(s) = \frac{RCs + 1}{Cs}$$

$$(b)\ I_N(s) = \frac{RI_A}{(Ls + R)(s + \alpha)} \qquad Z_N(s) = \frac{Ls + R}{LCs^2 + RCs + 1}$$

10–4 NODE-VOLTAGE ANALYSIS IN THE *s* DOMAIN

The previous sections deal with basic analysis methods using equivalence, reduction, and circuit theorems. These methods are valuable because we work directly with the element impedances and thereby gain insight into *s*-domain circuit behavior. We also need general methods to deal with more complicated circuits that these basic methods cannot easily handle.

FORMULATING NODE-VOLTAGE EQUATIONS

Formulating node-voltage equations involves selecting a reference node and assigning a node-to-datum voltage to each of the remaining nonreference nodes. Because of KVL, the voltage across any two-terminal element is equal to the difference of two node voltages. This fundamental property of node voltages, together with element impedances, allows us to write KCL constraints at each of the nonreference nodes.

For example, consider the *s*-domain circuit in Figure 10–28. The sum of currents leaving node A can be written as

$$I_{S2}(s) - I_{S1}(s) + \frac{V_A(s)}{Z_1(s)} + \frac{V_A(s) - V_B(s)}{Z_2(s)} + \frac{V_A(s) - V_C(s)}{Z_3(s)} = 0$$

Rewriting this equation with unknown node voltages grouped on the left and inputs on the right yields

$$\left[\frac{1}{Z_1(s)} + \frac{1}{Z_2(s)} + \frac{1}{Z_3(s)}\right] V_A(s) - \frac{1}{Z_2(s)} V_B(s) - \frac{1}{Z_3(s)} V_C(s) = I_{S1}(s) - I_{S2}(s)$$

Expressing this result in terms of admittances produces the following equation:

$$[Y_1(s) + Y_2(s) + Y_3(s)] V_A(s) - [Y_2(s)] V_B(s)$$
$$- [Y_3(s)]V_C(s) = I_{S1}(s) - I_{S2}(s)$$

This equation has a familiar pattern. The unknowns are the node-voltage transforms $V_A(s)$, $V_B(s)$, and $V_C(s)$. The coefficient $[Y_1(s) + Y_2(s) + Y_3(s)]$ of $V_A(s)$ is the sum of the admittances of the elements connected to node A. The coefficient $[Y_2(s)]$ of $V_B(s)$ is the admittance of the elements connected between nodes A and B, while $[Y_3(s)]$ is the admittance of the ele-

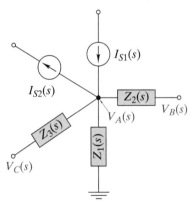

FIGURE   10 – 28 *An example node.*

ments connected between nodes A and C. Finally, $I_{S1}(s) - I_{S2}(s)$ is the sum of the source currents directed into node A. These observations suggest that we can write node-voltage equations for s-domain circuits by inspection, just as we did with resistive circuits.

The formulation method just outlined assumes that there are no voltage sources in the circuit. When transforming the circuit we can always select the current source models to represent the initial conditions. However, the circuit may contain dependent or independent voltage sources. If so, they can be treated using the following methods:

Method 1: If there is an impedance in series with the voltage source, use a source transformation to convert it into an equivalent current source.

Method 2: Select the reference node so that one terminal of one or more of the voltage sources is connected to ground. The source voltage then determines the node voltage at the other source terminal, thereby eliminating an unknown.

Method 3: Create a supernode surrounding any voltage source that cannot be handled by method 1 or 2.

Some circuits may require more than one of these methods.

Formulating a set of equilibrium equations in the s domain is a straightforward process involving concepts developed in Chapters 3 and 4 for resistance circuits. The following example illustrates the formulation process.

EXAMPLE 10-10

Formulate s-domain node-voltage equations for the circuit in Figure 10–29(a).

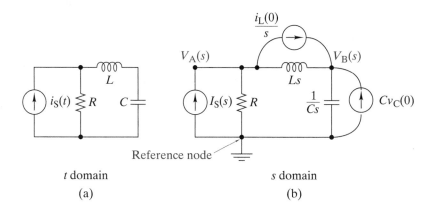

FIGURE 10-29

$i_L(0)$ over s

$V_A(s)$

L

Ls

$V_B(s)$

$i_S(t)$ R C

$I_S(s)$ R $\frac{1}{Cs}$ $Cv_C(0)$

Reference node

t domain

(a)

s domain

(b)

SOLUTION:

Figure 10–29(b) shows the circuit in the s domain. In transforming the circuit, we use current sources to represent the inductor and capacitor initial conditions. This choice facilitates writing node equations since the resulting s-domain circuit contains only current sources and fewer nodes. The sum of currents leaving nodes A and B can be written as

$$\text{Node A:} \frac{V_A(s)}{R} + \frac{V_A(s) - V_B(s)}{Ls} - I_S(s) + \frac{i_L(0)}{s} = 0$$

$$\text{Node B:} \frac{V_B(s)}{1/Cs} + \frac{V_B(s) - V_A(s)}{Ls} - \frac{i_L(0)}{s} - Cv_C(0) = 0$$

Rearranging these equations in the standard format with the unknowns on the left and the inputs on the right yields

$$\text{Node A:} \left(G + \frac{1}{Ls}\right) V_A(s) - \left(\frac{1}{Ls}\right) V_B(s) = I_S(s) - \frac{i_L(0)}{s}$$

$$\text{Node B:} -\left(\frac{1}{Ls}\right) V_A(s) + \left(\frac{1}{Ls} + Cs\right) V_B(s) = Cv_C(0) + \frac{i_L(0)}{s}$$

where $G = 1/R$ is the conductance of the resistor. Note that $(G + 1/Ls)$ is the sum of the admittance connected to node A, $(1/Ls + Cs)$ is the sum of the admittances connected to node B, and $1/Ls$ is the admittance connected between nodes A and B. The circuit is driven by an independent current source $I_S(s)$ and two initial condition sources $Cv_C(0)$ and $i_L(0)/s$. The terms on the right side of these equations are the sum of source currents directed into each node. With practice, we learn to write these equations by inspection. ■

SOLVING *s*-DOMAIN CIRCUIT EQUATIONS

Examples 10–10 and 10–11 show that node-voltage equations are linear algebraic equations in the unknown node voltages. In theory, solving these node equations is accomplished using techniques such as Cramer's rule or perhaps Gaussian reduction. In practice, we quickly lose interest in Gaussian reduction since coefficients in the equations are polynomials, making the algebra rather complicated. For hand analysis Cramer's rule is the better approach, especially when the element parameters are in symbolic form. For computer aided analysis we use the symbolic analysis capability of programs like Mathcad, MATLAB, and others to solve these linear equations.

We will illustrate the Cramer's rule solution process using an example from earlier in this section. In Example 10–10 we formulated the following node-voltage equations for the circuit in Figure 10–29:

$$\left(G + \frac{1}{Ls}\right) V_A(s) - \frac{1}{Ls} V_B(s) = I_S(s) - \frac{i_L(0)}{s}$$

$$-\frac{1}{Ls} V_A(s) + \left(Cs + \frac{1}{Ls}\right) V_B(s) = \frac{i_L(0)}{s} + Cv_C(0)$$

When using Cramer's rule, it is convenient first to find the determinant of these equations:

$$\Delta(s) = \begin{vmatrix} G + 1/Ls & -1/Ls \\ -1/Ls & Cs + 1/Ls \end{vmatrix}$$

$$= (G + 1/Ls)(Cs + 1/Ls) - (1/Ls)^2$$

$$= \frac{GLCs^2 + Cs + G}{Ls}$$

We call $\Delta(s)$ the **circuit determinant** because it only depends on the element parameters L, C, and $G = 1/R$. The determinant $\Delta(s)$ characterizes the circuit and does not depend on the input driving forces or initial conditions.

The node voltage $V_A(s)$ is found using Cramer's rule.

$$V_A(s) = \frac{\Delta_A(s)}{\Delta(s)} = \frac{\begin{vmatrix} I_S(s) - \dfrac{i_L(0)}{s} & -\dfrac{1}{Ls} \\ \dfrac{i_L(0)}{s} + Cv_C(0) & Cs + \dfrac{1}{Ls} \end{vmatrix}}{\Delta(s)} \tag{10–35}$$

$$= \underbrace{\frac{(LCs^2 + 1)\,I_S(s)}{GLCs^2 + Cs + G}}_{\text{Zero State}} + \underbrace{\frac{-LCsi_L(0) + Cv_C(0)}{GLCs^2 + Cs + G}}_{\text{Zero Input}}$$

Solving for the other node voltage $V_B(s)$ yields

$$V_B(s) = \frac{\Delta_B(s)}{\Delta(s)} = \frac{\begin{vmatrix} G + \dfrac{1}{Ls} & I_S(s) - i_L(0) \dfrac{}{s} \\ -\dfrac{1}{Ls} & \dfrac{i_L(0)}{s} + Cv_C(0) \end{vmatrix}}{\Delta(s)} \tag{10–36}$$

$$= \underbrace{\frac{I_S(s)}{GLCs^2 + Cs + G}}_{\text{Zero State}} + \underbrace{\frac{GLi_L(0) + (GLs + 1)Cv_C(0)}{GLCs^2 + Cs + G}}_{\text{Zero Input}}$$

Cramer's rule gives both the zero-input and zero-state components of the response transforms $V_A(s)$ and $V_B(s)$.

Cramer's rule yields the node voltages as a ratio of determinants of the form

$$V_X(s) = \frac{\Delta_X(s)}{\Delta(s)} \tag{10–37}$$

The response transform is a rational function of s whose poles are either zeros of the circuit determinant or poles of the determinant $\Delta_X(s)$. That is, $V_X(s)$ in Eq. (10–37) has poles when $\Delta(s) = 0$ or $\Delta_X(s) \to \infty$. The partial-fraction expansion of $V_X(s)$ will contain terms for each of these poles. We call the zeros of $\Delta(s)$ the **natural poles** because they depend only on the circuit and give rise to the natural response terms in the partial-fraction expansion. We call the poles of $\Delta_X(s)$ the **forced poles** because they depend on the form of the input signal and give rise to the forced response terms in the partial-fraction expansion.

EXAMPLE 10–11

The response transforms $V_A(s)$ and $V_B(s)$ in Eqs. (10-35) and (10-36) include the zero-state and zero-input components. The purpose of this example is to calculate the zero-state components of the waveforms $v_A(t)$ and $v_B(t)$ when $R = 1\ \text{k}\Omega$, $L = 0.5\ \text{H}$, $C = 0.2\ \mu\text{F}$, and $i_S(t) = 10u(t)\ \text{mA}$.

SOLUTION:

Inserting the specified numerical values into Eqs. (10-35) and (10-36) yields

$$V_A(s) = \left(\frac{10^{-7}s^2 + 1}{10^{-10}s^2 + 0.2 \times 10^{-6}s + 10^{-3}}\right)\frac{10^{-2}}{s}$$

$$= 10\frac{s^2 + 10^7}{s[(s + 1000)^2 + 3000^2]} \qquad \text{V-s}$$

$$V_B(s) = \left(\frac{1}{10^{-10}s^2 + 0.2 \times 10^{-6}s + 10^{-3}}\right)\frac{10^{-2}}{s}$$

$$= 10\frac{10^7}{s[(s + 1000)^2 + 3000^2]} \qquad \text{V-s}$$

Both response transforms have three poles: a forced pole at $s = 0$ and two natural poles at $s = -1000 \pm j3000$. The forced pole comes from the step function input, and the two natural poles are zeros of the circuit determinant. Expanding these rational functions as

$$V_A(s) = \frac{10}{s} - \frac{20}{3}\left(\frac{3000}{(s + 1000)^2 + 3000^2}\right)$$

$$V_B(s) = \frac{10}{s} - \frac{10}{3}\left(\frac{3000}{(s + 1000)^2 + 3000^2}\right) - 10\left(\frac{s + 1000}{(s + 1000)^2 + 3000^2}\right)$$

and taking the inverse transforms yields the required zero-state response waveforms:

$$v_A(t) = 10u(t) - 20e^{-1000t}\left[\frac{1}{3}\sin(3000\,t)\right]u(t) \qquad \text{V}$$

$$v_B(t) = 10u(t) - 10e^{-1000t}\left[\frac{1}{3}\sin(3000\,t) + \cos(3000\,t)\right]u(t) \qquad \text{V}$$

The step function in both responses is the forced response caused by the forced pole at $s = 0$. The damped sinusoids are natural responses determined by the natural poles.

Figure 10–30 shows part of an Excel spreadsheet that produces plots of $v_A(t)$ and $v_B(t)$. Spreadsheets are useful for generating graphs, especially when we wish to compare waveforms. The two plots show that the two response waveforms are different even though they have the same poles. In other words, the basic form of a response is determined by the forced and natural poles, but the relative amplitudes are influenced by the zeros as well. ∎

APPLICATION NOTE: EXAMPLE 10–12

In _s_-domain circuit analysis and design, the location of complex poles is often specified in terms of the undamped natural frequency (ω_0) and damping ratio (ζ) parameters introduced in our study of second-order circuits. Using these parameters, the standard form of a second-order factor is $s^2 + 2\zeta\omega_0 s + \omega_0^2$, which locates the poles at

$$s_{1,2} = \omega_0(-\zeta \pm \sqrt{\zeta^2 - 1})$$

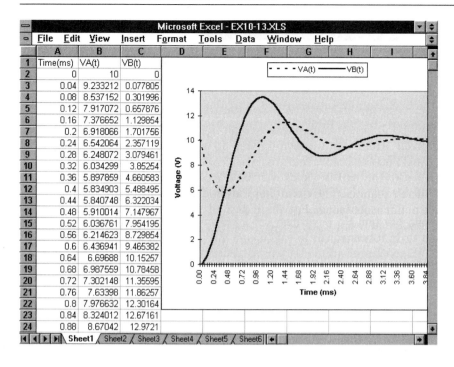

FIGURE 10–30

The quantity under the radical depends only on the damping ratio ζ. When $\zeta > 1$ the quantity is positive and the two poles are real and distinct, and the second-order factor becomes the product of two first-order terms. If $\zeta = 1$ the quantity under the radical vanishes and there is a double pole as $s = -\omega_0$. If $\zeta < 1$ the quantity under the radical is negative and the two roots are complex conjugates.

The location of complex poles is also defined in terms of the two natural frequency parameters α and β. Using these parameters, the poles are at $s_{1,2} = -\alpha \pm j\beta$, and the standard form of a second-order factor is $(s + \alpha)^2 + \beta^2$.

In *s*-domain circuit design, we often need to convert from one set of parameters to the other. First, equating their standard forms

$$s^2 + 2\alpha s + \alpha^2 + \beta^2 = s^2 + 2\zeta\omega_0 s + \omega_0^2$$

and then equating the coefficients of like powers of *s* yields

$$\omega_0 = \sqrt{\alpha^2 + \beta^2} \quad \text{and} \quad \zeta = \frac{\alpha}{\sqrt{\alpha^2 + \beta^2}}$$

and conversely

$$\alpha = \zeta\omega_0 \quad \text{and} \quad \beta = \omega_0\sqrt{1 - \zeta^2}$$

Figure 10–31 shows how these parameters define the locations of complex poles in the *s* plane. The natural frequency parameters α and β define the rectangular coordinates of the poles. In a sense, the parameters ω_0 and ζ define the corresponding polar coordinates. The parameter ω_0 is the radial distance from the origin to the poles. The angle θ is determined by the damping ratio ζ alone, since $\theta = \cos^{-1} \zeta$.

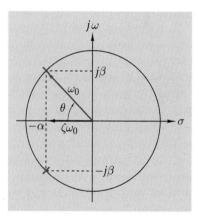

FIGURE 10–31 *s-plane geometry relating α and β to ζ and ω_0.*

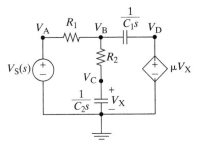

FIGURE 10-32

The *s*-domain circuit in Figure 10–32 is to be designed to produce a pair of complex poles defined by $\zeta = 0.5$ and $\omega_0 = 1000$ rad/s. To simplify production the design will use equal element values $R_1 = R_2 = R$ and $C_1 = C_2 = C$. Select the values of R, C, and the gain μ so that the circuit has the desired natural poles.

SOLUTION:

To locate the natural poles, we find the circuit determinant using node-voltage equations. The circuit has four nodes but only two of these involve independent variables. For the indicated reference node, the voltages at nodes A and D are $V_A(s) = V_S(s)$ and $V_D(s) = \mu V_X(s) = \mu V_C(s)$. That is, the two grounded voltage sources specify the voltages at nodes A and D. Consequently, we only need equations at nodes B and C. The sum of currents leaving these nodes are

$$\text{Node B:} \quad \frac{V_B(s) - V_S(s)}{R_1} + \frac{V_B(s) - V_C(s)}{R_2} + \frac{V_B(s) - \mu V_C(s)}{1/C_1 s} = 0$$

$$\text{Node C:} \quad \frac{V_C(s) - V_B(s)}{R_2} + \frac{V_C(s)}{1/C_2 s} = 0$$

Arranging these equations with unknown node voltages on the left and the source terms on the right yields

$$(C_1 s + G_1 + G_2)V_B(s) - (\mu C_1 s + G_2)V_C(s) = G_1 V_S(s)$$

$$- (G_2)V_B(s) + (C_2 s + G_2)V_C(s) = 0$$

where $G_1 = 1/R_1$ and $G_2 = 1/R_2$. The natural poles are zeros of the circuit determinant:

$$\Delta(s) = \begin{vmatrix} (C_1 s + G_1 + G_2) & -(\mu C_2 s + G_2) \\ -(G_2) & (C_2 s + G_2) \end{vmatrix}$$

$$= C_1 C_2 s^2 + (G_2 C_1 + G_1 C_2 + G_2 C_2 - \mu G_2 C_2)s + G_1 G_2$$

For equal resistances $R_1 = R_2 = R$ and equal capacitances $C_1 = C_2 = C$, the circuit determinant reduces to

$$\frac{\Delta(s)}{C^2} = s^2 + \left(\frac{3 - \mu}{RC}\right)s + \left(\frac{1}{RC}\right)^2$$

Comparing this second-order factor to the standard form $s^2 + 2\zeta \omega_0 s + \omega_0^2$ yields the following design constraints:

$$\omega_0 = \frac{1}{RC} = 1000 \quad \text{and} \quad \zeta = \frac{3 - \mu}{2} = 0.5$$

These constraints lead to the conditions $RC = 10^{-3}$ and $\mu = 3$. Selecting $R = 10$ kΩ makes $C = 100$ nF. For the specified conditions the natural poles are located at $s = \alpha \pm j\beta$, where

$$\alpha = -\zeta \omega_0 = -500 \text{ rad/s} \quad \text{and} \quad \beta = \omega_0 \sqrt{1 - \zeta^2} = 866 \text{ rad/s}$$

EXAMPLE 10-14

(a) For the s-domain circuit in Figure 10–33, solve for the zero-state output $V_O(s)$ in terms of a general input $V_S(s)$.
(b) Solve for the zero-state output $v_O(t)$ when the input is a unit step function $v_S(t) = u(t)$.

FIGURE 10–33

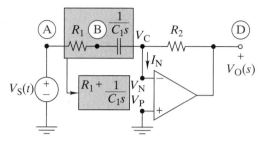

SOLUTION:

(a) We use the node-voltage method to find the OP AMP output. A node equation is not required at node A since the selected reference node makes $V_A(s) = V_S(s)$. Likewise, one is not needed at node D since it is the OP AMP output. Finally, we can avoid writing a node equation at node B by observing that the element impedances R_1 and $1/Cs$ are connected in series. We can treat this series combination as a single element with an equivalent impedance $R_1 + 1/Cs$. Using all of these observations, the sum of the currents leaving Node C is

$$\text{Node C:} \quad \frac{V_C(s) - V_S(s)}{R_1 + 1/C\,s} + \frac{V_C(s) - V_O(s)}{R_2} + I_N(s) = 0$$

In the s-domain the ideal OP AMP model in Eq. (10–4) requires $I_N(s) = 0$ and $V_P(s) = V_N(s)$. But $V_P(s) = 0$ since the noninverting input is grounded, hence $V_N(s) = V_C(s) = 0$. Inserting these conditions in the node C equation and solving for the output voltage yields

$$V_O(s) = \left[-\frac{R_2}{R_1 + 1/C\,s} \right] V_S(s)$$

$$= \left[-\frac{R_2}{R_1} \left(\frac{s}{s + 1/R_1 C} \right) \right] V_S(s) \quad \text{V-s}$$

This equation relates the zero-state output to a general input $V_S(s)$. The output transform is proportional to the input transform since the circuit is linear. The proportionality factor within the brackets is called a *network function*. In this case the network function has a natural pole at $s = -1/R_1 C$ and a zero at $s = 0$.

(b) A step function input $V_S(s) = 1/s$ produces a forced pole at $s = 0$. However, the zero in the network function cancels the forced pole so that

$$v_O(t) = \mathscr{L}^{-1} \left\{ -\frac{R_2}{R_1} \left(\frac{s}{s + 1/R_1 C} \right) \frac{1}{s} \right\} = \mathscr{L}^{-1} \left\{ -\frac{R_2}{R_1} \left(\frac{1}{s + 1/R_1 C} \right) \right\}$$

$$= \left(-\frac{R_2}{R_1} e^{-t/R_1 C} \right) u(t) \quad \text{V}$$

For a step function input the zero-state output has no forced pole, only a natural pole at $s = -1/R_1 C$. The general principle is that the forced response can be zero even when the input is not zero. In the s domain this occurs when the network function relating output to input has zeros at the same location as forced poles.

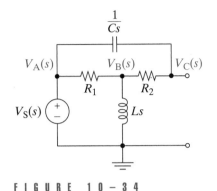

FIGURE 10–34

Exercise 10–7

Formulate node-voltage equations for the circuit in Figure 10–34 and find the circuit determinant. Assume that the initial conditions are zero.

Answer:
The node equations are:

$$\text{Node B} \quad \left(G_1 + G_2 + \frac{1}{Ls}\right)V_B(s) - G_2 V_C(s) = G_1 V_S(s)$$

$$\text{Node C} \quad -G_2 V_B(s) + (G_2 + Cs)V_C(s) = Cs V_S(s)$$

The circuit determinant is

$$\Delta(s) = \frac{(G_1 + G_2)LCs^2 + (G_1 G_2 L + C)s + G_2}{Ls}$$

Exercise 10–8

Formulate node-voltage equations for the circuit in Figure 10–34 when a resistor R_3 is connected between node C and ground. Assume that the initial conditions are zero.

Answer:

$$\text{Node B} \quad \left(G_1 + G_2 + \frac{1}{Ls}\right)V_B(s) - G_2 V_C(s) = G_1 V_S(s)$$

$$\text{Node C} \quad -G_2 V_B(s) + (G_2 + G_3 + Cs)V_C(s) = Cs V_S(s)$$

10–5 MESH-CURRENT ANALYSIS IN THE s DOMAIN

We can use the mesh-current method only when the circuit can be drawn on a flat surface without crossovers. Such planar circuits have special loops called meshes that are defined as closed paths that do not enclose any elements. The mesh-current variables are the loop currents assigned to each mesh in a planar circuit. Because of KCL the current through any two-terminal element can be expressed as the difference of two adjacent mesh currents. This fundamental property of mesh currents, together with the element impedances, allows us to write KVL constraints around each of the meshes.

For example, in Figure 10–35 the sum of voltages around mesh A can be written as

$$Z_1(s)I_A(s) + Z_3[I_A(s) - I_C(s)] - V_{S1}(s)$$
$$+ Z_2[I_A(s) - I_B(s)] + V_{S2}(s) = 0$$

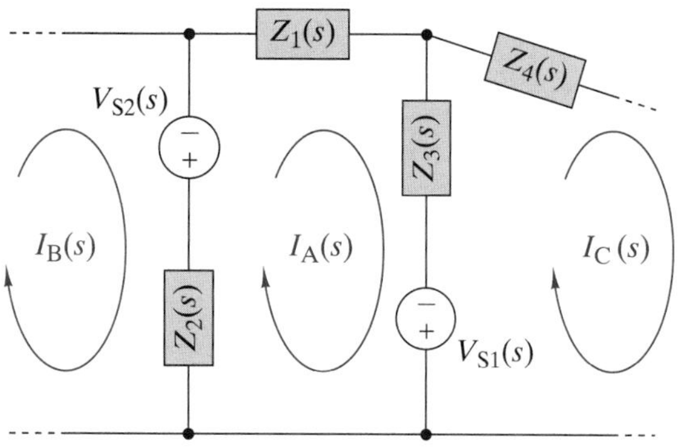

FIGURE 10–35   *An example mesh.*

Rewriting this equation with unknown mesh currents grouped on the left and inputs on the right yields

$$(Z_1 + Z_2 + Z_3)I_A(s) - Z_2 I_B(s) - Z_3 I_C(s) = V_{S1}(s) - V_{S2}(s)$$

This equation displays the following pattern. The unknowns are the mesh-current transforms $I_A(s)$, $I_B(s)$, and $I_C(s)$. The coefficient $[Z_1(s) + Z_2(s) + Z_3(s)]$ of $I_A(s)$ is the sum of the impedances of the elements in mesh A. The coefficients $[Z_2(s)]$ of $I_B(s)$ and $[Z_3(s)]$ of $I_C(s)$ are the impedances common to mesh A and the other meshes. Finally, $V_{S1}(s) - V_{S2}(s)$ is the sum of the source voltages around mesh A. These observations suggest that we can write node-voltage equations for *s*-domain circuits by inspection, just as we did with resistive circuits.

The formulation approach just outlined assumes that there are no current sources in the circuit. When writing mesh-current equations, we select the voltage source model to represent the initial conditions. If the circuit contains dependent or independent current sources, they can be treated using the following methods:

Method 1: If there is an admittance in parallel with the current source, use a source transformation to convert it into an equivalent voltage source.

Method 2: Draw the circuit diagram so that only one mesh current circulates through the current source. This mesh current is then determined by the source current.

Method 3: Create a supermesh for any current source that cannot be handled by methods 1 or 2.

Some circuits may require more than one of these methods.

The following examples illustrate the mesh-current method of *s*-domain circuit analysis.

EXAMPLE 10–15

(a) Formulate mesh-current equations for the circuit in Figure 10–36(a).
(b) Solve for the zero-input component of $I_A(s)$ and $I_B(s)$.
(c) Find the zero-input responses $i_A(t)$ and $i_B(t)$ for $R_1 = 200\ \Omega$, $R_2 = 300\ \Omega$, $L_1 = 50$ mH, and $L_2 = 100$ mH.

FIGURE 10 – 36

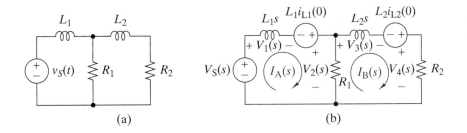

(a)   (b)

SOLUTION:

(a) Figure 10–36(b) shows the circuit transformed into the *s* domain. In transforming the circuit we used the voltage source model for the initial conditions. The net result is that the transformed circuit contains only voltage sources. The sum of voltages around meshes A and B can be written as

Mesh A: $-V_S(s) + L_1 s I_A(s) - L_1 i_{L_1}(0) + R_1[I_A(s) - I_B(s)] = 0$

Mesh B: $R_1[I_B(s) - I_A(s)] + L_2 s I_B(s) - L_2 i_{L_2}(0) + R_2 I_B(s) = 0$

Rearranging these equations in standard form yields

Mesh A: $(L_1 s + R_1) I_A(s) - R_1 I_B(s) = V_S(s) + L_1 i_{L_1}(0)$

Mesh B: $-R_1 I_A(s) + (L_2 s + R_1 + R_2) I_B(s) = L_2 i_{L_2}(0)$

These *s*-domain circuit equations are two linear algebraic equations in the two unknown mesh currents $I_A(s)$ and $I_B(s)$.

(b) To solve for the mesh equations, we first find the circuit determinant:

$$\Delta(s) = \begin{vmatrix} L_1 s + R_1 & -R_1 \\ -R_1 & L_2 s + R_1 + R_2 \end{vmatrix}$$

$$= L_1 L_2 s^2 + (R_1 L_2 + R_1 L_1 + R_2 L_1)s + R_1 R_2$$

To find the zero-input component of $I_A(s)$, we let $V_S(s) = 0$ and use Cramer's rule:

$$I_A(s) = \frac{\begin{vmatrix} L_1 i_{L_1}(0) & -R_1 \\ L_2 i_{L_2}(0) & L_2 s + R_1 + R_2 \end{vmatrix}}{\Delta(s)}$$

$$= \frac{(L_2 s + R_1 + R_2) L_1 i_{L_1}(0) + R_1 L_2 i_{L_2}(0)}{L_1 L_2 s^2 + (R_1 L_2 + R_1 L_1 + R_2 L_1)s + R_1 R_2}$$

Similarly, the zero-input component in $I_B(s)$ is

$$I_B(s) = \frac{\begin{vmatrix} L_1 s + R_1 & L_1 i_{L_1}(0) \\ -R_1 & L_2 i_{L_2}(0) \end{vmatrix}}{\Delta(s)}$$

$$= \frac{(L_1 s + R_1) L_2 i_{L_2}(0) + R_1 L_1 i_{L_1}(0)}{L_1 L_2 s^2 + (R_1 L_2 + R_1 L_1 + R_2 L_1)s + R_1 R_2}$$

(c) To find the time-domain response, we insert the numerical parameters into the preceding expressions to obtain

$$I_A(s) = \frac{(0.005s + 15)i_{L_1}(0) + 10i_{L_2}(0)}{0.005s^2 + 250s + 20{,}000}$$

$$= \frac{(s + 3000)i_{L_1}(0) + 2000i_{L_2}(0)}{(s + 1000)(s + 4000)}$$

$$I_B(s) = \frac{(0.005s + 100)i_{L_2}(0) + 5i_{L_1}(0)}{0.005s^2 + 250s + 20{,}000}$$

$$= \frac{(s + 2000)i_{L_2}(0) + 1000i_{L_1}(0)}{(s + 1000)(s + 4000)}$$

The circuit has natural poles at $s = -1000$ and -4000 rad/s. Expanding by partial fractions yields

$$I_A(s) = \frac{2}{3} \times \frac{i_{L_1}(0) + i_{L_2}(0)}{s + 1000} + \frac{1}{3} \times \frac{i_{L_1}(0) - 2i_{L_2}(0)}{s + 4000} \quad \text{A-s}$$

$$I_B(s) = \frac{1}{3} \times \frac{i_{L_1}(0) + i_{L_2}(0)}{s + 1000} - \frac{1}{3} \times \frac{i_{L_1}(0) - 2i_{L_2}(0)}{s + 4000} \quad \text{A-s}$$

The inverse transforms of these expansions are the required zero-state response waveforms:

$$i_A(t) = \left[\frac{2}{3}[i_{L_1}(0) + i_{L_2}(0)]e^{-1000t} + \frac{1}{3}[i_{L_1}(0) - 2i_{L_2}(0)]e^{-4000t}\right]u(t) \quad \text{A}$$

$$i_B(t) = \left[\frac{1}{3}[i_{L_1}(0) + i_{L_2}(0)]e^{-1000t} - \frac{1}{3}[i_{L_1}(0) - 2i_{L_2}(0)]e^{-4000t}\right]u(t) \quad \text{A}$$

Notice that if the initial conditions are $i_{L_1}(0) = -i_{L_2}(0)$, then both $I_A(s)$ and $I_B(s)$ have a zero at $s = -1000$. This zero effectively cancels the natural pole at $s = -1000$. As a result, this pole has zero residue in both partial fraction expansions, and the corresponding terms disappear from the time-domain responses. Likewise, if the initial conditions are $i_{L_1}(0) = 2i_{L_2}(0)$, then both $I_A(s)$ and $I_B(s)$ have a zero at $s = -4000$, and the natural pole at $s = -4000$ disappears in the s-domain responses. The general principle is that all of the circuit's natural poles may not be present in a given response. When this happens the response transform has a zero at the same location as a natural pole, and we say that the natural pole is not observable in the specified response. ∎

EXAMPLE 10–16

(a) Formulate mesh-current equations for the circuit in Figure 10–37(a).
(b) Solve for the zero-input component of $i_A(t)$ for $i_L(0) = 0$, $v_C(0) = 10$ V, $L = 250$ mH, $C = 1$ μF and $R = 1$ kΩ.

SOLUTION:
(a) Figure 10–37(a) is the s-domain circuit used in Example 10–10 to develop node equations. In this circuit each current source is connected in parallel with an impedance. Source transformations convert these current sources into the equivalent voltage sources shown in Figure 10–37(b). The circuit in Figure 10–37(b) is a series RLC circuit of the

FIGURE 10-37

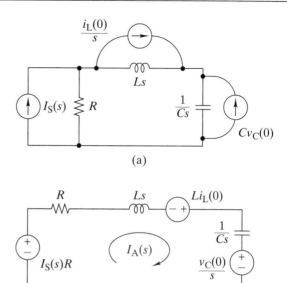

type treated in Chapter 8. By inspection, the KVL equation for the single mesh in this circuit is

$$\left[R + Ls + \frac{1}{Cs}\right] I_A(s) = RI_S(s) + Li_L(0) - \frac{v_C(0)}{s}$$

The circuit determinant is the factor $R + Ls + 1/Cs = (LCs^2 + RCs + 1)/Cs$. The zeros of the circuit determinant are roots of the quadratic equation $LCs^2 + RCs + 1 = 0$, which we recognize as the characteristic equation of a series RLC circuit. In our study of RLC circuits we called these roots natural frequencies. Thus, the natural poles of the circuit are its natural frequencies.

(b) Solving the mesh equation for the zero-input component yields

$$I_A(s) = \frac{LCsi_L(0) - Cv_C(0)}{LCs^2 + RCs + 1} \quad \text{A-s}$$

Inserting the given numerical values produces

$$I_A(s) = -\frac{10^{-5}}{0.25 \times 10^{-6}s^2 + 10^{-3}s + 1} = -\frac{40}{s^2 + 4 \times 10^3 s + 4 \times 10^6}$$

$$= -\frac{40}{(s + 2000)^2} \quad \text{A-s}$$

The zero-state response has two natural poles, both located at $s = -2000$. The inverse transform of $I_A(s)$ is a damped ramp waveform:

$$i_A(t) = -[40te^{-2000t}] u(t) \quad \text{A}$$

The damped ramp response indicates a critically damped second-order circuit. The minus sign means the direction of the actual current is opposite to the reference mark assigned to $I_A(s)$ in Figure 10–37. This sign makes sense physically since the capacitor initial condition source in Figure 10–37(b) tends to drive current in a direction opposite to the assigned reference mark. ∎

D **DESIGN EXAMPLE 10−17**

Select the element values in Figure 10–38 so that the circuit has a real natural pole at $s = -20$ krad/s and a pair of complex poles with $\zeta = 0.5$ and $\omega_0 = 20$ krad/s.

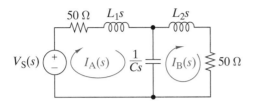

FIGURE 10−38

SOLUTION:
To locate the circuit poles, we first write two mesh equations:

$$\text{Mesh A:} \quad \left(50 + \frac{1}{Cs} + L_1 s\right) I_A(s) - \frac{1}{Cs} I_B(s) = V_S(s)$$

$$\text{Mesh B:} \quad -\frac{1}{Cs} I_A(s) + \left(50 + \frac{1}{Cs} + L_2 s\right) I_B(s) = 0$$

The circuit determinant is

$$\Delta(s) = \left(50 + \frac{1}{Cs} + L_1 s\right)\left(50 + \frac{1}{Cs} + L_2 s\right) - \left(\frac{1}{Cs}\right)^2$$

$$= \frac{L_1 L_2 C s^3 + (50L_1 C + 50\,L_2 C)s^2 + (L_1 + L_2 + 2500\,C)s + 100}{Cs}$$

The circuit characteristic equation in the numerator of $\Delta(s)$ can be placed in the following form:

$$q(s) = s^3 + \frac{(50L_1 + 50L_2)}{L_1 L_2}s^2 + \frac{(L_1 + L_2 + 2500C)}{L_1 L_2 C}s + \frac{100}{L_1 L_2 C}$$

To have the required pole positions, the circuit characteristic equation must have the form

$$q(s) = (s + \alpha)(s^2 + 2\zeta\omega_0 s + \omega_0^2)$$

where $\alpha = \omega_0 = 20$ krad/s and $\zeta = 0.5$. Hence,

$$q(s) = (s + 20{,}000)\,(s^2 + 20{,}000s + 20{,}000^2)$$

$$= s^3 + 4 \times 10^4 s^2 + 8 \times 10^8 s + 8 \times 10^{12}$$

Comparing the circuit's symbolic equation with the required polynomial leads to the following design constraints:

$$\frac{50L_1 + 50L_2}{L_1 L_2} = 4 \times 10^4 \qquad \frac{L_1 + L_2 + 2500C}{L_1 L_2 C} = 8 \times 10^8$$

$$\frac{100}{L_1 L_2 C} = 8 \times 10^{12}$$

The symbolic tools in MATLAB can solve equations like these. To use these tools the expressions need to be written in the form $f(x, y) = 0$. Accordingly, we rewrite the three design constraints as follows:

$$50L_1 + 50L_2 - 4 \times 10^4 L_1 L_2 = 0$$
$$L_1 + L_2 + 2500\, C - 8 \times 10^8 L_1 L_2 C = 0$$
$$100 - 8 \times 10^{12} L_1 L_2 C = 0$$

In the MATLAB command window we identify the symbolic variables and write the three design constraints as

```
syms  C  L1  L2
f1=50*L1+50*L2-4*10^4*L1*L2;
f2=L1+L2+2500*C-8*10^8*L1*L2*C;
f3=100-8*10^12*L1*L2*C;
```

and use the MATLAB solve function as

```
[C, L1, L2]=solve[f1, f2, f3]

C=

1/500000

L1=

1/400

L2=

1/400
```

Thus, the element values that produce the desired pole locations are

$$C = \frac{1}{500000} = 2\ \mu\text{F} \quad \text{and} \quad L_1 = L_2 = \frac{1}{400} = 2.5\ \text{mH}$$

In this case we were able to find a set of element values that meet the nonlinear design constraints. This may not always be possible. ∎

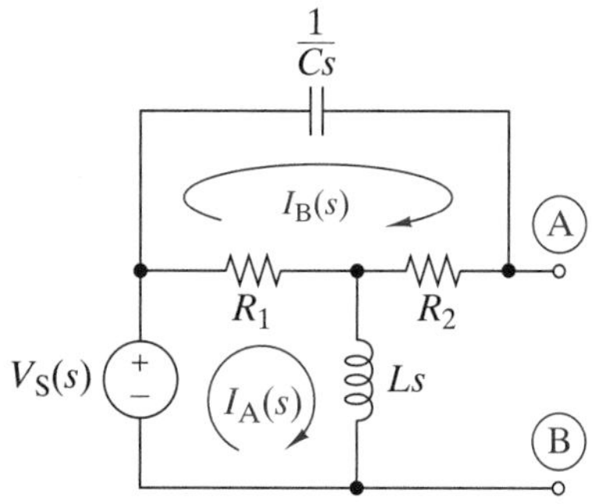

Exercise 10–9

(a) Formulate mesh-current equations for the circuit in Figure 10–39. Assume that the initial conditions are zero.
(b) Find the circuit determinant.
(c) Solve for the zero-state component of $I_B(s)$.

Answers:

(a) $\begin{cases} (R_1 + Ls)I_A(s) - R_1 I_B(s) = V_S(s) \\[2mm] -R_1 I_A(s) + (R_1 + R_2 + 1/Cs)I_B(s) = 0 \end{cases}$

(b) $\Delta(s) = \dfrac{(R_1 + R_2)LCs^2 + (R_1 R_2 C + L)s + R_1}{Cs}$

(c) $I_B(s) = \dfrac{R_1 Cs V_S(s)}{(R_1 + R_2)LCs^2 + (R_1 R_2 C + L)s + R_1}$

FIGURE 10 – 39

Formulate mesh-current equations for the circuit in Figure 10–39 when a resistor R_3 is connected between nodes A and B. Assume that the initial conditions are zero.

Answer:

$$(R_1 + Ls)I_A(s) - R_1I_B(s) - LsI_C(s) = V_S(s)$$

$$-R_1I_A(s) + \left(R_1 + R_2 + \frac{1}{Cs}\right)I_B(s) - R_2I_C(s) = 0$$

$$-LsI_A(s) - R_2I_B(s) + (R_2 + R_3 + Ls)I_C(s) = 0$$

10-6 SUMMARY OF S-DOMAIN CIRCUIT ANALYSIS

At this point we review our progress and put s-domain circuit analysis into perspective. We have shown that linear circuits can be transformed from the time domain into the s domain. In this domain KCL and KVL apply to transforms and the passive element i–v characteristics become impedances with series or parallel initial condition sources. In relatively simple circuits we can use basic analysis methods, such as reduction, superposition, and voltage/current division. For more complicated circuits we use systematic procedures, such as the node-voltage or mesh-current methods, to solve for the circuit response.

In theory, we can perform s-domain analysis on circuits of any complexity. In practice, the algebraic burden of hand computations gets out of hand for circuits with more than three or four nodes or meshes. Of what practical use is an analysis method that becomes impractical at such a modest level of circuit complexity? Why not just appeal to computer-aided analysis tools in the first place?

Unquestionably, large-scale circuits are best handled by computer-aided analysis. Computer-aided analysis is probably the right approach even for small-scale circuits when numerical values for all circuit parameters are known and the desired end product is a plot or numerical listing of the response waveform. Simply put, s-domain circuit analysis is not a particularly efficient algorithm for generating numerical response data.

The purpose of s-domain circuit analysis is to gain insight into circuit behavior, not to grind out particular response waveforms. In this regard s-domain circuit analysis complements programs like PSpice. It offers a way of characterizing circuits in very general terms. It provides guidelines that allow us to use computer-aided analysis tools intelligently. Some of the useful general principles derived in this chapter are the following.

The response transform $Y(s)$[1] is a rational function whose partial fraction expansion leads directly to a response waveform of the form

---

1 In this context $Y(s)$ is not an admittance but the Laplace transform of the circuit output $y(t)$.

$$y(t) = \sum_{j=1}^{\text{number of poles}} k_j e^{p_j t}$$

where k_j is the residue of the pole in $Y(s)$ located at $s = -p_j$. The location of the poles tells us a great deal about the form of the response. The pair of conjugate complex poles in Example 10–11 produced a damped sine waveform, the two distinct real poles in Example 10–15 produced exponential waveforms, and the double pole in Example 10–16 led to a damped ramp waveform. The general principle illustrated is as follows:

> *The poles of **Y(s)** are either real or complex conjugates. Simple real poles lead to exponentials, double real poles lead to a damped ramp, and complex conjugate poles lead to damped sinusoids.*

The poles in $Y(s)$ are introduced either by the circuit itself (natural poles) or by the input driving force (forced poles).

> *The natural poles are zeros of the circuit determinant and lead to the natural response. The forced poles are poles of the input **X(s)** and lead to the forced response.*

Stability is a key concept in circuit analysis and design. For our present purposes we say that a linear circuit is **stable** if its natural response decays to zero as $t \to \infty$. Figure 10–40 shows the waveforms of the natural modes corresponding to different pole locations in the s plane. Poles in the left-half plane give rise to waveforms that decay to zero as time increases,

FIGURE 10–40 *Form of the natural response corresponding to different pole locations.*

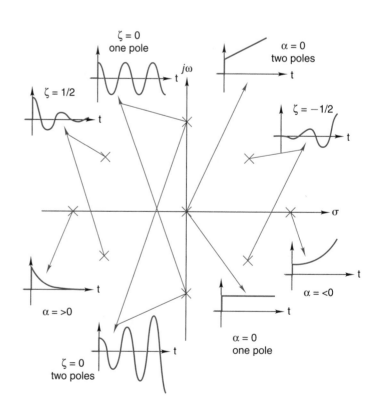

while those in the right-half plane increase without bound. As a result, we can say that

> *A circuit is stable if all of its natural poles are located in the left half of the s plane.*

Stability requires *all* of the natural poles to be in the left-half plane (LHP). The circuit is *unstable* if even one natural pole falls in the right-half plane (RHP).

In Figure 10–40 the $j\omega$-axis is the boundary between the LHP (stable circuits) and RHP (unstable circuits). Poles exactly on this boundary require further discussion. As Figure 10–40 shows, *simple j-axis poles at $s = 0$ and $s = \pm j\beta$* lead to natural modes like $u(t)$ and $\cos(\beta t)$ that neither decay to zero nor increase without bound. The figure also shows that *double poles on the j-axis* lead to natural modes like $tu(t)$ and $t\cos(\beta t)$ that increase without bound. Circuits with *simple* poles on the *j*-axis are sometimes said to be **marginally stable**,[2] while those with *multiple* poles on the *j*-axis are clearly *unstable*.

Circuit stability is determined by natural poles and not forced poles. For example, suppose an input $x(t) = e^{10t}$ produces an output transform

$$Y(s) = \frac{12s}{\underbrace{(s + 2)}_{\text{LHP}}\underbrace{(s - 10)}_{\text{RHP}}}$$

This transform has a left-half plane (LHP) pole and a right-half plane (RHP) pole. The corresponding waveform

$$y(t) = \underbrace{2e^{-2t}}_{\substack{\text{Natural} \\ \text{bounded}}} + \underbrace{10e^{10t}}_{\substack{\text{Forced} \\ \text{unbounded}}} \quad t > 0$$

has an unbounded term due to the RHP pole. Even with an unbounded response the circuit is still said to be stable because the natural pole at $s = -2$ is in the LHP and leads to a natural response that decays to zero. The unbounded part of the response waveform comes from the forced RHP pole caused by the unbounded input.

Since the natural response play a key role, we would like to predict the number of natural poles by simply examining the circuit. Figure 10–41 summarizes examples from this chapter and leads to the following observations. Circuits with only one energy storage element (inductor or capacitor) have only one pole, circuits with two independent elements have two poles, and Example 10–17 has three poles to go with its three energy storage elements. The conclusion appears to be that the number of natural poles is equal to the number of energy storage elements. While this rule is a useful guideline, there are exceptions (capacitors in parallel, for example). The best we can say is that

---

2 They could just as logically be called marginally unstable. The stability status of simple *j*-axis poles depends on the application. For example, electronic circuits with simple poles firmly rooted on the *j*-axis are called stable oscillators. On the other hand, *j*-axis poles in audio amplifiers cause "ringing," a dirty word among audiophiles.

FIGURE 10-41 *Summary of Chapter 10 examples.*

Example	Circuit Diagram	Natural Poles
10-2	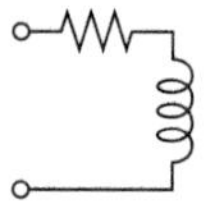 1 Inductor	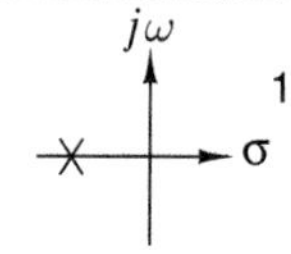 1 LHP Pole
10-4	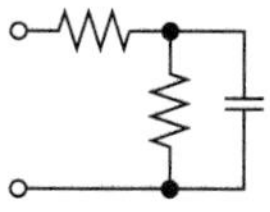 1 Capacitor	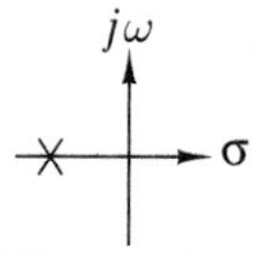 1 LHP Pole
10-6	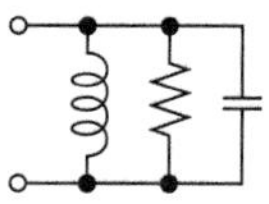 1 Capacitor / 1 Inductor	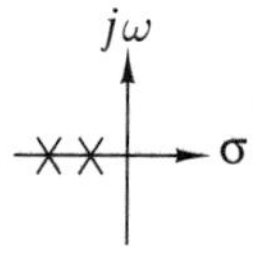 2 LHP Poles / Overdamped
10-11	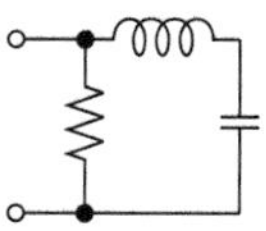 1 Capacitor / 1 Inductor	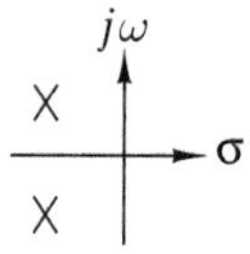 2 LHP Poles / Underdamped
10-16	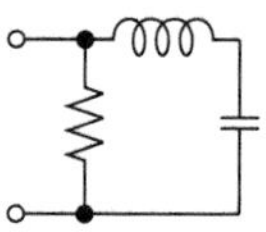 1 Capacitor / 1 Inductor	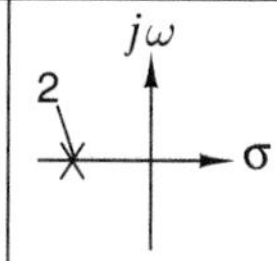 2 LHP Poles / Critically Damped
10-17	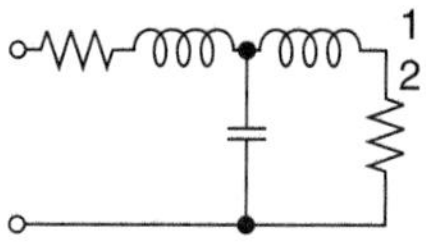 1 Capacitor / 2 Inductors	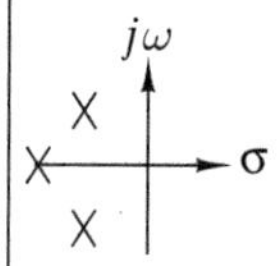 3 LHP Poles

The number of natural poles does not exceed the number of energy storage elements.

Another implication in Figure 10–41 comes from two additional observations. First, all of the natural poles are in the LHP; hence, all of the circuits are stable. Second, all of the circuits contain only passive resistors, capacitors, and/or inductors; that is, there are no active elements. These observations infer that

Circuits consisting of passive resistors, capacitors, and inductors are inherently stable.

This conclusion makes sense physically since the passive elements can only store or dissipate energy. They cannot produce the energy needed to sustain an unbounded response.

What about circuits with active elements like dependent sources or OP AMPs? Such circuits can be unstable as we can see by reviewing Example 10–13. In that example we analyzed the active *RC* circuit in Figure 10–32 and found the circuit determinant to be

$$\frac{\Delta(s)}{C^2} = s^2 + \left(\frac{3-\mu}{RC}\right)s + \left(\frac{1}{RC}\right)^2$$

and the two natural poles are defined by

$$\omega_0 = \frac{1}{RC} \quad \text{and} \quad \zeta = \frac{3-\mu}{2}$$

where μ is the gain of the dependent source, the active element in the circuit. For $\mu = 0$ (active element turned off), the damping ratio is $\zeta = 1.5 > 1$ and the circuit is overdamped. For $\mu = 1$, the damping ratio is $\zeta = 1$ and the circuit is critically damped. For $3 > \mu > 1$, the damping ratio is $1 > \zeta > 0$ and the circuit is underdamped. For $\mu = 3$, the damping ratio is $\zeta = 0$ and the circuit is undamped. Finally, for $\mu > 3$, the damping ratio is $\zeta < 0$ and the circuit is said to have negative damping, which is an unstable condition.

Figure 10-42 shows the locus of the natural poles as the gain increases. For $\mu > 3$, the poles move into the RHP and the circuit becomes unstable. This makes sense physically. When the gain is high enough the active element can produce the energy needed to sustain an unbounded output. Since instability is almost always undesirable we usually state the conclusion the other way around. That is, this active RC circuit is stable provided the gain $\mu < 3$. It is common for active circuits to be stable when circuit parameters are in one range and unstable when they are outside this range. For double-pole circuits the stable range can be found by relating the damping ratio to circuit parameters. For single-pole circuits the stable range ensures that the pole lies on the negative real axis. The following exercise involves a single-pole circuit.

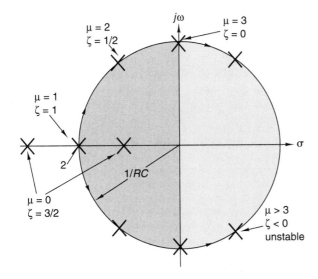

F I G U R E   1 0 – 4 2 *Complex pole locus on a circle of radius ω_0.*

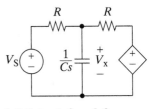

F I G U R E   1 0 – 4 3

Exercise 10–11

Find the range of the gain μ for which the circuit in Figure 10-43 is stable.

Answer: $\mu < 2$

SUMMARY

- Kirchhoff's laws apply to voltage and current waveforms in the time domain and to the corresponding transforms in the *s* domain.

- The *s*-domain models for the passive elements include initial condition sources and the element impedance or admittance. Impedance is the proportionality factor in the expression $V(s) = Z(s)I(s)$ relating the voltage and current transforms. Admittance is the reciprocal of impedance.

- The impedances of the three passive elements are $Z_R(s) = R$, $Z_L(s) = Ls$, and $Z_C(s) = 1/Cs$.

- The *s*-domain circuit analysis techniques closely parallel the analysis methods developed for resistance circuits. Basic analysis techniques, such as circuit reduction, Thévenin's and Norton's theorems, the unit output method, or superposition, can be used in simple circuits. More complicated networks require a general approach, such as the node-voltage or mesh-current methods.

- Response transforms are rational functions whose poles are zeros of the circuit determinant or poles of the transform of the input driving forces. Poles introduced by the circuit determinant are called natural poles and lead to the natural response. Poles introduced by the input are called forced poles and lead to the forced response.

- In linear circuits, response transforms and waveforms can be separated into zero-state and zero-input components. The zero-state component is found by setting the initial capacitor voltages and inductor currents to zero. The zero-input component is found by setting all input driving forces to zero.

- The main purpose of *s*-domain circuit analysis is to gain insight into circuit performance without necessarily finding the time-domain response. The natural poles reveal the form, stability, and observability of the circuit's response. The number of natural poles is never greater than the number of energy storage elements in the circuit.

- A circuit is stable if all of its natural poles are in the left half of the *s* plane. Passive circuits are inherently stable so their natural poles are in the left-half of the *s*-plane. Active circuits can be stable when circuit parameters are in one range and unstable for parameters outside this range.

PROBLEMS

ERO 10–1 EQUIVALENT IMPEDANCE (SECTS. 10–1, 10–2)

Given a linear circuit, use series and parallel equivalence to find the equivalent impedance at specified terminal pairs. Select element values to obtain specified pole locations. See Examples 10–2, 10–4, 10–5, 10–6, and Exercises 10–1, 10–2

10–1 Find the poles and zeros of the equivalent impedance between terminals 1 and 2 in Figure P10–1. Repeat for terminals 3 and 4.

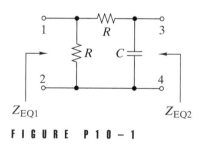

F I G U R E   P 1 0 – 1

10–2 A 2-kΩ resistor and a 0.1-μF capacitor are connected in series. The series combination is connected in parallel with a 0.5-H inductor. Find the equivalent impedance of the combination. Express the impedance as a rational function and locate its poles and zeros.

10–3 A 100-nF capacitor and a 100-mH inductor are connected in series. The series combination is connected in parallel with a 2000-Ω resistor. Find the equivalent impedance of the combination. Express the impedance as a rational function and locate its poles and zeros.

10–4 Find the poles and zeros of the equivalent impedance between terminals 1 and 2 in Figure P10–4 for $R = 2$ kΩ and $L = 10$ mH. Evaluate $Z_{EQ}(s)$ at $s = 0$ and $s = \infty$. Explain your results in terms of the inductor impedance.

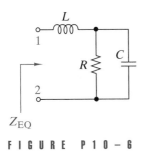

F I G U R E   P 1 0 – 4

10–5 A 1-kΩ resistor and a 100-mH inductor are connected in parallel. The parallel combination is connected in series with a 50-nF capacitor. Find the equivalent impedance of the combination. Express the impedance as a rational function and locate its poles and zeros.

10–6 Find the equivalent impedance between terminals 1 and 2 in the circuit of Figure P10–6. Express the impedance as a rational function and locate its poles and zeros. Evaluate $Z_{EQ}(s)$ at $s = 0$ and $s = \infty$. Explain your results in terms of the impedances of the inductor and capacitor.

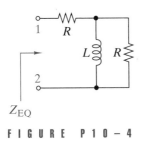

F I G U R E   P 1 0 – 6

10–7 Find the equivalent impedance between terminals 1 and 2 in the circuit of Figure P10–7. Express the impedance as a rational function and locate its poles and zeros.

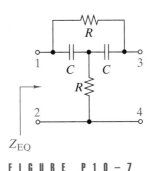

F I G U R E   P 1 0 – 7

10–8 Find the equivalent impedance between terminals 1 and 2 in the circuit of Figure P10–7 when a short circuit is connected between terminals 3 and 4. Express the impedance as a rational function and locate its poles and zeros.

10–9 (a) Find the equivalent impedance $Z_{EQ1}(s)$ in the circuit of Figure P10–9. Express the impedance as a rational function and locate its poles and zeros.
(b) Find the equivalent impedance $Z_{EQ2}(s)$ when a short circuit is connected between terminals 1 and 2. Express the impedance as a rational function and locate its poles and zeros.

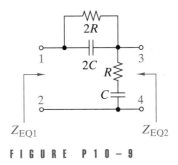

F I G U R E   P 1 0 – 9

10–10 Find the equivalent impedance between terminals 1 and 2 in the circuit of Figure P10–10. Select values of R and L so that $Z_{EQ}(s)$ has a pole at $s = -5000$ rad/s. Identify the location of the zeros of $Z_{EQ}(s)$ for your values of R and L.

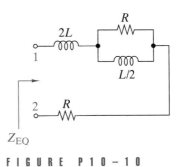

F I G U R E   P 1 0 – 1 0

ERO 10–2 BASIC CIRCUIT ANALYSIS TECHNIQUES (SECTS. 10–2, 10–3)

Given a linear circuit,

(a) Determine the initial conditions (if not given) and transform the circuit into the *s* domain.
(b) Solve for zero-state and/or zero-input response transforms and waveforms using basic analysis methods such as circuit reduction, unit output, Thévenin/Norton equivalent circuits, or superposition.
(c) Locate the forced and natural poles or select circuit parameters to place the natural poles at specified locations.

See Examples 10–2, 10–3, 10–4, 10–5, 10–6, 10–7, 10–8, and 10–9

10–11 The switch in Figure P10–11 has been open for a long time and is closed at $t = 0$. Transform the circuit into the *s* domain and solve for $I_L(s)$ and $i_L(t)$ in symbolic form.

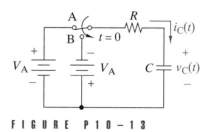

F I G U R E   P 1 0 – 1 1

10–12 The switch in Figure P10–11 has been closed for a long time and is opened at $t = 0$. Transform the circuit into the *s* domain and solve for $I_L(s)$ and $i_L(t)$ in symbolic form.

10–13 The switch in Figure P10–13 has been in position A for a long time and is moved to position B at $t = 0$. Transform the circuit into the *s* domain and solve for $I_C(s)$ and $i_C(t)$ in symbolic form.

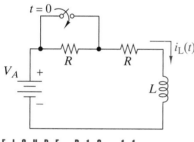

F I G U R E   P 1 0 – 1 3

10–14 The switch in Figure P10–13 has been in position B for a long time and is moved to position A at $t = 0$.

Transform the circuit into the *s* domain and solve for $V_C(s)$ and $v_C(t)$ in symbolic form.

10–15 The switch in Figure P10–15 has been closed for a long time and is opened at $t = 0$. Transform the circuit into the *s* domain and solve for $I_L(s)$ and $i_L(t)$.

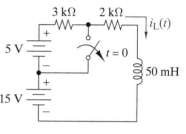

F I G U R E   P 1 0 – 1 5

10–16 The switch in Figure P10–15 has been open for a long time and is closed at $t = 0$. Transform the circuit into the *s* domain and solve for $I_L(s)$ and $i_L(t)$.

10–17 The switch in Figure P10–17 has been in position A for a long time and is moved to position B at $t = 0$.
   (a) Transform the circuit into the *s* domain and solve for $I_L(s)$ in symbolic form.
   (b) Find $i_L(t)$ for $R_1 = R_2 = 500$ Ω, $L = 250$ mH, $C = 4$ μF, and $V_A = 15$ V.

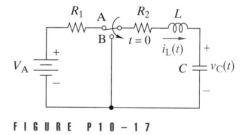

F I G U R E   P 1 0 – 1 7

10–18 The switch in Figure P10–17 has been in position B for a long time and is moved to position A at $t = 0$.
   (a) Transform the circuit into the *s* domain and solve for $V_C(s)$ in symbolic form.
   (b) Find $v_C(t)$ for $R_1 = R_2 = 500$ Ω, $L = 500$ mH, $C = 1$ μF, and $V_A = 15$ V.

10–19 The switch in Figure P10–19 has been in position A for a long time and is moved to position B at $t = 0$.
   (a) Transform the circuit into the *s* domain and solve for $I_L(s)$ in symbolic form.
   (b) Find $i_L(t)$ for $R_1 = 5$ kΩ, $R_2 = 1$ kΩ, $L = 40$ mH, $C = 0.25$ μF, $V_A = 10$ V, and $V_B = 5$ V.

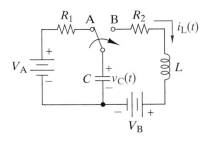

FIGURE P10-19

10-20 The switch in Figure P10-19 has been in position B for a long time and is moved to position A at $t = 0$.
  (a) Transform the circuit into the s domain and solve for $V_C(s)$ in symbolic form.
  (b) Find $v_C(t)$ for $R_1 = 5$ kΩ, $R_2 = 2$ kΩ, $L = 1$ H, $C = 10$ nF, $V_A = 10$ V, and $V_B = 5$ V

10-21 The initial conditions in Figure P10-21 are $v_C(0) = V_0$ and $i_L(0) = 0$. Transform the circuit into the s domain and use superposition and voltage division to find the zero-state and zero-input components of $V_C(s)$.

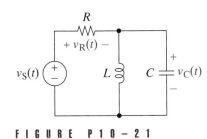

FIGURE P10-21

10-22 The initial conditions in Figure P10-21 are $v_C(0) = 0$ and $i_L(0) = I_0$. Transform the circuit into the s domain and use superposition and voltage division to find the zero-state and zero-input components of $V_R(s)$.

10-23 There is no energy stored in the capacitor in Figure P10-23 at $t = 0$. Transform the circuit into the s domain and use current division to find $v_O(t)$ when the input is $i_S(t) = 3.2\ e^{-100t}u(t)$ mA. Identify the forced and natural poles in $V_O(s)$.

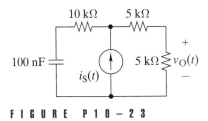

FIGURE P10-23

10-24 Repeat Problem 10-23 when $i_S(t) = 3.2\ e^{-1000t}u(t)$ mA.

10-25 The initial capacitor voltage in Figure P10-25 is V_0.
  (a) Transform the circuit into the s domain and find the Thévenin equivalent at the interface shown in the figure.
  (b) Use the Thévenin equivalent to determine the zero-state and zero-input components of $I(s)$ when $i_S(t) = I_A\ u(t)$.

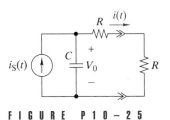

FIGURE P10-25

10-26 Repeat Problem 10-25 when the current source delivers $i_S(t) = I_A\ [\cos \beta t]u(t)$.

10-27 Select the value of C in Figure P10-27 so that the circuit has a natural pole at $s = -10$ Mrad/s when $R_S = 550\ \Omega$, $R_L = 2$ kΩ, and $g = 40$ mS.

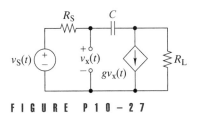

FIGURE P10-27

10-28 There is no initial energy stored in the circuit in Figure P10-28. Transform the circuit into the s domain and use superposition to find the zero-state component of $V(s)$. Identify the forced and natural poles in $V(s)$.

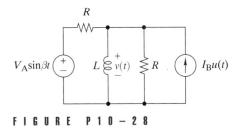

FIGURE P10-28

10-29 The equivalent impedance between a pair of terminals is

$$Z(s) = 1000 \left[\frac{s + 2000}{s} \right]$$

A voltage $v(t) = 10u(t)$ is applied across the terminals. Find the resulting current response $i(t)$.

10–30 There is no initial energy stored in the circuit in Figure P10–30. Use circuit reduction to find the output voltage $V_2(s)$ in terms of the input voltage $V_1(s)$.

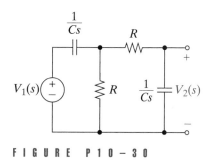

FIGURE P10–30

ERO 10–3 GENERAL CIRCUIT ANALYSIS (SECTS. 10–4, 10–5, 10–6)

Given a linear circuit,

(a) Determine the initial conditions (if not given) and transform the circuit into the s domain.

(b) Solve for zero-state and/or zero-input response transforms and waveforms using node-voltage or mesh-current methods.

(c) Identify the forced and natural poles or select circuit parameters to place the natural poles at specified locations.

See Examples 10–10, 10–11, 10–13, 10–14, 10–15, 10–16, and 10–17

10–31 There is no initial energy stored in the circuit in Figure P10–31.

(a) Transform the circuit into the s domain and formulate mesh-current equations.

(b) Solve these equations for $I_2(s)$ in symbolic form.

(c) Find $i_2(t)$ for $v_1(t) = 10u(t)$, $R_1 = R_2 = 2$ kΩ, $L = 250$ mH, and $C = 250$ nF.

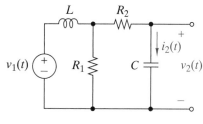

FIGURE P10–31

10–32 The initial conditions in the circuit in Figure P10–31 are $v_C(0) = V_0$ and $i_L(0) = 0$.

(a) Transform the circuit into the s domain and formulate node-voltage equations.

(b) Solve these equations for the zero-input components of $V_2(s)$ in symbolic form.

(c) Find $v_2(t)$ for $v_1(t) = 0$, $V_0 = 10$ V, $R_1 = R_2 = 50$ Ω, $L = 1.25$ mH, and $C = 2$ μF.

10–33 There is no initial energy stored in the circuit in Figure P10–33.

(a) Transform the circuit into the s domain and formulate node-voltage equations.

(b) Solve these equations for $V_2(s)$ in symbolic form.

(c) Find $v_2(t)$ for $v_1(t) = 10u(t)$, $R_1 = 10$ kΩ, $R_2 = 20$ kΩ, $C_1 = 200$ nF, and $C_2 = 100$ nF.

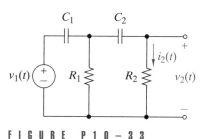

FIGURE P10–33

10–34 There is no initial energy stored in the circuit in Figure P10–33.

(a) Transform the circuit into the s domain and formulate mesh-current equations.

(b) Solve these equations for $I_2(s)$ in symbolic form.

(c) Find $i_2(t)$ for $v_1(t) = 10u(t)$, $R_1 = 1$ kΩ, $R_2 = 12.5$ kΩ, $C_1 = 6$ μF, and $C_2 = 2/3$ μF.

10–35 There is no initial energy stored in the bridged-T circuit in Figure P10–35.

(a) Transform the circuit into the s domain and formulate mesh-current equations.

(b) Solve these equations for input impedance seen by the voltage source .

(c) Use circuit reduction to check your answer in (b).

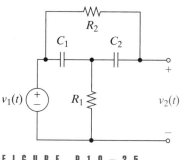

FIGURE P10–35

10–36 There is no initial energy stored in the bridged-T circuit in Figure P10–35.

(a) Transform the circuit into the s domain and formulate node-voltage equations.

(b) Find the poles and zeros of the zero-state component of $V_2(s)$ for $R_1 = 3$ kΩ, $R_2 = 25$ kΩ, $C_1 = 80$ nF, $C_2 = 20$ nF, and $v_1(t) = V_A e^{-2000t}u(t)$.

10–37 **D** With zero initial conditions transform the circuit in Figure P10–37 into the *s* domain and find the circuit determinant. For $R_1 = R_2 = 10 \text{ k}\Omega$, select values of C_1 and C_2 so that the circuit has natural poles defined by $\zeta = 0.5$ and $\omega_0 = 1000$ rad/s.

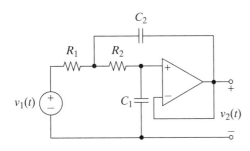

FIGURE P10–37

10–38 **D** With zero initial conditions transform the circuit in Figure P10–38 into the *s* domain and find the circuit determinant. For $R_1 = R_2 = R_3 = R$, and $C_1 = C_2 = C$, select values of R and C so that the circuit has natural poles at $s_{1,2} = -1000 \pm j1000$ rad/s.

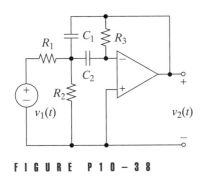

FIGURE P10–38

10–39 **D** With zero initial conditions transform the circuit in Figure P10–39 into the *s* domain and find the circuit determinant. Select values of R, C, and μ so that the circuit has natural poles at $s_{1,2} = \pm j5000$ rad/s.

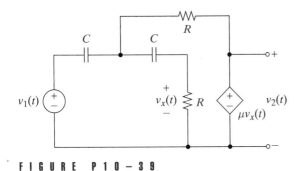

FIGURE P10–39

10–40 Find the natural poles of the circuit in Figure P10–40 for $R_1 = 10 \text{ k}\Omega$, $R_2 = 20 \text{ k}\Omega$, $C_1 = C_2 = 10$ nF, and $\mu = 5$.

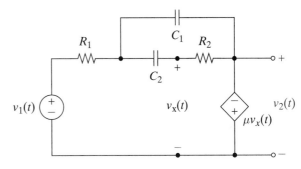

FIGURE P10–40

10–41 The zero-state output transform of the circuit in Figure P10–41 is

$$V_O(s) = \frac{500}{(s + 100)(s - 20)}$$

(a) Plot the pole-zero diagram of $V_O(s)$ and find $v_O(t)$.
(b) If $RC = 10^{-2}$, what is the input waveform $v_S(t)$?
(c) Is the circuit stable or unstable? Explain.

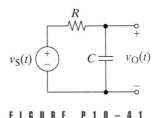

FIGURE P10–41

10–42 In order to match the Thévenin impedance of a source, the load impedance in Figure P10–42 must be

$$Z_L(s) = \frac{20(s + 5)}{s + 10}$$

(a) What impedance $Z_2(s)$ is required?
(b) How would you realize $Z_2(s)$ using only resistors, inductors, and/or capacitors?

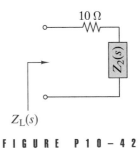

FIGURE P10–42

10–43 Figure P10–43 shows the three mesh currents in an *s*-domain circuit with a dependent source.

(a) Explain why only two of these mesh currents are independent.
(b) Write a set of two s-domain mesh currents in two independent mesh currents.

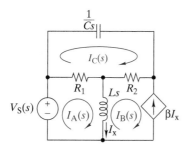

FIGURE P10-43

10-44 The switch in Figure P10–44 has been in position A for a long time and is moved to position B at $t = 0$. Solve for $V_C(s)$ and $v_C(t)$.

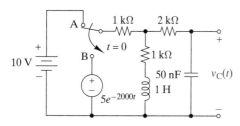

FIGURE P10-44

10-45 There is no energy stored in the circuit in Figure P10–45 at $t = 0$. Transform the circuit into the s domain and solve for $V_O(s)$ and $v_O(t)$.

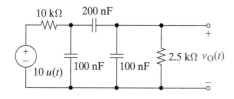

FIGURE P10-45

10-46 The switch in Figure P10–46 has been open for a long time and is closed at $t = 0$. Transform the circuit into the s domain and solve for $I_O(s)$ and $i_O(t)$.

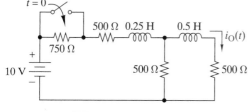

FIGURE P10-46

10-47 There is no initial energy stored in the circuit in Figure P10–47.
(a) Transform the circuit into the s domain and solve for $V_O(s)$ in symbolic form.
(b) Find $v_O(t)$ when $v_S(t) = 10^4 t u(t)$ V, $R_1 = R_2 = 200\ \Omega$, $C = 0.5\ \mu$F, and $L = 20$ mH.

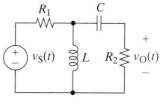

FIGURE P10-47

10-48 There is no initial energy stored in the circuit in Figure P10–48.
(a) Show that the circuit has natural poles at $s = -1000$ rad/s and $s = -2000$ rad/s when $R_1 = 10$ kΩ, $C_1 = 50$ nF, $R_2 = 10$ kΩ, and $C_2 = 100$ nF.
(b) When the input is $v_S(t) = 10u(t)$ V, show that $V_1(s)$ has a forced pole at $s = 0$ but only one natural pole at $s = -2000$ rad/s.
(c) For the input in (b), show that $V_2(s)$ has both of the circuit's natural poles but does not have a forced pole at $s = 0$.

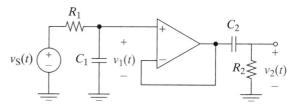

FIGURE P10-48

10-49 Show that when $8L = R^2C$ the circuit in Figure P10–49 has a simple natural pole $s = -8/RC$ and a double pole at $s = -4/RC$.

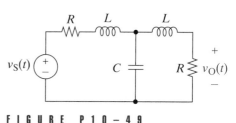

FIGURE P10-49

10-50 Select values of R_1, R_2, C_1, and C_2 so that the circuit in Figure P10–50 has natural poles located at $s = -200$ rad/s and $s = -1000$ rad/s.

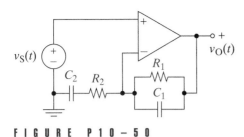

FIGURE P10 – 50

INTEGRATING PROBLEMS

10–51 **A** THÉVENIN'S THEOREM FROM TIME-DOMAIN DATA

A black box containing a linear circuit has an on-off switch and a pair of external terminals. When the switch is turned on, the open-circuit voltage between the external terminals is observed to be

$$v_{OC}(t) = (9e^{-10t} - 6e^{-40t})u(t)$$

When a 50-Ω load resistance is connected across the terminals and the switch again turned on, the voltage delivered to the load is observed to be

$$v_L(t) = (5e^{-10t} - 5e^{-40t})u(t)$$

What is the Thévenin's impedance looking into the box?

10–52 **A** DESIGN CONSTRAINTS IN THE s PLANE

A design specification requires a circuit to have a pair of complex conjugate poles with $\omega_0 \le 10$ and $\zeta \ge 0.5$. Sketch the allowable region of pole locations in the s plane.

10–53 **A D** RC CIRCUIT ANALYSIS AND DESIGN

The RC circuits in Figure P10–53 represent the situation at the input to an oscilloscope. The parallel combination of R_1 and C_1 represents the probe used to connect the oscilloscope to a test point. The parallel combination of R_2 and C_2 represents the input impedance of the oscilloscope.

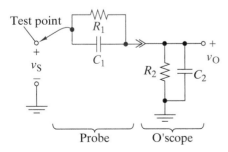

FIGURE P10 – 53

(a) **A** Assuming zero inital conditions, transform the circuit into the s-domain and find the relationship between the test point voltage $V_S(s)$ and the voltage $V_O(s)$ at the oscilloscope's input.

(b) **D** For $R_2 = 10$ MΩ and $C_2 = 2$ pF determine the values of R_1 and C_1 that make the input voltage a scaled duplicate of the test point voltage.

10–54 **A D** LIGHTNING PULSER DESIGN

The circuit in Figure P10–54 is a simplified circuit diagram of a pulser that is used to test equipment against lightning-induced transients. Select the values of V_0, R_1, R_2, C_1, and C_2 so the the pulser delivers an output pulse

$$v_O(t) = 500(e^{-2000t} - e^{-80000t}) \quad V$$

The pulser is completely discharged prior to delivering a pulse.

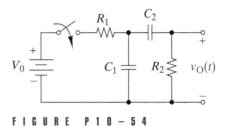

FIGURE P10 – 54

10–55 **E** PULSE CONVERSION CIRCUIT

The purpose of the test setup in Figure P10–55 is to deliver damped sine pulses to the test load. The excitation comes from a 1-Hz square wave generator. The pulse conversion circuit must deliver damped sine waveforms with $\zeta < 0.5$ and $\omega_0 > 10$ krad/s to 50-Ω and 600-Ω loads. The recommended values for the pulse conversion circuit are $L = 10$ mH and $C = 100$ nF. Verify that the test setup meets the specifications.

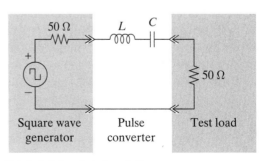

FIGURE P10 – 55

CHAPTER 11

NETWORK FUNCTIONS

The driving-point impedance of a network is the ratio of an impressed electromotive force at a point in a branch of the network to the resulting current at the same point.

Ronald M. Foster, 1924,
American Engineer

11–1 Definition of a Network Function

11–2 Network Functions of One- and Two-port Circuits

11–3 Network Functions and Impulse Response

11–4 Network Functions and Step Response

11–5 Network Functions and Sinusoidal Steady-state Response

11–6 Impulse Response and Convolution

11–7 Network Function Design

Summary

Problems

Integrating Problems

The concept of network functions emerged in the 1920s during the development of systematic methods of designing electric filters for long-distance telephone systems. The filter design effort eventually evolved into a formal realizability theory that came to be known as network synthesis. The objective of network synthesis is to design one or more networks that accomplish or realize a given network function. The foregoing quotation is taken from one of the first papers to approach network design from a synthesis viewpoint.[1] Ronald Foster, along with Sidney Darlington, Hendrik Bode, Wilhelm Cauer, and Otto Brune are generally considered the founders of the field of network synthesis. The classical synthesis methods they developed emphasized passive realizations using only resistors, capacitors, and inductors because the active elements of their era were costly and unreliable vacuum tubes. With the advent of semiconductor electronics, emphasis shifted to realizations using resistors, capacitors, and active devices such as transistors and OP AMPs.

In the last chapter we found that the transform of the zero-state output of a linear circuit is proportional to the transform of the input signal. In the *s* domain the proportionality factor is a rational function of *s* called a network function. The network function concept is used extensively in linear circuit analysis and design. This chapter begins a study of the properties of these functions that continues throughout the rest of this book.

The first two sections of this chapter show how to define and calculate network functions for different types of circuits. The next three sections discuss how to use network functions to obtain the impulse response, step response, and sinusoidal steady-state responses of a circuit. The final section treats some basic methods of synthesizing circuits that realize a specified network function.

11−1 DEFINITION OF A NETWORK FUNCTION

The proportionality property of linear circuits states that the output is proportional to the input. In Chapter 10 we noted that in the *s* domain the proportionality factor is a rational function of *s* called a network function. More formally, a network function is defined as the ratio of a zero-state response transform (output) to the excitation (input) transform.

$$\text{Network Function} = \frac{\text{Zero-state Response Transform}}{\text{Input Signal Transform}} \qquad (11\text{−}1)$$

Note carefully that this definition specifies zero initial conditions and implies only one input.

To study the role of network functions in determining circuit responses, we write the *s*-domain input-output relationship as

$$Y(s) = T(s)X(s) \qquad (11\text{−}2)$$

---

1 R. M. Foster, "A Reactance Theorem," *Bell System Technical Journal,* No. 3, pp. 259–267, 1924.

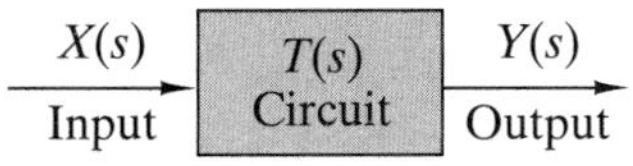

FIGURE 11–1 *Block diagram for an s-domain input-output relationship.*

where $T(s)$ is a network function, $X(s)$ is the input signal transform, and $Y(s)$ is a zero-state response or output.[2] Figure 11–1 shows a block diagram representation of the s-domain input-output relationship in Eq. (11–2).

In an analysis problem, the circuit and input [$X(s)$ or $x(t)$] are specified. We determine $T(s)$ from the circuit, use Eq. (11–2) to find the response transform $Y(s)$, and use the inverse transformation to obtain the response waveform $y(t)$. In a design problem the circuit is unknown. The input and output are specified, or their ratio $T(s) = Y(s)/X(s)$ is given. The objective is to devise a circuit that realizes the specified input-output relationship. A linear circuit analysis problem has a unique solution, but a design problem may have one, many, or even no solutions.

Equation (11–2) points out that the poles of the response $Y(s)$ come from either the network function $T(s)$ or the input signal $X(s)$. When there are no repeated poles, the partial-fraction expansion of the right side of Eq. (11–2) takes the form

$$Y(s) = \underbrace{\sum_{j=1}^{N} \frac{k_j}{s - p_j}}_{\substack{\text{Natural} \\ \text{Poles}}} + \underbrace{\sum_{\ell=1}^{M} \frac{k_\ell}{s - p_\ell}}_{\substack{\text{Forced} \\ \text{Poles}}} \tag{11–3}$$

where p_j ($j = 1, 2, \ldots N$) are the poles of $T(s)$ and $s = p_\ell$ ($\ell = 1, 2, \ldots M$) are the poles of $X(s)$. The inverse transform of this expansion is

$$y(t) = \underbrace{\sum_{j=1}^{N} k_j e^{p_j t}}_{\substack{\text{Natural} \\ \text{Response}}} + \underbrace{\sum_{\ell=1}^{M} k_\ell e^{p_\ell t}}_{\substack{\text{Forced} \\ \text{Response}}} \tag{11–4}$$

The poles of $T(s)$ lead to the natural response. In a stable circuit, the natural poles are all in the left half of the s plane, and all of the exponential terms in the natural response eventually decay to zero. The poles of $X(s)$ lead to the forced response. In a stable circuit, those elements in the forced response that do not decay to zero are called the **steady-state response**.

It is important to remember that the complex frequencies in the natural response are determined by the circuit and do not depend on input. Conversely, the complex frequencies in the forced response are determined by the input and do not depend on the circuit. However, the amplitude of its part of the response depends on the residues in the partial fraction expansion in Eq. (11–3). These residues are influenced by all of the poles and zeros, whether forced or natural. Thus, the amplitudes of the forced and natural responses depend on an interaction between the poles and zeros of $T(s)$ and $X(s)$.

The following example illustrates this discussion.

---

2 In this context $Y(s)$ is not an admittance, but the transform of the output waveform $y(t)$.

EXAMPLE 11-1

The transfer function of a circuit is

$$T(s) = \frac{V_2(s)}{V_1(s)} = \frac{2000\,(s + 2000)}{(s + 1000)\,(s + 4000)}$$

Find the zero-state response $v_2(t)$ when the input waveform is $v_1(t) = [20 + 15e^{-5000t}]u(t)$.

SOLUTION:

The transform of the input waveform is

$$V_1(s) = \frac{20}{s} + \frac{15}{s + 5000} = \frac{35\,s + 10^5}{s(s + 5000)}$$

Using the s-domain input-output relationship in Eq. (11-2), the transform of the response is

$$V_2(s) = \frac{10^4(s + 2000)\,(7s + 20000)}{(s + 1000)\,(s + 4000)s(s + 5000)}$$

Expanding by partial fractions,

$$V_2(s) = \underbrace{\frac{k_1}{s + 1000} + \frac{k_2}{s + 4000}}_{\text{Natural Poles}} + \underbrace{\frac{k_3}{s} + \frac{k_4}{s + 5000}}_{\text{Forced Poles}}$$

The two natural poles came from the circuit via the network function $T(s)$. The forced poles came from the step function and exponential inputs. Using the cover-up method to evaluate the residues yields

$$k_1 = \left.\frac{10^4(s + 2000)\,(7s + 20000)}{(s + 4000)s(s + 5000)}\right|_{s = -1000} = -\frac{65}{6}$$

$$k_2 = \left.\frac{10^4(s + 2000)\,(7s + 20000)}{(s + 1000)s(s + 5000)}\right|_{s = -4000} = \frac{40}{3}$$

$$k_3 = \left.\frac{10^4(s + 2000)\,(7s + 20000)}{(s + 1000)\,(s + 4000)\,(s + 5000)}\right|_{s = 0} = 20$$

$$k_4 = \left.\frac{10^4(s + 2000)\,(7s + 20000)}{(s + 1000)\,(s + 4000)s}\right|_{s = -5000} = -\frac{45}{2}$$

Collectively the residues depend on all of the poles and zeros. The inverse transform yields the zero-state response as

$$v_2(t) = \left[\underbrace{-\frac{65}{6}e^{-1000t} + \frac{40}{3}e^{-4000t}}_{\text{Natural Response}} + \underbrace{20 - \frac{45}{2}e^{-5000t}}_{\text{Forced Response}}\right]u(t)$$

The natural poles as $s = -1000$ and $s = -4000$ are in the left half of the s plane, so the natural response decays to zero. The forced pole as $s = -5000$

leads to an exponential term that also decays to zero, leaving a steady-state response of $20u(t)$.   ■

TEST SIGNALS

While the transfer function is a useful concept, it is clear that we cannot find the circuit response until we are given an input signal. Here, we encounter a central paradox of circuit analysis. In practice, the input signal is a carrier of information and is therefore unpredictable. We could spend a lifetime studying a circuit for various inputs and still not treat all possible signals that might be encountered in practice. What we must do is calculate the responses due to certain standard test signals. Although these test signals may never occur as real input signals, their responses tell us enough to understand the signal processing capabilities of a circuit.

The two premier test signals used are the pulse and the sinusoid. The study of pulse response divides into two extreme cases, short and long. When the pulse is very short compared to the circuit response time, the sudden injection of energy causes a circuit response long after the input returns to zero. The short pulse is modeled by an impulse, and the resulting *impulse response* is treated in Sec. 11–3. At the other extreme, the long pulse has a duration that greatly exceeds the circuit response time. In this case, the circuit has ample time to be driven from the zero state to a new steady-state condition. The step function is used to model the long pulse input, and the resulting *step response* is studied in Sec. 11–4.

The impulse response is of great importance because it contains all of the information needed to calculate the response due to any other input. The step response is important because it describes how a circuit response transitions from one state to another. The signal transition requirements for circuits and systems are often stated in terms of the step response using partial waveform descriptors such as rise time, fall time, propagation delay, and overshoot.

The unique properties of the sinusoid make it a useful input for characterizing the signal-processing capabilities of linear circuits and systems. When a stable linear circuit is driven by a sinusoidal input, the steady-state output is a sinusoid with the same frequency, but with a different phase angle and amplitude. The frequency-dependent relationship between the sinusoidal input and the steady-state output is called *frequency response,* a signal-processing description that is often used to specify the performance of circuits and systems. The relationship between network functions and the sinusoidal steady-state response is studied in Sec. 11–5.

11–2   NETWORK FUNCTIONS OF ONE- AND TWO-PORT CIRCUITS

The two major types of network functions are driving-point impedance and transfer functions. A **driving-point impedance** relates the voltage and current at a pair of terminals called a port. The driving-point impedance $Z(s)$ of the one-port circuit in Figure 11–2 is defined as

$$Z(s) = \frac{V(s)}{I(s)} \qquad (11–5)$$

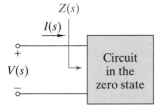

F I G U R E   1 1 – 2   *A one-port circuit.*

When the one port is driven by a current source the response is $V(s) = Z(s)I(s)$ and the natural frequencies in the response are the poles of impedance $Z(s)$. On the other hand, when the one port is driven by a voltage source the response is $I(s) = [Z(s)]^{-1}V(s)$ and the natural frequencies in the response are the poles of $1/Z(s)$; that is, the zeros of $Z(s)$. In other words, the driving-point impedance is a network function whether upside down or right side up.

The term *driving point* means that the circuit is driven at one port and the response is observed at the same port. The element impedances defined in Sec. 10–1 are elementary examples of driving-point impedances. The equivalent impedances found by combining elements in series and parallel are also driving-point impedances. Driving-point functions are the s-domain generalization of the concept of the input resistance. The terms *driving-point impedance*, *input impedance*, and *equivalent impedance* are synonymous.

The driving point impedance seen at a pair of terminals determines the loading effects that result when those terminals are connected to another circuit. When two circuits are connected together, these loading effects can profoundly alter the responses observed when the same two circuits operated in isolation. In an analysis situation it is important to be able to predict the response changes that occur when one circuit loads another. In design situations it is important to know when the circuits can be designed separately and then interconnected without encountering loading effects that alter their designed performance. The conditions under which loading can or cannot be ignored will be studied in this and subsequent chapters.

Transfer functions are usually of greater interest in signal-processing applications than driving-point impedances because they describe how a signal is modified by passing through a circuit. A **transfer function** relates an input and response (or output) at different ports in the circuit. Figure 11–3 shows the possible input-output configurations for a two-port circuit. Since the input and output signals can be either a current or a voltage, we can define four kinds of transfer function:

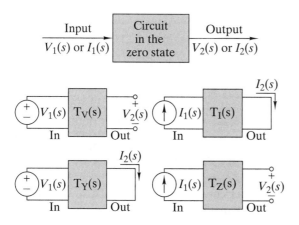

FIGURE 11–3 *Two-port circuits and transfer functions.*

$$T_V(s) = \text{Voltage Transfer Function} = \frac{V_2(s)}{V_1(s)}$$

$$T_I(s) = \text{Current Transfer Function} = \frac{I_2(s)}{I_1(s)}$$

$$T_Y(s) = \text{Transfer Admittance} = \frac{I_2(s)}{V_1(s)}$$

$$T_Z(s) = \text{Transfer Impedance} = \frac{V_2(s)}{I_1(s)}$$

(11–6)

The functions $T_V(s)$ and $T_I(s)$ are dimensionless since the input and output signals have the same units. The function $T_Z(s)$ has units of ohms and $T_Y(s)$ has units of siemens.

The functions in Eq. (11–6) are sometimes called forward transfer functions because they relate inputs applied at port 1 to outputs occurring at port 2. There are, of course, reverse transfer functions that relate inputs at port 2 to outputs at port 1. It is important to realize that a transfer function is only valid for a specified input port and output port. For example, the voltage transfer function $T_V(s) = V_2(s)/V_1(s)$ relates the input voltage applied at port 1 in Figure 11–3 to the voltage response observed at the output port. The reverse voltage transfer function for signal transmission from output to input is *not* $1/T_V(s)$. Unlike driving-point impedance, transfer functions are not network functions when they are turned upside down.

DETERMINING NETWORK FUNCTIONS

The rest of this section illustrates analysis techniques for deriving network functions. The application of network functions in circuit analysis and design begins in the next section and continues throughout the rest of this book. But first, we illustrate ways to find the network functions of a given circuit.

The divider circuits in Figure 11–4 occur so frequently that it is worth taking time to develop their transfer functions in general terms. Using s-domain voltage division in Figure 11–4(a), we can write

$$V_2(s) = \left[\frac{Z_2(s)}{Z_1(s) + Z_2(s)}\right] V_1(s)$$

Therefore, the voltage transfer function of a voltage divider circuit is

$$T_V(s) = \frac{V_2(s)}{V_1(s)} = \frac{Z_2(s)}{Z_1(s) + Z_2(s)}$$

(11–7)

Similarly, using s-domain current division in Figure 11–4(b) yields the transfer function of a current divider circuit as

$$T_I(s) = \frac{I_2(s)}{I_1(s)} = \frac{Y_2(s)}{Y_1(s) + Y_2(s)}$$

(11–8)

By series equivalence, the driving-point impedance at the input of the voltage divider is $Z_{EQ}(s) = Z_1(s) + Z_2(s)$. By parallel equivalence the driving-point impedance at the input of the current divider is $Z_{EQ}(s) = 1/(Y_1(s) + Y_2(s))$.

Two other useful circuits are the inverting and noninverting OP AMP configurations shown in Figure 11–5. To determine the voltage transfer

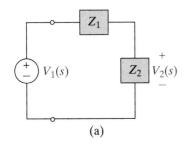

(a)

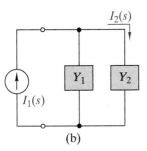

(b)

F I G U R E   1 1 – 4 *Basic divider circuits. (a) Voltage divider. (b) Current divider.*

function of the inverting circuit in Figure 11–5(a), we write the sum of currents leaving node B:

$$\frac{V_B(s) - V_A(s)}{Z_1(s)} + \frac{V_B(s) - V_C(s)}{Z_2(s)} + I_N(s) = 0$$

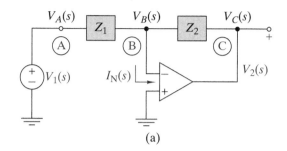

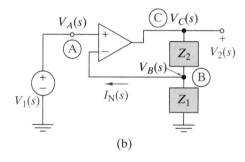

FIGURE 11–5 *Basic OP AMP circuits. (a) Inverting amplifier. (b) Noninverting amplifier.*

But the ideal OP AMP constraints require that $I_N(s) = 0$ and $V_B(s) = 0$ since the noninverting input is grounded. By definition, the output voltage $V_2(s)$ equals node voltage $V_C(s)$ and the voltage source forces $V_A(s)$ to equal the input voltage $V_1(s)$. Inserting all of these considerations into the node equations and solving for the voltage transfer function yields

$$T_V(s) = \frac{V_2(s)}{V_1(s)} = -\frac{Z_2(s)}{Z_1(s)} \tag{11–9}$$

From the study of OP AMP circuits in Chapter 4, you should recognize Eq. (11–9) as the s-domain generalization of the inverting OP AMP circuit gain equation, $K = -R_2/R_1$.

The driving-point impedance at the input to the inverting circuit is

$$Z_{IN}(s) = \frac{V_S(s)}{[V_A(s) - V_B(s)]/Z_1(s)}$$

But $V_A(s) = V_S(s)$ and $V_B(s) = 0$; hence the input impedance is $Z_{IN}(s) = Z_1(s)$.

For the noninverting circuit in Figure 11–5(b), the sum of currents leaving node B is

$$\frac{V_B(s) - V_C(s)}{Z_2(s)} + \frac{V_B(s)}{Z_1(s)} + I_N(s) = 0$$

In the noninverting configuration the ideal OP AMP constraints require that $I_N(s) = 0$ and $V_B(s) = V_1(s)$. By definition, the output voltage $V_2(s)$

equals node voltage $V_C(s)$. Combining all of these considerations and solving for the voltage transfer function yields

$$T_V(s) = \frac{V_2(s)}{V_1(s)} = \frac{Z_1(s) + Z_2(s)}{Z_1(s)} \tag{11-10}$$

Equation (11–10) is the s-domain version of the noninverting amplifier gain equation, $K = (R_1 + R_2)/R_1$. The transfer function of the noninverting configuration is the reciprocal of the transfer function of the voltage divider in the feedback path. The ideal OP AMP draws no current at its input terminals, so theoretically the input impedance of the noninverting circuit is infinite.

The transfer functions of divider circuits and the basic OP AMP configurations are useful analysis and design tools in many practical situations. However, a general method is needed to handle circuits of greater complexity. One general approach is to formulate either node-voltage or mesh-current equations with all initial conditions set to zero. These equations are then solved for network functions using Cramer's rule for hand calculations or symbolic math analysis programs such as Mathcad. The algebra involved can be a bit tedious at times, even with Mathcad. But the tedium is reduced somewhat because we only need the zero-state response for a single input source.

The following examples illustrate methods of calculating network functions.

EXAMPLE 11–2

(a) Find the transfer functions of the circuits in Figure 11–6.
(b) Find the driving-point impedances seen by the input sources in these circuits.

SOLUTION:

(a) These are all divider circuits, so the required transfer functions can be obtained using Eq. (11–7) or (11–8).

For circuit C1: $Z_1 = R$, $Z_2 = 1/Cs$, and $T_V(s) = 1/(RCs + 1)$

For circuit C2: $Z_1 = Ls$, $Z_2 = R$, and $T_V(s) = 1/(GLs + 1)$

For circuit C3: $Y_1 = Cs$, $Y_2 = G$, and $T_I(s) = 1/(RCs + 1)$

These transfer functions are all of the form $1/(\tau s + 1)$, where τ is the circuit time constant.

(b) The driving-point impedances are found by series or parallel equivalence.

For circuit C1: $Z(s) = Z_1 + Z_2 = (RCs + 1)/Cs$

For circuit C2: $Z(s) = Z_1 + Z_2 = Ls + R$

For circuit C3: $Z(s) = 1/(Y_1 + Y_2) = 1/(Cs + G) = R/(RCs + 1)$

The three circuits have different driving-point impedances even though they have the same transfer functions.

The general principle illustrated here is that several different circuits can have the same transfer function. Put differently, a desired transfer function can be realized by several different circuits. This fact is important in design because circuits that produce the same transfer function offer alternatives that may differ in other features. In this example, they all have different input impedances. ∎

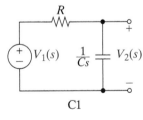

C1

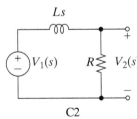

C2

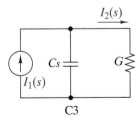

C3

FIGURE 11–6

EXAMPLE 11-3

(a) Find the input impedance seen by the voltage source in Figure 11–7.
(b) Find the voltage transfer function $T_V(s) = V_2(s)/V_1(s)$ of the circuit.
(c) Locate the poles and zeros of $T_V(s)$ for $R_1 = 10$ kΩ, $R_2 = 20$ kΩ, $C_1 = 100$ nF, and $C_2 = 50$ nF.

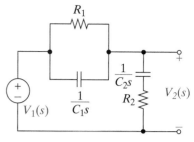

FIGURE 11-7

SOLUTION:

(a) The circuit is a voltage divider. We first calculate the equivalent impedances of the two legs of the divider. The two elements in parallel combine to produce the series leg impedance $Z_1(s)$ as

$$Z_1(s) = \frac{1}{C_1s + 1/R_1} = \frac{R_1}{R_1C_1s + 1}$$

The two elements in series combine to produce shunt leg impedance $Z_2(s)$:

$$Z_2(s) = R_2 + 1/C_2s = \frac{R_2C_2s + 1}{C_2s}$$

Using series equivalence, the driving-point impedance seen at the input is

$$Z_{EQ}(s) = Z_1(s) + Z_2(s)$$
$$= \frac{R_1C_1R_2C_2s^2 + (R_1C_1 + R_2C_2 + R_1C_2)s + 1}{C_2s(R_1C_1s + 1)}$$

(b) Using voltage division, the voltage transfer function is

$$T_V(s) = \frac{Z_2(s)}{Z_{EQ}(s)} = \frac{(R_1C_1s + 1)(R_2C_2s + 1)}{R_1C_1R_2C_2s^2 + (R_1C_1 + R_2C_2 + R_1C_2)s + 1}$$

(c) Inserting the specified numerical values into $T_V(s)$ yields

$$T_V(s) = \frac{(10^{-3}s + 1)(10^{-3}s + 1)}{10^{-6}s^2 + 2.5 \times 10^{-3}s + 1} = \frac{(s + 1000)^2}{(s + 500)(s + 2000)}$$

which indicates a double zero at $s = -1000$ rad/s, and simple poles at $s = -500$ rad/s and $s = -2000$ rad/s. ■

EXAMPLE 11-4

Find the driving-point impedance seen by the voltage source in Figure 11–8. Find the voltage transfer function $T_V(s) = V_2(s)/V_1(s)$ of the circuit. The poles of $T_V(s)$ are located at $p_1 = -1000$ rad/s and $p_2 = -5000$ rad/s. If $R_1 = R_2 = 20$ kΩ, what values of C_1 and C_2 are required?

FIGURE 11-8

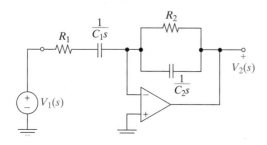

SOLUTION:

The circuit is an inverting OP AMP configuration of the form in Figure 11–5(a). The input impedance of this circuit is

$$Z_1(s) = R_1 + \frac{1}{C_1 s} = \frac{R_1 C_1 s + 1}{C_1 s}$$

The impedance Z_2 in the feedback path is

$$Z_2(s) = \frac{1}{C_2 s + 1/R_2} = \frac{R_2}{R_2 C_2 s + 1}$$

and the voltage transfer function is

$$T_V(s) = -\frac{Z_2(s)}{Z_1(s)} = -\frac{R_2 C_1 s}{(R_1 C_1 s + 1)(R_2 C_2 s + 1)}$$

The poles of $T_V(s)$ are located at $p_1 = -1/R_1 C_1 = -1000$ and $p_2 = -1/R_2 C_2 = -5000$. If $R_1 = R_2 = 20\,\mathrm{k\Omega}$, then $C_1 = 1/1000 R_1 = 50\,\mathrm{nF}$ and $C_2 = 1/5000 R_2 = 10\,\mathrm{nF}$. ∎

EXAMPLE 11–5

For the circuit in Figure 11–9 find the input impedance $Z(s) = V_1(s)/I_1(s)$, the transfer impedance $T_Z(s) = V_2(s)/I_1(s)$, and the voltage transfer function $T_V(s) = V_2(s)/V_1(s)$.

F I G U R E 1 1 – 9

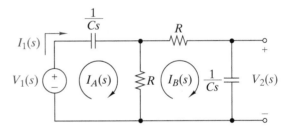

SOLUTION:

The circuit is not a simple voltage divider, so we use mesh-current equations to illustrate the general approach to finding network functions. By inspection, the mesh-current equations for this ladder circuit are

$$\left(R + \frac{1}{Cs}\right) I_A(s) - R I_B(s) = V_1(s)$$

$$-R I_A(s) + \left(2R + \frac{1}{Cs}\right) I_B(s) = 0$$

In terms of the mesh current, the input impedance is $Z(s) = V_1(s)/I_A(s)$. Using Cramer's rule to solve for $I_A(s)$ yields

$$I_A(s) = \frac{\Delta_A}{\Delta} = \frac{\begin{vmatrix} V_1(s) & -R \\ 0 & 2R + \dfrac{1}{Cs} \end{vmatrix}}{\begin{vmatrix} R + \dfrac{1}{Cs} & -R \\ -R & 2R + \dfrac{1}{Cs} \end{vmatrix}} = \frac{Cs(2RCs + 1)}{(RCs)^2 + 3RCs + 1} V_1(s)$$

The input impedance of the circuit is

$$Z(s) = \frac{V_1(s)}{I_A(s)} = \frac{(RCs)^2 + 3RCs + 1}{Cs(2RCs + 1)}$$

In terms of mesh current, the transfer impedance is $T_Z(s) = V_2(s)/I_A(s)$. The mesh-current equations do not yield the output voltage directly. But since $V_2(s) = I_B(s)Z_C(s)$, we can solve the second mesh equation for $I_B(s)$ in terms of as $I_A(s)$ as

$$I_B(s) = \frac{RCs}{2RCs + 1} I_A(s)$$

and obtain the specified transfer impedance as

$$T_Z(s) = \frac{I_B(s)[1/Cs]}{I_A(s)} = \frac{R}{2RCs + 1}$$

To obtain the specified voltage transfer function, we could use Cramer's rule to solve for $I_B(s)$ in terms of $V_1(s)$ and then use the fact that $V_2(s) = I_B(s)Z_C(s)$. But a moment's reflection reveals that

$$T_V(s) = \frac{V_2(s)}{V_1(s)} = \left[\frac{V_2(s)}{I_1(s)}\right]\left[\frac{I_1(s)}{V_1(s)}\right] = T_Z(s) \times \frac{1}{Z(s)}$$

Hence the specified voltage transfer function is

$$T_V(s) = \frac{R}{2RCs + 1} \times \frac{Cs(2RCs + 1)}{(RCs)^2 + 3RCs + 1}$$

$$= \frac{RCs}{(RCs)^2 + 3RCs + 1} \qquad \blacksquare$$

EXAMPLE 11–6

Find the voltage transfer function $T_V(s) = V_2(s)/V_1(s)$ of the circuit in Figure 11–10.

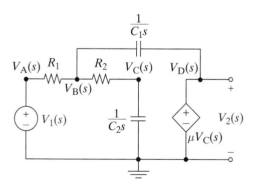

FIGURE 11 – 10

SOLUTION:
The voltage-controlled voltage source makes this an active RC circuit. We use node-voltage equations in this problem because the required output is a voltage. The circuit contains two voltage sources connected at a common node. Selecting this common node as the reference eliminates two un-knowns since $V_A(s) = V_1(s)$ and $V_D(s) = \mu V_C(s) = V_2(s)$. The sums of currents leaving nodes B and C are

$$\text{Node B:} \quad \frac{V_B(s) - V_1(s)}{R_1} + \frac{V_B(s) - V_C(s)}{R_2} + \frac{V_B(s) - \mu V_C(s)}{1/C_1 s} = 0$$

$$\text{Node C:} \quad \frac{V_C(s) - V_B(s)}{R_2} + \frac{V_C(s)}{1/C_2 s} = 0$$

Multiplying both equations by $R_1 R_2$ and rearranging terms produces

Node B: $(R_1 + R_2 + R_1 R_2 C_1 s)V_B(s) - (R_1 + \mu R_1 R_2 C_1 s)V_C(s) = R_2 V_1(s)$

Node C: $\qquad\qquad -V_B(s) + (1 + R_2 C_2 s)V_C(s) = 0$

Using the node C equation to eliminate $V_B(s)$ from the node B equation leaves

$$(R_1 + R_2 + R_1 R_2 C_1 s)\,(1 + R_2 C_2 s)V_C(s)$$
$$- (R_1 + \mu R_1 R_2 C_1 s)V_C(s) = R_2 V_1(s)$$

Since the output $V_2(s) = \mu V_C(s)$, the required transfer function is

$$T_V(s) = \frac{V_2(s)}{V_1(s)} = \frac{\mu}{R_1 R_2 C_1 C_2 s^2 + (R_1 C_1 + R_1 C_2 + R_2 C_2 - \mu R_1 C_1)s + 1}$$

This circuit is a member of the Sallen-Key family often used in filter design with $R_1 = R_2 = R$ and $C_1 = C_2 = C$, in which case the transfer function reduces to

$$T_V(s) = \frac{\mu}{(RCs)^2 + (3 - \mu)RCs + 1}$$

We will encounter this result again in later chapters. ■

Exercise 11–1

Find the driving-point impedance seen by the voltage source in Figure 11–11.

Answer:

$$Z(s) = \frac{RLCs^2 + Ls + R}{LCs^2 + 1}$$

Exercise 11–2

Find the voltage transfer function $T_V(s) = V_2(s)/V_1(s)$ in Figure 11–11.

Answer:

$$T_V(s) = \frac{LCs^2 + 1}{LCs^2 + GLs + 1}$$

THE CASCADE CONNECTION AND THE CHAIN RULE

Signal-processing circuits often involve a **cascade connection** in which the output voltage of one circuit serves as the input to the next stage. In some cases, the overall voltage transfer function of the cascade can be related to the transfer functions of the individual stages by a **chain rule**

F I G U R E   1 1 – 1 1

$$T_V(s) = T_{V1}(s)T_{V2}(s) \cdots T_{Vk}(s) \qquad (11-11)$$

where T_{V1}, T_{V2}, ... and T_{Vk} are the voltage transfer functions of the individual stages when operated separately. It is important to understand when the chain rule applies since it greatly simplifies the analysis and design of cascade circuits.

The chain rule in Eq. (11–11) applies if connecting the stages in cascade does not load (change) the output of any stage in the cascade. At any given stage, interface loading does not occur if (1) the Thévenin impedance of the source stage is zero or (2) the input impedance of the load stage is infinite. As a practical matter, however, the chain rule yields acceptable results when the Thévenin impedance of the source is simply much smaller than the input impedance of the load.

To illustrate the effect of loading, consider the two RC circuits or stages in Figure 11–12. When disconnected and operated separately, the transfer functions of each stage are easily found using voltage division as follows:

$$T_{V1}(s) = \frac{R}{R + 1/Cs} = \frac{RCs}{RCs + 1}$$

$$T_{V2}(s) = \frac{1/Cs}{R + 1/Cs} = \frac{1}{RCs + 1}$$

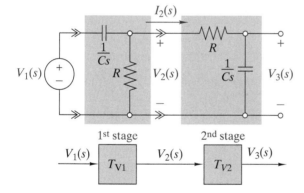

FIGURE 11−12 *Two-port circuits connected in cascade.*

When connected in cascade, the output of the first stage serves as the input to the second stage. If the chain rule applies, we would obtain the overall transfer function as

$$T_V(s) = \frac{V_3(s)}{V_1(s)} = \left(\frac{V_2(s)}{V_1(s)}\right)\left(\frac{V_3(s)}{V_2(s)}\right) = (T_{V1}(s))(T_{V2}(s)) \qquad (11-12)$$

$$= \underbrace{\left(\frac{RCs}{RCs + 1}\right)}_{\substack{\text{First} \\ \text{Stage}}} \underbrace{\left(\frac{1}{RCs + 1}\right)}_{\substack{\text{Second} \\ \text{Stage}}} = \underbrace{\frac{RCs}{(RCs)^2 + 2RCs + 1}}_{\text{Overall}}$$

However, in Example 11–5, the overall transfer function of this circuit was found to be

$$T_V(s) = \frac{RCs}{(RCs)^2 + 3RCs + 1} \qquad (11-13)$$

which disagrees with the chain rule result in Eq. (11–12).

The reason for the discrepancy is that when they are connected in cascade, the second circuit "loads" the first circuit. That is, the voltage-divider rule requires that the interface current $I_2(s)$ in Figure 11–12 be zero. The no-load condition $I_2(s) = 0$ applies when the stages operate separately, but when connected in cascade, the interface current is not zero. The chain rule does not apply here because loading caused by the second stage changes the transfer function of the first stage.

However, Figure 11–13 shows how the loading problem goes away when an OP AMP voltage follower is inserted between the RC circuit stages. The follower does not draw any current from the first RC circuit [$I_2(s) = 0$] and applies $V_2(s)$ directly across the input of the second RC circuit. With this modification the chain rule in Eq. (11–11) applies because the voltage follower isolates the two circuits, thereby solving the loading problem.

F I G U R E   1 1 – 1 3 *Cascade connection with voltage follower isolation.*

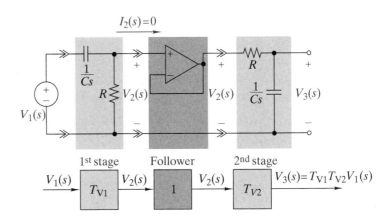

Thus, loading can be avoided by connecting an OP AMP voltage follower between stages, as in Figure 11–13. More important, loading does not occur if the output of the driving stage is the output of an OP AMP or controlled source. These elements act like ideal voltage sources whose outputs are unchanged by connecting the subsequent stage.

For example, the two circuits in Figure 11–14 contain the same two stages. The chain rule applies to the configuration in Figure 11–14(a) because the output of the first stage is an OP AMP. In Figure 11–14(b) the two stages are interchanged so that the output of the voltage divider drives the input to OP AMP stage. The chain rule does not apply to this configuration because the input impedance of OP AMP stage loads the output of the voltage divider stage.

EXAMPLE 11–7

Find the voltage transfer function of the circuit in Figure 11–14(a).

FIGURE 11-14

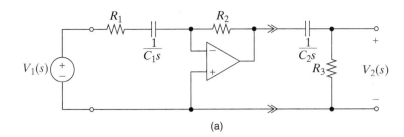

(a)

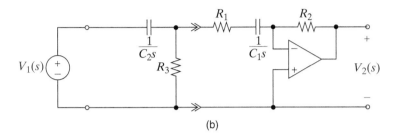

(b)

SOLUTION:

The chain rule applies to this circuit since the output of the first stage is an OP AMP. The transfer function of this inverting OP AMP stage is

$$T_{V1}(s) = -\frac{Z_2(s)}{Z_1(s)} = -\frac{R_2}{R_1 + 1/C_1 s} = -\frac{R_2 C_1 s}{R_1 C_1 s + 1}$$

The second stage is a voltage divider whose transfer function is

$$T_{V2}(s) = \frac{Z_2(s)}{Z_2(s) + Z_1(s)} = \frac{R_3}{R_3 + 1/C_2 s} = \frac{R_3 C_2 s}{R_3 C_2 s + 1}$$

and the chain rule yields the overall transfer function as

$$T_V(s) = T_{V1}(s) \times T_{V2}(s) = -\frac{R_2 C_1 R_3 C_2 s^2}{(R_1 C_1 s + 1)(R_3 C_2 s + 1)} \qquad \blacksquare$$

Exercise 11–3

Find the voltage transfer function of the circuit in Figure 11–14(b).

Answer:

$$T_V(s) = -\frac{R_2 C_1 R_3 C_2 s^2}{R_1 C_1 R_3 C_2 s^2 + (R_1 C_1 + R_3 C_2 + R_3 C_1)s + 1} \neq T_{V2}(s) \times T_{V1}(s)$$

Note that the overall transfer function is not the product of the individual stage transfer functions because second stage loads the first stage.

11-3   NETWORK FUNCTIONS AND IMPULSE RESPONSE

The **impulse response** is the zero-state response of a circuit when the driving force is a unit impulse applied at $t = 0$. When the input signal is $x(t) =$

$\delta(t)$ then $X(s) = \mathcal{L}\{\delta(t)\} = 1$ and the input-output relationship in Eq. (11–2) reduces to

$$Y(s) = T(s) \times 1 = T(s)$$

The impulse response transform equals the network function, and we could treat $T(s)$ as if it is a signal transform. However, to avoid possible confusion between a network function (description of a circuit) and a transform (description of a signal), we denote the impulse response transform as $H(s)$ and use $h(t)$ to denote the corresponding waveform.[3] That is,

Impulse Response

Transform   Waveform

$$H(s) = T(s) \times 1 \qquad h(t) = \mathcal{L}^{-1}\{H(s)\} \tag{11–14}$$

When there are no repeated poles, the partial fraction expansion of $H(s)$ is

$$H(s) = \underbrace{\frac{k_1}{s - p_1} + \frac{k_2}{s - p_2} + \cdots + \frac{k_N}{s - p_N}}_{\text{Natural Poles}}$$

where $p_1, p_2, \ldots p_N$ are the natural poles in the denominator of the transfer function $T(s)$. All of the poles of $H(s)$ are natural poles since the impulse excitation does not introduce any forced poles. The inverse transform gives the impulse response waveform as

$$h(t) = \underbrace{\left[k_1 e^{p_1 t} + k_2 e^{p_2 t} + \cdots + k_N e^{p_N t} \right]}_{\text{Natural Response}} u(t)$$

When the circuit is stable, all of the natural poles are in the left half plane and the impulse response waveform $h(t)$ decays to zero as $t \to \infty$. A linear circuit whose impulse response ultimately returns to zero is said to be **asymptotically stable**. Asymptotic stability means that the impulse response has a finite time duration. That is, for every $\varepsilon > 0$ there exists a finite time duration T_D such that $|h(t)| < \varepsilon$ for all $t > T_D$.

It is important to note that the impulse response $h(t)$ contains all the information needed to determine the circuit response to any other input. That is, since $\mathcal{L}\{h(t)\} = T(s)$, we can calculate the output $y(t)$ for any Laplace transformable input $x(t)$ as

$$y(t) = \mathcal{L}^{-1}\{H(s)X(s)\}$$

This expression, known as the *convolution theorem,* states that the impulse response can be used to relate the input and output of a linear circuit. Thus, the impulse response $h(t)$ or $H(s)$ can be considered as a mathematical model of a linear circuit. Obviously it is important to be able to find the impulse response and to know how to use the impulse response to find the output for other inputs. The following examples illustrate both of these issues.

---

3 Not all books make this distinction. Books on signals and circuits often use $H(s)$ to represent both a transfer function and the impulse response transform.

EXAMPLE 11–8

Find the response $v_2(t)$ in Figure 11–15 when the input is $v_1(t) = \delta(t)$. Use the element values $R_1 = 10 \text{ k}\Omega$, $R_2 = 12.5 \text{ k}\Omega$, $C_1 = 1 \text{ }\mu\text{F}$, and $C_2 = 2 \text{ }\mu\text{F}$.

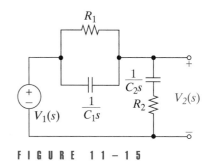

FIGURE 11–15

SOLUTION:

In Example 11–3, the transfer function of this circuit was found to be

$$T_V(s) = \frac{V_2(s)}{V_1(s)} = \frac{(R_1C_1s + 1)(R_2C_2s + 1)}{R_1C_1R_2C_2s^2 + (R_1C_1 + R_2C_2 + R_1C_2)s + 1}$$

For the given element values, the impulse response transform is

$$H(s) = \frac{(s + 100)(s + 40)}{s^2 + 220s + 4000} = \frac{(s + 100)(s + 40)}{(s + 20)(s + 200)}$$

This $H(s)$ is not a proper rational function, so we use one step of a long division plus a partial-fraction expansion to obtain

$$H(s) = 1 + \frac{80/9}{s + 20} - \frac{800/9}{s + 200}$$

and the impulse response is

$$h(t) = \delta(t) + \frac{80}{9}[e^{-20t} - 10e^{-200t}]u(t)$$

In this case, the impulse response contains an impulse because the network function is not a proper rational function. ∎

EXAMPLE 11–9

The impulse response of a linear circuit is $h(t) = 200e^{-100t}u(t)$. Find the output when the input is a unit ramp $r(t) = tu(t)$.

SOLUTION:

The circuit impulse response is

$$H(s) = \mathcal{L}\{h(t)\} = \mathcal{L}\{200e^{-100t}u(t)\} = \frac{200}{s + 100}$$

The Laplace transform of the unit ramp input is $1/s^2$, hence using the convolution theorem the response due to a ramp input is

$$y(t) = \mathcal{L}^{-1}\left\{ H(s)\frac{1}{s^2} \right\} = \mathcal{L}^{-1}\left\{ \frac{200}{(s + 100)s^2} \right\}$$

$$= \mathcal{L}^{-1}\left\{ \frac{1/50}{s + 100} + \frac{2}{s^2} - \frac{1/50}{s} \right\}$$

$$= \frac{1}{50}(e^{-100t} + 100t - 1)u(t)$$

This example illustrates that the impulse response $h(t)$ contains all the information needed to calculate the response due to another input. ∎

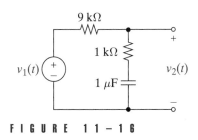

F I G U R E   1 1 – 1 6

Find the impulse response of the circuit in Figure 11–16.

Answer: $h(t) = 0.1\delta(t) + [90e^{-100t}]u(t)$

When the input to a linear circuit is $v_1(t) = \delta(t)$ the output is $v_2(t) = \delta(t) - 10e^{-20t}u(t)$ Find the poles and zeros of the network function $T_V(s) = V_2(s)/V_1(s)$.

Answer: A simple pole at $s = -20$ rad/s and a simple zero at $s = -10$ rad/s.

11–4   NETWORK FUNCTIONS AND STEP RESPONSE

The **step response** is the zero-state response of the circuit output when the driving force is a unit step function applied at $t = 0$. When the input is $x(t) = u(t)$, then $X(s) = \mathcal{L}\{u(t)\} = 1/s$ and the s-domain input-output relationship in Eq. (11–2) yields $Y(s) = T(s)/s$. The step response transform and waveform will be denoted by $G(s)$ and $g(t)$, respectively. That is,

Step Response

Transform   Waveform

$$G(s) = \frac{T(s)}{s} \qquad g(t) = \mathcal{L}^{-1}\{G(s)\} \qquad (11–15)$$

The poles of $G(s)$ are the natural poles contributed by the network function $T(s)$ and a forced pole at $s = 0$ introduced by the step function input. The partial fraction expansion of $G(s)$ takes the form

$$G(s) = \underbrace{\frac{k_0}{s}}_{\substack{\text{Forced} \\ \text{Pole}}} + \underbrace{\frac{k_1}{s - p_1} + \frac{k_2}{s - p_2} + \cdots + \frac{k_N}{s - p_N}}_{\substack{\text{Natural} \\ \text{Poles}}}$$

where $p_1, p_2, \ldots p_N$ are the natural poles in $T(s)$. The inverse transformation gives the step response waveform as

$$g(t) = \underbrace{k_0 u(t)}_{\substack{\text{Forced} \\ \text{Response}}} + \underbrace{[k_1 e^{p_1 t} + k_2 e^{p_2 t} + \cdots + k_N e^{p_N t}]u(t)}_{\substack{\text{Natural} \\ \text{Response}}}$$

When the circuit is stable, the natural response decays to zero, leaving a forced component called the **dc steady-state response**. The amplitude of the steady-state response is the residue in the partial-fraction expansion of the forced pole at $s = 0$. By the cover-up method, this residue is

$$k_0 = sG(s)\big|_{s=0} = T(0)$$

For a unit step input the amplitude of the dc steady-state response equals the value of the transfer function at $s = 0$. By linearity, the general princi-

ple is that an input $Au(t)$ produces a dc steady-state output whose amplitude is $AT(0)$.

We next show the relationship between the impulse and step responses.
First, combining Eqs. (11–14) and (11–15) gives

$$G(s) = \frac{H(s)}{s}$$

The step response transform is the impulse response transform divided by
s. The integration property of the Laplace transform tells us that division
by s in the s domain corresponds to integration in the time domain. Therefore, in the time domain, we can relate the impulse and step response
waveforms by integration:

$$g(t) = \int_0^t h(\tau)d\tau \qquad (11\text{–}16)$$

Using the fundamental theorem of calculus, the impulse response waveform is expressed in terms of the step response waveform

$$h(t)(=)\frac{dg(t)}{dt} \qquad (11\text{–}17)$$

where the symbol (=) means equal almost everywhere, a condition that excludes those points at which $g(t)$ has a discontinuity. In the time domain,
the step response waveform is the integral of the impulse response waveform. Conversely, the impulse response waveform is (almost everywhere)
the derivative of the step response waveform.

The key idea is that there are relationships between the network function $T(s)$ and the responses $H(s)$, $h(t)$, $G(s)$, and $g(t)$. If any one of these
quantities is known, we can obtain any of the other four using relatively
simple mathematical operations.

EXAMPLE 11–10

The element values for the circuit in Figure 11–17 are $R_1 = 10$ kΩ, $R_2 =$
100 kΩ, $C_1 = C_2 = 100$ nF, and $v_1(t) = u(t)$ V. Find the response $v_2(t)$.

FIGURE 11–17

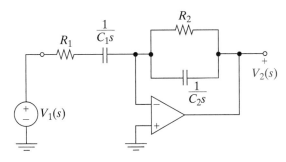

SOLUTION:
For a unit step input $V_1(s) = 1/s$. In Example 11–4 the transfer function of
this circuit is shown to be

$$T_V(s) = \frac{V_2(s)}{V_1(s)} = -\frac{R_2C_1s}{(R_1C_1s + 1)(R_2C_2s + 1)}$$

Substituting the numerical element values into this expression yields the step response transform as

$$V_2(s) = \frac{T_V(s)}{s} = -\frac{1000}{(s + 100)(s + 1000)}$$
$$= \frac{-10/9}{s + 100} + \frac{10/9}{s + 1000}$$

The inverse transform yields

$$v_2(t) = \frac{10}{9}(-e^{-100t} + e^{-1000t})u(t)$$

In this case the forced pole at $s = 0$ is cancelled by a zero of $T_V(s)$ and the final value of the step response is zero. Using the final value theorem, we have

$$g(\infty) = \underset{t \to \infty}{\text{Limit}}\, g(t) = \underset{s \to 0}{\text{Limit}}\, sG(s) = T(0)$$

the final steady-state value of the step response is equal to the value of the transfer function evaluated at $s = 0$. In the present case $T_V(0) = 0$, so the dc steady-state output of the circuit is zero. Recall from Chapter 6 that in the dc steady-state capacitors can be replaced by open circuits. Replacing C_1 in Figure 11–17 by an open circuit disconnects the input source from the OP AMP so no dc signals can be transferred through the circuit. A series capacitor that prevents the passage of dc signals is commonly called a *blocking capacitor*. ∎

EXAMPLE 11–11

The impulse response of a linear circuit is

$$h(t) = 5000\,(e^{-1000t} \sin 2000\ t)u(t)$$

Find the step response.

SOLUTION:
Using Eq. (11–16) we have

$$g(t) = \int_0^t h(\tau)d\tau = 5000\int_0^t e^{-1000\tau} \sin 2000\ \tau\ d\tau$$
$$= (-e^{-1000\tau} \sin 2000\ \tau - 2e^{-1000\tau} \cos 2000\ \tau)|_0^t$$
$$= (2 - e^{-1000t} \sin 2000\ t - 2e^{-1000t} \cos 2000\ t)u(t)$$

We could have found $g(t)$ by first finding the Laplace transform of the impulse response $H(s)$ and then calculating the step response as $g(t) = \mathcal{L}^{-1}\{H(s)/s\}$. However, we choose to use Eq. (11–16), which means that the step response calculation took place entirely in the time domain. This is a simple example of a general time-domain method based on the *convolution integral*, a procedure that will be covered in Sec. 11.6. ∎

APPLICATION NOTE: EXAMPLE 11−12

Three time-domain parameters often used to describe the step response are rise time, delay time, and overshoots. **Rise time (T_R)** is the time interval required for the step response to rise from 10% to 90% of its steady-state value $g(\infty)$. **Delay time (T_D)** is the time interval required for the step response to reach 50% of its steady-state value. **Overshoot** is the difference between the peak value of the step response and its steady-state value. Overshoot is usually expressed as a percentage of the steady-state value, namely

$$\text{Overshoot} = \frac{g_{\max} - g(\infty)}{g(\infty)} \times 100$$

Figure 11–18 illustrates these descriptors for a typical step response.

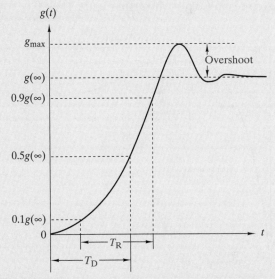

FIGURE 11 − 18 *Step response showing rise time (T_R), delay time (T_D), and overshoot.*

Step response descriptors are used to specify the performance of both analog and digital systems. Rise time governs how rapidly the system responds to an abrupt change in the input. Delay time controls the time between the application of an abrupt change and the appearance of a significant change in the output. Overshoot indicates the amount of damping present in the system. Lightly damped oscillations produce large overshoots and may cause erroneous state changes in digital systems.

Rise time, delay time, and overshoot can be determined experimentally or calculated using modern computer tools. For example, the MATLAB function `step` calculates and plots the step response of a transfer function of the form $T(s) = n(s)/d(s)$, where $n(s)$ is the numerator polynomial and $d(s)$ is the denominator polynomial. In the MATLAB command window these polynomials are defined by the row matrices `num` and `den` that contain the coefficients of the polynomials $n(s)$ and $d(s)$ in descending powers of s.

For example, suppose we need to evaluate the step response descriptors of a circuit whose transfer function is

$$T(s) = \frac{400(s + 100)}{(s + 100)^2 + 100^2}$$

The two coefficient matrices for this transfer function are entered in the MATLAB command window as

```
num=[400,400*100];
den=[1,2*100,100^2+100^2];
```

The first element in either matrix is the coefficient of the highest power of s, which is followed by the other coefficients in descending powers of s. The MATLAB statement

```
step(num,den)
```

produces the step response plot shown in Figure 11–19.

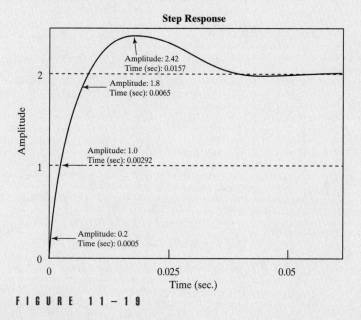

FIGURE 11 – 19

First note that the steady-state response is $g(\infty) = 2$. Using the cursor in the MATLAB graphics window, we can find the points at which certain amplitudes are reached. The key points shown in the figure are:

(1) The 10% rise point (amplitude = 0.2 and time = 0.0005).
(2) The 50% rise point (amplitude = 1 and time = 0.00292).
(3) The 90% rise point (amplitude = 1.8 and time = 0.0065).
(4) The maximum point (amplitude = 2.42 and time = 0.0157).

Hence the step response descriptors for the circuit are:

Rise time: $T_R = 0.0065 - 0.005 = 6$ ms

Delay time: $T_D = 0.00292 = 2.92$ ms

Overshoot = $(2.42 - 2)/2 = 21\%$

Exercise 11–6

The impulse response of a circuit is $h(t) = 0.1\delta(t) + 90e^{-100t}u(t)$. Find the step response.

Answer: $g(t) = [1 - 0.9e^{-1000t}]u(t)$

Exercise 11–7

The step response of a linear circuit is $g(t) = 5[e^{-1000t}\sin(2000t)]u(t)$. Find the circuit transfer function $T(s)$.

Answer:

$$T(s) = \frac{10^4 s}{s^2 + 2000s + 5 \times 10^6}$$

11–5   NETWORK FUNCTIONS AND SINUSOIDAL STEADY-STATE RESPONSE

When a stable, linear circuit is driven by a sinusoidal input, the output contains a steady-state component that is a sinusoid of the same frequency as the input. This section deals with using the circuit transfer function to find the amplitude and phase angle of the sinusoidal steady-state response. To begin, we write a general sinusoidal input in the form

$$x(t) = X_A \cos(\omega t + \phi) \tag{11–18}$$

which can be expanded as

$$x(t) = X_A(\cos \omega t \cos \phi - \sin \omega t \sin \phi)$$

The waveforms $\cos \omega t$ and $\sin \omega t$ are basic signals whose transforms are given in Table 9–2 as $\mathcal{L}\{\cos \omega t\} = s/(s^2 + \omega^2)$ and $\mathcal{L}\{\sin \omega t\} = \omega/(s^2 + \omega^2)$. Therefore, the input transform is

$$X(s) = X_A\left[\frac{s}{s^2 + \omega^2} \cos \phi - \frac{\omega}{s^2 + \omega^2} \sin \phi\right]$$

$$= X_A\left[\frac{s \cos \phi - \omega \sin \phi}{s^2 + \omega^2}\right] \tag{11–19}$$

Equation (11–19) is the Laplace transform of the general sinusoidal waveform in Eq. (11–18).

Using Eq. (11–2), we obtain the response transform for a general sinusoidal input:

$$Y(s) = X_A\left[\frac{s \cos \phi - \omega \sin \phi}{(s - j\omega)(s + j\omega)}\right] T(s) \tag{11–20}$$

The response transform contains forced poles at $s = \pm j\omega$ because the input is a sinusoid. Expanding Eq. (11–20) by partial fractions,

$$Y(s) = \underbrace{\frac{k}{s - j\omega} + \frac{k^*}{s + j\omega}}_{\text{Forced Poles}} + \underbrace{\frac{k_1}{s - p_1} + \frac{k_2}{s - p_2} + \cdots + \frac{k_N}{s - p_N}}_{\text{Natural Poles}}$$

where $p_1, p_2, \ldots p_N$ are the natural poles contributed by the transfer function $T(s)$. To obtain the response waveform, we perform the inverse transformation:

$$y(t) = \underbrace{ke^{j\omega t} + k^*e^{-j\omega t}}_{\text{Forced Response}} + \underbrace{k_1 e^{p_1 t} + k_2 e^{p_2 t} + \cdots + k_N e^{p_N t}}_{\text{Natural Response}}$$

When the circuit is stable, the natural response decays to zero, leaving a sinusoidal steady-state response due to the forced poles as $s = \pm j\omega$. The steady-state response is

$$y_{\text{SS}}(t) = ke^{j\omega t} + k^*e^{-j\omega t}$$

where the subscript SS identifies a steady-state condition.

To determine the amplitude and phase of the steady-state response, we must find the residue k. Using the cover-up method from Chapter 9, we find k to be

$$k = (s - j\omega)X_A \left[\frac{s\cos\phi - \omega\sin\phi}{(s - j\omega)(s + j\omega)}\right]T(s)\Big|_{s = j\omega}$$

$$= X_A \left[\frac{j\omega\cos\phi - \omega\sin\phi}{2j\omega}\right]T(j\omega)$$

$$= X_A \left[\frac{\cos\phi + j\sin\phi}{2}\right]T(j\omega) = \frac{1}{2}X_A e^{j\phi}T(j\omega)$$

The complex quantity $T(j\omega)$ can be written in magnitude and angle form as $|T(j\omega)|e^{j\theta}$. Using these results, the residue becomes

$$k = \left[\frac{1}{2}X_A e^{j\phi}\right]\left|T(j\omega)\right|e^{j\theta}$$

$$= \frac{1}{2}X_A \left|T(j\omega)\right|e^{j(\phi + \theta)}$$

The inverse transform yields the steady-state response in the form

$$y_{\text{SS}}(t) = 2|k|\cos(\omega t + \phi + \angle k)$$

$$= \underbrace{X_A|T(j\omega)|}_{\text{Amplitude}}\cos(\omega t + \underbrace{\phi + \theta}_{\text{Phase}}) \qquad (11\text{-}21)$$

In the development leading to Eq. (11–21), we treat frequency as a general variable where the symbol ω represents all possible input frequencies. In some cases the input frequency has a specific value, which we denote as ω_A. In this case the input is written as

$$x(t) = X_A \cos(\omega_A t + \phi)$$

To obtain the steady-state output, we evaluate the transfer function at the specific frequency (ω_A) of the input sinusoid, namely

$$T(j\omega_A) = |T(j\omega_A)| \angle T(j\omega_A)$$

and the steady-state output is expressed as

$$y_{SS}(t) = X_A |T(j\omega_A)| \cos[\omega_A t + \phi + \angle T(j\omega_A)]$$

This result emphasizes three things about the steady-state response:

(1) Output Frequency = Input Frequency = ω_A

(2) Output Amplitude = Input Amplitude $\times |T(j\omega_A)|$ (11–22)

(3) Output Phase = Input Phase + $\angle T(j\omega_A)$

   The next two examples illustrate sinusoidal steady-state response calculations. In the first example, frequency is treated as a general variable and we examine the steady-state response as frequency varies over a wide range. In the second example, we evaluate the steady-state response at two specific frequencies.

EXAMPLE 11–13

Find the steady-state output in Figure 11–20 for a general input $v_1(t) = V_A \cos(\omega t + \phi)$.

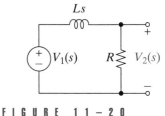

FIGURE 11–20

SOLUTION:

In Example 11–2, the circuit transfer function is shown to be

$$T(s) = \frac{R}{Ls + R}$$

The magnitude and angle of $T(j\omega)$ are

$$|T(j\omega)| = \frac{R}{\sqrt{R^2 + (\omega L)^2}}$$

$$\theta(\omega) = -\tan^{-1}\left(\frac{\omega L}{R}\right)$$

Using Eq. (11–21), the sinusoidal steady-state output is

$$v_{2SS}(t) = \frac{V_A R}{\sqrt{R^2 + (\omega L)^2}} \cos[\omega t + \phi - \tan^{-1}(\omega L/R)]$$

Note that both the amplitude and phase angle of the steady-state response depend on the frequency of the input sinusoid. In particular, at $\omega = 0$ the amplitude of the steady-state output reduces to V_A, which is the same as the amplitude of the input sinusoid. This makes sense because at dc the inductor in Figure 11–20 acts like a short circuit that directly connects the output port to the input port. At very high frequency the amplitude of the steady-state output approaches zero. This also makes sense because at very high frequency the inductor acts like an open circuit that disconnects the output port from the input port. In between these two extremes the output amplitude decreases as the frequency increases. ∎

EXAMPLE 11-14

The impulse response of a linear circuit is $h(t) = 5000[2e^{-1000t} \cos 2000t - e^{-1000t} \sin 2000t]u(t)$.
(a) Find the sinusoidal steady-state response when $x(t) = 5 \cos 1000t$.
(b) Repeat part (a) when $x(t) = 5 \cos 3000t$.

SOLUTION:
The transfer function corresponding to $h(t)$ is

$$T(s) = H(s) = \mathcal{L}\{h(t)\}$$

$$= 5000\left[\frac{2(s + 1000)}{(s + 1000)^2 + (2000)^2} - \frac{2000}{(s + 1000)^2 + (2000)^2}\right]$$

$$= \frac{10^4 s}{s^2 + 2000s + 5 \times 10^6}$$

(a) At $\omega_A = 1000$ rad/s, the value of $T(j\omega_A)$ is

$$T(j1000) = \frac{10^4(j1000)}{(j1000)^2 + 2000\,(j1000) + 5 \times 10^6}$$

$$= \frac{j10^7}{(5 \times 10^6 - 10^6) + j2 \times 10^6} = \frac{j10}{4 + j2}$$

$$= \frac{10\,e^{j90°}}{\sqrt{20}e^{j26.6°}} = 2.24\,e^{j63.4°}$$

and the steady-state response for $x(t) = 5 \cos 1000t$ is

$$y_{SS}(t) = 5 \times 2.24 \cos(1000t + 0° + 63.4°)$$

$$= 11.2 \cos(1000t + 63.4°)$$

(b) At $\omega_A = 3000$ rad/s, the value of $T(j\omega_A)$ is

$$T(j3000) = \frac{10^4(j3000)}{(j3000)^2 + 2000\,(j3000) + 5 \times 10^6}$$

$$= \frac{j3 \times 10^7}{5 \times 10^6 - 9 \times 10^6 + j6 \times 10^6}$$

$$= \frac{j30}{-4 + j6} = \frac{30\,e^{j90°}}{\sqrt{52}e^{j123.7°}} = 4.16\,e^{-j33.7°}$$

and the steady-state response for $x(t) = 5 \cos 3000t$ is

$$y_{SS}(t) = 5 \times 4.16 \cos(3000t + 0° - 33.7°)$$

$$= 20.8 \cos(3000t - 33.7°)$$

Again note that the amplitude and phase angle of the steady-state response depend on the input frequency. ∎

Exercise 11–8

The transfer function of a linear circuit is $T(s) = 5(s + 100)/(s + 500)$. Find the steady-state output for
(a) $x(t) = 3 \cos 100t$
(b) $x(t) = 2 \sin 500t$.

Answers:
(a) $y_{SS}(t) = 4.16 \cos(100t + 33.7°)$
(b) $y_{SS}(t) = 7.21 \cos(500t − 56.3°)$

Exercise 11–9

The impulse response of a linear circuit is $h(t) = \delta(t) − 100[e^{-100t}]u(t)$. Find the steady-state output for
(a) $x(t) = 25 \cos 100t$
(b) $x(t) = 50 \sin 100t$.

Answers:
(a) $y_{SS}(t) = 17.7 \cos(100t + 45°)$
(b) $y_{SS}(t) = 35.4 \cos(100t − 45°)$

NETWORK FUNCTIONS AND PHASOR CIRCUIT ANALYSIS

In this text we present two methods of finding the sinusoidal steady-state response of a linear circuit. Both methods depend on the fact that if the input is a sinusoid of frequency ω_A, then in the steady state every voltage and current will be a sinusoid of frequency ω_A. As a result, every method of steady-state analysis boils down to finding the amplitude and phase angle of response waveforms, since their frequency must be ω_A.

The method developed in this section involves finding the network function $T(s)$ relating the input and the desired output. We then form the complex quantity $T(j\omega_A)$, and use $|T(j\omega_A)|$ and $\angle T(j\omega_A)$ to find the amplitude and phase of steady-state waveforms via Eq. (11-21).

The other method is the phasor approach discussed in the chapter on sinusoidal steady-state analysis. With this method, input sinusoids are represented by complex numbers called phasors and the individual circuit elements described by their impedances at $s = j\omega_A$. Conventional circuit analysis is then used to find complex numbers representing phasor responses. The magnitude and angle of these complex numbers represent the amplitude and phase of the steady-state waveforms.

Under what conditions is one method preferable to the other?

Phasor circuit analysis works best when the circuit is driven at a single frequency and we need to find several voltages and currents, or perhaps average and maximum available power. The network function method works best when the circuit is driven at several frequencies and we only need to find a single response called the output. The network function is also the only method available when all we know about the circuit is its impulse or step response waveform. Thus, the preferred method depends on how the circuit is driven (single or multiple frequency) and what we need to find out (single or multiple response).

11–6 IMPULSE RESPONSE AND CONVOLUTION

In signal processing the term **convolution** refers to a process by which the impulse response of a linear system is used to determine the zero-state response due to other inputs. When the impulse response and input are given in the *s* domain as $H(s)$ and $X(s)$, the zero-state response is obtained from the inverse Laplace transform of their product.

$$y(t) = \mathcal{L}^{-1}\{H(s)X(s)\} \tag{11–23}$$

The purpose of this section is to introduce the notion that the convolution process can also be viewed and carried out entirely in time domain. Specifically, given a causal impulse response $h(t)$ and a causal input $x(t)$, the zero-state response is obtained from the **convolution integral**[1]

$$y(t) = \int_0^t h(t - \tau)x(\tau)d\tau \tag{11–24}$$

where τ is a dummy variable of integration. The shorthand notation $y(t) = h(t)*x(t)$ is used to represent the *t*-domain process, where the asterisk indicates a convolution integral, not a multiplication. That is, the expression $h(t)*x(t)$ reads "$h(t)$ convolved with $x(t)$," not "$h(t)$ times $x(t)$."

Figure 11–21 indicates the parallelism between the input-output relationships in the *s* domain and *t* domain. In the *s* domain the impulse response $H(s)$ multiplies the input transform to produce the output transform. In the *t* domain the impulse response $h(t)$ is convolved with the input waveform to produce the output waveform.

The following example illustrates that the same result is achieved whether convolution is carried out in the *s* domain or the *t* domain.

FIGURE 11–21 *Input-output relationships in the* t *domain and the* s *domain.*

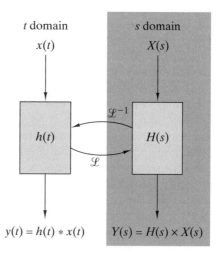

---

1 Recall from Chapter 5 that a waveform is causal if $f(t) = 0$ for all $t < 0$. If the impulse response $h(t)$ is not causal, the upper limit in the convolution integral becomes $+\infty$. If the input $x(t)$ is not casual, the lower limit becomes $-\infty$.

EXAMPLE 11-15

A linear circuit has an impulse response $h(t) = 2e^{-t}u(t)$ and an input $x(t) = e^{-2t}u(t)$. Find the zero state response using

(a) the s-domain process in Eq. (11–23) and
(b) the t-domain convolution integral in Eq. (11–24).

SOLUTION:
(a) Converting $h(t)$ and $x(t)$ to the s domain yields

$$H(s) = \mathcal{L}\{2e^{-t}\} = \frac{2}{s+1} \quad \text{and} \quad X(s) = \mathcal{L}\{e^{-2t}\} = \frac{1}{s+2}$$

Applying the Eq. (11–23) produces $y(t)$ as

$$y(t) = \mathcal{L}^{-1}\{H(s)X(s)\} = \mathcal{L}^{-1}\left\{\frac{1}{(s+1)(s+2)}\right\}$$

$$= \mathcal{L}^{-1}\left\{\frac{2}{s+1} - \frac{2}{s+2}\right\} = 2e^{-t} - 2e^{-2t} \quad \text{for } t > 0$$

(b) Meanwhile, back in the t domain the convolution integral in Eq. (11–24) yields

$$y(t) = \int_0^t h(t-\tau)x(\tau)d\tau = \int_0^t 2e^{-(t-\tau)}e^{-2\tau}d\tau$$

$$= 2e^{-t}\int_0^t e^{\tau}e^{-2\tau}d\tau = 2e^{-t}\int_0^t e^{-\tau}d\tau$$

$$= 2e^{-t}(1 - e^{-t}) = 2e^{-t} - 2e^{-2t} \quad \text{for } t > 0$$

The two methods produce the same result. The difference is that the convolution integral evaluation is carried out entirely in the time domain.

EQUIVALENCE OF S-DOMAIN AND T-DOMAIN CONVOLUTION

The approach used here starts out in the s domain with $Y(s) = H(s)X(s)$ and proceeds to show that $y(t) = h(t)*x(t)$. By beginning in the s domain we are assuming that the waveforms are causal. The s-domain input-output relationship can be written as

$$Y(s) = H(s)X(s) = H(s)\underbrace{\left[\int_0^\infty x(\tau)e^{-s\tau}d\tau\right]}_{\mathcal{L}\{x(t)\}}$$

where the bracketed term is the integral definition of $\mathcal{L}\{x(t)\}$. The impulse response can be moved inside the integration since $H(s)$ does not depend on the dummy variable τ.

$$Y(s) = \int_0^\infty [H(s)e^{-s\tau}]x(\tau)d\tau \qquad (11\text{–}25)$$

Using the time translation property from Chapter 9 and the integral definition of the Laplace transformation, the bracketed term can be written as

$$H(s)e^{-s\tau} = \mathcal{L}\{h(t-\tau)u(t-\tau)\}$$

$$= \int_0^\infty h(t-\tau)u(t-\tau)e^{-st}dt$$

When the last line in this result replaces the bracketed term in Eq. (11–25), we obtain

$$Y(s) = \int_0^\infty \left[\underbrace{\int_0^\infty h(t-\tau)u(t-\tau)e^{-st}\,dt}_{H(s)e^{-st}} \right]x(\tau)d\tau$$

Interchanging the order of integration produces

$$Y(s) = \int_0^\infty \left[\int_0^t h(t-\tau)\,x(\tau)\,d\tau \right]e^{-st}\,dt$$

The inner integration is now carried out with respect to the dummy variable τ. The upper limit on this integration need only extend to $\tau = t$ (rather than infinity) since $u(t-\tau) = 0$ for $\tau > t$. By definition the outer integration (now with respect to t) yields the Laplace transform of the quantity inside the bracket. In other words, this equation is equivalent to the statement

$$Y(s) = \mathcal{L} \left[\underbrace{\int_0^t h(t-\tau)\,x(\tau)\,d\tau}_{h(t)\,*\,x(t)} \right]$$

So finally we have

$$y(t) = \mathcal{L}^{-1}\{Y(s)\} = \mathcal{L}^{-1}\{H(s)Y(s)\} = h(t)*x(t)$$

which establishes the equivalence of Eqs. (11–23) and (11–24). It is important to remember that this equivalence only applies to casual waveforms.

APPLICATIONS OF THE CONVOLUTION INTEGRAL

Given that s-domain and t-domain convolutions are equivalent, why study both methods? The best answer the authors can give at this point will have an "eat your spinach" ring to it. Suffice it to say that in subsequent courses you will encounter signals for which Laplace transforms do not exist; hence, only the t-domain method convolution is possible. Examples of such application are the noncausal waveforms used in communication systems and the discrete-time signals used in digital signal processing. We cannot treat such applications here, but rather only introduce the student to the concept of viewing convolution as a t-domain process.

EXAMPLE 11–16

A linear circuit has an impulse response $h(t) = 2e^{-t}u(t)$. Use the convolution integral to find the zero-state response for $x(t) = tu(t)$.

SOLUTION:
Direct application of Eq. (11−24) produces the following results

$$y(t) = \int_0^t 2e^{-(t-\tau)}\tau \, d\tau = 2e^{-t}\int_0^t \tau \, e^\tau \, d\tau$$

$$= 2 \, e^{-t}[e^\tau(\tau - 1)]_0^t$$

$$= 2(t - 1 + e^{-t}) \quad \text{for} \quad t \geq 0$$

Although evaluation of the integral is straightforward in this case, a geometric interpretation of the process helps us proceed in more complicated situations. ∎

Figure 11−22 shows a geometric interpretation of the convolution in Example 11−16. In Figures 11−22 (a) and (b) the input and impulse response waveforms are plotted against the dummy variable τ. Forming $h(-\tau)$ *reflects* the impulse response across the $\tau = 0$ axis as shown in Figure 11−22(c). Forming $h(t - \tau)$ *shifts* the reflected impulse response to the right by t seconds as shown in Figure 11−22 (d). *Multiplying* the reflected/shifted impulse response by the input produces the product $h(t - \tau) \times x(\tau)$ shown in Figure 11−22(e). The *integrating* from $\tau = 0$ to $\tau = t$ yields the area under this product, which is the value of the zero-state output at time t. At a later instant of time the reflected impulse response $h(t - \tau)$ shifts further to the right, creating a new product $h(t - \tau) \times x(\tau)$ with a new area and a new value of $y(t)$.

Thus, the geometric interpretation of t-domain convolution involves four operations: reflecting, shifting, multiplying, and integrating. We visualize convolution as a process that reflects the impulse response across the origin and then progressively shifts it to the right as t increases. At any time t the output is the area under the product of the reflected/shifted impulse response and the input. Under this interpretation we can think of the impulse response as a weighting function. That is, when integrating the product $h(t - \tau) \times x(\tau)$, the impulse response tells us how much weight to assign to previous values of the input.

EXAMPLE 11-17

A certain circuit has an impulse response $h(t) = [2e^{-t}]u(t)$. Use the convolution integral to find the zero-state response for $x(t) = 5[u(t) - u(t-2)]$.

SOLUTION:
Evaluation of the convolution integral can be divided into the three situations shown in Figure 11−23. The situation for $t < 0$ is shown in Figure 11−23(a). For this case the reflected impulse response $h(t - \tau)$ and the input $x(\tau)$ do not overlap so the area under their product is zero. Hence $y(t) = 0$ for $t < 0$. This simply says that the zero-state response is causal when the impulse response and input are causal.

For $0 < t < 2$, the reflected impulse response and input overlap, as shown in Figure 11−23(b). In this situation the area under the product $h(t - \tau) \times x(\tau)$ is found by integrating $\tau = 0$ to $\tau = t$.

FIGURE 11 – 22 *Graphical interpretation of convolution.*

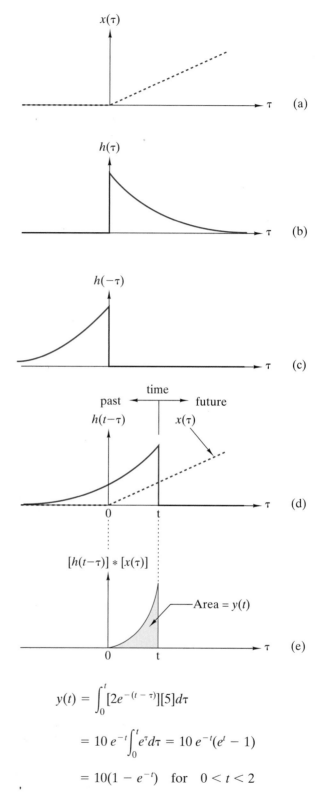

$$y(t) = \int_0^t [2e^{-(t-\tau)}][5]d\tau$$

$$= 10\,e^{-t}\int_0^t e^{\tau}d\tau = 10\,e^{-t}(e^t - 1)$$

$$= 10(1 - e^{-t}) \quad \text{for} \quad 0 < t < 2$$

For $t > 2$, the reflected impulse response and input overlap as shown in Figure 11–23(c). In this case the product $h(t - \tau) \times x(\tau)$ is confined to the

FIGURE 11-23

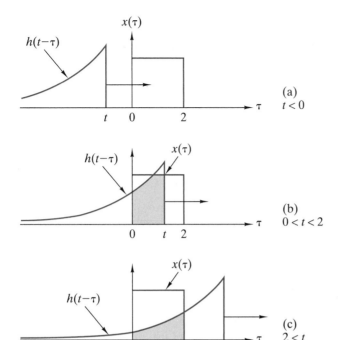

(a) $t < 0$

(b) $0 < t < 2$

(c) $2 < t$

interval $0 < \tau < 2$ since the input $x(\tau)$ is zero everywhere outside of this interval. In this situation the area under the product is found by integrating from $\tau = 0$ and $\tau = 2$.

$$y(t) = \int_0^2 [2e^{-(t-\tau)}][5]d\tau$$

$$= 10\, e^{-t} \int_0^2 e^\tau\, d\tau$$

$$= 10\, e^{-t}(e^2 - 1) \quad \text{for} \quad 2 \le t$$

Evaluation of the convolution integral was guided by the geometric interpretation in Figure 11-23 and leads to a zero-state response defined on three intervals.

$$y(t) = \begin{vmatrix} 0 & \text{for} \quad t < 0 \\ 10(1 - e^{-t}) & \text{for} \quad 0 \le t < 2 \\ 10e^{-t}(e^2 - 1) & \text{for} \quad 2 \le t \end{vmatrix}$$

Exercise 11-10

Use the convolution integral to find the zero-state response for $h(t) = 2u(t)$ and $x(t) = 5[u(t) - u(t-2)]$.

Answer:

$$y(t) = \begin{vmatrix} 0 & \text{for} \quad t < 0 \\ 10t & \text{for} \quad 0 \le t < 2 \\ 20 & \text{for} \quad 2 \le t \end{vmatrix}$$

APPLICATION NOTE: EXAMPLE 11–18

In mathematical context the convolution integral involves two general functions and is written

$$f_1(t) * f_2(t) = \int_{-\infty}^{+\infty} f_1(t - \tau)f_2(\tau)d\tau$$

If both functions are causal, the integration limits become $\tau = 0$ to $\tau = t$. In either form the convolution integral has the following mathematical properties.

$$f(t) * \delta(t) = f(t) \qquad\qquad\qquad\qquad\qquad\qquad \text{Identity}$$
$$f_1(t) * f_2(t) = f_2(t) * f_1(t) \qquad\qquad\qquad\qquad \text{Commutative}$$
$$f_1(t) * [f_2(t) * f_3(t)] = [f_1(t) * f_2(t)] * f_2(t) \qquad \text{Associative}$$
$$[f_1(t) + f_2(t)] * f_3(t) = f_1(t) * f_3(t) + f_2(t) * f_3(t) \quad \text{Distributive}$$

These properties may simplify convolution calculations in some cases. However, our interest here is to illustrate what these mathematical properties tell us about time-domain signal processing.

A circuit is said to be *memoryless* if its output at any time t depends only on the input at the same time; that is, $y(t) = Kx(t)$, where K is a scalar constant. The following steps use the commutative and identity properties to show that if $h(t) = K\delta(t)$ then the circuit is memoryless.

$$
\begin{aligned}
y(t) &= K\,\delta(t) * x(t) && \text{Given} \\
&= K[x(t) * \delta(t)] && \text{Commutative property} \\
&= Kx(t) && \text{Identity property}
\end{aligned}
$$

In sum, a circuit whose impulse response is an impulse is memoryless. The block diagram representation of this conclusion is shown in Figure 11–24(a), which is nothing more than the proportionality property of linear resistive circuits.

These properties can be used to prove the following time-domain signal processing property. If the output of a system $h_1(t)$ serves as the input to a second system $h_2(t)$, then the overall system impulse response is $h(t) = h_1(t) * h_2(t)$.

$$
\begin{aligned}
y(t) &= h_2(t) * [h_1(t) * x(t)] && \text{Given} \\
&= [h_2(t) * h_1(t)] * x(t) && \text{Associative property} \\
&= \underbrace{[h_1(t) * h_2(t)]}_{h(t)} * x(t) && \text{Commutative property}
\end{aligned}
$$

Figure 11–24(b) shows the block diagram representation of this result, which is none other than the time-domain version of the chain rule.[2]

The block diagram representation of the distributive property is shown in Figure 11–24(c). The parallel paths in this diagram superimpose the responses due to each term in the impulse response acting alone. The block diagrams of these signal processing properties are helpful because they show how systems operate.

2 This interpretation assumes that the cascade connection does not alter (load) the individual impulse responses.

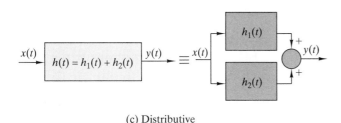

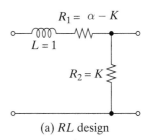

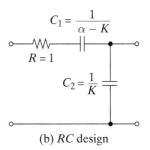

(a) Memoryless

(b) Associative

(c) Distributive

FIGURE 11–24 *Block diagram inter-pretation of the properties of convolution.*

11-7 NETWORK FUNCTION DESIGN

Finding and using a network function of a given circuit is an *s*-domain **analysis** problem. An *s*-domain **synthesis** problem involves finding a circuit that realizes a given network function. For linear circuits an analysis problem always has a unique solution. In contrast, a synthesis problem may have many solutions because different circuits can have the same network function. A transfer function design problem involves synthesizing several circuits that realize a given function and evaluating the alternative designs, using criteria such as input or output impedance, cost, and power consumption.

FIRST-ORDER VOLTAGE-DIVIDER CIRCUIT DESIGN

We begin our study of transfer function design by developing a voltage-divider realization of a first-order transfer function of the form $K/(s + \alpha)$. The impedances $Z_1(s)$ and $Z_2(s)$ are related to the given transfer function using the voltage-divider relationship.

$$T_V(s) = \frac{K}{s + \alpha} = \frac{Z_2(s)}{Z_1(s) + Z_2(s)} \qquad (11–26)$$

To obtain a circuit realization, we must assign part of the given $T_V(s)$ to $Z_2(s)$ and the remainder to $Z_1(s)$. There are many possible realizations of $Z_1(s)$ and $Z_2(s)$ because there is no unique way to make this assignment. For example, simply equating the numerators and denominators in Eq. (11–26) yields

$$Z_2(s) = K \quad \text{and} \quad Z_1(s) = s + \alpha - Z_2(s) = s + \alpha - K \qquad (11–27)$$

Inspecting this result, we see that $Z_2(s)$ is realizable as a resistance ($R_2 = K \ \Omega$) and $Z_1(s)$ as an inductance ($L_1 = 1$ H) in series with a resistance [$R_1 = (\alpha - K) \ \Omega$]. The resulting circuit diagram is shown in Figure 11–25(a). For $K = \alpha$ the resistance R_1 can be replaced by a short circuit because its resis-

FIGURE 11–25 *Circuit realizations of* $T(s) = K/(s + \alpha)$ *for K $\leq \alpha$.*

tance is zero. A gain restriction $K \leq \alpha$ is necessary because a negative R_1 is not physically realizable as a single component.

An alternative synthesis approach involves factoring s out of the denominator of the given transfer function. In this case Eq. (11–26) is rewritten in the form

$$T_V(s) = \frac{K/s}{1 + \alpha/s} = \frac{Z_2(s)}{Z_1(s) + Z_2(s)} \tag{11–28}$$

Equating numerators and denominators yields the branch impedances

$$Z_2(s) = \frac{K}{s} \quad \text{and} \quad Z_1(s) = 1 + \frac{\alpha}{s} - Z_2(s) = 1 + \frac{\alpha - K}{s} \tag{11–29}$$

In this case we see that $Z_2(s)$ is realizable as a capacitance ($C_2 = 1/K$ F) and $Z_1(s)$ as a resistance ($R_1 = 1 \, \Omega$) in series with a capacitance [$C_1 = 1/(\alpha - K)$ F]. The resulting circuit diagram is shown in Figure 11–25(b). For $K = \alpha$ the capacitance C_1 can be replaced by a short circuit because its capacitance is infinite. A gain restriction $K \leq \alpha$ is required to keep C_1 from being negative.

As a second design example, consider a voltage-divider realization of the transfer function $Ks/(s + \alpha)$. We can find two voltage-divider realizations by writing the specified transfer function in the following two ways:

$$T(s) = \frac{Ks}{s + \alpha} = \frac{Z_2(s)}{Z_1(s) + Z_2(s)} \tag{11–30a}$$

$$T(s) = \frac{K}{1 + \alpha/s} = \frac{Z_2(s)}{Z_1(s) + Z_2(s)} \tag{11–30b}$$

Equation (11–30a) uses the transfer function as given, while Eq. (11–30b) factors s out of the numerator and denominator. Equating the numerators and denominators in Eqs. (11–30a) and (11–30b) yields two possible impedance assignments:

Using Eq. (11– 30a): $Z_2 = Ks$ and $Z_1 = s + \alpha - Z_2 = (1 - K)s + \alpha$
$$\tag{11–31a}$$

Using Eq. (11– 30b): $Z_2 = K$ and $Z_1 = 1 + \frac{\alpha}{s} - Z_2 = (1 - K) + \frac{\alpha}{s}$
$$\tag{11–31b}$$

The assignment in Eq. (11–31a) yields $Z_2(s)$ as an inductance ($L_2 = K$ H) and $Z_1(s)$ as an inductance [$L_1 = (1 - K)$ H] in series with a resistance ($R_1 = \alpha \, \Omega$). The assignment in Eq. (11–31b) yields $Z_2(s)$ as a resistance ($R_2 = K$) and $Z_1(s)$ as a resistance [$R_1 = (1 - K) \, \Omega$] in series with a capacitance ($C_1 = 1/\alpha$ F). The two realizations are shown in Figure 11–26. Both realizations require $K \leq 1$ for the branch impedances to be realizable and both simplify when $K = 1$.

VOLTAGE-DIVIDER AND OP AMP CASCADE CIRCUIT DESIGN

The examples in Figures 11–25 and 11–26 illustrate an important feature of voltage-divider realizations. In general, we can write a transfer function as a quotient of polynomials $T(s) = r(s)/q(s)$. A voltage-divider realization re-

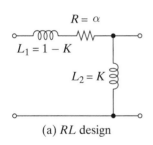

$R = \alpha$
$L_1 = 1 - K$
$L_2 = K$

(a) *RL* design

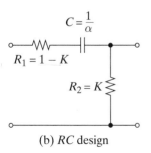

$C = \frac{1}{\alpha}$
$R_1 = 1 - K$
$R_2 = K$

(b) *RC* design

F I G U R E   1 1 – 2 6 *Circuit realizations of* $T(s) = Ks/(s + \alpha)$ *for* $K \leq 1$.

quires the impedances $Z_2(s) = r(s)$ and $Z_1(s) = q(s) - r(s)$ to be physically realizable. A voltage-divider circuit usually places limitations on the gain K. This gain limitation can be overcome by using an OP AMP circuit in cascade with divider circuit.

For example, a voltage-divider realization of the transfer function in Eq. (11–26) requires $K \leq \alpha$. When $K > \alpha$, then $T(s)$ is not realizable as a simple voltage divider, since $Z_2(s) = s + \alpha - K$ requires a negative resistance. However, the given transfer function can be written as a two-stage product:

$$T_V(s) = \frac{K}{s + \alpha} = \underbrace{\left[\frac{K}{\alpha}\right]}_{\substack{\text{First} \\ \text{Stage}}} \underbrace{\left[\frac{\alpha}{s + \alpha}\right]}_{\substack{\text{Second} \\ \text{Stage}}}$$

When $K > \alpha$, the first stage has a positive gain greater than unity. This stage can be realized using a noninverting OP AMP circuit with a gain of $(R_1 + R_2)/R_1$. The first stage design constraint is

$$\frac{K}{\alpha} = \frac{R_1 + R_2}{R_1}$$

Choosing $R_1 = 1\ \Omega$ requires that $R_2 = (K/\alpha) - 1$. An RC voltage divider realization of the second stage is obtained by factoring an s out of the stage transfer function. This leads to the second stage design constraint

$$\frac{\alpha/s}{1 + \alpha/s} = \frac{Z_2(s)}{Z_1(s) + Z_2(s)}$$

Equating numerators and denominators yields $Z_2(s) = \alpha/s$ and $Z_1(s) = 1$. Figure 11–27 shows a cascade connection of a noninverting first stage and the RC divider second stage. The chain rule applies to this circuit, since the first stage has an OP AMP output. The cascade circuit in Figure 11–27 realizes the first-order transfer function $K/(s + \alpha)$ for $K > \alpha$, a gain requirement that cannot be met by the divider circuit alone.

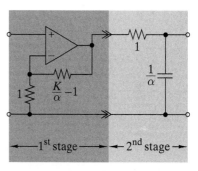

FIGURE 11–27 *Circuit realization of* $T(s) = K/(s + \alpha)$ *for* $K > \alpha$.

D DESIGN EXAMPLE: 11–19

Design a circuit to realize the following transfer function using only resistors, capacitors, and OP AMPs:

$$T_V(s) = \frac{3000s}{(s + 1000)(s + 4000)}$$

SOLUTION:
The given transfer function can be written as a three-stage product.

$$T_V(s) = \underbrace{\left[\frac{K_1}{s + 1000}\right]}_{\substack{\text{First} \\ \text{Stage}}} \underbrace{[K_2]}_{\substack{\text{Second} \\ \text{Stage}}} \underbrace{\left[\frac{K_3 s}{s + 4000}\right]}_{\substack{\text{Third} \\ \text{Stage}}}$$

where the stage gains K_1, K_2, and K_3 have yet to be selected. Factoring s out of the denominator of the first-stage transfer function leads to an RC divider realization:

$$\frac{K_1/s}{1 + 1000/s} = \frac{Z_2(s)}{Z_1(s) + Z_2(s)}$$

Equating numerators and denominators yields

$$Z_2(s) = K_1/s \quad \text{and} \quad Z_1(s) = 1 + (1000 - K_1)/s$$

The first stage $Z_1(s)$ is simpler when we select $K_1 = 1000$. Factoring s out of the denominator of the third-stage transfer function leads to an RC divider realization:

$$\frac{K_3}{1 + 4000/s} = \frac{Z_2(s)}{Z_1(s) + Z_2(s)}$$

Equating numerators and denominators yields

$$Z_2(s) = K_3 \text{ and } Z_1(s) = 1 - K_3 + 4000/s$$

The third stage $Z_1(s)$ is simpler when we select $K_3 = 1$. The stage gains must meet the constraint $K_1 \times K_2 \times K_3 = 3000$ since the overall gain of the given transfer function is 3000. We have selected $K_1 = 1000$ and $K_3 = 1$, which requires $K_2 = 3$. The second stage must have a positive gain greater than 1 and can be realized using a noninverting amplifier with $K_2 = (R_1 + R_2)/R_1 = 3$. Selecting $R_1 = 1 \, \Omega$ requires that $R_2 = 2 \, \Omega$.

Figure 11–28 shows the three stages connected in cascade. The chain rule applies to this cascade connection because the OP AMP in the second stage isolates the RC voltage-divider circuits in the first and third stages. The circuit in Figure 11–28 realizes the given transfer function but is not a realistic design because the values of resistance and capacitance are impractical. For this reason we call this circuit a **prototype** design. We will shortly discuss how to scale a prototype to obtain practical element values. ■

FIGURE 11–28

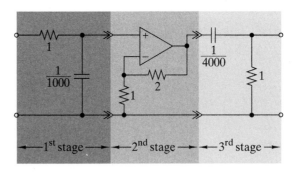

INVERTING OP AMP CIRCUIT DESIGN

The inverting OP AMP circuit places fewer restrictions on the form of the desired transfer function than does the basic voltage divider. To illustrate this, we will develop two inverting OP AMP designs for a general first-order transfer function of the form

$$T_V(s) = -K\frac{s+\gamma}{s+\alpha}$$

The general transfer function of the inverting OP AMP circuit is $-Z_2(s)/Z_1(s)$, which leads to the general design constraint

$$-K\frac{s+\gamma}{s+\alpha} = -\frac{Z_2(s)}{Z_1(s)} \qquad (11-32)$$

The first design is obtained by equating the numerators and denominators in Eq. (11–32) to obtain the OP AMP circuit impedances as $Z_2(s) = Ks + K\gamma$ and $Z_1(s) = s + \alpha$. Both of these impedances are of the form $Ls + R$ and can be realized by an inductance in series with a resistance, leading to the design realization in Figure 11–29(a).

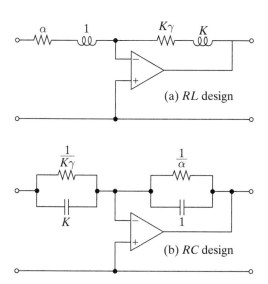

(a) RL design

(b) RC design

FIGURE 11−29 Inverting OP AMP circuit realizations of $T(s) = -K(s+\gamma)/(s+\alpha)$.

A second inverting OP AMP realization is obtained by equating $Z_2(s)$ in Eq. (11–32) to the reciprocal of the denominator and equating $Z_1(s)$ to the reciprocal of the numerator. This assignment yields the impedances $Z_1(s) = 1/(Ks + K\gamma)$ and $Z_2(s) = 1/(s + \alpha)$. Both of these impedances are of the form $1/(Cs + G)$, where Cs is the admittance of a capacitor and G is the admittance of a resistor. Both impedances can be realized by a capacitance in parallel with a resistance. These impedance identifications produce the RC circuit in Figure 11–29(b).

Because it has fewer restrictions, it is often easier to realize transfer functions using the inverting OP AMP circuit. To use inverting circuits, the given transfer function must require an inversion or be realized using an even number of inverting stages. In some cases, the sign in front of the transfer function is immaterial and the required transfer function is specified as $\pm T_V(s)$. *Caution:* The input impedance of an inverting OP AMP circuit may load the source circuit.

D DESIGN EXAMPLE: 11–20

Design a circuit to realize the transfer function given in Example 11–19 using inverting OP AMP circuits.

SOLUTION:
The given transfer function can be expressed as the product of two inverting transfer functions:

$$T_V(s) = \frac{3000s}{(s+1000)(s+4000)} = \underbrace{\left[-\frac{K_1}{s+1000}\right]}_{\text{First Stage}}\underbrace{\left[-\frac{K_2 s}{s+4000}\right]}_{\text{Second Stage}}$$

where the stage gains K_1 and K_2 have yet to be selected. The first stage can be realized in an inverting OP AMP circuit since

$$-\frac{K_1}{s+1000} = -\frac{K_1/1000}{1+s/1000} = -\frac{Z_2(s)}{Z_1(s)}$$

Equating the $Z_2(s)$ to the reciprocal of the denominator and $Z_1(s)$ to the reciprocal of the numerator yields

$$Z_2 = \frac{1}{1+s/1000} \quad \text{and} \quad Z_1 = 1000/K_1$$

The impedance $Z_2(s)$ is realizable as a capacitance ($C_2 = 1/1000$ F) in parallel with a resistance ($R_2 = 1\,\Omega$) and $Z_1(s)$ as a resistance ($R_1 = 1000/K_1\,\Omega$). We select $K_1 = 1000$ so that the two resistances in the first stage are equal. Since the overall gain requires $K_1 \times K_2 = 3000$, this means that $K_2 = 3$. The second-stage transfer function can also be produced using an inverting OP AMP circuit:

$$-\frac{3s}{s+4000} = -\frac{3}{1+4000/s} = -\frac{Z_2(s)}{Z_1(s)}$$

Equating numerators and denominators yields $Z_2(s) = R_2 = 3$ and $Z_1(s) = R_1 + 1/C_1 s = 1 + 4000/s$.

Figure 11–30 shows the cascade connection of the RC OP AMP circuits that realize each stage. The overall transfer function is noninverting because the cascade uses an even number of inverting stages. The chain rule applies here since the first stage has an OP AMP output. The circuit in Figure 11–30 is a prototype design because the values of resistance and capacitance are impractical. ∎

FIGURE 11–30

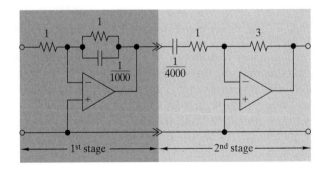

1ˢᵗ stage   2ⁿᵈ stage

MAGNITUDE SCALING

The circuits obtained in Examples 11–19 and 11–20 are called prototype designs because the element values are outside of practical ranges. The allowable ranges depend on the fabrication technology used to construct the circuits. For example, monolithic integrated circuit technology limits capacitances to a few hundred picofarads. An OP AMP circuit should have a feedback resistance greater than around 10 kΩ to keep the output current demand within the capabilities of general-purpose OP AMP devices. Other technologies and applications place different constraints on element values.

There are no hard and fast rules here, but, roughly speaking, a circuit is probably realizable by some means if its passive element values fall in the following ranges:

   Capacitors: 1 pF to 100 μF

   Inductors: 10 μH to 100 mH

   Resistors: 10 Ω to 100 MΩ.

The important idea here is that circuit designs like Figure 11–30 are impractical because 1-Ω resistors and 1-mF capacitors are unrealistic values.

It is often possible to scale the magnitude of circuit impedances so that the element values fall into practical ranges. The key is to scale the element values in a way that does not change the transfer function of the circuit. Multiplying the numerator and denominator of the transfer function of a voltage-divider circuit by a scale factor k_m yields

$$T_V(s) = \frac{k_m}{k_m} \frac{Z_2(s)}{Z_1(s) + Z_2(s)} = \frac{k_m Z_2(s)}{k_m Z_1(s) + k_m Z_2(s)} \qquad (11–33)$$

Clearly, this modification does not change the transfer function, but scales each impedance by a factor of k_m and changes the element values in the following way:

$$R_{after} = k_m R_{before} \quad L_{after} = k_m L_{before} \quad C_{after} = \frac{C_{before}}{k_m} \qquad (11–34)$$

Equation (11–34) was derived using the transfer function of a voltage-divider circuit. It is easy to show that we would reach the same conclusion if we had used the transfer functions of inverting or noninverting OP AMP circuits.

In general, a circuit is magnitude scaled by multiplying all resistances, multiplying all inductances, and dividing all capacitances by a scale factor k_m. The scale factor must be positive, but can be greater than or less than 1. Different scale factors can be used for each stage of a cascade design, but only one scale factor can be used for each stage. These scaling operations do not change the voltage transfer function realized by the circuit.

Our design strategy is first to create a prototype circuit whose element values may be unrealistically large or small. Applying magnitude scaling to the prototype produces a design with practical element values. Sometimes there may be no scale factor that brings the prototype element values into a practical range. When this happens, we must seek alternative realizations because the scaling process is telling us that the prototype is not a viable candidate.

EXAMPLE 11–21

Magnitude scale the circuit in Figure 11–30 so all resistances are at least 10 kΩ and all capacitances are less than 1 μF.

SOLUTION:

The resistance constraint requires $k_m R \geq 10^4$ Ω. The smallest resistance in the prototype circuit is 1 Ω; therefore, the resistance constraint requires $k_m \geq 10^4$. The capacitance constraint requires $C/k_m \leq 10^{-6}$ F. The largest capacitance in the prototype is 10^{-3} F; therefore, the capacitance constraint requires $k_m \geq 10^3$. The resistance condition on k_m dominates the two constraints. Selecting $k_m = 10^4$ produces the scaled design in Figure 11–31. This circuit realizes the same transfer function as the prototype in Figure 11–30, but uses practical element values.   ■

FIGURE   11 – 31

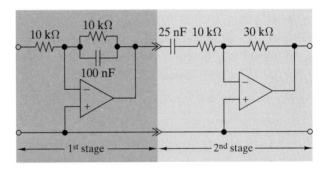

Exercise 11–11

Select a magnitude scale factor for each stage in Figure 11–28 so that both capacitances are 10 nF and all resistances are greater than 10 kΩ.

Answer: $k_m = 10^5$ for the first stage, $k_m = 10^4$ for the second stage, and $k_m = 0.25 \times 10^5$ for the third stage.

SECOND-ORDER CIRCUIT DESIGN

An *RLC* voltage divider can also be used to realize second-order transfer functions. For example, the transfer function

$$T_V(s) = \frac{K}{s^2 + 2\zeta\omega_0 s + \omega_0^2}$$

can be realized by factoring s out of the denominator and equating the result to the voltage-divider input-output relationship:

$$T_V(s) = \frac{K/s}{s + 2\zeta\omega_0 + \omega_0^2/s} = \frac{Z_2(s)}{Z_1(s) + Z_2(s)}$$

Equating numerators and denominators yields

$$Z_2(s) = \frac{K}{s} \quad \text{and} \quad Z_1(s) = s + 2\zeta\omega_0 + \frac{\omega_0^2 - K}{s}$$

The impedance $Z_2(s)$ is realizable as a capacitance ($C_2 = 1/K$ F) and $Z_1(s)$ as a series connection of an inductance ($L_1 = 1$ H), resistance ($R_1 = 2\zeta\omega_0$ Ω), and capacitance [$C_1 = 1/(\omega_0^2 - K)$ F]. The resulting voltage-divider circuit is shown in Figure 11–32(a). The impedances in this circuit are physically realizable when $K \le \omega_0^2$. Note that the resistance controls the damping ratio ζ because it is the element that dissipates energy in the circuit.

When $K > \omega_0^2$, we can partition the transfer function into a two-stage cascade of the form

$$T_V(s) = \underbrace{\left[\frac{K}{\omega_0^2}\right]}_{\substack{\text{First} \\ \text{Stage}}} \underbrace{\left[\frac{\omega_0^2/s}{s + 2\zeta\omega_0 + \omega_0^2/s}\right]}_{\substack{\text{Second} \\ \text{Stage}}}$$

The first stage requires a positive gain greater than unity and can be realized using a noninverting OP AMP circuit. The second stage can be realized as a voltage divider with $Z_2(s) = \omega_0^2/s$ and $Z_1(s) = s + 2\zeta\omega_0$. The resulting cascade circuit is shown in Figure 11–32(b).

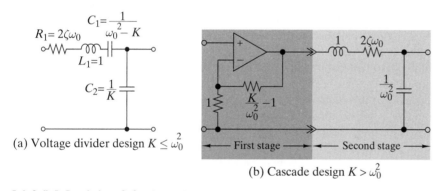

(a) Voltage divider design $K \le \omega_0^2$

(b) Cascade design $K > \omega_0^2$

FIGURE 11–32 *Second-order circuit realizations.*

D DESIGN EXAMPLE 11–22

Find a second-order realization of the transfer function given in Example 11–19.

SOLUTION:
The given transfer function can be written as

$$T_V(s) = \frac{3000s}{(s + 1000)(s + 4000)} = \frac{3000s}{s^2 + 5000s + 4 \times 10^6}$$

Factoring s out of the denominator and equating the result to the transfer function of a voltage divider gives

$$\frac{3000}{s + 5000 + 4 \times 10^6/s} = \frac{Z_2(s)}{Z_1(s) + Z_2(s)}$$

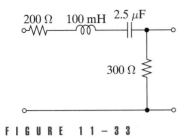

FIGURE 11 – 33

Equating the numerators and denominators yields

$$Z_2(s) = 3000 \quad \text{and} \quad Z_1(s) = s + 2000 + 4 \times 10^6/s$$

Both of these impedances are realizable, so a single-stage voltage-divider design is possible. The prototype impedance $Z_1(s)$ requires a 1-H inductor, which is a bit large. A more practical value is obtained using a scale factor of $k_m = 0.1$. The resulting scaled voltage divider circuit is shown in Figure 11–33. ∎

DESIGN EVALUATION SUMMARY

Examples 11–19, 11–20, and 11–22 show that there are several ways to realize a given transfer function, as summarized here:

EXAMPLE	FIGURE	DESCRIPTION	R	NUMBER OF L	C	OP AMP
11–19	11–28	RC voltage-divider cascade	4	0	2	1
11–20	11–30	RC inverting cascade	4	0	2	2
11–22	11–33	RLC voltage divider	2	1	1	0

Selecting a final design from among these alternatives involves evaluation using factors such as element count, element types, power requirements, and loading effects at the circuit input and output interfaces. For example, the *RLC* divider has the lowest element count but inductors can be heavy and lossy in low-frequency applications. The two *RC* OP AMP circuits have higher element counts but are preferred in these applications because they are "inductorless." The OP AMPs in these inductorless circuits place demands on the system dc power supply. The voltage divider cascade with only one OP AMP requires less dc power than the inverting cascade with two OP AMPs. On the other hand, the inverting cascade has an OP AMP output which means it is better suited to interfacing with a variety of output loads.

A design problem involves more than simply finding a prototype that realizes a given transfer function. In general, the first step in a design problem involves determining an acceptable transfer function, one that meets performance requirements such as the characteristics of the step or frequency response. In other words, we must first design the transfer function and then design several circuits that realize the transfer function. To deal with transfer function design we must understand how performance characteristics are related to transfer functions. The next two chapters provide some background on this issue.

D E DESIGN AND EVALUATION EXAMPLE: 11 – 23

Given the step response $g(t) = \pm[1 + 4e^{-500t}]u(t)$,

(a) Find the transfer function $T(s)$.
(b) Design two *RC* OP AMP circuits that realize the $T(s)$ found in part (a).
(c) Compare the two designs on the basis of element count, input impedance, output impedance.

SOLUTION:

(a) The transform of the step response is

$$G(s) = \pm\mathcal{L}\{[1 + 4e^{-500t}]u(t)\} = \pm\left[\frac{1}{s} + \frac{4}{s + 500}\right] = \pm\frac{5s + 500}{s(s + 500)}$$

and the required transfer function is

$$T(s) = H(s) = sG(s) = \pm\frac{5s + 500}{s + 500}$$

(b) The first design uses an inverting OP AMP configuration. Using the minus sign on the transfer function $T(s)$ and factoring an s out of the numerator and denominator yield

$$-T(s) = -\frac{5 + 500/s}{1 + 500/s} = -\frac{Z_2(s)}{Z_1(s)}$$

Equating numerators and denominators yields $Z_2(s) = 5 + 500/s$ and $Z_1(s) = 1 + 500/s$. The impedance $Z_2(s)$ is realizable as a resistance ($R_2 = 5\ \Omega$) in series with a capacitance ($C_2 = 1/500$ F) and $Z_1(s)$ as a resistance ($R_1 = 1\ \Omega$) in series with a capacitance ($C_1 = 1/500$ F). Using a magnitude scale factor $k_m = 10^5$ produces circuit C1 in Figure 11–34.

The second design uses a noninverting OP AMP configuration. Using the plus sign on the transfer function $T(s)$ and factoring an s out of the numerator and denominator yield

$$T(s) = \frac{5 + 500/s}{1 + 500/s} = \frac{Z_1(s) + Z_2(s)}{Z_1(s)}$$

Equating numerators and denominators yields

$$Z_1(s) = 1 + \frac{500}{s} \quad \text{and} \quad Z_2(s) = 5 + \frac{500}{s} - Z_1(s) = 4$$

The impedance $Z_1(s)$ is realizable as a resistance ($R_1 = 1\ \Omega$) in series with a capacitance ($C_1 = 1/500$ F) and $Z_2(s)$ as a resistance ($R_2 = 4\ \Omega$). Using a scale factor of $k_m = 10^4$ produces circuit C2 in Figure 11–34.

(c) Circuit C1 uses one more capacitor than circuit C2. The OP AMP output on both circuits means that they each have almost zero output impedance. The input impedance to circuit C2 is very large, because its input is the noninverting input of the OP AMP. The input impedance of circuit C1 is $Z_1(s) = k_m(1 + 500/s)$; hence, the scale factor must be selected to avoid loading the source circuit. The final design for circuit C1 in Figure 11–34 uses $k_m = 10^5$, which means that $|Z_1| > 100\ k\Omega$, which should be high enough to avoid loading the source circuit. ∎

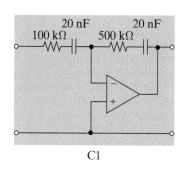

C1

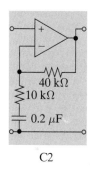

C2

FIGURE 11-34

D **DESIGN EXAMPLE 11-24**

Verify that circuit C2 in Figure 11–34 meets its design requirements.

SOLUTION:

One of the important uses of computer-aided analysis is to verify that a proposed design meets the performance specifications. The circuit C2 in Figure 11–34 is designed to produce a specified step response

$$g(t) = [1 + 4e^{-500t}]u(t) \quad \text{V}$$

This response jumps from zero to 5 V at $t = 0$ and then decays exponentially to 1 V at large t. The time constant of the exponential is $1/500 = 2$ ms, which means that the final value is effectively reached after about five time constants, or 10 ms.

Figure 11–35 shows the circuit diagram drawn in the Electronics Workbench circuit window. From the **Analysis** menu we bring up the **Transient Analysis** dialog box where we set the initial conditions to zero (to get a zero-state response), set the start time at zero, the end time at 10 ms (five time constants), and set node 4 (to OP AMP output) as the node for analysis. The step response calculated by Electronics Workbench is shown at the bottom of Figure 11–35. The response jumps to 5 V at $t = 0$ and then exponentially decays to 1 V after about 10 ms. The two cursors are set at $x_1 = 2.0079$ ms (about one time constant) where simulation predicts $y_1 = 2.4657$ and $x_2 = 4.0079$ ms (about two time constants) where simulation predicts $y_2 = 1.5392$. The theoretical values of $g(t)$ at these points are

$$g(t_1) = 1 + 4e^{-1} = 2.4715$$
$$g(t_2) = 1 + 4e^{-2} = 1.5413$$

FIGURE 11–35

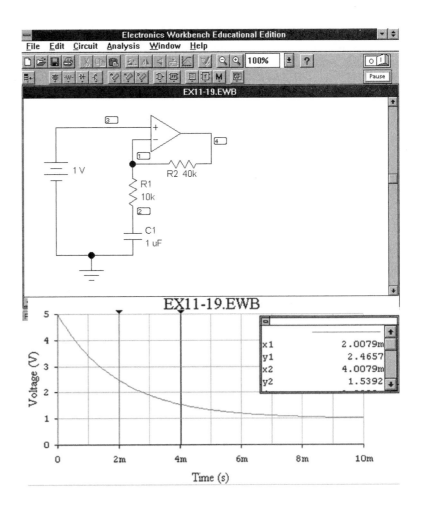

Thus, theory and simulation agree to three significant figures, confirming that circuit C2 meets the design requirement.   ∎

APPLICATION NOTE: EXAMPLE 11–25

The operation of a digital system is coordinated and controlled by a periodic waveform called a clock. The *clock waveform* provides a standard timing reference to maintain synchronization between signal processing results that are generated asynchronously. Because of differences in digital circuit delays, there must be agreed-upon instants of time at which circuit outputs can be treated as valid inputs to other circuits.

Figure 11–36 shows a section of the clock distribution network in an integrated circuit. In this network the clock waveform is generated at one point and distributed to other on-chip locations by interconnections that can be modeled as lumped resistors and capacitors. Clock distribution problems arise when the *RC* circuit delays at different locations are not the same. This delay dispersion is called *clock skew*, defined as the time difference between a clock edge at one location and the corresponding edge at another location.

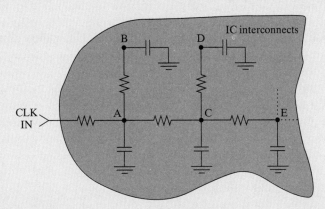

FIGURE 11–36 *Clock distribution network.*

To qualitatively calculate a clock skew, we will find the step responses in the *RC* circuit in Figure 11–37. The input $V_S(s)$ is a unit step function which simulates the leading edge of a clock pulse. The resulting step responses $V_A(s)$ and $V_B(s)$ represent the clock waveforms at points A and B in a clock distribution network. To find the step responses, we use the following *s*-domain node-voltage equations.

$$\text{Node A:} \quad \left(\frac{2}{R} + Cs\right)V_A(s) - \left(\frac{1}{R}\right)V_B(s) = \frac{V_S(s)}{R}$$

$$\text{Node B:} \quad -\left(\frac{1}{R}\right)V_A(s) + \left(\frac{1}{R} + Cs\right)V_B(s) = 0$$

The circuit determinant is

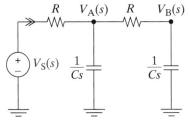

FIGURE 11–37 *Two-stage RC circuit model.*

$$\Delta(s) = \frac{(RCs)^2 + 3(RCs) + 1}{R^2} = \frac{(RCs + 0.382)(RCs + 2.618)}{R^2}$$

which indicates that the circuit has simple poles at $s = -0.382/RC$ and $s = -2.618/RC$. Using the circuit determinant and a unit step input, we can easily solve the node equations for $V_A(s)$ and $V_B(s)$ as

$$V_A(s) = \frac{RCs + 1}{s(RCs + 0.382)(RCs + 2.618)}$$

$$= \frac{1}{s} - \frac{0.7235}{s + 0.382/RC} - \frac{0.2764}{s + 2.618/RC}$$

$$V_B(s) = \frac{1}{s(RCs + 0.382)(RCs + 2.618)}$$

$$= \frac{1}{s} - \frac{1.171}{s + 0.382/RC} + \frac{0.1710}{s + 2.618/RC}$$

From which we obtain the time-domain step responses as

$$v_A(t) = 1 - 0.7235\,e^{-0.382t/RC} - 0.2764\,e^{-2.618t/RC}$$

$$v_B(t) = 1 - 1.171\,e^{-0.382t/RC} + 0.1710\,e^{-2.618t/RC} \quad \text{for} \quad t > 0$$

These two responses are plotted in Figure 11–38. For a unit step input, both responses have a final value of unity. Using the definition of step response *delay time* given in Example 11–12 (time required to reach 50% of the final value), we see that

$$T_{D_A} = 1.06/RC \quad \text{and} \quad T_{D_B} = 2.23/RC$$

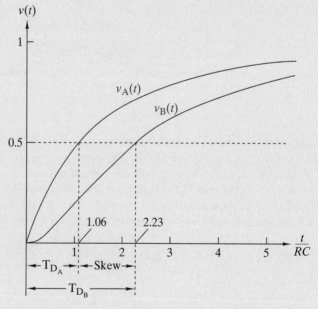

F I G U R E   1 1 – 3 8 *Step responses showing clock skew.*

The delay time skew is

$$\text{Delay skew} = T_{D_B} - T_{D_A} = 1.17/RC$$

The clock distribution problem is not that the RC elements representing the interconnects produce time delay, but that delays are not all the same. Ideally, digital devices at different locations should operate on their respective digital inputs at exactly the same instant of time. Erroneous results may occur when the clock pulse defining that instant does not arrive at all locations at the same time. Minimizing clock skew is one of the major constraints on the design of the clock distribution network in large-scale integrated circuits.

SUMMARY

- A network function is defined as the ratio of the zero-state response transform to the input transform. Network functions are either driving-point functions or transfer functions. Network functions are rational functions of s with real coefficients whose complex poles and zeros occur in conjugate pairs.

- Network functions for simple circuits like voltage and current dividers and inverting and noninverting OP AMPs are easy to derive and often useful. Node-voltage or mesh-current methods are used to find the network functions for more complicated circuits. The transfer function of a cascade connection obeys the chain rule when each stage does not load the preceding stage in the cascade.

- The impulse response is the zero-state response of a circuit for a unit impulse input. The transform of the impulse response is equal to the network function. The impulse response contains only natural poles and decays to zero in stable circuits. The impulse response of a linear, time-invariant circuit obeys the proportionality and time-shifting properties. The short pulse approximation is a useful way to simulate the impulse response in practical situations.

- The step response is the zero-state response of a circuit when the input is a unit step function. The transform of the step response transform is equal to the network function times $1/s$. The step response contains natural poles and a forced pole at $s = 0$ that leads to a dc steady-state response in stable circuits. The amplitude of the dc steady-state response can be found by evaluating the network function at $s = 0$. The step response waveform can also be found by integrating the impulse response waveform.

- The sinusoidal steady-state response is the forced response of a stable circuit for a sinusoidal input. With a sinusoidal input the response transform contains natural poles and forced poles at $s = \pm j\omega$ that lead to a sinusoidal steady-state response in stable circuits. The amplitude and phase angle of the sinusoidal steady-state response can be found by evaluating the network function at $s = j\omega$.

- The sinusoidal steady-state response can be found using phasor circuit analysis or directly from the transfer function. Phasor circuit analysis works best when the circuit is driven at only one frequency and several responses are needed. The transfer function method works best when the circuit is driven at several frequencies and only one response is needed.

- The convolution integral is a t-domain method relating the impulse response $h(t)$ and input waveform $x(t)$ to the zero-state response $y(t)$. Symbolically the convolution integral is represented by $y(t) = h(t)*x(t)$. Time-domain convolution and s-domain convolution are equivalent; that is, $y(t) = h(t)*x(t) = \mathscr{L}^{-1}\{H(s)X(s)\}$. The geometric interpretation of t-domain convolution involves four operations: reflecting, shifting, multiplying, and integrating.

- First- and second-order transfer functions can be designed using voltage dividers and inverting or noninverting OP AMP circuits. Higher-order transfer functions can be realized using a cascade connection of first- and second-order circuits. Prototype designs usually require magnitude scaling to obtain practical element values.

PROBLEMS

ERO 11–1 NETWORK FUNCTIONS (SECTS. 11–1, 11–2)

(a) Given a linear circuit, find specified network functions and locate their poles and zeros for specified element values.

(b) Given a linear circuit, find element values required to produce a given network function.

See Examples 11–2, 11–3, 11–4, 11–5, 11–6, 11–7 and Exercises 11–1, 11–2, 11–3

11–1 Connect a voltage $v_1(t)$ at the input port and an open circuit at the output port of the circuit in Figure P11–1.
  (a) Transform the circuit into the s domain and find the driving-point impedance $Z(s) = V_1(s)/I_1(s)$ and the transfer function $T_V(s) = V_2(s)/V_1(s)$.
  (b) Find the poles and zeros of $Z(s)$ and $T_V(s)$ for $R_1 = R_2 = 300\ \Omega$, $L = 50$ mH, $C = 4\ \mu$F.

11–2 Connect a current source $i_1(t)$ at the input port and an open circuit at the output port of the circuit in Figure P11–1.
  (a) Transform the circuit into the s domain and find the input impedance $Z(s) = V_1(s)/I_1(s)$ and the transfer function $T_Z(s) = V_2(s)/I_1(s)$.
  (b) Find the poles and zeros of $Z(s)$ and $T_Z(s)$ for $R_1 = R_2 = 1$ kΩ, $L = 100$ mH, $C = 50$ nF.

11–3 Connect a voltage source $v_1(t)$ at the input port and a short circuit at the output port of the circuit in Figure P11–3.
  (a) Transform the circuit into the s domain and find the driving-point impedance $Z(s) = V_1(s)/I_1(s)$ and the transfer function $T_Y(s) = I_2(s)/V_1(s)$.
  (b) Find the poles and zeros of $Z(s)$ and $T_Y(s)$ for $R_1 = 100\ \Omega$, $R_2 = 50\ \Omega$, $L_1 = 0.75$ H, and $L_2 = 0.25$ H.

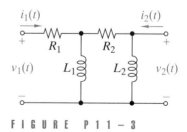

FIGURE P11–3

11–4 Connect a current source $i_1(t)$ at the input port and a short circuit at the output port of the circuit in Figure P11–3.

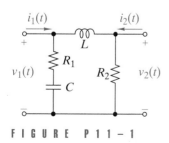

FIGURE P11–1

(a) Transform the circuit into the s domain and find the driving-point impedance $Z(s) = V_1(s)/I_1(s)$ and the transfer impedance $T_1(s) = I_2(s)/I_1(s)$.

(b) Find the poles and zeros of $Z(s)$ and $T_1(s)$ for $R_1 = 500\ \Omega$, $R_2 = 2000\ \Omega$, and $L_1 = 0.4$ H.

11–5 Transform the circuit in Figure P11-5 into the s domain and show that

$$T_V(s) = \frac{V_2(s)}{V_1(s)} = \frac{R_1C_1R_2C_2s^2 + (R_1C_1 + R_1C_2)s + 1}{R_1C_1R_2C_2s^2 + (R_1C_1 + R_2C_2 + R_1C_2)s + 1}$$

Locate the poles and zeros of $T_V(s)$ for $R_1 = 1$ kΩ, $R_2 = 4$ kΩ, $C_1 = 1.5\ \mu$F, and $C_2 = 0.5\ \mu$F.

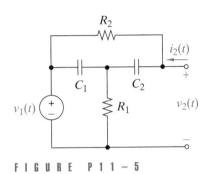

FIGURE P11–5

11–6 Show that the impedance seen by the voltage source in Figure P11–5 is

$$Z(s) = \frac{R_1C_1R_2C_2s^2 + (R_1C_1 + R_1C_2 + R_2C_2)s + 1}{s(R_2C_1C_2s + C_1 + C_2)}$$

Locate the poles and zeros of $Z(s)$ for $R_1 = 1$ kΩ, $R_2 = 2$ kΩ, $C_1 = 0.5\ \mu$F, and $C_2 = 0.25\ \mu$F.

11–7 The voltage transfer function of the circuit in Figure P11–7 is

$$T_V(s) = \frac{V_2(s)}{V_1(s)} = \frac{s + 400}{s + 200}$$

For $R = 10$ kΩ, find the value of C.

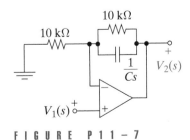

FIGURE P11–7

11–8 The current transfer function of the circuit in Figure P11–8 is

$$T_1(s) = \frac{I_2(s)}{I_1(s)} = \frac{200}{s + 400}$$

For $R = 100\ \Omega$, find the value of L and the input impedance $Z(s)$.

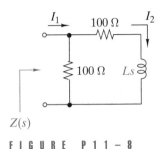

FIGURE P11–8

11–9 Transform the circuit in Figure P11–9 into the s domain and solve for the transfer function $T_V(s) = V_2(s)/V_1(s)$. Locate the poles and zeros of the transfer function.

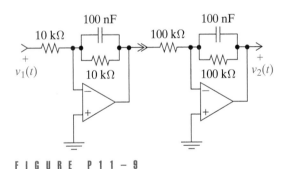

FIGURE P11–9

11–10 The input impedance of the circuit in Figure P11–10 is $Z(s) = 10^4(s + 200)/(s + 100)\ \Omega$. Find the value of C.

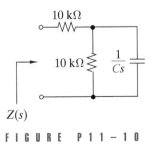

FIGURE P11–10

ERO 11–2 NETWORK FUNCTIONS AND IMPULSE AND STEP RESPONSES (SECT. 11–3, 11–4)

(a) Given a first- or second-order linear circuit, find its impulse or step response.
(b) Given the impulse or step response of a linear circuit, find the network functions.
(c) Given the impulse (step) response of a linear circuit, find the step (impulse) response.

See Examples 11–8, 11–9, 11–10, 11–11 and Exercises 11–4, 11–5, 11–6, 11–7

11–11   Find $v_2(t)$ in Figure P11–11 when $v_1(t) = \delta(t)$.

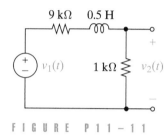

FIGURE   P 1 1 – 1 1

11–12   Find $v_2(t)$ in Figure P11–12 when $v_1(t) = \delta(t)$.

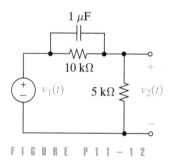

FIGURE   P 1 1 – 1 2

11–13   Find $v_2(t)$ in Figure P11–13 when $v_1(t) = u(t)$.

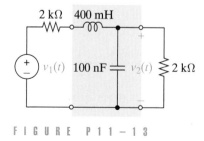

FIGURE   P 1 1 – 1 3

11–14   Find $v_2(t)$ in Figure P11–14 when $v_1(t) = u(t)$.

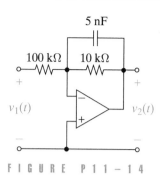

FIGURE   P 1 1 – 1 4

11–15   Find $v_2(t)$ in Figure P11–15 when $v_1(t) = \delta(t)$.

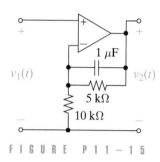

FIGURE   P 1 1 – 1 5

11–16   The impulse reponse of a linear circuit is $h(t) = 15(e^{-10t} - e^{-30t})u(t)$. Find the circuit step response $g(t)$.

11–17   Find the transfer function and the impulse-response waveform corresponding to each of the following step responses.
(a) $g(t) = [-e^{-2000t}]u(t)$
(b) $g(t) = [1 - e^{-2000t}]u(t)$

11–18   Find the step response waveforms corresponding to each of the following impulse responses.
(a) $h(t) = [-1000e^{-2000t}]u(t)$
(b) $h(t) = \delta(t) - [2000e^{-2000t}]u(t)$

11–19   The step response of a linear circuit is $g(t) = [e^{-2000t}\sin 5000t]u(t)$. Find the poles and zeros of the transfer function.

11–20   A one-port circuit is driven by a unit impulse of current and the voltage response is found to be $v(t) = \delta(t) - 200[e^{-200t}\sin(200t)]u(t)$ V. Find the current response when the one-port is driven by a unit impulse of voltage.

ERO 11–3 NETWORK FUNCTIONS AND SINUSOIDAL STEADY-STATE RESPONSE (SECT. 11–5)

(a) Given a first- or second-order linear circuit with a specified input sinusoid, find the sinusoidal steady-state response.
(b) Given the network function, impulse or step response, find the sinusoidal steady-state response for a specified input sinusoid.

See Examples 11–13, 11–14 and Exercises 11–8, 11–9

11–21 Find the amplitude and phase angle of the steady-state output $v_{2SS}(t)$ of the circuit in Figure P11–21 for $v_1(t) = 5 \cos 500t$ V. Repeat for $v_1(t) = 10 \cos 1000t$ V.

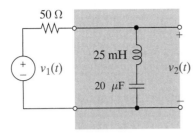

FIGURE   P11–21

11–22 Find the amplitude and phase angle of the steady-state output $v_{2SS}(t)$ of the circuit in Figure P11–22 for $v_1(t) = 10 \cos 2000t$ V. Repeat for $v_1(t) = 3 \cos 4000t$ V.

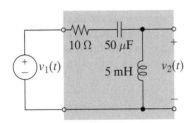

FIGURE   P11–22

11–23 For $R_1 = 100 \ \Omega$, $R_2 = 400 \ \Omega$, and L = 100 mH in Figure P11–23, find the amplitude and phase angle of the steady-state output $i_{2SS}(t)$ for $i_1(t) = 5 \cos 5000t$ mA. Repeat for $i_1(t) = 10 \cos 10^4 t$ mA.

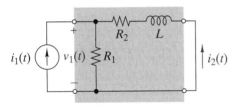

FIGURE   P11–23

11–24 For $R = 1 \ k\Omega$, L = 2 H, and C = 500 nF in Figure P11–24, find the amplitude and phase angle of the steady-state output $i_{2SS}(t)$ for $i_1(t) = 5 \cos 1000t$ mA. Repeat for $i_1(t) = 10 \sin 1000t$ mA.

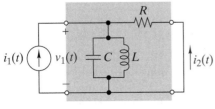

FIGURE   P11–24

11–25 Find the amplitude and phase angle of the steady-state output $v_{2SS}(t)$ of the circuit in Figure P11–25 for $i_1(t) = 10 \cos 5000t$ mA. Repeat for $i_1(t) = 5 \cos 2500t$ mA.

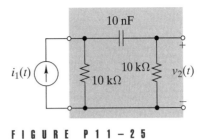

FIGURE   P11–25

11–26 The transfer function of a linear circuit is $T(s) = (s + 200)/(s + 50)$. Find the amplitude and phase angle of the sinusoidal steady-state output for the sinusoidal input $x(t) = 5 \cos 200t$.

11–27 The step response of a linear circuit is $g(t) = [e^{-1000t}]u(t)$. Find the amplitude and phase angle of the sinusoidal steady-state output for the sinusoidal input $x(t) = 5 \cos 2000t$.

11–28 The step response of a linear circuit is $g(t) = [2e^{-100t} - 1]u(t)$. Show that the amplitude of the sinusoidal steady-state output for $x(t) = 5 \cos \omega t$ is 5 V regardless of the frequency.

11–29 The impulse response of a linear circuit is $h(t) = [800e^{-1000t}]u(t) - \delta(t)$. Find the amplitude and phase angle of the sinusoidal steady-state output for the sinusoidal input $x(t) = 3 \sin 400t$.

11–30 The impulse response of a linear circuit is $h(t) = 400[e^{-100t} - e^{-5000t}]u(t)$. Find the amplitude and phase angle of the sinusoidal steady-state output for the sinusoidal input $x(t) = 5 \cos 700t$.

ERO 11–5 NETWORK FUNCTIONS AND CONVOLUTION (SECT. 11–6)

(a) Given the impulse response of a linear circuit, use the convolution integral to find the response to a specified input.

(b) Use the convolution integral to derive properties of linear circuits.

See Examples 11–15, 11–16, 11–17 and Exercise 11–10

11–31 The impulse response of a linear circuit is $h(t) = [u(t) - u(t - 1)]$. Use the convolution integral to find the response due to an input $x(t) = u(t - 1)$.

11–32 The impulse response of a linear circuit is $h(t) = t[u(t) - u(t - 1)]$. Use the convolution integral to find the response due to an input $x(t) = u(t - 1)$.

11–33 The impulse response of a linear circuit is $h(t) = e^{-t}u(t)$. Use the convolution integral to find the response due to an input $x(t) = u(t - 1)$.

11–34 Repeat Problem 11–33 for $x(t) = e^{-t}u(t)$.

11–35 The impulse response of a linear circuit is $h(t) = e^{-5t}u(t)$. Use the convolution integral to find the response due to an input $x(t) = tu(t)$.

11–36 The impulse response of a linear circuit is $H(s) = 1/(s + 5)$ and $x(t) = tu(t)$. Use s-domain convolution to find the zero-state response $y(t)$.

11–37 Use the convolution integral to find the $f(t)$ corresponding to

$$F(s) = \frac{s}{(s + 1)(s + 2)}$$

11–38 A circuit whose impulse response is a step function operates as an integrator. Use the convolution integral to show that if $h(t) = u(t)$ then the zero-state output for any input $x(t)$ is

$$y(t) = \int_0^t x(\tau)d\tau$$

11–39 In Section 11–4 we found that the step response is the integral of the impulse response. Use the convolution integral to show that if $x(t) = u(t)$ then the zero-state output is

$$y(t) = \int_0^t h(\tau)d\tau$$

11–40 The impulse responses of two linear circuits are $h_1(t) = e^{-2t}u(t)$ and $h_2(t) = 4e^{-4t}u(t)$. What is the impulse response of a cascade connection of these two circuits?

ERO 11–5 GENERAL RESPONSES (SECTS. 11–3, 11–4, 11–5, 11–6)

Given the impulse or step response of a linear circuit, find the response due to other specified inputs.

11–41 The impulse response of a linear circuit is $h(t) = 1000[e^{-1000t}]u(t)$. Find the output waveform when the input is $x(t) = 5tu(t)$ V.

11–42 The step response of a linear circuit is $g(t) = 0.5[1 - e^{-100t}]u(t)$. Find the output waveform when the input is $x(t) = [e^{-200t}]u(t)$.

11–43 The step response of a linear circuit is $g(t) = 20[tu(t) - (t - 5)u(t - 5)]$. Find the impulse response waveform.

11–44 The impulse response of a linear circuit is $H(s) = (s + 2000)/(s + 1000)$. Find the response waveform when the input is $x(t) = 5e^{-1000t}u(t)$.

11–45 The impulse response of a linear circuit is $h(t) = u(t) - u(t - 10)$. Find the step response.

11–46 The response of a linear circuit due to an input $x(t) = tu(t)$ is $y(t) = [e^{-200t}]u(t)$. Show that $g(t) = dy(t)/dt$. Explain.

11–47 The impulse response of a linear circuit is $H(s) = s/(s + 200)$. For a certain input the response waveform is $y(t) = [e^{-100t}]u(t)$. What is the input waveform?

11–48 The impulse response of a linear circuit is $h(t) = 200u(t) + \delta(t)$. Find the response transform when the input is $x(t) = [e^{-200t}]u(t)$.

11–49 The step response of a linear circuit is $g(t) = 10[e^{-1000t}\sin 2000t]u(t)$. Find the sinusoidal steady-state response for the input $x(t) = 10 \cos 2000t$.

11–50 The impulse response of a linear circuit is $h(t) = \delta(t - 0.002)$. Find the circuit response waveform when the input is $x(t) = 10[e^{-1000t}]u(t)$.

ERO 11–6 NETWORK FUNCTION DESIGN (SECT. 11–7)

(a) Design a circuit that realizes a given $T(s)$ and meets other stated constraints.
(b) Evaluate alternative designs using stated criteria.

11–51 **D** Design a circuit to realize the transfer function below using only resistors, capacitors, and OP AMPs. Scale the circuit so that all resistors are greater than 10 kΩ and all capacitors are less than 1 μF.

$$T_V(s) = \pm \frac{10^5}{(s + 200)(s + 2500)}$$

11–52 **D** Design a circuit to realize the transfer function below using only resistors, capacitors, and not more than one OP AMP. Scale the circuit so that all capacitors are exactly 100 pF.

$$T_V(s) = \pm \frac{100(s + 500)}{(s + 200)(s + 2500)}$$

11–53 **D** Design a circuit to realize the following transfer function using only resistors, capacitors, and no more than one OP AMP. Scale the circuit so that the final design uses only 10-kΩ resistors.

$$T_V(s) = \pm \frac{1000 s}{(s + 500)(s + 2000)}$$

11–54 **D** Design a circuit to realize the transfer function below using only resistors, capacitors, and inductors (no OP AMPs allowed). Scale the circuit so that all inductors are 10 mH or less.

$$T_V(s) = \frac{s^2}{(s + 1000)(s + 4000)}$$

11–55 **D** Design a circuit to realize the following transfer function using only resistors, capacitors, and not more than one OP AMP. Scale the circuit so that all resistors are greater than 10 kΩ and all capacitors are less than 1 μF.

$$T_V(s) = \pm \frac{(s + 100)(s + 1000)}{(s + 200)(s + 500)}$$

11–56 **D** Design a circuit to realize the transfer function below using only resistors, capacitors, and OP AMPs. Scale the circuit so that all capacitors are exactly 100 nF.

$$T_V(s) = - \frac{400 (s + 100)}{s (s + 200)}$$

11–57 **D** Design a circuit to realize the following transfer function using practical element values.

$$T_V(s) = \pm \frac{100 s + 10^6}{s^2 + 100 s + 10^6}$$

11–58 **E** It is claimed that both circuits in Figure P11–58 realize the transfer function

$$T_V(s) = K\left(\frac{s + 2000}{s + 1000}\right)$$

(a) Verify that both circuits realize the specified $T_V(s)$.
(b) Which design would you choose if the output must drive a 1 kΩ load?
(c) Which design would you choose if the input comes from a 50 Ω source?
(d) It is further claimed that connecting the two circuits in cascade produces an overall transfer function of $[T_V(s)]^2$ no matter which circuit is the first stage and which is the second stage. Do you agree or disagree? Explain.

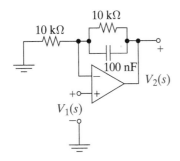

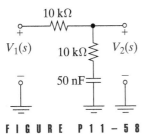

FIGURE P11–58

11–59 **E** It is claimed that both circuits in Figure P11–59 realize the transfer function

$$T_V(s) = \frac{\pm 1000 s}{(s + 1000)(s + 4000)}$$

(a) Verify that both circuits realize the specified $T_V(s)$.
(b) Which design would you choose if the output must drive a 1 kΩ load?
(c) Which design would you choose if the input comes from a 50 Ω source?
(d) It is further claimed that connecting the two circuits in cascade produces an overall transfer function of $[T_V(s)]^2$ no matter which circuit is the first stage and which is the second stage. Do you agree or disagree? Explain.

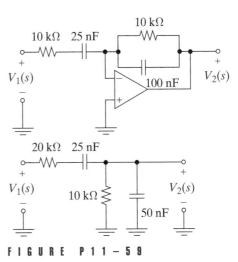

FIGURE P11–59

11–60 **D** Design a circuit that produces the following step response.

$$g(t) = [1 - e^{-50t} - 50 t e^{-50t}]u(t)$$

INTEGRATING PROBLEMS

11–61 **A D** FIRST-ORDER CIRCUIT IMPULSE AND STEP RESPONSES

Each row in the table shown in Figure P11–61 refers to a first-order circuit with an impulse response $h(t)$ and a step response $g(t)$. Fill in the missing entries in the table.

Circuit	$h(t)$	$g(t)$
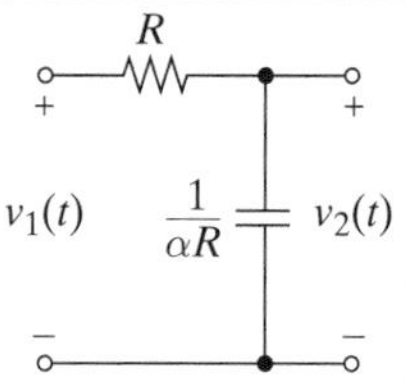 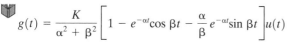		
	$\delta(t) - [\alpha\,e^{-\alpha t}]u(t)$	
		$\left(1 + \dfrac{e^{-\alpha t}}{2}\right)u(t)$

FIGURE   P11–61

11–62   **A D** SECOND-ORDER CIRCUIT STEP RESPONSE

The step response in Figure P11–62 is of the form

$$g(t) = \frac{K}{\alpha^2 + \beta^2}\left[1 - e^{-\alpha t}\cos\beta t - \frac{\alpha}{\beta}e^{-\alpha t}\sin\beta t\right]u(t)$$

(a) **A** Estimate values for the parameters K, α, and β from the plot of $g(t)$.

(b) **A** Find the transfer function $T(s)$ corresponding to $g(t)$.

(c) **D** Design a circuit that realizes the $T(s)$ found in (b).

(d) **D** Use computer-aided circuit analysis to verify your design.

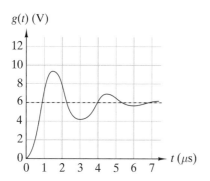

FIGURE   P11–62

11–63   **A** CIRCUIT STABILITY AND STEP RESPONSE

The step response of a circuit is $g(t) = [2e^{-\alpha t} - 1]u(t)$.

(a) Find the circuit transfer function. Is the circuit stable?

(b) For a certain input $x(t)$, the output is $y(t) = [e^{-2\alpha t}]u(t)$. Find one possible input.

(c) Is the input found in (b) bounded? If not, does this change your answer in part (a)?

11–64   **A** STEP RESPONSE AND FAN-OUT

The fan-out of a digital device is defined as the maximum number of inputs to similar devices that can be reliably driven by the device output. Figure P11–64 is a simplified diagram of a device output driving n identical capacitive inputs. To operate reliably, a 5-V step function at the device output must drive the capacitive inputs to 3.7 V in 10 ns or less. Determine the device fan-out for $R = 1\ \text{k}\Omega$ and $C = 3\ \text{pF}$.

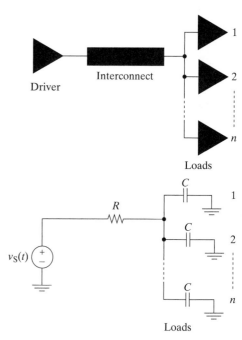

FIGURE   P11–64

11–65   **A** SINUSOIDAL STEADY-STATE RESPONSE

The input to the RL circuit in Figure P11–65 is $v_1(t) = V_A \cos\omega t$.

(a) Use the s-domain network function method in Chapter 11 to find the sinusoidal steady-state response $i_{SS}(t)$.

(b) Formulate the circuit differential equation and use the classical methods of Chapter 7 to find the forced component of $i(t)$.

(c) The responses found in (a) and (b) should be identical. Which method do you think is easiest to apply? Discuss.

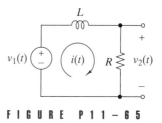

FIGURE   P11–65

CHAPTER 12

FREQUENCY RESPONSE

The advantage of the straight-line approximation is, of course, that it reduces the complete characteristics to a sum of elementary characteristics.

Hendrik W. Bode, 1945,
American Engineer

12-1 Frequency-response Descriptors

12-2 First-order Circuit Frequency Response

12-3 Second-order Circuit Frequency Response

12-4 The Frequency Response of *RLC* Circuits

12-5 Bode Diagrams

12-6 Bode Diagrams with Complex Critical Frequencies

12-7 Frequency Response and Step Response

Summary

Problems

Integrating Problems

In the sinusoidal steady state, both the input and the output of a stable linear circuit are sinusoids of the same frequency. The term *frequency response* refers to the frequency-dependent relation, in both gain and phase, between a sinusoidal input and the resulting sinusoidal steady-state output. The gain and phase relationships can be presented as equations or graphs. When presented as separate graphs of gain and phase versus logarithmic frequency, the result is commonly called a **Bode diagram**.

Hendrick Bode (pronounced Bow Dee) spent most of his distinguished career as a member of the technical staff of the Bell Telephone Laboratories. In the 1930s Bode made major contributions to circuit and feedback amplifier theory as part of the effort to develop the precise frequency-selective amplifiers required by long-distance telephone systems. In describing his work Bode used straight-line approximations as an aid to understanding and simplifying the slide rule calculations required in his day. With the advent of digital computers, the computational benefits of Bode diagrams are no longer important. His straight-line method is still useful today, however, because it allows us to quickly visualize how the poles and zeros of a transfer function affect the frequency response of a circuit. Thus, Bode diagrams are a conceptual tool that simplifies the analysis and design of circuits and systems, just as they did in Bode's day.

The first section of this chapter defines the parameters used to characterize the frequency response. The frequency response of first- and second-order circuits is discussed in the following two sections. The fourth section describes how the frequency-response characteristics of series and parallel *RLC* circuits can be traced to variation in the impedance of specific elements in the circuit. The formal treatment of Bode diagrams in the fifth and sixth sections generalizes the results obtained using basic first- and second-order circuits. The final section discusses the relationship between frequency response and the time-domain step response.

12–1 FREQUENCY-RESPONSE DESCRIPTORS

In Chapter 11, we found that replacing s by $j\omega$ allows us to find the steady-state response of a stable circuit directly from its transfer function. The circuit transfer function influences the sinusoidal steady-state response through the **gain** function $|T(j\omega)|$ and **phase** function $\theta(\omega)$.

$$\text{Output amplitude} = |T(j\omega)| \times (\text{Input amplitude})$$

$$\text{Output phase} = \text{Input phase} + \theta(\omega)$$

Taken together, the gain and phase functions show how the circuit modifies the input amplitude and phase angle to produce the output sinusoid. These two functions define the **frequency response** of the circuit since they are frequency-dependent functions that relate the sinusoidal steady-state input and output. The gain and phase functions can be expressed mathematically or presented graphically as the frequency response plots in Figure 12–1.

The terminology used to describe the frequency response of circuits and systems is based on the form of the gain plot. For example, at high fre-

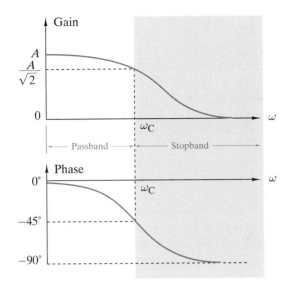

FIGURE   12 – 1   *Frequency   response plots.*

quencies the gain in Figure 12–1 falls off so that output signals in this frequency range are reduced in amplitude. The range of frequencies over which the output is significantly attenuated is called the **stopband**. At low frequencies the gain is essentially constant and there is relatively little attenuation. The frequency range over which there is little attenuation is called a **passband**. The frequency associated with the boundary between a passband and an adjacent stopband is called the **cutoff frequency** ($\omega_C = 2\pi f_C$). In general, the transition from the passband to the stopband is gradual, so the precise location of the cutoff frequency is a matter of definition. The most widely used definition specifies the cutoff frequency to be the frequency at which the gain has decreased by a factor of $1/\sqrt{2} = 0.707$ from its maximum value in the passband.

Again this definition is arbitrary, since there is no sharp boundary between a passband and an adjacent stopband. However, the definition is motivated by the fact that the power delivered to a resistance by a sinusoidal current or voltage waveform is proportional to the square of its amplitude. At a cutoff frequency the gain is reduced by a factor of $1/\sqrt{2}$ and the square of the output amplitude is reduced by a factor of one-half. For this reason the cutoff frequency is also called the **half-power frequency**.

When the gain response is plotted on log-log scales, we obtain the four prototype gain characteristics shown in Figure 12–2. A **low-pass** gain characteristic has a single passband extending from zero frequency (dc) to the cutoff frequency. A **high-pass** gain characteristic has a single passband extending from the cutoff frequency to infinite frequency. A **bandpass** gain has a single passband with two cutoff frequencies, neither of which is zero or infinite. Finally, the **bandstop** gain has a single stopband with two cutoff frequencies, neither of which is zero or infinite.

The **bandwidth** of a gain characteristic is defined as the frequency range spanned by its passband. The bandwidth of a low-pass circuit is equal to its cutoff frequency ($B = \omega_C$). The bandwidth of a high-pass characteristic is infinite since passband extends to infinity. For the bandpass and bandstop

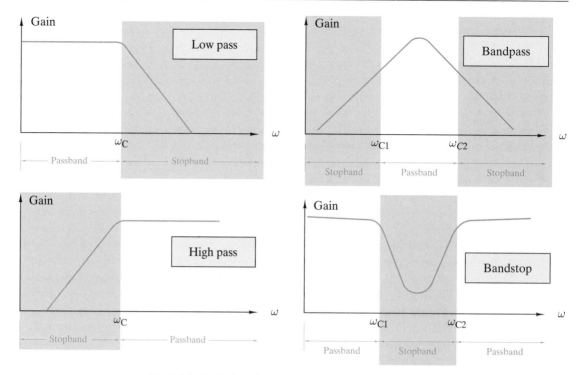

cases in Figure 12–2, the bandwidth is the difference in the two cutoff frequencies:

$$B = \omega_{C2} - \omega_{C1} \qquad (12\text{–}1)$$

For the bandstop case Eq. (12–1) defines the width of the stopband rather than the passbands.

The gain responses in Figure 12–2 have different characteristics at zero and infinite frequency—namely,

	GAIN AT	
	$\omega = 0$	$\omega = \infty$
Low pass	Finite	0
High pass	0	Finite
Bandpass	0	0
Bandstop	Finite	Finite

Since these extreme values form a unique pattern, the type of gain response can be inferred from the values of $|T(0)|$ and $|T(\infty)|$. These endpoint values, in turn, are usually determined by the impedance of capacitors and inductors in the circuit. In the sinusoidal steady state the impedances of these elements are

$$Z_C(j\omega) = \frac{1}{j\omega C} \quad \text{and} \quad Z_L(j\omega) = j\omega L \qquad (12\text{–}2)$$

These impedances form a unique pattern at zero and infinite frequency—namely,

	IMPEDANCE AT	
	$\omega = 0$	$\omega = \infty$
Capacitor	Infinite	0
Inductor	0	Infinite

Thus, the capacitor acts like an open circuit at zero frequency and a short circuit at infinite frequency, while the inductor acts like a short circuit at dc and an open circuit at infinite frequency. These observations often allow us to infer the type of gain response directly from the circuit itself without finding the transfer function.

When frequency-response plots use logarithmic scales for the frequency variable, they are normally referred to as **Bode diagrams**. Logarithmic scales are used to compress the data range because the frequency ranges of interest often span several orders of magnitude. The use of a logarithmic frequency scale involves some special terminology. Any frequency range whose end points have a 2:1 ratio is called an **octave**. Any range whose end points have a 10:1 ratio is called a **decade**. For example, the frequency range from 10 Hz to 20 Hz is one octave, as is the range from 20 MHz to 40 MHz. The standard UHF (ultra high frequency) band spans one decade from 0.3 to 3 GHz.

In Bode diagrams the gain $|T(j\omega)|$ is usually expressed in **decibels** (dB), defined as

$$|T(j\omega)|_{dB} = 20 \log_{10}|T(j\omega)| \qquad (12\text{–}3)$$

To construct and interpret frequency-response plots, you must have some familiarity with gains expressed in decibels. First note that the gain in dB can be either positive, negative, or zero. A gain of zero dB means that $|T(j\omega)| = 1$ (i.e., the input and output amplitudes are equal). When $|T(j\omega)| > 1$ the output amplitude exceeds the input and $|T(j\omega)|_{dB} > 0$. When $|T(j\omega)| < 1$ the output amplitude is less than the input and $|T(j\omega)|_{dB} < 0$. A cutoff frequency occurs when the gain is reduced from its maximum passband value by a factor $1/\sqrt{2}$. Expressed in dB, this is a gain reduction of

$$20 \log_{10}\left(\frac{1}{\sqrt{2}}|T|_{MAX}\right) = 20 \log_{10}|T_{MAX}| - 20 \log_{10}\sqrt{2}$$

$$\approx |T|_{MAX,\, dB} - 3 \text{ dB}$$

That is, cutoff occurs when the dB gain is reduced by about 3 dB. For this reason the cutoff is also called the **3-dB down frequency**.

The generalization of this idea is that multiplying $|T(j\omega)|$ by a factor K means $20 \times \log(K|T(j\omega)|) = |T(j\omega)|_{dB} + 20 \log(K)$. That is, multiplying the gain by K changes the gain in dB by an additive factor K_{dB}. Some multiplicative factors whose dB equivalents are worth remembering are as follows:

- A multiplicative factor $K = 1$ changes $|T(j\omega)|_{dB}$ by 0 dB.
- A multiplicative factor $K = \sqrt{2}$ $(1/\sqrt{2})$ changes $|T(j\omega)|_{dB}$ by about +3 dB (–3 dB).
- A multiplicative factor $K = 2$ $(1/2)$ changes $|T(j\omega)|_{dB}$ by about +6 dB (–6 dB).
- A multiplicative factor $K = 10$ $(1/10)$ changes $|T(j\omega)|_{dB}$ by +20 dB (–20 dB).
- A multiplicative factor $K = 30$ $(1/30)$ changes $|T(j\omega)|_{dB}$ by about +30 dB (–30 dB).
- A multiplicative factor $K = 100$ $(1/100)$ changes $|T(j\omega)|_{dB}$ by +40 dB (–40 dB).
- A multiplicative factor $K = 1000$ $(1/1000)$ changes $|T(j\omega)|_{dB}$ by +60 dB (–60 dB).

In summary, Bode diagrams use logarithmic frequency scales and linear scales for the gain (in dB) and phase (in radians or degrees). Both plots are semilog graphs, although in effect the gain plot is log/log because of the logarithmic definition of $|T(j\omega)|_{dB}$. To create frequency-response plots we must calculate $|T(j\omega)|$ and $\theta(\omega)$ at a sufficient number of frequencies to define adequately the stop and passband characteristics. The most efficient way to generate accurate plots is to use one of the many computer-aided analysis tools. Spreadsheets and math analysis programs like Mathcad can be used to calculate and plot frequency-response curves. Circuit analysis programs such as SPICE have ac analysis options that generate the frequency-response plots. We will illustrate the use of these tools in several examples in this chapter.

As always, some knowledge of the expected response is required to use computer-aided circuit analysis tools effectively. Using computer tools to generate frequency-response plots requires at least some preliminary analysis or a rough sketch of the expected response. To develop the required insight, we treat first- and second-order circuits and then use Bode plots to show how these building blocks combine to produce the frequency response of more complicated circuits.

APPLICATION NOTE: EXAMPLE 12–1

The use of the decibel as a measure of performance pervades the literature and folklore of electrical engineering. The decibel originally came from the definition of power ratios in **bels**.[1]

$$\text{Number of bels} = \log_{10}\frac{P_{\text{OUT}}}{P_{\text{IN}}}$$

---

1 The name of the unit honors Alexander Graham Bell (1847–1922), the inventor of the telephone.

The decibel (dB) is more commonly used in practice. The number of decibels is 10 times the number of bels:

$$\text{Number of dB} = 10 \times (\text{Number of bels}) = 10 \log_{10} \frac{P_{\text{OUT}}}{P_{\text{IN}}}$$

When the input and output powers are delivered to equal input and output resistances R, then the power ratio can be expressed in terms of voltages across the resistances.

$$\text{Number of dB} = 10 \log_{10} \frac{v_{\text{OUT}}^2/R}{v_{\text{IN}}^2/R} = 20 \log_{10} \frac{v_{\text{OUT}}}{v_{\text{IN}}}$$

or in terms of currents through the resistances:

$$\text{Number of dB} = 10 \log_{10} \frac{i_{\text{OUT}}^2 \times R}{i_{\text{IN}}^2 \times R} = 20 \log_{10} \frac{i_{\text{OUT}}}{i_{\text{IN}}}$$

The definition of gain in dB in Eq. (12–3) is consistent with these results, since in the sinusoidal steady state the transfer function equals the ratio of output amplitude to input amplitude. The preceding discussion is not a derivation of Eq. (12–3) but simply a summary of its historical origin. In practice Eq. (12–3) is applied when the input and output are not measured across resistances of equal value.

When the chain rule applies to a cascade connection, the overall transfer function is a product

$$T(j\omega) = T_1 \times T_2 \times \cdots T_N$$

where $T_1, T_2, \ldots T_N$ are the transfer functions of the individual stages in the cascade. Expressed in dB, the overall gain is

$$|T(j\omega)|_{\text{dB}} = 20 \log_{10}(|T_1| \times |T_2| \times \cdots |T_N|)$$
$$= 20 \log_{10}|T_1| + 20 \log_{10}|T_2| + \cdots + 20 \log_{10}|T_N|$$
$$= |T_1|_{\text{dB}} + |T_2|_{\text{dB}} + \cdots + |T_N|_{\text{dB}}$$

Because of the logarithmic definition, the overall gain (in dB) is the sum of the gains (in dB) of the individual stages in a cascade connection. The effect of altering a stage or adding an additional stage can be calculated by simply adding or subtracting the change in dB. Since summation is simpler than multiplication, the enduring popularity of the dB comes from its logarithmic definition and not its somewhat tenuous relationship to power ratios.

12–2   FIRST-ORDER CIRCUIT FREQUENCY RESPONSE

FIRST-ORDER LOW-PASS RESPONSE

We begin the study of frequency response with the first-order low-pass transfer function:

$$T(s) = \frac{K}{s + \alpha} \qquad\qquad (12\text{–}4)$$

The constants K and α are real. The constant K can be positive or negative, but α must be positive so that the natural pole at $s = -\alpha$ is in the left half of the s plane to ensure that the circuit is stable. Remember, the concepts of sinusoidal steady state and frequency response do not apply to unstable circuits that have poles in the right half of the s plane or on the j-axis.

To describe the frequency response of the low pass transfer function, we replace s by $j\omega$ in Eq. (12–4)

$$T(j\omega) = \frac{K}{\alpha + j\omega} \qquad (12\text{--}5)$$

and express the gain and phase functions as

$$|T(j\omega)| = \frac{|K|}{\sqrt{\omega^2 + \alpha^2}}$$

$$\theta(\omega) = \angle K - \tan^{-1}(\omega/\alpha) \qquad (12\text{--}6)$$

The gain function is a positive number. Since K is real, the angle of K ($\angle K$) is either $0°$ when $K > 0$ or $\pm 180°$ when $K < 0$. An example of a negative K occurs in an inverting OP AMP configuration where $T(s) = -Z_2(s)/Z_1(s)$.

Figure 12–3 shows the gain and phase functions plotted versus normalized frequency ω/α. In the gain plot the left vertical scale is a log scale for $|T(j\omega)|$ and the right scale is the equivalent linear scale for $|T(j\omega)|_{dB}$. The maximum passband gain occurs at $\omega = 0$ where $|T(0)| = |K|/\alpha$. As frequency increases the gain gradually decreases until at $\omega = \alpha$

$$|T(j\alpha)| = \frac{|K|}{\sqrt{\alpha^2 + \alpha^2}} = \frac{|K|/\alpha}{\sqrt{2}} = \frac{|T(0)|}{\sqrt{2}} \qquad (12\text{--}7)$$

That is, the cutoff frequency of the first-order low-pass transfer function is $\omega_C = \alpha$. The graph of the gain function in Figure 12–3 displays a low-pass characteristic with a finite dc gain and zero infinite frequency gain.

The low- and high-frequency gain asymptotes shown in Figure 12–3 are especially important. The low-frequency asymptote is the horizontal line and the high-frequency asymptote is the sloped line. At low frequencies ($\omega \ll \alpha$) the gain approaches $|T(j\omega)| \rightarrow |K|/\alpha$. At high frequencies ($\omega \gg \alpha$) the gain approaches $|T(j\omega)| \rightarrow |K|/\omega$. The intersection of the two asymptotes occurs when $|K|/\alpha = |K|/\omega$. The intersection forms a "corner" at $\omega = \alpha$, so the cutoff frequency is also called the **corner frequency**.

The high-frequency gain asymptote decreases by a factor of 10 (–20 dB) whenever the frequency increases by a factor of 10 (one decade). As a result, the high-frequency asymptote has a slope of –1 or –20 dB per decade and the low-frequency asymptote has a slope of 0 or 0 dB/decade. These two asymptotes provide a straight-line approximation to the gain response that differs from the true response by a maximum of 3 dB at the corner frequency.

The semilog plot of the phase shift of the first-order low-pass transfer function is shown in Figure 12–3. At $\omega = \alpha$ the phase angle in Eq. (12–12) is

$$\theta(\omega_C) = \angle K - \tan^{-1}\left(\frac{\alpha}{\alpha}\right)$$

$$= \angle K - 45°$$

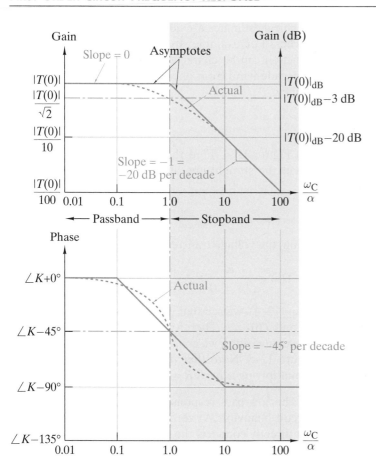

FIGURE 12–3 *First-order low-pass frequency response plots.*

At low frequency ($\omega \ll \alpha$) the phase angle approaches $\angle K$ and at high frequencies ($\omega \gg \alpha$) the phase approaches $\angle K - 90°$. Almost all of the $-90°$ phase change occurs in the two-decade range from $\omega/\alpha = 0.1$ to $\omega/\alpha = 10$. The straight-line segments in Figure 12–3 provide an approximation of the phase response. The phase approximation below $\omega/\alpha = 0.1$ is $\theta = \angle K$ and above $\omega/\alpha = 10$ is $\theta = \angle K - 90°$. Between these values the phase approximation is a straight line that begins at $\theta = \angle K$, passes through $\theta = \angle K - 45°$ at the cutoff frequency, and reaches $\theta = \angle K - 90°$ at $\omega/\alpha = 10$. The slope of this line segment is $-45°$/decade since the total phase change is $-90°$ over a two-decade range. The $-45°$/decade slope only applies over a frequency range from one decade below to one decade above the cutoff frequency. Outside of this range the straight-line approximation has zero slope.

To construct the straight-line approximations for a first-order low-pass transfer function, we need two parameters, the value of $T(0)$ and α. The parameter α defines the cutoff frequency and the value of $T(0)$ defines the passband gain $|T(0)|$ and the low-frequency phase $\angle T(0)$. The required quantities $T(0)$ and α can be determined directly from the transfer function $T(s)$ and can often be estimated by inspecting the circuit itself.

Using logarithmic scales in frequency-response plots allows us to make straight-line approximations to both the gain and phase responses. These approximations provide a useful way of visualizing a circuit's frequency response. Often such graphical estimates are adequate for developing analysis and design approaches. For example, the frequency response of the first-order low-pass function can be characterized by calculating the gain and phase over a two-decade band from one decade below to one decade above the cutoff frequency.

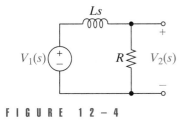

F I G U R E 1 2 – 4

EXAMPLE 12–2

Consider the circuit in Figure 12–4. Find the transfer function $T(s) = V_2(s)/V_1(s)$ and construct the straight-line approximations to the gain and phase responses.

SOLUTION:

Applying voltage division, the voltage transfer function for the circuit is

$$T(s) = \frac{R}{Ls + R} = \frac{R/L}{s + R/L}$$

Comparing this with Eq. (12–4), we see that the circuit has a low-pass gain response with $\alpha = R/L$ and $T(0) = 1$. Therefore, $|T(0)|_{dB} = 0$ dB, $\omega_C = R/L$, and $\angle K = 0°$. Given these quantities, we construct the straight-line approximations shown in Figure 12–5. Note that the frequency scale in Figure 12–5 is normalized by multiplying ω by $L/R = 1/\alpha$.

Circuit Interpretation: The low-pass response in Figure 12–5 can be explained in terms of circuit behavior. At zero frequency the inductor acts like a short circuit that directly connects the input port to the output port to produce a passband gain of 1 (or 0 dB). At infinite frequency the inductor acts like an open circuit, effectively disconnecting the input and output ports and leading to a gain of zero. Between these two extremes the impedance of the inductor gradually increases, causing the circuit gain to decrease. In particular, at the cutoff frequency we have $\omega L = R$, the impedance of the inductor is $j\omega L = jR$, and the transfer function reduces to

$$T(j\omega_C) = \frac{R}{R + jR} = \frac{1}{\sqrt{2}} \angle -45°$$

In other words, at the cutoff frequency the gain is –3 dB and the phase shift –45°. Obviously, the changing impedance of the inductor gives the circuit its low-pass gain features. ■

▶ DESIGN EXAMPLE 12–3

(a) Show that the transfer function $T(s) = V_2(s)/V_1(s)$ in Figure 12–6 has a low-pass gain characteristic.
(b) Select element values so the passband gain is 4 and the cutoff frequency is 100 rad/s.
(c) The design is to be tested using a 2-V sinusoidal input $v_S(t) = 2 \cos(\omega_A t)$ at $\omega_A = 5$ rad/s, 500 rad/s, and 50 krad/s. Use the asymptotic straight-line gain to estimate the amplitude of the steady-state output at each of these frequencies.

FIGURE 12-5

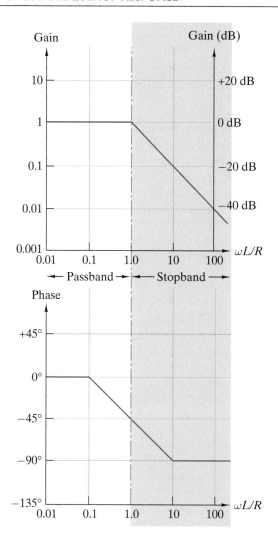

SOLUTION:

(a) The circuit is an inverting amplifier configuration with

$$Z_1(s) = R_1 \quad \text{and} \quad Z_2(s) = \frac{1}{C_2 s + \dfrac{1}{R_2}} = \frac{R_2}{R_2 C_2 s + 1}$$

The circuit transfer function is found as

$$T(s) = -\frac{Z_2(s)}{Z_1(s)} = -\frac{R_2}{R_1} \times \frac{1}{R_2 C_2 s + 1}$$

Rearranging the standard low-pass form in Eq. (12–4) as

$$T(s) = \frac{K/\alpha}{s/\alpha + 1}$$

shows that the circuit transfer function has a low-pass form with

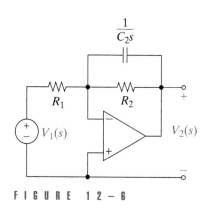

FIGURE 12-6

$$\omega_C = \alpha = \frac{1}{R_2 C_2} \quad \text{and} \quad T(0) = -\frac{R_2}{R_1}$$

This is an inverting circuit, so the $-90°$ phase swing of the low-pass form runs from $\angle T(0) = -180°$ to $\angle T(\infty) = -270°$, passing through $\angle T(j\omega_C) = -225°$ along the way.

Circuit Interpretation: The low-pass response is easily deduced from known circuit performance. At dc the capacitor acts like an open circuit and the circuit in Figure 12–6 reduces to a resistance inverting amplifier with $K = T(0) = -R_2/R_1$. At infinite frequency the capacitor acts like a short circuit that connects the OP AMP output directly to the inverting input. This connection results in zero output since the node voltage at the inverting input is necessarily zero. In between these two extremes the gain gradually decreases as the decreasing capacitor impedance gradually pulls the OP AMP output down to zero at infinite frequency.

(b) The design constraints require that $\omega_C = 1/R_2 C_2 = 100$ and $|T(0)| = R_2/R_1 = 4$. Selecting $R_1 = 10\ k\Omega$ implies that $R_2 = 40\ k\Omega$ and $C_2 = 250\ nF$.

(c) The input frequency $\omega_A = 5\ rad/s$ falls in the passband below the cutoff frequency at $\omega_C = 100\ rad/s$. In passband the asymptotic gain is $K = 4$; hence, the amplitude of the steady-state output is approximately $K \times$ input amplitude $= 4 \times 2 = 8\ V$.

The input frequencies at $\omega_A = 500\ rad/s$ and $50\ krad/s$ fall in the stopband above the cutoff frequency. In the stopband the asymptotic gain is

$$|T(j\omega)| = \underbrace{[K]}_{\substack{\text{Passband} \\ \text{Gain}}} \times \underbrace{\left[\frac{\omega_C}{\omega}\right]}_{\substack{\text{Stopband} \\ \text{Roll Off}}} = [4] \times \left[\frac{100}{\omega}\right] = \frac{400}{\omega}$$

Hence, at $\omega_A = 500\ rad/s$ the steady-state output would be approximately $(400/500) \times$ input amplitude $= 1.6\ V$. At $\omega_A = 50\ krad/s$ the steady-state output would be approximately $(400/50000) \times$ input amplitude $= 16\ mV$. ∎

APPLICATION NOTE: EXAMPLE 12–4

In terms of frequency response, the ideal OP AMP model introduced in Chapter 4 assumes that the device has an infinite gain and an infinite bandwidth. A more realistic model of the device is shown in Figure 12–7(a). The controlled source gain in Figure 12–7(a) is a low-pass transfer function with a dc gain of A and a cutoff frequency ω_C. The straight-line asymptotes of controlled source gain are shown in Figure 12–7(b). The **gain-bandwidth product** ($G = A\omega_C$) is the basic performance parameter of this model.

With no feedback the OP AMP transfer function is the same as the controlled-source transfer function. The gain-bandwidth product of the open-loop transfer function is

$$G = A\omega_C \quad \text{(Open Loop)}$$

FIGURE 12–7

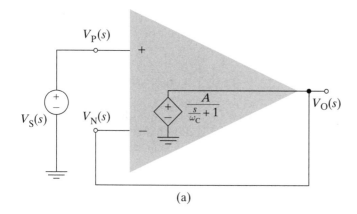

(a)

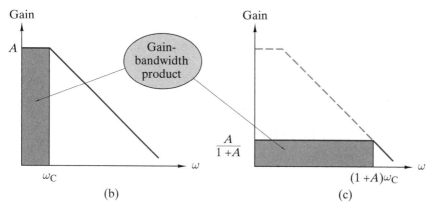

(b)   (c)

The closed-loop transfer function of the circuit in Figure 12–7(a) is found by writing the following device and connection equations:

$$\text{Device equation: } V_O = \frac{A}{s/\omega_C + 1}(V_P - V_N)$$

$$\text{Input connection: } V_P = V_S$$

$$\text{Feedback connection: } V_N = V_O$$

Substituting the connection equations into the OP AMP device equation yields

$$V_O(s) = \frac{A}{s/\omega_C + 1}(V_S(s) - V_O(s))$$

Solving for the closed-loop transfer function produces

$$T(s) = \frac{V_O(s)}{V_S(s)} = \frac{A}{A + 1}\left[\frac{1}{\dfrac{s}{(A + 1)\omega_C} + 1}\right]$$

The straight-line asymptotes of the closed-loop transfer function are shown in Figure 12–7(c).

The closed-loop circuit has a low-pass transfer function with a dc gain of $A/(A + 1)$ and a cutoff frequency of $(A + 1)\omega_C$. The gain-bandwidth product of the closed-loop circuit is

$$G = \left[\frac{A}{A + 1}\right][(A + 1)\omega_C] = A\omega_C \quad \text{(Closed Loop)}$$

which is the same as the open-loop case. In other words, the gain-bandwidth product is invariant and is not changed by feedback. It can be shown that this result is a general one that applies to all linear OP AMP circuits, regardless of the circuit configuration.

Gain-bandwidth product is a fundamental parameter that limits the frequency response of OP AMP circuits. For example, an OP AMP with a gain-bandwidth product of $G = 10^6$ Hz is connected as a noninverting amplifier with a closed–loop gain of 20. The frequency response of the resulting closed-loop circuit has a low-pass characteristic with a passband gain of 20 and a cutoff frequency of

$$f_C = \frac{10^6}{20} = 50\,\text{kHz}$$

FIRST-ORDER HIGH-PASS RESPONSE

We next treat the first-order high-pass transfer function

$$T(s) = \frac{Ks}{s + \alpha} \tag{12–8}$$

The high-pass function differs from the low-pass case by the introduction of a zero at $s = 0$. Replacing s by $j\omega$ in $T(s)$ and solving for the gain and phase functions yields

$$|T(j\omega)| = \frac{|K|\omega}{\sqrt{\omega^2 + \alpha^2}}$$

$$\theta(\omega) = \angle K + 90° - \tan^{-1}(\omega/\alpha) \tag{12–9}$$

Figure 12–8 shows the gain and phase functions versus normalized frequency ω/α. The maximum gain occurs at high frequency ($\omega \gg \alpha$) where $|T(j\omega)| \to |K|$. At low frequency ($\omega \ll \alpha$) the gain approaches $|K|\omega/\alpha$. At $\omega = \alpha$ the gain is

$$|T(j\alpha)| = \frac{|K|\alpha}{\sqrt{\alpha^2 + \alpha^2}} = \frac{|K|}{\sqrt{2}} \tag{12–10}$$

which means the cutoff frequency is $\omega_C = \alpha$. The gain response plot in Figure 12–8 displays a high-pass characteristic with a passband extending from $\omega = \alpha$ to infinity and a stopband between zero frequency and $\omega = \alpha$.

The low- and high-frequency gain asymptotes approximate the gain response in Figure 12–8. The high-frequency asymptote ($\omega \gg \alpha$) is the horizontal line whose ordinate is $|K|$ (slope = 0 or 0 dB/decade). The low-frequency asymptote ($\omega \ll \alpha$) is a line of the form $|K|\omega/\alpha$ (slope = +1 or +20 dB/decade). The intersection of these two asymptotes occurs when $|K| = |K|\omega/\alpha$, which defines a corner frequency at $\omega = \alpha$.

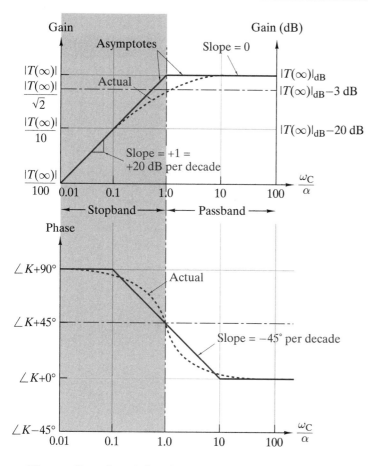

FIGURE   12-8   *First-order   high-pass frequency response plots.*

The semilog plot of the phase shift of the first-order high-pass function is shown in Figure 12–8. The phase shift approaches $\angle K$ at high frequency, passes through $\angle K + 45°$ at the cutoff frequency, and approaches $\angle K + 90°$ at low frequency. Most of the 90° phase change occurs over the two-decade range centered on the cutoff frequency. The phase shift can be approximated by the straight-line segments shown in the Figure 12–8. As in the low-pass case, $\angle K$ is 0° when K is positive and $\pm 180°$ when K is negative.

Like the low-pass function, the first-order high-pass frequency response can be approximated by straight-line segments. To construct these lines we need two parameters, $T(\infty)$ and α. The parameter α defines the cutoff frequency, and the quantity $T(\infty)$ gives the passband gain $|T(\infty)|$ and the high-frequency phase angle $\angle T(\infty)$. The quantities $T(\infty)$ and α can be determined directly from the transfer function or estimated directly from the circuit in some cases. The straight line shows that the first-order high-pass response can be characterized by calculating the gain and phase over a two-decade band from one decade below to one decade above the cutoff frequency.

EXAMPLE 12-5

Show that the transfer function $T(s) = V_2(s)/V_1(s)$ in Figure 12–9 has a high-pass gain characteristic. Construct the straight-line approximations to the gain and phase responses of the circuit.

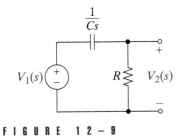

FIGURE 12 – 9

SOLUTION:
Applying voltage division, the voltage transfer function for the circuit is

$$T(s) = \frac{R}{R + 1/Cs} = \frac{RCs}{RCs + 1}$$

Rearranging Eq. (12–8) as

$$T(s) = \frac{K(s/\alpha)}{s/\alpha + 1}$$

shows that the circuit has a high-pass gain characteristic with $\alpha = 1/RC$ and $T(\infty) = 1$. Therefore, $|T(\infty)|_{dB} = 0$ dB, $\omega_C = 1/RC$, and $\angle T(\infty) = 0°$. Given these quantities, we construct the straight-line gain and phase approximations in Figure 12–10. The frequency scale in Figure 12–10 is normalized by multiplying ω by $RC = 1/\alpha$.

FIGURE 12 – 10

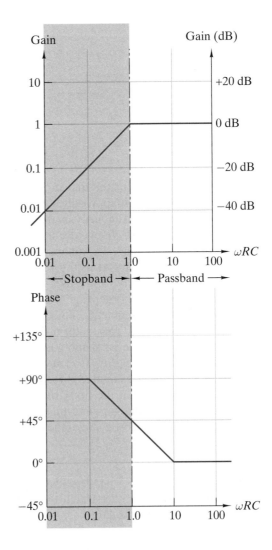

Circuit Interpretation: The high-pass response in Figure 12–10 can be understood in terms of known circuit behavior. At zero frequency the capacitor acts like an open circuit that effectively disconnects the input signal source, leading to zero gain. At infinite frequency the capacitor acts like a short circuit that directly connects the input to the output, leading to a passband gain of 1 (or 0 dB). Between these two extremes the impedance of the capacitor gradually decreases, causing the gain to increase. In particular, at the cutoff frequency we have $1/\omega C = R$, the impedance of the capacitor is $1/j\omega C = -jR$, and the transfer function is

$$T(j\omega_C) = \frac{R}{R - jR} = \frac{1}{\sqrt{2}} \angle + 45°$$

In other words, at the cutoff frequency the gain is –3 dB and the phase shift +45°. Obviously, the decreasing impedance of the capacitor gives the circuit its high-pass gain characteristics.   ■

DESIGN EXAMPLE 12–6

(a) Show that the transfer function $T(s) = V_2(s)/V_1(s)$ of the circuit in Figure 12–11 has a high-pass gain characteristic.

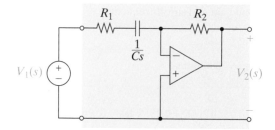

FIGURE 12–11

(b) Select the element values to produce a passband gain of 4 and a cutoff frequency of 40 krad/s.

SOLUTION:

(a) The branch impedances of the inverting OP AMP configuration in Figure 12–11 are

$$Z_1(s) = R_1 + \frac{1}{Cs} = \frac{R_1 Cs + 1}{Cs} \quad \text{and} \quad Z_2(s) = R_2$$

and the voltage transfer function is

$$T(s) = -\frac{Z_2(s)}{Z_1(s)} = -\frac{R_2 Cs}{R_1 Cs + 1} = \frac{(-R_2/R_1)s}{s + 1/R_1 C}$$

This results in a high-pass transfer function of the form $Ks/(s + \alpha)$ with $K = -R_2/R_1$ and $\alpha = \omega_C = 1/R_1 C$.

Circuit Interpretation: The high-pass response of this circuit is easily understood in terms of element impedances. At dc the capacitor in Figure 12–11

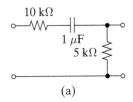

(a)

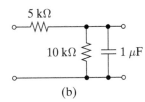

(b)

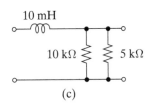

(c)

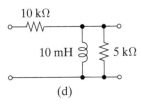

(d)

FIGURE 12–12

acts like an open circuit that effectively disconnects the input source, resulting in zero gain. At infinite frequency the capacitor acts like a short circuit that reduces the circuit to an inverting amplifier with $K = T(\infty) = -R_2/R_1$. As the frequency varies from zero to infinity, the gain gradually increases as the capacitor impedance decreases.

(b) The design requirements specify that $1/R_1C = 4 \times 10^4$ and $R_2/R_1 = 4$. Selecting $R_1 = 10$ kΩ requires $R_2 = 40$ kΩ and $C = 2.5$ nF. ∎

Exercise 12–1

For each circuit in Figure 12–12 identify whether the gain response has low-pass or high-pass characteristics and find the passband gain and cutoff frequency.

Answers:
(a) High pass, $|T(\infty)| = 1/3$, $\omega_C = 66.7$ rad/s
(b) Low pass, $|T(0)| = 2/3$, $\omega_C = 300$ rad/s
(c) Low pass, $|T(0)| = 1$, $\omega_C = 333$ krad/s
(d) High pass, $|T(\infty)| = 1/3$, $\omega_C = 333$ krad/s

Exercise 12–2

A first-order circuit has a step response $g(t) = 5[e^{-2000t}]u(t)$.
(a) Is the circuit frequency response low-pass or high-pass?
(b) Find the passband gain and cutoff frequency.

Answers:
(a) High-pass
(b) $|T(\infty)| = 5$ and $\omega_C = 2000$ rad/s

BANDPASS AND BANDSTOP RESPONSES USING FIRST-ORDER CIRCUITS

The first-order high-pass and low-pass circuits can be used in a building block fashion to produce circuits with bandpass or bandstop responses. Figure 12–13 shows a cascade connection of first-order high-pass and low-pass circuits. When the second stage does not load the first, the overall transfer function can be found by the chain rule:

$$T(s) = T_1(s) \times T_2(s) = \underbrace{\left(\frac{K_1 s}{s + \alpha_1}\right)}_{\text{High Pass}} \underbrace{\left(\frac{K_2}{s + \alpha_2}\right)}_{\text{Low Pass}} \tag{12–11}$$

Replacing s by $j\omega$ in Eq. (12–11) and solving for the gain response yields

$$|T(j\omega)| = \underbrace{\left(\frac{|K_1|\omega}{\sqrt{\omega^2 + \alpha_1^2}}\right)}_{\text{High Pass}} \underbrace{\left(\frac{|K_2|}{\sqrt{\omega^2 + \alpha_2^2}}\right)}_{\text{Low Pass}} \tag{12–12}$$

Note that the gain of the cascade is zero at $\omega = 0$ and at infinite frequency.

FIGURE 12–13 *Cascade connection of high-pass and low-pass circuits.*

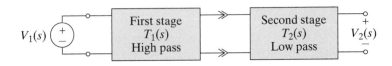

When $\alpha_1 \ll \alpha_2$ the high-pass cutoff frequency is much lower than the low-pass cutoff frequency, and the overall transfer function has a bandpass characteristic. At low frequencies ($\omega \ll \alpha_1 \ll \alpha_2$) the gain approaches $|T(j\omega)| \to |K_1 K_2| \omega / \alpha_1 \alpha_2$. At mid frequencies ($\alpha_1 \ll \omega \ll \alpha_2$) the gain approaches $|T(j\omega)| \to |K_1 K_2| / \alpha_2$. The low- and mid-frequency asymptotes intersect when $|K_1 K_2| \omega / \alpha_1 \alpha_2 = |K_1 K_2| / \alpha_2$ at $\omega = \alpha_1$ (that is, at the cutoff frequency of the high-pass stage). At high frequencies ($\alpha_1 \ll \alpha_2 \ll \omega$) the gain approaches $|T(j\omega)| \to |K_1 K_2| / \omega$. The high- and mid-frequency asymptotes intersect when $|K_1 K_2| / \omega = |K_1 K_2| / \alpha_2$ at $\omega = \alpha_2$ (that is, at the cutoff frequency of the low-pass stage). The plot of these asymptotes in Figure 12–14 shows that the asymptotic gain exhibits a passband between α_1 and α_2. Input sinusoids whose frequencies are outside of this range fall in one of the two stopbands.

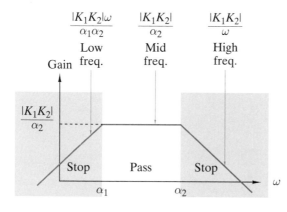

FIGURE 1 2 – 1 4   *Asymptotic gain of a bandpass characteristic.*

In the bandpass cascade connection the input signal must pass both a low- and a high-pass stage to reach the output. In the parallel connection in Figure 12–15 the input can reach the output via either a low- or a high-pass path. The overall transfer function is the sum of the low- and high-pass transfer functions

$$|T(j\omega)| = \underbrace{\left(\frac{|K_1| \omega}{\sqrt{\omega^2 + \alpha_1^2}} \right)}_{\text{High Pass}} + \underbrace{\left(\frac{|K_2|}{\sqrt{\omega^2 + \alpha_2^2}} \right)}_{\text{Low Pass}} \qquad (12\text{–}13)$$

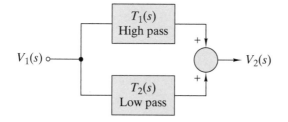

FIGURE 1 2 – 1 5   *Parallel connection of high-pass and low-pass circuits.*

Any sinusoid whose frequency falls in either passband will find its way to the output unscathed. An input sinusoid whose frequency falls in both stopbands will be attenuated.

When $\alpha_1 \gg \alpha_2$ the high-pass cutoff frequency is much higher than the low-pass cutoff frequency, and the overall transfer function has a bandstop gain response as shown in Figure 12–16. At low frequencies ($\omega \ll \alpha_2 \ll \alpha_1$) the high-pass gain is negligible and the over asymptote approaches the passband gain of the low-pass function $|T(j\omega)| \rightarrow |K_2|/\alpha_2$. At high frequencies ($\alpha_2 \ll \alpha_1 \ll \omega$) the low-pass gain is negligible and the overall gain approaches the bandpass gain of the high-pass function $|T(j\omega)| \rightarrow |K_1|$. With a bandstop function the two passbands normally have the same gain, hence $|K_1| = |K_2|/\alpha_2$. Between these two passbands there is a stopband. For $\omega > \alpha_2$ the low-pass asymptote is $|K_2|/\omega$ and for $\omega < \alpha_1$ the high-pass asymptote is $|K_1|\omega/\alpha_1$. The asymptotes intersect at $\omega^2 = \alpha_1|K_2|/|K_1|$. But equal gains in the two passband frequencies require $|K_1| = |K_2|/\alpha_2$, so the intersection frequency is $\omega = \sqrt{\alpha_1 \alpha_2}$. Below this frequency the stopband attenuation is determined by the low-pass function, and above this frequency the attenuation is governed by the high-pass function.

FIGURE 12–16 *Asymptotic gain of a bandstop characteristic.*

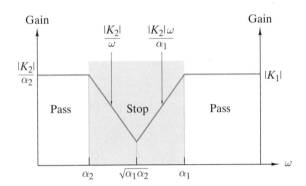

The analysis of the bandpass and bandstop transfer functions in Eqs. (12–12) and (12–13) illustrates that the asymptotic gain plots of first-order functions can help us understand and describe other types of gain response as well. The asymptotic responses in Figures 12–14 and 12–16 are a reasonably good approximation as long as the two first-order cutoff frequencies are widely separated. The straight-line approximation shows us that the frequency range of interest extends from a decade below the lowest cutoff frequency to a decade above the highest. This range could be a very wide indeed, since the two cutoff frequencies may be separated by several decades.

DESIGN EXAMPLE 12–7

(a) Design a bandpass circuit with a passband gain of 10 and cutoff frequencies at 20 Hz and 20 kHz.
(b) The circuit is to be tested using a 2-V sinusoidal input at $f_A = 6$ Hz, 600 Hz, and 60 kHz. Use the asymptotic straight-line gain to estimate the amplitude of the steady-state output at each of these frequencies.

SOLUTION:
(a) Our design uses a cascade connection of first-order low- and high-pass building blocks. The required transfer function has the form

$$T(s) = \left(\frac{K_1 s}{s + \alpha_1}\right)\left(\frac{K_2}{s + \alpha_2}\right)$$

with the following constraints:

Lower cutoff frequency: $\alpha_1 = 2\pi(20) = 40\pi$ rad/s

Upper cutoff frequency: $\alpha_2 = 2\pi(20 \times 10^3) = 4\pi \times 10^4$ rad/s

Midband Gain: $\dfrac{|K_1 K_2|}{\alpha_2} = 10$

With numerical values inserted, the required transfer function is

$$T(s) = \underbrace{\frac{s}{s + 40\pi}}_{\text{High Pass}} \times \underbrace{10}_{\text{Gain}} \times \underbrace{\frac{4\pi \times 10^4}{s + 4\pi \times 10^4}}_{\text{Low Pass}}$$

This transfer function can be realized using the high-pass/low-pass cascade circuit in Figure 12–17. The first stage is the RC high-pass circuit from Example 12–5 and the third stage is the RL low-pass circuit from Example 12–2. The noninverting OP AMP second stage serves two purposes: (1) It isolates the first and third stages, so the chain rule applies, and (2) it supplies the midband gain. Using the chain rule, the transfer function of this circuit is

$$T(s) = \underbrace{\left[\frac{s}{s + 1/R_C C}\right]}_{\text{High Pass}} \underbrace{\left[\frac{R_1 + R_2}{R_1}\right]}_{\text{Gain}} \underbrace{\left[\frac{R_L/L}{s + R_L/L}\right]}_{\text{Low Pass}}$$

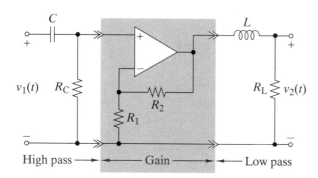

FIGURE 12-17

High pass ⟶ ⟵ Gain ⟶ ⟵ Low pass

Comparing this to the required transfer function leads to the following design constraints:

High-pass stage: $R_C C = 1/40\pi$ Let $R_C = 100$ kΩ Then $C = 79.6$ nF

Gain stage: $(R_1 + R_2)/R_1 = 10$ Let $R_1 = 10$ kΩ Then $R_2 = 90$ kΩ

Low-pass stage: $R_L/L = 40000\pi$ Let $R_L = 200$ kΩ Then $L = 0.628$ H

(b) The input frequency $f_A = 6$ Hz falls in the lower stopband below the lower cutoff frequency at $f_{C1} = 20$ Hz. In this region the circuit gain is controlled by the high-pass and gain stages. The asymptotic straight-line gain in this region is

$$|T(j\omega)| = \underbrace{\left[\frac{\omega}{2\pi f_{C1}}\right]}_{\substack{\text{High-pass}\\\text{Stage}}} \times \underbrace{[10]}_{\substack{\text{Gain}\\\text{Stage}}} \times \underbrace{[1]}_{\substack{\text{Low-pass}\\\text{Stage}}} = \frac{f}{f_{C1}} \times 10 = \frac{f}{2}$$

and the amplitude of the steady-state output is approximately $(f/2) \times$ input amplitude $= 3 \times 2 = 6$ V.

The input frequency $f_A = 600$ Hz falls in the passband between the lower cutoff frequency at $f_{C1} = 20$ Hz and the upper cutoff frequency at $f_{C2} = 20$ kHz. In this region the circuit gain is controlled by the gain stage and the amplitude of the steady-state output would be approximately $10 \times$ input amplitude $= 20$ V.

The input frequency $f_A = 60$ kHz falls in the upper stopband above the upper cutoff frequency at $f_{C2} = 20$ kHz. In this region the circuit gain is controlled by the low-pass and gain stages. The asymptotic straight-line gain in this region is

$$|T(j\omega)| = \underbrace{[1]}_{\substack{\text{High-pass}\\\text{Stage}}} \times \underbrace{[10]}_{\substack{\text{Gain}\\\text{Stage}}} \times \underbrace{\left[\frac{2\pi f_{C2}}{\omega}\right]}_{\substack{\text{Low-pass}\\\text{Stage}}} = [10] \times \left[\frac{f_{C2}}{f}\right] = \frac{2 \times 10^5}{f}$$

and the amplitude of the steady-state output is approximately $(2 \times 10^5/6 \times 10^4) \times$ input amplitude $= (10/3) \times 2 = 6.67$ V. ∎

EXAMPLE 12–8

Use computer-aided analysis to show that the circuit in Figure 12–18 implements a bandstop gain response.

FIGURE 12 – 18

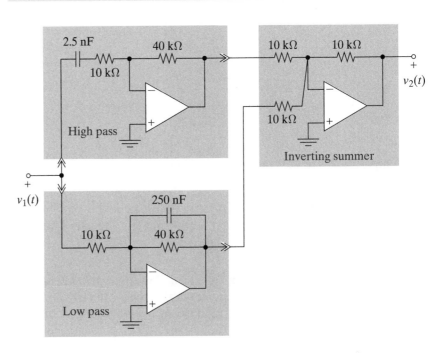

SOLUTION:

This circuit implements the block diagram in Figure 12–15. The two first-order circuits have the same input. The upper circuit is the high-pass OP AMP circuit designed in Example 12–6 to have a passband gain of 4 and a cutoff frequency $\alpha_1 = 40$ krad/s. The lower path is the low-pass OP AMP circuit designed in Example 12–3 to have a passband gain of 4 and a cutoff frequency $\alpha_2 = 100$ rad/s. The inverting summer at the far right implements the summing point function in Figure 12–15. Since the cutoff frequency of the low pass circuit is much lower than the cutoff frequency of the high-pass circuit, we expect to see two passbands on either side of a stopband centered at

$$\omega = \sqrt{\alpha_1\,\alpha_2} = \sqrt{100 \times 4 \times 10^4} = 2000 \ \text{rad/s}$$

Figure 12–19 shows the circuit as drawn in the Electronics Workbench circuit window. The input voltage source V_1 is set to produce 1-V ac, so the circuit gain is equal to the ac voltage at node 8 (the summer output). Selecting **AC Frequency** from the **Analysis** menu brings up the **AC Frequency Analysis** dialog box shown in the figure. Here we must specify the analysis frequency range. This range should extend from at least a decade below the lowest cutoff frequency ($\alpha_2/2\pi = 16$ Hz) to at least a decade above the highest cutoff frequency ($\alpha_1/2\pi = 6.4$ kHz). To span this range we select a **Start frequency (FSTART)** of 1 Hz and an **End frequency (FSTOP)** of 100 kHz. To span this range we choose a **Decade** sweep and select the summer output (node 8) as the **Node for analysis**. Clicking the simulate button launches an ac frequency analysis that produces the output voltage plot shown at the bottom of Figure 12–19.

As expected, the frequency response of the overall circuit has two passbands surrounding a stopband. In this plot one cursor is set in the lower passband and the other at the stopband minimum. The passband output voltage is $y_1 = 3.9921$ V (gain = 12 dB) at $x_1 = 1.0000 = 1$ Hz, which confirms that the two passbands have gain = 4 as straight-line analysis predicts. The minimum output is $y_2 = 20.1212$ mV (gain = –33.9 dB) at $x_2 = 316.2278$ Hz. This notch frequency corresponds to a radian frequency of $\omega = 2\pi x_2 = 1987$ rad/s, which is very close to the 2000 rad/s predicted by straight-line gain analysis. ■

12–3   SECOND-ORDER CIRCUIT FREQUENCY RESPONSE

We begin our study with the second-order bandpass transfer function:

$$T(s) = \frac{Ks}{s^2 + 2\zeta\omega_0 s + \omega_0^2} \tag{12–14}$$

The second-order polynomial in the denominator is expressed in terms of the damping ratio ζ and undamped natural frequency ω_0 parameters. These parameters play a key role in describing the frequency response of second-order circuits. When the damping ratio is less than 1, the complex poles of $T(s)$ dramatically influence the frequency response.

To gain a qualitative understanding of the effect of complex poles, we write Eq. (12–14) in factored form:

FIGURE 12-19

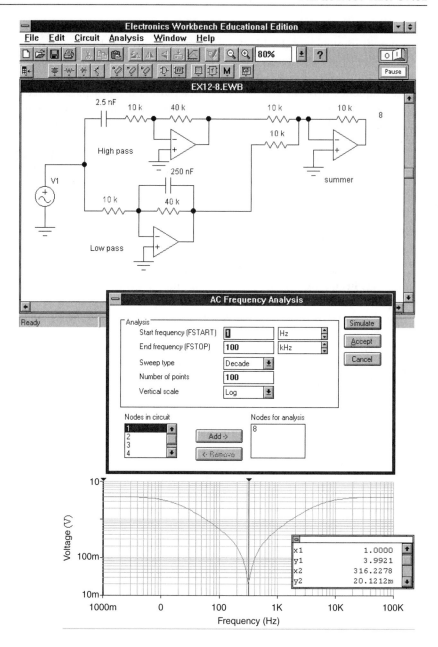

$$T(s) = \frac{Ks}{(s - p_1)(s - p_2)} \tag{12-15}$$

where p_1 and p_2 are the complex poles of $T(s)$. The frequency-response gain function is

$$|T(j\omega)| = \frac{|K|\omega}{|j\omega - p_1||j\omega - p_2|} = \frac{|K|\omega}{M_1 M_2} \tag{12-16}$$

As shown in Figure 12–20, the factors $(j\omega - p_1)$ and $(j\omega - p_2)$ can be interpreted as vectors from the natural poles at $s = p_1$ and $s = p_2$ to the point $s = j\omega$, where ω is the frequency at which the gain is being calculated. The lengths of these vectors are

$$M_1 = |j\omega - p_1|$$

$$M_2 = |j\omega - p_2|$$

The frequency response of a circuit depends on how the circuit responds to sinusoidal inputs of constant amplitude by changing frequencies. The sinusoidal input produces two forced poles on the j-axis whose location depends of the frequency of the input. Changing the input frequency causes the forced pole at $s = j\omega$ to slide up or down the imaginary axis. This, in turn, changes the lengths M_1 and M_2, since the tips of the vectors $j\omega - p_1$ and $j\omega - p_2$ move up or down the j-axis while the tails remain fixed at the location of the natural poles. The length M_1 reaches a minimum when the forced pole at $s = j\omega$ is close to the natural pole at $s = p_1$. Since M_1 appears in the denominator of Eq. (12–16), this minimum tends to produce a maximum or peak in the gain response.

Figure 12–21 shows how different input frequencies change the lengths M_1 and M_2 to produce the gain response curve. The lengths M_1 and M_2 are approximately equal at low frequencies ($\omega \ll \omega_1$) and again at high frequencies ($\omega \gg \omega_3$). At intermediate frequencies ($\omega_1 < \omega < \omega_3$) the lengths differ, especially in the neighborhood of ω_2, where length M_1 reaches a minimum and is very small compared with length M_2. The small value of the length M_1 produces a maximum in the gain response indicative of a phenomenon called resonance. **Resonance** is the increased response observed when the frequency of a periodic driving force like a sinusoid is close to a natural frequency of the circuit. In terms of s-plane geometry, resonance occurs when the point at $s = j\omega$ and the natural pole at $s = p_1$ are close together so the length M_1 is small.

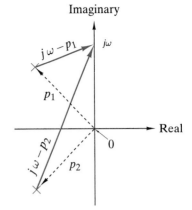

FIGURE   12–20   s-Plane vectors defining $|T(j\omega)|$.

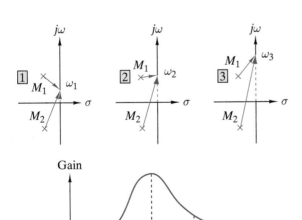

FIGURE   12–21   Gain response resonance from s-plane geometry.

The resonant peak decreases when p_1 and p_2 are shifted to the left in the s plane. Replacing s by $s + \alpha$ produces such a shift. The s-domain translation property

$$\mathcal{L}^{-1}\{F(s + \alpha)\} = e^{-\alpha t}f(t)$$

then shows that shifting the poles of $F(s)$ to the left increases the damping in the time domain. Qualitatively, we expect resonant peaks in the frequency response to be less pronounced when circuit damping increases and more pronounced when it decreases.

SECOND-ORDER BANDPASS CIRCUITS

To develop a quantitative understanding of resonance, we replace s by $j\omega$ in Eq. (12–14) to produce

$$T(j\omega) = \frac{Kj\omega}{-\omega^2 + 2\zeta\omega_0 j\omega + \omega_0^2} \tag{12–17}$$

At low frequencies ($\omega \ll \omega_0$) the gain approaches $|T(j\omega)| \rightarrow |K|\omega/\omega_0^2$. At high frequencies ($\omega \gg \omega_0$) the gain approaches $|T(j\omega)| \rightarrow |K|/\omega$. The low-frequency asymptote is directly proportional to frequency (slope = +1 or +20 dB/decade), while the high-frequency asymptote is inversely proportional to frequency (slope = –1 or –20 dB/decade). As shown in Figure 12–22, the two asymptotes intersect at $\omega = \omega_0$. At this intersection the gain is $|K|/\omega_0$. Note that the frequency scale in Figure 12–16 is logarithmic and normalized by dividing ω by the undamped natural frequency ω_0.

F I G U R E 1 2 – 2 2 *Asymptotic gain of a second-order bandpass response.*

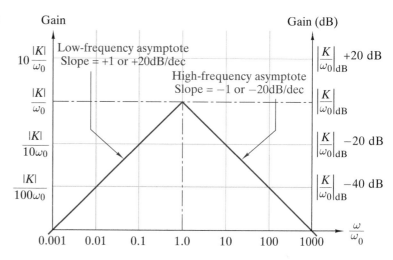

The two asymptotes display the characteristics of a bandpass response located around ω_0. For this reason, in a second-order bandpass circuit ω_0 is also called the **center frequency**. To examine the gain response in more detail, we must consider the effect of circuit damping. Factoring $j\omega\omega_0$ out of the denominator of Eq. (12–17) produces

$$T(j\omega) = \frac{K/\omega_0}{2\zeta + j\left(\dfrac{\omega}{\omega_0} - \dfrac{\omega_0}{\omega}\right)} \tag{12–18}$$

Arranging the transfer function in this form shows that only the imaginary part in the denominator varies with frequency. Clearly, the maximum gain occurs when the imaginary part in the denominator of Eq. (12–18) vanishes at $\omega = \omega_0$. The maximum gain is

$$|T(j\omega)|_{\text{MAX}} = \frac{|K|/\omega_0}{2\zeta} \qquad (12\text{–}19)$$

Since the maximum gain is inversely proportional to the damping ratio, the resonant peak will increase as circuit damping decreases, and vice versa. The quantity $|K|/\omega_0$ in the numerator of Eq. (12–19) is the gain at the point in Figure 12–22 where the low- and high-frequency asymptotes intersect.

Figure 12–23 shows plots of the gain of the bandpass function for several values of the damping ratio ζ. For $\zeta < 0.5$ the gain curve lies above the asymptotes and has a narrow resonant peak at ω_0. For $\zeta > 0.5$ the gain curve flattens out and lies below the asymptotes. When $\zeta = 0.5$ the gain curve remains fairly close to the asymptotes.

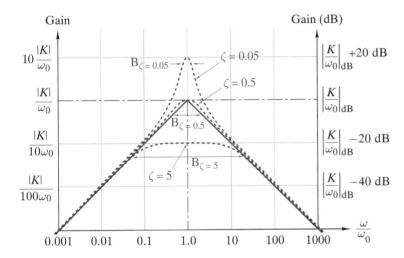

FIGURE 12–23 *Second-order bandpass gain responses.*

The gain response has a bandpass characteristic regardless of the value of the damping ratio. The passband is defined by a cutoff frequency on either side of the center frequency ω_0. To locate the two cutoff frequencies, we must find the values of ω at which the gain is $|T(j\omega)|_{\text{MAX}}/\sqrt{2}$, where the maximum gain is given in Eq. (12–19). We note that when the imaginary part in the denominator of Eq. (12–18) equals $\pm 2\zeta$, the gain is

$$|T(j\omega)| = \frac{|K|/\omega_0}{|2\zeta \pm j2\zeta|} = \frac{\dfrac{|K|/\omega_0}{2\zeta}}{\sqrt{2}} = \frac{|T(j\omega)|_{\text{MAX}}}{\sqrt{2}} \qquad (12\text{–}20)$$

The values of frequency ω that cause the imaginary part to be $\pm j2\zeta$ are the cutoff frequencies because at these points the gain is reduced by a factor of $1/\sqrt{2}$ from its maximum value at $\omega = \omega_0$.

To find the cutoff frequencies, we set the imaginary part in the denominator in Eq. (12–18) equal to $\pm 2\zeta$:

$$\frac{\omega}{\omega_0} - \frac{\omega_0}{\omega} = \pm 2\zeta \qquad (12\text{–}21)$$

which yields the quadratic equation

$$\omega^2 \mp 2\zeta\omega_0\omega - \omega_0 = 0 \tag{12–22}$$

Because of the $\mp$ sign this quadratic has four roots:

$$\omega_{C1}, \ \omega_{C2} = \omega_0(\pm \ \zeta \ \pm \ \sqrt{1 + \zeta^2}) \tag{12–23}$$

Only the two positive roots have physical significance. Since $\sqrt{1 + \zeta^2} > \zeta$, the two positive roots are

$$\omega_{C1} = \omega_0(-\zeta + \sqrt{1 + \zeta^2})$$

$$\omega_{C2} = \omega_0(+\zeta + \sqrt{1 + \zeta^2}) \tag{12–24}$$

Since $\zeta > 0$, these equations show that $\omega_{C1} < \omega_0$ is the lower cutoff frequency, while $\omega_{C2} > \omega_0$ is the upper cutoff frequency. Multiplying ω_{C1} times ω_{C2} produces

$$\omega_0^2 = \omega_{C1}\omega_{C2} \tag{12–25}$$

This result means that the center frequency ω_0 is the geometric mean of the two cutoff frequencies. The **bandwidth (B)** of the passband is found by subtracting ω_{C1} from ω_{C2}:

$$B = \omega_{C2} - \omega_{C1} = 2\zeta\omega_0 \tag{12–26}$$

The bandwidth is proportional to the product of the damping ratio and the natural frequency. When $\zeta < 0.5$ then $B < \omega_0$ and when $\zeta > 0.5$ then $B > \omega_0$.

We can think of $\zeta = 0.5$ as the boundary between two extreme cases. In the **narrowband** case ($\zeta \ll 0.5$) the complex natural poles are close to the j-axis and the gain response has a resonant peak that produces a bandwidth that is small compared to the circuit's natural frequency. In the **wideband** case ($\zeta \gg 0.5$) the gain response is relatively flat because the natural poles are farther from the j-axis because the circuit has more damping. The narrowband response is highly selective, passing only a very restricted range of frequencies, often less than one octave. In contrast, the wideband response usually encompasses several decades within its passband. Both cases are widely used in applications.

The narrow and broadband cases also show up in the phase response. From Eq. (12–18) the phase angle is

$$\theta(\omega) = \angle K - \tan^{-1}\left[\frac{\omega/\omega_0 - \omega_0/\omega}{2\zeta}\right] \tag{12–27}$$

At low frequency ($\omega \ll \omega_0$) the phase angle approaches $\theta(\omega) \rightarrow \angle K + 90°$. At the center frequency ($\omega = \omega_0$) the phase reduces to $\theta(\omega_0) = \angle K$. Finally, at high frequencies ($\omega \gg \omega_0$) the phase approaches $\theta(\omega) \rightarrow \angle K - 90°$. Graphs of the phase response for several values of damping are shown in Figure 12–24. The phase shift at the center frequency is 90° and the total phase change is −180° regardless of the damping ratio. However, the transition from +90° to −90° is very abrupt for the narrowband case ($\zeta = 0.05$) and more gradual for the wideband case ($\zeta = 5$).

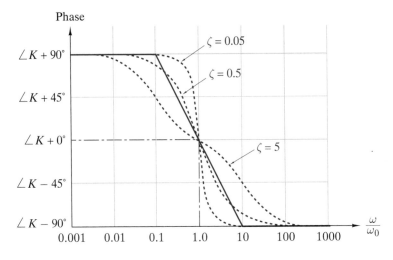

Phase

$\angle K + 90°$

$\angle K + 45°$

$\angle K + 0°$

$\angle K - 45°$

$\angle K - 90°$

0.001   0.01   0.1   1.0   10   100   1000 $\dfrac{\omega}{\omega_0}$

$\zeta = 0.05$

$\zeta = 0.5$

$\zeta = 5$

FIGURE 12–24 *Second-order band-pass phase responses.*

EXAMPLE 12–9

(a) Show that the transfer function $T(s) = V_2(s)/V_1(s)$ of the circuit in Figure 12–25 has a bandpass characteristic.

(b) Select the element values to obtain a center frequency of 455 kHz and a bandwidth of 10 kHz.[2]

(c) Calculate the cutoff frequencies for the values selected in (b).

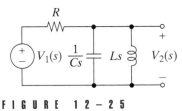

FIGURE 12–25

SOLUTION:

(a) The given circuit is a voltage divider with branch impedances

$$Z_1 = R$$

$$Z_2 = \frac{1}{Cs + \dfrac{1}{Ls}} = \frac{Ls}{LCs^2 + 1}$$

By voltage division, the transfer function is

$$T(s) = \frac{Z_2}{Z_1 + Z_2} = \frac{Ls}{RLCs^2 + Ls + R}$$

$$= \frac{s/RC}{s^2 + s/RC + 1/LC}$$

This transfer function is of the standard bandpass form $Ks/(s^2 + 2\zeta\omega_0 s + \omega_0^2)$.

Circuit Interpretation: The circuit's bandpass response is easily explained by the element impedances at zero and infinite frequency. At zero frequency the inductor acts like a short circuit that forces the output down to zero. At infinite frequency the capacitor is a short circuit again, forcing the

---

2 These values are typical of the bandpass filters in a conventional AM radio receiver.

output to zero. Thus, the circuit has zero gain at dc and zero gain at infinite frequency, which is the characteristic pattern of a bandpass response.

(b) Comparing the circuit's transfer function with the standard form yields the following relationships:

$$\omega_0^2 = \frac{1}{LC}$$

$$\zeta = \frac{\sqrt{L/C}}{2R}$$

The center frequency and bandwidth requirements impose the following numerical values:

$$\omega_0 = 2\pi \times 455 \times 10^3 = 2.86 \times 10^6 \text{ rad/s}$$

$$\zeta = \frac{B}{2\omega_0} = \frac{2\pi \times 10 \times 10^3}{2(2\pi \times 455 \times 10^3)} = 0.0110$$

and the resulting design constraints on the element values are

$$LC = \frac{1}{\omega_0^2} = 1.22 \times 10^{-13} \quad \text{and} \quad \frac{\sqrt{L/C}}{2R} = \zeta = 0.011$$

Selecting $C = 10$ nF yields $L = 12.2$ μH and $R = 1.59$ kΩ.

(c) Using Eq. (12–24), the cutoff frequencies are

$$f_{C1} = \frac{\omega_0(-\zeta + \sqrt{1 + \zeta^2})}{2\pi} = 450.2 \text{ kHz}$$

$$f_{C2} = \frac{\omega_0(+\zeta + \sqrt{1 + \zeta^2})}{2\pi} = 460.2 \text{ kHz}$$

Since the damping ratio is very small, the natural poles are quite close to the j-axis, producing a very narrow resonant peak at 455 kHz. ■

Exercise 12–3

Find the natural frequency, damping ratio, cutoff frequencies, bandwidth, and maximum gain of the following bandpass transfer functions:

$$\text{(a) } T(s) = \frac{500s}{s^2 + 200s + 10^6}$$

$$\text{(b) } T(s) = \frac{5 \times 10^{-3} s}{2.5 \times 10^{-5}s^2 + 5 \times 10^{-3}s + 1}$$

Answers:

	ω_0(rad/s)	ζ	ω_{C1}(rad/s)	ω_{C2}(rad/s)	B(rad/s)	$|T(j\omega)|_{MAX}$
(a)	1000	0.1	905	1105	200	2.5
(b)	200	0.5	123.6	323.6	200	1

Exercise 12–4

A bandpass circuit has a center frequency of $\omega_0 = 1$ krad/s and a bandwidth of $B = 10$ krad/s.

Find the damping ratio ζ and cutoff frequencies. Is this a narrowband or broadband circuit?

Answers:

$\zeta = 5$, $\omega_{C1} = 99.02$ rad/s, $\omega_{C2} = 10.099$ krad/s, broadband

---

Exercise 12–5

A bandpass circuit has cutoff frequencies at $\omega_{C1} = 20$ krad/s and $\omega_{C2} = 50$ krad/s. Find the center frequency ω_0, damping ratio ζ, and bandwidth B.

Answers:

$\omega_0 = 31.6$ krad/s, $\zeta = 0.474$, $B = 30$ krad/s

SECOND-ORDER LOW-PASS CIRCUITS

The transfer function of a second-order low-pass prototype has the form

$$T(s) = \frac{K}{s^2 + 2\zeta\omega_0 s + \omega_0^2} \tag{12–28}$$

The frequency response is found by replacing s by $j\omega$.

$$T(j\omega) = \frac{K}{\omega_0^2 - \omega^2 + j2\zeta\omega_0\omega} \tag{12–29}$$

At low frequencies ($\omega \ll \omega_0$) the gain approaches $|T(j\omega)| \to |K|/\omega_0^2 = |T(0)|$. At high frequencies ($\omega \gg \omega_0$) the gain approaches $|T(j\omega)| \to |K|/\omega^2$. The low-frequency asymptote is a constant equal to the zero frequency or dc gain. The high-frequency asymptote is inversely proportional to the square of the frequency. As a result, the high-frequency gain asymptote has a slope of –40 dB/decade (slope = –2). The two asymptotes intersect to form a corner at $\omega = \omega_0$, as indicated in Figure 12–26. The two asymptotes display a low-pass gain characteristic with a passband below ω_0, and a stopband above ω_0. The passband gain is the dc gain $|T(0)|$ and the slope of the gain asymptote in the stopband is –40 dB/decade.

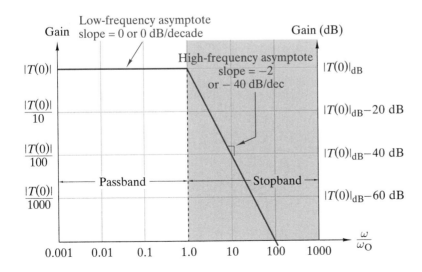

FIGURE 12–26 *Asymptotic gain of a second-order low-pass response.*

The influence of the damping ratio on the gain response can be illustrated by evaluating the gain at $\omega = \omega_0$:

$$\left|T(j\omega_0)\right| = \frac{|K|/\omega_0^2}{2\zeta} = \frac{|T(0)|}{2\zeta} \qquad (12\text{–}30)$$

This result and the plots in Figure 12–27 show that the actual gain curve lies above the asymptotes when $\zeta < 0.5$, below the asymptotes when $\zeta > 0.5$, and close to the asymptotes when $\zeta = 0.5$. When $\zeta = 1/\sqrt{2}$ then the gain at $\omega = \omega_0$ is

$$\left|T(j\omega_0)\right| = \frac{|T(0)|}{\sqrt{2}} \qquad (\zeta = 1/\sqrt{2}) \qquad (12\text{–}31)$$

FIGURE 12–27 *Second-order low-pass gain responses.*

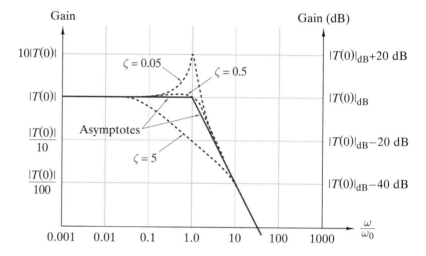

Since the passband gain is $|T(0)|$, this result means that the cutoff frequency is equal to ω_0 when $\zeta = 1/\sqrt{2}$

The distinctive feature of the low-pass case is a narrow resonant peak in the neighborhood of ω_0 for lightly damped circuits. The actual maximum gain of a second-order low-pass response is

$$\left|T(j\omega)\right|_{\text{MAX}} = \frac{|T(0)|}{2\zeta\sqrt{1 - \zeta^2}} \qquad (12\text{–}32)$$

and occurs at

$$\omega_{\text{MAX}} = \omega_0\sqrt{1 - 2\zeta^2} \qquad (12\text{–}33)$$

Note that the ω_{MAX} is always less than the natural frequency ω_0. Deriving Eqs. (12–32) and (12–33) is straightforward and is left as a problem at the end of this chapter (Problem 12–20).

Figure 12–28 compares the response of the first-order low-pass prototype with the second-order gain response for $\zeta = 1/\sqrt{2}$. The important difference is that the gain slope in the stopband is –2 or –40 dB/decade for the second-order response and only –20 dB/decade for the first-order circuit. The two poles in the second-order case produce a steeper slope because the high-frequency response decreases as $1/\omega^2$, rather than $1/\omega$ as in the first-order case. The generalization is that the high-frequency gain asymptote of an n-pole low-pass function has a slope of $-n$ or $-20n$ dB/decade.

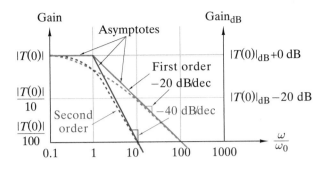

FIGURE 12–28 *Comparison of first- and second-order gain responses.*

EXAMPLE 12–10

(a) Show that the transfer function $T(s) = V_2(s)/V_1(s)$ of the circuit in Figure 12–29 has a low-pass gain characteristic.

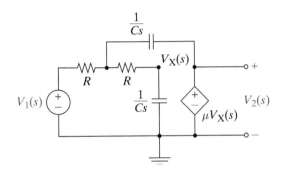

FIGURE 12–29

(b) Derive expressions relating the damping ratio ζ and undamped natural frequency ω_0 to the circuit element values. Use these relationships to select the element values to produce $\zeta = 0.1$ and $\omega_0 = 5$ krad/s.

SOLUTION:

(a) The transfer function of this circuit is found in Example 11–6 using node-voltage analysis.

$$T(s) = \frac{V_2(s)}{V_1(s)} = \frac{\mu}{(RCs)^2 + (3 - \mu)RCs + 1}$$

This is a second-order low-pass transfer function of the form $K/(s^2/\omega_0^2 + 2\zeta s/\omega_0 + 1)$.

Circuit Interpretation: At zero frequency both capacitors in Figure 12–29 act like open circuits and the control voltage $V_X(s) = V_1(s)$ since there is no voltage drop across either resistor. As a result, the dc output voltage is $V_2(s) = \mu V_X(s) = \mu V_1$. In other words, the dc gain of the circuit is $T(0) = \mu$. At infinite frequency both capacitors act like short circuits. The shunt capacitor shorts the control node X to ground so that $V_X = 0$, the dependent voltage source μV_X is effectively turned off, and the output voltage is zero. The circuit has a finite gain at dc and zero gain at infinite frequency, which is the characteristic pattern of a low-pass response.

(b) Comparing the circuit transfer function with the standard form yields the following relationships:

$$\omega_0 = \frac{1}{RC} \quad \text{and} \quad \zeta = \frac{1}{2}(3 - \mu)$$

For the specified values the design constraints on the element values are

$$RC = \frac{1}{\omega_0} = 2 \times 10^{-4} \quad \text{Let } R = 10 \text{ k}\Omega \quad \text{Then } C = 20 \text{ nF}$$

$$\mu = 3 - 2\zeta = 2.8$$

The maximum gain for these conditions is

$$|T(j\omega)|_{\text{MAX}} = \frac{|T(0)|}{2\zeta\sqrt{1 - \zeta^2}} = 5.03 \times |T(0)|$$

The peak gain is five times as large as the dc gain. Thus, lightly damped second-order circuits of all types have strong resonant peaks in their gain response. ∎

Exercise 12–6

Find the natural frequency ω_0 damping ratio ζ, dc gain $T(0)$, and maximum gain of the following second-order low-pass transfer functions:

(a) $T(s) = \dfrac{(4000)^2}{(s + 3000)^2 + (4000)^2}$

(b) $T(s) = \dfrac{5}{0.25 \times 10^{-4}s^2 + 10^{-2}s + 1}$

Answers:

(a) $\omega_0 = 5$ krad/s, $\zeta = 0.6$, $T(0) = 0.64$, $T_{\text{MAX}} = 0.667$
(b) $\omega_0 = 200$ rad/s, $\zeta = 1.00$, $T(0) = 5$, $T_{\text{MAX}} = T(0)$

Exercise 12–7

The impulse response of a second-order low-pass circuit is

$$h(t) = 2000[e^{-1000t} \sin 1000t]u(t)$$

Find the natural frequency ω_0, damping ratio ζ, dc gain $T(0)$, and maximum gain.

Answers:
$\omega_0 = 1414$ rad/s, $\zeta = 0.707$, $T(0) = T_{\text{MAX}} = 1$

SECOND-ORDER HIGH-PASS CIRCUITS

To complete the study of second-order circuits, we treat the high-pass case.

$$T(s) = \frac{Ks^2}{s^2 + 2\zeta\omega_0 s + \omega_0^2} \tag{12-34}$$

The high-pass frequency response is a mirror image of the low-pass case reflected about $\omega = \omega_0$. At low frequencies ($\omega \ll \omega_0$) the gain approaches $|T(j\omega)| \to |K|\omega^2/\omega_0^2$. At high frequencies ($\omega \gg \omega_0$) the gain approaches $|T(j\omega)| \to |K| = |T(\infty)|$. The high-frequency gain is a constant $|T(\infty)|$ called the infinite frequency gain. The low-frequency asymptote is inversely proportional to the square of the frequency (slope = +2 or +40 dB/decade). The two asymptotes intersect at the natural frequency ω_0, as shown in Figure 12–30. The asymptotes display a high-pass gain response with a passband above ω_0 and a stopband below ω_0. The passband gain is $|T(\infty)|$ and the slope of the gain asymptote in the stopband is +2 or +40 dB/decade.

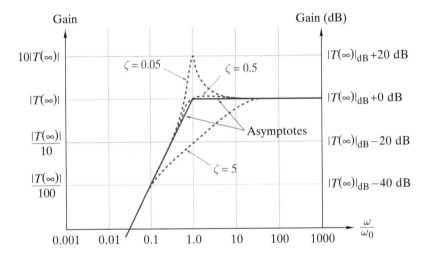

FIGURE 12–30 Second-order high-pass gain responses.

Figure 12–30 also shows the effect of the damping ratio on the actual gain curve. Evaluating the gain response at $\omega = \omega_0$ produces

$$\left|T(j\omega_0)\right| = \frac{|K|}{2\zeta} = \frac{|T(\infty)|}{2\zeta} \tag{12-35}$$

This result and the plots in Figure 12–30 show that the gain response will be above the asymptotes for $\zeta < 0.5$, below for $\zeta > 0.5$, and close to the asymptotes for $\zeta = 0.5$. When $\zeta = 1/\sqrt{2}$ the gain in Eq. (12–35) is $|T(\infty)|/\sqrt{2}$. That is, ω_0 is the cutoff frequency when $\zeta = 1/\sqrt{2}$.

Again, the distinctive feature of the second-order response is the peak in the neighborhood of ω_0 for lightly damped circuits with $\zeta \ll 0.5$. For the second-order high-pass function the maximum gain is

$$\left|T(j\omega)\right|_{\text{MAX}} = \frac{|T(\infty)|}{2\zeta\sqrt{1 - \zeta^2}} \tag{12-36}$$

and occurs at

$$\omega_{\text{MAX}} = \frac{\omega_0}{\sqrt{1 - 2\zeta^2}} \tag{12-37}$$

The frequency at which the gain maximum occurs is always above ω_0. Deriving Eqs. (12–36) and (12–37) is straightforward and is similar to the low-pass case.

EXAMPLE 12–11

(a) Show that the transfer function $T(s) = V_2(s)/V_1(s)$ of the circuit in Figure 12–31 has a high-pass characteristic.

FIGURE   12 – 31

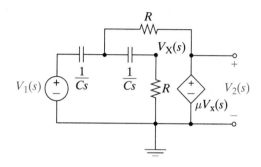

(b) Derive expressions relating the damping ratio ζ and undamped natural frequency ω_0 to the circuit element values.

SOLUTION:

This high-pass circuit is obtained from the low-pass circuit in Figure 12–29 by interchanging the positions of the resistors and capacitors. This interchange is an example of a transformation that converts any low-pass RC circuit into a high-pass RC circuit. The CR-RC transformation is carried out by:

1. Replacing every impedance R by an impedance $1/Cs$, and
2. Replacing every impedance $1/Cs$ by an impedance R.

Applying this transformation at the circuit level converts the low-pass circuit in Figure 12–29 into the high-pass circuit in Figure 12–31. The transformation can also be applied at the transfer function level. The transfer function of the low-pass prototype is given in Example 12–10 as

$$T_{LP}(s) = \frac{\mu}{(RCs)^2 + (3 - \mu)RCs + 1}$$

$$= \frac{\mu}{(R)^2\left(\dfrac{Cs}{1}\right)^2 + (3 - \mu)(R)\left(\dfrac{Cs}{1}\right) + 1}$$

Performing the CR-RC transformation defined previously on the second expression yields

$$T_{HP}(s) = \frac{\mu}{\left(\dfrac{1}{Cs}\right)^2\left(\dfrac{1}{R}\right)^2 + (3 - \mu)\left(\dfrac{1}{Cs}\right)\left(\dfrac{1}{R}\right) + 1}$$

$$= \frac{\mu(RCs)^2}{(RCs)^2 + (3 - \mu)RCs + 1}$$

The low-pass transfer function is converted into a high-pass form with an infinite frequency gain of $|T(\infty)| = \mu$. Comparing the denominator of $T_{HP}(s)$ with the standard second-order form yields

$$\omega_0 = \frac{1}{RC} \quad \text{and} \quad \zeta = \frac{3 - \mu}{2}$$

These are the same relationships we obtained for the low-pass prototype in Example 12–10. Note that the *CR-RC* transformation only affects the resistors and capacitors and does not change the dependent source.

Circuit Interpretation: It is not hard to see why the circuit in Figure 12–31 has a high-pass response with zero gain at dc and finite gain at infinite frequency. At zero frequency both capacitors act like open circuits that disconnect the input source, leading to zero output and zero gain. At infinite frequency the capacitors act like short circuits that directly connect the input source to the control voltage node, forcing the condition $V_X(s) = V_1(s)$. As a result, the output is $V_2(s) = \mu V_1(s)$, which means the infinite frequency gain is μ.   ■

The transfer function of a high-pass circuit is $T(s) = 0.2s^2/(s^2 + 10^4 s + 10^6)$. Find the undamped natural frequency ω_0, damping ratio ζ, high-frequency gain $T(\infty)$, and maximum gain of the transfer function.

Answers:
$\omega_0 = 10^3$ rad/s, $\zeta = 5$, $T(\infty) = 0.2$, $T_{MAX} = T(\infty)$

The step response of a second-order high-pass circuit is

$$g(t) = [e^{-1000t} \cos 2000t - 0.5e^{-1000t} \sin 2000t]u(t)$$

Find the natural frequency ω_0, damping ratio ζ, high-frequency gain $T(\infty)$, and maximum gain of the circuit.

Answers:
$\omega_0 = 2236$ rad/s, $\zeta = 0.447$, $T(\infty) = 1$, $T_{MAX} = 1.25$

12–4   THE FREQUENCY RESPONSE OF *RLC* CIRCUITS

The series *RLC* circuit and the parallel *RLC* circuit are canonical examples traditionally used in electrical engineering to illustrate the response of second-order circuits. These circuits give us a special insight because the form of a response is easily related to the circuit elements. We first encountered these circuits in Chapter 8, where they served as vehicles for studying the solution of second-order differential equations. In this section we concentrate on the frequency response of the traditional *RLC* circuits starting with the series circuit in Figure 12–32.

By voltage division, the transfer function of the series *RLC* circuit in Figure 12–32 is

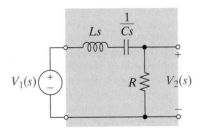

F I G U R E   1 2 – 3 2 *Series RLC bandpass circuit.*

$$T(s) = \frac{R}{Ls + 1/Cs + R} = \frac{\dfrac{R}{L}s}{s^2 + \dfrac{R}{L}s + \dfrac{1}{LC}} \qquad (12\text{-}38)$$

Comparing this result with the standard bandpass in Eq. (12–14), we conclude that the circuit has a bandpass response with

$$K = \frac{R}{L}$$

$$\omega_0 = \frac{1}{\sqrt{LC}} \qquad (12\text{-}39)$$

$$2\zeta\omega_0 = \frac{R}{L}$$

Inserting these conclusions into Eqs. (12–19), (12–24), and (12–26), we find the descriptive parameters of the passband to be

$$|T(j\omega)|_{\text{MAX}} = \frac{K}{2\zeta\omega_0} = 1$$

$$B = 2\zeta\omega_0 = \frac{R}{L} \qquad (12\text{-}40)$$

and

$$\omega_{C1} = -\frac{R}{2L} + \sqrt{\left(\frac{R}{2L}\right)^2 + \frac{1}{LC}}$$

$$\omega_{C2} = +\frac{R}{2L} + \sqrt{\left(\frac{R}{2L}\right)^2 + \frac{1}{LC}} \qquad (12\text{-}41)$$

For historical reasons, it is traditional to add a fifth descriptive parameter called the **quality factor** Q.[3] We can define the quality factor of any bandpass filter as $Q = \omega_0/B$. In general, this means that for a second-order circuit, $Q = 1/2\zeta$. In the particular case of the series RLC circuit, this requires

$$Q = \frac{1/\sqrt{LC}}{R/L} = \frac{\sqrt{L/C}}{R} \qquad (12\text{-}42)$$

The consequences of these results are illustrated graphically in Figure 12–33. The maximum gain is always 1 (or 0 dB) regardless of the bandwidth. The bandwidth is directly proportional to the resistance, so ω_{C1} decreases and ω_{C2} increases as R increases. The boundary between broadband and narrowband occurs when $\zeta = 0.5$, which requires that $Q = 1$. Figure 12–33 shows the response for values of Q one decade above and one decade

---

3 Q was originally defined as a measure of the quality of lossy inductors. In modern usage it is a substitute parameter for the damping ratio of a second-order factor. The relationship between the two parameters is $2\zeta = 1/Q$. Since the two are reciprocally related, their comparative adjectives are reversed; that is, high Q means low damping and low Q means high damping.

below this boundary value. Increasing Q causes the passband to shrink into a high-Q, narrowband response. Conversely, decreasing Q expands the passband into a low-Q, broadband response. But regardless of the value of Q, the maximum gain remains anchored at $\omega = \omega_0$ with a value of 1.

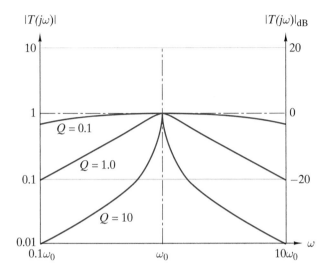

F I G U R E   1 2 – 3 3 *Effect of Q on the series RLC bandpass response.*

What is interesting, and worth remembering, about this circuit is the way the impedances of the inductor and capacitor combine to produce the bandpass response. Again, using voltage division in Figure 12–32, we write the transfer function in the form

$$T(j\omega) = \frac{R}{R + Z_{LC}(j\omega)}$$

where

$$Z_{LC}(j\omega) = j\omega L + \frac{1}{j\omega C} = j\left(\omega L - \frac{1}{\omega C}\right)$$

is the impedance of inductor and capacitor in the series leg of the voltage divider. The variation of this impedance affects the gain response as follows:

- At $\omega = 0$ the impedance $Z_{LC}(j\omega)$ is infinite since the capacitor acts like an open circuit. The open circuit disconnects the input source, leading to zero output and $|T(0)| = 0$.

- At $\omega = \infty$ the impedance $Z_{LC}(j\omega)$ is infinite since the inductor acts like an open circuit. The open circuit disconnects the input source, leading to zero output and $|T(\infty)| = 0$.

- At $\omega = \omega_0$ the impedance $Z_{LC}(j\omega)$ is zero since

$$Z_{LC}(j\omega_0) = j\left(\omega_0 L - \frac{1}{\omega_0 C}\right) = j\left(\sqrt{\frac{L}{C}} - \sqrt{\frac{L}{C}}\right) = 0$$

The inductor and capacitor together act like a short circuit that directly connects the input across the output, leading to $|T(j\omega_0)| = 1$.

- At $\omega = \omega_{C1}$ the impedance $Z_{LC}(j\omega) = -jR$ since this value yields

$$T(j\omega_{C1}) = \frac{R}{R - jR} = \frac{1}{\sqrt{2}} \angle + 45°$$

The inductor and capacitor together act like an impedance $-jR$ with the result that the the gain at the lower cutoff frequency is $|T(j\omega_{C1})| = 1/\sqrt{2}$ as required.

- At $\omega = \omega_{C2}$ the impedance $Z_{LC}(j\omega) = +jR$ since this value yields

$$|T(j\omega_{C2})| = \frac{R}{R + jR} = \frac{1}{\sqrt{2}} \angle - 45°$$

The inductor and capacitor together act like an impedance $+jR$ with the result that the the gain at the upper cutoff frequency is $|T(j\omega_{C2})| = 1/\sqrt{2}$ as required.

Figure 12–34 summarizes these observations graphically and shows how the impedance $Z_{LC}(j\omega)$ controls the gain response of the circuit. Knowing

F I G U R E   1 2 – 3 4 *Effect of Z_{LC} on the series RLC bandpass response.*

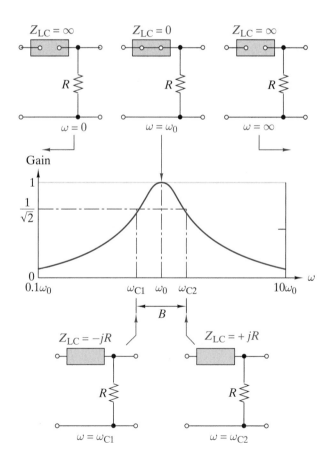

this impedance at five frequencies is enough to describe the bandpass characteristics.

EXAMPLE 12–12

A series *RLC* circuit has a resonant frequency of $\omega_0 = 20$ krad/s, $Q = 5$, and a resistance of 50 Ω. Find the values of L, C, B, ω_{C1}, and ω_{C2}.

SOLUTION:
From the definition of Q, we have $B = \omega_0/Q = 4000$ rad/s. For the series *RLC* circuit, $B = R/L$; hence we can calculate the inductance as $L = R/B = 12.5$ mH and the capacitance as $C = 1/\omega_0^2 L = 0.2$ µF. Inserting these results into Eq. (12–41) yields the cutoff frequencies as

$$\omega_{C1} = -\frac{4000}{2} + \sqrt{\left(\frac{4000}{2}\right)^2 + 20{,}000^2} = 18.1 \text{ krad/s}$$

$$\omega_{C2} = +\frac{4000}{2} + \sqrt{\left(\frac{4000}{2}\right)^2 + 20{,}000^2} = 22.1 \text{ krad/s} \qquad \blacksquare$$

EXAMPLE 12–13

(a) Show that the current transfer function $T(s) = I_2(s)/I_1(s)$ in Figure 12–35 has a bandpass gain response with a center frequency at $\omega_0 = 1/\sqrt{LC}$.
(b) Derive expression relating the descriptive parameters B, Q, ω_{C1}, and ω_{C2} to the circuit parameters R, L, and C.

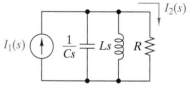

FIGURE   12–35   *Parallel RLC bandpass circuit.*

SOLUTION:
(a) The current transfer function is derived using current division as

$$T(s) = \frac{I_2(s)}{I_1(s)} = \frac{Y_R}{Y_L + Y_C + Y_R} = \frac{1/R}{1/Ls + Cs + 1/R}$$

$$= \frac{\dfrac{s}{RC}}{s^2 + \dfrac{s}{RC} + \dfrac{1}{LC}}$$

Comparing this transfer function to the standard bandpass form in Eq. (12–14), we see that this is a bandpass response with

$$K = \frac{1}{RC} \qquad \omega_0 = \frac{1}{\sqrt{LC}} \qquad 2\zeta\omega_0 = \frac{1}{RC}$$

(b) In general, the bandwidth of a second-order bandpass response is $B = 2\zeta\omega_0$. Hence for the parallel *RLC* circuit the bandwidth and Q are

$$B = \frac{1}{RC} \quad \text{and} \quad Q = \frac{\omega_0}{B} = \frac{R}{\sqrt{L/C}}$$

Inserting these results into the Eq. (12–24) leads to the following expressions:

$$\omega_{C1} = -\frac{1}{2RC} + \sqrt{\left(\frac{1}{2RC}\right)^2 + \frac{1}{LC}}$$

$$\omega_{C2} = +\frac{1}{2RC} + \sqrt{\left(\frac{1}{2RC}\right)^2 + \frac{1}{LC}}$$ ∎

Exercise 12–10

A series RLC circuit has cutoff frequencies of ω_{C1} = 100 rad/s and ω_{C2} = 10 krad/s. Find the values of B, ω_0, and Q. Does the circuit have a broadband or narrowband response?

Answers:
B = 9.9 krad/s, ω_0 = 1 krad/s, Q = 0.101, broadband

Exercise 12–11

A parallel RLC bandpass circuit has a center frequency at ω_0 = 200 krad/s, bandwidth of B = 10 krad/s, and a resistance of R = 10 kΩ. Find ω_0 and B when the resistance is increased to 40 kΩ.

Answers:
ω_0 = 200 krad/s, B = 2.5 krad/s

THE SERIES *RLC* BANDSTOP CIRCUIT

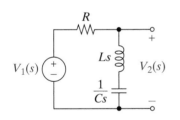

FIGURE 12–36 *Series RLC bandstop circuit.*

The transfer function of the series RLC circuit in Figure 12–36 can be derived by voltage division as

$$T(s) = \frac{Ls + 1/Cs}{Ls + 1/Cs + R} = \frac{s^2 + \dfrac{1}{LC}}{s^2 + \dfrac{R}{L}s + \dfrac{1}{LC}}$$

If we define $\omega_0 = 1/\sqrt{LC}$ then $|T(j\omega_0)| = 0$ because the transfer function has zeros on the j-axis at $s = \pm j\omega_0$. The dc gain is $|T(0)| = 1$ and the infinite frequency gain is $|T(\infty)| = 1$. Thus, this transfer function has finite gain at dc and infinite frequency, and zero gain at an intermediate frequency. These features are indicative of a bandstop response.

The bandstop circuit in Figure 12–36 and the bandpass circuit in Figure 12–32 are both series RLC circuits. The difference is that the bandstop circuit takes its output across the inductor and capacitor in series while the bandpass circuit defines its output across the resistor. In either case, the shape of the frequency response is controlled by the impedance $Z_{LC}(j\omega) = j\omega L + 1/j\omega C$.

Figure 12–37 shows how this impedance controls the shape of the bandstop response. At dc and infinite frequency Z_{LC} acts like an open circuit so the output voltage equals the input voltage and the passband gains are both one. At $\omega = \omega_0$ the Z_{LC} acts like a short circuit so the output voltage and the gain are both zero. At the cutoff frequencies $Z_{LC} = \pm jR$, so the passband gains are reduced to $1/\sqrt{2}$. In sum, the factors defining the shape of the bandstop response are the same as those defining the shape of the bandpass response.

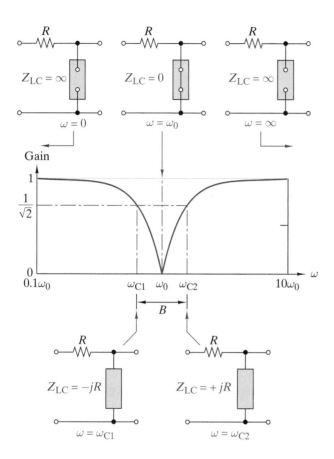

FIGURE 12−37 *Effect of* Z_{LC} *on the series RLC bandstop response.*

The net result is that the equations relating circuit parameters and the stopband descriptive parameters are the same as the equations for the bandpass case. That is, we can use Eqs. (12–39), (12–40), and (12–41) for both the bandpass and bandstop series *RLC* circuits. The difference is that B, ω_{C1}, and ω_{C2} describe the passband in the former and the stopband in the latter.

D **DESIGN EXAMPLE 12–14**

A series connection of an inductor and capacitor are to be connected across the output of a source whose Thévenin resistance is 50 Ω. The goal is to create a bandstop filter to eliminate a spurious (undesirable) source output at 25 krad/s. Select the values of L and C that eliminate the spurious response and produce a stopband bandwidth less than 1 krad/s.

SOLUTION:

Using Eq. (12–40), we have $B = R/L < 1000$. Since $R = 50\ \Omega$, this constraint requires $L > 50$ mH. Using Eq. (12–39), the bandstop notch must be located at $\omega_0 = 1/\sqrt{LC} = 25,000$. Selecting $L = 100$ mH to meet the bandwidth constraint means that $C = 1/(\omega_0^2 L) = 16$ nF.   ∎

Exercise 12–12

A series RLC circuit has a bandstop response with a B of 2 krad/s and a notch at $\omega_0 = 10$ krad/s. Find the circuit gain at $\omega = 9$ krad/s and $\omega = 11$ krad/s.

A n s w e r s : 0.726, 0.690

12–5   **B**ODE **D**IAGRAMS

Bode diagrams are separate graphs of the gain $|T(j\omega)|_{dB}$ and phase $\theta(j\omega)$ versus log-frequency scales. The frequency-response performance of devices, circuits, and systems is often presented in this format. The poles and zeros of the transfer functions can be estimated from Bode plots of experimental frequency-response data. Thus, the Bode plot format is a very useful way to develop an understanding of frequency response of circuits and systems. Any of our computer-aided analysis tools can precisely calculate and plot Bode diagrams. However, often a simple hand-calculated approximation will tell us what we want to know about the frequency response of a circuit.

The purpose of this section is to present a method of quickly drawing straight-line approximations to Bode plot of transfer functions with real poles and zeros. In an analysis situation the straight-line approximations tell us how the poles and zeros of $T(s)$ affect frequency response. In circuit design these straight-line plots serve as a shorthand notation for outlining design approaches or developing design requirements. The straight-line plots help us use computer-aided analysis effectively because they tell us what frequency ranges are important and what features of the response need to be investigated in greater detail.

The straight-line versions of Bode plots are particularly useful when the poles and zeros are located on the real axis in the s plane. As a starting place, consider the transfer function

$$T(s) = \frac{Ks(s + \alpha_1)}{(s + \alpha_2)(s + \alpha_3)} \tag{12–43}$$

where K, α_1, α_2, and α_3 are real. This function has zeros at $s = 0$ and $s = -\alpha_1$, and poles at $s = -\alpha_2$ and $s = -\alpha_3$. All of these critical frequencies lie on the real axis in the s plane. When making Bode plots we put $T(j\omega)$ in a standard format obtained by factoring out α_1, α_2, and α_3:

$$T(j\omega) = \left(\frac{K\alpha_1}{\alpha_2\alpha_3}\right) \frac{j\omega(1 + j\omega/\alpha_1)}{(1 + j\omega/\alpha_2)(1 + j\omega/\alpha_3)} \tag{12–44}$$

Using the following notation

$$\text{Magnitude} = M = |1 + j\omega/\alpha| = \sqrt{1 + (\omega/\alpha)^2}$$

$$\text{Angle} = \theta = \angle(1 + j\omega/\alpha) = \tan^{-1}(\omega/\alpha) \qquad (12\text{–}45)$$

$$\text{Scale Factor} = K_0 = \frac{K\alpha_1}{\alpha_2\alpha_3}$$

we can write the transfer function in Eq. (12–44) in the form

$$T(j\omega) = K_0 \frac{(\omega\, e^{j90°})(M_1\, e^{j\theta_1})}{(M_2\, e^{j\theta_2})(M_3\, e^{j\theta_3})} = \frac{|K_0|\omega M_1}{M_2 M_3} e^{j(\angle K_0 + 90° + \theta_1 - \theta_2 - \theta_3)} \qquad (12\text{–}46)$$

The gain (in dB) and phase responses are

$$|T(j\omega)|_{dB} = \underbrace{20\log_{10}|K_0|}_{\text{Scale Factor}} + \underbrace{20\log_{10}\omega}_{\text{Zero}} + \underbrace{20\log_{10} M_1}_{\text{Zero}} - \underbrace{20\log_{10} M_2}_{\text{Pole}} - \underbrace{20\log_{10} M_3}_{\text{Pole}}$$

$$\theta(\omega) = \qquad \angle K_0 \quad + \quad 90° \quad + \quad \theta_1 \quad - \quad \theta_2 \quad - \quad \theta_3$$

$$(12\text{–}47)$$

The terms in Eq. (12–47) caused by zeros have positive signs and increase the gain and phase angle, while the pole terms have negative signs and decrease the gain and phase.

The summations in Eq. (12–47) illustrate a general principle. In a Bode plot, the gain and phase responses are determined by the following types of factors:

1. The scale factor K_0

2. A factor of the form $j\omega$ due to a zero or a pole at the origin

3. Factors of the form $(1 + j\omega/\alpha)$ caused by a zero or pole at $s = -\alpha$.

We can construct Bode plots by considering the contributions of these three factors.

The Scale Factor. The gain and phase contributions of the scale factor are constants that are independent of frequency. The gain contribution $20\log_{10}|K_0|$ is positive when $|K_0| > 1$ and negative when $|K_0| < 1$. The phase contribution $\angle K_0$ is $0°$ when $K_0 > 0$ and $\pm 180°$ when $K_0 < 0$.

The Factor $j\omega$. A simple zero or pole at the origin contributes $\pm 20\log_{10}\omega$ to the gain and $\pm 90°$ to the phase, where the plus sign applies to a zero and the minus to a pole. When $T(s)$ has a factor s^n in the numerator (denominator), it has a zero (pole) of order n at the origin. Multiple zeros or poles at $s = 0$ contribute $\pm 20n\log_{10}\omega$ to the gain and $\pm n90°$ to the phase. Figure 12–38 shows that the gain factors contributed by zeros and poles at the origin are straight lines that pass through a gain of 1 (0 dB) at $\omega = 1$.[4]

The Factor $1 + j\omega/a$. The gain contributions of first-order zeros and poles are shown in Figure 12–39. Like the first-order transfer functions studied earlier in this chapter, these factors produce straight-line gain

---

4 Strictly speaking, a circuit with a natural pole at the origin is unstable and does not have a sinusoidal steady-state response. Nevertheless, it is traditional to treat poles at the origin in Bode diagrams because there are practical applications in which such poles are important considerations.

FIGURE **1 2 – 3 8** *Gain responses of poles and zeros at* s = 0.

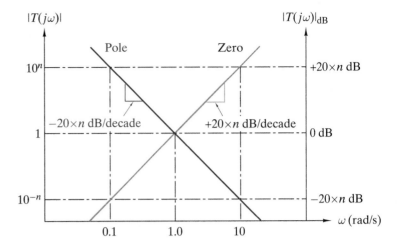

FIGURE **1 2 – 3 9** *Gain responses of real poles and zeros at* s = α.

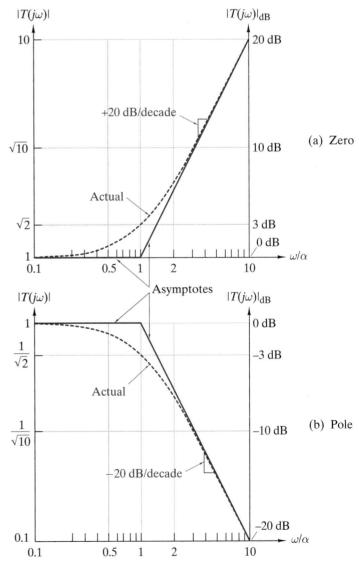

asymptotes at low and high frequency. In a Bode plot the low-frequency ($\omega \ll \alpha$) asymptotes are horizontal lines at with gain = 1 (0 dB). The high-frequency ($\omega \gg \alpha$) asymptotes are straight lines of the form $\pm \omega/\alpha$ ($\pm 20 \log(\omega/\alpha)$ dB), where the plus sign applies to a zero and the minus to a pole. The high-frequency gain asymptote is proportional to the frequency ω (slope = +1 or +20 dB/decade) for a zero and proportional to $1/\omega$ (slope = −1 or −20 dB/decade) for a pole. In either case, the low- and high-frequency asymptotes intersect at the corner frequency $\omega = \alpha$.

To construct straight-line (SL) gain approximations, we develop a piecewise linear function $|T(j\omega)|_{SL}$ defined by the asymptotes of each of the factors in $T(j\omega)$. The function $|T(j\omega)|_{SL}$ has a corner frequency at each of the critical frequencies of the transfer function. At frequencies below a corner ($\omega < \alpha$) a first-order factor is represented by a gain of 1. Above the corner frequency ($\omega > \alpha$) the factor is represented by its high-frequency asymptote ω/α. To generate $|T(j\omega)|_{SL}$ we start out below the lowest critical frequency with a low-frequency baseline that accounts for the scale factor K_0 and any poles or zeros at the origin. We increase frequency and change the form of $|T(j\omega)|_{SL}$ whenever we pass a corner frequency. We proceed upward in frequency until we have gone beyond the highest critical frequency, at which point we have a complete expression for $|T(j\omega)|_{SL}$.

The following example illustrates the process.

EXAMPLE 12–15

(a) Construct the Bode plot of the straight-line approximation of the gain of the transfer function

$$T(s) = \frac{12{,}500\,(s + 10)}{(s + 50)\,(s + 500)}$$

(b) Find the point at which the high-frequency gain falls below the dc gain.

SOLUTION:

(a) Written in the standard form for a Bode plot, the transfer function is

$$T(j\omega) = \frac{5(1 + j\omega/10)}{(1 + j\omega/50)\,(1 + j\omega/500)}$$

The scale factor is $K_0 = 5$ and the corner frequencies are at $\omega_C = 10$ (zero), 50 (pole), and 500 (pole) rad/s. At low frequency ($\omega < 10$ rad/s) all of the first-order factors are represented by their low-frequency asymptotes. As a result $|T(j\omega)| \approx 5(1)/[(1)(1)] = 5$, so the low-frequency baseline is $|T(j\omega)|_{SL} = 5$ for $\omega \le 10$. At $\omega = 10$ rad/s we encounter the first critical frequency. Beginning at this point the factor $(1 + j\omega/10)$ is represented by its high-frequency asymptote $\omega/10$ and the straight-line gain becomes $|T(j\omega)|_{SL} = 5(\omega/10) = \omega/2$. This expression applies until we pass the critical frequency at $\omega_C = 50$ due to the pole at $s = -50$. After this point the gain contribution due to the pole factor $1/(1 + j\omega/50)$ is represented by its high-frequency asymptote $\omega/50$ and $|T(j\omega)|_{SL} = 5(\omega/10)/(\omega/50) = 25$. This version applies until we pass the final critical frequency at $\omega_C = 500$ beyond which the gain contribution of last pole is approximated by $\omega/500$ and the high-frequency gain rolls off as $|T(j\omega)|_{SL} = 5(\omega/10)/[(\omega/50)(\omega/500)] = 12500/\omega$. In summary, the straight-line approximation to the gain is

$$|T(j\omega)|_{SL} = \begin{cases} 5 & \text{if} \quad 0 < \omega \leq 10 \\ \omega/2 & \text{if} \quad 10 < \omega \leq 50 \\ 25 & \text{if} \quad 50 < \omega \leq 500 \\ 12{,}500/\omega & \text{if} \quad 500 < \omega \end{cases}$$

Given this function, we can easily plot the straight-line gain response in Figure 12–40. At low frequency ($\omega < 10$) the gain is flat at value of 5 (14 dB). At $\omega = 10$ the zero causes the gain to increase as ω (slope = +1 or +20 dB/decade). This increasing gain continues until $\omega = 50$, where the first pole cancels the effect of the zero and the gain is flat at a value of 25 (28 dB). The gain remains flat until the final pole causes a corner at $\omega = 500$. Thereafter the gain falls off as $1/\omega$ (slope = −1 or −20 dB/decade).

(b) The dc gain is 5. A quick look at the sketch in Figure 12–40 shows that the high-frequency gain falls below the dc gain in the region above $\omega = 500$, where the straight-line gain is $12{,}500/\omega$. Hence we estimate the required frequency to be $\omega = 12{,}500/5 = 2500$ rad/s. ∎

FIGURE   1 2 – 4 0

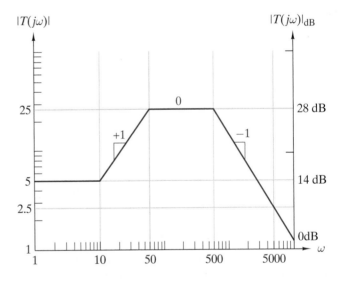

EXAMPLE 12–16

(a) Construct the Bode plot of the straight-line approximation of the gain of the transfer function:

$$T(s) = \frac{10s^2}{(s + 40)(s + 200)}$$

(b) Find the point at which the low-frequency gain falls −40 dB below the passband gain.

SOLUTION:

(a) Writing $T(j\omega)$ in standard form produces

$$T(j\omega) = \frac{1}{800}\left[\frac{(j\omega)^2}{(1 + j\omega/40)(1 + j\omega/200)}\right]$$

In this form $T(j\omega)$ has a scale factor of $K_0 = 1/800$, a double zero ($n = 2$) at the origin, and finite critical frequencies at $\omega = 40$ and 200, both due to poles. At low frequency ($\omega < 40$) the two first-order factors are represented by their low-frequency asymptote. As a result, the low-frequency gain baseline is $|T(j\omega)|_{SL} = \omega^2/800$. This trend continues until we pass the critical frequency at $\omega = 40$. After this point the straight-line gain is

$$\left|T(j\omega)\right|_{SL} = (\omega^2/800)/(\omega/40) = \omega/20$$

The gain continues to increase at a reduced rate until we pass the final critical frequency at $\omega = 200$. Beyond this point the straight-line gain is constant at the high frequency of

$$\left|T(j\omega)\right|_{SL} = (\omega^2/800)/[(\omega/40)(\omega/200)] = 10$$

In summary, the straight-line approximation to the gain response is

$$\left|T(j\omega)\right|_{SL} = \begin{vmatrix} \omega^2/800 & \text{if} & 0 < \omega \leq 40 \\ \omega/20 & \text{if} & 40 < \omega \leq 200 \\ 10 & \text{if} & 200 < \omega \end{vmatrix}$$

We can easily plot the straight-line gain response shown in Figure 12–41. At low frequency ($\omega < 20$) the gain is increasing as ω^2 (slope = +2 or +40 dB/decade). This upward trend lasts until the critical frequency (a pole) at $\omega = 40$, at which point the gain is $(40)^2/800 = 2$. Thereafter the gain increases as ω (slope = +1 or +20 dB/decade) until the final critical frequency at $\omega = 200$ reduces the slope to zero at a flat gain of 10 (20 dB).

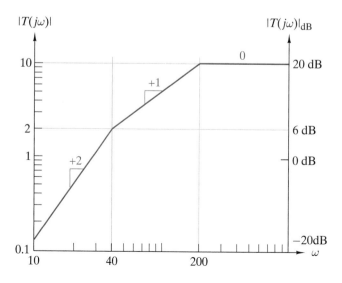

FIGURE 12 – 41

(b) A quick look at the plot in Figure 12–41 shows that the gain response has a high-pass characteristic with a passband above $\omega = 200$, where the passband gain is 20 dB. The low-frequency gain will be –40 dB below this level in the frequency range below $\omega = 40$, where the straight-line gain is $\omega^2/800$. The actual gain must be 20 dB – 40 dB = –20 dB or 0.1. This gain occurs when $\omega^2/800 = 0.1$ at $\omega = 8.9$ rad/s. ■

Exercise 12–13

(a) Derive an expression for the straight-line approximation to the gain response of the following transfer function:

$$T(s) = \frac{500\,(s + 50)}{(s + 20)(s + 500)}$$

(b) Find the straight-line gains at $\omega = 10, 30$, and 100 rad/s.
(c) Find the frequency at which the high-frequency gain asymptote falls below –20 dB.

Answers:

(a) $|T(j\omega)|_{SL} = \begin{cases} 2.5 & \text{if} & 0 < \omega \le 2 \\ 50/\omega & \text{if} & 20 < \omega \le 50 \\ 1 & \text{if} & 50 < \omega \le 500 \\ 500/\omega & \text{if} & 500 < \omega \end{cases}$

(b) 8 dB, 4.4 dB, 0 dB
(c) 5 krad/s

Exercise 12–14

Given the following transfer function

$$T(s) = \frac{4000s}{(s + 100)(s + 2000)}$$

(a) Find the straight-line approximation to the gain at $\omega = 50, 100, 500, 2000$, and 4000 rad/s.
(b) Estimate the actual gain at the frequencies in (a).

Answers:
(a) 0 dB, 6 dB, 6 dB, 6 dB, 0 dB
(b) –1 dB, 3 dB, 6 dB, 3 dB, –1 dB

In many situations the straight-line Bode plot tells us all we need to know. When greater accuracy is needed the straight-line gain plot can be refined by adding gain corrections in the neighborhood of the corner frequency. Figure 12–39 shows that the actual gains and the straight-line approximations differ by ±3 dB at the corner frequency. They differ by roughly ±1 dB an octave above or below the corner frequency. When these "corrections" are included, we can sketch the actual response and achieve somewhat greater accuracy. However, making the graphical gain corrections is usually not worth the trouble. First, the gain corrections overlap unless the corner frequencies are separated by more than two octaves. More important, the purpose of a straight-line gain analysis is to provide insight, not to generate accurate frequency-response data. The straight-line plots are useful in preliminary analysis and in the early stages of design. At some point accurate response data will be needed, in which case it is better to use computer-aided analysis rather than trying to "correct the errors" graphically in a straight-line plot.

EXAMPLE 12–17

Use computer-aided analysis to compare the actual and straight-line approximation to the gain of the transfer function in Example 12–15.

SOLUTION:

Figure 12–42 shows a Mathcad worksheet that calculates both the straight-line gain and the actual gain. In Mathcad we can write the transfer function $T(s)$ exactly as it is given in Example 12–15 and then have Mathcad calculate and plot the gain $|T(j\omega)|$. Similarly, we can use an extended conditional statement to write the straight-line gain $|T(j\omega)|_{SL}$ exactly as it appears in Example 12–15 and then have Mathcad calculate and plot this function as well. The frequency range of interest extends from one decade below the lowest critical frequency ($\omega = 10$) to the first decade point above the highest critical frequency ($\omega = 500$). The resulting graph shows that the straight-line gain (the solid line) is a reasonable approximation that captures the major features of the actual gain (the dashed curve). When numerical accuracy is important it is best to have a program like Mathcad calculate values of $|T(j\omega)|$ rather than try graphically to correct the plot of $|T(j\omega)|_{SL}$. ■

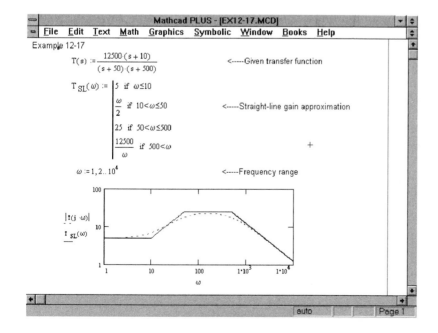

FIGURE 12 – 4 2

STRAIGHT-LINE PHASE ANGLE PLOTS

Figure 12–43 shows the phase contributions of first-order zeros and poles. The straight-line approximations are similar to the gain asymptotes except that there are two slope changes. The first occurs a decade below the gain corner frequency, and the second occurs a decade above. The total phase changes by 90° over this two-decade range, so the straight-line approximations have slopes of ±45° per decade, where the plus sign applies to a zero and the minus to a pole. Poles and zeros at the origin contribute a constant phase angle of ±n90°, where n is the order of the critical frequency and the plus (minus) sign applies to zeros (poles).

To generate a straight-line phase plot, we begin with the low-frequency phase asymptote. This low-frequency baseline accounts for the effect of the scale factor K_0 and any poles or zeros at the origin. We account for the effect of other critical frequencies by introducing a slope change of ±45°/decade one decade below and one decade above each gain corner frequency. These slope changes generate a straight-line phase plot as we proceed from the low-frequency baseline to a high frequency that is at least a decade above the highest gain corner frequency.

It is important to remember that a decade above the highest corner frequency the phase asymptote is a constant value with zero slope. That is, at high frequency the straight-line phase plot is a horizontal line at $\theta(j\omega) = (m - n)$90°, where m is the number of finite zeros and n is the number of poles.

EXAMPLE 12–18

Find the straight-line approximation to the phase response of the transfer function in Example 12–15.

SOLUTION:

In Example 12–15 the standard from of $T(j\omega)$ is shown to be

$$T(j\omega) = \frac{5(1 + j\omega/10)}{(1 + j\omega/50)(1 + j\omega/500)}$$

The scale factor is $K_0 = 5$ and the corner frequencies are $\omega_C = 10$ (zero), 50 (pole), and 500 (pole) rad/s. At low frequency $T(j\omega) \rightarrow K_0 = 5$, so the low-frequency phase asymptote is $\theta(\omega) \rightarrow \angle K_0 = 0°$. Proceeding from one decade below the lowest corner frequency (1 rad/s) to one decade above the highest corner frequency (5000 rad/s), we encounter the following slope changes:

FREQUENCY	CAUSED BY	SLOPE CHANGE	NET SLOPE
1	zero at $s = -10$	+45°/decade	+45°/decade
5	pole at $s = -50$	−45°/decade	0°/decade
50	pole at $s = -500$	−45°/decade	−45°/decade
100	zero at $s = -10$	−45°/decade	−90°/decade
500	pole at $s = -50$	+45°/decade	−45°/decade
5000	pole at $s = -500$	+45°/decade	0°/decade

FIGURE 12–43 *Phase responses of real poles and zeros.*

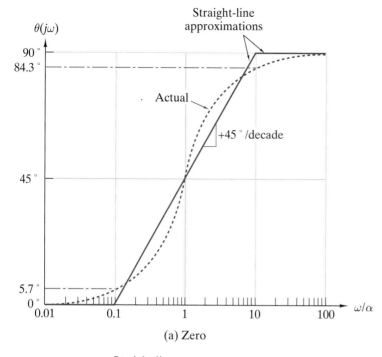

(a) Zero

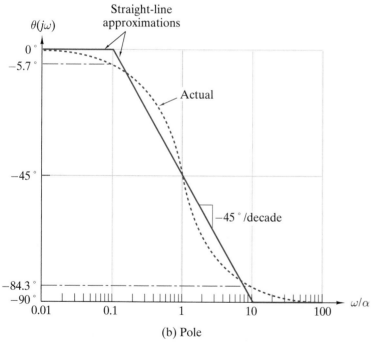

(b) Pole

FIGURE   12 – 44

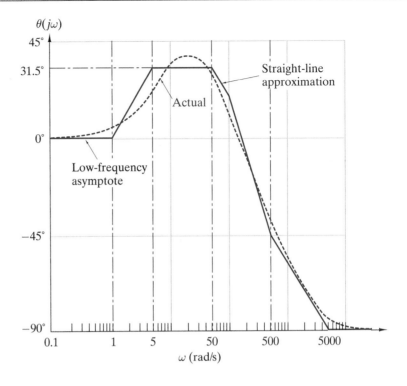

Figure 12–44 shows the straight-line approximation and the actual phase response. ∎

Exercise 12–15

Construct a Bode plot of the straight-line approximation to the phase response of the transfer function in Exercise 12–13. Use the plot to estimate the phase angles at $\omega = 1, 15, 300,$ and 10^4 rad/s.

Answers: $0°, -18°, -45°, -90°$

12–6 BODE DIAGRAMS WITH COMPLEX CRITICAL FREQUENCIES

The frequency response of transfer functions with complex poles and zeros can be analyzed using straight-line gain plots. However, complex critical frequencies may produce resonant peaks (or valleys) where the actual gain response departs significantly from the straight-line approximation. Straight-line plots can be used to define a starting place for describing the frequency response of these highly resonant circuits.

Complex poles and zeros occur in conjugate pairs that appear as quadratic factors of the form

$$s^2 + 2\zeta\omega_0 s + \omega_0^2 \tag{12–48}$$

where ζ and ω_0 are the damping ratio and undamped natural frequency. In a Bode diagram the appropriate standard form of the quadratic factor is obtained by factoring out ω_0^2 and replacing s by $j\omega$ to obtain

$$1 - (\omega/\omega_0)^2 + j2\zeta(\omega/\omega_0) \tag{12–49}$$

In a Bode diagram this quadratic factor introduces gain and phase terms of the following form:

$$|T(j\omega)|_{dB} = \pm 20 \log_{10} \sqrt{[1 - (\omega/\omega_0)^2]^2 + (2\zeta\omega/\omega_0)^2} \tag{12–50a}$$

$$\theta(\omega) = \pm \tan^{-1} \frac{2\zeta\omega/\omega_0}{1 - (\omega/\omega_0)^2} \tag{12–50b}$$

where the plus sign applies to complex zeros of $T(s)$ and the minus sign to complex poles.

Figure 12–45 shows the gain contribution of complex poles and zeros for several values of the damping ratio ζ. The low-frequency ($\omega \ll \omega_0$) asymp-

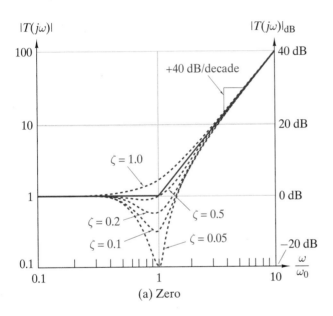

(a) Zero

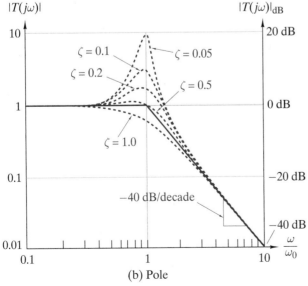

(b) Pole

F I G U R E   1 2 – 4 5 *Gain responses of complex poles and zeros.*

totes for these plots is a gain of unity (0 dB). The high-frequency ($\omega \gg \omega_0$) gain asymptotes are of the form $(\omega/\omega_0)^{\pm 2}$. Expressed in dB the high-frequency asymptote is $\pm 40 \log_{10}(\omega/\omega_0)$, which in a Bode diagram is a straight line with a slope of ± 2 or ± 40 dB/decade, where again the plus sign applies to zeros and the minus to poles.

These asymptotes intersect at a corner frequency of $\omega_C = \omega_0$. The gains in the neighborhood of the corner frequency are strong functions of the damping ratio. From our study of second-order low-pass transfer functions we can infer the following. For $\zeta > 1/\sqrt{2}$ the actual gain lies entirely above the asymptotes for complex zeros and entirely below the asymptotes for complex poles. For $\zeta < 1/\sqrt{2}$ the gain is a minimum at $\omega = \omega_0 \sqrt{1 - 2\zeta^2}$ for complex zeros and a maximum for complex poles. These valleys (for zeros) and peaks (for poles) are not particularly conspicuous until $\zeta < 0.5$.

To develop a straight-line gain plot for complex critical frequencies we insert a corner frequency at $\omega = \omega_0$. Below this corner frequency we use the low-frequency asymptote to approximate the gain and above the corner we use the high-frequency asymptote. The actual gain around the corner frequency depends on ζ. But generally speaking, the straight-line gain is within ± 3 dB of the actual gain for ζ in the range from about 0.3 to about 0.7. When ζ falls outside this range we can calculate the actual gain at the corner frequency and perhaps a few points on either side of the corner frequency. These gains may give us a better picture of the gain plot in the vicinity of the corner frequency.

However, we should keep in mind that the purpose of straight-line gain analysis is insight into the major features of a circuit's frequency response. If greater accuracy is required, then computer-aided analysis is the best approach. The straight-line gain gives useful results when the resonant peaks and valleys are not too abrupt. When a circuit has lightly damped critical frequencies, the straight-line approach may not be particularly helpful. The following examples illustrate both of these cases.

EXAMPLE 12–19

(a) Construct the straight-line gain plot for the transfer function

$$T(s) = \frac{5000(s + 100)}{s^2 + 400s + (500)^2}$$

(b) Use the straight-line plot to estimate the maximum gain and the frequency at which it occurs.

SOLUTION:

(a) The transfer function has a real zero at $s = -100$ rad/s and a pair of complex poles with $\zeta = 0.4$ and $\omega_0 = 500$ rad/s. This damping ratio falls in the range (0.3 to 0.7) in which the resonant peak due to the complex poles is not too pronounced. Hence we expect the straight-line gain to give a useful approximation. Written in standard form, $T(j\omega)$ is

$$T(j\omega) = 2\left(\frac{1 + j\omega/100}{1 - (\omega/500)^2 + j400\omega/500}\right)$$

The scale factor is $K_0 = 2$, and there are corner frequencies at $\omega = 100$ rad/s due to the zero and $\omega = 500$ rad/s due to the pair of complex poles. At low frequency $T(j\omega) \rightarrow 2$, so the low-frequency ($\omega < 100$) baseline is $|T(j\omega)|_{SL} = 2$. This gain applies until we pass the first critical frequency at $\omega = 100$. Beginning at that point, the zero is represented by its high-frequency asymptote ($\omega/100$) and the straight-line gain becomes

$$|T(j\omega)|_{SL} = 2(\omega/100) = \omega/50$$

This linearly increasing gain applies until we pass the critical frequency at $\omega_0 = 500$. Thereafter the complex poles are represented by their high-frequency asymptote ($500^2/\omega^2$). After this point the gain rolls off as

$$|T(j\omega)|_{SL} = 2(\omega/100)(500^2/\omega^2) = 5000/\omega$$

In summary, the straight-line gain function is

$$|T(j\omega)|_{SL} = \begin{vmatrix} 2 & \text{if} & 0 < \omega \leq 100 \\ \omega/50 & \text{if} & 100 < \omega \leq 500 \\ 5000/\omega & \text{if} & 500 < \omega \end{vmatrix}$$

Figure 12–46 shows a plot of the straight-line gain. We expect to see a gain peak around $\omega = 500$ rad/s due to the complex poles. The plot in Figure 12–46 shows that the zero at $s = -100$ rad/s causes the gain to bend upward prior to the corner frequency at $\omega = 500$ rad/s. This upward bend enhances the height of the resonant peak caused by the complex poles.

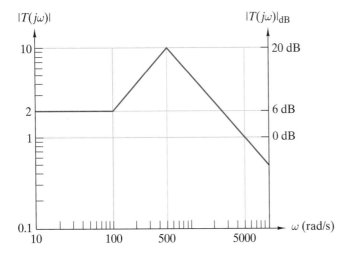

FIGURE 12-46

(b) The straight-line gain predicts a maximum gain of 10 (20 dB) at $\omega = 500$ rad/s. The actual gain at that point is

$$|T(j500)| = 2\left|\frac{1 + j5}{j0.8}\right| = 12.7 \ (22 \ \text{dB})$$

In this case the straight-line plot gives us a useful approximation of the gain in the vicinity of the resonant peak. ∎

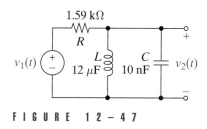

FIGURE   12 – 47

EXAMPLE 12–20

Calculate the straight-line and the actual gain response of the *RLC* circuit shown in Figure 12–47.

SOLUTION:

The analysis of this circuit is in Example 12–9, where its transfer function is shown to be

$$T(s) = \frac{s/RC}{s^2 + s/RC + 1/LC}$$

This is a second-order bandpass transfer function with a center frequency at $\omega_0 = 1/\sqrt{LC}$. The appropriate standard form for straight-line gain analysis is

$$|T(j\omega)| = \left| \frac{j\omega L/R}{1 - LC\omega^2 + j\omega L/R} \right|$$

At low frequency ($\omega \ll \omega_0$) the gain asymptote is $\omega L/R$. At high frequency ($\omega \gg \omega_0$) the gain asymptote is $1/\omega RC$. These asymptotes intersect when $\omega L/R = 1/\omega RC$, which occurs at $\omega^2 = 1/LC$; that is, at the center frequency $\omega_0 = 1/\sqrt{LC}$. The straight-line gain for this circuit is

$$|T(j\omega)|_{\text{SL}} = \left| \begin{array}{ll} \omega L/R & \text{if } 0 < \omega \leq \omega_0 \\ 1/\omega RC & \text{if } \omega_0 < \omega \end{array} \right.$$

Figure 12–48 shows a Mathcad worksheet that includes the circuit parameters, the circuit transfer function, the straight-line gain function, and a frequency range from a decade below to a decade above the center frequency. The Mathcad worksheet includes a plot of the actual gain (the

FIGURE   12 – 48

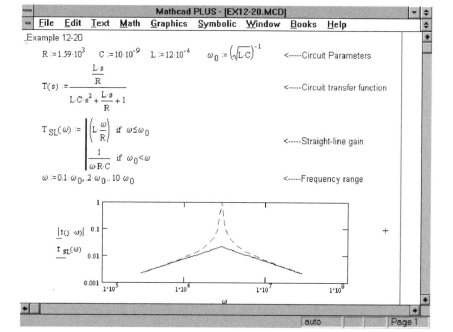

dashed curve) and the straight-line gain (the solid lines). In this case the straight-line gain plot indicates a bandpass response but does not tell us the bandwidth.

In Example 12–9 the circuit element values used are selected to produce a center frequency at 455 kHz and a bandwidth of 10 kHz. The damping ratio of this circuit is

$$\zeta = \frac{B}{2\omega_0} = \frac{10 \times 10^3}{2 \times 455 \times 10^3} = 0.011$$

Since $\zeta << 0.5$, the circuit has a very narrow resonant peak at 455 kHz. For highly resonant circuits like this one the straight-line gain is not particularly helpful. ∎

Exercise 12–16

(a) Construct the straight-line gain function for the transfer function

$$T(s) = \frac{10^5}{[s^2 + 100s + 10^4](s + 10)}$$

(b) Find the straight-line gain at $\omega = 1, 10, 100, 1000$ rad/s.

Answers:

(a) $|T(j\omega)|_{SL} = \begin{cases} 1 & \text{if} \quad 0 < \omega < 10 \\ 10/\omega & \text{if} \quad 10 < \omega \leq 100 \\ 10^5/\omega^3 & \text{if} \quad 100 < \omega \end{cases}$

(b) 0 dB, 0 dB, –20 dB, –80 dB

PHASE PLOTS FOR COMPLEX CRITICAL FREQUENCIES

Figure 12–49 shows the phase contribution from complex poles or zeros for several values of ζ. The low-frequency phase asymptotes are $0°$ and the high-frequency limits are $\pm 180°$. The phase is always $\pm 90°$ at $\omega = \omega_0$ regardless of the value of the damping ratio. The total phase change is $\pm 180°$, and most of this change occurs in a two-decade range from $\omega_0/10$ to $10\omega_0$. As a result, the straight lines in Figure 12–49 offer crude approximations to the phase shift. The shape of the phase curves change radically with the damping ratio, so these straight-line approximations are of little use except at ω_0 and around the end points.

The net result is that phase angle plots for complex critical frequencies are best generated by computer-aided analysis. In practical applications we can often derive useful information from the gain plot alone without generating phase response. However, the converse is not true. When we need phase response we usually need gain as well. The next example shows how to obtain both using a computer-aided analysis tool.

EXAMPLE 12–21

Plot the gain and phase response of the transfer function

$$T(s) = \frac{5000(s + 100)}{s^2 + 400s + (500)^2}$$

FIGURE   12–49   *Phase responses of complex poles and zeros.*

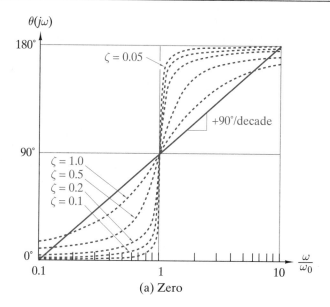

(a) Zero

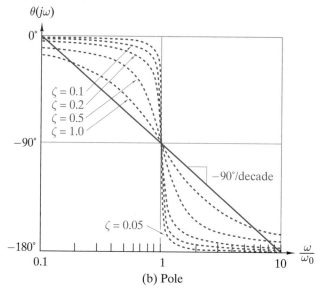

(b) Pole

SOLUTION:

The MATLAB function `bode` calculates and plots the gain and phase responses of transfer functions of the form $T(s) = n(s)/d(s)$, where $n(s)$ is the numerator polynomial and $d(s)$ is the denominator polynomial. In the MATLAB command window these polynomials are defined by the row matrices `num` and `den` that contain the coefficients of the polynomials $n(s)$ and $d(s)$ in descending powers of s. In the present case these matrices are

```
num=[5000,5000*100];
```

```
den=[1,400,500^2];
```

The first element in both matrices is the coefficient of the highest power of s followed by the other coefficients in descending powers of s. The order of the polynomial is determined by the total number of entries in the matrix. This means that every polynomial coefficient must be entered in the matrix even if it is zero. The MATLAB statement

```
bode (num,den)
```

produces the Bode plots shown in Figure 12–50.

FIGURE   12-50

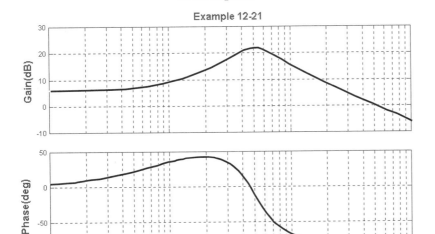

Bode Diagrams

Example 12-21

At low frequencies the phase angle gradually increases under the influence of the zero at $s = -100$ rad/s. As the input frequency approaches the natural frequency of the complex poles, the phase begins to decrease. As the frequency passes natural frequency at $\omega = 500$ rad/s, the phase rapidly decreases and eventually approaches $-90°$ at high frequency. The net effect of the zero and the two poles is that the phase initially increases and then decreases as the poles come into play. The result is that the phase has a maximum of about $42°$ at around 200 rad/s. ∎

12–7   FREQUENCY RESPONSE AND STEP RESPONSE

Frequency and step reponses are often used to describe or specify the performance of devices, circuits, and systems. In Chapter 11 we found that a transfer function contains all of the data needed to construct the step response. In the present chapter we found that a transfer function also gives us the frequency response in the form of gain $|T(j\omega)|$ and phase $\angle T(j\omega)$. This section explores the relationship between the frequency-domain response of a circuit and its time-domain step response.

If we are given a transfer function then obviously we can calculate either the step or the frequency response. If we are given the time-domain step response, we can obtain the transfer function as $T(s) = s\mathcal{L}\{g(t)\}$ and then calculate the gain and phase reponses. In other words, it is a conceptually easy matter to calculate the gain and phase responses when given $g(t)$, or $T(s)$.

But, can we determine the time-domain step response given only the frequency-domain gain response $|T(j\omega)|$? The answer is a qualified yes, provided we can determine the circuit transfer function. It may not always be possible to construct a unique transfer function because gain alone does not necessarily determine a unique phase response. Nonetheless it is true that the transfer function serves as an intermediary that allows us to calculate the response due to any input, including the step response and gain response. It is important to be able to go from one domain to the other because design requirements and device performance are often stated in both the time and frequency domains.

The next series of examples illustrates problems in which we must operate in both domains.

EXAMPLE 12–22

Figure 12–51 shows the straight-line gain response of a low-pass circuit. Find the rise time of the circuit step response.

FIGURE 12–51

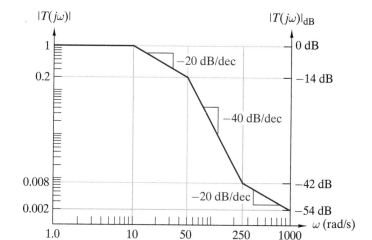

SOLUTION:

The gain plot shows corner frequencies at ω = 10, 50, and 250 rad/s. The following observations determine whether the critical frequencies are poles or zeros:

- The gain slope between ω = 10 and 50 is m = (−14 − 0)/(log 50 − log 10) = −20 dB/decade, so the critical frequency at ω = 10 is a pole.

- The slope between ω = 50 and 250 is m = [−42 −(−14)]/(log 250 − log 50) = −40 dB/dec, so the critical frequency at ω = 50 is a pole.

- The gain slope between ω = 250 and 1000 is m = [−54 − (−42)]/(log 1000 − log 250) = −20 dB/dec, so the critical frequency at ω = 250 is a zero.

Combining these observations, the transfer function must have the form

$$T(s) = \frac{K(s + 250)}{(s + 10)(s + 50)}$$

The dc gain of this transfer function is $T(0) = K/2$. The dc gain in Figure 12–51 is one; hence, $K = 2$ and the required transfer function $T(s)$ is

$$T(s) = \frac{2(s + 250)}{(s + 10)(s + 50)} = \frac{2s + 500}{s^2 + 60s + 500}$$

The MATLAB function `step` calculates and plots the step response of a transfer function of the form $T(s) = n(s)/d(s)$, where $n(s)$ is the numerator polynomial and $d(s)$ is the denominator polynomial. In the MATLAB command window these polynomials are defined by the row matrices `num` and `den` that contain the coefficients of the polynomials $n(s)$ and $d(s)$ in descending powers of s. In the present case these matrices are

```
num=[2,500];
```

```
den=[1,60,500];
```

The first element in both matrices is the coefficient of the highest power of s followed by the other coefficients in descending powers of s. The MATLAB statement

```
step(num,den);
```

produces the step response plot show in Figure 12–52.

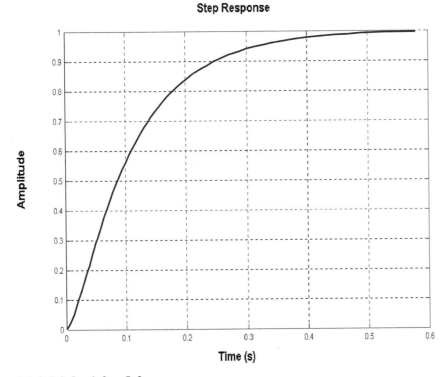

Step Response

FIGURE 12 – 52

The plot shows that the final value of the step response $g(\infty) = 1$. The rise time is the time required for the step response to rise from 10% to 90% of its final value. The plot in Figure 12–52 shows that the step response rises to 0.1 (10% rise) at $t = 0.02$ s and to 0.9 (90% rise) at $t = 0.25$ s. Hence, the step response rise time is about $T_R = 0.25 - 0.02 = 0.23$ s. ∎

EXAMPLE 12–23

Characterize the time-domain step response of a second-order bandpass circuit with a maximum gain of 12 dB, a center frequency of 2000 rad/s, and a bandwidth of 100 rad/s.

SOLUTION:
The bandpass circuit transfer function is of the form

$$T(s) = \frac{Ks}{s^2 + 2\zeta\omega_0 s + \omega_0^2}$$

The bandwidth and center frequency values require that $2\zeta\omega_0 = B = 100$ and $\omega_0 = 2000$. The maximum gain occurs at the center frequency. A gain requirement of 12 dB (factor of 4) means that $|T(j\omega_0)| = K/(2\zeta\omega_0) = 4$, or $K = 4B = 400$. Hence the circuit transfer function is

$$T(s) = \frac{400s}{s^2 + 100s + 2000^2}$$

The circuit step response is

$$g(t) = \mathcal{L}^{-1}\left\{\frac{T(s)}{s}\right\} = \mathcal{L}^{-1}\left\{\frac{400}{1999.4} \times \frac{1999.4}{(s + 50)^2 + 1999.4^2}\right\}$$

$$= 0.200\, e^{-50t} \sin(1999.4t) \qquad t > 0$$

The bandpass transfer function has zero at the origin that cancels the forced pole due to the step function input. As a result, the steady-state value of the step response is zero. The standard definition of rise time does not work here because the initial value and the final value are both zero. In this case we use a time-domain parameter called settling time. **Settling time (T_S)** is defined as the time required for the response to enter and remain within specified limits centered about its steady-state value. For step response the limits are often specified as ±2%.

Figure 12–53 shows a graph of the step response. The narrowband frequency response requires a small damping ratio ($\zeta = 0.025$, in this example) so the step response is a slowly decaying damped sine. The decay rate is controlled by an exponential envelope $\pm 0.2e^{-50t}$. This envelope decays to ±2% of its initial value when $e^{-50t} = 0.02$, which means that $T_S = 0.0782$ s. ∎

FIGURE 12-53

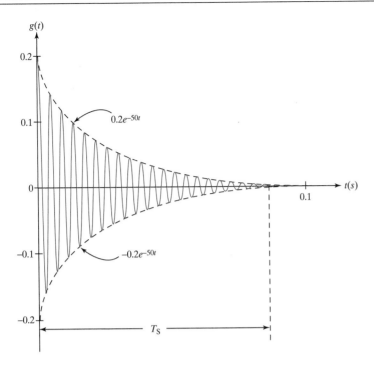

EXAMPLE 12-24

A class of piezoelectric pressure transducers have step reponses of the form

$$v_O(t) = K_V e^{-t/T_C} u(t)$$

where K_V is called the voltage sensitivity of the transducer and T_C is its time constant. In a certain application a transducer is needed to sense sinusoidal pressure waves in the range from 2 Hz to 2000 Hz. The following transducers are available for use in this application.

MODEL NUMBER	VOLTAGE SENSITIVITY (MV/PSI)	TIME CONSTANT (MS)
2612-50	50	600
2612-20	20	240
2612-10	10	120
2612-5	5	60
2612-2	2	24
2612-1	1	12

Select the transducer with the largest sensitivity that meets the frequency response requirement.

SOLUTION:

Because $v_O(t)$ is the step response of the transducer, its transfer function from pressure to output voltage is of the form

$$T(s) = s\mathcal{L}\{K_V e^{-t/T_C}\} = \frac{K_V s}{s + 1/T_C}$$

This transfer function has a first-order high-pass characteristic with a cut-off frequency of $\omega_C = 1/T_C$. The sinusoidal pressure waves range from 2 hz to 2000 Hz. This frequency range must fall in the passband of the transducer. Since the transducer frequency response is high pass, the transducer cutoff frequency must fall below 2 Hz. In equation form this requires that

$$\omega_C = \frac{1}{T_C} < 2\pi \times 2 \qquad \text{or} \qquad T_C > \frac{1}{4\pi} = 79.6 \text{ ms}$$

The first three transducers in the table meet this requirement. The model with the largest sensitivity is 2612-50 with $K_V = 50$ mV/psi. ∎

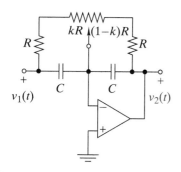

FIGURE 12–54 *Bass tone control circuit.*

APPLICATION NOTE: EXAMPLE 12-25

Figure 12–54 is an active *RC* circuit that functions as a bass tone control in an audio system. The purpose of the tone control is to emphasize or de-emphasize low-frequency audio signals. Qualitatively we see that this is accomplished by adjusting the potentiometer, which in turn changes the gain response of the inverting amplifier.

To proceed quantitatively we first find the input and feedback impedances of the noninverting circuit

$$Z_1(s) = \frac{1}{Cs + \dfrac{1}{R + kR}} = \frac{C}{s + \dfrac{1}{(1 + k)RC}}$$

$$Z_2(s) = \frac{1}{Cs + \dfrac{1}{R + (1 - k)R}} = \frac{C}{s + \dfrac{1}{(2 - k)RC}}$$

where the parameter k defines the position of the potentiometer wiper $(0 \le k \le 1)$. Given these impedances the voltage transfer function of the circuit is found to be

$$T_V(s) = -\frac{Z_2(s)}{Z_1(s)} = -\frac{s + \dfrac{1}{(1 + k)RC}}{s + \dfrac{1}{(2 - k)RC}}$$

Note the following features of this transfer function:

1. The dc gain ($s = 0$) is $|T_V(0)| = (2 - k)/(1 + k)$, which ranges between $|T_V(0)| = 2$ (+6 dB) when $k = 0$ and $|T_V(0)| = 0.5$ (−6 dB) when $k = 1$. Thus, varying the potentiometer changes the dc gain by up to ±6 dB.
2. The high-frequency ($s \to \infty$) gain is $|T_V(\infty)| = 1$ (0 dB) regardless of the value of k. In other words, varying the potentiometer does not affect the response at high frequency.
3. When $k = 0.5$, the gain is $|T_V(j\omega)| = 1$ (0 dB) since the numerator and denominator are both equal to $(s + 1/1.5RC)$ and the poles and zero cancel. In other words, when the potentiometer is set at its

midpoint ($k = 0.5$) the circuit does not affect the gain at any frequency.

For $k \neq 0.5$, the shape of the gain response depends on the relative locations of the pole at $s = -1/(2 - k)RC$ and the zero at $s = -1/(1 + k)RC$. Figure 12–55 shows how these critical frequencies migrate as k varies from zero to one. The pole is closest to the origin when $k = 0$ and moves to the left as k increases. Conversely, the zero is farthest from the origin when $k = 0$ and moves to the right as k increases. The important point is that the pole and zero move in opposite directions. The pole is closest to the origin for $0 \leq k < 0.5$, and the zero is closest for $0.5 < k \leq 1$. Thus, when k increases from zero to one the relative pole/zero positions reverse at $k = 0.5$ where the two coincide.

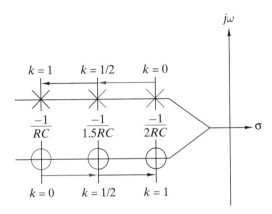

FIGURE 12–55 *Bass tone control pole-zero diagram.*

When $k = 0$, the dc gain is +6 dB, the pole has a corner frequency at $1/2RC$, and the zero has a corner frequency at $1/RC$. This set of Bode plot parameters produces the upper straight-line gain response in Figure 12–56. Conversely, when $k = 1$ the dc gain is –6 dB, the zero has a corner frequency at $1/2RC$, and the pole has a corner frequency at $1/RC$. These parameters produce the lower straight-line gain response in Figure 12–56. The straight-line gain responses for other values of k fall between the upper and lower bounds defined by the $k = 0$ and $k = 1$.

The straight-line responses indicate that the circuit emphasizes (amplifies) signal components below $\omega = 1/RC$ for potentiometer settings in the range $0 \leq k < 0.5$. It de-emphasizes (attenuates) these components for settings in the range $0.5 < k \leq 1$. There is a reciprocal relationship between emphasis and de-emphasis gain responses. For example, for $k = 0.25$ (emphasis) and $k = 0.75$ (de-emphasis), we have

$$T_V(s)\big|_{k=0.25} = -\frac{s + \dfrac{1}{1.25\,R\,C}}{s + \dfrac{1}{1.75\,R\,C}} = \frac{1}{T_V(s)\big|_{k=0.75}}$$

This reciprocal relationship comes from the fact that the pole and zero move in opposite directions when k changes. The net result is that the

FIGURE 12-56 *Bass tone control Bode diagram.*

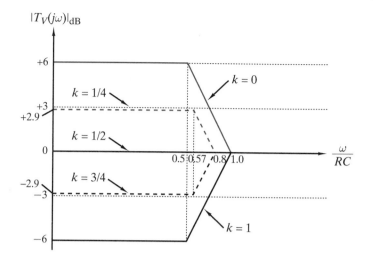

emphasis and de-emphasis gain responses (in dB) are mirror images across the 0 dB gain axis (see Figure 12–56).

Numerical values of R and C are determined by the frequency range to be affected by the circuit. For example, to affect frequencies below 300 Hz requires $RC = 1/(2\pi 300) = 5.3 \times 10^{-4}$. Selecting $R = 10$ kΩ requires $C = 53$ nF.

SUMMARY

- The frequency response of a circuit is defined by the variation of the gain $|T(j\omega)|$ and phase $\angle T(j\omega)$ with frequency. The gain function is usually expressed in dB in frequency-response plots. Logarithmic frequency scales are used on frequency-response plots of the gain and phase functions.

- A passband is a range of frequencies over which the steady-state output is essentially constant with very little attenuation. A stopband is a range of frequencies over which the steady-state output is significantly attenuated. The cutoff frequency is the boundary between a passband and the adjacent stopband.

- Circuit gain responses are classified as low pass, high pass, bandpass, and bandstop depending on the number and location of the stop and passbands. The performance of devices and circuits is often specified in terms of frequency-response descriptors such as bandwidth, passband gain, and cutoff frequency.

- The low- and high-frequency gain asymptotes of a first-order circuit intersect at a corner frequency determined by the location of its pole. The total phase change from low to high frequency is ±90°. First-order circuits can be connected to produce bandpass and bandstop responses.

- The low- and high-frequency gain asymptotes of second-order circuits intersect at a corner frequency determined by the natural frequency of the poles. The total phase change from low to high frequency is 180°.

Second-order circuits with complex critical frequencies may exhibit narrow resonant peaks and valleys.

- Series and parallel *RLC* circuits provide bandpass and bandstop gain characteristics that are easily related to the circuit parameters.

- Bode plots are graphs of the gain (in dB) and phase angle (in degrees) versus log-frequency scales. Straight-line approximations to the gain and phase can be constructed using the corner frequencies defined by the poles and zeros of $T(s)$. The purpose of the straight-line approximations is to develop a conceptual understanding of frequency response. The straight-line plots do not necessarily provide accurate data at all frequencies, especially for circuits with complex poles.

- Computer-aided circuit analysis programs can accurately generate and plot frequency-response data. The user must have a rough idea of the gain and frequency ranges of interest to use these tools intelligently.

- It is often necessary to relate the time-domain characteristics of a circuit to its frequency response, and vice versa. The transfer function provides a link between the frequency-domain and time-domain responses.

PROBLEMS

ERO 12–1 FIRST-ORDER CIRCUIT FREQUENCY RESPONSE (SECTS. 12–1, 12–2)

Given a first-order circuit or transfer function,

(a) Determine frequency response descriptors and classify the response.
(b) Draw the straight-line approximations of the gain and phase responses.
(c) Use the straight-line approximations to estimate the gain, phase, or steady-state output at specified frequencies.
(d) Select circuit parameters to produce specified descriptors.

See Examples 12–2, 12–3, 12–4, 12–5, 12–6, 12–7, 12–8 and Exercise 12–1

12–1 Find the transfer function $T_V(s) = V_2(s)/V_1(s)$ of the circuit in Figure P12–1.
(a) Find the dc gain and cutoff frequency. What kind of gain response is this?
(b) Draw the straight-line approximations of the gain and phase responses.
(c) Use the straight-line approximation to estimate the gain at $\omega = 0.5\omega_C$, ω_C, and $2\omega_C$.

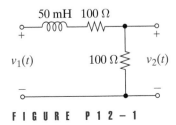

FIGURE P12–1

12–2 Find the transfer function $T_V(s) = V_2(s)/V_1(s)$ of the circuit in Figure P12–2.
(a) Find the dc gain, infinite frequency gain, and cutoff frequency. What kind of gain response is this?
(b) Use the straight-line approximation to estimate the gain at $\omega = 0.5\omega_C$, ω_C, and $2\omega_C$.

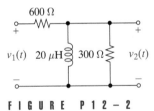

FIGURE P12–2

12–3 Find the transfer function $T_V(s) = V_2(s)/V_1(s)$ of the circuit in Figure P12–3.
(a) Find the dc gain and cutoff frequency. What kind of gain response is this?
(b) Draw the straight-line approximations of the gain and phase responses.
(c) Use the straight-line approximation to estimate the phase response at $\omega = 0.5\omega_C$, ω_C, and $2\omega_C$.

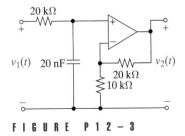

FIGURE P12–3

12–4 Find the transfer function $T_V(s) = V_2(s)/V_1(s)$ of the circuit in Figure P12–4.

(a) Find the dc gain and cutoff frequency. What kind of gain response is this?

(b) Draw the straight-line approximations of the gain and phase of $T_V(j\omega)$.

(c) Use the straight-line gain to estimate the amplitude of the steady-state output for a 5-V sinusoidal input with $\omega = 0.5\omega_C$, ω_C, and $2\omega_C$.

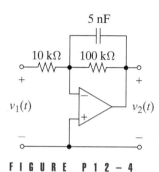

FIGURE P12-4

12–5 (a) Find the dc gain, infinite frequency gain, and cutoff frequency of the circuit in Figure P12–5.

(b) What type of gain response does this circuit have?

(c) Draw the straight-line approximations of the gain and phase responses.

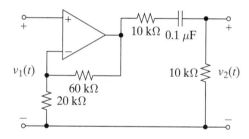

FIGURE P12-5

12–6 Find the transfer function $T_V(s) = V_2(s)/V_1(s)$ of the circuit in Figure P12–6. What type of gain response does the circuit have? What is the passband gain? Select values of R and C to produce a cutoff frequency of 500 Hz. What is the phase shift at the cutoff frequency?

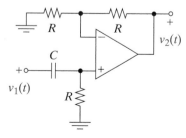

FIGURE P12-6

12–7 A first-order low-pass circuit has a passband gain of 20 dB and a cutoff frequency of 350 rad/s. Find the straight-line approximations of the gain (in dB) at $\omega = 200$, 400, and 800 rad/s?

12–8 A first-order high-pass circuit has a passband gain of 5 dB and a cutoff frequency of 2 krad/s. Find the straight-line approximations to the phase response at $\omega = 0.5\omega_C$, ω_C, and $2\omega_C$.

12–9 The transfer function of a first-order circuit is

$$T(s) = \frac{0.1}{10^{-2} + \dfrac{20}{s}}$$

(a) What type of gain response does this circuit have? What is the cutoff frequency and the passband gain?

(b) Draw the straight-line approximations of the gain and phase responses.

(c) Use the straight-line approximation to estimate the gain at $\omega = 0.5\omega_C$, ω_C, and $2\omega_C$.

12–10 Repeat Problem 12–9 for

$$T(s) = \frac{10}{10^2 + \dfrac{s}{20}}$$

ERO 12–2 SECOND-ORDER CIRCUIT FREQUENCY RESPONSE (SECT. 12–3)

Given a second-order linear circuit or its transfer function,

(a) Determine frequency response descriptors and classify the circuit response.

(b) Draw the straight-line approximations of the gain and phase responses.

(c) Use the straight-line approximations to estimate the gain, phase, or steady-state output at specified frequencies.

(d) Select circuit parameters to produce specified descriptors.

See Examples 12–9, 12–10, 12–11 and Exercises 12–3, 12–4, 12–5, 12–6, 12–8.

12–11 Find the transfer function $T_V(s) = V_2(s)/V_1(s)$ of the circuit in Figure P12–11.

(a) Find the dc gain, infinite-frequency gain, the damping ratio ζ, and the undamped natural frequency ω_0. What type of gain response does this circuit have?

(b) Draw the straight-line approximation of the gain.

(c) Use the straight-line gain to estimate the gain $\omega = 0.5\omega_0$, ω_0, and $2\omega_0$.

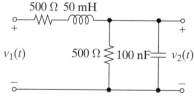

FIGURE P12-11

12–12 Find the transfer function $T_V(s) = V_2(s)/V_1(s)$ of the circuit in Figure P12–12.

(a) Determine the dc gain, infinite-frequency gain, the damping ratio ζ, and the undamped natural frequency ω_0. What type of gain response does this circuit have?

(b) Draw the straight-line approximation of the gain.

(c) Use the straight-line gain to estimate the gain $\omega = 0.5\omega_0$, ω_0, and $2\omega_0$.

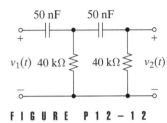

FIGURE P12–12

12–13 Find the transfer function $T_V(s) = V_2(s)/V_1(s)$ of the circuit in Figure P12–13.

(a) Determine the dc gain, infinite-frequency gain, the damping ratio ζ, and the undamped natural frequency ω_0. What type of gain response does this circuit have?

(b) Draw the straight-line approximation of the gain.

(c) Compare the straight-line gain and the actual gain at $\omega = 0.5\omega_0$, ω_0, and $2\omega_0$.

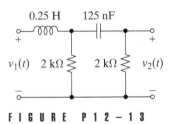

FIGURE P12–13

12–14 Find the transfer function $T_V(s) = V_2(s)/V_1(s)$ of the circuit in Figure P12–14.

(a) Determine the dc gain, infinite-frequency gain, the damping ratio ζ, and the undamped natural frequency ω_0. What type of gain response does this circuit have?

(b) Use the straight-line approximation to estimate the gain at $\omega = 0.5\omega_0$, ω_0, and $2\omega_0$.

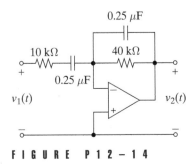

FIGURE P12–14

12–15 Find the transfer function $T_V(s) = V_2(s)/V_1(s)$ of the circuit in Figure P12–15.

(a) Determine the dc gain, infinite-frequency gain, the damping ratio ζ, and the undamped natural frequency ω_0. What type of gain response does this circuit have?

(b) Draw the straight-line approximation of the gain.

(c) Compare the straight-line gain and the actual gain at $\omega = 0.5\omega_0$, ω_0, and $2\omega_0$.

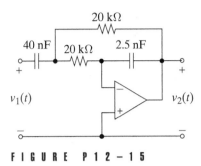

FIGURE P12–15

12–16 (a) Find the transfer function $T_V(s) = V_2(s)/V_1(s)$ of the circuit in Figure P12–16 for $R_1 = 10$ kΩ, $R_2 = 40$ kΩ, and $C = 20$ nF.

(b) Determine the dc gain, infinite-frequency gain, the damping ratio ζ, and the undamped natural frequency ω_0. What type of gain response does this circuit have?

(c) Draw the straight-line approximation of the gain.

(d) Use the straight-line gain to estimate the amplitude of the steady-state output for a 0.5-V sinusoidal input with $\omega = 0.5\omega_C$, ω_C, and $2\omega_C$.

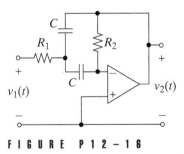

FIGURE P12–16

12–17 Find the transfer function $T_V(s) = V_2(s)/V_1(s)$ of the circuit in Figure P12–17. What type of gain response does this circuit have? Derive expressions for the damping ratio ζ and the undamped natural frequency ω_0 in terms of circuit parameters. Select values of the circuit parameters so that $\zeta = 2$ and $\omega_0 = 5$ krad/s. What is the passband gain for your choice?

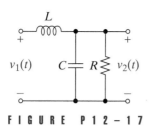

FIGURE P12-17

12-18 Find the transfer function $T_V(s) = V_2(s)/V_1(s)$ of the circuit in Figure P12-18. What type of gain response does this circuit have? Derive expressions for the damping ratio ζ and the undamped natural frequency ω_0 in terms of circuit parameters. Select values of the circuit parameters so that $\zeta = 0.7$ and $\omega_0 = 2500$ rad/s. What is the passband gain for your choice?

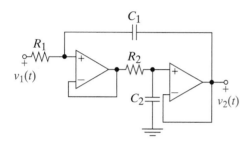

FIGURE P12-18

12-19 The circuit in Figure P12-19 produces a bandpass response. With $R_1 = R_2 = R_4 = 10$ kΩ, select the values of R_3, C_1, and C_2 to produce a passband gain of 5 with cutoff frequencies at 500 rad/s and 50 krad/s.

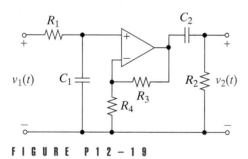

FIGURE P12-19

12-20 The transfer function of a bandpass circuit can be written as

$$T(s) = \frac{2.5 \times 10^4}{s + 5.5 \times 10^3 + \dfrac{2.5 \times 10^6}{s}}$$

(a) What are the cutoff frequencies and the passband gain?

(b) Draw the straight-line approximations of the gain and phase responses.

ERO 12-3 THE FREQUENCY RESPONSE OF *RLC* CIRCUITS (SECT. 12-4)

Given a series or parallel *RLC* circuit connected as a bandpass or bandstop filter,

(a) Find the circuit parameters or frequency response descriptors.

(b) Select the circuit parameters to achieve specified filter characteristics.

(c) Derive expressions for the frequency response descriptors.

See Examples 12-12, 12-13, 12-14 and Exercises 12-10, 12-11, 12-12

12-21 A series *RLC* circuit is designed to have a bandwidth of 8 Mrad/s and an input impedance of 50 Ω at its resonant frequency of 40 Mrad/s. Determine the values L, C, Q, and the upper and lower cutoff frequencies.

12-22 A parallel *RLC* circuit with $C = 20$ pF and $Q = 10$ has a resonant frequency of 100 Mrad/s. Find the values of R, L, ω_{C1}, and ω_{C2}.

12-23 A series *RLC* circuit with $R = 100$ Ω, $L = 20$ mH, and $C = 200$ pF is driven by a sinusoidal voltage source with a peak amplitude of 10 V.

(a) Calculate the circuit bandwidth and the upper and lower cutoff frequencies.

(b) Calculate the amplitude of the steady-state voltage across the capacitor and the inductor at the resonant frequency.

12-24 A series *RLC* circuit has a resonant frequency of 400 krad/s and an upper cutoff frequency of 420 krad/s. Find the bandwidth, Q, and the lower cutoff frequency.

12-25 A parallel *RLC* circuit with $R = 40$ kΩ is to have a resonant frequency of 100 MHz. Calculate the values of L and C required to produce a bandwidth of 100 kHz.

12-26 **D** A series *RLC* bandpass filter is required to have a resonance at $f_0 = 200$ kHz. The series connected L and C are to be driven by a sinusoidal source with a Thévenin resistance of 50 Ω. The following standard capacitors are available in the stock room: 1 μF, 680 nF, 470 nF, 330 nF, 200 nF, and 120 nF. The inductor will be custom-designed to match the capacitor used. Select from available capacitors the one that minimizes the circuit bandwidth.

12-27 **D** A series *RLC* circuit is to be used as a notch filter to eliminate a bothersome 60-Hz hum in an audio channel. The signal source has a Thévenin resistance of 600 Ω. Select values of L and C so the upper cutoff frequency is below 200 Hz.

12-28 Find the transfer function $T_V(s) = V_2(s)/V_1(s)$ for the bandpass circuit in Figure P12-28. Derive expressions for the parameters B, Q, ω_{C1}, and ω_{C2} in terms of the circuit parameters R, L, and C.

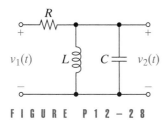

FIGURE P12-28

12–29 The transfer function $T_V(s) = V_2(s)/V_1(s)$ of the circuit in Figure P12–29 has a bandstop filter characteristic. Without solving for the transfer function, use the element impedances to

(a) Explain why the low- and high-frequency passband gains are unity.

(b) Explain why the bandstop notch occurs at $\omega_0 = 1/\sqrt{LC}$.

(c) Derive expressions ω_{C1} and ω_{C2} in terms of circuit parameters R, L, and C.

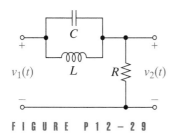

FIGURE P12-29

12–30 Find the transfer impedance $T_Z(s) = V_2(s)/I_1(s)$ for the bandpass circuit in Figure P12–30. Derive expressions for the parameters B, Q, ω_{C1}, and ω_{C2} in terms of the circuit parameters R, L, and C.

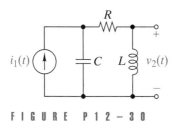

FIGURE P12-30

ERO 12–4 BODE PLOTS (SECTS. 12–5, 12–6)

(a) Construct Bode plots of the straight-line approximations of gain and phase responses of a given circuit or transfer function.

(b) Construct the transfer function corresponding to a given straight-line gain plot.

(c) Use the straight-line gain plots to estimate frequency response descriptors or steady-state outputs.

See Examples 12–15, 12–16, 12–17, 12–18, 12–19, 12–20, 12–21 and Exercises 12–13, 12–14, 12–15, 12–16

12–31 Construct Bode plots of the straight-line approximations of gain response of the circuit in Figure P12–31. Use the straight-line gain plot to estimate the amplitude of the steady-state output for an input $v_1(t) = 10 \sin 500t$ V. Calculate the actual amplitude of the steady-state output for this input and compare the two results.

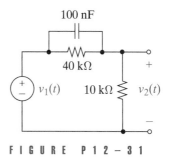

FIGURE P12-31

12–32 Repeat Problem 12–31 using the circuit in Figure P12–32.

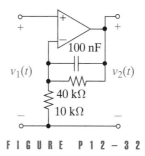

FIGURE P12-32

12–33 Construct Bode plots of the straight-line approximations of gain and phase responses of the following transfer function. Is this a low-pass, high-pass, bandpass, or bandstop function? Estimate the cutoff frequency and passband gain.

$$T(s) = \frac{(s + 25)}{(5s + 5)(s + 5)}$$

12–34 Construct Bode plots of the straight-line approximations to the gain and phase responses of the following transfer function. Is this a low-pass, high-pass, bandpass, or bandstop function? Estimate the cutoff frequency and passband gain.

$$T(s) = \frac{5(s + 5)}{(s + 1)(s + 25)}$$

12–35 Construct Bode plots of the straight-line approximations to the gain and phase responses of the following transfer function. Is this a low-pass, high-pass, band-pass, or bandstop function? Estimate the cutoff frequency and passband gain.

$$T(s) = \frac{25\,s(s + 100)}{(s + 10)(s + 25)}$$

12–36 Construct Bode plots of the straight-line approximations to the gain and phase responses of the following transfer function. Is this a low-pass, high-pass, band-pass, or bandstop function? Estimate the cutoff frequency and passband gain.

$$T(s) = \frac{2.5s^2}{(0.05\,s + 1)^2}$$

12–37 Construct Bode plots of the straight-line approximations to the gain and phase responses of the following transfer function. Is this a low-pass, high-pass, band-pass, or bandstop function? Estimate the cutoff frequency and passband gain.

$$T(s) = \frac{2000}{s^2 + 5\,s + 100}$$

12–38 Construct Bode plots of the straight-line approximations to the gain and phase responses of the following transfer function. Is this a low-pass, high-pass, band-pass, or bandstop function? Estimate the cutoff frequency and passband gain.

$$T(s) = \frac{100\,s^2}{s^2 + 5\,s + 25}$$

12–39 Find the transfer function corresponding to the straight-line gain plot in Figure P12–39. Compare the straight-line gain and the actual gain at the corner frequencies shown in the figure.

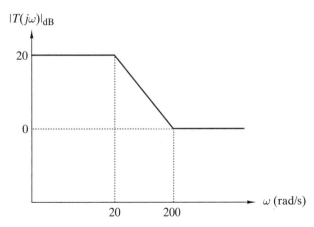

F I G U R E   P 1 2 – 3 9

12–40 Find the transfer function corresponding to the straight-line gain plot in Figure P12–40. Compare the straight-line gain and the actual gain at the corner frequencies shown in the figure.

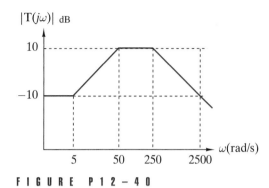

F I G U R E   P 1 2 – 4 0

ERO 12–5 FREQUENCY RESPONSE AND STEP RESPONSE (SECT. 12–7)

(a) Find the step response corresponding to a given straight-line gain response.
(b) Find the straight-line approximations to the gain and phase responses corresponding to a given step response.
(c) Construct a transfer function that meets constraints on both the frequency and step responses.

See Examples 12–22, 12–23, 12–24 and Exercises 12–2, 12–7, 12–9

12–41 Find the step response corresponding to the straight-line gain response in Figure P12–41.

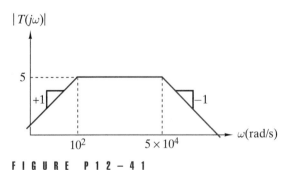

F I G U R E   P 1 2 – 4 1

12–42 Repeat Problem 12–41 for the gain response in Figure P12–39.

12–43 Repeat Problem 12–41 for the gain response in Figure P12–40

12–44 The step response of a linear circuit is

$$g(t) = 12 - 10\,e^{-10t} - 2\,e^{-100t} \quad t > 0$$

Is the circuit a low-pass, high-pass, bandpass, or band-stop filter? Construct Bode plots of the straight-line approximations to the gain and phase responses.

12–45 The step response of a linear circuit is

$$g(t) = 5 - 5\,e^{-100t}\sin 200\,t \quad t > 0$$

Is the circuit a low-pass, high-pass, bandpass, or band-stop filter? Construct Bode plots of the straight-line approximations to the gain and phase responses.

12–46 The step response of a linear circuit is

$$g(t) = 10\,e^{-100t}\sin 200t \quad t > 0$$

Is the circuit a low-pass, high-pass, bandpass, or band-stop filter? Construct a Bode plot of the straight-line approximation to the gain response.

12–47 The step response of a linear circuit is

$$g(t) = e^{-100t} + 2\,e^{-2000t} \quad t > 0$$

Is the circuit a low-pass, high-pass, bandpass, or band-stop filter? Construct Bode plots of the straight-line approximations to the gain and phase responses.

12–48 Construct a first-order low-pass transfer function with a dc gain of 10, a bandwidth less than 250 rad/s, and a step response that rises to 50% of its final value in less than 4 ms.

12–49 Construct a first-order high-pass transfer function whose step response decays to 5% of its peak value in less than 10 μs and whose cutoff frequency is less than 100 kHz.

12–50 Construct a second-order bandpass transfer function with a midband gain of 10, a center frequency of 100 kHz, a bandwidth less than 50 kHz, and a step response that decays to less thanw 20% of its peak value in less than 5 μs.

INTEGRATING PROBLEMS

12–51 A DC, AC, and Impulse Responses

A linear circuit is driven by an input $v_1(t) = 2e^{-1000t}$. The zero-state response for this input is observed to be $v_2(t) = 5e^{-1000t}[1 - \cos(2000t)]$.
(a) Find the steady-state output when the input is $v_1(t) = 4 \cos(2000t)$.
(b) Find the steady-state output when the input is 15 V dc.
(c) Construct the Bode plot of the circuit gain as a function of frequency.
(d) Find the circuit impulse response.

12–52 A E Bandpass Filter Step Responses

The step responses of the two second-order bandpass filters are

$$g_1(t) = 2e^{-60t}\sin(600t)u(t) \text{ and } g_2(t) = 10(e^{-30t} - e^{-12000t})u(t)$$

(a) A Find the transfer function, center frequency, and bandwidth of each filter. Sketch the Bode plot of the gain response of each filter.
(b) E A filter is required for an input signal

$$v_1(t) = 8 \cos (100t) + 6 \cos(600t) + 10 \cos(3000t)$$

that will pass the desired sinusoid at $\omega = 600$ with about 20 dB gain, and keep amplitudes of the undesired sinusoids at $\omega = 100$ and $\omega = 3000$ at least 20 dB below the amplitude of the desired sinsuoid. Which filter would you use?
(c) E A filter is required for an input signal

$$v_1(t) = 6 \cos(400t) + 6 \cos(800t) + 6 \cos(1600t)$$

that will pass all three sinusoids with about 20 dB gain. Which filter would you use?

12–53 D Filter Design with Input Impedance Specified

Design a circuit whose gain response lies entirely within the unshaded region in Figure P12–53 and whose input impedance is around 50 Ω for most of the frequencies in the passband.

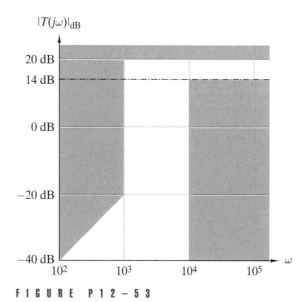

F I G U R E   P 1 2 – 5 3

12–54 A High-Frequency Model of a Resistor

Figure P12–54 shows a circuit model of a resistor R that includes parasitic capacitance C and parasitic lead inductance L. The purpose of this problem is to investigate the effect of these parasitic elements on the high-frequency characteristics of the resistor device.

(a) Derive an expression for the impedance $Z(s)$ of the circuit model in Figure P12–54 in terms of the circuit parameters R, L, and C.

(b) For $R = 10$ kΩ, $L = 2$ μH, and $C = 4$ pF, find the poles and zeros of $Z(s)$. Construct a Bode plot of the straight-line asymptotes of the impedance magnitude $|Z(j\omega)|$ expressed in ohms. Do not express $|Z(j\omega)|$ in dB. Identify the corner frequencies and slopes (in ohms per decade) in this plot.

(c) Use the straight-line plot in (b) to identify the frequency range (in Hz) over which the resistor device appears to be: (1) a 10-kΩ resistance, (2) a 4-pF capacitance, and (3) a 2-μH inductor.

(d) Over what frequency range can this device be treated as a 10-kΩ resistor?

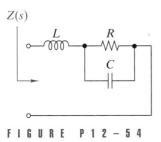

$Z(s)$

F I G U R E   P 1 2 – 5 4

12–55   A E D COMPENSATOR DESIGN

A modification is required to improve the performance of an existing output transducer. The compensator/transducer combination must meet the following performance specification: *dc gain 0 dB, gain = 0 ±5 dB for ω between 20 rad/s and 200 rad/s, gain slope of –40 dB/dec for ω > 1000 rad/s*. The two designs in Figure P12–55 have been proposed.

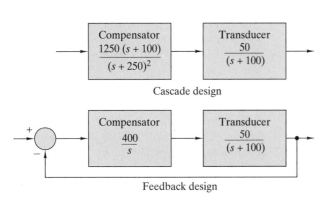

Cascade design

Feedback design

F I G U R E   P 1 2 – 5 5

(a) **A**   Verify that both designs meet the performance specification.

(b) **E**   The compensator is to be built using an *RC* OP AMP circuit with no more than one OP AMP. Which of the two designs would you recommend and why?

(c) **D**   Design an OP AMP circuit to implement your recommendation in (b).

12–56   A D USING A BODE PLOT

Figure P12–56 shows the gain response of a second-order filter. The purpose of this problem is to infer things about the performance of the filter with various inputs. The only additional data is that the filter output is an OP AMP with $V_{CC} = \pm15$ V.

(a) What is the maximum allowable dc input voltage?

(b) Estimate the maximum allowable sinusoidal input voltage at $f = 20$ Hz.

(c) A short rectangular pulse is to be used to evaluate the impulse response of the filter. Estimate the required pulse duration.

(d) What is the final value of the impulse response of the filter? About how long does it take to settle within 1% of this value?

(e) A 0.5-volt step function is to be applied at the input. Estimate the final value of the output response and about how long it takes the response to reach 50% of this value.

(f) The signal $v(t) = 0.5 \sin(100t) + 0.25 \sin(1000t)$ V is to be applied at the input. About how long will it take for a steady-state condition to be reached? What will be the ratio of the steady-state output amplitude at $\omega = 100$ rad/s to the amplitude of the output at $\omega = 1000$ rad/s? Express the ratio in dB.

(g) Your boss needs to know the frequency at which the phase shift is –90°. What is your estimate?

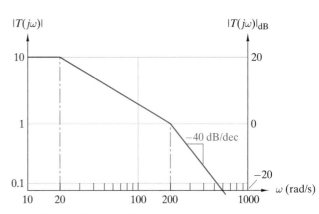

F I G U R E   P 1 2 – 5 6

CHAPTER 13

FOURIER SERIES

The series formed of sines or cosines of multiple arcs are therefore adapted to represent between definite limits all possible functions, and the ordinates of lines or surfaces whose form is discontinuous.

Jean Baptiste Joseph Fourier, 1822,
French Mathematician

13–1 Overview of Fourier Analysis

13–2 Fourier Coefficients

13–3 Waveform Symmetries

13–4 Circuit Analysis Using Fourier Series

13–5 RMS Value and Average Power

Summary

Problems

Integrating Problems

The analysis techniques in this chapter have their roots in the works of the French mathematician Joseph Fourier (1768–1830). While studying heat transfer, Fourier found that he could represent discontinuous functions by an infinite series of harmonic sinusoids, or sines of multiple arcs as he called them. Fourier was guided by his physical intuition and presented experimental evidence to support his claim, but he left the question of series convergence unanswered. As a result, his theory was incomplete and his 1807 treatise was initially rejected by the scientific community of his day. Some years later an acceptable proof of the "Fourier theorem" was supplied by the German mathematician P. G. L. Dirichlet (1805–1859). The basic ideas behind Fourier analysis were eventually accepted during his lifetime and are now firmly entrenched in all branches of science and engineering.

The purpose of circuit analysis is to describe how a network modifies the input to produce the output. Fourier analysis resolves a periodic input into an infinite sum of harmonic sinusoids. In linear circuit we treat each of these sinusoids as a separate input and find the response due to each input acting alone. The total response is found by adding up these individual responses. The result is that the circuit inputs and outputs are expressed as infinite series. To carry out this approach we must first learn how to construct the Fourier series of periodic inputs. That process is the subject of the first three sections. The last two sections treat the process of finding the Fourier series of the output.

13–1 OVERVIEW OF FOURIER ANALYSIS

In this chapter we develop a method of finding the steady-state response of circuits to periodic inputs such as the waveforms in Figure 13–1. These periodic waveforms can be written as a Fourier series consisting of an infinite sum of harmonically related sinusoids. More specifically, if $f(t)$ is periodic with period T_0 and is reasonably well behaved, then $f(t)$ can be expressed as a **Fourier series** of the form

$$f(t) = a_0 + a_1 \cos(2\pi f_0 t) + a_2 \cos(2\pi 2 f_0 t) + \ldots + a_n \cos(2\pi n f_0 t) + \ldots$$
$$+ b_1 \sin(2\pi f_0 t) + b_2 \sin(2\pi 2 f_0 t) + \ldots + b_n \sin(2\pi n f_0 t) + \ldots \qquad \text{(13–1)}$$

or, more compactly,

$$f(t) = \underbrace{a_0}_{\text{dc}} + \underbrace{\sum_{n=1}^{\infty} [a_n \cos(2\pi n f_0 t) + b_n \sin(2\pi n f_0 t)]}_{\text{ac}} \qquad \text{(13–2)}$$

The coefficient a_0 is the dc component or average value of $f(t)$. The constants a_n and b_n ($n = 1, 2, 3, \ldots$) are the **Fourier coefficients** of the sinusoids in the ac component. The lowest frequency in the ac component occurs for $n = 1$ and is called the **fundamental frequency** defined as $f_0 = 1/T_0$. The other frequencies are integer multiples of the fundamental called the second harmonic ($2f_0$), third harmonic ($3f_0$) and, in general, the nth harmonic (nf_0).

Since Eq. (13–2) is an infinite series, there is always a question of convergence. We have said that the series converges as long as $f(t)$ is reason-

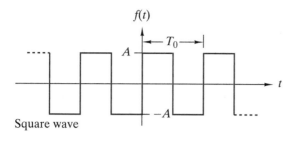

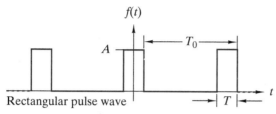

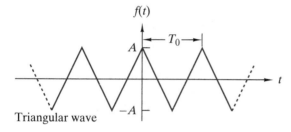

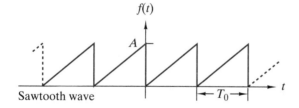

FIGURE 13–1 *Examples of periodic waveforms.*

ably well behaved. Basically, this means that $f(t)$ is single valued, the integral of $|f(t)|$ over a period is finite, and $f(t)$ has a finite number of discontinuities in any one period. These requirements, called the **Dirichlet conditions**, are sufficient to assure convergence. Every periodic waveform that meets the Dirichlet conditions has a convergent Fourier series. However, there are waveforms that do not meet the Dirichlet conditions that also have convergent Fourier series. That is, while the Dirichlet conditions are sufficient, they are not necessary and sufficient. This limitation does not present a serious problem because the Dirichlet conditions are satisfied by the waveforms generated in physical systems. All of the periodic waveforms in Figure 13–1 meet the Dirichlet requirements.

It is important to have an overview of how a Fourier series is used in circuit analysis. In our previous study we learned how to find the steady-state responses due to a dc or an ac input. The Fourier series resolves a periodic input into a dc component and an infinite sum of ac components. We treat each of the terms in the Fourier series as a separate input source and use our circuit analysis tools to find the steady-state responses due to each term acting alone. The complete steady-state response is found by superposition, that is, adding up the responses due to each term acting alone. The net result is that the output is a modified version of the Fourier series of the periodic input.

At first glance it may seem complicated to find the complete response by finding the individual responses due to infinitely many inputs. However, each step involves simply circuit analysis tools we have already mastered. The end result tells us how the circuit transforms the input series into the output series. The distribution of amplitudes and phase angles in a Fourier series is called the **spectrum** of a periodic waveform. The frequency-domain signal processing involves modifying a given input spectrum to produce a desired output spectrum. Thus, finding the Fourier series of periodic waveforms is not an end in itself, but the first step in the study of frequency-domain signal processing.

13–2 FOURIER COEFFICIENTS

The Fourier coefficients for any periodic waveform $f(t)$ satisfying the Dirichlet conditions can be obtained from the equations

$$a_0 = \frac{1}{T_0}\int_{-T_0/2}^{+T_0/2} f(t)dt$$

$$a_n = \frac{2}{T_0}\int_{-T_0/2}^{+T_0/2} f(t)\cos(2\pi nt/T_0)dt \qquad (13\text{--}3)$$

$$b_n = \frac{2}{T_0}\int_{-T_0/2}^{+T_0/2} f(t)\sin(2\pi nt/T_0)dt$$

The integration limits in these equations extend from $-T_0/2$ to $+T_0/2$. However, the limits can span any convenient interval as long as it is exactly one period. For example, the limits could be from 0 to T_0 or $-T_0/4$ to $3T_0/4$. We will show where Eq. (13–3) comes from in a moment, but first we use these equations to obtain the Fourier coefficients of the sawtooth wave.

EXAMPLE 13–1

Find the Fourier coefficients for the sawtooth wave in Figure 13–1.

SOLUTION:
An expression for a sawtooth wave on the interval $0 \le t \le T_0$ is

$$f(t) = \frac{At}{T_0} \qquad 0 \le t < T_0$$

For this definition of $f(t)$ we use 0 and T_0 as the limits in Eq. (13–3). The first expression in Eq. (13–3) yields a_0 as

$$a_0 = \frac{1}{T_0}\int_0^{T_0}\frac{At}{T_0}dt = \frac{At^2}{2T_0^2}\bigg|_0^{T_0} = \frac{A}{2}$$

This result states that the average or dc value is $A/2$, which is easy to see because the area under one cycle of the sawtooth wave is $AT_0/2$. The second expression in Eq. (13–3) yields a_n as

$$a_n = \frac{2}{T_0} \int_0^{T_0} \frac{At}{T_0} \cos(2\pi nt/T_0)dt$$

$$= \frac{2A}{T_0^2} \left[\frac{\cos(2\pi nt/T_0)}{(2\pi n/T_0)^2} + \frac{t \times \sin(2\pi nt/T_0)}{(2\pi n/T_0)} \right]_0^{T_0}$$

$$= \frac{2A}{T_0^2} \left[\frac{\cos(2\pi n) - \cos(0)}{(2\pi n/T_0)^2} \right] = 0 \quad \text{for all } n$$

Since $a_n = 0$ for all n, there are no cosine terms in the series. The b_n coefficients are found using the third expression in Eq. (13–3):

$$b_n = \frac{2}{T_0} \int_0^{T_0} \frac{At}{T_0} \sin(2\pi nt/T_0)dt$$

$$= \frac{2A}{T_0^2} \left[\frac{\sin(2\pi nt/T_0)}{(2\pi n/T_0)^2} - \frac{t \times \cos(2\pi nt/T_0)}{(2\pi n/T_0)} \right]_0^{T_0}$$

$$= \frac{2A}{T_0^2} \left[\frac{T_0 \cos(2\pi n)}{(2\pi n/T_0)} \right] = -\frac{A}{n\pi} \quad \text{for all } n$$

Given the coefficients a_n and b_n found above, the Fourier series for the sawtooth wave is

$$f(t) = \frac{A}{2} + \sum_{n=1}^{\infty} \left[-\frac{A}{n\pi} \right] \sin(2\pi nf_0 t) \qquad ∎$$

EXAMPLE 13–2

In this example we use a computer tool to show that a truncated Fourier series approximates a periodic waveform. The waveform is a sawtooth with $A = 10$ and $T_0 = 2$ ms. Calculate the Fourier coefficients of the first 20 harmonics and plot the truncated series representation of the waveform using the first 5 harmonics and the first 10 harmonics.

SOLUTION:

From Example 13–1 the Fourier coefficients for the sawtooth wave are

$$a_0 = A/2 \quad a_n = 0 \quad \text{and} \quad b_n = -\frac{A}{n\pi} \quad \text{for all } n$$

Figure 13–2 shows a Mathcad worksheet that generates truncated Fourier series. The first line defines the waveform amplitude and period. The Fourier coefficients of the first 20 harmonics are calculated in the second line using the index variable $n = 1,2, \ldots 20$. In Mathcad syntax $n = I, J, \ldots$

K means FOR $n = I$ TO K STEP $J - I$ DO. The next line defines a truncated Fourier series $f(k,t)$ consisting of the dc component a_0 plus the sum of the first k harmonics. Specifically, the function $f(5, t)$ is the sum of the dc component plus the first 5 harmonics and $f(10, t)$ is the sum of the dc component plus the first 10 harmonics. The plots of these two functions at the bottom of the worksheet show how function $f(k, t)$ approaches the sawtooth wave as more harmonics are added. Infinitely many harmonics are needed to represent a sawtooth wave exactly. However, the plots suggest that a relatively small number, say 5 or 10, provides a reasonable approximation of the major features of the sawtooth wave.

FIGURE 13–2

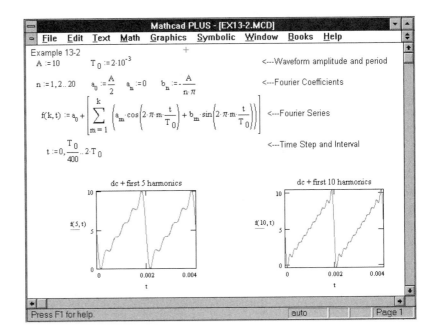

The Mathcad worksheet in Figure 13–2 is a template for generating truncated Fourier series. Changing the amplitude and period in the first line and the Fourier coefficients in the second line generates plots of the truncated Fourier series for other periodic waveforms. The index for n in the second line may need to be changed as well. For example, if only odd harmonics are present in the series, then the index should be $n = 1, 3, \ldots 21$. ∎

DERIVING EQUATIONS FOR a_n AND b_n

The sawtooth wave example shows how to calculate the Fourier coefficients using Eq. (13–3). We now turn to the derivation of these equations. An equation for a Fourier coefficient is derived by multiplying both sides of Eq. (13–2) by the sinusoid associated with the coefficient and then integrating the result over one period. This multiply and integrate process isolates one coefficient because it turns out that all of the integrations produce zero except one.

The following derivation makes use of the fact that the area under a sine or cosine wave over an integer number of cycles is zero. That is,

$$\int_{-T_0/2}^{+T_0/2} \sin(2\pi k f_0 t)dt = 0 \qquad \text{for all } k$$

$$\int_{-T_0/2}^{+T_0/2} \cos(2\pi k f_0 t)dt = 0 \qquad \text{for } k \neq 0 \qquad (13\text{–}4)$$

$$= T_0 \qquad \text{for } k = 0$$

where k is an integer. These equations state that integrating a sinusoid over $k \neq 0$ cycles produces zero, since the areas under successive half cycles cancel. The single exception occurs when $k = 0$, in which case the cosine function reduces to one and the net area for one period is T_0.

We derive the equation for the amplitude of the dc component a_0 by integrating both sides of Eq. (13–2):

$$\int_{-T_0/2}^{+T_0/2} f(t)dt = \qquad\qquad\qquad (13\text{–}5)$$

$$a_0 \int_{-T_0/2}^{+T_0/2} dt + \sum_{n=1}^{\infty}\left[a_n \int_{-T_0/2}^{+T_0/2} \cos(2\pi n f_0 t)dt + b_n \int_{-T_0/2}^{+T_0/2} \sin(2\pi n f_0 t)dt \right]$$

$$= a_0 T_0 + \qquad\qquad 0 \qquad + \qquad\qquad 0$$

The integrals of the ac components vanish because of the properties in Eq. (13–4), and the right side of this expression reduces to $a_0 T_0$. Solving for a_0 yields the first expression in Eq. (13–3).

To derive the expression for a_n we multiply Eq. (13–2) by $\cos(2\pi m f_0 t)$ and integrate over the interval from $-T_0/2$ to $+T_0/2$:

$$\int_{-T_0/2}^{+T_0/2} f(t)\cos(2\pi m f_0 t)dt = a_0 \int_{-T_0/2}^{+T_0/2} \cos(2\pi m f_0 t)dt$$

$$+ \sum_{n=1}^{\infty}\left[a_n \int_{-T_0/2}^{+T_0/2} \cos(2\pi m f_0 t)\cos(2\pi n f_0 t)dt + \right. \qquad (13\text{–}6)$$

$$\left. b_n \int_{-T_0/2}^{+T_0/2} \cos(2\pi m f_0 t)\sin(2\pi n f_0 t)dt \right]$$

All of the integrals on the right side of this equation are zero except one. To show this we use identities

$$\cos(x)\cos(y) = \frac{1}{2}\cos(x - y) + \frac{1}{2}\cos(x + y)$$

$$\cos(x)\sin(y) = \frac{1}{2}\sin(x - y) + \frac{1}{2}\sin(x + y)$$

to change Eq. (13–6) into the following form:

$$\int_{-T_0/2}^{+T_0/2} f(t)\cos(2\pi m f_0 t)dt = a_0 \int_{-T_0/2}^{+T_0/2} \cos(2\pi m f_0 t)dt$$

$$+ \sum_{n=1}^{\infty} \left\{ \frac{a_n}{2} \left[\int_{-T_0/2}^{+T_0/2} \cos[2\pi(m-n)f_0 t]dt + \int_{-T_0/2}^{+T_0/2} \cos[2\pi(m+n)f_0 t]dt \right] \right\}$$

$$+ \sum_{n=1}^{\infty} \left\{ \frac{b_n}{2} \left[\int_{-T_0/2}^{+T_0/2} \sin[2\pi(m-n)f_0 t]dt + \int_{-T_0/2}^{+T_0/2} \sin[2\pi(m+n)f_0 t]dt \right] \right\}$$

(13–7)

All of the integrals are now in the form of expressions in Eq. (13–4). Consequently, we see all of the integrals on the right side of Eq. (13–7) vanish, except for one cosine integral when $m = n$. This one survivor corresponds to the $k = 0$ case in Eq. (13–4), and the right side of Eq. (13–7) reduces to

$$\int_{-T_0/2}^{T_0/2} f(t)\cos(2\pi n f_0 t)dt = \frac{a_n}{2} \int_{-T_0/2}^{T_0/2} \cos[2\pi(n-n)f_0 t]dt$$

$$= \frac{a_n}{2} T_0$$

Solving Eq. (13–7) for a_n yields the second expression in Eq. (13–3).

To obtain the expression for b_n we multiply Eq. (13–2) by $\sin(2\pi m f_0 t)$ and integrate over the interval $t = -T_0/2$ to $+T_0/2$. The derivation steps then parallel the approach used to find a_n. The end result is that the dc component integral vanishes and the ac component integrals reduce to $b_n T_0/2$, which yields the expression for b_n in Eq. (13–3).

The derivation of Eq. (13–3) focuses on the problem of finding the Fourier coefficients of a given periodic waveform. Some experience and practice are necessary to understand the implications of this procedure. On the other hand, it is not necessary to go through these mechanics for every newly encountered periodic waveform because tables of Fourier series expansions are available. For our purposes the listing in Figure 13–3 will suffice. For each waveform defined graphically, the figure lists the expressions for a_0, a_n, and b_n as well as restrictions on the integer n.

EXAMPLE 13–3

Verify the Fourier coefficients given for the square wave in Figure 13–3 and write the first three nonzero terms in its Fourier series.

SOLUTION:
An expression for a square wave on the interval $0 < t <$ to T_0 is

$$f(t) = \begin{cases} A & 0 < t < T_0/2 \\ -A & T_0/2 < t < T_0 \end{cases}$$

Using the first expression in Eq. (13–3) to find a_0 yields

Waveform	Fourier Coefficients	Waveform	Fourier Coefficients
Constant (dc) 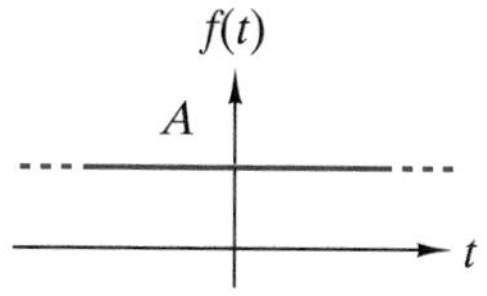	$a_0 = A$ $a_n = 0$ all n $b_n = 0$ all n	Sawtooth wave 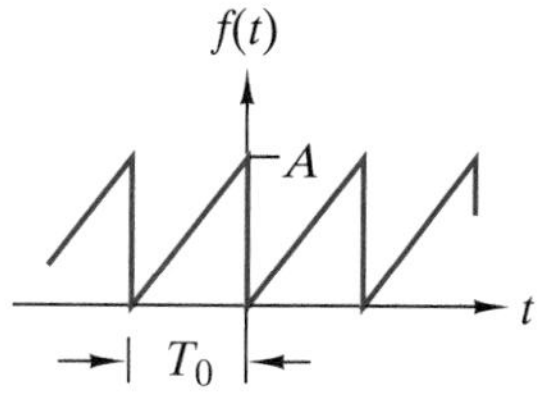	$a_0 = \dfrac{A}{2}$ $a_n = 0$ all n $b_n = -\dfrac{A}{n\pi}$ all n
Cosine wave 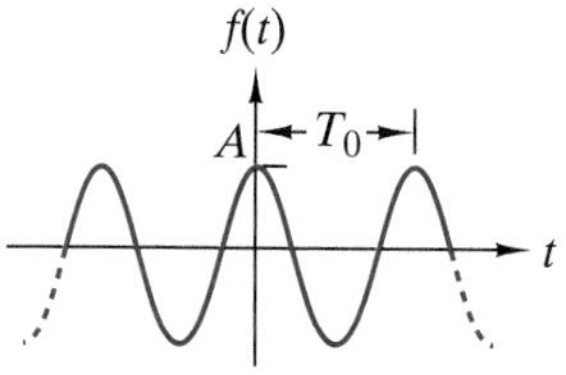	$a_0 = 0$ $a_1 = A$ $a_n = 0$ $n \neq 1$ $b_n = 0$ all n	Triangular wave 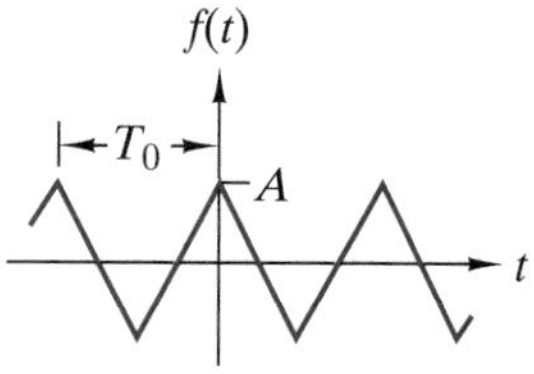	$a_0 = 0$ $a_n = \dfrac{8A}{(n\pi)^2}$ n odd $a_n = 0$ n even $b_n = 0$ all n
Sine wave 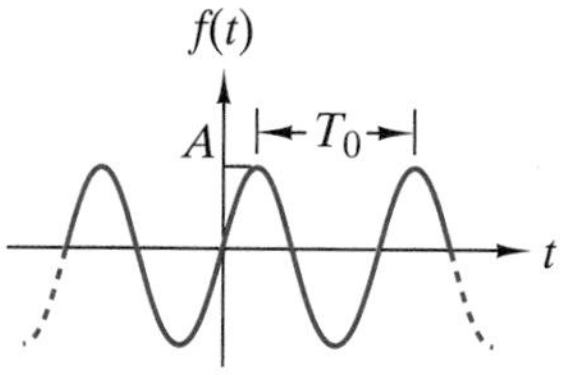	$a_0 = 0$ $a_n = 0$ all n $b_1 = A$ $b_n = 0$ $n \neq 1$	Half-wave rectified sine wave 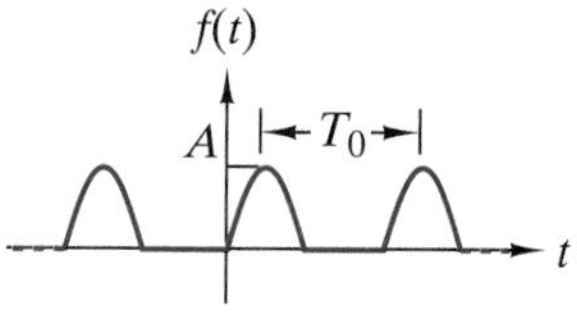	$a_0 = \dfrac{A}{\pi}$ $a_n = \dfrac{2A/\pi}{1-n^2}$ n even $a_n = 0$ n odd $b_1 = \dfrac{A}{2}$ $n = 1$ $b_n = 0$ $n \neq 1$
Square wave 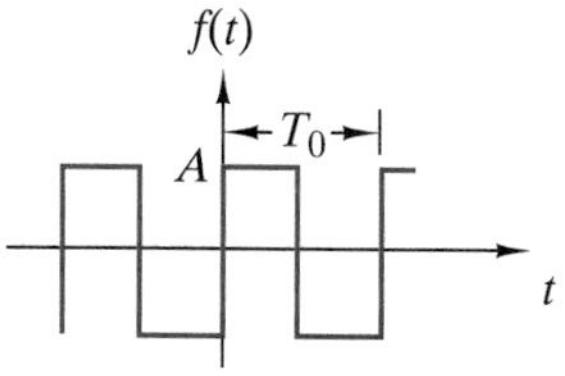	$a_0 = 0$ $a_n = 0$ all n $b_n = \dfrac{4A}{n\pi}$ n odd $b_n = 0$ n even	Full-wave rectified sine wave 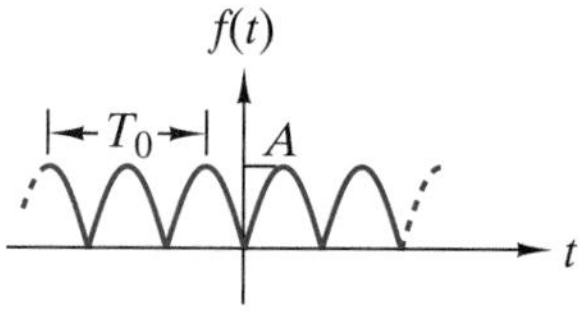	$a_0 = 2A/\pi$ $a_n = \dfrac{4A/\pi}{1-n^2}$ n even $a_n = 0$ n odd $b_n = 0$ all n
Rectangular pulse 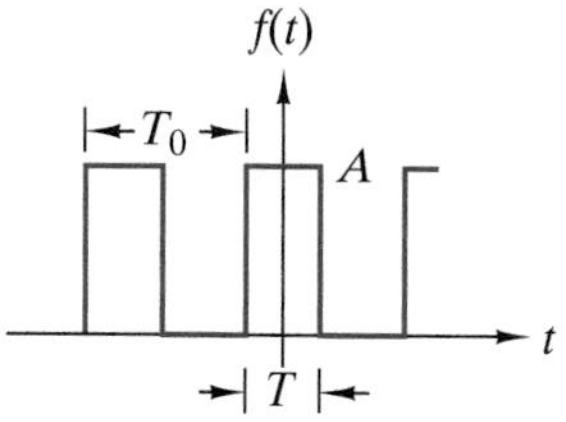	$a_0 = \dfrac{AT}{T_0}$ $a_n = \dfrac{2A}{n\pi}\sin\left(\dfrac{n\pi T}{T_0}\right)$ $b_n = 0$ all n	Parabolic wave 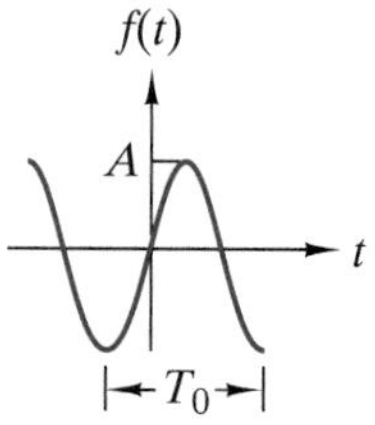	$a_0 = 0$ $a_n = 0$ all n $b_n = \dfrac{32A}{(n\pi)^3}$ n odd $b_n = 0$ n even

FIGURE 13-3 *Fourier coefficients for some periodic waveforms.*

$$a_0 = \frac{1}{T_0}\int_0^{T_0/2} A\, dt + \frac{1}{T_0}\int_{T_0/2}^{T_0} (-A)dt$$

$$= \frac{A}{T_0}\left[\frac{T_0}{2} - 0 - T_0 + \frac{T_0}{2}\right] = 0$$

The result $a_0 = 0$ means that the dc value of the square wave is zero, which is easy to see because the area under a positive half cycle cancels the area under a negative half cycle. Using the second expression in Eq. (13–3) to find a_n produces

$$a_n = \frac{2}{T_0}\int_0^{T_0/2} A\cos(2\pi nt/T_0)dt + \frac{2}{T_0}\int_{T_0/2}^{T_0} (-A)\cos(2\pi nt/T_0)dt$$

$$= \frac{2A}{T_0}\left[\frac{\sin(2\pi nt/T_0)}{2\pi n/T_0}\right]_0^{T_0/2} - \frac{2A}{T_0}\left[\frac{\sin(2\pi nt/T_0)}{2\pi n/T_0}\right]_{T_0/2}^{T_0}$$

$$= \frac{A}{n\pi}[\sin(n\pi) - \sin(0) - \sin(2n\pi) + \sin(n\pi)] = 0$$

Since $a_n = 0$ for all n there are no cosine terms in the series. This makes some intuitive sense because a sinewave with the same fundamental frequency as the square wave fits nicely inside the square wave with zeros crossing at the same points, whereas a cosine with the same frequency does not fit at all. The b_n coefficients for the sine terms are found using the third expression in Eq. (13–3):

$$b_n = \frac{2}{T_0}\int_0^{T_0/2} A\sin(2\pi nt/T_0)dt + \frac{2}{T_0}\int_{T_0/2}^{T_0} (-A)\sin(2\pi nt/T_0)dt$$

$$= \frac{2A}{T_0}\left[-\frac{\cos(2\pi nt/T_0)}{2\pi n/T_0}\right]_0^{T_0/2} - \frac{2A}{T_0}\left[-\frac{\cos(2\pi nt/T_0)}{2\pi n/T_0}\right]_{T_0/2}^{T_0}$$

$$= \frac{A}{n\pi}[-\cos(n\pi) + \cos(0) + \cos(2n\pi) - \cos(n\pi)]$$

$$= \frac{2A}{n\pi}[1 - \cos(n\pi)]$$

The term $[1 - \cos(n\pi)] = 2$ if n is odd and zero if n is even. Hence b_n can be written as

$$b_n = \begin{cases} \dfrac{4A}{n\pi} & n \text{ odd} \\ 0 & n \text{ even} \end{cases}$$

The first three nonzero terms in the Fourier series of the square wave are

$$f(t) = \frac{4A}{\pi}\left[\sin 2\pi f_0 t + \frac{1}{3}\sin 2\pi 3f_0 t + \frac{1}{5}\sin 2\pi 5f_0 t + \ldots\right]$$

Note that this series contains only odd harmonic terms.   ■

Exercise 13–1

The triangular wave in Figure 13–3 has a peak amplitude of $A = 10$ and $T_0 = 2$ ms. Calculate the Fourier coefficients of the first nine harmonics.

Answers:
$a_1 = 8.11$, $a_2 = 0$, $a_3 = 0.901$, $a_4 = 0$, $a_5 = 0.324$, $a_6 = 0$, $a_7 = 0.165$, $a_8 = 0$, $a_9 = 0.100$, $b_n = 0$ for all n

ALTERNATIVE FORM OF THE FOURIER SERIES

The series in Eq. (13–1) can be written in several alternative yet equivalent forms. From our study of sinusoids in Chapter 5, we recall that the Fourier coefficients determine the amplitude and phase angle of the general sinusoid. Thus, we can write a general Fourier series in the form

$$f(t) = A_0 + A_1 \cos(2\pi f_0 t + \phi_1) + A_2 \cos(2\pi 2 f_0 t + \phi_2) + \ldots$$
$$+ A_n \cos(2\pi n f_0 t + \phi_n) + \ldots \qquad (13\text{–}8)$$

where

$$A_n = \sqrt{a_n^2 + b_n^2} \quad \text{and} \quad \phi_n = \tan^{-1}\frac{-b_n}{a_n} \qquad (13\text{–}9)$$

The coefficient A_n is the amplitude of the nth harmonic and ϕ_n is its phase angle.[1]

Note that the amplitude A_n and phase angle ϕ_n contain all of the information needed to construct the Fourier series in the form of Eq. (13–8). Figure 13–4 shows how plots of this information are used to display the spectral content of a periodic waveform $f(t)$. The plot of A_n versus $n f_0$ (or $n\omega_0$) is called the **amplitude spectrum**, while the plot of ϕ_n versus $n f_0$ (or $n\omega_0$) is called the **phase spectrum**. Both plots are **line spectra** because spectral content can be represented as a line at discrete frequencies.

In theory, a Fourier series includes infinitely many harmonics, although the harmonics tend to decrease in amplitude at high frequency. For example, the summary in Figure 13–3 shows that the amplitudes of the square wave decrease as $1/n$, the triangular wave as $1/n^2$, and the parabolic wave as $1/n^3$. The $1/n^3$ dependence means that the amplitude of the fifth harmonic in a parabolic wave is less than 1% of the amplitude of the fundamental (actually 1/125th of the fundamental). In practical signals the harmonic amplitudes decrease at high frequency so that at some point the higher-order components become negligibly small. This means that we can truncate the series at some finite frequency and still retain the important features of the signal. This is an important consideration in systems with finite bandwidth.

---

1 There is a 180° ambiguity in the value returned by the inverse tangent function in most computational tools. The ambiguity is resolved by the following rule: $b_n < 0$ implies that the angle is in the range 0 to –180°, while $b_n > 0$ implies the 0 to +180° range.

FIGURE 13−4 *Amplitude and phase spectra.*

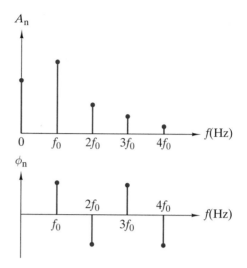

EXAMPLE 13−4

Derive expressions for the amplitude A_n and phase angle ϕ_n of the Fourier series of the sawtooth wave in Figure 13–3. Sketch the amplitude and phase spectra of a sawtooth wave with $A = 5$ and $T_0 = 4$ ms.

SOLUTION:

Figure 13–3 gives the Fourier coefficients of the sawtooth wave as

$$a_0 = \frac{A}{2} \qquad a_n = 0 \qquad b_n = -\frac{A}{n\pi} \qquad \text{for all } n$$

Using Eq. (13–9) yields

$$A_n = \sqrt{a_n^2 + b_n^2} = \begin{cases} \dfrac{A}{2} & n = 0 \\[2mm] \dfrac{A}{n\pi} & n > 0 \end{cases}$$

and

$$\phi_n = \tan^{-1}\frac{-b_n}{a_n} = \begin{cases} \text{undefined} & n = 0 \\ 90° & n > 0 \end{cases}$$

For $A = 5$ and $f_0 = 1/T_0 = 250$ Hz the first four nonzero terms in the series are

$$f(t) = 2.5 + 1.59 \cos(2\pi 250t + 90°)$$

$$+ \, 0.796 \cos(2\pi 500t + 90°) + 0.531 \cos(2\pi 750t + 90°) + \cdots$$

Figure 13–5 shows the amplitude and phase spectra for this signal. Note that the lines in the amplitude spectrum are inversely proportional to frequency. As frequency increases the amplitudes decrease so that the high-frequency components become negligible. ∎

FIGURE 13–5

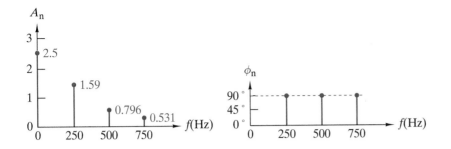

Exercise 13–2

Derive expressions for the amplitude A_n and phase angle ϕ_n for the triangular wave in Figure 13–3 and write an expression for the first three nonzero terms in the Fourier series with $A = \pi^2/8$ and $T_0 = 2\pi/5000$ s.

Answers:

$$A_n = \frac{8A}{(n\pi)^2} \quad \phi_n = 0° \quad n \text{ odd}$$

$$A_n = 0 \quad \phi_n \text{ undefined} \quad n \text{ even}$$

$$f(t) = \cos(5000t) + \frac{1}{9}\cos(15{,}000t) + \frac{1}{25}\cos(25{,}000t) + \cdots$$

13–3 WAVEFORM SYMMETRIES

Many of the Fourier coefficients are zero when a periodic waveform has certain types of symmetries. It is helpful to recognize these symmetries, since they may simplify the calculation of the Fourier coefficients.

The first expression in Eq. (13–3) shows that the amplitude of the dc component a_0 is the average value of the periodic waveform $f(t)$. If the waveform has equal area above and below the time axis, then the integral over one cycle vanishes, the average value is zero, and $a_0 = 0$. The square wave, triangular wave, and parabolic wave in Figure 13–3 are examples of periodic waveforms with zero average value.

A waveform is said to have **even symmetry** if $f(-t) = f(t)$. The cosine wave, rectangular pulse, and triangular wave in Figure 13–3 are examples of waveforms with even symmetry. The Fourier series of an even waveform is made up entirely of cosine terms: that is, all of the b_n coefficients are zero. To show this we write the Fourier series for $f(t)$ in the form

$$f(t) = a_0 + \sum_{n=1}^{\infty}[a_n \cos(2\pi nf_0t) + b_n \sin(2\pi nf_0t)] \qquad (13\text{–}10)$$

Given the Fourier series for $f(t)$, we use the identities $\cos(-x) = \cos(x)$ and $\sin(-x) = -\sin(x)$ to write the Fourier series for $f(-t)$ as follows:

$$f(-t) = a_0 + \sum_{n=1}^{\infty}[a_n \cos(2\pi nf_0t) - b_n \sin(2\pi nf_0t)] \qquad (13\text{–}11)$$

For even symmetry $f(t) = f(-t)$ and the right sides of Eqs. (13–10) and (13–11) must be equal. Comparing the Fourier coefficients term by term, we find that $f(t) = f(-t)$ requires $b_n = -b_n$. The only way this can happen is for $b_n = 0$ for all n.

A waveform is said to have **odd symmetry** if $-f(-t) = f(t)$. The sine wave, square wave, and parabolic wave in Figure 13–3 are examples of waveforms with this type of symmetry. The Fourier series of odd waveforms are made up entirely of sine terms: that is, all of the a_n coefficients are zero. Given the Fourier series for $f(t)$ in Eq. (13–10), we use the identities $\cos(-x) = \cos(x)$ and $\sin(-x) = -\sin(x)$ to write the Fourier series for $-f(-t)$ in the form

$$-f(-t) = -a_0 + \sum_{n=1}^{\infty}[-a_n \cos(2\pi nf_0 t) + b_n \sin(2\pi nf_0 t)] \qquad (13\text{–}12)$$

With odd symmetry $f(t) = -f(-t)$ and the right sides of Eq. (13–10) and (13–12) must be equal. Comparing the Fourier coefficients term by term, we find that odd symmetry requires $a_0 = -a_0$ and $a_n = -a_n$. The only way this can happen is for $a_n = 0$ for all n, including $n = 0$.

A waveform is said to have **half-wave symmetry** if $-f(t - T_0/2) = f(t)$. This requirement states that inverting the waveform $[-f(t)]$ and then time shifting by half a cycle $(T_0/2)$ must produce the same waveform. Basically, this means that successive half cycles have the same waveshape but opposite polarities. In Figure 13–3 the sine wave, cosine wave, square wave, triangular wave, and parabolic wave have half-wave symmetry. The sawtooth wave, half-wave sine, rectangular pulse train, and full-wave sine do not have this symmetry.

With half-wave symmetry the amplitudes of all even harmonics are zero. To show this we use the identities $\cos(x - n\pi) = (-1)^n \cos(x)$ and $\sin(x - n\pi) = (-1)^n \sin(x)$ to write the Fourier series of $-f(t - T_0/2)$ in the form

$$-f(t - T_0/2) = -a_0 + \sum_{n=1}^{\infty}[-(-1)_n\, a_n \cos(2\pi nf_0 t) -$$

$$(-1)^n\, b_n \sin(2\pi nf_0 t)] \qquad (13\text{–}13)$$

For half-wave symmetry the right sides of Eqs. (13–10) and (13–13) must be equal. Comparing the coefficients term by term, we find that equality requires $a_0 = -a_0$, $a_n = -(-1)^n a_n$, and $b_n = -(-1)^n b_n$. The only way this can happen is for $a_0 = 0$ and for $a_n = b_n = 0$ when n is even. In other words, the only nonzero Fourier coefficients occur when n is odd.

A waveform may have more than one symmetry. For example, the triangular wave in Figure 13–3 has even symmetry and half-wave symmetry, while the square wave has both odd and half-wave symmetries. The sawtooth wave in Figure 13–3 is an example where an underlying odd symmetry is masked by a dc component. A symmetry that is not apparent until the dc component is removed is sometimes called a **hidden symmetry**.

Finally, whether a waveform has even or odd symmetry (or neither) depends on where we choose to define $t = 0$. For example, the triangular wave in Figure 13–3 has even symmetry because the $t = 0$ vertical axis is located at a local maximum. If the axis is shifted to a zero crossing, the waveform has odd symmetry and the cosine terms in the series are replaced by sine terms. If the vertical axis is shifted to a point between a zero cross and a maximum, then the resulting waveform is neither even nor odd and its Fourier series contains both sine and cosine terms.

EXAMPLE 13-5

Given that $f(t)$ is a square wave of amplitude A and period T_0, use the Fourier coefficients in Figure 13-3 to find the Fourier coefficients of $g(t) = f(t + T_0/4)$.

SOLUTION:
Figure 13-6 compares the square waves $f(t)$ and $g(t) = f(t + T_0/4)$. The square wave $f(t)$ has odd symmetry (sine terms only) and half-wave symmetry (odd harmonics only). Using the coefficients in Figure 13-3, the Fourier series for $f(t)$ is

$$f(t) = \sum \frac{4A}{n\pi} \sin(2\pi nt/T_0) \quad n \text{ odd}$$

The Fourier series for $g(t) = f(t + T_0/4)$ can be written in the form

$$g(t) = f(t + T_0/4) = \sum \frac{4A}{n\pi} \sin[2\pi n(t + T_0/4)/T_0] \quad n \text{ odd}$$

$$= \sum \frac{4A}{n\pi} \sin(2\pi nt/T_0 + n\pi/2)$$

$$= \sum \frac{4A}{n\pi} \cos(2\pi nt/T_0)\sin(n\pi/2)$$

$$= \sum \frac{4A}{n\pi} \cos(2\pi nt/T_0)(-1)^{\frac{n-1}{2}} \quad n \text{ odd}$$

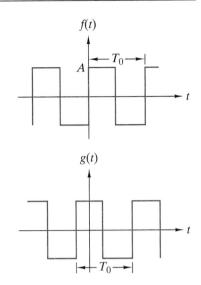

FIGURE   13 - 6

Figure 13-6 shows that $g(t)$ has even and half-wave symmetry so its Fourier series has only cosine terms and odd harmonics. The Fourier coefficients for $g(t)$ are

$$a_0 = 0 \quad a_n = \begin{cases} 0 & n \text{ even} \\ \left[\dfrac{4A}{n\pi}\right](-1)^{\frac{2n-1}{2}} & n \text{ odd} \end{cases}$$

$$b_n = 0 \quad \text{all } n$$

Shifting the time origin alters the even or odd symmetry properties of a periodic waveform because these symmetries depend on values of $f(t)$ on opposite sides of the vertical axis at $t = 0$. The half-wave symmetry of a waveform is not changed by time shifting because this symmetry only requires successive half cycles to have the same form but opposite polarities. ■

Exercise 13-3

(a) Identify the symmetries in the waveform $f(t)$ whose Fourier series is

$$f(t) = \frac{2\sqrt{3}A}{\pi}\left[\cos(\omega_0 t) - \frac{1}{5}\cos(5\,\omega_0 t) + \frac{1}{7}\cos(7\,\omega_0 t)\right.$$

$$\left. - \frac{1}{11}\cos(11\,\omega_0 t) + \frac{1}{13}\cos(13\,\omega_0 t) + \ldots\right]$$

(b) Write the corresponding terms of the function $g(t) = f(t - T_0/4)$.

Answers:
(a) Even symmetry, half-wave symmetry, zero average value.

(b) $g(t) = \dfrac{2\sqrt{3}A}{\pi}\left[\sin(\omega_0 t) - \dfrac{1}{5}\sin(5\,\omega_0 t) - \dfrac{1}{7}\sin(7\,\omega_0 t)\right.$

$\left. + \dfrac{1}{11}\sin(11\,\omega_0 t) + \dfrac{1}{13}\sin(13\,\omega_0 t) + \ldots\right]$

13–4   CIRCUIT ANALYSIS USING FOURIER SERIES

Up to this point we have concentrated on finding the Fourier series description of periodic waveforms. We are now in a position to address circuit analysis problems of the type illustrated in Figure 13–7. This first-order *RL* circuit is driven by a periodic sawtooth voltage, and the objective is to find the steady-state current $i(t)$.

FIGURE 13–7 *Linear circuit with a periodic input.*

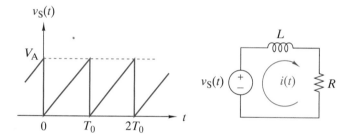

We begin by using the results in Example 13–4 to express the input voltage as a Fourier series in the form

$$v_S(t) = \underbrace{\frac{V_A}{2}}_{\text{dc}} + \underbrace{\sum_{n=1}^{\infty} \frac{V_A}{n\pi}\cos(n\omega_0 t + 90°)}_{\text{ac}} \qquad (13\text{–}14)$$

This result expresses the input driving force as the sum of a dc component plus ac components at harmonic frequencies $n\omega_0 = 2\pi n f_0$, $n = 1, 2, 3, \ldots$. Since the circuit is linear, we find the steady-state response caused by each component acting alone and then obtain the total response by superposition.

In the dc steady state the inductor acts like a short circuit, so the steady-state current due to the dc input $V_A/2$ is simply $I_0 = V_A/(2R)$. The amplitude and phase angle of the nth harmonic of the sawtooth input are

$$V_n = \frac{V_A}{n\,\pi} \quad \text{and} \quad \phi_n = 90°$$

In Chapter 11 we found that the sinusoidal steady-state response at frequency ω_A can be found directly from a network function $T(s)$ as

$$\text{Output amplitude} = \text{Input amplitude} \times |T(j\,\omega_A)|$$

$$\text{Output phase} = \text{Input phase} + \angle T(j\,\omega_A)$$

In the present case the input is the nth harmonic at $\omega_A = n\omega_0$, the output is the current $I(s) = V(s)/Z(s)$, and hence the network function of interest is $T(s) = 1/(Ls + R)$. Applying the Chapter 11 method here yields the amplitude and phase of the nth harmonic current as

$$\text{Amplitude} = V_n \times \left| \frac{1}{jn\omega_0 L + R} \right| = \frac{V_A}{n\pi R} \frac{1}{\sqrt{1 + jn\omega_0 L/R}}$$

$$\text{Phase} = \phi_n + \text{angle}\left(\frac{1}{jn\omega_0 L + R}\right) = 90° - \tan^{-1}(n\omega_0 L/R)$$

Defining $\theta_n = \tan^{-1}(n\omega_0 L/R)$, we get the waveform of the steady-state response to the nth harmonic input.

$$i_n(t) = \underbrace{\frac{V_A}{n\pi R} \frac{1}{\sqrt{1 + (n\omega_0 L/R)^2}}}_{\text{Amplitude}} \cos\underbrace{(n\omega_0 t + 90° - \theta_n)}_{\text{Phase}}$$

We have now found the steady-state response of the circuit due to the dc component acting alone and the nth harmonic ac component acting alone. Since the circuit is linear, superposition applies and we find the steady-state response caused the sawtooth input triangular summing the contributions of each of these sources:

$$i(t) = I_0 + \sum_{n=1}^{\infty} i_n(t)$$

$$= \frac{V_A}{2R} + \frac{V_A}{R}\sum_{n=1}^{\infty} \frac{1}{n\pi\sqrt{1 + (n\omega_0 L/R)^2}} \cos(n\omega_0 t + 90° - \theta_n) \qquad (13\text{--}15)$$

The Fourier series in Eq. (13–15) represents the steady-state current due to a sawtooth driving force whose Fourier series is in Eq. (13–14).

EXAMPLE 13–6

Find the first four nonzero terms of the Fourier series in Eq. (13–15) for $V_A = 25$ V, $R = 50\ \Omega$, $L = 40\ \mu$H, $\omega_0 = 1$ Mrad/s.

SOLUTION:
Equation (13–15) gives a Fourier series of the form

$$i(t) = I_0 + \sum_{n=1}^{\infty} I_n \cos(n\omega_0 t + \psi_n)$$

where

$$I_0 = \frac{V_A}{2R} \qquad I_n = \frac{V_A}{2R} \frac{1}{n\pi\sqrt{1 + (n\omega_0 L/R)^2}} \qquad \psi_n = \frac{\pi}{2} - \tan^{-1}(n\omega_0 L/R)$$

Inserting the numerical values leads to

$$I_0 = \frac{1}{4} \qquad I_n = \frac{1}{2n\pi\sqrt{1 + (n \times 0.8)^2}} \qquad \psi_n = \frac{\pi}{2} - \tan^{-1}(n \times 0.8)$$

The first four nonzero terms in the Fourier series include the dc component plus the first three harmonics. For $n = 0, 1, 2, 3$, we have

$$I_0 = 0.25$$

$$I_1 = 0.124 \quad \psi_1 = 0.896 \ \text{rad} = 51.3°$$

$$I_2 = 0.0422 \quad \psi_2 = 0.559 \ \text{rad} = 32.0°$$

$$I_3 = 0.0204 \quad \psi_3 = 0.395 \ \text{rad} = 22.6°$$

hence, the desired expression is

$$i(t) = 0.25 + 0.124 \cos(10^6 t + 51.3°)$$

$$+ 0.0422 \cos(2 \times 10^6 t + 32°) + 0.0204 \cos(3 \times 10^6 t + 22.6°) \ldots$$

Given the Fourier series of a periodic input, it is a straightforward procedure to obtain the Fourier series of the steady-state response. But interpreting the analysis result requires some thought since the response is presented as an infinite series. In practice the series converges rather rapidly, so we can calculate specific values or generate plots using computer tools to sum a truncated version of the infinite series. Before doing so let us look closely at Eq. (13–15) to see what we can infer about the response.

First note that if $L = 0$ Eq. (13–15) reduces to

$$i(t) = \frac{V_A/R}{2} + \sum_{n=1}^{\infty} \frac{V_A/R}{n\pi} \cos(n\omega_0 t + 90°)$$

which is the Fourier series of a sawtooth wave of amplitude V_A/R. This makes sense because without the inductor the circuit in Figure 13–7 is a simple resistive circuit in which $i(t) = v_S(t)/R$, so the input and response must have the same waveform.

When $L \neq 0$ the response is not a sawtooth, but we can infer some features of its waveform if we examine the amplitude spectrum. At high frequency $[(n\omega_0 L/R) \gg 1]$ Eq. (13–15) points out that the amplitudes of the ac components are approximately

$$I_n \approx \frac{V_A}{R} \frac{1}{n^2 \pi \omega_0 L/R} \tag{13–16}$$

In the steady-state response the amplitudes of the high-frequency ac components decrease as $1/n^2$, whereas the ac components in the input sawtooth decrease as $1/n$. In other words, the relative amplitudes of the high-frequency components are much smaller in the response than in the input. This makes sense because the inductor's impedance increases with frequency and thereby reduces the amplitudes of the high-frequency ac currents. We would expect the circuit to filter out the high-frequency components in the input and produce a response without the sharp corners and discontinuities in the input sawtooth.

The next example examines this thought for a specific set of parameters.

EXAMPLE 13–7

The parameters of the steady-state waveform in Eq. (13–15) are $V_A = 25$ V, $T_0 = 5$ μs, $L = 40$ μH, and $R = 50$ Ω. Calculate and plot a truncated Fourier series representation of the steady-state current using the first 5 harmonics and the first 10 harmonics.

SOLUTION:

Figure 13–8 shows a Mathcad worksheet that calculates truncated Fourier series of the steady-state current in Eq. (13–15). The first two lines define the waveform and circuit parameters. The Fourier parameters for the first 20 harmonics are calculated in the third line using the index variable $n = 1, 2, \ldots 20$. The next line defines a truncated Fourier series $i(k, t)$ consisting of the dc component I_0 plus the sum of the first k harmonics. The responses $I(5, t)$ and $I(10, t)$ are plotted at the bottom of the worksheet. There is not much change between the two plots, suggesting that a truncated series converges rather rapidly and gives a reasonable approximation to the steady-state current.

The plots in Figure 13–8 show that the steady-state current is indeed smoother than the input driving force. The reason is that the inductor suppresses the high-frequency components that are important contributors to the discontinuities in the sawtooth waveform. As a result, the response does not have the abrupt changes and sharp corners present in the sawtooth input. ■

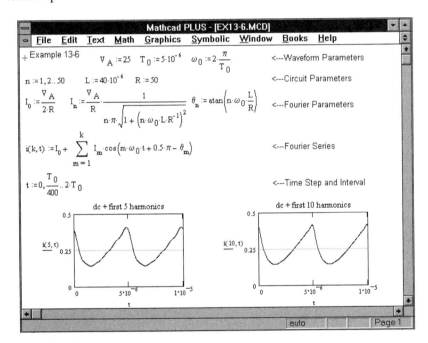

FIGURE 13 – 8

Exercise 13-4

Derive an expression for the steady-state current of the RL circuit in Figure 13–7 when the input voltage is a triangular wave with amplitude V_A and frequency ω_0.

Answer:

$$i(t) = \sum_{n=1}^{\infty} \frac{8V_A/R}{(n\pi)^2\sqrt{1 + (n\omega_0L/R)^2}} \cos(n\omega_0 t - \theta_n) \quad n \text{ odd}$$

where

$$\theta_n = \tan^{-1}(n\omega_0 L/R)$$

EXAMPLE 13–8

A sawtooth wave with $V_A = 12$ V and $\omega_0 = 40$ rad/s drives a circuit with a transfer function $T(s) = 100/(s + 50)$. Find the amplitude of the first four nonzero terms in the Fourier series of the steady-state output.

SOLUTION:

The Fourier coefficients of the input are

$$a_0 = \frac{V_A}{2} \qquad a_n = 0 \qquad b_n = -\frac{V_A}{n\pi} \qquad \text{for all } n$$

The amplitudes of the Fourier series for the steady-state output are found from

$$A_0 = |a_0 T(0)|$$

$$A_n = |b_n T(jn\omega_0)| = \left| \frac{12}{n\pi} \frac{100}{\sqrt{50^2 + (n\,40)^2}} \right|$$

Which yields $A_0 = 12$, $A_1 = 5.97$, $A_2 = 2.02$, and $A_3 = 0.979$. ∎

EXAMPLE 13–9

Figure 13–9 shows a block diagram of a dc power supply. The ac input is a sinusoid that is converted to a full-wave sine by the rectifier. The filter passes the dc component in the rectified sine and suppresses the ac components. The result is an output consisting of a small residual ac ripple riding on top of a much larger dc signal.

   Calculate and plot the first 10 harmonics in amplitude spectra of the filter input and output for $V_A = 23.6$ V, $T_0 = 1/60$ s, and a lowpass filter transfer function of

$$T(s) = \frac{(200)^2}{s^2 + 70s + (200)^2}$$

FIGURE 13 – 9

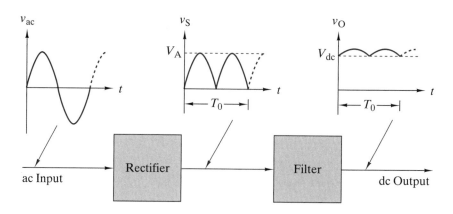

SOLUTION:

The amplitude spectrum of the filter input is obtained using the Fourier coefficients for the full-wave rectified sine in Figure 13–3:

$$V_0 = 2V_A/\pi = 15.02 \text{ V}$$

$$V_n = \begin{cases} 0 & n \text{ odd} \\ \left|\dfrac{4V_A/\pi}{1 - n^2}\right| = \dfrac{30.04}{n^2 - 1} & n \text{ even} \end{cases} \qquad (13\text{-}17)$$

The magnitude of the transfer function at each of these discrete frequencies is

$$|T(jn\omega_0)| = \frac{(200)^2}{\sqrt{[(200)^2 - (n\omega_0)^2]^2 + (280\,n\omega_0)^2}} \qquad (13\text{-}18)$$

To obtain the specified output spectrum, we must generate the product of the input amplitude times the transfer function magnitude for $n = 0, 1, 2, 3, \ldots 10$.

Spreadsheets are ideally suited to making repetitive calculations of this type. Figure 13-10 shows an Excel spreadsheet that implements the required calculations. Column A gives the index n and column B gives the corresponding frequencies in Hz. The input amplitudes in column C are calculated using Eq. (13-17), while the entries in column D are based on Eq. (13-18). Finally, the entries in column E are the product of those in columns C and D. Since the lowpass filter has unity gain at zero frequency, the dc components in the input and output are equal. The first nonzero ac component is the second harmonic, which has an amplitude of 10 V in the input but less than 1 V in the output. By the time we get to the next nonzero harmonic at 240 Hz, the ac amplitudes in the output are entirely negligible. ■

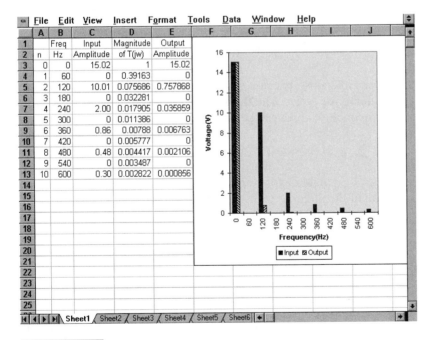

FIGURE 13-10

Exercise 13-5

Derive an expression for the first three nonzero terms in the Fourier series of the steady-state output voltage in Example 13-9.

Answer:
$$v_O(t) = 15.02 + 0.758 \cos(2\pi\,120t + 5.7°) + 0.036 \cos(2\pi\,240t + 2.7°) \text{ V}$$

13–5 R M S V A L U E A N D A V E R A G E P O W E R

In Chapter 5 we introduced the rms value of a periodic waveform as a descriptor of the average power carried by a signal. In this section we relate the rms value of the waveform to the amplitudes of the dc and ac components in its Fourier series. The rms value of a periodic waveform is defined as

$$F_{rms} = \sqrt{\frac{1}{T_0}\int_0^{T_0} [f(t)]^2 \, dt} \tag{13–19}$$

The waveform $f(t)$ can be expressed as a Fourier series in the amplitude and phase form as

$$f(t) = A_0 + \sum_{n=1}^{\infty} A_n \cos(n\omega_0 t + \phi_n)$$

Substituting this expression into Eq. (13–19), we can write F^2_{rms} as

$$F^2_{rms} = \frac{1}{T_0}\int_0^{T_0} \left[A_0 + \sum_{n=1}^{\infty} A_n \cos(n\omega_0 t + \phi_n) \right]^2 dt \tag{13–20}$$

Squaring and expanding the integrand on the right side of this equation produces three types of terms. The first is the square of the dc component:

$$\frac{1}{T_0}\int_0^{T_0} [A_0]^2 \, dt = A_0^2 \tag{13–21}$$

The second is the cross product of the dc and ac components, which takes the form

$$\frac{1}{T_0} \sum_{n=1}^{\infty} 2A_0 \int_0^{T_0} A_n \cos(n\omega_0 t + \phi_n) dt = 0 \tag{13–22}$$

These terms all vanish because they involve integrals of sinusoids over an integer number of cycles. The third and final type of term is the square of the ac components, which can be written as

$$\frac{1}{T_0} \sum_{n=1}^{\infty} \sum_{m=1}^{\infty} \int_0^{T_0} A_n \cos(n\omega_0 t + \phi_n) A_m \cos(m\omega_0 t + \phi_m) dt = \frac{1}{2} \sum_{n=1}^{\infty} A_n^2 \tag{13–23}$$

This rather formidable expression boils down to a simple sum of squares because all of the integrals vanish except when $m = n$.

Combining Eqs. (13–19) through (13–23), we obtain the rms value as

$$F_{rms} = \sqrt{A_0^2 + \sum_{n=1}^{\infty} \frac{A_n^2}{2}}$$

$$= \sqrt{A_0^2 + \sum_{n=1}^{\infty} \left(\frac{A_n}{\sqrt{2}}\right)^2} \tag{13–24}$$

Since the rms value of a sinusoid of amplitude A is $A/\sqrt{2}$, we conclude that

The rms value of a periodic waveform is equal to the square root of the sum of the square of the dc value and the square of the rms value of each of the ac components.

In Chapter 5 we found that the average power delivered to a resistor is related to its rms voltage or current as

$$P = \frac{V^2_{rms}}{R} = I^2_{rms}R$$

Combining these expressions with the result in Eq. (13–24), we can write the average power delivered by a periodic waveform to the average power delivered by each of its Fourier components:

$$P = \frac{V^2_0}{R} + \sum_{n=1}^{\infty} \frac{V^2_n}{2R} = I^2_0 R + \sum_{n=1}^{\infty} \frac{I^2_n}{2}R$$

$$= P_0 + \sum_{n=1}^{\infty} P_n \qquad (13\text{--}25)$$

where P_0 is the average power delivered by the dc component and P_n is the average power delivered by the nth ac component. This additive feature is important because it means we can find the total average power by adding the average power carried by the dc plus that carried by each of the ac components.

Caution: In general, we cannot find the total power by adding the power delivered by each component acting alone because the superposition principle does not apply to power. However, the average power carried by harmonic sinusoids is additive because they belong to a special class call orthogonal signals.

EXAMPLE 13-10

Derive an expression for the average power delivered to a resistor by a sawtooth voltage of amplitude V_A and period T_0. Then calculate the fraction of the average power carried by the dc component plus the first three ac components.

SOLUTION:
An equation for the sawtooth voltage is $v(t) = V_A t/T_0$ for the range $0 < t < T_0$. The square of the rms value of a sawtooth is

$$V^2_{rms} = \frac{1}{T_0}\int_0^{T_0}\left(\frac{V_A t}{T_0}\right)^2 dt = V^2_A\left[\frac{t^3}{3T_0^3}\right]_0^{T_0} = \frac{V^2_A}{3}$$

The average power delivered to a resistor is

$$P = \frac{V^2_{rms}}{R} = \frac{V^2_A}{3R} = 0.333\,\frac{V^2_A}{R}$$

This result is obtained directly from the sawtooth waveform without having to sum an infinite series. The same answer could be obtained by summing an infinite series. The question this example asks is, How much of the average power is carried by the first four components in the Fourier series of the sawtooth wave? The amplitude spectrum of the sawtooth wave (see Example 13–4) is

$$V_n = \begin{cases} \dfrac{V_A}{2} & n = 0 \\[2mm] \dfrac{V_A}{n\pi} & n > 0 \end{cases}$$

From Eq. (13–25) the average power in terms of amplitude spectrum is

$$P = \frac{(V_A/2)^2}{R} + \sum_{n=1}^{\infty} \frac{(V_A/n\pi)^2}{2R}$$

which can be arranged in the form

$$P = \frac{V_A^2}{R}\left\{\underbrace{\underbrace{\frac{1}{(2)^2} + \frac{1}{2(\pi)^2} + \frac{1}{2(2\pi)^2}}_{0.319(96\%)} + \frac{1}{2(3\pi)^2} + \frac{1}{2(4\pi)^2} + \cdots}_{0.333}\right\}$$

The infinite series within the braces must sum to 0.333 to match the average power we calculated directly from the waveform itself. The dc component plus the first three ac terms contribute 0.319 to the infinite sum. In other words, these four components alone deliver 96% of the average power carried by the sawtooth wave.   ∎

Exercise 13–6

The full-wave sine shown in Figure 13–3 has an rms value of $A/\sqrt{2}$. What fraction of the average power that the waveform delivers to a resistor is carried by the first two nonzero terms in its Fourier series?

Answer:

$$\text{Fraction} = 88/9\pi^2 \text{ or } 99.07\%.$$

EXAMPLE 13–11

Figure 13–7 shows a series RL circuit driven by a sawtooth voltage source. Estimate the average power delivered to the resistor for $V_A = 25$ V, $R = 50\ \Omega$, $L = 40\ \mu$H, $T_0 = 5\ \mu$s, and $\omega_0 = 2\pi/T_0 = 1.26$ Mrad/s.

SOLUTION:

The Fourier series of the current in a series RL circuit is given in Eq. (13–15) as

$$i(t) = I_0 + \sum_{n=1}^{\infty} I_n \cos(n\omega_0 t + \psi_n)$$

where

$$I_0 = \frac{V_A}{2R} \qquad I_n = \frac{V_A}{R}\frac{1}{n\,\pi\sqrt{1 + (n\omega_0 L/R)^2}}$$

We cannot directly calculate the rms value of this current without a closed form expression for its waveform. However, we can get an estimate of the average power from the Fourier series. In terms of its Fourier series, the average power delivered by the current is

$$P = I_0^2 R + \sum_{n=1}^{\infty} \frac{I_n^2}{2} R$$

When $n\omega_0 L/R \gg 1$, the amplitude I_n decreases as $1/n^2$ and its contribution to the average power falls off as $1/n^4$. In other words, the infinite series for the average power converges very rapidly. To show how rapidly the series converges, we calculate the first few terms using the specified numerical values

$$P = I_0^2 R + \frac{I_1^2}{2} R + \frac{I_2^2}{2} R + \frac{I_3^2}{2} R + \ldots$$

$$= 3.125 + 0.315 + 0.0314 + 0.00697 + \ldots$$

$$\longleftarrow\!\!-3.44\longrightarrow$$
$$\longleftarrow\!\!\!-\!\!-3.471\!-\!\!-\longrightarrow$$
$$\longleftarrow\!\!\!-\!\!\!-3.478\!-\!\!\!-\!\!\longrightarrow$$

These results indicate that $P = 3.48$ W is a reasonable estimate of the average power, an estimate obtained using only four terms in the infinite series. The important point is that the high-frequency ac components are not important contributors to the average power carried by a signal. ■

APPLICATION NOTE: EXAMPLE 13-12

Digital signal processing treats samples of an analog waveform, as contrasted with analog processing which operates on the entire waveform. Sampling refers to the process of selecting discrete values of a time-varying analog waveform for further processing. By far the most common method of sampling is to record the waveform amplitudes at equally spaced time intervals. The set of samples v_k $\{k = 1, 2, 3, \ldots\}$ of an analog waveform $v(t)$ is defined as

$$v_k = v(k T_S)$$

where k is an integer and T_S is the time interval between successive samples. Figure 13-11 shows an example of sampling an analog waveform.

The digitized samples can be stored in a computer or some other medium such as a compact disk. These samples become the only means of describing the original analog signal. How good of a description can they be? From our experience we know that a reasonable facsimile can be obtained if the samples are closely spaced. Intuitively it might seem that exact reconstruction of the analog waveform would require the time between samples to approach zero. Surprisingly, exact reproduction is possible from samples taken at finite intervals.

An analog waveform $v(t)$ whose spectral content falls below f_{max} can be reproduced exactly from sample $v_k = v(kT_S)$, if the sampling rate $f_S = 1/T_S$ is greater than $2f_{max}$.

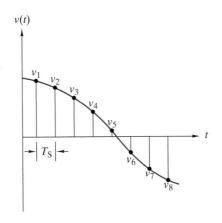

FIGURE 13-11 *Sampled Signal.*

This statement, known as the *sampling theorem*, is one of the key principles of signal processing. The theorem states that exact recovery requires a minimum sampling rate of $2f_{max}$.[2] Analog waveforms are usually sampled at a rate higher than the minimum. For example, the industry standard sampling rate for recording digital audio signals is $f_S = 44.1$ kHz, which is slightly more than twice the generally accepted upper limit on human hearing at $f_{max} = 20$ kHz.

The important point is that the minimum sampling rate is determined by the spectral content of the original analog waveform. Waveforms defined as a sum of sinusoids whose frequencies all fall in the band $0 \le f \le f_{max}$ are said to be *bandlimited*. For a simple bandlimited waveform such as

$$v(t) = 8 + 5\cos(2\pi 200\, t) + 3\cos(2\pi 400\, t) + 1.5\cos(2\pi 1200\, t)$$

it is easy to see that $f_{max} = 1.2$ kHz and hence $f_S > 2.4$ kHz. In general, a Fourier series contains infinitely many terms, which means that periodic waveforms are not strictly bandlimited. However, we have seen that the amplitudes of the high-frequency harmonic become negligibly small and make little contribution to the waveform attributes such as energy content. Hence, a bandlimited approximation to a periodic waveform can be made by terminating its Fourier series at $f_{max} = nf_0$, in which case we refer to nf_0 as the *waveform bandwidth*.

One simple way to define waveform bandwidth is to compare the amplitudes of the high-frequency harmonics with the amplitude of the largest ac component, usually the fundamental. Harmonics whose amplitudes are less than some specified fraction (say 5%) of the fundamental are ignored. Applying a 5% rule to waveforms whose Fourier coefficients decrease as $1/n$ (square wave, rectangular pulse, sawtooth wave; see Figure 13–3) leads to bandwidths of about $20f_0$. Waveforms whose harmonics fall off as $1/n^2$ (triangular wave, rectified sine waves) have 5% bandwidths of about $4f_0$. Regardless of the criteria used, the basic idea is that signals can be thought of as bandlimited by ignoring the frequency range in which the spectral content is negligibly small.

But we cannot simply ignore the out-of-band spectral content. It turns out that sampling signals that are not strictly bandlimited leads to *aliasing*, a process by which seemingly negligible out-of-band spectral content reappears as in-band distortion. The answer to the aliasing problem is to filter the analog signal prior to sampling. These *anti-aliasing filters* must pass spectral content up to the *highest frequency of interest* f_{max}, and suppress the spectral content above the *lowest aliasing frequency* at $f_S - f_{max}$, where f_S is the sampling frequency.

Figure 13–12 shows the gain response of an anti-aliasing filter used in telecommunication. This lowpass filter has 0 dB gain at $f_{max} = 3.3$ kHz and less than -60 dB gain at $2f_{max}$. This type of gain response cannot be achieved by the first- and second-order filters studied in Chapter 12. It calls for higher-order filters of the type discussed in Chapter 14.

---

2 This minimum sampling rate is called the *Nyquist rate*. The name honors Harry Nyquist, who along with Claude Shannon and others made several key breakthroughs in signal processing in the era from 1920 to 1950 at the Bell Telephone Laboratories.

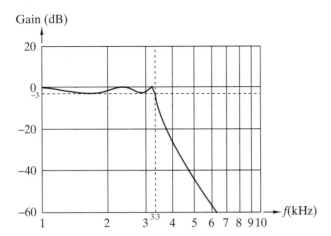

F I G U R E   1 3 – 1 2 *Anti-Aliasing Filter.*

SUMMARY

- The Fourier series resolves a periodic waveform into a dc component plus an ac component containing an infinite sum of harmonic sinusoids. The dc component is equal to the average value of the waveform. The amplitudes of the sine and cosine terms in the ac component are called Fourier coefficients.

- The fundamental frequency of the ac component is determined by the period T_0 of the waveform ($f_0 = 1/T_0$). The harmonic frequencies in the ac component are integer multiples of the fundamental frequency.

- Waveform symmetries cause the amplitudes of some terms in a Fourier series to be zero. Even symmetry causes all of the sine terms in the ac component to be zero. Odd symmetry causes all of the cosine terms to be zero. Half-wave symmetry causes all of the even harmonics to be zero.

- An alternative form of the Fourier series represents each harmonic in the ac component by its amplitude and phase angle. A plot of amplitudes versus frequency is called the amplitude spectrum. A plot of phase angles versus frequency is called the phase spectrum. A periodic waveform has spectral components at the discrete frequencies present in its Fourier series.

- The steady-state response of a linear circuit for a periodic driving force can be found by first finding the steady-state response due each term in the Fourier series of the input. The Fourier series of the steady-state response is then found by adding (superposing) responses due to each term acting alone. The individual responses can be found using either phasor or *s*-domain analysis.

- The rms value of a periodic waveform is equal to the square root of the sum of the square of the dc value and the square of the rms value of each of the ac components. The average power delivered by a periodic waveform is equal to the average power delivered by the dc component plus the sum of average power delivered by each of the ac components.

PROBLEMS

ERO 13–1 THE FOURIER SERIES (SECTS. 13–1, 13–2, 13–3)

(a) Given an equation or graph of a periodic waveform, derive expressions for the Fourier coefficients.
(b) Given an equation or graph of a periodic waveform in Figure 13–3, calculate the Fourier coefficients and plot its amplitude and phase spectrum.
(c) Given a Fourier series of a periodic waveform, determine properties of the waveform, plot its amplitude and phase spectrum, and identify the waveform.

See Examples 13–1, 13–3, 13–4, 13–5 and Exercises 13–1, 13–2, 13–3

13–1 The equation for the first cycle ($0 \leq t \leq T_0$) of a periodic waveform is $v(t) = V_A(t/T_0 - 1)$.
  (a) Sketch the first two cycles of the waveform.
  (b) Derive expressions for the Fourier coefficients a_n and b_n.

13–2 The equation for the first cycle ($0 \leq t \leq T_0$) of a periodic pulse train is

$$v(t) = V_A[u(t) - u(t - T_0/2)]$$

  (a) Sketch the first two cycles of the waveform.
  (b) Derive expressions for the Fourier coefficients a_n and b_n.

13–3 Derive expressions for the Fourier coefficients of the periodic waveform in Figure P13–3.

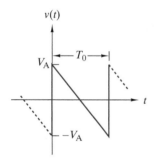

FIGURE P13–3

13–4 Derive expressions for the Fourier coefficients of the periodic waveform in Figure P13–4.

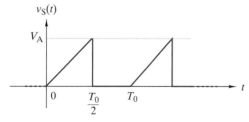

FIGURE P13–4

13–5 Use the expressions in Figure 13–3 to calculate Fourier coefficients of the square wave in Figure P13–5. Write an expression for the first four nonzero terms in the Fourier series.

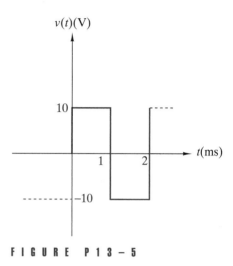

FIGURE P13–5

13–6 Use the expressions in Figure 13–3 to calculate the Fourier coefficients of the shifted triangular wave in Figure P13–6. Write an expression for the first four nonzero terms in the Fourier series.

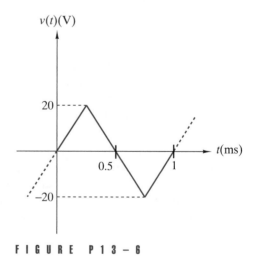

FIGURE P13–6

13–7 Use the expressions in Figure 13–3 to calculate the Fourier coefficients of the full-wave rectified sine wave in Figure P13–7. Write an expression for the first four nonzero terms in the Fourier series.

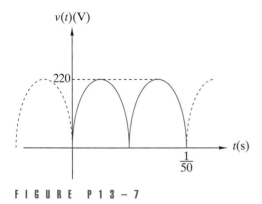

$v(t)$(V)

220

$\dfrac{1}{50}$

t(s)

F I G U R E   P 1 3 - 7

13–8 A half-wave rectified sine wave has an amplitude of 50 V and a fundamental frequency of 60 Hz. Use the results in Figure 13–3 to write an expression for the first four nonzero terms in the Fourier series and plot the amplitude spectrum of the signal.

13–9 The waveform $g(t) = 10 + f(t)$, where $f(t)$ is a 100-kHz square wave with a *peak-to-peak* amplitude of 10 V. Use the results in Figure 13–3 to write an expression for the first four nonzero terms in the Fourier series of $g(t)$ and plot its amplitude spectrum.

13–10 A parabolic wave has a *peak-to-peak* amplitude of 25 V and a fundamental frequency of 2 kHz. Use the results in Figure 13–3 to write an expression for the first four nonzero terms in the Fourier series and plot the amplitude spectrum of the signal.

13–11 The Fourier series of a periodic waveform is

$$f(t) = 10 + 8.11 \cos(500\,\pi\,t - 45°)$$
$$+ 0.901 \cos(1500\,\pi\,t - 135°)$$

  (a) Find the period and fundamental frequency in rad/s and Hz.
  (b) Plot the amplitude and phase spectrum of the signal.
  (c) Express $f(t)$ as a sum of cosine and sine terms.
  (d) Does the waveform have even or odd symmetry?

13–12 The first four terms in the Fourier series of a periodic waveform are

$$v(t) = 25\left[\sin(100\,\pi\,t) - \frac{1}{9}\sin(300\,\pi\,t)\right.$$
$$\left. + \frac{1}{25}\sin(500\,\pi\,t) - \frac{1}{49}\sin(700\,\pi\,t) + \dots \right]$$

  (a) Find the period and fundamental frequency in rad/s and Hz.
  (b) Does the waveform have even or odd symmetry?
  (c) Plot the waveform using the truncated series above to confirm your answer in (b). Can you identify the waveform from this plot?

13–13 The first four terms in the Fourier series of a periodic waveform are

$$v(t) = 30 + 20\left[\cos(400\,t) - \frac{1}{5}\cos(800\,t)\right.$$
$$\left. + \frac{1}{7}\cos(1200\,t) - \frac{1}{21}\cos(1600\,t) + \dots \right]$$

  (a) Find the period and fundamental frequency in rad/s and Hz.
  (b) Does the waveform have even or odd symmetry?
  (c) Plot the waveform using the truncated series above to confirm your answer in (b). Can you identify the waveform from this plot?

13–14 The Fourier series of a periodic waveform is

$$f(t) = 15\cos(2000\,\pi\,t) + 5\cos(6000\,\pi\,t) + 2\cos(10^4\pi\,t)$$

  (a) Find the period and fundamental frequency in rad/s and Hz.
  (b) Plot the amplitude and phase spectrum of the signal.
  (c) Does the waveform have even or odd symmetry?
  (d) A composite signal is formed as $g(t) = f(t) + 10 + 3\sin(4000\pi t)$. Plot the amplitude spectrum of $g(t)$. Does $g(t)$ have even or odd symmetry?

13–15 The Fourier series of a periodic waveform is

$$f(t) = \sum_{n=1}^{\infty} \frac{10}{n\pi}\cos(40\,\pi\,n\,t)$$

  (a) Find the period and fundamental frequency in rad/s and Hz.
  (b) Does the waveform have even or odd symmetry?
  (c) Plot the amplitude and phase spectrum of the signal.
  (d) A composite signal is formed as $g(t) = 2f(t - 0.005)$. Does $g(t)$ have even or odd symmetry?

ERO 13–2 FOURIER SERIES AND CIRCUIT ANALYSIS (SECT. 13–4)

  (a) Given a linear circuit with a periodic input waveform, find the Fourier series representation of a steady-state response.
  (b) Given a network function with a periodic input, find the amplitude and/or phase spectrum of the steady-state output.

See Examples 13–6, 13–7, 13–8, 13–9 and Exercises 13–4, 13–5

13–16 The periodic pulse train in Figure P13–16 is applied to the RL circuit shown in the figure.
  (a) Use the results in Figure 13–3 to find the Fourier coefficients of the input for $V_A = 5$ V, $T_0 = \pi/2$ ms, and $T = T_0/5$.
  (b) Find the first four nonzero terms in the Fourier series of $v_O(t)$ for $R = 200\ \Omega$, and $L = 100$ mH.

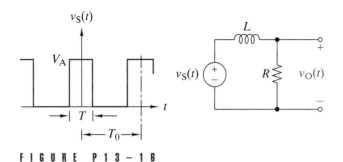

FIGURE P13-16

13–17 The periodic triangular wave in Figure P13–17 is applied to the *RC* circuit shown in the figure. The Fourier coefficients of the input are

$$a_0 = 0 \qquad a_n = 0 \qquad b_n = \frac{8 V_A}{(n \pi)^2} \sin\left(n \frac{\pi}{2}\right)$$

If $V_A = 20$ V and $T_0 = 2\pi$ ms, find the first four nonzero terms in the Fourier series of $v_O(t)$ for $R = 10$ kΩ, and $C = 50$ nF.

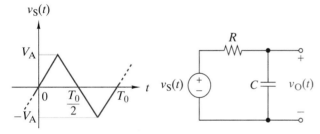

FIGURE P13-17

13–18 The periodic sawtooth wave in Figure P13–18 drives the OP AMP circuit shown in the figure.

(a) Use the results in Figure 13–3 to find the Fourier coefficients of the input for $V_A = 5$ V and $T_0 = 400\pi$ μs.

(b) Find the first four nonzero terms in the Fourier series of $v_O(t)$ for $R_1 = 10$ kΩ, $R_2 = 50$ kΩ, and $C = 10$ nF.

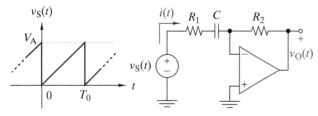

FIGURE P13-18

13–19 The periodic sawtooth wave in Figure P13–18 drives the OP AMP circuit shown in the figure.

(a) Use the results in Figure 13–3 to find the Fourier coefficients of the input for $V_A = 4$ V and $T_0 = 800\pi$ μs.

(b) Find the first four nonzero terms in the Fourier series of $i(t)$ for $R_1 = 10$ kΩ, $R_2 = 50$ kΩ, and $C = 400$ nF.

13–20 The periodic triangular wave in Figure P13–20 is applied to the *RLC* circuit shown in the figure.

(a) Use the results in Figure 13–3 to find the Fourier coefficients of the input for $V_A = 5$ V and $T_0 = 400\pi$ μs.

(b) Find the amplitude of the first five nonzero terms in the Fourier series for $v_O(t)$ when $R = 1$ Ω, $L = 2$ mH, and $C = 800$ nF. What term in the Fourier series tends to dominate the response? Explain.

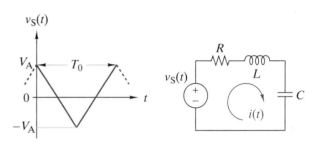

FIGURE P13-20

13–21 A square wave with $V_A = 15$ V and $T_0 = 40\pi$ ms drives a circuit with a transfer function $T(s) = s/(s + 200)$. Find the amplitude of the first four nonzero terms in the Fourier series of the steady-state output. Compare the amplitudes of the harmonics in the input and output waveforms and comment on the differences.

13–22 Repeat Problem 13–21 for $T(s) = 100/(s + 100)$.

13–23 The voltage across a 1-nF capacitor is a triangular wave with $V_A = 12$ V and $T_0 = 2$ ms. Construct plots of the amplitude spectrum of the capacitor voltage and current. Discuss any differences in spectral content.

13–24 An ideal time delay is a signal processor whose output is $v_O(t) = v_{IN}(t - T_D)$. Construct spectral plots of the input and output for $T_D = 1$ ms and

$$v_{IN}(t) = 10 + 10 \cos(2\pi 500 t)$$
$$+ 3 \cos(2\pi 1000 t) + 2 \cos(2\pi 4000 t)$$

Discuss the spectral changes caused by the time delay.

13–25 A square wave with $V_A = 5\pi$ V and $T_0 = 20\pi$ ms drives a circuit whose transfer function is

$$T(s) = \frac{100 \, s}{(s + 50)^2 + 300^2}$$

Find the amplitude of the first four nonzero terms in the Fourier series of the steady-state output. What term in the Fourier series tends to dominate the response? Explain.

ERO 13–3 RMS Value and Average Power (Sect. 13–5)

(a) Given a periodic waveform, find the rms value of the waveform and the average power delivered to a specified load.

(b) Given the Fourier series of a periodic waveform, find the fraction of the average power carried by specified components and estimate the average power delivered to a specified load.

See Examples 13–10, 13–11 and Exercise 13–6

13–26 A rectangular pulse train has an amplitude V_A, a period of T_0, and a duration $T < T_0$. Find the rms value of the rectangular pulse train and the average power the waveform delivers to a resistor. Find the dc component of the pulse train. The ratio T/T_0 is called the pulse *duty cycle*. For 50% duty cycle, calculate the fraction of the total average power carried by the dc component. What fraction is carried by the ac components? Repeat for 10% duty cycle. Comment on the effect of the duty cycle on the waveform amplitude spectrum.

13–27 The voltage

$$v(t) = 10 + 15 \cos(300\,\pi\,t) - 20 \sin(900\,\pi\,t)$$

appears across a 50-Ω resistor. Find the power delivered to the resistor and the rms value of the voltage.

13–28 Find the rms value of a square wave and the fraction of the total average power carried by the first three nonzero harmonics in the Fourier series of the waveform.

13–29 The Fourier coefficients of the periodic waveform in Figure P13–29 are

$$a_0 = \frac{V_A}{4} \qquad a_n = \frac{V_A}{(n\,\pi)^2}\,[\cos(n\pi) - 1] \qquad b_n = \frac{-V_A}{n\,\pi}\cos(n\pi)$$

Find the rms value of the waveform and the average power the waveform delivers to a resistor. What fraction of the total average power is carried by the dc component plus the first three nonzero ac components in the Fourier series of the waveform?

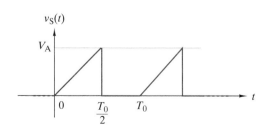

FIGURE P13–29

13–30 Find the rms value of the periodic waveform in Figure P13–30 and the average power the waveform delivers to a resistor. Find the dc component of the waveform. What fraction of the total average power is carried by the dc component? What fraction is carried by the ac components?

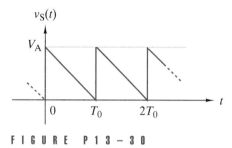

FIGURE P13–30

13–31 A first-order low-pass filter has a cutoff frequency of 500 rad/s and a passband gain of 20 dB. The input to the filter is $x(t) = 6 \sin 1000t + 2 \cos 3000t$ V. Find the rms value of the steady-state output.

13–32 The current through a 10-Ω resistor is

$$i(t) = \frac{2}{\pi}\left[\sum_{n=1}^{\infty} \frac{(-1)^n}{n} \sin n\omega_0 t\right]$$

Use the first three harmonics to estimate the average power delivered to the resistor.

13–33 Estimate the rms value of the periodic voltage

$$v(t) = V_A\left[\frac{\pi^2}{12} - \cos \omega_0 t + \frac{1}{4}\cos 2\,\omega_0 t - \frac{1}{9}\cos 3\,\omega_0 t + \ldots\right]$$

13–34 The input to the circuit in Figure P13–34 is the voltage

$$v(t) = 40 \cos(2\,\pi\,3000\,t) + 15 \sin(2\,\pi\,9000\,t)$$

Calculate the average power delivered to the 50-Ω resistor.

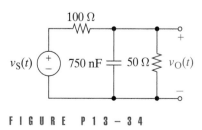

FIGURE P13–34

13–35 The input to the circuit in Figure P13–34 is the voltage given in Problem 13-33 with $V_A = 4$ V and $T_0 = 50\pi\,\mu$s. Estimate the average power delivered to the 50-Ω resistor.

INTEGRATING PROBLEMS

13–36 ⓐ FOURIER SERIES FROM A BODE PLOT

Figure P13–36 shows the straight-line Bode plots of the magnitude and phase angle of a transfer function. The first four terms in the Fourier series of a periodic input to the circuit are

$$x(t) = 14\cos(10\,t) + 4.667\cos(30\,t)$$
$$+ 2.8\cos(50\,t) + 2\cos(70\,t)$$

(a) Use the straight-line Bode Plot and the input Fourier series to estimate the first four terms in the Fourier series of the steady-state output.

(b) Find a function $T(s)$ that matches the Bode plots shown in the figure. Use $T(j\omega)$ and the Fourier series of the input found to calculate the exact values of the steady-state output.

(c) Compare your results in (a) and (b) and discuss the differences.

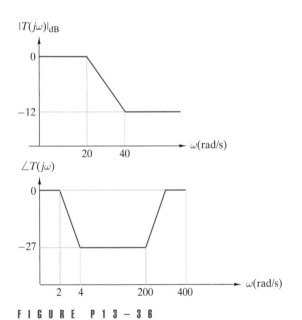

FIGURE P13-36

13–37 ⓐ STEADY-STATE RESPONSE FOR A PERIODIC IMPULSE TRAIN

A periodic impulse train can approximate a pulse train when the individual pulse durations are very short compared with the circuit response time. This example explores the response of a first-order circuit to a periodic impulse train. An impulse response of a linear circuit is

$$h(t) = \frac{1}{T_0} e^{-t/T_0} u(t)$$

The circuit is driven by a periodic impulse train:

$$x(t) = T_0 \sum_{n=-\infty}^{\infty} \delta(t - nT_0)$$

(a) Use Eq. (13–3) to find the Fourier coefficients of $x(t)$. Write a general expression for the Fourier series of the input $x(t)$.

(b) Find expressions for the Fourier coefficients of the steady-state output $y(t)$ when the input is the periodic impulse train $x(t)$.

(c) Write a general expression for the Fourier series of the steady-state output $y(t)$.

(d) Use a computer tool to generate a plot of a truncated Fourier series that gives a reasonable approximation of the steady-state response found in (c).

13–38 ⓓ LOW-PASS FILTER DESIGN REQUIREMENTS

The gain of an mth-order Butterworth low-pass filter is

$$|T(j\omega)| = \frac{1}{\sqrt{1 + \left(\dfrac{\omega}{\omega_C}\right)^{2m}}}$$

where ω_C is the cutoff frequency. The input is a periodic waveform with a fundamental frequency ω_0. The filter must pass all harmonics up to the 3rd with a gain of at least –3 dB and have a gain less than –20 dB for all harmonics above the 3rd. Find values of m and $\omega_C = k\omega_0$ that meet this requirement. *Hint*: The Butterworth gain decreases monotonically with frequency.

13–39 ⓓ BANDPASS FILTER DESIGN REQUIREMENTS

The transfer function of a 2nd-order bandpass filter is of the form

$$T(s) = \frac{\dfrac{\omega_C}{Q} s}{s^2 + \dfrac{\omega_C}{Q} s + \omega_C^{\,2}}$$

where Q is the quality factor and ω_C is the center frequency. The filter input is a periodic waveform with even symmetry and $T_0 = 2\pi\,\mu s$. The filter must pass the 7th harmonic with a gain of 0 dB and have a gain less than -20 dB at the two adjacent harmonic frequencies. Find Q and ω_C.

13–40 ⓔ SPECTRUM ANALYZER CALIBRATION

A certain spectrum analyzer measures the average power delivered to a calibrated resistor by the individual harmonics of periodic waveforms. The calibration of the analyzer has been checked by applying a 1-MHz square wave and the following results reported

f(MHz)	1	3	5	7	9	11
P(dBm)	12.1	2.56	–1.88	–4.80	–6.98	–8.73

The reported power in dBm is $P = 10\,\text{Log}(P_n)$, where P_n is the average power delivered by the nth harmonic in mW. Is the spectrum analyzer correctly calibrated?

CHAPTER 14

ANALOG FILTER DESIGN

In its usual form the electric wave-filter transmits currents of all frequencies lying within one or more specified ranges, and excludes currents of all other frequencies.

George A. Campbell, 1922,
American Engineer

14–1 Frequency-domain Signal Processing

14–2 Design with First-order Circuits

14–3 Design with Second-order Circuits

14–4 Low-pass Filter Design

14–5 High-pass Filter Design

14–6 Bandpass and Bandstop Filter Design

Summary

Problems

Integrating Problems

The electric filter was independently invented during World War I by George Campbell in the United States and by K. W. Wagner in Germany. Electric filters and vacuum tube amplifiers were key technologies that triggered the growth of telephone and radio communication systems in the 1920s and 1930s. The emergence of semiconductor electronics in the 1960s, especially the integrated circuit OP AMP, allowed the functions of filtering and amplification to be combined into what are now called active filters.

This chapter treats an important type of electric wave filter made possible by operational amplifiers. These circuits are often called active filters, although the term *analog filter* is also common. Modern integrated circuit technology allows signal-processing systems to use both analog and digital filters, sometimes on the same chip. Although analog and digital filters have some common attributes, the subject of digital filtering is covered in courses in discrete-time signal processing.

Some signal-processing functions performed by analog filters are described in the first section of this chapter. The second section introduces filter design techniques using a cascade connection of first-order circuits. The third section treats the design of second-order low-pass, high-pass, and bandpass circuits. The fourth section develops cascade realizations of low-pass transfer functions using the first- and second-order building blocks. The fifth section shows how the low-pass filter serves as a prototype for designing high-pass filters. The final section illustrates that low-pass and high-pass filters are the building blocks for bandpass and bandstop filters.

14–1 FREQUENCY-DOMAIN SIGNAL PROCESSING

A signal spectrum describes a signal in terms of the frequencies it contains. Signal processing can be described in terms of how different input frequencies are treated. The idea that signals can be described and processed in terms of their frequency content is a fundamental concept in electrical engineering.

The Touch-Tone™ telephone provides an example of frequency-domain signal processing. The Touch-Tone™ keypad is shown in Figure 14–1. Pressing a keypad button transmits a signal that is the sum of two sinusoids. One of the sinusoids is from the low group(697, 770, 852, or 941 Hz) and the other from the high group (1209, 1336, or 1477 Hz). For instance, pressing the no. 3 button transmits a sum of sinusoids at 697 and 1447 Hz. The relative amplitudes of these dual-frequency waveforms are not important. What matters is that each of the twelve keypad characters is encoded as a unique combination of two frequencies. To decode the data carried by the signal, we must determine what two frequencies are present.

FIGURE 14–1 *Touch-tone dialing frequencies.*

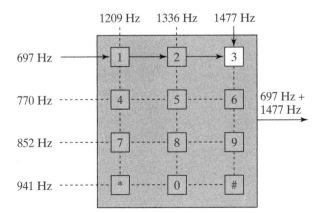

A **filter** is a signal processor that separates signals on the basis of their frequency. It does this by passing signals in some frequency bands and rejecting (or attenuating) signals in other bands. Figure 14–2 shows a filter-based decoding tree for Touch-Tone™ signals. During any one transmission we don't know the exact waveform of the input signal. We only know that it contains two frequencies, one from the low group and one from the high group.

The first step in decoding involves separating the input into two bands using a low-pass filter and a high-pass filter. A *low-pass filter* passes all frequencies below its cutoff frequency and attenuates the frequencies above the cutoff frequency. The cutoff frequency of the low-pass filter in Figure 14–2 is set at 941 Hz, the maximum frequency in the low group. A *high-pass filter* passes all frequencies above its cutoff frequency and attenuates those below cutoff. In the present case, the high-pass cutoff is at 1209 Hz, the minimum frequency in the high group. The result is that the input is split into two signals in different frequency bands. One band contains the low-group frequencies and the other the high-group frequencies.

The two split-band signals are applied to a bank of seven bandpass filters. A *bandpass* filter passes all frequencies in a band around its center frequency and attenuates all frequencies above or below this band. The bandpass filters in Figure 14–2 are each centered at one of the seven unique frequencies generated by the Touch-Tone™ keypad. For any given transmission only one of the low-group filters and one of the high-group

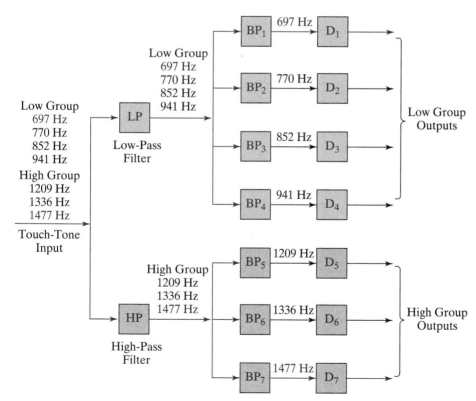

FIGURE 14 – 2 *Touch-tone signal decoding.*

filters will have a large output. The tone-detection filter outputs drive threshold detectors (D_k, $k = 1, 2, \ldots 7$), whose digital outputs go high when their input exceeds a certain level. For example, the sinusoids at 697 and 1447 Hz representing the keypad character no. 3 cause the outputs of detectors D_1 and D_7 to go high. Since each keypad character is encoded by a unique combination of two frequencies, the two threshold detectors with high outputs represent a unique digital code for the character.

Note the underlying viewpoint in frequency-domain signal processing. We do not know exactly what the Touch-Tone™ signal looks like. We only have a partial description of its spectrum in terms of the frequencies it contains. This knowledge, though incomplete, is enough to design filters to process the signal. In an application note at the end of this chapter we will consider the design of the band-splitting low-pass and high-pass filters in Figure 14–2.

FILTER DESIGN

This chapter introduces methods of designing filter circuits with prescribed frequency-domain characteristics. A filter design problem is defined by specifying a required gain response $|T(j\omega)|$. We then devise one or more circuits that produce the desired gain response. In general, a given gain response can be produced by several different circuits, including passive circuits containing only resistors, capacitors, and inductors. We limit our study to active filters involving resistors, capacitors, and OP AMPs. These *active RC filters* offer the following advantages:

1. They combine amplifier gain with the frequency response characteristics of passive *RLC* filters.

2. The transfer function can be divided into stages that can be designed independently and then connected in cascade to realize the required gain function.

3. They are often smaller and less expensive than *RLC* filters because they do not require inductors that can be quite large in low-frequency applications.

In our discussion we concentrate on producing a given gain response. The filter phase response is not directly controlled. Designing circuits to produce a specified phase response will not be treated here.

14–2 DESIGN WITH FIRST-ORDER CIRCUITS

In this section we begin the study of filter design problems in which characteristics of the gain response $|T(j\omega)|$ are specified. Our objective is to obtain a circuit whose transfer function has the specified gain response and whose element values are in practical ranges. The design process begins by constructing a transfer function whose gain response meets the specification. Except in the case of simple first- and second-order filters, we partition the transfer function into a product of the form

$$T(s) = T_1(s) \times T_2(s) \times T_3(s) \times \ldots T_n(s) \tag{14–1}$$

where each $T_k(s)$ in this equation is a first-order transfer function with one pole and no more than one finite zero. We then design prototype first-

First Order Circuits	Transfer functions
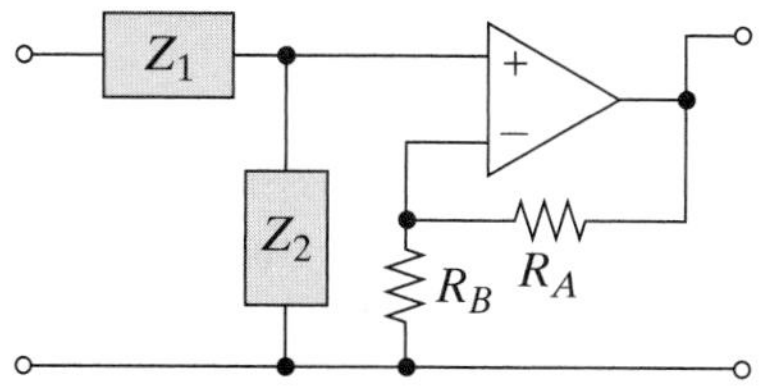 **Voltage divider with gain**	$T(s) = K \dfrac{Z_2}{Z_1 + Z_2} = K \dfrac{Y_1}{Y_1 + Y_2}$ where $K = \dfrac{R_A + R_B}{R_B}$
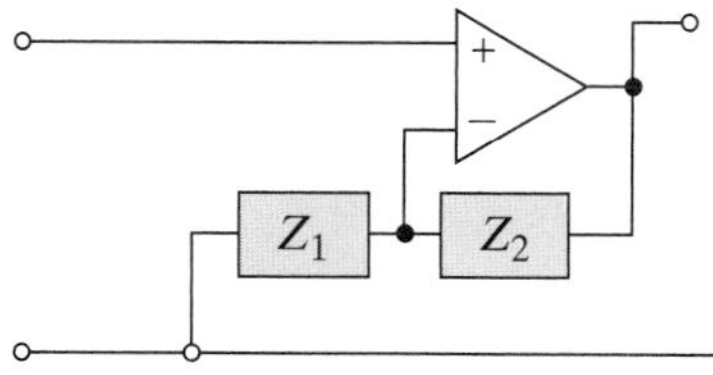 **Noninverting amplifier**	$T(s) = \dfrac{Z_1 + Z_2}{Z_1} = \dfrac{Y_1 + Y_2}{Y_2}$
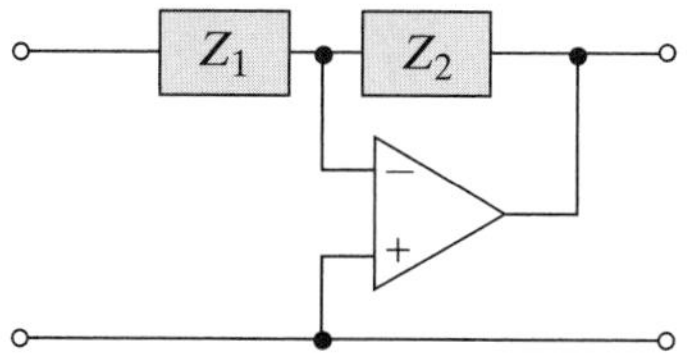 **Inverting amplifier**	$T(s) = -\dfrac{Z_2}{Z_1} = -\dfrac{Y_1}{Y_2}$

FIGURE 14–3 *First-order circuit building blocks.*

order circuits using the voltage divider, noninverting amplifier, and the inverting amplifier building blocks in Figure 14–3. The circuit element values in the prototype may be unrealistically large or small, so we use a magnitude scale factor to bring the element values into practical ranges.

The required transfer function $T(s)$ is realized by connecting the scaled circuit for each stage in cascade, as shown in Figure 14–4. The cascade connection produces the required $T(s)$ because the building blocks in Figure 14–3 have OP AMP outputs. As a result, loading does not change the stage transfer functions and the chain rule in Eq. (14–1) applies. The noninverting and inverting amplifier circuits in Figure 14–3 inherently have low-impedance outputs. Adding a noninverting amplifier to the output of the basic voltage-divider circuit in Figure 14–3 produces a circuit that can be connected in cascade without changing the stage transfer function.

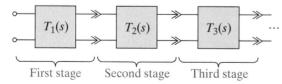

FIGURE 14–4 *A cascade connection.*

The end result is a cascade circuit whose gain response has the attributes required by the specification. The aforementioned procedure does not produce a unique solution for the following reasons:

1. Several transfer functions may exist that adequately approximate the required gain characteristics.

2. Partitioning the selected transfer function into a product of first-order transfer functions can be carried out in several different ways.

3. Any of the three building blocks in Figure 14–3 can be used to realize the first-order transfer functions in the selected partitioning.

4. There usually are several ways to assign numerical values to circuit elements in the selected building block for each stage.

5. Each stage normally requires a different magnitude scale factor to bring the assigned numerical values into practical ranges.

In summary, since there are many choices, there are no unique answers to the circuit design problem. The following examples illustrate the design of active *RC* circuits.

D DESIGN EXAMPLE 14–1

Design two active *RC* circuits that realize a first-order low-pass filter with a cutoff frequency of 2 krad/s and a passband gain of 5.

SOLUTION:

In Chapter 12 we found that a first-order low-pass transfer function has a pole at $s = -\alpha$, cutoff frequency of $\omega_C = \alpha$, and passband gain of $T(0)$. For design puposes it is convenient to write the first-order lowpass transfer function in the Bode plot standard format

$$T(s) = \frac{K}{s/\alpha + 1}$$

Written in this form, we see that the design specification requires that $\omega_C = \alpha = 2000$ and that passband gain $T(0) = K = 5$. Thus, the required transfer function is

$$T(s) = \frac{\pm 5}{s/2000 + 1}$$

where the ± sign indicates that the phase response is not specified.

First Design: Using the voltage divider circuit in Figure 14–3, we use the plus sign and partition $T(s)$ as

$$T(s) = [K]\left[\frac{Y_1}{Y_1 + Y_2}\right] = [+5]\left[\frac{1}{s/2000 + 1}\right]$$

from which we identify $K = 5$, $R_A = (5 - 1)R_B$, $Y_1 = 1$, and $Y_2 = s/2000$. Selecting $R_B = 1\ \Omega$, we obtain $R_A = 4\ \Omega$. The admittance Y_1 can be realized by a resistor $R_1 = 1\ \Omega$, and admittance Y_2 by a capacitor $C_2 = 1/2000$ F. These results lead to the prototype design in Figure 14–5(a). To obtain practical element values, we apply a magnitude scale factor of $k_m = 10^4$ to produce the final design shown in Figure 14–5(b).

Second Design: Using the inverting amplifier circuit in Figure 14–3, we use the minus sign and write $T(s)$ as

$$T(s) = \left[-\frac{Y_1}{Y_2}\right] = \left[-\frac{5}{s/2000 + 1}\right]$$

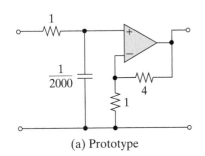

(a) Prototype

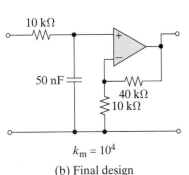

$k_m = 10^4$

(b) Final design

FIGURE 14–5

from which we identify $Y_1 = 5$ and $Y_2 = s/2000 + 1$. The admittance Y_1 can be realized by a resistor $R_1 = 1/5$ Ω. The admittance Y_2 is of the form $C_2 s + G_2$ and hence can be realized by capacitor $C_2 = 1/2000$ F connected in parallel with a resistor $R_2 = 1$ Ω. These results lead to the prototype design in Figure 14–6(a). To obtain practical element values, we apply a magnitude scale factor of $k_m = 5 \times 10^4$ to produce the final design shown in Figure 14–6(b). ■

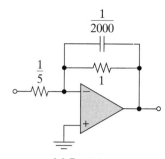

(a) Prototype

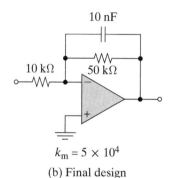

$k_m = 5 \times 10^4$

(b) Final design

FIGURE 14–6

D DESIGN EXAMPLE 14–2

(a) Construct a transfer function $T(s)$ with the straight-line gain response shown in Figure 14–7.
(b) Design an active RC circuit to realize the $T(s)$ found in (a).

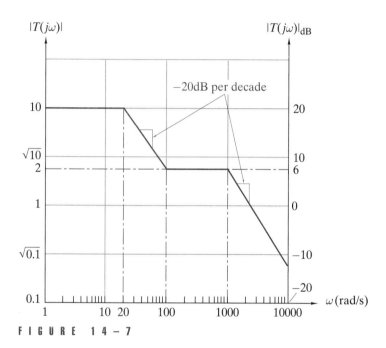

FIGURE 14–7

SOLUTION:
(a) The gain response shows slope changes at $\omega = 20$ rad/s, 100 rad/s, and 1000 rad/s. Since the net slope changes at $\omega = 20$ rad/s and 1000 rad/s are both –20 dB/decade, the denominator of $T(s)$ must have poles at $s = -20$ rad/s and $s = -1000$ rad/s. Since the net slope change at $\omega = 100$ rad/s is +20 dB/decade, the numerator of $T(s)$ must have a zero at $s = -100$ rad/s. These critical frequencies account for all of the slope changes, so the required transfer function has the form

$$T(s) = K \frac{s/100 + 1}{(s/20 + 1)(s/1000 + 1)}$$

where the gain K is yet to be determined. The dc gain of the proposed transfer function is $T(0) = K$ and the specified dc gain in Figure 14–7 is +20 dB; hence $K = 10$.

(b) To design first-order circuits, we partition $T(s)$ into a product of first-order transfer functions:

$$T(s) = T_1(s)T_2(s) = \left(\frac{s/100 + 1}{s/20 + 1}\right)\left(\frac{10}{s/1000 + 1}\right)$$

This partitioning is not unique because the overall gain $K = K_1 \times K_2 = 10$ can be distributed between the two functions in any number of ways (for example, $K_1 = 2$ and $K_2 = 5$). We could move the zero from the first function to the second or reverse the order in which the functions are assigned to stages in the cascade design. Selecting the inverting amplifier building block in Figure 14–3 for both stages leads to the design sequence shown in Figure 14–8.

The transfer function of the inverting amplifier is $-Z_2(s)/Z_1(s)$, so the selected building block places constraints on Z_1 and Z_2, as indicated in the second row in Figure 14–8. Equating $Z_1(s)$ to the reciprocal of the transfer function numerator and $Z_2(s)$ to the reciprocal of the denominator yields impedances of the form $1/(Cs + G)$. An impedance of this form is a resistor in parallel with a capacitor, as shown in the third and fourth rows in Figure 14–8. The element values of the prototype designs shown in the fifth row are not within practical ranges. To make all resistors 10 kΩ or greater, we use a magnitude scale factor of $k_m = 10^4$ in the first stage and $k_m = 10^5$ in the second stage. Applying these scale factors to the prototype designs produces the final designs shown in the last row of Figure 14–8.   ∎

▷ ⬤ D E S I G N   E X A M P L E   1 4 – 3 ⬤

(a) Construct a transfer function $T(s)$ with the straight-line gain response shown in Figure 14–9.
(b) Design an active RC circuit to realize the $T(s)$ found in (a).

S O L U T I O N :
(a) The straight-line gain response shows slope changes at $\omega = 10$ rad/s, 100 rad/s, 500 rad/s, and 5000 rad/s. The net slope changes at $\omega = 10$ rad/s and 5000 rad/s are −20 dB/decade. The net slope changes at $\omega = 100$ rad/s and 500 rad/s are +20 dB/decade. Therefore, $T(s)$ must have poles at $s = -10$ rad/s and $s = -5000$ rad/s and zeros at $s = -100$ rad/s and $s = -500$ rad/s. The required transfer function has the form

$$T(s) = K\frac{(s/100 + 1)(s/500 + 1)}{(s/10 + 1)(s/5000 + 1)}$$

where the gain K is yet to be determined. The gain plot in Figure 14–9 indicates that the required dc gain is 0 dB. The dc gain of the proposed transfer function is $T(0) = K$, so we conclude that $K = 1$.

Item	First Stage	Second Stage
Transfer function	$T_1(s) = -\dfrac{(s/100) + 1}{(s/20) + 1}$	$T_2(s) = -\dfrac{10}{(s/1000) + 1}$
Design constraints	Inverting Amplifier $\dfrac{Z_2}{Z_1} = \dfrac{(s/100) + 1}{(s/20) + 1}$	Inverting Amplifier $\dfrac{Z_2}{Z_1} = \dfrac{10}{(s/1000) + 1}$
Z_1	$\dfrac{1}{(s/100) + 1}$ 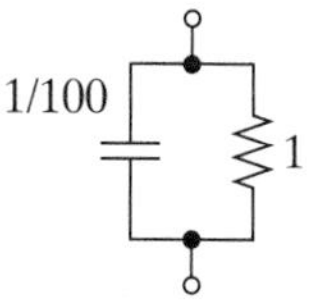	$\dfrac{1}{10}$ 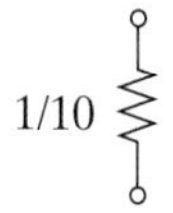
Z_2	$\dfrac{1}{(s/20) + 1}$ 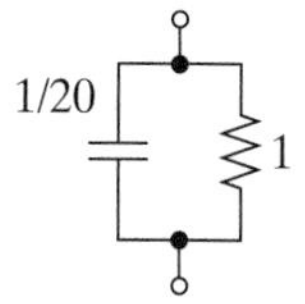	$\dfrac{1}{(s/1000) + 1}$ 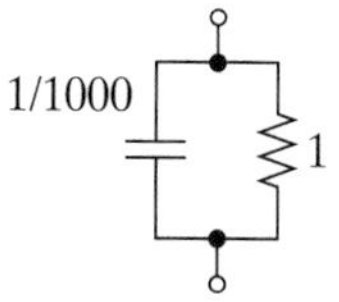
Prototype designs	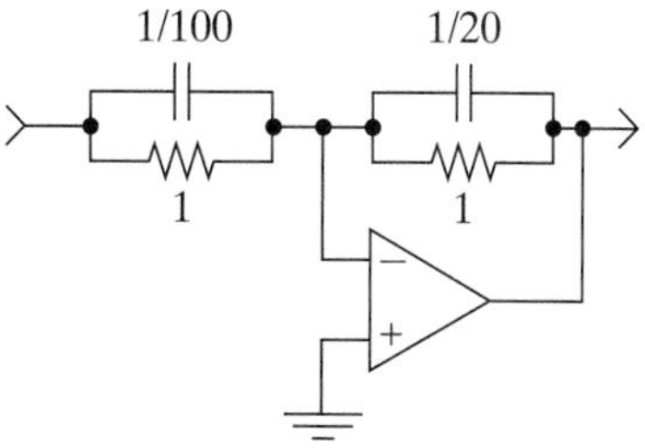	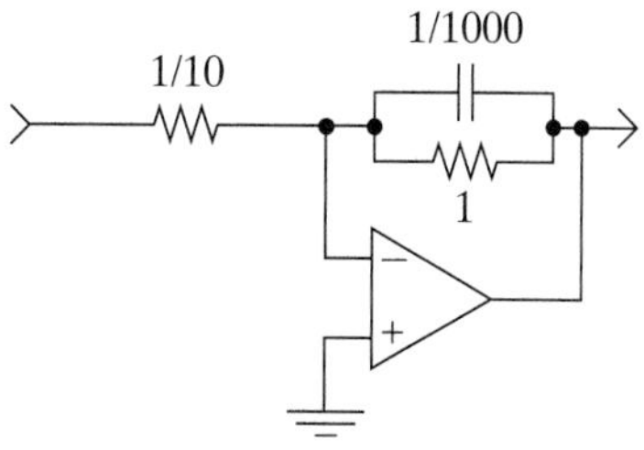
Final designs	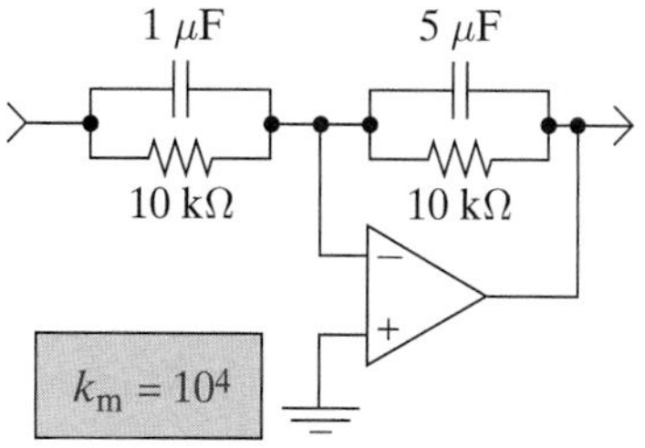 $k_m = 10^4$	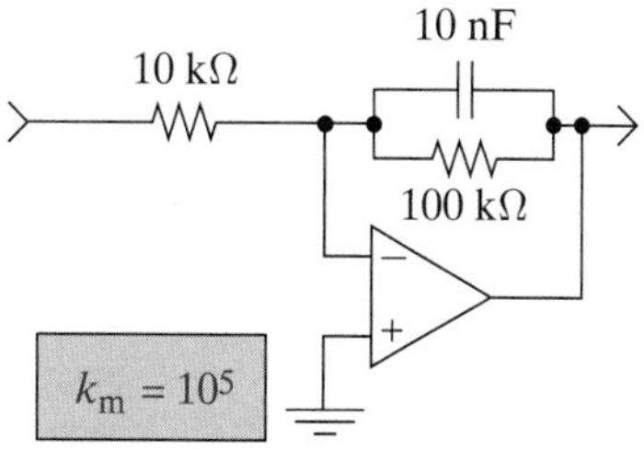 $k_m = 10^5$

FIGURE 14–8 *Design sequence for Example 14–2.*

FIGURE 14-9

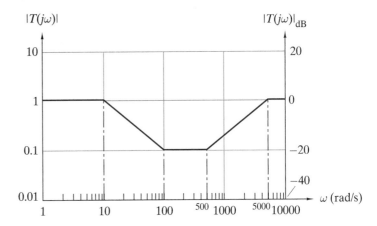

(b) To begin the circuit design, we partition $T(s)$ into the following product of first-order transfer functions:

$$T(s) = T_1(s)T_2(s) = \left(\frac{s/100 + 1}{s/10 + 1}\right)\left(\frac{s/500 + 1}{s/5000 + 1}\right)$$

This partitioning of $T(s)$ is not unique, since we could, for example, interchange the zeros or reverse the order in which the stages are listed. The design sequence shown in Figure 14–10 uses a voltage-divider circuit for the first stage and a noninverting amplifier for the second stage.

Factoring an s out of the numerator and denominator of the stage transfer functions produces the design constraints in the second row of Figure 14–10. Equating numerators and denominators of the design constraint equations leads to the RC circuits shown in the fourth and fifth rows of Figure 14–10. The gain of the voltage-divider stage is unity ($K = 1$) so the OP AMP in the first stage is a voltage follower, as indicated in the prototype design. However, the output of the first stage drives the noninverting input of the second-stage OP AMP. Since OP AMP inputs draw negligible current, no loading occurs when the voltage follower in the first stage is eliminated. The final design in the last row of Figure 14–10 eliminates the first-stage OP AMP and uses magnitude scale factors of $k_m = 10^6$ and $k_m = 5 \times 10^7$ to make all resistors greater than 10 kΩ. ∎

D DESIGN EXERCISE: 14–1

Design an active RC circuit that realizes a first-order high-pass filter with a cutoff frequency of 500 rad/s and a passband gain of 10.

Answers:

There are no unique answers. Two possible unscaled prototypes are shown in Figure 14–11.

Item	First Stage	Second Stage
Transfer function	$T_1(s) = \dfrac{(s/100) + 1}{(s/10) + 1}$	$T_2(s) = \dfrac{(s/500) + 1}{(s/5000) + 1}$
Design constraints	Voltage Divider $K\dfrac{Z_2}{Z_1 + Z_2} = \dfrac{1/100 + 1/s}{(1/10) + 1/s}$	Noninverting Amplifier $\dfrac{Z_1 + Z_2}{Z_1} = \dfrac{1/500 + 1/s}{(1/5000) + 1/s}$
K	1	Not Applicable
Z_1	$\dfrac{9}{100}$ 9/100	$\dfrac{1}{s} + \dfrac{1}{5000}$ 1/5000 1
Z_2	$\dfrac{1}{s} + \dfrac{1}{100}$ 1/100 1	$\dfrac{9}{5000}$ 9/5000
Prototype designs	9/100 1/100 1 $K = 1$	1/5000 9/5000 1
Final designs	90 kΩ 10 kΩ 1 μF $k_m = 10^6$	10 kΩ 90 kΩ 20 nF $k_m = 5 \times 10^7$

FIGURE 14–10 *Design sequence for Example 14–3.*

FIGURE 14–11

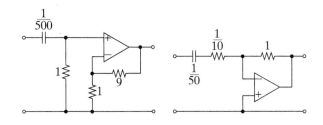

D DESIGN EXERCISE: 14–2

(a) Construct a transfer function $T(s)$ with the asymptotic gain response shown in Figure 14–12.

(b) Design an active RC circuit to realize the $T(s)$ found in (a).

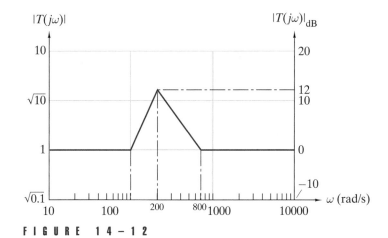

FIGURE 14–12

Answers:

(a) $T(s) = \dfrac{(s/100 + 1)^2\,(s/800 + 1)}{(s/200 + 1)^3}$

(b) There is no unique answer. One possible unscaled prototype design is shown in Figure 14–13.

FIGURE 14–13

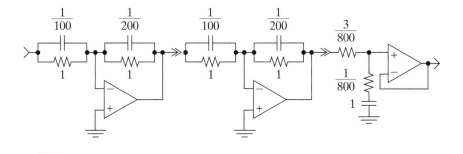

14–3 DESIGN WITH SECOND-ORDER CIRCUITS

Cascade connections of first-order circuit can only produce transfer functions with poles on the negative real axis in the s plane. The complex poles available in second-order circuits produce highly selective filter characteristics that cannot be realized using only real poles. While passive RLC circuits can produce complex poles, our interest centers on active RC circuits. In this regard we need second-order building block circuits analogous to the first-order circuits in Figure 14–3.

SECOND-ORDER LOW-PASS CIRCUITS

The active RC circuit in Figure 14–14(a) is analyzed in Chapter 11 (Example 11–6), where its voltage transfer function is shown to be

$$T(s) = \frac{V_2(s)}{V_1(s)} = \frac{\mu}{R_1R_2C_1C_2s^2 + (R_1C_1 + R_2C_2 + R_1C_2 - \mu R_1C_1)s + 1} \quad (14\text{-}2)$$

In circuit applications the dependent source in Figure 14–14(a) is replaced by the noninverting OP AMP circuit in Figure 14–14(b). The OP AMP circuit simulates the dependent source because it draws negligible input current and has very low output resistance. To serve as a replacement for the dependent source, the gain of the noninverting OP AMP circuit must be

$$\frac{R_A + R_B}{R_B} = \mu \quad (14\text{-}3)$$

where μ is the gain of the dependent source in Figure 14–14(a).

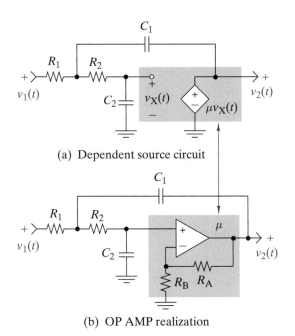

(a) Dependent source circuit

(b) OP AMP realization

FIGURE 14–14 *A second-order low-pass circuit.*

In second-order circuit design we select element values to produce specified values of the natural frequency ω_0 and damping ratio ζ. Comparing the denominator of Eq. (14–2) with the standard form $(s/\omega_0)^2 + 2\zeta(s/\omega_0) + 1$ gives

$$\omega_0 = \frac{1}{\sqrt{R_1R_2C_1C_2}} \text{ and } 2\zeta = \sqrt{\frac{R_2C_2}{R_1C_1}} + \sqrt{\frac{R_1C_2}{R_2C_1}} + (1 - \mu)\sqrt{\frac{R_1C_1}{R_2C_2}}$$

$$(14\text{-}4)$$

Since this yields two design constraints and five unknown element values, the designer has more options than are actually needed. Two common methods of selecting element values are discussed next.

The equal R and equal C method requires that $R_1 = R_2 = R$ and $C_1 = C_2 = C$. Using the equal element constraints together with Eqs. (14–3) and (14–4) yields the conditions

$$RC = \frac{1}{\omega_0} \quad \text{and} \quad \frac{R_A}{R_B} = 2(1 - \zeta) \qquad (14\text{–}5)$$

The unity gain method requires that $R_1 = R_2 = R$ and $\mu = 1$. Using these constraints together with Eqs. (14–3) and (14–4) yields the conditions

$$R_A = 0, \quad R\sqrt{C_1 C_2} = \frac{1}{\omega_0}, \quad \text{and} \quad \frac{C_2}{C_1} = \zeta^2 \qquad (14\text{–}6)$$

Since $R_A = 0$, the noninverting OP AMP circuit in Figure 14–14(b) can be a voltage follower, so neither feedback resistor in the OP AMP circuit is needed.

Finally, combining Eqs. (14–2) and (14–3), we find that the dc gain is $T(0) = (R_A + R_B)/R_B$. That is, the dc gain of the transfer function equals the gain of the OP AMP circuit.

Note: Under the equal element or unity gain approaches, the designer does not control the dc gain since the OP AMP circuit gain is determined by the damping ratio or is set to unity. In some cases the design conditions may not specify the dc gain because what really matters in a filter is the relative gains in passband and stopband. An additional gain stage can be added if the dc gain is specified.

D DESIGN EXAMPLE 14–4

(a) Construct a second-order low-pass transfer function with a natural frequency of $\omega_0 = 1000$ rad/s and with $|T(0)| = |T(j\omega_0)|$.
(b) Design second-order circuits that realize the transfer function found in (a).

SOLUTION:
(a) The transfer function has the form

$$T(s) = \frac{K}{(s/1000)^2 + 2\zeta(s/1000) + 1}$$

where K and ζ are to be determined. The dc gain of this function is $|T(0)| = |K|$. The gain at the natural frequency is $|T(j1000)| = |K|/2\zeta$. The condition $|T(0)| = |T(j\omega_0)|$ requires $\zeta = 0.5$. Hence a transfer function meeting the design statement is

$$T(s) = \frac{K}{(s/1000)^2 + (s/1000) + 1}$$

Note that the design conditions do not specify a value for the dc gain $|T(0)| = K$.
(b) We use both methods of assigning component values to obtain alternative designs.

Equal R and Equal C Design: Inserting the design constraints $\omega_0 = 1000$ rad/s and $\zeta = 0.5$ into Eq. (14–5) yields $RC = 10^{-3}$ s and $R_A = R_B$. Selecting $R = R_A = 10$ kΩ requires $C = 100$ nF and $R_B = 10$ kΩ and produces the circuit in Figure 14–15(a). The transfer function realized by the equal element value design has a dc gain of $K = (R_A + R_B)/R_B = 2$.

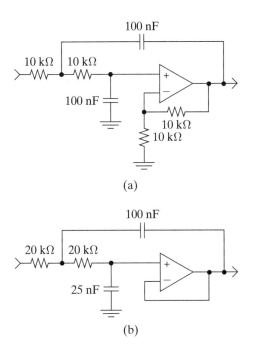

F I G U R E   1 4 – 1 5

(a)

(b)

Unity Gain Design: Inserting the design constraints $\omega_0 = 1000$ rad/s and $\zeta = 0.5$ into Eq. (14–6) yields $R\sqrt{C_1 C_2} = 10^{-3}$ rad/s and $C_2 = 0.25C_1$. Selecting $C_1 = 100$ nF dictates that $C_2 = 25$ nF and $R = 20$ kΩ and yields the circuit in Figure 14–15(b). The unity gain design uses two fewer resistors since the OP AMP is connected as a voltage follower. The transfer function realized by the method has a dc gain $K = 1$, which is less than our equal R and equal C design. ∎

SECOND-ORDER HIGH-PASS CIRCUITS

The active *RC* circuit in Figure 14–16(a) has a second-order high-pass voltage transfer function of the form

$$T(s) = \frac{V_2(s)}{V_1(s)} = \frac{\mu R_1 R_2 C_1 C_2 s^2}{R_1 R_2 C_1 C_2 s^2 + (R_2 C_2 + R_1 C_1 + R_1 C_2 - \mu R_2 C_2)s + 1} \quad (14\text{–}7)$$

This high-pass circuit is obtained from the low-pass circuit in Figure 14–14(a) by interchanging the locations of the resistors and capacitors.[1]

---

1 Both circuits belong to a family of circuits originally proposed by R. P. Sallen and E. L. Key in "A Practical Method of Designing RC Active Filters," *IRE Transactions on Circuit Theory*, Vol. CT-2, pp. 74–85, 1955. In 1955 the controlled source was derived using vacuum tubes.

F I G U R E   1 4 – 1 6   *A second-order high-pass circuit.*

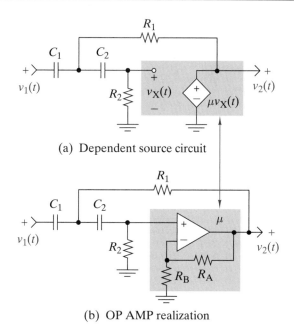

(a) Dependent source circuit

(b) OP AMP realization

Derivation of the transfer function in Eq. (14–7) is left as an exercise (see Problem 14–26).

In design applications the dependent source in Figure 14–16(a) is replaced by the noninverting OP AMP circuit in Figure 14–16(b). To serve as a replacement for the dependent source, the gain of the noninverting OP AMP circuit must be

$$\frac{R_A + R_B}{R_B} = \mu \tag{14–8}$$

where μ is the gain of the noninverting OP AMP circuit that simulates the dependent source in Figure 14–16(a). Comparing the denominator of Eq. (14–7) with the standard form $(s/\omega_0)^2 + 2\zeta\,(s/\omega_0) + 1$ yields the following results:

$$\omega_0 = \frac{1}{\sqrt{R_1 R_2 C_1 C_2}} \text{ and } 2\zeta = \sqrt{\frac{R_1 C_1}{R_2 C_2}} + \sqrt{\frac{R_1 C_2}{R_2 C_1}} + (1 - \mu)\sqrt{\frac{R_2 C_2}{R_1 C_1}} \tag{14–9}$$

As is the case with the low-pass circuit, we have two design constraints and five unknown element values. Two often used methods of defining element values are discussed next.

The equal R and equal C method requires that $R_1 = R_2 = R$ and $C_1 = C_2 = C$. Inserting these constraints into Eqs. (14–8) and (14–9) yields the conditions

$$RC = \frac{1}{\omega_0} \text{ and } \frac{R_A}{R_B} = 2(1 - \zeta) \tag{14–10}$$

which is the same as the result in Eq. (14–5) for the low-pass circuit. The unity gain method requires that $C_1 = C_2 = C$ and $\mu = 1$. Inserting these constraints into Eqs. (14–8) and (14–9) yields the conditions

$$R_A = 0, \quad C\sqrt{R_1 R_2} = \frac{1}{\omega_0}, \quad \text{and} \quad \frac{R_1}{R_2} = \zeta^2 \qquad (14\text{--}11)$$

Since $R_A = 0$, the noninverting OP AMP circuit in Figure 14–16(b) can be a voltage follower, so neither feedback resistor is needed.

Combining Eqs. (14–7) and (14–8) shows that the passband gain of the high-pass transfer function is $|T(\infty)| = (R_A + R_B)/R_B$. As is the case for the low-pass circuit, the passband gain equals the gain of the OP AMP circuit.

Note: Under the equal element or unity gain approaches, the designer does not control the passband gain since the OP AMP circuit gain is determined by the damping ratio or is set to unity. An additional gain stage can be added if a different value of gain is specified.

The next example illustrates the problem of achieving both a specified damping ratio and a passband gain.

D DESIGN EXAMPLE 14–5

(a) Design a second-order high-pass transfer function with a passband gain of 0 dB and a cutoff frequency of $\omega_C = \omega_0 = 20$ krad/s.
(b) Design active RC circuits that realize the transfer function.

SOLUTION:

(a) A high-pass transfer function with a natural frequency of $\omega_0 = 20$ krad/s has the form

$$T(s) = \frac{K(s/20{,}000)^2}{(s/20{,}000)^2 + 2\zeta(s/20{,}000) + 1}$$

where K and ζ are to be determined. The passband gain of this transfer function is $|T(\infty)| = |K|$. The specification calls for a passband gain of 0 dB so $|K| = 1$. For the natural frequency also to be the cutoff frequency requires that $|T(j\omega_0)| = |K|/\sqrt{2}$. This, in turn, means that $|T(j20{,}000)| = |K|/\sqrt{2\zeta} = |K|/\sqrt{2}$, or $2\zeta = \sqrt{2}$. A transfer function that meets the specification is

$$T(s) = \frac{(s/20{,}000)^2}{(s/20{,}000)^2 + \sqrt{2}(s/20{,}000) + 1}$$

(b) We use both methods of assigning component values to obtain alternative designs.

Equal R and Equal C Design: Substituting the conditions $\omega_0 = 20$ krad/s and $\zeta = 1/\sqrt{2}$ into Eq. (14–10) yields $RC = 5 \times 10^{-5}$ s and $R_A = (2 - \sqrt{2})R_B = 0.586R_B$. Selecting $C = 5$ nF and $R_B = 50$ kΩ requires $R = 10$ kΩ and $R_A = 29.3$ kΩ. The passband gain of the resulting second-order circuit is $|T(\infty)| = K = (R_A + R_B)/R_B = 1.586$ and not unity (0 dB), as given in the specification. To overcome this problem we partition the transfer function as

$$T(s) = \underbrace{\left(\frac{1}{1.586}\right)}_{\text{First stage}} \underbrace{\left(\frac{1.586(s/20{,}000)^2}{(s/20{,}000)^2 + \sqrt{2}(s/20{,}000) + 1}\right)}_{\text{Second stage}}$$

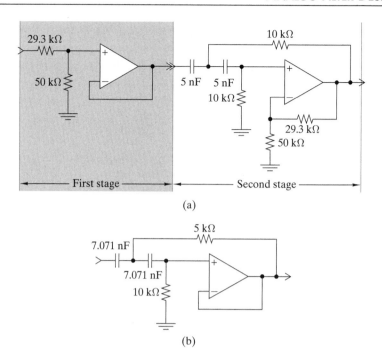

(a)

(b)

The second stage has a passband gain of $K = 1.586$ and can be realized using the second-order high-pass circuit designed previously. The first stage is a voltage divider providing an attenuation of $1/1.586$ to bring the overall passband gain down to the 0 dB level specified. Figure 14–17(a) shows the resulting two-stage cascade design.

Unity Gain Design: Substituting the conditions $\omega_0 = 20$ krad/s and $\zeta = 1/\sqrt{2}$ into Eq. (14–11) yields $R_A = 0$, $C\sqrt{R_1 R_2} = 5 \times 10^{-5}$ rad/s, and $R_1 = 0.5 R_2$. Selecting $R_2 = 10$ kΩ dictates that $R_1 = 5$ kΩ and $C = 7.071$ nF. The transfer function realized by the method has a passband gain $|T(\infty)| = (R_A + R_B)/R_B = 1$, which matches the 0 dB condition in the specification. Since no gain correction is needed, the single stage design in Figure 14–17(b) suffices. ∎

SECOND-ORDER BANDPASS CIRCUITS

The active *RC* circuit in Figure 14–18 has a second-order bandpass transfer function of the form

$$T(s) = \frac{V_2(s)}{V_1(s)} = \frac{-R_2 C_2 s}{R_1 R_2 C_1 C_2 s^2 + (R_1 C_1 + R_1 C_2)s + 1} \tag{14–12}$$

Derivation of the transfer function in Eq. (14–12) is left as an exercise (see Problem 14–27). This bandpass circuit has two negative feedback paths, one provided by the resistor R_2 and the other by the capacitor C_2. This arrangement identifies this circuit as a member of a class of active *RC* circuits called multiple feedback circuits.[2]

---

2 For an extensive discussion of this family of circuits, see Wai-Kai Chen, Ed., *The Circuits and Filters Handbook*, Boca Raton, Fla., CRC Press, 1995, Chapter 76, p. 2372ff. The bandpass circuit in Figure 14–18 is sometimes called the Delyannis-Friend circuit. See M. E. Van Valkenburg, *Analog Filter Design*, New York, Holt, Rinehart and Winston, 1982, p. 203.

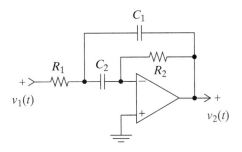

FIGURE 14–18 *A second-order band-pass circuit.*

The key descriptors of a bandpass filter are its center frequency ω_0 and its bandwidth $2\zeta\omega_0$. Comparing the denominator of Eq. (14–12) with the standard form $(s/\omega_0)^2 + 2\zeta\,(s/\omega_0) + 1$ yields the following results:

$$\omega_0 = \frac{1}{\sqrt{R_1R_2C_1C_2}} \quad \text{and} \quad 2\zeta = \sqrt{\frac{R_1C_1}{R_2C_2}} + \sqrt{\frac{R_1C_2}{R_2C_1}} \qquad (14\text{–}13)$$

In the multiple feedback bandpass case we have two design constraints and four unknown element values. One way to assign element values is to use equal capacitors ($C_1 = C_2 = C$), in which case Eq. (14–13) yields the conditions

$$C\sqrt{R_1R_2} = \frac{1}{\omega_0} \quad \text{and} \quad \frac{R_1}{R_2} = \zeta^2 \qquad (14\text{–}14)$$

For this choice of element values the gain at the center frequency is $|T(j\omega_0)| = R_2/2R_1 = 1/2\zeta$. As is the case for the low-pass and high-pass circuits, we cannot independently choose the damping ratio and the passband gain.

Bandpass filters are traditionally described in terms of center frequency ω_0 and a quality factor $Q = 1/2\zeta$. For second-order functions the filter bandwidth is

$$B = 2\zeta\omega_0 = \omega_0/Q$$

When $Q > 1$ ($\zeta < 0.5$) the filter is said to be narrowband because the bandwidth is less than the center frequency. Conversely, $Q < 1$ ($\zeta > 0.5$) describes a broadband filter. The active RC circuit shown in Figure 14–18 is best suited to narrowband applications. The design of broadband filters is discussed in Sect. 14–6.

D DESIGN EXAMPLE 14–6

Design an active RC bandpass circuit that has a center frequency of 10 kHz and a bandwidth of 2 kHz.

SOLUTION:
From the definitions of center frequency and bandwidth we have

$$\omega_0 = 2\pi \times 10^4 = 6.283 \times 10^4 \text{ rad/s} \quad \text{and} \quad Q = \frac{\omega_0}{B} = 5$$

Selecting $R_2 = 100$ kΩ and using $\zeta = 1/2Q = 0.1$ in Eq. (14–14) yields $R_1 = 1$ kΩ and $C_1 = C_2 = 1.592$ nF. The passband gain produced by this design is $|T(j\omega_0)| = 1/2 \zeta = 5$. ■

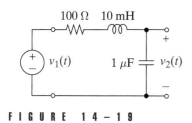

FIGURE 14–19

DESIGN EXERCISE: 14–3

Design an active RC circuit that has the same transfer function $T(s) = V_2(s)/V_1(s)$ as the RLC circuit in Figure 14–19.

Answer: One possible solution is shown in Figure 14–20.

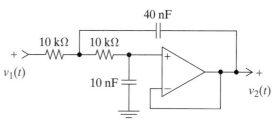

FIGURE 14–20

DISCUSSION: *At first glance the RLC filter in Figure 14–19 seems simpler than the active RC circuit in Figure 14–20. However, active RC filters offer the advantages discussed in the first section of this chapter. Briefly, the advantages are that they produce both gain and frequency selectivity, they can be connected in cascade to produce high-order filters, they can drive a variety of loads, and they are often smaller and less expensive in low-frequency applications.*

DESIGN EXERCISE: 14–4

Design a second-order active RC high-pass circuit with $\omega_0 = 500$ rad/s and $Q = 0.5$ using only 10-kΩ resistors.

Answers: Two possible solutions are shown in Figure 14–21.

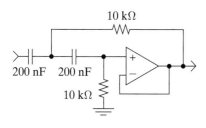

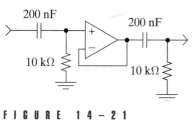

FIGURE 14–21

14–4 LOW-PASS FILTER DESIGN

The gain of the ideal low-pass filter response in Figure 14–22 is unity (0 dB) in the passband and zero ($-\infty$ dB) in the stopband. The transfer functions of physically realizable (real) low-pass filters are rational functions that can only approximate the ideal response. These rational functions have asymptotic high-frequency responses that roll off as ω^{-n} ($-20n$ dB/decade), where n is the number of poles in $T(s)$. Increasing the number of poles improves the approximation, as indicated in Figure 14–22. On the other hand, increasing the number of poles makes the circuit realizing $T(s)$ more complicated. Since cost and circuit complexity are usually related, there is an important trade-off between circuit complexity and how closely different filters approximate the ideal response.

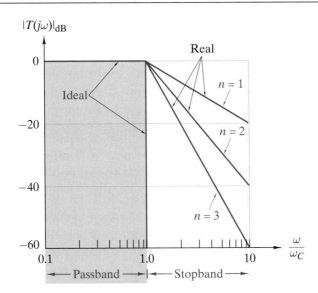

Ideal and real low-pass filter asymptotic responses.

Since we can only approximate the ideal response filter, design requirements specify how closely we must approach the ideal. A commonly used way to specify this is to require that the gain response fall within the unshaded region in Figure 14–23. Many different transfer functions can meet this restriction, as illustrated by two responses shown in the figure. The allowable region in Figure 14–23 is defined by four parameters. The parameter T_{MAX} is the maximum gain in the passband, called simply the **passband gain**. Within the passband the gain $|T(j\omega)|$ must remain in the range

$$\frac{T_{\text{MAX}}}{\sqrt{2}} \le |T(j\omega)| \le T_{\text{MAX}}$$

and must equal $T_{\text{MAX}}/\sqrt{2}$ at the **cutoff frequency** ω_{C}. In the stopband the gain must decrease to and remain below T_{MIN} for all $\omega \ge \omega_{\text{MIN}}$. In sum, the parameters T_{MAX} and ω_{C} specify the passband response and the parameters T_{MIN} and ω_{MIN} specify how rapidly the stopband response must decrease.

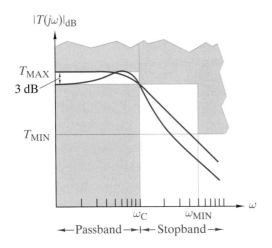

Gain responses meeting a low-pass filter specification.

To design a low-pass filter we must construct a transfer function whose gain $|T(j\omega)|$ approximates the ideal filter response within the tolerances allowed by the four parameters T_{MAX}, T_{MIN}, ω_C, and ω_{MIN}. Such a low-pass transfer function has the form

$$T_n(s) = \frac{K}{q_n(s)}$$

where $q_n(s)$ is an nth-order polynomial defining the poles. Thus, filter design involves two major tasks: (1) Construct a transfer function $T_n(s)$ whose gain response meets the filter performance specification and (2) devise one or more circuits that produce the transfer function $T_n(s)$.

We will consider three approaches to the first task: first-order cascade responses, Butterworth responses, and Chebychev responses. For the second task we partition the transfer function into a product of first- and second-order functions each of which can be realized using the building block circuits described in the preceding sections. A circuit realizing the transfer function is obtained by a cascade connection of these first- and second-order stages.

FIRST-ORDER CASCADE RESPONSES

A simple way to produce an nth-order low-pass filter is to connect n identical first-order low-pass filters in cascade. When we connect n identical filters in cascade the overall transfer function is

$$T_n(s) = \underbrace{\left[\frac{K}{s/\alpha + 1}\right] \times \left[\frac{K}{s/\alpha + 1}\right] \times \cdots \left[\frac{K}{s/\alpha + 1}\right]}_{n \text{ Stages}} = \frac{K^n}{(s/\alpha + 1)^n}$$

(14–15)

In a design problem we must select K, α, and n such that the gain response of this transfer function meets the performance requirements defined by T_{MAX}, T_{MIN}, ω_C, and ω_{MIN}.

The transfer function in Eq. (14–15) produces a gain response of

$$|T_n(j\omega)| = \frac{|K|^n}{\sqrt{1 + (\omega/\alpha)^2}^{\,n}}$$

(14–16)

Note that the low-frequency gain asymptote is $|K|^n$ and the high-frequency asymptote is $(|K|\alpha/\omega)^n$. These asymptotes intersect (i.e., are equal) at a corner frequency located at $\omega = \alpha$. However, when $n > 1$ this corner frequency is *not* the filter cutoff frequency.

The maximum gain in Eq. (14–16) occurs at $\omega = 0$, where gain is $|T(0)| = |K|^n$. To meet the passband gain requirement, we select $|K|^n = T_{MAX}$. The passband specification also requires the gain at the cutoff frequency to be

$$|T_n(j\omega_C)| = \frac{|K|^n}{\sqrt{1 + (\omega_C/\alpha)^2}^{\,n}} = \frac{T_{MAX}}{\sqrt{2}}$$

With $|K|^n = T_{MAX}$ we can equate the denominators on the right side of this expression and solve for α as

$$\alpha = \frac{\omega_C}{\sqrt{2^{1/n} - 1}}$$

(14–17)

Each first-order function in the cascade has a corner frequency at $\omega = \alpha$ whose value depends on ω_C and the number of poles n. Since $2^{1/n} - 1 \leq 1$ it follows that $\alpha \geq \omega_C$; that is, the corner frequency is higher than the cutoff frequency for all $n > 1$. Said another way, increasing the order of a cascade filter requires that we increase the corner frequency to maintain the same bandwidth.

Figure 14–24 shows normalized gain responses of first-order low-pass cascades for $n = 1$ to $n = 8$. All of these responses meet the passband requirements. As expected, the stopband attenuation increases as the number of poles n increases since the high-frequency asymptote is proportional to $(\alpha/\omega)^n$. We can estimate the number of poles needed to meet the stopband requirements from the graphs in Figure 14–24. For example, suppose the stopband requirement is $T_{MIN}/T_{MAX} = 0.01$ (–40 dB) at $\omega_{MIN} = 10\omega_C$. In Figure 14–24 we see that at $\omega/\omega_C = 10$ the normalized gain $|T(j\omega)|/T_{MAX}$ is –32 dB for $n = 2$ and –42 dB for $n = 3$. Hence $n = 3$ is the smallest number of poles in a first-order cascade response that meets these requirements.

In summary, to construct a first-order cascade transfer function, we proceed as follows. Given values of T_{MAX}, T_{MIN}, ω_C, and ω_{MIN}, we set $|K|^n = T_{MAX}$ and use Figure 14–24 (or trial and error) to find the smallest integer n that meets the stopband requirements. Given n and ω_C, we calculate α using Eq. (14–17) and obtain the required transfer function $T_n(s)$ from Eq. (14–15). The transfer function is partitioned into a product of identical first-order functions, each of which can be obtained using a first-order circuit. The design process is relatively simple and yields identical stages, which may reduce manufacturing and logistical support costs. The next example illustrates these features.

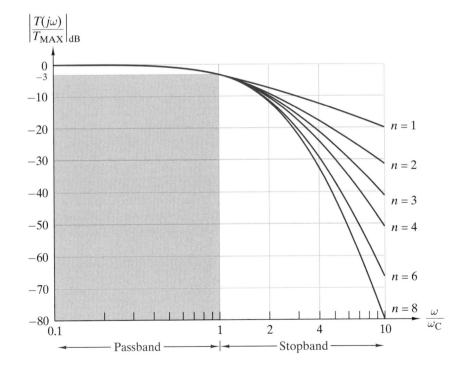

FIGURE 14–24 *First-order low-pass cascade filter responses.*

D   DESIGN EXAMPLE 14–7

(a) Construct a first-order cascade transfer function that meets the following requirements: $T_{MAX} = 10$ dB, $\omega_C = 200$ rad/s, $T_{MIN} = -20$ dB, and $\omega_{MIN} = 800$ rad/s.
(b) Design a cascade of active RC circuits that realizes the transfer function developed in (a).

SOLUTION:

(a) The stopband requirement is that the normalized gain be less than -30 dB at $\omega/\omega_C = 800/200 = 4$. Figure 14–24 suggests that $n = 7$ might work, but we cannot be sure since the curve for $n = 7$ is not shown. To resolve this question, we use trial and error. Using Eq. (14–17), we calculate corner frequencies for $n = 7$ and $n = 8$.

$$\text{For } n = 7, \quad \alpha = \frac{200}{\sqrt{2^{1/7} - 1}} = 620 \text{ rad/s}$$

$$\text{For } n = 8, \quad \alpha = \frac{200}{\sqrt{2^{1/8} - 1}} = 665 \text{ rad/s}$$

Next we use Eq. (14–16) to calculate the corresponding normalized stopband gains:

$$\text{For } n = 7, \quad \left| \frac{T_7(j\omega_{MIN})}{T_{MAX}} \right| = 20 \log \left(\frac{1}{\left[\sqrt{1 + (800/620)^2} \right]^7} \right) = -29.8 \text{ dB}$$

$$\text{For } n = 8, \quad \left| \frac{T_8(j\omega_{MIN})}{T_{MAX}} \right| = 20 \log \left(\frac{1}{\left[\sqrt{1 + (800/665)^2} \right]^8} \right) = -31.1 \text{ dB}$$

Strictly speaking, $n = 8$ is the smallest integer for which the stopband gain meets the design requirement, although $n = 7$ comes within 0.2 dB (roughly 2.3%). Using $n = 8$, we solve for K. Since $T_{MAX} = 10$ dB (factor of $10^{1/2}$), we can write the passband requirement as $|K|^8 = 10^{1/2}$; hence $K = 1.155$. So, finally, the required first-order cascade transfer function is

$$T_8(s) = \left(\frac{1.155}{s/665 + 1} \right)^8$$

Note that the corner frequency is 665 rad/s while the cutoff frequency is 200 rad/s.

(b) The transfer function developed in (a) can be partitioned into a product of eight identical first-order functions of the form $1.155/(s/665 + 1)$. Using the voltage-divider building block in Figure 14–3 produces the following design constraints:

$$K \frac{Z_2}{Z_1 + Z_2} = 1.155 \times \frac{1/s}{1/665 + 1/s}$$

These conditions yield $K = (R_A + R_B)/R_B = 1.155$, $Z_2(s) = 1/s$, and $Z_1(s) = 1/665$. Selecting $R_B = 100$ kΩ leads to $R_A = 15.5$ kΩ. Using the magnitude scale factor of $k_m = 10^7$ on the RC voltage divider circuit produces the first-order low-pass circuit in Figure 14–25. A cascade connection of eight such circuits produces a low-pass filter that meets the design requirements. The gain of each stage is the 8th root of the overall passband gain since the overall gain is the product of the gain in each of the identical eight stages. The OP AMP at the output of each stage also provides isolation so that the cascade connection of eight such stages does not cause loading effects that change the frequency response of the individual stages. ∎

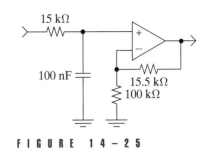

FIGURE 14–25

BUTTERWORTH LOW-PASS RESPONSES

All Butterworth low-pass filters produce a gain response of the form

$$|T_n(j\omega)| = \frac{|K|}{\sqrt{1 + (\omega/\omega_C)^{2n}}} \tag{14–18}$$

where ω_C is the cutoff frequency and n is the number of poles in $T_n(s)$. By inspection, the maximum passband gain occurs at $\omega = 0$, where the gain is $|T(0)| = |K| = T_{MAX}$. At the cutoff frequency the gain is $|T_n(j\omega_C)| = |K|/\sqrt{2}$ for all n. Thus, selecting $|K| = T_{MAX}$ ensures that the Butterworth gain response in Eq. (14–18) satisfies the passband requirements for all values of n.

The low-frequency gain asymptote in Eq. (14–18) is $|K|$, while the high-frequency asymptote is $|K| (\omega_C/\omega)^n$. These two asymptotes intersect at a corner frequency of $\omega = \omega_C$. In a Butterworth response the corner frequency is always equal to the cutoff frequency, which means that it is independent of the order of the filter. This feature is different from the first-order cascade response, where the corner frequency increases with filter order.

Figure 14–26 compares normalized first-order cascade and Butterworth gain responses for $n = 4$. Both responses meet the passband requirements,

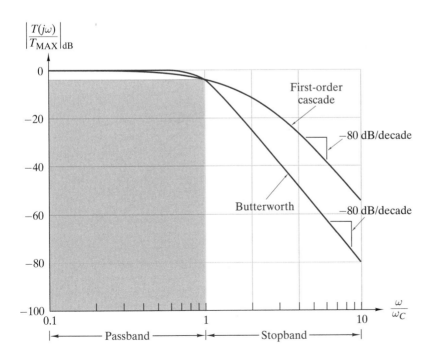

FIGURE 14–26 *First-order cascade and Butterworth low-pass filter responses for* n = 4.

and both have high-frequency asymptotic slopes of $-20n = -80$ dB/decade. However, the Butterworth response provides greater attenuation in the stopband because its corner frequency is lower and hence it approaches its high-frequency asymptote at lower frequencies than does the first-order cascade response.

With the Butterworth response we can analytically solve for the required number of poles. Since $|K| = T_{MAX}$, we insert the stopband requirements in Eq. (14–18) to obtain

$$|T(j\omega_{MIN})| = \frac{T_{MAX}}{\sqrt{1 + (\omega_{MIN}/\omega_C)^{2n}}} \leq T_{MIN} \qquad (14\text{–}19)$$

Solving this equation for n yields the constraint

$$n \geq \frac{1}{2} \frac{\ln[(T_{MAX}/T_{MIN})^2 - 1]}{\ln[\omega_{MIN}/\omega_C]} \qquad (14\text{–}20)$$

For example, the stopband requirement of $T_{MAX}/T_{MIN} = 10^{3/2}$ (30 dB) at $\omega_{MIN} = 4\omega_C$ yields

$$n \geq \frac{1}{2} \frac{\ln[(10^{3/2})^2 - 1]}{\ln[4]} = 2.49$$

The smallest integer meeting this constraint is $n = 3$.

Figure 14–27 shows normalized plots of the Butterworth gain response in Eq. (14–18) for $n = 1$ to $n = 6$. These graphs can be used to estimate the number of poles required to meet the stopband requirements. For example, consider the requirements used previously, namely, $|T_{MIN}/T_{MAX}|_{dB} = -30$ dB at $\omega_{MIN} = 4\omega_C$. In Figure 14–27 we see that a $\omega/\omega_C = 4$, $|T(j\omega)/T_{MAX}|_{dB}$ is -24 dB for $n = 2$, and is -36 dB for $n = 3$. Hence $n = 3$ is the smallest number

F I G U R E   1 4 – 2 7 *Butterworth low-pass filter responses.*

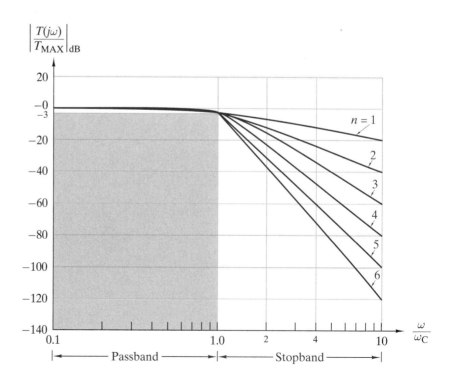

of poles in a Butterworth response that meets these stopband requirements. Thus, either Eq. (14–20) or Figure 14–27 can be used to determine n.

To design low-pass filters we need polynomials $q_n(s)$ such that the transfer function

$$T_n(s) = \frac{K}{q_n(s)}$$

produces the Butterworth gain response in Eq. (14–18). We begin with a first-order polynomial of the form $q_1(s) = s/\omega_C + 1$, which produces the gain response

$$|T_1(j\omega)| = \frac{|K|}{\sqrt{1 + (\omega/\omega_C)^2}} \qquad (14\text{–}21)$$

This response has the Butterworth form for $n = 1$. Next consider a second-order polynomial of the form $q_2(s) = (s/\omega_C)^2 + 2\zeta(s/\omega_C) + 1$, which produces the gain response

$$\begin{aligned}|T_2(j\omega)| &= \frac{|K|}{\sqrt{[1 - (\omega/\omega_C)^2]^2 + (2\zeta\omega/\omega_C)^2}} \\ &= \frac{|K|}{\sqrt{1 + (4\zeta^2 - 2)(\omega/\omega_C)^2 + (\omega/\omega_C)^4}} \qquad (14\text{–}22)\end{aligned}$$

When $4\zeta^2 = 2$ this response has the Butterworth form for $n = 2$. As a final example, consider a third-order polynomial of the form $q_3(s) = (s/\omega_C + 1)[(s/\omega_C)^2 + 2\zeta(s/\omega_C) + 1]$, which produces a gain response

$$|T_3(j\omega)| = \frac{|K|}{\sqrt{1 + (\omega/\omega_C)^2}\sqrt{[1 - (\omega/\omega_C)^2]^2 + 2\zeta(\omega/\omega_C)^2}}$$

$$= \frac{|K|}{\sqrt{1 + (4\zeta^2 - 1)(\omega/\omega_C)^2 + (4\zeta^2 - 1)(\omega/\omega_C)^4 + (\omega/\omega_C)^6}} \qquad (14\text{–}23)$$

When $4\zeta^2 = 1$ this response has the Butterworth form of Eq. (14–18) for $n = 3$.

The preceding analysis shows that the polynomials

$$\begin{aligned}q_1(s) &= s/\omega_C + 1 \\ q_2(s) &= (s/\omega_C)^2 + \sqrt{2}(s/\omega_C) + 1 \\ q_3(s) &= (s/\omega_C + 1)[(s/\omega_C)^2 + (s/\omega_C) + 1]\end{aligned}$$

produce the Butterworth gain response in Eq. (14–18) for $n = 1, 2,$ and 3. Proceeding in this fashion, we can generate the normalized ($\omega_C = 1$) polynomials in Table 14–1 for values of n up to 6.

A general Butterworth low-pass transfer function with a cutoff frequency of ω_C is

$$T_n(s) = \frac{K}{q_n(s/\omega_C)} \qquad (14\text{–}24)$$

where $q_n(s)$ is the nth-order normalized polynomial in Table 14–1. With $|K| = T_{MAX}$ this transfer function meets the passband conditions because the normalized polynomials produce Butterworth gain responses. The order of the polynomial, and hence the number of poles, is determined from the stopband requirement using Figure 14–27 or Eq. (14–20).

T A B L E   14–1   NORMALIZED POLYNOMIALS THAT PRODUCE BUTTERWORTH RESPONSES

ORDER	NORMALIZED DENOMINATOR POLYNOMIALS
1	$(s + 1)$
2	$(s^2 + 1.414s + 1)$
3	$(s + 1)(s^2 + s + 1)$
4	$(s^2 + 0.7654s + 1)(s^2 + 1.848s + 1)$
5	$(s + 1)(s^2 + 0.6180s + 1)(s^2 + 1.618s + 1)$
6	$(s^2 + 0.5176s + 1)(s^2 + 1.414s + 1)(s^2 + 1.932s + 1)$

Once we have constructed the required $T_n(s)$, we partition it into a product of first- and second-order functions and realize each function using the building blocks developed in the previous sections. The next example illustrates the design procedure for a Butterworth low-pass filter.

D   DESIGN EXAMPLE 14–8

The spectrum of a transducer signal is concentrated in the frequency ranges from dc to $f_{MAX} = 1000/2\pi$ Hz. The signal is to be converted into digital form by sampling at a frequency of $f_S = 5000/2\pi$ Hz. Prior to sampling, an anti-aliasing filter is required with a passband gain of 20 dB and a stopband gain less than −20 db in the aliasing frequency range.

(a) Construct a Butterworth low-pass transfer function that meets these requirements.
(b) Design a cascade of active RC circuits that produces the transfer function found in (a).

SOLUTION:
(a) The required passband gain and low-pass cutoff frequency are $T_{MAX} = 20$ dB and $\omega_C = 2\pi f_{MAX} = 1000$ rad/s. The aliasing frequencies all fall above $f_S - f_{MAX} = 4000/2\pi$ Hz. The stopband gain at these frequencies must be below −20 dB. Hence, the stopband requirements are $T_{MIN} = -20$ dB at $\omega_{MIN} = 2\pi(f_S - f_{MAX}) = 4000$ rad/s. The relative frequency range is $\omega_{MIN}/\omega_C = 4$. The 40 dB difference in between passband and stopband gains means that $T_{MAX}/T_{MIN} = 100$. Applying Eq. (14–20) yields the required filter order.

$$n \geq \frac{1}{2} \frac{\ln[(100)^2 - 1]}{\ln[4]} = 3.32$$

from which we see that $n = 4$ is the lowest-order polynomial that meets the stopband requirements. Using the fourth-order polynomial from Table 14–1, the required Butterworth low-pass transfer function is

$$T(s) = \frac{K}{q_4(s/\omega_C)}$$

$$= \frac{10}{\left[\left(\frac{s}{1000}\right)^2 + 0.7654\left(\frac{s}{1000}\right) + 1\right]\left[\left(\frac{s}{1000}\right)^2 + 1.848\left(\frac{s}{1000}\right) + 1\right]}$$

(b) The transfer function developed in (a) can be partitioned as follows:

$$\frac{T(s)}{10} = \left[\frac{1}{\left(\dfrac{s}{1000}\right)^2 + 0.7654\left(\dfrac{s}{1000}\right) + 1}\right]\left[\frac{1}{\left(\dfrac{s}{1000}\right)^2 + 1.848\left(\dfrac{s}{1000}\right) + 1}\right]$$

Figure 14–28 shows a design sequence for these second-order transfer functions using the Sallen-Key low-pass circuit shown in Figure 14–14. The first and second rows in the figure give the transfer functions for each stage and the stage parameters. Using the equal R and equal C design method [see Eq. (14–5)] leads to the design constraints in the

Item	First Stage	Second Stage
Prototype transfer function	$\dfrac{1}{(s/1000)^2 + 0.7654(s/1000) + 1}$	$\dfrac{1}{(s/1000)^2 + 1.848(s/1000) + 1}$
Stage parameters	$\omega_0 = 1000 \quad \zeta = 0.7654/2 = 0.3827$	$\omega_0 = 1000 \quad \zeta = 1.848/2 = 0.924$
Stage prototype	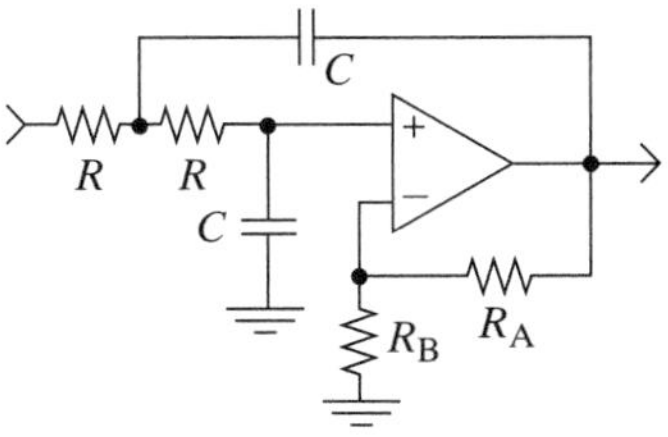	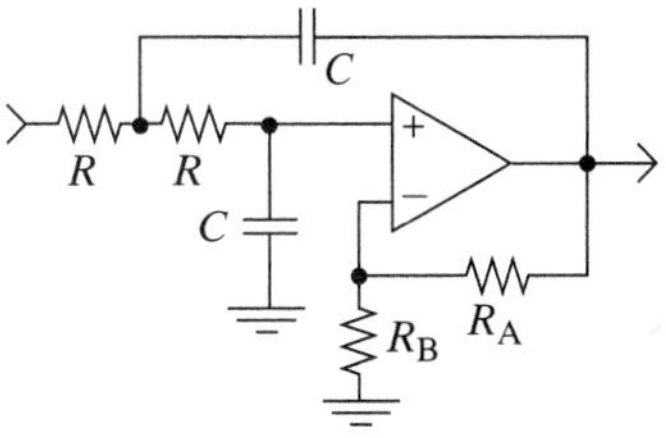
Design constraints	$RC = \dfrac{1}{\omega_0} = 0.001$ $\dfrac{R_A}{R_B} = 2(1 - \zeta) = 1.23$	$RC = \dfrac{1}{\omega_0} = 0.001$ $\dfrac{R_A}{R_B} = 2(1 - \zeta) = 0.152$
Element values	Let $R = 100$ kΩ, then $C = 10$ nF Let $R_B = 100$ kΩ, then $R_A = 123$ kΩ	Let $R = 100$ kΩ, then $C = 10$ nF Let $R_B = 100$ kΩ, then $R_A = 15.2$ kΩ
Final designs	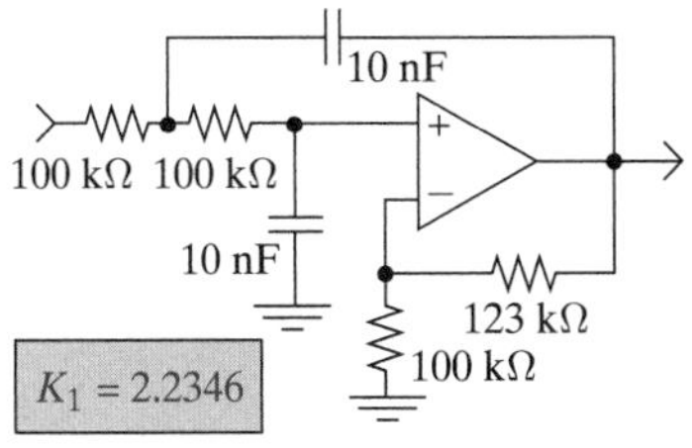 $K_1 = 2.2346$	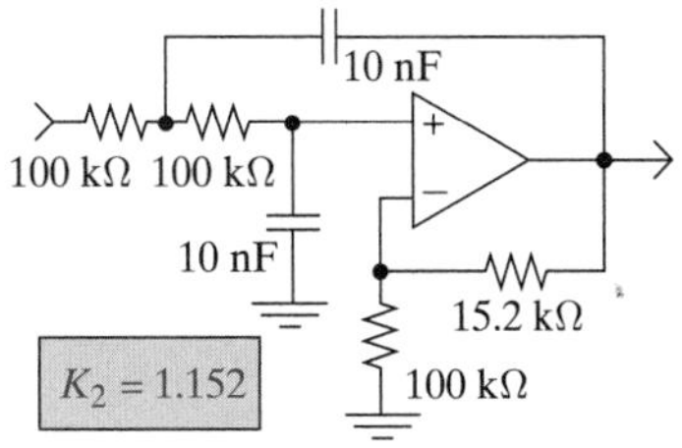 $K_2 = 1.152$

FIGURE 14–28 Design sequence for Example 14–8.

fourth row. The element values selected in the fifth row produce the final design shown in the last row of Figure 14–28.

With the Sallen-Key circuit the gain of each stage is determined by the stage damping ratio. The final designs in Figure 14–28 produce an overall passband gain of

$$K = K_1 K_2 = 2.23 \times 1.152 = 2.57$$

which is less than the requirement of $K = 10$. To meet the gain requirement, we add a third stage with a gain of $K_3 = 10/2.57 = 3.89$. Figure 14–29 shows a final three-stage cascade design that meets all of the design requirements. ∎

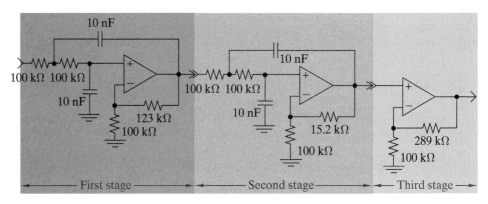

FIGURE 14-29

CHEBYCHEV LOW-PASS RESPONSES

All Chebychev low-pass filters produce a gain response of the form

$$|T_n(j\omega)| = \frac{|K|}{\sqrt{1 + C_n^2(\omega/\omega_C)}} \tag{14–25}$$

where $C_n(x)$ is an nth-order Chebychev polynomial defined by

$$C_n(x) = \cos[n \times \cos^{-1}(x)] \quad \text{for} \quad x \leq 1 \tag{14–26a}$$

and

$$C_n(x) = \cosh[n \times \cosh^{-1}(x)] \quad \text{for} \quad x > 1 \tag{14–26b}$$

In the passband ($x = \omega/\omega_C \leq 1$) the function $C_n(x)$ in Eq. (14–26a) is a cosine function that varies between –1 and +1. Hence $C_n^2(x)$ varies between 0 and 1 with the result that the passband gain of the Chebychev response in Eq. (14–25) varies over the range

$$\frac{|K|}{\sqrt{2}} \leq |T_n(j\omega)| \leq |K| \quad \text{for} \quad \omega \leq \omega_C$$

and, in particular at $\omega = \omega_C$,

$$|T_n(j\omega_C)| = \frac{|K|}{\sqrt{1 + C_n^2(1)}} = \frac{|K|}{\sqrt{2}}$$

for all values of n. By selecting $|K| = T_{\text{MAX}}$ we ensure that the Chebychev response in Eq. (14–25) meets the passband gain requirements for all n.

Like Butterworth responses, the Chebychev order is determined by the stopband requirements. Inserting these requirements into Eq. (14–25) produces

$$|T_n(j\omega_{MIN})| = \frac{|T_{MAX}|}{\sqrt{1 + C_n^2(\omega_{MIN}/\omega_C)}} \leq T_{MIN} \tag{14–27}$$

Solving this constraint for $C_n(\omega_{MIN}/\omega_C)$ yields

$$C_n(\omega_{MIN}/\omega_C) \geq \sqrt{\left(\frac{T_{MAX}}{T_{MIN}}\right)^2 - 1}$$

In the stopband ($x = \omega/\omega_C \geq 1$) the function $C_n(x)$ is defined by the hyperbolic cosine function in Eq. (14–26b). Inserting this definition into the preceding constraint and then solving for n yields

$$n \geq \frac{\cosh^{-1}(\sqrt{(T_{MAX}/T_{MIN})^2 - 1})}{\cosh^{-1}(\omega_{MIN}/\omega_C)} \tag{14–28}$$

In Example 14–8 the Butterworth response requires $n = 4$ to meet the stopband conditions $T_{MAX}/T_{MIN} = 100$ (40 dB) at $\omega_{MIN} = 4\omega_C$. Inserting these conditions into Eq. (14–28) yields

$$n \geq \frac{\cosh^{-1}(\sqrt{(100)^2 - 1})}{\cosh^{-1}(4)} = 2.57$$

The Chebychev response can meet the same stopband conditions using $n = 3$.

Figure 14–30 compares the Butterworth and Chebychev gain responses for $n = 4$. Both responses have stopband asymptotic slopes of –80 dB/decade. The Butterworth response is relatively flat in the passband and has a smooth transition to its high-frequency asymptote. In contrast, the Chebychev passband gain displays a succession of resonant peaks and valleys with a more abrupt transition to the stopband asymptote. Because

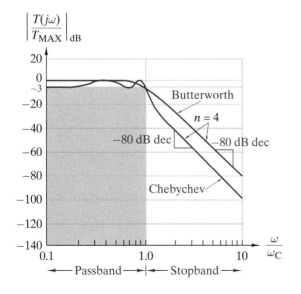

FIGURE 14–30 Butterworth and Chebychev low-pass filter responses for n = 4.

TABLE 14–2 NORMALIZED POLYNOMIALS THAT PRODUCE CHEBYCHEV RESPONSES

ORDER	NORMALIZED DENOMINATOR POLYNOMIALS
1	$(s + 1)$
2	$[(s/0.8409)^2 + 0.7654(s/0.8409) + 1]$
3	$[(s/0.2980) + 1][(s/0.9159)^2 + 0.3254(s/0.9159) + 1]$
4	$[(s/0.9502)^2 + 0.1789(s/0.9502) + 1)][(s/0.4425)^2 + 0.9276(s/0.4425) + 1]$
5	$[(s/0.1772) + 1][(s/0.9674)^2 + 0.1132(s/0.9674) + 1][(s/0.6139)^2 + 0.4670(s/0.6139) + 1]$
6	$[(s/0.9771)^2 + 0.0781(s/0.9771) + 1][(s/0.7223)^2 + 0.2886(s/0.7223) + 1][(s/0.2978)^2 + 0.9562(s/0.2978) + 1]$

$C_n(\omega)$ varies as cosine function in the passband, the resonant peaks in the Chebychev gain characteristic are all equal to T_{MAX} and the valleys all equal to $T_{MAX}/\sqrt{2}$. For this reason the Chebychev gain response in Eq. (14–25) is called the **equal-ripple response**. The net result of these resonances is that the Chebychev response transitions to its high-frequency asymptote rather abruptly.

To design a low-pass filter we need polynomials $q_n(s)$ such that the transfer function

$$T_n(s) = \frac{K}{q_n(s)}$$

produces a Chebychev gain response. Derivation of these polynomials involves complex variable theory beyond the scope of our study.[3] The analysis carried out in the reference allows us to write the normalized ($\omega_C = 1$) polynomials in Table 14–2. The transfer function of an nth-order Chebychev low-pass filter with a cutoff frequency of ω_C is then written as follows

$$T_n(s) = \frac{K}{q_n(s/\omega_C)} \qquad n \text{ odd} \qquad (14\text{–}29a)$$

or

$$T_n(s) = \frac{K/\sqrt{2}}{q_n(s/\omega_C)} \qquad n \text{ even} \qquad (14\text{–}29b)$$

where $q_n(s)$ is the nth-order normalized polynomial in Table 14–2. When we select $|K| = T_{MAX}$ the scale factor adjustment of $1/\sqrt{2}$ in Eq. (14–29b) is needed to ensure that $|T_n(j\omega)|$ meets the passband requirements when n is even. As with the Butterworth response, the order of the denominator polynomial can be found analytically from Eq. (14–28) or graphically using the normalized gain plots in Figure 14–31.

---

3 For example, see M. E. Van Valkenburg, *Analog Filter Design*, Chicago: Holt, Rinehart, and Winston, 1982, pp. 233–241.

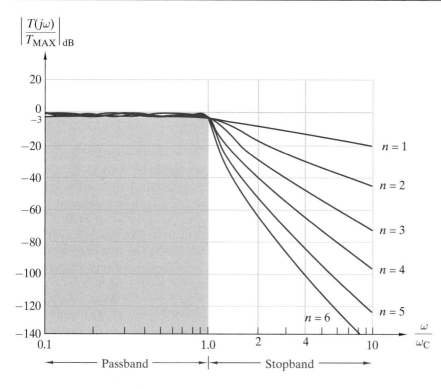

FIGURE   14-31 *Chebychev low-pass filter responses.*

DESIGN EXAMPLE 14-9

(a) Construct a Chebychev low-pass transfer function that meets a passband requirement of $T_{MAX} = 20$ dB and $\omega_C = 10$ rad/s, and a stopband requirement of $T_{MIN} = -30$ dB at $\omega_{MIN} = 50$ rad/s.

(b) Design a cascade of active *RC* circuits that produces the transfer function found in part (a).

(c) The filter is to be driven by a voltage source with a 600-Ω Thévenin resistance. Design the filter to avoid loading of the input source.

SOLUTION:

(a) The passband gain condition requires $K = 10$. The stopband requirement specifies that the stopband gain at $\omega_{MIN}/\omega_C = 5$ must be 50 dB (factor of $10^{5/2}$) less than the passband gain. Using Eq. (14–28) yields

$$n \geq \frac{\cosh^{-1}(\sqrt{(10^{2.5})^2 - 1})}{\cosh^{-1}(5)} = 2.81$$

Hence $n = 3$ is the smallest integer meeting the stopband requirement. Using the third-order polynomial from Table 14–2, we construct the required Chebychev low-pass transfer function:

$$T(s) = \frac{K}{q_3(s/\omega_C)}$$

$$= \frac{10}{\left[\left(\dfrac{s}{0.298 \times 10}\right) + 1\right]\left[\left(\dfrac{s}{0.9159 \times 10}\right)^2 + 0.3254\left(\dfrac{s}{0.9159 \times 10}\right) + 1\right]}$$

(b) The Chebychev transfer function developed in (a) can be partitioned as follows:

$$\frac{T(s)}{10} = \left[\frac{1}{\left(\dfrac{s}{0.298 \times 10}\right) + 1} \right]\left[\frac{1}{\left(\dfrac{s}{0.9159 \times 10}\right)^2 + 0.3254\left(\dfrac{s}{0.9159 \times 10}\right) + 1} \right]$$

realizing this partition requires a first-order stage in cascade with a second-order stage.

The design sequence in Figure 14–32 begins with the required transfer functions in the first row. The next three rows show the stage parameters, stage prototype, and design constraints for the selected prototypes. The constraints are then used to select element values that produce the final designs shown in the last row of Figure 14–32. The second stage is a Sallen-Key circuit, so its gain ($K_2 = 2.675$) is determined by the second-stage damping ratio. The first stage is a voltage divider circuit whose gain can be adjusted without changing its pole location. Thus, we can adjust the first-stage gain at $K_1 = 10/K_2$ so that the overall passband gain is $K_1 \times K_2 = K = 10$. A cascade connection of the final designs in the last two of Figure 14–32 meets all design requirements without introducing an additional gain stage.

(c) The input to the first stage in Figure 14–32 is an RC voltage divider circuit. The input impedance of this circuit is at least 100 kΩ, which is high enough to avoid loading of the 600-Ω input source.   ■

Exercise 14–5

What is the minimum order of the first-order cascade, Butterworth, and Chebychev transfer functions that meet the following stopband conditions?

(a) $T_{MIN}/T_{MAX} = -20$ dB at $4\omega_C$
(b) $T_{MIN}/T_{MAX} = -30$ dB at $5\omega_C$
(c) $T_{MIN}/T_{MAX} = -40$ dB at $6\omega_C$
(d) $T_{MIN}/T_{MAX} = -60$ dB at $8\omega_C$

Answers:

	First Order	Butterworth	Chebychev
(a)	$n = 3$	$n = 2$	$n = 2$
(b)	$n = 4$	$n = 3$	$n = 2$
(c)	$n = 5$	$n = 3$	$n = 3$
(d)	$n = 7$	$n = 4$	$n = 3$

COMPARISON OF LOW-PASS FILTER RESPONSES

We have described low-pass filter design methods for three responses, the first-order cascade, the Butterworth, and the Chebychev. At this point we want to compare the methods and discuss how we might choose between them. In filter applications the gain response is obviously important. Figure 14–33 shows the three straight-line asymptotes (solid lines) and the actual gain responses (dashed curves) for $n = 4$. All three responses meet the same passband requirements, have the same cutoff frequency, and have high-frequency asymptotes with slopes of $-20n = -80$ dB/decade. However, the three responses have different corner frequencies. At one extreme the

Item	First Stage	Second Stage
Prototype transfer function	$\dfrac{1}{(s/2.98)+1}$	$\dfrac{1}{(s/9.159)^2 + 0.3254(s/9.159)+1}$
Stage parameters	$\omega_0 = 2.980$	$\omega_0 = 9.159 \quad \zeta = 0.3254/2 = 0.1627$
Stage prototype	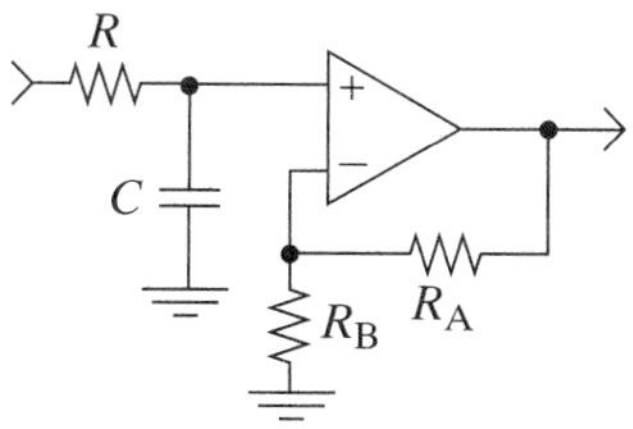 *(circuit diagram: R, C, R_B, R_A, op-amp)*	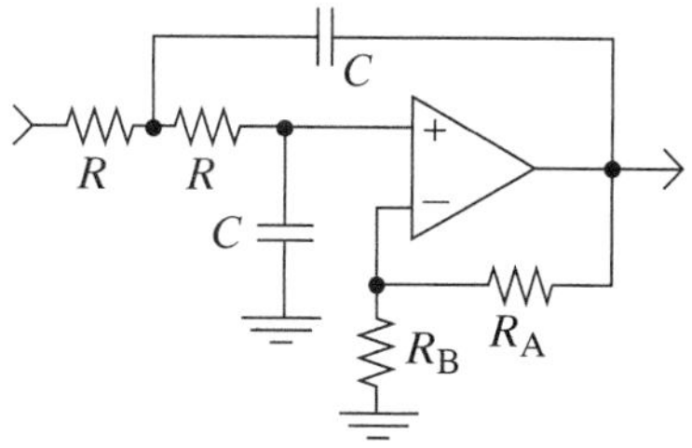 *(circuit diagram: R, R, C, C, R_B, R_A, op-amp)*
Design constraints	$RC = \dfrac{1}{\omega_0} = 0.3357$ $\dfrac{R_A}{R_B} = K_1 - 1 = 2.74$	$RC = \dfrac{1}{\omega_0} = 0.1092$ $\dfrac{R_A}{R_B} = 2(1-\zeta) = 1.67$
Element values	Let $R = 100\ \text{k}\Omega$, then $C = 3.36\ \mu\text{F}$ Let $R_B = 100\ \text{k}\Omega$, then $R_A = 274\ \text{k}\Omega$	Let $R = 100\ \text{k}\Omega$, then $C = 1.092\ \mu\text{F}$ Let $R_B = 100\ \text{k}\Omega$, then $R_A = 167\ \text{k}\Omega$
Final designs	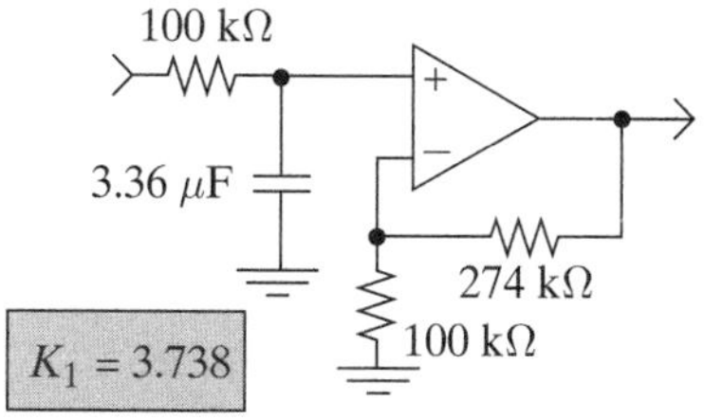 *(final circuit: 100 kΩ, 3.36 μF, 274 kΩ, 100 kΩ)* $K_1 = 3.738$	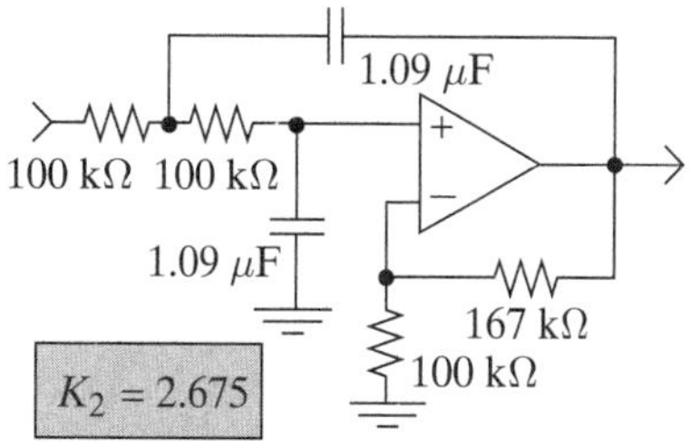 *(final circuit: 100 kΩ, 100 kΩ, 1.09 μF, 1.09 μF, 167 kΩ, 100 kΩ)* $K_2 = 2.675$

FIGURE 14 – 3 2   *Design sequence for Example 14–9.*

corner frequency of the first-order cascade response is above the cutoff frequency, so its actual gain response approaches its asymptote very gradually. At the other extreme the Chebychev corner frequency lies below the cutoff frequency and the actual response has a resonant peak that fills the gap between the corner frequency and the cutoff frequency. This resonance causes the Chebychev response to decrease rapidly in the neighborhood of the cutoff frequency. The Butterworth response has its corner frequency at the cutoff frequency, so its gain response falls between these two extremes.

FIGURE 14–33 *First-order cascade, Butterworth, and Chebychev low-pass responses for* n = 4.

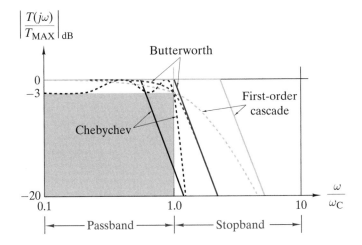

The differences in gain response can be understood by examining the pole-zero diagram in Figure 14–34. The Butterworth poles are evenly distributed on a circle of radius ω_C. The Chebychev poles lie on an ellipse whose minor axis is much smaller than ω_C. As a result, the Chebychev poles are closer to the *j*-axis, have lower damping ratios, and produce a gain response with pronounced resonant peaks. These resonant peaks lead to the equal-ripple response in the passband and the steep gain slope in the neighborhood of the cutoff frequency. At the other extreme the first-order cascade response has a fourth-order pole (quadruple pole) located on the negative real axis. The distance from the *j*-axis to the first-order cascade poles is much larger than ω_C, which explains the rather leisurely way its gain response transitions from the passband to stopband asymptote. As might be expected, the Butterworth poles fall between these two extremes.

This discussion illustrates the following principle. For any given value of *n*, the Chebychev response produces more stopband attenuation than the Butterworth response, which, in turn, produces more than the first-order

FIGURE 14–34 *First-order cascade, Butterworth, and Chebychev pole locations for* n = 4.

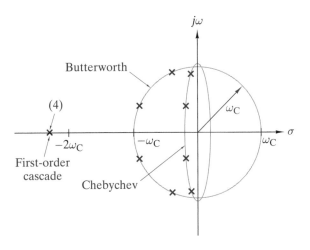

cascade response. If stopband performance is the only consideration, then we should choose the Chebychev response. However, the Chebychev response comes at a price.

Figure 14–35 shows the step response of these three low-pass filters for $n = 4$. The step response of the Chebychev filter has lightly damped oscillations that produce a large overshoot and a long settling time. These undesirable features of the Chebychev step response are a direct result of the low-damping-ratio complex poles that produce the desirable features of its gain response. At the other extreme the step response of the first-order cascade filter rises rapidly to its final value without overshooting. This result should not be surprising since the remote poles of the first-order cascade produce exponential waveforms that have relatively short durations. In other words, the desirable features of the first-order cascade step response are a direct result of the remote real poles that produce the undesirable features of its gain response. Not surprisingly, the step response of the Butterworth filter lies between these two extremes.

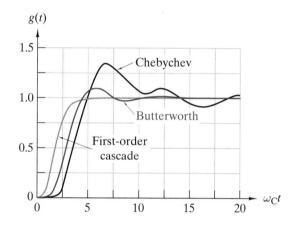

FIGURE 14–35 *First-order cascade, Butterworth, and Chebychev step responses for n = 4.*

Finally, consider the element values in the circuit realizations of these filters. Examination of Table 14–2 reveals that each pair of complex poles in a Chebychev filter has a different ω_0 and a different ζ. These parameters define the constraints on the element values for each stage in the filter. As a result, each stage in a cascade realization of a Chebychev filter has a different set of element values. In contrast, the stages in a first-order cascade filter can be exactly the same. From a manufacturing point of view, it may be better to produce and stock identical circuits rather than uniquely different circuits.

The essential point is that stopband attenuation is important, but it does not tell the whole story. Filter design, and indeed all design, involves trade-offs between conflicting requirements. The choice of a design approach is driven by the weight assigned to conflicting requirements.

14–5 HIGH-PASS FILTER DESIGN

High-pass transfer functions can be derived from low-pass prototypes using a transformation of the complex frequency variable s. If a low-pass transfer function $T_{LP}(s)$ has a passband gain of T_{MAX} and cutoff frequency of ω_C, then the transfer function $T_{HP}(s)$ defined as

$$T_{HP}(s) = T_{LP}(\omega_C^2/s) \qquad \text{(14–30)}$$

has high-pass characteristics with the same cutoff frequency and passband gain. On a logarithmic frequency scale this transformation amounts to a horizontal reflection of the gain response about ω_C. In other words, replacing s by ω_C^2/s changes a low-pass transfer function into a high-pass function without changing the parameters defining passband and stopband performances.

The first-order low-pass transfer function provides a simple example of the low-pass to high-pass transformation. Given a first-order low-pass function

$$T_{LP}(s) = \frac{T_{MAX}}{\dfrac{s}{\omega_C} + 1}$$

replacing s by ω_C^2/s yields a first-order high-pass transfer function

$$T_{HP}(s) = \frac{T_{MAX}}{\dfrac{\omega_C^2/s}{\omega_C} + 1} = \frac{T_{MAX}s}{s + \omega_C}$$

Note that the cutoff frequency and passband gain of the high-pass function are the same as those of the low-pass function from which it was derived.

Figure 14–36 shows normalized Butterworth high-pass responses obtained by applying the low-pass to high-pass transformation to the Butterworth low-pass responses shown in Figure 14–27. The plots confirm that the transformation interchanges the locations of the passband and stopband while leaving the cutoff frequency unchanged. The maximally flat Butterworth gain characteristic (or equal ripple in the case of Chebychev) is relocated to a passband *above* ω_C. The stopband is relocated *below* ω_C where the gain response decreases with filter order.

A high-pass filter design problem is defined by specifying four parameters: T_{MAX}, ω_C, T_{MIN}, and ω_{MIN}. With a high-pass filter the stopband gain T_{MIN} is specified at a frequency ω_{MIN} that falls *below* ω_C. The number of Butterworth poles needed to meet a high-pass specification can be estimated from the normalized plots in Figure 14–36. Or, we can use the high-pass version of the bound Eq. (14–20) for Butterworth low-pass functions. Applying the low-pass to-high-pass transformation to Eq. (14–20) leads to the following bound for Butterworth high-pass filters:

$$n \geq \frac{1}{2} \frac{\ln\left[(T_{MAX}/T_{MIN})^2 - 1\right]}{\ln\left[\omega_C/\omega_{MIN}\right]} \qquad \text{High-pass order} \qquad \text{(14–31)}$$

For example, consider a high-pass filter requirement of $T_{MIN} = T_{MAX}/100$ (–40 dB) at $\omega = \omega_C/4$. Figure 14–36 shows that a stopband requirement of –40 dB at $\omega/\omega_C = 0.25$ cannot be met by any order less than $n = 4$. Using $T_{MAX}/T_{MIN} = 100$ and $\omega_C/\omega_{MIN} = 4$ in Eq. (14–31) yields

$$n \geq \frac{1}{2} \frac{\ln\left[(10^2)^2 - 1\right]}{\ln[4]} = 3.32$$

Either method shows that a fourth-order Butterworth high-pass function is needed to meet this requirement.

Once the filter order is determined, the Butterworth high-pass function is obtained by first creating a low-pass prototype. Using Table 14–1, we

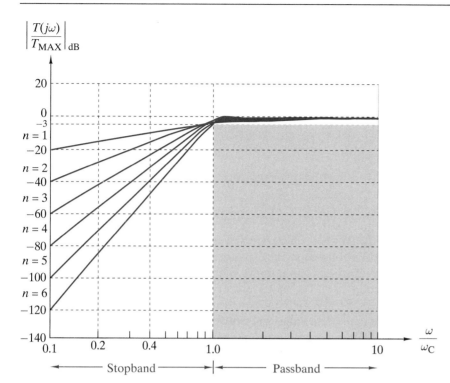

FIGURE   14-36 *Butterworth high-pass filter responses.*

create a Butterworth low-pass function whose cutoff frequency, passband gain, and order are the same as the desired high-pass function. Applying the low-pass to high-pass transformation changes this low-pass prototype into the desired high-pass function. The high-pass function is then partitioned into first- and second-order factors, and realized using high-pass building blocks discussed in previous sections.

The next example illustrates the procedure.

D DESIGN EXAMPLE 14-10

(a) Construct a Butterworth high-pass transfer function with $T_{MAX} =$ 20 dB, $\omega_C = 10$ rad/s, and a stopband requirement of $T_{MIN} = -10$ dB at $\omega_{MIN} = 3$ rad/s.
(b) Design a cascade of active RC circuits that produces the transfer function found in part (a).

SOLUTION:

(a) The difference between T_{MAX} and T_{MIN} is 30 dB; hence $T_{MAX}/T_{MIN} = 10^{1.5}$. Using Eq. (14-31) yields

$$n \geq \frac{1}{2} \frac{\ln\left[(10^{3/2})^2 - 1\right]}{\ln(10/3)} = 2.87$$

The lowest-order Butterworth response that meets the high-pass requirements is $n = 3$. Using the third-order polynomial in Table 14-1 with $\omega_C = 10$ and $K = 10$ ($T_{MAX} = 20$ dB) produces the following low-pass prototype.

$$T_{\text{LP}}(s) = \frac{K}{q_3(s/10)} = \frac{10}{\left[\left(\dfrac{s}{10}\right)^2 + \left(\dfrac{s}{10}\right) + 1\right]\left[\left(\dfrac{s}{10}\right) + 1\right]}$$

The high-pass transfer function is obtained by replacing s with $\omega_C^2/s = 100/s$:

$$T_{\text{HP}}(s) = \frac{10}{\left[\left(\dfrac{100/s}{10}\right)^2 + \left(\dfrac{100/s}{10}\right) + 1\right]\left[\left(\dfrac{100/s}{10}\right) + 1\right]}$$

$$= \left[\frac{(s/10)^2}{(s/10)^2 + (s/10) + 1}\right]\left[\frac{10(s/10)}{(s/10) + 1}\right]$$

The function $T_{\text{HP}}(s)$ is a high-pass transfer function that meets the filter design requirements.

(b) Figure 14–37 shows a design sequence for realizing $T_{\text{HP}}(s)$ using the first- and second-order high-pass active *RC* building blocks. The stage transfer functions given in the first row yield the stage parameters in the second row. The stage prototypes are a second-order high-pass section with unity gain and a first-order high-pass section with an adjustable gain. The stage prototypes together with the stage parameters yield the design constraints in the fourth row. Selecting element values within these constraints produces the final stage designs shown in the last row of Figure 14–37. The required passband gain is included in the first-order circuit, so no gain correction stage is required in this example. In other words, connecting the two stages in Figure 14–37 in cascade meets all design requirements, including the passband gain of 20 dB. ∎

EXAMPLE 14–11

Use Electronics Workbench to verify the high-pass design developed in Example 14–10.

SOLUTION:

One of the most important uses of computer-aided analysis is to simulate a proposed circuit design and confirm that it meets the performance requirements. To verify the design in Example 14–10, we must confirm that the frequency response of the two-stage circuit in the last row of Figure 14–37 has a high-pass characteristic with the following features:

1. A passband gain of 20 dB (gain = 10).
2. A cutoff frequency of 10 rad/s (gain = $10/\sqrt{2} = 7.07$ at $f = 10/2\pi = 1.59$ Hz).
3. Stopband gain less than -10 dB (gain < 0.316) below 3 rad/s (0.477 Hz).

Figure 14–38 shows the two-stage circuit as constructed in the Electronics Workbench circuit window. The input source $\mathbf{V_1}$ is set to produce a 1 V ac

Item	First Stage	Second Stage
Prototype transfer function	$\dfrac{(s/10)^2}{(s/10)^2 + (s/10) + 1}$	$\dfrac{10\,(s/10)}{(s/10) + 1}$
Stage parameters	$\omega_0 = 10 \quad \zeta = 0.5 \quad K_1 = 1$	$\omega_0 = 10 \quad K_2 = 10$
Stage prototype	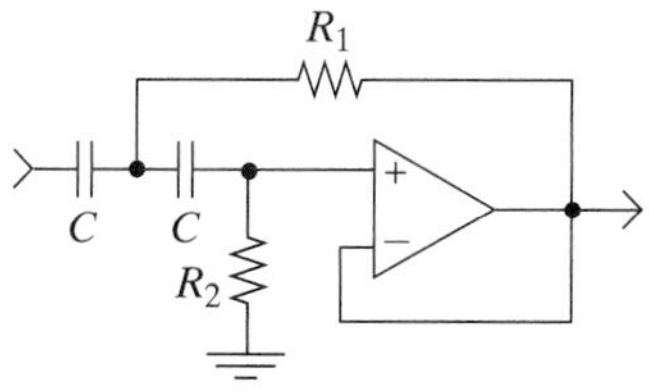	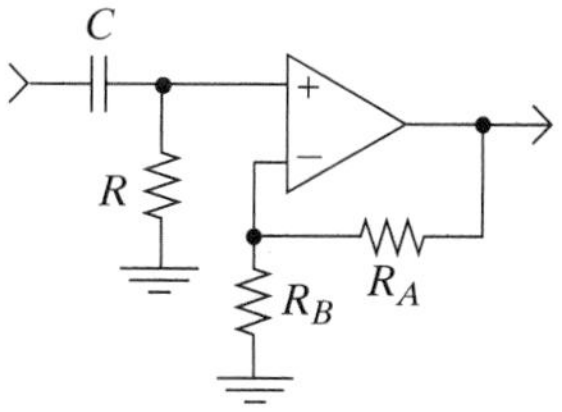
Design constraints	$\sqrt{R_1 R_2}\,C = 1/\omega_0 = 0.1$ $R_1/R_2 = \zeta^2 = 0.25$	$RC = 1/\omega_0 = 0.1$ $R_A/R_B = K_2 - 1 = 9$
Element values	Let $R_2 = 100 \text{ k}\Omega$, then $R_1 = 25 \text{ k}\Omega$ and $C = 2\ \mu\text{F}$	Let $R = 50 \text{ k}\Omega$, then $C = 2\ \mu\text{F}$ Let $R_B = 10 \text{ k}\Omega$, then $R_A = 90 \text{ k}\Omega$
Final designs	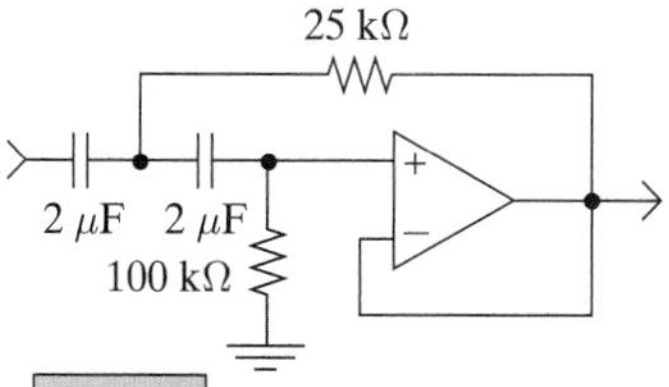	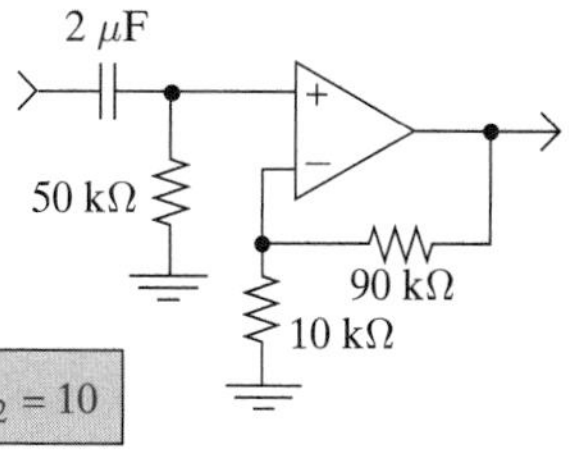

FIGURE 14 – 37 *Design sequence for Example 14–10.*

input, so the ac voltage at node 9 (the second OP AMP output) is the overall filter circuit gain response. Using the **AC Analysis** command from the **Analysis** menu, we sweep the frequency of the source $\mathbf{V_1}$ from 0.3 Hz to 4 Hz, and select node 9 as the node for analysis. The **Simulate** command produces an ac analysis run leading to the gain response shown in Figure 14–38. The gain plot shows that

1. The circuit has a high-pass response with a passband of 10.

2. The x2 cursor is set at x2 = 1.59 Hz where the gain is found to be y2 = 7.07 confirming that the cutoff frequency is 10 rad/s.

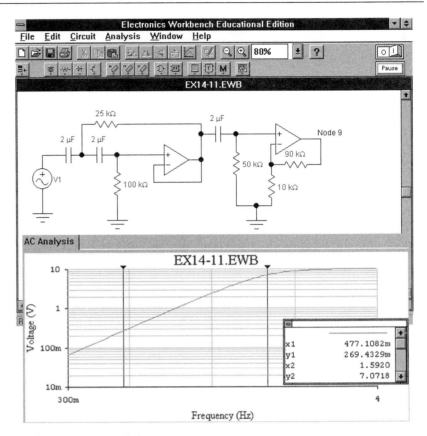

FIGURE 14 – 38

3. The x1 cursor is set at x1 = 0.477 Hz where the gain is found to be y1 = 0.269 (–11.4 dB) confirming that the stopband gain is below –10 dB at 3 rad/sec.

The plot confirms that the circuit meets the design requirements. ■

Exercise 14–6

Construct first-order cascade and Butterworth high-pass transfer functions that have passband gains of 0 dB, cutoff frequencies of 50 rad/s, and $T_{\text{MIN}} = -40$ dB at $\omega_{\text{MIN}} = 10$ rad/s.

Answers:

$$T_{\text{FO}}(s) = \frac{s^8}{(s + 15)^8}, \; T_{\text{BU}}(s) = \frac{s^3}{(s + 50)(s^2 + 50s + 2500)}$$

14–6 BANDPASS AND BANDSTOP FILTER DESIGN

In Chapter 12 we found that the cascade connection in Figure 14–39 can produce a bandpass filter. When the cutoff frequency of the low-pass filter (ω_{CLP}) is higher than the cutoff frequency of the high-pass filter (ω_{CHP}), the

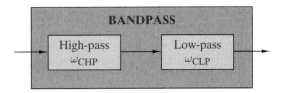

FIGURE 14–39 *Cascade connection of high-pass and low-pass filters.*

interval between the two frequencies is a passband separating two stop-bands. The low-pass filter provides the high-frequency stopband and the high-pass filter the low-frequency stopband. Frequencies between the two cutoffs fall in the passband of both filters and are transmitted through the cascade connection, producing the passband of the resulting bandpass filter.

When $\omega_{CLP} \gg \omega_{CHP}$, two bandpass cutoff frequencies are approximately $\omega_{C1} \approx \omega_{CHP}$ and $\omega_{C2} \approx \omega_{CLP}$. Under these conditions the center frequency and bandwidth of the bandpass filter are

$$\omega_0 = \sqrt{\omega_{CHP}\omega_{CLP}} \text{ and } B = \omega_{CLP} - \omega_{CHP}$$

and the ratio of the center frequency over the bandwidth is approximately

$$\frac{\omega_0}{B} = Q \approx \sqrt{\frac{\omega_{CHP}}{\omega_{CLP}}} \ll 1$$

Since the quality factor is less than 1, this method of bandpass filter design produces a broadband filter, as contrasted with the narrowband response ($Q > 1$) produced by the active *RC* bandpass circuit studied earlier in Sect. 14–3.

D DESIGN EXAMPLE 14–12

Use second-order Butterworth low-pass and high-pass functions to obtain a fourth-order bandpass function with a passband gain of 0 dB and cutoff frequencies at $\omega_{C1} = 10$ rad/s and $\omega_{C2} = 50$ rad/s.

SOLUTION:
The upper cutoff frequency at 50 rad/s is produced by a low-pass function. Using second-order Butterworth poles the required low-pass function is

$$T_{LP}(s) = \frac{1}{(s/50)^2 + \sqrt{2}(s/50) + 1}$$

The lower cutoff frequency at 10 rad/s is to be produced by a second-order high-pass function. Using second-order Butterworth poles, the required high-pass function is

$$T_{HP}(s) = \frac{(s/10)^2}{(s/10)^2 + \sqrt{2}(s/10) + 1}$$

When circuits realizing these two transfer functions are connected in cascade the overall transfer function is

$$T_{HP}(s) \times T_{LP}(s) = \left[\frac{(s/10)^2}{(s/10)^2 + \sqrt{2}(s/10) + 1}\right] \times \left[\frac{1}{(s/50)^2 + \sqrt{2}(s/50) + 1}\right]$$

$$= \frac{2500 \, s^2}{s^4 + 60\sqrt{2}s^3 + 3600s^2 + 30000 \sqrt{2}s + 250000}$$

With the transfer function expressed as a quotient of polynomials, we use MATLAB to generate Bode plots of the frequency response. In the MATLAB command window, we define the numerator and denominator polynomials using the row matrices num and den.

```
num=[2500 0 0];

den=[1 60*sqrt(2) 3600 30000*sqrt(2) 250000];
```

These matrices list the polynomial coefficients in descending order. The MATLAB command

```
bode(num,den)
```

produces the Bode plots shown in Figure 14–40. The gain response displays a bandpass characteristic with a lower cutoff frequency at 10 rad/s and an upper cutoff frequency at 50 rad/s. These two cutoff frequencies come from the high-pass and low-pass functions, respectively. The phase shift swings from +180° at low frequency to −180° at high frequency passing through zero at the passband center frequency of $\sqrt{10 \times 50} =$ 22.36 rad/s. ∎

FIGURE 14-40

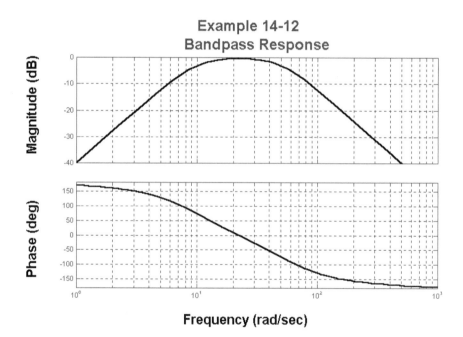

**Example 14-12
Bandpass Response**

Frequency (rad/sec)

Figure 14–41 shows the dual situation in which a high-pass and a low-pass filter are connected in parallel to produce a bandstop filter. When $\omega_{CLP} \ll \omega_{CHP}$, the region between the two cutoff frequencies is a stopband separating two passbands. The low-pass filter provides the low-frequency passband via the lower path and the high-pass filter the high-frequency passband via the upper path. Frequencies between the two cutoffs fall in the stopband of both filters and are not transmitted through either path in the parallel connection. As a result, the two filters produce the stopband of the resulting bandstop filter. When $\omega_{CHP} \gg \omega_{CLP}$, two cutoff frequencies are approximately $\omega_{C1} \approx \omega_{CLP}$ and $\omega_{C2} \approx \omega_{CHP}$.

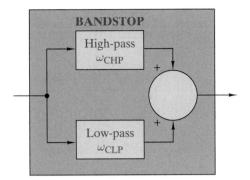

BANDSTOP

High-pass
ω_{CHP}

Low-pass
ω_{CLP}

FIGURE 14–41 *Parallel connection of high-pass and low-pass filters.*

D **DESIGN EXAMPLE 14–13**

Use second-order Butterworth low-pass and high-pass functions to obtain a fourth-order bandstop function with passband gains of 0 dB and cutoff frequencies at $\omega_{C1} = 10$ rad/s and $\omega_{C2} = 50$ rad/s.

SOLUTION:

In Example 14–12 we obtained a Butterworth bandpass response using a cascade connection of a second-order high-pass function and a low-pass function with $\omega_{CLP} = 50 >> \omega_{CHP} = 10$. To obtain a bandstop response, we interchange the cutoff frequencies and write the two functions as

$$T_{LP}(s) = \frac{1}{(s/10)^2 + \sqrt{2}(s/10) + 1}$$

$$T_{HP}(s) = \frac{(s/50)^2}{(s/50)^2 + \sqrt{2}(s/50) + 1}$$

We now have $\omega_{CLP} = 10 << \omega_{CHP} = 50$ which leads to a bandstop response. For the parallel connection the overall transfer function is the sum of the two transfer functions.

$$T_{LP}(s) + T_{HP}(s) = \frac{1}{(s/10)^2 + \sqrt{2}(s/10) + 1} + \frac{(s/50)^2}{(s/50)^2 + \sqrt{2}(s/50) + 1}$$

$$= \frac{s^4 + 10\sqrt{2}s^3 + 200s^2 + 5000\sqrt{2}s + 250000}{s^4 + 60\sqrt{2}s^3 + 3600s^2 + 30000\sqrt{2}s + 250000}$$

With the transfer function expressed as a quotient of polynomials, we use MATLAB to generate Bode plots of the frequency response. In the MATLAB command window, we define the numerator and denominator polynomials using the row matrices num and den.

```
num=[1 10*sqrt(2) 200 5000*sqrt(2) 250000];

den=[1 60*sqrt(2) 3600 30000*sqrt(2) 250000];
```

These matrices list the polynomial coefficients in descending order. The MATLAB command

```
bode(num,den)
```

produces the Bode plots shown in Figure 14–42. The gain displays a bandstop response with a lower cutoff frequency at 10 rad/s and an upper cutoff frequency at 50 rad/s. These two cutoff frequencies come from the low-

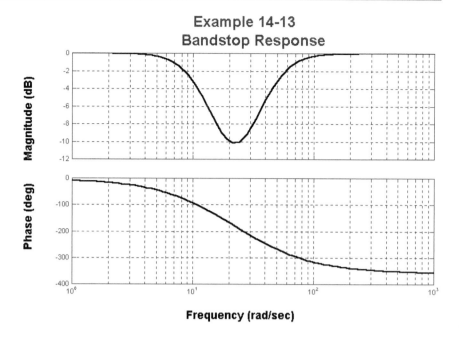

Example 14-13
Bandstop Response

Frequency (rad/sec)

pass and high-pass functions, respectively. The phase shift changes from +0° at low frequency to −360° at high frequency passing through −180° at the stopband center frequency of $\sqrt{10 \times 50} = 22.36$ rad/s. ∎

When the two cutoff frequencies are widely separated, we can realize broadband bandpass and bandstop filters using a cascade or parallel connection of low-pass and high-pass filters. The design problem reduces to designing separate low-pass and high-pass filters and then connecting them in cascade or parallel to obtain the required overall response. In the bandpass case the active *RC* circuits building blocks can be connected in cascade without violating the chain rule. The bandstop case requires a summation, which can be implemented using an OP AMP summer.

Exercise 14–7

Construct Butterworth low-pass and high-pass transfer functions whose cascade connection produces a bandpass function with cutoff frequencies at 20 rad/s and 500 rad/s, a passband gain of 0 dB, and a stopband gain less than −20 dB at 5 rad/s and 2000 rad/s.

Answer:

$$T(s) = \left[\frac{500^2}{s^2 + 707s + 500^2}\right]\left[\frac{s^2}{s^2 + 28.3s + 400}\right]$$

Exercise 14–8

Develop Butterworth low-pass and high-pass transfer functions whose parallel connection produces a bandstop filter with cutoff frequencies at 2 rad/s and 800 rad/s, passband gains of 20 dB, and stopband gains less than −30 dB at 20 rad/s and 80 rad/s.

Answer:

$$T_{LP}(s) = \frac{80}{(s^2 + 2s + 4)(s + 2)}$$

$$T_{HP}(s) = \frac{10s^3}{(s^2 + 800s + 800^2)(s + 800)}$$

APPLICATION NOTE: EXAMPLE 14–14

At the start of this chapter, we described the dual-tone signals generated by the Touch-Tone™ key pad in Figure 14–1. Briefly, each signal is a sum of two sinusoids with one frequency from a low group (697, 770, 852, 941 Hz) and the other from a high group (1209, 1336, 1477 Hz). At the receiving end a low-pass filter and a high-pass filter split the input into two signals, one containing the low-group sinusoid and the other the high-group sinusoid. The purpose of this example is to explore the design of these band-splitting filters.

The basic design requirement for both the low- and the high-pass filters is that the stopband gain be at least 40 dB below the passband gain. The low-pass filter passes all low-group sinusoids and attenuates all high-group sinusoids. The passband cutoff frequency is set at $f_C = 941$ Hz, which is the maximum frequency in the low group. The stopband begins at $f_{MIN} = 1209$ Hz, which is the minimum frequency in the high group. The 40-dB requirement then means that $T_{MAX}/T_{MIN} \geq 100$ for $f \geq f_{MIN}$. Using Eq. (14–20), the lowest order Butterworth filter meeting these conditions turns out to be

$$n \geq \frac{1}{2} \frac{\ln(T_{MAX}/T_{MIN})}{\ln(f_{MIN}/f_C)} = \frac{1}{2} \frac{\ln(100)}{\ln(1209/941)} = 18.4$$

A 19th-order Butterworth low-pass filter is required, which would be a considerable design challenge. Things get better using a Chebychev filter. Equation (14–28) gives the lowest Chebychev order as

$$n \geq \frac{\cosh^{-1}(\sqrt{T_{MAX}/T_{MIN} - 1})}{\cosh^{-1}(f_{MIN}/f_C)} = \frac{\cosh^{-1}(\sqrt{99})}{\cosh^{-1}(1209/941)} = 7.18$$

Thus, an eighth-order Chebychev low-pass filter is necessary.

The other band-splitting filter is high pass. This filter passes all frequencies in the high group and attenuates all frequencies in the low group. The passband cutoff is set at $f_C = 1209$ Hz, which is the minimum frequency in the high group. The stopband begins at $f_{MIN} = 941$ Hz, which is the maximum frequency in the low group. As we have seen in Sect. 14–5, designing a high-pass filter involves a low-pass prototype found by reversing the roles of f_C and f_{MIN} in the high-pass specification. Applying this rule here means using $f_C = 941$ Hz and $f_{MIN} = 1209$ Hz, which leads us back to the preceding computations for the low-pass filter. In other words, it takes a 19th-order Butterworth or an eighth-order Chebychev to meet the high-pass requirements.

The Butterworth approach would require two 19th-order filters, each involving nine second-order stages and one first-order stage. The circuit complexity is significantly reduced by using two eighth-order Chebychev filters, each involving four second-order stages. On the basis of circuit complexity, the Chebychev filters are the best of the available options.[4]

We do not discuss the circuit design for these Chebychev filters because this would involve multiple repetitions of procedures illustrated in several previous examples. Instead, we discuss the actual performance these filters would deliver. Figure 14–43 shows the gain responses of the two band-splitting Chebychev filters. Note that the gain and frequency axes use linear scales rather than logarithmic scales.

Both filters deliver a very steep transition from the passband to the stopband. This performance is achieved by having complex poles close to the *j*-axis in the s plane. The result is a sequence of resonant peaks and valleys in the passbands. This passband ripple means that inband sinusoids are not amplified equally. For instance, in the low-pass filter 697 Hz is near a valley while 852 Hz is near a peak. This unequal handling of inband signals is an unavoidable consequence of using Chebychev filters. For the Touch-Tone™ signals this is not a significant problem, the reason being that the data are not encoded in the analog amplitudes but encoded digitally in the presence or absence of different frequencies. ∎

FIGURE 14–43 *Touch-Tone™ band-splitting filter responses.*

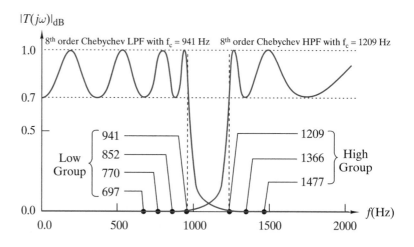

4 The perceptive reader may wonder why we have not included first-order cascade poles as an option. Unlike Butterworth or Chebychev, the first-order cascade gain roll off approaches a finite limit as *n* increases. The filter requirements here cannot be met by a first-order cascade, no matter how large the order.

SUMMARY

- A filter design problem is defined by specifying attributes of the gain response such as a straight-line gain plot, cutoff frequency, passband gain, and stopband attenuation. The first step in the design process is to construct a transfer function $T(s)$ whose gain response meets the specification requirements.

- In the cascade design approach, the required transfer function is partitioned into a product of first- and second-order transfer functions, which can be independently realized using basic active RC building blocks.

- Transfer functions with real poles and zeros can be realized using the voltage divider, noninverting amplifier, or inverting amplifier building blocks. Transfer functions with complex poles can be realized using second-order active RC circuits.

- Transfer functions meeting low-pass filter specifications can be constructed using first-order cascade, Butterworth, or Chebychev poles. First-order cascade filters are easy to design but have poor stopband performance. Butterworth responses produce maximally flat passband responses and more stopband attenuation than a first-order cascade with the same number of poles. The Chebychev responses produce equal-ripple passband responses and more stopband attenuation than the Butterworth response with the same number of poles.

- A high-pass transfer function can be constructed from a low-pass prototype by replacing s with ω_C^2/s. Bandpass (bandstop) filters can be constructed using a cascade (parallel) connection of a low-pass and a high-pass filter.

PROBLEMS

ERO 14–1 DESIGN WITH FIRST-ORDER CIRCUITS (SECT. 14–2)

Given a filter specification or a Bode plot of a straight-line gain,

(a) Construct a transfer function $T(s)$ that has the specified characteristics.
(b) Design a cascade of first-order circuits that realizes $T(s)$.

See Examples 14–1, 14–2, 14–3 and Exercises 14–1, 14–2

14–1 **D** Design a first-order low-pass filter with a cutoff frequency of 1500 Hz and a passband gain of 26 dB.

14–2 **D** Design a first-order high-pass filter with a cutoff frequency of 500 Hz and passband gain of 10 dB. Check the passband gain and cutoff frequency of your design.

14–3 **D** Construct a first-order low-pass transfer function that has at least 10 dB gain at 5 kHz and no more than −10 dB at 2 MHz.

14–4 **D** The open-circuit voltage of the source in Figure P14–4 is

$$v_S(t) = 5 + 0.02 \sin (4000 \, \pi \, t) \quad V$$

The purpose of the bypass capacitor is to reduce the ac voltage delivered to the 20-Ω load below 10 mV peak. Select a value of C.

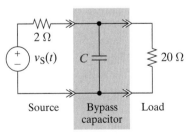

Source   Bypass   Load
           capacitor

F I G U R E   P 1 4 – 4

14–5 **D** Design a first-order low-pass filter that has a dc gain of 5 and a gain of less than −20 dB at 20 kHz.

14–6 **D** Design a cascade of first-order active *RC* circuits to realize the transfer function

$$T(s) = \frac{8 \times 10^6}{(s + 500)(s + 2000)}$$

14–7 **D** The straight-line gain plot in Figure P14–7 emphasizes the frequencies below 20 rad/s and de-emphasizes the frequencies above 200 rad/s.

(a) Construct a transfer function that has this gain response using only real poles and zeros.

(b) Design a cascade of first-order active *RC* circuits to realize the *T(s)* found in (a). Scale the circuits so that the element values are in practical ranges.

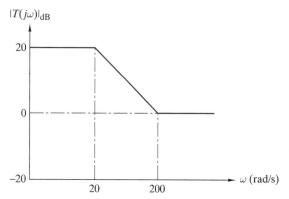

FIGURE P14–7

14–8 **D** The straight-line plot in Figure P14–8 has a low-pass characteristic.

(a) Construct a transfer function *T(s)* that has this gain response using only real poles and zeros.

(b) Design a cascade of first-order active *RC* circuits to realize the transfer function found in (a). Scale the circuits so that the element values are in practical ranges.

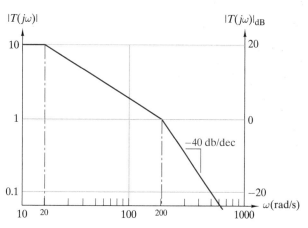

FIGURE P14–8

14–9 **D** The straight-line gain response in Figure P14–9 emphasizes the frequencies between 200 rad/s and 1000 rad/s.

(a) Construct a transfer function *T(s)* that has this gain response using only real poles and zeros.

(b) Design a cascade of first-order active *RC* circuits to realize the *T(s)* found in (a). Scale the circuits so that the element values are in practical ranges.

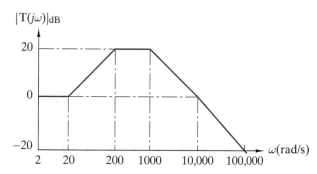

FIGURE P14–9

14–10 **D** The gain response in Figure P14–10 has a passband between 40 Hz and 2 kHz.

(a) Construct a transfer function *T(s)* that has this gain response using only real poles and zeros.

(b) Design a cascade of first-order active *RC* circuits to realize the transfer function found in (a). Scale the circuits so that the element values are in practical ranges.

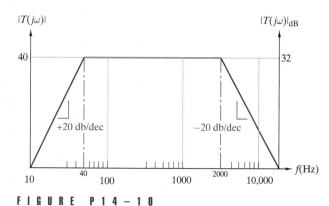

FIGURE P14–10

14–11 **D** The high-pass straight-line gain response in Figure P14–11 provides a two-pole stopband roll off at low frequencies.

(a) Construct a transfer function *T(s)* that has this gain response using only real poles and zeros.

(b) Design an active *RC* circuit to realize the *T(s)* found in (a) using only one OP AMP. Scale the circuit so that the element values are in practical ranges.

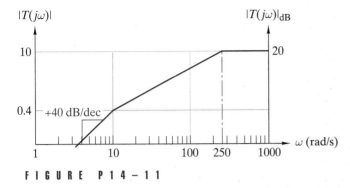

FIGURE P14-11

14-12 **D** The straight-line gain response in Figure P14-12 is the design specification for an audio preamplifier. Note that frequencies are specified in Hz.
  (a) Construct a transfer function $T(s)$ that has this gain response using only real poles and zeros.
  (b) Design an active RC circuit to realize the $T(s)$ found in (a) using only one OP AMP. Scale the circuit so that the element values are in practical ranges.

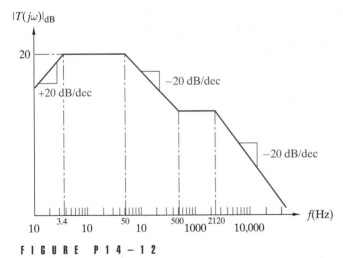

FIGURE P14-12

14-13 **D** A first-order low-pass power amplifier has a dc gain of 7 dB and a cutoff frequency of 2 kHz. Design an active RC preamp to connect in cascade with the power amplifier so that the overall circuit has a first-order low-pass response with a dc gain of 20 dB and a cutoff frequency of 5 kHz. Scale the circuit so that the element values are in practical ranges.

14-14 **D** A first-order low-pass amplifier has a dc gain of 10 dB and a cutoff frequency of 10 kHz. Design an active RC preamp to connect in cascade with the amplifier so that the resulting combination has a bandpass characteristic with cutoff frequencies at 10 Hz and 10 kHz and a passband gain of 20 dB. Scale the circuit so that the element values are in practical ranges.

14-15 **D** Your boss asks you to design a cost-effective RC circuit that realizes the transfer function given below. You are cautioned that here "cost effective" means using as few devices as possible and using standard values wherever possible. Also, your boss would really like you to use some of those 5.6-nF capacitors left over from your last project. There is an extra bonus if you can design the circuit without using any OP AMPs.

$$T(s) = \pm \frac{K\,s}{(s + 10^2)(s + 10^5)}$$

ERO 14-2 DESIGN WITH SECOND-ORDER CIRCUITS (SECT. 14-3)

(a) Given a circuit that realizes a second-order transfer function, develop a method of selecting circuit parameters to achieve specified filter characteristics.
(b) Given a filter specification,
    (1) Develop a second-order transfer function that has the required characteristics.
    (2) Design an active RC circuit that realizes the transfer function.

See Examples 14-4, 14-5, 14-6 and Exercises 14-3, 14-4

Design second-order active RC circuits to meet the following requirements:

Problem		Type	ω_0(rad/s)	ζ	Constraints
14-16	**D**	Low pass	2000	0.5	Use 10-kΩ resistors.
14-17	**D**	Low pass	2000	0.25	Gain of 20 dB at dc.
14-18	**D**	Low pass	2000	?	Cutoff frequency at 2 krad/s.
14-19	**D**	High pass	2000	?	20 dB gain at corner frequency.
14-20	**D**	High pass	200	0.5	Use 200-nF capacitors.
14-21	**D**	High pass	1000	0.75	High-frequency gain of 40 dB.
14-22	**D**	Bandpass	5000	?	$Q = 5$.
14-23	**D**	Bandpass	5000	?	Bandwidth of 400 rad/s.

14–24 🅳 A digital data system transmits a tone at 950 Hz to indicate a zero and 1050 Hz to indicate a one. A second-order bandpass filter is needed to eliminate unwanted frequencies above and below the two tones. Using the tone frequencies as the cutoff frequencies, what values of ω_0 and ζ are needed? Design an active RC circuit that meets the design requirements.

14–25 🅳 A second-order filter is needed with a center frequency at 5 kHz and a one-octave bandwidth, i.e., $f_{C2} = 2f_{C1}$. What value of ζ is required? Design an active RC circuit that meets the design requirements.

14–26 🅳 Design a second-order low-pass filter to meet the following requirements: (1) 20 dB passband gain, (2) high-frequency slope = –40 dB/dec, and (3) cutoff frequency at 1 kHz.

14–27 🅳 Design a second-order high-pass filter to meet the following requirements: (1) 20 dB passband gain, (2) low-frequency slope = +40 dB/dec, and (3) cutoff frequency at 1 kHz.

14–28 Another approach to designing the low-pass circuit in Figure 14–14 is to specify the conditions $\mu = 2$ and $R_1C_1 = R_2C_2$. Develop a method of selecting R_1, R_2, C_1, and C_2 to achieve specified values of ω_0 and ζ using this approach.

14–29 Figure P14–29 shows an active RC low-pass circuit with two voltage followers. Show that the circuit has a transfer function

$$T(s) = \frac{V_2(s)}{V_1(s)} = \frac{1}{R^2C_1C_2s^2 + RC_2s + 1}$$

Develop a method of selecting R, C_1, and C_2 to achieve specified values of ω_0 and ζ.

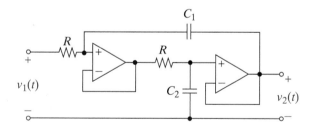

FIGURE P14–29

14–30 Figure P14–30 shows an active RC high-pass circuit with two voltage followers. Show that the circuit has a transfer function

$$T(s) = \frac{V_2(s)}{V_1(s)} = \frac{R^2C_1C_2s^2}{R^2C_1C_2s^2 + RC_1s + 1}$$

Develop a method of selecting R, C_1, and C_2 to achieve specified values of ω_0 and ζ.

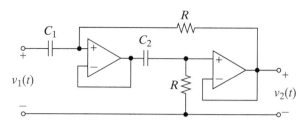

FIGURE P14–30

ERO 14–3 LOW-PASS FILTER DESIGN (SECT. 14–4)

Given a low-pass filter specification:

(a) Construct a transfer function $T(s)$ that meets the specification using first-order cascade, Butterworth, or Chebychev poles.

(b) Design a cascade of first- and second-order circuits that realizes $T(s)$.

See Examples 14–7, 14–8, 14–9 and Exercise 14–5

14–31 Construct a transfer function with a first-order cascade low-pass gain response for $n = 4$, $\omega_C = 1000$ rad/s, and a dc gain of 20 dB. Find the gain (in dB) at $2\omega_C$, $5\omega_C$, and $10\omega_C$. Plot the straight-line gain response on the range $0.1\omega_C \le \omega \le 10\omega_C$ and sketch the actual response.

14–32 Construct a transfer function with a third-order Butterworth low-pass response with $\omega_C = 500$ rad/s and a passband gain of 20 dB. Find the gain (in dB) at $2\omega_C$, $5\omega_C$, and $10\omega_C$. Plot the straight-line gain response on the range $0.1\omega_C \le \omega \le 10\omega_C$ and sketch the actual response.

14–33 Construct a transfer function with a fourth-order Chebychev low-pass response with $\omega_C = 500$ rad/s and a passband gain of 0 dB. Find the gain (in dB) at $2\omega_C$, $5\omega_C$, and $10\omega_C$. Plot the straight-line gain response on the range $0.1\omega_C \le \omega \le 10\omega_C$ and sketch the actual response.

14–34 🅳 A low-pass filter specification requires $\omega_C = 2$ krad/s, a passband gain of 0 dB, and a stopband gain less than –30 dB at 10 krad/s.

(a) Find the lowest order (n) of a first-order cascade response that meets these requirements.

(b) Repeat (a) using a Butterworth response and a Chebychev response.

14–35 🅳 Design a fourth-order low-pass filter with $\omega_C = 1500$ rad/s and a passband gain of 10 dB using first-order cascade poles.

14–36 🅳 Design a third-order Butterworth low-pass filter with $\omega_C = 300$ rad/s and a passband gain of 10 dB.

14–37 🅳 Design a third-order Chebychev low-pass filter with $\omega_C = 8$ krad/s and a passband gain of 10.

14–38 D A low-pass filter is needed to suppress the harmonics in a periodic waveform with $f_0 = 1$ kHz. The filter must have unity passband gain, less than –25 dB gain at the 3rd harmonic, and less than –40 dB gain at the 5th harmonic. Design a Butterworth filter that meets these requirements.

14–39 D Design a low-pass filter with 6 dB passband gain, a cutoff frequency of 3.2 kHz, and stopband gains less than –20 dB at 6.4 kHz and –40 dB at 12.8 kHz. Calculate the gains realized by your design at 3.2 kHz, 6.4 kHz, and 12.8 kHz.

14–40 D A pesky signal at 200 kHz is interfering with a desired signal at 25 kHz. A careful analysis suggests that reducing the interfering signal by 30 dB will eliminate the problem, provided the desired signal is not reduced by more than 3 dB. Design an active RC filter that meets these requirements.

14–41 D A low-pass filter is required with a cutoff frequency of 500 Hz and a steep roll off in the stopband. A summer intern working on this problem claims the transfer function

$$T(s) = \frac{7.751 \times 10^6}{[(s + 468.2)^2 + 2839^2](s + 936.2)}$$

is a Chebychev function with a really cool roll off. Is this a Chebychev low pass?

14–42 D The circuit in Figure P14–42 has been designed as a third-order Butterworth low-pass filter with a cutoff frequency of 2000 rad/s. Verify the circuit design.

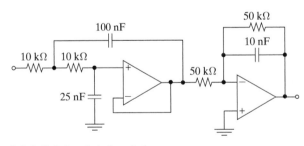

F I G U R E   P 1 4 – 4 2

ERO 14–4 HIGH-PASS, BANDPASS AND BANDSTOP FILTER DESIGN (SECTS. 14–5, 14–6)

Given a high-pass, broadband bandpass, or bandstop filter specification:

(a) Construct a transfer function $T(s)$ that meets the specification using first-order cascade, Butterworth, or Chebychev poles.

(b) Design a cascade or parallel connection of first- and second-order circuits that realizes $T(s)$.

See Examples 14–10, 14–12, 14–13 and Exercises 14–6, 14–7, 14–8

14–43 D A high-pass filter specification requires $\omega_C = 20$ krad/s, a passband gain of 20 dB, and a stopband gain less than –20 dB at 5 krad/s.

(a) Construct a high-pass transfer function $T(s)$ that meets the specification using first-order cascade poles.

(b) Repeat (a) using Butterworth and Chebychev poles.

(c) Which of these responses would you use and why?

14–44 D A high-pass filter specification requires $\omega_C = 25$ krad/s, a passband gain of 0 dB, a stopband gain less than –20 dB at 10 krad/s, and a stopband gain less than –40 dB at 5 krad/s.

(a) Construct a high-pass transfer function $T(s)$ that meets the specification using Chebychev poles.

(b) Calculate the actual stopband gain of $T(s)$ at $\omega = 5$ krad/s, 10 krad/s, and 25 krad/s. Which stopband requirement determines the number of poles in $T(s)$?

14–45 D A bandpass filter specification requires cutoff frequencies at 100 rad/s and 3 krad/s, a passband gain of 0 dB, and stopband gains of –40 dB or less at 10 rad/s and at 40 krad/s.

(a) Construct a transfer function $T(s)$ that meets the specification.

(b) Calculate the actual gain of $T(s)$ at the cutoff frequencies and the stopband gains at $\omega = 10$ rad/s and 40 krad/s.

14–46 D A bandpass filter specification requires cutoff frequencies at 100 rad/s and 2 krad/s, a passband gain of 0 dB, and stopband gains less than –20 dB at 25 krad/s and 8 krad/s.

(a) Construct a bandpass transfer function $T(s)$ that meets the specification using Butterworth poles.

(b) Repeat (a) using Chebychev poles.

(c) Calculate the stopband gain at 8 and 25 krad/s for the transfer functions found in (a) and (b).

(d) Which one of these responses would you choose and why?

14–47 D Design a third-order Butterworth high-pass filter with $\omega_C = 2$ krad/s and a passband gain of 17 dB.

14–48 D Design a Butterworth high-pass filter with a cutoff frequency of 2 kHz, 0 dB passband gain, and a gain no greater than –25 dB at 400 Hz.

14–49 D Design a fourth-order Butterworth bandpass filter with unity passband gain and cutoff frequencies at 400 and 2500 rad/s.

14–50 D A bandstop filter specification requires cutoff frequencies at 3 krad/s and 40 krad/s, passband gains of 0 dB, and a stopband gain less than –15 dB at 10 krad/s.

(a) Construct a transfer function $T(s)$ using a second-order Butterworth low-pass and a second-order Butterworth high-pass.

(b) Calculate the actual gain of $T(s)$ at $\omega = 3$, 10, and 40 krad/s.

INTEGRATING PROBLEMS

14–51   D EEG SIGNAL PROCESSING

Electroencephalography (EEG) measures electrical signals generated by the human brain. For diagnostic purposes, the useful part of the spectrum of these signals generally falls in a frequency range below 30 Hz. This range is sometimes divided into four bands called delta (0-4 Hz), theta (5-7 Hz), alpha (8-13 Hz), and beta (14-30 Hz). Each band is an indicator of different states of awareness and consciousness. Reduced or elevated spectral content in one or more of these bands provides a diagnostic indicator to a trained clinician. The purpose of this problem is to design a low-pass filter to separate EEG signals in the delta band from the other bands. The filter must have a cutoff frequency of 4 Hz, unity passband gain, and a gain no greater than −10 dB in any of the other bands. Calculate the gains realized by your design at the center frequency of the other three bands.

14–52   D MODIFYING AN EXISTING DESIGN

An existing digital data channel uses a second-order Butterworth low-pass filter with a cutoff frequency of 10 kHz. Field experience reveals that noise is causing the equipment performance to fall below advertised levels. Analysis by the engineering department suggests that decreasing the filter gain by 20 dB at 30 kHz will solve the problem, provided the cutoff frequency does not change. The manufacturing department reports that adding anything more than one second-order stage will cause a major redesign and an unacceptable slip in scheduled deliveries. Design a second-order low-pass filter to connect in cascade with the existing second-order Butterworth filter. The frequency response of the circuit consisting of your second-order circuit in cascade with the original second-order Butterworth circuit must have a cutoff frequency of 10 kHz and reduce the gain at 30 kHz by at least 20 dB.

14–53   A E SECOND-ORDER CASCADE RESPONSE

The transfer function of a second-order Butterworth low-pass filter is

$$T_2(s) = \frac{K}{\left(\dfrac{s}{\omega_0}\right)^2 + \sqrt{2}\left(\dfrac{s}{\omega_0}\right) + 1}$$

where ω_0 is the cutoff frequency of the second-order response. A second-order cascade response is obtained by a cascade connection of n stages each with the transfer function given above. The overall transfer function of the n-stage cascade is $T_2(s) \times T_2(s) \times \ldots T_2(s) = [T_2(s)]^n$.

(a) Derive an expression relating the cutoff frequency (ω_C) of an n-stage second-order cascade to the cutoff frequency (ω_0) of the individual stages. *Hint*: Review the derivation of the first-order cascade response.

(b) Derive an expression relating the corner frequency (ω_{CORNER}) to the cutoff frequency (ω_C) of an n-stage second-order cascade.

(c) Plot the normalized gain response of a two-stage second-order cascade response on the range $0.1\omega_C \leq \omega \leq 10\omega_C$. Plot the normalized gain response of a first-order cascade response with the same number of poles. Which response produces the most stopband attenuation and why?

(d) Discuss the advantages of the second-order cascade over other low-pass responses such as Butterworth or Chebychev.

14–54   D SEISMIC TRANSDUCER EQUALIZER

A seismic transducer converts mechanical motion into an electrical signal. At low frequency these transducers behave as a first-order high-pass filter of the form

$$T(s) = \frac{K_T s}{s + \omega_C}$$

where K_T is the transducer sensitivity. In some applications, signals of interest have the frequency components below ω_C. In such cases an equalizer is used to extend the measurements range below ω_C without changing the transducer sensitivity. The equalizer and transducer are connected in cascade, with transducer output as the equalizer input. The overall transfer function of the combination has the form

$$T(s)G(s) = \frac{K_T s}{s + \lambda \omega_C}$$

where $G(s)$ is the equalizer transfer function and λ is less than one. Suppose that the measurement range of a transducer with a cutoff frequency of 0.1 Hz is to be extended down to 0.01 Hz.

(a) Determine the required equalizer transfer function $G(s)$.

(b) Design an active RC circuit that implements the equalizer transfer function.

14–55 A D E THIRD-ORDER BUTTERWORTH CIRCUIT

(a) **A** Show that the circuit in Figure P14–55 produces a third-order Butterworth low-pass filter with a cutoff frequency of $\omega_C = 1/RC$ and a pass-band gain of $K = 4$.

(b) **D** Use the cascade design method in Sec. 14–4 to produce a two-stage circuit with the same third-order Butterworth response.

(c) **E** Rank order the two circuits using each of the following criteria: (1) smallest number of components required, (2) smallest OP AMP gain bandwidth required, (3) smallest number of different component values required, and (4) ease of fine tuning the cutoff frequency in production.

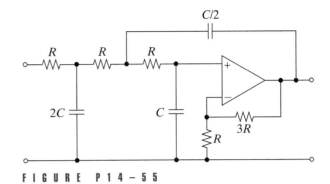

FIGURE P14–55

CHAPTER 15

SINUSOIDAL STEADY-STATE RESPONSE

The vector diagram of sine waves gives the best insight into the mutual relationships of alternating currents and emf's.

Charles P. Steinmetz, 1893,
American Engineer

15–1 Sinusoids and Phasors

15–2 Phasor Circuit Analysis

15–3 Basic Circuit Analysis with Phasors

15–4 Circuit Theorems with Phasors

15–5 General Circuit Analysis with Phasors

15–6 Energy and Power

Summary

Problems

Integrating Problems

In this chapter we introduce the vector representation of sinusoids as a tool for finding the response of a linear circuit. The vector model was first discussed in detail by Charles Steinmetz (1865–1923) at the International Electric Congress of 1893. Although there is some evidence of earlier use by Oliver Heaviside, Steinmetz is generally credited with popularizing the vector approach by demonstrating its many applications to alternating current devices and systems. By the turn of the century, the concept was well established in engineering practice and education. In the 1950s the Steinmetz vector representation of sinusoids came to be called *phasors* to avoid possible confusion with the space vectors used to describe electromagnetic fields.

The response of a linear circuit consists of a natural component and a forced component. In a stable circuit the natural response eventually vanishes, leaving a sustained response caused by the input driving force. When the input is a sinusoid, the forced response is a sinusoid, with the same frequency as the input but with a different amplitude and phase angle. In electrical engineering the forced sinusoidal component remaining after the natural component disappears is called the **sinusoidal steady-state response**. The sinusoidal steady-state response is also called the **ac response** since the driving force is an alternating current signal.

The phasor defined in the first section of this chapter is the key concept in sinusoidal steady-state analysis. The second section examines the form taken by the device and connection constraints when sinusoidal voltages and currents are represented by phasors. It turns out that the device and connection constraints have the same form as *s*-domain circuits. As a result, we can analyze circuits in the sinusoidal steady state using the analysis tools developed in previous chapters. The basic circuit analysis techniques, such as series equivalence and voltage division, are treated in the third section. The phasor domain versions of circuit theorems, such as superposition and Thévenin's theorem, are treated in the fourth section. The fifth section treats general phasor analysis methods using node voltages and mesh currents. The last section treats power transfer, an important application of phasor circuit analysis.

15-1 SINUSOIDS AND PHASORS

The phasor concept is the foundation for the analysis of linear circuits in the sinusoidal steady state. Simply put, a **phasor** is a complex number representing the amplitude and phase angle of a sinusoidal voltage or current. The connection between sinewaves and complex numbers is provided by Euler's relationship:

$$e^{j\theta} = \cos \theta + j \sin \theta \qquad (15\text{--}1)$$

Equation (15–1) relates the sine and cosine functions to the complex exponential $e^{j\theta}$. To develop the phasor concept, it is necessary to adopt the point of view that the cosine and sine functions can be written in the form

$$\cos \theta = \text{Re}\{e^{j\theta}\} \qquad (15\text{--}2)$$

and

$$\sin \theta = \text{Im}\{e^{j\theta}\} \qquad (15\text{--}3)$$

where Re stands for the "real part of" and Im for the "imaginary part of." Development of the phasor concept can begin with either Eq. (15–2) or (15–3). The choice between the two involves deciding whether to describe the eternal sinewave using a sine or cosine function. In Chapter 5 we chose the cosine, so we will reference phasors to the cosine function.

When Eq. (15–2) is applied to the general sinusoid defined in Chapter 5, we obtain

$$v(t) = V_A \cos(\omega t + \phi)$$
$$= V_A \text{Re}\{e^{j(\omega t + \phi)}\} = V_A \text{Re}\{e^{j\omega t} e^{j\phi}\} \quad (15\text{–}4)$$
$$= \text{Re}\{(V_A e^{j\phi}) e^{j\omega t}\}$$

In the last line of Eq. (15–4), moving the amplitude V_A inside the real part operation does not change the final result because it is a real constant.

By definition, the quantity $V_A e^{j\phi}$ in the last line of Eq. (15–4) is the **phasor representation** of the sinusoid $v(t)$. The phasor **V** is written as

$$\mathbf{V} = V_A e^{j\phi} = V_A \cos\phi + j V_A \sin\phi \quad (15\text{–}5)$$

Note that **V** is a complex number determined by the amplitude and phase angle of the sinusoid. Figure 15–1 shows a graphical representation commonly called a phasor diagram.

The phasor is a complex number that can be written in either polar or rectangular form. An alternative way to write the polar form is to replace the exponential $e^{j\phi}$ by the shorthand notation $\angle\phi$. In subsequent discussions, we will often express phasors as $\mathbf{V} = V_A \angle\phi$, which is equivalent to the polar form in Eq. (15–5).

Two features of the phasor concept need emphasis:

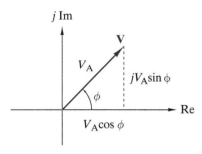

FIGURE 15–1 *Phasor diagram.*

1. Phasors are written in boldface type like **V** or $\mathbf{I}_1$ to distinguish them from signal waveforms such as $v(t)$ and $i_1(\text{t})$.

2. A phasor is determined by amplitude and phase angle and does not contain any information about the frequency of the sinusoid.

The first feature points out that signals can be described in different ways. Although the phasor **V** and waveform $v(t)$ are related concepts, they have different physical interpretations and our notation must clearly distinguish between them. The absence of frequency information in the phasors results from the fact that in the sinusoidal steady state, all currents and voltages are sinusoids with the same frequency. Carrying frequency information in the phasor would be redundant, since it is the same for all phasors in any given steady-state circuit problem.

In summary, given a sinusoidal signal $v(t) = V_A \cos(\omega t + \phi)$, the corresponding phasor representation is $\mathbf{V} = V_A e^{j\phi}$. Conversely, given the phasor $\mathbf{V} = V_A e^{j\phi}$, the corresponding sinusoid is found by multiplying the phasor by $e^{j\omega t}$ and reversing the steps in Eq. (15–4) as follows:

$$v(t) = \text{Re}\{\mathbf{V} e^{j\omega t}\} = \text{Re}\{(V_A e^{j\phi}) e^{j\omega t}\}$$
$$= V_A \text{Re}\{e^{j(\omega t + \phi)}\} = V_A \text{Re}\{\cos(\omega t + \phi) + j\sin(\omega t + \phi)\} \quad (15\text{–}6)$$
$$= V_A \cos(\omega t + \phi)$$

The frequency ω in the complex exponential $\mathbf{V}e^{j\omega t}$ in Eq. (15–6) must be expressed or implied in a problem statement, since by definition it is not contained in the phasor. Figure 15–2 shows a geometric interpretation of the complex exponential $\mathbf{V}e^{j\omega t}$ as a vector in the complex plane of length V_A, which rotates counterclockwise with a constant angular velocity of ω. The real part operation projects the rotating vector onto the horizontal (real) axis and thereby generates $v(t) = V_A \cos(\omega t + \phi)$. The complex exponential is sometimes called a **rotating phasor**, and the phasor $\mathbf{V}$ is viewed as a snapshot of the situation at $t = 0$.

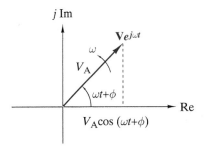

FIGURE   15 – 2 *Complex exponential $Ve^{j\omega t}$.*

PROPERTIES OF PHASORS

The utility of the phasor concept is based on its additive and network function properties. The **additive property** states that the phasor representing the sum sinusoidal waveforms of the same frequency is found by adding the phasors representing the individual sinusoids. To establish this property, we write the sum of sinusoids as

$$v(t) = v_1(t) + v_2(t) + \ldots + v_N(t)$$
$$= \text{Re}\{\mathbf{V}_1 e^{j\omega t}\} + \text{Re}\{\mathbf{V}_2 e^{j\omega t}\} + \ldots + \text{Re}\{\mathbf{V}_N e^{j\omega t}\} \qquad (15\text{–}7)$$

where $v_1(t), v_2(t), \ldots$ and $v_N(t)$ are sinusoids of the same frequency whose phasor representations are $\mathbf{V}_1, \mathbf{V}_2, \ldots$ and $\mathbf{V}_N$. The real part operation is additive, so the sum of real parts equals the real part of the sum. Consequently, Eq. (15–7) can be written in the form

$$v(t) = \text{Re}\{\mathbf{V}_1 e^{j\omega t} + \mathbf{V}_2 e^{j\omega t} + \ldots + \mathbf{V}_N e^{j\omega t}\}$$
$$= \text{Re}\{(\mathbf{V}_1 + \mathbf{V}_2 + \ldots + \mathbf{V}_N)e^{j\omega t}\} \qquad (15\text{–}8)$$

Comparing the last line in Eq. (15–8) with the definition of a phasor, we conclude that the phasor $\mathbf{V}$ representing $v(t)$ is

$$\mathbf{V} = \mathbf{V}_1 + \mathbf{V}_2 + \ldots + \mathbf{V}_N \qquad (15\text{–}9)$$

The result in Eq. (15–9) applies only if the component sinusoids all have the same frequency so that $e^{j\omega t}$ can be factored out as shown in the last line in Eq. (15–8).

The **network function property** of phasors provides a way to find phasor responses using s-domain network functions. Specifically, given an input phasor $\mathbf{V}_1$ and a network function $T(s) = V_2(s)/V_1(s)$, the output phasor is found as $\mathbf{V}_2 = T(j\omega)\mathbf{V}_1$. To establish this property we use network function properties from Chapter 11. Given a sinusoidal input of the form

$$v_1(t) = V_A \cos(\omega t + \phi)$$

The steady-state output is found as

$$v_{2SS}(t) = V_A|T(j\omega)| \cos(\omega t + \phi + \theta)$$

where $T(j\omega) = |T(j\omega)|e^{j\theta}$ is the network function $T(s)$ evaluated at $s = j\omega$. Using Euler's relationship we can rewrite the steady-state output in the form

$$v_{2SS}(t) = \text{Re}\{V_A|T(j\omega)| \, e^{j(\omega t + \phi + \theta)}\}$$
$$= \text{Re}\{(\underbrace{|T(j\omega)| \, e^{j\theta}}_{T(j\omega)})(\underbrace{V_A \, e^{j\phi}}_{\mathbf{V}_1})e^{j\omega t}\}$$

Using the identifications in the last line yields

$$v_{2SS}(t) = \text{Re}\{[T(j\omega)\mathbf{V}_1]e^{j\omega t}\} \qquad (15\text{--}10)$$

We can also write the steady-state response in terms of its phasor as

$$v_{2SS}(t) = \text{Re}\{\mathbf{V}_2\, e^{j\omega t}\}$$

where $\mathbf{V}_2$ is the phasor representation $v_{2SS}(t)$. Comparing this expression with Eq. (15–10) we conclude that

$$\mathbf{V}_2 = T(j\omega)\,\mathbf{V}_1 \qquad (15\text{--}11)$$

That is, the input and output phasors are related by the network function $T(s)$ evaluated at $s = j\omega$, where ω is the frequency of the input sinusoid.

In summary, the additive property indicates that adding phasors is equivalent to adding sinusoidal waveforms of the same frequency. The network function property allows us to use the s-domain concept of a network function to find the phasors corresponding to the sinusoidal steady-state responses. The following examples show applications of these two properties.

EXAMPLE 15–1

(a) Construct the phasors for the following signals:

$$v_1(t) = 10 \cos(1000t - 45°) \text{ V}$$

$$v_2(t) = 5 \cos(1000t + 30°) \text{ V}$$

(b) Use the additive property of phasors and the phasors found in (a) to find $v(t) = v_1(t) + v_2(t)$.

SOLUTION:

(a) The phasor representations of $v_1(t)$ and $v_2(t)$ are

$$\mathbf{V}_1 = 10e^{-j45°} = 10 \cos(-45°) + j10 \sin(-45°)$$

$$= 7.07 - j7.07$$

$$\mathbf{V}_2 = 5\, e^{+j30°} = 5 \cos(30°) + j5 \sin(30°)$$

$$= 4.33 + j2.5$$

(b) The two sinusoids have the same frequency so the additive property of phasors can be used to obtain their sum:

$$\mathbf{V} = \mathbf{V}_1 + \mathbf{V}_2 = 11.4 - j4.57 = 12.3e^{-j21.8°}$$

The waveform corresponding to this phasor sum is

$$v(t) = \text{Re}\{(12.3e^{-j21.8°})e^{j1000t}\}$$

$$= 12.3 \cos(1000t - 21.8°) \text{ V}$$

The phasor diagram in Figure 15–3 shows that summing sinusoids can be viewed geometrically in terms of phasors. ∎

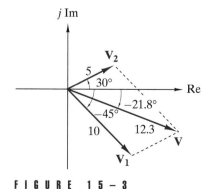

FIGURE 15 – 3

EXAMPLE 15–2

(a) Construct the phasors representing the following signals:

$$i_A(t) = 5 \cos(377t + 50°) \text{ A}$$

$$i_B(t) = 5 \cos(377t + 170°) \text{ A}$$

$$i_C(t) = 5 \cos(377t - 70°) \text{ A}$$

(b) Use the additive property of phasors and the phasors found in (a) to find the sum of these waveforms.

SOLUTION:

(a) The phasor representation of the three sinusoidal currents are

$$\mathbf{I}_A = 5e^{j50°} = 5 \cos(50°) + j5 \sin(50°) = 3.21 + j3.83 \text{ A}$$

$$\mathbf{I}_B = 5e^{j170°} = 5 \cos(170°) + j5 \sin(170°) = -4.92 + j0.87 \text{ A}$$

$$\mathbf{I}_C = 5e^{-j70} = 5 \cos(-70°) + j5 \sin(-70°) = 1.71 - j4.70 \text{ A}$$

(b) The currents have the same frequency, so the additive property of phasors applies. The phasor representing the sum of these currents is

$$\mathbf{I}_A + \mathbf{I}_B + \mathbf{I}_C = (3.21 - 4.92 + 1.71) + j(3.83 + 0.87 - 4.70)$$

$$= 0 + j0 \text{ A}$$

It is not obvious by examining the waveforms that these three currents add to zero. However, the phasor diagram in Figure 15–4 makes this fact clear, since the sum of any two phasors is equal and opposite to the third. Phasors of this type occur in balanced three-phase power systems, which we study in Chapter 17. The balanced condition occurs when three equal-amplitude phasors are displaced in phase by exactly 120°. ∎

EXAMPLE 15–3

A linear circuit has a transfer function $T(s) = V_2(s)/V_1(s) = s/(s + 100)$. Find the input and output phasors for $v_1 = 10 \cos(200t + 30°)$.

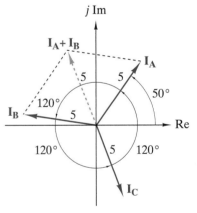

FIGURE   15–4

SOLUTION:

The phasor for the input sinusoid is $\mathbf{V}_1 = 10\angle30°$. Evaluating $T(s)$ at $s = j200$ yields

$$T(j200) = \frac{j200}{j200 + 100} = \frac{200 \angle 90°}{224 \angle 63.4°} = 0.893 \angle 26.6°$$

Using the network function property the phasor output is

$$\mathbf{V}_2 = \mathbf{V}_1 T(j200) = (10 \, e^{j30°}) \times (0.894 \, e^{j26.6°})$$

$$= 8.94 \, e^{j56.6°}$$

The sinusoidal waveform corresponding to the phasor $\mathbf{V}_2$ is

$$v_2(t) = \text{Re} \{(8.94 \, e^{j56.6°})e^{j200t}\} = 8.94 \cos(200 \, t + 56.6°) \quad ∎$$

EXAMPLE 15-4

(a) Convert the following phasors into sinusoidal waveforms.

$$\mathbf{V}_1 = 20 + j20 \text{ V}, \quad \omega = 500 \text{ rad/s}$$

$$\mathbf{V}_2 = 10\sqrt{2}\, e^{-j45°} \text{ V}, \quad \omega = 500 \text{ rad/s}$$

(b) Use phasors addition to find the sinusoidal waveform $v_3(t) = v_1(t) + v_2(t)$.

SOLUTION:

(a) Since $\mathbf{V}_1 = 20 + j20 = 20\sqrt{2}\, e^{j45°}$, the waveforms corresponding to the phasors $\mathbf{V}_1$ and $\mathbf{V}_2$ are

$$v_1(t) = \text{Re}\,\{20\sqrt{2}\, e^{j45°}\, e^{j500t}\} = 20\sqrt{2}\cos(500\,t + 45°) \quad \text{V}$$

$$v_2(t) = \text{Re}\,\{10\sqrt{2}\, e^{-j45°}\, e^{j500t}\} = 10\sqrt{2}\cos(500\,t - 45°) \quad \text{V}$$

(b) Since $\mathbf{V}_2 = 10\sqrt{2}\, e^{-j45°} = 10 - j10$, the additive property of phasors yields

$$\mathbf{V}_3 = \mathbf{V}_1 + \mathbf{V}_2 = 20 + j20 + 10 - j10$$

$$= 30 + j10 = 31.6\, e^{j18.4°}$$

Hence,

$$v_3(t) = \text{Re}\,\{31.6\, e^{j18.4°}\, e^{j500t}\} = 31.6\cos(500\,t + 18.4°) \quad \text{V} \qquad ■$$

Exercise 15-1

Convert the following sinusoids to phasors in polar and rectangular form:

(a) $v(t) = 20\cos(150t - 60°)$ V
(b) $v(t) = 10\cos(1000t + 180°)$ V
(c) $i(t) = -4\cos 3t + 3\cos(3t - 90°)$ A

Answers:

(a) $\mathbf{V} = 20\angle -60° = 10 - j17.3$ V
(b) $\mathbf{V} = 10\angle 180° = -10 + j0$ V
(c) $\mathbf{I} = 5\angle -143° = -4 - j3$ A

Exercise 15-2

Convert the following phasors to sinusoids:

(a) $\mathbf{V} = 169\angle -45°$ V at $f = 60$ Hz
(b) $\mathbf{V} = 10\angle 90° + 66 - j10$ V at $\omega = 10$ krad/s
(c) $\mathbf{I} = 15 + j5 + 10\angle 180°$ mA at $\omega = 1000$ rad/s

Answers:

(a) $v(t) = 169\cos(377t - 45°)$ V
(b) $v(t) = 66\cos 10^4 t$ V
(c) $i(t) = 7.07\cos(1000t + 45°)$ mA

Exercise 15–3

Find the phasor corresponding to the waveform $v(t) = V_A \cos(\omega t) + 2V_A \sin(\omega t)$.

Answer:

$$\mathbf{V} = V_A - j2V_A$$

Exercise 15–4

The transfer function of a linear circuit is $T(s) = V_2(s)/V_1(s) = 500/(s + 200)$. Find the phasor output when the input is $v_1(t) = 20 \cos(50t)$.

Answer:

$$\mathbf{V}_2 = 47.06 - j11.76 = 48.51 \angle -14.04°$$

15–2 PHASOR CIRCUIT ANALYSIS

Phasor circuit analysis is a method of finding sinusoidal steady-state responses directly from the circuit without using differential equations. How do we perform phasor circuit analysis? At several points in our study we have seen that circuit analysis is based on two kinds of constraints: (1) connection constraints (Kirchhoff's laws), (2) device constraints (element equations). To analyze phasor circuits, we must see how these constraints are expressed in phasor form.

CONNECTION CONSTRAINTS IN PHASOR FORM

The sinusoidal steady-state condition is reached after the circuit's natural response decays to zero. In the steady state all of the voltages and currents are sinusoids with the same frequency as the driving force. Under these conditions, the application of KVL around a loop could take the form

$$V_1 \cos(\omega t + \phi_1) + V_2 \cos(\omega t + \phi_2) \ldots + V_N \cos(\omega t + \phi_N) = 0$$

These sinusoids have the same frequency but have different amplitudes and phase angles. The additive property of phasors discussed in the preceding section shows that there is a one-to-one correspondence between waveform sums and phasor sums. Therefore, if the sum of the waveforms is zero, then the corresponding phasors must also sum to zero.

$$\mathbf{V}_1 + \mathbf{V}_2 \ldots + \mathbf{V}_N = 0$$

Clearly the same result applies to phasor currents and KCL. In other words, we can state Kirchhoff's laws in phasor form as follows:

> *KVL: The algebraic sum of phasor voltages around a loop is zero.*

> *KCL: The algebraic sum of phasor currents at a node is zero.*

DEVICE CONSTRAINTS IN PHASOR FORM

In the *s* domain the device constraints for the three passive elements are expressed in terms of their impedances.

Resistor $\quad V_R(s) = Z_R(s)I_R(s) = R\,I_R(s)$

Inductor $\quad V_L(s) = Z_L(s)I_L(s) = Ls\,I_L(s)$ $\qquad$ (15–12a)

Capacitor $\quad V_C(s) = Z_C(s)I_C(s) = \dfrac{1}{Cs}\,I_C(s)$

The element impedances are network functions relating the current through to the voltage across the elements. Applying the network function property of phasors to the element impedances yields the device constraints in phasor form.

Resistor $\quad \mathbf{V}_R = R\,\mathbf{I}_R$

Inductor $\quad \mathbf{V}_L = j\,\omega\,L\,\mathbf{I}_L$ $\qquad$ (15–12b)

Capacitor $\quad \mathbf{V}_C = \dfrac{1}{j\,\omega\,C}\,\mathbf{I}_C$

The phasor form is derived from the *s*-domain form by simply replacing *s* replaced by $j\omega$ in the element impedances.

Although the *s*-domain and phasor-domain device equations look similar, they have quite different interpretations. The *s*-domain versions in Eq. (15–12a) relate voltage and current transforms and applies to signals in general. The phasor domain versions in Eq. (15–12b) relate complex numbers representing the amplitude and phase angles of sinusoidal steady-state voltages and currents.

To explore the phasor device constraints in more detail, we assume the current through a resistor is $i_R(t) = I_M \cos(\omega t + \phi)$. In phasor form this current is $\mathbf{I}_R = I_M e^{j\phi}$ and phasor voltage $\mathbf{V}_R$ is found by using Eq. (15–12b).

$$\mathbf{V}_R = R\mathbf{I}_R = RI_M e^{j\phi} \qquad (15\text{–}13)$$

The phasor diagram in Figure 15–5 shows that the phasors $\mathbf{V}_R$ and $\mathbf{I}_R$ have the same phase angle ϕ. Phasors with the same phase angle are said to be **in phase**. Phasors with unequal phase angles are said to be **out of phase**.

We next assume that the current through an inductor is $i_L(t) = I_M \cos(\omega t + \phi)$, in which case the phasor current is $\mathbf{I}_L = I_M e^{j\phi}$. The phasor voltage $\mathbf{V}_L$ is found by using Eq. (15–12b).

$$\mathbf{V}_L = j\,\omega\,L\,\mathbf{I}_L = \left(\omega\,L\,e^{j90°}\right)\left(I_M e^{j\phi}\right)$$
$$= \omega\,L\,I_M e^{j(\phi\,+\,90°)} \qquad (15\text{–}14)$$

The phasor diagram in Figure 15–6 shows that the phasors $\mathbf{V}_L$ and $\mathbf{I}_L$ are out of phase. The phase angle of phasor voltage is $\phi + 90°$, which shifts $\mathbf{V}_L$ ahead of $\mathbf{I}_L$ by 90°. In ac circuit terminology we say that the voltage phasor *leads* the current phasor by 90° or equivalently the current *lags* the voltage 90°.

Finally, we assume that the current through a capacitor is $i_C(t) = I_M \cos(\omega t + \phi)$, in which case the phasor current is $\mathbf{I}_C = I_M e^{j\phi}$. The phasor voltage $\mathbf{V}_C$ is found by using Eq. (15–12b).

$$\mathbf{V}_C = \frac{1}{j\omega C}\,\mathbf{I}_C = \left(\frac{1}{\omega C}\,e^{-j90°}\right)\left(I_M e^{j\phi}\right)$$
$$= \frac{I_M}{\omega C}\,e^{j(\phi\,-\,90°)} \qquad (15\text{–}15)$$

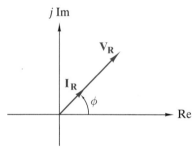

FIGURE 15–5 *Phasor i–v characteristics of the resistor.*

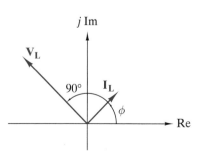

FIGURE 15–6 *Phasor i–v characteristics of the inductor.*

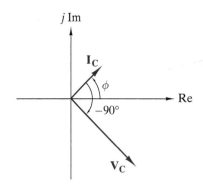

FIGURE 15 – 7 *Phasor i–v characteristics of the capacitor.*

The phasor diagram in Figure 15–7 shows that the phasors $\mathbf{V}_C$ and $\mathbf{I}_C$ are out of phase. The phase angle of phasor voltage is $\phi - 90°$, which shifts $\mathbf{V}_C$ behind $\mathbf{I}_C$ by 90°. In this case we say that the voltage phasor *lags* the current phasor by 90° or equivalently the current *leads* the voltage 90°.

PHASOR DOMAIN IMPEDANCE

The device constraints Eq. (15–12b) are all of the form

$$\mathbf{V} = Z\mathbf{I} \tag{15–16}$$

where Z is the phasor-domain impedance of the element. The phasor-domain impedances of the three passive elements are

$$\text{Resistor} \quad Z_R = R$$

$$\text{Inductor} \quad Z_L = j\omega L \tag{15–17}$$

$$\text{Capacitor} \quad Z_C = \frac{1}{j\omega C} = \frac{-j}{\omega C}$$

In general, phasor impedance can be defined in terms of the phasor ratio $Z = \mathbf{V}/\mathbf{I}$. From this it is clear that phasor impedance is a complex quantity whose units are ohms. Although impedance can be a complex number, it is not a phasor. Phasors describe the amplitudes and phase angles of sinusoidal voltages and currents, whereas impedances characterize devices and circuits in the sinusoidal steady state.

EXAMPLE 15–5

The circuit in Figure 15–8 is operating in the sinusoidal steady state with $i(t) = 4\cos(5000t)$ A. Find the steady-state voltage $v(t)$.

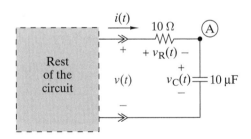

FIGURE 15 – 8

SOLUTION:
The sinusoidal frequency is $\omega = 5000$; hence, the resistor and the capacitor have impedances

$$Z_R = R = 10 + j0 \quad \text{and} \quad Z_C = \frac{1}{j\omega C} = \frac{1}{j5000 \times 10^{-5}} = 0 - j20$$

Applying KCL at node A shows that $i(t) = 4\cos(5000t)$ is the current through both the resistor and the capacitor. The corresponding phasor current is $\mathbf{I} = 4 + j0$. Using the element impedances, the phasor voltages across the resistor and the capacitor are

$$\mathbf{V_R} = Z_R\mathbf{I} = 10 \times (4 + j0) = 40 + j0$$

$$\mathbf{V_C} = Z_C\mathbf{I} = (-j20) \times (4 + j0) = 0 - j80$$

Applying KVL around the loop yields $\mathbf{V} = \mathbf{V_R} + \mathbf{V_C}$; hence,

$$\mathbf{V} = 40 - j80 = 89.4 \angle -63.4°$$

and the steady-state voltage waveform is

$$v(t) = \mathrm{Re}\ \{89.4\ e^{-j63.4°}e^{j5000t}\}$$

$$= 89.4 \cos(5000\ t - 63.4°)\quad \mathrm{V}$$

Exercise 15–5

The circuit in Figure 15–9 is operating in the sinusoidal steady state with $v(t) =$ 50 cos(500t) V and $i(t) = 4$ cos(500t – 60°) A. Find the impedance of the elements in the box.

Answer:

$$Z = 6.25 + j10.8\ \Omega$$

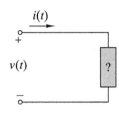

$i(t)$

$v(t)$

?

FIGURE 15–9

Exercise 15–6

The current through a 12-mH inductor is $i_L(t) = 20$ cos(10^6t) mA. Determine

(a) The impedance of the inductor
(b) The phasor voltage across the inductor
(c) The waveform of the voltage across the inductor.

Answers:
(a) $Z_L = j12\ k\Omega$
(b) $\mathbf{V_L} = 240\angle90°$ V
(c) $v_L = 240 \cos(10^6t + 90°)$ V

Exercise 15–7

The current through a 20-pF capacitor is $i_C(t) = 0.3$ cos(10^6t) mA.

(a) Find the impedance of the capacitor.
(b) Find the phasor voltage across the capacitor.
(c) Find the waveform of the voltage across the capacitor.

Answers:
(a) $Z_C = -j50\ k\Omega$
(b) $\mathbf{V_C} = 15\angle-90°$ V
(c) $v_C = 15 \cos(10^6t - 90°)$ V

15–3 BASIC CIRCUIT ANALYSIS WITH PHASORS

Functions of time like $v(t) = V_A \cos(\omega t + \phi_V)$ and $i(t) = I_A \cos(\omega t + \phi_I)$ are time-domain representations of sinusoidal signals. Producing the corresponding phasors can be thought of as a transformation that carries $v(t)$

and $i(t)$ into a complex-number domain where signals are represented as phasors **V** and **I**. We call this complex-number domain the **phasor domain**. When we analyze circuits in this phasor domain, we obtain sinusoidal steady-state responses in terms of phasors like **V** and **I**. Performing the inverse phasor transformation as $v(t) = \text{Re}\{\mathbf{V}e^{j\omega t}\}$ and $i(t) = \text{Re}\{\mathbf{I}e^{j\omega t}\}$ carries the responses back into the time domain. To perform ac circuit analysis in this way, we obviously need to develop methods of analyzing circuits in the phasor domain.

In the preceding section, we showed that KVL and KCL apply in the phasor domain and that the phasor element constraints all have the form $\mathbf{V} = Z\mathbf{I}$. These element and connection constraints have the same format as the underlying constraints for resistance and s-domain circuit analysis. Therefore, familiar algebraic circuit analysis tools, such as series and parallel equivalence, voltage and current division, proportionality and superposition, and Thévenin and Norton equivalent circuits, are applicable in the phasor domain. In other words, we do not need new analysis techniques to handle circuits in the phasor domain.

We can think of phasor domain circuit analysis in terms of the flow diagram in Figure 15–10. The analysis begins in the time domain with a linear circuit operating in the sinusoidal steady state and involves three major steps:

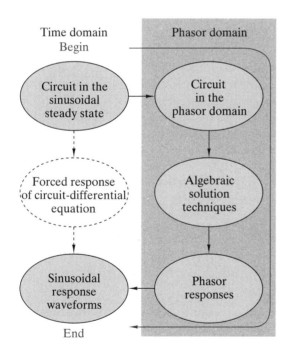

FIGURE 15 – 10 *Flow diagram for phasor circuit analysis.*

STEP 1. The circuit is transformed into the phasor domain by representing the input and response sinusoids as phasors and the passive circuit elements by their impedances.

STEP 2. Standard algebraic circuit analysis techniques are applied to solve the phasor domain circuit for the desired unknown phasor responses.

STEP 3. The phasor responses are inverse transformed back into time-domain sinusoids to obtain the response waveforms.

The third step assumes that the required end product is a time-domain waveform. However, a phasor is just another representation of a sinusoid. With some experience, we learn to think of the response as a phasor without converting it back into a time-domain waveform.

SERIES EQUIVALENCE AND VOLTAGE DIVISION

We begin the study of phasor-domain analysis with two basic analysis tools—series equivalence and voltage division. In Figure 15–11 the two-terminal elements are connected in series, so by KCL, the same phasor current $\mathbf{I}$ exists in impedances $Z_1, Z_2, \ldots Z_N$. Using KVL and the element constraints, the voltage across the series connection can be written as

$$
\begin{aligned}
\mathbf{V} &= \mathbf{V}_1 + \mathbf{V}_2 + \ldots + \mathbf{V}_N \\
&= Z_1\mathbf{I} + Z_2\mathbf{I} + \ldots + Z_N\mathbf{I} \\
&= (Z_1 + Z_2 + \ldots + Z_N)\mathbf{I}
\end{aligned}
\tag{15–18}
$$

The last line in this equation points out that the phasor responses $\mathbf{V}$ and $\mathbf{I}$ do not change when the series connected elements are replaced by an equivalent impedance:

$$
Z_{EQ} = Z_1 + Z_2 + \ldots + Z_N \tag{15–19}
$$

In general, the equivalent impedance Z_{EQ} is a complex quantity of the form

$$
Z_{EQ} = R + jX
$$

where R is the real part and X is the imaginary part. The real part of Z is called **resistance** and the imaginary part (X, not jX) is called **reactance**. Both resistance and reactance are expressed in ohms (Ω), and both can be functions of frequency (ω). For passive circuits, resistance is always positive, while reactance X can be either positive or negative. A positive X is called an **inductive** reactance because the reactance of an inductor is ωL, which is always positive. A negative X is called a **capacitive** reactance because the reactance of a capacitor is $-1/\omega C$, which is always negative.

Combining Eqs. (15–18) and (15–19), we can write the phasor voltage across the kth element in the series connection as

$$
\mathbf{V}_k = Z_k\mathbf{I} = \frac{Z_k}{Z_{EQ}}\mathbf{V} \tag{15–20}
$$

Equation (15–20) is the phasor version of the voltage division principle. The phasor voltage across any element in a series connection is equal to the ratio of its impedance to the equivalent impedance of the connection times the total phasor voltage across the connection.

FIGURE 15–11 *A series connection of impedances.*

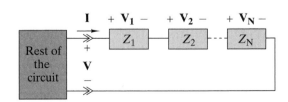

EXAMPLE 15–6

The circuit in Figure 15–12 is operating in the sinusoidal steady state with $v_S(t) = 35 \cos 1000t$ V.

(a) Transform the circuit into the phasor domain.
(b) Solve for the phasor current $\mathbf{I}$.
(c) Solve for the phasor voltage across each element.
(d) Construct the waveforms corresponding to the phasors found in (b) and (c).

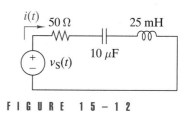

FIGURE   15 – 12

SOLUTION:
(a) The phasor representing the input source voltage is $\mathbf{V}_S = 35\angle 0°$. The impedances of the three passive elements are

$$Z_R = R = 50 \ \Omega$$

$$Z_L = j\omega L = j1000 \times 25 \times 10^{-3} = j25 \ \Omega$$

$$Z_C = \frac{1}{j\omega C} = \frac{1}{j1000 \times 10^{-5}} = -j100 \ \Omega$$

Using these results, we obtain the phasor-domain circuit in Figure 15–13.
(b) The equivalent impedance of the series connection is

$$Z_{EQ} = 50 + j25 - j100 = 50 - j75 = 90.1\angle -56.3° \ \Omega$$

The current in the series circuit is

$$\mathbf{I} = \frac{\mathbf{V}_S}{Z_{EQ}} = \frac{35\angle 0°}{90.1\angle -56.3°} = 0.388\angle 56.3° \ \text{A}$$

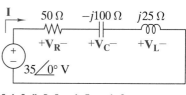

FIGURE   15 – 13

(c) The current $\mathbf{I}$ exists in all three series elements, so the voltage across each passive element is

$$\mathbf{V}_R = Z_R\mathbf{I} = 50 \times 0.388\angle 56.3° = 19.4\angle 56.3° \ \text{V}$$

$$\mathbf{V}_L = Z_L\mathbf{I} = j25 \times 0.388\angle 56.3° = 9.70\angle 146.3° \ \text{V}$$

$$\mathbf{V}_C = Z_C\mathbf{I} = -j100 \times 0.388\angle 56.3° = 38.8\angle -33.7° \ \text{V}$$

Note that the voltage across the resistor is in phase with the current, the voltage across the inductor leads the current by 90°, and the voltage across the capacitor lags the current by 90°.
(d) The sinusoidal steady-state waveforms corresponding to the phasors in (b) and (c) are

$$i(t) = \text{Re}\{0.388e^{j56.3°}e^{j1000t}\} = 0.388 \cos(1000t + 56.3°) \ \text{A}$$

$$v_R(t) = \text{Re}\{19.4e^{j56.3°}e^{j1000t}\} = 19.4 \cos(1000t + 56.3°) \ \text{V}$$

$$v_L(t) = \text{Re}\{9.70e^{j146.3°}e^{j1000t}\} = 9.70 \cos(1000t + 146.3°) \ \text{V}$$

$$v_C(t) = \text{Re}\{38.8e^{-j33.7°}e^{j1000t}\} = 38.8 \cos(1000t - 33.7°) \ \text{V} \qquad ■$$

D DESIGN EXAMPLE 15–7

Design the voltage divider in Figure 15–14 so that an input $v_S = 15 \cos 2000t$ V produces a steady-state output $v_O(t) = 2 \sin 2000t$ V.

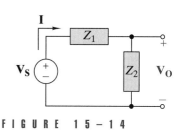

FIGURE   15 – 14

SOLUTION:

Using voltage division, we can relate the input and output phasor as follows:

$$\mathbf{V}_O = \frac{Z_2}{Z_1 + Z_2}\mathbf{V}_S$$

The phasor representation of the input voltage is $\mathbf{V}_S = 15\angle 0 = 15 + j0$. Using the identity $\cos(x - 90°) = \sin x$, we write the required output phasor as $\mathbf{V}_O = 2\angle{-90°} = 0 - j2$. The design problem is to select the impedances Z_1 and Z_2 so that

$$0 - j2 = \frac{Z_2}{Z_1 + Z_2}(15 + j0)$$

Solving this design constraint for Z_1 yields

$$Z_1 = \frac{15 + j2}{-j2}Z_2$$

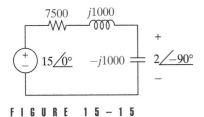

7500 j1000

$15\angle 0°$ $-j$1000 $2\angle{-90°}$

FIGURE 15–15

Evidently, we can choose Z_2 and then solve for Z_1. In making this choice, we must keep some physical realizability conditions in mind. In general, an impedance has the form $Z = R + jX$. The reactance X can be either positive (an inductor) or negative (a capacitor), but the resistance R must be positive. With these constraints in mind, we select $Z_2 = -j1000$ (a capacitor) and solve for $Z_1 = 7500 + j1000$ (a resistor in series with an inductor). Figure 15–15 shows the resulting phasor circuit. To find the values of L and C, we note that the input is $v_S = 15 \cos 2000t$ hence, the frequency is $\omega = 2000$. The inductive reactance $\omega L = 1000$ requires $L = 0.5$ H, while the capacitive reactance requires $-(\omega C)^{-1} = -1000$ or $C = 0.5$ μF. Other possible designs are obtained by selecting different values of Z_2. To be physically realizable, the selected value of Z_2 must produce $R \geq 0$ for Z_1 and Z_2. ∎

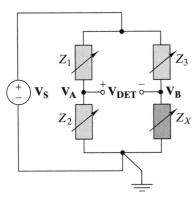

FIGURE 15–16 *Impedance bridge.*

APPLICATION NOTE: EXAMPLE 15–8

The purpose of the impedance bridge in Figure 15–16 is to measure the unknown impedance Z_X by adjusting known impedances Z_1, Z_2, and Z_3 until the detector voltage $\mathbf{V}_{DET}$ is zero. The circuit consists of a sinusoidal source $\mathbf{V}_S$ driving two voltage dividers connected in parallel. Using the voltage division principle, we find that the detector voltage is

$$\mathbf{V}_{DET} = \mathbf{V}_A - \mathbf{V}_B = \frac{Z_2}{Z_1 + Z_2}\mathbf{V}_S - \frac{Z_X}{Z_3 + Z_X}\mathbf{V}_S$$

$$= \left[\frac{Z_2 Z_3 - Z_1 Z_X}{(Z_1 + Z_2)(Z_3 + Z_X)}\right]\mathbf{V}_S$$

This equation shows that the detector voltage will be zero when $Z_2 Z_3 = Z_1 Z_X$. When the branch impedances are adjusted so that the detector voltage is zero, the unknown impedance can be written in terms of the known impedances as follows:

$$Z_X = R_X + jX_X = \frac{Z_2 Z_3}{Z_1}$$

This equation is called the bridge balance condition. Since the equality involves complex quantities, at least two of the known impedances must be adjustable to balance both the resistance R_X and the reactance X_X of the unknown impedance. In practice, bridges are designed assuming that the sign of the unknown reactance is known. Bridges that measure only positive reactance are called inductance bridges, while those that measure only negative reactance are called capacitance bridges.

The Maxwell inductance bridge in Figure 15–17 is used to measure the resistance R_X and inductance L_X of an inductive device by alternately adjusting resistances R_1 and R_2 to balance the bridge circuit. The impedances of the legs of this bridge are

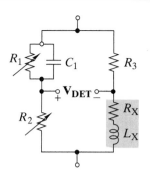

FIGURE   15 – 17 *Maxwell bridge.*

$$Z_1 = \frac{1}{j\omega C_1 + \dfrac{1}{R_1}}$$

$$Z_2 = R_2 \quad Z_3 = R_3$$

For the Maxwell bridge, the balance condition $Z_X = Z_2 Z_3 / Z_1$ yields

$$R_X + j\omega L_X = \frac{R_2 R_3}{R_1} + j\omega C_1 R_2 R_3$$

Equating the real and imaginary parts on each side of this equation yields the parameters of the unknown impedance in terms of the known impedances:

$$R_X = \frac{R_2 R_3}{R_1} \quad \text{and} \quad L_X = R_2 R_3 C_1$$

Note that adjusting R_1 only affects R_X. The Maxwell bridge measures inductance by balancing the positive reactance of an unknown inductive device with a calibrated fraction of negative reactance of the known capacitor C_1. If the reactance of the unknown device is actually capacitive (negative), then the Maxwell bridge cannot be balanced.

PARALLEL EQUIVALENCE AND CURRENT DIVISION

In Figure 15–18 the two-terminal elements are connected in parallel, so the same phasor voltage $\mathbf{V}$ appears across the impedances $Z_1, Z_2, \ldots Z_N$. Using the phasor element constraints, the current through each impedance is $\mathbf{I}_k = \mathbf{V}/Z_k$. Next, using KCL, the total current entering the parallel connection is

$$\mathbf{I} = \mathbf{I}_1 + \mathbf{I}_2 + \ldots + \mathbf{I}_N$$

$$= \frac{\mathbf{V}}{Z_1} + \frac{\mathbf{V}}{Z_2} + \ldots + \frac{\mathbf{V}}{Z_N} \tag{15–21}$$

$$= \left(\frac{1}{Z_1} + \frac{1}{Z_2} + \ldots + \frac{1}{Z_N} \right) \mathbf{V}$$

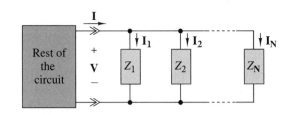

The same phasor responses **V** and **I** exist when the parallel connected elements are replaced by an equivalent impedance.

$$\frac{1}{Z_{EQ}} = \frac{\mathbf{I}}{\mathbf{V}} = \frac{1}{Z_1} + \frac{1}{Z_2} + \ldots + \frac{1}{Z_N} \qquad (15\text{–}22)$$

These results can also be written in terms of admittance Y, which is defined as the reciprocal of impedance:

$$Y = \frac{1}{Z} = G + jB$$

The real part of Y is called **conductance** and the imaginary part B is called **susceptance,** both of which are expressed in units of siemens (S).

Using admittances to rewrite Eq. (15–21) yields

$$\mathbf{I} = \mathbf{I}_1 + \mathbf{I}_2 + \ldots + \mathbf{I}_N$$
$$= Y_1\mathbf{V} + Y_2\mathbf{V} + \ldots + Y_N\mathbf{V} \qquad (15\text{–}23)$$
$$= (Y_1 + Y_2 + \ldots + Y_N)\mathbf{V}$$

Hence, the equivalent admittance of the parallel connection is

$$Y_{EQ} = \frac{\mathbf{I}}{\mathbf{V}} = Y_1 + Y_2 + \ldots + Y_N \qquad (15\text{–}24)$$

Combining Eqs. (15–23) and (15–24), we find that the phasor current through the kth element in the parallel connection is

$$\mathbf{I}_k = Y_k\mathbf{V} = \frac{Y_k}{Y_{EQ}}\mathbf{I} \qquad (15\text{–}25)$$

Equation (15–25) is the phasor version of the current division principle. The phasor current through any element in a parallel connection is equal to the ratio of its admittance to the equivalent admittance of the connection times the total phasor current entering the connection.

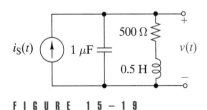

F I G U R E 1 5 – 1 9

EXAMPLE 15–9

The circuit in Figure 15–19 is operating in the sinusoidal steady state with $i_S(t) = 50 \cos 2000t$ mA.

(a) Transform the circuit into the phasor domain.
(b) Solve for the phasor voltage **V**.
(c) Solve for the phasor current through each element.
(d) Construct the waveforms corresponding to the phasors found in (b) and (c).

SOLUTION:

(a) The phasor representing the input source current is $\mathbf{I}_S = 0.05\angle0°$ A.
    The impedances of the three passive elements are

$$Z_R = R = 500 \ \Omega$$

$$Z_L = j\omega L = j2000 \times 0.5 = j1000 \ \Omega$$

$$Z_C = \frac{1}{j\omega C} = \frac{1}{j2000 \times 10^{-6}} = -j500 \ \Omega$$

Using these results, we obtain the phasor-domain circuit in Figure 15–20.

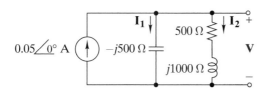

FIGURE 1 5 – 2 0

(b) The admittances of the two parallel branches are

$$Y_1 = \frac{1}{-j500} = j2 \times 10^{-3} \ \text{S}$$

$$Y_2 = \frac{1}{500 + j1000} = 4 \times 10^{-4} - j8 \times 10^{-4} \ \text{S}$$

The equivalent admittance of the parallel connection is

$$Y_{EQ} = Y_1 + Y_2 = 4 \times 10^{-4} + j12 \times 10^{-4}$$

$$= 12.6 \times 10^{-4}\angle71.6° \ \text{S}$$

and the voltage across the parallel circuit is

$$\mathbf{V} = \frac{\mathbf{I}_S}{Y_{EQ}} = \frac{0.05\angle0°}{12.6 \times 10^{-4}\angle71.6°}$$

$$= 39.7\angle-71.6° \ \text{V}$$

(c) The current through each parallel branch is

$$\mathbf{I}_1 = Y_1\mathbf{V} = j2 \times 10^{-3} \times 39.7\angle-71.6° = 79.4\angle18.4° \ \text{mA}$$

$$\mathbf{I}_2 = Y_2\mathbf{V} = (4 \times 10^{-4} - j8 \times 10^{-4}) \times 39.7\angle-71.6°$$

$$= 35.5 \angle-135° \ \text{mA}$$

(d) The sinusoidal steady-state waveforms corresponding to the phasors in
    (b) and (c) are

$$v(t) = \text{Re}\{39.7e^{-j71.6°}e^{j2000t}\} = 39.7 \cos(2000t - 71.6°) \ \text{V}$$

$$i_1(t) = \text{Re}\{79.4e^{j18.4°}e^{j2000t}\} = 79.4 \cos(2000t + 18.4°) \ \text{mA}$$

$$i_2(t) = \text{Re}\{35.5e^{-j135°}e^{j2000t}\} = 35.5 \cos(2000t - 135°) \ \text{mA} \quad \blacksquare$$

EXAMPLE 15–10

Find the steady-state currents $i(t)$, $i_C(t)$, and $i_R(t)$ in the circuit of Figure
15–21 for $v_S = 100 \cos 2000t$ V, $L = 250$ mH, $C = 0.5 \ \mu$F, and $R = 3 \ \text{k}\Omega$.

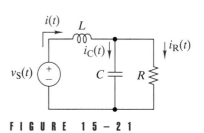

FIGURE 1 5 – 2 1

SOLUTION:

The phasor representation of the input voltage is $100\angle0°$. The impedances of the passive elements are

$$Z_L = j500 \ \Omega \quad Z_C = -j1000 \ \Omega \quad Z_R = 3000 \ \Omega$$

Figure 15–22(a) shows the phasor-domain circuit.

To solve for the required phasor responses, we reduce the circuit using a combination of series and parallel equivalence. Using parallel equivalence, we find that the capacitor and resistor can be replaced by an equivalent impedance

$$Z_{EQ1} = \frac{1}{Y_{EQ1}} = \frac{1}{\dfrac{1}{-j1000} + \dfrac{1}{3000}}$$

$$= 300 - j900 \ \Omega$$

The resulting circuit reduction is shown in Figure 15–22(b). The equivalent impedance Z_{EQ1} is connected in series with the impedance $Z_L = j500$. This series combination can be replaced by an equivalent impedance

$$Z_{EQ2} = j500 + Z_{EQ1} = 300 - j400 \ \Omega$$

This step reduces the circuit to the equivalent input impedance shown in Figure 15–22(c). The phasor input current in Figure 15–22(c) is

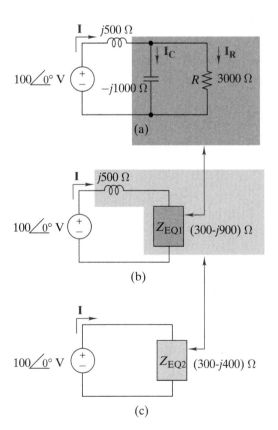

$$I = \frac{100\angle 0°}{Z_{EQ2}} = \frac{100\angle 0°}{300 - j400} = 0.12 + j0.16 = 0.2\angle 53.1° \text{ A}$$

Given the phasor current I, we use current division to find I_C

$$I_C = \frac{Y_C}{Y_C + Y_R}I = \frac{\dfrac{1}{-j1000}}{\dfrac{1}{-j1000} + \dfrac{1}{3000}}0.2\angle 53.1°$$

$$= 0.06 + j0.18 = 0.19\angle 71.6° \text{ A}$$

By KCL, $I = I_C + I_R$, so the remaining unknown current is

$$I_R = I - I_C = 0.06 - j0.02 = 0.0632\angle -18.4° \text{ A}$$

The waveforms corresponding to the phasor currents are

$$i(t) = \text{Re}\{Ie^{j2000t}\} = 0.2\cos(2000t + 53.1°) \text{ A}$$
$$i_C(t) = \text{Re}\{I_Ce^{j2000t}\} = 0.19\cos(2000t + 71.6°) \text{ A}$$
$$i_R(t) = \text{Re}\{I_Re^{j2000t}\} = 0.0632\cos(2000t - 18.4°) \text{ A} \qquad \blacksquare$$

APPLICATION NOTE: EXAMPLE 15–11

In general, the equivalent impedance seen at any pair of terminals can be written in rectangular form as

$$Z_{EQ} = R_{EQ} + jX_{EQ} \qquad (15\text{–}26)$$

In a passive circuit the equivalent resistance R_{EQ} must always be non-negative, that is, $R_{EQ} \geq 0$. However, the equivalent reactance X_{EQ} can be either positive (inductive) or negative (capacitive). When inductance and capacitance are both present, their reactances may exactly cancel at certain frequencies. When $X_{EQ} = 0$, the impedance is purely resistive and the circuit is said to be in **resonance**. The frequency at which this occurs is called a **resonant frequency**, denoted by ω_0.

For example, suppose we want to find the resonant frequency of the circuit in Figure 15–23. We first find the equivalent impedance of the parallel resistor and capacitor:

$$Z_{RC} = \frac{1}{Y_R + Y_C} = \frac{1}{\dfrac{1}{R} + j\omega C} = \frac{R}{1 + j\omega RC}$$

This expression can be put into rectangular form by multiplying and dividing by the conjugate of the denominator:

$$Z_{RC} = \frac{R}{1 + j\omega RC}\frac{1 - j\omega RC}{1 - j\omega RC} = \frac{R}{1 + (\omega RC)^2} - j\frac{\omega R^2 C}{1 + (\omega RC)^2}$$

The impedance Z_{RC} is connected in series with the inductor. Therefore, the overall equivalent impedance Z_{EQ} is

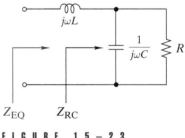

$$Z_{EQ} \qquad Z_{RC}$$

FIGURE 1 5 – 2 3

$$Z_{EQ} = Z_L + Z_{RC}$$

$$= \underbrace{\frac{R}{1 + (\omega RC)^2}}_{R_{EQ}} + \underbrace{j\left[\omega L - \frac{\omega R^2 C}{1 + (\omega RC)^2}\right]}_{jX_{EQ}}$$

$$= \quad R_{EQ} \quad + \quad jX_{EQ}$$

Note that the equivalent resistance R_{EQ} is positive for all ω. However, the equivalent reactance X_{EQ} can be positive or negative. The resonant frequency is found by setting the reactance to zero

$$X_{EQ}(\omega_0) = \omega_0 L - \frac{\omega_0 R^2 C}{1 + (\omega_0 RC)^2} = 0$$

and solving for the resonant frequency:

$$\omega_0 = \sqrt{\frac{1}{LC} - \frac{1}{(RC)^2}} \tag{15–27}$$

Note the reactance X_{EQ} is inductive (positive) when $\omega > \omega_0$ and capacitive (negative) with $\omega < \omega_0$.

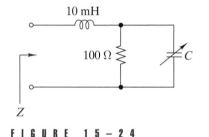

FIGURE 15 – 24

EXAMPLE 15–12

The circuit in Figure 15–24 is operating in the sinusoidal steady state with $\omega = 5$ krad/s.

(a) Find the value of capacitance C that causes the input impedance Z to be purely resistive.

(b) Find the real part of the input impedance for this value of C.

SOLUTION:

(a) At the specified frequency the impedance is

$$Z = j5000 \times 0.01 + \frac{1}{\dfrac{1}{100} + j5000\,C}$$

which can be written as

$$Z = j50 + \frac{100}{1 + j5 \times 10^5\,C} \times \frac{1 - j5 \times 10^5\,C}{1 - j5 \times 10^5\,C}$$

$$= j50 + \frac{100}{1 + (5 \times 10^5\,C)^2} - j\,\frac{5 \times 10^7\,C}{1 + (5 \times 10^5\,C)^2}$$

A purely resistive input impedance means that the imaginary part of Z is zero, which requires that

$$50 - \frac{5 \times 10^7\,C}{1 + (5 \times 10^5\,C)^2} = 0 \quad \text{or} \quad 25 \times 10^{10}\,C^2 - 10^6\,C + 1 = 0$$

This quadratic has a double root at $C = 2 \times 10^{-6}$ F.

(b) For this value of capacitance the real part of Z is

$$Re\{Z\} = R_{EQ} = \frac{100}{1 + (5 \times 10^5\, C)^2} = 50\ \Omega$$

Exercise 15–8

(a) Find the equivalent impedance Z in Figure 15–25 at $\omega = 1$ krad/s.
(b) Repeat (a) with $\omega = 4$ krad/s.

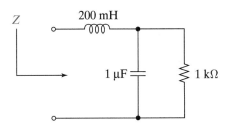

F I G U R E   1 5 – 2 5

Answers:
(a) $Z = 500 - j300\ \Omega$
(b) $Z = 58.8 + j565\ \Omega$

Exercise 15–9

A voltage source $v_S(t) = 15 \cos 2000t$ V is applied at the input in Figure 15–25.

(a) Find the steady-state current through the inductor.
(b) Find the steady-state voltage across the 1-kΩ resistor.

Answers:
(a) $75 \cos 2000t$ mA
(b) $33.5 \cos (2000t - 63.4°)$ V

Exercise 15–10

The circuit in Figure 15–26 is operating at $\omega = 10$ krad/s.

(a) Find the equivalent impedance Z.
(b) What element should be connected in series with Z to make the total reactance zero?

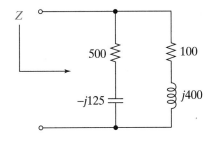

F I G U R E   1 5 – 2 6

Answers:
(a) $Z = 256 + j195 = 322\angle37.3°\ \Omega$
(b) A capacitor with $C = 513$ nF

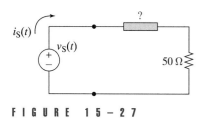

In Figure 15–27 $v_S(t) = 12.5 \cos 1000t$ V and $i_S(t) = 0.2 \cos (1000t - 36.9°)$ A. What is the impedance seen by the voltage source and what element is in the box?

Answer:
$Z = 50 + j37.5 \ \Omega$ and the element is a 37.5-mH inductor.

15–4 CIRCUIT THEOREMS WITH PHASORS

In this section we treat basic properties of phasor circuits that parallel the circuit theorems discussed in previous chapters. Circuit linearity is the foundation for all of these properties. The proportionality and superposition properties are two fundamental consequences of linearity.

PROPORTIONALITY

The **proportionality** property states that phasor output responses are proportional to the input phasor. Mathematically, proportionality means that

$$\mathbf{Y} = K\mathbf{X} \tag{15–28}$$

where $\mathbf{X}$ is the input phasor, $\mathbf{Y}$ is the output phasor, and K is the proportionality constant. In phasor circuit analysis, the proportionality constant is generally a complex number.

The unit output method is based on the proportionality property and is applicable to phasors. To apply the unit output method in the phasor domain, we assume that the output is a unit phasor $\mathbf{Y} = 1\angle 0°$. By successive application of KCL, KVL, and the element impedances, we solve for the input phasor required to produce the unit output. Because the circuit is linear, the proportionality constant relating input and output is

$$K = \frac{\text{Output}}{\text{Input}} = \frac{1\angle 0°}{\text{Input phasor for unit output}}$$

Once we have the constant K, we can find the output for any input or the input required to produce any specified output.

The next example illustrates the unit output method for phasor circuits.

EXAMPLE 15–13

Use the unit output method to find the input impedance, current $\mathbf{I}_1$, output voltage $\mathbf{V}_C$, and current $\mathbf{I}_3$ of the circuit in Figure 15–28 for $\mathbf{V}_S = 10\angle 0°$ V.

SOLUTION:
The following steps implement the unit output method for the circuit in Figure 15–28:

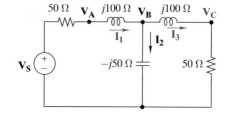

1. Assume a unit output voltage $\mathbf{V}_C = 1 + j0$ V.

2. By Ohm's law, $\mathbf{I}_3 = \mathbf{V}_C/50 = 0.02 + j0$ A.

3. By KVL, $\mathbf{V}_B = \mathbf{V}_C + (j100)\mathbf{I}_3 = 1 + j2$ V.

4. By Ohm's law, $\mathbf{I}_2 = \mathbf{V}_B/(-j50) = -0.04 + j0.02$ A.

5. By KCL, $\mathbf{I}_1 = \mathbf{I}_2 + \mathbf{I}_3 = -0.02 + j0.02$ A.

6. By KVL, $\mathbf{V}_S = (50 + j100)\mathbf{I}_1 + \mathbf{V}_B = -2 + j1$ V.

Given $\mathbf{V}_S$ and $\mathbf{I}_1$, the input impedance is

$$Z_{IN} = \frac{\mathbf{V}_S}{\mathbf{I}_1} = \frac{-2 + j1}{-0.02 + j0.02} = 75 + j25 \ \Omega$$

The proportionality factor between the input $\mathbf{V}_S$ and output voltage $\mathbf{V}_C$ is

$$K = \frac{1}{\mathbf{V}_S} = \frac{1}{-2 + j} = -0.4 - j0.2$$

Given K and Z_{IN}, we can now calculate the required responses for an input $\mathbf{V}_S = 10\angle 0°$:

$$\mathbf{V}_C = K\mathbf{V}_S = -4 - j2 = 4.47\angle -153° \text{ V}$$

$$\mathbf{I}_1 = \frac{\mathbf{V}_S}{Z_{IN}} = 0.12 - j0.04 = 0.126\angle -18.4° \text{ A}$$

$$\mathbf{I}_3 = \frac{\mathbf{V}_C}{50} = -0.08 - j0.04 = 0.0894\angle -153° \text{ A} \qquad ■$$

SUPERPOSITION

The superposition principle applies to phasor responses only if all of the independent sources driving the circuit have the *same frequency*. That is, when the input sources have the same frequency, we can find the phasor response due to each source acting alone and obtain the total response by adding the individual phasors. If the sources have different frequencies, then superposition can still be used but its application is different. With different frequency sources, each source must be treated in a separate steady-state analysis because the element impedances change with frequency. The phasor response for each source must be changed into waveforms and then superposition applied in the time domain. In other words, the superposition principle always applies in the time domain. It also applies in the phasor domain when all independent sources have the same frequency. The following examples illustrate both cases.

EXAMPLE 15-14

Use superposition to find the steady-state voltage $v_R(t)$ in Figure 15-29 for $R = 20 \ \Omega$, $L_1 = 2$ mH, $L_2 = 6$ mH, $C = 20 \ \mu$F, $v_{S1} = 100 \cos 5000t$ V, and $v_{S2} = 120 \cos(5000t + 30°)$ V.

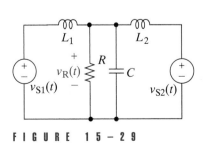

FIGURE 15-29

SOLUTION:

In this example, the two sources operate at the same frequency. Figure 15–30(a) shows the phasor domain circuit with source no. 2 turned off and replaced by a short circuit. The three elements in parallel in Figure 15–30(a) produce an equivalent impedance of

$$Z_{EQ} = \frac{1}{\dfrac{1}{20} + \dfrac{1}{-j10} + \dfrac{1}{j30}} = 7.20 - j9.60 \ \Omega$$

By voltage division, the phasor response $\mathbf{V}_{R1}$ is

$$\mathbf{V}_{R1} = \frac{Z_{EQ1}}{j10 + Z_{EQ1}} 100\angle 0°$$

$$= 92.3 - j138 = 166\angle -56.3° \ \text{V}$$

FIGURE 15–30

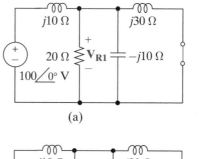

(a)

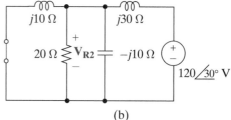

(b)

Figure 15–30(b) shows the phasor-domain circuit with source no. 1 turned off and source no. 2 on. The three elements in parallel in Figure 15–30(b) produce an equivalent impedance of

$$Z_{EQ2} = \frac{1}{\dfrac{1}{20} + \dfrac{1}{-j10} + \dfrac{1}{j10}} = 20 - j0 \ \Omega$$

By voltage division, the response $\mathbf{V}_{R2}$ is

$$\mathbf{V}_{R2} = \frac{Z_{EQ2}}{j30 + Z_{EQ2}} 120\angle 30°$$

$$= 59.7 - j29.5 = 66.6\angle -26.3° \ \text{V}$$

Since the sources have the same frequency, the total response can be found by adding the individual phasor responses $\mathbf{V}_{R1}$ and $\mathbf{V}_{R2}$:

$$\mathbf{V}_R = \mathbf{V}_{R1} + \mathbf{V}_{R2} = 152 - j167 = 226\angle -47.8° \text{ V}$$

The time-domain function corresponding to the phasor sum is

$$v_R(t) = \text{Re}\{\mathbf{V}_R e^{j5000t}\} = 226 \cos(5000t - 47.8°) \text{ V}$$

The overall response can also be obtained by adding the time-domain functions corresponding to the individual phasor responses $\mathbf{V}_{R1}$ and $\mathbf{V}_{R2}$:

$$v_R(t) = \text{Re}\{\mathbf{V}_{R1} e^{j5000t}\} + \text{Re}\{\mathbf{V}_{R2} e^{j5000t}\}$$

$$= 166 \cos(5000t - 56.3°) + 66.6 \cos(5000t - 26.3°) \text{ V}$$

You are encouraged to show that the two expressions for $v_R(t)$ are equivalent using the additive property of sinusoids. ∎

EXAMPLE 15–15

Use superposition to find the steady-state current $i(t)$ in Figure 15–31 for $R = 10$ kΩ, $L = 200$ mH, $v_{S1} = 24 \cos 20000t$ V, and $v_{S2} = 8 \cos(60000t + 30°)$ V.

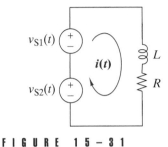

FIGURE 15–31

SOLUTION:
In this example the two sources operate at different frequencies. With source no. 2 off, the input phasor is $\mathbf{V}_{S1} = 24\angle 0°$ V at a frequency of $\omega = 20$ krad/s. At this frequency the equivalent impedance of the inductor and resistor is

$$Z_{EQ1} = R + j\omega L = 10 + j4 \text{ kΩ}$$

The phasor current due to source no. 1 is

$$\mathbf{I}_1 = \frac{\mathbf{V}_{S1}}{Z_{EQ1}} = \frac{24\angle 0°}{10000 + j4000} = 2.23\angle -21.8° \text{ mA}$$

With source no. 1 off and source no. 2 on, the input phasor $\mathbf{V}_{S2} = 8\angle 30°$ V at a frequency of $\omega = 60$ krad/s. At this frequency the equivalent impedance of the inductor and resistor is

$$Z_{EQ2} = R + j\omega L = 10 + j12 \text{ kΩ}$$

The phasor current due to source no. 2 is

$$\mathbf{I}_2 = \frac{\mathbf{V}_{S2}}{Z_{EQ2}} = \frac{8\angle 30°}{10000 + j12000} = 0.512\angle -20.2° \text{ mA}$$

The two input sources operate at different frequencies, so the phasors responses $\mathbf{I}_1$ and $\mathbf{I}_2$ cannot be added to obtain the overall response. In this case the overall response is obtained by adding the corresponding time-domain functions.

$$i(t) = \text{Re}\{\mathbf{I}_1 e^{j20000t}\} + \text{Re}\{\mathbf{I}_2 e^{j60000t}\}$$

$$= 2.23 \cos(20000t - 21.8°) + 0.512 \cos(60000t - 20.2°) \text{ mA} ∎$$

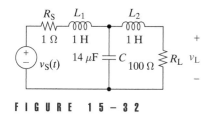

FIGURE 15 – 32

EXAMPLE 15–16

The voltage source in Fig. 15–32 produces a 60-Hz sinusoid with a peak amplitude of 200 V plus a 180-Hz third harmonic with a peak amplitude of 10 V. The purpose of the *LC* circuit is to reduce the relative size of the third harmonic component delivered to the 100-Ω load resistor. Use Orcad Capture to calculate the amplitude of the 60-Hz fundamental and 180-Hz third harmonic in the voltage across the load.

SOLUTION:

Figure 15–33 shows the circuit diagram as constructed in Orcad Capture. The value ACMAG = 1V and ACPHASE = 0 means that $\mathbf{V_S}$ is an ac voltage source with an amplitude of 1 V and a phase angle of 0°. The printer symbol[1] attached to Node N00576 causes Orcad PSpice to write the magnitude and phase of the node voltage $\mathbf{V(N00576)}$ in the output file. This example involves ac circuit analysis with input sinusoids at 60 Hz and 180 Hz. Selecting **PSpice/Simulations Settings/AC Sweep/Noise** from the Schematics menu bar brings up the AC sweep menu also shown in Figure 15–33. We select a two-point linear sweep that starts at 60 Hz and ends at 180 Hz. Returning to the main menu and selecting **PSpice/Run** cause Orcad PSpice to make two AC analysis runs, first at $f = 60$ Hz and then again at $f = 180$ Hz. The input phasor for each run is $\mathbf{V_S} = 1\angle 0°$. The relevant portion of the

FIGURE 15 – 33 *Orcad Capture circuit diagram, ac sweep analysis set up, and analysis output.*

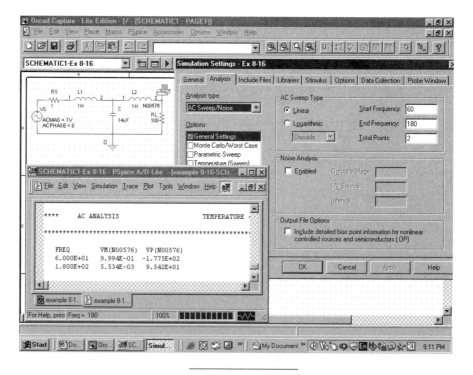

---

1 The printer is found under the "Place Part" menu. Choose **Add Library** and select **Special**. Scroll down and select **VPRINT1**. Connect the printer to the desired node. In order to set the parameters that the printer will record, it is necessary to double-click on the printer symbol. This brings up the Property Editor. Click on the **AC** box and place a "Y" for "Yes." Repeat this for the boxes labeled "MAG" and "PHASE." Before closing the Property Editor, click on the **Apply** button. After running the simulation, the Printer records the desired results in the Output File.

output file is shown at the bottom in Figure 15–33. The symbol **VM(N00576)** is the magnitude of the output voltage phasor in volts, and **VP(N00576)** is the phase angle in degrees. These values are for the normalized input $\mathbf{V_S} = 1\angle 0°$. To obtain the actual output, we must scale the normalized results by the actual input amplitudes at each frequency.

Component	Freq (Hz)	Input(V)	Output(V)
Fundamental	60	200	$200 \times 0.9994 = 199.9$
3rd Harmonic	180	10	$10 \times 0.005534 = 0.05534$

The amplitude of the 60-Hz component is virtually unchanged while the amplitude of the 180-Hz component is reduced from 10 V to 55.3 mV. This frequency selectivity or filtering occurs because the impedances of the inductor and capacitor change with frequency. ■

Exercise 15–12

The two sources in Figure 15–34 have the same frequency. Use superposition to find the phasor current $\mathbf{I_X}$.

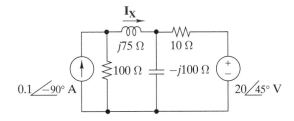

FIGURE 15–34

Answer:

$$\mathbf{I_X} = 0.206\angle -158° \text{ A}$$

THÉVENIN AND NORTON EQUIVALENT CIRCUITS

In the phasor domain, a two-terminal circuit containing linear elements and sources can be replaced by the Thévenin or Norton equivalent circuits shown in Figure 15–35. The general concept of Thévenin's and Norton's theorems and their restrictions are the same as in a resistive circuit. The important difference here is that the signals $\mathbf{V_T}$, $\mathbf{I_N}$, $\mathbf{V}$, and $\mathbf{I}$ are phasors, and $Z_T = 1/Y_N$ and Z_L are complex numbers representing the source and load impedances.

Finding the Thévenin or Norton equivalent of a phasor circuit involves the same process as for resistance circuits, except that now we must manipulate complex numbers. The Thévenin and Norton circuits are equivalent to each other, so their circuit parameters are related as follows:

$$\mathbf{V}_{OC} = \mathbf{V}_T = \mathbf{I}_N Z_T$$

$$\mathbf{I}_{SC} = \frac{\mathbf{V}_T}{Z_T} = \mathbf{I}_N$$ (15–29)

$$Z_T = \frac{1}{Y_N} = \frac{\mathbf{V}_{OC}}{\mathbf{I}_{SC}}$$

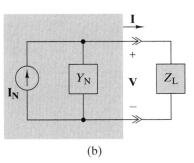

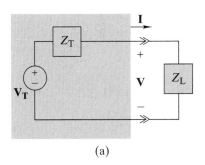

(a)

(b)

FIGURE 15–35 *Thévenin and Norton equivalent circuits in the phasor analysis.*

Algebraically, the results in Eq. (15–29) are identical to the corresponding equations for resistance circuits. The important difference is that these equations involve phasors and impedances rather than waveforms and resistances. These equations point out that we can determine a Thévenin or Norton equivalent by finding any two of the following quantities: (1) the open-circuit voltage $\mathbf{V}_{OC}$, (2) the short-circuit current $\mathbf{I}_{SC}$, and (when there are no dependent sources) (3) the impedance Z_T looking back into the source circuit with all independent sources turned off.

The relationships in Eq. (15–29) define source transformations that allow us to convert a voltage source in series with an impedance into a current source in parallel with the same impedance, or vice versa. Phasor-domain source transformations simplify circuits and are useful in formulating general node-voltage or mesh-current equations, discussed in the next section.

The next two examples illustrate applications of source transformation and Thévenin equivalent circuits.

EXAMPLE 15–17

Both sources in Figure 15–36(a) operate at a frequency of $\omega = 5000$ rad/s. Find the steady-state voltage $v_R(t)$ using source transformations.

SOLUTION:

Example 15–14 solves this problem using superposition. In this example we use source transformations. We observe that the voltage sources in Figure 15–36(a) are connected in series with an impedance and can be converted into the following equivalent current sources:

$$\mathbf{I}_{EQ1} = \frac{100\angle 0°}{j10} = 0 - j10 \text{ A}$$

$$\mathbf{I}_{EQ2} = \frac{120\angle 30°}{j30} = 2 - j3.46 \text{ A}$$

FIGURE 15–36

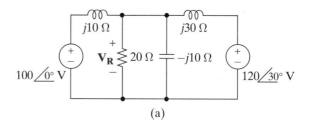

(a)

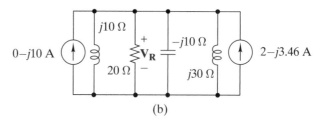

(b)

Figure 15–36(b) shows the circuit after these two source transformations. The two current sources are connected in parallel and can be replaced by a single equivalent current source:

$$\mathbf{I}_{EQ} = \mathbf{I}_{EQ1} + \mathbf{I}_{EQ2} = 2 - j13.46 = 13.6\angle-81.5° \text{ A}$$

The four passive elements are connected in parallel and can be replaced by an equivalent impedance:

$$Z_{EQ} = \cfrac{1}{\cfrac{1}{20} + \cfrac{1}{-j10} + \cfrac{1}{j10} + \cfrac{1}{j30}} = 16.6\angle33.7° \text{ }\Omega$$

The voltage across this equivalent impedance equals $\mathbf{V}_R$, since one of the parallel elements is the resistor R. Therefore, the unknown phasor voltage is

$$\mathbf{V}_R = \mathbf{I}_{EQ}Z_{EQ} = (13.6\angle-81.5°) \times (16.6\angle33.7°) = 226\angle-47.8° \text{ V}$$

The value of $\mathbf{V}_R$ is the same as found in Example 15–14 using superposition. The corresponding time-domain function is

$$v_R(t) = \text{Re}\{\mathbf{V}_R e^{j5000t}\} = 226 \cos(5000t - 47.8°) \text{ V} \qquad \blacksquare$$

EXAMPLE 15–18

Use Thévenin's theorem to find the current $\mathbf{I}_X$ in the bridge circuit shown in Figure 15–37.

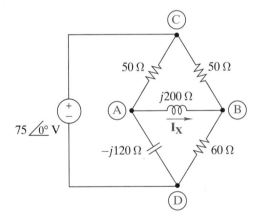

FIGURE 15 – 37

SOLUTION:
Disconnecting the impedance $j200$ from the circuit in Figure 15–37 produces the circuit shown in Figure 15–38(a). The voltage between nodes A and B is the Thévenin voltage since removing the impedance $j200$ leaves an open circuit. The voltages at nodes A and B can each be found by voltage division. Since the open-circuit voltage is the difference between these node voltages, we have

$$\mathbf{V}_T = \mathbf{V}_A - \mathbf{V}_B$$

$$= \frac{-j120}{50 - j120}75\angle0° - \frac{60}{60 + 50}75\angle0°$$

$$= 23.0 - j26.6 \text{ V}$$

FIGURE 15–38

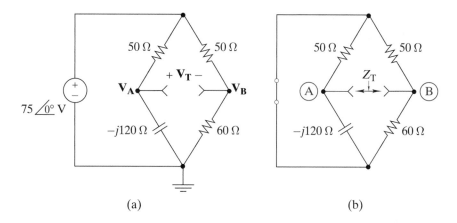

(a)   (b)

Turning off the voltage source in Figure 15–38(a) and replacing it by a short circuit produces the situation shown in Figure 15–38(b). The look-back impedance seen at the interface is a series connection of two pairs of elements connected in parallel. The equivalent impedance of the series/parallel combination is

$$Z_T = \cfrac{1}{\cfrac{1}{50} + \cfrac{1}{-j120}} + \cfrac{1}{\cfrac{1}{50} + \cfrac{1}{60}} = 69.9 - j17.8 \ \Omega$$

Given the Thévenin equivalent circuit, we treat the impedance $j200$ as a load connected at the interface and calculate the resulting load current $\mathbf{I}_X$ as

$$\mathbf{I}_X = \frac{\mathbf{V}_T}{Z_T + j200} = \frac{23.0 - j26.6}{69.9 + j182} = 0.180\angle -118° \ \text{A} \qquad \blacksquare$$

EXAMPLE 15–19

In the steady state, the open-circuit voltage at an interface is observed to be

$$v_{OC}(t) = 12 \cos 2000t \quad \text{V}$$

When a 50-mH inductor is connected across the interface, the interface voltage is observed to be

$$v(t) = 17 \cos(2000t + 45°) \quad \text{V}$$

Find the Thévenin equivalent circuit at the interface.

SOLUTION:

The phasors for $v_{OC}(t)$ and $v(t)$ are $\mathbf{V}_{OC} = 12\angle 0°$ and $\mathbf{V} = 17\angle 45°$. The pha-sor Thévenin voltage at the interface is $\mathbf{V}_T = \mathbf{V}_{OC} = 12\angle 0°$. The impedance of the inductor is $Z_L = j\omega L = j2000 \times 0.050 = j100 \ \Omega$. When the inductor load is connected across the interface, we use voltage division to express the interface voltage as

$$\mathbf{V} = \frac{Z_L}{Z_T + Z_L}\mathbf{V}_T$$

Inserting the known numerical values yields

$$17\angle 45° = \frac{j100}{Z_T + j100}12\angle 0°$$

Solving for Z_T, we have

$$Z_T = j100 \times \frac{12\angle 0°}{17\angle 45°} - j100 = 49.9 - j50.1 \ \Omega$$

The Thévenin equivalent circuit at the interface is defined by $\mathbf{V}_T = 12\angle 0°$ and $Z_T = 49.9 - j50.1 \ \Omega$. ∎

Exercise 15-13

(a) Find the Thévenin equivalent circuit seen by the inductor in Figure 15–34.
(b) Use the Thévenin equivalent to calculate the current $\mathbf{I}_X$.

Answers:
(a) $\mathbf{V}_T = -15.4 - j22.6$ V, $Z_T = 109.9 - j0.990 \ \Omega$
(b) $\mathbf{I}_X = 0.206\angle -158°$ A

Exercise 15-14

By inspection, determine the Thévenin equivalent circuit seen by the capacitor in Figure 15–28 for $\mathbf{V}_S = 10\angle 0°$ V.

Answer:
$\mathbf{V}_T = 5\angle 0°$ V, $Z_T = 25 + j50 \ \Omega$

Exercise 15-15

In the steady state the short-circuit current at an interface is observed to be

$$i_{SC}(t) = 0.75 \sin \omega t \ \text{A}$$

When a 150-Ω resistor is connected across the interface, the interface current is observed to be

$$i(t) = 0.6 \cos(\omega t - 53.1°) \ \text{A}$$

Find the Norton equivalent phasor circuit at the interface.

Answer:

$\mathbf{I}_N = 0 - j0.75$ A, $Z_N = 0 + j200 \ \Omega$

15-5 GENERAL CIRCUIT ANALYSIS WITH PHASORS

The previous sections discuss basic analysis methods based on equivalence, reduction, and circuit theorems. These methods are valuable because we work directly with element impedances and thereby gain insight into steady-state circuit behavior. We also need general methods, such as node and mesh analysis, to deal with more complicated circuits than the basic methods can easily handle. These general methods use node-voltage or

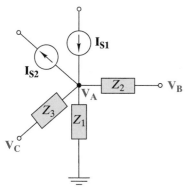

FIGURE 15–39 *An example node.*

mesh-current variables to reduce the number of equations that must be solved simultaneously.

Node-voltage equations involve selecting a reference node and assigning a node-to-datum voltage to each of the remaining nonreference nodes. Because of KVL, the voltage between any two nodes equals the difference of the two node voltages. This fundamental property of node voltages plus the element impedances allow us to write KCL constraints at each of the nonreference nodes.

For example, consider node A in Figure 15–39. The sum of currents leaving this node can be written as

$$\mathbf{I}_{S2} - \mathbf{I}_{S1} + \frac{\mathbf{V}_A}{Z_1} + \frac{\mathbf{V}_A - \mathbf{V}_B}{Z_2} + \frac{\mathbf{V}_A - \mathbf{V}_C}{Z_3} = 0$$

Rewriting this equation with unknowns grouped on the left and known inputs on the right yields

$$\left[\frac{1}{Z_1} + \frac{1}{Z_2} + \frac{1}{Z_3}\right]\mathbf{V}_A - \frac{1}{Z_2}\mathbf{V}_B - \frac{1}{Z_3}\mathbf{V}_C = \mathbf{I}_{S1} - \mathbf{I}_{S2}$$

Expressing this result in terms of admittances produces the following equation.

$$[Y_1 + Y_2 + Y_3]\mathbf{V}_A - [Y_2]\mathbf{V}_B - [Y_3]\mathbf{V}_C = \mathbf{I}_{S1} - \mathbf{I}_{S2}$$

This equation has a familiar pattern. The unknowns $\mathbf{V}_A$, $\mathbf{V}_B$, and $\mathbf{V}_C$ are the node-voltage phasors. The coefficient $[Y_1 + Y_2 + Y_3]$ of $\mathbf{V}_A$ is the sum of the admittances of all of the elements connected to node A. The coefficient $[Y_2]$ of $\mathbf{V}_B$ is admittance of the elements connected between nodes A and B, while $[Y_3]$ is the admittance of the elements connected between nodes A and C. Finally, $\mathbf{I}_{S1}$ and $\mathbf{I}_{S2}$ are the phasor current sources connected to node A, with $\mathbf{I}_{S1}$ directed into and $\mathbf{I}_{S2}$ directed away from the node. These observations suggest that we can write node-voltage equations for phasor circuits by inspection, just as we did with resistive circuits.

Circuits that can be drawn on a flat surface with no crossovers are called **planar** circuits. The mesh-current variables are the loop currents assigned to each mesh in a planar circuit. Because of KCL, the current through any two-terminal element is equal to the difference of the two adjacent meshes. This fundamental property of mesh currents together with the element impedances allow us to write KVL constraints around each of the meshes.

For example, the sum of voltages around mesh A in Figure 15–40 is

$$Z_1\mathbf{I}_A + Z_2[\mathbf{I}_A - \mathbf{I}_B] + Z_3[\mathbf{I}_A - \mathbf{I}_C] - \mathbf{V}_{S1} + \mathbf{V}_{S2} = 0$$

The mesh A equation is obtained by equating this sum to the sum of the source voltages produced in mesh A. Arranging this equation in standard form yields

$$[Z_1 + Z_2 + Z_3]\mathbf{I}_A - [Z_2]\mathbf{I}_B - [Z_3]\mathbf{I}_C = \mathbf{V}_{S1} - \mathbf{V}_{S2}$$

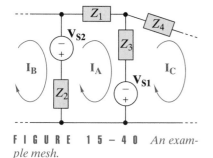

FIGURE 15–40 *An example mesh.*

This equation also displays a familiar pattern. The unknowns $\mathbf{I}_A$, $\mathbf{I}_B$, and $\mathbf{I}_C$ are mesh-current phasors. The coefficient $[Z_1 + Z_2 + Z_3]$ of $\mathbf{I}_A$ is the sum of the impedances in mesh A. The coefficient $[Z_2]$ of $\mathbf{I}_B$ is the impedance in both mesh A and mesh B, while $[Z_3]$ is the impedance common to meshes A and C. Finally, $\mathbf{V}_{S1}$ and $\mathbf{V}_{S2}$ are the phasor voltage sources in mesh A. These observations allow us to write mesh-current equations for phasor circuits by inspection.

The preceding discussion assumes that the circuit contains only current sources in the case of node analysis and voltage sources in mesh analysis. If there is a mixture of sources, we may be able to use the source transformations discussed in Sect. 15–4 to convert from voltage to current sources, or vice versa. A source transformation is possible only when there is an impedance connected in series with a voltage source or an admittance in parallel with a current source. When a source transformation is not possible, we use the phasor version of the modified node- and mesh-analysis methods.

Formulating a set of equilibrium equations in phasor form is a straightforward process involving concepts that we have used before. Once formulated, we use Cramer's rule or Gaussian reduction to solve these equations for phasor responses, although this requires manipulating linear equations with complex coefficients. In principle, the solution process can be done by hand, but as a practical matter circuits with more than three nodes or meshes are best handled using computer tools. Modern hand-held scientific calculators and math analysis programs like Mathcad can deal with sets of linear equations with complex coefficients. Circuit analysis programs such as SPICE have ac analysis options that handle steady-state circuit analysis problems.

If computer tools are required for all but the simplest circuits, why bother with the hand solution at all? Why not always use SPICE or Mathcad? The answer is that hand analysis and computer-aided analysis are complementary rather than competitive. Computer-aided analysis excels at generating numerical responses when numerical values of circuit parameters are given. With hand analysis we can generate responses in symbolic form for all possible element values and input frequencies. Solving simple circuits in symbolic form gives insight as to how various circuit elements affect the steady-state response which helps us intelligently use the numerical profusion generated by computer-based "what if" analysis of large-scale circuits.

EXAMPLE 15−20

Use node analysis to find the node voltages $\mathbf{V}_A$ and $\mathbf{V}_B$ in Figure 15–41(a).

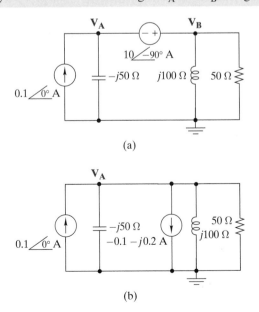

(a)

(b)

FIGURE   15 − 41

SOLUTION:

The voltage source in Figure 15–41(a) is connected in series with an impedance consisting of a resistor and inductor connected in parallel. The equivalent impedance of this parallel combination is

$$Z_{EQ} = \frac{1}{\dfrac{1}{50} + \dfrac{1}{j100}} = 40 + j20 \ \Omega$$

Applying a source transformation produces an equivalent current source of

$$\mathbf{I}_{EQ} = \frac{10\angle{-90°}}{40 + j20} = -0.1 - j0.2 \ \text{A}$$

Figure 15–41(b) shows the circuit produced by the source transformation. Note that the transformation eliminates node B. The node-voltage equation at the remaining nonreference node in Figure 15–41(b) is

$$\left(\frac{1}{-j50} + \frac{1}{j100} + \frac{1}{50}\right)\mathbf{V}_A = 0.1\angle{0°} - (-0.1 - j0.2)$$

Solving for $\mathbf{V}_A$ yields

$$\mathbf{V}_A = \frac{0.2 + j0.2}{0.02 + j0.01} = 12 + j4 = 12.6\angle{18.4°} \ \text{V}$$

Referring to Figure 15–41(a), we see that KVL requires $\mathbf{V}_B = \mathbf{V}_A + 10\angle{-90°}$. Therefore, $\mathbf{V}_B$ is found to be

$$\mathbf{V}_B = (12 + j4) + 10\angle{-90°} = 12 - j6 = 13.4\angle{-26.6°} \ \text{V} \quad ■$$

EXAMPLE 15–21

Use node analysis to find the current $\mathbf{I}_X$ in Figure 15–42.

F I G U R E   1 5 – 4 2

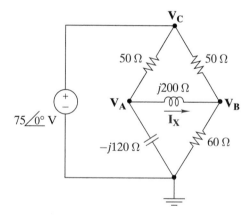

SOLUTION:

In this example we use node analysis on a problem solved in Example 15–18 using a Thévenin equivalent circuit. The voltage source cannot be replaced by source transformation because it is not connected in series with an impedance. By inspection, the node equations at nodes A and B are

Node A: $\dfrac{\mathbf{V}_A}{-j120} + \dfrac{\mathbf{V}_A - \mathbf{V}_B}{j200} + \dfrac{\mathbf{V}_A - \mathbf{V}_C}{50} = 0$

Node B: $\dfrac{\mathbf{V}_B}{60} + \dfrac{\mathbf{V}_B - \mathbf{V}_A}{j200} + \dfrac{\mathbf{V}_B - \mathbf{V}_C}{50} = 0$

A node equation at node C is not required because the voltage source forces the condition $\mathbf{V}_C = 75\angle 0°$. Substituting this constraint into the equations of nodes A and B and arranging the equations in standard form yields two equations in two unknowns:

Node A: $\left(\dfrac{1}{50} + \dfrac{1}{-j120} + \dfrac{1}{j200}\right)\mathbf{V}_A - \left(\dfrac{1}{j200}\right)\mathbf{V}_B = \left(\dfrac{75\angle 0°}{50}\right)$

Node B: $-\left(\dfrac{1}{j200}\right)\mathbf{V}_A + \left(\dfrac{1}{50} + \dfrac{1}{60} + \dfrac{1}{j200}\right)\mathbf{V}_B = \left(\dfrac{75\angle 0°}{50}\right)$

Solving these equations for $\mathbf{V}_A$ and $\mathbf{V}_B$ yields

$$\mathbf{V}_A = 70.4 - j21.4 \text{ V}$$

$$\mathbf{V}_B = 38.6 - j4.33 \text{ V}$$

Using these values for $\mathbf{V}_A$ and $\mathbf{V}_B$, the unknown current is found to be

$$\mathbf{I}_X = \dfrac{\mathbf{V}_A - \mathbf{V}_B}{j200} = \dfrac{31.8 - j17.1}{j200} = 0.180\angle -118° \text{ A}$$

This value of $\mathbf{I}_X$ is the same as the answer obtained in Example 15–18. ∎

EXAMPLE 15–22

Use node-voltage analysis to determine the phasor input-output relationship of the OP AMP circuit in Figure 15–43.

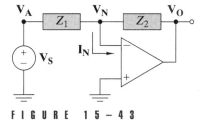

FIGURE 15–43

SOLUTION:
In the sinusoidal steady state the sum of currents leaving the inverting input node is

$$\dfrac{\mathbf{V}_N - \mathbf{V}_S}{Z_1} + \dfrac{\mathbf{V}_N - \mathbf{V}_O}{Z_2} + \mathbf{I}_N = 0$$

This is the only required node equation, since input source forces the condition $\mathbf{V}_A = \mathbf{V}_S$ and no node equation is ever required at an OP AMP output. In the time domain the i–v relationships of an ideal OP AMP are $v_P(t) = v_N(t)$ and $i_P(t) = i_N(t) = 0$. In the sinusoidal steady state these equations are written in phasor form as $\mathbf{V}_P = \mathbf{V}_N$ and $\mathbf{I}_P = \mathbf{I}_N = 0$. In the present case this means $\mathbf{V}_N = 0$ since the noninverting input is grounded. When the ideal OP AMP constraints are inserted in the node equation, we can solve for the OP AMP input-output relationship as

$$\mathbf{V}_O = -\dfrac{Z_2}{Z_1}\mathbf{V}_S$$

This result is the phasor domain version of the inverting amplifier configuration. In the phasor domain, the "gain" $K = -Z_2/Z_1$ is determined by a ratio of impedances rather than resistances. Thus, the gain affects both the amplitude and the phase angle of the steady-state output. ∎

Use node-voltage analysis to determine the phasor input-output relationship of the OP AMP circuit in Figure 15–44.

FIGURE 15–44

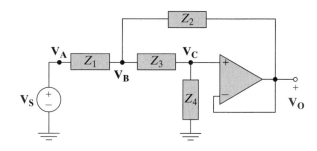

SOLUTION:
The input source forces the condition $\mathbf{V}_A = \mathbf{V}_S$. The OP AMP is connected as a voltage follower so the inverting input voltage is $\mathbf{V}_N = \mathbf{V}_O$. The voltage at the noninverting input is $\mathbf{V}_P = \mathbf{V}_C = \mathbf{V}_O$ since OP AMP voltage constraint requires $\mathbf{V}_P = \mathbf{V}_N$. Taking all these constraints into account, we can write the sum of currents leaving nodes B and C as

$$\text{Node B:} \quad \frac{\mathbf{V}_B - \mathbf{V}_S}{Z_1} + \frac{\mathbf{V}_B - \mathbf{V}_O}{Z_2} + \frac{\mathbf{V}_B - \mathbf{V}_O}{Z_3} = 0$$

$$\text{Node C:} \quad \frac{\mathbf{V}_O - \mathbf{V}_B}{Z_3} + \frac{\mathbf{V}_O}{Z_4} = 0$$

Solving the node C equation for $\mathbf{V}_B$ yields

$$\mathbf{V}_B = \frac{Z_3 + Z_4}{Z_4} \mathbf{V}_O$$

Using this result to eliminate $\mathbf{V}_B$ from the node B equation and then solving for the output $\mathbf{V}_O$ produces

$$\mathbf{V}_O = \frac{Z_2 Z_4}{Z_1 Z_2 + Z_1 Z_3 + Z_2 Z_3 + Z_2 Z_4} \mathbf{V}_S$$ ∎

The circuit in Figure 15–45 is an equivalent circuit of an ac induction motor. The current $\mathbf{I}_S$ is called the stator current, $\mathbf{I}_R$ the rotor current, and $\mathbf{I}_M$ the magnetizing current. Use the mesh-current method to solve for the branch currents $\mathbf{I}_S$, $\mathbf{I}_R$, and $\mathbf{I}_M$.

FIGURE 15-45

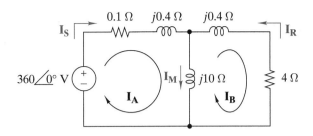

SOLUTION:

Applying KVL to the sum of voltages around each mesh in Figure 15–45 yields

Mesh A: $-360\angle0° + (0.1 + j0.4)\mathbf{I}_A + j10(\mathbf{I}_A - \mathbf{I}_B) = 0$

Mesh B: $j10(\mathbf{I}_B - \mathbf{I}_A) + (4 + j0.4)\mathbf{I}_B \qquad = 0$

Arranging these equations in standard form yields

$$(0.1 + j10.4)\mathbf{I}_A - (j10)\mathbf{I}_B = 360\angle0°$$

$$-(j10)\mathbf{I}_A + (4 + j10.4)\mathbf{I}_B = 0$$

Solving these equations for $\mathbf{I}_A$ and $\mathbf{I}_B$ produces

$$\mathbf{I}_A = 79.0 - j48.2 \text{ A}$$

$$\mathbf{I}_B = 81.7 - j14.9 \text{ A}$$

The required stator, rotor, and magnetizing currents are related to these mesh currents, as follows:

$$\mathbf{I}_S = \mathbf{I}_A = 92.5\angle-31.4° \text{ A}$$

$$\mathbf{I}_R = -\mathbf{I}_B = -81.8 + j14.7 = 83.0\angle170° \text{ A}$$

$$\mathbf{I}_M = \mathbf{I}_A - \mathbf{I}_B = -2.68 - j33.3 = 33.4\angle-94.6° \text{ A} \qquad ∎$$

EXAMPLE 15-25

Use the mesh-current method to solve for output voltage $\mathbf{V}_2$ and input impedance Z_{IN} of the circuit in Figure 15–46.

FIGURE 15-46

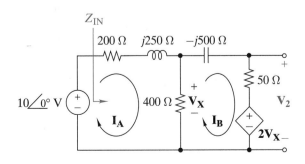

SOLUTION:
The circuit contains a voltage-controlled voltage source. We initially treat the dependent source as an independent source and use KCL to write the sum of voltages around each mesh:

$$\text{Mesh A:} \quad -10\angle 0° + (200 + j250)\mathbf{I}_A + 400(\mathbf{I}_A - \mathbf{I}_B) = 0$$

$$\text{Mesh B:} \quad 400(\mathbf{I}_B - \mathbf{I}_A) + (50 - j500)\mathbf{I}_B + 2\mathbf{V}_X = 0$$

Arranging these equations in standard form produces

$$\text{Mesh A:} \; (600 + j250)\mathbf{I}_A - 400\mathbf{I}_B = 10\angle 0°$$

$$\text{Mesh B:} \; -400\mathbf{I}_A + (450 - j500)\mathbf{I}_B = -2\mathbf{V}_X$$

Using Ohm's law, the control voltage $\mathbf{V}_X$ is

$$\mathbf{V}_X = 400(\mathbf{I}_A - \mathbf{I}_B)$$

Eliminating $\mathbf{V}_X$ from the mesh equations yields

$$\text{Mesh A:} \; (600 + j250)\mathbf{I}_A - 400\mathbf{I}_B = 10\angle 0°$$

$$\text{Mesh B:} \; 400\mathbf{I}_A + (-350 - j500)\mathbf{I}_B = 0$$

Solving for the two mesh currents produces

$$\mathbf{I}_A = 10.8 - j11.1 \text{ mA}$$

$$\mathbf{I}_B = -1.93 - j9.95 \text{ mA}$$

Using these values of the mesh currents, the output voltage and input impedance are

$$\mathbf{V}_2 = 2\mathbf{V}_X + 50\mathbf{I}_B = 800(\mathbf{I}_A - \mathbf{I}_B) + 50\mathbf{I}_B$$

$$= 800\mathbf{I}_A - 750\mathbf{I}_B = 10.1 - j1.42$$

$$= 10.2\angle -8.00° \text{ V}$$

$$Z_{IN} = \frac{10\angle 0°}{\mathbf{I}_A} = \frac{10\angle 0°}{0.0108 - j0.0111} = 450 + j463 \; \Omega \qquad \blacksquare$$

EXAMPLE 15–26

In the circuit in Figure 15–47 the input voltage is $v_S(t) = 10 \cos 10^5 t$ V. Use mesh equations and MATLAB to find the input impedance at the input interface and the proportionality constant relating the input voltage phasor to the phasor voltage across the 50-Ω load resistor.

SOLUTION:
In matrix form, the mesh-current equations for this circuit are

$$
\begin{bmatrix}
R_S + Z_{C1} + Z_{L2} & -Z_{L2} & -Z_{C1} \\
-Z_{L2} & Z_{L2} + Z_{C2} + R_L & -Z_{C2} \\
-Z_{C1} & -Z_{C2} & Z_{C1} + Z_{C2} + Z_{L1}
\end{bmatrix}
\begin{bmatrix}
\mathbf{I}_A \\
\mathbf{I}_B \\
\mathbf{I}_C
\end{bmatrix}
=
\begin{bmatrix}
\mathbf{V}_S \\
0 \\
0
\end{bmatrix}
$$

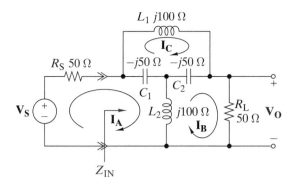

This matrix equation is of the form $\mathbf{ZI} = \mathbf{V}$, where $\mathbf{Z}$ is a 3×3 matrix of element impedance, $\mathbf{I}$ is the 3×1 mesh current vector, and $\mathbf{V}$ is the 3×1 input voltage vector. The entries on the main diagonal of the $\mathbf{Z}$ matrix are the total impedances in each mesh. The $\mathbf{Z}$ matrix is symmetrical because the off diagonal entries are the negatives of the impedances common to two meshes. For example, the entry in the 2nd row and 3rd column $(-Z_{C2})$ is equal to the entry in the 3rd row and 2nd column.

With the circuit equations formulated, we can use MATLAB to solve for the required circuit descriptors. To solve for the mesh current vector, we need the matrix product $\mathbf{Z}^{-1}\mathbf{V}$. We have used MATLAB for this purpose in previous examples. The only thing new here is that the matrices contain complex numbers.

One of the advantages of MATLAB is the ease with which it handles complex numbers. Complex inputs are identified using either the mathematical convention $(i = \sqrt{-1})$ or the electrical engineering convention $(j = \sqrt{-1})$. That is, MATLAB accepts complex number inputs written as z=x+yi or z=x+yj. Accordingly, in the MATLAB command window, we enter the element parameters as

```
VS=10; RS=50; RL=50;
ZC1=-50j; ZC2=-50j;
ZL1=100j; ZL2=100j;
```

the $\mathbf{Z}$ and $\mathbf{V}$ matrices as

```
Z=[RS+ZC1+ZL2      -ZL2        -ZC1
       -ZL2      ZL2+ZC2+RL    -ZC2
       -ZC1        -ZC2     ZC1+ZC2+ZL1];
V=[VS ; 0 ; 0];
```

and solve for the mesh currents as

```
I=inv(Z)*V
I=

      0.0100 - 0.0300i
     -0.0100 + 0.0300i
           0 - 0.1000i
```

The form of the output points out that MATLAB accepts inputs in either convention, but it reports complex number outputs using the mathematical convention. In MATLAB matrix notation, the mesh currents are $\mathbf{I}_A = I(1)$, $\mathbf{I}_B = I(2)$, and $\mathbf{I}_C = I(3)$. We can now calculate the required parameters as

```
ZIN=(VS-I(1)*RS)/I(1)
ZIN =
    5.0000e+001 + 3.0000e+002i
K=I(2)*RL/VS
K=
    -0.05000 + 0.15000i
```

That is, in electrical engineering notation, $Z_{IN} = 50 + j300 \ \Omega$ and $K = -0.05 + j0.15$. ■

Exercise 15–16

Use the mesh-current or node-voltage method to find the branch currents $\mathbf{I}_1$, $\mathbf{I}_2$, and $\mathbf{I}_3$ in Figure 15–48.

FIGURE 15 – 48

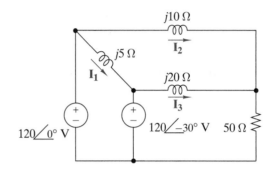

Answer:

$$\mathbf{I}_1 = 12.4\angle{-15°} \ \text{A}, \ \mathbf{I}_2 = 3.61\angle{-16.1°} \ \text{A}, \ \mathbf{I}_3 = 1.31\angle{166°} \ \text{A}$$

Exercise 15–17

Use the mesh-current or node-voltage method to find the output voltage $\mathbf{V}_2$ and input impedance Z_{IN} in Figure 15–49.

FIGURE 15 – 49

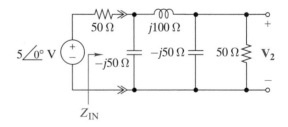

Answer:

$$\mathbf{V}_2 = 1.77\angle{-135°} \ \text{V}, \ Z_{IN} = 100 - j100 \ \Omega$$

Use the mesh-current or node-voltage method to find the current $\mathbf{I}_X$ in Figure 15–50.

Answer:

$\mathbf{I}_X = 1.44 \angle 171° \text{ mA}$

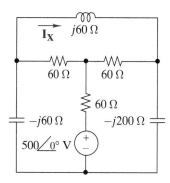

F I G U R E   1 5 – 5 0

15–6   E NERGY   AND   P OWER

In the sinusoidal steady state, ac power is transferred from sources to various loads. To study the transfer process, we must calculate the power delivered in the sinusoidal steady state to any specified load. It turns out that there is an upper bound on the available load power; hence, we need to understand how to adjust the load to extract the maximum power from the rest of the circuit. In this section the load is assumed to be made up of passive resistance, inductance, and capacitance. To reach our objectives, we must first study the power and energy delivered to these passive elements in the sinusoidal steady state.

In the sinusoidal steady state the current through a resistor can be expressed as $i_R(t) = I_A \cos(\omega t)$. The instantaneous power delivered to the resistor is

$$p_R(t) = R i_R^2(t) = R I_A^2 \cos^2(\omega t) \qquad (15\text{–}30)$$

$$= \frac{R I_A^2}{2}[1 + \cos(2\omega t)]$$

where the identity $\cos^2(x) = \frac{1}{2}[1 + \cos(2x)]$ is used to obtain the last line in Eq. (15–30). The energy delivered for $t \geq 0$ is found to be

$$w_R(t) = \int_0^t p_R(x)dx = \frac{R I_A^2}{2}\int_0^t dx + \frac{R I_A^2}{2}\int_0^t \cos 2\omega x \, dx$$

$$= \frac{R I_A^2}{2}t + \frac{R I_A^2}{4\omega}\sin 2\omega t$$

Figure 15–51 shows the time variation of $p_R(t)$ and $w_R(t)$. Note that the power is a periodic function with twice the frequency of the current, that both $p_R(t)$ and $w_R(t)$ are always positive, and that $w_R(t)$ increases without bound. These observations remind us that a resistor is a passive element that dissipates energy.

In the sinusoidal steady state an inductor operates with a current $i_L(t) = I_A \cos(\omega t)$. The corresponding energy stored in the element is

$$w_L(t) = \frac{1}{2} L i_L^2(t) = \frac{1}{2} L I_A^2 \cos^2 \omega t$$

$$= \frac{1}{4} L I_A^2 (1 + \cos 2\omega t)$$

where the identity $\cos^2(x) = \frac{1}{2}[1 + \cos(2x)]$ is again used to produce the last line. The instantaneous power delivered to the inductor is

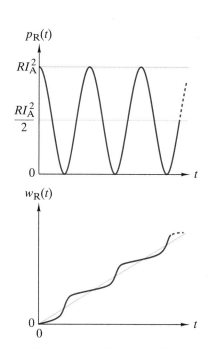

F I G U R E   1 5 – 5 1   *Resistor power and energy in the sinusoidal steady state.*

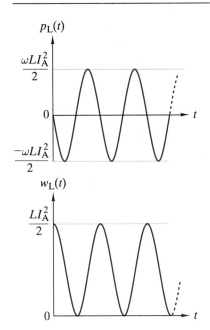

FIGURE 15 – 52 *Inductor power and energy in the sinusoidal steady state.*

$$p_L(t) = \frac{dw_L}{dt} = -\frac{\omega L I_A^2}{2}\sin(2\omega t) \qquad (15\text{–}31)$$

Figure 15–52 shows the time variation of $p_L(t)$ and $w_L(t)$. Observe that both $p_L(t)$ and $w_L(t)$ are periodic functions at twice the frequency of the ac current, that $p_L(t)$ is alternately positive and negative, and that $w_L(t)$ is never negative. Since $w_L(t) \geq 0$, the inductor does not deliver net energy to the rest of the circuit. Unlike the resistor's energy in Figure 15–51, the energy in the inductor is bounded by $\frac{1}{2} L I_A^2 \geq w_L(t)$, which means that the inductor does not dissipate energy. Finally, since $p_L(t)$ alternates signs, we see that the inductor stores energy during a positive half cycle and then returns the energy undiminished during the next negative half cycle. Thus, in the sinusoidal steady state there is a lossless interchange of energy between an inductor and the rest of the circuit.

In the sinusoidal steady state the voltage across a capacitor is $v_C(t) = V_A \cos(\omega t)$. The energy stored in the element is

$$w_C(t) = \frac{1}{2}Cv_C^2(t) = \frac{1}{2}CV_A^2 \cos^2 \omega t$$

$$= \frac{1}{4}CV_A^2(1 + \cos 2\omega t)$$

The instantaneous power delivered to the capacitor is

$$p_C(t) = \frac{dw_C}{dt} = -\frac{\omega C V_A^2}{2}\sin(2\omega t) \qquad (15\text{–}32)$$

Figure 15–53 shows the time variation of $p_C(t)$ and $w_C(t)$. Observe that these relationships are the duals of those found for the inductor. Thus, in the sinusoidal steady state the element power is sinusoidal and there is a lossless interchange of energy between the capacitor and the rest of the circuit.

AVERAGE POWER

We are now in a position to calculate the average power delivered to various loads. The instantaneous power delivered to any of the three passive elements is a periodic function that can be described by an average value. The **average power** in the sinusoidal steady state is defined as

$$P = \frac{1}{T_0}\int_0^{T_0} p(t)dt$$

The power variation of the inductor in Eq. (15–31) and capacitor in Eq. (15–32) have the same sinusoidal form. The average value of any sinusoid is zero since the areas under alternate cycles cancel. Hence, the average power delivered to an inductor or capacitor is zero:

Inductor: $P_L = 0$

Capacitor: $P_C = 0$

The resistor power in Eq. (15–30) has both a sinusoidal ac component and a constant dc component $\frac{1}{2} R I_A^2$. The average value of the ac component is zero, but the dc component yields

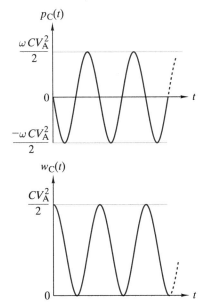

FIGURE 15 – 53 *Capacitor power and energy in the sinusoidal steady state.*

$$\text{Resistor: } P_R = \frac{1}{2}RI_A^2$$

To calculate the average power delivered to an arbitrary load $Z_L = R_L + jX_L$, we use phasor circuit analysis to find the phasor current $\mathbf{I}_L$ through Z_L. The average power delivered to the load is dissipated in R_L, since the reactance X_L represents the net inductance or capacitance of the load. Hence the average power to the load is

$$P = \frac{1}{2}R_L|\mathbf{I}_L|^2 \qquad (15\text{–}33)$$

Caution: When a circuit contains two or more sources, superposition applies only to the total load current and not to the total load power. You cannot find the total power to the load by summing the power delivered by each source acting alone.

The following example illustrates a power transfer calculation.

EXAMPLE 15–27

Find the average power delivered to the load to the right of the interface in Figure 15–54.

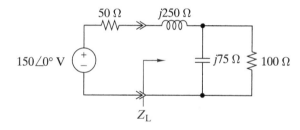

FIGURE 15 – 54

SOLUTION:
The equivalent impedance to the right of the interface is

$$Z_L = j250 + \cfrac{1}{\cfrac{1}{-j75} + \cfrac{1}{100}} = 36 + j202 \ \ \Omega$$

The current delivered to the load is

$$\mathbf{I}_L = \frac{150\angle 0^\circ}{50 + Z_L} = 0.683\angle -66.9^\circ \ \text{A}$$

Hence the average power delivered across the interface is

$$P = \frac{1}{2}R_L|\mathbf{I}_L|^2 = \frac{36}{2}|0.683|^2 = 8.40 \ \ \text{W}$$

Note: All of this power goes into the 100-Ω resistor since the inductor and capacitor do not absorb average power.   ■

Exercise 15–19

Find the average power delivered to the 25-Ω load resistor in Figure 15–55.

FIGURE 15 – 55

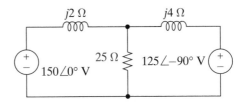

Answer: 234 W

MAXIMUM POWER

To address the maximum power transfer problem, we model the source/load interface as shown in Figure 15–56. The source circuit is represented by a Thévenin equivalent circuit with source voltage $\mathbf{V}_T$ and source impedance $Z_T = R_T + jX_T$. The load circuit is represented by an equivalent impedance $Z_L = R_L + jX_L$. In the maximum power transfer problem the source parameters $\mathbf{V}_T$, R_T, and X_T are given, and the objective is to adjust the load impedance R_L and X_L so that average power to the load is a maximum.

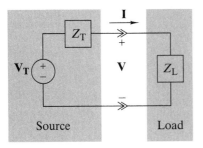

FIGURE 15 – 56 *A source-load interface in the sinusoidal steady state.*

The average power to the load is expressed in terms of the phasor current and load resistance:

$$P = \frac{1}{2}R_L|\mathbf{I}|^2$$

Then, using series equivalence, we express the magnitude of the interface current as

$$|\mathbf{I}| = \left|\frac{\mathbf{V}_T}{Z_T + Z_L}\right| = \frac{|\mathbf{V}_T|}{|(R_T + R_L) + j(X_T + X_L)|}$$

$$= \frac{|\mathbf{V}_T|}{\sqrt{(R_T + R_L)^2 + (X_T + X_L)^2}}$$

Combining the last two equations yields the average power delivered across the interface:

$$P = \frac{1}{2}\frac{R_L|\mathbf{V}_T|^2}{(R_T + R_L)^2 + (X_T + X_L)^2} \tag{15–34}$$

The quantities $|\mathbf{V}_T|$, R_T, and X_T in Eq. (15–34) are fixed. Our problem is to select R_L and X_L to maximize P.

Clearly, for every value of R_L the denominator in Eq. (15–34) is minimized and P maximized when $X_L = -X_T$. This choice of X_L is possible because a reactance can be positive or negative. When the source Thévenin equivalent has an inductive reactance ($X_T > 0$), we modify the load to have a capacitive reactance of the same magnitude, and vice versa. This step reduces the net reactance of the series connection in Figure 15–56 to zero, creating a condition in which the net impedance seen by the Thévenin voltage source is purely resistive.

When the source and load reactances cancel out, the expression for average power in Eq. (15–34) reduces to

$$P = \frac{1}{2} \frac{R_L |\mathbf{V}_T|^2}{(R_T + R_L)^2} \qquad (15\text{–}35)$$

This equation has the same form encountered in Chapter 3 in dealing with maximum power transfer in resistive circuits. From the derivation in Sect. 3–5, we know P is maximized when $R_L = R_T$. In summary, to obtain maximum power transfer in the sinusoidal steady state, we select the load resistance and reactance so that

$$R_L = R_T \quad \text{and} \quad X_L = -X_T \qquad (15\text{–}36)$$

These conditions can be compactly expressed in the following way:

$$Z_L = Z_T^* \qquad (15\text{–}37)$$

The condition for maximum power transfer is called a **conjugate match**, since the load impedance is the conjugate of the source impedance. When the conjugate-match conditions are inserted into Eq. (15–34), we find that the maximum average power available from the source circuit is

$$P_{\text{MAX}} = \frac{|\mathbf{V}_T|^2}{8R_T} \qquad (15\text{–}38)$$

where $|\mathbf{V}_T|$ is the peak amplitude of the Thévenin equivalent voltage.

It is important to remember that conjugate matching applies when the source is fixed and the load is adjustable. These conditions arise frequently in power-limited communication systems. However, as we will see in Chapter 17, conjugate matching does not apply to electrical power systems because the power transfer constraints are different.

EXAMPLE 15–28

(a) Calculate the average power delivered to the load in the circuit shown in Figure 15–57 for $v_S(t) = 5 \cos 10^6 t$ V, $R = 200$ Ω, $R_L = 200$ Ω, and $C = 10$ nF.
(b) Calculate the maximum average power available at the interface and specify the load required to draw the maximum power.

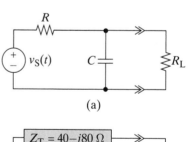

(a)

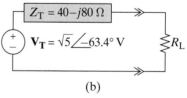

(b)

FIGURE   15 – 57

SOLUTION:

(a) To find the power delivered to the 200-Ω load resistor, we use a Thévenin equivalent circuit. By voltage division, the open-circuit voltage at the interface is

$$\mathbf{V}_T = \frac{Z_C}{Z_R + Z_C}\mathbf{V}_S = \frac{-j100}{200 - j100}5\angle 0°$$

$$= 1 - j2 = \sqrt{5}\,\angle -63.4° \text{ V}$$

By inspection, the short-circuit current at the interface is

$$\mathbf{I}_N = \frac{5\angle 0°}{200} = 0.025 + j0 \text{ A}$$

Given $\mathbf{V}_T$ and $\mathbf{I}_N$, we calculate the Thévenin source impedance.

$$Z_T = \frac{\mathbf{V}_T}{\mathbf{I}_N} = \frac{1 - j2}{0.025} = 40 - j80 \quad \Omega$$

Using the Thévenin equivalent shown in Figure 15–57(b), we find that the current through the 200-Ω resistor is

$$\mathbf{I} = \frac{\mathbf{V}_T}{Z_T + Z_L} = \frac{\sqrt{5}\angle -63.4°}{40 - j80 + 200} = 8.84\angle -45° \quad \text{mA}$$

and the average power delivered to the load resistor is

$$P = \frac{1}{2}R_L|\mathbf{I}|^2 = 100(8.84 \times 10^{-3})^2 = 7.81 \quad \text{mW}$$

(b) Using Eq. (15–38), the maximum average power available at the interface is

$$P_{\text{MAX}} = \frac{|\mathbf{V}_T|^2}{8R_T} = \frac{(\sqrt{5})^2}{(8)(40)} = 15.6 \quad \text{mW}$$

The 200-Ω load resistor in part (a) draws about half of the maximum available power. To extract maximum power, the load impedance must be

$$Z_L = Z_T^* = 40 + j80 \quad \Omega$$

This impedance can be obtained using a 40-Ω resistor in series with a reactance of +80 Ω. The required reactance is inductive (positive) and can be produced by an inductance of

$$L = \frac{|X_T|}{\omega} = \frac{80}{10^6} = 80 \ \mu\text{H} \qquad \blacksquare$$

Exercise 15–20

Calculate the maximum average power available at the interface in Figure 15–58.

Answer: 125 mW.

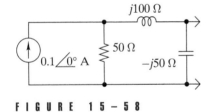

FIGURE 15-58

SUMMARY

- A phasor is a complex number representing a sinusoidal waveform. The magnitude and angle of the phasor correspond to the amplitude and phase angle of the sinusoid. The phasor does not provide frequency information.

- In the sinusoidal steady state, phasor currents and voltages obey Kirchhoff's laws and the element i–v relationships are written in terms of impedances. Impedance can be defined as the ratio of phasor voltage over phasor current. The device and connection constraints for phasor circuit analysis have the same form as resistance circuits.

- Phasor circuit analysis techniques include series equivalence, parallel equivalence, circuit reduction, Thévenin's and Norton's theorems, unit output method, superposition, node-voltage analysis, and mesh-current analysis.

- In the sinusoidal steady state the equivalent impedance at a pair of terminals is $Z(j\omega) = R(\omega) + jX(\omega)$, where $R(\omega)$ is called resistance and $X(\omega)$ is called reactance. A frequency at which an equivalent impedance is purely real is called a resonant frequency. Admittance is the reciprocal of impedance.

- In the sinusoidal steady state the instantaneous power to a passive element is a periodic function at twice the frequency of the driving force. The average power delivered to an inductor or capacitor is zero. The average power delivered to a resistor is $\frac{1}{2} R|\mathbf{I}_R|^2$. The maximum average power is delivered by a fixed source to an adjustable load when the source and load impedances are conjugates.

PROBLEMS

ERO 15–1 SINUSOIDS AND PHASORS (SECT. 15–1)

Use the additive properties of phasors to convert sinusoidal waveforms into phasors and vice versa.
See Examples 15–1, 15–2, 15–3, 15–4 and Exercises 15–1, 15–2, 15–3, 15–4

15–1 Transform the following sinusoids into phasor form and draw a phasor diagram. Use the additive property of phasors to find $v_1(t) + v_2(t)$.

    (a) $v_1(t) = 250 \cos(\omega t + 45°)$ V

    (b) $v_2(t) = 150 \cos \omega t + 100 \sin \omega t$ V

15–2 Transform the following sinusoids into phasor form and draw a phasor diagram. Use the additive property of phasors to find $i_1(t) + i_2(t)$.

    (a) $i_1(t) = 6 \cos \omega t$ A

    (b) $i_2(t) = 3 \cos(\omega t - 90°)$ A

15–3 Convert the following phasors into sinusoidal waveforms.

    (a) $\mathbf{V}_1 = 10\,e^{-j30°}$, $\omega = 10^4$

    (b) $\mathbf{V}_2 = 60\,e^{-j220°}$, $\omega = 10^4$

    (c) $\mathbf{I}_1 = 5\,e^{j90°}$, $\omega = 200$

    (d) $\mathbf{I}_2 = 2\,e^{j270°}$, $\omega = 200$

15–4 Use the phasors in Problem 15–3 and the additive property to find $2v_1(t) + v_2(t)$ and $i_1(t) + 3i_2(t)$.

15–5 A sinusoid with $\omega = 20$ rad/s has a phasor representation $\mathbf{V} = 20 + j5$ V. Find the time derivative of the waveform $v(t) = \text{Re}\{\mathbf{V}e^{j\omega t}\}$.

15–6 Convert the following phasors into sinusoids.

    (a) $\mathbf{V}_1 = 10 + j40$, $\omega = 10$

    (b) $\mathbf{V}_2 = (8 - j3)5\,e^{-j60°}$, $\omega = 20$

    (c) $\mathbf{I}_1 = 8 - j3 + \dfrac{3}{j}$, $\omega = 300$

    (d) $\mathbf{I}_2 = \dfrac{3 + j1}{1 - j3}$, $\omega = 50$

15–7 The input–output relationship of a signal processor is

$$v_2(t) = \frac{1}{200}\frac{dv_1(t)}{dt} + v_1(t)$$

Use phasors to find the output $v_2(t)$ when the input is $v_1(t) = 10 \cos(500t + 45°)$.

15–8 Given the sinusoids

$$v_1(t) = 50 \cos(\omega t - 45°) \quad \text{and} \quad v_2(t) = 25 \sin \omega t$$

use the additive property of phasors to find $v_3(t)$ such that $v_1 + v_2 + v_3 = 0$.

15–9 Find the phasor corresponding to $v(t) = 12\sqrt{2} \cos [1000\pi(t - 2.5 \times 10^{-4})]$.

15–10 Given a phasor $\mathbf{V}_1 = -3 + j4$, use phasor methods to find a voltage $v_2(t)$ that leads $v_1(t)$ by 90° and has an amplitude of 10 V.

ERO 15–2 EQUIVALENT IMPEDANCE (SECTS. 15–2, 15–3)

Given a linear circuit, use series and parallel equivalence to find the equivalent impedance at a specified pair of terminals.
See Examples 15–5, 15–6, 15–9, 15–10, 15–12 and Exercises 15–8, 15–10, 15–11

15–11 Express the equivalent impedance of the following circuits in rectangular and polar form:
    (a) A 25-Ω resistor in series with a 20-mH inductor at $\omega = 1000$ rad/s.

(b) A 25-Ω resistor in parallel with a 20-μF capacitor at $\omega = 1000$ rad/s.

(c) The circuit formed by connecting the circuits in (a) and (b) in parallel.

(d) Repeat (c) at $\omega = 4000$ rad/s.

15–12 Find the equivalent impedance Z in Figure P15–12. Express the result in both polar and rectangular form.

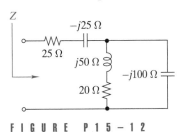

FIGURE P15–12

15–13 Express the equivalent impedance of the following circuits in rectangular and polar form:

(a) A 100-mH inductor in series with a 10-μF capacitor at $\omega = 2000$ rad/s.

(b) A 30-mH inductor in parallel with 60-Ω resistor at $\omega = 2000$ rad/s.

(c) The circuit formed by connecting the circuits in (a) and (b) in series.

(d) Repeat (c) at $\omega = 1000$ rad/s.

15–14 Find the equivalent impedance Z in Figure P15–14. Express the result in both polar and rectangular form.

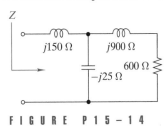

FIGURE P15–14

15–15 The voltage applied at the input to a linear circuit is $v(t) = 200 \cos(1000t - 60°)$ V. In the sinusoidal steady state the input current is observed to be $i(t) = 20 \cos 1000t$ mA.

(a) Find the equivalent impedance at the input.

(b) Find the steady-state current $i(t)$ for $v(t) = 150 \cos(1000t - 270°)$ V.

15–16 The circuit in Figure P15–16 is operating in the sinusoidal steady state with $\omega = 20$ krad/s. Find the equivalent impedance Z.

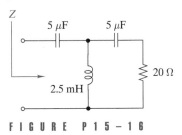

FIGURE P15–16

15–17   An inductor L is connected in parallel with a 600-Ω resistor. The parallel combination is then connected in series with a capacitor C. Select the values of L and C so that the equivalent impedance of the combination is $100 + j0$ Ω at $\omega = 100$ krad/s.

15–18 A 200-Ω resistor is connected in series with a capacitor. When the current through the series combination is $i(t) = 10 \cos(1000t)$ mA, the voltage is $v(t) = 2\sqrt{2} \cos(1000 t - 45°)$. Find the value of C.

15–19   A capacitor C is connected in parallel with a resistor R. Select values of R and C so that the equivalent impedance of the combination is $600 - j600$ Ω at $\omega = 1$ Mrad/s.

15–20 A relay coil can be modeled as a resistor and inductor in series. When a certain coil is connected across a 110-V, 60-Hz source the amplitude of the relay current is observed to be 0.25 A. When the frequency of the 110-V source is increased to 400 Hz, the amplitude drops to 0.12 A. Find the resistance and inductance of the coil.

ERO 15–3 BASIC PHASOR CIRCUIT ANALYSIS (SECTS. 15-3, 15–4)

Given a linear circuit operating in the sinusoidal steady state, find phasor responses using basic analysis methods such as series and parallel equivalence, voltage and current division, circuit reduction, Thévenin or Norton equivalent circuits, and proportionality or superposition.
See Examples 15–6, 15–8, 15–9, 15–10, 15–13, 15–14, 15–15, 15–17, 15–18, 15–19 and Exercises 15–9, 15–13, 15–14, 15–15

15–21 A voltage $v(t) = 10 \cos 500t$ V is applied across a series connection of a 100-Ω resistor and 40-mH inductor. Find the steady-state current $i(t)$ through the series connection.

15–22 The circuit in Figure P15–22 is operating in the sinusoidal steady state with $v_S(t) = V_A \cos \omega t$ V. Use circuit reduction to derive a general expression for the phasor response $\mathbf{I}_L$.

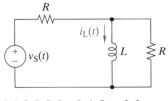

FIGURE P15–22

15–23 A current source delivering $i(t) = 300 \cos 2000t$ mA is connected across a parallel combination of a 10-kΩ resistor and a 50-nF capacitor. Find the steady-state current $i_R(t)$ through the resistor and the steady-state current $i_C(t)$ through the capacitor. Draw a phasor diagram showing $\mathbf{I}$, $\mathbf{I}_C$, and $\mathbf{I}_R$.

15–24 The circuit in Figure P15–24 is operating in the sinusoidal steady state with $i_S(t) = I_A \cos \omega t$ A. Use phasor circuit reduction to derive a general expression for the steady-state response $\mathbf{V}_R$.

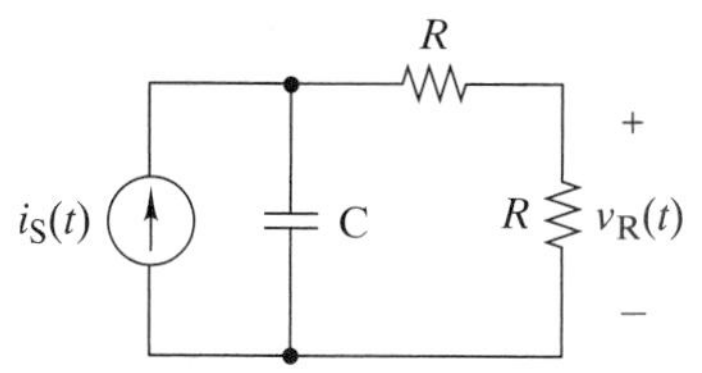

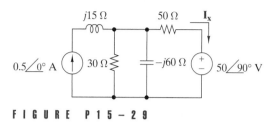

FIGURE P15–24

15–25 A voltage $v(t) = 50\cos(1000t + 45°)$ V is applied across a parallel connection of a 5-kΩ resistor and a 200-nF capacitor. Find the steady-state current $i_C(t)$ through the capacitor and the steady-state current $i_R(t)$ through the resistor. Draw a phasor diagram showing $\mathbf{V}$, $\mathbf{I}_C$, and $\mathbf{I}_R$.

15–26 The circuit in Figure P15–26 is operating in the sinusoidal steady state. Use circuit reduction to find the input impedance seen by the voltage source and the steady-state response $v_x(t)$.

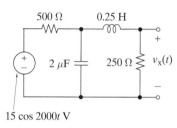

FIGURE P15–26

15–27 The circuit in Figure P15–27 is operating in the sinusoidal steady state. Use circuit reduction to find the input impedance seen by the current source and steady-state response $v_x(t)$.

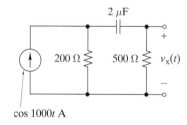

cos 1000t A

FIGURE P15–27

15–28 The circuit in Figure P15–28 is operating in the sinusoidal steady state. Use circuit reduction to find the input impedance seen by the voltage source and the steady-state phasor response $\mathbf{V}_x$.

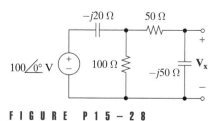

FIGURE P15–28

15–29 The circuit in Figure P15–29 is operating in the sinusoidal steady state. Use superposition to find the phasor response $\mathbf{I}_x$.

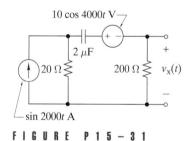

FIGURE P15–29

15–30 The circuit in Figure P15–30 is operating in the sinusoidal steady state. Use superposition to find the response $v_x(t)$. *Note:* The sources do not have the same frequency.

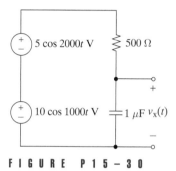

FIGURE P15–30

15–31 The circuit in Figure P15–31 is operating in the sinusoidal steady state. Use superposition to find the response $v_x(t)$. *Note:* The sources do not have the same frequency.

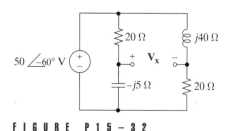

FIGURE P15–31

15–32 The circuit in Figure P15–32 is operating in the sinusoidal steady state. Find the input impedance seen by the voltage source and the phasor response $\mathbf{V}_x$.

FIGURE P15–32

15–33 The circuit in Figure P15–33 is operating in the sinusoidal steady state. Use the unit output method to find the input impedance seen by the voltage source and the phasor response $\mathbf{V}_x$.

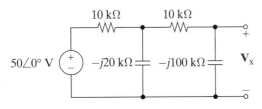

FIGURE P15-33

15–34 Find the phasor Thévenin equivalent of the source circuit to the left of the interface in Figure P15–34. Then use the equivalent circuit to find the steady-state voltage $v(t)$ and current $i(t)$ delivered to the load.

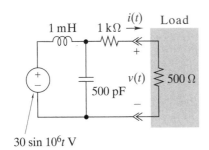

FIGURE P15-34

15–35 Find the phasor Thévenin equivalent of the source circuit to the left of the interface in Figure P15–35. Then use the equivalent circuit to find the phasor voltage $\mathbf{V}$ and current $\mathbf{I}$ delivered to the load.

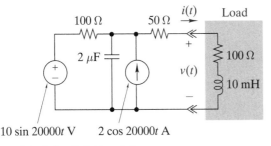

FIGURE P15-35

15–36 The circuit in Figure P15–36 is operating in the sinusoidal steady state. When $Z_L = 0$, the phasor current at the interface is $\mathbf{I} = 4.8 - j3.6$ mA. When $Z_L = -j20$ kΩ, the phasor interface current is $\mathbf{I} = 10 + j0$ mA. Find the Thévenin equivalent of the source circuit.

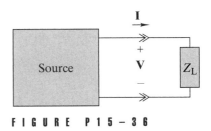

FIGURE P15-36

15–37 The circuit in Figure P15–36 is operating in the sinusoidal steady state with $\omega = 2$ krad/s. When $Z_L = 0$ the phasor short-circuit current is $\mathbf{I} = \mathbf{I}_{SC} = 0.8 - j0.4$ A. When a 1-nF capacitor is connected across the interface, the current delivered is $\mathbf{I} = \mathbf{I}_{LOAD} = 3 + j0$ A. Note that the amplitude of the load current is greater than the amplitude of the short-circuit current. Find the Thévenin equivalent of the source circuit and then explain why $|\mathbf{I}_{LOAD}| > |\mathbf{I}_{SC}|$.

15–38 Use a Thévenin equivalent circuit to find the phasor response $\mathbf{V}_x$ in Figure P15–33.

15–39 **D** Design a two-port circuit so that an input voltage $v_S(t) = 100 \cos(10^4 t)$ V delivers a steady-state output current of $i_O(t) = 10 \cos(10^4 t - 35°)$ mA to a 50-Ω resistive load.

15–40 **D** Design a two-port circuit so that an input voltage $v_S(t) = 50 \sin(1000t)$ V delivers a steady-state output voltage of $v_O(t) = 50 \cos(1000t - 45°)$ V.

ERO 15–4 GENERAL CIRCUIT ANALYSIS (SECT. 15–5)

Given a linear circuit operating in the sinusoidal steady state, find equivalent impedances and phasor responses using node-voltage or mesh-current analysis.
See Examples 15–20, 15–21, 15–22, 15–23, 15–24, 15–25, 15–26 and Exercises 15–16, 15–17, 15–18

15–41 Use node-voltage analysis to find the sinusoidal steady-state response $v_x(t)$ in the circuit shown in Figure P15–41.

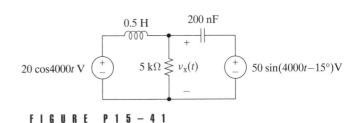

FIGURE P15-41

15–42 Use node-voltage analysis to find the steady-state phasor response $\mathbf{V}_O$ in the circuit shown in Figure P15–42.

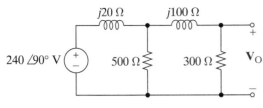

FIGURE P15–42

15–43 Use node-voltage analysis to find the input impedance Z_{IN} and phasor gain $K = \mathbf{V}_O/\mathbf{V}_S$ of the circuit shown in Figure P15–43 with $\mu = 100$.

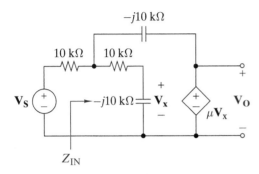

FIGURE P15–43

15–44 Use mesh-current analysis to find the phasor branch currents $\mathbf{I}_1$, $\mathbf{I}_2$, and $\mathbf{I}_3$ in the circuit shown in Figure P15–44.

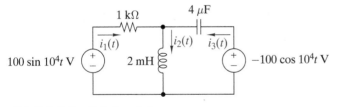

FIGURE P15–44

15–45 Use mesh-current analysis to find the phasor response $\mathbf{V}_y$ and $\mathbf{I}_x$ in the circuit shown in Figure P15–45.

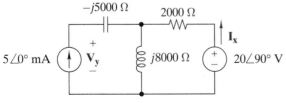

FIGURE P15–45

15–46 Use mesh-current analysis to find the input impedance Z_{IN} and phasor gain $K = \mathbf{V}_O/\mathbf{V}_S$ of the circuit shown in Figure P15–46.

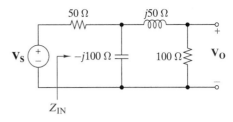

FIGURE P15–46

15–47 Find the input impedance Z_{IN} and phasor gain $K = \mathbf{V}_O/\mathbf{V}_S$ of the circuit shown in Figure P15–47.

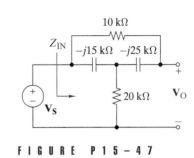

FIGURE P15–47

15–48 Find the sinusoidal steady state response $v_O(t)$ in the circuit shown in Figure P15–48. The element values are $\mathbf{V}_S = 80$ mV, $R_1 = 10$ kΩ, $R_P = 5$ kΩ, $R_F = 1$ MΩ, $R_C = 10$ kΩ, $R_L = 100$ kΩ, $C = 0.25$ nF, and $\beta = 50$. The frequency is $\omega = 500$ rad/s.

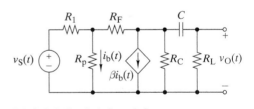

FIGURE P15–48

15–49 Find the phasor response $\mathbf{I}_{IN}$ and $\mathbf{V}_O$ in the circuit shown in Figure P15–49.

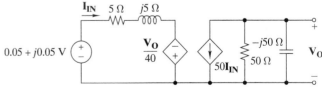

FIGURE P15–49

15–50 The OP AMP circuit in Figure P15–50 is operating in the sinusoidal steady state with $\omega = 100$ krad/s. Find the input phasor $\mathbf{V}_S$ when the output phasor is $\mathbf{V}_O = 10 + j0$ V.

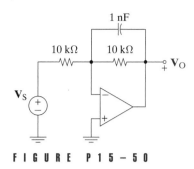

FIGURE P15-50

ERO 15-5 AVERAGE POWER AND MAXIMUM POWER TRANSFER (SECT. 15-6)

Given a linear circuit operating in the sinusoidal steady state,

(a) Find the average power delivered at a specified interface.
(b) Find the maximum average power available at a specified interface.
(c) Find the load impedance required to draw the maximum available power.

See Examples 15-27, 15-28 and Exercises 15-19, 15-20

15-51 A load consisting of a 50-Ω resistor in series with a 100-mH inductor is connected across a voltage source $v_S(t) = 35 \cos 500t$ V. Find the phasor voltage, current, and average power delivered to the load.

15-52 A load consisting of a 50-Ω resistor in series with an 8-μF capacitor is connected across a voltage source $v_S(t) = 50 \cos 2500t$ V. Find the phasor voltage, current, and average power delivered to the load.

15-53 A load consisting of a resistor and capacitor connected in parallel draws an average power of 100 mW and a peak current of 35 mA when connected to voltage source $v_S(t) = 15 \cos 2500t$ V. Find the values of R and C.

15-54 A load consists of a 800-Ω resistor in parallel with an inductor with a reactance of 400 Ω is connected to a current source whose peak amplitude is 0.2 A. Find the average power delivered to the load.

15-55 (a) Find the average power delivered to the load in Figure P15-55.
(b) Find the maximum available average power at the interface shown in Figure P15-55.
(c) Specify the load required to extract the maximum average power.

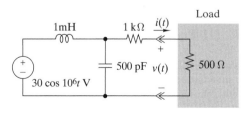

FIGURE P15-55

15-56 A 200-V (rms) source delivers 100 W to an inductive load whose series resistance is 8 Ω. What is the load impedance?

15-57 (a) Find the maximum average power available at the interface in Figure P15-57.
(b) Specify the values of R and C that will extract the maximum power from the source circuit.

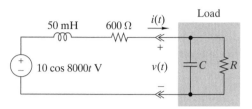

FIGURE P15-57

15-58 The steady-state open-circuit voltage at the output interface of a source circuit is $10\angle0°$ V. When a 100-Ω resistor is connected across the interface, the steady-state output voltage is $4\angle-60°$ V. Find the maximum average power available at the interface and the load impedance required to extract the maximum power.

15-59 The phasor Thévenin voltage of a certain source is $\mathbf{V}_T = 10 + j0$ V, and the Thévenin impedance is $Z_T = 100 + j50$ Ω. The problem is to extract as much average power as possible from this source using a single standard value resistor as the load. The values of resistance available are 1.0, 1.5, 2.2, 3.3, 4.7, 6.8, and multiples of 10 times these values.
(a) Find the maximum available average power from the source.
(b) Find the standard load resistor that extracts the most average power.

15-60 The 220-V (rms) power delivery circuit can be modeled as shown in Figure P15-60. Find the average power delivered by the source and absorbed by the load.

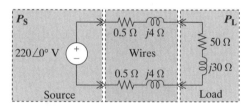

FIGURE P15-60

INTEGRATING PROBLEMS

15-61 AC VOLTAGE MEASUREMENT

An ac voltmeter measurement indicates the amplitude of a sinusoid and not its phase angle. The magnitude and phase can be inferred by making several measurements and using KVL. For example, Figure P15-61 shows a relay coil of unknown resistance and inductance. The following ac voltmeter readings are taken with the circuit operating in the sinusoidal steady state at $f = 1$ kHz: $|\mathbf{V}_S| = 10$ V, $|\mathbf{V}_1| = 4$ V, and $|\mathbf{V}_2| = 8$ V.

(a) Use these voltage magnitude measurements to solve for R and L.

(b) Determine the phasor voltage across each element and show that they satisfy KVL.

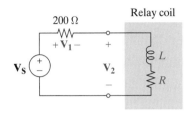

F I G U R E   P 1 5 – 6 1

15–62   🛦 OP AMP Circuits

The characteristics of resistive OP AMP circuits can be extended to dynamic circuits operating in the sinusoidal steady state using the concept of impedance. For the circuit in Figure P15–62,

(a) Find the phasor gain $K = \mathbf{V}_O/\mathbf{V}_S$ in terms of the impedances Z_1 and Z_2.

(b) Find the magnitude and angle of K when $Z_1 = 100\,\text{k}\Omega$ and $Z_2 = (10 - j20)\,\text{k}\Omega$.

(c) Select values for Z_1 and Z_2 so that $K = 2\angle{-60^\circ}$.

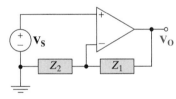

F I G U R E   P 1 5 – 6 2

15–63   🛦 Circuit Forced Response

The purpose of this problem is to demonstrate that the steady-state response obtained using phasor circuit analysis is the forced component of the solution of the circuit differential equation.

(a) Transform the circuit in Figure P15–63 into the phasor domain and use voltage division to solve for phasor output $\mathbf{V}$.

(b) Convert the phasor found in (a) into a sinusoid of the form $v(t) = a \cos \omega t + b \sin \omega t$.

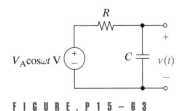

F I G U R E . P 1 5 – 6 3

(c) Formulate the circuit differential equation in the capacitor voltage $v(t)$.

(d) Show that the $v(t)$ found in part (b) satisfies the differential equation in (c).

15–64   🛦 Average Stored Energy

In the sinusoidal steady state the phasor voltage across a capacitor is $\mathbf{V}_C = V_A + j0$. Derive an expression for the average energy stored in the capacitor.

15–65   🅓 AC Circuit Design

Select values for the reactances L and C in Figure P15–65 so that the input impedance seen by the voltage source is $75 + j0\ \Omega$ when the frequency is $\omega = 10^6$ rad/s. For these values of L and C, find the Thévenin impedance seen by the 600-Ω load resistor.

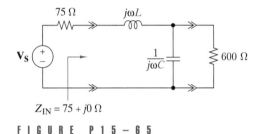

$Z_{IN} = 75 + j0\ \Omega$

F I G U R E   P 1 5 – 6 5

15–66   🛦 AC Circuit Analysis

Ten years after graduating with a BSEE, you decide to go to graduate school for a masters degree. In desperate need of income, you agree to sign on as a grader in the basic circuit analysis course. One of the problems asks the students to find $v(t)$ in Figure P15–66 when the circuit operates in the sinusoidal steady state. One of the students offers the following solution:

$$v(t) = (R + j\omega L) \times i(t)$$
$$= (20 + j20) \times 0.5 \cos 200t$$
$$= 10 \cos 200\,t + j10 \cos 200\,t$$
$$= 10\sqrt{2} \cos (200\,t + 45^\circ)$$

Is the answer correct? If not, what grade would you give the student? If correct, what comments would you give the student about the method of solution?

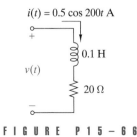

$i(t) = 0.5 \cos 200t$ A

F I G U R E   P 1 5 – 6 6

CHAPTER 16

MUTUAL INDUCTANCE

The results which I had by this time obtained with magnets led me to believe that the battery current through one wire, did, in reality, induce a similar current through the other wire, but that it continued for an instance only . . .

Michael Faraday, 1831
British Physicist

16–1 Coupled Inductors

16–2 The Dot Convention

16–3 Energy Analysis

16–4 The Ideal Transformer

16–5 Transformers in the Sinusoidal Steady State

16–6 Transformer Equivalent Circuits

Summary

Problems

Integrating Problems

Ampere's discovery that a force existed between two adjacent current-carrying wires led Micheal Faraday to investigate whether the current in one wire would affect the current in the other. He wound two separate insulted wires around an iron ring. One winding was connected to a battery by a switch and the other to a galvanometer. He observed that current in one wire did indeed induce a current in the other, but only when the switch was opened or closed. He correctly deduced that it was the change in the magnetic field that created the inductive effect, and he thereby discovered a basic principle that has come to be known as Faraday's law.

Faraday's iron ring with its two separate windings was actually a crude forerunner of an electrical device called a transformer. The development of practical transformers in 1882 by the British engineers Lucien Gauland and Josiah Gibbs gave the ac electrical power system one of its major advantages. Thomas Edison had pioneered the development of electrical power generation and distribution with his 110-V dc system. The low system voltage severely limited the distances over which dc electrical power could be transferred without encountering unacceptable losses. The transformer allows ac voltages to be increased to high levels for improved transmission efficiency and then reduced to safer levels at the load. Early ac systems used 1000-V lines while current-day systems use much higher levels. The development of the transformer was one of the key technology advances that lead to widespread electrical power systems of today.

The first section in this chapter develops the i–v relationships for a pair of windings called coupled inductors. The central feature of these relationships is that the time rate of change of current in one winding induces a voltage in the other, and vice versa. Because it is necessary to keep track of the polarity of these induced voltages, the second section introduces the dot convention—a special set of reference marks associated with coupled inductors. Energy calculations are used in the third section to define a parameter that indicates the degree of coupling between windings. The concept of perfect coupling is then used in the fourth section to develop an ideal model of the transformer device. The final two sections deal with phasor analysis of transformer circuits operating in the sinusoidal steady state.

16–1 COUPLED INDUCTORS

The i–v characteristic of the inductor results from the magnetic field produced by current through a coil of wire. A constant current produces a constant magnetic field that forms closed loops of magnetic flux lines in the vicinity of the inductor. A changing current causes these closed loops to expand or contract, thereby cutting the turns in the winding that makes up the inductor. Faraday's law states that voltage across the inductor is equal to the time rate of change of the total flux linkage. In earlier chapters we expressed this relationship between a time-varying current and an induced voltage in terms of an circuit parameter called inductance L.

Now suppose that a second inductor is brought close to the first so that the flux from the first inductor links with the turns of the second inductor. If the current in the first inductor is changing, then this flux linkage will generate a voltage in the second inductor. The magnetic coupling between the changing current in one inductor and the voltage generated in a second inductor produces **mutual inductance**.

I–V CHARACTERISTICS

The i–v characteristics of coupled inductors unavoidably involve describing the effects observed in one inductor due to causes occurring in the other. We will use a double subscript notation because it clearly identifies the various cause-and-effect relationships. The first subscript indicates the inductor in which the effect takes place, and the second identifies the inductor in which the cause occurs. For example, $v_{12}(t)$ is the voltage across inductor no. 1 due to causes occurring in inductor no. 2, whereas $v_{11}(t)$ is the voltage across inductor no. 1 due to causes occurring in inductor no. 1 itself.

We begin by assuming that the inductors are far apart as shown in Figure 16–1(a). Under these circumstances there is no magnetic coupling between the two. A current $i_1(t)$ passes through the N_1 turns of the first inductor and $i_2(t)$ through N_2 turns in the second. Each inductor produces a flux

$$\text{Inductor 1} \qquad \text{Inductor 2}$$
$$\phi_1(t) = k_1 N_1 i_1(t) \qquad \phi_2(t) = k_2 N_2 i_2(t) \tag{16–1}$$

where k_1 and k_2 are proportionality constants. The flux linkage in each inductor is proportional to the number of turns:

$$\text{Inductor 1} \qquad \text{Inductor 2}$$
$$\lambda_{11}(t) = N_1 \phi_1(t) \qquad \lambda_{22}(t) = N_2 \phi_2(t) \tag{16–2}$$

By Faraday's law the voltage across an inductor is equal to the time rate of change of the flux linkage. Using Eqs. (16–1) and (16–2) together with relationship between voltage and time rate of change of flux linkage gives

$$\text{Inductor 1:} \quad v_{11}(t) = \frac{d\lambda_{11}(t)}{dt} = N_1 \frac{d\phi_1(t)}{dt} = [k_1 N_1^2]\frac{di_1(t)}{dt}$$
$$\text{Inductor 2:} \quad v_{22}(t) = \frac{d\lambda_{22}(t)}{dt} = N_2 \frac{d\phi_2(t)}{dt} = [k_2 N_2^2]\frac{di_2(t)}{dt} \tag{16–3}$$

Equations (16–3) provide the i–v relationships for the inductors when there is no mutual coupling. These results are the same as previously found in Chapter 6.

Now suppose the inductors are brought close together so that part of the flux produced by each inductor intercepts the other, as indicated in Figure 16–1(b). That is, part (but not necessarily all) of the fluxes $\phi_1(t)$ and $\phi_2(t)$ in Eq. (16–1) intercept the opposite inductor. We describe the cross coupling using the double subscript notation:

$$\text{Inductor 1} \qquad \text{Inductor 2}$$
$$\phi_{12}(t) = k_{12} N_2 i_2(t) \qquad \phi_{21}(t) = k_{21} N_1 i_1(t) \tag{16–4}$$

The quantity $\phi_{12}(t)$ is the flux intercepting inductor 1 due to the current in inductor 2, and $\phi_{21}(t)$ is the flux intercepting inductor 2 due to the current in inductor 1. The total flux linkage in each inductor is proportional to the number of turns:

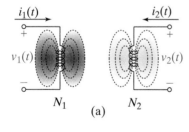

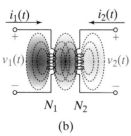

FIGURE 16–1 (a) Inductors separated, only self-inductance present. (b) Inductors coupled, both self- and mutual inductance present.

$$\text{Inductor 1} \qquad\qquad \text{Inductor 2}$$
$$\lambda_{12}(t) = N_1\phi_{12}(t) \qquad \lambda_{21}(t) = N_2\phi_{21}(t) \qquad\qquad (16\text{–}5)$$

By Faraday's law, the voltage across a winding is equal to the time rate of change of the flux linkage. Using Eqs. (16–4) and (16–5) together with derivative relationship, the time rate of change of flux linkages and voltages gives

$$\text{Inductor 1:} \quad v_{12}(t) = \frac{d\lambda_{12}(t)}{dt} = N_1\frac{d\phi_{12}(t)}{dt} = [k_{12}N_1N_2]\frac{di_2(t)}{dt}$$

$$\text{Inductor 2:} \quad v_{21}(t) = \frac{d\lambda_{21}(t)}{dt} = N_2\frac{d\phi_{21}(t)}{dt} = [k_{21}N_1N_2]\frac{di_1(t)}{dt} \qquad (16\text{–}6)$$

The expressions in Eq. (16–6) are the i–v relationships describing the cross coupling between inductors when there is mutual coupling.

When the magnetic medium supporting the fluxes is linear, the superposition principle applies, and the total voltage across the inductors is the sum of the results in Eqs. (16–3) and (16–6):

$$\text{Inductor 1:} \quad v_1(t) = v_{11}(t) + v_{12}(t)$$

$$= [k_1N_1^2]\frac{di_1(t)}{dt} + [k_{12}N_1N_2]\frac{di_2(t)}{dt}$$

$$\text{Inductor 2:} \quad v_2(t) = v_{21}(t) + v_{22}(t) \qquad\qquad (16\text{–}7)$$

$$= [k_{21}N_1N_2]\frac{di_1(t)}{dt} + [k_2N_2^2]\frac{di_2(t)}{dt}$$

We can identify four inductance parameters in these equations:

$$L_1 = k_1N_1^2 \qquad\qquad L_2 = k_2N_2^2 \qquad\qquad (16\text{–}8)$$

and

$$M_{12} = k_{12}N_1N_2 \qquad\qquad M_{21} = k_{21}N_2N_1 \qquad\qquad (16\text{–}9)$$

The two inductance parameters in Eq. (16–8) are the **self-inductance** of the inductors. The two parameters in Eq. (16–9) are the **mutual inductances** between the two inductors. In a linear magnetic medium, $k_{12} = k_{21} = k_{\mathrm{M}}$. As a result, we can define a single mutual inductance parameter M as

$$M = M_{12} = M_{21} = k_{\mathrm{M}}N_1N_2 \qquad\qquad (16\text{–}10)$$

Using the definitions in Eqs. (16–8) and (16–10), the i–v characteristics of two coupled inductors are

$$\text{Inductor 1:} \quad v_1(t) = L_1\frac{di_1(t)}{dt} + M\frac{di_2(t)}{dt}$$

$$\text{Inductor 2:} \quad v_2(t) = M\frac{di_1(t)}{dt} + L_2\frac{di_2(t)}{dt} \qquad\qquad (16\text{–}11)$$

Coupled inductors involve three inductance parameters, the two self-inductances L_1 and L_2 and the mutual inductance M.

The preceding development assumes that the cross coupling is additive. Additive coupling means that a positive rate of change of current in inductor 2 induces a positive voltage in inductor 1, and vice versa. The additive assumption produces the positive sign on the mutual inductance terms in Eq. (16–11). Unhappily it is possible for a positive rate of change of current in one inductor to induce a negative voltage in the other. To account for additive and subtractive coupling, the general form of the coupled inductor i–v characteristics includes a $\pm$ sign on the mutual inductance terms:

$$\text{Inductor 1:} \quad v_1(t) = L_1 \frac{di_1(t)}{dt} \pm M \frac{di_2(t)}{dt}$$

$$\text{Inductor 2:} \quad v_2(t) = \pm M \frac{di_1(t)}{dt} + L_2 \frac{di_2(t)}{dt}$$

(16–12)

When applying these element equations, it is necessary to know when to use a plus sign and when to use a minus sign.

16–2 THE DOT CONVENTION

The parameter M is positive, so the question is, what sign should be placed in front of this positive parameter in the i–v relationships in Eq. (16–12)? The correct sign depends on two things: (1) the spatial orientation of the two windings, and (2) the reference marks given to the currents and voltages.

Figure 16–2 shows the additive and subtractive spatial orientation of two coupled inductors. In either case, the direction of the flux produced by a current is found using the right-hand rule treated in physics courses. In the additive case, currents i_1 and i_2 both produce clockwise fluxes ϕ_1 and ϕ_2. In the subtractive case, the currents produce opposing fluxes because ϕ_1 is clockwise and ϕ_2 is counterclockwise. The sign for the mutual inductance term is positive for the additive orientation and negative for the subtractive case.

In general, it is awkward to show the spatial features of the windings in circuit diagrams. The dots shown near one terminal of each winding in Figure 16–2 are special reference marks indicating the relative orientation of the windings. The reference directions for the currents and voltages are arbitrary. They can be changed as long as we follow the passive sign convention. However, the dots indicate physical attributes of the windings that make up the coupled inductors. They are not arbitrary. They cannot be changed.

The correct sign for the mutual inductance term hinges on how the reference marks for currents and voltages are assigned relative to the dots. For a given winding orientation, Figure 16–3 shows all four possible current and voltage reference assignments under the passive sign convention. In cases A and B the fluxes are additive, so the mutual inductance term is positive. In cases C and D the fluxes are subtractive and the mutual inductance term is negative. From these results we derive the following rule:

Mutual inductance is additive when both current reference directions point toward or both point away from dotted terminals; otherwise, it is subtractive.

Because we always use the passive sign convention, the rule can be stated in terms of voltages as follows:

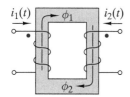

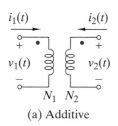

(a) Additive

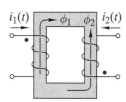

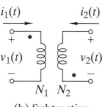

(b) Subtractive

FIGURE 16–2 *Winding orientations and corresponding reference dots. (a) Additive. (b) Subtractive.*

Mutual inductance is additive when the voltage reference marks are both positive or both negative at the dotted terminals; otherwise, it is subtractive.

Because the current reference directions or voltage polarity marks can be changed, a corollary of this rule is that we can always assign reference directions so that the positive sign applies to the mutual inductance. This corollary is important because a positive sign is built into the mutual inductance models in circuit analysis programs like SPICE.

It may seem that all of this discussion about signs and dots is much ado about nothing. Not so. First, selecting the wrong sign can have nontrivial consequences because the polarity of the output signal is reversed. If the signal is a command to your car's autopilot, then you really need to know whether stepping on the brake pedal will slow the car down or speed it up. Another problem is that the two coupled inductors may appear in different parts of a circuit diagram and may be assigned voltage and current reference marks for other reasons. In such circumstances it is important to understand the underlying principle to select the correct sign for the mutual inductance term.

The following examples and exercises illustrate selecting the correct sign and applying the *i–v* characteristics in Eq. (16–12)

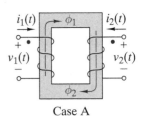

Case A

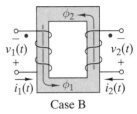

Case B

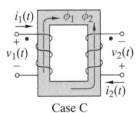

Case C

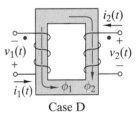

Case D

FIGURE 16-3 *All possible current and voltage reference marks for a fixed winding orientation.*

EXAMPLE 16-1

The source voltage in Figure 16–4 is $v_S(t) = 10 \cos 100t$ V. Find the output voltage $v_2(t)$.

SOLUTION:
The sign of the mutual inductance term is positive because both current reference directions are toward the dots. Since the load connected across inductor 2 is an open circuit, $i_2(t) = 0$ and the *i–v* equations of the coupled inductors reduce to

$$\text{Inductor 1:} \quad 10 \cos 100t = 0.01 \frac{di_1(t)}{dt} + 0$$

$$\text{Inductor 2:} \quad v_2(t) = 0.002 \frac{di_1(t)}{dt} + 0$$

Solving the first equation for di_1/dt yields

$$\frac{di_1(t)}{dt} = 1000 \cos 100t$$

Substituting this equation for $i_1(t)$ into the second equation yields

$$v_2(t) = 2 \cos 100t \text{ V (correct)}$$

If we had incorrectly chosen a minus sign, the mutual inductance would have produced a voltage of

$$v_2(t) = -2 \cos 100t \text{ V (incorrect)}$$

which differs from the correct answer by a signal inversion.　■

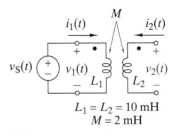

$L_1 = L_2 = 10$ mH
$M = 2$ mH

FIGURE 16-4

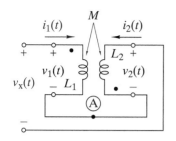

F I G U R E   1 6 – 5

EXAMPLE 16–2

Solve for $v_x(t)$ in terms of $i_1(t)$ for the coupled inductors in Figure 16–5.

SOLUTION:

In this case the signs of mutual inductance terms are negative because the reference direction for $i_1(t)$ points toward the dot for inductor 1 and the reference direction for $i_2(t)$ points away from the dot for inductor 2. The coupled inductor i–v equations are

$$\text{Inductor 1:} \quad v_1(t) = L_1 \frac{di_1(t)}{dt} - M \frac{di_2(t)}{dt}$$

$$\text{Inductor 2:} \quad v_2(t) = -M \frac{di_1(t)}{dt} + L_2 \frac{di_2(t)}{dt}$$

A KCL constraint at node A requires that $i_1(t) = -i_2(t)$. Therefore, these i–v equations can be written in the form

$$\text{Inductor 1:} \quad v_1(t) = L_1 \frac{di_1(t)}{dt} - M \frac{d[-i_1(t)]}{dt}$$

$$\text{Inductor 2:} \quad v_2(t) = -M \frac{di_1(t)}{dt} + L_2 \frac{d[-i_1(t)]}{dt}$$

A KVL constraint around the loop requires that $v_x(t) = v_1(t) - v_2(t)$. Subtracting the second equation from the first yields

$$v_x(t) = L_1 \frac{di_1(t)}{dt} + M \frac{di_1(t)}{dt} + M \frac{di_1(t)}{dt} + L_2 \frac{di_1(t)}{dt}$$

$$= (L_1 + L_2 + 2M) \frac{di_1(t)}{dt}$$

In this case the two mutual inductance terms add to produce an equivalent inductance of

$$L_{EQ} = L_1 + L_2 + 2M.$$

If the plus sign in the coupled inductor i–v relationships was appropriate, the mutual inductance terms would subtract to produce an equivalent inductance of $L_{EQ} = L_1 + L_2 - 2M$. Clearly it is important to have the right sign in the i–v relationships. ∎

Exercise 16–1

Find $v_1(t)$ and $v_2(t)$ for the circuit in Figure 16–6.

F I G U R E   1 6 – 6

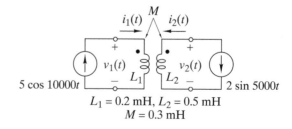

$L_1 = 0.2$ mH, $L_2 = 0.5$ mH
$M = 0.3$ mH

16-3 ENERGY ANALYSIS

Calculating the total energy stored in a pair of coupled inductors reveals a fundamental limitation on allowable values of the self- and mutual inductances. To uncover this limitation, we first calculate the total power absorbed. Multiplying the first equation in Eq. (16–12) by $i_1(t)$ and the second equation by $i_2(t)$ produces

$$p_1(t) = v_1(t)i_1(t) = L_1 i_1(t) \frac{di_1(t)}{dt} \pm M i_1(t) \frac{di_2(t)}{dt}$$

$$p_2(t) = v_2(t)i_2(t) = \pm M i_2(t) \frac{di_1(t)}{dt} + L_2 i_2(t) \frac{di_2(t)}{dt}$$

$$(16\text{--}13)$$

The quantities $p_1(t)$ and $p_2(t)$ are the powers absorbed with inductors 1 and 2. The total power is the sum of the individual inductor powers:

$$p(t) = p_1(t) + p_2(t) \tag{16-14}$$

$$= L_1 \left[i_1(t) \frac{di_1(t)}{dt} \right] \pm M \left[i_1(t) \frac{di_2(t)}{dt} + i_2(t) \frac{di_1(t)}{dt} \right] + L_2 \left[i_2(t) \frac{di_2(t)}{dt} \right]$$

Each of the bracketed terms in Eq. (16–14) is a perfect derivative. Specifically,

$$i_1(t) \frac{di_1(t)}{dt} = \frac{1}{2} \frac{di_1^2(t)}{dt}$$

$$i_2(t) \frac{di_2(t)}{dt} = \frac{1}{2} \frac{di_2^2(t)}{dt} \tag{16-15}$$

$$i_1(t) \frac{di_2(t)}{dt} + i_2(t) \frac{di_1(t)}{dt} = \frac{di_1(t)i_2(t)}{dt}$$

Therefore, the total power in Eq. (16–14) is

$$p(t) = \frac{d}{dt} \left[\frac{1}{2} L_1 i_1^2(t) \pm M i_1(t)i_2(t) + \frac{1}{2} L_2 i_2^2(t) \right] \tag{16-16}$$

Because power is the time rate of change of energy, the quantity inside the brackets in Eq. (16–16) is the total energy stored in the two inductors. That is,

$$w(t) = \frac{1}{2} L_1 i_1^2(t) \pm M i_1(t)i_2(t) + \frac{1}{2} L_2 i_2^2(t) \tag{16-17}$$

In Eq. (16–17) the self-inductance terms are always positive. However, the mutual-inductance term can be either positive or negative. At first glance it appears that the total energy could be negative. But the total energy must be positive: Otherwise the coupled inductors could deliver net energy to the rest of the circuit.

The condition $w(t) \geq 0$ places a constraint on the values of the self- and mutual inductances. First, if $i_2(t) = 0$, then $w(t) \geq 0$ in Eq. (16–17) requires $L_1 > 0$. Next, if $i_1(t) = 0$, then $w(t) \geq 0$ in Eq. (16–17) requires $L_2 > 0$. Finally, if $i_1(t) \neq 0$ and $i_2(t) \neq 0$, then we divide Eq. (16–17) by $[i_2(t)]^2$ and defined a variable $x = i_1/i_2$. With these changes, the energy constraint $w(t) > 0$ becomes

$$\frac{w(t)}{i_2^2(t)} = f(x) = \frac{1}{2}L_1 x^2 \pm Mx + \frac{1}{2}L_2 \geq 0 \tag{16–18}$$

The minimum value of $f(x)$ occurs when

$$\frac{df(x)}{dx} = L_1 x \pm M = 0 \quad \text{hence} \quad x_{\min} = \mp \frac{M}{L_1} \tag{16–19}$$

The value $x_{\min}$ yields the minimum of $f(x)$ because the second derivative of $f(x)$ is positive. Substituting $x_{\min}$ back into Eq. (16–18) yields the condition

$$f(x_{\min}) = \frac{1}{2}L_1 \frac{M^2}{L_1^2} - \frac{M^2}{L_1} + \frac{1}{2}L_2 = \frac{1}{2}\left[-\frac{M^2}{L_1} + L_2\right] \geq 0 \tag{16–20}$$

The constraint in Eq. (16–20) means that the stored energy in a pair of coupled inductors is positive if

$$L_1 L_2 \geq M^2 \tag{16–21}$$

Energy considerations dictate that in any pair of coupled inductors, the product of the self-inductances must exceed the square of the mutual inductance.

The constraint in Eq. (16–21) is usually written in terms of a new parameter called the **coupling coefficient** k:

$$k = \frac{M}{\sqrt{L_1 L_2}} \leq 1 \tag{16–22}$$

The parameter k ranges from 0 to 1. If $M = 0$ then $k = 0$ and the coupling between the inductors is zero. The condition $k = 1$ requires **perfect coupling** in which all of the flux produced by one inductor links the other. Perfect coupling is physically impossible, although careful design can produce coupling coefficients of 0.99 and higher.

The next section discusses a transformer model that assumes perfect coupling ($k = 1$). Computer-aided circuit analysis programs like SPICE specify the parameters of a pair of coupled inductors in terms of the self-inductances L_1 and L_2 and the coupling coefficient k.

Exercise 16–2

What is the coupling coefficient of the coupled inductors in Figure 16–6.

Answer:
$k = 0.949$

16-4 THE IDEAL TRANSFORMER

A **transformer** is an electrical device that utilizes magnetic coupling be-tween two inductors. Transformers find application in virtually every type of electrical system, but especially in power supplies and commercial power grids. Some example devices from these applications are shown in Figure 16-7.

In Figure 16-8 the transformer is shown as an interface device between a source and a load. The winding connected to the source is called the **primary winding**, and the one connected to the load is called the **secondary winding**. In most applications the transformer is a coupling device that transfers signals (especially power) from the source to the load. The basic purpose of the device is to change voltage and current levels so that the conditions at the source and load are compatible.

Transformer design involves two primary goals: (1) to maximize the magnetic coupling between the two windings, and (2) to minimize the power loss in the windings. The first goal produces nearly perfect coupling ($k \approx 1$) so that almost all of the flux in one winding links the other. The second goal produces nearly zero power loss so that almost all of the power delivered to the primary winding transfers to the load. The **ideal transformer** is a circuit element in which coupled inductors are assumed to have perfect coupling and zero power loss. Using these two idealizations, we can derive the $i–v$ characteristics of an ideal transformer.

PERFECT COUPLING

Perfect coupling means that all of the flux in the first winding links the sec-ond, and vice versa. Equation (16-1) defines the total flux in each wind-ing as

$$
\begin{array}{cc}
\text{Winding 1} & \text{Winding 2} \\
\phi_1(t) = k_1 N_1 i_1(t) & \phi_2(t) = k_2 N_2 i_2(t)
\end{array}
\tag{16-23}
$$

where k_1 and k_2 are proportionality constants. Equation (16-4) defines the cross coupling using the double subscript notation

$$
\begin{array}{cc}
\text{Winding 1} & \text{Winding 2} \\
\phi_{12}(t) = k_{12} N_2 i_2(t) & \phi_{21}(t) = k_{21} N_1 i_1(t)
\end{array}
\tag{16-24}
$$

In this equation $\phi_{12}(t)$ is the flux intercepting winding 1 due to the current in winding 2, and $\phi_{21}(t)$ is the flux intercepting winding 2 due to the cur-rent in winding 1. Perfect coupling means that

$$
\phi_{21}(t) = \phi_1(t) \quad \text{and} \quad \phi_{12}(t) = \phi_2(t)
\tag{16-25}
$$

Comparing Eqs. (16-23) and (16-24) shows that perfect coupling requires $k_1 = k_{21}$ and $k_2 = k_{12}$. But in a linear magnetic medium $k_{12} = k_{21} = k_M$, so perfect coupling implies

$$
k_1 = k_2 = k_{12} = k_{21} = k_M
\tag{16-26}
$$

Substituting the perfect coupling conditions in Eq. (16-26) into the $i–v$ characteristics in Eq. (16-7) gives

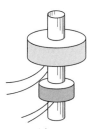

Iron core

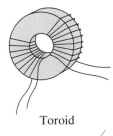

Air core

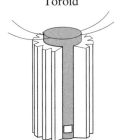

Toroid

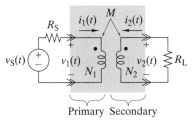

Powerline

FIGURE   16-7 *Examples of transformer devices.*

FIGURE   16-8 *Transformer connected at a source/load inter-face.*

$$v_1(t) = [k_M N_1^2] \frac{di_1(t)}{dt} \pm [k_M N_1 N_2] \frac{di_2(t)}{dt}$$

$$v_2(t) = \pm [k_M N_1 N_2] \frac{di_1(t)}{dt} + [k_M N_2^2] \frac{di_2(t)}{dt}$$

(16–27)

Factoring N_1 out of the first equation and $\pm N_2$ out of the second produces

$$v_1(t) = N_1 \left([k_M N_1] \frac{di_1(t)}{dt} \pm [k_M N_2] \frac{di_2(t)}{dt} \right)$$

$$v_2(t) = \pm N_2 \left([k_M N_1] \frac{di_1(t)}{dt} \pm [k_M N_2] \frac{di_2(t)}{dt} \right)$$

(16–28)

Dividing the second equation by the first shows that perfect coupling implies

$$\frac{v_2(t)}{v_1(t)} = \pm \frac{N_2}{N_1} = \pm n$$

(16–29)

where the parameter n is called the **turns ratio**.

With perfect coupling the secondary voltage is proportional to the primary voltage, so they have the same waveshape. For example, when the primary voltage is $v_1(t) = V_A \sin \omega t$, the secondary voltage is $v_2(t) = \pm n V_A \sin \omega t$. When the turns ratio $n > 1$, the secondary voltage amplitude is larger than the primary and the device is called a **step-up transformer**. Conversely, when $n < 1$, the secondary voltage is smaller than the primary and the device is called a **step-down transformer**. The ability to increase or decrease ac voltages is a basic feature of transformers. Commercial power systems use transmission voltages of several hundred kilovolts. For residential applications the transmission voltage is reduced to safer levels (typically 220/110 V_{rms}) using step-down transformers.

The $\pm$ sign in Eq. (16–29) reminds us that mutual inductance can be additive or subtractive. Selecting the correct sign is important because of signal inversion. The sign depends on the reference marks given the primary and secondary currents relative to the dots indicating the relative winding orientations. The rule for the ideal transformer is a corollary of the rule for selecting the sign of the mutual inductance term for coupled inductors.

> *The sign in Eq. (16–29) is positive when the reference directions for both currents point toward or both point away from a dotted terminal; otherwise it is negative.*

Exercise 16–3

The transformer in Figure 16–9 has perfect coupling and a turns ratio of $n = 0.1$. The input voltage is $v_S(t) = 120 \sin 377t$ V.

(a) What is the secondary voltage?
(b) What is the secondary current for a 50-Ω load?
(c) Is this a step-up or step-down transformer?

Answers:
(a) $v_2(t) = +12 \sin 377t$ V
(b) $i_2(t) = -0.24 \sin 377t$ A
(c) Step down

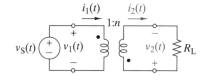

F I G U R E   1 6 – 9

ZERO POWER LOSS

The ideal transformer model also assumes that there is no power loss in the transformer. With the passive sign convention, the quantity $v_1(t)i_1(t)$ is the power in the primary winding and $v_2(t)i_2(t)$ is the power in the secondary winding. Zero power loss requires

$$v_1(t)i_1(t) + v_2(t)i_2(t) = 0 \qquad (16\text{--}30)$$

This equation states that whatever power enters the transformer on one winding immediately leaves via the other winding. This not only means that there is no energy lost in the ideal transformer, but also that there is no energy stored within the element. The zero-power loss constraint can be rearranged as

$$\frac{i_2(t)}{i_1(t)} = -\frac{v_1(t)}{v_2(t)} \qquad (16\text{--}31)$$

But under the perfect coupling assumption $v_2(t)/v_1(t) = \pm n$. With zero power loss and perfect coupling the primary and secondary currents are related as

$$\frac{i_2(t)}{i_1(t)} = \mp \frac{1}{n} \qquad (16\text{--}32)$$

The correct sign in this equation depends on the orientation of the current reference directions relative to the dots describing the transformer structure.

With both perfect coupling and zero power loss, the secondary current is inversely proportional to the turns ratio. A step-up transformer ($n > 1$) increases the voltage and decreases the current, which improves transmission line efficiency because the i^2R losses in the conductors are smaller.

i--v CHARACTERISTICS

Equations (16–29) and (16–32) define the i--v characteristics of the ideal transformer circuit element.

$$v_2(t) = \pm n\, v_1(t)$$
$$i_2(t) = \mp \frac{1}{n} i_1(t) \qquad (16\text{--}33)$$

where $n = N_2/N_1$ is the turns ratio. The correct sign in these equations depends on the assigned reference directions and transformer dots, as previously discussed.

Because the turns ratio is a key transformer parameter, it is worth relating it to the inductance parameters. Looking back at Eq. (16–27), we see that in a linear magnetic medium with perfect coupling the inductance parameters are $L_1 = k_M N_1^2$, $L_2 = k_M N_2^2$, and $M = k_M N_1 N_2$. As a result, the turns ratio can be expressed in terms of the inductance parameters as

$$n = \frac{N_2}{N_1} = \sqrt{\frac{L_2}{L_1}} = \frac{M}{L_1} = \frac{L_2}{M} \qquad (16\text{--}34)$$

Although actual transformer devices are not ideal, they do approach perfect coupling. Hence, these relations are useful approximations with transformers that are tightly coupled ($k \approx 1$).

Using the ideal transformer model requires some caution. The relationships in Eq. (16–33) state that the voltages are independent of the currents. For example, if the secondary winding is connected to an open circuit then $i_2(t) = 0$. The ideal transformer characteristics require that $i_1(t) = 0$ regardless of the value of $v_1(t)$. In effect this means that the inductance parameters are all infinite, but infinite in such a way that their ratios in Eq. (16–34) are still equal to the turns ratio. Equally important is the fact that the element equations appear to apply to all signals, including constant (dc) signals. In theory, a transformer with perfect coupling and zero power loss would pass dc signals. In practice, some transformers approach these conditions but no real transformer actually achieves this state of perfection. Remember that the element equations of the ideal transformer are an idealization of mutual coupling between inductors, and that time-varying signals are necessary; otherwise, the coupling terms $M di_1/dt$ and $M di_2/dt$ are zero.

EXAMPLE 16–3

The input to the primary of an ideal transformer with a turns ratio of $n = 10$ is $v_1(t) = 120 \sin 377t$ V. The load connected to the secondary winding is a 50-Ω resistor. Assuming additive coupling, find the secondary voltage, the secondary current, the secondary power, the primary current, and the primary power.

SOLUTION:
With additive coupling the secondary voltage is

$$v_2(t) = +nv_1(t) = 10 \times 120 \sin 377\,t = 1200 \sin 377t \text{ V}$$

From Ohm's law the secondary current is

$$i_2(t) = -\frac{v_2(t)}{R_L} = -\frac{1200}{50} \sin 377\,t = -24 \sin 377t \text{ A}$$

The minus sign in this expression comes about because the reference marks for $v_2(t)$ and $i_2(t)$ follow the passive sign convention with respect to the secondary winding and not the load resistance. The power in the secondary winding is

$$p_2(t) = v_2(t)i_2(t) = -28.8 \sin^2 377t \text{ kW}$$

The minus sign here means that the secondary delivers power to the load resistor. With additive coupling the primary current is

$$i_1 = -ni_2(t) = -10 \times (-24 \sin 377\,t) = 240 \sin 377t \text{ A}$$

and the primary power is

$$p_1(t) = v_1(t)i_1(t) = 28.8 \sin^2 377t \text{ kW}$$

The plus sign here means that the primary winding is absorbing power from the source. Note that $p_1(t) + p_2(t) = 0$ as required by the zero power loss assumption. ∎

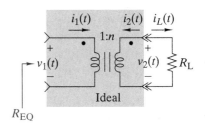

EQUIVALENT INPUT RESISTANCE

Because a transformer changes the voltage and current levels, it effectively changes the load resistance seen by a source in the primary circuit. To derive the equivalent input resistance, we write the device equations for the ideal transformer shown in Figure 16–10.

Resistor: $v_2(t) = R_L i_L(t)$

Transformer: $v_2(t) = n\, v_1(t)$ and $i_2(t) = -\dfrac{1}{n} i_1(t)$

FIGURE 16−10 *Equivalent resistance seen on the primary winding.*

Dividing the first transformer equation by the second and inserting the load resistance constraint yields

$$\frac{v_2(t)}{i_2(t)} = \frac{i_L(t)R_L}{i_2(t)} = -n^2 \frac{v_1(t)}{i_1(t)}$$

Applying KCL at the output interface tells us that $i_L(t) = -i_2(t)$. Therefore, the equivalent resistance seen on the primary side is

$$R_{EQ} = \frac{v_1(t)}{i_1(t)} = \frac{1}{n^2} R_L \qquad (16{-}35)$$

The equivalent load resistance seen on the primary side depends on the turns ratio and the load resistance. Adjusting the turns ratio can make R_{EQ} equal to the source resistance. Transformer coupling can produce the conditions for maximum power transfer when the source and load resistances are not equal.

The derivation leading to Eq. (16–35) used the ideal transformer with the dot markings and reference directions in Figure 16–10. However, the final result does not depend on the location of the dot marks relative to voltage and current reference directions. In other words, Eq. (16–35) yields the input resistance for any ideal transformer with a turns ratio of n and a load R_L.

EXAMPLE 16−4

A stereo amplifier has an output resistance of 600 Ω. The input resistance of the speaker (the load) is 8 Ω. Select the turns ratio of a transformer to obtain maximum power transfer.

SOLUTION:

The maximum power transfer theorem in Chapter 3 states that the source and load resistance must be matched (equal) to achieve maximum power. Directly connecting the amplifier (600 Ω) to the speaker (8 Ω) produces a mismatch. If a transformer is inserted as shown in Figure 16–11, then the equivalent load resistance seen by the amplifier is

$$R_{EQ} = \frac{1}{n^2} R_L = \frac{1}{n^2} 8$$

To produce a resistance match we need $R_{EQ} = 600 = 8/n^2$ or a turns ratio of $n = 1/8.66$. ∎

FIGURE 16–11

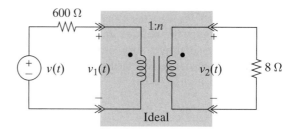

The load resistor in Figure 16–10 is 50 Ω, and the number of turns on the primary and secondary windings are $N_1 = 540$ and $N_2 = 108$. Find the equivalent resistance on the primary side.

Answer:
$R_{EQ} = 1250 \ \Omega$

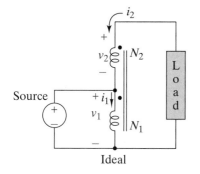

FIGURE 16–12 *The autotransformer connection.*

APPLICATION NOTE: EXAMPLE 16–5

In a transformer the primary and secondary windings are magnetically coupled but are usually electrically isolated. Transformer performance in some applications can be improved by electrically connecting the two magnetically coupled windings in a configuration called an **autotransformer**. Figure 16–12 shows a two-winding ideal transformer connected in an autotransformer step-up configuration.

The rating of the ideal transformer when connected in the usual electrically isolated two-winding configuration is $P_{rated} = v_1 i_1 = -v_2 i_2$. Connected in the autotransformer configuration in Figure 16–12, the power delivered to the load is

$$P_{load} = (v_1 + v_2)(-i_2)$$

For an ideal transformer $v_2 = nv_1$ and $i_2 = -i_1/n$, where $n = N_2/N_1$. Hence the power delivered to the load is

$$P_{load} = (v_1 + nv_1)(i_1/n) = \left(1 + \frac{1}{n}\right) v_1 i_1$$

$$= \left(1 + \frac{1}{n}\right) P_{rated}$$

The autotransformer configuration delivers more power to the load than the power rating of the two-winding transformer. Put differently, the autotransformer can supply a specified load power using a transformer with a lower power rating. Autotransformers are normally used when the turns ratio is less than 3:1, so this advantage can be significant. A disadvantage is that the electrical isolation provided by the usual transformer configuration is lost.

16–5 Transformers in the Sinusoidal Steady State

By far the most common application of transformers occurs in electric power systems where they operate in the sinusoidal steady state. In this context we describe transformers in terms of phasors and impedances and deal with average power transfer. The ac analysis of a transformer begins with the time-domain element equations for a pair of coupled inductors

$$v_1(t) = L_1 \frac{di_1(t)}{dt} \pm M \frac{di_2(t)}{dt}$$

$$v_2(t) = \pm M \frac{di_1(t)}{dt} + L_2 \frac{di_2(t)}{dt}$$

To transform these equations into the phasor domain, we first transform them into the s domain with zero initial conditions

$$V_1(s) = L_1 s I_1(s) + M s I_2(s)$$

$$V_2(s) = M s I_1(s) + L_2 s I_2(s)$$

Then, using the network function property of phasor, we replace the transforms $V_1(s)$, $V_2(s)$, $I_1(s)$, and $I_2(s)$ by the corresponding phasors and set $s = j\omega$.

$$\mathbf{V}_1 = j\omega L_1 \mathbf{I}_1 \pm j\omega M \mathbf{I}_2$$
$$\mathbf{V}_2 = \pm j\omega M \mathbf{I}_1 + j\omega L_2 \mathbf{I}_2 \tag{16–36}$$

where

1. $\mathbf{V}_1$ and $\mathbf{I}_1$ are the phasors representing ac voltage and current of the first winding, and $j\omega L_1$ is the impedance of the self-inductance of first winding.

2. $\mathbf{V}_2$ and $\mathbf{I}_2$ are the phasors representing ac voltage and current of the second winding, and $j\omega L_2$ is the impedance of the self-inductance of second winding.

3. $j\omega M$ is the impedance of the mutual inductance between the two windings.

The self-inductance impedances relate phasor voltage and current at the same pair of terminals, while the mutual inductance impedance relates the phasor voltage at one pair of terminals to the phasor current at the other pair. The phasor domain equations can also be written in terms of reactances as

$$\mathbf{V}_1 = jX_1 \mathbf{I}_1 \pm jX_M \mathbf{I}_2$$
$$\mathbf{V}_2 = \pm jX_M \mathbf{I}_1 + jX_2 \mathbf{I}_2$$

where the three reactances are $X_1 = \omega L_1$, $X_2 = \omega L_2$, and $X_M = \omega M$. The degree of coupling between windings is indicated by the coupling coefficient, which can be written in terms of reactances as

$$k = \frac{M}{\sqrt{L_1 L_2}} = \frac{X_M}{\sqrt{X_1 X_2}}$$

In either form, energy considerations dictate that $0 \leq k \leq 1$.

FIGURE 16-13 *Phasor circuit model of the two-winding transformer.*

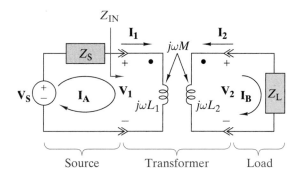

Figure 16–13 shows the phasor-domain version of transformer coupling between a source and a load. Following our previous notation, the winding connected to the source is called the **primary,** and the winding connected to the load is called the **secondary.** Although the transformer is a bilateral device, we normally think of signal and power transfer as passing from the primary to the secondary winding.

Our immediate objective is to write circuit equations for the transformer using the mesh currents I_A and I_B in Figure 16–13. Applying KVL around the primary circuit (mesh A) and secondary circuit (mesh B), we obtain the following equations:

$$\text{Mesh A:} \quad Z_S I_A + V_1 = V_S$$
$$\text{Mesh B:} \quad -V_2 + Z_L I_B = 0 \tag{16–37}$$

The reference directions for the inductor currents in Figure 16–13 are both directed in at dotted terminals, so the mutual inductance coupling is additive and the plus signs in Eq. (16–36) apply. Using KCL we see that the reference directions for the currents lead to the relations $I_1 = I_A$ and $I_2 = -I_B$. The **I–V** relationships of the coupled inductors in terms of the mesh currents are

$$V_1 = +j\omega L_1 I_A + j\omega M(-I_B)$$
$$V_2 = +j\omega M I_A + j\omega L_2(-I_B) \tag{16–38}$$

Substituting the inductor voltages from Eq. (16–38) into the KVL equations in Eq. (16–37) yields

$$\text{Mesh A:} \quad (Z_S + j\omega L_1)I_A - j\omega M I_B = V_S$$
$$\text{Mesh B:} \quad -j\omega M I_A + (Z_L + j\omega L_2)I_B = 0 \tag{16–39}$$

This set of mesh equations provides a complete description of the circuit ac response. Once we solve for the mesh currents, we can calculate every phasor voltage and current using Kirchhoff's laws and element equations.

EXAMPLE 16–6

The source circuit in Figure 16–13 has $Z_S = 0 + j20\ \Omega$ and $V_S = 2500\angle 0°$ V at $\omega = 377$ rad/s. The transformer has $L_1 = 2$ H, $L_2 = 0.2$ H, and $M = 0.6$ H. The load impedance is $Z_L = 25 + j15\ \Omega$. Find I_A, I_B, V_1, V_2, Z_{IN}, and the average power delivered by the source at input interface.

SOLUTION:
The transformer impedances are

$$j\omega L_1 = j377 \times 2 = j754 \ \Omega$$

$$j\omega L_2 = j377 \times 0.2 = j75.4 \ \Omega$$

$$j\omega M = j377 \times 0.6 = j226 \ \Omega$$

Using these impedances in Eq. (16–39) yields the following mesh equations:

Mesh A: $(j20 + j754)\mathbf{I}_A - j226\mathbf{I}_B = 2500 \angle 0°$

Mesh B: $-j226\mathbf{I}_A + (25 + j15 + j75.4)\mathbf{I}_B = 0$

Solving these equations for the mesh currents yields

$$\mathbf{I}_A = 4.36 - j7.49 = 8.67 \angle -59.8° \quad A$$

$$\mathbf{I}_B = 15.0 - j14.6 = 20.9 \angle -44.2° \quad A$$

The winding voltages are found to be

$$\mathbf{V}_1 = \mathbf{V}_S - Z_S\mathbf{I}_A = 2350 - j87.2 = 2350 \angle -2.1° \ V$$

$$\mathbf{V}_2 = Z_L\mathbf{I}_B = 594 - j140 = 610 \angle -13.2° \ V$$

The input impedance seen by the source circuit is

$$Z_{IN} = \frac{\mathbf{V}_1}{\mathbf{I}_A} = R_{IN} + jX_{IN} = 145 + j229 \ \Omega$$

The average power delivered by the source at the input interface is

$$P_{IN} = \frac{|\mathbf{I}_A|^2}{2} R_{IN} = \frac{(8.67)^2}{2} 145 = 5.45 \ kW$$

Exercise 16–5

Repeat Example 16–6 when the dot on the secondary winding in Figure 16–13 is moved to the bottom terminal and all other reference marks stay the same.

Answers:

$\mathbf{I}_A = 4.36 - j7.49 \ A; \ \mathbf{I}_B = -15.0 + j14.6 \ A$

$\mathbf{V}_1 = 2350 - j87.2 \ V; \ \mathbf{V}_2 = -594 + j140 \ V; \ Z_{IN} = 145 + j \ 229 \ \Omega; \ P_{IN} = 5.45 \ kW$

The method used in the preceding example illustrates a general approach to the analysis of transformer circuits. The steps in the method are as follows:

STEP 1. Write KVL equations around the primary and secondary circuits using assigned mesh currents, source voltages, and inductor voltages.

STEP 2. Write the **I–V** characteristics of the coupled inductors in terms of the mesh currents using the dot convention to determine whether the coupling is additive or subtractive.

STEP 3. Use the **I–V** relationships from step 2 to eliminate the inductor voltages from the KVL equations obtained in step 1 to obtain mesh-current equations.

The next two examples illustrate this method of formulating mesh equations.

EXAMPLE 16–7

Find $\mathbf{I}_A$, $\mathbf{I}_B$, $\mathbf{V}_1$, $\mathbf{V}_2$, and the impedance seen by the voltage source in Figure 16–14.

FIGURE 16 – 14

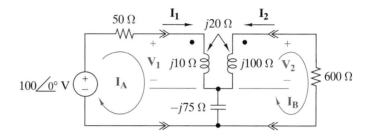

SOLUTION:

STEP 1. The KVL equations around meshes A and B are

Mesh A: $(50 - j75)\mathbf{I}_A - (-j75)\mathbf{I}_B + \mathbf{V}_1 = 100 \angle 0°$
Mesh B: $- (-j75)\mathbf{I}_A + (600 - j75)\mathbf{I}_B - \mathbf{V}_2 = 0$

STEP 2. For the assigned reference directions, the coupling is additive. By KCL we have $\mathbf{I}_1 = \mathbf{I}_A$ and $\mathbf{I}_2 = -\mathbf{I}_B$. Hence, the element equations for the coupled inductors in terms of the mesh currents are

$$\mathbf{V}_1 = j10\mathbf{I}_A + j20(-\mathbf{I}_B)$$
$$\mathbf{V}_2 = j20\mathbf{I}_A + j100(-\mathbf{I}_B)$$

STEP 3. Using these equations to eliminate the inductor voltages from the KVL equations found in step 1 yields the following mesh equations:

Mesh A: $(50 - j75 + j10)\mathbf{I}_A + (j75 - j20)\mathbf{I}_B = 100 \angle 0°$
Mesh B: $(j75 - j20)\mathbf{I}_A + (600 - j75 + j100)\mathbf{I}_B = 0$

Solving these equations for the mesh currents produces

$$\mathbf{I}_A = 0.756 + j0.896 = 1.17 \angle 49.8° \text{ A}$$
$$\mathbf{I}_B = 0.0791 - j0.0726 = 0.107 \angle -42.5° \text{ A}$$

Given the mesh currents, we find the inductor voltages from the I–V relations:

$$\mathbf{V}_1 = j10\mathbf{I}_A - j20\mathbf{I}_B = -10.4 + j5.98 = 12.0 \angle 150° \text{ V}$$
$$\mathbf{V}_2 = j20\mathbf{I}_A - j100\mathbf{I}_B = -25.2 + j7.21 = 26.2 \angle 164° \text{ V}$$

Finally, the impedance seen by the input voltage source is

$$Z_{IN} = \frac{\mathbf{V}_S}{\mathbf{I}_A} = 55.0 - j65.2 \ \Omega$$

EXAMPLE 16–8

Figure 16–15 shows an ideal transformer connected as an autotransformer. Find the voltage and average power delivered to the load for V_S = 500∠0° V, $Z_S = j10 \ \Omega$, $Z_L = 50 + j0 \ \Omega$, $N_1 = 200$, and $N_1 = 280$.

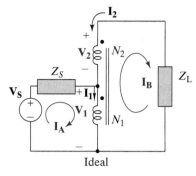

FIGURE 16–15

SOLUTION:

The three-step method of writing mesh equations can be used here with a modification to the second step:

STEP 1. The KVL equations around meshes A and B are

$$\text{Mesh A:} \quad Z_S\mathbf{I}_A + \mathbf{V}_1 = \mathbf{V}_S$$
$$\text{Mesh B:} \quad -\mathbf{V}_1 - \mathbf{V}_2 + Z_L\mathbf{I}_B = 0$$

STEP 2. For the assigned reference directions, the coupling is additive so the ideal transformer voltages and currents are related as

$$\mathbf{V}_2 = n\mathbf{V}_1 \quad \text{and} \quad \mathbf{I}_1 = -n\mathbf{I}_2$$

By KCL we have $\mathbf{I}_1 = \mathbf{I}_A - \mathbf{I}_B$ and $\mathbf{I}_2 = -\mathbf{I}_B$. Hence, the ideal transformer constrains the two mesh currents as

$$\mathbf{I}_A = (n + 1)\mathbf{I}_B$$

STEP 3. Using these results to eliminate $\mathbf{I}_A$ and $\mathbf{V}_2$ from the KVL equation in step 1 yields

$$\text{Mesh A:} \quad (n + 1)Z_S\mathbf{I}_B + \mathbf{V}_1 = \mathbf{V}_S$$
$$\text{Mesh B:} \quad -(n + 1)\mathbf{V}_1 + Z_L\mathbf{I}_B = 0$$

Notice that we cannot eliminate both $\mathbf{V}_1$ and $\mathbf{V}_2$ from the KVL equations since voltage and current are independent in an ideal transformer. Substituting the numerical values produces

$$\text{Mesh A:} \quad j24\mathbf{I}_B + \mathbf{V}_1 = 500 \angle 0°$$
$$\text{Mesh B:} \quad -2.4\mathbf{V}_1 + 50\mathbf{I}_B = 0$$

Solving these equations for $\mathbf{I}_B$ and $\mathbf{V}_1$ yields

$$\mathbf{I}_B = 10.3 - j11.9 = 15.7 \angle -49.1° \text{ A}$$
$$\mathbf{V}_1 = 215 - j247 = 328 \angle -49.0° \text{ A}$$

Given the $\mathbf{I}_B$, we find the output quantities as

$$\mathbf{V}_L = Z_L\mathbf{I}_B = 515 - j595 = 787 \angle -49.1° \text{ A}$$
$$P_L = \frac{|\mathbf{I}_B|^2}{2} R_L = \frac{15.7^2}{2} 50 = 6.16 \text{ kW}$$

Exercise 16–6

Find $\mathbf{V}_L$, $\mathbf{V}_1$, $\mathbf{V}_2$, and Z_{IN} in the circuit shown in Figure 16–16.

FIGURE 16–16

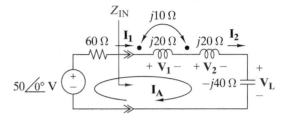

Answers: $\mathbf{V}_L = 31.6 \angle{-108°}$ V, $\mathbf{V}_1 = 23.7 \angle 71.6°$ V, $\mathbf{V}_2 = 23.7 \angle 71.6°$ V, $Z_{IN} = 0 + j20\ \Omega$.

16–6 TRANSFORMER EQUIVALENT CIRCUITS

The circuit analysis may be simplified when coupled inductors are replaced by an equivalent circuit made up of uncoupled inductors. The inductive T-circuit in Figure 16–17 is one example of such a circuit. By definition two circuits are equivalent if they have the same **I-V** characteristics at specified terminals. Equation (16–36) gives the **I-V** relations for coupled inductors at the input and output ports. To establish an equivalence, we must find the values of L_A, L_B, and L_C in Figure 16–17 that produce the same **I-V** properties as Eq. (16–36).

Applying KCL in Figure 16–17 shows that the current through L_C is $\mathbf{I}_1 + \mathbf{I}_2$. As a result, applying KVL around the circuit's input and output loops yields

$$-\mathbf{V}_1 + j\omega L_A \mathbf{I}_1 + j\omega L_C(\mathbf{I}_1 + \mathbf{I}_2) = 0$$

$$-j\omega L_C(\mathbf{I}_1 + \mathbf{I}_2) - j\omega L_B \mathbf{I}_2 + \mathbf{V}_2 = 0$$

These equations can be rearranged in the form

$$\mathbf{V}_1 = j\omega(L_A + L_C)\mathbf{I}_1 + j\omega L_C \mathbf{I}_2$$

$$\mathbf{V}_2 = j\omega L_C \mathbf{I}_1 + j\omega(L_B + L_C)\mathbf{I}_2$$

Comparing these equations to the **I-V** relations in Eq. (16–36), we conclude that equivalence requires

$$L_A + L_C = L_1$$

$$L_B + L_C = L_2$$

$$L_C = \pm M$$

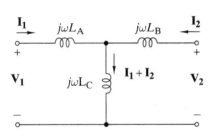

FIGURE 16–17 *T-equivalent circuit of a two-winding transformer.*

Using the third equation to eliminate L_C from the first two equations yields the required values of L_A, L_B, and L_C.

$$L_A = L_1 \mp M$$

$$L_B = L_2 \mp M$$

$$L_C = \pm M$$

where the upper sign applies for additive coupling and the lower for subtractive.

For additive coupling, the equivalent T-circuit involves three uncoupled inductances whose values are $L_1 - M$, $L_2 - M$, and M. For subtractive coupling, the three inductances are $L_1 + M$, $L_2 + M$, and $-M$. In either case, one of the inductances is negative. A negative inductance is not a passive element, which means we can't actually build an equivalent T-circuit out of ordinary inductors.

The fact that we cannot actually build the equivalent T-circuit in the laboratory does not invalidate the equivalence. Equivalence simply means that a pair of coupled inductors and the T-circuit defined by Eq. (16–40) have the same **I-V** properties at the input and output ports. Coupled inductors interact with the rest of the circuit only via these ports. The equivalent T-circuit will correctly predict external interaction even though one of its inductances is negative.

One final comment is that the T-circuit in Figure 16–17 shows why coupled inductors do not pass steady-state dc signals. Recall that in the dc steady state, inductors act like short circuits. Applying this idea in Figure 16–17 shows that the input and output ports are short-circuited, making it impossible for dc signals to pass from input to output.

EXAMPLE 16–9

Use the equivalent T-circuit in Figure 16–17 to find the input impedance of the transformer in Example 16–6.

SOLUTION:
The transformer in Example 16–6 had additive coupling with $L_1 = 2$ H, $L_2 = 0.2$ H, $M = 0.6$ H, $\omega = 377$ rad/s, and $Z_L = 25 + j15\ \Omega$. Applying Eqs. (16–40) using the upper sign yields the impedances of the equivalent T-circuit as

$$j\omega L_A = j\omega (L_1 - M) = j528\quad \Omega$$

$$j\omega L_B = j\omega (L_2 - M) = -j151\quad \Omega$$

$$j\omega L_C = j\omega M = j226\quad \Omega$$

In this case the inductance $L_B = L_2 - M = -0.4$ H is negative. Figure 16–18 shows the equivalent T-circuit together with the load impedance. The input impedance of this equivalent circuit is easily calculated by circuit reduction.

$$Z_{IN} = j528 + j226 \parallel (-j151 + Z_L)$$

$$= j528 + \cfrac{1}{\cfrac{1}{j226} + \cfrac{1}{25 - j136}}$$

$$= 145 + j228 \ \Omega$$

This result agrees with the answer in Example 16–6, where the input impedance was found using mesh-current analysis. ∎

FIGURE 16–18

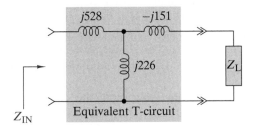

Equivalent T-circuit

Z_{IN}

Exercise 16–7

Use an equivalent T-circuit to find the input impedance of a transformer with subtractive coupling and $L_1 = 20$ mH, $L_2 = 250$ mH, $M = 70$ mH, $\omega = 2000$ rad/s, and $Z_L = 500 + j100 \ \Omega$.

Answer:
$Z_{IN} = 16.1 + j20.7 \ \Omega$

SUMMARY

- Mutual inductance describes magnetic coupling between two inductors. The mutual inductance parameter relates the voltage induced in one inductor to the rate of change of current in the other inductor. The induced voltage can be either positive (additive coupling) or negative (subtractive coupling).

- The dot convention describes the physical orientation of the two magnetically coupled inductors. Mutual inductance coupling is additive when both current reference arrows point toward or away from dotted terminals; otherwise, it is subtractive. The reference directions for the currents can always be selected so that the coupling is additive.

- The degree of coupling is indicated by the coupling coefficient. Energy analysis shows that the coupling coefficient must lie between zero and one. A coupling coefficient of unity is called perfect coupling and means that all of the flux produced by first winding links with a second winding, and vice versa. A coupling coefficient of zero indicates no magnetic linkage between the windings.

- A transformer is an electrical device based on the mutual inductance coupling between two windings. Transformers find applications in almost all electrical systems, especially in power supplies and the electrical power grid. The transformer winding connected to the power source is called the primary winding and the winding connected to the load is called the secondary winding.

- The ideal transformer is a circuit element in which the primary and secondary windings are assumed to be perfectly coupled and to have no power loss. In an ideal transformer the voltages and currents in the primary and secondary windings are related by the turns ratio, which is the ratio of the number of turns in the secondary winding to that in the primary winding.

- In the sinusoidal steady state, transformers can be analyzed using phasors to determine steady-state currents, voltages, impedances, and average power transfer. The phasor analysis of transformers is carried out using a modified mesh-current approach.

- Transformers have many different equivalent circuits. These equivalent circuits provide insight into the performance of transformers and may simplify certain types of analysis.

PROBLEMS

ERO 16–1 MUTUAL INDUCTANCE (SECT. 16–1, 16–2, 16–3)

(a) Given the current through or voltage across two coupled inductors, find the unspecified currents or voltages.
(b) Find characteristics of coupled inductors when given some device parameters and circuit connections.

See Examples 16–1, 16–2 and Exercises 16–1, 16–2

16–1 The input to the coupled inductors in Figure P16–1 is a voltage source $v_S(t) = 10 \sin 2000t$ V.
    (a) Write the i–v relationships for the coupled inductors using the reference marks in the figure.
    (b) Solve for $v_2(t)$ when the output terminals are open circuited ($i_2 = 0$).

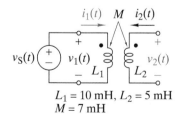

$$L_1 = 10 \text{ mH}, L_2 = 5 \text{ mH}$$
$$M = 7 \text{ mH}$$

FIGURE P16–1

16–2 The input to the coupled inductors in Figure P16–1 is a voltage source $v_S(t) = 5 \sin 500t$ V.

    (a) Write the i–v relationships for the coupled inductors using the reference marks in the figure.
    (b) Solve for $i_1(t)$ and $i_2(t)$ when the output terminals are short circuited ($v_2 = 0$).

16–3 In Figure P16–1 the outputs are $v_2(t) = 0$ and $i_2(t) = 35 \sin 500t$ A.
    (a) Write the i–v relationships for the coupled inductors using the reference marks given.
    (b) Solve for the source voltage $v_S(t)$.

16–4 The input to the coupled inductors in Figure P16–4 is a current source $i_S(t) = 10 \sin 500t$ A. Solve for $v_1(t)$ and $v_2(t)$ when the output terminals are open circuited ($i_2 = 0$).

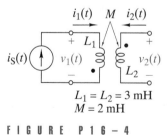

$$L_1 = L_2 = 3 \text{ mH}$$
$$M = 2 \text{ mH}$$

FIGURE P16–4

16–5 The output terminals in Figure P16–4 are short circuited and the short-circuit current is found to be $i_2(t) = 0.5 \sin 1000t$ A. Solve for $v_1(t)$ and $i_1(t)$.

16–6 A pair of coupled inductors have $L_1 = 1.2$ H and $L_2 = 2.4$ H. With the output open circuited ($i_2(t) = 0$), the coil voltages are observed to be $v_1(t) = 3 \sin 1000t$ and $v_2(t) = 4 \sin 1000t$. Find the mutual inductance and the coupling coefficient. Is the coupling additive or subtractive?

16–7 In Figure P16–7, $L_1 = L_2 = 10$ mH, $M = 9$ mH, and $i_1(t) = 2 \cos(1000t)$ A. Find the input voltage $v_x(t)$.

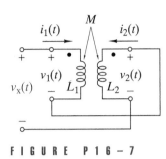

FIGURE P16-7

16–8 In Figure P16–8, $L_1 = 75$ mH, $L_2 = 400$ mH, and $k = 0.97$. Find the equivalent inductance L_{EQ}.

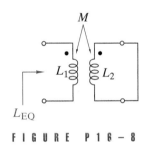

FIGURE P16-8

16–9 The self-inductances of two coupled inductors are $L_1 = 40$ mH and $L_2 = 60$ mH. What is the maximum value M can have?

16–10 The self-inductances of two coupled inductors are found to be 110 mH and 266 mH. When the two coupled inductors are connected in series, the total inductance is found to be 237 mH. What is the mutual inductance and the coupling coefficient?

ERO 16–2 THE IDEAL TRANSFORMER (SECT. 16–4)

Given a circuit containing ideal transformers,

(a) Find specified voltages, currents, powers, and equivalent circuits.
(b) Select the turns ratio to meet prescribed conditions.

See Examples 16–3, 16–4, 16–5 and Exercises 16–3, 16–4

16–11 The primary voltage in an ideal transformer is a 120-V, 60-Hz sinusoid. The secondary voltage is a 480-V, 60-Hz sinusoid. The secondary terminals are connected to an 800-Ω resistive load.
   (a) Find the transformer turns ratio.
   (b) Write expressions for the primary current and voltage.

16–12 The number of turns in the primary and secondary of an ideal transformer are $N_1 = 50$ and $N_2 = 500$. The primary is connected to a 120-V, 60-Hz source with a source resistance of 4 Ω. The secondary is connected to a 600-Ω load. Find the primary and secondary currents.

16–13 The turns ratio of the second ideal transformer in Figure P16–13 is $n = 1/3$. Find the equivalent resistance indicated in the figure.

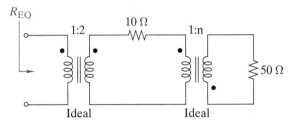

FIGURE P16-13

16–14 Select the turns ratio of the second ideal transformer in Figure P16–13 so that the same power is delivered to each resistor.

16–15 Figure P16–15 shows an ideal transformer connected as an autotransformer. Find $i_L(t)$ and $i_S(t)$ when $v_S(t) = 200 \sin 400t$ and $R_L = 50$ Ω.

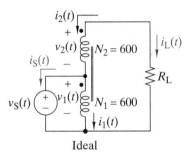

FIGURE P16-15

16–16 The primary of an ideal transformer with $n = 5$ is connected to a voltage source with a source resistance of 100 Ω. What value of load resistance in the secondary circuit will draw maximum power from the source?

16–17 Select the turns ratio of an ideal transformer in the interface circuit shown in Figure P16–17 so that the maximum available power is delivered to the 1-kΩ output resistor.

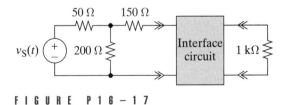

FIGURE P16-17

16–18 A voltage source with Thévenin parameters $v_T = 5 \sin 1000t$ V and $R_T = 50\ \Omega$ drives the primary winding of an ideal transformer with $n = 5$. Find the power delivered to a 300-Ω load connected across the secondary winding.

16–19 A voltage source with $v_S = 12 \sin 377t$ V drives the primary windings of the ideal transformer in Figure P16–19. Find the source current.

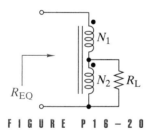

FIGURE P16–19

16–20 Show that the equivalent resistance in Figure P16–20 is

$$R_{EQ} = \left(\frac{N_1 + N_2}{N_2}\right)^2 R_L$$

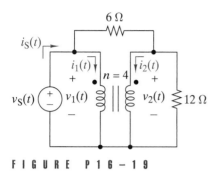

FIGURE P16–20

ERO 16–3 TRANSFORMERS IN THE SINUSOIDAL STEADY STATE (SECTS. 16–5, 16–6)

Given a linear transformer in a linear circuit operating in the sinusoidal steady state, find phasor voltages and currents, average powers, and equivalent impedances.
See Examples 16–6, 16–7, 16–8, 16–9 and Exercises 16–5, 16–6, 16–7

16–21 The input voltage to the transformer in Figure P16–21 is a sinusoid $v_1(t) = 20 \cos 1000t$. With the circuit operating in the sinusoidal steady state, use mesh-current analysis to find the phasor output voltage $\mathbf{V}_2$ and the equivalent impedance seen by the input source.

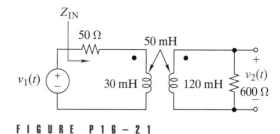

FIGURE P16–21

16–22 A transformer operating in the sinusoidal steady state has self-inductances $L_1 = 200$ mH, $L_2 = 450$ mH, and $k = 0.98$. The source has a peak amplitude of 100 V and a frequency of 377 rad/s. The load connected to the secondary of the transformer is a 75-Ω resistor. Assume additive coupling.
(a) Find the phasor responses $\mathbf{I}_1$, $\mathbf{I}_2$, $\mathbf{V}_1$, and $\mathbf{V}_2$.
(b) Find the input impedance seen by the source.
(c) Find the average power delivered by the source.

16–23 The circuit in Figure P16–23 is operating in the sinusoidal steady state with $v_S(t) = 20 \cos 20000t$. Find the steady-state response $i(t)$.

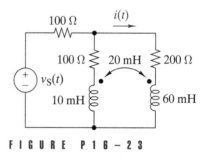

FIGURE P16–23

16–24 Repeat Problem 16–23 when the input is $v_S(t) = 20 \cos 10000t$.

16–25 The circuit in Figure P16–25 is operating in the sinusoidal steady state with $v_S(t) = 150 \sin 2000t$ V. The resistors in series with the coupled inductors simulate winding resistances. Find input impedance Z_{IN}, and the phasor responses $\mathbf{V}_1$ and $\mathbf{V}_L$.

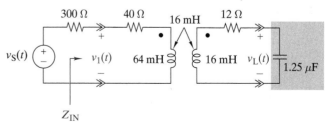

FIGURE P16–25

16–26 The frequency of the voltage source in Figure P16–25 is adjustable. Find the frequency at which the input impedance Z_{IN} is purely real.

16–27 The circuit in Figure P16–27 is operating in the sinusoidal steady state with $v_S(t) = 10 \cos 4000t$. Find the impedance seen by the voltage source, input current $i(t)$, and the average power supplied by the voltage source.

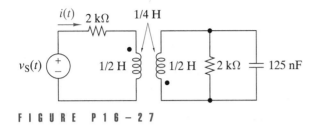

FIGURE P16–27

16–28 Find the equivalent input impedance in Figure P16–28.

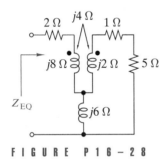

FIGURE P16–28

16–29 Find $\mathbf{V}_1$ and $\mathbf{V}_2$ in Figure P16–29.

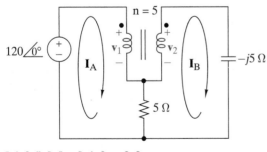

FIGURE P16–29

16–30 A transformer operating in the sinusoidal steady state has inductances $L_1 = 5$ mH, $L_2 = 20$ mH, and $M = 10$ mH. The load connected to the secondary is $Z_L = 200 + j100\ \Omega$. The voltage source connected to the primary side has a frequency of $\omega = 10$ krad/s. Find the impedance seen by the source.

16–31 The self- and mutual-inductances of a transformer can be calculated from measurements of the steady-state ac voltages and currents with the secondary winding open circuited and short circuited. Suppose the measurements are $|\mathbf{V}_1| = 100$ V, $|\mathbf{I}_1| = 120$ mA, and $|\mathbf{V}_2| = 220$ V when the secondary is open and $|\mathbf{I}_1| = 10$ A and $|\mathbf{I}_2| = 2.2$ A when the secondary is shorted. All measurements were made at $f = 400$ Hz. Find L_1, L_2, and M.

16–32 The secondary winding of an ideal transformer with a turns ratio of $n = 0.1$ is connected to a load $Z_L = 25 + j10\ \Omega$. The primary is connected to a voltage source with a peak amplitude of 2.5 kV and an internal impedance of $Z_S = j2\ \Omega$. Find the average power delivered to the load.

16–33 A certain resistive load requires 25 kW at a voltage of 480 V rms. The available supply is 2400 V rms. A transformer is to be used to reduce the supply voltage to 480 V, and a circuit breaker is to be installed on the primary side to protect the equipment. The available circuit breakers have tripping currents of 5 A, 10 A, 15 A, and 20 A. Which breaker would you choose in this application?

16–34 A 60-Hz, 7200-V(rms) to 480-V(rms) step-down transformer delivers 50 kW to a resistive load. Treat the transformer as ideal and estimate the rms primary and secondary currents.

16–35 A transformer that can be treated as ideal has 480 turns in the primary winding and 120 turns in the secondary winding. The primary is connected to a 60-Hz source with a peak amplitude of 480 V. The secondary delivers 5 kW to a resistive load. Find the primary and secondary currents and the impedance seen by the source.

INTEGRATING PROBLEMS

16–36 Ⓐ **COUPLED INDUCTOR STEP RESPONSE**

Figure P16–36 shows a coupled inductor circuit. Assuming zero initial conditions,
  (a) Write s-domain mesh-current equations for the circuit and solve for $V_2(s)$ in term of $V_S(s)$.
  (b) Solve for the unit step response of $v_2(t)$ for $L_1 = 0.1$ H, $L_2 = 0.5$ H, $M = 0.1$ H, $R_1 = 10\ \Omega$, and $R_2 = 40\ \Omega$.

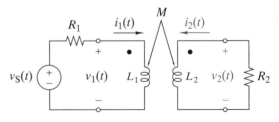

FIGURE P16–36

16–37 TRANSFER FUNCTION FOR COUPLED INDUCTORS

Figure P16–36 shows a coupled inductor circuit. Assuming zero initial conditions,

(a) Write s-domain mesh-current equations for the circuit.

(b) Solve the mesh equations for the transfer function $V_2(s)/V_S(s)$. Show that the transfer function has a second-order bandpass characteristic.

(c) Find the center frequency, upper and lower cutoff frequencies, and the bandwidth when $L_1 = 0.25$ H, $L_2 = 1$ H, $M = 0.499$ H, $R_1 = 50\ \Omega$, and $R_2 = 200\ \Omega$. Does the transformer have a wide band or narrow band bandpass frequency response?

(d) Find the step response of the transformer using the element values in (c).

16–38 THREE-WINDING IDEAL TRANSFORMER

The i–v characteristics of the three-winding ideal transformer in Figure P16–38 are

$$\frac{v_1(t)}{N_1} = \frac{v_2(t)}{N_2} = \frac{v_3(t)}{N_3}$$

and

$$N_1 i_1(t) + N_2 i_2(t) + N_3 i_3(t) = 0$$

The total instantaneous power delivered to the transformer is

$$p(t) = v_1(t)i_1(t) + v_2(t)i_2(t) + v_3(t)i_3(t)$$

Show that the i–v characteristics imply $p(t) = 0$, that is, zero power loss.

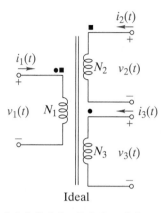

FIGURE P16-38

CHAPTER 17

POWER IN THE SINUSOIDAL STEADY STATE

George Westinghouse was, in my opinion, the only man on the globe who could take my alternating current (power) system under the circumstances then existing and win the battle against prejudice and money power.

Nikola Tesla, 1932,
American Engineer

17–1 Average and Reactive Power

17–2 Complex Power

17–3 AC Power Analysis

17–4 Load-flow Analysis

17–5 Three-phase Circuits

17–6 Three-phase AC Power Analysis

Summary

Problems

Integrating Problems

In the last decade of the nineteenth century there were two competing electrical power systems in this country. The direct current approach was developed by Thomas Edison, who installed the first of his dc power systems at the famous Pearl Street Station in New York City in 1882. By 1884 there were more than 20 such power stations operating in the United States. The competing alternating current technology was hampered by the lack of practical motors for the single-phase systems initially produced. The three-phase ac induction motor that met this need was the product of the mind of the brilliant Serbian immigrant Nikola Tesla (1856–1943). Tesla was educated in Europe and in 1884 emigrated to the United States, where he initially worked for Edison. In 1887 Tesla founded his own company to develop his inventions, eventually producing some 40 patents on three-phase equipment and controls. The importance of Tesla's work was recognized by George Westinghouse, a hard-driving electrical pioneer who purchased the rights to Tesla's patents. In the early 1890s the competition between the older dc system championed by Edison and the newer ac technology sponsored by Westinghouse developed into a heated controversy called the "war of the currents." The showdown came over the equipment to be installed in a large electric power station at Niagara Falls, New York. The choice of an ac system for Niagara Falls established a trend that led to large interconnected ac systems that form the backbone of modern industry today.

In this chapter we study the flow of electrical power in the sinusoidal steady state using phasors. The first section shows that the instantaneous power at an interface can be divided into two components called real and reactive power. The unidirectional real component produces a net transfer of energy from the source to the load. The oscillatory reactive power represents an interchange between the source and load with no net transfer of energy. The concept of complex power developed in the second section combines these two components into a single complex entity relating power flow to the interface voltage and current phasors. The third and fourth sections develop the basic tools for using complex power to analyze power flow in ac circuits. The final two sections discuss the three-phase circuits and systems that deal with the large blocks of electrical power required in a modern industrial society.

17–1   AVERAGE AND REACTIVE POWER

We begin our study of electric power circuits with the two-terminal interface in Figure 17–1. In power applications we normally think of one circuit as the source and the other as the load. Our objective is to describe the flow of power across the interface when the circuit is operating in the sinusoidal steady state. To this end, we write the interface voltage and current in the time domain as sinusoids of the form

$$v(t) = V_A \cos(\omega t + \theta)$$

$$i(t) = I_A \cos \omega t \qquad (17\text{–}1)$$

In Eq. (17–1) V_A and I_A are real, positive numbers representing the peak amplitudes of the voltage and current, respectively.

In Eq. (17–1) we have selected the $t = 0$ reference at the positive maximum of the current $i(t)$ and assigned a phase angle to $v(t)$ to account for

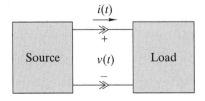

FIGURE   17–1   *A   two-terminal interface.*

the fact that the voltage maximum may not occur at the same time. In the phasor domain the angle $\theta = \phi_V - \phi_I$ is the angle between the phasors $\mathbf{V} = V_A \angle \phi_V$ and $\mathbf{I} = I_A \angle \phi_I$. In effect, choosing $t = 0$ at the current maximum shifts the phase reference by an amount $-\phi_I$ so that the voltage and current phasors become $\mathbf{V} = V_A \angle \theta$ and $\mathbf{I} = I_A \angle 0$.

A method of relating power to phasor voltage and current will be presented in the next section, but at the moment we write the instantaneous power in the time domain.

$$p(t) = v(t) \times i(t)$$
$$= V_A I_A \cos(\omega t + \theta) \cos \omega t \qquad (17\text{–}2)$$

This expression for instantaneous power contains dc and ac components. To separate the components, we first use the identity $\cos(x + y) = \cos x \cos y - \sin x \sin y$ to write $p(t)$ in the form

$$p(t) = V_A I_A [\cos \omega t \cos \theta - \sin \omega t \sin \theta] \cos \omega t$$
$$= [V_A I_A \cos \theta]\cos^2 \omega t - [V_A I_A \sin \theta]\cos \omega t \sin \omega t \qquad (17\text{–}3)$$

Using the identities $\cos^2 x = \frac{1}{2}(1 + \cos 2x)$ and $\cos x \sin x = \frac{1}{2} \sin 2x$, we write $p(t)$ in the form

$$p(t) = \underbrace{\left[\frac{V_A I_A}{2} \cos \theta\right]}_{\text{dc Component}} + \underbrace{\left[\frac{V_A I_A}{2} \cos \theta\right] \cos 2\omega t - \left[\frac{V_A I_A}{2} \sin \theta\right] \sin 2\omega t}_{\text{ac Component}} \qquad (17\text{–}4)$$

Written this way, we see that the instantaneous power is the sum of a dc component and a double-frequency ac component. That is, the instantaneous power is the sum of a constant plus a sinusoid whose frequency is 2ω, which is twice the angular frequency of the voltage and current in Eq. (17–1).

Note that instantaneous power in Eq. (17–4) is periodic. In Chapter 5 we defined the average value of a periodic waveform as

$$P = \frac{1}{T}\int_0^T p(t)dt$$

where $T = 2\pi/2\omega$ is the period of $p(t)$. In Chapter 5 we also showed that the average value of a sinusoid is zero, since the area under the waveform during a positive half cycle is canceled by the area under a subsequent negative half cycle. Therefore, the **average value** of $p(t)$, denoted as P, is equal to the constant or dc term in Eq. (17–4):

$$P = \frac{V_A I_A}{2} \cos \theta \qquad (17\text{–}5)$$

The amplitude of the $\sin 2\omega t$ term in Eq. (17–4) has a form much like the average power in Eq. (17–5), except it involves $\sin \theta$ rather than $\cos \theta$. This amplitude factor is called the **reactive power** of $p(t)$, where reactive power Q is defined as

$$Q = \frac{V_A I_A}{2} \sin \theta \qquad (17\text{–}6)$$

Substituting Eqs. (17-5) and (17-6) into Eq. (17-4) yields the instantaneous power in terms of the average power and reactive power:

$$p(t) = \underbrace{P(1 + \cos 2\omega t)}_{\text{Unipolar}} - \underbrace{Q \sin 2\omega t}_{\text{Bipolar}} \qquad (17\text{-}7)$$

The first term in Eq. (17-7) is said to be unipolar because the factor $1 + \cos 2\omega t$ never changes sign. As a result, the first term is either always positive or always negative depending on the sign of P. The second term is said to be bipolar because the factor $\sin 2\omega t$ alternates signs every half cycle.

The energy transferred across the interface during one cycle $T = 2\pi/2\omega$ of $p(t)$ is

$$\begin{aligned} W &= \int_0^T p(t)dt \\ &= P\underbrace{\int_0^T (1 + \cos 2\omega t)dt}_{\text{Net Energy}} - Q\underbrace{\int_0^T \sin 2\omega t\, dt}_{\text{No Net Energy}} \qquad (17\text{-}8) \\ &= \qquad P \times T \qquad - \qquad 0 \end{aligned}$$

Only the unipolar term in Eq. (17-7) provides any net energy transfer, and that energy is proportional to the average power P. With the passive sign convention the energy flows from source to load when $W > 0$. Equation (17-8) shows that the net energy will be positive if the average power $P > 0$. Equation (17-5) points out that the average power P is positive when $\cos \theta > 0$, which in turn means $|\theta| < 90°$. We conclude that

> *The net energy flow in Figure 17-1 is from source to load when the angle between the interface voltage and current is bounded by $-90° < \theta < 90°$; otherwise the net energy flow is from load to source.*

The bipolar term in Eq. (17-7) is a power oscillation that transfers no net energy across the interface. In the sinusoidal steady state the load in Figure 17-1 borrows energy from the source circuit during part of a cycle and temporarily stores it in the load inductance or capacitance. In another part of the cycle the borrowed energy is returned to the source unscathed. The amplitude of the power oscillation is called reactive power because it involves periodic energy storage and retrieval from the reactive elements of the load. The reactive power can be either positive or negative depending on the sign of $\sin \theta$. However, the sign of Q tells us nothing about the net energy transfer, which is controlled by the sign of P.

We are obviously interested in average power, since this component carries net energy from source to load. For most power system customers the basic cost of electrical service is proportional to the net energy delivered to the load. Large industrial users may also pay a service charge for their reactive power. This may seem unfair, since reactive power transfers no net energy. However, the electric energy borrowed and returned by the load is generated within a power system that has losses. From a power company's viewpoint, the reactive power is not free because there are losses in the system connecting the generators in the power plant to source/load interface at which the lossless interchange of energy occurs.

In ac power circuit analysis, it is necessary to keep track of both the average power and reactive power. These two components of power have the same dimensions, but because they represent quite different effects they traditionally are given different units. The average power is expressed in watts (W) while reactive power is expressed in VARs, which is an acronym for "volt-amperes reactive."

Exercise 17–1

Using the reference marks in Figure 17–1, calculate the average and reactive power for the following voltages and currents. State whether the load is absorbing or delivering net energy.

(a) $v(t) = 168 \cos(377t + 45°)$ V, $i(t) = 0.88 \cos 377t$ A
(b) $v(t) = 285 \cos(2500t – 68°)$ V, $i(t) = 0.66 \cos 2500t$ A
(c) $v(t) = 168 \cos(377t + 45°)$ V, $i(t) = 0.88 \cos(377t – 60°)$ A
(d) $v(t) = 285 \cos(2500t – 68°)$ V, $i(t) = 0.66 \sin 2500t$ A

Answers:
(a) $P = +52.3$ W, $Q = +52.3$ VAR, absorbing
(b) $P = +35.2$ W, $Q = -87.2$ VAR, absorbing
(c) $P = -19.1$ W, $Q = +71.4$ VAR, delivering
(d) $P = +87.2$ W, $Q = +35.2$ VAR, absorbing

17–2 COMPLEX POWER

It is important to relate average and reactive power to phasor quantities because steady-state analysis is conveniently carried out using phasors. In our previous work the magnitude of a phasor represented the peak amplitude of a sinusoid. However, in power circuit analysis it is convenient to express phasor magnitudes in rms (root mean square) values. In this chapter phasor voltages and currents are expressed as

$$\mathbf{V} = V_{rms}e^{j\phi_V} \quad \text{and} \quad \mathbf{I} = I_{rms}e^{j\phi_I} \tag{17–9}$$

Notice that the phasor magnitudes are the rms amplitude of the corresponding sinusoid.

Equations (17–5) and (17–6) express average and reactive power in terms of peak amplitudes V_A and I_A. In Chapter 5 we showed that the peak and rms values of a sinusoid are related by $V_{rms} = V_A/\sqrt{2}$. The expression for average power easily can be converted to rms amplitudes since we can write Eq. (17–5) as

$$P = \frac{V_A I_A}{2} \cos \theta = \frac{V_A}{\sqrt{2}} \frac{I_A}{\sqrt{2}} \cos \theta$$

$$= V_{rms} I_{rms} \cos \theta \tag{17–10}$$

where $\theta = \phi_V – \phi_I$ is the angle between the voltage and current phasors. By similar reasoning, Eq. (17–6) becomes

$$Q = V_{rms} I_{rms} \sin \theta \tag{17–11}$$

Using rms phasors the **complex power** (*S*) at a two-terminal interface is defined as follows:

$$S = \mathbf{VI}^* \qquad (17\text{--}12)$$

That is, the complex power at an interface is the product of the voltage phasor times the conjugate of the current phasor. Substituting Eq. (17–9) into this definition yields

$$S = \mathbf{VI}^* = V_{rms}e^{j\phi_V}I_{rms}e^{-j\phi_I}$$
$$= [V_{rms}\,I_{rms}]e^{j(\phi_V - \phi_I)} \qquad (17\text{--}13)$$

Using Euler's relationship and the fact that the angle is $\theta = \phi_V - \phi_I$, we can write complex power as

$$S = [V_{rms}I_{rms}]e^{j\theta}$$
$$= [V_{rms}I_{rms}]\cos\theta + j[V_{rms}I_{rms}]\sin\theta \qquad (17\text{--}14)$$
$$= P + jQ$$

The real part of the complex power S is the average power, and the imaginary part is the reactive power. Although S is a complex number, it is not a phasor. However, it is a convenient variable for keeping track of the two components of power when the voltage and currents are expressed as phasors.

The power triangles in Figure 17–2 provide a convenient way to remember complex power relationships and terminology. We confine our study to cases in which net energy is transferred from source to load. In such cases $P > 0$ and the power triangles fall in the first or fourth quadrant, as indicated in Figure 17–2.

The magnitude $|S| = V_{rms}I_{rms}$ is called **apparent power** and is expressed using the unit volt-ampere (VA). The ratio of the average power to the apparent power is called the **power factor** (pf). Using Eq. (17–10), we see that the power factor is

$$\text{pf} = \frac{P}{|S|} = \frac{V_{rms}I_{rms}\cos\theta}{V_{rms}I_{rms}} = \cos\theta \qquad (17\text{--}15)$$

Since pf = $\cos\theta$, the angle θ is called the **power factor angle**.

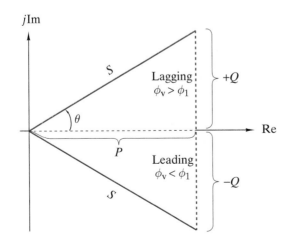

FIGURE 17-2 *Power triangles.*

When the power factor is unity the phasors $\mathbf{V}$ and $\mathbf{I}$ are in phase ($\theta = 0°$) and the reactive power is zero since $\sin \theta = 0$. When the power factor is less than unity, the reactive power is not zero and its sign is indicated by the modifiers lagging or leading. The term *lagging power factor* means the current phasor lags the voltage phasor so that $\theta = \phi_V - \phi_I > 0$. For a lagging power factor S falls in the first quadrant in Figure 17–2 and the reactive power is positive since $\sin \theta > 0$. The term *leading power factor* means the current phasor leads the voltage phasor so that $\theta = \phi_V - \phi_I < 0$. In this case S falls in the fourth quadrant in Figure 17–2 and the reactive power is negative since $\sin \theta < 0$. Most industrial and residential loads have lagging power factors.

The apparent power rating of electrical power equipment is an important design consideration. The ratings of generators, transfomers, and transmission lines are normally stated in kVA. The ratings of most loads are stated in kW and power factor. The wiring must be large enough to carry the required current and insulated well enough to withstand the rated voltage. However, only the average power is potentially available as useful output, since the reactive power represents a lossless interchange between the source and device. Because reactive power increases the apparent power rating without increasing the net energy output, it is desirable for electrical devices to operate with zero reative power or, equivalently, at unity power factor.

Exercise 17–2

Determine the average power, reactive power, and apparent power for the following voltage and current phasors. State whether the power factor is lagging or leading.

(a) $\mathbf{V} = 208\angle{-90°}$ V (rms), $\mathbf{I} = 1.75\angle{-75°}$ A (rms)
(b) $\mathbf{V} = 277\angle{+90°}$ V (rms), $\mathbf{I} = 11.3\angle{0°}$ A (rms)
(c) $\mathbf{V} = 120\angle{-30°}$ V (rms), $\mathbf{I} = 0.30\angle{-90°}$ A (rms)
(d) $\mathbf{V} = 480\angle{+75°}$ V (rms), $\mathbf{I} = 8.75\angle{+105°}$ A (rms)

Answers:
(a) $P = 352$ W, $Q = -94.2$ VAR, $|S| = 364$ VA, leading
(b) $P = 0$ W, $Q = +3.13$ kVAR, $|S| = 3.13$ kVA, lagging
(c) $P = 18$ W, $Q = +31.2$ VAR, $|S| = 36$ VA, lagging
(d) $P = 3.64$ kW, $Q = -2.1$ kVAR, $|S| = 4.20$ kVA, leading

COMPLEX POWER AND LOAD IMPEDANCE

In many cases power circuit loads are described in terms of their power ratings at a specified voltage or current level. To find voltages and current elsewhere in the circuit, it is necessary to know the load impedance. For this reason we need to relate complex power and load impedance.

Figure 17–3 shows the general case for a two-terminal load. For the assigned reference directions the load produces the element constraint $\mathbf{V} = Z\mathbf{I}$. Using this constraint in Eq. (17–12), we write the complex power of the load as

$$S = \mathbf{V} \times \mathbf{I}^* = Z\mathbf{I} \times \mathbf{I}^* = Z|\mathbf{I}|^2$$
$$= (R + jX)\,I^2_{rms}$$

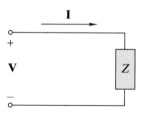

FIGURE 17–3 *A two-terminal impedance.*

where R and X are the resistance and reactance of the load, respectively. Since $S = P + jQ$, we conclude that

$$R = \frac{P}{I_{rms}^2} \quad \text{and} \quad X = \frac{Q}{I_{rms}^2} \tag{17–16}$$

The load resistance and reactance are proportional to the average and reactive power of the load, respectively.

The first condition in Eq. (17–16) demonstrates that resistance cannot be negative, since P cannot be negative for a passive circuit. That is, in the sinusoidal steady state a passive circuit cannot produce average power; otherwise, perpetual motion would be possible and the energy crisis would be a small footnote in the great sweep of human history. The second condition in Eq. (17–16) points out that when the reactive power is positive the load is inductive, since $X_L = \omega L$ is positive. Conversely, when the reactive power is negative the load is capacitive, since $X_C = -1/\omega C$ is negative. The terms *inductive load, lagging power factor,* and *positive reactive power* are synonymous, as are the terms *capacitive load, leading power factor,* and *negative reactive power.*

EXAMPLE 17–1

At 440 V (rms) a two-terminal load draws 3 kVA of apparent power at a lagging power factor of 0.9. Find

(a) I_{rms}
(b) P
(c) Q
(d) the load impedance.

Draw the power triangle for the load.

SOLUTION:
(a) $I_{rms} = |S|/V_{rms} = 3000/440 = 6.82$ A (rms)
(b) $P = V_{rms}I_{rms} \cos \theta = 3000 \times 0.9 = 2.7$ kW
(c) For $\cos \theta = 0.9$ lagging, $\sin \theta = 0.436$ and $Q = V_{rms}I_{rms} \sin \theta = 1.31$ kVAR
(d) $Z = (P + j Q)/(I_{rms})^2 = (2700 + j1310)/46.5 = 58.0 + j28.2$ Ω.

Figure 17–4 shows the power triangle for this load. ∎

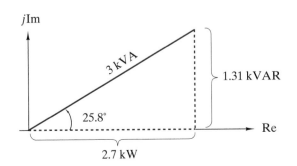

FIGURE 17–4

jIm

3 kVA

1.31 kVAR

25.8°

Re

2.7 kW

Find the impedance of a two-terminal load under the following conditions.
(a) $\mathbf{V} = 120\angle 30°$ V (rms) and $\mathbf{I} = 20\angle 75°$ A (rms)
(b) $|S| = 3.3$ kVA, $Q = -1.8$ kVAR, and $I_{rms} = 7.5$ A
(c) $P = 3$ kW, $Q = 4$ kVAR, and $V_{rms} = 880$ V
(d) $V_{rms} = 208$ V, $I_{rms} = 17.8$ A, and $P = 3$ kW

Answers:
(a) $Z = 4.24 - j4.24$ Ω
(b) $Z = 49.2 - j32$ Ω
(c) $Z = 92.9 + j124$ Ω
(d) $Z = 9.47 \pm j6.85$ Ω

17–3 AC POWER ANALYSIS

The nature of ac power analysis can be modeled in terms of the phasor voltage, phasor current, and complex power at the source/load interface in Figure 17–1. Two different types of problems are treated using this model. In the **direct analysis** problem the source and load circuit are given and we are required to find the steady-state responses at one or more interfaces. This type of problem is essentially the same as the phasor circuit analysis problems, except that we calculate complex powers as well as phasor responses. In the **load-flow** problem we are required to adjust the source so that a prescribed complex power is delivered to the load at a specified interface voltage magnitude. This type of problem arises in electrical power systems where the objective is to supply changing energy demands at a fixed voltage level.

The following examples are direct analysis problems that illustrate the computational tools needed to deal with the load-flow problems discussed in the next section. One of the useful tools is the principle of the **conservation of complex power**, which can be stated as follows:

> *In a linear circuit operating in the sinusoidal steady state, the sum of the complex powers produced by each independent source is equal to the sum of the complex power absorbed by all other two-terminal elements in the circuit.*

To apply this principle, it is important to distinguish between the complex power "produced by" and "absorbed by" a two-terminal element. To do so, we modify our practice of using the passive sign convention. We continue to use the passive convention (current reference directed in at the + voltage mark) for loads, but use the active convention (current reference directed out at the + voltage mark) for sources. In either case, the complex power is always calculated as $S = \mathbf{VI}^* = P + jQ$. The net result is that the average power produced by a source is positive and the average power absorbed by a load is positive.

EXAMPLE 17–2

(a) Calculate the complex power absorbed by each parallel branch in Figure 17–5.

FIGURE 17–5

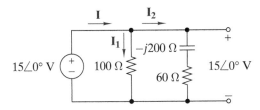

(b) Calculate the complex power produced by the source and the power factor of the load seen by the source.

SOLUTION:

(a) The voltage across each branch is $15\angle0°$ V and the branch impedances are $Z_1 = 100$ and $Z_2 = 60 - j200$. Therefore, the branch currents are

$$\mathbf{I}_1 = \frac{15\angle0°}{100} = 0.15\angle0° \text{ A}$$

$$\mathbf{I}_2 = \frac{15\angle0°}{60 - j200} = 0.0718\angle73.3° \text{ A}$$

The reference marks for the voltage across and currents through these loads follow the passive sign convention. Hence the complex powers absorbed by the loads are

$$S_1 = (15\angle0°)\mathbf{I}_1^* = (15\angle0°)(0.15\angle - 0°) = 2.25\angle0° \text{ VA}$$

$$S_2 = (15\angle0°)\mathbf{I}_2^* = (15\angle0°)(0.0718\angle - 73.3°) = 1.08\angle - 73.3° \text{ VA}$$

(b) Using KCL, the source current $\mathbf{I}$ is

$$\mathbf{I} = (\mathbf{I}_1 + \mathbf{I}_2) = (0.15\angle0° + 0.0718\angle73.3°)$$

$$= 0.171 + j0.0688 = 0.184\angle21.9° \text{ A}$$

The reference marks for the source current $\mathbf{I}$ and source voltage $15\angle0°$ conform to the active sign convention, and the complex power produced by the source is

$$S = (15\angle0°)\mathbf{I}^* = (15\angle0°)(0.184\angle - 21.9°)$$

$$= 2.76\angle - 21.9° \text{ VA}$$

The power factor is $\cos(-21.9°) = 0.928$ leading. Alternatively, we can use the conservation of complex power to obtain the complex power produced by the source as the sum of the complex powers delivered to the passive elements:

$$S_1 + S_2 = 2.25\angle0° + 1.08\angle - 73.3°$$

$$= 2.25 + j0 + 0.310 - j1.03$$

$$= 2.56 - j1.03 = 2.76\angle - 21.9° \text{ VA}$$

This result is the same as the answer obtained previously.   ■

EXAMPLE 17–3

Find the complex power produced by each source in Figure 17–6 when $\mathbf{V}_{S1} = 440\angle0°$ V (rms) and $\mathbf{V}_{S2} = 500\angle0°$ V (rms).

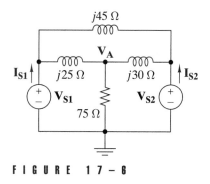

FIGURE 17-6

SOLUTION:

Since both voltage sources are connected to ground, the voltage at node A is the only unknown node voltage in the circuit. By inspection, the node equation at node A is

$$\left[\frac{1}{75} + \frac{1}{j25} + \frac{1}{j30}\right]\mathbf{V}_A = \frac{\mathbf{V}_{S1}}{j25} + \frac{\mathbf{V}_{S2}}{j30} = -j34.27$$

which yields $\mathbf{V}_A = 452 - j82.2$ V. The current supplied by source no. 1 is

$$\mathbf{I}_{S1} = \frac{\mathbf{V}_{S1} - \mathbf{V}_A}{j25} + \frac{\mathbf{V}_{S1} - \mathbf{V}_{S2}}{j45} = 3.29 + j1.81 \text{ A (rms)}$$

The complex power supplied by source no. 1 is

$$S_{S1} = \mathbf{V}_{S1}\mathbf{I}_{S1}^* = 440 \times (3.29 - j1.83)$$
$$= 1450 - j796 \quad \text{VA}$$

The current supplied by source no. 2 is

$$\mathbf{I}_{S2} = \frac{\mathbf{V}_{S2} - \mathbf{V}_A}{j30} + \frac{\mathbf{V}_{S2} - \mathbf{V}_{S1}}{j45} = 2.74 - j2.93 \quad \text{A (rms)}$$

and the complex power supplied by source no. 2 is

$$S_{S2} = \mathbf{V}_{S2}\mathbf{I}_{S2}^* = 500 \times (2.74 + j2.93)$$
$$= 1370 + j1460 \quad \text{VA} \qquad \blacksquare$$

EXAMPLE 17-4

Given the transformer circuit in Figure 17–7, find the complex power produced by the input voltage source and the power absorbed by the 100-Ω load resistor.

FIGURE 17-7

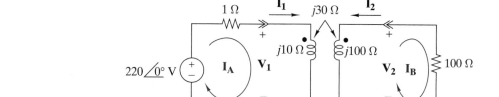

SOLUTION:

Using the voltages and mesh currents in Figure 17–7, the KVL equations around the primary and secondary circuits are

$$1 \times \mathbf{I}_A + \mathbf{V}_1 = 220\angle 0°$$
$$100\,\mathbf{I}_B - \mathbf{V}_2 = 0$$

The coil currents are related to the mesh currents as $\mathbf{I}_1 = \mathbf{I}_A$ and $\mathbf{I}_2 = -\mathbf{I}_B$. Both coil currents are directed inward at the winding dots, so the mutual inductive coupling is additive and the i–v characteristics of the transformer in terms of the mesh currents are

$$\mathbf{V}_1 = j10\mathbf{I}_A + j30(-\mathbf{I}_B)$$

$$\mathbf{V}_2 = j30\mathbf{I}_A + j100(-\mathbf{I}_B)$$

Substituting these i–v characteristics into the KVL equation produces the two mesh current equations for the transformer:

$$(1 + j10)\mathbf{I}_A - j30\mathbf{I}_B = 220$$

$$-j30\mathbf{I}_A + (100 + j100)\mathbf{I}_B = 0$$

Solving for the two mesh current produces

$$\mathbf{I}_A = 20 - j20 \text{ A (rms)}$$

$$\mathbf{I}_B = 6 + j0 \text{ A (rms)}$$

Using mesh current $\mathbf{I}_A$, the complex power produced by the source is

$$S_{IN} = (220\angle 0°)\mathbf{I}_A^* = 4400 + j4400 \quad \text{VA}$$

and the power absorbed by the 100-Ω output resistor is

$$S_{OUT} = (100\mathbf{I}_B)\mathbf{I}_B^* = 3600 + j0 \text{ VA} \qquad ■$$

Exercise 17–4

Calculate the complex power delivered by each source in Figure 17–8.

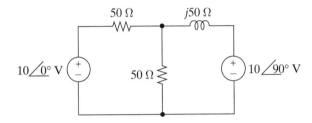

FIGURE 17 − 8

Answers: $S_1 = 0.4 + j0.8$ VA, $S_2 = 1.6 + j1.2$ VA

17–4 LOAD-FLOW ANALYSIS

The analysis of ac electrical power systems is one of the major applications of phasor circuit analysis. Although the loads on power systems change throughout the day, these variations are extremely slow compared with the period of the 50/60-Hz sinusoid involved.[1] Consequently, electrical power system analysis can be carried out using steady-state concepts and phasors. In fact, it was the study of the steady-state performance of ac power equipment that led Charles Steinmetz to advocate phasors in the first place.

In this section we treat ac power analysis using the simple model of an electrical power system in Figure 17–9. This model is a series circuit with

---

1 In the United States commercial ac power systems operate at 60 Hz. In most of the rest of the world the standard operating frequency is 50 Hz.

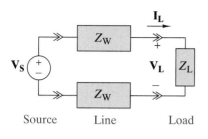

F I G U R E 1 7 – 9 *A simple electrical power system.*

an ac source connected to a load via power lines whose wire impedances are Z_W. In a **load-flow problem** the complex power delivered to the load is specified and we are asked to find either the source voltage for a given load voltage or the load voltage for a given source voltage. The analysis approach is similar to the unit output method. That is, we begin with conditions at the load and work backward through the circuit to establish the required source voltage. The load-flow problem is different from the maximum power transfer problem studied in Chapter 15. In a maximum power transfer problem the source is fixed and the load is adjusted to achieve a conjugate match. Conjugate matching does not apply to electrical power systems because the load power is fixed and the source is adjusted to meet customer demands.

Large industrial customers are charged for their reactive power, so in some cases it is desirable to reduce the load reactance. Since power system loads are normally inductive, the net reactive power of the load can be reduced by adding a capacitor in parallel with the load. The amount of the negative reactive power taken by the capacitor is selected to cancel some or all of the positive reactance power drawn by the inductive load. Physically, this means that the oscillatory interchange of energy represented by reactive power takes place between the capacitor and inductance in the load, rather than between the load inductance and the lossy power system.

Adding parallel capacitance is called **power factor correction**, since the net power factor of the composite load is increased. If the power factor is increased to unity, then the net reactance is zero and the load is in resonance. Power factor correction reduces the reactive power drain on the power system but does not change the average power delivered to the load.

The following examples illustrate ac power analysis problems, including load flow and power factor correction.

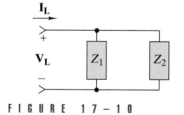

F I G U R E 1 7 – 1 0

EXAMPLE 17–5

In this problem the two parallel connected impedances in Figure 17–10 are the load in the power system model in Figure 17–9. With $V_L = 480$ V (rms), load Z_1 draws an average power of 10 kW at a lagging power factor of 0.8 and load Z_2 draws 12 kW at a lagging power factor of 0.75. The line impedances shown in Figure 17–9 are $Z_W = 0.35 + j1.5$ Ω.

(a) Find the total complex power delivered to the composite load.
(b) Find the apparent power delivered by the source and the source voltage phasor $\mathbf{V}_S$.
(c) Calculate the transmission efficiency of the system.

SOLUTION:

(a) In this example we are given the power factor (pf) and the average power (P) for each load. Because the power factor angle is $\theta = \cos^{-1}(\text{pf})$ and $Q = P \tan \theta$, we can find the complex power for each load as $S = P + P \tan(\cos^{-1} \text{pf})$, hence

$$S_{L1} = 10 + j10 \tan(\cos^{-1} 0.8) = 10 + j7.5 \text{ kVA}$$

$$S_{L2} = 12 + j12 \tan(\cos^{-1} 0.75) = 12 + j10.6 \text{ kVA}$$

The total complex power delivered to the composite load is

$$S_L = S_{L1} + S_{L2} = 22 + j18.1 = 28.5\angle 39.4° \text{ kVA}$$

(b) Using the load voltage as the phase reference, we find the load current to be

$$\mathbf{I}_L^* = \frac{S_L}{\mathbf{V}_L} = \frac{28500\angle 39.4°}{480\angle 0°} = 59.4\angle 39.4° \text{ A (rms)}$$

or $\mathbf{I}_L = 59.4\angle -39.4° \text{ A (rms)}$. To find the source power, we need to find the complex power lost in the transmission to the line:

$$S_W = 2|\mathbf{I}_L|^2(R_W + jX_W) = 2 \times (59.4)^2 (0.35 + j1.5)$$

$$= 2.47 + j10.6 \text{ kVA}$$

Using the conservation of complex power, we find that the source must produce

$$S_S = S_L + S_W = 24.5 + j28.7 = 37.7\angle 49.5° \text{ kVA}$$

For the series model in Figure 17–9 the source and load currents are equal so the source power is $S_S = \mathbf{V}_S(\mathbf{I}_L)^*$. Given the source power and load current, we find that the required source voltage is

$$\mathbf{V}_S = \frac{S_S}{\mathbf{I}_L^*} = \frac{37700\angle 49.5°}{59.4\angle 39.4°} = 635\angle 10.1° \text{ V (rms)}$$

(c) The transmission efficiency is defined in terms of source and load average power as

$$\eta = \frac{P_L}{P_S} \times 100\% = \frac{22}{24.5} \times 100 = 89.8\% \qquad \blacksquare$$

EXAMPLE 17–6

A power system modeled by the circuit in Figure 17–9 delivers a load power of $S_L = 25 + j10 \text{ kVA}$ when the source voltage is $\mathbf{V}_S = 600\angle 0°$. The line impedances are $Z_W = 0.35 + j1.5 \text{ }\Omega$. Find the load current, load voltage, and transmission efficiency.

SOLUTION:
In the preceding example the known quantities are the load power and load voltage. In this example the load power and the source voltage are given. As a result both the load current and the load voltage are unknowns. We can write two constraints on these unknowns. First, the specified complex power delivered to the load requires that

$$S_L = 25000 + j10000 = \mathbf{V}_L\mathbf{I}_L^*$$

Next, writing a KVL equation around the loop in Figure 17–9 yields $\mathbf{V}_S - \mathbf{V}_L - 2Z_W\mathbf{I}_L = 0$. Inserting the known quantities and solving for $\mathbf{V}_L$ produces

$$\mathbf{V}_L = 600 - (0.7 + j3)\mathbf{I}_L$$

Substituting this result into the complex power constraint yields

$$25000 + j1000 - 600\,\mathbf{I}_L^* + (0.7 + j3)\mathbf{I}_L\mathbf{I}_L^* = 0$$

We have a single complex constraint in which the only unknown is the load current phasor $\mathbf{I}_L$. This condition requires both the real part and the imaginary part of the constraint to vanish. Hence, the constraint can be written in terms of two real value functions. Writing the load current phasor as $\mathbf{I}_L = I_1 + jI_2$, hence $\mathbf{I}_L\mathbf{I}_L^* = I_1^2 + I_2^2$, the constraints on the real part and the imaginary part are

$$\text{Real part:} \quad 25000 - 600\,I_1 + 0.7(I_1^2 + I_2^2) = 0$$
$$\text{Imag. part:} \quad 10000 + 600\,I_2 + 3(I_1^2 + I_2^2) = 0$$

We have now set up the equations so that MATLAB can be used to solve the complex constraints. In the MATLAB command window we write the following statements:

```
syms I1 I2

FR=25000-600*I1+0.7*(I1^2+I^2);

FI=10000+600*I2+3*(I1^2+I2^2);
```

The first statement declares I1 and I2 to be symbolic variables. The next two statements define the real part constraint FR and the imaginary part constraint FI as functions of the form $f(x, y) = 0$. The MATLAB solve command is set up to solve equations of this type. Accordingly, we write the command

```
[I1,I2]=solve(FR,FI)
```

and MATLAB responds with the values of I1 and I2 that satisfy the conditions FR=0 and FI=0, namely,

```
I1=

[55000/949-700/2847*2669^(1/2)]
[55000/949+700/2847*2669^(1/2)]


I2=

[-246200/2847+1000/949*2669^(1/2)]
[-246200/2847-1000/949*2669^(1/2)]
```

MATLAB returns two possible solutions for the load current. The first solution is

$$\mathbf{I}_L = I_1 + jI_2$$
$$= \left(\frac{55000}{949} - \frac{700}{2847}\sqrt{2669}\right) + j\left(\frac{-246200}{2847} + \frac{1000}{949}\sqrt{2669}\right)$$
$$= 45.253 - j32.083 \ \text{A}$$

which produces a load voltage phasor of

$$\mathbf{V}_L = 600 - (0.7 + j3)\mathbf{I}_L = 472.208 - j113.333 \ \text{V}$$

The apparent power delivered by the source is

$$S_S = \mathbf{V}_S\mathbf{I}_L^* = (600 + j0)(45.253 + j32.038)$$
$$= 27.2 + j19.2 \ \text{kVA}$$

and the transmission efficiency for the first solution is

$$\eta = \frac{P_L}{P_S} \times 100\% = \frac{25}{27.2} \times 100 = 91.9\%$$

The second MATLAB solution is

$$\mathbf{I}_L = I_1 + jI_2$$
$$= \left(\frac{55000}{949} + \frac{700}{2847}\sqrt{2669}\right) + j\left(\frac{-246200}{2847} - \frac{1000}{949}\sqrt{2669}\right)$$
$$= 70.658 - j140.916 \ \text{A}$$

which produces a load voltage phasor of

$$\mathbf{V}_L = 600 - (0.7 + j3)\mathbf{I}_L = 127.790 - j113.333 \ \text{V}$$

The apparent power delivered by the source is

$$S_S = \mathbf{V}_S\mathbf{I}_L^* = (600 + j0)(70.658 + j140.915)$$
$$= 42.39 + j84.55 \ \text{kVA}$$

and the transmission efficiency for the second solution is

$$\eta = \frac{P_L}{P_S} \times 100\% = \frac{25}{42.395} \times 100 = 59.0\%$$

Several comments are in order. This type of load flow problem may have several solutions, two in this case. On the basis of transmission efficiency we would choose the first solution over the second. This simple example illustrates some features of the general load flow problem in large power systems. Such systems have a number of generating stations with different efficiencies and many geographically separated load centers. The load flow problem involves controlling the output of the different generating stations to meet the total load requirements in a way that maximizes the system efficiency. ∎

EXAMPLE 17-7

Using the loads and line impedances defined in Example 17–5,

(a) Find the parallel capacitance needed so that the load power factor is at least 0.95. The power system frequency is 60 Hz.
(b) Find the transmission efficiency with the capacitor connected.

SOLUTION:
(a) To determine the capacitance, it is necessary to relate the reactive power of a capacitor to the load voltage. Since the capacitor is in parallel with the load, the current through it is $\mathbf{I}_C = j\omega C\mathbf{V}_L$. Therefore, the reactive power of the capacitor can be written as

$$Q_C = |\mathbf{I}_C|^2 X_C = |j\omega C \mathbf{V}_L|^2 \left(\frac{-1}{\omega C}\right)$$
$$= |\mathbf{V}_L|^2(-\omega C)$$

Note that Q_C is negative. In Example 17–5 the complex power of the two parallel loads is found to be

$$S_L = P_L + jQ_L = 22 + j18.1 \quad \text{kVA}$$

When the capacitor is connected the complex power delivered to the composite load becomes

$$S_L = P_L + j(Q_L + Q_C)$$

Since the capacitor is a reactive element, it only changes the reactive power and not the average power. To correct the power factor to at least 0.95 requires that

$$\cos\theta = \frac{P_L}{\sqrt{P_L^2 + (Q_L + Q_C)^2}} \geq 0.95$$

Solving for $-Q_C$ yields

$$-Q_C \geq Q_L - P_L\sqrt{1/(0.95)^2 - 1} = 10.9 \quad \text{kVAR}$$

which means that a lower bound on the capacitance is

$$C = \frac{-Q_C}{V_L^2\omega} \geq \frac{10,900}{(480)^2(2\pi60)} = 125 \ \mu\text{F}$$

This yields a lower bound on the required capacitance. An upper bound is found by increasing the load-power factor to unity ($\cos\theta = 1$). The required capacitive reactance is

$$-Q_C = Q_L = 18.1 \quad \text{kVAR}$$

which yields an upper bound on the capacitance of

$$C = \frac{-Q_C}{V_L^2\omega} \leq \frac{18,100}{(480)^2(2\pi60)} = 208 \ \mu\text{F}$$

Thus, the design requirement can be met by any capacitance in the range from 125 μF to 208 μF.

(b) With the minimum capacitance of $C = 125 \ \mu$F connected in parallel with the load, the apparent power delivered to the composite load is $S_L = 22 + j7.2$ kVA. Using the load voltage for the phase reference, we find that the load current is

$$\mathbf{I}_L^* = \frac{S_L}{\mathbf{V}_L} = \frac{22,000 + j7200}{480\angle0°}$$
$$= 48.2\angle18.1° \quad \text{A (rms)}$$

The apparent power lost in the line is

$$S_W = |\mathbf{I}_L|^2 \, 2(R_W + jX_W) = 1.62 + j6.97 \quad \text{kVA}$$

Hence, with power factor correction, the source produces

$$S_S = S_L + S_W = 23.6 + j14.2 \quad \text{kVA}$$

and the transmission efficiency is

$$\eta = \frac{P_L}{P_S} \times 100\% = \frac{22}{23.6} \times 100 = 93.2\%$$

In Example 17–5 we found the transmission efficiency without power factor correction to be 89.8%. With power factor correction the source delivers the same average power to the load with an increase in efficiency. Reactive power is a burden to a power system even though it represents a lossless interchange of energy at the terminals of the load.

■

Exercise 17–5

Find the source voltage and apparent power required to deliver 2400 V (rms) to a load that draws 25 kVA at a 0.85 lagging power factor from a line with a total line impedance of $2Z_W = 4 + j20 \ \Omega$.

Answers: 2.55 kV (rms) and 26.6 kVA at a lagging power factor of 0.82

___

APPLICATION NOTE: **EXAMPLE 17—8**

The electrical power for most residential customers in the United States is supplied by the 60-Hz, 110/220-V (rms) single-phase, three-wire system modeled in Figure 17–11. The term *single phase* means that the phasors representing the two source voltages are in phase. The three lines connecting the sources and loads are labeled A, B, and N (for neutral). The impedances Z_W and Z_N are small compared with the load impedances, so the load voltages differ from the source voltages by only a few percent. The impedances Z_1 and Z_2 connected from Lines A or B to neutral represent small appliance and lighting loads which require 110 V (rms) service. The impedances Z_3 connected between lines A and B are heavier loads that require 220 V (rms) service, such as water heaters or clothes dryers.

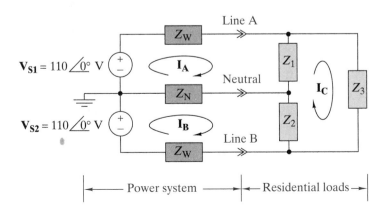

FIGURE 17–11 *Residential power distribution circuit.*

When the two source voltages are exactly equal and $Z_1 = Z_2$ the system is said to be balanced. Under balanced conditions the current in the neutral wire is zero. To show why the neutral current is zero, we write two mesh-current equations:

$$\text{Mesh A: } (Z_W + Z_1 + Z_N)\mathbf{I}_A - Z_N\mathbf{I}_B - Z_1\mathbf{I}_C = \mathbf{V}_{S1}$$

$$\text{Mesh B: } -Z_N\mathbf{I}_A + (Z_W + Z_2 + Z_N)\mathbf{I}_B - Z_2\mathbf{I}_C = \mathbf{V}_{S2}$$

For balanced conditions $\mathbf{V}_{S1} = \mathbf{V}_{S2}$ and $Z_1 = Z_2 = Z_L$. Subtracting the mesh B equation from the mesh A equation yields the condition

$$(Z_W + Z_L + 2Z_N)(\mathbf{I}_A - \mathbf{I}_B) = 0$$

This condition requires $\mathbf{I}_A - \mathbf{I}_B = 0$, since the impedance sum cannot be zero for all loads. Therefore, the net current in the neutral line is zero and theoretically the neutral wire can be disconnected. In practice, the balance is never perfect and the neutral line is included for safety reasons. But even so, the current in the neutral is usually less than the line currents, so losses in the feeder lines are reduced.

17–5 THREE-PHASE CIRCUITS

The three-phase system shown in Figure 17–12 is the predominant method of generating and distributing ac electrical power. The system uses four lines (A, B, C, N) to transmit power from the source to the loads. The symbols stand for the three phases A, B, and C, and a neutral line labeled N. The three-phase generator in Figure 17–12 is modeled as three independent sources, although the physical hardware is a single unit with three separate windings. Similarly, the loads are modeled as three separate impedances, although the actual equipment may be housed within a single container.

FIGURE 17–12 *A three-phase ac electrical power system.*

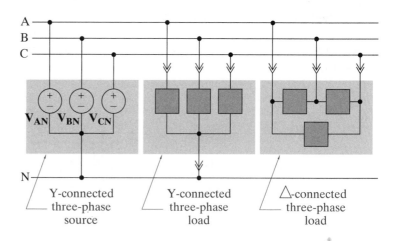

A three-phase system involves three voltages and three currents. In a balanced three-phase system the generator produces phasor voltages that are equal in magnitude and symmetically displaced in phase at 120° intervals. When the three-phase load is balanced (equal impedances), the resulting phasor currents have equal magnitude and are symmetically displaced in phase at 120° intervals. Thus, a **balanced three-phase** system is one in which the phasor currents and voltages have equal magnitudes and phase differences of 120°.

The terminology Y-connected and Δ-connected refers to the two ways the source and loads can be electrically connected. Figure 17–13 shows the same electrical arrangement as Figure 17–12 with the elements rearranged to show the Y and Δ nature of the connections (the Δ is upside down in the figure). The circuit diagrams in the two figures are electrically equivalent, but we will use the form in Figure 17–12 because it highlights the purpose of the system. You need only remember that in a Y-connection the three elements are connected from line to neutral, while in the Δ-connection they are connected from line to line. In most systems the source is Y-connected while the loads can be either Y or Δ, although the latter is more common.

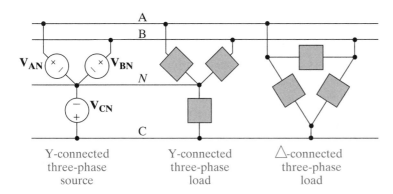

FIGURE 17 – 13 *A three-phase power system with the loads rearranged.*

Y-connected
three-phase
source

Y-connected
three-phase
load

△-connected
three-phase
load

Three-phase sources usually are Y-connected because the Δ-connection involves a loop of voltage sources. Large currents may circulate in this loop if the three voltages do not exactly sum to zero. In analysis situations, a Δ-connection of ideal voltage sources is awkward because it is impossible to determine the current in each source.

We use a double subscript notation to identify voltages in the system. The reason is that there are at least six voltages to deal with: three line-to-line voltages and three line-to-neutral voltages. If we use the usual plus and minus reference marks to define all of these voltages, our circuit diagram would be hopelessly cluttered and confusing. Hence we use two subscripts to define the points across which a voltage is defined. For example, $\mathbf{V}_{XY}$ means the voltage between points X and Y, with an implied plus reference mark at the first subscript (X) and an implied minus at the second subscript (Y).

The three line-to-neutral voltages are called the **phase voltages** and are written in double subscript notation as $\mathbf{V}_{AN}$, $\mathbf{V}_{BN}$, and $\mathbf{V}_{CN}$. Similarly, the three line-to-line voltages, called simply the **line voltages**, are identified as $\mathbf{V}_{AB}$, $\mathbf{V}_{BC}$, and $\mathbf{V}_{CA}$. From the definition of the double subscript notation it follows that $\mathbf{V}_{XY} = -\mathbf{V}_{YX}$. Using this result and KVL, we derive the relationships between the line voltages and phase voltages:

$$\mathbf{V}_{AB} = \mathbf{V}_{AN} + \mathbf{V}_{NB} = \mathbf{V}_{AN} - \mathbf{V}_{BN}$$

$$\mathbf{V}_{BC} = \mathbf{V}_{BN} + \mathbf{V}_{NC} = \mathbf{V}_{BN} - \mathbf{V}_{CN} \qquad (17\text{--}17)$$

$$\mathbf{V}_{CA} = \mathbf{V}_{CN} + \mathbf{V}_{NA} = \mathbf{V}_{CN} - \mathbf{V}_{AN}$$

A balanced three-phase source produces phase voltages that obey the following two constraints:

$$|\mathbf{V}_{AN}| = |\mathbf{V}_{BN}| = |\mathbf{V}_{CN}| = V_P$$

$$\mathbf{V}_{AN} + \mathbf{V}_{BN} + \mathbf{V}_{CN} = 0 + j0$$

That is, the phase voltages have equal amplitudes (V_P) and sum to zero. There are two ways to satisfy these constraints:

Positive Phase Sequence	Negative Phase Sequence	
$\mathbf{V}_{AN} = V_P\angle 0°$	$\mathbf{V}_{AN} = V_P\angle 0°$	
$\mathbf{V}_{BN} = V_P\angle{-}120°$	$\mathbf{V}_{BN} = V_P\angle{-}240°$	(17–18)
$\mathbf{V}_{CN} = V_P\angle{-}240°$	$\mathbf{V}_{CN} = V_P\angle{-}120°$	

Figure 17–14 shows the phasor diagrams for the positive and negative phase sequences. It is apparent that both sequences involve three equal-length phasors that are separated by angles of 120°. As a result, the sum of any two phasors cancels the third. In the positive sequence the phase B voltage lags the phase A voltage by 120°. In the negative sequence phase B lags by 240°. It also is apparent that we can convert one phase sequence into the other by simply interchanging the labels on lines B and C. From a circuit analysis viewpoint, there is no conceptual difference between the two sequences. Consequently, in analysis problems we will use the positive phase sequence unless otherwise stated.

However, the phrase *no conceptual difference* does not mean that phase sequence is unimportant. It turns out that three-phase motors run in one direction when the positive sequence is applied, and in the opposite direction for the negative sequence. This could be a matter of some importance if the motor is driving a conveyor belt at a sewage treatment facility. In practice, it is essential that there be no confusion about which is line A, B, and C and whether the source phase sequence is positive or negative.

A simple relationship between the line and phase voltages is obtained by substituting the positive phase sequence voltages from Eq. (17–18) into the phasor sums in Eq. (17–17). For the first sum

$$\begin{aligned}
\mathbf{V}_{AB} &= \mathbf{V}_{AN} - \mathbf{V}_{BN} \\
&= V_P\angle 0° - V_P\angle{-}120° \\
&= V_P(1 + j0) - V_P(-1/2 - j\sqrt{3}/2) \qquad (17\text{--}19) \\
&= V_P(3/2 + j\sqrt{3}/2) \\
&= \sqrt{3}V_P\angle 30°
\end{aligned}$$

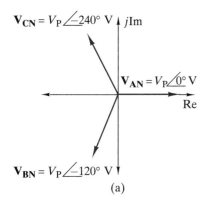

(a)

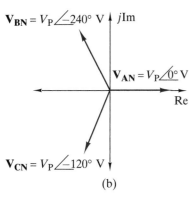

(b)

FIGURE 17–14 *Two possible phase sequences: (a) Positive. (b) Negative.*

Using the other sums, we find the other two positive sequence line voltages as

$$\mathbf{V}_{BC} = \sqrt{3}V_P\angle -90°$$
$$\mathbf{V}_{CA} = \sqrt{3}V_P\angle -210° \tag{17-20}$$

Figure 17–15 shows the phasor diagram of these results. The line voltage phasors have the same amplitude and are displaced from each other by 120°. Hence they obey equal-amplitude and zero-sum constraints like the phase voltages.

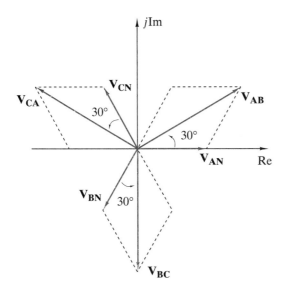

FIGURE 17-15 *Phasor diagram showing phase and line voltages for the positive phase sequence.*

If we denote the amplitude of the line voltages as V_L, then

$$V_L = \sqrt{3}V_P \tag{17-21}$$

In a balanced three-phase system the line voltage amplitude is $\sqrt{3}$ times the phase voltage amplitude. This ratio appears in equipment descriptions such as 277/480 V three phase, where 277 is the phase voltage and 480 the line voltage.

It is necessary to choose one of the phasors as the zero-phase reference when defining three-phase voltages and currents. Usually the reference is the line A phase voltage (i.e., $\mathbf{V}_{AN} = V_P\angle 0°$), as illustrated in Figures 17–14 and 17–15. Unless otherwise stated, $\mathbf{V}_{AN}$ will be used as the phase reference in this chapter.

Exercise 17–6

A balanced Y-connected three-phase source produces positive sequence phase voltages with 2400-V (rms) amplitudes. Write expressions for the phase and line voltage phasors.

Answers:

$\mathbf{V}_{AN} = 2400\angle 0°$	$\mathbf{V}_{BN} = 2400\angle -120°$	$\mathbf{V}_{CN} = 2400\angle -240°$
$\mathbf{V}_{AB} = 4160\angle +30°$	$\mathbf{V}_{BC} = 4160\angle -90°$	$\mathbf{V}_{CA} = 4160\angle -210°$

Given that $\mathbf{V}_{BC} = 480\angle+135°$ in a balanced, positive sequence, three-phase system, write expressions for the three phase-voltage phasors.

Answer:

$$\mathbf{V}_{AN} = 277\angle-135° \qquad\qquad \mathbf{V}_{BN} = 277\angle+105° \qquad\qquad \mathbf{V}_{CN} = 277\angle-15°$$

17–6 THREE-PHASE AC POWER ANALYSIS

This section treats the analysis of balanced three-phase circuits. We first treat the direct analysis problem beginning with the Y-connected source and load shown in Figure 17–16. In a direct analysis problem we are given the source phase voltages $\mathbf{V}_{AN}$, $\mathbf{V}_{BN}$, and $\mathbf{V}_{CN}$ and the load impedances Z. Our objective is to determine the three line currents $\mathbf{I}_A$, $\mathbf{I}_B$, and $\mathbf{I}_C$ and the total complex power delivered to the load.

FIGURE 17–16 *A balanced three-phase system with a Y-connected source and load.*

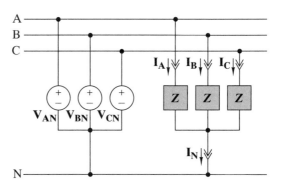

Y-CONNECTED SOURCE AND Y-CONNECTED LOAD

The load in Figure 17–16 is balanced because the phase impedances in the legs of the Y are equal. With the neutral point at the load connected, the voltage across each phase impedance is a phase voltage. Using $\mathbf{V}_{AN}$ as the phase reference, we find that the line currents are

$$\mathbf{I}_A = \frac{\mathbf{V}_{AN}}{Z} = \frac{V_P\angle 0°}{|Z|\angle\theta} = \frac{V_P}{|Z|}\angle-\theta$$

$$\mathbf{I}_B = \frac{\mathbf{V}_{BN}}{Z} = \frac{V_P\angle-120°}{|Z|\angle\theta} = \frac{V_P}{|Z|}\angle-120°-\theta \qquad (17\text{–}22)$$

$$\mathbf{I}_C = \frac{\mathbf{V}_{CN}}{Z} = \frac{V_P\angle-240°}{|Z|\angle\theta} = \frac{V_P}{|Z|}\angle-240°-\theta$$

Figure 17–17 shows the phasor diagram of the line currents and phase voltages.

The line current phasors in Eq. (17–22) and Figure 17–17 have the same amplitude I_L, where

$$I_L = \frac{V_P}{|Z|} \text{ (Y-connected Load)} \qquad (17\text{–}23)$$

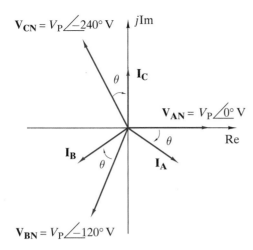

The line currents have equal amplitudes and are symmetrically disposed at $120°$ intervals, so they obey the zero-sum condition $\mathbf{I}_A + \mathbf{I}_B + \mathbf{I}_C = 0$. Applying KCL at the neutral point of the load in Figure 17–16, we find that $\mathbf{I}_N = \mathbf{I}_A + \mathbf{I}_B + \mathbf{I}_C = 0$.

Thus, in a balanced Y-Y circuit there is no current in the neutral line. The neutral connection could be replaced by any impedance whatsoever, including infinity, without affecting the power delivered to the load. In other words, the neutral wire can be disconnected without changing the circuit response. Real systems may or may not have a neutral wire, but in solving three-phase problems it is helpful to draw the neutral line because it serves as a reference point for the phase voltages.

The total complex power delivered to the load is

$$S_L = \mathbf{V}_{AN}\mathbf{I}_A^* + \mathbf{V}_{BN}\mathbf{I}_B^* + \mathbf{V}_{CN}\mathbf{I}_C^*$$

$$= (V_P\angle 0)(I_L\angle\theta) + (V_P\angle -120°)(I_L\angle 120° +\theta) +$$

$$(V_P\angle -240°)(I_L\angle 240° +\theta)$$

$$= 3V_P I_L\angle\theta \qquad (17\text{--}24)$$

Since $V_P = V_L/\sqrt{3}$, the expression for complex power can also be written using the line voltage:

$$S_L = \sqrt{3}V_L I_L\angle\theta \qquad (17\text{--}25)$$

In either Eq. (17–24) or (17–25) the power factor angle θ is the angle of the per-phase impedance of the Y-connected load.

EXAMPLE 17-9

When the line voltage is 480 V (rms), a balanced Y-connected load draws an apparent power of 40 kVA at a lagging power factor of 0.9.

(a) Find the per-phase impedance of the load.
(b) Find the three line current phasors using $\mathbf{V}_{AN}$ as the phase reference.

SOLUTION:

(a) For the given conditions the phase voltage, line current, and power factor angle are

$$V_P = \frac{V_L}{\sqrt{3}} = \frac{480}{\sqrt{3}} = 277 \text{ V (rms)}$$

$$I_L = \frac{|S_L|}{\sqrt{3}V_L} = \frac{4 \times 10^4}{\sqrt{3} \times 480} = 48.1 \text{ A (rms)}$$

$$\theta = \cos^{-1}(0.9) = 25.8°$$

and the per-phase impedance of the Y-connected load is

$$Z_L = \frac{V_P}{I_L}\angle\theta = \frac{277}{48.1}\angle 25.8°$$

$$= 5.18 + j2.51 \quad \Omega$$

(b) With $\mathbf{V}_{AN} = 277\angle 0°$ as the phase reference, we calculate the line current $\mathbf{I}_A$ as

$$\mathbf{I}_A = \frac{\mathbf{V}_{AN}}{Z_L} = I_L\angle -\theta = 48.1\angle -25.8° \text{ A (rms)}$$

Hence for a balanced load the other two line current phasors are

$$\mathbf{I}_B = I_L\angle -120° -\theta = 48.1\angle -145.8° \text{ A (rms)}$$

$$\mathbf{I}_C = I_L\angle -240° -\theta = 48.1\angle -265.8° \text{ A (rms)} \quad \blacksquare$$

Exercise 17–8

A Y-connected load with $Z = 10 + j4$ Ω/phase is driven by a balanced, positive sequence three-phase generator with $V_L = 4.16$ kV (rms). Using $\mathbf{V}_{AN}$ as the phase reference,

(a) Find the line currents.
(b) Find the average and reactive power delivered to the load.

Answers:
(a) $\mathbf{I}_A = 223\angle -21.8°$ A $\mathbf{I}_B = 223\angle -141.8°$ A $\mathbf{I}_C = 223\angle -261.8°$ A
(b) $P_L = 1.49$ MW $Q_L = 0.597$ MVAR

Y-CONNECTED SOURCE AND Δ-CONNECTED LOAD

We now turn to the balanced Δ-connected load shown in Figure 17–18. When the objective is to determine the line currents and total complex power, it is convenient to replace the Δ-connected load by an equivalent Y-connected load. Using Figure 17–19, we can easily determine the required transformation. Looking between any two terminals in the Δ-connected load we see an impedance $Z \| 2Z$. Similarly, looking between any two terminals in the Y-connected load we see an impedance $2Z_Y$. For these two circuits to be equivalent we must see the same equivalent impedance between any two terminals. Hence, equivalence requires

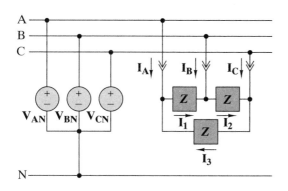

FIGURE   17-18 *A balanced three-phase system with a Y-connected source and a Δ-connected load.*

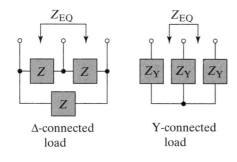

Δ-connected
load

Y-connected
load

FIGURE   17-19 *Equivalent Y-connected and Δ-connected loads.*

$$Z \| 2Z = \frac{2Z^2}{Z + 2Z} = \frac{2}{3}Z = 2Z_\text{Y}$$

or

$$Z_\text{Y} = \frac{Z}{3} \qquad\qquad (17\text{--}26)$$

Using the Δ-to-Y transformation, we reduce the problem to a circuit in which the source and load are Y-connected. The resulting Y-Y configuration can then be analyzed to determine line currents and total power, as discussed previously.

However, when we need to know the current or power delivered to each leg of the delta load, we must determine the phase currents $\mathbf{I}_1$, $\mathbf{I}_2$, and $\mathbf{I}_3$ shown in Figure 17–18. The phase currents can be expressed in terms of the phase impedances and the line voltage. Assuming the positive phase sequence and using $\mathbf{V}_\text{AN}$ as the phase reference, these expressions are

$$\mathbf{I}_1 = \frac{\mathbf{V}_\text{AB}}{Z} = \frac{V_\text{L}\angle 30°}{|Z|\angle\theta} = \frac{V_\text{L}}{|Z|}\angle 30° - \theta$$

$$\mathbf{I}_2 = \frac{\mathbf{V}_\text{BC}}{Z} = \frac{V_\text{L}\angle -90°}{|Z|\angle\theta} = \frac{V_\text{L}}{|Z|}\angle -90° - \theta \qquad (17\text{--}27)$$

$$\mathbf{I}_3 = \frac{\mathbf{V}_\text{CA}}{Z} = \frac{V_\text{L}\angle -210°}{|Z|\angle\theta} = \frac{V_\text{L}}{|Z|}\angle -210° - \theta$$

The phase currents have the same amplitude I_P defined as

$$I_\text{P} = \frac{V_\text{L}}{|Z|} \ (\Delta\text{- connected Load}) \qquad (17\text{--}28)$$

Although of no immediate physical importance, note that the phase currents sum to zero because they have equal amplitudes and are symmetrically disposed at 120° intervals.

Using the results in Eq. (17–27), the complex power delivered to each leg of the delta load is

$$S_1 = \mathbf{V}_{AB} \times \mathbf{I}_1^* = (V_L \angle 30°)\,(I_P \angle \theta - 30°) = V_L I_P \angle \theta$$

$$S_2 = \mathbf{V}_{BC} \times \mathbf{I}_2^* = (V_L \angle -90°)\,(I_P \angle \theta + 90°) = V_L I_P \angle \theta \qquad (17\text{–}29)$$

$$S_3 = \mathbf{V}_{CA} \times \mathbf{I}_3^* = (V_L \angle -210°)\,(I_P \angle \theta + 210°) = V_L I_P \angle \theta$$

The total complex power delivered to the Δ-connected load is

$$S_L = S_1 + S_2 + S_3 = 3V_L I_P \angle \theta \qquad (17\text{–}30)$$

where the power factor angle θ is the angle of the per-phase impedance of the Δ-connected load.

The result in Eq. (17–30) can be put in the same form as Eq. (17–25) by replacing the phase current I_P by the line current I_L. As noted previously, the line currents for a balanced Δ-connected load can be calculated using a Δ-to-Y transformation. In the transformed circuit the line current amplitude is $I_L = V_P/|Z_Y|$, where Z_Y is the impedance in each leg of the equivalent Y-connected load. But in a balanced system $V_P = V_L/\sqrt{3}$, and according to Eq. (17–26) $Z_Y = Z/3$. Therefore, the amplitudes of the line and phase currents in a Δ-connected load are related as follows:

$$I_L = \frac{V_L/\sqrt{3}}{|Z/3|} = \sqrt{3}\frac{V_L}{|Z|} = \sqrt{3}I_P \qquad (17\text{–}31)$$

When $I_P = I_L/\sqrt{3}$ is substituted into Eq. (17–30), we obtain

$$S_L = \sqrt{3}V_L I_L \angle \theta \qquad (17\text{–}32)$$

Equations (17–25) and (17–32) are identical, which means that the relationship applies to balanced three-phase loads whether Y- or Δ-connected. In either case the power factor angle θ is the angle of the per-phase impedance of the load because the transformation $Z_Y = Z/3$ does not alter the phase angle of the phase impedance.

EXAMPLE 17–10

A Δ-connected load with $Z = 40 + j30\ \Omega$/phase is driven by a balanced, positive sequence three-phase generator with $V_L = 2400$ V (rms). Using $\mathbf{V}_{AN}$ as the phase reference,

(a) Find the phase currents.
(b) Find the line currents.
(c) Find the average and reactive power delivered to the load.

SOLUTION:
(a) The first phase current is

$$\mathbf{I}_1 = \frac{\mathbf{V}_{AB}}{Z} = \frac{2400 \angle 30°}{40 + j30} = 48.0 \angle -6.87°\ \text{A (rms)}$$

Since the circuit is balanced, the other phase currents all have amplitudes of $I_P = 48.0$ A and are displaced at 120° intervals.

$$\mathbf{I}_2 = 48.0\angle -126.87° \text{ A (rms)}$$

$$\mathbf{I}_3 = 48.0\angle -246.87° \text{ A (rms)}$$

(b) Using $V_P = 2400/\sqrt{3}$ and Δ-Y transformation with $Z_Y = (40 + j30)/3$, we find that the phase A line current is

$$\mathbf{I}_A = \frac{\mathbf{V}_{AN}}{Z_Y} = \frac{2400/\sqrt{3}}{(40 + j30)/3} = 83.1\angle -36.9° \text{ A (rms)}$$

Since the circuit is balanced, the other line currents have amplitudes of $I_L = 83.1$ A and are displaced at 120° intervals.

$$\mathbf{I}_B = 83.1\angle -156.9° \text{ A (rms)}$$

$$\mathbf{I}_C = 83.1\angle -276.9° \text{ A (rms)}$$

(c) The complex power delivered to the load is

$$S_L = \sqrt{3}V_L I_L \angle\theta = \sqrt{3} \times 2400 \times 83.1\angle 36.9°$$

$$= 345\angle 36.9° \text{ kVA}$$

Therefore, $P_L = 276$ kW and $Q_L = 207$ kVAR. ■

Exercise 17–9

Exercise 17–8 involved a balanced Y-connected load with $Z = 10 + j4$ Ω/phase and a line voltage of $V_L = 4.16$ kV (rms). In this exercise these same parameters apply to a Δ-connected load. Using $\mathbf{V}_{AN}$ as the phase reference,

(a) Find phase currents.
(b) Find the line currents.
(c) Find the average and reactive power delivered to the load.

Answers:
(a) $\mathbf{I}_1 = 386\angle +8.20°$ A, $\mathbf{I}_2 = 386\angle -111.8°$ A, $\mathbf{I}_3 = 386\angle -231.8°$ A
(b) $\mathbf{I}_A = 669\angle -21.8°$ A, $\mathbf{I}_B = 669\angle -141.8°$ A, $\mathbf{I}_C = 669\angle -261.8°$ A
(c) $P_L = 4.47$ MW, $Q_L = 1.79$ MVAR

THREE-PHASE LOAD-FLOW ANALYSIS

The analysis and examples given thus far treat direct analysis problems in which the source parameters are given and the power flow is unknown. In a load-flow problem we are required to find the source or load voltages when the power flow is given.

A simple model of the three-phase circuit for the load-flow problem is shown in Figure 17–20. The impedance Z_W represents the wire impedances of the power lines connecting the source and load. It is clear even in this very simple case that including all three phases in a circuit diagram is unwieldy. In more complicated situations, including all three phases in circuit diagrams tends to obscure the working of the system. In a balanced three-phase system we have seen that once we find one of the line currents or voltages, the others are easily derived by shifting the known response at

FIGURE 17-20 *A balanced three-phase system with line impedances.*

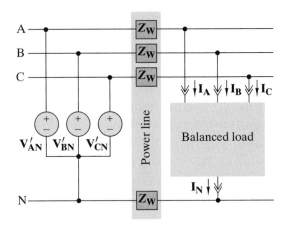

120° intervals. Thus, in effect, we really do not need all three phases in the circuit diagram to analyze balanced three-phase systems.

A simpler representation of a balanced three-phase system is obtained by omitting the neutral and using one line to represent all three phases. Figure 17–21 is a single-line representation of the circuit in Figure 17–20. In power system terminology a **bus** is a group of conductors that serve as a common connection for two or more circuits. In the single-line diagram in Figure 17–21 the buses are represented by short horizontal lines. Bus no. 1 is a generator bus connecting a three-phase source represented by the circle and a transmission line represented by Z_W. Bus no. 2 is a load bus connecting the transmission line to a balanced three-phase load represented by an arrow indicating the delivery of complex power. The line voltages and complex power flow are written beside the buses using a subscript that identifies the bus.

FIGURE 17-21 *Single-line representation of a three-phase system.*

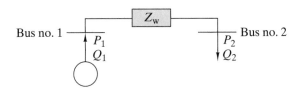

Using single-line diagrams makes three-phase load-flow analysis similar to the single-phase two-wire load-flow problems in Sect. 17–5. The only complication here is that we must account for the power in all three phases and we must distinguish between the line and phase voltages. In a three-phase load-flow problem the bus voltages and line currents are represented by their phasor magnitudes without reference to their phase angles. A load-flow problem does not require that we find these phase angles since the required phase information is carried by the specified complex power.

EXAMPLE 17-11

The single-line diagram in Figure 17–22 shows a power system with a generator bus connected to a load bus via power lines with $Z_W = 4 + j16\ \Omega$. The line voltage at the load bus is $V_{L2} = 4800$ V (rms) and the load draws $P_2 = 100$ kW at 0.8 lagging power factor. Find the complex power produced by the source, the line voltage at the source bus, and the transmission efficiency.

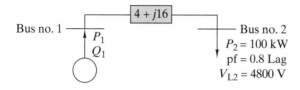

FIGURE 17-22

SOLUTION:

The total complex power delivered to the three-phase load connected to bus no. 2 is

$$S_2 = 100 + j100 \tan(\cos^{-1} 0.8) = 100 + j75$$

$$= 125\angle 36.9°\ \text{kVA}$$

Using Eq. (17–32) to calculate the line current at the load bus yields

$$I_L = \frac{|S_2|}{\sqrt{3}\ V_{L2}} = \frac{125000}{\sqrt{3} \times 4800} = 15.04\ \text{A (rms)}$$

The total complex power absorbed by the power line is

$$S_W = 3I_L^2(R_W + jX_W) = 3(15.04)^2(4 + j16)$$

$$= 2.71 + j10.8\ \text{kVA}$$

The preceding computation includes the three lines for phases A, B, and C but does not include the neutral wire since the circuit is balanced and there is no current in the neutral line. Using the conservation of complex power yields the source power as

$$S_1 = S_2 + S_W = 102.7 + j85.8$$

$$= 133.8\angle 39.9°\ \text{kVA}$$

We now use Eq. (17–32) again to find the line voltage at the source bus as

$$V_{L1} = \frac{|S_1|}{\sqrt{3}I_L} = \frac{133800}{\sqrt{3} \times 15.04}$$

$$= 5.14\ \text{kV (rms)}$$

The transmission efficiency is

$$\eta = \frac{P_2}{P_1} \times 100\% = \frac{100}{102.7} \times 100 = 97.4\%$$

∎

EXAMPLE 17–12

The transmission line in Figure 17–23 has a maximum rated capacity of 250 kVA at line voltage of 4160 V (rms). When operating at rated capacity, the resistive and reactive voltage drops in the line are 2.5% and 6% of the rated voltage, respectively.

FIGURE 17–23

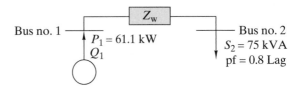

Bus no. 1 — $P_1 = 61.1$ kW Q_1 — Bus no. 2 Z_w $S_2 = 75$ kVA pf = 0.8 Lag

(a) Find the wire impedance Z_W.
(b) The system is operating with $P_1 = 61.1$ kW and $S_2 = 75$ kVA at pf = 0.8 lagging. Find the line voltage at the load bus.

SOLUTION:

(a) When the power line is operating at maximum rated capacity, the line current is

$$I_L = \frac{|S_L|}{\sqrt{3}\,V_L} = \frac{250{,}000}{\sqrt{3} \times 4160} = 34.7 \text{ A (rms)}$$

The magnitudes of the voltage across line resistance and reactance are

$$I_L R_W = 4160 \times 0.025 = 104 \text{ V}$$

$$I_L X_W = 4160 \times 0.060 = 250 \text{ V}$$

Hence, the wire impedance of the line is

$$Z_W = \frac{104 + j250}{I_L} = 3 + j7.2\ \Omega$$

(b) When the system operates with $P_1 = 61.1$ kW and $S_2 = 75$ kVA at pf = 0.8 lagging, the average power lost in the line is $P_W = P_1 - S_2 \times 0.8 = 1.1$ kW. The line current is

$$I_L = \sqrt{\frac{P_W}{3\,R_W}} = \sqrt{\frac{1100}{3 \times 3}} = 11.06 \text{ A (rms)}$$

and the line voltage at load bus is

$$V_{L2} = \frac{|S_2|}{\sqrt{3}I_L} = 3.92 \text{ kV (rms)} \quad \blacksquare$$

Exercise 17–10

A balanced three-phase load draws 8 kVA at lagging power factor of 0.9 when the phase voltage is 277 V (rms). The load is fed by a balanced three-phase source via lines with $Z_W = 0.3 + j5\ \Omega$.

(a) Find the line current.
(b) Find the line voltage at the source.

Answers:
(a) $I_L = 9.63$ A
(b) $V_L = 526$ V

INSTANTANEOUS THREE-PHASE POWER

We began this chapter by showing that the instantaneous power in a single-phase circuit consists of a constant dc component (the average power) plus a double-frequency ac component that makes no net contribution to energy transfer. One of the advantages of three-phase operation is that the net ac component is zero. In other words, the total instantaneous power in a balanced three-phase circuit is constant.

To show that $p_T(t)$ is constant, we write the instantaneous power in each phase of the balanced three-phase circuit:

$$p_A(t) = v_{AN}(t) \times i_A(t) = [\sqrt{2}V_P\cos(\omega t)] \times [\sqrt{2}\,I_L\cos(\omega t - \theta)]$$

$$p_B(t) = v_{BN}(t) \times i_B(t) = [\sqrt{2}V_P\cos(\omega t - 120°)] \times [\sqrt{2}I_L\cos(\omega t - 120° - \theta)]$$

$$p_C(t) = v_{CN}(t) \times i_C(t) = [\sqrt{2}V_P\cos(\omega t - 240°)] \times [\sqrt{2}I_L\cos(\omega t - 240° - \theta)]$$

where $\sqrt{2}V_P$ is the peak amplitude of each line-to-neutral voltage and $\sqrt{2}I_L$ is the peak amplitude of each line current. Using the trigonometric identity,

$$[\cos x] \times [\cos y] = \frac{1}{2}\cos(x - y) + \frac{1}{2}\cos(x + y)$$

the individual phase powers can be put into the form

$$p_A(t) = V_PI_L \cos\theta + V_PI_L \cos(2\omega t - \theta)$$

$$p_B(t) = V_PI_L \cos\theta + V_PI_L \cos(2\omega t - 240° - \theta)$$

$$p_C(t) = V_PI_L \cos\theta + V_PI_L \cos(2\omega t - 480° - \theta)$$

Each phase power has a constant dc term $V_PI_L\cos\theta$ plus a double-frequency ac term. The double-frequency terms all have the same amplitude V_PI_L and are symmetrically disposed at 120° intervals because –480° is the same as –120°. When viewed as phasors it is easy to see that the double-frequency sinusoidal terms sum to zero and the total instantaneous power is

$$p_T(t) = p_A(t) + p_B(t) + p_C(t) = 3\,V_PI_L \cos\theta$$

The fact that the total instantaneous power is constant means, among other things, that three-phase motors produce constant mechanical output and the three-phase generators require constant mechanical input. As a result, there is smoother operation with less vibration at the electromechanical interfaces of the system.

APPLICATION NOTE: **EXAMPLE 17–13**

The purpose of this example is to use conventional phasor circuit analysis to analyze a three-phase power system. Figure 17–24 shows a single-line diagram of a power system in which a three-phase source with $V_L = 250$ kV supplies power to two power distribution centers through a radial network of transmission lines. The two transmission lines can be modeled by series impedances of

$$Z_{W1} = 6.5 + j14 \ \ \Omega \quad \text{and} \quad Z_{W2} = 8.6 + j19 \ \ \Omega$$

The two load centers can be modeled as Y-connected loads with per phase impedance of

$$Z_{Y1} = 1000 \ \angle \ 10° \ \ \Omega \quad \text{and} \quad Z_{Y2} = 2400 \ \angle \ 20° \ \ \Omega$$

Find the complex power delivered by the source and the magnitude of the line voltage delivered to each distribution center.

FIGURE 17–24

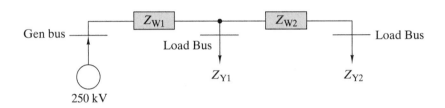

SOLUTION:

We can analyze a balanced three-phase system on a per phase basis. If we look at just one phase, say phase A, we get the single-phase equivalent circuit in Figure 17–25. We analyze this circuit to get a solution for phase A. If need be, we use the phase sequence to shift the solution to get phase B and phase C results. The single-phase equivalent circuit includes a neutral connection. We may do this even if the system does not have a neutral wire. In a balanced system, neutral points in Y-connections have the same voltage. Thus, for analysis purposes we can tie these points together with a "virtual" neutral wire without altering the response of the system.

The source voltage in Figure 17–25 is the phase A voltage of the 250-kV source.

$$\mathbf{V}_{AN} = \frac{V_L}{\sqrt{3}} = \frac{250}{\sqrt{3}} = 144.3 \ \angle \ 0° \ \ \text{kV}$$

In matrix format the mesh-current equations for the equivalent circuit are

$$\begin{bmatrix} Z_{W1} + Z_{Y1} & -Z_{Y1} \\ -Z_{Y1} & Z_{W2} + Z_{Y1} + Z_{Y2} \end{bmatrix} \begin{bmatrix} \mathbf{I}_A \\ \mathbf{I}_B \end{bmatrix} = \begin{bmatrix} \mathbf{V}_{AN} \\ 0 \end{bmatrix}$$

Note that in this context $\mathbf{I}_A$ and $\mathbf{I}_B$ are the mesh currents in Figure 17–25, not line currents. Inserting the preceding numerical values and solving for these mesh currents yield

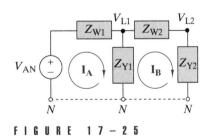

FIGURE 17–25

$$\mathbf{I}_A = 194.9 - j48.5 \quad \text{A} \quad \text{and} \quad \mathbf{I}_B = 55.06 - j21.4 \quad \text{A}$$

We can now calculate the total complex power delivered by the 25-kV source as

$$S = 3\,\mathbf{S}_A = 3\,\mathbf{V}_{AN}\mathbf{I}_A^* = 82.32 + j24.89 \quad \text{kVA}$$

The $3\,\mathbf{S}_A$ in this equation accounts for the complex power delivered by all three phases. The phase A voltage at each load center is

$$\mathbf{V}_{AN1} = \mathbf{V}_{AN} - \mathbf{I}_A Z_{W1} = 142.4 - j2.41 \quad \text{kV}$$

$$\mathbf{V}_{AN2} = \mathbf{V}_{AN1} - \mathbf{I}_B Z_{W2} = 141.5 - j3.24 \quad \text{kV}$$

From which we get the line voltage magnitude at each load center.

$$V_{L1} = \sqrt{3}\,|\mathbf{V}_{AN1}| = 246.6 \quad \text{kV}$$

$$V_{L2} = \sqrt{3}\,|\mathbf{V}_{AN2}| = 245.5 \quad \text{kV}$$

SUMMARY

- In the sinusoidal steady state the instantaneous power at a two-terminal interface contains average and reactive power components. The average power represents a unidirectional transfer of energy from source to load. The reactive power represents a lossless interchange of energy between the source and load.

- In ac power analysis problems amplitudes of voltage and current phasors are expressed in rms values. Complex power is defined as the product of the voltage phasor and the conjugate of the current phasor. The real part of the complex power is the average power in watts (W), the imaginary part is the reactive power in volt-amperes reactive (VAR), and the magnitude is the apparent power in volt-amperes (VA).

- In a direct analysis problem the load impedance, line impedances, and source voltage are given, and the load voltage, current, and power are the unknowns. In a load-flow problem the source and load powers are given, and the unknowns are line voltages and currents at different points.

- Three-phase systems transmit power from source to load on four lines labeled A, B, C, and N. The three line-to-neutral voltages $\mathbf{V}_{AN}$, $\mathbf{V}_{BN}$, and $\mathbf{V}_{CN}$ are called the phase voltages. The three line-to-line voltages $\mathbf{V}_{AB}$, $\mathbf{V}_{BC}$, and $\mathbf{V}_{CA}$ are called line voltages.

- In a balanced three-phase system, (1) the neutral wire carries no current, (2) the line-voltage amplitude V_L is related to the phase voltage amplitude V_P by $V_L = \sqrt{3}\,V_P$, (3) the three line currents $\mathbf{I}_A$, $\mathbf{I}_B$, and $\mathbf{I}_C$ have the same amplitude I_L, and (4) the total complex power delivered to a Y- or a Δ-connected load is $\sqrt{3}V_L I_L \angle\theta$, where θ is the angle of the phase impedance.

PROBLEMS

ERO 17-1 COMPLEX POWER (SECTS. 17-1, 17-2)

Given a linear circuit operating in the sinusoidal steady state,

(a) Find the complex power delivered at a specified interface.
(b) Find the phasor voltages or currents required to deliver a specified complex power at an interface.
(c) Find the load impedance required to draw a specified complex power at a given voltage level.

See Example 17-1 and Exercises 17-1, 17-2

17-1 Calculate the average power, reactive power, and power factor for the circuit in Figure P17-1 using the following voltages and currents. State whether the circuit is absorbing or delivering net energy.
  (a) $v(t) = 1200 \cos(\omega t - 145°)$ V, $i(t) = 2 \cos(\omega t + 50°)$ A
  (b) $v(t) = 280 \cos(\omega t + 60°)$ V, $i(t) = 15 \cos(\omega t - 20°)$ A

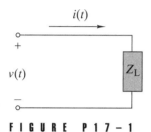

FIGURE P17-1

17-2 Calculate the average power, reactive power, and power factor for the circuit in Figure P17-1 using the following voltages and currents. State whether the circuit is absorbing or delivering net energy.
  (a) $v(t) = 135 \sin(\omega t)$ V, $i(t) = 1.5 \cos(\omega t + 30°)$ A
  (b) $v(t) = 370 \cos(\omega t)$ V, $i(t) = 11.5 \cos(\omega t + 130°)$ A

17-3 Calculate the average power, reactive power, and power factor for the circuit in Figure P17-1 using the following conditions.
  (a) $\mathbf{V} = 250 \angle 0°$ V (rms), $\mathbf{I} = 0.25 \angle -15°$ A (rms)
  (b) $\mathbf{V} = 120 \angle 135°$ V (rms), $\mathbf{I} = 12.5 \angle 165°$ A (rms)

17-4 Calculate the average power, reactive power, and power factor for the circuit in Figure P17-1 using the following conditions.
  (a) $\mathbf{V} = 2.4 \angle 45°$ kV (rms), $Z_L = 250 \angle -10.5°$ Ω
  (b) $Z_L = 300 - j400$ Ω, $\mathbf{I} = 120 \angle 25°$ mA (rms)

17-5 Find the power factor of the circuit in Figure P17-1 under the following conditions. State whether the power factor is lagging or leading.
  (a) $S = 800 + j250$ kVA
  (b) $|S| = 12$ kVA, $Q = -9$ kVA, $\cos \theta > 0$

17-6 An inductive load draws 10 kW and 12 kVA from a 240-V (rms) source.
  (a) Find Q, I_{rms}, and the power factor.
  (b) Find the load impedance.

17-7 A load draws 20 kVA at a power factor of 0.75 lagging from a 2400-V (rms) source.
  (a) Find P, Q, and I_{rms}.
  (b) Find the load impedance.

17-8 A load impedance with $Z_L = 4 - j1.5$ Ω draws an apparent power of 10 kVA. Find P, Q, I_{rms}, and V_{rms}.

17-9 The load draws 12 A (rms), 4.8 kW, and 2.4 kVARS (lagging) from a 60-Hz source. Find the load power factor and impedance.

17-10 An electrical load is rated at 440 V (rms), 10 A (rms), and 4 kW. Find the load impedance.

ERO 17-2 AC POWER ANALYSIS (SECT. 17-3)

Given a linear circuit operating in the sinusoidal steady state,

(a) Find the complex power associated with any element.
(b) Find the load impedance required to draw a given complex power from a source.

See Examples 17-2, 17-3, 17-4 and Exercise 17-3

17-11 A load consisting of a 100-ohm resistor in parallel with a 150-mH inductor is connected across a 60-Hz voltage source that delivers 240 V (rms). Find the complex power delivered to the load.

17-12 A load consisting of a 50-Ω resistor in series with a 100-mH inductor is connected across a 133-Hz voltage source that delivers 110 V (rms). Find the complex power delivered to the load.

17-13 A load consisting of a resistor and capacitor connected in parallel absorbs a complex power $S = 10 - j12$ VA when connected to a 440-V (rms) 60-Hz line. Find the values of R and C.

17-14 The load in Figure P17-14 consists of a 200-Ω resistor in parallel with an inductor whose reactance is 50 Ω. The load draws 12 A (rms) at 60 Hz from the source, and the wire impedance of the line is $Z_W = 1 + j10$ Ω.
  (a) Find the rms source voltage.
  (b) Find the complex power absorbed by the load and line.
  (c) Calculate the transmission efficiency (η).

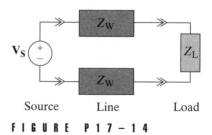

FIGURE P17-14

17-15 The load in Figure P17-14 consists of a 60-Ω resistor in series with a capacitor whose reactance is 40 Ω. The voltage source produces 220 V (rms) at 60 Hz, and the wire impedance of the line is $Z_W = 2 + j5$ Ω.

(a) Find the rms current in the line.

(b) Find the complex power absorbed by the load and line.

(c) Calculate the transmission efficiency (η).

17–16 The transmission line in Figure P17–14 has a maximum rated capacity of 50 kVA at 2400 V (rms). When operated at rated capacity, the resistive voltage drop in the line is 3% and the reactive drop is 12% of the rated voltage. Find the line impedance. Find the transmission efficiency when the load is 200-Ω resistor and the source voltage is 2400 V (rms).

17–17 The power line in Figure 17–14 has a transmission efficiency of $\eta = 90\%$ when it delivers 25 kW to the load. To improve efficiency, a second identical transmission line is added in parallel with this line. Find the transmission efficiency when the two lines in parallel deliver 25 kW to the load.

17–18 The three load impedances in Figure P17–18 are: $Z_1 = 25 + j6\ \Omega$, $Z_2 = 16 + j8\ \Omega$, and $Z_3 = 10 + j2\ \Omega$.

(a) Find the current in lines A, B, and N.

(b) Find the complex power produced by each source.

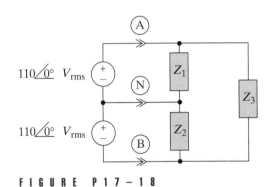

FIGURE P17–18

17–19 The three loads in Figure P17–18 are absorbing the following complex powers:

$$S_1 = 1250 + j500\ \text{VA},\ S_2 = 800 + j0\ \text{VA},$$

$$S_3 = 2000 + j400\ \text{VA}.$$

(a) Find the current in lines A, B, and N.

(b) Find the complex power produced by each source.

17–20 A load is rated at 440 V(rms), 30 A(rms), and 10 kW. The load is supplied by a 440-V source via a two-wire line with wire impedances of $0.3 + j2\ \Omega$. Find the voltage and average power actually delivered to the load.

ERO 17–3 LOAD FLOW ANALYSIS (SECT. 17–4)

Find unknown voltages or currents required to produce a specified power flow.
See Examples 17–5, 17–6, 17–7 and Exercise 17–5

17–21 The load in Figure P17–14 draws 25 kW at 2400 V(rms) with a power factor of 0.85 lagging. The wire impedance is $2 + j8\ \Omega$. Find the complex power produced by the source and the transmission efficiency (η).

17–22 The apparent power delivered to the load in Figure P17–14 is 30 kVA at a power factor of 0.75 lagging when the load voltage is 2400 V(rms). The wire impedance is $2 + j\,10\ \Omega$. Find the source voltage and power factor.

17–23 The source in Figure P17–14 delivers 37 kW when the apparent power delivered to the load is $35 + j20$ kVA. The wire impedance is $2.1 + j12\ \Omega$. Find the load and source voltages.

17–24 The two loads in Figure P17–24 absorb complex powers of $S_1 = 12 + j6$ kVA and $|S_2| = 25$ kVA at a lagging power factor of 0.7. The load voltage is $|\mathbf{V}_L| = 2400$ V (rms), and the line impedances are $Z_W = 2 + j8\ \Omega$.

(a) Find the line current and source voltage.

(b) Find the complex power produced by the source.

(c) Calculate the transmission efficiency (η).

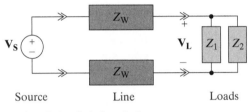

Source   Line   Loads

FIGURE P17–24

17–25 The two loads in Figure P17–24 absorb complex powers of $|S_1| = 16$ kVA at a lagging power factor of 0.75 and $|S_2| = 25$ kVA at unity power factor. The source delivers 50 kVA at a lagging power factor of 0.8. The line impedances are $Z_W = 0.1 + j0.5\ \Omega$. Find the load voltage and source voltage.

17–26 A load draws 15 A (rms) and 5 kW at a power factor 0.75 (lagging) from a 60-Hz source. Find the capacitance needed in parallel with the load to raise the power factor to unity.

17–27 The load in Figure P17–27 operates at 440 V (rms), 60 Hz, and draws 33 kVA at a power factor of 0.75 lagging.

(a) Calculate the value of C required to produce an overall power factor of at least 0.95.

(b) Calculate the transmission efficiency with and without the capacitor found in (a).

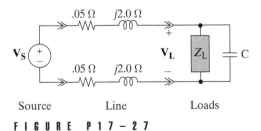

Source   Line   Loads

FIGURE P17–27

17–28 A load draws an apparent power of 25 kVA at a lagging power factor 0.6 from a 60-Hz voltage source whose output is 440 V (rms). Find the capacitance to be connected in parallel with the load so that the combined load has a power factor of 0.9.

17–29 The load in Figure P17–29 operates at 60 Hz. With $\mathbf{V}_L = 12\angle 0°$ kV (rms), the load draws 1.2 MVA at a lagging power factor of 0.8. The first source voltage is $\mathbf{V}_{S1} = 12.7 + j0.96$ kV (rms). Find the complex power supplied by each of the sources.

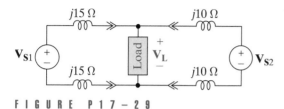

FIGURE P17–29

17–30 A two-wire power line with wire impedances of $Z_W = 1.2 + j9$ Ω connects a source to two loads connected in parallel. The line current is 8 A (rms) when one load draws 20 kW at a lagging power factor of 0.8, and the other load draws 12 kW at a lagging power factor of 0.9. Find the source power factor.

ERO 17–4 THREE-PHASE POWER (SECTS. 17–5, 17–6)

Given a balanced three-phase circuit operating in the sinusoidal steady state,

(a) Find the line (phase) voltages or currents when given the phase sequence and the phase (line) voltages or currents.
(b) Find the line current and total complex power (load impedance) when given the line or phase voltage and the load impedance (total complex power).
(c) Find unknown voltages or currents required to produce a specified power flow.

See Examples 17–9, 17–10, 17–11, 17–12, 17–13 and Exercises 17–7, 17–8, 17–9, 17–10

17–31 In a balanced Y-connected three-phase circuit, the magnitude of the line voltage is 480 V (rms) and the phase sequence is positive.
   (a) Write the line and phase-voltage phasors in polar form using $\mathbf{V}_{AN}$ as the reference phasor.
   (b) Draw a phasor diagram of the line and phase voltages.

17–32 In a balanced Δ-connected three-phase circuit the magnitude of the line voltage is 2400 V (rms) and the phase sequence is positive.
   (a) Write the line and phase voltage phasors in polar form using $\mathbf{V}_{AB}$ as the reference phasor.

   (b) Draw a phasor diagram of the line and phase voltages.

17–33 In a balanced Δ-connected three-phase circuit, the magnitude of the phase current is $I_P = 3.5$ A (rms) and the phase sequence is negative.
   (a) Write the line and phase current phasors in polar form using $\mathbf{I}_A$ as the reference phasor.
   (b) Draw a phasor diagram of the line and phase currents.

17–34 A balanced Y-connected three-phase load with a per-phase impedance of $20 + j15$ Ω operates with a line voltage magnitude of 480 V (rms) using a positive phase sequence. Using $\mathbf{V}_{AN}$ as the reference phasor,
   (a) Find the line current phasors in rectangular form.
   (b) Calculate the total complex power delivered to the load.
   (c) Draw a phasor diagram showing the line currents and phase voltages.
   (c) Draw a phasor diagram showing the line currents.

17–35 A balanced Δ-connected three-phase load with a per-phase impedance of $15 + j12$ Ω operates with a line voltage magnitude of 480 V (rms) using a negative phase sequence. Using $\mathbf{V}_{AB}$ as the reference,
   (a) Find the line current phasors in rectangular form.
   (b) Calculate the total complex power delivered to the load.
   (c) Draw a phasor diagram showing the line currents and voltages.

17–36 A balanced Δ-connected three-phase load with a per-phase impedance of $50 + j15$ Ω operates with a line voltage magnitude of 480 V (rms) using a positive phase sequence. Using $\mathbf{I}_A$ as the reference,
   (a) Find the line current phasors in rectangular form.
   (b) Calculate the total complex power delivered to the load.

17–37 A balanced Δ-connected three-phase load with a per-phase impedance of $8 - j6$ Ω is connected in parallel with a balanced Y-connected three-phase load with a per-phase impedance of $12\angle 40°$ Ω. The line voltage is $V_L = 480$ V (rms). Find the magnitude of the line current and the power factor of the combined loads.

17–38 In a balanced Δ-connected three-phase load, the phase A line current is $\mathbf{I}_A = 25\angle -40°$ A (rms) with a line voltage of $\mathbf{V}_{AB} = 480\angle 30°$ V(rms). Find the phase impedance of the load assuming a positive phase sequence.

17–39 A balanced Y-connected three-phase load absorbs 30 kVA at a power factor of 0.75 lagging when the line voltage magnitude is 480 V (rms).
   (a) Find the magnitude of the line current.
   (b) Calculate the resistance and reactance of the per-phase impedance.

17–40 A balanced Y-connected three-phase load absorbs 20 kW when the line current magnitude is 32 A (rms) and the line voltage magnitude is 480 V (rms). Find the resistance and reactance of the per-phase impedance assuming a lagging power factor.

17–41 A balanced Δ-connected three-phase load absorbs 30 kVA at a power factor of 0.75 lagging when the line voltage magnitude is 2400 V (rms).

(a) Find the magnitude of the line current.

(b) Calculate the resistance and reactance of the per-phase impedance.

17–42 Two three-phase loads are connected in parallel. The first is a balanced Δ-connected circuit absorbing 30 kVA at a power factor of 0.85 lagging. The second is a balanced Y-connected load with a per-phase impedance of $20 + j15\ \Omega$. The magnitude of the line voltage at the loads is 480 V (rms). Find the magnitude of the total line current and the total complex power delivered to the loads.

17–43 The wire impedances in the single-line diagram of Figure P17–43 are $Z_W = 0.6 + j4\ \Omega$ per phase. The balanced load connected to Bus no. 2 is rated at $S_2 = 16 + j12$ kVA when the line voltage is 480 V(rms). The generator connected to Bus no. 1 produces a line voltage of 440 V (rms). Find the line voltage at Bus no. 2 and the complex power produced by the source.

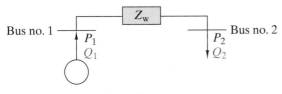

FIGURE P17–43

17–44 The source at Bus no. 1 in Figure P17–43 is a Y-connected generator with an internal impedance of $j2.2\ \Omega$/phase and a Thévenin voltage of 4160 V/phase. The load connected to Bus no. 2 is a Y-connected three-phase load with a phase impedance of $80 + j66\ \Omega$/phase. The power lines connecting the two buses have wire impedances of $1.2 + j5\ \Omega$/phase.

(a) Find the line current.

(b) Find the complex power delivered to the load and phase voltage at Bus no. 2.

(c) Find the total complex power produced by the source.

(d) Find the transmission efficiency (η).

17–45 The three-phase line in Figure P17–43 has wire impedances of $Z_W = 0 + j8\ \Omega$ per phase. The line voltage at Bus no. 2 is 7200 V (rms), and the load absorbs 800 kVA at a power factor of 0.8 lagging. Find the line current, the line voltage at Bus no. 1, and the complex power produced by the source.

17–46 The three-phase line in Figure P17–43 has wire impedances of $Z_W = 3 + j6\ \Omega$ per phase. The line voltage

at Bus no. 1 is 7.2 kV (rms), and the load at Bus no. 2 absorbs 300 kW at a power factor of 0.8 lagging. Find the line current and the line voltage at Bus no. 2.

17–47 The three-phase line in Figure P17–43 has wire impedances of $Z_W = 0.5 + j2\ \Omega$ per phase. The load at Bus no. 2 absorbs an average power of $P_2 = 4.5$ kW at a lagging power factor of $1/\sqrt{2}$. The line voltage at Bus no. 2 is 440 V(rms). Find the line current and the line voltage at Bus no. 1.

17–48 The three-phase load connected to Bus no. 2 in Figure P17–48 draws an average power of $P_2 = 100$ MW at a power factor of 0.8 lagging. At Bus no. 2 the Phase A voltage phasor is $90 \angle 0°$ kV (rms). At the Bus no. 3 the Phase A voltage phasor is $90 \angle 12°$ kV (rms). Find the complex power produced by each source.

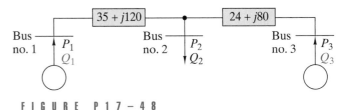

FIGURE P17–48

17–49 Three balanced three-phase loads are connected in parallel and connected to a source by lines with $Z_W = 1 + j10\ \Omega$. The first load is Y-connected with an impedance of $200 + j100\ \Omega$/phase. The second load is Δ-connected with an impedance of $2700 - j1200\ \Omega$/phase. The third load draws a complex power of $110 + j95$ kVA. The phase voltage at the load is 7.2 kV(rms). Find the complex power delivered by the source.

17–50 A balanced Δ-connected three-phase load has per-phase impedances of $60 + j25\ \Omega$. The magnitude of the line voltage at the load is 480 V(rms).

(a) The line connecting the load to the source has wire impedances of $Z_W = 5 + j15\ \Omega$ per phase. Find the transmission efficiency (η).

(b) A second line in parallel with the first has been proposed as a way to improve efficiency. The wire impedances of the proposed new line are $Z_W = 2 + j15\ \Omega$ per phase. Find the transmission efficiency (η) with the two lines connected in parallel.

INTEGRATING PROBLEMS

17–51 ▲ UNBALANCED THREE-PHASE LOAD

A balanced three-phase source with $V_P = 208$ V (rms) drives a Y-connected three-phase load with phase impedances $Z_A = 100\ \Omega$, $Z_B = 100\ \Omega$, and $Z_C = 50 + j100\ \Omega$.

(a) Calculate the line currents, line voltages, and total complex power delivered to the load.

(b) Repeat part (a) when a zero-resistance neutral wire is connected between the source and load.

17–52   **A D** THREE-PHASE POWER FACTOR CORRECTIONS

The line voltage in Figure P17–52 is 480 V(rms). The balanced three-phase load operates 8 hours per day and draws 105 kVA at a lagging power factor of 0.8. The electric power supplier charges 7¢/kW-hr when the load power factor is greater than 0.95 and 9¢/kW-hr when the power factor is less than 0.95. The following three-phase capacitor banks are commercially available.

480-V THREE-PHASE CAPACITOR EQUIPMENT					
kVAR	PART NUMBER	UNIT PRICE	kVAR	PART NUMBER	UNIT PRICE
10	1N0240A05	$500	40	1N0240A17	$900
20	1N0240A09	$600	50	1N0240A19	$1100
30	1N0240A11	$700	60	1N0240A23	$1500

If one or more capacitor banks are purchased to increase the power factor above 0.95, estimate the time it will take for the accumulated savings in operating costs to equal the equipment capital investment.

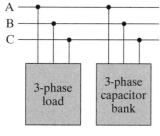

F I G U R E   P 1 7 – 5 2

17–53   **A** THREE-PHASE TRANSFORMER CALCULATIONS

The line voltage and line current at the input to a three-phase transformer are $2540 \angle -30°$ V (rms) and $43.3 \angle -66°$ A (rms). The line voltage and line current at the transformer output are $240 \angle 0°$ V (rms) and $417 \angle -30°$ A (rms).

(a) What are the complex power and power factor at the input?

(b) What are the complex power and power factor at the output?

(c) Is this an ideal transformer?

(d) What is the efficiency of the transformer?

17–54   **E** COMPARISON OF THREE-PHASE AND SINGLE-PHASE SYSTEMS

The three-phase and single-phase power systems in Figure P17–54 operate under the following conditions:

(a) The two systems deliver the same total complex power to the load.

(b) The two systems have the same line-to-line voltage.

(c) The two systems have the same transmission efficiency.

(d) The distance from source to load is the same.

(e) The resistance of a line is proportional to its length divided by the cross-sectional area of the wire.

Show that the transmission line in a three-phase system requires 25% less copper than the single-phase system.

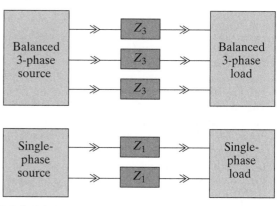

F I G U R E   P 1 7 – 5 4

APPENDIX A

STANDARD VALUES

To reduce inventory costs the electrical industry has agreed on standard values and tolerances for commonly used discrete components such as resistors, capacitors, and zener diodes. The standard values for resistors and capacitors with 5%, 10%, and 20% tolerances are shown in Table A–1. Standard resistances are obtained by multiplying the values in the table by different powers of 10. For example, multiplying the values for $\pm20\%$ tolerance by 10^4 yields a decade range of standard resistances of 100 kΩ, 150 kΩ, 220 kΩ, 330 kΩ, 470 kΩ, and 680 kΩ. Standard values for tolerances down to $\pm0.1\%$ are defined, although the resulting proliferation of values tends to defeat the purpose of standardization.

TABLE A–1   **STANDARD VALUES FOR RESISTORS AND CAPACITORS**

VALUE	TOLERANCES	VALUE	TOLERANCES	VALUE	TOLERANCES
10	±5%, ±10%, ±20%	22	±5%, ±10%, ±20%	47	±5%, ±10%, ±20%
11	±5%	24	±5%	51	±5%
12	±5%, ±10%	27	±5%, ±10%	56	±5%, ±10%
13	±5%	30	±5%	62	±5%
15	±5%, ±10%, ±20%	33	±5%, ±10%, ±20%	68	±5%, ±10%, ±20%
16	±5%	36	±5%	75	±5%
18	±5%, ±10%	39	±5%, ±10%	82	±5%, ±10%
20	±5%	43	±5%	91	±5%

APPENDIX B

SOLUTION OF LINEAR EQUATIONS

The purpose of this appendix is to review the methods of solving systems of linear algebraic equations. Circuit analysis often requires solving linear algebraic equations of the type

$$5x_1 - 2x_2 - 3x_3 = 4$$
$$-5x_1 + 7x_2 - 2x_3 = -10 \qquad \text{(B–1)}$$
$$-3x_1 - 3x_2 + 8x_3 = 6$$

where x_1, x_2, and x_3 are unknown voltages or currents. Often some of the unknowns may be missing from one or more of the equation. For example, the equations

$$5x_1 - 2x_2 = 5$$
$$-4x_1 + 7x_2 = 0$$
$$-3x_2 + 8x_3 = 0$$

involve three unknowns with one variable missing in each equation. Such equations can always be put in the standard square form by inserting the missing unknowns with a coefficient of zero.

$$5x_1 - 2x_2 - 0x_3 = 5$$
$$-4x_1 + 7x_2 - 0x_3 = 0 \qquad \text{(B–2)}$$
$$0x_1 - 3x_2 + 8x_3 = 0$$

Equations (B–1) and (B–2) will be used to illustrate the different methods of solving linear equations.

CRAMER'S RULE

Cramer's rule states that the solution of a system of linear equations for any unknown x_k is found as the ratio of two determinants

$$x_k = \frac{\Delta_k}{\Delta} \tag{B-3}$$

where Δ and Δ_k are determinants derived from the given set of equations. A **determinant** is a square array of numbers or symbols called **elements**. The elements are arranged in horizontal rows and vertical columns and are bordered by two vertical straight lines. In general, a determinant contains n^2 elements arranged in n rows and n columns. The value of the determinant is a function of the value and position of its n^2 elements.

The **system determinant** Δ in Eq. (B–3) is made up of the coefficients of the unknowns in the given system of equations. For example, the system determinant for Eq. (B–1) is

$$\Delta = \begin{vmatrix} 5 & -2 & -3 \\ -5 & 7 & -2 \\ -3 & -3 & 8 \end{vmatrix}$$

and for Eq. (B–2) is

$$\Delta = \begin{vmatrix} 5 & -2 & 0 \\ -4 & 7 & 0 \\ 0 & -3 & 8 \end{vmatrix}$$

These two equations are examples of the general 3×3 determinant

$$\Delta = \begin{vmatrix} a_{11} & a_{12} & a_{13} \\ a_{21} & a_{22} & a_{23} \\ a_{31} & a_{32} & a_{33} \end{vmatrix} \tag{B-4}$$

where a_{ij} is the element in the ith row and jth column.

The determinant Δ_k in Eq. (B–3) is derived from the system determinant by replacing the kth column by the numbers on the right side of the system of equations. For example, Δ_1 for Eq. (B–1) is

$$\Delta_1 = \begin{vmatrix} 4 & -2 & -3 \\ -10 & 7 & -2 \\ 6 & -3 & 8 \end{vmatrix}$$

and Δ_3 for Eq. (B–2) is

$$\Delta_3 = \begin{vmatrix} 5 & -2 & 5 \\ -4 & 7 & 0 \\ 0 & -3 & 0 \end{vmatrix}$$

These examples are 3×3 determinants because the system determinants from which they are derived are 3×3.

In summary, using Cramer's rule to solve linear equations boils down to evaluating the determinants formed using the coefficients of the unknowns and the right side of the system of equations.

EVALUATING DETERMINANTS

The **diagonal rule** gives the value of a 2×2 determinant as the difference in the product of the elements on the main diagonal $(a_{11}a_{22})$ and the product of the elements on the off diagonal $(a_{21}a_{12})$. That is, for a 2×2 determinant

$$\Delta = \begin{vmatrix} a_{11} & a_{12} \\ a_{21} & a_{22} \end{vmatrix} = a_{11}a_{22} - a_{21}a_{12} \tag{B-5}$$

The value of 3×3 and higher-order determinants can be found using the method of minors. Every element a_{ij} has a **minor** M_{ij}, which is formed by deleting the row and column containing a_{ij}. For example, the minor M_{21} of the general 3×3 determinant in Eq. (B–4) is

$$M_{21} = \begin{vmatrix} a_{12} & a_{13} \\ a_{32} & a_{33} \end{vmatrix} = a_{12}a_{33} - a_{32}a_{13}$$

The **cofactor** C_{ij} of the element a_{ij} is its minor M_{ij} multiplied by $(-1)^{i+j}$.

$$C_{ij} = (-1)^{i+j}M_{ij}$$

The signs of the cofactors alternate along any row or column. The appropriate sign for cofactor C_{ij} is found by starting in position a_{11} and counting plus, minus, plus, minus . . . along any combination of rows or columns leading to the position a_{ij}.

To use the **method of minors** we select one (and only one) row or column. The determinant is the sum of the products of the elements in the selected row or column and their cofactors. For example, selecting the first column in Eq. (B–4), we obtain Δ as follows:

$$\Delta = a_{11}C_{11} + a_{21}C_{21} + a_{31}C_{31}$$

$$= a_{11}(-1)^2\begin{vmatrix} a_{22} & a_{23} \\ a_{32} & a_{33} \end{vmatrix} + a_{21}(-1)^3\begin{vmatrix} a_{12} & a_{13} \\ a_{32} & a_{33} \end{vmatrix} + a_{31}(-1)^4\begin{vmatrix} a_{12} & a_{13} \\ a_{22} & a_{23} \end{vmatrix}$$

$$= a_{11}(a_{22}a_{33} - a_{32}a_{23}) - a_{21}(a_{12}a_{33} - a_{32}a_{13}) + a_{31}(a_{12}a_{23} - a_{22}a_{13})$$

An identical expression for Δ is obtained using any other row or column. For determinants greater than 3×3 the minors themselves can be evaluated using this approach. However, a system of equations leading to determinants larger than 3×3 is probably better handled using computer tools.

EXAMPLE B–1

Solve for the three unknowns in Eq. (B–1) using Cramer's rule.

SOLUTION:

Expanding the system determinant about the first column yields

$$\Delta = \begin{vmatrix} 5 & -2 & -3 \\ -5 & 7 & -2 \\ -3 & -3 & 8 \end{vmatrix} = 5\begin{vmatrix} 7 & -2 \\ -3 & 8 \end{vmatrix} - (-5)\begin{vmatrix} -2 & -3 \\ -3 & 8 \end{vmatrix} + (-3)\begin{vmatrix} -2 & -3 \\ 7 & -2 \end{vmatrix}$$

$$= 5[7 \times 8 - (-2)(-3)] - (-5)[(-2) \times 8 - (-3)(-3)] + (-3)[(-2)(-2) - (7)(-3)]$$

$$= 250 - 125 - 75 = 50$$

Expanding Δ_1 about the first column yields

$$\Delta_1 = \begin{vmatrix} 4 & -2 & -3 \\ -10 & 7 & -2 \\ 6 & -3 & 8 \end{vmatrix} = 4\begin{vmatrix} 7 & -2 \\ -3 & 8 \end{vmatrix} - (-10)\begin{vmatrix} -2 & -3 \\ -3 & 8 \end{vmatrix} + (6)\begin{vmatrix} -2 & -3 \\ 7 & -2 \end{vmatrix}$$

$$= 200 - 250 + 150 = 100$$

Expanding Δ_2 about the first column yields

$$\Delta_2 = \begin{vmatrix} 5 & 4 & -3 \\ -5 & -10 & -2 \\ -3 & 6 & 8 \end{vmatrix} = 5\begin{vmatrix} -10 & -2 \\ 6 & 8 \end{vmatrix} - (-5)\begin{vmatrix} 4 & -3 \\ 6 & 8 \end{vmatrix} + (-3)\begin{vmatrix} 4 & -3 \\ -10 & -2 \end{vmatrix}$$

$$= -340 + 250 + 114 = 24$$

Expanding Δ_3 about the first column yields

$$\Delta_3 = \begin{vmatrix} 5 & -2 & 4 \\ -5 & 7 & -10 \\ -3 & -3 & 6 \end{vmatrix} = 5\begin{vmatrix} 7 & -10 \\ -3 & 6 \end{vmatrix} - (-5)\begin{vmatrix} -2 & 4 \\ -3 & 6 \end{vmatrix} + (-3)\begin{vmatrix} -2 & 4 \\ 7 & -10 \end{vmatrix}$$

$$= 60 - 0 + 24 = 84$$

Now, applying Cramer's rule, we solve for the three unknowns.

$$x_1 = \frac{\Delta_1}{\Delta} = \frac{100}{50} = 2$$

$$x_2 = \frac{\Delta_2}{\Delta} = \frac{24}{50} = 0.48$$

$$x_3 = \frac{\Delta_3}{\Delta} = \frac{84}{50} = 1.68 \qquad\blacksquare$$

Exercise B-1

Evaluate Δ, Δ_1, Δ_2, and Δ_3 for Eq. (B-2).

Answer: 216, 280, 160, and 60

MATRICES AND LINEAR EQUATIONS

Circuit equations can be formulated and solved in matrix format. By definition, a **matrix** is a rectangular array written as

$$\mathbf{A} = \begin{bmatrix} a_{11} & a_{12} & a_{13} & \ldots & a_{1n} \\ a_{21} & a_{22} & a_{23} & \ldots & a_{2n} \\ \ldots & \ldots & \ldots & \ldots & \ldots \\ a_{m1} & a_{m2} & a_{m3} & \ldots & a_{mn} \end{bmatrix} \qquad (B-6)$$

The matrix $\mathbf{A}$ in Eq. (B-6) contains m rows and n columns and is said to be of order m by n (or $m \times n$). The matrix notation in Eq. (B-6) can be abbreviated as follows:

$$\mathbf{A} = [a_{ij}]_{mn} \tag{B–7}$$

where a_{ij} is the element in the ith row and jth column.

SOME DEFINITIONS

Different types of matrices have special names. A **row matrix** has only one row ($m = 1$) and any number of columns. A **column matrix** has only one column ($n = 1$) and any number of rows. A **square matrix** has the same number of rows as columns ($m = n$). A **diagonal matrix** is a square matrix in which all elements not on the main diagonal are zero ($a_{ij} = 0$ for $i \neq j$). An **identity matrix** is a diagonal matrix for which the main diagonal elements are all unity ($a_{ii} = 1$).

For example, given

$$\mathbf{A} = \begin{bmatrix} 1 & -2 & 0 & 4 \end{bmatrix} \quad \mathbf{B} = \begin{bmatrix} 3 \\ -2 \\ 6 \\ 0 \end{bmatrix} \quad \mathbf{C} = \begin{bmatrix} 1 & 0 & -7 \\ -3 & 12 & 0 \\ 0 & 0 & -4 \end{bmatrix} \quad \mathbf{U} = \begin{bmatrix} 1 & 0 & 0 & 0 \\ 0 & 1 & 0 & 0 \\ 0 & 0 & 1 & 0 \\ 0 & 0 & 0 & 1 \end{bmatrix}$$

we say that $\mathbf{A}$ is a 1×4 row matrix, $\mathbf{B}$ is a 4×1 column matrix, $\mathbf{C}$ is a 3×3 square matrix, and $\mathbf{U}$ is a 4×4 identity matrix.

The **determinant** of a square matrix $\mathbf{A}$ (denoted det $\mathbf{A}$) has the same elements as the matrix itself. For example, given

$$\mathbf{A} = \begin{bmatrix} 4 & -6 \\ 1 & -2 \end{bmatrix} \quad \text{then} \quad \det \mathbf{A} = \begin{vmatrix} 4 & -6 \\ 1 & -2 \end{vmatrix} = -8 + 6 = -2$$

The **transpose** of a matrix $\mathbf{A}$ (denoted $\mathbf{A}^T$) is formed by interchanging the rows and columns. For example, given

$$\mathbf{A} = \begin{bmatrix} 1 & 2 & 0 & 8 \\ 4 & 7 & -1 & -3 \end{bmatrix} \quad \text{then} \quad \mathbf{A}^T = \begin{bmatrix} 1 & 4 \\ 2 & 7 \\ 0 & -1 \\ 8 & -3 \end{bmatrix}$$

The **adjoint** of a square matrix $\mathbf{A}$ (denoted adj $\mathbf{A}$) is formed by replacing each element a_{ij} by its cofactor C_{ij} and then transposing.

$$\text{adj } \mathbf{A} = [C_{ij}]^T \tag{B–8}$$

For example, if

$$\mathbf{A} = \begin{bmatrix} -3 & 2 \\ 0 & 5 \end{bmatrix} \quad \text{then} \quad C_{11} = 5 \quad C_{12} = 0 \quad C_{21} = -2 \quad C_{22} = -3$$

and therefore

$$\text{adj } \mathbf{A} = \begin{bmatrix} 5 & 0 \\ -2 & -3 \end{bmatrix}^T = \begin{bmatrix} 5 & -2 \\ 0 & -3 \end{bmatrix}$$

MATRIX ALGEBRA

The matrices $\mathbf{A}$ and $\mathbf{B}$ are equal if and only if they have the same number of rows and columns, and $a_{ij} = b_{ij}$ for all i and j. Matrix addition is only possible when two matrices have the same number of rows and columns. When two matrices are of the same order, their sum is obtained by adding the corresponding elements: that is,

$$\text{If } \mathbf{C} = \mathbf{A} + \mathbf{B} \text{ then } c_{ij} = a_{ij} + b_{ij} \qquad \text{(B–9)}$$

For example, given

$$\mathbf{A} = \begin{bmatrix} -1 & 4 \\ -3 & -2 \end{bmatrix} \text{ and } \mathbf{B} = \begin{bmatrix} 3 & 0 \\ 2 & -4 \end{bmatrix} \text{ then } \mathbf{C} = \mathbf{A} + \mathbf{B} = \begin{bmatrix} 2 & 4 \\ -1 & -6 \end{bmatrix}$$

Multiplying a matrix $\mathbf{A}$ by a scalar constant k is accomplished by multiplying every element by k; that is, $k\mathbf{A} = [ka_{ij}]$. In particular, if $k = -1$ then $-\mathbf{B} = [-b_{ij}]$, and applying the matrix addition rule yields matrix **subtraction**.

$$\text{If } \mathbf{C} = \mathbf{A} - \mathbf{B} \text{ then } c_{ij} = a_{ij} - b_{ij} \qquad \text{(B–10)}$$

Multiplication of two matrices $\mathbf{AB}$ is defined only if the number of columns in $\mathbf{A}$ equals the number of rows in $\mathbf{B}$. In general, if $\mathbf{A}$ is of order $m \times n$ and $\mathbf{B}$ is of order $n \times r$, then the product $\mathbf{C} = \mathbf{AB}$ is a matrix of order $m \times r$. The element c_{ij} is found by summing the products of the elements in the ith row of $\mathbf{A}$ and the jth column of $\mathbf{B}$.

$$c_{ij} = [a_{i1} \, a_{i2} \, \ldots \, a_{in}] \begin{bmatrix} b_{1j} \\ b_{2j} \\ .. \\ .. \\ .. \\ b_{nj} \end{bmatrix} = a_{i1}b_{1j} + a_{i2}b_{2j} + \cdots a_{in}b_{nj}$$

$$= \sum_{k=1}^{n} a_{ik}b_{kj} \qquad \text{(B–11)}$$

In other words, matrix multiplication is a row by column operation.

Matrix multiplication is not commutative, so usually $\mathbf{AB} \neq \mathbf{BA}$. Two important exceptions are (1) the product of a square matrix $\mathbf{A}$ and an identity matrix $\mathbf{U}$ for which $\mathbf{UA} = \mathbf{AU} = \mathbf{A}$, and (2) the product of a square matrix $\mathbf{A}$ and its **inverse** (denoted $\mathbf{A}^{-1}$) for which $\mathbf{A}^{-1}\mathbf{A} = \mathbf{AA}^{-1} = \mathbf{U}$. A closed-form formula for the inverse of a square matrix is

$$\mathbf{A}^{-1} = \frac{\text{adj } \mathbf{A}}{\det \mathbf{A}} \qquad \text{(B–12)}$$

That is, the inverse can be found by multiplying the adjoint matrix of $\mathbf{A}$ by the scalar $1/\det \mathbf{A}$. If $\det \mathbf{A} = 0$ then $\mathbf{A}$ is said to be **singular** and $\mathbf{A}^{-1}$ does not exist. Equation (B–12) is useful for deriving properties of the inverse of a matrix. It is not, however, a very efficient way to calculate the inverse of a matrix of order greater than 3×3.

Exercise B–2

Given:

$$\mathbf{A} = \begin{bmatrix} -5 & 7 \\ 7 & 11 \end{bmatrix} \text{ and } \mathbf{B} = \begin{bmatrix} 3 & -1 \\ 6 & -2 \end{bmatrix}$$

Calculate AB, BA, $\mathbf{A}^{-1}$, and $\mathbf{B}^{-1}$.

Answers:

$$\mathbf{AB} = \begin{bmatrix} 27 & -9 \\ 87 & -29 \end{bmatrix} \quad \mathbf{BA} = \begin{bmatrix} -22 & 10 \\ -44 & 20 \end{bmatrix}$$

$$\mathbf{A}^{-1} = \frac{1}{104}\begin{bmatrix} -11 & 7 \\ 7 & 5 \end{bmatrix} \quad \mathbf{B}^{-1} \text{ does not exist}$$

MATRIX SOLUTION OF LINEAR EQUATIONS

The three linear equations in Eq. (B–1) are

$$5x_1 - 2x_2 - 3x_3 = 4$$
$$-5x_1 + 7x_2 - 2x_3 = -10$$
$$-3x_1 - 3x_2 + 8x_3 = 6$$

These equations are expressed in matrix form as follows:

$$\begin{bmatrix} 5 & -2 & -3 \\ -5 & 7 & -2 \\ -3 & -3 & 8 \end{bmatrix}\begin{bmatrix} x_1 \\ x_2 \\ x_3 \end{bmatrix} = \begin{bmatrix} 4 \\ -10 \\ 6 \end{bmatrix} \qquad \text{(B–13)}$$

The left side of Eq. (B–13) is the product of a 3×3 square matrix and a 3×1 column matrix of unknowns. The elements in the square matrix are the coefficients of the unknown in the given equations. The matrix product on the left side in Eq. (B–13) produces a 3×1 matrix, which equals the 3×1 column matrix on the right side. The elements of the 3×1 on the right side are the constants on the right sides of the given equations.

In symbolic form we write the matrix equation in Eq. (B–13) as

$$\mathbf{AX} = \mathbf{B} \qquad \text{(B–14)}$$

where

$$\mathbf{A} = \begin{bmatrix} 5 & -2 & -3 \\ -5 & 7 & -2 \\ -3 & -3 & 8 \end{bmatrix}, \quad \mathbf{X} = \begin{bmatrix} x_1 \\ x_2 \\ x_3 \end{bmatrix} \text{ and } \mathbf{B} = \begin{bmatrix} 4 \\ -10 \\ 6 \end{bmatrix}$$

Left multiplying Eq. (B–14) by $\mathbf{A}^{-1}$ yields

$$\mathbf{A}^{-1}\mathbf{AX} = \mathbf{A}^{-1}\mathbf{B}$$

But by definition $\mathbf{A}^{-1}\mathbf{A} = \mathbf{U}$ and $\mathbf{UX} = \mathbf{X}$; therefore

$$\mathbf{X} = \mathbf{A}^{-1}\mathbf{B} \qquad \text{(B–15)}$$

To solve linear equations by matrix methods we calculate the product $\mathbf{A}^{-1}\mathbf{B}$.

To implement the matrix approach we must first find $\mathbf{A}^{-1}$ using Eq. (B–12). The determinant of the coefficient matrix is

$$\det \mathbf{A} = \begin{vmatrix} 5 & -2 & -3 \\ -5 & 7 & -2 \\ -3 & -3 & 8 \end{vmatrix} = 50$$

The cofactors of the first row of the coefficient matrix are

$$C_{11} = -\begin{vmatrix} +7 & -2 \\ -3 & 8 \end{vmatrix} = 50 \quad C_{12} = \begin{vmatrix} -5 & -2 \\ -3 & 8 \end{vmatrix} = 46$$

$$C_{13} = -\begin{vmatrix} -5 & +7 \\ -3 & -3 \end{vmatrix} = 36$$

The cofactors for the second and third rows are

$$C_{21} = 25 \quad C_{22} = 31 \quad C_{23} = 21$$

$$C_{31} = 25 \quad C_{32} = 25 \quad C_{33} = 25$$

Now, using Eq. (B–12), we obtain $\mathbf{A}^{-1}$ as

$$\mathbf{A}^{-1} = \frac{\text{adj } \mathbf{A}}{\det \mathbf{A}} = \frac{1}{50} \begin{bmatrix} 50 & 46 & 36 \\ 25 & 31 & 21 \\ 25 & 25 & 25 \end{bmatrix}^T = \frac{1}{50} \begin{bmatrix} 50 & 25 & 25 \\ 46 & 31 & 25 \\ 36 & 21 & 25 \end{bmatrix}$$

Using Eq. (B–15), we solve for the column matrix of unknowns as

$$\begin{bmatrix} x_1 \\ x_2 \\ x_3 \end{bmatrix} = \mathbf{X} = \mathbf{A}^{-1}\mathbf{B} = \frac{1}{50} \begin{bmatrix} 50 & 25 & 25 \\ 46 & 31 & 25 \\ 36 & 21 & 25 \end{bmatrix} \begin{bmatrix} 4 \\ -10 \\ 6 \end{bmatrix} = \frac{1}{50} \begin{bmatrix} 100 \\ 24 \\ 84 \end{bmatrix}$$

which yields $x_1 = 2$, $x_2 = 24/50$, and $x_3 = 84/50$. These are, of course, the same results previously obtained using Cramer's rule.

USING COMPUTER TOOLS

Computer tools for solving linear equations range from inexpensive hand-held calculators to sophisticated software packages capable of solving hundreds of equations. At the intermediate level are math analysis software packages such as MATLAB and Mathcad. Under what circumstances should you consider using these computer tools in linear circuit analysis?

There are no hard and fast rules here. Somewhere around three or four equations the burden of hand calculations becomes mildly excruciating. Sets of equations with N up to 20 or 30 are routinely solved using computer tools, except when the equations are ill conditioned (several equations are almost linearly dependent). Well-conditioned systems of equations with $N = 50$ or more can be solved using sophisticated numerical methods. On the other hand, these sophisticated computer tools probably don't buy you very much in linear circuit applications. If you encounter a problem that requires solving (say) 20 or more linear equations, you should redefine the problem so that it can be partitioned into smaller pieces.

MATLAB is a software package for matrix-based computations and data analysis. The original MATLAB was a main frame resident software written in the 1970s to provide a "matrix laboratory" for linear algebra and matrix theory courses. Since then, personal computer versions have been developed that fit on modest computer platforms. In its present form MATLAB is an interactive system and programming language whose capabilities extend far beyond the original matrix laboratory application.

To solve the matrix equation in Eq. (B–13) using MATLAB, we first enter the **A** and **B** matrices in the MATLAB command window. In this window the command line prompt is either the character string << or EDU<<, depending on the version in use. This prompt indicates that MATLAB is ready to accept data and commands entered via the keyboard. For example, a matrix can be entered by typing its elements one row at a time. Elements in the same row are separated by spaces. The end of a row is indicated by a semicolon or an ↵enter keystroke. The row-by-row entry is enclosed by a left bracket [at the beginning and a right bracket] at the end of the last row.

For example, the **A** matrix in Eq. (B–13) can be entered by typing

```
A=[5  -2  -3
  -5   7  -2
  -3  -3   8]
```

After the last ↵enter keystroke, MATLAB responds by listing the elements of **A.**

```
A=
    5  -2  -3
   -5   7  -2
   -3  -3   8
```

This echo check allows you to verify that the elements of **A** have been entered correctly. The **A** can also be entered by typing

```
A=[5  -2  -3;-5  7  -2;-3  -3  8];
```

The semicolon following the closing right bracket suppresses the Matlab echo check if you have supreme confidence that you have entered the matrix elements without error. An individual matrix element can be referenced by enclosing its subscripts in parenthesis. For instance, the command line query

```
A(3,1)
```

asks a_{31} = ? and produces the MATLAB response

```
A(3,1) =
       -3
```

The **B** matrix for Eq. (B–13) can be entered as

```
B = [4; -10; 6];
```

The matrix equations in Eq. (B–13) can be solved for the unknown column matrix **X** by forming the matrix product $\mathbf{A}^{-1}\mathbf{B}.$ In the MATLAB command window we enter the statement

```
X=inv(A)*B
```

to which MATLAB responds with

```
X=
    2.0000
    0.4800
    1.6800
```

We can verify this result by checking to see that the matrix product **AX** is equal to the matrix **B**. In the command window we type

```
A*X
```

to which MATLAB responds with

```
ans=
    4
   -10
    6
```

which is the matrix **B** on the right side of Eq. (B–13).

Figure B–1 shows a Mathcad worksheet that performs the same matrix operations. The format and appearance of the mathematical operations in the worksheet are virtually self-defining. Again we have checked the answer by forming the matrix product **AX** in final step.

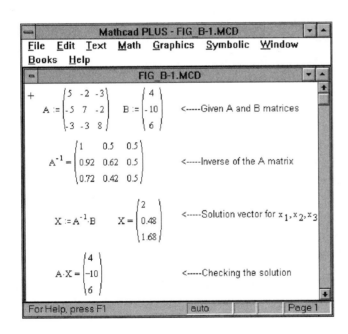

FIGURE B – 1 *A Mathcad worksheet.*

APPENDIX C

COMPLEX NUMBERS

Using complex numbers to represent signals and circuits is a fundamental tool in electrical engineering. This appendix reviews complex-number representations and arithmetic operations. These procedures, though rudimentary, must be second nature to all who aspire to be electrical engineers. Exercises are provided to confirm your mastery of these basic skills.

COMPLEX-NUMBER REPRESENTATIONS

A complex number z can be written in rectangular form as

$$z = x + jy \qquad (C-1)$$

where j represents $\sqrt{-1}$. Mathematicians customarily use i to represent $\sqrt{-1}$, but i represents current in electrical engineering so we use the symbol j instead.

The quantity z is a two-dimensional number represented as a point in the complex plane, as shown in Figure C–1. The x component is called the **real part** and y (not jy) the **imaginary part** of z. A special notation is sometimes used to indicate these two components:

$$x = \text{Re}\{z\} \quad \text{and} \quad y = \text{Im}\{z\} \qquad (C-2)$$

where $\text{Re}\{z\}$ means the real part and $\text{Im}\{z\}$ the imaginary part of z.

Figure C–1 also shows the polar representation of the complex number z. In polar form a complex number is written

$$z = M\angle\theta \qquad (C-3)$$

where M is called the **magnitude** and θ the **angle** of z. A special notation is also used to indicate these two components.

$$|z| = M \quad \text{and} \quad \angle z = \theta \qquad \text{(C–4)}$$

where $|z|$ means the magnitude and $\angle z$ the angle of z.

The real and imaginary parts and magnitude and angle of z are all shown geometrically in Figure C–1. The relationships between the rectangular and polar forms are easily derived from the geometry in Figure C–1:

Rectangular to polar $\quad M = \sqrt{x^2 + y^2} \quad \theta = \tan^{-1}\dfrac{y}{x}$

Polar to rectangular $\quad x = M \cos \theta \quad y = M \sin \theta \qquad \text{(C–5)}$

The inverse tangent relation for θ involves an ambiguity that can be resolved by identifying the correct quadrant in the z-plane using the signs of the two rectangular components. [See Exercise C–1(b) and (c).]

Another version of the polar form is obtained using Euler's relationship:

$$e^{j\theta} = \cos \theta + j \sin \theta \qquad \text{(C–6)}$$

We can write the polar form as

$$z = Me^{j\theta} = M \cos \theta + jM \sin \theta \qquad \text{(C–7)}$$

This polar form is equivalent to Eq. (C–3), since the right side yields the same polar-to-rectangular relationships as Eq. (C–5). Thus, a complex number can be represented in three ways:

$$z = x + jy \quad z = M\angle\theta \quad z = Me^{j\theta} \qquad \text{(C–8)}$$

The relationships between these forms are given in Eq. (C–5).

The quantity z^* is called the conjugate of the complex number z. The asterisk indicates the **conjugate** of a complex number formed by reversing the sign of the imaginary component. In rectangular form the conjugate of $z = x + jy$ is written as $z^* = x - jy$. In polar form the conjugate is obtained by reversing the sign of the angle of z, $z^* = Me^{-j\theta}$. The geometric interpretation in Figure C–2 shows that conjugation simply reflects a complex number across the real axis in the complex plane.

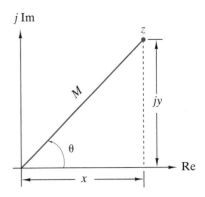

FIGURE C–1 *Graphical representation of complex numbers.*

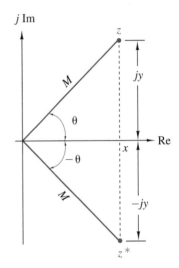

FIGURE C–2 *Graphical representation of conjugate complex numbers.*

Exercise C–1

Convert the following complex numbers to polar form.
(a) $1 + j\sqrt{3}$ (b) $-10 + j20$ (c) $-2000 - j8000$ (d) $60 - j80$

Answers: (a) $2e^{j60°}$ (b) $22.4e^{j117°}$ (c) $8246e^{j256°}$ (d) $100e^{j307°}$

Exercise C–2

Convert the following complex numbers to rectangular form.
(a) $12e^{j90°}$ (b) $3e^{j45°}$ (c) $400\angle\pi$ (d) $8e^{-j60°}$ (e) $15e^{j\pi/6}$

Answers: (a) $0 + j12$ (b) $2.12 + j2.12$ (c) $-400 + j0$ (d) $4 - j6.93$ (e) $13 + j7.5$

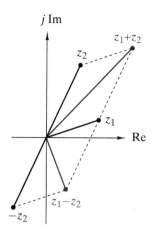

FIGURE C – 3 *Graphical representation of addition and subtraction of complex numbers.*

Exercise C–3

Evaluate the following expressions.

(a) $\text{Re}(12e^{j\pi})$ (b) $\text{Im}(100\angle 60°)$ (c) $\angle(-2 + j6)$ (d) $\text{Im}[(4e^{j\frac{\pi}{4}})^*]$

Answers:
(a) -12 (b) 86.6 (c) $108.4°$ (d) -2.83

ARITHMETIC OPERATIONS: ADDITION AND SUBTRACTION

Addition and subtraction are defined in terms of complex numbers in rectangular form. Two complex numbers

$$z_1 = x_1 + jy_1 \quad \text{and} \quad z_2 = x_2 + jy_2 \tag{C–9}$$

are added by separately adding the real parts and imaginary parts. The sum $z_1 + z_2$ is defined as

$$z_1 + z_2 = (x_1 + x_2) + j(y_1 + y_2) \tag{C–10}$$

Subtraction follows the same pattern except that the components are subtracted:

$$z_1 - z_2 = (x_1 - x_2) + j(y_1 - y_2) \tag{C–11}$$

Figure C–3 shows a geometric interpretation of addition and subtraction. In particular, note that $z + z^* = 2x$ and $z - z^* = j2y$.

MULTIPLICATION AND DIVISION

Multiplication and division of complex numbers can be accomplished with the numbers in either rectangular or polar form. For complex numbers in rectangular form the multiplication operation yields

$$\begin{aligned} z_1 z_2 &= (x_1 + jy_1)(x_2 + jy_2) \\ &= (x_1 x_2 + j^2 y_1 y_2) + j(x_1 y_2 + x_2 y_1) \\ &= (x_1 x_2 - y_1 y_2) + j(x_1 y_2 + x_2 y_1) \end{aligned} \tag{C–12}$$

For numbers in polar form the product is

$$\begin{aligned} z_1 z_2 &= (M_1 e^{j\theta_1})(M_2 e^{j\theta_2}) \\ &= (M_1 M_2)e^{j(\theta_1 + \theta_2)} \end{aligned} \tag{C–13}$$

Multiplication is somewhat easier to carry out with the numbers in polar form, although both methods should be understood. In particular, the product of a complex number z and it conjugate z^* is the square of its magnitude, which is always positive.

$$zz^* = (Me^{j\theta})(Me^{-j\theta}) = M^2 \tag{C–14}$$

For complex numbers in polar form the division operation yields

$$\frac{z_1}{z_2} = \frac{Me^{j\theta_1}}{M_2 e^{j\theta_2}}$$

$$= \left(\frac{M_1}{M_2}\right) e^{j(\theta_1 - \theta_2)} \tag{C-15}$$

When the numbers are in rectangular form the numerator and denominator of the quotient are multiplied by the conjugate of the denominator.

$$\frac{z_1 \, z_2^*}{z_2 \, z_2^*} = \frac{(x_1 + jy_1)(x_2 - jy_2)}{(x_2 + jy_2)(x_2 - jy_2)}$$

Applying the multiplication rule from Eq. (C–12) to the numerator and denominator yields

$$\frac{z_1}{z_2} = \frac{(x_1 x_2 + y_1 y_2) + j(x_2 y_1 - x_1 y_2)}{x_2^2 + y_2^2} \tag{C-16}$$

Complex division is easier to carry out with the numbers in polar form, although both methods should be understood.

Exercise C–4

Evaluate the following expressions using $z_1 = 3 + j4$, $z_2 = 5 - j7$, $z_3 = -2 + j3$, and $z_4 = 5\angle -30°$:

(a) $z_1 z_2$ (b) $z_3 + z_4$ (c) $z_2 z_3 / z_4$ (d) $z_1{}^* + z_3 z_1$ (e) $z_2 + (z_1 z_4)^*$

Answers:
(a) $43 - j$ (b) $2.33 + j0.5$ (c) $-0.995 + j6.12$ (d) $-15 - j3$ (e) $28 - j16.8$

Exercise C–5

Given $z = x + jy = Me^{j\theta}$, evaluate the following statements:

(a) $z + z^*$ (b) $z - z^*$ (c) z/z^* (d) z^2 (e) $(z^*)^2$ (f) zz^*

Answers:
(a) $2x$ (b) $j2y$ (c) $e^{j2\theta}$ (d) $x^2 - y^2 + j2xy$ (e) $x^2 - y^2 - j2xy$ (f) $x^2 + y^2$

Exercise C–6

Given $z_1 = 1$, $z_2 = -1$, $z_3 = j$, and $z_4 = -j$, evaluate (a) z_1/z_3 (b) z_1/z_4 (c) $z_3 z_4$ (d) $z_3 z_3$ (e) $z_4 z_4$ (f) $z_2 z_3^*$.

Answers:
(a) $-j$ (b) $+j$ (c) 1 (d) -1 (e) -1 (f) j

Exercise C–7

Evaluate the expression $T(\omega) = j\omega/(j\omega + 10)$ at $\omega = 5, 10, 20, 50, 100$.

Answers: $0.447\angle 63.4°, 0.707\angle 45°, 0.894\angle 26.6°, 0.981\angle 11.3°, 0.995\angle 5.71°$

ANSWERS

TO SELECTED PROBLEMS

CHAPTER ONE

1–3 1.188×10^7 C

1–5   (a) 10^6;   (b) 10^{-3};   (c) 10^3;   (d) 10^{-6}

1–7 $i(t) = -60 \, e^{-3t} \, \mu$A

1–9 $Q = 8$ C

1–11 $Q = 75$ C

1–13 (a) 4.167 A; (b) 48 hr

1–15   2.16 MJ

1–17   at $v = -3$ V, $p = 29.8$ W, absorbing;
          at $v = 1.5$ V, $p = -8.28$ W, delivering;
          at $v = 3$ V, $p = 30.3$ W, absorbing

1–19 $i_{max} = 5$ mA

1–21 (a) $p = -72.6$ W, transfer from B to A;
      (b) $p = 14.4$ mW, transfer from A to B;
      (c) $p = 1.5$ W, transfer from A to B;
      (d) $p = 645$ mW, transfer from A to B;

CHAPTER TWO

2–1   (a) $i = 10$ mA; $p = 500$ mW
      (b) $v = -3$ V

2–3 $G = 0.25$ mS

2–5   at $v = -15$ V, $i = -7.5$ mA; device is linear

2–7 $i = 1.095$ mA

2–11 (a) Nodes A, B, C; Loops 1,2;2,3,4;1,3,4

(b) Series 3 & 4; Parallel 1 & 2

(c) KCL: Node A $- i_1 - i_2 - i_3 = 0$;

Node B $i_3 - i_4 = 0$;

Node C $i_1 + i_2 + i_4 = 0$

KVL: Loop 1,2 $-v_1 + v_2 = 0$;

Loop 2,3,4 $-v_2 + v_3 + v_4 = 0$;

Loop 1,3,4 $-v_1 + v_3 + v_4 = 0$

(d) $i_3 = -2$ mA; $i_4 = -2$ mA

2–13 (a) Nodes A, B, C, D;

Loops 1,3,2;2,4,5;3,6,4;1,5,6;2,3,6,5

(b) Series none; Parallel none

(c) KCL: Node A $- i_2 - i_3 - i_4 = 0$;

Node B $- i_1 + i_3 - i_6 = 0$;

Node C $i_1 + i_2 + i_5 = 0$

Node D $i_4 - i_5 + i_6 = 0$

KVL: Loop 1,3,2 $- v_1 + v_2 - v_3 = 0$;

Loop 2,4,5 $- v_2 + v_4 + v_5 = 0$;

Loop 3,6,4 $v_3 - v_4 + v_6 = 0$

(d) $v_1 = 9$ V; $v_2 = 1$ V; $v_6 = 0$ V

2–16 $v_2 = 15$ V; $v_5 = -20$ V

2–17 $i_4 = 11$ A; $i_5 = -7$ A

2–21 $v_x = -10$ V; $i_x = -2$ A

2–23 $v_x = 60$ V

2–27 $R_{EQ} = 110 \ \Omega$

2–29 Open, $R_{EQ} = 167 \ \Omega$; Closed, $R_{EQ} = 100 \ \Omega$

2–31 $R_{AB} = 90 \ \Omega$; $R_{AC} = 60 \ \Omega$; $R_{AD} = 40 \ \Omega$; $R_{BC} = 130 \ \Omega$; $R_{BD} = 110 \ \Omega$; $R_{CD} = 40 \ \Omega$

2–33 $i_S = 0.3$ A; $R_S = 50 \ \Omega$

2–35 $R_{AB} = 8R/5$ cannot be obtained

2–41 $v_L = \dfrac{R_L v_S}{R + 2R_L}$

2–43 $i_x = 0.667$ A

2–45 For $v_G = 5$ V, $v_O = 2.14$ V; for $v_G = 0.5$ V, $v_O = 1.5 \ \mu$V

2–47 $R_1 = 40/19 \ \Omega$; $R_2 = 10/19 \ \Omega$

2–51 $i_x = -0.625$ A; $v_x = -6.25$ V

2–53 $i_x = 2.5$ A; $v_x = 5$ V

2–55 $v_{AB} = 7.27$ V

CHAPTER THREE

3–1 (b) $v_x = 15.8$ V; $i_x = 0.183$ A

3–3 (b) $v_x = -15$ V; $i_x = 0$

3–5 (b) $v_x = 2$ V; $i_x = -1.2$ mA

3–7 (b) $v_x = 11.4$ V; $i_x = 37.1$ mA;
(c) total power delivered is 1.86 W

3–9 (b) $v_x = -1.6$ V; $i_x = 1.28$ mA

3–11 (b) $v_x = 6.25$ V; $i_x = 0.375$ A

3–13 (b) $v_x = -3$ V; $i_x = 3.35$ mA;
(c) power delivered by S1 is 40.2 mW

3–15 (b) $v_x = 25.7$ V; $i_x = -5.71$ mA;
(c) total power delivered is 529 mW

3–21 $$K = \frac{R_1 R_3}{R_1 + R_2 + R_3}$$

3–23 $$K = \frac{R_2 R_3 - R_1 R_4}{R_1 + R_2 + R_3 + R_4}$$

3–25 $i_O = 81.7$ mA

3–27 $K = 3/20$

3–29 $i_O = 923$ μA

3–31 $v_O = R(i_1 - 3i_2)/5$

3–37 (a) $R_T = 100$ Ω; $v_T = 25$ V;
(b) $p_L = 1.39$ W for $R_L = 50$ Ω; $p_L = 1.56$ W for $R_L = 100$ Ω;
$p_L = 0.868$ W for $R_L = 500$ Ω;

3–39 (a) $R_T = 12$ kΩ; $v_T = 184$ V;
(b) $i_L = 10.2$ mA for $R_L = 6$ kΩ; $i_L = 7.67$ mA for $R_L = 12$ kΩ;
$i_L = 5.11$ mA for $R_L = 24$ kΩ; $i_L = 3.07$ mA for $R_L = 48$ kΩ;

3–41 (a) $R_T = 5R/3$; $v_T = 2v_1/3 + v_2/3$;
(b) $v_O = v_1/4 + v_2/8$

3–43 $i_L = 20$ mA

3–45 $R_L \geq 350$ Ω

3–47 (a) $R_T = 100$ kΩ; $v_T = 9$ V;
(b) $R_L = 200$ kΩ

3–51 (a) $p_{MAX} = 200$ mW for $R_L = 20$ Ω;
(b) $v_{MAX} = 4$ V for $R_L = \infty$;
(c) $i_{MAX} = 200$ mA for $R_L = 0$

3–53 $p_{MAX} = 801$ mW for $R_L = 21.7$ Ω

3–55 $p_{MAX} = 1.125$ mW for $R_L = 2$ kΩ

Chapter Four

4–1 (a) $K_V = -6.67$; (b) $K_I = -100$

4–3 $v_3 = 2.27$ V; $K_I = 3.33 \times 10^5$

4–5 (a) $i_2 = -0.262$ mA; $K_V = -2.211$
(b) $R_{IN} = 90$ kΩ

4–7 $K_V = 8.18$

4–9 $K_V = \dfrac{g\,R_S R_O + R_O}{g\,R_S R_O + R_O + R_S}$

4–11 $v_O = V_{CC} - \dfrac{\beta\,R_C}{\beta + 1}\,i_S;\quad i_B = \dfrac{i_S}{\beta + 1}$

4–13 $v_S = 2$ V; $i_C = 2.6$ mA; $v_{CE} = 4.8$ V;
$v_S = 5$ V; $i_C = 5$ mA; $v_{CE} = 0$ V

4–15 $v_S \geq 4.96$ V

4–17 $v_S = 0.5$ V; $i_C = 0$; $v_{CE} = 6.67$ V;
$v_S = 1$ V; $i_C = 1$ mA; $v_{CE} = 0$ V

4–19 $i_C = 1.68$ mA; $v_{CE} = 1.54$ V

4–21 $v_O = -1.5\,v_S$

4–23 (a) $v_O = 3\,v_S$; (b) $i_O = 0.9$ mA

4–25 (a) $R_1 = 75$ kΩ; $R_2 = 25$ kΩ; $R_3 = 15$ kΩ;
(b) $-11.5 < v_1 < 18.5$

4–27 $v_O = \dfrac{R_3 + R_4}{R_3}\left(\dfrac{R_2 v_{S1} + R_1 v_{S2}}{R_1 + R_2}\right)$

4–29 $1 \leq K \leq 2$

4–31 $v_2 = 6\,v_1 + 9$

4–33 $v_O = -0.3\,v_S$

4–35 $v_O = 15\,v_1 - 30\,v_2$

4–47 (a) $v_O = V_{OH}$ when $v_S > -2$ V; $v_O = V_{OL}$ when $v_S < -2$ V

4–49 (a) $v_O = V_{OH}$ when $v_S > 5$ V; $v_O = V_{OL}$ when $v_S < 5$ V

CHAPTER FIVE

5–3 $v(t) = V_A\,[u(t + T/2) - u(t - T/2)]$

5–4 $v(t) = -5\,u(0.003 - t) + 5\,u(t - 0.003) - 10\,u(t - 0.005)$ V

5–5 $v(t) = 2500\,r(t)\,[u(t) - u(t - 0.002)]$ V

5–6 (a) $V_A = 10$ V; $T_C = 0.5$ s;
(b) $V_A = 10$ V; $T_C = 2$ s;
(c) $V_A = -10$ V; $T_C = 0.05$ s;
(d) $V_A = -10$ V; $T_C = 20$ s;

5–8 $V_A = 128$ V; $T_C = 1.443$ ms

5–9   8.54 V

5–11 (a) $T_0 = 1$ ms; $f_0 = 1$ kHz; $V_A = 14.1$ V; $T_S = 0.125$ ms; $\phi = -45°$;
(b) $T_0 = 1$ ms; $f_0 = 1$ kHz; $V_A = 36.1$ V; $T_S = -0.406$ ms; $\phi = 146°$;
(c) $T_0 = 10$ s; $f_0 = 0.1$ Hz; $V_A = 14.1$ V; $T_S = -1.25$ s; $\phi = 45°$;
(d) $T_0 = 2.5$ ms; $f_0 = 400$ Hz; $V_A = 36.1$ V; $T_S = 0.859$ ms; $\phi = -124°$

5–12 $T_0 = 1$ ms; $f_0 = 1$ kHz; $V_A = 22.4$ V; $T_S = -0.426$ ms; $\phi = 153°$

5–13 $a = 10$ V; $b = 11.2$ V

5–15 (a) $f_0 = 2$ kHz; $T_0 = 0.5$ ms; $a = -20$ V; $b = 0$ V;
(b) $f_0 = 2$ kHz; $T_0 = 0.5$ ms; $a = 0$ V; $b = 20$ V;
(c) $f_0 = 0.0025$ Hz; $T_0 = 400$ s; $a = 21.2$ V; $b = 21.2$ V;
(d) $f_0 = 1$ kHz; $T_0 = 1$ ms; $a = 42.4$ V; $b = 42.4$ V;

5–19 $v_{max} = 0.368\ V_A$ at $t = 1/\alpha$

5–21 $V_A = 9.94$ V; $\alpha = 994$ rad/s; $\beta = 1257$ rad/s

5–23 $v(t) = 2\ u(t) - 2\ u(t-1) - u(t-2) + u(t-3)$ mV

5–25 $v(t) = -1.5 - 3.5\cos(1000\pi t)$ V

5–29 $V_A = 24.5$ V; $\alpha = 235$ rad/s; $\beta = 1257$ rad/s

5–31 (a) aperiodic, causal; $V_p = 230$ V; $V_{pp} = 160$ V;
(b) aperiodic, causal; $V_p = 40$ V; $V_{pp} = 80$ V;
(c) periodic, noncausal; $V_p = 18$ V; $V_{pp} = 36$ V; $V_{avg} = 0$ V; $V_{rms} = 12.7$ V;
(d) aperiodic, causal; $V_p = 10$ V; $V_{pp} = 5$ V

5–33 (a) $V_p = 20$ V; $V_{pp} = 40$ V; $V_{avg} = 0$ V; $V_{rms} = 14.1$ V;
(b) $V_p = 20$ V; $V_{pp} = 40$ V; $V_{avg} = 0$ V; $V_{rms} = 14.1$ V;
(c) $V_p = 30$ V; $V_{pp} = 60$ V; $V_{avg} = 0$ V; $V_{rms} = 21.2$ V;
(d) $V_p = 60$ V; $V_{pp} = 120$ V; $V_{avg} = 0$ V; $V_{rms} = 42.4$ V

5–35 $V_p = 2$ mV; $V_{pp} = 3$ mV; $V_{avg} = 0.25$ mV; $V_{rms} = 1.118$ mV

5–37 $5 \geq V_0 \geq -5$ V

CHAPTER SIX

6–1 $i_C(t) = -24\ e^{-4000\,t}\ u(t)$ mA; $p_C(t) = -72\ e^{-8000\,t}\ u(t)$ mW; $w_C = 9\ e^{-8000\,t}\ u(t)$ μJ; delivering

6–3 $v_L(t) = -12\ e^{-4000\,t}\ u(t)$ V; $p_L(t) = -360\ e^{-8000\,t}\ u(t)$ μW; $w_L(t) = 45\ e^{-8000\,t}\ u(t)$ μJ; delivering

6–5 $v_C(t) = [20\sin(10^5\,t) + 30]\ u(t)$ V

6–7 0 V; 5 V; 5 V

6–9 $v_L(t) = 5\ e^{-2000\,t}[-2\sin(1000t) + \cos(1000t)]\ u(t)$ V; $p_L(t) = 0.05\ e^{-4000\,t}[-2\sin^2(1000t) + \sin(1000t)\cos(1000t)]\ u(t)$ W; $w_L(t) = 25\ e^{-4000\,t}\sin^2(1000t)\ u(t)$ μJ; both

6–11 $i_C(t) = 5\ e^{-1000\,t}\ u(t)$ mA; delivering

6–13 $i_L(0) = -125$ mA

6–15 $i_C(t) = 0.1\ e^{-1000\,t}\ u(t)$ A; $p_C(t) = [e^{-1000\,t} - e^{-2000\,t}]u(t)$ W; absorbing

6–21 $v_O(t) = v_S(t) + \dfrac{1}{RC}\displaystyle\int_0^t v_S(x)\,dx + v_O(0) - v_S(0)$

6–23 $v_O(t) = -RC_1\dfrac{dv_{S1}(t)}{dt} - RC_2\dfrac{dv_{S2}(t)}{dt}$

6–25 $T \leq 6$ ms

6–27 $-15 \leq V_A \leq 15$ V

6–29 $v_O(t) = 5 \times 10^{-6} \, \alpha \, e^{-\alpha t} \, u(t)$ V; $\alpha \le 3$ Mrad/s

6–36 $C_{EQ} = 3 \, \mu$F; $L_{EQ} = 1.0493$ H

6–38 $L_{EQ} = 2.464$ mH

6–40 $C_{EQ} = 4.128 \, \mu$F

6–42   Four strings in parallel each consisting of four 20 μF capacitors in series

CHAPTER SEVEN

7–1 $v(t) = -15[\, e^{-1500 \, t}\,]u(t)$ V

7–3   C1: $T_C = 0.5$ ms; C2: $T_C = 0.418$ ms

7–5 $v_C(t) = 15 \, [e^{-3000 \, t}]u(t)$ V; $i_O(t) = 1.5 \, [e^{-3000 \, t}]u(t)$ mA

7–7 $v_C(t) = V_A \, [-1 + 2e^{-t/RC}]$; $i_L(t) = V_A \, [-1 + 2e^{-R \, t/L}]/R$

7–9 $v_O(t) = I_A R \, [1 - e^{-t/RC}]$

7–11 $v_C(t) = [6 - 3 \, e^{-200 \, t}]u(t)$ V

7–13 $v_C(t) = 11.538 \, [\, 1 - e^{-10400 \, t}\,]u(t)$ V

7–15 $v(t) = [- \, 100 \, e^{-200 \, t} + 100 \cos (100t) + 50 \sin (100t)]u(t)$ mV

7–17 $i_L(t) = [- \, 300 \, e^{-15 \, t} + 300 \cos (5t) + 100 \sin (5t)\,]u(t)$ mA

7–23 (a) $T_C = 1$ ms; (b) $v_C(0) = 10$ V;
    (c) $R = 500 \, \Omega$; $C = 2 \, \mu$F; (d) 13.5 μJ

7–25 (a) $T_C = 0.5$ ms; (b) $v_C(0) = 5$ V;
    (c) $v_S(t) = 15$ V; $R = 1$ kΩ; $C = 0.5 \, \mu$F; (d) 46.6 μJ

7–27 (a) $T_C = 1$ ms; (b) $i_L(0) = -5$ mA; $i_L(\infty) = 5$ mA;
    (c) $v_S(t) = 0.5$ V; $R = 100 \, \Omega$; $L = 0.1$ H;
    (d) $w_L(0) = w_L(\infty) = 1.25 \, \mu$J

CHAPTER EIGHT

8–1 $v(t) = 12[\, t \, e^{-8 \, t}\,]u(t)$

8–3   C1: $s_1, s_2 = [-2.5 \pm j49.94] \times 10^4$; underdamped;
    C2: $s_1 = -1.01 \times 10^5$; $s_2 = -9.899 \times 10^6$, overdamped

8–5 $v_C(t) = e^{-1000 \, t} [10 \cos (2000 \, t) + 5 \sin (2000 \, t)]$ V;
    $i_L(t) = -5 \, e^{-1000 \, t} \sin (2000 \, t)$ mA; underdamped

8–7 $v_C(t) = 15 - e^{-1000 \, t} [5 \cos (2000 \, t) + 2.5 \sin (2000 \, t)]$ V;
    $i_L(t) = 5 \, e^{-1000 \, t} \sin (2000 \, t)$ mA; underdamped

8–9 $i_L(t) = e^{-1000 \, t} [3 \cos (7000 \, t) + 0.4286 \sin (7000 \, t)]$ mA;
    $v_C(t) = -1.714 \, e^{-1000 \, t} \sin (7000 \, t)$ V; underdamped

8–11 $i_L(t) = 5 - 6.667e^{-2000 \, t} + 1.667e^{-8000 \, t}$ mA;
    $v_C(t) = 16.667e^{-2000 \, t} - 16.667e^{-8000 \, t}$ V; overdamped

8–13 $i_L(t) = 0.24 - 0.54e^{-100 \, t} - 54te^{-100 \, t}$ A;
    $v_C(t) = 5400t \, e^{-100 \, t}$ V; critically damped

8–15 $v_C(t) = 5 - 5e^{-2000 \, t} - 10^4te^{-2000 \, t}$ V;

$$i_L(t) = 20t\,e^{-2000\,t} \text{ A; critically damped}$$

8–21 (a) $v_T = 10$ V; $R = 450\ \Omega$; $L = 50$ mH;
(b) $i_L(t) = 0.2\,e^{-4000\,t} - 0.2\,e^{-5000\,t}$ A

8–23 $R = 2600\ \Omega$; $L = 2$ H; $C = 1.923\ \mu$F

8–25 $v_C(t) = -10\,e^{-10\,t}\,[\cos(20\,t) + 2\sin(20\,t)]$ V;

CHAPTER NINE

9–1 $F(s) = \dfrac{-A(s + 2\,\alpha)}{s(s + \alpha)}$
zero at $s = -2\alpha$; poles at $s = -\alpha$ and $s = 0$

9–3 $F(s) = \dfrac{A(s + \beta)^2}{s(s^2 + \beta^2)}$
double zero at $s = -\beta$; poles at $s = \pm j\beta$ and $s = 0$

9–5 $F(s) = \dfrac{A(s + \alpha)^2}{(s + \alpha)^2 + \beta^2}$
double zero at $s = -\alpha$; poles at $s = -\alpha \pm j\beta$

9–7 (a) $F(s) = \dfrac{-5s}{(s + 5)(s + 10)}$

(b) $F(s) = \dfrac{-6s(s^2 + 30^2)}{(s^2 + 10^2)(s^2 + 20^2)}$

9–11 $F(s) = \dfrac{A\,\beta\,e^{-Ts}}{(s + \alpha)^2 + \beta^2}$

9–17 (a) $f_1(t) = [2 - e^{-5\,t}]u(t)$;

(b) $f_2(t) = [2\,e^{-10\,t} - e^{-5\,t}]u(t)$;

9–19 $f(t) = \delta(t) - \beta\,[\cos(\beta\,t) + \sin(\beta\,t)]u(t)$;

9–21 (a) $f_1(t) = 36\,[e^{-20t} - e^{-45t}]u(t)$;
(b) $f_2(t) = [900\,t\,e^{-30t}]u(t)$;
(c) $f_3(t) = 37.5\,[e^{-18t}\sin(24\,t)]u(t)$

9–23 (a) $f_1(t) = \left[\dfrac{3}{8} + \dfrac{1}{4}e^{-4t} + \dfrac{3}{8}e^{-8t}\right]u(t)$;

(b) $f_2(t) = \left[3 - \dfrac{15}{8}\cos(\sqrt{20}\,t) - \dfrac{1}{8}\cos(\sqrt{60}t)\right]u(t)$

9–31 (a) $y(t) = [5\,e^{-20\,t}]u(t)$;

(b) $y(t) = [10 - 20\,e^{-100\,t}]u(t)$

9–33 (a) $2 \times 10^{-3}\dfrac{di_L(t)}{dt} + i_L(t) = 0.1\,u(t)$ $i_L(0) = 0$

(b) $i_L(t) = 0.1\,[1 - e^{-500\,t}]u(t)$ A

9–35 (a) $\dfrac{1}{3750}\dfrac{dv_C(t)}{dt} + v_C(t) = \dfrac{2}{3}v_S(t)$ $v_C(0) = 0$;

(b) $v_O(t) = 18.75\,[e^{-2500\,t} - e^{-3750\,t}]u(t)$ V

9–37 $y(t) = [15 e^{-5t} - 5 e^{-15t}]u(t)$

9–39 (a) $2 \times 10^{-7} \dfrac{d^2 i_L(t)}{dt^2} + 4 \times 10^{-4} \dfrac{d\, i_L(t)}{dt} + i_L(t) = 0$

$i_L(0) = 5 \times 10^{-3};\quad \dfrac{d\, i_L(0)}{dt} = 0$

(b) $i_L(t) = 5 e^{-1000\,t}[\cos(2000\,t) + 0.5\sin(2000\,t)]u(t)$ mA

9–45 (a) $f_1(0) = 0$; $f_1(\infty) = 0$;

(b) $f_2(0) = 1$; $F_2(s)$ has j-axis poles

9–47 (a) $f_1(0) = 2$; $f_1(\infty) = 0$;

(b) $f_2(0) = 10$; $F_2(s)$ has right half plane poles

9–49 (a) $f_1(0) = 0$; $f_1(\infty) = 2$;

(b) $f_2(0) = 1$; $f_2(\infty) = 0$

CHAPTER TEN

10–1 $Z_{12}(s) = R\left[\dfrac{RCs+1}{2RCs+1}\right]$; pole at $s = -1/2RC$ zero at $s = -1/RC$;

$Z_{34}(s) = \left[\dfrac{2R}{2RCs+1}\right]$;   pole at $s = -1/2RC$

10–3 $Z(s) = 2000\,\dfrac{(s+j10^4)(s-j10^4)}{(s+10^4)(s+10^4)}$

10–5 $Z(s) = 1000\,\dfrac{(s+10^4+j10^4)(s+10^4-j10^4)}{s(s+10^4)}$

10–7 $Z(s) = \dfrac{(RCs+0.382)(RCs+2.618)}{Cs\,(RCs+2)}$

10–11 $I_L(s) = \dfrac{V_A}{2R}\dfrac{Ls+2R}{s(Ls+R)};\ i_L(t) = \dfrac{V_A}{R}[1 - 0.5\, e^{-Rt/L}]u(t)$

10–13 $I_C(s) = \dfrac{-2CV_A}{RCs+1};\ i_C(t) = -\dfrac{2V_A}{R}[e^{-t/RC}]u(t)$

10–15 $I_L(s) = 2.5 \times 10^{-3}\left[\dfrac{3s+16\times10^4}{s(s+10^5)}\right]$;

$i_L(t) = [4 + 3.5\, e^{-10^4 t}]u(t)$ mA

10–17 (a) $I_L(s) = \dfrac{-V_A C}{LC s^2 + R_2 C s + 1}$;

(b) $i_L(t) = -[60\, t\, e^{-1000t}]u(t)$ A

10–19 (a) $I_L(s) = \dfrac{(V_A - V_B)C}{LC s^2 + R_2 C s + 1}$;

(b) $i_L(t) = 8.333\,[e^{-5000t} - e^{-20000t}]u(t)$ mA

10–21 $V_{Czs}(s) = \dfrac{L s\, V_S(s)}{RLC s^2 + L s + R}$

$V_{Czi}(s) = \dfrac{RLC s\, V_0}{RLC s^2 + L s + R}$

10–23 $v_O(t) = [-10\,e^{-500t} + 18\,e^{-100t}]u(t)$ V;
Forced pole as $s = -100$; natural pole at $s = -500$

10–31 (b) $I_2(s) = \dfrac{R_1 C\, s\, V_1(s)}{(R_1 + R_2)LC\,s^2 + (L + R_1 R_2 C)s + R_1};$
(c) $i_2(t) = [15.1\,e^{-2500t}\sin(1323\,t)]u(t)$ mA

10–33 (b) $V_2(s) = \dfrac{R_1 R_2 C_1 C_2 s^2 V_1(s)}{R_1 R_2 C_1 C_2 s^2 + (R_1 C_1 + R_2 C_2 + R_1 C_2)s + 1}$
(c) $v_2(t) = [-3.33\,e^{-250t} + 13.3\,e^{-1000t}]u(t)$ V

10–35 $Z(s) = \dfrac{R_1 R_2 C_1 C_2 s^2 + (R_1 C_1 + R_2 C_2 + R_1 C_2)s + 1}{R_2 C_1 C_2 s^2 + (C_1 + C_2)s}$

10–37 $C_1 = 50$ nF; $C_2 = 200$ nF

10–39 (a) $\Delta(s) = [(RCs)^2 + (3 - \mu)RCs + 1]/R^2$;
(b) $\mu = 3$; $RC = 2 \times 10^{-4}$

10–41 (a) $v_C(t) = [-4.17\,e^{-100t} + 4.17\,e^{20t}]u(t)$ V;
(b) $v_S(t) = [5\,e^{20t}]u(t)$ V;
(c) Circuit is stable.

10–43 (a) Mesh A and B are dependent $I_B(s) = \dfrac{\beta}{\beta - 1}I_A(s);$
(b) $[R_1 + Ls(1 - \beta)^{-1}]I_A(s) - R_1 I_C(s) = V_S(s);$

$-[R_1 + \beta\,R_2(\beta - 1)^{-1}]I_A(s) + \left[R_1 + R_2 + \dfrac{1}{C\,s}\right]I_C(s) = 0$

CHAPTER ELEVEN

11–1 (a) $Z(s) = \dfrac{(Ls + R_2)(R_1 Cs + 1)}{LCs^2 + (R_1 + R_2)Cs + 1}$; $T_V(s) = \dfrac{R_2}{Ls + R_2}$;
(b) $Z(s) = \dfrac{300\,(s + 6000)(s + 833)}{(s + 432)(s + 11570)}$; $T_V(s) = \dfrac{6000}{s + 6000}$

11–3 (a) $Z(s) = \dfrac{(R_1 + R_2)L_1 s + R_1 R_2}{L_1 s + R_2}$; $T_Y(s) = \dfrac{-L_1 s}{(R_1 + R_2)L_1 s + R_1 R_2}$;
(b) $Z(s) = 150\,\dfrac{s + 44.4}{s + 66.7}$; $T_Y(s) = \dfrac{1}{150}\dfrac{-s}{s + 44.4}$

11–5 $T_V(s) = \dfrac{(s + 333 + j471)(s + 333 - j471)}{(s + 333)(s + 1000)}$;

11–7 $C = 0.5\ \mu$F

11–11 $v_2(t) = [2000e^{-20000\,t}]u(t)$ V

11–13 $v_2(t) = \left[\dfrac{1}{2} + \dfrac{1}{\sqrt{2}}\,e^{-5000t}\cos(5000\,t - 225°)\right]u(t)$ V

11–15 $v_2(t) = \delta(t) + [100e^{-200t}]u(t)$ V

11–17 (a) $h(t) = -\delta(t) + [2000e^{-2000t}]u(t);$
(b) $h(t) = [2000e^{-2000t}]u(t);$

11–21 $v_{2SS}(t) = 4.34\cos(500t - 29.7°)$ V; $v_{2SS}(t) = 4.47\cos(1000t - 63.4°)$ V

11–23 $i_{2SS}(t) = 0.707\cos(5000t + 135°)$ mA; $i_{2SS}(t) = 0.894\cos(10000t + 117°)$ mA

11–25 $v_{2SS}(t) = 35.4 \cos(5000t + 45°)$ V; $v_{2SS}(t) = 11.2 \cos(2500t + 63.4°)$ V

11–27 $y_{SS}(t) = 4.47 \cos(2000t + 26.6°)$

11–31 $y(t) = \begin{vmatrix} 0 & t < 1 \\ t - 1 & 1 \le t \le 2 \\ 1 & 2 \le t \end{vmatrix}$

11–33 $y(t) = [1 - e^{-(t-1)}]u(t-1)$

11–35 $y(t) = \left[\dfrac{1}{25} (e^{-5t} - 1) + \dfrac{t}{5} \right]u(t)$

11–37 $f(t) = [2e^{-2t} - e^{-t}]u(t)$

11–41 $y(t) = [5t - 5 \times 10^{-3} (1 - e^{-1000t})]u(t)$

11–42 $y(t) = 0.5 [e^{-100t} - e^{-200t}]u(t)$

11–43 $h(t) = 20 [u(t) - u(t-5)]$

11–44 $y(t) = [(5000t + 5) e^{-1000t}]u(t)$

Chapter Twelve

12–1 $T_V(s) = \dfrac{2000}{s + 4000}$ (a) $|T_V(0)| = 0.5$; $\omega_C = 4$ krad/s; low pass

(c) $T_{SL}(0.5\omega_C) = 0.5$; $T_{SL}(\omega_C) = 0.5$; $T_{SL}(2\omega_C) = 0.25$

12–3 $T_V(s) = \dfrac{7500}{s + 2500}$ (a) $|T_V(0)| = 3$; $\omega_C = 2500$ rad/s; low pass

(c) $\phi(0.5\omega_C) = -31.5°$; $\phi(\omega_C) = -45°$; $\phi(2\omega_C) = -58.5°$;

12–5 $T_V(s) = \dfrac{2s}{s + 500}$ (a) $|T_V(0)| = 0$; $|T_V(\infty)| = 2$; $\omega_C = 500$ rad/s;

(b) high pass

12–7   20 dB @ $\omega = 200$ rad/s ; 18.8 dB @ $\omega = 400$ rad/s ; 12.8 dB @ $\omega = 800$ rad/s

12–11 $T_V(s) = \dfrac{2 \times 10^8}{s^2 + 3 \times 10^4 s + 4 \times 10^8}$

(a) $\omega_0 = 20$ krad/s; $\zeta = 0.75$; $|T_V(0)| = 0.5$; $|T_V(\infty)| = 0$; low pass;
(c) $|T_{SL}(0.5\omega_0)| = 0.5$; $|T_{SL}(\omega_0)| = 0.5$; $|T_{SL}(2\omega_0)| = 0.125$

12–13 (a) $T_V(s) = \dfrac{4000 s}{s^2 + 6000 s + (4000)^2}$;

(b) $\omega_0 = 4$ krad/s; $\zeta = 0.75$; $|T_V(0)| = 0$; $|T_V(\infty)| = 0$; bandpass;
(c) $|T_{SL}(0.5\omega_0)| = 0.5$; $|T_{SL}(\omega_0)| = 1$; $|T_{SL}(2\omega_0)| = 0.5$
$|T_V(0.5\omega_0)| = 0.471$; $|T_V(\omega_0)| = 0.667$; $|T_V(2\omega_0)| = 0.471$

12–14 $T_V(s) = \dfrac{-400 s}{s^2 + 500 s + (200)^2}$

(a) $\omega_0 = 200$ rad/s; $\zeta = 1.25$; $|T_V(0)| = 0$; $|T_V(\infty)| = 0$; bandpass;
(b) $|T_{SL}(0.5\omega_0)| = 1$; $|T_{SL}(\omega_0)| = 2$; $|T_{SL}(2\omega_0)| = 1$

12–17 $T_V(s) = \dfrac{R}{RLC s^2 + L s + R}$

$$\omega_0 = \frac{1}{\sqrt{LC}}; \ \zeta = \frac{\sqrt{L/C}}{2R}; \ |T_V(0)| = 1; \ |T_V(\infty)| = 0; \text{ low pass;}$$

For $L = 0.1$ H; $C = 0.4$ μF and $R = 125$ Ω

12–21 $L = 6.25$ μH; $C = 100$ pF; $Q = 5$; $\omega_{C1} = 36.2$ Mrad/s; $\omega_{C2} = 44.2$ Mrad/s

12–23 (a) $B = 5$ krad/s ; $\omega_{C1} = 497.5$ krad/s; $\omega_{C2} = 502.5$ krad/s;

(b) $|V_C(\omega_0)| = |V_L(\omega_0)| = 1$ kV

12–25 $L = 63.7$ nH; $C = 39.8$ pF

12–27 $L = 0.555$ H; $C = 12.7$ μF

12–31 $T_V(s) = (s + 250)/(s + 1250)$; $A_{SL} = 4$ V; $A_{ACTUAL} = 4.15$ V

12–32 $T_V(s) = (s + 1250)/(s + 250)$; $A_{SL} = 25$ V; $A_{ACTUAL} = 24.1$ V

12–33 low pass; $\omega_C = 1$ rad/s; passband gain = 0 dB

12–35 bandpass; $\omega_{C1} = 10$ rad/s; $\omega_{C1} = 25$ rad/s; passband gain = 40 dB

12–41 $g(t) = 5.10(e^{-100t} - e^{-5000t})$

12–43 $g(t) = 0.316 + 3.56e^{-50t} - 3.87 \ e^{-250t}$

12–44 $T_V(s) = \dfrac{300 \ (s + 40)}{(s + 10)(s + 100)}$; low pass

CHAPTER THIRTEEN

13–1 $a_0 = -V_A/2$; $a_n = 0$; $b_n = -V_A/n\pi$; for all n

13–3 $a_0 = 0$; $a_n = 0$; $b_n = 2 \ V_A/n\pi$; for all n

13–5 $v(t) = 12.7 \sin(2\pi500t) + 4.24 \sin(2\pi1500t) + 2.55 \sin(2\pi2500t) + 1.82 \sin(2\pi3500t)$

13–7 $v(t) = 140 - 93.4 \cos(2\pi100t) - 18.7 \cos(2\pi200t) - 8.0 \sin(2\pi300t) - 4.44 \sin(2\pi400t)$

13–9 $f_0 = 10^5$ Hz, $g(t) = 10 + 6.37 \sin(2\pi f_0 \ t) + 2.12 \sin(2\pi3f_0 \ t) + 1.27 \sin(2\pi5f_0 \ t)$

13–17 $v_O(t) = 14.5 \cos(1000t - 117°) + 0.999 \cos(3000t + 34°) + 0.241 \cos(5000t - 168°) + 0.091 \cos(7000t + 16°)$ V

13–19 $i(t) = 127 \cos(2500t - 264°) + 63.6 \cos(5000t - 267°) + 42.4 \cos(7500t - 268°) + 31.8 \cos(10000t - 269°)$ μA

13–21 The high-pass filter reduces the fundamental more than the harmonics.

$n\omega_0$	INPUT AMPLITUDE	OUTPUT AMPLITUDE
50	19.1	4.63
150	6.37	3.82
250	3.82	2.98
350	2.73	2.37

13–27 $P = 8.25$ W; $V_{rms} = 20.31$ V

13–29 $V_{rms} = V_A/\sqrt{6}$; $P = V_A^2/6R$;   91.3%

13–31 $Y_{rms} = 19.1$

13–33   First three terms give $V_{rms} = 1.165 V_A$; First four terms give $V_{rms} = 1.178 V_A$;
       Both are within 5% of true value of $V_{rms} = 1.224 V_A$

CHAPTER FOURTEEN

14–1 $T(s) = \dfrac{19.95}{s/9425 + 1}$

14–3 $T(s) = \dfrac{K}{s/\omega_C + 1}$; Many solutions with $K > 3.15$ and $\omega_C < 1.26$ Mrad/s

14–5 $T(s) = \dfrac{K}{s/\omega_C + 1}$; Many solutions with $K = 5$ and $\omega_C < 2.51$ krad/s

14–7 (a) $T(s) = 10\left[\dfrac{s/200 + 1}{s/20 + 1}\right]$

14–24 $\omega_0 = 6275$ rad/s: $\zeta = 0.05$

14–25 $\zeta = 0.354$

14–28   Select C_1 then $C_2 = 2\zeta C_1$, $R_1 = (\omega_0 C_1)^{-1}$ and $R_2 = (\omega_0 C_2)^{-1}$

14–30   Select R then $C_1 = 2\zeta/\omega_0 R$ and $C_2 = 1/2\zeta\omega_0 R$

14–31 $T(s) = \dfrac{10}{(s/2299 + 1)^4}$

14–32 $T(s) = \dfrac{10}{[s/500 + 1][(s/500)^2 + s/500 + 1]}$

14–34 (a) First-order cascade $n = 4$; (b) Butterworth $n = 3$; Chebychev $n = 2$

14–35 $T(s) = \left(\dfrac{1.334}{s/3448 + 1}\right)^4$

14–36 $T(s) = \dfrac{\sqrt{10}}{[s/300 + 1][(s/300)^2 + s/300 + 1]}$

14–38   Butterworth with $n = 3$, $T_{MAX} = 1$, and $\omega_C = 2\pi 1000$ rad/s

14–43 (a) First-order cascade $n = 24$; (b) Butterworth $n = 4$; Chebychev $n = 3$

14–48 $T(s) = \dfrac{s^2}{s^2 + 17770\,s + (12570)^2}$

CHAPTER FIFTEEN

15–1 $\mathbf{V}_1 = 177 + j177$ V; $\mathbf{V}_2 = 150 - j100$ V;
       $v_1(t) + v_2(t) = 336\cos(\omega t + 13.2°)$ V

15-3 (a) $v_1(t) = 10 \cos(10^4 t - 30°)$ V;
  (b) $v_2(t) = 60 \cos(10^4 t - 220°)$ V;
  (c) $i_1(t) = 5 \cos(200t + 90°)$ A;
  (d) $i_2(t) = 2 \cos(200t + 270°)$ A

15-5 $412 \cos(20t + 104°)$ V/s

15-7 $v_2(t) = 26.9 \cos(500t + 113°)$ V

15-11 (a) $Z_{EQ} = 25 + j20 = 32.0 \angle 38.7°$ Ω;
  (b) $Z_{EQ} = 20 - j10 = 22.4 \angle -26.6°$ Ω;
  (c) $Z_{EQ} = 15.5 - j0.118 = 15.5 \angle -0.434°$ Ω;
  (d) $Z_{EQ} = 6.59 - j10.4 = 12.3 \angle -57.6°$ Ω;

15-13 (a) $Z_{EQ} = j150 = 150 \angle 90°$ Ω;
  (b) $Z_{EQ} = 30 + j30 = 42.4 \angle 45°$ Ω;
  (c) $Z_{EQ} = 30 + j180 = 183 \angle 80.5°$ Ω;
  (d) $Z_{EQ} = 12 + j24 = 26.8 \angle 63.4°$ Ω;

15-15 (a) $Z_{EQ} = 5 - j8.66$ kΩ;
  (b) $i(t) = 15 \cos(1000t + 150°)$ mA

15-17 $L = 2.68$ mH; $C = 44.7$ nF

15-21 $i(t) = 98.1 \cos(500t - 11.3°)$ mA

15-23 $i_R(t) = 212 \cos(2000t - 45°)$ mA;
  $i_C(t) = 212 \cos(2000t + 45°)$ mA

15-25 $i_R(t) = 10 \cos(1000t + 45°)$ mA;
  $i_C(t) = 10 \cos(1000t + 135°)$ mA

15-27 $Z = 162 - j27.0$ Ω; $v_x(t) = 116 \cos(1000t + 35.5°)$ V

15-29 $\mathbf{I_x} = 0.278 - j0.712$ A

15-31 $v_x(t) = 12.0 \cos(2000t - 41.4°) + 7.90 \cos(4000t - 150°)$ V

15-33 $Z = 10.3 - j16.7$ kΩ; $\mathbf{V_x} = 34.1 - j25.1$ V

15-41 $v_x(t) = 121.2 \cos(4000t - 84.1°)$ V

15-43 $Z_{IN} = 9897 + j101$ Ω; $K = j1.031$

15-45 $\mathbf{V_y} = 4.71 - j3.82$ V; $\mathbf{I_x} = -2.35 - j0.588$ mA

15-47 $Z_{IN} = 21.3 - j9.71$ kΩ; $K = 0.847 - j0.0281$

15-51 $\mathbf{V_L} = 35 + j0$ V; $\mathbf{I_x} = 0.35 - j0.35$ A; $P_L = 6.125$ W

15-53 $R = 1125$ Ω; $C = 0.863$ μF

15-55 (a) $P_L = 144$ mW;
  (b) $P_{MAX} = 450$ mW;
  (c) $Z_L = 1 - j2$ kΩ

CHAPTER SIXTEEN

16-1 (a) $v_1(t) = 10 \times 10^{-3} \dfrac{di_1(t)}{dt} + 7 \times 10^{-3} \dfrac{di_2(t)}{dt}$

  $v_2(t) = 7 \times 10^{-3} \dfrac{di_1(t)}{dt} + 5 \times 10^{-3} \dfrac{di_2(t)}{dt}$

  (b) $v_2(t) = 7 \sin(2000t)$ V

16–3 (a) See 16–1 (a) above.

(b) $v_S(t) = -2.5 \sin (500t)$ V

16–5 $v_1(t) = 1.25 \cos (1000t)$ V; $i_1(t) = 0.75 \sin (1000t)$ A

16–7 $v_x(t) = -76 \sin (1000t)$ V

16–11 (a) $n = 4$; (b) $v_1(t) = 120 \cos (2\pi60t)$ V; $i_1(t) = 2.4 \cos (2\pi60t)$ A

16–13 $R_{EQ} = 115 \ \Omega$

16–15 $i_S(t) = -16 \sin (400t)$ A; $i_L(t) = 8 \sin (400t)$ A

16–17 $n = 2.29$

16–21 $\mathbf{V}_2 = 10.2 + j12.3 = 16.0 \angle 50.3° $ V; $Z_{IN} = 54.0 + j29.2 \ \Omega$

16–23 $i(t) = 20.1 \cos (20000t - 161.2°)$ mA

16–25 $\mathbf{V}_1 = 44.3 - j34.7$ V; $\mathbf{V}_L = 13.2 - j5.57$ V; $Z_{IN} = 40.1 + j131 \ \Omega$

16–27 $Z_{IN} = 2.5 + j1.5$ kΩ; $i(t) = 3.43 \cos (4000t - 31.0°)$ mA; $P_{IN} = 14.7$ mW

16–29 $\mathbf{V}_1 = -29.4 - j9.34$ V; $\mathbf{V}_2 = -147 - j46.7$ V

16–31 $L_1 = 0.332$ H; $L_2 = 3.32$ H; $M = 0.729$ H;

CHAPTER SEVENTEEN

17–1 (a) $P = -1159$ W; $Q = 311$ VAR: pf $= -0.966$; delivering;

(b) $P = 364.7$ W; $Q = 2068$ VAR: pf $= 0.174$; absorbing

17–3 (a) $P = 60.37$ W; $Q = 16.2$ VAR: pf $= 0.966$; absorbing;

(b) $P = 1299$ W; $Q = -750$ VAR: pf $= 0.866$; absorbing

17–5 (a) pf $= 0.954$; lagging; (b) pf $= 0.661$; leading

17–7 (a) $P = 15$ kW; $Q = 13.2$ kVAR; $I_{rms} = 8.33$ A; (b) $Z_L = 216 + j190 \ \Omega$

17–9   pf $= 0.894$; $Z_L = 33.3 + j16.7 \ \Omega$

17–11 $S = 576 + j1019$ VA

17–13 $R_L = 19.4$ kΩ; $C_L = 0.164 \ \mu$F

17–15 (a) $I_{rms} = 3.11$ A; (b) $S_L = 581 - j388$ VA; $\eta = 93.75\%$

17–17 $\eta = 94.74\%$

17–21 $S_S = 25.6 + j17.9$ kVA; $\eta = 97.65\%$

17–22 $\mathbf{V}_S = 2.6 + j0.154$ kV; $pf_S = 0.71$

17–25 $\mathbf{V}_L = 266 + j76.2$ V; $\mathbf{V}_S = 294 + j215$ V

17–31 (a) $\mathbf{V}_{AN} = 277 \angle 0°$ V; $\mathbf{V}_{BN} = 277 \angle -120°$ V; $\mathbf{V}_{CN} = 277 \angle -240°$ V;

$\mathbf{V}_{AB} = 480 \angle 30°$ V; $\mathbf{V}_{BC} = 480 \angle -90°$ V; $\mathbf{V}_{CA} = 480 \angle 150°$ V

17–33 (a) $\mathbf{I}_A = 6.06 \angle 0°$ A; $\mathbf{I}_B = 6.06 \angle 120°$ A; $\mathbf{I}_C = 6.06 \angle -120°$ A;

$\mathbf{I}_1 = 3.5 \angle -30°$ A; $\mathbf{I}_2 = 3.5 \angle 90°$ A; $\mathbf{I}_3 = 3.5 \angle -150°$ A

17–35 (a) $\mathbf{I}_A = 42.8 - j6.5$ A; $\mathbf{I}_B = -15.7 + j40.3$ A; $\mathbf{I}_C = -27 - j33.8$ A;

(b) $S_L = 28.1 + j22.5$ kVA

17–37 $|\mathbf{I}_L| = 91.2$ A; pf $= 0.923$

17–39 (a) $|\mathbf{I_L}| = 36.1$ A; (b) $Z = 5.76 + j5.08$ Ω

17–41 (a) $|\mathbf{I_L}| = 7.22$ A; (b) $Z = 432 + j381$ Ω

APPENDIX W1

W1–1 $\quad F(\omega) = \dfrac{A}{j\omega}\left(1 - e^{-j\omega}\right)$

W1–3 $\quad F(\omega) = A\left[-4\,\pi\,\dfrac{\cos(\omega)}{4\omega^2 - \pi^2}\right]$

W1–5 $\quad f(t) = 10\left[\dfrac{\cos(t) - 1}{t}\right]$

W1–7 (a) $f_1(t) = 20\left[e^{-20t} - e^{-40t}\right]u(t)$; (b) $f_2(t) = \left[-e^{-20t} + 2\,e^{-40t}\right]u(t)$

W1–13 (a) $F_1(\omega) = \dfrac{2j\omega + 4}{-\omega^2 + 4j\omega + 20}$; (b) $F_2(\omega) = \dfrac{j\omega + 6}{(j\omega + 2)(j\omega + 4)}$

W1–15 (a) $f_1(t) = 2 + 2\cos(2t)$; (b) $f_2(t) = 2 + \operatorname{sgn}(t)$; (c) $f_3(t) = 2u(t)$

W1–17 (a) $f_1(t) = 2 + \operatorname{sgn}(t - 2)$; (b) $f_2(t) = (2e^{-2(t-2)})u(t - 2)$:
(c) $f_3(t) = \operatorname{sgn}(t + 2) + \operatorname{sgn}(t - 2)$

W1–19 $\quad g(t) = 0.25\,(1 - 2e^{-2t} + e^{-4t})u(t)$

W1–27 $\quad v_2(t) = 8[e^{5t}\,u(-t) + e^{-20t}\,u(t)]$

W1–29 $\quad v_2(t) = 10\,e^{-100t}\,u(t)$

W1–31 $\quad y(t) = 0.5[u(-t) + e^{-2t}\,u(t)]$

W1–33 $\quad y(t) = 4\,e^{-t}\,u(t) - \operatorname{sgn}(t)$

W1–35 $\quad$ System is an ideal high-pass filter with $\omega_C = \pm\beta$.

W1–41 $\quad W_{1\Omega} = A^2/2\alpha$

W1–43 $\quad W_{1\Omega} = A^2/4\alpha$; $W_\alpha = 0.182\,W_{1\Omega}$

APPENDIX W2

W2–1 $\quad z_{11} = 400\ \Omega$; $z_{12} = z_{21} = 100\ \Omega$; $z_{22} = 167\ \Omega$

W2–2 $\quad y_{11} = 2.94$ mS; $y_{12} = y_{21} = -1.77$ mS; $y_{22} = 7.06$ mS

W2–7 $\quad I_1 = 18$ mA; $I_2 = -12$ mA

W2–8 $\quad T_V = -4.17$

W2–9 $\quad V_2 = 9$ V; $I_1 = 30$ mA; $I_2 = -6$ mA

W2–10 $\quad Y_{IN} = 25 + j20$ mS

W2–13 $\quad h_{11} = 0$; $h_{12} = 1$; $h_{21} = -1$; $h_{22} = Y$

W2–15 $\quad h_{11} = R_1 + R_2$; $h_{12} = 0$; $h_{21} = \beta$; $h_{22} = G_3$

W2–18 $\quad A = 1$; $B = R_2$; $C = G_1$; $D = R_2\,G_1$

W2–19 $\quad V_T = 6$ V; $R_T = 250\ \Omega$

W2–21 (a) $R_{IN} = 800\ \Omega$; (b) $R_{IN} = 400\ \Omega$; (c) $R_{IN} = 600\ \Omega$

INDEX

ac (alternating current) signal, 234
Active circuit, 141
Active device, 141
   OP AMP, 161
   transistor, 156
Active *RC* filter, 616
Admittance, 401, 684
   parallel connection, 406, 684
   transfer, 454
Aliasing, 606
Ammeter, 46
Ampere (unit), 5
Amplifier:
   differential, 175
   noninverting, 164
   summing, 172
   voltage follower, 169
Amplitude, 211, 215, 220
Amplitude spectrum, 591, W-1
Analog-to-Digital Converter, 195, 267
Angular frequency, 222
Apparent power, 755
Autotransformer, 736
Average power, 710, 752
   of a periodic signal, 602
Average value, 234

Balanced three-phase, 769
   Y-Δ connection, 774
   Y-Y connection, 772

Bandlimited signal, 606
Bandpass filter, 507
   design, 654
   with first-order circuits, 522
   narrowband, 532
   second order, 527, 630
   wideband, 532
Bandstop filter, 507
   design, 654
   with first-order circuits, 523
Bandwidth, 507, 532, 606
Bilateral element, 16
Bode diagrams, 509, 548
   complex poles and zeros, 558
   using Mathcad, 555
   using MATLAB, 563
Bridge circuits:
   impedance, 682
   Maxwell, 683
   Wheatstone, 191
Bus, 778
Butterworth:
   high-pass response, 650
   low-pass response, 637
   polynomial, 640

Capacitance, 248
Capacitive reactance, 680
Capacitor, 247
   average power, 710

Capacitor (*cont.*)
   dc response, 250
   energy, 249
   impedance, 401, 677
   i–v relationship, 248, 249
   parallel connection, 268
   series connection, 269
   state variable, 250
Cascade connection, 460
Cascade design, 616
Causal waveform, 232, 353
Center frequency, 530
Chain rule, 460
Characteristic equation:
   first order, 282
   second order, 320
Charge, 5
Chebychev:
   low-pass response, 642
   polynomials, 644
Circuit, 21
Circuit analysis, 3
   computer aided, 52
   phasor domain, 675
   s domain, 395
   resistive, 28
Circuit design, 3
   evaluation, 3, 492, 646
   filters, 616
   interface circuits, 117
   network functions, 483
   using Mathcad, 125
   using MATLAB, 434
   verification, 493, 652
   with OP AMPs, 183
Circuit determinant, 423
Circuit reduction, 46
Circuit simulation, 52
Circuit theorems:
   maximum power transfer, 115, 712
   proportionality, 94, 409, 690
   superposition, 98, 410, 691
   Thévenin/Norton, 101, 415, 695
Clock, 237, 495
   detection, 296
   duty cycle, 238
   edges, 238
   frequency, 238
   skew, 495
   waveform, 238

Comparator, 193
Complex frequency, 325
Complex numbers, A-12
   Arithmetic operations, A-14
   conjugate, A-13
   exponential form, A-13
   imaginary part, A-12
   real part, A-12
Complex power, 754
   and load impedance, 756
   conservation of, 758
Computer tools, 52
   Electronics Workbench, 181, 336, 526, 652
   Excel, 216, 308, 424, 600
   MATLAB, 89, 364, 433, 563, 763
   Mathcad, 125, 258, 555, 585
   Orcad Capture, 53, 153, 286, 694
Conductance, 16, 684
Connection constraints, 20
   phasor, 675
   s domain, 396
Convolution, 464, 476, W-22
   graphical interpretation, 480
   integral, 476, W-23
   properties of, 482
Corner frequency, 512
Coulomb (unit), 5
Coupled inductors, 723
   energy, 729
   i–v relationships, 725
Coupling coefficient, 730
Cramer's rule, A-3
Critically damped, 326
Current, 5
   mesh, 84
   short circuit, 13
Current division, 43, 406, 684
Current gain, 142
Current source, 19, 397
   dependent, 142
   Norton equivalent, 101, 415, 695
Current transfer function, 454
Cutoff frequency, 507, 633

Δ-to-Y transformation, 775

Damped ramp, 229
Damped sine, 229
Damping, 326
Damping ratio, 333

d'Arsonval meter, 46
dc (direct current) signal, 209
Decade, 509
Decibel, 509
Delay time, 469
Dependent sources, 141, 397
Determinant, A-3
Device, 15
Differential amplifier, 175
Differential equation:
    first order, 280
    second order, 318, 329
    solution by Laplace transforms, 376
Differentiator, 263
Digital-to-analog converter, 186
Dirichlet conditions, 583, W-2
Dot convention, 726
Driving point impedance, 452
Duality, 260, 281
Duty cycle, 238

Electronics Workbench, 181, 336, 526, 652
Element, 15
Energy, 5
    capacitor, 249
    coupled inductors, 729
    inductor, 256
Energy spectral density, W-28
En route objectives, 2
Equivalent circuits, 33
    capacitors, 268
    dc, 270
    inductors, 269
    resistance, 33, 34
    sources, 36
    summary of, 39
    Thévenin/Norton, 101, 415, 695
Euler's relationship, A-13
Evaluation, 3
Even symmetry, 593
Excel, 216, 308, 424, 600
Exponential waveform, 215
    double, 231
    properties of, 216

Farad (unit), 248
Filter, 615
Filter design, 616
Final conditions, 298

Final-value property, 384
First-order circuits, 281
    design with, 616
    differential equation, 280
    frequency response, 511
    RC and RL circuits, 279
    sinusoidal response, 304
    step response, 289
    zero-input response, 281, 296
    zero-state response, 296
Forced pole, 423
Forced response, 289, 436
Fourier series, 582
    alternative form, 591
    coefficients, 584
    in circuit analysis, 596
    table of, 589
Fourier transforms, W-1
    and Laplace transforms, W-7
    in circuit analysis, W-18
    inverse, W-5
    table of pairs, W-17
    table of properties, W-18
    uniqueness property, W-2
    Parseval's theorem, W-27
Frequency:
    angular, 222
    center, 530
    complex, 325
    corner, 512
    critical, 362
    cutoff, 507, 633
    fundamental, 231, 582
    half power, 507
    harmonic, 231, 582
    natural, 325
    radian, 325
    resonant, 687
Frequency response, 506
    and step response, 565
    Bode diagrams, 509, 548
    descriptors, 507
    first order, 511
    RLC circuits, 541
    second order, 527
Fundamental frequency, 231, 582

Gain-bandwidth product, 516
Gain function, 506
Ground, 9

Half-power frequency, 507
Half-wave symmetry, 594
Harmonic frequency, 231, 582
Henry (unit), 255
Hertz (unit), 222
High-pass filter, 508
    design, 649
    first order, 518
    second order, 539, 627

Ideal models:
    current source, 18, 397
    low-pass filter, W-25
    OP AMP, 164
    switch, 17
    voltage source, 18, 397
Ideal transformer, 731
    i–v characteristics, 733
    input resistance, 735
    turns ratio, 732
Impedance, 400, 677
    driving point, 452
    equivalent, 405, 680
    input, 453
    magnitude scaling, 489
    matching, 713
    transfer, 454
Impulse, 212
Impulse response, 463
    and convolution, 476, W-22
    from transfer function, 464
    from step response, 467
Inductance, 255
    mutual, 725
    self, 725
Inductive reactance, 680
Inductor, 254
    average power, 710
    dc response, 257
    energy, 256
    impedance, 401, 677
    i–v relationship, 255, 256
    parallel connection, 269
    series connection, 269
    state variable, 257
Initial conditions, 298
Initial-value property, 384
Input, 93
Input impedance, 453
Instantaneous power, 752

three phase, 781
Instrumentation systems, 189
Integrator, 261
Integrodifferential equation, 319, 380
Interface, 2
Interface circuit, 118
Inverting amplifier, 170
Inverter:
    CMOS, 302
    Digital, 122
Inverse transform:
    Fourier, W-5
    Laplace, 352, 365

Joule (unit), 6

Kirchhoff's laws, 20
    current (KCL), 21
    phasor domain, 675
    s domain, 396
    voltage (KVL), 23

Ladder circuit, 46
Lagging power factor, 756
Laplace transforms, 351
    inverse, 352, 365
    table of pairs, 360
    table of properties, 359
    uniqueness property, 353
    using MATLAB, 364
Leading power factor, 756
Linear circuit, 93
Linear element, 16
Line current, 772
Line spectrum, 591
Line voltage, 769
Load flow, 758
    single phase, 761
    three phase, 777
    using MATLAB, 763
Load line, 112
Loading, 462
Lookback resistance, 109
Loop, 21
Low-pass filter, 509
    Butterworth, 637
    Chebychev, 642
    design, 632
    first order, 511

first-order cascade, 634
ideal, W-25
second order, 535, 625

Magnitude scaling, 489
Mathcad, 125, 258, 555, 585
MATLAB, 89, 364, 433, 563, 763
Matrix, A-5
    adjoint, A-6
    and linear equations, A-8
    inverse, A-7
Matrix algebra, A-6
Maximum signal transfer, 114, 712
Mesh current, 84
Mesh-current analysis, 84
    by inspection, 87
    fundamental property, 85
    phasor domain, 700
    s domain, 428
    summary of, 93
    using MATLAB, 89, 707
    with current sources, 89, 429
    with dependent sources, 150
Mutual inductance, 725

Natural frequency, 325
Natural pole, 423
Natural response, 289, 436
Netlist, 55
Network function, 409, 449
    design of, 483
    driving-point impedance, 452
    transfer functions, 453
Node, 21
Node voltage, 70
Node-voltage analysis, 69
    by inspection, 73
    fundamental property, 70
    phasor domain, 700
    s domain, 420
    summary of, 83
    with dependent sources, 146
    with OP AMPs, 179
    with voltage sources, 78
Noninverting amplifier, 164
Nonlinear element, 16, 111
Norton equivalent circuit, 101
    phasor domain, 695
    s domain, 415
Nyquist rate, 606

Octave, 509
Odd symmetry, 594
Ohm (unit), 16
Ohm's law, 16
OP AMP, 161
    dependent source model, 163
    effect of finite gain, 166
    gain–bandwidth product, 516
    ideal model, 163
    in the s domain, 398
    notation, 161
    operating modes, 162
OP AMP circuits:
    bandpass filter, 630
    differential amplifier, 175
    differentiator, 263
    high-pass filter, 627
    integrator, 261
    inverting amplifier, 170
    low-pass filter, 625
    noninverting amplifier, 164
    noninverting summer, 185
    second order, 339
    subtractor, 175
    summary of, 177, 262
    summing amplifier, 172
    voltage follower, 169
Open circuit, 17
Open-circuit voltage, 103
Operational amplifier. see OP AMP
Orcad Capture, 53, 153, 286, 694
Output, 93
Overdamped, 326
Overshoot, 469

Parallel connection, 25
    admittances, 406, 684
    capacitors, 268
    current division in, 43, 406, 684
    inductors, 269
    resistors, 34
Parseval's theorem, W-26
Partial fraction expansion, 366
    of improper rational functions, 371
    with complex poles, 368
    with multiple poles, 373
Passband, 507
Passband gain, 633
Passive circuit, 438
Passive sign convention, 8

Peak to peak value, 233
Peak value, 233
Period, 220
Periodic waveform, 232
    rms value of, 235, 602
Phase angle, 220
Phase current, 775
Phase function, 506
Phase sequence, 770
Phase spectrum, 591, W-1
Phase voltage, 769
Phasor, 669
    diagram, 670
    domain, 679
    properties, 671
    rotating, 671
Phasor circuit analysis, 675
    current division, 684
    mesh current, 700
    node voltage, 700
    parallel equivalence, 679
    proportionality, 690
    series equivalence, 680
    superposition, 691
    Thévenin/Norton, 695
    using MATLAB, 706
    voltage division, 680
Planar circuit, 84
Pole, 362
    forced, 423
    multiple, 373
    observable, 431
    natural, 423
    number of, 438
    simple, 366
    stable, 437
Pole-zero diagram, 362
Port, 117, W-37
Potentiometer, 42
Power, 6
    apparent, 755
    average, 602, 710, 752
    complex, 754
    maximum transfer, 115, 712
    reactive, 752
Power factor, 755
Power factor angle, 755
Power factor correction, 762
Power triangle, 755
Primary winding, 731

Proportionality, 94, 409, 690
PSpice, 53

Quality factor (Q), 542

Radian frequency, 222
Ramp, 213
Rational function, 366
RC and *RL* circuits, 279
Reactance, 680
Reactive power, 752
Residue, 366
Resistance, 16, 680
    lookback, 109
    standard values, A-1
Resistor, 15
    average power, 711
    impedance, 401, 677
    i–v relationship, 16
    power, 16
Resonance, 529, 687
Resonant frequency, 687
Response:
    ac steady-state, 306, 669
    dc steady-state, 306, 466
    critically damped, 326
    forced, 289, 436
    frequency, 506
    impulse, 453
    natural, 289, 436
    overdamped, 326
    sinusoidal steady-state, 306, 471, 669
    steady-state, 306, 450
    step, 289, 333, 466, 565
    underdamped, 326
    with periodic inputs, 596
    zero input, 281, 319, 411
    zero state, 295, 411
Rise time, 469
RLC circuits, 317
    bandpass, 541
    bandstop, 546
    design, 490
    critically damped, 326
    overdamped, 326
    step response, 335
    underdamped, 326
    zero-input response, 319
Root-mean-square (rms) value, 235, 602
Rotating phasor, 671

s domain, 350
s plane, 363
Sample-and-hold, 253
Sampling, 605
   aliasing, 606
   Nyquist rate, 606
   theorem, 605
Secondary winding, 731
Second-order circuits, 317
   design with, 624
   frequency response, 527
   OP AMP, 339
   parallel *RLC*, 328
   series *RLC*, 317
   step response, 333
   zero-input response, 319
Self-inductance, 725
Series connection, 26
   capacitors, 269
   impedances, 405, 680
   inductors, 269
   resistors, 33
   voltage division in, 40, 406, 680
Settling time, 568
Short circuit, 17
Short-circuit current, 103
Siemens (unit), 16
Signal, 29
   ac, 234
   bandlimited, 606
   conditioning, 189
   dc, 209
   composite, 227
   exponential, 215
   impulse, 212
   ramp, 213
   sampling, 604
   signum, 227
   sinusoidal, 219
   step function, 211
Single phase, 767
Sinusoidal steady-state, 306, 669
   from transfer function, 471
Sinusoidal waveform, 219
   damped, 229
   period, 220
   properties of, 224
   phasor representation, 670
Source:
   current, 19, 397

   dependent, 142
   equivalent, 36
   practical, 20
   Thévenin/Norton, 101, 415, 695
   three phase, 769
   voltage, 18, 397
Source transformation, 37
   phasor domain, 696
   s domain, 415
Spectrum, 584, W-1
Stable, 436
   Asymptotic, 464
   Marginally, 437
State variable, 250, 256, 280
Steady-state response, 306, 450
Step function, 211
Step response:
   descriptors, 469
   first-order circuit, 289
   from frequency response, 565
   from impulse response, 467
   from transfer function, 466
   second-order circuit, 333
   using MATLAB, 470
Stopband, 507
Summing amplifier, 172
Super capacitor, 277, 315
Supermesh, 90
Supernode, 79
Superposition, 98, 410, 691
Susceptance, 684
Switch, 17
   analog, 18
   ideal, 17

Thévenin equivalent circuit, 101
   phasor domain, 695
   s domain, 415
   with dependent sources, 154
   with nonlinear loads, 111
Three phase power analysis, 772
   load flow, 777
   single-line diagram, 778
   Y- Δ connection, 774
   Y-Y connection, 772
Time constant, 215
   RC circuit, 282
   RL circuit, 283
Time shift, 211
Touch-tone telephone, 614

Transconductance, 142
Transducer, 189
Transfer admittance, 454
Transfer function, 453
    dividers, 454
    Fourier domain, W-24
    inverting amplifier, 455
    noninverting amplifier, 455
Transfer impedance, 454
Transformer, 731
    autotransformer, 736
    equivalent circuit, 742
    ideal, 733
    sinusoidal steady-state, 737
    turns ratio, 732
Transform pair, 350
Transistor, 151, 153, 156
    dependent source model, 157
    operating modes, 157
Transresistance, 142
Turns ratio, 732
Two-port network, 117, W-37
    connection, W-50
    reciprocal, W-39
Two-port parameters, W-38
    admittance, W-41
    conversions, W-49
    hybrid, W-43
    impedance, W-39
    transmission, W-46
Undamped natural frequency, 333
Underdamped, 326

Unit output method, 95, 690
Unit impulse, 212
Unit ramp, 213
Units, table of, 4

VA (unit), 755
VARs (unit), 754
Volt (unit), 6
Voltage, 6
    line, 769
    node, 70
    open circuit, 103
    phase, 764
    threshold, 156
Voltage division, 40, 406, 680
Voltage follower, 169
Voltage gain, 142
Voltage source, 18
    dependent, 142
    Thévenin equivalent, 101, 415, 695
Voltage transfer function, 454

Watt (unit), 6
Waveform, 209
Waveform bandwidth, 606
Waveform symmetries, 593
Weber (unit), 255
Wheatstone bridge, 191

Zero, 362
Zero–input response, 281, 319, 411
Zero–state response, 295, 411

BASIC LAPLACE TRANSFORMATION PROPERTIES

PROPERTIES	TIME DOMAIN	FREQUENCY DOMAIN
Independent Variable	t	s
Signal Representation	$f(t)$	$F(s)$
Uniqueness	$\mathscr{L}^{-1}\{F(s)\}(=)[f(t)]u(t)$	$\mathscr{L}\{f(t)\} = F(s)$
Linearity	$Af_1(t) + Bf_2(t)$	$AF_1(s) + BF_2(s)$
Integration	$\int_0^t f(\tau)d\tau$	$\dfrac{F(s)}{s}$
Differentiation	$\dfrac{df(t)}{dt}$	$sF(s) - f(0-)$
	$\dfrac{d^2 f(t)}{dt^2}$	$s^2 F(s) - sf(0-) - f'(0-)$
	$\dfrac{d^3 f(t)}{dt^3}$	$s^3 F(s) - s^2 f(0-) - sf'(0-) - f''(0-)$
t-Translation	$[f(t-a)]u(t-a)$	$e^{-as} F(s)$
s-Translation	$e^{-\alpha t} f(t)$	$F(s + \alpha)$
Scaling	$f(at)$	$\dfrac{1}{a} F\left(\dfrac{s}{a}\right)$
Final Value	$\lim\limits_{t \to \infty} f(t)$	$\lim\limits_{s \to 0} sF(s)$
Initial Value	$\lim\limits_{t \to 0+} f(t)$	$\lim\limits_{s \to \infty} sF(s)$

The Analysis and Design of Linear Circuits: Laplace Early

150601

BASIC OP AMP MODULES

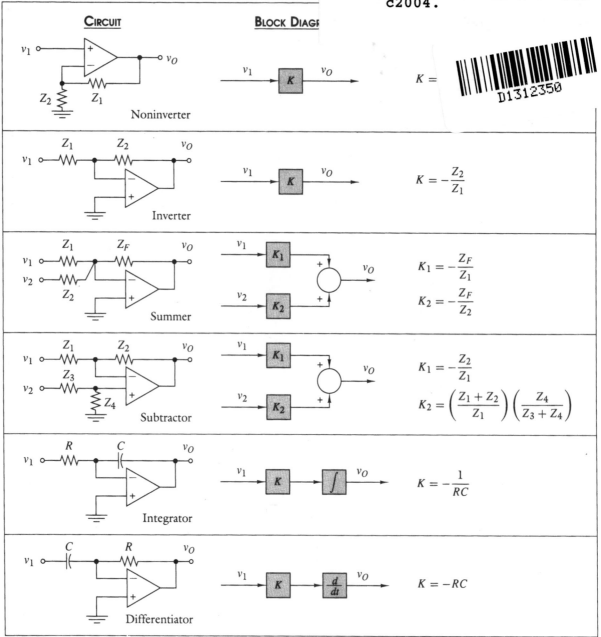

$$K =$$

$$K = -\frac{Z_2}{Z_1}$$

$$K_1 = -\frac{Z_F}{Z_1}$$

$$K_2 = -\frac{Z_F}{Z_2}$$

$$K_1 = -\frac{Z_2}{Z_1}$$

$$K_2 = \left(\frac{Z_1 + Z_2}{Z_1}\right)\left(\frac{Z_4}{Z_3 + Z_4}\right)$$

$$K = -\frac{1}{RC}$$

$$K = -RC$$

Figures and Charts

Figure 2.1. Variations of Čoček Rhythmic Patterns   29
Chart 9.1. Intersecting Circles: Chalga, Wedding Music,
      Romani Music   179

Acknowledgments

This book has benefited from several decades of assistance from many individuals, families, communities, institutions, and granting agencies. I would like to acknowledge funding from the International Research and Exchanges Board, the Open Society Institute, the National Council for Eurasian and East European Research, and the National Endowment for the Humanities. At the University of Oregon, I was supported by a Summer Research Grant and grants from the Oregon Humanities Center and the Center for the Study of Women in Society. I would also like to thank the Institut za Folklor "Marko Cepenkov" in Skopje, Macedonia, for serving as my academic home in 1990.

Above all, I owe tremendous gratitude to the Romani community members who generously hosted me in Macedonia, Bulgaria, Toronto, Melbourne, and New York and who invited me to their homes and guided me through cultural events. They include Yuri and Lidia Yunakov; Pera, Kjani, Binas, Rafet, Erhan, Sevgul, and Shengul Redžeposki; Zada, Zekir, Ferhan, Selviana, Rechko, Cindy, Redžep, Perijan, Šeman, Nuri, Zejnep, Idris, Gjulfa, Zulfikjar, and Bajramša Ismail; Afrodita Salievska and her family; Sonya and Jašar Jašaroski; Sadet, Seido, and Sanela Mamudoski; Mevlude, Sazija and Ferat Arifovi; Seido, Nimet, and Isa Salifoski; Lahorka and Ali Jašar; Tair, Selfija, Sabuhan, Severdžan, and Turkijana Azirovi; Mirka and Firus Redžeposki; Nešo Ajvazi; Ismail Lumanovski; Erhan, Gjulče, Husamedin, Mikrema, Jusuf, Sevim, Turan, and Uska Umer; Esengul Edipova; Muren and Ajten Ibraimovi; Sevim, Nurije, and Sal Mamudoski; Sebihana and David Neziroski; Imer and Gjula Sulemanoski; Mizka, Ruse, and Bajram Amzoski; Sebihana, Kaimet, and Šeno Ademoski; Ajša Sefuloska and Ferdi Memedoski; Virgil and Dalip Asanovi; Gjulten and Šaban Dervisoski; Perijana and Nedžat Useinoski; Ramiz Islami; Romeo and Kurte Kurtali; Kujtim and Muamed Ismaili; Ilmi and Bisa Teraski; Sevda and Marem Bajramovski; Ali, Muzo, Kenedi, and Altan Zekiroski; Memet Dželoski; Trajče Džemaloski; Šani Rifati; Abdula Durak and his extended family; Esma Redžepova, Stevo Teodosievski, and Simeon Atanasovski; Zahir Ramadanov; Sami Zekiroski; Mustafa Gjuneš; Trajko and Sabo Petrovski; Muharem Serbezovski; Muzafer and Altan Mahmut; Gjulizar Dželjadin; Bajsa Arifovska; Adžerka and Sukri Arifoski; Snezhana Gocheva and her extended family; Yashko Argirov; Yordan and Vera Kenderov; Ivo Papazov and Maria Karafezieva; Neshko Neshev; Salif Ali; Dobri and Matyo Dobrev; Hristo Kyuchukov; Mihail and Dimitŭr Georgiev; and Anzhelo Malikov.

In Bulgaria and Macedonia, I was also graciously hosted by Petŭr Ralchev, Ivan Milev, Georgi Yanev, Ahmed Yunakov, Vergiili and Nadya Atanasov, Radost Ivanova, Aleksandar and Olga Džukeski, Vladimir and Olivera Cvetkovski, and many other friends and colleagues. I would also like to thank the many families who invited me to their family celebrations.

For help with translation, I owe thanks to Afrodita Salievska, Šani and Dževrija Rifati, Dušan Ristić, Zada and Ferhan Ismail, Rachel MacFarlane, and especially Victor Friedman, who read the entire manuscript. I would like to acknowledge editing assistance from Angela Montague, from the staff of Oxford University Press, and from series editor Mark Slobin.

I have greatly benefited over the years from fruitful intellectual exchanges with an inspiring group of colleagues, among them Jane Sugarman, Amy Shuman, Donna Buchanan, Gail Kligman, Steven Feld, Timothy Rice, Svanibor Pettan, Martin Stokes, Judith Okely, Brana Mijatovič, Margaret Beissinger, Elsie Dunin, Victor Friedman, Regina Bendix, Dorothy Noyes, Mark Slobin, Sonia Seeman, Michael Beckerman, Petra Gelbart, and Mirjana Lausević. Scholars from the Balkans, including Lozanka Peycheva, Ventsislav Dimov, Elena Marushiakova, Vesselin Popov, Trajko Petrovski, Claire Levy, Radost Ivanova, Vergiili Atanasov, Tsenka Iordanova, and Speranta Radulescu, all generously shared their ideas with me. Ventsislav Dimov and Lozanka Peycheva helped greatly with permissions in Bulgaria.

I would especially like to express my appreciation to Seido Salifoski, Šani Rifati, Afrodita Salievska, Kalin Kirilov, Mark Levy, Garth Cartwright, and Nick Nasev for their helpful comments on specific chapters. Francis Fung, Traci Lindsey, Henry Ernst, Helmut Neumann, Victor Friedman, Rumen Shopov, Šani Rifati, and Villie Shumanov helped with musical materials. Kalin Kirilov was a valuable video editor. Ian Hancock generously provided encouragement in difficult moments. My heartfelt thanks go to Jane Sugarman for reading the entire manuscript and offering many insightful suggestions. I would also like to thank the anonymous readers from Oxford University Press for their detailed comments.

In the course of researching and writing, several community members and colleagues passed away; they are sorely missed and will be deeply remembered in my heart. Finally, I owe much gratitude to my family, Mark and Nesa Levy, for their unwavering moral support.

The title of this book, *Romani Routes,* was inspired by the NGO (nongovernmental organization) Voice of Roma and its Romani Routes touring program (www.voiceofroma.com/culture/romani-routes.shtml). I would like to thank the officers of VOR for their permission to use the phrase.

Portions of Chapter 6 were reprinted from "Transnational Chochek: Gender and the Politics of Balkan Romani Dance," in *Balkan Dance:*

Essays on Characteristics, Performance, and Teaching, ed A. Shay, 2008, with permission from McFarland Publishers. Portions of Chapters 7 and 8 were revised from "Bulgarian Wedding Music Between Folk and Chalga: Politics, Markets, and Current Directions," in *Musicology* 7(2007): 69–97 with permission of the Serbian Academy of Sciences, Institute of Musicology. Portions of Chapter 10 were revised from "The Gender of the Profession: Music, Dance and Reputation Among Balkan Muslim Romani (Gypsy) Women," in *Gender and Music in the Mediterranean*, ed. Tullia Magrini, 2003,with permission from the University of Chicago Press. Portions of Chapter 11 were reprinted from "Music and Transnational Identity: The Life of Romani Saxophonist Yuri Yunakov," in *Džaniben* (Czech Journal of Romani Studies), Winter 2009: 59–84, with permission of the publisher. Portions of Chapter 12 were reprinted from "Trafficking in the Exotic with Gypsy Music," in *Balkan Popular Culture and the Ottoman Ecumeme*, ed. D. Buchanan, 2007, with permission from Scarecrow Press, a division of Rowman & Littlefield Publishing Group.

This book is dedicated to the memory of my parents, Evelyn and Larry Silverman, who eagerly followed my research and live in my heart.

Notes on Transliteration

This book deals with four Balkan languages—Romani, Bulgarian, Macedonian, and Serbian (Bosnian/Croatian/Serbian)—each with its own conventions of transliteration. Bulgarian, Macedonian, and Serbian belong to the family of South Slavic languages. Romani is an Indo-Aryan language related to other languages spoken in northern India; its orthography differs from country to country. Bulgarian and Macedonian are written in Cyrillic; Bosnian/Croatian/Serbian may be written in either the Cyrillic or Latin alphabet. A few Turkish words are also used in this book.

I have provided translations of all foreign words; when these words are not Romani I have indicated the language used, as in *"Južni Ekspres* (Southern express [Macedonian])." If no language is indicated, the language is Romani, as in *"Bijav Geljum te Bašalav* (I went to a wedding to play)." I note that many words in Macedonian, Bulgarian, and Romani are derived from Turkish; for example, the Macedonian and Romani term *bovčalok* (gifts for the groom sewn on a sheet) comes from the Turkish *bohça*. However, there are too many words from Turkish for me to indicate this connection in this book.

In effort to make it easier for the reader, I have modified existing transliteration practices according to this system:

Romani, Serbian, and Macedonian words are written with Bosnian/Croatian/Serbian Latin orthography and pronounced as follows:

> a = as in art
> e = as in met
> i = as in machine
> o = as in port
> u = as in lunar
> ' = short u (schwa) as in but
> š= sh as in shop
> ž = zh as s in pleasure
> c = ts as in hats
> č = ch as in change
> ć = ch as t in nature

dž = dzh as j in jazz
j = y as in yes

Bulgarian words are written according to one widely accepted scholarly system with h's; for example:

The sound sh as in shop is written sh
dzh as j in jazz, etc.
ŭ = short u (schwa) as in but

In Romani, four aspirated consonants are written as čh, ph, kh, and th

Most Turkish words used follow standard Turkish orthography, for example, *Laço Tayfa,* where ç is ch.

Because I am dealing with two transliteration systems and several cultural systems, the same word or concept may appear two ways; for example *čoček* (Romani, Serbian, and Macedonian) and *kyucheck* (Bulgarian); *surla* or *zurla* (Macedonian) and *zurna* (Bulgarian).

For previously published materials and for names already transliterated with Latin letters, I have retained the previous forms. There will thus inevitably be some inconsistencies in the text.

About the Companion Website

www.oup.com/us/romaniroutes

The author and Oxford have created a password-protected website to accompany this book. The website contains video examples, audio examples, photographs, and text supplements (including song texts and historical information). Users may access the website with the username Music1 and password Book5983.

All examples were used with duplication permission or are fair use. When not indicated, the source is the author.

The website materials are explained below.

CHAPTER 1

Photographs

1.1 Working-class home, Šutka, Skopje, Macedonia, 1994
1.2 Poor home, Šutka, 1994
1.3 Arthur Ave. street, 2009
1.4 Arthur Ave. market, 2009
1.5 Arthur Ave. burek and pizza store, 2009

CHAPTER 2

Photographs

2.1 Zurla and tapan, Šutka, 1990
2.2 Bear trainer, Bulgaria, 1980
2.3 Monkey trainer, Bulgaria, 1980
2.4 Ferus Mustafov plays at a celebration for the birth of Muamet Čun's granddaughter, Šutka, 1990

Video Examples

2.1 Zurla and tapan, men's heavy crossing dance, wedding, Šutka, 1990
2.2 Zurna player Samir Kurtov, wedding, Gotse Delchev, Bulgaria, 2004
2.3 Ferus Mustafov plays at a party for the birth of a girl, Šutka, 1990
2.4 Ferus Mustafov, Macedonian TV show, 2/4 čoček, Belly Dances, Maestro (YU Video 5046, 1985)
2.5 Džipsi Aver, Stara Zagora Romfest, 2005
2.6 Amza Tairov performing at a wedding, 2003

Audio Examples

2.1 Shalvar Kyuchek, Nezhniya Tsumani (The Delicate Tidal Wave), Fekata, Filip Simeonov (Crystal Records, 2005)
2.2 E Sitakoro Oro, Ora i Čočeci, Euro Čoček 2002, Titanik (Voice of Roma)
2.3 Bijav Geljum me Bašalav, Orkestŭr Knezha: Ko Džamije Me Bešav, Iliya Marinov (Lazarov Records, 1990s) with text supplement
2.4 Me Phirava, Orkestŭr Sever: O Dzhumaya, Dancho Panov (Folkton, 1990s)
2.5 Mirveta, music Ilir Karimani, Dae (Mother), Safet Ibrahimi (Chrom 002, 1990s) with text supplement
2.6 Džansever, Astargja o Horo, 1990s recording, with text supplement
2.7 Džansever, Astargja o Horo, Gypsy Queens (Frankfurt, Germany, Network 32.843, 1999)
2.8 Džansever, Astargja o Horo with Kristali at wedding in Bujanovac, Serbia, 2002 with text supplement
2.9 Čita performs Germanija, Germany (live) on Čita (Milena Records MR 200513-2, 2005), with text supplement

CHAPTER 3

Photographs

3.1 Romani flag
3.2 Esma Redžepova singing Dželem Dželem to American Kalderash Roma, private party sponsored by Macedonian Roma, New York City, 1996

Video Examples

3.1 Muharem Serbezovski, performing Ramajana at a private Romani New Year's Eve party, with Bilhan Mačev (clarinet), Trajče

Džemaloski (keyboard), Ilhan Rahmanovski (guitar), Kujtim Ismaili (bass), and Severdžan Azirov (drums), Yonkers, New York, 1997

3.2 Esma Redžepova performing Dželem Dželem at Šutkafest, accompanied by Stevo Teodosievski, 1993

3.3 Esma Redžepova singing Dželem Dželem (Serbian text) to American Kalderash Roma, private party sponsored by Macedonian Roma, New York, 1996

3.4 Esma Redžepova performing Dželem Dželem at the Macedonian church in Garfield, New Jersey, 2004

Audio Examples

3.1 Celo Dive Mangasa (All Day We Beg), Ciganske Pemse Pevaju, Muharem Čizmoli (Beograd Disk EBD 0207, 1970s), with text supplement

3.2 Ramajama, Muharem Serbezovski with Medo Čun (RTB EP 11 191, 1970s), with text supplement

3.3 Stranci (Strangers), Holivud, Zvonko Demirovič with Južni Vetar, 1990s, with text supplement

3.4 Kemano Bašal (The violin plays), Bašal Kemano/ Violino Sviri, Džansever, music Ferus Mustafov/Ahmed Rasimov; text Neždet Mustafa; arrangement Ahmed Rasimov (Sokoj MP 21102, 1992), with text supplement

Text Supplements (not attached to other media)

3.1 Dželem Dželem
3.2 Cosmopolitanism

CHAPTER 4

Photographs

4.1 Album cover, Amanet, Rome (sic) Songs, 1995
4.2 Musa Mosque and Islamic Center in Belmont, 2009
4.3 Wedding sheet displayed in tray, Šutka, 1990
4.4 Groom with gifts dances with the bride's mother, Šutka, 1990

Video examples

4.1 Bride's mother pinning gifts on the groom, blaga rakija, Šutka, 1990
4.2 Macedonian Romani woman in Melbourne, Australia, sends greetings in Romani to her relatives in New York, 1998

Audio Examples

4.1 Gurbeti, sung by Ferhan Ismail, Rome [sic] Songs with Amanet: Ramiz Islami (clarinet), Erhan Umer (keyboard), Ilhan Rahmanovski (guitar), Seido Salifoski (dumbek), New York, 1990s, with text supplement

4.2 Gurbetluko sung by Ramadan Bislim (Ramko), Najšužo Kilibari (Ramko Produkcija, 1990s), with text supplement

4.3 To Phurano Bunari, Abas Muzafer on Alen; Adžiker te Bajrovav, n.d., with text supplement

Text Supplement (not attached to other media)

4.1 Education and Gender

CHAPTER 5

Photographs

5.1 Women making stuffed grape leaves, Šutka wedding, 1994
5.2 Women making breads, circumcision party, Šutka, 1990
5.3 Men preparing meat for a wedding, Šutka, 1994
5.4 Women and girls in fancy šalvari, dance line, Šutka wedding, 1990
5.5 Gifts on trays, Šutka wedding, 1990
5.6 Gifts on trays, Šutka wedding, 1994
5.7 Bride led out of her house for henna ceremony wearing tel, silver streamers, Šutka, 1990
5.8 Bride wearing tel, silver streamers, Šutka, 1990
5.9 Bride wearing tel, silver streamers, Šutka, 1990
5.10 Bride's female relatives dance at henna ceremony, Šutka, 1990
5.11 Preparing the sieve with greenery, a red scarf, and popcorn, Šutka, 1994
5.12 Fancy outfits of girls, Šutka wedding, 1990
5.13 Leading the dance line with a decorated sieve, Šutka wedding, 1990
5.14 Leading the dance line with a decorated sieve, Šutka wedding, 1990
5.15 Leading the dance line with a decorated sieve, Šutka wedding, 1994
5.16 Leading the dance line with a decorated sieve, Šutka wedding, 1994
5.17 Groom's female relatives with a tray of henna with candle for the second henna ceremony, Šutka, 1990
5.18 Bride with henna on her hands and feet, second henna ceremony, Šutka, 1990
5.19 Bride's male relative leads her to groom's family, holding her head down, Šutka, 1994
5.20 Bride gazing downwards, Šutka, 1994
5.21 Grooms' mother leads the dance line at double wedding, New York, 1995

5.22   Muharem Serbezovski, private Romani New Year's Eve party, Bilhan Mačev (clarinet), Trajče Džemaloski (keyboard), Ilhan Rahmanovski (guitar), Kujtim Ismaili (bass), and Severdžan Azirov (drums), Yonkers, New York, 1997

5.23   Ramiz Islami, circumcision party, New York, 1988

5.24   Album cover, Ramiz Islami and Grupi Sazet E Ohrit, New York, 1995

5.25   Ismail Lumanovski performing with Ilhan Rahmanovski and Šaban Dervisoski, Maia Meyhane, New York, 2006

5.26   Seido Salifoski, circumcision, New York, 1988

Video Examples

5.1   Henna ceremony: groom's female relatives arriving with trays of gifts, Šutka, 1990

5.2   Bride's father plays zurla at her henna ceremony, Šutka, 1990

5.3   Groom's female relative announces gifts for the bride, henna ceremony, Šutka, 1990

5.4   Bride led out for henna application, Šutka, 1990

5.5   Applying henna to the bride's hair, Šutka, 1990

5.6   Dancing in the street, henna ceremony, Šutka, 1990

5.7   Bride led out from the bath to greet groom's female relatives; she kisses their hands, Šutka, 1990

5.8   Young girl dances as her female relatives instruct her, Šutka wedding, 1990

5.9   Bride's female relatives lead the dance line with a decorated sieve, Šutka, 1990

5.10   Line dance at igranka, with solo čoček dancers in the center, Šutka, 1990

5.11   Father of a gifted young female dancer beckons her to dance solo čoček at the front of the line, Šutka wedding, 1990

5.12   Afet Dude, dance in 9/8 with solo dancing in the center, Šutka wedding, 1990

5.13   Elder woman leads slow line dance, Šutka wedding, 1990

5.14   Elder woman leads crossing line dance, Šutka wedding,1990

5.15   Feta's band on stage; young boys play makeshift drums under the stage, Šutka wedding, 1990

5.16   Night henna ceremony for bride's hands and feet, Šutka, 1990

5.17   Banquet sponsored by the bride's relatives, Šutka, 1990

5.18   Muamet Čun, clarinet, solo dancer, Šutka wedding,1990

5.19   Young solo dancer, Šutka wedding, 1990

5.20   Bride dances sadly with her family before she is transferred to the groom's family, Šutka, 1990

5.21   Bride is led by her brother from her family to the groom's family, Šutka, 1990

5.22   Bride's relatives arrive for the džumaluk, village near Šutka, 1990

5.23 Bride dances with her relatives at the džumaluk, village near Šutka, 1990

5.24 Bride kisses her aunt, mother, and father as they give her money at the džumaluk, village near Šutka, 1990

5.25 Bride leading dance line, henna party, New York, 2004

5.26 Bride led out for henna ceremony, New York, 2004, with Oj Borije text supplement

5.27 Henna ceremony, New York, 2004

5.28 Bridal couple emerging from the mosque, New York, 2004

5.29 Dancing in front of the mosque, New York wedding, 2004, with Zapevala Sojka Ptica text supplement

5.30 Bride led out of her house, takes temana, New York wedding, 1995

5.31 Groom's parents bargain for the bride with bride's parents, New York wedding, 1995

5.32 Bride takes temana, New York, 1995

5.33 Bride's incorporation rituals, New York, 1995

5.34 Grooms' mother leads the first dance line, New York wedding, 1995

5.35 Grooms' mother leads; bride's mother leads a crossing dance, New York wedding, 1995

5.36 Ramiz Islami (clarinet), his son Romeo (clarinet), Erhan Umer (synthesizer and vocals) and his father Husamedin (drum set and vocals), Trajče Džemaloski (synthesizer), Kujtim Ismaili (guitar), New York wedding, 1995

5.37 Ramiz Islami's band playing a 9/8 dance, New York wedding, 1995

5.38 Husamedin Umer playing tapan for line dance in 7/8, New York wedding, 1995

5.39 Bride's parents lead the dance line, New York, 2004

5.40 Ramiz's band at blaga rakija, New York, 1995

5.41 Groom and bride receive gifts at blaga rakija, New York, 1995

5.42 Men lead line čoček at blaga rakija, New York, 1995

5.43 Bride's mother lead line dance Beranče in 12/8, blaga rakija, New York, 1995

5.44 Men lead slow 7/8 line dance, blaga rakija, New York, 1995

5.45 Groom leads slow 2/4 line dance, blaga rakija, New York, 1995

5.46 Slow 2/4 line dance speeds up, blaga rakija, New York, 1995

5.47 Muharem Serbezovski, private Romani New Year's Eve party, with Bilhan Mačev (clarinet), Trajče Džemaloski (keyboard), Ilhan Rahmanovski (guitar), Kujtim Ismaili (bass), and Severdžan Azirov (drums), Yonkers, New York, 1997

5.48 Šadan Sakip sings Geljan Dade accompanied by Bilhan Mačev; dancers do Jeni Jol, Romani party, New York, 1996

5.49 Šadan Sakip sings, accompanied by Bilhan Mačev, Romani party, New York, 1996

5.50 Solo čoček, Romani party, New York, 1996

5.51 Erhan Umer (keyboard and vocals), Yuri Yunakov (saxophone), Sal Mamudoski (clarinet), Rumen Sali Shopov (drums), California Herdeljezi festival, 2008

5.52　Erhan Umer (keyboard and vocals), Rumen Sali Shopov (tambura, tapan and vocals), performing Red Bul, with Yuri Yunakov (saxophone) and Seido Salifoski (dumbek), who then takes a solo with Rumen, California Herdeljezi festival, 2008

5.53　Uska Umer sings with Amanet: Erhan Umer (keyboard and vocals), Turan Umer (guitar), Sevim Umer (drums), and Ismail Lumanovski (clarinet), Herdeljezi festival, California, 2007

5.54　Uska Umer sings Red Bul with Amanet: Erhan Umer (keyboard and vocals), Turan Umer (guitar), Sevim Umer (drums), Muren Ibraimov (dumbek), and Ismail Lumanovski (clarinet), Herdeljezi festival, California, 2007

5.55　Džengis Rahmanovski (dumbek), Ilhan Rahmanovksi (guitar), Seido Salifoski (dumbek), Šaban Dervisoski (accordion) and Ismail Lumanovski (clarinet), Maia Meyhane, New York, 2006

5.56　Sal Mamudoski (clarinet), Yuri Yunakov (saxophone), and Alfred Popaj (keyboard), Hungaria House, New York, 2007

5.57　Sal Mamudoski (clarinet), Yuri Yunakov (saxophone), and Alfred Popaj (keyboard) performing Kjuperlika, Hungaria House, New York, 2007

5.58　Menderes Azirov leading Čačak, Yuri Yunakov (saxophone), wedding, New York, 1996

5.59　Nešo Ajvazi performs talava, accompanied by Seido Salifoski (dumbek), Toni Jankuloski (keyboard), and Ismail Lumanovski (clarinet), Balkan Music and Dance Workshop, Iroquois Springs, New York, 2005

5.60　Nešo Ajvazi sings Red Bul, accompanied by Seido Salifoski (dumbek), Toni Jankuloski (keyboard), and Ismail Lumanovski (clarinet), Balkan Music and Dance Workshop, Iroquois Springs, New York, 2005

5.61　Ismail Lumanovski (clarinet) performs Gaida, Seido Salifoski (dumbek) and Toni Jankuloski (keyboard), Balkan Music and Dance Workshop, Iroquois Springs, New York, 2005

5.62　Ismail Lumanovski (clarinet) performs an improvisatory čoček, Seido Salifoski (dumbek) and Toni Jankuloski (keyboard), Balkan Music and Dance Workshop, Iroquois Springs, New York, 2005

5.63　Seido Salifoski (dumbek) performs an improvisatory solo, Ismail Lumanovski (clarinet) and Toni Jankuloski (keyboard), Balkan Music and Dance Workshop, Iroquois Springs, New York, 2005

Audio Example

5.1　Muharem Serbezovski, Gilaven Romalen, Zaljubih Se, Ans. Crni Dijamanti (Diskoton DTK 9430, 1987), with text supplement

CHAPTER 6

Photographs

6.1 Solo dancer on top of a car, circumcision procession, Šutka, 1990
6.2 Solo dancer receives tips, Bulgaria, 1984
6.3 Frula ensemble, Tsigane, 1986

Video Examples

6.1 Solo čoček dancers (one getting tips), Orkestŭr Orfei, Sofia, 1994
6.2 Male and female solo čoček dancers in the middle of the line, celebration for the birth of a girl, Šutka, 1990
6.3 Solo čoček dancer, recorded music, family gathering, Šutka, 1990
6.4 7/8 line dance, Yuri Yunakov (saxophone), Hasan Isakut (kanun), Trajče Džemaloski (keyboard), Kujtim Ismaili (guitar), Severdžan Azirov (drums), wedding, New York, 1996
6.5 Beranče (12/8) danced at celebration for the birth of a girl, Šutka, 1990
6.6 Crossing dance, celebration for the birth of a girl, Šutka, 1990
6.7 Opening, Šutkafest 1993: Esma, dumbek players, Bitolska Gaida and solo čoček, Šutkafest 1993 (MRT Sokom 1994)

Audio example

6.1 Romani Čhaj Sijum, Džansever, Kemano Bašal/Violino Sviri, text Džansever, music and arrangement Ferus Mustafov (Sokoj 21102,1992), with text supplement

Text Supplement (not attached to other media)

6.1 History of Romani Dance

CHAPTER 7

Photographs

7.1 Zurna players, Pomak wedding, Avramovo, 1980
7.2 Tŭpan players, Pomak wedding, Avramovo, 1980
7.3 Dancing, zurna/tŭpan, field above Pirin Pee festival, 1985
7.4 Mancho Kamburov, Pirin Pee, 1985
7.5 Ivo Papazov playing saxophone and clarinet, 1980s
7.6 Ivo Papazov and Ali Garzhev (accordion), wedding procession, Iskra, 1980

7.7 Musicians' market, Sofia, 1984

7.8 Anzhelo Malikov playing cimbalom at a Hungarian restaurant, Sofia, 1984

7.9 View of Stambolovo festival, 1988, cover of journal *Bŭlgarska Muzika*, 1989

Video Examples

7.1 Ivo Papazov removing pieces of his clarinet and playing saxophone and clarinet simultaneously, Bulgarian National Television, 1987

7.2 Nedyalka Keranova sings, Karadzha Duma Rusanke with Akademikus, Zvezdite na Trakiya Folk 1994 (Payner 96001, 1995)

7.3 Ivo Papazov (clarinet) and Yuri Yunakov (saxophone) improvise a pravo horo (2/4), Bulgarian National Television, 1987

7.4 Neshko Neshev (accordion) and Ivo Papazov (clarinet) improvise a rŭchenitsa (7/16), Bulgarian National Television, 1987

7.5 Kyuchek in 2/4, Ivo Papazov (clarinet), Neshko Neshev (accordion), Radi Kazakov (guitar), Vasil Denev (keyboard), Salif Ali (drums), Matyo Dobrev (kaval), and Ahmed Yunakov (saxophone), wedding banquet, Thrace, 1994

7.6 Filips kyuchek (9/8), Ivo Papazov (clarinet), Neshko Neshev (accordion), Radi Kazakov (guitar), Vasil Denev (keyboard), Salif Ali (drums), Matyo Dobrev (kaval), and Ahmed Yunakov (saxophone), wedding banquet, Thrace, 1994

7.7 Improvisation by Ahmed Yunakov (saxophone), with Trakiya, wedding banquet, Thrace, 1994

7.8 Improvisations by Ivo Papazov (clarinet), Neshko Neshev (accordion), Yuri Yunakov (saxophone) in pravo horo (2/4), with Salif Ali (drums) and Kalin Kirilov (guitar), Seattle, 2005

7.9 Filip Simeonov improvising rŭchenitsa, Orkestŭr Trŭstenik, Zvezdite na Trakiya Folk 1994 (Payner 96001, 1995)

Audio Examples

7.1 Trakiya, Kŭrdzhaliisko Horo, wedding, Iskra, Bulgaria 1980

7.2 Trakiya, Kŭrdzhaliisko Horo, arranged by Dimitŭr Trifonov (Balkanton BHA 11330), 1970s

Text Supplement (not attached to other media)

7.1 Heritage, Nationalism, and Socialism

7.2 Stambolovo Festivals during Bulgarian Socialism

CHAPTER 8

Photographs

8.1 Publicity shot, Trakiya, Ivo Papazov (clarinet), Neshko Neshev (accordion), Maria Karafezieva (vocals), Yuri Yunakov (saxophone), Radi Kazakov (guitar), Salif Ali (drums), Ryko, 1990
8.2 Romani music concert, Sofia circus arena, 1990

Video Examples

8.1 Orkestŭr Kanari: Nie Bŭlgarite, Kanari 25 Godini, Horovodna Broenitsa, Plovdiv amphitheater, opening (Payner, 2000)
8.2 Gloria sings Moma v Zandani with Orkestŭr Kanari: Nie Bŭlgarite, Kanari 25 Godini (Payner, 2000)

CHAPTER 9

Video Examples

9.1 Toni Dacheva and Orkestŭr Kristal perform Chudesen Sŭn (Vsichko e Lyubov, Payner, 1998), with text supplement
9.2 Toni Dacheva and Orkestŭr Kristal perform Svadba (Vsichko e Lyubov, Payner, 1998), with text supplement
9.3 Amet, Belgiiski Vecheri/Dzhamovete (Payner DVD Collection 5, 2004), with text supplement
9.4 Emiliya, Zabravi! Hitove na Planeta Payner 3 (Payner, 2005), with text supplement
9.5 Ballads MegaMix by DJ Jerry (Payner DVD Collection 5, 2004)
9.6 Antigeroi, Azis: The Best Video Clips (Sunny, 2004)
9.7 Nyama, Azis: The Best Video Clips (Sunny, 2004), with text supplement
9.8 Azis and Sofi Marinova, Edin Zhivot Ne Stiga, Azis: The Best Video Clips (Sunny, 2004), with text supplement

Audio Examples

9.1 Sladka Rabota, Toni Dacheva i Orkestŭr Kristal, Vsichko e Lyubov (Payner, 1998), with text supplement
9.2 Bŭlgarina v Evropa, Magiya: Orkestŭr Kristal s Mariana Kalcheva (Payner, 2001), with text supplement
9.3 Danyova Mama, Sofi Marinova: Studen Plamŭk (Ara 266, n.d.), with text supplement

CHAPTER 10

Photographs

10.1 Esma Redžepova Ansambl Steve Teodosievskog RTB EP 12725, early 1970s
10.2 Esma Redžepova, Jugoton EPY 3736, early 1970s
10.3 Esma Redžepova Ansambl Stevo Teodosievski, late 1970s
10.4 Esma Redžepova in Slovenian, Croatian, and Bosnian costumes, late 1970s
10.5 Esma Redžepova in Romani, Indian, and Spanish costumes, late 1970s
10.6 Esma Redžepova in modern clothing, late 1970s
10.7 Esma Redžepova Ansambl Stevo Teodosievski publicity shot, late 1970s

Video Examples

10.1 Bašal Seljadin, Putevima Pesme Esma Ansambl Teodosievski (MP 31005) 1988, with text supplement
10.2 Hajri Mate (sic), Putevima Pesme Esma Ansambl Teodosievski (MP 31005) 1988, with text supplement
10.3 Čhaje Šukarije, Putevima Pesme Esma Ansambl Teodosievski (from Zapej Makedonija 1968, MP 31004) 1979, with text supplement
10.4 Ciganski Čoček, Putevima Pesme Esma Ansambl Teodosievski (from Zapej Makedonija 1968, MP 31004, 1979
10.5 Čhaje Šukarije, Esma Ansambl Teodosievski, 1965, Austrian Public Broadcasting, rebroadcast on Macedonian National Television, with text supplement
10.6 Romano Horo, Esma Ansambl Teodosievski, 1965, Austrian Public Broadcasting, rebroadcast on Macedonian National Television, with text supplement
10.7 Šadan Sakip, tarabuka, Ibro Demir "Kec", vocals Lenorije Čhaj, Putevima Pesma Esma Ansambl Teodosievski, MP 31004, 1979, with text supplement
10.8 Čini (Magija), Toše Proeski and Esma Redžepova, 2002, with text supplement

Audio Example

10.1 Bašal Seljadin, Anka Gieva and Dragica Mavrovska, Jugoton SY 1090, 1960s, with text supplement

Text Supplement (not attached to other media)

10.1 Female Singers and Sexuality in Historical Perspective

CHAPTER 11

Photographs

11.1 Yuri Yunakov (saxophone), Sunaj Saraçi (violin), Severdžan Azirov (drums), Ilhan Rahmanovski (guitar), Kujtim Ismaili (guitar), Trajče Džemaloski (keyboard), wedding, New York, 1997
11.2 Publicity photograph of Yuri Yunakov and Ivo Papazov, 2005, courtesy Traditional Crossroads

Video Example

11.1 Yuri Yunakov (saxophone), Hasan Isakut (kanun), Trajče Džemaloski (keyboard), Kujtim Ismaili (guitar), Severdžan Azirov (drums), wedding, New York, 1996

CHAPTER 12

Photographs

12.1 Poster advertising the 1999 Gypsy Caravan from Bass Hall, Fort Worth, Texas
12.2 Cover of CD Band of Gypsies: Taraf de Haidouks (Nonesuch 79641, 2001)

Video Example

12.1 Finale, Gypsy Caravan: A Celebration of Roma Music and Dance, 1999, filmed by Jasmine Delall

CHAPTER 13

Video Example

13.1 Godzila, Alyosha and Orkestŭr Kristali, Folk Kasino 3 (Payner, 2005), with text supplement

Audio Examples

13.1 Godzila, Jony Iliev and Fanfare Ciocarlia on Gili Garabdi Ancient Secrets of Gypsy Brass (Asphalt Tango ATR 0605, 2005), with text supplement

13.2 Godzila, Jony Iliev and Band, Ma Maren Ma (Asphalt Tango ATR 0102, 2002), with text supplement

13.3 Godzila, Alyosha, Džansever and Orkestŭr Kristali, wedding, Bujanovac, Serbia, 2002, with text supplement

13.4 Lake Bul, Ajgara, wedding, Šutka, 2000, with text supplement

13.5 Red Bul, Džansever and Orkestŭr Kristali, wedding, Bujanovac, Serbia, 2002

13.6 Red Bula, Mahala Rai Banda (CRAW 31 Crammed Discs 2004)

13.7 Red Bula, Balkan Beat Box vs. Mahala Rai Banda, Electric Gypsyland 2 (CRAW 37 Crammed Discs 2008)

Text Supplement (not attached to other media)

13.1 Herdeljezi song text

Romani Routes

PART I
INTRODUCTION

1

Balkan Roma

History, Politics, and Performance

In the last fifteen years, as the fusion music terms Gypsy[1] Punk and Balkan Beats have proliferated and Gypsy motifs in clothing have become fashionable, Gypsy music has become a staple at world music festivals and dance clubs in the United States and Western Europe.[2] Moreover, Gypsy style seems to be simultaneously familiar and exotic. Many consumers profess to know who and what Gypsies are, and what Gypsy music is. Some audience members repeat stereotypical generalizations drawing on a plethora of written, visual, and oral formulations from the last few centuries: Gypsies are innately talented, artistic, embodying their wildness in their music; they are consummate musical technicians; they magically sense the desires of their patrons; but in the end, they can't be trusted. Indeed, the fictional Gypsy musician is a ubiquitous exotic fantasy figure in Western literature, art, and oral tradition (Trumpener 1992; Van de Port 1998).

How does music mediate between these poles of fascination and rejection? Since the fall of socialism in 1989, thousands of Roma have emigrated westward because of deteriorating living conditions in Eastern Europe; as a result, fear of "Gypsy hordes" and entrenched stereotypes of thievery and trickery are being revived. In this heightened atmosphere of xenophobia, Roma are paradoxically revered as musicians and reviled as people. Underlying this phenomenon are the dichotomous emotions of fear and admiration.

Two contrasting phenomena encapsulate the dichotomy of how most North Americans and Europeans think about Roma: the warning about Gypsy beggars in European cities, and the craze for Gypsy music in American and West European clubs. When Madonna performed a fusion of East European Romani (Gypsy) music on her summer 2009 tour, she epitomized how celebrity patrons appropriate the music of marginal groups. But when she was booed by 60,000 Romanian fans after she bemoaned the plight of Gypsies, she further exposed the dichotomy that Roma, loved for their music, are hated as people.

3

Romani Routes deliberately positions the recent popularity of Gypsy music alongside the recent refugee flow of Eastern European Roma westward, contrasting the discrimination faced by the majority of Roma with the new commercial ventures of a small group of successful Romani musicians. I further contrast both the poverty-stricken majority and a few rich musicians with the Balkan Romani community in New York City, dating from the 1960s, where working-class refugees and immigrants toil for a better life for their children while cultivating music as a vital communicative link. The placement of this book in Oxford's American Music-spheres series reflects its ethnographic grounding in the United States while underlining the connections that American Balkan Roma have with both Eastern Europe and world music markets.

The book combines a transnational approach with an ethnography of community life in relation to music. My community-based fieldwork focuses on two diasporic Macedonian Romani communities: Belmont, located in the Bronx, New York; and Šuto Orizari (known as Šutka), located outside of Skopje, with comparative materials from several Bulgarian Romani communities. In Šutka and Belmont (and in most Balkan Romani communities), music and dance are emblematic of Romani identity and embedded in numerous and elaborate ritual displays. Weddings are the main focus of families, and marriage is a transnational public event, often negotiated over long distances. Music is the vehicle for enacting social relationships and enhancing status. It is also a commodity to sell to non-Roma and other Roma.

Situating music in relation to individuals, communities, states, policy, and world music markets, I confound the simplistic assumption that music starts out "pure" or "authentic" in bounded communities and becomes hybrid only when it moves to non-Romani markets. I show how innovation, hybridity, and market forces all operate within communities, and between communities in the diaspora, and how Romani musicians move among these sites. I also examine how hybridity is recast in transnational sites and commercial venues by managers and producers. Furthermore, I confound the assumption that music starts out as noncommercial in Romani contexts and becomes commercial for the world music market. Balkan Romani musicians have been professionals for hundreds of years, marketing their product and tailoring their performances to Romani as well as non-Romani patrons. The interplay among economic necessity, marginalization, identity formation, and symbolic display via music is the subject of this book.

A performance framework highlights the dramatic and processual quality of music, of discourse about music, and of identity making. Following Bauman (1975), Hymes (1975), Abrahams (1977) and Goffman (1974) I define *performance* as a marked mode of communication with specific generic features signaled by various "frames." Performers assume responsibility to display communicative competence and to be judged by audiences (Bauman 1975:293).[3] Kapchan underscores that performance "not only fabricates meanings in highly condensed symbols . . . but comments on those meanings, interpreting them for the larger community and often critiquing and subverting them as well" (1996b:480).[4] This book, then, discusses the multiple meanings of Romani music that are interpreted processually through performance by various actors, including musicians, their varied audiences, their communities, their marketers, and state and local officials.

Whereas most folklorists have used a performance framework to study bounded events, some scholars have fruitfully expanded the concept to embrace identity construction and gender management. Kirshenblatt-Gimblett's phrase "the political economy of showing" (1998) is useful in reminding us that performances of identity are always embedded in hierarchies of power and class. For Roma, displaying or hiding one's Romani identity is both historically informed and negotiated on the spot. Particularly revealing of identity management are debates about musical authenticity that take place among Romani musicians, non-Romani audiences, managers and marketers, and scholars. Like Povinelli (2002), I investigate the challenges Roma face in inhabiting various "spaces of recognition," such as community member, authentic musician, world music star, European minority, American minority, and activist.

Judith Butler's work on gender performativity (1990, 1993) can help to frame representational issues among Roma. Butler claims that people dramatically perform conventions of maleness and femaleness according to implicit heteronormativity. Gender parody, such as drag, may be transgressive, but it also cites and may even reinscribe gender norms.[5] Similarly, when Roma play the part of Gypsy musicians, that is, deliver the stereotype that is expected, are they reinscribing ethnic and racial norms or subverting them?[6] To begin to answer this question, we must ask what choices Roma have and how they maneuver within them. Can and do they perform outside the stereotype? If so, what are the results? Judith Okely's work is relevant here; in an aptly titled article, "Trading Stereotypes," she underlines that in dealing with non-Roma, identities are "exoticized, concealed, degraded, or neutralized" (1996:52). We must also consider the transformative power of performance to create new subjectivities. As Diamond writes, "In performance . . . signifying (meaning-ful) acts may enable new subject positions and new perspectives to emerge, even as the performative present contests the conventions and assumptions of oppressive cultural habits" (1996:6).[7] For Roma, musical performance has been one of the positively coded arenas in a long history of exclusion, and thus it charts a potential site of transformation.

Romani Routes investigates the cultural politics and the political economy of Balkan Romani music making embedded in changing historical inequalities. Moving from American and Balkan communities to policy and states, I examine how the socialist governments in Bulgaria and Macedonia positioned Romani culture in relation to categories of folklore, and how the postsocialist state repositioned it in relation to political and economic agendas.[8] Since 1989, privatization has opened up new capitalist markets that promote Gypsy music, but these commercial ventures are usually managed by non-Roma; Romania is a notable exception. Romani music has been appropriated by non-Roma into fusion genres such as Gypsy Punk and into remixes by international DJs. Although a small number of Romani musicians have been catapulted into fame, the vast majority of professional Romani musicians struggle to maintain their trade amid economic crisis and political instability.

In addition to entrepreneurs and fans, nation/states have also become interested in displays of Romani culture in relation to political agendas. In 2007 Bulgaria and Romania became members of the European Union, and Macedonia is in the initial stage of negotiations. Accession criteria sometimes link human rights and economic development to the visibility of Romani culture, for example, in music festivals. In Europe as well as in the United States, Balkan Romani musicians respond to state policies at the same time they are dependent on commercial forces and a volatile market. How Romani musicians negotiate the relationship between politics and music in the context of neoliberal privatization is one aspect of this book.

Despite the celebration of Romani culture, anti-Romani xenophobic sentiments are growing all over Europe. According to the European Roma Rights Centre, Roma remain to date the most persecuted people of Europe (www.errc.org); in fact, *The Economist* titled an article on European Roma (2008) "Bottom of the Heap." Their fundamental human rights are threatened in many locations, and racist violence has increased since 1989, reaching an alarming rate in 2009–10. Racism is no longer merely the purview of extremists; rather, anti-immigrant and anti-Romani sentiment is becoming more mainstream in Europe. For example, in 2008, the Italian government fingerprinted Gypsies living in camps in an effort to crack down on crime; in 2009 numerous violent incidents such as fire bombings occurred in Hungary and the Czech Republic, and armed militias began patrolling "against Gypsies"; and in 2010 the French government evicted and deported Roma back to Romania and Bulgaria. All over Europe, nationalist parties are on the rise (often under the guise of populism) and the population is growing more polarized.[9] In the United States, racism against Roma is less pronounced but nevertheless exists in many realms; for example, "Gypsy Crime" units are found in police departments, and discrimination in housing and employment persists (Hancock 1987; Becerra 2006).

In addition to focusing on states and politics, *Romani Routes* also highlights several Romani individuals, communities, and genres of music from Macedonia and Bulgaria to ethnographically document their diasporic routes. Examining musicians in their Balkan communities and following them to their North American neighborhoods and on their tours, I explore how, through performance, they grapple with representational issues and enact multiple positions in transnational contexts. A wider political and economic context frames how musicians negotiate viable performances for various audiences, including their own communities, other ethnic communities of the Balkans, and non-Romani world music audiences in the United States. I highlight the Balkans, specifically Macedonia and Bulgaria, because they are home to populous European Romani communities and because many of the most famous Romani musicians have come from

Muslim Balkan Roma display many similarities to other Muslims of the region in terms of culture, ritual, and music. This shared history mitigates against the tendency in both scholarly and lay writings to see Roma as exceptional or unique. True, Roma have their particular historical trajectory of marginalization; but a healthy dose of comparison to other Balkan peoples, especially Balkan Muslims, shows many commonalities in family life, gender roles, ritual, custom, and music. In fact, Balkan Roma often share more cultural patterns with their Balkan neighbors than with other Romani groups, such as American Kalderash. Throughout this book, I aim de-exoticize Roma by suggesting comparisons that are historically and ethnographically emplaced.

In the Balkans today, approximately half of the Roma are Muslim and half are Eastern Orthodox, with a small percentage of Catholics and a new rising percentage of Pentecostals.[16] In practice, the religion of Muslim Balkan Roma is quite syncretic, incorporating elements of paganism and Eastern Orthodoxy. Approximately half of Balkan Roma have lost the Romani language; in the southern Balkans many of those who have done so speak Turkish as their first language. Multilingualism is the norm among Balkan Roma, with the older generation sometimes speaking four or five languages.[17]

Petrova suggests that negative stereotypes of Roma blossomed in fifteenth-century Western Europe and spread eastward (2003:128). Roma were viewed as intruders probably because of their dark skin, non-European physical features, foreign customs, and association with both magic and invading Turks. She asserts the rising tide of the Protestant work ethic condemned vagrancy, idleness, and lenience as well as alms for wanderers and beggars (125). Perhaps most important was the late arrival of Roma into Europe, plus their lack of roots in terms of land and property: "Ultimately the main difference that set the Roma apart was that they were the only ethnically distinct nomadic communities in a civilization that had been non-nomadic for centuries" (126).

The positive yet dangerous coding of Romani otherness hinges on their romanticization, on the part of non-Roma, as free souls (outside the rules and boundaries of European society); their association with the arts, especially music and the occult; and their proximity to nature and sexuality. Using Said's concept, we can claim that Roma are "orientalized" and exoticized (1978).[18] Trumpener emphasizes the association of Roma with an ahistoric, timeless nostalgia: "Nomadic and illiterate, they wander down an endless road, without a social contract or country to bind them, carrying their home with them, crossing borders at will" (1992:853). Simultaneously they are reviled as unreformable and untrustworthy, liars, and rejected from civilization. This contrast expresses the "ideology of Gypsy alterity—feared as deviance, idealized as autonomy" (854). Roma, then, serve as Europe's quintessential others.

The most tragic period of Romani history was perhaps World War II. With the Nazi rise to power, Roma faced an extermination campaign that has only recently been documented. According to various authors, from

500,000 to 1.5 million Roma were murdered, representing between one-fourth and one-fifth of their total population (Lewy 2002; Hancock 2002; Kenrick and Puxon 1972). After the war, Roma received neither compensation nor recognition as victims, and only recently have several claims to property and assets been filed.

The post–World War II communist regimes in Eastern Europe officially downplayed ethnicity but nevertheless defined Roma as a social problem. Targeted for integration into the planned economy, Roma were sometimes forced to give up their traditional occupations, and assigned to the lowest-skilled and lowest-paid industrial and agricultural state jobs (e.g., street cleaners). Nomadic Roma were forcibly settled; settled Roma were sometimes forcibly moved; and sometimes aspects of their culture, such as music, were outlawed. Specific policies varied by country; for example, forced sterilization was common in Czechoslovakia in the 1970s. Cheap housing was nominally provided, but segregated neighborhoods were commonplace. On the positive side, during socialism Romani school attendance grew (despite inferior and segregated schools), violence was rare, and Roma held steady employment and received the benefits of the paternalistic state (Verdery 1996; Silverman 1988).

The situation of Bulgarian Roma during socialism, such as forced changing of Muslim names to Slavic names and prohibitions against Romani and Turkish musical genres, is discussed in Chapter 7. In contrast to Bulgaria, Tito's brand of Yugoslav socialism emphasized (at least theoretically) *bratstvo i jedinstvo* (brotherhood and unity; Bosnian-Croatian-Serbian), a policy that promoted acceptance of different ethnicities. However, there was an official hierarchy: *narodi* (nations), *nardonosti* (nationalities), and *etničeski grupi* (ethnic groups, where Roma fell).[19] Culture, especially music, was an area where the groups could acceptably display their distinctiveness (Maners 2006). Hundreds of soccer clubs, amateur music groups, well-funded professional ensembles, and a well-coordinated network of festivals all served as a public manifestation for Yugoslav multiculturalism. Romani music thus was somewhat visible in official contexts, but this did not diminish engrained discrimination in employment, housing, health care, and education.

As in Bulgaria, Roma in Yugoslavia for the most part were ignored in scholarly folklore research because the discipline focused on rural peasants who were assumed to constitute the "pure" national culture. Roma were seen as "others" and excluded from the rubric "folk" (see Chapter 7 for a discussion of heritage); instead, they appeared as exotic, erotic, wild figures in films, literature, and children's stories (Pettan 2001). When Roma were scrutinized by ethnographers, they were found to be disorderly primitives, existing on the borders of civilization and lacking a unified culture (Van de Port 1998:137–159). In fact, the Serbian anthropologist Bajraktarović predicted they would disappear: they "do not have any prospects for remaining a separate . . . element of our society" (1970:747). In Macedonia, there is one Romani folklorist, Trajko Petrovski;[20] however, all over the Balkans young Romani scholars are currently emerging.

In the postsocialist period, harassment and violence toward the Roma of Eastern Europe have increased, along with marginalization and poverty. They are the largest minority in Europe and have the lowest standard of living in every country, with unemployment reaching 80 percent in some regions. Census statistics are unreliable because states are reluctant to report true numbers, but activists may by contrast overestimate numbers; in addition, some Roma report themselves as other ethnic groups to avoid the stigma of being Gypsy. Scholars agree there are about ten to twelve million European Roma, with the largest numbers of Roma in Spain, Bulgaria, and Romania (Petrova 2003; Ringold, Orenstein, and Wilkins 2004; Barany 2002).

Today East European Roma face inferior and segregated housing and education, including tracking of children into special schools for the disabled. Poor health conditions, specifically higher infant mortality and morbidity, shorter life expectancy, and higher frequency of chronic diseases, all plague Roma. Discrimination is widespread in employment and the legal system, and even educated people routinely express disdain for Gypsies. Hate speech and racial profiling are common in the media. Perhaps most troubling are the hundreds of incidents of physical violence against Roma perpetrated by ordinary citizens and also by the police.[21]

In response to historic discrimination and recent abuses, a Romani human rights movement has mobilized in the last twenty years via a network of activists and NGOs (nongovernmental organizations) such as the European Roma Rights Centre (ERRC), the European Roma Information Office (ERIO), the International Romani Union, the Roma National Congress, the European Roma and Travellers Forum, and European bodies such as the European Union, the Council of Europe, and the Organization for Security and Cooperation in Europe (Klimova-Alexander 2005; Acton and Klimova 2001; Cahn 2001a; Petrova 2003; Barany 2002; Guy 2001; Vermeersch 2006). This movement has drawn much public attention and funding to the plight of Roma, but material conditions have hardly improved in some areas. The "Decade of Roma Inclusion," inaugurated in 2005 by the Open Society Institute and the World Bank and currently endorsed by twelve European governments, aims to ensure that Roma have equal access to education, housing, employment, and health care.[22]

Macedonian Roma, numbering 130,000–200,000,[23] are currently represented by four Romani political parties and more than thirty active NGOs.[24] Although the prime minister claimed in 2003 that "I am proud of being representative of the country in which the Roma have perhaps the highest level of rights compared to all the other European countries" (Plaut and Memedova 2005:15), many disagree. For example, the European Romani Rights Centre titled its 1998 human rights report on Macedonian Roma "A Pleasant Fiction." On the one hand, Roma are now a "nationality," the 1991 constitution mentions full equality, the Romani language is spoken by 80 percent of Macedonian Roma, and there are several Romani-language radio programs and two television stations. On the other hand, there is widespread police brutality and discrimination in hiring, education, service in public establishments, and the legal system; moreover, surveys show 59–80 percent of non-Roma have negative feelings toward Roma (Kanev 1996: 24; Plaut and Memedova 2005:16).

The municipality of Šuto Orizari (Šutka), outside Skopje, is home to more than 40,000 Romani inhabitants; as the largest population concentration of Macedonian Roma, it has become a cultural center for music, dance, and politics (Silverman 1995b). Šutka Roma occupy every class sector, from poverty-stricken to rich, but most inhabitants are poor (see photographs 1.1 and 1.2, and see Chapter 4). Šutka was settled in 1963 when the government offered Roma housing after a devastating earthquake; it has grown steadily with migrants from other parts of Yugoslavia, and more recently with several thousand Kosovo Romani refugees from the 1990s Yugoslav wars.

Bulgarian Roma, numbering approximately 800,000 or 10 percent of the total population, do not have their own political parties because ethnic-based parties are legally prohibited.[25] Thus lobbying is done via NGOs; in Sofia alone there were more than 150 Romani NGOs in 2004 (Mihaylova 2005:48). Marushiakova and Popov observe that the NGO "Gypsy industry" often perpetuates itself rather than seeking solutions to problems (2005). Moreover, as elsewhere in Eastern Europe, a strong NGO sector is often an excuse for the state to do less. In Bulgaria, for example, funding and infrastructure for desegregation of schools comes primarily from NGOs.

In 1999 the Bulgarian government signed the Framework Program for Equal Integration of the Roma in Bulgarian Society, but very little has been done since then (Rechel 2008). In fact the framework expired in 2009, and the National Council for Cooperation on Ethnic and Demographic Issues (NCCEDI) has not prepared a new document. Attention to the human rights of Roma by the multiple coalition governments that have ruled Bulgaria since 1989 was clearly motivated by the prospect of joining the European Union in 2007; for accession, progress needed to be demonstrated.[26] Since accession, motivation has decreased. In spite of an "Anti-Discrimination Act," there are numerous cases of discrimination in local labor bureaus, social welfare offices, and health and education institutions (Mihaylova 2005:65), and some are being contested in the courts. The 2008 U.S. State Department report on human rights in Bulgaria claimed that unemployment was 65 percent among Roma, reaching 80 percent in some regions.[27] According to the World Bank, 13 percent of Roma completed secondary education, in comparison to 90 percent of ethnic Bulgarians.[28] In November 2009, the EU human rights commissioner noted that the situation of the Roma community is of particular concern; after viewing a settlement in Sofia, he stated, "No one should live in these conditions in today's Europe."[29]

In 2005, Ataka (Attack), an extreme nationalist party that openly proselytizes against Roma, won more than 8 percent representation in Parliament; in 2006 it won 26 percent of the presidential vote. Ataka's leaders have characterized Roma as criminals and as a threat to Bulgarians because of their high birth rate; one of their slogans is "No to Gypsifica-tion, No to Turkification" (Kanev 2005; Cohen 2005).[30] Although Ataka has recently lost popularity, some of its ideas have been adopted by more mainstream parties (Ciobanu 2008; Ghodsee 2008). It has also gained allies among Western European xenophobic parties in the European Parliament.

As a result of these inequalities of postsocialism, Romani refugees and emigrants can now be found in every Western European nation and in the United States and Canada. A profound refugee crisis has occurred in Kosovo, from which the vast majority of the Roma have fled as a result of the Serbian-Albanian conflict.[31] Because of inferior living conditions, many Balkan Roma would like to emigrate to the west, but immigration has become extremely difficult. Western European nations are deporting Roma, nationalist parties are on the rise, and xenophobia is growing. With their racial taint, their low class stigma, and their baggage of historic ste-reotypes, East European Roma are among the least desirable immigrants; Muslim Roma are even more suspect.[32]

To recap the complexity of Romani migration: I emphasize that Europe rather than India has been home to Roma since the fifteenth century and that multiple rediasporizations within and from Europe have occurred. For example, Roma migrated out of southern Romania after slavery was abolished, Eastern European Roma migrated to Western Europe after the fall of Communism, and Roma have migrated from Europe to North America since colonial times.

Migration to the United States

The first trickle of Romani travel to the United States occurred with the colonists, followed by waves from England in the 1850s. The largest numbers came during the second wave of immigration, from 1880 to World War I, along with eastern and southern Europeans (Lockwood and Salo 1994). The current Kalderash Romani population in the United States, numbering close to one million, can be traced to this last period of immigration. The United States hosts Roma from every group and sub-group, but they do not coalesce as a viable community. The New York Macedonian Romani community that is the subject of this book dates from the 1960s; these Roma interact with neither Kalderash nor other Romani groups, although occasionally intermarriage with other Balkan Roma takes place.

The center of Macedonian Romani life in the United States is located in the New York City borough of the Bronx, in the Belmont neighborhood. Belmont is a historic Italian neighborhood known as the "Little Italy of the Bronx." Italian groceries, restaurants, and bakeries, most of them family-owned, still line the main shopping street, Arthur Avenue, but in the past forty years Hispanics, Albanians, Bosnian Muslims, Montene-grin Muslims, south Serbian Muslims (from Sandžak), and Balkan Roma have moved into the area while the Italian population has declined.[33] Now Balkan (mostly Albanian) restaurants, groceries, photography studios, and pizza/*burek* (a doughy pie with feta cheese, spinach, or meat) parlors are interspersed with older Italian businesses (see photographs 1.3, 1,4, and 1.5).

Macedonian Roma began moving to New York City in the late 1960s,[34] specifically from the city of Prilep, but also from Bitola and Skopje. At the time, the Yugoslav government supported sending "guest workers" to Western Europe (especially Germany),[35] the United States, and Australia because of hard currency remittances. Emigrants saw working abroad as a way to make good money, move up the social scale, and help out relatives at home. After the guest worker policy ended, sponsorship of relatives and the need for spouses continued. The wars in Yugoslavia 1991–1995 brought economic crisis to the entire region, causing another wave of emigration. Although Macedonia was peaceful during the war and declared independence in 1991, its economy was in ruins during the entire decade. Push factors (out of Yugoslavia) in the 1960s were mostly economic, but now they include lack of hope, absence of a political future, and fear of police brutality and other forms of discrimination. Pull factors (to the United States) include the need for spouses, better employment possibilities, and the upwardly mobile models that migrant relatives have set.

"Chain" immigration, that is, one family sponsoring another, is the common pattern. This was the case until the mid-1990s, when American laws became more restrictive. Now spouses are the most numerous migrants. Women very rarely migrate without a relative (e.g. husband, father, son, brother) sponsoring them. Once in the United States, Roma face a new set of challenges. The majority are working-class, but some families have reached the middle class. First-generation Roma had poor English skills, and they lacked the legal connections needed to apply for documents. Many arrived with tourist visas, tried to regularize their status, and were sometimes exploited by lawyers. Many Roma simply overstayed their visas and became irregular migrants.

Some Macedonian Roma have applied for refugee status since 1991, but they have usually been unsuccessful in part because the persecution against Roma is underdocumented and unrelated to conflicts between nation-states. Refugee status has historically been easier for claimants fleeing wars; thus during the Yugoslav wars Bosnian Roma could more easily became refugees. In addition to Bosnian Roma, Bulgarian Roma are more often granted refugee status than Macedonian Roma because there are more numerous reports of human rights abuses against them. Recently it has become harder to receive asylum. Immigration judges have claimed that some applicants are merely posing as Roma; ironically, Roma now have to prove in court that they are Roma, while most of their lives they have had to pass as non-Roma to avoid discrimination.

Virtually no country wants Gypsy refugees, and many Roma remain undocumented or mired in legal battles for years. According to Xenos, "For the

Gypsies, assimilation into the world of nations appears to be impossible—they are perpetual refugees" (1996:240). The situation in Western Europe is more highly charged than in America because the numbers of Roma are larger and xenophobic parties advocate anti-immigrant policies (Castle-Kanerova 2001: Bilefsky and Fisher 2006). Although asylum in Western Europe was more liberally granted in the early 1990s, a decade later the trend reversed and Roma are being deported while social welfare is being dismantled. In the Macedonian Romani community in the United States, few distinguish between refugees and migrants: "displacement is a process that is not limited to those who meet the legal criteria for refugee status" (Lubkemann 2002:1). It is clear that a vibrant community life (see Chapters 4 and 5) helps diasporic Macedonian Roma feel at home in New York.

Issues of Representation in Fieldwork and Writing

Although this book distills many years of fieldwork, it is still only a "partial truth" in many senses (Clifford and Marcus 1986). My interpretation is not only one among many but also situated in specific places at certain times. My access to resources, my non-Romani "outsider" status, my gender, and my training have certainly affected my perceptions. Much of the postmodern discussion of ethnography rests on acknowledgment of multiple views; my account, then, has become the occasion for my Romani collaborators to discuss their interpretations of my interpretations. Heeding Lassiter's call for collaborative ethnography (2005) and specifically employing Elaine Lawless's concept of "reciprocal ethnography" (1992), I asked a number of Romani activists, musicians, numerous members of the New York Macedonian Romani community, and several non-Romani managers of Romani bands to comment on portions of the book; their reactions and our discussions have been incorporated. Romani scholar and University of Texas Professor Ian Hancock reminds us that until recently all representations of Roma were constructed by non-Roma, and Roma exercised no control over these descriptions and images, whether scientific, artistic, or literary (1997:39–40). This is finally changing, and the ethnographer is either obsolete or must delicately negotiate her place.

Studying a minority during the socialist and postsocialist periods highlights many issues of ethics, the role of the fieldworker, the power differential between fieldworker and informants, and the give-and-take in relationships.[36] As I accepted hospitality and knowledge from Roma, I continually asked myself, What is my relationship to these people? What am I doing for those who so generously taught me? How can I best discuss my own positionality in this research? As the Romani human rights movement emerged in the 1990s, I struggled to combine activism and scholarship and was alternately accused of neglecting one for the other. Whereas one Romani activist said I should concentrate on documenting human rights abuses and forget about analyzing music, some of my colleagues in academia said I was spending too much time on activism (which some regarded as "service," not "scholarship"). Some Roma said I should forgo a music focus because that would promote the stereotypical connection of Roma with music; some said I should focus on middle-class educated Roma to counteract the ubiquitous image of poor, begging Gypsies.

I have frequently interrogated myself as to the role of a non-Romani scholar. What right do I have to speak about a group that is trying to define its own voice?[37] While one activist questioned my right and ability to speak about Roma, other activists defended my commitment to Roma. I believe I have a role among non-Roma in education and advocacy, but Roma have their own organizing to do among themselves; thus I learned to withdraw when a context required my exclusion. Among Roma, non-Roma such as myself can facilitate, mediate, and provide resources for various cultural, economic, and political projects while eschewing paternalistic and colonizing stances (Smith 1999). Along these lines, I have been active with the nongovernmental organization Voice of Roma (www.voiceofroma.com). As Kamala Visweswaran writes, "If we have learned anything about anthropology's encounter with colonialism, the question is not really whether anthropologists can represent people better, but whether we can be accountable to people's struggles for self-representation and self-determinism" (1988:39). Because Roma are currently engaged in precisely the struggle for self-determinism, my research turned to the use of music as symbolic currency in self-presentation.

My Romani research started in the United States in 1975 when I became a volunteer teacher in a Romani alternative school in Philadelphia. My dissertation research (1975–1979) with the largest Romani groups in the United States, Kalderash and Machwaya, dealt with ethnic identity, gender, and pollution and taboo systems (Silverman 1981, 1982, 1988; also see Sutherland 1975; Gropper 1975). Having migrated to the United States from various parts of Eastern Europe about a hundred years ago, many Kalderash know very little about other Romani groups in Europe and the United States. Among the few tangible things I was able to give to American Kalderash Roma were historical information and cassette tapes of East European Romani music. After immersing myself in American Kalderash culture and gaining some fluency in their Vlach dialects of the Romani language, I was anxious to pursue Romani fieldwork in Bulgaria, a country I had visited regularly since 1972.[38]

I first worked with Roma in Bulgaria in the 1980s, in the context of research on wedding music, a fusion genre that was prohibited by the government (see Chapter 7). I met the stars of the wedding music, including Ivo Papazov, Yuri Yunakov, Neshko Neshev, Matyo Dobrev, and Salif Ali, and many others; as a fan, I tagged along their performance trail. Working with Roma in socialist Bulgaria was challenging because by 1984 they did not officially exist.[39] Despite government policy, I managed to circumvent prohibitions and spend considerable time in several Romani settlements. I documented ritual events such as weddings, baptisms, soldier send-off celebrations, and house warmings; I recorded, photographed, and videotaped where possible, given the constraints of socialism. Families viewed videotapes with me and offered valuable interpretations. Often their interpretations centered around the "Romani way of doing things," in light of their being both Bulgarian and Romani. Discussions frequently turned to the role of the state in their lives.

The idea of my working with Roma Macedonia was suggested by Aiše,[40] a Muslim Macedonian Romani woman whom I met in New York in 1988 when she was visiting her brother Osman (see Chapters 4 and 5). They lived in Belmont, located close to the neighborhood where I was born and my parents lived. From the beginning of my research, then, I approached Macedonian Roma from at least two locations. Aiše's family arranged my living arrangements in Šuto Orizari, the largest neighborhood of Roma in Skopje, Macedonia, where I resided for six months in 1990 and one month in 1994. I have continued to work with Roma in Macedonia and New York until the present, with long stays in New York and short trips to Macedonia as well as to Macedonian Romani communities in Australia and Toronto.

My Bulgarian and Macedonian Romani connections merged in the Bronx in 1994 when the Bulgarian wedding music star Yuri Yunakov emigrated to the United States and took up residence in the Macedonian Romani community. Yuri formed a new band, asked me to join, and reluctantly I accepted. I had sung Balkan music for more than twenty years and Romani music for about five years, but mostly with Americans. In addition to bestowing on me a great honor and challenge, performing with the Yuri Yunakov Ensemble gave me both backstage and front-stage perspectives. In 1999 I toured for two months with "The Gypsy Caravan: A Festival of Roma Music and Dance" as a performer and as the "education coordinator," delivering lectures and leading panel discussions for the general public. Dozens of concerts with the Yunakov Ensemble, including tours in 2003 and 2005 with Ivo Papazov, Neshko Neshev, and Salif Ali, have yielded invaluable information and professional experience. Recording with these artists on several CDs, writing liner notes for their albums, and arranging tours for them has given me a chance to help their music reach wider audiences and also to study the roles of audiences, marketers, managers, and producers.

Finally, my most recent ethnographic tool is YouTube, through which I have explored the transnational flow of music and dance. Romani materials that are posted on YouTube (by both Roma and non-Roma) include not only commercial videos and television programs from the Balkans but also excerpts of family and community celebrations from the Balkans and the Western European and American diasporas. I have followed the commentaries posted on YouTube and interviewed several prolific posters via the internet. *Romani Routes*, then, takes account of my non-Romani identity, my transnational fieldwork, and my multiple roles of ethnographer, performer, and educational activist.

Chapter Overview

Romani Routes is divided into four parts to reflect the transnational underpinning of the materials and the dynamic movement of people and music between the Balkans and America. I aim to underline the dialogue between homes and migration, between states and capitalist markets, and between communities and individuals. Please see the website for information on video examples, audio examples, and photographs as well as supplementary textual material including song texts and historical information (a guide is found in the front of the book).

Part I, "Introduction," discusses the basic analytical questions and the theoretical framework, plus general background information on Romani musical genres and styles.

Chapter 1, "Balkan Roma: History, Politics, and Performance," presents the issues raised in the book and a historical overview of Roma and their diasporic migration to the Balkans, as well as their rediasporization to North America. I trace the legacy of discrimination and discuss the status of Roma during socialism and postsocialism, highlighting Macedonia and Bulgaria. I chronicle the challenges of being a non-Romani researcher and introduce the issue of representation: how have Roma been depicted, and by whom?

Chapter 2, "Musical Styles and Genres," explores music as a historic Romani profession and offers insight into patron-client relationships and stylistic change. I look at Balkan Romani music in terms of rhythm, melody, genre, style, text, improvisation, and variation, highlighting the emergence of new and revived styles. Balkan Roma have been extremely influential in many of the fusions of the last forty years, and the cross-pollination of regional styles fosters innovation. I also profile several important Romani artists.

Chapter 3, "Dilemmas of Diaspora, Hybridity, and Identity," theoretically explores the interdisciplinary scholarship on diaspora, transnationalism, hybridity, and cosmopolitanism and asks how Roma interrogate these concepts. How have music and essentialism been used in multicultural discourse and in identity politics? As a motley group of disparate peoples lacking unity in territory, language, and religion, how have Roma united around their history of exclusion to build a pan-Romani ethnic movement? I explore activists' attempts to construct nationalist symbols of Romani culture, such as the Indian homeland, the flag, the literary language, and the anthem.

Part II, "Music in Diasporic Homes," ethnographically profiles Roma in their multiple homes in the Balkans and the United States and explores how and why they travel among diasporic homes. I illustrate the significance of music and dance by analyzing the complex relationship between social relations and family and community rituals. I explore how the New York Macedonian community cements its ties to Macedonia and other diasporic locations through marriage, language, and ritual, all enacted performatively via cultural markers such as music and dance.

Chapter 4, "Transnational Families," traces the history of the Macedonian Romani community in New York via kin networks, occupational trajectories, migration of brides and grooms, and the movement of musicians and media products. I look at the multiple ways Roma in the diaspora constitute their identities and their gendered roles, and how changes are occurring, especially for educated youths.

Chapter 5, "Transnational Celebrations," focuses on family celebrations that are the symbolic focus and the glue binding the Romani community together. I analyze how celebrations display and interpret values through music and dance and also reveal conflict. I compare weddings in Macedonia with weddings in New York, analyzing their structure, rituals, music, dance, costume, and economic and social implications. I profile several key New York musicians and describe their repertoires and training.

Chapter 6, "Transnational Dance," compares and discusses Romani dance in numerous contexts, emphasizing its stylistic, social, and power dimensions in relation to the marginality of Roma in wider society and the ambivalent position of women. I historically trace the dance genre *čoček* from Ottoman times until the present. I explore how women negotiate dance performances within Romani diasporic communities and how professional dance becomes symbolic capital to negotiate in the commercial marketplace. I compare Gypsy dance performances in several ensembles to show the range of representational styles and the use of stereotypes.

Part III, "Music, States, and Markets," examines the legacy of socialism in Balkan Romani music and traces economic, social, artistic, and political changes through the postsocialist period. Examining the exclusionary practices of states and the constraining forces of the market, I investigate how Roma have resisted, collaborated, and adapted.

Chapter 7, "Dilemmas of Heritage and the Bulgarian Socialist State," examines how and why Roma were excluded from the rubrics of "nation" and "folk." I explore the trajectory of the fusion genre Bulgarian "wedding music," which was prohibited by the socialist government and became a countercultural phenomenon. I document how wedding musicians negotiated and resisted the state, and how the state responded.

Chapter 8, "Cultural Politics of Postsocialist Markets and Festivals," deals with music in the postsocialist period in relation to new contexts, markets, and political configurations and a rising tide of xenophobia. As Roma are squeezed between states and markets, how do they respond? I examine Bulgarian wedding music, Romani music festivals, popular music contests, and a Macedonian UNESCO application in relation to nationalism, multiculturalism, and public negotiation of Romani identity.

Chapter 9, " Bulgarian Pop/Folk: Chalga," investigates the rise of *chalga* in the 1990s as a pan-Balkan fusion of Romani, folk, and popular musics. I trace the depictions of the orient, formulaic packaging, and the recent signs of audience fatigue. I also examine the challenges that Romani performers face, using case studies of Sofi Marinova and the transgendered diva Azis. The controversy surrounding chalga illuminates the debate regarding definitions of what it means to be Balkan, European, and "modern."

Part IV, "Musicians in Transit," widens the focus of the book to international audiences and revisits integration of American and Balkan viewpoints. I illustrate how two Romani stars, one male and one female, navigated transnational border crossings and strategized to expand their careers. Looking at tours and commercial enterprises managed by non-Roma, I discuss representational dilemmas from the point of view of managers, producers, audiences, and Romani musicians. I interrogate the political economy of collaborations and appropriations by examining recent DJ remixes and the issue of who represents whom, who benefits, and why.

Chapter 10, "Esma Redžepova: Queen of Gypsy Music," explores the life history of the Macedonian superstar in the context of Yugoslav multiculturalism and as a bridge between Roma and non-Roma. She resisted gender norms, and her husband skillfully crafted her image as a respectful singer. I examine her recent collaborations and explore her humanitarian efforts.

Chapter 11, "Yuri Yunakov: Saxophonist, Refugee, Citizen," illustrates, through life history, how musical performance is a strategy in personal identity politics. Emigrating to New York in 1994, Yunakov plays regularly for Roma, Macedonians, Turks, Armenians, Albanians, Bulgarians, and Americans. I show how through music he mediates the tension between such supposed binaries as official-unofficial, traditional-modern, authentic-hybrid, inclusion-exclusion, and local-global.

Chapter 12, "Romani Music as World Music," discusses what happens when community performers achieve international fame and when the local becomes the global. I chart the relationship among festival producers and managers of Romani music acts (who provide a saleable item), audience members (who claim to support a liberal multicultural agenda), the press (eager to exoticize), and Romani musicians (trying to eke out a living). Debate about authenticity and the emergence of new fusions such as Gypsy Punk reveal the strategies of marketers and performers.

Chapter 13, "Collaboration, Appropriation, and Transnational Flows," ties together threads from previous chapters to discuss collaboration, appropriation, and the transnational movement of music in relation to political and economic matrices. I interrogate who is producing and marketing Romani music and how power relationships are implicated in these exchanges. I examine issues of ownership and compensation through case studies of DJ remixes, Balkan Beats dance clubs, and the movie *Borat*.

2

Musical Styles and Genres

What Is Romani Music?

Music is one of the oldest Romani occupations, a fact corroborated by historical documentation dating from the fifteenth century.[1] This traditional link may be one reason music has a deep symbolic connection for Roma and the terms *music* and *Roma* are almost synonymous for non-Roma. We should not forget, however, that music has been a viable occupation for professional Roma for more than 600 years, and that in the current period of postsocialist transnational mobility it remains viable. Although *Gypsy* and *music* are commonly paired terms, the nature of Romani music is only now receiving the scholarly attention it deserves.[2] Historically, writers have assumed more than they have proved; exaggerations run the gamut from the position that Roma are "sponges," that is to say, they merely borrow and have no music of their own (Starkie 1933; Bhattacharya 1965), to the position that they are the staunchest preservers of tradition. For example, the CD *Rromano Suno 2* (Gypsy Dream, B92, Serbia, 2006) asserts that "they found themselves in a strange place where their repertoire is described both as the deepest repository of tradition and a generator of irresponsible innovation."

As early as 1910, Serbian music scholar Tihomir Djordjević disparaged Gypsies because they failed to preserve their own music and, when adopting Serbian music, they "gypsified" it.[3] He wrote that when Gypsies perform Serbian folk music, "they decharacterize and gypsify it. They change primitive folk music as they choose, and they interfere with the most essential aspects of this music; they change the details as they see fit, or if these details are already attractive, they overemphasize them or add decorations which sound gentle and beautiful at first, but have no place in that music. That is the Gypsy Quality of folk music" (1984 [1910]:38). In 1977 Gojković wrote that Gypsies "corrupt not only national music of various countries but also new music, for instance, jazz" (1977:48). Thus the typical older Balkan scholarly attitude toward Romani musical innovation was one of contempt.

In Hungary, Roma have either been hailed as the most authentic preservers of peasant music (Vekerdi 1976) or assailed as corrupters and distorters of peasant music (Bartók 1931).[4] There are many levels of this controversy, which continues to the present day (Frigyesi 1994; J. Brown 2000; Hooker 2007) and spills over from Hungary into the Balkans. The core of the conflict lies in varied interpretations of the concept of creativity and in the vain search for origins and "authenticity." Roma are neither a "primitive folk which has no authentic music of its own, either vocal or instrumental" (Spur 1947:114) nor merely sponges. What Frigyesi points out for Hungary is also true for the Balkans: Gypsy music not only was stylistically at the "crossroads of folk, popular and high art" but also was "the common ground between the 'rich' and the 'poor'" (Frigyesi 1994:267; Peycheva 1999a). Because of their professional niche, Roma creatively molded the popular repertoire and interacted dynamically with local musics. Examples of how this happens in the Balkans are discussed throughout this book.

This debate shows that the historical nexus of Romani music is quite complex. For centuries, Romani groups in Eastern Europe have been professional musicians, playing for non-Romani peasants and city dwellers of many classes for remuneration in cafes and taverns and at events such as weddings, baptisms, circumcisions, fairs, and village dances. This professional niche, primarily male and instrumental,[5] requires Roma to know expertly the co-territorial repertoire and interact with it creatively. A nomadic way of life, often forced on Roma from harassment and prejudice, gave them opportunity to enlarge their repertoires and become multimusical as well as multilingual.

In additional to nomadic Roma, large groups of sedentary Roma in major European cities became professionals who performed urban folk, classical, and popular music. In Hungary, Spain, and Russia, certain forms of Romani music became national music, veritable emblems of the country (Frigyesi 1994; Leblon 1994; Lemon 2000). The music played by professional Romani musicians in in-group contexts may or may not differ, depending on the historical situation, from the music played for other ethnic groups. Finally, there are many groups of Roma who are not professional musicians but have their own music. Furthermore, all these groups have migrated within Europe to varying degrees, and also to the Americas and Australia.

It should be clear by now that there is neither one worldwide nor one pan-European Romani music. Roma constitute a rich mosaic of groups that distinguish among themselves musically. This is not to deny that there is an emerging ethnic awareness of unity and a scholarly basis for comparison. A Bulgarian Romani song may have more in common with an ethnic Bulgarian song than with a Polish Romani song, reflecting centuries of co-territorial traffic in music. Are there stylistic elements common to all European Romani musics? In answering negatively, I explore in

Chapter 12 why this question is so urgent for many music producers. Certainly the professional niche continues to exist (in a wide area) and can generate comparable data on repertoire and performance. Over and over again in Eastern Europe, we hear of virtuosic performances of Roma that move people to tears, of seemingly endless variations in melody, of the capturing of emotion in music. Proverbs attest that "a wedding without a Gypsy isn't worth anything" (Bulgarian) and "give a Hungarian a glass of water and a Gypsy fiddler and he will become completely drunk" (Hungarian). Although the prominence of Roma in Balkan folk music cannot be denied, facile searches for a unifying style must be met with suspicion.

In the quest for the universal and unique in Romani music, some scholars have turned to the Indian homeland and claimed to find musical links with present-day groups (Bhattacharya 1965; Fonseca 1995; Hancock 2002:71; Acković 1989). This work has been highly speculative and remains unproven.[6] Rather than seeking the unique or the pure, I seek to explore Romani music as it exists in Balkan diaspora Romani communities and for non-Romani local and international patrons. I start by examining the music Balkan Roma play and sing in New York, Macedonia, and Bulgaria, making note of what is shared with co-territorial peoples and what travels in which direction. Balkan Romani music, then, rather than being a unified whole, can be considered a constellation varying regionally and historically.

Balkan Historical Threads

As mentioned in Chapter 1, the Balkans are home to several million Roma divided into many groups; in their migrations to North America, Balkan Roma have retained music as a focus of their community life, but that music has changed over time. Historically, Roma have been professionally involved in many musical forms, both folk and popular; have had a virtual monopoly of some forms; and have been virtually absent from other forms (such as Istrian, Dalmatian, and Slovenian music and some shepherd's flute and bagpipe genres). The probable explanation of their absence in these musics is that there simply wasn't a steady income to be earned from them. For example, Macedonian Roma have never played rural instruments such as *gaida* (bagpipe) and *kaval* (end-blown flute). In Bulgaria, however, Roma played these instruments because there was a market for them.

For centuries all over the southern Balkans, Roma have had a virtual monopoly of professional ensembles consisting of one or two *zurli* (*zurla* and *surla* in Macedonian, *zurna* in Bulgarian and Turkish, *pipiza* in Greek), double-reed conical-bore instruments, plus one or two *tapani* (*tŭpan* in Bulgarian, *tapan* in Macedonian, *davuli* and *daouli* in Greek), double-headed cylindrical drums (see photograph 2.1 and video examples 2.1 and 2.2). A few ethnic Slavic zurna players have been reported in the literature (Peycheva 1993:50). Citing evidence from a fourteenth-century fresco in

Ohrid, some ethnomusicologists believe that the zurla and tapan were brought to the Balkans by the Roma before the arrival of the Ottoman Turks, but there is debate on this topic (Ilnitchi 2007).[7]

It cannot be doubted that for hundreds of years zurla and tapan ensembles have played professionally for many ethnic groups in southern Serbia, Kosovo, Macedonia, the Macedonian province of Greece, the Pirin region of Bulgaria (southwest), and some parts of Balkan Thrace. At large public events such as fairs, zurla and tapan ensembles were (and are) hired by dancers or picnicking families. They were also associated with wrestling matches, found among many Muslims of the southern Balkans. Among Muslim Macedonian Roma, zurla and tapan music is essential for ritual moments such as application of henna to the bride's hair, hands, and feet (see Chapter 5, and video examples 5.1, 5.2, 5.4, 5.5, 5.6, and 5.16); the act of male circumcision; and the slaughter of the lamb on Erdelezi (Gjurgjovden, St. George's Day; see Chapter 5 and Dunin 1998). Today zurla and tapan ensembles coexist with amplified modern bands because of the ritual function of the zurla and tapan, their role in playing traditional dance music, and their symbolic association with Romani identity (Blau, Keil, Keil, and Feld 2002; Silverman 1996a; Peycheva and Dimov 2002). In New York, there are no zurla and tapan ensembles—but there are a few Romani zurla players in the Bridgeport, Connecticut, region and many tapan players as well.

In the 1980s, the zurna was prohibited in Bulgaria because of its Romani and Muslim associations. In Chapter 7 I discuss socialist state policy and how Roma resisted it. Since 1989, Pirin zurna and tŭpan music has emerged as a vital force; players now serve patrons of all ethnicities and play for a variety of events; they have also been rehired (albeit part-time and for low wages) by some local ensembles. Famous Pirin zurna players such as Samir Kurtov, who was born in 1971 in Kavrakirovo, Petrich region, are highly paid and well respected, and they now perform at weddings outside their region; Kurtov now performs often in Greece, Serbia, Macedonia, and Turkey (Hunt 2009; Peycheva 2009). Video example 2.2 shows Kurtov playing at a wedding in Gotse Delchev, Bulgaria, in 2004; a slow melody changes into a 2/4 dance. In postsocialist Bulgaria and Macedonia, the instruments have also been used at political events for their role in outdoor announcement at parades and rallies (Peycheva and Dimov 2002:183).

Zurla and tapan playing, like all instrumental performance, is exclusively male, transmitted along kin lines. In some communities, zurla players are from a single family of Roma. Training takes place from elder to younger, and repertoire and technique are learned by listening and watching. This principle can be generalized to transmission of all Balkan Romani music: oral immersion without written notation. (For example, video example 5.15 shows how young boys sit under the stage at a wedding and follow along with makeshift drums.) For zurla training, typically the learner drones while the master plays melody. Once mastery is reached, the parts may alternate. Occasionally, the two zurlas will play in unison, in octaves, or more recently in parallel thirds. In addition to play-ing the dance or song melody, the lead zurla player does free rhythmic improvisations, known as *mane*, and also metric improvisations. Size of repertoire and technical virtuosity distinguish good zurla players. Orna-mentation consists of rapid and even finger trills, mordents, and grace notes (Peycheva and Dimov 2002; Rice 1982). Master tapan players impro-vise rhythmically and texturally, creatively using the differing sounds of the two drumheads.

Roma have also been active in the realm of brass bands, in both rural and urban environments. Adopting brass instruments from central Europe about a hundred years ago, Roma became especially prominent in brass bands in southern Serbia, Macedonia, and Bulgaria. Serbian peasants also play in brass bands, but Roma tend to be professionals, and they per-form a more Turkish-influenced repertoire. Serbian festivals, such as those at Guča, have given wider visibility to this tradition and introduced a sense of hierarchy through awarding prizes. The brass band has become a Serbian national symbol, and bands such as Boban Marković are pop-ular on the world music circuit.[8]

Professional male Romani bear leaders have been found throughout the Balkans since the sixteenth century, often traveling with their families and teaching their bears to perform to tambourine and voice accompaniment (photograph 2.2). In the nineteenth century their centers were Romania, Serbia, Bosnia, and North Bulgaria. They still entertain peasants at fairs and in courtyards; according to a Bulgarian proverb, "a festival without a bear trainer is a waste of time." Bears can also heal various illnesses (this being related to the power of the bear in folk belief) and perform tricks such as dancing on the hind legs and passing the tambourine around to collect money. Since the nineteenth century, Romani monkey leaders have also been common (photograph 2.3). Bear and monkey trainers in Bul-garia play a vertically held, three-stringed, pear-shaped, bowed lute called *gŭdulka* (*kemene* in Macedonia). They identify as *Kopanari* (part of the Rudari group) and speak Romanian. Many make their own instruments, since they are usually woodworkers as well as animal trainers. In addition to playing dance music to which the animal performs, they also play and sing improvised historical ballads or humorous songs, sometimes pro-viding social commentary (Silverman 1986:55). The Bulgarian socialist government strictly regulated and even prohibited animal trainers, but since 1989 the restrictions have been eased, and they can now be found in parks and playgrounds of major cities.

From Ottoman times there has been a trafficking of musical styles fa-cilitated, in part by Balkan Roma (Peycheva 1999a; see discussion of the historical roots of wedding bands later in this chapter). For example, in the nineteenth century the Romani fiddlers of Negotin transmitted Romanian music to the Vlachs of east Serbia (Vukanović 1962:48). Today, not only do Romani musicians travel, but also there is trafficking in media products. In examining the interplay among "Oriental" (Turkish-influenced) style, marketing, and Romani identity, Rasmussen notes that

Serbian Roma played a vital role in facilitating interaction among several distinct musical genres: village folk music, urban folk music, popular music, and *novokomponovana narodna muzika* (newly composed folk music; 1991, 2002, 2007). Similarly, in Macedonia and Bulgaria Roma played eclectic repertoires in urban ensembles that developed into contemporary wedding bands.

Kovalcsik, using Hungarian materials, claims Romani music can be distinguished from co-territorial folk music by improvisation and a readiness to adopt new influences, especially commercial popular genres (1987). Using material from Kosovo, Pettan also cites improvisation and the value of change as specifically Romani features (1992); Kertesz-Wilkinson uses Hungarian materials to analyze how changes in performance are incorporated *selectively*, according to a Romani aesthetic system (1992); and Rasmussen analyzes Romani openness to popular music (1991). Dimov similarly comments that the presence of Gypsies is "the clearest example of the polyethnic characteristic of Balkan music, and its negotiation, translation and integration" (1995:14). Peycheva also discusses the "polylingualism" of Gypsy musicians in text and style (1995). On the other hand, I note as well that recently other ethnic groups in the Balkans are adopting new and varied styles.

It is worth remembering that professional Balkan Romani musicians regularly serve patrons from many ethnic groups, and thus their repertoires tend to be large and varied.[9] Although the economic patron-client relationship is often the framework within which Roma perform, the artistic imperative is the creative engine behind exchange of services. For centuries, Roma have been one of the main forces of innovation in Balkan music (Pettan 1992; Peycheva 1999a). Their role as innovators can be partially traced to the motivation of generating new material to sell to patrons, but it can't be solely reduced to economic imperative. Musicians also value innovation for its artistry; they carefully listen to and evaluate each other, detecting what precisely is new and worthwhile in technique, melody, harmony, improvisation, genre, text, and form. But innovation isn't enough to win admiration, novelty must be accompanied by superior technique and soulful passion.[10]

Innovation is accomplished in myriad ways: sometimes by looking to other local genres (e.g., using local instruments or folk or pop styles such as African-American rap or electronic music styles), sometimes by looking across regional borders (e.g., using Albanian, Serbian, or other Balkan styles), and sometimes by looking toward distant Romani musics (e.g., Spanish or Indian styles). The guitar style of the Gipsy Kings (primarily rumba and flamenco) was appropriated in the early 1990s after the group made their first Eastern European tour. Rap in the Romani language was employed in Macedonia and Bulgaria first in the early 1990s and is now a growing style; the vocalist Ševćet (who lives in Germany) created the genre "Roma Reggaeton Hip Hop," which masterfully fuses hip hop and Romani elements into his "Gio Style" shows (see www.myspace.com/sevcet).

Another way of innovating is to revive and reinterpret older repertoires. In Bulgaria, for example, Ibro Lolov and Anzhelo Malikov rerecorded songs that Yashar Malikov (Anzhelo Malikov's father) collected or composed in the 1950s. The elder Malikov was a prolific arranger credited with dozens of songs (Peycheva 1999a). The 2005 CD *Romane Merikle/Roma Beads* is dedicated to Yashar Malikov, Hasan Chinchiri, and several other deceased Romani composers; it features remakes of Chinchiri's songs as well as other repertoire from the 1950s.[11] Although these examples could be termed "covers" or remakes, I would counter that this is more than mere borrowing. In some cases (for example, wedding music; see Chapter 7), Roma have created new genres from existing elements.

Malvinni represents Gypsy music with an equation:

$$I + V = E$$

where I is improvisation, V is virtuosity, and E is emotion (2004:10). I agree that these three elements are significant in the Balkans, but I think Malvinni's equation is too narrow and too mathematical a formulation. Furthermore, I can envision another possible equation: $V = I + E$. Moreover, Malvinni's equation does not hold for all Romani musics, and it is also applicable to some non-Romani musics; improvisation is not very important in American Kalderash music but definitely is in (non-Romani) jazz. In the Balkans, it is nevertheless true that improvisation holds an almost sacred place for Romani instrumentalists; it is the core of Bulgarian wedding music, to give just one illustration. For singers, emotion and technique tend to be more important than improvisation, and vocal melodic improvisation is usually confined to variation in ornamentation. In contrast, among instrumentalists improvisation is a conscious item of practice and discourse.

Čoček/Kyucheck

In the Balkans the most characteristic Romani musical genre is called *čoček*, or *čuček* in Macedonia, Serbia, and Kosovo and *kyuchek* in Bulgaria. Note that čoček also refers to the solo dance genre associated with this music, discussed in Chapter 6. In Chapter 9, I discuss how the genre has traveled north since 1989 to Romania (and even Hungary and Slovakia). Whereas before 1989 čoček was found in Yugoslavia, Bulgaria, Greece, Albania, and Turkey, today it is shared across much of Eastern Europe. It has also spread with the Romani diaspora to Western Europe, North America, and Australia and is the mainstay of Balkan Romani celebrations in New York.

Čočeks use Turkish-derived scales (*makams*) that sometimes employ microtones and sometimes Westernized pitches. There is no one typical scale or typical makam for Balkan Romani music, and indeed the term "Gypsy scale"[12] is a misnomer. A variety of scales are used, including major, minor, phrygian (similar to the Turkish makam *kurd*), and other modes and Turkish makams (modal scalar patterns) such as *hicaz, nihavent,* etc. Čočeks typically have precomposed sections plus solo sections distinguished by *taksim* or mane,[13] a highly improvised free-rhythm or metric exploration of the scale or makam, often using stock motives and figures, played over a metric ostinato. In the mane musicians display their improvisatory virtuosity. Čočeks are associated with characteristic rhythms, some of which are displayed in Figure 2.1. Actually there are many variations on the rhythms in this figure, each imparting a distinct style that sometimes indicates to dancers what should be danced. Number 6, for example, is known as *čifteteli*.

The 9/8 meter is associated with Turkish-speaking Roma in the Balkans, and in fact in Turkey it is the characteristic rhythm of Roma.[14] The meter can be played fast and light or slow and syncopated, as in the Bulgarian Romani rhythm known as *kaba zurna* (low-pitched zurna), whose name suggests it was a zurna style adopted by the clarinet. Kaba zurna is rhythm number 14 in Figure 2.1. Bulgarian wedding musician Filip Simeonov's (Fekata's) "Shalvar Kyuchek" is a kaba zurna, see audio example 2.1 (he can also be seen in video example 7.9, playing a Bulgarian rǔchenitsa in 7/16); the whole piece is an improvisation—there is no precomposed part. He explores the various timbres of the clarinet while playing in a makam similar to a minor scale with a neutral second degree (between a major and minor second). The neutral second is characteristic of solo improvisation in both Bulgaria and Macedonia; in almost all the taksims discussed in this chapter, regardless of the scale or makam, the second degree has this microtonal element.

Another important rhythm is 7/8, especially in Macedonia and southwest Bulgaria, where it takes a number of forms. Fast and light forms of 7/8 (sometimes known as *lesno*, light or easy in Macedonian) are used in many songs from Macedonia such as "Oj Borije" (Oh Bride!) and "Zapevala Sojka Ptica" (The Jaybird Began to Sing [Macedonian]), sung during a wedding (see Chapter 5, video examples 5.26 and 5.32) and Esma Redžepova's song "Naktareja Mo Ilo Phanlja" (With a Key He Closed My Heart), discussed in Chapter 10. A slower 7/8 is illustrated by "Sitakoro Oro" (Sieve Dance), in rhythm number 10 in Figure 2.1, performed by Orkestar *Titanik* (Titanic), a well-known band from Šutka (audio example 2.2); note that the sieve is used in the wedding ritual in reference to fertility (see Chapter 5). Again there is no precomposed tune; rather, the whole piece is improvised around familiar motives by the saxophonist, followed by the synthesizer player. The phrygian mode is used, and the second degree of the scale tends to be microtonally high. If the piece were longer, it would characteristically end with a short fast section, typical of Macedonian dances. Titanik's repertoire also includes several Albanian-influenced pieces, reflecting the fact that Skopje Roma have regularly played for Albanian speakers (both Romani and non-Romani). This sets them squarely apart from Bulgarian Roma, who have only recently imported Albanian-influenced styles.

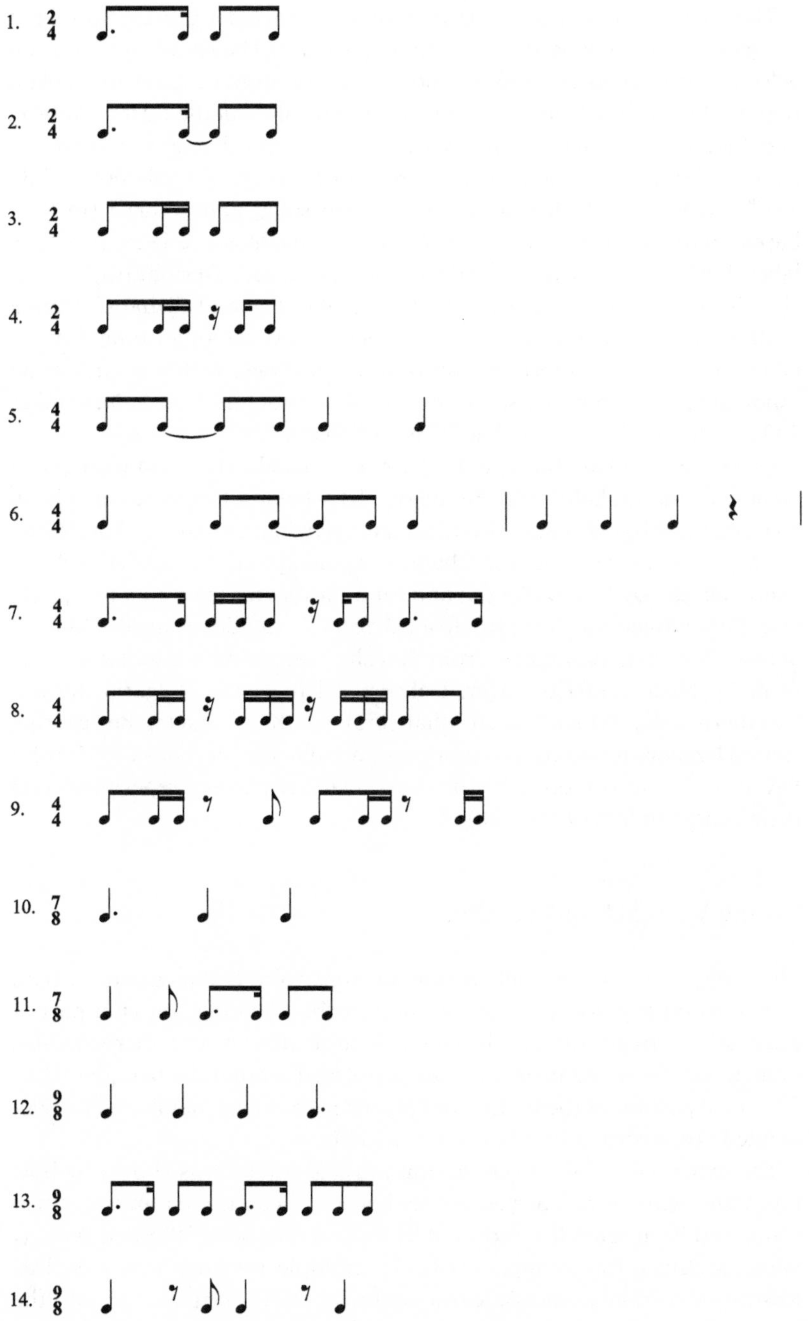

Figure 2.1. Variations of Čoček Rhythmic Patterns

The tunes for čočeks are sometimes drawn from older Romani tunes but are more often composed by wedding musicians. They are inspired by an eclectic array of sources: folk and popular music from neighboring Balkan regions, film scores from the West, cartoon music, Middle Eastern music, and Indian film music. Kyuchek titles in Bulgaria during the 1980s included "Sarajevo '84" and "Olimpiada," in honor of the Olympics; "Alo Taxi" (Hello Taxi [Bulgarian]), from a pop song; and "Pinko" (in 9/8), based on the musical theme from *The Pink Panther*. Since 1989, inventive labels for kyucheks have been added, such as "Evro," "Germaniya," "Arabski," "Bingo," "Hazart" (Hazard/Risk, a gambling game), "Isuara" (from a Latin soap opera), and more. As mentioned above, among Romani musicians there is a cross-fertilization of musical styles, with a premium on innovation. For example, soon after the American election in November 2008, "Barack Obama Kyuchek" was composed.[15]

Peycheva divided Bulgarian clarinet kyuchek styles into regional "schools": the Makalov and Pamukov clans belong to the Kotel school, characterized by energetic, sharp, staccato playing; Osman Zhekov, Nesho Neshev, and Ivo Papazov (see Chapter 7) belong to the Kŭrdzhali school, characterized by legato playing and fluid movement from pitch to pitch; and Filip Simeonov (Fekata) from Trŭstenik (audio example 2.1, see above), Marin Dzhambazov from Knezha (see below), Dimitŭr Paskov from Vŭrbitsa, and Kuti from Dobrich, all in North Bulgaria, have a "northern style," which is somewhat similar to Kotel playing but characterized by more tonguing and more staccato playing (Peycheva 1995:16).[16] Peycheva has also thoroughly analyzed clarinet players' repertoires and their hybrid styles (1999a; 2008b).

Song Variants and Versions

The largest part of the Balkan Romani repertoire is dance music, both instrumental and vocal, reflecting the fact that dancing is a vital part of celebrations. In addition to dance music in regular meters, there are also unmetered songs and instrumentals performed around the banquet table. The vocal portion of the unmetered repertoire has been neither well documented nor recorded (but see talava below).

The dynamism of the Romani oral musical tradition is shown by how tunes and texts have traveled across borders, been traded among musicians, and been remade or covered by singers. The song "Phirava Daje" (I Went, Mother), for example, exists in multiple variants across Serbia, Kosovo, Macedonia, and Bulgaria, and also in Western Europe and the United States. The text of this song and analysis of three variants are found in audio examples with text supplement 2.3–2.5. Another illustration of the breadth of variation comes from comparing versions of one song by the same singer performed or recorded at different times. In audio examples with text supplement 2.6–2.8, I compare three variations of a song by Macedonian singer, Džansever, and discuss her life. Note that any good piece of Romani music, vocal or instrumental, tends to exist in multiple variants; that is a mark of its excellence. If fellow musicians embrace a piece, they do not hesitate to change it; Pettan's research in Kosovo, for example, presents several variations of songs and instrumentals (2002:251–276). One of the most common paths of dissemination is from Serbia to Bulgaria or Macedonia, and another path is from Greece northward; however, all directions are operational.[17]

Stylistic Trends

The current Romani wedding bands in Macedonia are heirs to the urban professional *čalgija* tradition of the early twentieth century, which flourished until World War II. The word comes from the Turkish root *çalg*, meaning instrumental music or a musical instrument. Čalgija ensembles played Ottoman-derived multi-ethnic vocal and instrumental music in a heterophonic style based on the makam system, emphasizing innovation and improvisation. Roma were the major performers, joined by Macedonians, Armenians, and Jews (and, though rarely, Turks). Seeman speculates that the absence of Turks was due to the association of čalgija music with the lower social classes (1990a). In addition, secular music had a somewhat ambiguous status in Islam; thus musicians were often non-Turkish. Families of Roma such as the venerated Čun family have played čalgija for generations; Roma may have played a significant role in importing this genre from Turkey (Seeman 1990a:17–19). The Čuns lived in Kosovo before moving to Macedonia, and their repertoire also draws from Kosovo styles. Muamet Čun played in the Radio Skopje čalgija band and also for community events (see video example 5.18). In audio example 3.2, Muamet's brother, Medo Čun, performs the song "Ramajana" with Muharem Serbezovski on vocals (this song is discussed in Chapter 3, video example 3.1), and on video examples 10.3 and 10.5 he plays "Čhaje Šukarije" (which he claims to have composed) with Esma Redžepova and the Teodosievski Ansambl (this song is discussed in Chapter 10).

Early čalgija ensembles in Macedonian cities consisted of violin, ud (plucked, short-necked, fretless lute), *kanun* (plucked zither), *dajre* (frame drum with jingles), and voice, but they grew to feature *džumbuš* (fretless, plucked lute with a metal resonator and skin face), clarinet, *truba* (trumpet or flugelhorn), accordion, and *tarabuka* (Seeman 1990a:13; Džimrevski 1985). Čalgija repertoire included light Turkish classical pieces, rural folk music, and urban popular songs in the many languages of the Ottoman city: Turkish, Albanian, Vlach, Macedonian, and Romani. Čalgija music flourished in the Ottoman period in contexts such as the coffee house, weddings and other life-cycle celebrations, fairs, and saints' day celebrations. Note that Macedonian Roma have never played rural instruments such as gaida (bagpipe) and kaval (end-blown flute); nor have they sung the ritual songs of Slavic Macedonian villagers. Profound changes in the 1960s, such as migration of rural populations into urban centers, the spread of Western harmony and instruments, and introduction of amplification, affected the style and texture of čalgija (Seeman 1990a). Wedding bands in Macedonia, Bulgaria, and Kosovo were updated to include saxophone, keyboard, and drum set.

In comparing Skopje Romani celebrations of Herdelezi (St. George's Day) over a ten-year period, for example, Dunin notes that in 1967 unamplified music with no singer was the rule, but by 1977 amplified music including synthesizer and vocalist was more common. Amplification necessitated a fixed location for the band: a raised stage that is now ubiquitous (1985). At events such as weddings, the amplified band plays for the large dance gatherings, but either zurla and tapan or a smaller, unamplified, portable version of the amplified band plays for the rituals and processionals.

Through the Yugoslav socialist period, Romani music was available on recordings and on radio and television in Macedonia, as part of Tito's multicultural agenda. By contrast, the state music industry and institutions were themselves discriminatory to Roma; in Chapter 10 I trace the challenges of musicians in this period through the life history of Esma Redžepova. In Chapter 6 I describe how the amateur Romani dance ensemble Phralipe was formed in Skopje in 1949 and toured widely. I also describe how Yugoslav ensembles danced Gypsy suites using gross stereotypical movements and excluding Roma performers. Alongside the ensembles, however, there was a relatively healthy commercial Romani music industry.

Coming from a long line of male wedding musicians, Ferus Mustafov was one of the first Macedonian Romani instrumentalists to regularly appear on Western recordings. The son of the famous saxophonist Ilmi Jašarov, Ferus was born in 1950 in Štip, a center for brass bands. Ilmi issued some of the earliest LPs of čočeks in Yugoslavia in the 1970s; Ferus's mother was also a saxophonist (but not a professional), a highly unusual role for a woman. He went to music school until he was thirteen years old and then became a professional (Cartwright 2005b:125; Burton 1995). For several years, he left Macedonia to live in Sarajevo in order to work for the Bosnian television orchestra, and he later worked for Macedonian television. Ferus developed his father's style into a tighter, slicker sound and was more influenced by Bulgarian wedding music. In Chapter 7 I discuss how he traded tunes with Bulgarian wedding musician Ivo Papazov on the telephone; for example, his signature "Tikino" (Small) is claimed by Ivo. His son, Ilmi Mustafov, is now a respected musician. In video example 2.3 and photograph 2.4, Ferus plays a 2/4 čoček as a guest musician at the *babina* (birth party) of Muamet Čun's granddaughter.

In video example 2.4, from a 1985 Macedonian television show, Ferus and his band perform a 2/4 čoček (rhythm number 1 in Figure 2.1) dressed in pseudo-Turkish costumes, including turbans (illustrating the orientalizing and self-stereotyping that I discuss in Chapter 12). Also note the staging: in addition to two solo belly dancers who have much exposed skin but use characteristic Romani stomach movements, there is a group of non-Romani performers (the Ballet Troupe of Macedonian Television) doing unsubtle modern dance choreographies that have little in common with Romani dance (see Chapter 6).

In Macedonia today, a viable but somewhat unstable commercial recording industry regularly produces Romani artists (although it is smaller than in Bulgaria). Political rallies often feature music, and Romani radio stations are often aligned with politicians. In 1992 the private television station BTR began programming; in 1994 the private television station Šutel began; and since 1992 a national station, MTV 2, has produced the Romani language program *Bijandipe* (Renaissance). There are also several Romani radio stations in Skopje and others in smaller cities that feature music; they are financed by local advertisements and paid "greetings," mostly in the Romani language (for example birthday, wedding, or anniversary messages) that patrons write.

Roma in the Macedonian diaspora keep abreast of music through recordings, visits of performers for events, and most recently YouTube and Facebook postings. Popular Macedonian Romani bands of the last decade include Versace (cf. the Italian fashion house), Mladi Talenti (Young Talents [Macedonian]), Veseli Momci (Jolly Boys [Macedonian]), Bistijani, Ongeni Momčinja (Fiery Boys [Macedonian]); Gazoza (Shpritzer), and Titanik. Performers listen widely across Balkan borders. The older repertoire of Feta Šakir's band from the 1990s, is featured in video examples 5.9–5.16.

I discuss Romani music festivals in Chapter 8, but here I want to note that other institutionalized events in Macedonia (e.g., Romani calendrical holidays, beauty contests, and film festivals) sometimes feature music.[18] For example, the spring holiday of Herdelezi and International Roma Day (April 8) are often celebrated with concerts sponsored by NGOs. In January 1998, a combined celebration of Romska Vasilica (Romani St. Basil's Day) and Romska Ubavica (Most Beautiful Romani Woman contest) featured a singing contest with cash prizes. It drew singers from the diaspora, was attended by more than 1,200 people even with little advertising, and was held in the largest theater in Skopje. It was sponsored by one of the two Skopje Romani television stations, which is affiliated with a Romani political party.[19]

The history of Bulgarian Romani music has been documented in great detail by Peycheva (1999a, 2008) and Dimov (2009a, 2009b). Ibro Lolov (accordionist), Yashar and Anzhelo Malikov (composers), Hasan Chinchiri (composer and violinist), and the Takev brothers (violinists and guitarists) all represent a strain of Sofia-based Romani music that can be contrasted with the more Turkish-influenced styles in Thrace and north Bulgaria. As mentioned earlier, Yashar Malikov was a prolific song writer, composer, and arranger who (along with his son and Lolov) was involved in several pioneering recordings in the 1980s featuring songs in the Romani language. Since the 1990s, Lolov has issued new recordings and remakes of older songs.[20]

After the fall of socialism, Bulgarian Romani music burst forth in the public domain after years of government suppression.[21] The excitement at the time was palpable; in fact, the first concerts were tied to political mobilization, as Romani nongovernmental organizations sprang up everywhere. Anzhelo Malikov organized a Sofia Romani dance and music ensemble, but it disbanded after a few years for lack of financing and infrastructure (Peycheva 1999a). New commercial recording companies such as Lazarov Records, Sunny Music, Payner, Ara/Diapason BMK (Bulgarian Music Company), Unison Stars, and Milena began issuing scores of cassettes of Romani music (in addition to other genres); bands often self-produced master tapes and sold them to these companies, which functioned as distributors (Peycheva 1995: Dimov 1995). In the 1990s piracy was rampant, production standards were low, and the mafia controlled parts of the music business.[22]

In addition to ubiquitous cassettes, other media venues such as private radio and television channels sprang up in the early 1990s (Dimov 1995). Radio Signal Plyus and Radio Veselina offered pop/folk music, wedding music, rock music, folk music, and liberal amounts of Serbian, Macedonian, and Greek genres, which were the new rage,[23] financed by paid greetings and advertisements. Radio Veselina, for example, broadcast *Muabet Bez Parsa* (Dinner at the Table Without Tips [Bulgarian]), which accepted free requests. Music videos made inroads into homes in the 1990s, at first through videocassettes and later through private cable television stations. The Payner company started Planeta TV for pop/folk in 2001; Ara TV, Fen, Veselina TV, and other channels followed. (See Chapter 9 for discussion of Balkanika TV and folk/pop fusions.)

Today practically every Bulgarian family has cable, as prices have dropped to just a few dollars per month. In 2008, a corporately financed channel, www.gypsytv.tv started to broadcast sporadically from Sofia. Its programming includes clips from Romfest (see Chapter 8), Ivo Papazov (Chapter 7), and Goran Bregović (Chapter 13), as well as Gypsy jazz and belly dance instruction (Chapter 6). Romani music is widely available via privately owned media, but government channels rarely feature Romani music. Lack of government support is nothing new for Roma, and thus they do not rely on it. New talent quickly becomes known via dissemination of clips, and more recently via YouTube. For example, since 2008 the young clarinetist Sali Okka has become widely known in Romani communities even though he hasn't cut an album (Peycheva 2008a); similarly, the clarinetist/vocalist Alyosha (formerly with Orkestŭr Kristali) has received acclaim with Orkestŭr Universal, which has limited recordings. Musicians comment that they make their money from weddings, not from albums. With rampant sharing of music files and clips, fewer Roma buy music. On the other hand, the most popular Romani bands, such as Nasmi'ler, Kristal, and Kristali, have healthy album sales.

In the early 1990s, the band Džipsi Aver (Another Gypsy) from Sofia captured the spotlight for their innovations, for example with rap in the Romani language; they won the grand prize at the Stara Zagora Romfest in 1995 (see video example 2.5). Their name came from the English term Gypsy, previously unknown in Bulgaria but probably taken from the Gipsy

Kings. They also adopted stylistic elements from the Gipsy Kings, notably guitars, Flamenco-style clapping (synthesized), and dancers dressed in flared pseudo-Gitano skirts and scarves. On their video *Imam li Dobŭr Kŭsmet* (Do I Have Good Luck? [Bulgarian], Video Total) vocalist Dzhago Traykov performs a rap version of Esma Redžepova's song "Čhaje Šukarije" (Beautiful Girl; see Chapter 10). This video, in contrast to the slick videos of a decade later, depicted the band casually shopping in an open-air market; it has no back-up dancers and no set choreography, but there are two kyuchek dancers (with exposed midriffs) and a Michael Jackson imitator. The rap element links Roma to African Americans; this is a resonant tie, because both Roma and African Americans are minorities of color who have had a strong influence in popular culture but face discrimination (see Levy 2002; Currid 2000; Marian-Bălașa 2004).[24]

Amza Tairov is a relatively new but extremely popular artist in the Balkan Romani scene. Around 2003, a legend started circulating among musicians about a young Rom from Vinica, Macedonia, who shut himself up in his room for several years to learn the synthesizer. Although the legend is untrue, it illustrates the aura surrounding "Amzata." Synthesizer players from all over the Balkans have tried to imitate him, including his tendency to use only three fingers, and his equipment: a small Casio keyboard mounted above a larger Korg Triton that he uses for sampling. Note that Amza's reputation was established without the aid of commercial media recordings; it happened among musicians via traded live-performance recordings, much in the same manner that wedding music was transmitted in Bulgaria the 1970s and 1980s via unofficial cassettes (see Chapter 7).

Because Amza plays the synthesizer, he is able to innovate in multiple ways: melodically, harmonically, rhythmically, texturally, and timbrally (video example 2.6). Rhythmically he pushes the limits by syncopating inventively. In addition, he is a master of melody, especially improvisation and ornamentation, distinguished by his pronounced use of the pitch bender. Using a Korg Pandora effect processor, he also is a master of timbre; he samples folk instruments such as zurla for timbral and textural variation. He is a composer, arranger, and performer all in one. Not surprisingly, Amza's reputation has become trans-Balkan; he is now featured on several albums produced in Bulgaria, has been a guest on Bulgarian cable television, and is hired to play for Romani weddings all over Bulgaria. He is a pan-European star with multiple engagements in western Europe in the Romani diaspora and with multiple videos on YouTube (Peycheva 2008a).

Amza is much influenced by Romani styles from Kosovo. He imports an Albanian style into Macedonian music, and this has been adopted by many musicians and even exported to Bulgaria. Another significant musician in this Macedonian/Kosovo trajectory is the singer Ćita, from Mitrovica, Kosovo, living in Germany. Ćita is a master of the *talava* or *telava* Kosovo style. Talava is believed to be a contraction of the Romani phrase *tel o vas*, literally under the hand, referring to the solo women's Romani dance čoček where the hands are waved delicately.[25] Until the 1980s, talava was performed in segregated female gatherings by singers who were presumed to be homosexual (Pettan 1996a and 2003). Talava style is characterized by highly ornamented and syncopated vocal lines, often against a drone or constant chord, with *amanes*, vocal improvisations often sung on the syllables of the word *aman*, in which names of guests, commentary, and greetings are inserted. As a Macedonian Rom in Toronto stated, "They don't have a specific melody; they make it up on the spot, commenting on the bride, the groom, and their families." A New York Romani musician commented that "it has free style words about the people involved." These textual improvisations are especially evocative in the diaspora; Roma react emotionally when they hear personalized greetings from distant family members. This emotional and personal way of singing, with masterful technique, is admired by Roma all over the Balkans and in the diaspora. Even in Bulgaria, where it was never done before, Romani singers now improvise over a drone in talava style.

In audio example 2.9 with text supplement, Ćita performs a talava-style song over a repeated chord (with rhythm number 9 in Figure 2.1) at a Kosovo Romani event in Germany. Ćita's song comments on the perils of life as a refugee, raising themes of domestic violence, alcohol, and poverty. Other talava singers now popular are Ševćet (see above), Babuš (from Kosovo), Erdžan (from Kumanovo, Macedonia), Muharem Ahmeti (from Tetovo, Macedonia), and Džemailj Gaši (from Mitrovica, Kosovo, now living in Italy); Džemailj is the son of Nehat Gaši, popular in the 1970s. Many of these performers now live in Western Europe and command high fees, sometimes several thousand euros (see YouTube videos from western Europe and the Balkans). These singers employ the ornamental virtuosity characteristic of talava in their song repertoire; as one Rom stated, "They take an older style and put it in modern songs." It is ironic that Kosovo Romani music is currently such a strong influence among Balkan Romani musicians precisely at a time when Kosovo Roma lead a precarious existence; they are displaced, dispersed, and unwelcome in Kosovo.[26]

In concluding this chapter, I want to emphasize the vitality of Romani music in Macedonia and Bulgaria, as well as the huge number of performing musicians and the popularity of music in the diaspora. Although economic times are very bad, there are still celebrations among various ethnic groups, even if these events are shorter than they were in the 1980s. In Macedonia and Pirin, Bulgaria, zurla and tapan are typically used for rituals and processions, with an amplified band on stage for the banquets and evening dance. On any summer weekend in large Romani neighborhoods such as Šutka, there are still numerous weddings. There is also a prominent trend, especially in the Western European Romani diaspora, of hiring Romani musical stars from various countries. For a Romani wedding in Hamburg, Germany, in 2003, for example, the musicians included Husnu Senlendirici (a Turkish Romani clarinetist who leads the band Laço Tayfa, or Good Band), Vasillis Saleas (a Greek Romani clarinetist), Bilhan (a clarinetist from Mitrovica, Kosovo), and the Kosovo Romani singer Babuš. A 2007 wedding in Bujanovac, southern Serbia, featured

Orkestŭr Universal from Bulgaria (with Sasho Bikov on drums, and Alyosha, clarinet, saxophone, and vocals), along with Macedonian guests Amza and saxophonist Džafer.[27] A celebration in Dusseldorf, Germany, that I attended in 2011 featured the band Južni Kovači from Šutka, the violinist Sunaj, and talava singers Džemailj Gaši and Tarkan (living in Belgium). The New York community also hosts many talented musicians; in Chapter 5 I describe how musicians from the Balkans visit New York regularly to perform at community celebrations.

3

e⍟⍟

Dilemmas of Diaspora, Hybridity, and Identity

Although the title of my book solidly invokes diaspora, in this chapter I interrogate its theoretical provenance by exploring its applicability to Roma. Do Roma fit received definitions of diaspora? If not, what does this say about the model, and about Roma? How can the Romani case help interrogate related concepts of transnationalism, hybridity, and cosmopolitanism? Finally, I explore how these themes highlight identity issues in relation to music and the current Romani rights movement. Although the scholarly literature on diaspora is vast, only recently have ethnomusicologists tackled this concept (Slobin 2003; Ramnarine 2007a).[1]

William Safran's classic definition of *diaspora*, in the inaugural issue of the journal of the same title, relies on several core factors: migration from a singular historic homeland, vivid memories of the homeland postulating an eventual return, and belief by migrants that "they are not—and perhaps cannot be—fully accepted by their host country" (1991:83–84; also see Clifford 1994).[2] The Jewish and Armenian cases are usually posed as examples, but Daniel and Jonathan Boyarin (1993) have illustrated how variable the Jewish case is, and this may provide some parallels with Roma. Safran's definition fits Roma in terms of historic migration from India but does not fit in terms of migrants' consciousness of a homeland or their desire to return. Many Roma have learned about their Indian origins (and their victimization during the Holocaust) from scholars and activists, not from oral history. Activists are trying to inform them, as I discuss below. Note that although Roma do not wish to return to India, it figures symbolically in activist agendas and musical motifs. Rather than having a primary consciousness of an Indian homeland, Balkan Roma relate to more recent homelands in their historical rediasporization from the Balkans to Western Europe, North America, and Australia.[3] Macedonian Roma, for example, vividly refer to Macedonia as "home" (*doma*; Macedonian); those few who are informed of their Indian origins see no contradiction.

39

Homelands may be multiple and invoked strategically depending on context. For example, in the performance and marketing of Balkan Romani music, both the Indian homeland and the new homelands are invoked. The challenge remains to resist essentializing diasporas by attaching them to particular places of origin, i.e., homelands. A second challenge is to resist equating all diasporic subjects merely because they are related to a posited homeland; a third is to resist diluting the concept so much as to equate it with all migration. Mark Slobin writes that since the concept of diaspora has grown to embrace myriad forms of movement, it is overwhelmed by complexity and multiplicity (2003:290–291). Perhaps it is more practical to see diaspora as a special kind of migration involving some kinds of homelands, but not necessarily fixed ones. The Romani case, like the Jewish case, evinces multiple rediasporizations "which do not necessarily succeed each other in historical memory but echo back and forth" (Boyarin and Boyarin 1993, cited in Clifford 1994:305).

For diasporic Macedonian and Bulgarian Roma, I posit that Macedonia and Bulgaria, not India, are the more relevant "homelands." But even so, they do not function like iconic homelands in that Roma do not seek to return; rather, they make new homes in which they invest physically and emotionally but that they might leave. The very notion of home has to be reconceived (Malkki 1995:509). By asking "what does it mean to be emplaced" (515), we can approach the relationship between displacement and emplacement. The goal, then, is to study how ties to various homelands "are conceived and articulated and whether or not they erase significant historical differences . . . in different locations" (Dirlik 2000:177).[4] For Balkan Roma, migration—whether forced or voluntary—has become a way of life and a mode of adaptation; it is prevalent and valued because it is often necessary and irreversible. As Massey et al. write, "As migration grows in prevalence within a community it changes values and cultural perceptions in ways that increase the probability of future migration. . . . Migration becomes deeply ingrained into the repertoire of people's behaviors, and values associated with migration become part of the community's values" (1993:452–453).

Arif Dirlik, in analyzing the "Chinese overseas," offers a useful critique of how the concept of diaspora tends to level disparate peoples into one diasporic unity. This may lead to a "cultural reification" that erases the particulars of history and class and furthermore racializes the group (2002:95–99).[5] Similarly, Roma constitute a multiplicity of cultures that neither intermarry nor identify as one group; this variation is erased by conceiving of the Romani diaspora as a unified cultural unit. Notions of Romani identity that are based on Indian origins and homogeneous culture, then, may racialize Roma by emphasizing their non-European origins. Activists can sometimes capitalize on these notions and use them in pursuit of political agendas; similarly, musicians and music producers may use the Indian homeland concept in their art. But there may be a risk of exotification, which I will discuss below.

Transnationalism and Hybridity

In part to overcome the diasporic emphasis on a singular homeland, Basch, Glick Schiller, and Blanc promoted transnationalism as "the processes by which immigrants forge and sustain multi-stranded social relations that link together their societies of origin and settlement" (1997:7; also see Glick Schiller, Basch, and Blanc Szanton 1992 and Glick Schiller 1995).[6] This concept easily applies to Roma because it sidesteps the issue of origins and focuses on people and communities. According to Roger Rouse, diasporic groups "find that their most important kin and friends are as likely to be living hundreds of miles away as immediately around them. More significantly, they are able to maintain these spatially extended relationships as actively and effectively as the ties that link them with their neighbors" (1991:13). In Chapters 4, and 5, I detail how Roma in New York and Macedonia communicate via telephone, internet, and videos of music and ritual, as well as trips for new spouses. Here I emphasize the agency of transnational actors in enabling an "active display of identification in the making of diaspora" (Werbner 2002b:11); enactment of identity via performative genres, especially music, is a visible, audible symbol in the Balkan Romani diaspora. Through performance, identity is conceptualized: "the imagination of diaspora is constituted . . . by a compelling sense of moral co-responsibility and embodied performance" (11; also see Ramnarine 2007b). Some genres of music and dance play have become veritable emblems of identity in the diaspora.

At the same time, we must remember that diasporas are not homogeneous; a diaspora is a "site of multiple consciousness" (Toloyan 1996:28). Diasporas are "lived and relived though multiple modalities," as "differentiated, heterogeneous and contested spaces, even as they are implicated in the construction of a common 'we'" (Brah 1996:184). The tension among modalities applies to Balkan Roma; sometimes they identify as Roma and sometimes they adopt other labels, as I discuss in Chapter 4. Sometimes they unite with Roma from other places and other religions, and sometimes they reject other Roma. Finally, wherever they are, all of their negotiations are informed by historical discrimination and stereotypification.

Barbara Kirshenblatt-Gimblett reminds us that historically diaspora has had a negative, almost pathological connotation: "The terms diaspora and ghetto form a linked pair. What is not blamed on one is attributed to (and often entailed by) the other—stranger and marginal man flow from them" (1994:340) It is no accident that in America Jews, African Americans, and Gypsies iconically define the diaspora/ghetto mold in terms of where they lived and how they were conceived and rejected as "others." The classic American social science literature sees all three groups as problematic and deficient, needing to be assimilated and acculturated.[7] For refugees in particular, according to Malkki, "The bare fact of movement or displacement is often assumed a priori to entail not a transformation but a loss of culture and/or identity" (1995:508). Roma, for example, are often assumed to have no culture (especially music) of their own and to be inveterate borrowers (see Chapter 1). Kirshenblatt-Gimblett's statement "This is not a site of privilege" (1994:340) is echoed by Ong and Nonini: "There is nothing intrinsically liberating about diasporic cultures" (1997:325).

Current diaspora studies reject this pathology in favor of reclaiming hybridities, routes, mixings, and border crossings, but these concepts may be too celebratory and too ahistoric. The concept of hybridity, first popularized by cultural studies and postcolonial studies and then adopted by anthropology and other disciplines, is useful to destabilize binaries and bounded notions of culture that permeate the classic social science literature. As part of the postcolonial project, hybridity challenges Eurocentric master narratives, homogenization of ethnic identities, and the assumption that nations are composed of singular nationalities.[8] Perhaps its radical potential lies in its implicit rejection of those who have had the power to label and classify: "Hybridity in contemporary culture is in a fundamental sense a rebellion of those who are, or feel, culturally disposed . . . who challenge the claims of the centers of power" (Dirlik 2000:182).

The fact that Roma embrace hybridity and that Romani music is hybrid is perhaps obvious, but if the concept is to have any validity we must show how this hybridity works, why it exists, and how it differs from other explanatory models. Indeed, the fluidity of Romani music grows from the multiple diasporas of Roma, their openness to adopting non-Romani and multiple Romani styles, and their outsider status. For centuries Roma have had neither a singular state nor a national language, territory, religion, or culture. Historically, the professional marginal musician must be a hybrid to survive; multiple patrons require multiple musical repertoires.

Hybridity, however, can be a problematic concept because of its vagueness and its theoretical positioning. Hybridity is now so fashionable and applied to so many situations that it has begun to lose its specificity. As Dirlik writes: "If hybridity is indeed a condition of everyday life, what is radical about it?" (2002:189). Hybridity also brings up the problem of antecedent purity: "The idea of hybridity, of intermixture, presupposes two anterior purities. . . . I think there isn't any . . . anterior purity . . . that's why I try not to use the word hybrid. . . . Cultural production is not like mixing cocktails" (Gilroy 1994:54–55).[9] Either hybridity is everywhere, thus losing its theoretical force, or else it exists in specific places and is contrasted with the nonhybrid. I align with the latter position but do not subscribe to Paul Gilroy's view that hybridity always implies prior purities. True, no cultures are pure or bounded, which is to say, nothing is nonhybrid; but some interactions are more hybrid than others. I believe we can usefully recover the concept if we keep it grounded in historical specificities and resist its vague discursive seduction.

This brings up another criticism of hybridity, its abstractness, or more precisely its location in discourse rather in specific socioeconomic conditions. In Edward Soja's and Homi Bhabha's writings, hybridity is claimed to be a mode of consciousness that "releases the imagination to conceive of the world in new ways" (Dirlik 2000:182). Soja's term "thirdspace"

(1996) and Bhabha's term "inbetweenness" (1994) are similar to Appadurai's term "global imaginaries" (1996) in that they emphasize the realm of thought and creativity rather than on-the ground realities. Although this realm can fruitfully lead us to performance and style, it can also become too abstract, losing sight of precisely the material realities that inform the imaginary. Dirlik has been a vocal critic of valorization of the hybrid and the diasporic because these concepts can easily elide specific histories, structures, and power inequalities (1997, 2000, 2002). He points out that hybridity as an abstract concept may actually blur "in the name of difference, significant distinctions between differences . . . as if the specific character of what is being mixed (from class to gender to ethnicity and race) did not matter" (2000:184). Dirlik reminds us that hybridity may serve "not to illuminate but to disguise social inequality and exploitation by reducing to a state of hybridity all those who may be considered 'marginal,' covering up the fact that there is great deal of difference between marginalities" (184). Specific histories must always be examined.

Another danger in glibly using a term like hybridity is that it takes on a life of its own in identity discourse and loses political mooring.[10] Furthermore, the power of hybridity can be harnessed by reactionary as well as progressive causes: "Hybridity in and of itself is not a marker of any kind of politics but a deconstructive strategy that may be used for different political ends" (Dirlik 2000:187). Rey Chow elaborates the position that hybridity, though valorizing difference and disjuncture, may acquiesce to and support the status quo of global capitalism: "The enormous seductiveness of the postmodern hybridite's discourse lies . . . in its invitation to join the power of global capitalism by flattening out past injustices" (1998:156). John Hutnyk similarly writes that there is no problem with creative trading of cultures, but rather we must investigate the terms of the trade: "To think that a celebration of the trade is sufficient is the problem. Celebration of multicultural diversity and fragmentation is exactly the logic of the mass market" (2000:135). Along these lines, in Chapter 13 I investigate appropriation of Romani music by non-Roma for commercial transactions.

Dirlik points out that hybridity means different things to different class constituencies. To business investors it means internationalizing consumption markets, but to postcolonial scholars such as Bhabha and Soja it means a new kind of radical politics (Dirlik 2002). Concepts of multiculturalism, transnationalism, and globalism have been successfully used by corporations to recruit wider markets (Dirlik 1997:94–95). Gilroy similarly points out that hybridity has been annexed by corporate culture (2004:xix), and Žižek underlines that multiculturalism is manipulated by commerce (1997). Indeed, music marketers and producers have played an important role in the proliferation of hybrid Gypsy fusion genres such as Gypsy Punk, Balkan Beats, and DJ remixes under the rubric "world music" (see Chapter 13).

These critiques are useful for showing that celebrating hybridity may mask underlying inequalities. Similarly, generalizing all Roma as hybrid flattens them into one homogeneous group and obscures on-the-ground differences. Roma are divided by class, religion, language, sense of homeland, and identity label. Just as Dirlik's Chinese colleague felt "silenced by a concept such as hybridity which erases his differences from other Chinese" (2000:198), Roma, both activists and ordinary citizens, often reject pan-Romani scholarly labels such as *hybrid* in favor of historically informed particularistic labels and positions. The label *Roma* is a case in point because it is sometimes rejected as an outside imposition; the historical reasons for adoption or rejection of the label reveal much about social positioning. Communities of Muslim Turkish-speaking Roma in eastern Bulgaria, for example, label themselves *Turks*. They "became" Turkish during the Ottoman Empire, when Turkish culture and language were the marks of civilization and they could ascend the social scale by adopting them.[11] Today these communities are not so easily convinced to join the pan-European Romani rights movement; a change of identity would require deep reevaluation of selfhood and political awareness. Note that they persist in calling themselves Turks in spite of the fact that the local population (including Turks) refer to them as Turkish Gypsies. The label *Gypsy* is, of course, pejorative, and they do not see the term Roma as an improvement. On the other hand, in Chapter 11 I show how one musician from this community, Yuri Yunakov, employed multiple identification labels throughout his life (including Bulgarian, Gypsy, Romani, and Turkish) depending on context.

The concepts of cosmopolitanism and modernity may also help to interrogate Romani identities precisely because Roma have often been excluded from these categories; Roma are usually presented by non-Romani scholars, music producers, and sometimes themselves as "traditional," "premodern," bound by kin, custom, and conservatism (Van de Port 1998). On the other hand, Roma may also be viewed as the epitome of global postmodern European citizens: motley, diasporic, urban, transnational, with ultimate loyalty to no one state (or to many states), having no common religion, language, or territory. Their cultural traits sometimes resemble those of their non-Romani neighbors more than other Romani groups, and their music is innovative and open to various generic influences. They are always multilingual, multicultural, and multioccupational: in short; they are multisited and cosmopolitan, at home everywhere in a Europe that despises them. (See text supplement 3.2 for a fuller discussion of cosmopolitanism and modernity.)

Hybridity and World Music

In his 2000 book, John Hutnyk, a Marxist anthropologist specializing in popular fusion musics of England, provides an insightful critique of hybridity in relation to world music.[12] World music is above all a marketing label; it emerged in the 1980s and converted the conservative "international folk" section of record stores into a hip site of fusion and hybridity (Taylor 1997). But what does global capitalism's embrace of heterogeneity really mean? Bringing the musics of marginal peoples into the mainstream may yield visibility and even hard cash for formerly impoverished performers if they have fair contracts. But valorization of hybridity rarely changes the structures of inequality. For Roma it is true that some performers have become rich (even supporting whole villages in the Balkans) and Gypsy styles have been appropriated by mainstream non-Romani artists (see Chapter 13). But the overall structural domination of Roma has not changed. On the other hand, as I illustrate in several chapters, there have been many emancipatory artistic moments and even movements that could count as resistance.

Hutnyk shows how certain cultural forms become "the flavour of the month . . . the seasoning for transnational commerce. . . . Hybridity sells difference as the logic of multiplicity" (2000:4–5). In its meekest form, hybridity is not too far from the Disney version of multiculturalism: watered down, safe, distant. Liberals can feel good when buying a hybrid product like a world music CD because of the imputed connection to the dispossessed. In fact, marginality can becomes a kind of asset, a type of political cache, because of the assumption that marginal folks make good music, and we owe it to them to buy their products. It is certainly no accident that African Americans and Roma occupy similar positions vis-à-vis race and music. Hutnyk writes that "other love (anti-racism, esotericism, anthropology) can turn out to be its opposite" (2000:6). This is reminiscent of Renato Rosaldo's concept of "imperialist nostalgia," whereby the powerful destroy a form of life and then yearn for it aesthetically: "Imperialist nostalgia uses the pose of 'innocent yearning' both to capture people's imaginations and to conceal its complicity with often brutal domination" (1989:69–70). Thus Roma (or African Americans or Native Americans) suffer discrimination for years, and then white folks idolize and appropriate their music (or spirituality) as a means to erase this history and feel good.

Although marginality may be an attraction in music, it also may be erased by the illusion of success on stage. Part of the deceptive seductiveness of hybridity for audiences is the assumption that in art there is a level playing field. Hybridity, especially in music, comes with an aura of equality. The logic goes something like this: if Africans or Gypsies use Western harmony and electric guitars and appear in large festivals, they must be already integrated into the West and successful; and if they are successful, we assume they are compensated fairly and accepted fully by the mainstream as musicians and people. Of course, these are all false presumptions.

Hutnyk writes: "Difference within the system is a condition and stimulus of the market—and this necessarily comes with an illusion of equality, . . . 'crossed' cultural forms merely competing for a fair share" (2000:33). Few audience members bother to find out what performers are paid, what Western styles and instruments mean to performers, or how performers are treated once the show is over. Romani musicians relate many stories of being idolized on stage but being suspect in walking down the street (see Chapter 12). Furthermore, successful performers are unrepresentative of the vast majority of poverty-stricken Roma. Neither can we presume Romani world musicians to be representative of all Roma; nor can we presume that they have solved the problems of marginality.

The role of the exotic in representation of many world music styles has been noted by numerous scholars (Taylor 1997 and 2007; Stokes 2004; Radano and Bohlman 2000; Kapchan 2007). As I explore in several chapters, Gypsies are iconically pictured as sexual, eastern, passionate, genetically musical, and defiant of rules and regulations. It is precisely their outsiderness, their otherness, that makes them a valuable marketing commodity; many performers know this very well and capitalize on it. But performers are always negotiating the fine line between exoticism and rejection, between being a Gypsy on stage and passing as an ordinary citizen so as to avoid discrimination. Esma Redžepova's life history (Chapter 10), for example, illustrates how she embodied the multiethnic agenda of socialist Yugoslavia through her music at the same time she presented herself as an authentic Gypsy star on stage; simultaneously she faced prejudice in the Yugoslav recording industry.

In an ironic twist, the hybrid often becomes a mark of authenticity (even purity), and the two terms can even be found side by side in music marketing. In Chapter 12 I show how Gypsies are pictured as Europe's last bastion of authenticity by Europeans who are mourning their own loss of authenticity. Perhaps the authentic emanates magically from the hybrid because it is enacted by marginal artists: folks who look like they come from real communities with real rituals, songs, and dances, the very things most Europeans and Americans have lost, or think they have lost. Although the label "authentic" may valorize Romani music, it can also serve as a straightjacket, limiting choices of performers. In Chapter 13 I illustrate how West European audience members prefer acoustic instruments for Romani bands because then they can be sure they are getting "the real thing." As Hutnyk states, "The ghettoization of purity and authenticity serves only to corral the 'ethnically' marked performer yet again" (2000:31).

One of Hutnyk's most important points is that the celebration of hybridity by both postcolonial scholars and marketers is occurring precisely at a time when identities are becoming more political and battles are being waged for representation and turf:

> Why is it that cultural celebration rarely translates into political transformation? . . . At a time when class politics in the West seems blocked, does the shift to identity, hybridity and the postcolonial express a decline in aspirations (to transform the entire system) and an accommodation to things as they seem now and forever to be? Importing culturally "hybrid" styles via the mass media that sanitises and decontextualizes the political context for those styles . . . might be recognized as a danger [2000:119].

Hutnyk calls for engaged cultural studies where hybridity is not merely celebrated aesthetically and discursively but enmeshed in political struggles. The challenge I accept from Dirlik and Gilroy is to keep a focus on representation, performance, and aesthetics while still maintaining a solid connection to material conditions and history. Thus I turn to the relationship of music to identity issues and the current struggle of Roma for political rights.

Identity Politics and the Romani Rights Movement

As discussed earlier, Roma constitute a rich mosaic of groups that distinguish among themselves culturally and do not usually intermarry. The diversity of Romani groups is in part due to their diaspora; some Roma became sedentary, some are nomadic to varying degrees, some assimilated more than others linguistically and culturally, some adopted the religious beliefs of their neighbors (Hancock 2002; Guy 2001). Discrimination is sometimes the only thing that seems to unify Roma, and this is precisely what Roma seek to eliminate. Activists recognize this diversity:

> While East European administrators tend to look for the "uniqueness" and unity of a people's culture as a prerequisite for promoting distinct cultural entities . . . the Romani people is presenting itself as a huge diaspora embracing five continents, sharing the citizenship of a multitude of states, while lacking a territory of its own. The Gypsy "archipelago" is formed by a mosaic of various groups speaking both different dialects of Romani as an oral language and a variety of languages of the surrounding societies. The Romani communities share a number of religions . . .; they maintain cultural boundaries not only between themselves and the surrounding environment, but also between various Romani groups themselves [Gheorghe and Acton 2001:55–56].

Will Guy similarly asserts that "in view of the diversity of Romani experience, it would be more accurate to talk of a constellation of Romani cultures and . . . a cluster of varying and related identities rather than a homogeneous identity" (Guy 2001:28; also see Marushiakova and Popov 2001). Gheorghe and Acton also realize that the "multiculturality" of Roma can be a drawback to political mobilization: "it is still difficult to imagine how multiculturality and multi-territoriality could become the basis for the cultural affirmation and development of a people . . . which strive to identify themselves . . . in terms of unity and specificity" (2001:56). Although Mirga and Gheorghe suggest adopting the term "transnational minority,"[13] other activists use the terminology "ethnogenesis" (Guy 2001:19) or "nation." The International Romani Union's[14] Declaration of a Nation, in 2000, states: "Individuals belonging to the Roma Nation call for representation of their Nation which does not want to become a state. . . . We share the same tradition, the same culture, the same origin, we are a nation" (Acton and Klimova 2001:216).

Like indigenous rights movements that have used symbols for unification (such as the powwow of Native Americans), the Romani rights movement has created national symbols. They include a unifying label (Roma), a singular narrative of Indian origin, the Holocaust as a symbol of oppression, a flag, a literary language, and an anthem. Each is a trope that inscribes the legitimate historical place of Roma in the world; each corresponds to the dominant European tropes of defining the heritage of a singular nation. This is no accident, as the Romani movement seeks to legitimize the place of Roma in European politics. As discussed earlier in this chapter, the term *Roma* is used as an in-group label largely by Roma who speak Romani, but these Roma constitute only about half the world's Roma. Other designations, such as Gypsy, Sinti, Gitano, and Tsigan, have regional provenance but are sometimes contested in pan-European activist forums. Roma has emerged as the unifying term even in regions where it was never used.

Marushiakova and Popov claim a significant part of Romani nationalist ideology is "a fresh approach to Romani history emphasizing the Holocaust" (2001:49). Indeed, the Romani word for the Holocaust, *porrajmos,* is now widely used in Romani circles. The Holocaust has become a symbol of Romani oppression for several reasons: close to a million Roma perished at the hands of the Nazis; the facts are still not widely known and more research is needed; few Roma received compensation; and most important, Romani scholars and activists have had to fight to be included in Holocaust museums, memorials, and commemorations, both in Europe and America (Hancock 1987 and 2002). Unlike the Holocaust, which is a badge of suffering, the Romani anthem and flag are positive affirmations of Romani heritage and identity. Both were adopted at the First World Romani Congress, which took place in London in 1971. The flag is composed of a green lower portion, a blue upper portion, and a red wheel in the middle (photograph 3.1). Common explanations assert that green is the earth, blue is the sky, and the wheel is migration, but I have also heard activists claim the wheel is a spiritual sign, a mandala, signifying Indian ties.

Formation of a singular Romani literary language and production of a Romani dictionary were mandated several years ago by the International Romani Union, but progress has been slow. Many Romani dialects exist, and the language has changed and continues to change in relation to surrounding languages (V. Friedman 1985, Matras 2002). Deciding which dialect of Romani to elevate to the literary language is problematic, as well as deciding which orthography to use. According to V. Friedman (2005) and Matras (2005), a multiplicity of forms of literary Romani are emerging, which have national or regional provenance rather than international provenance. The challenge is how to network among these forms. According to Matras, the web of Romani language varieties (rather than a single form) "fits the specific Romani situation of a trans-national minority with dispersed regional centres of cultural and public life" (2005:31).

Music in the Romani Rights Movement:
Origins and Anthems

The Indian origin of Roma is supported by historical linguistics, but the precise time, location, and nature of the Indian exodus is contested (Matras 2002; Hancock 2002).[15] Activists, however, sometimes use dubious historical connections to prove cultural ties to India, such as asserting that certain Romani musical scales or dance steps come from India (see Chapter 2). Furthermore, Western European Gypsy music festivals are usually modeled on the documentary film *Latcho Drom*, which depicts linear nomadic migration, starting in Rajasthan, India, and ending in Spain (see Chapter 12). This may convey a misleading message that Rajasthani music today is what Romani music sounded like a thousand years ago. Furthermore, as discussed in Chapter 1, not all Roma are nomadic.

In the 1970s, Macedonian and Serbian Romani musicians embraced Indian-inspired melodies and songs, reflecting the growing diaspora consciousness of Roma. In Macedonia there was a veritable craze for Indian culture; parents gave their children Indian names such as Rajiv and Indira, and one famous singer made pilgrimages to India (see Chapter 10 on Esma Redžepova).[16] Movies from India were widely viewed by Roma (who could understand them because Hindi, like Romani, is related to Sanskrit), and movie tunes were turned into čočeks. For example, in 1990 the snake theme from the Indian movie *The Cobra* became "Sapeskiri Čoček" (the snake's čoček; Pettan 2003).

Muharem Serbezovski's Serbian song "Ramo Ramo," a tune inspired by an Indian film, became a hit in the 1970s.[17] Many versions were released in Yugoslavia. A Romani version appeared in Serbia/Kosovo in the 1970s as "Celo Dive Mangasa" (All Day We Beg; audio example with text supplement 3.1). This version emphasizes the themes of poverty and loss of a friend. Note the older Kosovo style of acoustic instrumental accompaniment, comprising clarinet, accordion, *džumbuš* (long-necked, fretless, plucked lute with a skin face), and tarabuka (the synthesizer had not entered the scene yet).

Another 1970s song invoking India is Serbezovski's "Ramajana," whose title refers both to the Hindu epic and his daughter's name. Like many musicians, Serbezovski (born in Topana, Skopje, in 1950) moved to Sarajevo in the 1970s to further his career; there he married a non-Romani Bosnian woman and faced issues around assimilation of his children.[18] This chapter of his life is chronicled in the song: when he asks his daughter if she speaks Romani, a chorus of children answer, "No I don't know Romani" (audio example with text supplement 3.2, video example 3.1). The song displays Indian elements not only in the text but also in the music.[19]

In the mid-1990s, the Serbian Romani singer Zvonko Demirović released "Stranci" (Strangers; Serbian) which deals with Indian origins and the tragic fate of Roma (audio example with text supplement 3.3). The song has a pop/jazz introduction followed by a slow 4/4 meter (number 5 in Figure 2.1). The text is more about the suffering of Roma than about Indian origins, but it does mention the posited origins of those in Macedonia who call themselves Egyptians. Hundreds of other songs deal with poverty, suffering, orphaned children, death, and lack of work—in general, the hard lives of Roma. In fact, these are the some of the most common themes of Balkan Romani song texts.

A few songs deal with resistance to suffering, that is, fighting back or rising up. One such song is "Kemano Bašal" (The violin plays; by the singer Džansever, discussed in Chapter 2; on *Bašal Kemano/ Violino Sviri*, Sokoj MP 21102), which metaphorically calls for the unity of all Roma (audio example with text supplement 3.4). Some Macedonian Roma refer to this song as their "anthem."[20] Besides Kemano Bašal, there are other local "anthems" such as "Ciganyhimnusz" (The Gypsy anthem), which is widespread in Hungary (Lange 1999). Recently, a group of Romani musicians including Esma Redžepova, Kal, and rapper R Point composed versions of the European Union anthem for the antiracism campaign *Dosta* (Enough), funded by the Council of Europe (http://www.coe.int/T/DG3/RomaTravellers/dosta_en.asp, accessed June 17, 2011).

The song that is claimed most widely as the Romani anthem is "Dželem Dželem" (I traveled and traveled), whose trajectory is a good illustration of the interplay between politics and music. The melody of the song (in a minor scale) was in oral circulation in multiple variants in the Balkans at least since the late nineteenth century. It is possibly a Romanian song adopted into Serbia (Marushiakova and Popov 1995:20–21). It became popular when it was featured in the 1960s in the Yugoslav feature film *Skupljači Perja* (literally, the feather buyers, known in English as *I Even Met Happy Gypsies*, 1967), directed by Aleksandar Petrovič.

The song's more recent political import is tied to the April 1971 meeting of the Comité Internationale Tsigane in London. Romani activists gathered and reconstituted the Comité as the First World Romani Congress, which was eventually constituted as the International Romani Union. Serbian singer Jarko Jovanović embellished the song for the Congress by taking several verses from oral tradition and writing several new ones. The British sociologist Thomas Acton reported that "Jarko composed them on a bus, the day people from the congress formed a delegation and went to Walsall, where three little Romani children had been burned to death in a trailer while the police were towing [it] away" (Gelbart 2004:1). Donald Kenrick, a British linguist and educator who was also there, remembers that Dr. Jan Cibula, a Romani activist from Slovakia, contributed to the text. It was adopted as the official song of the Congress and sung at its closing (Marushiakova and Popov 1995:21). It eventually became the "Romani anthem" through use at numerous political gatherings, and it now frequently opens or closes major Romani political events in the diaspora such as International Day of Roma, April 8. For example, Esma Redžepova sang "Dželem Dželem" at the opening of the Skopje Romani music festival Šutkafest in 1993 (video example 3.2 and Chapter 10).

The lyrics in text supplement 3.1 are taken from Gelbart (2004) but modified in translation and orthography. Gelbart took them from the Albanian

Romani group Rromani Dives (Romani day).[21] The text exhibits a strong and indicting reference to the Holocaust, not common in Romani songs. However, as I noted above, since 1989 activists have mobilized the Holocaust as an organizing symbol for Romani unity and resistance. I believe the song has a more general emotional appeal than a specific historical appeal. There are hundreds of variants in circulation, and many versions have been commercially recorded in myriad styles, although its popularity is greatest in Serbia and Macedonia.[22]

Most of the variants now in circulation neither have overtly political texts nor mention the Holocaust. Serbian Romani singer Šaban Bajramović's 1980s version about love has been countlessly emulated both in text and melodic contour. According to Gelbart, activist Valery Novoselsky claimed the song "is important not only for our politicians and representation but for ordinary people also. . . . [When non-Roma hear it] they can understand more of who we are" (2004:3). Activists point out its political function, but ordinary Roma often become teary when they hear the song. It sometimes helps to bridge the gap between Roma and non-Roma.

When Macedonian Romani singer Esma Redžepova performed the song in Serbian at a private New York City party in 1996, the audience consisted of Macedonian Roma, Serbian Roma, and American Kalderash. These groups do not normally socialize, and there is little camaraderie among them. Esma directed her performance of the song (in Serbian) to the Serbian Roma and the American Kalderash (see photograph 3.2 and video example 3.3); the rest of her program was directed to the Macedonian Roma. But at the moment of performance, there was a palpable feeling of unity in the room. On the other hand, sometimes the song fails to achieve this unifying function. In Chapter 12 I recount how "Dželem Dželem" was rejected as a finale piece by most of the musicians in the 1999 Gypsy Caravan tour because they didn't relate to it. Finally, video example 3.4 shows Esma singing "Dželem Dželem" to a mixed audience of Macedonians and Roma in 2004 at a Macedonian church in Garfield, New Jersey. The song thus reveals a complex web of identity politics and charts how Roma choose to represent themselves.

The Crucible of Identity

In examining use of the anthem, the flag, and the quest for a literary language, we see that although Roma have been excluded from the dominant tropes of national folklore and cultural heritage (see Chapter 7) they have constructed their own symbols of heritage as part of a strategizing process in European politics. Herzfeld points out that "states AND citizens both depend on the semiotic illusion—that identity is consistent; they both create or constitute homogeneity and produce iconicities" (1997:31). Although "essentialism is not exclusively a state activity . . . states do have a rich variety of devices [and I would add institutions] for essentializing. . . . It seems like common sense" (31). Marginalized ethnic groups such as Roma often engage in what Gayatri Spivak has termed "strategic essentialism" (1988) in the cause of mobilization. Herzfeld reminds us that "powerful state agents and humble social actors all engage in the strategy of essentialism to the same degree." In fact, "social poetics is the analysis of essentialism in everyday life" (1997:31).

Herzfeld rightly draws our attention back to essentialism, a concept that has been so demonized in cultural theory that Werbner called it "the bogey word of the human sciences" (1997:226). Perhaps the concept of hybridity became so fashionable because it seemed the perfect antidote to essentialism. Demonization of essentialism is quite unfortunate because we can never understand identity politics without it. Furthermore, as scholars we remove ourselves from the trenches of political struggle when we point fingers and assign accusatory labels. As Dirlik writes: "It seems that any admission of identity, including the identity that may be necessary to any articulated form of collective political action, is open to charges of essentialism" (2000:188; also see Dirlik 1997).[23] Similarly, bell hooks welcomes a critique of essentialism but warns:

> This critique should not become a means to dismiss differences or an excuse for the ignoring of experience. It is often evoked in a manner which suggests that all the ways black people think of ourselves as 'different' from whites are really essentialist, and therefore without concrete grounding. This way of thinking threatens the very foundation that makes resistance to domination possible [1990:130].

Both hooks and Dirlik remind us of the irony that postmodern/postcolonial intellectuals have the luxury to repudiate essentialized identities. These scholars construct "identities and histories almost at will in those 'in-between' places that are immune to the burden of the past," whereas those who suffer "the sentence of history" are supposedly too caught up in the past and thus misguided in their collective claims (Dirlik 1997:221).

Too often academics intellectualize an unequal playing field into an abstract argument. Nicholas Thomas states:

> Clifford writes as though the problem were merely intellectual: difference and hybridity are more appropriate analytically to the contemporary scene of global cultural transposition than claims about human sameness or bounded types. I would agree, but this does not bear upon the uses that essential discourses may have for people whose projects involve mobilization rather than analysis. . . . Nativist consciousness cannot be deemed undesirable merely because it is ahistorical. . . . The main problem is not that this imposes academic (and arguably ethnocentric) standards on non-academic and non-Western representation, but that it paradoxically essentializes nativism by taking its politics to be uniform [1994:176].

Both Thomas and Dirlik encourage scholars not to dismiss cultural and historical claims to collective identity as mere essentialisms, but to analyze them as works in progress in a hierarchical political playing field.

For Dirlik history is critical; he titles a chapter in his book The Postcolonial Aura "The Past as Legacy and Project: Postcolonial Criticism of Indigenous Historicism" to differentiate history as static heritage from history as a political project. Using indigenous cultural politics as a case study, he writes that its political significance lies "in its claims to a different historicity that challenges not just postcolonial denials of collective identity but the structure of power that contains it. To criticize indigenous ideology for its reification of culture is to give it at best an incomplete reading" (1997:228). Thus the use of cultural and historical symbols in political struggles of marginal peoples cannot be merely explained away as "social constructions."

With the case of Roma, although we may be tempted to label their nationalist symbols "invented traditions" because they are newly created, we fall into several traps by employing the term *invented*. Hobsbawm and Ranger (1983) first used the term to refer to symbols and practices that figured prominently in European nationalist discourse but were of recent historical provenance. They therefore implied that some traditions are real or authentic (meaning old) while others were invented, hence made-up and inauthentic. Handler and Linnekin (1984), Wagner (1979) and Hanson (1989) broadened the argument to claim that all traditions (and for Wagner, culture itself) are invented in the sense that they are social constructions. Thus authenticity is itself a social construction. This constructivist position fit nicely into the 1980s postmodernist critique of bounded notions of culture but couldn't have been more ill-timed in terms of world politics.

Indeed, the 1980s were precisely the era of the emergence of identity politics, when marginal groups were finally taking center stage and defining their own histories and symbols. As Clifford states: "For just at the moment the radical post-structuralisms became popular in the US academy, a whole range of formerly marginal and excluded peoples and perspectives were fighting for recognition: women, racial, and ethnic minorities, new immigrants. These groups, for the first time entering the public sphere, often felt the sophisticated cultural critics to be, in effect, telling them, 'Oh yes, we understand your gender, race, culture and identity are important to you, but you know, you're just essentializing'" (2003:64). Indigenous scholar/activists such as Haunani-Kay Trask (1991) rejected "the implication that dynamic traditions were merely politically contrived for current purposes" (Clifford 2004:156) and criticized constructivists as neocolonial outsiders who were thwarting the legitimate political agendas of marginalized people. Other scholars analyzed the confrontation between these two sides, arguing that we should simultaneously abandon the loaded language of "invention" and interrogate all positions as to motivations, agendas, and funding (Briggs 1996).

Taking Briggs's suggestion, I aim to elucidate how the concept of heritage/tradition can be pried from its narrow historical moorings so we may understand the symbols of the Romani rights movement as historically placed responses to marginality. At the same time, an expanded notion of heritage can help us widen bounded notions of national culture to embrace multicultural and hybrid forms. Indigenous heritage movements, such as those of various Native American groups, can serve as useful comparisons. Dirlik writes: "Contrary to critics . . . who see in every affirmation of cultural identity an ahistorical cultural essentialism, indigenous voices are quite open to change; what they insist on is not cultural purity or persistence, but the preservation of a particular historical trajectory of their own" (1997:223).

As Clifford notes, indigenous leaders are simultaneously loosening and reclaiming the notion of authenticity; sometimes authenticity can be "a straightjacket, making every engagement with modernity (religions, technologies, knowledges, markets, or media) a contamination, a 'loss' of true selfhood" (2004:156). Rejecting their emplacement in the past, native leaders are asserting their legitimate place in modernity through global displays of media, technology, and legality. Simultaneously they are claiming land, reviving languages and rituals, reclaiming sacred objects and burials from collections, building cultural centers, and representing themselves in museums. Similarly, Roma are starting to establish cultural centers, design exhibits, produce films, publish histories, and produce their own music festivals and albums (see Chapter 12). As Clifford writes, these are "zones of contact" (1997:188–219), "whereby authenticity thus becomes a process—the open-ended work of preservation and transformation. Living traditions must be selectively pure: mixing, matching, remembering, forgetting, sustaining, transforming their senses of communal continuity" (2004:156). To examine what Roma and other marginalized groups are doing is to implicitly interrogate and rethink received notions of tradition and authenticity.

Clifford claims that "what is at stake is the power to define tradition and authenticity, to determine the relationships though which . . . identity is negotiated in a changing world" (2004:157). The challenge is to reject both a pro- and anti-essentialist position and to embrace an anti-anti-essentialist position. As Clifford writes:

The two negatives do not, of course, add up to a positive, and so the anti-anti-essentialist position is not simply a return to essentialism. It recognizes that a rigorously anti-essentialist attitude, with respect to things like identity, culture, tradition, gender . . . is not really a position one can sustain in a consistent way. . . . Certainly one can't sustain a social movement or a community without certain apparently stable criteria for distinguishing us from them. These may be . . . articulated in connections and disconnections, but as they are expressed and become meaningful to people, they establish accepted truths. Certain key symbols come to define the we against the they; certain core elements . . . come to be separated out, venerated, fetishized, defended. This is the normal process, the politics, by which groups form themselves into identities [2003:62].

Stuart Hall makes the point that identity politics arises precisely around issues of representation (also see Hancock 1997): "Though they seem to invoke an origin in a historical past . . ., actually identities are about . . . using the resources of history, language and culture in the process of be-coming . . .; not 'who we are' or 'where we came from' so much as who we might become, how we have been represented, and how that bears on how we represent ourselves. Identities are, therefore constituted within, not without representation" (1996a:4). Hall's concept of identity rejects an un-changing traditional core; it "does *not* signal that stable core of the self, unfolding from beginning to end through all the vicissitudes of history without change. . . . Nor . . . is it that 'collective or true self hiding inside the many other, more superficial or artificially imposed "selves," which a people with a shared history . . . hold in common' and which can stabilize, fix, or guarantee an unchanging 'oneness' or cultural belongingness under-lying all the other superficial differences." Rather, identities are "never unified, and . . . increasingly fragmented and fractured, never singular but multiply constructed across . . . intersecting and antagonistic discourses, practices and positions" (Hall 1996a:3–4). For Roma, identity has always been construed in relation to hegemonic powers such as patrons of the arts, socialist ideologues, European Union officials, and NGO funders.

According to Clifford, "tradition is not a wholesale return to past ways, but a practical selection and critical reweaving of roots" whereby "some essentialisms are embraced while others are rejected (2004:157). Tradition should not be read as "endless reiteration but as 'the changing same,' not the so-called return to roots but a coming-to-terms with our routes" (Hall 1996a:4). Here Hall is referencing Paul Gilroy's useful formulation of tra-dition as the "changing same" (1993:101). Gilroy advocates that the term *tradition* be used "neither to identify a lost past nor to name a culture of compensation which would restore access to it" (198). The "lost past" is sometimes conceived by African-American writers and activists as the African homeland, whereby "Africa is retained as one special measure of their authenticity" (191). But, according to Gilroy, this ignores the impor-tant place of the diaspora in forging African-American identities. Simi-larly for Roma, Indian origins, whether historical, linguistic, or cultural, are valorized but diasporic flows and cultural circulations define the Romani experience.

Rather than standing in opposition to modernity, tradition indicates a specific relation to it: "We struggle to comprehend the reproduction of cul-tural traditions not in the unproblematic transmission of a fixed essence through time but in the breaks and interruptions which suggest that the invocation of tradition may itself be a distinct, though covert response to the post-contemporary world" (Gilroy 1993:101). Gilroy, Clifford, Hall, and Briggs all urge us to analyze specific identity projects in their historical contexts, paying special attention to inequalities and hierarchies. Just as the project of African-American identity making was forged in the crucible of slavery and diaspora (Gilroy 1993) and the project of Native American identity making was forged in the crucible of genocide and displacement, similarly the project of Romani identity making was forged during centuries of discrimination and diaspora. The marginal position of these groups has led to an urgency of cultural matters tied to human rights and global entitlements. And music in diaspora contexts assumes an especially important place in this process, as I illustrate in the chapters that follow.

PART II

MUSIC IN DIASPORIC HOMES

4

Transnational Families

This chapter examines the issue of transnationalism from the point of view of Romani communities. Romani families are "transnational," defined by Bryceson and Vuorela as those "that live some or most of the time separated from each other, yet hold together and create something that can be seen as a feeling of collective welfare and unity, namely 'familyhood,' even across national borders" (2002:3). Not only does Romani music travel in transnational circuits; Romani musicians and community members travel and communicate in a diasporic network.

As mentioned in Chapter 1, the center of Macedonian Romani life in the United States is located in the Belmont neighborhood of the Bronx. Nermin, a middle-aged woman living in the Bronx, summarized: "We stay in touch with the relatives at home [in Macedonia]—we speak on the phone, we send music, we send videos of our weddings. People come here." Almost all Roma in Macedonia have relatives abroad in many countries of Western Europe, in the United States, and in Australia. In this chapter, as well as the following two, I explore how the New York community cements its ties to Macedonia and other diasporic locations through marriage, language, and ritual, all enacted performatively via cultural markers such as music and dance.

Migration Narrated in Song

Since the 1960s, Balkan singers in general and Roma in particular have used the theme of *gurbet* or *pečalba* (working abroad)[1] to lament the separation of loved ones. For example, after he arrived in New York in 1992, Ferhan Ismail composed the text of the song "Gurbeti" to a Turkish melody and recorded it (audio example with text supplement 4.1, photograph 4.1;). Ferhan wrote this precisely when he had emigrated from Skopje, and its text voices the pain of separation. Another song, "O Gurbetluko," composed by the Macedonian Romani singer Ramko (Ramadan Bislim), tells about a dying father whose son went abroad to work (audio example with text supplement 4.2). The father's bitterness from illness and separation causes him to curse his son in his last living moments. This is a very grave utterance in Romani culture, as children are ideally sacred.

Similarly, the song "To Phurano Bunari" (Your old well; audio example with text supplement 4.3), composed by Abas Muzafer of Šuto Orizari, laments separation by way of his brother's wedding in Germany, which he cannot attend. In this text as well as the previous two, note first that money becomes worthless or "cursed" when compared to the ordeal of separation from family. Second, family rather than place is missed, underlining the person-oriented rather than place-oriented values of Roma. Money is blamed for the pain of loss, and, strikingly (in "To Phurano Bunari") it becomes bloodied money. Third, the guest worker is depicted as suffering abroad while his relatives suffer at home. He is pictured as lonely and isolated, reduced to a prisoner begging for bread, even wanting to die. Finally, in "To Bunaro" home is described as an "an old well." Muzafer's home, indeed, had a well in the courtyard, so this is a personal vision. These texts provide an artistic view of immigration stories; when the songs are performed in the diaspora, Romani audiences are visibly moved. Similar sentiments are kindled when talava singers improvise greetings to relatives abroad (see Chapter 2).

The movement of people, things, and ideas occurs among several sites in the diaspora, occasionally even without reference to Macedonia. Although Macedonia is the nominal "home," Roma often prefer to travel to other diasporic locations. A woman in Toronto, for example, saved money to visit her sister in Melbourne whom she had not seen for twenty-five years; this was more important than a cheaper trip to Macedonia, where she has many more relatives. Travel is contingent on having proper documents and substantial money for tickets and expensive gifts; visitors are expected to treat relatives to meals and sponsor banquets; these practices are often the reason families cannot afford to travel.

The trajectory of one family clearly illustrates transnational migratory patterns. In the late 1980s, Osman lived in Belmont; his natal family consisted of a brother who lived with their aging mother in their hometown, Prilep; a brother in Germany; and a sister, Aiše, who married Ali and lived in Skopje. He invited me to meet his wife, Jasmin, and his sister, Aiše, who was visiting in order to attend Osman's son's circumcision party and to earn some money.[2] Short work visits were a common occurrence in the 1980s. Aiše invited me to visit her in Macedonia, and she helped me arrange my living quarters in Šuto Orizari in 1990. After Aiše returned to Macedonia, one of her brothers came to Belmont to visit and work. His son also came to visit and eventually emigrated. Another brother emigrated from Germany with his wife and children. In 1992 Aiše's oldest son, Ramo (born 1970), arrived in Belmont, and then Aiše came again in 1994. During all these extended visits, Osman and Jasmin were their hosts, housing and feeding them in their small two-bedroom apartment, and helping them find work and social connections. Their generosity was boundless. I especially remember the care that Aiše gave Ramo when he had an accident, was incapacitated, and had to be nursed back to health.

Aiše and Ali also had a younger son, Rifat, who had remained home in Skopje with Ali. I remember a moment in August 1994 when Aiše and Ramo in the Bronx telephoned Ali and Rifat in Skopje and everyone started crying, all thinking of loved ones in the diaspora. Since family means everything, people suffer when family members disperse; yet they must, for economic and sometimes political reasons. Note that the two sons were at the age when marriage becomes a factor. In 1995, Ramo planned to marry Metola, who lived in the Bronx and was born in Prilep. Meanwhile, Rifat (living in Skopje with his father) became engaged to marry Fatima, a Romani woman born in 1973 in Toronto whose parents were from Prilep (see the story of their marriage later in this chapter). The parents wanted a double wedding for their two sons, but the celebration couldn't take place because Ali could not get a visa for the United States. Rifat could travel freely because of his wife's legal Canadian status; Aiše and Ramo, however, could not leave the States, so they couldn't go to Toronto to see Rifat. The only solution was to wait and keep applying for an American visa for Ali. After more than a year, Ali finally received a visa and the wedding was held (see next chapter).

Viewing this situation from both American and Balkan viewpoints helped me understand how Roma negotiate across distances. When I traveled, I was given gifts to bring to Skopje and was instructed to take copious photographs and videos to show to relatives. Whereas Aiše hadn't seen her husband in more than a year, other relatives who traveled saw him, brought gifts, and gave each side pictures and videotapes (some of which I had filmed). I even served as part of a large communication network consisting of relatives traveling back and forth visiting, working temporarily, and looking for spouses. After Ali emigrated to America, his family's transnational ties multiplied significantly. Almost all of the children of his brothers and sisters in Šutka have married Macedonian Roma in the diaspora. Rifat and Ramo now have cousins in Germany, Austria, Switzerland, Belgium, and Holland.

Early Emigration Stories

Nermin was one of the first Macedonian Roma to emigrate to America in the 1960s.[3] She narrated:

Before I came to America I worked in a state job in Macedonia. I had four kids. . . . A cousin of mine came to America via Vienna. . . . My husband was also in Vienna working. My cousin's wife was here six months and she encouraged me to emigrate. She said, "You have to come over here, Nermin, I can't live here alone." They sent us documents, guarantees. Roma didn't go to America then—perhaps to Austria but not any farther. We came to America in 1968. There weren't any Roma here then.

Leila, who was ten years old when her family arrived from Prilep in 1971, described why her family migrated:

My dad [Zahir] was one of the first few Roma in the town to become educated. That in itself is an accomplishment. He had the opportunity and the desire to . . . go to school. My uncle supported him—he put him through school. He became a veterinarian, got married, and had me and my sister. But, as an educated Rom he realized that opportunities were very limited for us, his children. It was super hard for him to get to where he was, and he didn't want it to be that hard for us. So, he came to the States. We came through Vienna, through some friends. Once we got established my dad brought his nephew here. And then my mom brought her sister and her brother, then my grandparents—we've extended the family.

Whether arriving in Australia, Western Europe, or New York, newcomers lived with relatives who furnished food, housing, clothing, child care, and work contacts for them until they could venture out on their own. Note that a substantial burden of providing for new arrivals falls on the women, since they are in charge of the domestic sphere. When the hosts themselves are newly arrived, the strain can be intense. Nermin illustrated:

After we were here for a year a brother of my husband came with his eight children. They stayed with us for three weeks. After that, three more families came. We took care of them for five-six months—it was a very harsh winter. Then five of my brothers came, then my sister, then two more brothers, then my mother, then another sister. Gradually the whole family came and now everyone is here. . . . We lived for three years as a super[4]; then we bought a building on Belmont Ave. More and more came, each one helping the next. We have over three hundred houses here.

Nermin's comments illustrate that kin ties are activated in the female as well as the male line. She sponsored more relatives than her husband, despite the fact that Romani society is (ideally) patrilineal and patrilocal.

Many of Nermin's relatives migrated to Western Europe and to Australia as well as New York. These families are transnational and multisited, although they label Macedonia home (Bryceson and Vuorela 2002; Al-Ali and Koser 2002). Many family members have lived in at least three countries. This experience means that multilingualism is the norm. In Macedonia, Roma were and still are multilingual; the older generation spoke Turkish in addition to Romani and Macedonian. Today, in Prilep and Skopje, Roma tend to speak Romani, while Roma in Bitola speak Turkish. The Romani language is, however, on the decline in Belmont; the trend is to retain Macedonian-English bilingualism. One counter trend, however, is the constant trickle of new Romani-speaking spouses and visitors coming from Macedonia.

Work and Family Life

All Belmont families started their immigrant experiences with virtually no material resources; rather, they relied on human resources. Leila illustrates: "When we came, we only had two hundred dollars on us. My uncle took us in. He helped my dad and my mom get a job. He helped us find an apartment. He took us under his wing, so to speak. He was a friend of my dad from back home. My dad bought a house within three years after coming here." Both men and women are expected to work outside the home; this was true in Macedonia and remains true in the diaspora, as Nermin narrates:

> There was work to be found even without the language, but for the best work you had to know the language. The kids were learning the language, but the parents? My husband's brother found work downtown as a janitor. One of my brothers was a tailor—we found him work; another brother was an electrician so we found him work. We worked for $1 or $1.50 an hour, $30 a week [in the 1960s]. My husband made $60 a week. Our salaries together were $90 a week. We worked at night—we had to leave the kids at home alone. But things weren't as dangerous then as they are now. Life was pretty good and we saved money even though we made so little.

There is a strong work ethic in this community; having a job is the norm and laziness is condemned. Everyone believes there is work to be had, even if it is unskilled or menial. Unlike in Western Europe, few Macedonian Roma in America are officially refugees, so they are not entitled to social services. They view welfare as somewhat of a stigma and prefer to support themselves.

The occupations in which Belmont Roma engage are coded by gender, just as they were in Macedonia. Some common male occupations, such as electrician, construction worker, car mechanic, and tailor, have transferred well to America. Other occupations such as metalworker do not as readily transfer. Ali, born in 1948 to a family of *kovači* (blacksmiths), had a Skopje home workshop where he crafted metal objects, plus a stall in the open market where he sold these objects and traded clothing. A creative combination of trades is very characteristic of Balkan Roma (Silverman 1986). In 1990, I observed that Ali was marketing his metal work in five languages. When he emigrated to the United States, he crafted fences and ornamental wrought iron for private homes, but his earnings suffered and he did not want his sons to continue his profession. Many electricians, on the other hand, have successfully trained their American-born sons to take over the family business.

Some males have to take any job available, such as factory work; security guarding; bread, pizza, and meat delivery; and janitorial work in schools, nursing homes, hospitals, and office buildings. One family opened a hot dog booth, sharing hours among male and female members and making a modest income. Ali's son, who was well educated, worked his way up from a meat deliveryman to a manager and eventually established his own meat distribution company. Several professional musicians combine music with a day job. For example, drummer Severdžan Azirov worked as a delivery van driver, and singer Nešo Ajvazi worked as a janitor (see Chapter 5). No females are professional musicians because of the stigma of performing in public (see Chapters 6 and 10, and Silverman 2003).

Female employment is a necessity in almost all families, although the ideal is a sole male breadwinner.[5] If there are small children and no older females to care for them, mothers stay home; day care centers are rarely used. The middle and older generation of women work as cleaning ladies in office buildings, as cutters in the clothing industry, as sales clerks in neighborhood shops, as caretakers for the elderly in their homes, as food managers in nursing homes, and as hair stylists and cosmeticians. These jobs are similar to those in Macedonia, with the exception that in the Balkans they were state jobs with stable pensions and vacations. In the United States, there is little security in terms of employment and benefits. Much depends on legal status. Those who are undocumented, male or female, are extremely limited in their jobs. In the 1980s, for example, Roma worked in a neighborhood plastics factory for $4.00 per hour. Undocumented workers have no job security, no vacations, no pension plans, and no medical insurance; they are constantly afraid their employers will report them.

Belmont is a multigenerational community. Despite the youth orientation of American culture, elders occupy a venerated position in Romani families. Female elders sometimes work, but they also do child care and visit. Typical Belmont households are multigenerational vertically, but not horizontally (via brothers), as is more common in Macedonia. Ideally, in one dwelling live a son, his parents (and perhaps his grandparents), his wife, and his children. Girls live at home until they marry, when they move in with their husband, whereas boys rarely move out. In Macedonia the *zadruga* was a patrilineal, patrilocal, extended familial residential unit that communally owned resources. In its classic form, all brothers with their families lived together and pooled income. Although this is rare today in Macedonia, the value of living together in a large unit persists in the diaspora. Tasks can be divided among available and skilled men and women, child care is easier for women, and emotional ties ensue.

On the other hand, living in close quarters generates conflict. For daughters-in- law, who are the least powerful members of the family because they are female outsiders, living with their husband's relatives is especially challenging. The mother-in-law, who supervises and trains the daughter-in-law in domestic and ritual tasks, can be very critical. Young people currently crave privacy, and if they can't get it at home they escape to the streets, especially if they are males. Monetary conflict may also erupt, especially when finances are tight. In spite of these challenges, children rarely move out before marriage not only because of family bonds but also because they can't afford it. Musician Seido Salifoski told me his mother simply fainted when he told her he was moving out. For her, Seido moving out signaled disrespect for the family and for Romani culture.

The membership in a residential unit, typically a small apartment, is quite variable not only because of migration but also from a belief that a variety of related women can raise children. Mothers do not hesitate to give their children to their mothers or sisters for a few weeks, or a few months if necessary. For example, one girl was raised by her grandparents because her mother migrated to Australia after a divorce. The community is extremely close-knit; everyone knows one another face-to-face, sees one another often at celebrations, and socializes within the community. Elders rarely have friends outside the community. For women and children, socializing takes place in homes or in front of buildings where people gather after work or on weekends. Men, on the other hand, congregate in several "clubs," which are community centers or bars in the basement of apartment buildings, where card playing, drinking, and recorded music are found. When misfortune or illness strikes, families rely on each other. If someone is ill, the extended family plus friends and neighbors keep vigil at home or in the hospital. The highlights of community life are life cycle and calendrical celebrations (see Chapter 5).

Leila illustrated: "Family values . . . are very important. My family was everything to me. . . . This was always the main issue growing up." She saw family as defense against the hostile outside world:

> Not to feel alone in the world, like many Americans, that is the main reason I stayed within the family. I could not imagine going against the family and the tradition, and being out there on my own and being ostracized from everything I knew from the time I opened my eyes. Your family is who you are, and it is there forever. The family is a positive thing, and it is our only defense. We have no choice, especially in Europe. If you go and you try to become a part of somebody else's community as a Rom, they don't want you. So you have to make the best of it. The family is so strong because we are not accepted anywhere. It has become almost an obsession.

Leila points out that kin orientation is an adaptive mechanism in a world filled with hostility against Roma. Relying on one's own family has been a way to ensure trust to counter the threats from a mistrustful environment. What defines the field of social relations is a "very high level of interpersonal and intercommunal investment and trust—economic, social, emotional and moral" (Werbner 2002a:272). Community members "define their subjectivities as moral individuals through long term relations of sociality such as marriage, family, and community" (272–273).

The Muslim religion is a strong cultural identification point, but the level of practice varies tremendously. In general, Macedonian Roma in the past were not very observant; even today, most eat pork, drink alcohol, and do not pray. During the 1980s, I rarely heard of anyone going to a mosque except for a funeral. But in the 1990s, the Musa Mosque was built in the heart of Belmont, financed by a rich community member. Subsequently, the mosque became a focal point; funerals, for example, were very crowded. Women and young adults became more involved, and some Romani male children began attending Arabic language classes. About a decade ago, the community center next to the mosque reorganized into another mosque. The Islamic Center (see photograph 4.2) is now a vital community center, and many young Roma have become quite religious. The marriages of several Romani couples, for example, have featured a mosque ceremony, and two nonalcoholic weddings took place recently.[6]

Identity Issues

As mentioned, New York is home to Roma from every group, but they neither socialize nor intermarry. If, as the anthropological literature suggests, identity is always configured in opposition to others (Barth 1969; Appiah and Gates 1995), then the boundary between Roma and non-Roma is definitive, and one is either in or out (Hancock 2002; Sutherland 1975). This division, however, applies more to Kalderash Roma, who are much less integrated into American society, than Balkan Roma.[7] The school system, as an institution for integration into American life, is viewed positively by most Macedonian Roma, especially the younger generation. Whereas Kalderash Roma tend to be distrustful of schools because of drugs and sex, Macedonian Roma are not. Given their history of compulsory education in socialist Macedonia, they see it is as very useful for work advancement. Some Roma voice concerns about drugs and sex, but they do not pull children out of school at the same rate as Kalderash Roma do. Most Belmont families educate their children through high school; higher education is not the norm, although a few families have stressed it.

Belmont Roma feel different not only from majority Americans but also from other Muslim Balkan ethnicities and from other Roma. When speaking Romani, they call themselves Roma, when speaking Macedonian they call themselves *Gjupci*, and when speaking English they call themselves Gypsies.[8] For Belmont Roma, identity issues arise in part because the dangers of assimilation are ever-present. They are well aware of the tension between American individualistic ethics and the collective family ethics of their community (Ong 2003:7–8). Living, working, and going to school alongside outsiders makes them aware of what they claim distinguishes them from others: their family orientation, their ties to Macedonia, and their culture, including customs, music, and languages. Note that this list does not include all the usual features of ethnic identity (Romanucci-Ross and De Vos 1995): shared territory, history, and language. Territory and history are missing. Belmont Roma know little about their origins from India; rather, as mentioned earlier, they relate to Macedonia as home. Home, however, is a discursive trope, a reference point, not a fundamental unchanging value. Home is wherever their community is;

this diasporic attitude minimizes a singular homeland. Because they are people-oriented more than place-oriented, they take their home with them wherever they are. For example, when people speak of wanting to travel to Macedonia, it is always to see people, never "to be there." I was surprised to hear how often people said they didn't need to visit Macedonia because "everyone was here."

Community members are proud to be Roma, but exactly what that means may be contested. For the older generation it may be language and customs, and for the middle generation it may be finding appropriate spouses for their children. Second-generation Roma, who were born here, are often challenged to define themselves. A twenty-one-year-old unmarried girl told me she doesn't really know what or who she is: "I've never been to Macedonia, so am I Macedonian?" When someone hasn't seen a homeland, indeed, it may seem very remote. Furthermore, how one constructs one's ethnicity for others is often a different issue from how one feels within one's own community.[9] Certainly this is true for musicians, who are forced to deal with marketing images that are usually created by non-Roma. But all Roma are forced to deal with their public identity whether they want to or not, because of the stigma associated with it. In the Balkans, non-Roma readily identify (and often stigmatize) Roma by where they live, what language they speak, how they dress, or the color of their skin. Of course there are Balkan Roma who have successfully passed as non-Roma, but this requires cutting off ties to one's community so as to avoid detection. In Europe, passing is extremely difficult.

For American immigrants, however, there are more choices available because America is a more mobile environment, and Americans tend to pry less than Europeans; privacy is valued. The American government is less intrusive into family life than in Eastern European states. In America, you can hide your family history, you don't live in an exclusively Romani neighborhood, few can pinpoint your foreign language, and there are other dark-skinned people around. As a result, in Belmont Roma exhibit a diversity of self-presentational attitudes. Many community members do not readily reveal their ethnicity to non-Roma because of discrimination. In the United States, stereotypes about Gypsies center on criminality. Police forces in several cities have divisions specializing in "Gypsy crime" (Becerra 2006); there have been several "exposés" about con schemes of American Kalderash families on television. The Peter Maas film *King of the Gypsies* was a hit in the 1970s; and in 2007 the FX cable network inaugurated the series *The Riches* about a Gypsy/Traveler family engaged in pickpocketing, robbery, and credit card and identity theft.

Belmont Roma, then, need to be cautious about their ethnicity. Leila said that her parents often claimed, "We're Turkish, to avoid not being able to find an apartment, a job. To avoid the whole issue." A Belmont resident who moved to Australia narrated:

> We are very cautious. If you say you're a Gypsy people begin to look at you. They think you steal, you can't be trusted. We would lose our

jobs. We say we are Muslim Macedonians. Australians don't know the difference—they just think we are Muslims; but at work if there are Christian Macedonians and Serbs, they begin to suspect. Then they hear our last names and begin to figure it out. Then they distance themselves. My cousin, on the other hand, does the opposite—she doesn't hide she is Romani. On her locker at work she wrote "Gypsy." She's not afraid like us.

One Rom neither volunteers he is Romani nor denies it, but if someone says something against Gypsies he will bring up his ethnicity. One woman specifically asked me not to tell the proprietors of a banquet hall her family was renting that they were Roma. She explained: "Gypsies are considered the lowest level of person by Americans. Blacks, they've come up, but we are still down. I don't tell people I'm Romani—they don't have to know. Once at work I told my co-workers I was a Gypsy and they didn't believe me. They said, 'But you've been at this job for over three years—you don't live in a tent!' My husband —he tells everyone, but not me." Occasionally, Roma raised the question of my role as a researcher in relation to their adaptive strategy of passing. One woman told me, "So when you come along, saying you are studying us, that you teach about Romani culture, we are suspicious of you. We pull back. We are always hiding who we are to non-Roma. We hear you say you take photos of Roma. We want to know why." One community member did not want me to identify Belmont as a Romani neighborhood.[10]

Leila's older daughter tells non-Roma that she is Macedonian.[11] According to Leila: "If they question further, she'll say she's Gypsy. And that's what I teach her. You can tell them we're Macedonian because we are. We were born there. We're citizens of that state. Our boys died in the war, too. If we're not Macedonians, why do you draft our boys?" Fatima, a college-educated married woman, explained, "Whether we say we are Roma depends on whom you talk to. You really have to pick carefully who you tell because they can throw it back at you. Some of my friends and co-workers know and some don't." Similarly, Ramo told me: "I'm proud that I'm a Rom, but others hide, they say they're Turkish, whatever. I hate that. A lot of people think that we steal, that we don't work. Where I used to work, I told them I am Gypsy and they didn't believe it—they said that is impossible, you can't be Gypsy. Most people think we live in tents." In 2007, in the Islamic Center, some women reported hearing disparaging comments about Roma from other Balkan Muslims.

Roma sometimes hide their ethnicity by refusing to publicly identify with symbols of their culture. In the 1990s, two brothers from Dračevo (a village near Skopje), Severdžan and Menderes Azirov, tired to organize a Romani dance group in Belmont. They are excellent dancers and had performed in several groups in Macedonia such as Kočo Racin, Orce Nikolov, and the Romani KUD Phralipe (brotherhood).[12] Parents, however, were reluctant to let their children attend, especially the girls (see discussion later in this chapter, and Chapters 6, and 10). Severdžan said, "When they reach fourteen or fifteen years old they don't want to let the girls out. Also, they don't want people to know they are Roma. This is art—they should be proud of Romani folklore. We gave up—these people just don't understand." Similarly, several Macedonian Romani musicians refused to play in a prestigious concert for non-Roma when they learned that they were identified as Roma in the program notes. These same Roma are extremely proud of and involved in their music when it occurs in all-Romani contexts, but they shy away from the Gypsy label in mainstream American contexts because of the stigma.

Leila explained that those who experienced racism in Macedonia or other diasporic locations were the most afraid to admit their ethnicity in America. She narrated: "In the 1970s I was afraid because my parents were afraid; I would say I'm Macedonian or Yugoslavian. If someone would ask how come you're so dark, I'd say we're Turkish. While we were in Vienna . . . the people we were living with said to us, 'Don't speak the Romani language because if people find out we're Gypsies, they'll deport us.' So we had to keep a low profile. And when we came here, my parents carried that through."

After a while, however, Leila realized that she could not reject her ethnicity:

I can't deny what I am. Maybe I can deny it to the world, but I can't deny it to the mirror. . . . It'll always stare right back at me. You may tell everybody you're Yugoslavian, or Macedonian, or Turkish but I know you're Gypsy. You carry your shadow everywhere you go, so that's the main reason why. . . . I'm not going to deny it. I never really felt racism here, growing up in the States. Once I started school and became unafraid, I would tell my teachers, I would tell my friends what I am. And I didn't feel the rejection and the racism like we do in Europe. So, once I started working, I would tell my manager, and she would make a comment like, "Oh, but Yugoslavians are so light-skinned—you're so dark." "Well, that's because I'm a Gypsy." "Oh, what is that—those people that fortune-tell?" So, I've become open about what I am, and I haven't felt the racism.

The ignorance of Americans is sometimes contrasted with the blatant discrimination back home. "Battle" stories are told and retold, almost as parables, legitimating why they emigrated. The most striking narratives are told when relatives visit or when Belmont residents return home from trips. Leila narrated this story about her 2003 trip home to Prilep (a story I heard repeated many times by members of her family):

I hadn't been home in seventeen years. I heard about going to places where they wouldn't let you in because you're a Gypsy. I thought "Yeah, right" [incredulous]. My daughter and I, we were walking by this coffee shop. She says, "Ma, I want some pizza." I went in, I sat down. The restaurant was empty. There were three tables occupied and another twenty-eight empty. The guy says, "Sorry, you can't sit here. These tables are all reserved." I said, "Reserved for what?" He goes, "For the tourists." I said, "But the place is empty." "Well, you can't sit here." And, I had to get up and leave. Very blatant! I didn't want to expose my daughter to that—she was only eight years old. She's never been told she can't sit here because she's a Gypsy. And I didn't want to create a scene in front of her. . . . I felt prejudice. I felt it very strongly. My daughter was very uncomfortable in town. She didn't want to go any place outside the Gypsy environment. She felt the stares and the comments. And it made her uncomfortable; it made her unhappy. She said, "Why should I go there and, and have them look at me like that?" And now I can understand a little bit easier, why the Gypsies in Europe tend to keep a low profile.

Musician Erhan Umer (see next chapter) narrated what happened when he took his family home to Bitola in 2002: "I was so excited to visit the city swimming pool that I had seen under construction years earlier. When I arrived with my family, an Albanian guy was selling entrance tickets. He said 'Ne zemame Gjupci' [we don't allow Gypsies; Macedonian]. I answered, 'You can't tell me that—this is my city, I was born here I have every right you have. In fact, I'm American.' Things are very bad. I would never go back to live there."

Leila also encountered racism via the internet. A few Belmont Roma participate in diasporic chat rooms with Balkan or Romani themes. She explained:

I chose a nick [nickname] that says exactly who I am. *Romani čhaj* [Romani girl]. When I first went in with that nick, I used to get bounced right away. Macedonians would throw me out of the room just for walking in. Because they don't want Gypsies in their room. And then they would start making comments. And I fight like crazy. They know me. They know when Romani walks into the room and if they make a Gypsy comment, she will start. That's the only time I create problems in the room. Otherwise, I don't argue with anybody. I just sit there and I play my music. I play Romani music. It's video-audio chat. And, I use Romani music as a statement—I put on a Romani song. In the beginning, they would bounce me right away. "No Gypsy music allowed in this room!" "Why not? It's the Internet. The Internet is free." "Oh, but it's a Macedonian room." "So what? I go to Macedonia and I hear Gypsy music in the cafés, in the stores. I hear it everywhere; it's on TV, on the radio. Who the freak are you to tell me I can't play my music on the internet, on my computer? If you don't want to listen to it, leave the room." That's when I encountered the racism.

There are, then, a range of responses among Belmont Roma regarding identifying their ethnicity to non-Roma; while some hide, others boast, and still others strategically pass. Gropper and Miller's concept of "selective multiculturalism" (2001:107) illuminates that Belmont Roma negotiate multiple ethnicities (e.g., American, European, Romani, Turkish, Macedonian). Their choices resonate with the choices musicians make in their diasporic encounters. Within the Belmont community, on the other hand, there is strong pride in being Romani; although the meaning of being Romani varies, a core value is marrying a Romani person.

Marriage

Marriage underscores the significance of the family and demonstrates transnational ties via the network for spouses. Everyone is expected to marry, and those who don't are more or less stigmatized.[13] Musician Seido Salifoski told me he was derided by his family when he turned twenty eight years old and still wasn't married: "I got married because of all the pressure my parents put on me." In Belmont, young people congregate in the streets, at ritual events, on dance lines, at the mosque, and at school. Officially, there is no "dating," but rather young people are supposed to socialize in groups with the elder generation supervising. In reality, however, young people do sometimes meet surreptitiously. At weddings, for example, teenagers meet outside for one-to-one conversations. This mirrors the situation in Macedonia, where people meet on the evening walk (korzo), at gatherings, in school, on public transportation, and in shopping areas (Silverman 1996b). Today communicating via cell phone and Facebook is common.

The ideal is an arranged marriage within the diasporic community. Traditionally, the parents looked for appropriate spouses and the children acquiesced. However, there are myriad variations to this process: at one extreme, parents do indeed pressure children to marry, and at the other parents may acquiesce to a match entirely orchestrated by the children. Sometimes surrogate parents such as aunts and uncles enact the role of parents if birth parents are absent. Parents rarely force a particular spouse on a child, although they put pressure on children to marry by their mid-twenties. If two people want to marry, the groom's parents visit the bride's parents to ask for her; the bride and groom supposedly agree before the deal is sealed. Parents get involved because they claim they can see beyond romantic love—they check out the reputation not only of the prospective spouse but also of the entire family. They obtain information such as economic standing, level of respectability, and how the family has treated its brides in the past. Women have "people knowledge": when they socialize, they discuss people and their reputations. Thus, although men may be the public face of the family in marriage negotiations, they rely on women precisely for the information that makes marriage negotiations possible.[14]

Marrying their children to Macedonian Roma is the goal of parents in Belmont; they believe this ensures continuity of the culture. Intermarriage, though discouraged, happens in a minority of cases. Leila explained: "We hope they'll stay within the community. As a mother, if my daughter falls in love with some American guy who is going to make her happy, and he's a good person, I have nothing against it. Because the most important thing to me is her safety, her happiness." In the 1980s, a woman married a Hispanic male against her family's wishes; although she still attends weddings and other large family celebrations, she is not immersed in the fabric of the community, and her children do not speak Macedonian or see their cousins regularly. On the other hand, there are several cases of men marrying Italian or Hispanic women. Although the parents disapproved, the children eloped and eventually the parents acquiesced. An Italian wife and several Hispanic wives have even learned some Macedonian language. Despite his parents' disapproval, Seido Salifoski married a Japanese woman, and she helped him raise his daughter from his first marriage. His wedding ceremony creatively combined customs from both cultures.

If a young man or woman can't find a suitable spouse in New York, usually the family takes a trip to Macedonia to "look around." Of course, only people who have legal status can travel abroad. Every summer, families embark on this ritualized journey. Word goes out "back home," in Prilep, for example, or in a diasporic location such as Vienna, that certain family members are coming, and their Macedonian relatives network to arrange meetings with prospective spouses. These trips can be very stressful, considering the short time period. Leila, who met her husband on a three-month trip to Prilep, commented:

> I met my husband through relatives. It was a group choice. I hadn't met anybody that I felt would be somebody I could work with. When you don't know people, you don't want to take a risk. But eventually, as time neared for me to come back, I had to take a risk. I have known people who were in love for five, six years, then got divorced. There are no guarantees in life, even if you know somebody. When I met my husband, he presented himself really well, and I thought this is someone I can work with. . . . I was honest with him about who I am; I'm too honest! I told him: "I'm not going to be a typical Romani wife. I go to school, I have a mind of my own, and I'm not afraid to express it. I'll be working with men; I may have to go on trips. Sometimes you may have to clean the house. You may have to pitch in and be an equal partner, and if you can deal with that, fine. If not, it's not going to work." We kind of agreed. And here we are, still married seventeen and a half years later.

Leila is somewhat of an exceptional case because she was twenty-five years old and in college. In spite of her family's insistence on education (see discussion later in this chapter), she knew she needed to find a spouse. She had an aura of self-confidence and honesty that was rare for young women of the 1980s. Her philosophy has carried through for her children:

I won't force my daughter to go back home for a husband. It's so stressful, especially if you don't have the support from home. Everybody is telling you what you should do. They don't approve of anybody, and you don't know anybody. How are you supposed to make a decision? One relative says—you can't pick this one—his family did something 250 years ago! I hope my daughter finds someone here in the community that she will be happy with. If not, if she chooses to go home, she'll have my support.

A woman from Skopje met her husband when he made a trip home to find a bride. She narrated:

I was sixteen years old when he came for me. I saw him twice. I really didn't want to get married but my parents arranged it. They made a small wedding but when we arrived in the Bronx my in-laws made a big party. I cried for weeks to go home, but I stayed, learned the language, and got used to it. My parents and siblings went to Germany as refugees, and I haven't seen them for years. The U.S. embassy turned down a visa for my mother to visit me.

Fatima, met her husband, Rifat, on a 1994 trip to Prilep. Note in Fatima's narrative that the couple is given some time alone together, plus the option of refusal on either side:

We were on vacation. My aunt is an in-law of Rifat's aunt. The two women arranged for Rifat and me to meet without us knowing. My aunt woke me up and said, "Get dressed up, we're having guests." He came over with his cousin and we started talking. We liked each other immediately but we weren't thinking of marriage at first. Only when we were leaving, at the airport, did we know that our parents were involved. At the airport, Rifat and I took a walk and when we returned to the table where everyone was sitting, we saw everyone was shaking hands. We asked, "What's going on?" and they told us the marriage was approved. We were very happy.

Aiše, Rifat's mother, who was in Belmont at the time of the summer trip, told her version of this story: "Rifat got engaged this summer to a wonderful Romani girl. She came to see her relatives and they met each other. At first, Rifat was reluctant. Ali called me to ask what we should do; we were thinking North America was too far away and Rifat was too young. But Rifat said, 'We are in love.' So we said, 'Since they are in love how could we separate them?' So we gave our blessing. She got him papers. Fatima is modest, a very good wife."

There are failed trips, but not many. Seido Salifoski told me of the reluctant trip he took with his parents; he didn't care for anyone in Prilep, so they went to Turkey to visit his relatives and "look around." He agreed to marry a Turkish woman, but the marriage lasted only a few years. Most marriages, however, are successful. Some parents of Belmont sons prefer a Macedonian bride because the girls in America are spoiled; as several people claimed, "They don't want to cook, clean, care for children and do domestic chores." They reason that if a bride is brought over from Macedonia, she is more likely to accept traditional roles because she wouldn't know English and her legal status would depend on them.

Bringing grooms to America is more complicated in cultural terms than bringing brides because it contradicts the patrilocal residence expectation; nevertheless, it is done regularly out of necessity, as with the case of Fatima and Rifat above. Given the patriarchal nature of the family, it is awkward for a man to move in with the bride's family and depend on them for language, employment, and legal status. He is known as a *domazet*, meaning a live-in son-in-law, in Macedonian, which has a pejorative connotation. Leila explained: "When he's a *zet* in the house, they are made aware of it from the moment the marriage is announced. They'll get the comments, *Sega kje bideš domazet. Žena kje ti se komandva* (Now you'll be a live-in husband. Your wife will command you [Macedonian]). And they'll get that cruel stare." The stigma, however, is balanced out by the opportunity to emigrate.[15]

Parents often will not agree to a match because of objections regarding the family. One family in Šutka refused to give their daughter because the man had a child with another woman. The bride's parents usually use euphemistic terms of refusal, saying the child is "too young," or "not ready," rather than the real reason, which may be related to character or economics. Elopement is a possibility when parents won't agree. In fact, it is quite common for a young woman and man "to run away." The bride is then called a *našli čhaj* (runaway girl) rather than a *manglardi čhaj* (asked-for girl). What this actually means is that they go to the home of a friend or relative, consummate the marriage, and then wait for the reactions. Sexual consummation is basically an irreversible act, since it signals the termination of the woman's virginity (see more on this later). People often refer to this situation after elopement with the terms "It's all over" (*gotovo*, Macedonian). The parents will typically relent and agree to the match at a ceremony known as *smiruvanje* (Macedonian, reconciliation). Some parents, however, never agree to their child's choice. One young woman ran away with a married man, and—despite the fact that he obtained a divorce to marry her, and that they are very happy together, and that his ex-wife had been having adulterous affairs—the woman's parents cut off relations with their daughter. The birth of a child often leads to reconciliation.

The Bride's Reputation

Among Roma, the test of the bride's virginity is an extremely significant custom, both in the Balkans and the United States. It presents visible manifestation of a girl's reputation and the honor of her family. Moreover, it is symbol of Romani identity in terms of keeping the proper order of things in a changing world. Until the 1960s, this custom was practiced among virtually all ethnic groups in the Balkans regardless of religion, but today it has declined. Many Macedonian and Bulgarian Roma, however, still practice it, whether they are Eastern Orthodox Christian or Muslim, and in the United States it is part of many marriage rituals. Gjulizar Dželjadin of Šutka commented: "The bride must be honest and honorable. On Monday we want to see the stained sheet. . . . If the bride had brought us all of Europe's wealth, it would not have been worth as much as what she gave us, her honor. That was the most beautiful gift to us."

Theoretically, in Macedonia the consummation of the marriage takes place during the wedding at the groom's house (see next chapter). The mother-in-law looks for blood stains on the wedding sheet, and if she finds them she publicly announces "the good news." Gjulizar explained that after the couple consummates the marriage, "we send word to her father's house that she is honest, that she was worth the expense. . . . The test happens at night without music. We Roma only accept blood. Even a doctor's note is not enough. If she is a virgin we send news to the bride's mother right away. If the mother-in-law doesn't see blood, she will send the bride home riding on a donkey with pots and pans tied on clattering. All gifts are returned." When I asked her if she ever witnessed this, she says she heard it did happen. Obviously, the threat is enough for most young girls to make them conform to sexual restraint before marriage.

Leila confirmed that the sheet is shown during the wedding in Macedonia, and I witnessed it may times: "If they haven't eloped then they show the sheet during the wedding. . . . They bring the sheet, that night, over to the mother's house, and then everybody celebrates. Technically, that's how it's supposed to be done." I asked her if she knew of instances where the woman is a virgin but doesn't bleed; would the sheet then be more of a symbolic object?

L: No, they want to see the blood.
C: They really want to see it?
L: Darn the symbolism.
C: Really? It's that literal?
L: They want to see blood.
C: But not everyone bleeds.
L: She's going have a tough time proving that she was a virgin.
C: I've heard all stories about a little chicken blood, whatever.
L: They do what they've got to do.

In Macedonia, after the test, the sheet is placed on a metal tray (*tepsija*), covered with gauze (see photograph 4.3), and the wedding party (which includes the groom but excludes the bride) processes to the bride's house to bring the good news to the mother of the bride. This is a very important moment because it vindicates not only the bride's reputation but also the family's; it is the job of the mother to raise her daughter in preparation for this very test. The entire ritual dramatizes transmission from mother to daughter of proper control of sexuality. Termed *blaga rakija* (sweet brandy, Macedonian), the ritual features a procession with zurla and tapan, led by the groom's women carrying a brandy bottle decorated with flowers, greenery, and red ribbons (fertility symbols). The mother is required to tip the groom and feed him feminine foods (sometimes literally placing a spoon in his mouth), most notably eggs. His friends play tricks on him, such as offering him cigarettes but pulling them away three times, then finally letting him smoke.

Elvis Huna, the keyboardist with Esma Redžepova's band, described his wedding night: "Normally the morning after the wedding I would go to my wife's family house and eat eggs. . . . It's a Gypsy tradition. Eggs signify birth and so I eat eggs to signal that we have good births" (Cartwright 2005b:118). The groom, in other words, eats fertility foods to display the transference of the bride's reproductive potential from her family to his. In addition he receives gifts from the bride's family (*bovčalok*, gifts sewn on a sheet) such as shirts and handkerchiefs, which are draped over him (see photograph 4.4 and video example 4.1 from a wedding in Šutka).

Despite the pride in ritual elaboration of the test of the bride's virginity, there is a recent campaign in Macedonia to eradicate the custom. It is based on the human rights dictum that every person has inalienable personal rights, regardless of culture.[16] Activists claim that "the test" is a form of subjugation of women (since only women need to be virgins), is humiliating for both men and women, and often leads to psychological trauma. The campaign was spurred by Romani activist Enisa Eminova, who in 2001 conducted a survey of 660 Roma (parents and children fourteen to twenty-five years old, from ten Macedonian Romani communities) funded by the Open Society Institute. Surprisingly, most Roma agreed to participate in the survey, and the older generation did not uniformly express traditional views. Nearly half of the parents said they would accept brides if they were not virgins, but 70 percent replied they were not sure whether their sons would. Many respondents saw no need to maintain the custom. In short, the survey revealed much uncertainty on the issue and opened up an avenue of debate.[17]

In Belmont, the custom of checking the sheet is simply called *adet* (the general word for custom in Macedonian; of Turkish origin) and is widely practiced. However, it is virtually never done during the wedding because the timing of rituals has been altered in the American context. Blaga rakiya has been removed from the test and is celebrated whether the test is done or not. It has morphed into a separate party in a banquet hall, put on by the bride's side a few days or up to a week after the wedding. The ritual brandy bottle is still decorated, the ritual foods are still consumed, and the ritual gifts are still given, but the setting may be at home or a rented hall, and the bride's virginity is not the issue (see Chapter 5).

Given Leila's liberal views on education and marriage, I was surprised to learn that she and other educated younger women approved of the custom.[18] She said: "It is oppressive. I have mixed feelings about that issue. I had to do it. And if I could do it, everybody should be able to do it. But, it's not necessarily a good thing. I was able to sacrifice, and remain a virgin and go through it, and it was a demand I had to meet. And if we're going to expect these girls to stay in the culture and in the community, then yes. It is still very important in this community. Very important." Leila explored the changes in her attitude as she aged:

When I was growing up, I was against it. I felt, "Why do I have to prove it to everybody?" "Why does it have to be done so publicly?" "Why can't just I bring the sheet out after I do whatever I do with my husband?" I've learned to accept that it is part of the culture, that it is part of the tradition, part of proving you are what you are. And, if you can't fight 'em, you join 'em. So, I've kind of learned to join them. I mean, a lot of the younger ones are against it, that is, until they become women and have children, and they have sons. And, their sons are expected to bring home a virgin and then all of sudden it becomes, you know, a major issue.

I asked Leila how the test works if the couple elopes, and she answered: "They're supposed to save the sheet, yes. Now, if they've run away, or if they've eloped, they'll pick a date. If she's menstruating, they'll wait until she's clean. Or, they may wait if the families aren't in agreement about the marriage. Some people will wait to see, 'Well, are they going to take her back?' And if the girl says, 'I'm not going back. I'm here to stay,' and the families are in agreement, they'll do it that night." Elvis Huna described how his elopement in Skopje dovetailed with the adet: "I had to steal her. My family went to ask for her. . . . Then my wife tells me that her family thinks it is better next year for the wedding, so I take her to my home and you know it's important that the Gypsy girl is a virgin . . . and I take her virginity. So we do it and we show the . . . sheet! Now her mother cannot . . . take her home. . . . This wedding tradition stretches back through my ancestors" (Cartwright 2005b:118).

One couple "ran away because the parents of the girl wouldn't give her. So they'll elope and then it's over. They will do the adet. We're going to have good news tomorrow." I was also told that one couple "waited to do the adet" until the groom's relatives drove to their town. Proper timing indicates respect. Finally, in one instance a mother who was against her daughter's marriage and still refuses to speak to her participated in the test because this was the respectful thing to do. The test of virginity, then, is not only about the bride but also about the bride's family's honor and reputation.

The Question of Women's Power

Romani culture is patriarchal, but the various forms of gendered power need to be dissected. Education is one important factor that may mitigate gendered power (see text supplement 4.1 for a discussion of education and gender). First-generation males have much more freedom of movement than women, and they expect to be respected and be served in any home. Women do all domestic tasks: they cook, shop, clean, and take care of children. Because there is age as well as gender hierarchy, new brides have the lowest status. This is the standard pattern among all ethnic groups in the Balkans, but Roma adhere to it very strictly. New brides sometimes will kiss the hand of and bow before older relatives. Women's sexuality is especially restricted; clothing, dance styles (see Chapter 6), and mobility are closely monitored, and brides also must endure the test of virginity, as has just been discussed.

Men are the nominal heads of the family and occupy positions of authority; for example, they represent the family in ritual occasions, such as arranging marriages, even if the knowledge on which it is based is obtained by women. Males, then, occupy the public sphere of Romani life, while women occupy the domestic.[19] This observation, however, obscures the fact that women influence the public realm from their position in the domestic realm (Nelson 1974). They are the links between the two families, and their reproductive abilities perpetuate the family. In addition, women provide substantial income; they may keep their own salary, and in fact some manage their husband's income. Family budgets, then, are sometimes in female hands.

Claiming I was giving too much credit to women, Leila insisted that large financial decisions are routinely made by men. She said, "A woman has a budget to run the house. But when the big things like weddings come up, when they go to rent the hall, a woman won't do that. The men do that. Or they go together. She won't go alone. Because that involves a large amount of money, and that has to be a mutual decision. Some men won't even take their wives. They'll just go alone." Leila took issue with an article I wrote about Šutka (Silverman 1996b, 2000b) where I claimed that women exercised substantial power. She saw more sexism than I did, as this conversation shows:

> L: A male-female relationship, it's never equal. It'll never be equal. There's always times when you have to be the woman and stay quiet. And there are times when he has to be the man and step away, and let the woman do what she has to do.
> C: I think in many Balkan cultures, but especially in Romani culture, that women have a great deal of power in the home in terms of money and raising the children. The men are in the public realm, for example, when guests come, they're in the living room, but day-to-day decisions about money, don't you think the women run much of it?
> L: Not the money. I read that in your article and it kind of upset me. I'm like, "No. Not really!" Maybe she has a say in the shopping, but major decisions about buying a home or a business, or putting on a wedding, those kinds of things, the man has the final say. If he's a decent person, he'll take the woman's view into consideration.

My 1996 article dealt with the myriad roles women occupy in celebrations, from leading dance lines to deciding and directing which rituals should be performed, to managing details of their execution and their budget (see Chapters 5 and 6). I argued that this leads to respect for ritual knowledge and a female sphere of ritual power. Leila, however, argued that men had more monetary power than I allowed: "The women buy the bride the dress and the gold, etc. But the man gives her the money."

Precisely because Leila disagreed with me, I think it is worth exploring how women negotiate power in the family. Leila may be reluctant to admit the extent of the power she wields. Or her interpretations may reflect the fact that in New York women are less mobile than men; perhaps they have lost some ritual and financial power. On the other hand, several female East European Romani activists have written about the domestic as a site of oppression rather than power (European Roma Rights Centre 2000). In Belmont I observed the entire gamut, from supportive husbands to abusive husbands. Supportive men share power and help with child care and shopping, usually by driving their female relatives to the supermarket (many women do not drive) and carrying bags; abusive husbands engage in domestic violence.

The literature on gender and migration has yielded mixed results in terms of women's power. According to Brettel, "In some cases scholars have documented greater independence of women and more equity in the family. . . . By contrast, other scholars have argued that even when immigrant women earn more than their spouses do, this does not necessarily result in greater decision-making power within the household or greater autonomy outside it" (2003:147). In the case of Roma, we must keep in mind that women regularly worked outside the home in Macedonia, and thus work is not an arena of great change in the diaspora. With the flowering of female education in the younger generation (see text supplement 4.1), however, it will be interesting to chart future changes in gender roles.

Video Diaspora

As I have emphasized, face-to-face communication is highly valued among Roma, and thus communication across distance poses challenges. Roma deliberately videotape in order to show relatives in the diaspora what is happening in their communities. Sometimes I was the conduit for video exchanges, and sometimes I made videos to facilitate communication between families. Roma also hire professional videographers to document their celebrations. Videos, then, figure as valuable gifts in a global network of reciprocity.

Videotaping in New York typically documents life cycle and calendrical events, such as circumcisions, baby showers, birthday parties, New Year's parties, religious holidays, and weddings. When documenting their celebrations, Roma focus on two intertwined subjects: people and music. There are few video frames that do not include people; location does not command attention.[20] Place is literally absent from Romani videos, supporting the notion that Romani communities exist wherever there are Roma, regardless of location.

Persons depicted in the videos are usually actively performing, e.g., speaking, playing an instrument, dancing, or singing. People enjoy being the object of the camera; there is neither shyness on the part of performers nor hesitation on the part of the people behind the camera. This reflects the positive coding of performance in Romani life. Parents, for example, encourage children to sing, play an instrument, and dance for relatives at celebrations. Video subjects either already know the audience for whom the video is intended, or else they ask and tailor their performance for it. They often face directly into the camera and offer greetings to the intended viewers. In video example 4.2, an elderly Macedonian Romani woman in Melbourne, Australia, sends greetings in Romani to her relatives in New York in 1998.

Along with personal greetings, music and dance are ubiquitous features in Romani videos. As I have explained, at Romani celebrations live music is the medium for hours or even days of dancing, and music also provides accompaniment for rituals. Time analyses of videos evince long sequences of dancing, often lasting for a few hours. As will be discussed in Chapter 6, dance is the site of displaying social relations. The video, then, is a guide to figuring out who is related to whom, who has married whom, who has children now, who has grown up, who looks ill, who has passed away, etc. This information is very important in the diaspora, where people rely on videos to evaluate information. It was also important to me, the ethnographer, as a graphic guide to who was who in the diaspora community.

Home videos of celebrations are a visual window into the aesthetic system of the community. The aesthetic system displays stylistic markers, which serve as badges of identity for the group, and compose a system of style, related to consumption and economic class (Bourdieu 1984). These markers surface most obviously in performances, where symbols (objects, genres, and behaviors) are elevated to representational icons for the group (Leuthold 1998:18). For Macedonian Roma, these include the musical genre and dance form čoček, female clothing, the instrument zurla, and certain ritual acts (such as temana, or bride's greeting), and gift giving (see Chapter 5). Videos feature these icons prominently, and they are evaluated most thoroughly. For example, not only is costume coded as Romani but also the cost of the fabric and of the seamstress is evaluated. Similarly, food displays (banquet tables, wedding cakes) and gift exchanges (e.g., jewelry) are the object of the camera's gaze as indices to class. In fact, there are moments in rituals when the cost of gifts is publicly announced. Videos, then, capture the verbal and visual dramas of style and class.

If possible, Roma begin watching the videos immediately. If the wedding is a three-day event, the family and guests might return home the first night and watch the video, no matter how late. Communal viewing elicits evaluative comments; input is generated as to how the rest of the event should unfold. Evaluations debate the aesthetic system; for example, viewers might comment on the costumes (are the hemlines too short for the women?), the music (how could the hosts afford to hire those famous musicians?), the dancers (what a great dancer she is!), the number of guests (few people came to their wedding because they aren't speaking to most of their relatives), the money spent on the event (the limousine cost $200 an hour!), the manner in which rituals are performed (the parents of the bride really didn't want to let her go!), or their omission (she forgot the sieve when she led the dance). In recent years, YouTube has become a forum for sharing family videos; I am currently studying this phenomenon.

In 1994, I gathered in the Bronx with Aiše and her son, Ramo, to watch a video of a double wedding of Aiše's husband Ali's brother's sons. I had shot this video in Šutka a few weeks earlier; it featured Ali and their younger son, Rifat, who played a prominent role in the wedding as the flag bearer (*bajraktar*; see next chapter). As we watched, Aiše caught up on all the news about her husband's family, whom she hadn't seen in several months. She commented on the clothing, the hair, the food, the music, who drank too much, and who was engaged to whom. She also used the occasion to explain the wedding rituals to Ramo, who was at a marriage-able age and "needed to know these things."

Videos illustrate and initiate conversations about the display of symbols of Romani ethnicity. They are the visual and social medium to concretize and reconfigure various subjectivities, including representations of iden-tity. Moreover, they convey information about people and situations across the distance of diaspora. Following Appadurai, I suggest that videos both create and reflect a Romani diasporic public sphere transcending the boundaries of the nation-state (1996:4). Appadurai asserts that a constitu-tive feature of modern subjectivity is the effect of media and migration on the work of the imagination (3). For Roma, electronic communication "impels the work of imagination" (4), which is, above all, the performance of identity: "People not only position themselves vis-à-vis modernity through multifarious practices but also struggle to *reposition* themselves, sometimes through deploying the very codes of the modern that have framed them as its others" (Schein 1999:364; and see Chapter 3). Through collective readings of images and words, videos create "communities of sentiment" (Appadurai 1996:8; also see Kapchan 2007), which supply meaning for Roma. The next chapter delves into the layered meanings of performative celebrations.

5

⚜

Transnational Celebrations

Celebrations are the glue that binds Roma to their families and communities. Both in the Balkans and in the diaspora, community members not only gather regularly for events, most of which include music and dance, but also plan them well ahead of time and discuss them long afterward; events thus have a long symbolic life. They figure clearly in how Roma performatively conceive of their identity and how they distinguish themselves from both non-Roma and non-Balkan Roma. Moreover, celebrations are motivations for and manifestations of diasporic migration; Roma plan travel to coincide with celebrations (e.g., attending a relative's wedding), and trips are sometimes the cause of events (visiting relatives sponsoring a farewell banquet). At any event, participants typically hail from several diasporic locations, as at the wedding of two brothers, Bilhan and Irfan in Šutka in 1994, which relatives and friends from Germany, Austria, Belgium, Australia, and the United States attended.

In the United States, life-cycle events are celebrated more regularly than calendrical events. The wedding (*bijav*) and the circumcision (for Muslim Romani boys, *sunet*; Romani and Macedonian) are the two most important celebrations; in Macedonia, some families also sponsor a babina (party for a newborn baby, especially a girl, since she won't have a circumcision) or a soldier-send off celebration; all families arrange Muslim funerals. The most important Balkan Romani calendrical celebrations are Herdelezi/Erdelezi/Herdeljezi (St. George's Day, early May celebration of spring renewal), *Vasilica* (St. Basil's Day, early January), Ramazan (fasting month), and the two Muslim *Bajram*s. *Baro Bajrami* or *Šeker Bajram* (big or sweet festival; Arabic *Eid-ul-Fitr*) falls at the end of Ramazan, and *Kurban Bajram* (festival of sacrifice; Arabic *Eid-al-Adha*) falls seventy days after Ramazan (Petrovski 1993 and 2002). Note that some Muslims of the Balkans, including Roma, enact rituals that are related to Eastern Orthodoxy such as dyeing eggs in the spring (related to the fertility concept that predates both Islam and Christianity); this underlines Balkan religious syncretism.

In New York, Herdelezi is still sometimes celebrated with an outdoor picnic of lamb, the traditional food. Muslim funerals are common and both bajrams are celebrated in the home or mosque setting. A New Year's dance is regularly held by Bronx Roma, and often musicians from the diaspora are invited to perform.

Celebrations display cultural values and serve as markers or organizing principles of the year and the life cycle; they are complex events entailing multiple genres, e.g., music, dance, costume, food, and ritual. Many dramatic roles are enacted, much economic planning is necessary, reputations are established or questioned, and individual and family power is negotiated. Furthermore, via celebration, Romani life evinces a conscious and heightened performative dimension. Video and photographic documentation has been common since the 1970s, and the resulting tapes become treasured historical documents. In emphasizing community, however, I mean to imply neither a conflict-free atmosphere nor a functional explanation of a system in balance. To the contrary, celebrations often lead to conflict or reenact prior arguments and schisms over resources and reputations.[1]

By far, weddings are the most frequent celebratory event and the focus of community attention. They are the ubiquitous subject of evaluative talk, ranging from the availability and suitability of spouses to future and recent marriages. Wedding and circumcisions often involve hundreds of guests, numerous meals, and lavish presents. Of course, poorer Roma put on more modest events,[2] but there is some truth to the claim that Roma spend their money on weddings.[3] For example, in Asen Balicki's Bulgarian film compilation *Roma Portraits*,[4] a young Romani director, Mincho Stambolov, explained that he chose weddings for his filmic portrait because they are so significant: "They [the family] have been saving money for five to ten years but when . . . it is time for the wedding they are ready to spend everything. After the wedding they might not have a cent left but they really want a big feast." In the same film, a family member explained:

> It's a big celebration for us. No matter how much money you don't have, you have to make a wedding. You remember my aunt's wedding? They didn't have any money—they sold their animals to put on the wedding and now they have no animals! They have to buy animals with the money they collected [at the wedding]. But they had to make a wedding! . . . There was no other way!

Weddings are sometimes delayed because of insufficient finances, even into the bride's pregnancy. The same family member explained: "They waited and waited. The bride gave birth two days after the wedding." Similarly, several Belmont weddings were delayed because of the lack of fit between relatives' schedules and the availability of a banquet hall; it is not uncommon for the bride to be several months pregnant at the wedding and "almost showing."

Weddings are the key to the growth of the family, and the Belmont community has indeed been growing thanks to spouses relocating from Macedonia to America. Because the bride is ideally brought into the groom's family, weddings affirm her reproductive importance to patrilocal residency and patriarchal relations. The bride is thus the ritual focus and the most symbolically endowed personage in the event. She is the person undergoing the most marked transition, and because the alliance between two families depends on her she is the most precarious person.[5] It is not surprising that most wedding customs and song texts involve the bride.

Song texts are a guide to the importance of weddings in general, and specifically of brides. *Bori* means bride in Romani (plural *borja*), but its meaning expands to "a woman married into our family," dramatizing that a bride belongs to everyone on the groom's side, not just the husband. Bori thus means daughter-in-law to the groom's parents, sister-in-law to the groom's sisters and brothers, etc. The groom's relatives tend to use the term bori when addressing her or talking about her; using her name might signal too much intimacy. Furthermore, a woman remains a bori her whole life; only perhaps in old age will she outgrow being called bori. Along with the label comes the expectation of service to the groom's family (not only to the men but also to the elder women). "First up in the morning and last to go asleep" is a proverb found in every Balkan language about the role of the new bride, but it is especially resonant among Roma. As a woman has children and matures, however, her status increases and younger borja serve her. Thus gender hierarchy is mediated by age hierarchy.

The majority of Romani songs, indeed, are about weddings or specifically borja.[6] In Chapter 2, I discussed "Astargja o horo" from the repertoire of Džansever (see audio examples and text supplement 2.6–2.8). I have heard this song performed at several weddings in Macedonia, Bulgaria, Serbia, and New York. The text reveals the significance of elaborate marriage rituals and illustrates how the bride's family gives her away and the groom's family welcomes her. Most songs in praise of the bride are expressed from the view of the groom's father. As the senior male in the sponsoring family, he represents the voice of authority. In addition, the bride herself has almost magical powers in that she brings luck and happiness to the entire family; also note that she is displayed to the Romani public through dance (see Chapter 6).

Weddings in Šutka

Contemporary Romani weddings in Macedonia are typically three-to-six-day events; their length distinguishes them from ethnic Macedonian weddings, which are typically one day or one evening. Several decades ago, however, Macedonian villagers also had weeklong weddings (Kličkova and Georgieva 1996; Silverman and Wixman 1983).[7] The entire Romani wedding conforms to a pan-Balkan structural pattern that was common fifty years ago regardless of region, religion, or ethnicity. Pan-Balkan themes include transference of the bride from her natal family to the groom's family and emphasis on her virginity and fertility.[8] Structurally, the wedding illustrates

Van Gennep's tripartite division of separation, transition, and incorporation, from the bride's point of view (1961). Muslim Romani Macedonian and Bulgarian weddings differ from Eastern Orthodox Romani weddings in costume, use of henna, and more recently in New York the mosque ceremony (see discussion later).

In Šuto Orizari, Macedonia, perhaps the largest Romani settlement in Europe, music and dance are the community's expressive focus (Silverman 1996b). Weddings can be found every summer weekend, although in the postsocialist period the size and duration of celebrations have declined because of economic constraints. Indeed, from June to September in Šutka on any weekend evening one can find five to ten weddings on the streets. The outdoor dance portions of the weddings are regularly viewed by scores of uninvited onlookers, and there are times when uninvited people may dance. Dance-crazy Šutka teenagers regularly make the rounds looking for the best music for dancing.

The Romani expression for putting on a wedding is *kerava bijav*, which means literally I make a wedding. Note that bijav is also used as a general term for a celebration; thus the party for a sunet (circumcision) is also a bijav. Also note that making and working are represented by the same word, as in *kerava buti* (I work), which implies that making a wedding is a type of work. Furthermore, ritual is a particular type of gendered work that charts the relationship between a family and the community via the aesthetic dimensions of music, dance, costume, and foodways. Female identity is thus constructed by the relationship of economics to kinship and is expressed aesthetically in a ritual and symbol system. I am inspired by Micaela di Leonardo's 1987 article in which she coins the term "kinwork" to describe female work other than wage work and domestic work. Unlike domestic work, which occurs within a household, kinwork cuts across households, and it mobilizes women across households. Kinwork also creates obligations and reciprocal work for the whole household, including men. The term nicely describes the kind of work Romani women do in planning, organizing, managing, performing (including dancing), and evaluating ritual celebrations (Silverman 1996b). Older women direct rituals much more than men and younger women; in fact, many men and younger women are quite ignorant about what needs to be done and when.

Music and dance are required at Balkan Romani weddings; music tends to be a male realm whereas dance is female (see Chapter 6 and Silverman 2008b). Through dance, participants enact some of the most important rituals in the wedding (Sugarman 1997; Cowan 1990). For example, guest families are called up one by one to lead dance lines, in the order of closeness to the sponsoring family; moreover, dance lines are usually led by women. Before the family begins dancing, someone requests a tune from the musicians, and a male family member tips them. Families are called up to lead by a "speaker," a man who is eloquent, is a good organizer, and knows the proper order. He must not insult people by omitting them or calling them in the wrong order (the sequence must be *ko redo*, in order). Dance, then, is a performative display of social structure. One common speaker's formula is *Akana ka khela* . . . (now So-and-So will dance). Often a speaker is instructed about the proper order by knowledgeable family members, but I have attended weddings where guests were furious at the order.

Weddings are occasions for parents to scrutinize potential spouses for their children. People discuss who is dancing next to whom and who is wearing what outfit; most important, parents of marriageable children ask *kaske* (whose) that son or daughter is, meaning to what family he or she belongs. Because dance lines are sexually integrated, they serve as a meeting place for young people. Since dating is not practiced and arranged marriages are still the ideal, young Roma look one another over on the dance line and exchange glances. Young men and women sometimes dance next to each other (while friends and relatives watch), and conversations are initiated. The seeds of future matches, then, are planted at weddings.

As I explain in Chapter 6, women of the sponsoring family are expected to dance for hours at a time at weddings. Because women have so many obligations, such as ritual enactments, food preparation, and dancing, men sometimes end up taking care of children—something that rarely happens outside of rituals. Male dancing is more optional than female dancing, although there are some ritual moments requiring male dancing (as when his family is called up to lead the dance line). The males of the bride's family must also solemnly dance with her just before she is transferred to the groom's family (see later discussion).

I argue that in addition to dance, food preparation and presentation are performative because they are public behaviors with a marked aesthetic dimension that others evaluate according to shared criteria (Bauman 1975). Both women and men participate in food preparation, but in a segregated manner. Women, for example, are mobilized across households to prepare foods, such as baking hundreds of bread products, making salads, and stuffing grape leaves (photographs 5.1 and 5.2). For one Šutka wedding, it took five women eight hours to prepare 1,200 *sarma* (stuffed grape leaves). Men slaughter animals, prepare meats, and transport ready-made foods from bakeries and warehouses (photograph 5.3). Serving food at banquets is done mostly by men who activate kin networks to recruit the necessary laborers; this is a significant reversal of the normal division of labor and is necessary in part because women need to be free to dance. Washing dishes and pots is done by women.

Costume is also an important performance arena under women's direction. The most widespread form of clothing worn by Balkan Romani women (whether they are Muslim or not) is *šalvari,* also called *čintijani* or *dimije,* wide billowing pants (often 10 feet), matched with vests or jackets for festive events (Dunin 1984). Women are expected to wear numerous and appropriately styled outfits during the course of the wedding. Knowing they are on display on dance lines, females dress up, and sometimes young unmarried women change their clothes several times (borrowing their friends' and relatives' outfits; see photograph 5.4). Clothing also figures significantly as wedding gifts. For example, at various rituals during the wedding (such as the henna party), the bride is given clothing by female members of the groom's family, who have tastefully arranged it on *tepsii* (metal trays, photographs 5.5 and 5.6). Women shop and sew (or hire someone to sew) the outfits they wear and give as gifts; they also financially manage all of the tasks mentioned here, sometimes quite independently of men. Does this female ritual knowledge represent power? In the last chapter, I discussed my conversation with Leila about this topic; whereas she focused on the underlying patriarchal nature of the Romani family, I noticed the arenas of female competence.

Owing to space limitations I will not describe the prewedding *manglaribe*, "asking for the bride," involving visits and bargaining sessions; the *angrustik*, the period of engagement; or the postwedding *prvič*, the first visit of the bride to her family.[9] Note that sometimes lavish gifts are given by the groom's family to the bride's family at the engagement ceremony. Gjulizar Dželjadin described the engagement gifts of Amdi Bajram's son[10] as follows: "twelve meters of fabric for šalvari for the mother or the grandmother, a lamb, fifteen beers, two liters of brandy, ten pairs of women's slippers," plus much jewelry (including the ring). She was careful to point out that "Our girls . . . have never been sold for money, only for a gift (*bakšiš*). But there are Roma that sell them for money, but not ours."[11] All of these reciprocal exchanges trace the alliance between the two families.

The order of the wedding week in the 1990s was as follows[12]:

Kana (henna, Romani; *kına*, Turkish), Wednesday

Banja (bath, Macedonian), Thursday

Igranka (dance parties at bride's house and at groom's house, Macedonian) and second henna ceremony, *Kana gedže* (Romani; Turkish *gece*, evening), Saturday

Zemane getting (or taking), transferring, and incorporating the bride (Macedonian), Sunday

Blaga rakija (sweet brandy ceremony, celebrating the virginity of the bride, Macedonian), Monday

Džumaluk (bride's relatives visit the bride at the groom's house), Monday

Henna and Bath Ceremonies

The first henna ceremony takes place at the bride's house during late Wednesday afternoon; henna is a vegetable dye used for beautification on women's hair, hands, and feet.[13] The groom's female relatives dance while processing through the streets toward the bride's house. The groom's women carry decorated metal trays (tepsii) laden with bridal gifts such as šalvari, jewelry, shoes, underwear, sometimes the white wedding gown, and items of clothing for other family members. Often chickpeas and candies are put on the trays to ensure fertility. One tray contains the henna paste and is covered with a red cloth. Like many

Romani rituals in Macedonia, the event takes place mostly outdoors; as a result, neighbors and other unrelated people watch. Romani homes in the Balkans are typically modest and poorly ventilated, so, weather permitting, people congregate on porches and balconies, or in courtyards, or spill out into the street. Weddings typically take place June through September in Macedonia and Bulgaria, a period called the "wedding season." Summer weddings allow relatives in the diaspora to attend, and the hot weather favors outdoor banquets and rituals. The outdoor location (often blocking traffic on a street) and the loud music both mark the event as public and performative.

Note that in my analysis of the spatial aspect of the henna ceremony, I am avoiding use of the dichotomy domestic-public.[14] For Roma there are many publics. First, there is the public sphere of macro society that is dominated by non-Roma but in which Roma work. Second, there is the sphere of the larger Romani public, which I term the Romani community. Third, there is the sphere of the extended family. When I speak of community, I mean the sphere of local public life that is visible to other Roma, be they kin or neighbors, as in the henna ceremony. The non-Roma public is irrelevant here, and as a rule non-Roma do not have access to these local settings. For the henna ceremony, we may speak of a specifically female public sphere. I deliberately avoid the term *domestic* to define this female space because it is too narrow. The configuration domestic-public obscures rather than illuminates because the domestic arena is not always private and subordinate but is instead part of community life.

The henna procession is accompanied by a zurla and tapan band (video example 5.1; see Chapter 2), while other dance events use a modern wedding orchestra. The two musical styles are markedly different. Zurla/tapan music is the oldest musical formation, and it evinces a more intense volume and texture (Blau et al. 2002; Peycheva and Dimov 2002). Note that it is also used for the actual cutting in the circumcision ceremony, another intense moment. Also, practically speaking, zurla/tapan bands are mobile, require no amplification, and are cheaper because they consist of only three or four people. Note that fifty years ago zurla and tapan bands were the often the only professional music available and were used for all musical aspects of the wedding. Today only the best zurla players know the appropriate ritual melodies for the kana and sunet.[15]

For the henna party of seventeen-year-old Ramisa, who married Hasan in 1990, the groom's female relatives traveled from their village outside Skopje to the bride's house in Šutka. The six sisters of the bride's mother, Nazlija, had all helped prepare for the henna party, including making the *tel*, silver streamers that the bride wears on either side of her face (see photographs 5.7, 5.8, and 5.9). Tel is usually prepared by the groom's females and brought to the bride's house for the kana, but in Ramisa's case her relatives made the tel, perhaps because the groom's mother was impoverished. Typically the only males present at the henna ceremony are the hired zurla/tapan players; male relatives, if present, stand at a distance and look from afar. At Ramisa's henna ritual, however, her father, Muzafer

Mahmud, a very famous zurla player,[16] took the zurla from the hired musician and played briefly at the ceremony (video example 5.2).

Outside Ramisa's house, the bride's women greeted the groom's women; they danced together and were led into the courtyard, where they sat down and were served *lokum* (Turkish delight). A representative of the groom's women opened the trays and announced each gift with formulaic language wishing the bride health and happiness: who it is from, where it is from, and often what it cost; then she handed it over to the bride's side. At Ramisa's ceremony, Hasan's sister was the "speaker" and Ramisa's sister gathered the gifts on a white sheet (see video example 5.3). Since the groom was poor (Hasan's father had died), the gifts were very modest. Wealthy families may give five or six pairs of šalvari, the white dress, and many gold necklaces.

After the gifts were announced, the music started again and the bride was led to the courtyard (photograph 5.7) as the women screamed, ululated, crowded around her, and threw candies and chickpeas. This was a very loud and intense moment, as it was first time she emerged publicly (video example 5.4). In 1990 Ramisa wore tel, šalvari, and a white veil (photographs 5.7, 5.8, and 5.9), but it is more typical for brides to have their white veil covered with a red veil (red is a powerful color, invoking blood and life). A series of rituals ensued, enacted by one woman from the bride's side (her sister) and another from the groom's side (his sister); both had to be nursing mothers. They put sugar in each other's mouths (for future sweetness) and squirted breast milk onto the bride's hair (for future fertility). Then they dipped a gold coin into the henna and stuck it onto the bride's hair (for future wealth); finally, they applied henna to her hair (video example 5.5). At this moment, they made Ramisa cry by telling her she would be moving far away and would never see her parents again. These remarks were formulaic and did not necessarily represent the bride's actual situation; they did, however, structurally show the bride's transition and the uncertainty of her future.[17] After her henna was applied, dancing resumed with lines led by her relatives (photograph 5.10, video example 5.6). After the groom's women departed with the musicians, we watched the videotape of the whole ceremony at Ramisa's house. Ramisa's relatives evaluated who wore what and how the rituals were enacted.

In intense ritual moments the bride generally assumes a passive stance, evinced by her downward gaze and how she is led around and told what to do. Only selected elderly women have the knowledge to direct the ritual, and they shout out instructions (and sometimes argue among themselves) as it transpires. At several henna ceremonies, I heard criticism regarding the lack of a red scarf for the bride. Decisions about the ritual unfold during the moment of performance. Ritual objects may be prepared ahead of time, but there is neither rehearsal nor instruction of the bride in advance. One display that is prepared in advance is the bride's trousseau, *čeiz*, which is arranged in her house and inspected by all females. A parallel display may occur in the groom's house, consisting of clothing (*borjana šeja*) and household gift items.

The next morning Ramisa had her *banja*, bath ceremony, where the henna is washed out. Fifty years ago the bride would go to the public bathhouse,[18] but almost all bathhouses were torn down in Macedonia and Bulgaria as part of an effort by the socialist governments to remove Muslim cultural elements and modernize. In 1990 Ramisa bathed in a neighbor's newly remodeled bathroom, accompanied by young female relatives from both families. Meantime, her aunts prepared *pita* (salty cheese pastry baked in a tepsija) and other foods for the groom's women, who arrived around noon with sweets, rakija, and trays of meat pastry. Ramisa was led out from her bath wearing tel and the šalvari and jewelry the groom's family gave her the day before, and she performed the ritual kissing of the hands of her future female relatives, receiving small monetary gifts. Among Macedonian Roma, kissing elders is performed in a specific manner: you take their right hand in your right hand, kiss it, and touch it to your chin and forehead in a bowing manner (video excerpt 5.7). Because this was a *ženski muabet* (an all-women party; Macedonian), the dancing was solo čoček and more sensual than it might be in male company. Video example 5.8 shows a young girl at this event being instructed to dance by a female relative. At another bath party, an older female relative did humorous suggestive dances. Women sometimes get drunk at these bath parties; they laugh, sing, and cry over close relatives they miss who are abroad in the diaspora.

Igranka

On Saturday, an *igranka* (dance party; Macedonian) is held at both the bride's house and the groom's house. For the bride's side, this is the single most important event; the groom's igranka is optional and depends on economic capacity. The term is actually shorthand for a series of rituals occurring on Saturday. During the day the women decorate a sita (sieve) with greenery and red ribbons (and sometimes tel plus a grain product such as popcorn), all symbols of fertility and prosperity (see photograph 5.11); this sieve will be used to lead dance lines. Costume for the igranka is typically the fanciest šalvari sets (pants and jacket or vest) and western dresses that women own (photograph 5.12). Often new šalvari are purchased specially for the igranka, and fabric styles change every few years. Women are quite critical of old-fashioned styles and fabrics, and indeed I received quite a few gifts of outdated šalvari.[19]

The igranka is a decidedly female event even though men are present, and dance lines are usually led by women. The first line dance of the event is led by the most respected female elder, who holds a sieve decorated with a grain product, greenery, and a red scarf; this known as *sitakoro* (the sieve dance). The sieve symbolically links the fertility of the land (wheat and flour) to the fertility of the bride (photographs 5.13 through 5.16; audio example 2.2; video example 5.9). At Ramisa's igranka, the bride's mother, Nazlija, led the first dance, followed by her brother's wife, followed by

Nazlija's sisters; eventually Ramisa led, dressed in the šalvari she had received at the henna party a few days earlier. Women were called up to lead the line one by one, in the order of age and closeness to the sponsoring family. Čoček as a female solo dance has an important place in ritual (see Chapter 6). It is danced in the middle of the area near the front of the dance line; simultaneously, the line snakes around (video examples 5.10 and 5.11). For example, at Ramisa's igranka the bride's close female kin danced čoček in the middle.

The igranka takes place in the late afternoon, often on the street, accompanied by acoustic instruments such as clarinet, saxophone, *dumbek* (hour-shaped hand drum), and accordion (and sometimes džumbuš, the plucked string instrument with skin face discussed in Chapter 3). These instruments are all portable; they can be played walking, unlike synthesizers and drum sets. For Ramisa's wedding, clarinetist/saxophonist Feta Šakir's band was hired; he led one of the most popular bands in Šutka in the early 1990s (video example 5.10 shows Feta and his son on drums). Toward dusk the musicians climbed up to the stage and plugged in their instruments, adding a synthesizer, *džez* (drum set), and singer (video examples 5.9 through 5.15 and 5.17 through 5.21 feature Feta's band; in video example 5.15 note how young boys under the stage played makeshift drums). As more relatives and friends arrived at the igranka, the dance line grew to fill the street. At first, only those who were "invited"[20] danced, but many others came to observe. A favorite activity in Romani neighborhoods in Macedonia and Bulgaria is strolling around to watch weddings. Observers tend to wait to dance until all the important relatives have been called up to lead.

As it gets dark and more people gather to dance, the groom's female relatives prepare a tray of henna with candles for the second henna ceremony. They leave their family's festivities and process to the bride's house with zurla and tapan music. At Ramisa's wedding, they arrived after dark with their music and their lighted candles, causing a loud cacophony and a visual glow amidst the dancing, thus increasing the intensity of the ritual (video example 5.16). Ramisa's female relatives led her into her courtyard, and the groom's women put one of his shirts on her head and then applied henna to her hands and her feet, which were encased in special blue (in other cases, red) silken cases. The color red, as mentioned earlier, brings luck, and blue wards off the evil eye, a force to which brides are especially vulnerable because they are beautiful and happy (photograph 5.18). The women threw chickpeas and candies over her. While she sat immobile with the cases on her hands and feet, the bride was made to cry again, as in the first henna ceremony, with warnings about her bleak future. At this point, the zurla melodies imitated the girl crying. The groom's women then departed with their musicians.

On Saturday night a meal is served at the igranka. During the banquet, a speaker calls up close relatives, in a respectful order, to lead the dance line, and the male head of each family tips the musicians. Later, gifts from each family to the couple are exchanged and announced and recorded by a speaker. Gifts combine soft goods like blankets and fabric with jewelry and money. Remembering who gave what and how much is essential because of reciprocity; at future weddings guests are expected to reciprocate with similar amounts.

At Ramisa's wedding, however, the groom did not host his own Saturday night banquet because of his family's economic condition; thus he and his relatives joined the bride's banquet (video example 5.17). Many famous musicians, notably the Romani clarinetist Muamet Čun (video examples 5.18 and 5.19), Macedonian singers Jagoda Filipovska and Jonče Hristovski, and the group Tavče Gravče, performed because they were colleagues of the bride's father at Radio Skopje. A modest hall was rented and the food was catered; in the 1990s, many banquets were set up on the street with rented tables and chairs. Conversely, when a rich groom marries a poorer bride, the bride and her relatives may join the groom's banquet. Such an arrangement occurred at Amdi Bajram's son's wedding in 1992. Amdi's banquet took place in a large Skopje convention center, and several hundred guests attended. Tables of roasted lamb were displayed, and the bride wore a heavy necklace of gold coins. The music consisted of several local groups, plus Esma Redžepova and her ensemble and Slobodan Salijević's brass band from southern Serbia. Hiring nonlocal performers is a mark of status for the sponsors. Similarly, the sponsors of a circumcision I attended in Šutka in 1990 gained status because Greek Roma were hired. This is also true in the diaspora; for example, in Western Europe and New York Roma invite and pay for the transportation of guest musicians from the Balkans (see Chapter 2).

Getting the Bride

Sunday is the day of transference of the bride from her natal home to the groom's home. It is enacted whether the couple are virtual strangers or have been living together, because it is the symbolic dramatization of the patrilocal principle. The bajraktar (flag bearer; Macedonian) leads the street procession to get the bride.[21] The bride's parents, if they can afford it, engage musicians, and all the relatives dance with the bride in front of her house (see video example 5.20). She is dressed in a white gown and her relatives demonstrably show their sadness by crying. As soon as they hear word that the groom's party is approaching (typically without the groom), they seclude the bride inside the house, and the men set up a barricade in the street. The bride's men wield sticks, knives, shovels, and axes and act very threatening. As the groom's party approaches, the cacophony of the two musical bands becomes more intense. The groom's men try to get inside the barricade, while the bride's men resist. The groom's men jokingly buy their way in with bribes and cases of beer. At many Šutka weddings, the bribe had to be paid in western currency! Note that this ritual dramatizes the close bond between the bride and her extended family, and their reluctance to see her leave. The ambivalence between their happiness to have her married and their pain of separation is performatively enacted.

The drama continues as the bride is led out from her house by an elder male relative, who holds her head down (photograph 5.19). In Ramisa's case, her brother fulfilled this role (video example 5.21). The bride is expected to gaze down demurely (photograph 5.20) and she sometimes performs temana (the expression used is *zema temana* [Macedonian] she takes temana), a slow arching movement done by one hand, then the other, then both (see New York video examples discussed later). Temana is done primarily in Skopje, Tetovo, and Gostivar and demonstrates respect.[22] The bride is then led into the street and transferred to the groom's males after more bargaining (see the end of video example 5.21). This is the saddest moment for the bride's side, and all her relatives cry. If they can afford it, the groom's family rents a *pajton* (horse driven cart, Macedonian), to transport the bride. Amdi Bajram rented an airplane to transfer the bride; it accommodated about fifty wedding guests and briefly circled above Skopje.[23] In any case, the path taken to bring the bride to the groom's house must be different from the path taken to get her; this is to confuse the evil eye.

Note that many rituals express traditional patriarchal values, e.g., the stance of the bride in which her eyes are lowered and she acts modestly, which seem to contradict both the powerful position of women in ritual management and the display of her sexuality through dancing. This paradox questions and sets into tension some of the traditional patriarchal tenants that the rituals themselves enact. Taking the ritual of a male elder leading the bride out of her house with her head lowered, for example, we would certainly be correct in assuming this was a symbol of female subordination. Yet we cannot assume a singular interpretation of this symbol. The fact that the ritual is directed by women who may have alternative views, and the fact that it is embedded in a complex set of female-centered performances and economic roles, mediates the patriarchal message.

The song that was performed during the transference of Ramisa from her natal family to her affine family was "Sine Moj" (My Son [Serbian], popularized by Muharem Serbezovski), which reflected the theme of the passing of childhood (video example 5.21). More typical for this moment, however, are the two songs "Oj Borije" and "Kote Isi Amalalen"; the former extols the beauty of the bride via a dialogue between her and her father-in-law (video example 5.26 with text supplement and video example 5.32 from a New York wedding). The latter extols the bride's pedigree (see audio example 5.1 with text supplement and video example 5.23). Another song often performed at this moment is "Ustaj Kato," a Serbian song in 7/8 telling of a jaybird's conversation with a young girl about her arranged marriage. See video example 5.29 with text supplement from a New York wedding.

The bride is incorporated into her husband's family with a series of rituals supervised by her mother-in-law (whom she often calls mother, *daj*). She is the most significant female to her; she will spend much time with her and has to follow the rules of her house.[24] The groom puts a belt around his bride's neck and pulls her slowly into the house as she carries a loaf of bread under each arm; a third loaf is held above her head. She smears honeyed water on the three walls of the threshold and kicks over (or steps on) a glass of sweetened water on the floor. As noted above, sweets and wheat ensure happiness and fertility.

Depending on the economic standing of the family, there may be a banquet on Sunday night, similar to the Saturday night banquet. If the couple have not run away, then at some point on Sunday night the marriage is consummated and the "good news" of the bride's virginity is announced. As described in the previous chapter, the blaga rakija ritual requires the groom and his relatives to process outdoors with music to the bride's house to show the blood-stained sheet to her mother. There they receive gifts and are fed eggs (video example 4.1 shows Ramisa's groom at the blaga rakija).

Džumaluk[25] refers to the first visit of the bride's relatives to the home of the groom. It usually occurs on Monday, the day immediately after the taking of the bride. Ramisa's džumaluk is illustrated in video example 5.22, where the bride's relatives arrived with decorated trays filled with food and gifts, and the groom's musicians met them. Ramisa, wearing her white gown (by now a bit faded) and looking forlorn, greeted her relatives and everyone cried as they took turns leading dance lines with her (video example 5.23; note the song "Kote Isi Amalalen," discussed earlier). In video example 5.24, Ramisa respectfully kissed her aunt, mother, and father as they gave her money. After several hours of dancing and feasting, Ramisa tearfully bade her relatives goodbye. Although Monday signaled the end of the wedding proper, visiting continued for several weeks, including prvič, the first visit of the bride to her natal home.

Weddings in Belmont

In the American context, many traditional rituals have been eliminated while a few new customs have been introduced. Rather than seeing diasporic weddings as lacking, however, I focus on how and why weddings change, that is, the decisions made regarding inclusion, exclusion, and innovation. For example, in Chapter 4 I described how in Macedonia invitations to a wedding are typically delivered formulaically face-to-face. In Belmont, on the other hand, invitations are professionally printed in the Macedonian language but with Latin letters, because many young Roma cannot read Cyrillic; but they are still delivered face-to face if possible. Belmont Roma usually confine their weddings to one or two weekend days. The rigid American work schedule stands in marked contrast to more flexible Macedonian work schedules.[26]

In the diaspora, sponsorship of weddings often departs from the patricentered norm. If a groom is brought over from Macedonia, for example, the bride's family, especially if they are poor, may feel no need to put on an event. The groom's side may have already sponsored an event in Macedonia.

On the other hand, if the bride's family can afford it and if they have many relatives here in America, they might sponsor a wedding. When a groom brings over a bride from Macedonia, there is usually a wedding. Because of migration, key family members are sometimes absent, and thus substitutes enact ritual roles. For example, people are assigned to be the bride's parents and a Belmont apartment may be designated as the bride's house. Of course, this adjustment also happens in Macedonia. For example, the parents of Bilhan's bride could not attend her wedding in Šutka because they lived abroad. To dramatize the transference of the bride to the groom's side, some friends of the groom acted the role of the bride's parents and used their Šutka house as her home.

As discussed in Chapter 4, Belmont families are involved in choosing spouses. The prewedding arrangements, visits and bargaining sessions, and gift giving and engagements are very similar to those in Macedonia, if both sets of parents are present. Sometimes engagement parties are held in banquet halls. The six-day wedding, however, is typically shortened to one to three days, but they are not necessarily consecutive. The henna party is sometimes eliminated, but if held it is combined with the second henna ceremony and may take place at any time prior to the wedding banquet. Surprisingly, in Belmont it is sometimes sponsored by the groom's side rather than the bride's side.

Ramo and Rifat's double wedding in 1995 (see the previous chapter) included a midweek kana in their parents' tiny three-room apartment. Approximately twenty-five women occupied the living room (which was nicely decorated with gifts for the two brides), and the few men were relegated to the kitchen. There was recorded music. At Samir and Lebabet's wedding in 2004, the groom's parents sponsored a large henna party on the night before the banquet in the courtyard of their apartment building in Belmont. They beautifully displayed clothing and gifts for the bride. Musicians used a powerful sound system, and guests danced outdoors until 3:00 AM. The clarinetist and singer, who were born in Macedonia and were relatives of the bride's father, were flown in from Germany. Not only is it very prestigious to hire musicians from abroad; they also bring news of relatives and return with gifts and videos. For this wedding the two guest musicians performed with local musicians. They all knew the same repertoire of songs and dances because every performer's musical reference point is Macedonia.

Lebadet led the first dance with a fancy handkerchief (not a decorated sieve, as in Macedonia; video example 5.25). One by one, her female relatives as well as the groom's led the line. Before each woman led, she (not a man) tipped the musicians. Lebadet wore a gown, then šalvari, then another gown. Šalvari are infrequently worn in the United States, but a few elders insist that brides wear them for rituals. As this was a female-centered party, the men stayed on the sidelines. The dancing became bawdy as the women loosened up, and the mothers of the bride and groom climbed on chairs and mimed sexually suggestive movements in a humorous way. In terms of the intense female presence, this event resembled the henna parties I had attended in Macedonia despite the absence of some of the specific customs.

Henna was brought out on a tepsija (with the song "Oj Borije," video example with text supplement 5.26) and, amidst loud screaming, was applied by an elder from the bride's side to the hands and feet, which were then covered in satin bags; a small amount was put on Lebadet's hair (video example 5.27). Most brides in Belmont do not want their hands to look odd at work, so they request just a dab of henna. After a while Lebadet washed off the henna; there was no bath ceremony. The next day (Saturday) included a ceremony in the mosque (this custom is becoming common in Belmont) and the formal banquet, but it did not include getting the bride. As the couple emerged from the mosque, they danced to a band of acoustic instruments, including clarinet, accordion, and tarabuka; several local musicians took turns performing (video example 5.28). Many relatives congregated, and although the dancing in the street blocked city traffic (video example 5.29) everyone was very polite; eventually a limousine arrived to take the couple to the banquet hall.

Other American weddings do feature "getting the bride." At Ramo's and Rifat's double wedding, both grooms and their relatives first went to Ramo's bride Metola's house (by car and limousine) and bargained their way into the courtyard, accompanied by an acoustic band. Metola emerged, eyes downcast and with her head held down by male relatives, and she "took the temana" (video example 5.30); this is done very rarely in the United States (Aiše, the grooms' mother, had previously instructed the brides in the temana). Metola was ushered into the limousine, and her trousseau was piled into another car. Then the limousine headed for Rifat's bride's (Fatima's) house, which was actually out of town, but a relative's Bronx apartment was used as a substitute. The same ritual ensued, with the bargaining (video example 5.31), the leading out, and the temana with the song "Oj Borije" (video example with text supplement 5.32). Then they drove to the grooms' apartment, where, Aiše, the grooms' mother, took charge of the incorporation rituals; many people remarked that these rituals are rarely done in the Untied States. Video example 5.33 shows Fatima as she walked up the stairs with a bread over her head and rubbed the threshold with sugar water; Rifat led her in with a belt around her neck and (gently) knocked her head on the walls (note the bridal gifts pinned to the walls); then he teased her with a knife and everyone kissed each other. All this was done with Aiše's formulaic blessings for luck, happiness, and "a male child next year."

The highlight of the wedding was the evening banquet, whose structure mirrored American weddings in that it took place in a rented hall with a dais, catered food, and a wedding cake. Typically, Romani weddings include a buffet plus cocktails followed by a sit-down dinner with Mediterranean foods. Other American customs include throwing the bouquet, throwing the bride's garter, drinking champagne, and cutting the first piece of cake. Bride's maids and ushers are other American adoptions; they are usually chosen from the younger unmarried relatives of both sides and wear matching gowns and suits. They may enter the hall in pairs, hold an arch decorated in greenery, and lead the bridal couple under it. At Ramo and Rifat's double wedding, the first dance line of the evening was led by Aiše (photograph 5.21), followed by her female relatives (video example 5.34); later Aiše led again, followed by Fatima's mother leading a slow crossing dance (video example 5.35). As in Macedonia, several close relatives danced solo inside the curve of the line. Then came the formal entrances into the hall, announced by the speaker, who was Osman (the grooms' mother's brother).[27] When the bridal couples entered, the brides again took the temana.

The band consisted of Ramiz Islami (clarinet and saxophone), his son Romeo (clarinet), Erhan Umer (synthesizer and vocals) and his father Husamedin (drum set and vocals), Trajče Džemaloski (synthesizer), Kujtim Ismaili (guitar), and several other drummers (video examples 5.36 and 5.37). I discuss these musicians later, but here I note that because Ramo was a singer, his colleagues were glad to play for him. Husamedin sometimes played dumbek and sometimes tapan in the center of the dancers (see video example 5.38).

After the two bridal couples entered, they did a slow American couple dance to a Romani song that Ramo sang, which morphed into a free-form čoček and became more intense as the three fathers threw money over the couples. As in Macedonia, the speaker called up the relatives to lead dance lines in order of closeness. This is an entirely constant element in the diaspora and, as mentioned earlier, is a visual interpretation of social structure of the extended family. For example, at Lebadet and Samir's 2004 wedding, the speaker announced, "The bride's mother and father will now dance." The father requested a tune and tipped the musicians, while the mother led the line and close females danced solo in the curve of the line. The father left the line to tip again so the dance would be extended. The momentum tangibly built up at the front of the line, until there was a visceral intensity (video example 5.39).

As in the Balkans, women are the primary dancers; they dance for hours while men dance sporadically. But in New York there is also a small, strong group of young male dancers. As in the Balkans, the solo čoček is considered a female specialty, and talented women are surrounded and encouraged by their relatives. Toward the end of Ramo and Rifat's wedding, after everyone had loosened up, the two couples danced solo čoček standing on chairs. At another wedding in 2004, the sponsoring family innovated by hiring a non-Romani American belly dancer. Note that, as will be explored in Chapter 6, although Romani čoček shares some movements with belly dancing, the latter is more overtly sexual, is costumed with naked flesh showing, and is danced by professionals for strangers for money. Wedding guests had mixed reactions to the belly dancer; elders for the most part disliked it (because it was not part of their tradition), and younger guests either liked the novelty or criticized it for taking the focus off the couple.

Another important part of weddings is the procession of the bridal couple around the banquet hall to every table. This can take several hours and is accompanied by slow songs, often in Turkish. The couple greets each elder family member with the customary kiss on the hand, and they receive money. The largest cash gifts, however, are given in envelopes at the end of the banquet to the sponsoring family, as each guest family lines up to bid farewell.

Common but not ubiquitous in New York is a blaga rakija banquet. It is usually sponsored by the bride's side, and held in a rented hall anywhere from a day to a few weeks after the banquet.[28] The day after Ramo and Rifat's banquet, for example, Ramo's bride's parents sponsored a blaga rakija party. Roughly the same musicians were hired as the night before, with the addition of Kurte Islami, Ramiz's son (video example 5.40). The traditional rituals were enacted, including the groom receiving gifts sewn to a sheet of fabric (bovčalok; see Chapter 4, and video example 5.41), eating eggs and doughy products, and joking with his male friends. Guests also brought gifts for the couple, and everyone danced for hours (see video examples 5.42 through 5.46).

Note that in New York weddings the financial outlay of gifts on all sides is substantially larger than in Macedonia. In New York, a guest (even distant kin) has to provide gifts for the engagement, the kana, the wedding banquet (in fact, several gifts during the wedding, plus tipping the musicians), and the blaga rakija. The main gift costs approximately $100 per adult. This represents a financial strain for most Belmont families, and they struggle to give honorably in spite of limited resources. They do, of course, expect to reciprocally receive gifts when they sponsor events. Putting on a wedding in New York is a huge financial commitment; if several hundred guests are invited, the cost of the rental hall, the caterer, and the musicians runs over $20,000, not including the gowns, gifts, limousine, and tips. According to *Condé Nast Bridal Magazine* the average national cost of an American wedding in 2005 was $28,000. This is precisely why families save up money.

Belmont Musicians

As I have emphasized, music in Romani communities is an important conduit through which social obligations are enacted. There are more than twenty regularly performing Macedonian Romani musicians in the New York area, and a younger generation is now emerging. They are all male, including singers, because it is considered immodest for a woman to be a professional and perform for strangers (see Chapter 6). Belmont Roma engage musicians for weddings from the local or diasporic Romani community because a successful performer needs to be familiar with the language, the rituals, and the dance repertoire. Families typically contract a lead player, who then assembles the rest of the band and secures a sound system (usually owned by one of the musicians). A fee is negotiated, but the musicians also expect to receive tips. Typical fees vary from $100 to $400 per person. For example, at one Belmont wedding in 1996 the saxophonist received $300, the keyboardist $250, one singer $200, another singer $400 (because he also owned the sound system), the guitarist $200, and the drummer $250; tips totaled $1,500, which were divided among the six musicians. These fees increased until 2008, when the economy declined and events became more sporadic. This is not a small income for one event, but Romani weddings are sporadic, and thus most musicians need to service other ethnic groups and have day jobs.

Some Romani musicians play for Bosnians, Serbs, Montenegrins, Macedonians, and Albanians as well as Roma; in Chapter 11 I discuss Yuri Yunakov in depth, who also plays for Turks, Armenians, and Bulgarians. Patrons hire musicians to perform for weddings, circumcisions, baptisms, graduations, and New Year's parties. Musicians also play in restaurants and ethnic nightclubs in Astoria, Ridgewood, Staten Island, and the Bronx, but these establishments frequently go out of business; for example, a Bosnian club in Clifton, New Jersey, employed Roma for many years as well as Bosnians, Serbs, and Bulgarians. For each patron group, musicians must know the appropriate dances and songs. Thus for non-Romani events musicians usually engage a singer from the ethnic group of the patron.

In the diaspora, good Romani musicians try to cast a wide net for patrons. Once someone is known to be talented and trustworthy, he is asked to perform by many colleagues. On a good weekend, for example, a musician may play in a Bosnian club on Friday, at an Albanian wedding on Saturday, and at a Romani circumcision on Sunday. Local musicians often combine with visiting Macedonian Romani artists, as at Samir and Lebadet's wedding discussed earlier. In the last fifteen years, visiting artists have included Esma Redžepova (see photograph 3.2 and Chapter 10), Bilhan Mačev (clarinetist who trained with Esma and Stevo Teodosievski), Tunan Kurtiš (clarinetist, now living in Germany), Ferus Mustafov (clarinetist and saxophonist see chapter 2), Severdžan Amzoski (clarinetist, known as Klepača from Bitola), Gardjian (singer), Vehbi Mefailov (clarinetist, from Bitola), Rifat Demirov (clarinetist now living in Vienna), and Muharem Serbezovski (singer now living in Germany; see Chapters 2, and 3).

When Serbezovski performed for a New Year's party in 1997, the crowd was especially large since he is a well-known artist and was accompanied by Bilhan Mačev (photograph 5.22, video examples 3.1 and 5.47). The older generation especially liked his hits from the 1970s, while the younger Roma preferred Mačev's contemporary repertoire. When Esma Redžepova visited New York in 1996, a group of Bronx Roma organized a dance party in a rented hall. Esma performed a miniconcert in which she sang her hits, and she then took requests by walking from table to table (video example 3.3); afterward, her drummer and gifted vocal protégé, Šadan Sakip, sang for a dance party accompanied by contemporary Skopje-style dance music played by Bilhan Mačev (video examples 5.48, 5.49, and 5.50).

Because Esma had such a broad fan base and because this event was not a private family event, Serbian Roma, Kalderash Roma, Slavic Macedonians, and American folk dancers attended as well as Macedonian Roma.

It is quite unusual to have these groups together in the same space because they do not normally socialize. On the dance lines, ethnicities clustered but everyone felt welcome. On the other hand, when Roma venture out beyond their community to musical events, there is sometimes tension; because of prejudice, they are not readily accepted by other Balkan ethnic groups. In 2003 I attended a New Year's party at a Macedonian Orthodox Church in New Jersey with about fifteen Macedonian Roma from Belmont (and with Šani Rifati, a Romani activist from Kosovo). The featured vocalist was Blagica Pavlovska, who sings in Romani style; the featured band were the Struškite Svadbari (Struga Wedding Musicians). Roma were not welcomed by most Macedonians at this event (especially when nationalistic songs were sung) and as a result were ill at ease, did not dance freely, and departed early. A similar experience happened when Esma played at a Macedonian Church in 2004 and several Romani families attended (video example 3.4). On the other hand, Romani musicians are much more used to interacting with Macedonians and other non-Roma, and they are immune to stares and hostile glances. They adapt readily and are much more adept at crossing borders thanks to their professional experience playing for patrons of varying ethnicities.

Ramiz Islami was a seminal musical figure in the New York Romani community in the 1980s; he lived in Brooklyn and died in 2004 at the young age of forty-eight. He was the favorite clarinetist/saxophonist for events through the 1990s, and he played at Ramo and Rifat's 1995 wedding. Ramiz is from a family of Albanian speakers from the Prespa (Resen/ Ohrid) area, and his uncle, Dule, the most famous clarinetist of the region, played regularly for Albanians and Macedonians (Leibman 1974).[29] Various incarnations of his band, Grupi Sazet E Ohrit (Ensemble from Ohrid; Albanian), included Kujtim Ismaili on guitar and vocals (born in Ohrid in 1965, married to Ramiz's sister); his brother Redžep on clarinet/saxophone; his two sons, Romeo on clarinet and Kurte on drum set; Erhan Umer on accordion/keyboard; Muren Ibraimov from Bitola on violin; and Seido Salifoski on dumbek. Ramiz trained Nešo Ajvazi (vocals) in addition to most of these musicians.

Ramiz had a sweet tone, great mastery of technique, and a varied repertoire. Photographs 5.23 and 5.24 and video examples 5.34 through 5.38 and 5.40 through 5.46 feature Ramiz and his band 1988–1995. Today his two talented sons, who moved to the Philadelphia area, continue his legacy, although Romeo has somewhat departed from his father's older Albanian Prespa style and repertoire in favor of a technically flashier Skopje Romani style (he is also influenced by Ivo Papazov's Bulgarian wedding style). [30] Kujtim' s son Muamed (born 1990) is also emerging as a talented keyboard player.

Erhan Umer (known as "Rambo") was born in Bitola, Macedonia, in 1973 and emigrated to the United States in 1986 with his parents; at home his family speaks Turkish and Macedonian. He plays accordion and keyboard, sings, and is a very popular musician. He comes from a long line of professional male musicians: his father, Husamedin, was invited to play tapan with Tanec, the national professional folk music ensemble of Macedonia; and his uncle, Jusuf, also played tapan; both brothers are excellent dancers (Jusuf leads a dance line in video example 5.44). Erhan's older brother Sevim narrated: "My father was invited to play with Tanec in Skopje 1959. He joined for one month but my grandfather objected, saying 'How can you leave me alone in Bitola?' So Husamedin had to drop out! My grandfather was also a tapan player but he wanted his youngest son near him at home. Those were the traditions back then." As a young man Husamedin played in the Bitola KUD (Kulturno Umetničko Društvo, amateur ensemble) Ilinden. Eran recalls how his father endured the stressful life of a professional musician even as a child: "Husamedin was eleven years old and had just received a new tapan from his father—he was playing it with the father for the first time at a wedding, and was so proud. In the middle of the wedding someone approached the father about buying the tapan, for a good price. My grandfather took the drum away from Husamedin and sold it right then and there. Husamedin felt like crying. Money counted, not a child's feelings."

Husamedin switched from tapan to drum set and currently plays regularly for Albanian events in New York (he speaks Albanian fluently). He still plays tapan when the music and the atmosphere require it (see video example 5.38). Most of Erhan's other male relatives are musicians: Vebi (first cousin) is a singer, Turan (brother, born 1968) plays guitar (he and his family lived in Germany for several years before coming to New York), Dževat (brother living Switzerland, born 1960) plays drums and keyboard, and Sevim (brother, born 1965, speaks Albanian) regularly plays drums and tours with Albanian stars Merita Halili and Raif Hyseni.

In addition to family celebrations, Erhan performs regularly in clubs that cater to Albanians, Bosnians, Serbs, and Macedonians. Erhan was trained by Ramiz as well as his father; video examples 5.34 through 5.38 and 5.40 through 5.46 feature Erhan playing with Ramiz at Ramo and Rifat's 1995 wedding. As young man, Erhan formed his own band, Amanet (Testament), which has had varying personnel and produced several cassettes, one of which features Ferhan Ismail (his cousin) on vocals, Ramiz Islami (clarinet), Ilhan Rahmanovski (guitar), Seido Salifoski (dumbek; photographs 4.1 and 5.26 and audio example and text supplement 4.1 of the song "Gurbeti," discussed in Chapter 4). Erhan recorded the master tape in New York and then, to save money, sent it to Macedonia for duplication. On several of Erhan's albums, Yuri Yunakov was contracted to play solo improvisations. In recent years Yuri, Erhan, and Sevim have collaborated extensively, and Yuri has included them on tours for American audiences (see Chapter 11). Video examples 5.51 and 5.52 feature Erhan and Yuri at the 2008 California Herdeljezi festival, sponsored by NGO Voice of Roma.

Erhan's teenage son Husamedin (nicknamed Uska, born 1989) is emerging as a talented vocalist and has begun to sing with his father. In May 2007, he performed with his family band (and guest Ismail Lumanovski) for Americans at the Voice of Roma Herdeljezi festival (see video examples 5.53 and 5.54). Like many other young Macedonian Roma in the diaspora, Uska keeps abreast of the newest musical developments in the Balkans via the internet. He has thousands of Romani songs on his computer and trades them with his cousins in the diaspora. He graciously shared his extensive knowledge of new Macedonian Romani music with me; like many young musicians, he is also knowledgeable about Bulgarian Romani music.

Ilhan Rahmanovski, who emigrated in 1993, is a guitar player who plays professionally, as well as a handyman for several buildings (photographs 4.1, 5.22, 5.25, and 11.1). He narrated:

> I didn't play music at home in Prilep, I learned here. I was living in the same apartment building as Erhan who was learning keyboard. He encouraged me to learn guitar. I play regularly in clubs and weddings for Bosnians and Serbs. The clubs were more popular in the 1990s. In those years Balkan men came over *na pečalba,* to work, to earn money, and they patronized clubs. They had no families, low rent, few bills, and no family life. They went out to clubs to meet each other, to hear music from back home. I used to play four to five nights a week. Now there are only a few clubs left. Today they are family men with bills, rent, and large families. Plus the war in Yugoslavia brought tension to these clubs.[31] But I still play fairly regularly.

Ilhan's son, Džengis, is a teenage tarabuka player who grew up surrounded by his father's musical friends. He watched Seido Salifoski perform for years, was invited to play with his father at several gigs, and played with other Belmont teenagers. Video example 5.55 shows Džengis performing at the East Village club Maia Meyhane with his father, Seido, Šaban Dervisoski (accordion), and Ismail Lumanovski.

Another young musician, Sal Mamudoski, a cousin of Seido born in 1988, is a promising clarinetist being trained by Yuri Yunakov. Sal debuted with Yuri for American audiences at the 2006 Zlatne Uste Golden Festival, in a band that included Ilhan, Džengis, Kujtim, and Muamed. In 2007 Sal toured nationally with Yuri, Erhan, and Bulgarian Romani drummer/tambura player Rumen Sali Shopov for American audiences. Currently Sal plays regularly with Yuri in his ensemble at the downtown club Mehanata with Alfred Popaj, an Albanian keyboardist. Video examples 5.56 and 5.57 feature them at Hungaria House in 2007; video example 5.51 features Sal, Yuri, Erhan, and Rumen at VOR's Herdeljezi festival in 2008. Seido and Yuri are thus training the younger generation and exposing them to varied American audiences.

Severdžan Azirov (born 1967) and his brother Menderes left their birthplace, Dračevo, a suburb near Skopje, in 1983, lived in Ljubljana, and emigrated to Belmont in 1990. When they resided in Macedonia, the two brothers were dancers in several KUDs, including Phralipe (brotherhood) in Skopje (see Chapter 6), and in ensembles such as Kočo Racin and Orce Nikolov. They know a large repertoire of ethnic Macedonian dances plus the Romani repertoire. At Belmont events, they often lead men's lines, and their parents are also strong dancers (video example 5.58 features

Menderes leading Čačak and 5.42 features his father leading a line čoček).[32] Severdžan is also an excellent drummer (photographs 5.22 and 11.1) and performs regularly in clubs and at weddings for Bosnians, Macedonians, Serbs, and Albanians, as well as Roma (he is playing in many of the video examples from Lebadet and Samir's wedding, and also in video examples 6.4 and 11.1). Almost every New Year's, for example, he plays at a Macedonian Church in Garfield, New Jersey. Severdžan's son, Sabuhan, in his early twenties, learned keyboard and production skills for hip hop and for some time played Romani fusion rap music with Yuri Yunakov's son, Danko.

Nešat (Nešo) Ajvazi is one of the few professional singers in the community, and he also has a janitorial job in an office building. He was born in Belgrade in 1970 but regularly spent his summers in Priština, visiting relatives from Kosovo; thus Serbian is the language he speaks most fluently, although he knows Albanian, Romani, Macedonian, and English as well. His father, a theater director, was from the Aškalija group (Albanian-speaking Roma) in Priština, and his grandfather was a džumbuš player. His mother, from Gnjilanje, Kosovo, died when he was very young and his father remarried a woman from Skopje, so he has always been close to Macedonian Roma. He emigrated in 1985 when he was fifteen years old and lives in Staten Island. Nešo learned to sing in the United States as an adult under the tutelage of Ramiz Islami and later with Yuri Yunakov. In video examples 5.59 and 5.60, he sings at the 2005 Balkan Music and Dance Workshop, sponsored by the East European Folklife Center, in Iroquois Springs, New York, accompanied by Ismail, Seido, and ethnic Macedonian keyboardist Toni Jankuloski.

Ismail Lumanovski (known as Smajko), born in Bitola in 1984 to a Turkish-speaking family, is a masterful addition to the pool of young Romani musicians; he is unusual in that he received western classical training. He started playing the clarinet at the age of nine and made his debut in 1998 with the Macedonian Philharmonic. He came to the United States as a teenager as a result of winning a stiff competition to study at Interlochen Academy in Michigan, making his American debut in 2002 at the Interlochen Arts Camp with the World Youth Symphony Orchestra and becoming its principal clarinetist. He graduated from the Julliard Conservatory in 2008 and is a member of the Juilliard Symphony Orchestra; in 2010 he obtained his master's from the Julliard graduate program. He has won numerous competitions and awards, including first prize at the 1998 Folk Music Competition in Macedonia and first prize at the twenty-third, twenty-fourth, and twenty-fifth National Clarinet Competitions for classical music in Macedonia. He has played in numerous classical music concerts and competitions in France, England, Australia, Belgium, Germany, Bulgaria, China, Turkey, and the United States.

Thus Ismail is one of the few New York musicians who read notation and play Western classical music as well as folk music. His father, Remzi, sang for Macedonian Radio in the Turkish, Macedonian, Serbian, and Albanian languages. Ismail is a brilliant master of a wide variety of styles, including Macedonian village repertoire. He is well connected to ethnic Macedonian musicians and often plays for the Macedonian church in New Jersey. He is very highly respected by Macedonian Roma and was the youngest performer to participate in the Clarinet All Stars at the first New York Gypsy Festivals in 2005 (see Chapter 13). His own group, the NY Gypsy All-Stars (www.myspace.com/thenygypsyallstarsband), regularly performs at the downtown club Drom (road). In photograph 5.25 he performs with Ilhan Rahmanovski and Šaban Dervisoski (from Prilep) at Maia Meyhane, New York, in 2006. Video example 5.61 shows Ismail playing "Gaida" (the clarinet imitates a gaida) in 2005 accompanied by Seido Salifoski on tapan and Tony Jankuloski keyboard in New York; video example 5.2 shows a masterful solo from 2005; video examples 5.53 and 5.54 are from California Herdeljezi 2007.

Seido Salifoski is one of the most versatile of the New York Macedonian Romani musicians (photograph 5.26). A brilliant dumbek player, Seido is the son of Nermin, who was one of the first émigrés to Belmont from Prilep (see Chapter 4); Seido emigrated when he was six years old. He has remained within the Romani community in terms of kinship ties but has moved away from it to forge an unusual musical career, live alone for many years, perform with non-Roma, travel extensively, and marry a Japanese woman. Surmounting many challenges, he lives in two worlds by balancing familial loyalties with personal goals. The fact that he is an only son makes him particularly obligated to his parents. However, he has had some conflicts with his relatives and can be quite critical of his community.

Musically, Seido bridges several worlds. From a young age he listened to Romani music and other folk and popular musics. Growing up in Belmont, he befriended a variety of performers: "At first I got together with the Black and Spanish musicians in my neighborhood. I was playing Latin music, Santana, all those styles. I hung out with them, they were my friends. I was really into that stuff. I brought these guys into my home in Laurelwood and we played music in the basement. My parents didn't mind—I guess that was pretty open of them." In his younger years, Seido played regularly at community events with Ramiz and the musicians mentioned above, and he recorded several albums with them (audio example 4.1). But he also sought other styles and wider audiences. He has played regularly with Bulgarian Romani wedding musicians Yuri Yunakov and Ivo Papasov, Turkish artists Tarkan and Omar Faruk Tekbilek, and Greek and Arab performers. He is equally at home in the jazz fusion world, and since 1990 he has been a member of the Paradox Trio, with whom he recorded four albums. The trio, made up of Matt Darriau, Brad Shepik, and Rufus Cappadocia, performs Balkan/jazz fusion and tours regularly in Western Europe.

Seido is one of the few Romani artists, along with Yuri Yunakov, who perform with and socialize with Americans who play Balkan music. He is a member of the Zlatne Uste Brass Band (composed of Americans) and has taught at the East European Folklife Center's Balkan Music and Dance

Workshops, and many other folk dance camps. In 2005 he formed his own band, Romski Boji (Romani Colors [Macedonian]), with various Romani musicians. The band performs in downtown clubs and for American folk dance events (see video example 5.55). Video examples 5.59 through 5.63 feature 2005 performances; 5.52 and 5.63 show dazzling solos. Seido is one of the only Romani musicians to offer music lessons to non-Romani students and maintain his own website (www.seidoism.com), designed by his wife. He also sells T-shirts (also designed by his wife) with his web logo and the words "Honorary Rom" printed on a field of blue and green (with the wheel as the letter O). The packaging of these shirts includes a quote from scholar/activist Ian Hancock explaining the symbolism of the colors and wheel as related to the Romani flag (see Chapter 3), and a pledge that a portion of the proceeds will be donated to Romani political causes. Seido is thus the most activist of Belmont musicians; he enthusiastically helped me by furnishing extensive commentary on chapters related to Belmont in this book.

The musicians described here are all respected members of the Belmont community; the music they offer is the means through which performative Romani identities are enacted. Diasporic Macedonian Roma, whether in New York or the Balkans, readily opine that music and dance help define them and set them apart from others. Now I turn more closely to Balkan Romani dance to explore diasporic styles and gender relationships.

6

ⸯⸯⸯⸯ

Transnational Dance

A s mentioned in previous chapters, dance is closely imbedded in the social life of Balkan Romani communities and is especially tied to music, gender, and status. Dance mediates female sexuality and reputation; its practice is governed by community ideas of propriety, context, and talent. As a solo dance, čoček has a long history rooted in Ottoman professional genres and lies in a continuum to contemporary forms of belly dance. In the last fifty years, čoček has traveled in the Balkan Romani diaspora to Western Europe and the United States and has also been appropriated into new settings, including professional and amateur ensembles, Romani music festivals, world music events, and Slavic, Albanian, and Romanian community dance events. This chapter compares and discusses Romani dance in all these locations, investigating its stylistic, social, and power dimensions in relation to the marginality of Roma in the wider society and the ambivalent position of women.[1]

Professional Dancers of the Balkans: Ottoman Roots

According to Ottoman sources, from the eighteenth to the early twentieth century Romani women and men were professional dancers, hired in aristocratic, courtly, and military as well as tavern settings (And 1959, 1963–64:26–8). A professional male dancer was known as *köçek* (or *raqq*) and a female dancer was known as *çengi*, although both sexes were often called çengi (And 1976:138). The appellation köçek also became associated with the style of music played for accompaniment and later for the sensuous solo dance form known all over the Balkans today as čoček (Bulgarian: kyuchek). See text supplement 6.1 for a history of Ottoman dance.

It is obvious that older çengi dancing informs contemporary female professional belly dancing in Egypt, Turkey, and the Balkans, as well as solo čoček dancing in various Balkan communities. In Turkey, Seeman reports that professional çengis adopted Egyptian style movements and tighter choreographies, resulting in a style known as "oriental." In Istanbul Romani neighborhoods such as Sulukule, çengis were hired for weddings and for tourist shows (Seeman 1998:3–5 and 2002; Potuoğlu-Cook 2006 and 2007). Seeman also reports that in the 1980s there was one professional Romani dancer in Skopje who was hired for men's celebrations (personal communication). Pettan remarks that in Kosovo in the early 1980s, "in the area of Peč it is customary that female musicians perform for a short while for male guests, and one of the musicians even dances" (1996a:317). In Bulgaria, the budget for a north Bulgarian *panair* (gathering) in 1884 included income from the kyuchecks of female performers (Peycheva 1999a:41). And more recently, in the early 1970s, Turkish-speaking Romani clarinetist Ivo Papazov partnered with Zvezda Salieva, a professional Romani dancer who performed at weddings. She and her sisters were part of a "dance dynasty" (214 and 247).

On the one hand, çengis were admired for their musicality and beauty, while on the other hand they were criticized for their licentiousness and abandon, and many were assumed to be prostitutes. In the early years of the Turkish Republic (1920s), belly dancing was "a despised genre" associated with "fallen women." It was rehabilitated in 1980 when featured on television for the first time. Now belly dancers regularly grace tourist brochures and furnish a steady income for restaurants and cafes (Öztürkmen 2001:143). Sugarman points out that the late-nineteenth-century nationalist movements of the Eastern Orthodox southern Slavs mobilized specifically against the perceived decadence of Muslim culture (as symbolized by çengis; 2003:101–102). In this emerging nationalist discourse, Roma had two strikes against them: they were Muslim and they were Gypsy. According to Sugarman:

> Their identity as Roma was yet another factor contributing to their poor reputation, leading to a highly ironic situation for them: having taken up the role of entertainer in part because it was one of the few economic niches available to them as a marginal social group, they were then further marginalized by the profession itself. Their perceived indecency could then be ascribed by non-Roma to the moral character of their ethnic groups, rather than to the particular social and economic conditions and gender arrangements that prevailed within late Ottoman society [101].

Later in this chapter I explore how the market has been a constant factor in determining the place of professional dance, but first we need to examine čoček in contemporary Romani communities and the historical, religious, and cultural baggage of ambivalence that surrounds it.

Sexuality and Dance

Condemnation of sensuous dancing is grounded in an ideology of female modesty and decorum that was historically shared by all Balkan peoples regardless of religion; today this ideology appears stronger among Muslims. As Cowan writes (on the basis of Eastern Orthodox Greek Macedonian materials), "Dance is a problem for women because in the dance site 'ambivalent attitudes about female sexuality as both pleasurable and threatening are juxtaposed'" (1990:190). The embodied nature of dance highlights its association with female sexuality. For Muslims "the female body is the embodiment of seductive power and its open expression is therefore strongly condemned in moral-religious discourses" (Nieuwkerk 2003:268).

The literature on honor and shame in the Mediterranean region is useful in that it identifies the honor of the family with control of female sexuality. But this literature must be criticized for reducing complex and variable systems to a rigid dichotomy. Various authors have shown that the supposed pan-Mediterranean concept of honor via music and dance means different things to different cultural groups (Magrini 2003). The Balkan Romani moral system contrasts *pativ* (*pačiv, pakiv*; Romani, respect) with *ladž* (Romani; shame). In the South Slavic languages, Roma speak of these concepts as *čest* (honor) and *sram* (shame). A bride who is a virgin is *čestna* (honest, pure). A professional belly dancer *nema sram* (has no shame). A family's reputation is expressed by offering hospitality to guests, respecting elders, and caring for family members in gender-specific ways. A man works and provides for his family; women work too, but they also cook, clean, and take care of children, and they serve men. In public, women are expected to cater to and defer to men, as the latter are nominally "heads of household."

This association of women with sexuality bears directly on the stigma of the female professional dancer, for it is both the commercial relationship with a paying audience and the display of the body to strange males that threaten female modesty. For this reason dancing professionally is regarded as far more immoral than singing professionally (Silverman 2003). In Chapter 10 I chronicle the challenges faced by Esma Redžepova in carving out a respectable niche as a vocalist whose performances often included dance. Yuri Yunakov, a Bulgarian Muslim Romani musician (see Chapter 11), remarked that he would never let his daughter (who is a very talented dancer) become a professional dancer, as it was a degrading profession. Dancing for money involves performing for strange men, marketing one's sexuality, and thereby devaluing it. Dancing nonprofessionally in the Romani community also has its challenges (more on this later), although they are mitigated by the high value placed on dance as a female art form.

Although men are the "heads" of families, the ideology of patriarchy is contradicted by realms of female power and influence. The female role in income-producing activities, budget decisions, marriage decisions, information networks, and ritual all mitigate her subordination (see Chapters

5 and 6 and Silverman 1996b). In the realm of sexuality, however, women theoretically must conform to ideal behavior precisely because sexuality poses the greatest danger of ladž or sram. Women are scrutinized by other women as to their bodily appearance and deportment. Clothing (especially hem lines and bodices), makeup, eye contact, socializing patterns, company kept, time spent outdoors . . . all are noted and evaluated for violations of modesty. The most highly charged symbol of the proper deportment of female sexuality is the test for the virginity of the bride, still performed today in many Romani families in the Balkans and in the diaspora (see Chapter 4).

The common social structural argument explaining the potency of female sexuality argues that in patrilineal and patrilocal societies the possibility of a woman having a child with a man who isn't her husband disrupts the patriarchal system and poses a problem of affiliation of the child. Other views argue that it is the female body itself that is inherently sexual, in contrast to the "productive" body of men (Nieuwkerk 1995:154). A third view interprets Islam as conceiving of women as more sexual than men, thus needing to be constrained (Mernissi 1975). These views are somewhat relevant for the Romani case, but they are insufficient explanations. Balkan Roma talk constantly about the problems of a child who isn't rightfully attached to an extended family; children conceived in adulterous relationships are pitied and their mothers are rebuked. But a woman's deviant sexual behavior is seen as part of her intrinsic immoral character. Roma seem to view sexuality as inherent to females, but not in contrast to the "productive bodies" of males. True, males have to worry less about public scrutiny of sexuality, but on the other hand Roma view females as productive bodies. In fact women are often viewed as more productive than men. Most Roma agree that women hold the family together emotionally and culturally, and in addition many families survive on women's incomes.

An important manifestation of the proper deportment of sexuality is monitoring where and for whom dance is performed. Because dance is so sexual, it should, ideally, be performed only among one's own sex. According to Dunin's pioneering research in Macedonian Romani communities, segregated male and female dancing was the norm until the 1970s. Women danced in private home settings to the accompaniment of a female dajre player and women's singing; women dancing for men was considered crude (1971:324–325; 1973:195; 2000; 2006). Note, however, that this was also true for Eastern Orthodox Christians and for non-Romani Muslims of the Balkans (Rice 1994; Sugarman 1997). Esma Redžepova, speaking of her childhood in the 1950s, remembered, "Women used to be in a separate room, men separate, and they used to celebrate segregated at weddings." During the women-only bathing-the-bride ritual at Esma's wedding in 1968, there was a female orchestra composed of one violin and two daires (Teodosievski and Redžepova 1984:108).

Some women conceived of older weddings as two simultaneous events: a women's party and a men's party. An elder woman remembered the

1950s as follows: "During the Saturday celebration of the wedding at the bride's house, there would be a professional female orchestra—two violins, a daire player, and the singer, usually the daire player." The spatial segregation during celebrations was often described in terms of the "inside" women's world and the "outside" men's world. This concept of space is shared with non-Romani Muslims of the region (Sugarman 1997; Ellis 2003). In the henna ceremony, for example, Esma recalls that "the women were inside, the men were outside with the zurlas and tapans." Pettan reminds us not to take the words "outside" and "inside" too literally. In the Balkans, many courtyards have high walls; thus a women's courtyard performance is outside, though not as public as the street. The courtyard is sharply distinguished from the street, where men perform (Pettan 1996a:316; 2003). Henna celebrations in Šutka, for example, take place either "inside" the house or "inside the courtyard" of the bride. As I described in Chapter 5, women from the groom's family dance through the streets and then make the bride's courtyard into women's space (video examples 5.1 through 5.8). Similarly, at Lebabet's Belmont henna ceremony, described in Chapter 5, an urban courtyard became women's space (video examples 5.25, 5.26, and 5.27); in both cases, men looked on from the periphery (except for musicians).

These descriptions do not imply that Romani women are confined to the domestic sphere. Although they are associated with the domestic, both historical references and ethnographic observations show that Romani women regularly occupy the public sphere, primarily for economic activities such as music, dance, seasonal agricultural work, and selling at markets. Pettan observes greater freedom of movement of Romani women in comparison to other non-Romani Muslims of Kosovo: "Similarly to non-Gypsy ethnic groups and musicians in Kosovo, Gypsy men are oriented towards the public domain while Gypsy women primarily towards the private domain. Their private domain, is however, extended in comparison to most non-Gypsy women" (Pettan 1996a:316). Pettan further explains that although this is true for sedentary Roma, nomadic Romani women are even more exposed in the public realm, through fortune telling, selling herbal medicines, and begging (Pettan 1996a:316; also see Okely 1983).

In the 1930s, freedom of movement of Romani women was noted by Catherine Brown, a British traveler: "One of the most striking features of these gypsy women is their great freedom and independence of bearing as compared with other Mohamedan women in Macedonia. Although among orthodox Mohamedans [non-Romani] one may occasionally see on feast days groups of men strolling about the village together, tinkling gently and rather halfheartedly on small stringed instruments, no women are ever to be seen with them, the women's festivities being invariably quite separate and confined to the harem. Here [among the Roma] men and women joined freely together in whole-hearted enjoyment. The whole scene resembled an enormous ballet" (C. Brown 1933:307). Unlike upper-class non-Romani Muslims, Romani women have historically worked outside the home among non-Romani women and men.

To return to the theme of propriety, Dunin's Macedonian research in the 1970s showed that line and processional dancing were sexually integrated while solo dancing (which is more sexually suggestive) was segregated. Dunin remarked that "whenever the dancing began during segregated parties, the curtains or drapes were secured so that no one could look indoors. If a child playfully pulled the curtains from outside, he was sent scurrying for fear of being punished. . . . This dance was meant to be performed by women for women and not in mixed situations" (1973:195). By the 1980s, however, thanks to relaxation of gender divisions in many areas of life, solo dances could be found in mixed company. Women now dance solo in the presence of men; women also continue to dance in sexually segregated events such as henna parties. How women negotiate varied contexts is discussed later.

Čoček in Diaspora Balkan Romani Communities

Čoček, or čuček in Macedonia and Kosovo (kyçek, Albanian) and kyuchek in Bulgaria, is the most characteristic Romani solo dance form. Note that čoček can also refer to the musical genre used for this dance, in 2/4, 7/8, and 9/8 (see Figure 2.1 and Chapter 2); the term thus serves a double function. Note too that čoček can also refer to the line dance performed to this musical genre. As a solo dance, čoček is improvised, using hand movements, contractions of the abdomen and pelvis, shoulder shakes, movement of isolated body parts (such as hips and head), and small footwork patterns. Men as well as women perform it, but it is overwhelmingly associated with women. Čoček is clearly an heir to the dances of the Ottoman çengis (the term comes from köçek), but in Romani communities its subtlety and restraint distinguish it from contemporary belly dancing. I conceive of solo čoček dancing as a continuum, with subtlety and a covered body (as found at Romani community events) on one end and belly dancing and exposed skin on the other.

Čoček is embedded in ritual events, which are numerous and obligatory in Romani communities, as discussed in Chapter 5. Close kin women are expected, even obliged, to dance for hours at weddings, sometimes for three or four days in a row, no matter how tired they are. The only excuse not to dance is illness or mourning. Women who do not dance well or are mentally or physically disabled also dance and even lead dance lines. Male dancing is more optional than female dancing. As I explained in Chapter 5, there are some moments where a man's dancing is required. Men dance for entertainment too, and some men dance a great deal, but they are rarely obliged to dance.

On the dance floor one finds both children learning by immersion and seasoned elders. A typical wedding dance line, whether located in the Balkans or in the diaspora, has a ratio of approximately three to one women to men. Men often dance together, put a great deal of energy into the dance for a short while, and then sit down. In Belmont, there are groups of young men who always dance together; they look for each other at celebrations and try out complicated steps. In fact, men seem to demonstrate their masculinity more in line dances rather than solo dances. Women and girls also tend to dance with their relatives and friends; they too join the dance line in pairs or groups, rarely alone. But unlike men, women and girls are on the dance floor for practically the whole event.

Dunin's research describes ordinary Romani women looking "very comfortable and confident of their movements probably due to the frequency of dancing, which occurs almost every week" (1971:323–324; also see 1973; 1977; 1985; 1997; 2008; 2009). Indeed, čoček as a solo dance has an important place within all rituals. It is danced in the middle of the floor by important females near the front of the dance line; simultaneously, the line snakes around (see video examples 5.6, 5.11, 5.12, 5.25, and many others in Chapter 5). For example, in Šutka, at an igranka, the bride's close female kin will dance čoček in the center as relatives are called up to lead the dance line. A few female members of the beckoned relatives (rarely men) might also join in the center. The style changes as new tunes are played and new family members are summoned. Even though women may be ostensibly doing the same dance for hours, its texture migrates, for example, from fast and bouncy to slow and heavy. Dancers may show their exuberance by climbing on tables to perform; photograph 6.1 shows a solo dancer on top of a car during a circumcision procession.

Ritual contexts of dance are obligatory, but dancing for entertainment is also common, for example, during the less ritualized parts of celebrations. Video examples 5.48, 5.49, and 5.50 show a party in New York in honor of Esma Redžepova with line and solo dances; example 5.50 shows a talented solo dancer. A good čoček dancer has the admiration of the entire community, and her family proudly displays her talents. At a wedding in Šutka, the father of an excellent teenage dancer was very angry with her because she was nowhere to be found when the family was called up to lead the dance line (video example 5.11). His family's artistic competence depended partly on his daughter, who possessed a valuable female asset. Women and girls squarely take center stage as excellent dancers, and people crowd around them to watch. Esma Redžepova remarked, "That was the most beautiful, to show dignity. A mother-in-law might say to another mother-in-law, 'my daughter-in-law raises her hand [while dancing] as if she could take everyone's life!' This would show how delicately she danced; this was the realization of Romani tradition." Girls more than boys are coached by family members to dance (see video example 5.8). At home, taped music is played as experienced female dancers demonstrate techniques. At dance events, mothers "put up" their daughters to dance on tables (video example 5.19).

The female art of dancing čoček is chronicled in hundreds of songs; for example, see audio example with text supplement 6.1. In "Romani Čhaj Sijum" Macedonian Romani singer Džansever (see Chapter 2) sings of "throwing" the stomach, implying that the movement is quite sharp and rhythmic; this is a characteristic move for Turkish Roma, and the phrase describing it is shared by Turks and Roma (see video examples 6.1 from Bulgaria and 6.2 and 6.3 from Šutka). Peycheva's Bulgarian Romani informants all speak of the stomach flick as an essential part of kyuchek (1999a:244–245). Seeman also documents the importance of the stomach flick (2002 and 2007); it becomes especially dramatic in the 9/8 rhythm that has become emblematic for Turkish Roma. In fact, at New Year's dances sponsored by Macedonian Roma in New York, 9/8 tunes are played exactly at midnight.

Although family members seek to show off the dancing of unmarried girls, they must delicately negotiate the propriety of the display. Some displays are crass and transgressive while others are appropriate, depending on context and audience. At Ramisa's wedding banquet in Šutka, for example, a mother put her sixteen-year-old daughter on a table to dance while she, her sister, her husband, and the dancer's sister and brother all danced in front of her (on the floor), encouraging her with shouts of appreciation and even with monetary tips (video example 5.19). Similarly, female relatives of good dancers often stop dancing and clap for the talented performer. In spatial terms, the nearest audience for proper female čoček dancers is composed of relatives. Strangers, however, do watch from afar. Ironically, it is precisely for strangers that the girl's talent needs to be shown (for marriage purposes). The physical proximity of relatives is not only a permeable wall—a shield of protection against claims of sexual immorality—but also a transparent screen through which to view female bodily displays.

Also note that Roma (especially in Macedonia) do line dances to čoček music that vary in step and style by region, age, and subgroup of Roma. Rhythms of 9/8 and 7/8 are less common than 2/4, and 9/8 tunes are often played later in the evening and induce increased intensity. The most common 2/4 line dance in Macedonia, Kosovo, and southern Serbia is a three-measure dance, sometimes called *oro*,[2] with versions that vary by rhythm and footwork (see video examples in Chapter 5); it is also danced to 7/8. Note too that in addition to čoček Balkan Roma have always adopted some of the dance repertoire of the non-Roma in their region. Line dances often start slow and speed up at the end, in typical Macedonian style. The New York Macedonian Romani repertoire of line dances includes *Bugarsko* (Bulgarian), *Lesno* (light), in 7/8 danced slow or fast (see video examples 5.38 and 5.44 from New York and 6.4 from Šutka); *čačak* (video example 5.58); *Bitolska or Romska Gaida* (Bitola or Romani bagpipe); *Eleno Mome* (Oh, Elena, girl); *Jeni Jol* (New way, Turkish; video example 5.48); and several 9/8 dances (including *Afe Dude*, video example 5.12 from Šutka). *Beranče or Ibraim Odža* is another popular line dance, especially loved by Roma from Bitola (it is often in 12/8, 3+2+2+3+2; see video examples 5.43 from New York and 6.5 from Šutka). Elder Roma or talented young Roma are often called on to lead the older, slow, heavy line dances (*Pharo/Teško*, heavy; Romani/Macedonian) in 2/4 or 7/8 and crossing dances that many young people do not know (see video examples 5.35 [second part] from New York and 2.1 and 6.6 from Šutka). Several older women in the New York community are excellent dancers. An exceptional solo dancer can even receive bakšiš (tips) on her forehead (see video example 6.1 and photograph 6.2).

Čoček as Social Dance Among Non-Roma in the Balkans and the Diaspora

As solo dance, čoček can currently be found at community events not only among Roma but also among Bulgarians, Albanians, Macedonians, Serbs, Romanians, Greeks, and Turks—that is, among virtually all the ethnic groups of the Balkans. The three-measure line version of čoček known among Roma in Macedonia is found among Macedonians and Albanians from Macedonia, but not among Bulgarians. As a variant of *lesno/pravo/oro*, it probably disseminated from Skopje and was picked up by Macedonians and Albanians in the 1970s (Dunin 2008); by the 1980s it had spread to Albanians in the Prespa region of Macedonia, and by the 1990s to Albanians in the North American diaspora, according to Sugarman (2003 and personal communication).

As a solo dance, čoček encodes a number of meanings for non-Roma, who to varying degrees may be aware of its sexual associations and its ties to Roma. Sugarman thoughtfully explores how contemporary young Muslim Albanian women from Macedonia redefine aspects of their own sexuality and their own modernity when they dance čoček with other women. They still condemn professional female dancers, but "the genres once associated with them have been adopted by 'respectable' women and even men" (2003:112). Furthermore, they relate čoček to Turkish urban culture, thereby placing it in the realm of art and "civilization."

I observed Bulgarians dancing solo kyuchek at community events in the 1970s (up to the present) when wedding bands included them in their repertoire (despite prohibitions against them). The typical pattern among Bulgarians is for guests to dance kyuchek at the middle or end of a wedding—at a moment of abandon and release. This may be a time for enacting the perceived freedom and unbridled sexuality of "the other" in the form of the internal Gypsy or Turk. For some Eastern Orthodox Bulgarians, Muslim and Romani cultures are coded as unbounded by the constraints of civilization. As I discussed in Chapter 1, the Muslim cultural and political issue has a long history and is still sensitive in Bulgaria today (Neuburger 2004; Ghodsee 2008 and 2009). For example, in Chapter 9 I explore how the pop/folk genre chalga is criticized by many Bulgarians for being too uncivilized, which means too Muslim and too Gypsy.

Van de Port reports a similar phenomenon among Serbs who frequent cafes with Romani music in Novi Sad, Vojvodina (1998). Neither Serbs nor the Romani musicians who play for them are Muslims (both are Eastern Orthodox); nevertheless, Roma function as the internal uncivilized "other." In the cafes, Serbs dress like Gypsies, dance čoček, and drink with abandon, as if enacting what they perceive as the culture of Roma: "Within the Gypsy bar the door is opened to all those forbidden and hidden things which were deposited in the figure of the Gypsy. . . . As would-be Gypsies the visitors gain access to what is labeled as primitive and Balkan in the civilization debate" (188).

Similarly, in Romania since the late 1980s a form of solo čoček known as *mahala* or *manele* has been adopted into popular youth culture.[3] Drawing models from the southern Balkans, manele does have historical roots in urban Romani Romanian dance of the Ottoman period (Garfias 1984). Critics associate it with commercialism, sex, and Roma—all marks of the uncivilized. Despite public condemnation by intellectuals and folk music scholars and performers, it is widely danced and heard and has a growing fan base among Romanians as well as Roma (Beissinger 2001, 2005, and 2007). For example, the Romani brass band Fanfare Ciocarlia from the northern Romanian region of Moldavia currently tours with two female manele dancers performing čoček with bare midriffs. Upholding the tradition of performing with family members, these women are the Romani wives of the German managers of the band.[4]

In the last decade, čoček in its belly dance incarnation has become popular in American and West European clubs where young urban hipsters congregate to dance to Balkan music, either live or spun by DJs. I explore this phenomenon in Chapter 13, but I note here that the atmosphere in these clubs is close to wild abandon. In my preliminary interviews with clubbers, I found that they viscerally identify with the unbridled frenzy of sexualized belly dance, which they attribute to Gypsies, thus enacting "the exotic other."

Čoček as Professional Dance: Ensembles, Festivals, and Music Videos

Since World War II čoček has been incorporated into professional and amateur Balkan ensemble choreographies, some of which feature romantic and orientalist images of Roma (Dunin 2008). According to Shay and Sellers-Young, "Belly dance contexts . . . negotiate a transnational discourse of exoticism" (2006:25). Similarly, in her research on tango, Savigliano notes that "exoticism is a way of establishing order in an unknown world through fantasy" (1995:169). Dance choreographies are effective visual communication about what constitutes civilization, the nation, and the folk versus "others" (Shay 2002).

In Yugoslavia, ensembles incorporated Gypsy suites into their repertoire to illustrate Tito's ideal of "brotherhood and unity" of the nation's ethnicities. Note that Roma did not typically dance in these companies. The Kolo (Serbian State Ensemble) suite, titled *Vranje* (a city in southern Serbia), choreographed by Branko Marković in 1949, has become a classic for many amateur and professional ensembles. According to Dunin, it depicts Gypsies nonrealistically: tapping tambourines on their hips, elbows, and shoulders; dancing solo steps in unison; and doing intricate footwork (including spinning) over large distances (2008:118).[5]

Shay describes another Gypsy suite with dances from Vojvodina, performed by Kolo in 1987 and displaying Roma "as childlike, irresponsible, sexually lax individuals who dance, sing and fornicate the night away. . . . All the visual clues—the campfire, the gypsy wagon, the false mustaches are present" (2002:8). The men pull knives and carry the women off, and the women wear costumes revealing their legs and breasts, uncharacteristic of Serbia's Romani communities (8):

> The Gypsies are shown as childlike, indolent, oversexed . . . people. The choreographies featured stereotypical props [such] as a Gypsy wagon, camp fire and the clothes were covered in patches. . . . In the Serbian folk dances the women were portrayed as demure. . . . The Gypsies, on the other hand, have their hair free and disordered to signify "sexual looseness," and the blouses are off the shoulder and they show bare legs (none of this is what actual Gypsies would do). At the end of the dance, a man runs his hand up the woman's leg under her skirt until the lights fade out, indicating a night of unbridled passion ahead (171–172).

I viewed a suite with similar images in 1989 at the Ohrid Folk Festival in Macedonia, performed by a visiting Dutch group. The men were bare-chested and had whips, and the women had flare skirts and off-the-shoulder blouses. While I was offended at the stereotypification, a Romani journalist accompanying me clapped wildly. When I asked him how he could approve, he answered that it was wonderful that the Dutch performers sang in Romani; he was pleased at *any* public recognition of Roma. Later in this chapter, I discuss the implications of stereotyping for marginal minorities.

The Serbian dance company Frula (which broke off from Kolo and is known for acrobatic stylizations) had an entire show titled *Tzigane* (photograph 6.3). Its 1986 press release states:

> Lacking any national folk heritage of their own, the Gypsies have adopted the cultural traits of the localities in which they have settled . . . and have mysteriously made it their own. In addition to the many songs and dances, the program will feature performances of hitherto secret tribal rites celebrating marriage, birth, and death, as they have been practiced since time immemorial in Gypsy encampments all over the world. For centuries Gypsies have been the objects of curiosity, fascination and persecution among the world's people. Their carefree, nomadic life style has inspired envy in the hearts of some, suspicion and disgust in the hearts of others. Their caravans and campfires have sung of the open road. Their flashing eyes, unbridled zest for living, and their passion for singing and dancing have made them popular attractions wherever they have settled or roamed. Though often identified with the

supernatural and the occult, Gypsies generally will adapt to their environment and are happy carefree people.

Although I do not have the space to analyze all the implications of this text, note that all the major Gypsy fantasy themes are present: mystery, secrecy, the occult, rootlessness, freedom, music, wildness, passion, and sex. The message is: these are people are *not* like us. Note also the alternation of the dual polarities of fascination and repulsion.

Up to 1960 Tanec, the Macedonian State Ensemble, included one traditional Romani line dance in its repertoire, "Čuperlika/Kjuperlika," performed to a well-known 7/8 melody. The 1950 Tanec program lists Čuperlika as a Turkish women's dance, but it is also widely done among Macedonian Roma. The dance was collected by the Janković sisters in 1939 in Skopje (1939:75–77). Tanec's 1950s line dance choreography was changed to add a solo čoček, and for the 1956 United States tour "costumes were changed from Turkish style shalvare to translucent and narrower type pantaloons because Americans like to see more of the legs" (Višinski and Dunin 1995:127).

In 2004 the Budapest Ensemble (composed almost entirely of non-Romani Hungarian dancers) presented an international tour of *Gypsy Spirit*, sponsored by Columbia Artists. Most of the show featured tasteful Hungarian Romani dance; one scene set in the Balkans aimed to capture the grace of čoček but slipped into the stereotypical trap of exaggerated belly dancing and flimsy šalvari. In Chapter 12, I explore the problematic staging of this show, but here I note that a Hungarian Romani female audience member was appalled that in one scene a male dancer put his head onto a female dancer's lap. For her, the proximity of his head near her crotch was a violation of public sexual modesty, and she wrote the management a letter of complaint.

When ensembles are composed of Roma, they too must constantly negotiate how to present "Gypsy" dance. For example, government-sponsored amateur Romani ensembles in Yugoslavia were encouraged to present their folklore and that of neighboring ethnicities at festivals. A Romani KUD (Kulturno Umetničko Društvo, Cultural Artistic Group) in Serbia was founded in Priština, Kosovo, in 1969, followed by others in Serbia (Dunin 1977:14). In the mid-1970s, a festival for Romani KUDs from all over Serbia was organized; it was an important moment in the non-Romani public's recognition of Romani musical talent. Not surprising, groups followed the typical ensemble model of presenting complicated choreographies, most unknown in Romani communities. Although subtle čočeks were danced in some Romani KUDs, other "Gypsy suites" in Romani KUDs imitated the gross erotic movements done in state and amateur ensembles (15).[6] This brings up the question of self-stereotyping, which I discuss shortly and in Chapters 12, and 13.

Known as Phralipe (brotherhood), the Macedonian Romani KUD founded in 1949 in Skopje was very popular, won prizes at Yugoslav folk festivals, and even traveled outside the country to France, Poland, Bulgaria,

Italy, and other locations (Džimrevski 1983:216). According to Dunin (2008), the group performed mostly Macedonian line dances (plus non-Romani dances from all over Yugoslavia, as was common for all KUDs) but also included the stereotypical Vranje choreography mentioned earlier. After 1969, a wedding scene was introduced and solo čočeks were incorporated (Dunin 2008). Dunin counted three Romani KUDs in 1988 (KUD Phralipe in Skopje, and Tetovo and Folklorna Grupa Trabotiviste from a village near Delčevo; Dunin 2009). By 2008 she counted twelve, five (from Kumanovo, Bitola, Veles, Kočani, and Delčevo) of which performed for International Romani day in Kumanovo in April 2008. According to Dunin's research, the young female čoček dancers in KUD Ternipe (Youth) from Delčevo learned their orientalized movements from the Brazilian television soap opera *O Clone*, which features scenes of belly dancing in Morocco (Dunin 2009).

In the 1970s, Skopje Phralipe members told Dunin that "it was difficult to maintain a repertoire of Rom dances, because the girls did not continue in the group beyond marriage (usually between the ages fourteen and seventeen)" (1977:13). Similarly, in 1990 I learned from former Phralipe members that the group had problems recruiting girls and had to disband for a while in the 1980s. As sites of male-female socializing, ensembles might compromise the morals of unprotected females. Pettan writes of Kosovo Roma: "Engagement of Gypsies with music and dance within the school or amateur ensembles ends with marriage. This is more strict with the female part of the Gypsy population than with its male counterpart" (1996a:316). This very same problem of female reputations plagued Severdžan Azirov in New York when he tried to start a Romani performing group. As I described in Chapter 4, parents were reluctant to let their daughters attend dance rehearsals.

The performance of Romani dance by ensembles in Yugoslavia can be contrasted sharply with its virtual absence in Bulgaria during the socialist period. In the 1980s, the genre kyuchek (both dance and music) was prohibited in the official media because the state claimed it was not "purely Bulgarian" (see Chapter 7). Weddings were sometimes raided by the police if musicians played and guests danced kyuchetsi. Of course, Roma, Turks, and Bulgarians found ways to resist, and the dance thrived in private settings, eventually emerging in vital form after 1989.

During the 1980s and early 1990s in Yugoslavia, gala television programs (for example for New Year's Eve) regularly featured Romani musicians and dancers performing čoček. Some Romani performers such as Esma Redžepova (see Chapter 10) danced modestly; others enacted orientalized versions of čoček with writhing, scantily clad belly dancers. In Chapter 2, I discussed video example 2.4, which features Ferus Mustafov and his band with two solo belly dancers who have exposed skin but employ characteristic Romani stomach movements; in addition there is a group of non-Romani performers (the Ballet Troupe of Macedonian Television) doing modern dance choreographies that have little in common with Romani dance. Watching these programs with my

Romani friends, I heard them remark how the solo dancers had virtually nothing in common with Romani community dancers; plus they gave Romani women a bad name. On the other hand, it is clear these programs exposed non-Roma to Romani music and dance, albeit an orientalized version.

The television program *Maestro*, with Ferus Mustafov from 1987, for example, featured him as a doctor in uniform playing to his bed-ridden female patients who shed their hospital sheets and emerge as belly dancers. Singer Esma Redžepova commented:

> In recent times, . . . there has appeared . . . with Ferus Mustafov a Macedonian woman (she isn't Romani) who does belly dance—and they show this as if it were Romani. This isn't Romani, it is Turkish. That is Ferus' mistake. He makes a profit—money—from this. . . . A Romani woman would never be undressed to show her belly button. . . . Women used to be in a separate room, men separate, and they used to celebrate segregated at weddings. At our weddings our women used to be dressed in beautiful dimije, beautiful shoes . . . nothing at all bare— beautiful vests, underdresses, handkerchiefs at their hands, blouses with handmade lace. When they got up to dance, two-by-two . . . all of the elders . . . would cheer whomever danced better. Among us, we didn't do any mixed [sex] dances—we only danced čoček. You dance čoček with your stomach, you don't dance (with your hips) in a circle, you don't dance it with moans; we didn't have any of the new things with which people now deceive people.

Esma's modest sensibility prevailed at Šutkafest, the 1993 Romani sponsored festival in Skopje (see Chapter 8); the performers were fully clothed in šalvari and danced modest solo or line čočeks. Video example 6.7 from Šutkafest shows the line dance Bitolska Gaida and solo čoček.

The tension between modesty and overt exhibition of female sexuality is juxtaposed in Romani beauty contests in Macedonia; they have emerged since 1991 as a forum for music, dance, and costume display, and sometimes music and dance contests are imbedded in them. Given the scrutiny of female behavior found in Romani communities, it may seem surprising that Romani beauty contests have been held annually in Macedonia with great success since 1991. With the typical walkway passes, panel of judges, and bathing suit and gown competitions, beauty pageants seem to embrace the opposite of feminine modesty and instead promote Western objectification of femininity. Closer examination reveals a more nuanced reading of these events. Beauty contests are framed to address intertwining issues of the modern world: the political struggle of Roma, the significance of Romani culture, the role of Romani women, and the importance of Romani music. Either music is liberally interspersed as entertainment during the contest (as in the 1997 International Miss Roma Contest) or music contests are piggybacked onto the beauty contest (as in the Romska Ubavica Contest of 1998).

Promoters clearly describe their motives related to women. A news release for the 1996 international contest reads: "Since 1991 this event has been extremely successful in strengthening the Gypsy Community of Central Europe as well as aiding in the cause of emancipation of Romani women. This event is a showcase for Romani culture and the talents and beauty of the Gypsy people, their songs, dances, folklore, art and fashion. . . . The pageant is open to young women between the ages of 18 and 24, preferably women who can represent Romani communities and organizations in their part of the world." Similarly, the producer of the 1997 contest said his vision was "to demonstrate the emancipation of the Rom woman and to remove the stereotype of the Roma as a backward peoples" (Dunin 1997:1). We should not, of course, naïvely believe the words of promoters, for they are primarily businessmen; rather we need to interrogate how and why displays are produced. Beauty contests, as well as pornography, have invaded all of postsocialist Eastern Europe, signaling objectification and commodification of all women and demonstrated by awarding "feminine" prizes such as jewelry, clothing, and cosmetics. We must also remember that the marketing of Romani dance and the trafficking of Romani images have always capitalized on the sexuality of females. In postsocialism, beauty contests have become a mark of modernity and progress for some women. In spite of Western feminist critiques of objectification of women in these contests, they do bring women more squarely into the public realm. However, it remains to been seen at what price.

It is clear that commercial dance images are directly related to the high market value of the hypersexualized female Gypsy body, a phenomenon with a long history.[7] A cursory glance at the graphic design of cassettes, CDs, videos, and DVDs with Romani music produced in Macedonia and Bulgaria since 1990 reveals that many of them feature seminude belly dancers, and some are explicitly pornographic.[8] The rise of this music/dance imagery is related to the spread of pornography throughout the Balkans after the fall of socialism, which is in turn related to reconfiguration of female roles and to economic insecurity (Daskalova 2000; Gal and Kligman 2000a, 2000b). In Bulgaria the genre chalga (a fusion of pop and folk with predominantly kyuchek rhythms) capitalizes on association of kyuchek with erotic belly dance. In Chapter, 9 I analyze several examples of how the feminized oriental is produced via belly dance.

Although we may view most of these examples as the product of non-Romani marketing, Roma themselves are not immune to these stereotypes. Activists condemn these images, but entertainers often capitalize on them. For example, even Esma Redžepova, whose eloquent protest against belly dancing we have just read, made videos in the 1970s and 1980s with veiled belly dancers;[9] Esma claimed that the dancers in her videos were Macedonian, not Romani, and she did not have full artistic control over the staging. Her videos also featured campfires, tents, and other stereotypic symbols that have nothing to do with the actual history of Esma's urban-based music. She explained to me that she thought the scenarios were staged beautifully, even though she knew they were not representative of her culture (see Chapter 10). Similarly, Lemon discusses Kelderara Roma in Russia who embraced stereotypic dancing around a campfire for a documentary film about them (2000:156–157).[10]

The postsocialist mania for belly dancing in Bulgaria illustrates the interplay of gender stereotypes and politics. During postsocialism Bulgarian kyuchek became more "orientalized" and was influenced by trends in Turkey and Yugoslavia. In 1990, at one of the first public concerts to be labeled with the words *Tsiganska Muzika* (Gypsy Music), the all-Romani audience was ecstatic to hear and see the formerly prohibited kyuchek. The dancing, however, was not the subtle kyuchek that Roma do at their in-group events but rather belly dancing with bare midriffs and bodily contortions. Belly dancers (some non-Romani, some Romani) now regularly appear on commercial videos with Romani singers in romanticized stagings. Furthermore, since the early 1990s in Bulgaria, belly dancers have been appearing with bands at festivals, creating a virtual craze. At the 1995 Romani Music Festival in Stara Zagora (see Chapter 8), the winning band Džipsi Aver (Gypsy Friend) appeared with five *kyuchekinyas* (kyuchek dancers) with bare midriffs and oriental-style outfits (video example 2.5). There is now a prize for the best kyuchekinya.

Russian Romani dance and costumes (flared skirts and shawls for women, wide shirts and boots for men) are also becoming more popular in Bulgarian Romani dance groups (Peycheva and Dimov 2005:21). In 2000 a new award category was created at the Stara Zagora Romani Festival, for best dance ensemble (10). As a result, many new dance groups, such as Ansel from Vidin, Šukaripe from Sofia, Romska Veseliya from Septemvri, and Romaniya from Sredets, have been formed and have wide Romani repertoires, including Russian, Hungarian, and Indian dance; these groups make a statement about a pan-Romani consciousness (see Chapter 3). The groups showcase teenagers, both girls and boys, and bring a welcome degree of pride and gender integration to young Roma. The dance groups also serve as community centers and gathering places for young Roma.[11]

The group Džipsi Aver also performed in the 1990s with another type of dancer: a Michael Jackson imitator. Partially clothed kyuchek dancers and Michael Jackson imitators may seem to have nothing in common, but I believe they are both viewed as symbols of a modern Romani identity that goes beyond the local Balkan region. Recall that in Chapter 2 I mentioned the emergence of Romani rap at this time, and in Chapter 9 I describe the use of rap in pop/folk videos. Indeed, Roma (as well as non-Romani Bulgarians) are experimenting with African-American hip-hop styles; Roma may have a special affinity for American-American pop culture because they are similarly configured in musical and racial terms (Levy 2002).

In Bulgaria, the belly dance boom has complex connotations embedded in ethnic and political displays of the postsocialist period. First, it is predominantly a youth phenomenon. In Chapter 8 I describe how the annual Stara Zagora festival, for example, has grown tremendously since its inception in 1993 and regularly attracts several thousand Roma, most under thirty years of age, who actively dance while watching the performances. There is a party atmosphere, and it is one of the places where Roma feel safe in congregating. Second, the festival is tied to the cultural and political mobilization of Roma. Romani festivals often feature speeches by politicians and are sponsored by political organizations or nongovernmental organizations.[12] Thus professionalization of čoček into belly dance is embedded in economic and political projects, all propelled by the precarious position of Roma in Bulgaria society.

Returning to the issue of dance stereotypes, we can begin to tackle the thorny question of why professional Romani productions of čoček now resemble non-Romani productions. Indeed, with the orientalization of čoček it is sometimes hard to distinguish anymore what is produced by Roma. Roma engage in self-stereotypification (or mimesis of other's projections of them) in part because it is economically profitable. Romani dancers, like Romani musicians, have never been in control of their own imagery, and they are quite used to being made (and making themselves) into "exotic others" (see Okely 1996; Lemon 2000; and Chapter 12).

We must also remember that the commercial success of belly dance performances and videos is one of the only positive economic niches in an otherwise bleak economy. Yet here too Roma remain marginal—they do not profit nearly as much as non-Romani performers, managers, and producers. Throughout history Roma have had to rely on outside patrons and the trade in outsider imagery for work. Some observers, even Romani activists, have criticized Roma for "cashing in" on outsider stereotypes. This position ignores the tremendous power inequalities between Roma and the non-Romani world of promoters and media producers. In truth, Roma have historically had few choices about their work and their images, and even today they lack access to image-creating mechanisms (Hancock 1997). Few Roma produce their own music and dance, and most are subject to the marketing decisions of others.

Female belly dance performances sell precisely because they fit the image non-Roma have of Romani women: sexually alluring, promiscuous, dangerous, provocative, and musically talented. The historical information about Ottoman çengis can be interpreted from this angle: çengis were selling not only their musicality but also their perceived (and often actual) sexuality. This is in contrast to Romani community čoček performers, whose sexuality is muted. A community dancer is monitored for modesty but must also display the potentially sensual fluidity of body movement that defines a talented dancer. Traditional social arrangements, such as where and when she dances, shield the čoček dancer from criticism, but the ambivalence about the female dancer remains. For Roma, female Romani professionals are suspect but necessary. Because they embody commodification of sexuality, they can disrupt the social system from the inside. On the other hand, their performances in the marketplace underline the paradox of economic necessity versus ideal modesty.

PART III

MUSIC, STATES, AND MARKETS

7

᥯᥍ᥐᥗᥱᥕ

Dilemmas of Heritage and the Bulgarian
Socialist State

This chapter examines the relationship of Roma to the nation/state via
music, taking Bulgaria as a case study. Although diasporas are usually
defined in contrast to the nation and state, dispersed peoples often are ideo-
logically, culturally, and historically connected to states (Werbner 2002a,
2002b; Lemon 2000). Bulgaria is a particularly illuminating case because
the socialist state consciously targeted Romani music in its ethnonational-
ist cultural project of "Bulgarization." This chapter traces the historical
trajectory of definitions of heritage and authenticity through the socialist
period in Bulgaria to show that Roma pose the question of belonging.
Roma raise the issue of exclusion versus inclusion in the nation/state; they
interrogate the framework of heritage by exposing its monoethnic framework. Inspired by Herzfeld's concept of cultural intimacy (1997), I investigate the complex performative relationship between Roma and the socialist
state through analysis of the politics of zurna and tŭpan ensembles and
Bulgarian wedding music.[1] I also explore the issue of resistance to state
policy, noting that resistance is often paired with collaboration. I continue
this discussion in Chapter 8, where I deal with the postsocialist period.

The terms *heritage, tradition,* and *folk* had great weight in nineteenth-
century East European nation-building projects; indeed, heritage and tradition were used to culturally define the nation as a community composed
of homogeneous "folk," thereby excluding Roma as well as other minorities. See text supplement 7.1 for discussion of the relationship among
Bulgarian nation building, socialism, and folklore policy. Because folk
music became a politicized symbol of the Bulgarian nation, its definitional
borders were carefully patrolled, and Romani music was clearly outside
those borders. Romani music, then, has never been performed in ensembles, festivals, or music schools under the rubric *folk.*

Zurna and the 1980s Anti-Muslim Campaign in Bulgaria

As mentioned, Balkan Roma have had a historical monopoly of ensembles consisting of zurna and tǔpan. Today in Bulgaria this ensemble plays the traditional dance music of the southwest (Pirin) region. In contrast to socialist Bulgaria, where zurna and tǔpan ensembles were regulated and eventually prohibited, in Macedonia zurla and tapan players were regularly hired by state-sponsored radio and ensembles. Dissemination of recordings gave wide media, festival, and concert visibility to zurla and tapan music in Macedonia. In Bulgaria, on the other hand, zurna was excluded from most official settings, including folk music schools. However, even in Bulgaria there were long periods during socialism when zurna players performed with ensembles; from 1964 to 1969, for example, Mancho Kamburov from Razlog was employed by the Pirin Ensemble (Peycheva and Dimov 2002:179; also see Buchanan 2006:267). Sometimes regional and village ensembles had their gaida (bagpipe) players (typically Bulgarians, not Roma) learn enough zurna to perform it (Peycheva and Dimov 2002:184). Romani tǔpan players were likewise sometimes employed by Bulgarian ensembles; for example, Angel Krǔstev was employed by the Yambol Ensemble from 1973 until his death in 2010. Although the tǔpan is regularly played with traditional village instruments, it is not formally taught in schools.

In 1984 the zurna was officially banned from all contexts, including festivals, media, urban and village celebrations, and private parties. Even earlier, however, it was prohibited in certain localities (Peycheva and Dimov 2002:213–214). In 1980 I attended a Pomak wedding in the village of Avramovo (Velingrad district, southwest region), where Romani zurna and tǔpan players were hired despite the local ban (photographs 7.1 and 7.2). Family members served as guards, watching from the roof of a house to warn if officials were approaching. This underscores how both Roma and their patrons resisted prohibition. When the zurna was prohibited from the 1985 *Pirin Pee* (Pirin Sings) folk festival, government officials substituted *svirki* (flutes) to accompany village dance groups. Svirki are much softer in volume and lighter in tone quality than zurni. Audiences failed to show up at the stages where dances were performed to svirki, and when they did they found the dancing boring and uninspired, lacking the vitality and loudness of zurna and tǔpan.

Despite the ban, Romani zurna and tǔpan players arrived at Pirin Pee and played for dancing in a meadow above the festival. They attracted a large crowd (photograph 7.3), and dancers tipped them generously; people of all ethnicities danced vigorously until the musicians were chased away by the police. Several zurna players, among them Mancho Kamburov of Razlog, managed to perform surreptitiously and even teach his son (photograph 7.4) despite prohibitions. In the mid-1980s, Kamburov had to accept a state job as a gardener for a hospital, which actually served as a cover for his music. These examples show how Roma and their patrons subverted the socialist system of musical management. This resistance allowed zurna and tŭpan bands to survive until 1989, when prohibitions were lifted and they emerged as a vital tradition (see video example 2.2 and Chapter 2).

The official reason for the ban was that zurna was a foreign (specifically Turkish) instrument and thus had no place in Bulgarian folk music. In actuality, zurna-type instruments are found from India to Spain, and until World War II they provided much of the outdoor dance music of the southern Balkans. The Bulgarian state was itself contradictory about the official performance of zurna; although it was banned at the Pirin Pee festival, the National Ensemble of Folk Music and Dance permitted its gaida players to play the zurna for its Pirin suite. Buchanan points out that the instrument was legitimated by being incorporated into the ensemble's stylized spectacle of the nation (2006:267). This illustrates Aretxaga's point that we need to "rethink the notion of the state in a new light as a contradictory ensemble of practices and processes" (2003:395).

The rhetoric about purity is directly related to the 1980s state policy of monoethnism and Bulgarization and its concomitant regulation of display of Muslim ethnicity (Poulton 1991; Neuburger 2004; Rechel 2008). In fact, the official *Vŭzroditelen Protses* (regeneration process) dictated that there were no minorities; everyone was Bulgarian (thus Roma didn't exist). Roma were referred to in official contexts as *grazhdani s novo-bŭlgarski proizhod* (citizens with new or modern Bulgarian ancestry). This policy included name changes and prohibition of religious and cultural observances among the country's Muslim minorities—Turks, Pomaks, and Roma. The policy was enacted among the three groups of Muslims at different times and with different consequences (Rechel 2008:138–141; Neuburger 2004).

As early as the 1960s, Roma were targeted with name changes; that is, their Muslim names were forcibly changed to Slavic ones (Pomak name changes were initiated even earlier). Several Roma satirized this process by chosing the names of famous Bulgarians; for example, one man chose the name Filip Kutev (after the head of the National Ensemble of Folk Music and Dance). Another well-known television director in Sofia chose a name that sounded Slavic but was actually part Romani: Manush Romanov (*manush* means man in Romani; Romanov is a hybrid of a Slavic ending (ov) and the word Roma). Many Muslims never fully abandoned their Muslim names; they used them at home among their family and used their Slavic names in official contexts.[2] When I asked Muslims in private "What is your name?" they would answer, "Which one?"

It was rare for Roma to overtly resist name changes;[3] they superficially went along with the process but resisted in more covert ways, as with using prohibited names and playing their music in private, which illustrates Herzfeld's concept of "cultural intimacy" with the state (1997). Many Turks, on the other hand, resisted overtly, some arming themselves, going to jail, and losing their lives in the process. Because the Turkish minority is very large and because the country of Turkey carefully monitors the

Turks of Bulgaria, the name-changing campaign drew international outrage. Human rights organizations reported numerous violations of civil rights. Perhaps Roma did not resist like the Turks precisely because they knew no one would come to their defense; no outside country represents the interests of Roma. Turks, on the other hand, had more security in their high numbers and in having Turkey next door. In spite of international outrage, the Bulgarian government forged ahead with the name-changing campaign. The climax of this process occurred in the summer of 1989, when 370,000 Muslims (mostly ethnic Turks) departed Bulgaria for Turkey (Poulton 1991; Eminov 1997). Although it was cast by the government as a voluntary move (dubbed *Golyamata Ekskursiya*, "The Great Excursion"), it was de facto an expulsion. Some observers have suggested that the fall of Bulgarian socialism was due in large part to this misguided policy (Buchanan 1996:221).

In addition to name changing, Muslim clothing (such as shalvari), customs (such as circumcision), the speaking of Turkish and Romani languages, production and distribution of Muslim literature, and listening to and performing Muslim music (see later discussion in this chapter) were all prohibited in the 1980s.[4] School administrators strip-searched male students, bus drivers refused to pick up women in Muslim garb, and police officers imposed fines for speaking Romani and Turkish in public. Many of these prohibited practices did not disappear but rather were driven underground or into the private sphere, which became a refuge from state regulation.[5] At home women would wear shalvari or distinctive Pomak or Romani aprons, but when they went out they would remove their aprons and substitute pants worn under a skirt for shalvari.

In the state's effort to rid Bulgarian culture of all "foreign" elements, music played a decisive role. From 1984 to 1989 kyuchek, the main musical genre among Roma and Turks, as well as songs in Romani and Turkish were eliminated from all Balkanton (the state-sponsored record label) recordings and from all official performances. Before 1984 there were a few albums with kyucheks, but the euphemism *tanc* (dance) was used. The last record of Romani music that was released in the socialist period was a 1983 collection of Anzhelo Malikov's songs and arrangements.[6] Anzhelo, son of Yashar Malikov (1922–1994), was a composer, arranger, and collector of Romani music (Peycheva 1999a:56 and 141). When I interviewed him in 1986, he was quite pessimistic about Romani music; at that time he was employed playing the cimbalom in a Hungarian restaurant and was not allowed to perform Bulgarian Romani music. But he played Hungarian urban music, some of which is influenced by Romani music. In the early 1980s, he performed a program of Bulgarian, Spanish, and Russian Gypsy music for the restaurant *Ogneni Ritmi* (Fiery Rhythms) in Sofia. In 1984, everything was censored from his program except the Russian Gypsy part, which was in the Russian language. He couldn't compose or perform anything in Romani. To understand these prohibitions, we must examine the rise of "wedding music."

Bulgarian Wedding Music, 1970s–1989: Instrumentation, Style, and Repertoire

In the 1970s and 1980s the genre wedding music (*svatbarska muzika*) cat-apulted to fame, causing "mass hysteria," according to one journalist (Gadjev 1987:10). Roma were prime innovators in the wedding music scene, and this fact fueled the controversy around the genre. Labeled "kitsch" and "corrupt" by purists and excluded from folk music festivals, wedding music was the most popular music of the 1970s and 1980s, with the most fans. During the 1980s, the socialist government prohibited wed-ding music from recordings, radio, television, and private settings; note that Serbian music, as well as western jazz and rock, were also prohibited. The absence of wedding music from state media ironically promoted its success in unofficial media. Fundamentally a grassroots pan-ethnic youth movement, wedding music struggled against state censorship and became a mass underground cultural phenomenon.

The rubric *wedding music* is somewhat of a misnomer because it en-compasses music played not only at weddings but also at baptisms, house warmings, and soldier send-off celebrations[7]—in short, at major ritual events in village and urban contexts, for both Bulgarians and Roma. Although its history reaches back to urban ensembles of the nineteenth century that were composed mostly of Roma,[8] wedding music as a distinct genre began to crystallize in the late 1960s and early 1970s, when amplifi-cation was introduced to folk music in village settings. The loudness of electric amplification and its affinity to rock music became a symbol of modernity and the West.

Hiring a band with a sound system enhanced a family's status in the village; the bigger the speakers and the louder the sound, the higher the status. Japanese sound systems were preferred. Every band had an *ured-badzhiya* (sound man), who provided, transported, and monitored the system and received a fee similar to that of the musicians. The loudness affected the texture of the music. As effects such as reverb and delay were introduced, an intentional, slightly overloaded distortion became desir-able.

What defines wedding music is a combination of instrumentation, rep-ertoire, and style. Instrumentation typically consists of clarinet, saxo-phone, accordion, electric guitar, electric bass guitar, and drum set, plus a vocalist.[9] In the late 1980s synthesizers were added, sometimes replac-ing guitar, bass, and drums. These instruments have a greater range and versatility than Bulgarian village instruments. Occasionally violin or trumpet or village folk instruments appropriate to the region, such as gŭdulka, gaida, or kava, are added. Note that the core instruments were outside the socialist rubric of folk. True, they were imports from Western Europe, but clarinet and accordion have been used in Bulgarian folk music by both villagers and urbanites since the early part of the twentieth century.

Even today, these instruments are not taught in folk music schools and are taught instead in schools for classical music. Ironically, if a student wishes to learn folk music on clarinet, he or she must attend a school for classical music and learn folk music on the side; or else a student can attend a folk music school with a different instrument, and then switch to clarinet. Such situations happened countless times. Kalin Kirilov, for example, a talented musician born in 1975 near Vidin, played folk music at home on accordion and studied classical music on accordion in Pleven; when he was six years old he was told by a teacher that if he wanted to compete for admission to the Kotel high school he would have to play "a folk instrument" such as tambura, and he should not mention that he played the accordion. Later he was accepted at the Plovdiv Academy of Folk Music on tambura but played accordion covertly.[10]

The repertoire of wedding music can be divided into two main categories: Bulgarian music and Romani music (kyucheks). Bulgarian music is divided into slow songs (*bavni pesni*) and dance music. In wedding music, the most common dance meters are *pravo horo* and *rŭchenitsa*, with an occasional *paidushko*, *krivo horo*, or other dance meter.[11] Tunes are either local or drawn from the standardized Thracian wedding repertoire created by famous wedding musicians. This repertoire has a Thracian emphasis because the most famous bands are from Thrace.

Instrumental wedding music is highly structured in some ways and highly unstructured in others; there are set passages played in unison or thirds that alternate with individual improvisations on the melody instruments. The set passages are composed by musicians, sometimes based on folk melodies; but they often have melodic and rhythmic surprises. Eclecticism is the preferred mode of creation. In the middle of a horo one may find the "Flight of the Bumble Bee," the "Can Can" (from Offenbach), a quote from an advertising jingle, a popular rock-and-roll song, or phrases more reminiscent of jazz and rock than folk music. The emphasis is on originality and cleverness. Versatility is also prized. Clarinetist Ivo Papazov composed "A Musical Stroll Around Bulgaria" to display his regional diversity. He also does an imitation of a gaida on his clarinet, removes pieces from his clarinet (down to the mouthpiece), and plays clarinet and saxophone at the same time while the tune morphs to swing (photograph 7.5, video example 7.1). The theatrical element is definitely present. Moreover, audience members, who are often musicians themselves, listen carefully for what is new and interesting; they are highly critical listeners and relentlessly compare musicians and performances.

Above all, ability to improvise is valued by both performers and audience. Each melody instrument in turn departs from the unison phrases and shows its virtuosity in original ways. Dazzling technique is displayed by complicated rhythmic syncopation, daring key changes, arpeggio passages, chromaticism, and extremely fast tempi. One journalist wrote, "Rhythms are frantic and unbridled showing a rare virtuosity as if playing were a question of life or death" (Gadjev 1987). Timothy Rice quotes the phrase *s hus* (with gusto) to illustrate how proponents differentiated wedding music from folk music, which they found *prosto* (simple) (1996:193). Indeed, musicians contrasted the *svobodno svirane* (free playing) of wedding music with the *shkoluvano svirane* (schooled playing) of folk music.[12]

From the 1970s to 1989, wedding music was inextricably tied to large, opulent life-cycle events that were the pride of Bulgarians of all ethnicities. Weddings were a status symbol; villagers saved for years to invite hundreds of guests for a three-day event. Despite totalitarianism, this period was the apex of community celebration and display. Ignoring government warnings about "bourgeois conspicuous consumerism," villagers insisted on abundant food and drink, expensive gifts, and good-quality music. Wedding music was central to the rituals (such as *daruvane*, public reciprocal gift giving), the banquets, and the dancing that occurred for many hours. Unmetered slow songs and slow instrumental tunes accompanied the rituals and the meals served at long banquet tables, and metric songs and instrumentals encouraged guests to dance (Silverman 1992). Thus many Romani wedding musicians had steady professional work in that era.

Wedding songs are either from the local folk corpus or composed by the singers and instrumentalists. Songs performed by the most famous singers in the 1980s are still sung. The vocal style emphasizes rhythmic vibrato and extensive ornamentation, imitating the melodic instruments and showcasing technique. The style is based on eastern Thracian models and was developed by Nedyalka Keranova, born in a village near Haskovo, Thrace; many musicians assert she was of Romani descent, from an Eastern Orthodox group locally known as *sivi gŭlŭbi*, grey doves (see Pamporovo 2009). Keranova was the leading wedding singer until her death in 1996, and her vocal style is widely imitated (Bakalov 1992:229–238). Video example 7.2 is an excerpt of a signature Bulgarian slow song from the Trakiya Folk festival in 1994.[13]

The second category of repertoire consists of kyucheks, comprising both instrumental music and songs. The tunes for kyucheks are sometimes drawn from older Romani tunes but are more often composed by wedding musicians. They are inspired by eclectic sources: folk and popular music from Serbia, Macedonia, Greece, and Turkey; film scores from the West; cartoon music; Middle Eastern music; and Indian film music. Kyuchek titles in the 1980s included "Sarajevo '84" and "Olimpiada," in honor of the Olympics; "Alo Taxi" (Hello Taxi), from a pop song; and "Pinko," in 9/8, based on the musical theme from the Pink Panther. There are also covers of Macedonian, Greek, Turkish, and Serbian Romani kyucheks. Ferus Mustafov, a noted Macedonian Romani musician, performs several pieces based on melodies composed by Ivo Papazov. As I have emphasized, among Romani musicians there is cross-fertilization of musical styles, with a premium on innovation. Papazov confirmed that he and Mustafov would trade tunes over the telephone in the 1980s because travel to Yugoslavia was impossible.

The Ivo Papazov Phenomenon

The unquestioned guru of wedding music was and still is Ivo Papazov. Born Ibryam Hapazov[14] in Kŭrdzhali in 1952 of Turkish Romani ancestry, he is a founder (with his cousin, the accordionist Neshko Neshev) of the band Trakiya. In the 1970s and 1980s he was the highest-paid wedding musician in the country and was in such demand that people waited months and years to engage him. People even married on midweek nights rather than the usual Sunday to accommodate his busy schedule. He narrated: "Some people came to see me about moving their wedding earlier, but I already had engagements. I offered them other musicians—but they wouldn't hear of a replacement. Only later I found out that the bride was quite pregnant and she had aborted the child so that we, only we, could play for her" (Sŭrnev 1988:23).

The family that hired Papazov achieved high social status not only for their monetary expenditure but also for being the guaranteed focus of attention. Whenever Trakiya played at a village event, uninvited people showed up from miles around to dance in the public parts of the event, or merely to crowd outside the tent or banquet hall to listen to the music and catch a glimpse of the stars. In 1980 I attended a wedding in Iskra, a village near Haskovo, where about 200 uninvited fans showed up, some from several hundred miles away, to hear Ivo play (photograph 7.6). An added attraction was a simultaneous wedding in the same village, where Nikola Iliev, a renowned clarinetist and leader of the Konushenska Grupa, was hired. People viewed the event as a contest between Ivo and Nikola in terms of stamina, technique, and number of fans. The two wedding bands set up their sound systems at opposite ends of the village square, and hundreds of people joined the dance line. The music went on continuously for five hours and resumed after dinner for another four hours at indoor locations.

Admired for both his technical and his creative talents, Ivo is known for masterful improvisations, creativity, stamina, daringly fast tempi, forays into jazz, numerous compositions, and charisma: "A virtuoso in the instinctive meaning of the word, improviser of the highest class, he quite freely led the horo into jazz, built on a Bulgarian musical foundation— something, which elsewhere we didn't find done with such mastery and strength. . . . He has set the tone for a large musical movement with hundreds of thousands of followers" (N. Kaufman 1987:79). When a journalist asked Yuri Yunakov, the saxophone player in Trakiya, why no one in the orchestra looks at the audience, he replied, "There's no time. Have you ever seen how a hunted wild rabbit runs? It runs zig-zag, stops, returns, does 8s, 16s. . . . That's how Ivo plays. And we chase him like hounds with our tongues hanging out" (Sŭrnev 1988:25).

In the 1980s, Papasov's popularity was enormous: "The concert hall literally exploded when Ivo Papazov, the uncontested king, got on stage. It was the apotheosis. I compared it in spirit to Alan Stivell, Joan Baez, . . . the modern bards I respect deeply. I thought about Art Pepper's[15] commentary after listening to an Ivo Papasov cassette: 'A man can't play like that.' I also thought of the beginning of the century in the slums of New Orleans when jazz was beginning" (Gadjev 1987). Numerous fans have testified to his superstar status:

> I have 100 cassettes of Ivo Papazov. When he plays I feel weak in the knees. His compositions are unending. For at least 40–50 years there will be no one who can surpass him. . . . He is a magician! A master! A phenomenon in folk music that we won't see repeated soon. . . . When Ivo Papazov plays I stop breathing. I can't explain why. Can you explain love? (Sŭrnev 1988:23).

Legends circulated about Ivo and the early emergence of his talent. His mother, for example, supposedly tied his umbilical cord with a thread from his father's clarinet.[16] In truth Ivo comes from several generations of zurna players. Referring to his colleagues from Kŭrdzhali, that is, Salif Ali (drummer) and Neshko Neshev (accordion), he remarked that "all our grandfathers were zurna players." Ivo stated that his elder male relatives were some of the first musicians to switch to clarinet; before World War II one of them "traded a cow for a clarinet on a trip to Greece. That's how the clarinet was introduced to my family." Clarinets were valued over zurnas because of their newness, versatility, greater range, ease of playing in different keys, and chromatic possibilities. For a period of time, zurnas and clarinets were combined in bands in the Kŭrdzhali region; Ivo showed me a photograph of his father performing at a wedding in the 1950s that had a clarinet player, a zurna player, and a tŭpan player.

Ivo played music from a young age; at nine, he switched from accordion to clarinet and was said to "play like a man." In truth, he was exposed to many fine musicians from the older generation who played the Turkish Romani style of Kŭrdzhali, such as Halil Dzhamgyoz (Peycheva 1999:136–137); he also listened widely, especially to jazz, which was prohibited: "In those years we learned the old style from the older performers. But even then we listened to jazz on illegal cassettes of Charlie Parker and Benny Goodman." One legend relates that when he was in his teens he went to a local restaurant to eat and was invited to play outdoors when the resident orchestra took a break. Even though it was raining, the diners came outside, wrapping themselves in tablecloths, and for a half hour they didn't budge. Years after that they were still asking "Isn't that boy coming to play again?" (Sŭrnev 1988:23). A second legend tells of Milcho Leviev, a noted Bulgarian-American jazz musician and composer, who was given a tape of Ivo to listen to on his return flight to the United States. He forced the pilot to turn around in midflight because he insisted on meeting the musician. When I spoke with Leviev, he confirmed that he was very impressed by Ivo's playing, but the rest of the story is conjecture.

Another legend relates that Ivo owns a solid gold clarinet. Perhaps this claim was inspired by the popular exaggerations of his wealth, or by infusing his instrument with magical qualities. Indeed, fans lifted their children to touch Ivo "for good luck." One story relates how a wedding scheduled in Istanbul was moved to Bulgaria when Ivo couldn't travel to Istanbul. Another tells of five boys who showed up late for their induction into the army but were willing to take the consequences because Ivo had played for their soldier send-off celebration. When asked by the commanding officer why they were negligent, they answered, "You haven't heard how Ivo Papazov plays!" (23).

Ivo claimed, "I can eat the same dish twenty times, but I can't play the same thing the same way twice" (25). In 2005 he embellished, "Wedding music existed for many years, but I modernized it with a new style, modern chords, modern accompaniment, a contemporary musicality with more improvisation. And the young generation liked it; from them we received our popularity." When a journalist remarked to Ivo that it was hard to figure out what style he plays, he answered: "I play in Papazov style. I play in Balkan style, I play in the style of ethnojazz. I play our jazz. . . . I really get angry when people say they can't categorize me. . . . One time Lyudmil Georgiev said 'I can't tell you if Ivo Papazov plays jazz, but he plays incredible music.' And Georgiev . . . was one of the greatest jazz players."[17] Ivo narrated: "Neshko and I changed the style. I just can't stay in one place. I have to develop. The old ways didn't please us."

Video examples 7.1, 7.3, and 7.4 feature Trakiya's dazzling improvisations playing Bulgarian music on television in 1987 (later I discuss the filming of this show). In video example 7.3 Ivo and Yuri improvise in a pravo horo, and in 7.4 Neshko and Ivo improvise in a rŭchenitsa. Video examples 7.5, 7.6, and 7.7 were filmed at a Romani wedding banquet in a tent 1994 where Ivo and Neshko performed with Radi Kazakov (guitar), Vasil Denev (keyboard), Salif Ali (drums), and guests Matyo Dobrev (kaval player from Straldzha, Yambol region) and Ahmed Yunakov (Yuri's son, on saxophone). Romani kyucheks in 2/4 and 9/8 predominated. In video example 7.8 Trakiya is featured in on their 2005 American reunion tour (with guitarist Kalin Kirilov) playing pravo horo and improvising. In Chapter 11, I discuss this tour.

Besides Ivo, there are also many other fine musicians. I discuss Yuri Yunakov in Chapter 11 and have mentioned Kŭrdzhali accordionist Neshko Neshev and clarinetist Nikola Iliev, from the Plovdiv region. The roster of wedding musicians is too long to list here,[18] but some of the most famous veterans of the 1980s who still perform are Romani clarinetists Mladen Malakov and Orlin Pamukov (from Kotel, who performed with Orfei for many years; video example 6.1), Filip Simeonov (see Chapter 13, video example 7.9, rŭchenitsa, 1994),[19] Boril Iliev (from Lyaskovets, North Bulgaria), Dimitŭr Paskov (from Sofia), Nesho Neshev (from Kŭrdzhali), and Yashko Argirov (from Brestovitsa, Pazardzhik); Turkish clarinetist Osman Žekov (from Kŭrdzhali); Bulgarian violinist Georgi Yanev (from Asenovgrad, Plovdiv region, founder of the band Orfei); Romani accordionist Traicho Sinapov (from Sofia); and Bulgarian accordionists Petŭr Ralchev (from north Bulgaria) and Ivan Milev (from Haskovo, founder of the band Mladost; youth).

This list is not exhaustive, but it illustrates the point that although Roma have had decisive roles in creating wedding music, Bulgarians also masterfully perform it and bands are often mixed. Second, the majority of well-known musicians are from Thrace. And third, virtually all instrumentalists are male. When women perform, they are singers, usually spouses of musicians. Female singers who perform in bands with no male relatives are considered by some to be "loose" because of the late-night work and uncertain lodging arrangements.[20] Note too that the musical background of wedding musicians varies considerably, from a few conservatory graduates who can read music (e.g., ethnic Bulgarians Nikola Iliev Petŭr Ralchev, and Ivan Milev) to the majority who play by ear. Romani musicians tend to play by ear and acquire skills informally within a family context, the way most ethnic Bulgarians learned before the 1960s. Ethnic Bulgarians are more tied to the ensembles and the folk music schools, which emphasize musical literacy. The wedding music tradition, however, is strictly oral.

Economics: The Free Market and State Control

Understanding the economic framework of wedding music helps us in understanding attempts in the 1980s at state intervention. Because of the phenomenal popularity of some bands, the market for them became grossly inflated. When a family hired a famous band, the family not only gained in social status but also displayed financial prosperity to neighbors and kin. At the high end of the scale, Trakiya charged approximately 2,000 leva or $1,000 in 1984, not counting tips, for a two-day wedding.[21] This computed to about 300 leva apiece. If we take 200 leva as a good monthly salary for a factory worker in the mid–1980s, it becomes obvious that the stars earned in two days what most Bulgarians earned in six weeks. Remember, however, that the majority of wedding musicians were not stars[22] and that a more typical salary for a two-day event was forty to fifty leva. Though nowhere equivalent to a star's fee, this sum was still roughly equivalent to a week's salary in a factory. It is not surprising that in some Romani neighborhoods almost every male played an instrument. In fact, among Roma and non-Roma alike wedding music became a viable economic niche in the 1970s and 1980s.

Hiring music was always located in the realm of the free market, even during the socialist period. Someone from the family would approach the band leader, and they would bargain. Musicians waited to be contacted at specific places and times, such as *pazari za muzikanti* (musicians' markets; Bulgarian; Peycheva 1999a:236–237). In Plovdiv, for example, at the *chetvŭrtŭk pazar* (Thursday market; Bulgarian) wedding musicians gathered over lunch and were contacted by clients. In Sofia, musicians met Monday through Friday at noon at a restaurant in the center. As Tome Chinciri, a noted Romani singer and son of the venerated violinist Hasan

Chinchiri from Sofia, narrated: "People pick up musicians for their bands—I need a singer, you need a guitar player. You have to be careful to watch out for musicians who take more than their share. It's better to work with people you know." On Fridays the pazar in Sofia was crowded with thirty to forty musicians (see photograph 7.7), the majority Roma. Besides securing work, the pazar also functioned as a place to socialize. Trends in musical style, fees, and sources for buying instruments were all discussed, and albums from Turkey, Yugoslavia, and Greece were traded.

When a client approaches a wedding musician (in person or by telephone) the tone of conversation becomes more formal and a bargaining mode ensues. Usually half the money is paid ahead of time (known as *kaparo*) and the other half is paid at the end of the event. Nothing is written down formally; rather, a handshake or verbal agreement seals the deal. In addition to the fee, the beginning and ending times for playing are fixed. If they are asked to play beyond the fixed time, musicians require additional fees.[23] Another reliable source of income is from tips, that is, from requests for particular songs. This money is called *parsa* (collection) or *bakshish* (tips). Patrons often line up at weddings to tell the master of ceremonies which songs and dances they and their families request. They pay by sticking bills onto a musician's forehead or in his instrument, by handing them to the singer, or by throwing them onto the stage, sometimes ostentatiously. Tips are also given when a dancer wants a particular piece of music to continue, or when someone is particularly moved by the music.[24] Tipping is illustrated in the videos of Romani weddings in Chapter 5. Tips can generate up to 100 percent more than the contracted fee, and they too are divided (Peycheva 1999a:238); in fact, their division can generate conflict. Wedding musicians tell many stories about tipping. One famous story tells of a guitar player who used gum on the bottom of his shoe to gather bills for himself!

In the 1970s and 1980s, most musicians had state-sponsored jobs in addition to wedding work. Many wedding musicians and singers worked in professional folk music ensembles; to put it conversely, most ensemble musicians played weddings on weekends. Some musicians preferred restaurant jobs to ensemble jobs because restaurant work usually took place Monday through Friday evenings, leaving the weekends free for weddings. In Sofia in the mid-1980s, Tome Chinchiri and violinist Ventsislav Takev did restaurant work; Anzhelo Malikov played cimbalom at a Hungarian restaurant on weekdays and played guitar at weddings on weekends (photograph 7.8)

In the 1980s, the salary for restaurant and ensemble work was approximately 150–200 leva ($75–100) a month, relatively low compared to wedding work. Having regular state work, however, entitled a musician to a pension, medical benefits, and vacation packages. These amenities were denied to full-time wedding musicians, who were also denied the right to join the musician's union (Buchanan 1991:538 and 1996:207; Rice 1994:247–250). Moreover, in ideological terms doing wage labor made you into a "worker," thereby affirming your place as a productive member of society. Still, many wedding musicians, such as Nikola Yankov (founder of the Lenovska Grupa) and Ivo, resisted wage labor and played only at weddings and concerts. They were permitted to do so, but they were very heavily taxed (Rice 1994:247). Bulgarian clarinetist Nikola Iliev, founder of the Konushenska Grupa, explained: "It became really bad for musicians. The government started collecting high taxes from us. Because I was from a 'fascist'[25] family, they targeted me first; I had to pay back taxes and fines for five years. The first time I paid over 2,000 leva, an enormous sum, equivalent to fourteen weddings." Similarly, in 1985 Nikola Yankov was fined 2,000 leva in back taxes.

The state, concerned about "conspicuous consumption," began more vigorously to regulate the earnings of musicians. In 1985, in a few targeted regions such as Stara Zagora and Sliven, a state commission assigned each band a category (*kategoria*) that dictated how much it could charge. The system was administered by the concert division of the *Dŭrzhavno Obedinenie Muzika* (State Music Society), which was responsible for all categories of professional music and dance, including classical, popular, and folk. Before 1984, Balkanturist, the state tourist bureau, ran a commission to assign categories for its own establishments, but after 1985 the system was applied all over the country. *The Dŭrzhavna Atestatsionna Komisiya* (State Certifying Commission), comprising government-decorated musicians and professors, traveled to every regional capitol twice a year to test musicians through a short performance. They assigned a category according to the level of expertise and mastery of "pure" Bulgarian music.

Each band also had to submit a repertory list, which was approved or amended by the commission, to ensure that only pure Bulgarian music was played (Rice 1994:249–250; Buchanan 1991:538–539, 1996). Finally the category system also regulated where a band worked. Singer Dinka Ruseva, for example, told me she was singing at a wedding in the Plovdiv district when the police arrived and stopped the music; they said she could sing only in the Stara Zagora district! Wedding musicians were extremely upset over the imposition of the category system, as my journal entry for September 24, 1985, shows:

The musicians are all talking about kategorii. Two days ago the commission (headed by Manol Todorov, a professor of music) came to Sliven to assign kategorii. In Sliven alone eighty groups auditioned, attesting to the vitality of the wedding scene here. Each gave a list of their repertoire to the commission and then played for fifteen minutes. From now on, every group has to receive a kategoria and a musician has to play regularly with the same band in order to be hired for events. The government wants to get rid of free-market bargaining in part because musicians are making too much money. One Romani musician commented: "They want to have a bureau where you would go to get musicians for your wedding. The pay would be 52 leva a musician for two days of work, very little. I'm very willing to pay taxes, but we need a free market for weddings. They've started fining people 200–300 leva in Stara Zagora for violations, so we came here to Sliven."

The category system was enforced only selectively. Almost immediately after it was implemented, musicians began to circumvent the system by charging the official fee over the table but requiring more money under the table. During the 1980s wedding music thus stubbornly clung to the free-market domain.

The Official Rhetoric of Purity

Despite its popularity, wedding music was excluded from official government-sponsored media channels such as recordings, radio, and television. It was also either neglected by scholars or else condescendingly labeled as "clichéd" or "kitsch." Manol Todorov, professor at the Music Conservatory in Sofia, wrote: "The harmonic language is modest and when it is complicated it is unconvincing. . . . Often they master clichés that are imitative and chaotic. . . . The repertoire [of the singers] is not carefully chosen, they do not perform the best folk songs. Very often pieces of doubtful Bulgarian ancestry are performed, songs made up 'especially' for weddings. These pieces, devoid of artistic value, are quickly disseminated" (1985:31). Another scholar referred to wedding music as stateless, impetuous, and out of control, like "cosmopolitan water" where "Bulgarian music is only a glaze-like covering." He further laments that no one has told wedding musicians which influences are good and which are bad (K. Georgiev 1986:90). Music professor Nikolai Kaufman wrote: "Recently it has been pointed out that these wedding bands are the illegitimate children of the music profession. The basis of this attitude was that the bands were not successful in performing Bulgarian and foreign music and lacked professional ability in harmony, construction of form, and maintaining pure Bulgarian style" (1987:78–79).

The most common criticism leveled against wedding music was that it incorporated foreign elements and did not retain the "purity" of Bulgarian folk music. It was, ironically, simultaneously too Western (like jazz and rock) and too Eastern (like Romani, Turkish, and Middle Eastern music). Manol Todorov espoused this position to me in 1985: "No one is playing pure folk material. We must keep Bulgarian music Bulgarian. Foreign elements—Spanish, Indian, Turkish—don't belong. You wouldn't throw foreign words in the middle of a sentence. A Spanish motif doesn't belong in Bulgarian folk music." In print, Todorov reiterated: "We heard harmonic stamps, clichéd in rhythmic treatments, which are foreign to the melodic tenor of Bulgarian folk music. In essence, the basic task should be the war against the foreign and clichéd in melody, harmony, and rhythm, and the search for contemporary musical thought resting on the great richness of Bulgarian national musical folklore" (1985:31).

This rhetoric about musical purity is directly related to the 1980s' state policy of monoethnism and concomitant regulation of the display of Muslim ethnicity. Earlier I discussed the forced name changes, banning of zurnas, prohibition against kyucheks, and the mission of scholars to prove that Bulgarian folk music had no foreign influences. Wedding music became a primary target; its Romani and Turkish manifestations (i.e., kyuchek) were banned entirely, and the jazz, rock, and non-Bulgarian elements in the Bulgarian repertoire were cleansed. Playing and dancing kyuchek was officially prohibited, and fines and jail sentences were threatened for lack of compliance. In 1985 members of the Lenovska Grupa told me that the punishment was a 200–300 leva fine or a prison sentence.

Ivo Papazov remembers these difficult years with bitterness:

We played in spite of the fact that many composers did not like our style. At that time there were people who were in charge of the style, the order, the framework of the music. They didn't like our style because we crossed the boundaries. We had more freedom, more improvisation. They didn't want us to experiment with authentic music— my music was prohibited in folk music schools so the students wouldn't forget authentic music. On the contrary, we used the authentic, but combined with the modern. The critics didn't like us until 1989, when democracy came and our music was no longer illegal.

He describes the development of wedding style:

We started to create a new style into which we mixed Romani elements. Even though it was forbidden, we put it in. And for that reason we were not recognized for so many years. We mixed styles and we saw that it enriched Bulgarian folklore. There is nothing at all wrong with mixing two folklore styles into one. And there was an incredible resonance between the styles, Turkish, Romani and Bulgarian. It was very beautiful; there were more possibilities for improvisation. The people loved precisely this, but the government officials in charge of culture started to follow us around, to harass us, to prohibit us from playing. This was the reason they didn't let us appear on radio, even though we really wanted to record our pieces. They chased us; they fined us.

I questioned Ivo further about the relationship between his music and his identity. Growing up, he heard predominantly Turkish and Romani music, but he was also exposed to Bulgarian music because musicians serviced patrons from all ethnic groups. Ivo's Romani consciousness actually emerged later in his life. He narrated:

I was raised thinking I was Turkish. To this day, my sister argues with me that we are Turkish even though she is very dark and I am one of the few light-skinned people in my family. But I knew I was Romani

deep down inside, we just didn't face it—it was an insult. On our passports, it said "Tsgani" but we said we were Turks. My grandparents
were basket makers, sieve makers; they sold these items, they showed
me how to make baskets. I realized I was Romani from language,
history, mannerisms, culture. I knew it inside, but to accept it is another thing. Some members of the older generation spoke Romani; in
my dialect of Turkish there are Romani words.

In Chapter 11 I chronicle Yuri Yunakov's identity shifts, but here I note
that both Yuri and Ivo were raised as Turks yet later realized they were
Roma.

What mattered more than ethnicity was religion: because they were
Muslims (though not practicing), the names of Yuri (formerly Husein
Husein) and Ivo (formerly Ibryam Hapazov) had to be changed. Yuri's
ordeal is described in Chapter 11. Ivo bitterly recalls: "My mother's mother
was Pomak from the Rhodopes so we witnessed their name changes in the
1970s. The police were on my trail for a long time, but I was constantly
traveling. Finally they caught up with me and said, 'We have orders to take
you to headquarters. If you won't go voluntarily, we'll handcuff you.' So I
had to go and my name was changed." Similarly, Ivo's cousin Neshko
Neshev, born Nedyatin Ibryamov, had to change his name when he married a Bulgarian woman (Statelova 2005). And Ivo's drummer Salif Ali
became Aleksandŭr Mihailov.

By the mid-1980s, wedding musicians faced a coordinated program of
prohibition, harassment, fines, and imprisonment. As the top musicians,
members of Trakiya were especially targeted by officials to display them as
examples for other musicians. Ivo stated: "In sum, they wanted to slap the
hand of Romani and Turkish folklore to show that, 'Look, the greatest
artists are in jail—the rest of you, be careful.' They wanted to warn people
not to make weddings like that. It was a horrible time." Trakiya members'
cars (or license plates) were confiscated, and they were fined, beaten, and
jailed; in prison their heads were shaved and they were forced to do menial
work such as breaking rock and digging canals. Ivo narrated:

Thank God we were saved—we survived—we only served forty-five days.
I had a white uniform and had to break cement. By the fifteenth day
everyone was my friend and they all gave me Marlboros to smoke. Some
of my friends from the army saved us, otherwise we would have served
longer or been sent to a labor camp, and when you are sent to a camp,
you never return. A police officer warned me that they would send me to
a camp to get rid of me—only me—the others were being released. I got
in touch with someone I knew from the army who loved music, and he
saved me. Actually he came at 3:00 A.M. the very morning they were
supposed to send me to a camp, and he arranged for my release.

Ivo vividly remembered that legal charges of "hooliganism" had to be
fabricated because no official law existed about kyucheks: "There was no

evidence—they had nothing to charge me with! I hadn't broken a law—there was no law about music I had broken! They charged me with political propaganda, that I didn't respect their laws, that I was spreading propaganda—as if I were a terrorist! It was humorous!" Also note that regardless of ethnicity and religion, a musician was guilty by playing wedding music; for example, ethnic Bulgarian accordionist Petŭr Ralchev was arrested. Ivo's ethnic Bulgarian wife, Maria Karafezieva, was also incarcerated. Ivo narrated: "Maria was inside too—she was arrested but they couldn't charge her because she only sang Bulgarian songs. They had to let her go. She yelled at them: 'We get Roma to listen to Bulgarian music—how many times did I sing about Hadzhi Dimitŭr [a famous Bulgarian hero]?"

Ivo explained how musicians tried to avoid the prohibitions but ultimately faced them. If they couldn't play in the official media, they concentrated on weddings:

So we started to play illegally. We played at weddings because these are private and nobody could tell you what to play. People would record us at weddings and sell these tapes, and we became very famous. We were approached for weddings because people wanted to hear this music live. We wanted to work in restaurants but they wouldn't let us. We still played Romani weddings even though they prohibited us from playing Romani music. It is absurd not to play kyucheks at a Romani wedding. So they hounded us; they wouldn't let us play that type of music, but it is impossible to omit this type of music. . . . And after we were in jail we weren't allowed to play at festivals. They followed us everywhere so we had to stop playing weddings for a while. I didn't want to be arrested a second time. There were so many weddings that we couldn't play—we bargained for weddings three years in advance!

Along with musicians, wedding sponsors were also arrested; all were enraged that the government intruded in the domestic sphere to ruin the events for which they had prepared for years. Ivo remembers: "My patrons protested while I was in jail. Our incarcerations ruined their weddings, their celebrations. You know when Bulgarians celebrate how many people gather; the sponsors prepare food and drink. You know how much money they had already spent preparing! People came from Plovdiv to protest because I cancelled so many weddings from the Plovdiv area. It was reported in the Radio Free Europe press, but not in the Bulgarian press."

Musicians and sponsors developed creative tactics for avoiding incarceration; at village events, family members kept watch (often from the roof) for approaching police officers. An obvious tactic was to hide when the police approached, as Yuri Yunakov describes in Chapter 11. Yuri recalled that Ivo was smart enough to hide his car in private garages during weddings. Yuri admits, "I wasn't so smart; my car was parked next to the stage, so even though I hid, the police confiscated my car." If it was too late to hide, a common tactic was to morph a kyuchek in progress into a traditional Bulgarian pravo horo (musicians illustrated this to me by converting a kyuchek to the popular song *Kara Kolyo Sedeshe* [Dark Kolyo was sitting down]). Yuri admitted that despite lookouts, running was sometimes the only alternative: "As soon as the police approached, most of us started running. It was humorous to see Ivo, as heavy as he is, running into the forest behind the stage. The worst thing was to run from the police. That was the highest insult. You were supposed to stay and face the consequences."

Here Yuri alludes to the complicated issue of resistance, suggesting that the bravest response would have been to continue playing kyucheks and face the harsh consequences. But resistance is never simple. Musicians, though brave, were survivors; they did not seek to become heroes because of lofty antigovernment principles. They defied the state because of economic rather than moral imperatives. Music was their profession, and they made a living by serving their patrons, who requested kyucheks. At the same time, moral outrage accompanied economic motives. Musicians did not shy away from critiquing the absurdity of the policy and its racist message.

Resistance to prohibition was also found among young musician fans. Ripe breeding grounds for wedding musicians were the folk music high schools in Shiroka Lŭka and Kotel and the Plovdiv Academy. Although playing wedding music was strictly forbidden at the schools, students would regularly sneak out on weekends to play or listen to famous musicians at weddings. After speaking with students at the Shiroka Lŭka school in October 1985, I made this journal entry:

> All the students talk about is wedding music. They are infatuated with it, and they test us to see what we know: "Who is the accordionist with Ivo now?" They live for this music but they are not allowed to listen to it or perform it. Playing weddings is strictly prohibited. The administration recently issued uniforms and confiscated all of their "civilian" clothing so they can't sneak off and pass unrecognized. Some students have no warm clothing now. We met a vocal student from Thrace who does weddings on weekends, but she has to sneak off or take sick leave.

Dragiya Enev (Bulgarian singer Dinka Ruseva's son) told me that he had to securely hide his accordion in his room because playing it at the Kotel school was forbidden. He wanted to move from the dormitory into an apartment so it would be easier to play weddings, but school officials locked him in his dormitory and refused to let him move. He managed to sneak out anyway.

Nikolai Kolev, a Thracian gŭdulka player living in New York, recalled:

> We students at Shiroka Lŭka were forbidden to play wedding music even in our dormitory rooms. We could be dropped from the school if we were found at weddings. In fact, a friend of mine was kicked out of

the Plovdiv Academy because he went to Varna to play in a restaurant. In spite of this, my friends and I would slip out at night and somehow get to weddings to hear Ivo or Nikola Iliev, and then sneak back in, or sleep on a bench somewhere. We were crazy for the new music. The atmosphere of Shiroka Lŭka was very enriching—not just the classes, but outside of class. We played and listened to wedding music all the time even though it was prohibited.

Kalin Kirilov described how students struggled in secret to learn wedding music from cassettes that had been poorly recorded and copied many times. These sentiments were repeated to me by countless other musicians. Many told the legendary story of being threatened about wedding music by their music teachers, of ignoring them, of sneaking out to a wedding, and of seeing their teachers at these weddings!

Resistance was located in many sites, even the most official. As described earlier, the teachers at the schools lectured their students about the evils of wedding music but sometimes broke rules to patronize it. Ivo recalls that some of his most ardent fans were police officers, and he even played at their private events. He claims that when he was arrested, the judge loved his music and so he received a soft sentence (Cartwright 2006c). In 1985, I attended the baptism of Romani kaval player Matyo Dobrev's son at his home in Straldzha, near Yambol, Thrace. One of the guests of honor was a local police officer, who danced kyucheck with abandon. Similarly, when I told folklore scholars that I was studying Roma, they responded with the official line, "They don't exist," but there was always an ironic smile.

These examples amplify Herzfeld's point that cultural intimacy with the state is highly nuanced (1997). Herzfeld commented on my last example above by pointing out, "For a brief instant we see the official representatives of state ideology as human beings capable of wincing at the absurdity of what they must nevertheless proclaim" (2000:226). He further explained that despite the external formality of states, they can be viewed in social terms as "intimate apparatuses." The state embodies "potentially disreputable but familiar cultural matter," which is "the very substance of what holds people together. . . . Some of that substance even includes resistance to the state itself" (224). On both sides, the official and the unofficial, there were cracks in dogma. Police officers arrested musicians but secretly loved kyuchek; wedding musicians not only resisted but also accommodated to the state. In the cracks in official ideology, then, wedding music thrived.

State Ambivalence

I have asserted that resistance is neither singular nor pure; as Ortner (1995) points out, it is always paired with collaboration. More precisely, resistance often involves accommodation to the state. Moreover, the state is not monolithic. I now discuss cracks within the official sphere, and its relationship to black and gray musical markets. In the 1980s, life was filled

with much ambivalence. Although it was illegal, most Bulgarians pro-
cured western currency on the black market, receiving a rate that was four
times the official rate of exchange. Although it was illegal, most Bulgarians obtained western goods. Although it was illegal, most Bulgarians listened to kyucheks. Verdery explicates how the socialist state permitted the
unofficial sphere to operate, so rebellion would not erupt (1996). The government, then, simultaneously prohibited wedding music, accommodated
to it, sold it, and tried to control it from within.

In the mid-1980s, for example, the state recording company Balkanton
released several official versions of wedding music that were sanitized of
foreign melodies, jazz, and kyucheks.[26] Manol Todorov wrote for the liner
notes of Papazov's Balkanton record (BHA 11330): "All this is based upon
the sound instrumental tradition of Bulgarian folklore, without the introduction of foreign elements, motifs, or manner of performance." Todorov
told me that he instructed Ivo not to play anything foreign at the recording
session, or else it wouldn't be pressed. On these albums, wedding music
was not only censored of foreign influences but also arranged by state
composers. In the process of *obrabotka* (arrangement), much of the wild,
spontaneous, improvisatory style was lost.

Furthermore, an ensemble-type orchestra was added as backup to the
band, further distancing the music from its typical format. A cassette featuring the winners of the Stambolovo 1986 festival, Trakiya and Mladost
(BHMC 7265), credits Dimitŭr Trifonov and Todor Prashtakov as arrangers
and directors. Even the album *S Orkestŭr Na Kanarite Na Svatba* (With the
Canaries at a Wedding, BHA 1111, 1982), which is supposed to simulate a
real wedding, has orchestral accompaniment. Musicians greatly resented
this obrabotka, claiming it detracted from the music and merely filled the
pockets of arrangers with money.[27] One musician complained, "We got
paid very little for our record. But the composer who did the obrabotka
got paid much more. He only added a few violins and contrabass and got
his name on the record as 'arranged by. . . .'" Neither wedding musicians
nor their fans accepted these Balkanton releases as representative.

Audio examples 7.1 and 7.2 contrast the same piece, "Kŭrdzhaliisko
Horo," played by Trakiya at a wedding and arranged and sanitized for a
Balkanton album. Mark Levy and I recorded audio example 7.1 in 1980 at
a wedding in the village of Iskra (described earlier). There is a wild, edgy,
improvisatory quality, and the energy is visceral. On the other hand, audio
example 7.2 from Balkanton BHA 11330 (1983) adds a string orchestra
arranged and led by Dimitŭr Trifonov. Musical phrases are squared off and
harmonic chord progressions typical of the socialist era are introduced.
Instead of improvisations, there are composed solo phrases with little energy; changes in timbre have also been eliminated; finally, electric bass
and guitar and drum set have been replaced by acoustic bass and guitar
and no drums.

Like Balkanton, the other official media channels of radio and television
permitted only censored versions of wedding music to air. The few times
in the late 1980s when famous wedding bands were allowed to play on

television without backup orchestras, the viewer turnout was enormous. Fans were glued to their home television set, or they crowded around televisions in hotel lobbies. One such event was the 1987 televised performance of a concert of the winners of the 1986 Stambolovo festival (more discussion on this in a moment). Video examples 7.3 and 7.4 show Trakiya performing a pravo horo and a rŭchenitsa in this show. I viewed this video with Ivo, Neshko, Yuri, and Salif in 2005 and asked them if someone in the government spoke to them beforehand about omitting kyucheks. Ivo answered: "They didn't need to speak to me. I had just been in jail for playing kyucheks. A few years earlier they made me change my name. It was absurd to think of playing kyucheks. They would have hung me."

This brings up the issue of self-censorship. Wedding musicians developed the ability to sense when they could push the limits of the state and when they had to toe the party line. This may help to explain the apparent puzzle of why musicians recorded these censored versions. Economics, not lofty moral principles, was the main motive guiding musicians. They reasoned that official versions would increase circulation of their music and even enhance the value of their live performances. In addition, they did not want to incite the government against them by refusing to cooperate.

James Scott's work on "everyday protest" (1985, 1990) suggests that analyzing resistance always requires analyzing power and its effects on the weak. The hegemony of the state depends not on brainwashing but on how public discourse triggers shifts in consciousness. Both wedding musicians and the state may have perceived "the advantage of avoiding open confrontation" (Sivaramakrishnan 2005:350). In addition, we can't assume that musicians had full agency; nor can we assume the state had total hegemony: "On the contrary, at times social structures, roles, statuses . . . modify agency and its consequences. . . . Actors may engage in everyday acts of resistance or desist from them under structural pressures" (351). Wedding musicians, then, strategically alternated between accommodation and resistance to the state.

In addition, the state itself was not monolithic, and indeed "different levels of the state may work at cross-purposes" (2005:351). Aretxaga reminds us that we need to "rethink the notion of the state in a new light as a contradictory ensemble of practices and processes" (2003:395). The state was ambivalent about a phenomenon that was fast becoming a mass movement. Policy was contradictory, and at times the state even cashed in on the popularity of wedding music, again illustrating Herzfeld's point about cultural intimacy and Verdery's point about gray markets.

In the early 1970s, when wedding music was first becoming popular, fans would record at events and then copy the tapes for friends or sell them at exorbitant rates on the black market; young people prized these unofficial recordings. In the 1980s, in an effort to undercut the black market in wedding tapes, the state established studios for selling wedding music and other cassette recordings made outside the auspices of Balkanton. At a *stereo zapis studio* (literally a tape recording studio),[28] one found for sale a selection of rock, funk, disco, "authentic" folk music, and wedding music. The largest seller was wedding music. The studios were, in effect, sites where popular taste was paramount and where official prohibitions were relaxed. When kyucheks were banned from records, they could still be found at studios; in fact, they were the best sellers among Roma. Similarly, when zurna music was banned it could still be found at studios. Although a printed notice posted in one studio read, "This studio is for copying tapes of Bulgarian music and music from other socialist countries," I regularly saw tapes of groups from Italy, Greece, and Serbia.

With the studios, the state simultaneously maintained its official folk music policy and also catered to public taste. More important, the studios were a means for the government to gain access to the inflated market of wedding music. The price of a studio cassette was high, 15 leva for a sixty-minute tape. Although this was equivalent to more than a day's wages, the demand was very high. Fans were willing to pay dearly for the music they loved but couldn't find on official Balkanton recordings, which cost a fraction of the studio tapes (about 2.5 leva).[29]

Given the popularity of wedding music, it was perhaps inevitable that the state would take a more direct hand. The form of state participation, the Sambolovo festivals (1985–1988), involved both promotion and regulation. Within a few years, scholars began lauding the talent of wedding musicians while policies dictated what could be played at the festival. Note the panoramic view of the huge 1988 Stambolovo audience, which appeared on the cover of the scholarly journal *Bŭlgarska Muzika* (photograph 7.9). See text supplement 7.2 for a discussion of this festival.

In sum, wedding music erupted as a mass youth phenomenon that eventually caused a fundamental shift in the official rhetoric of the state during late socialism; this resonated with cracks in socialist doctrine in other arenas of life, such as the emergence of limited private enterprise. The significant role of Roma in contesting the state via wedding music cannot be ignored, but we must also remember that non-Roma were jailed as well for playing wedding music. The next chapter continues my analysis of the role of the state, moving into the postsocialist period where the capitalist market dominates.

8

⌒⊙⊙⌒

Cultural Politics of Postsocialist Markets and Festivals

Turning to the postsocialist period, this chapter examines the challenges Bulgarian wedding musicians and other Balkan Romani performers face vis-à-vis capitalism, changing state policies, and polarizing world politics. As the state becomes weaker, private forces take its place. For Roma, professional music has always been about business, but now it is about big business, often with structural exclusions. How do Romani musicians negotiate this complicated terrain between state and commercial forces? Music may have touristic value as UNESCO-sponsored "world heritage," or musicians may be ignored by states. On the other hand, music idol contests and Romani music festivals have emerged as sites of negotiating national and transnational identity politics.

Bulgarian Wedding Music in the 1990s

Ironically, in the 1990s wedding music garnered effusive praise internationally while at home in Bulgaria it faced severe economic woes. It was "discovered" in the west by British impresario Joe Boyd of Hannibal records, who visited Bulgaria in 1987. Asking his guide, the music producer Rumyana Tsintsarska, to show him some contemporary folk music, he recalled: "I was taken to hear the Plovdiv Folk Jazz Band,[1] which I found dull, except for a brilliant guest solo by Papazov. I took Ivo aside and asked him if he had his own band. Ivo answered, 'Of course!' and invited me to a Romani wedding." Boyd was so taken with Trakiya that he planned an album and a tour.

The tour fell through when the government withheld the visas. Boyd arranged a tour of the village music group Balkana instead, and the government sent Trakiya to Moscow to perform. Boyd recalled, "The visas for Trakiya were never denied, but they weren't granted. The officers kept asking for more documents." Papazov bitterly remembers:

149

My career in music changed in 1987 when Joe Boyd came to Bulgaria. He had heard of us. He was Pink Floyd's first manager. He went around with us for a whole week to Romani weddings. He listened to our music. Then he proposed a tour to us. The government hassled him for a year with contract problems but he made a CD of us in Bulgaria. He was ready, the contract was sent to us, but the government wouldn't let us go, and they dragged it out for a year. It was a huge mockery in 1988 when I was supposed to leave the country. I had to go from bureau to bureau, to Todor Zhivkov's adviser, and in Stara Zagora to the administrative division for minorities. Three times a day I had to go for interviews. I said, "I want to travel, I don't want to emigrate." They said "You are this, you are that—a Turkish Rom—America will easily assimilate you." Joe Boyd had the tickets and everything ready but at the last moment they wouldn't let us go. Actually I had the right to apply for political asylum because of mistreatment. If it hadn't been for my two kids I might have thought about emigrating. I can live anywhere. . . . How many years did those guys from Internal Security follow me around? Now they all emigrated to America and I still live in Bulgaria!

It is clear that the state did not want Roma representing Bulgaria abroad. Indeed, many Bulgarians agreed with the sentiment that "we can't have a Gypsy or a Turk represent us internationally."[2] Most Bulgarians felt more comfortable with the international success of groups like Balkana and Le Mystère des Voix Bulgares, which played clearly sanctioned "folk music," even if it was highly arranged (Buchanan 1996:220–226).

Boyd persisted in his advocacy of Trakiya; as described earlier, he released the album *Orpheus Ascending* (HNCD 1346, 1989) to international acclaim (see publicity shot, photograph 8.1). But he was already planning the next album. In 1988, in a taxi in New York City on the way to hear Balkana, Boyd and I discussed whether including kyucheks on a second album, *Balkanology*, would hurt Trakiya's chances of receiving visas. I stressed how important kyucheks were in its repertoire, and to omit them would misrepresent its artistry. Boyd excluded Romani and Turkish music on *Orpheus Ascending* because he was guided by state representatives; he was reluctant to alienate the socialist authorities who were his co-producers. His liner notes are vague about ethnicity: "Bulgaria is sensitive to questions of racial or national origin, so accurate information is hard to come by, but Ivo and his group seem to be at least partly gypsy and much of their music is related as much to gypsy styles as to Bulgarian traditions" (1989).

Boyd decided to include Romani, Greek, Romanian, Macedonian, and Turkish repertoire on *Balkanology*, and he asked me to write the liner notes. They emphasize the Romani/Turkish ethnic dimension of Trakiya's music, but Boyd refused to label any tracks kyucheks and he did not want me to write about politics. Despite my protests, Boyd insisted on employing euphemistic names that the Bulgarian state had used in the 1970s, e.g.

Mladeshki Tants (young person's dance) for kyuchek. In fact, the marketing for Boyd's three American tours of Trakiya in 1989, 1990, and 1992 did not emphasize a Romani connection. Remember, this occurred before the craze for "Gypsy music" was initiated by the documentary film *Latcho Drom* (see Chapter 13); however, it was precisely at the time when "world music" became a viable marketing category, and in fact Joe Boyd was one of the key people in Britain who coined the term. *Balkanology* appeared in 1991 (HNCD 1363) to rave international reviews.

Trakiya members were successful in receiving their visas in autumn 1989, right before the fall of the Berlin Wall, November 9. The musicians heard about the fall of Bulgarian communism on November 10 from abroad, where they were awash in media adoration. Ivo recalled: "We started our tour in September 1989 in London at Ronnie's Club, where the Beatles started. The first shock was the opulence of London and the second was New York—because at the time there was nothing in the stores in Bulgaria." In September 1989 Bulgarian Radio broadcast an interview with Papazov from the United States in which he described the outpouring of praise that wedding music was receiving (Buchanan 1991:554). Ironically, wedding musicians received the recognition they craved from the West, not from their own government. From 1989 to 1994, Trakiya toured frequently in Europe and also traveled to America and Australia. The musicians made their mark on the international folk and jazz scenes; on the one hand, this increased their stature in Bulgaria, but on the other hand it made them less available for local weddings and concerts.

The transition to capitalism in postsocialist Bulgaria affected wedding musicians in contradictory ways: there were new freedoms, but the economy suffered greatly. Socialist restrictions related to purity were totally removed, allowing performance of kyucheks along with jazz, rock, and foreign musics. In spring 1990, for example, I attended the first post-socialist state-sponsored concert of Romani music, held in a Sofia theater (photograph 8.2). The audience was 99 percent Roma, and the excitement was palpable. Organized by Anzhelo Malikov, the musicians included Sofia-based Romani wedding musicians and dancers, and the master of ceremonies spoke both Romani and Bulgarian. The program, however, featured orientalized versions of kyuchek with half-naked women in synchronized choreographies, unlike what happened at Romani events, and instead appropriating Turkish belly dance (see Chapter 6). The Bulgarian public, meanwhile, enthusiastically embraced Serbian, Macedonian, and Greek musics and pop/folk fusions, which became the rage in restaurants and taverns. Wedding bands broadened their repertoire to include these musics. The opening of the borders permitted musicians to travel, and the best wedding bands went to Yugoslavia, Greece, and Western Europe.

Unfortunately, the euphoria of transition was short-lived and the reality of unfettered capitalism soon soured the populace. Economic crisis gripped Bulgaria in the early 1990s, negatively affecting work, health care, education, and sociability (Engelbrecht 1993). State enterprises closed and private companies struggled to operate, but they were poorly managed and heavily taxed. There were shortages of goods; thousands of people tried to emigrate. Corruption flourished in everyday transactions and also in the process of legal restitution of land and property. A tiny class of "new rich" emerged, flaunting their cars and jewelry, while the middle class sank closer to poverty and the rate of unemployment rose. Discrimination against Roma increased, violent crimes against them rose, and their unemployment reached 90 percent.

At first, wedding musicians embraced capitalism boldly, as most of them had experience in the free-market realm and had never relied on the state for security. Many bands released cassettes on newly formed private labels (none run by musicians) such as Payner, Lazarov, and Unison Stars (Peycheva and Dimov 1994). Folk ensemble musicians, on the other hand, suffered as bread-and-butter government support for the arts diminished (Buchanan 2006:426-478). Stereo zapis studios closed and Balkanton curtailed most of its production. Everyone, including state ensembles, looked for private sponsorship, either local or foreign.

The Stambolovo festivals of wedding music in 1990, 1992, 1994, and 1996 were financed mostly by private sponsors. Attendance dwindled because people had less disposable cash.[3] Wedding musicians, however, remember Stambolovo with fondness, and they regret its cessation. When a journalist asked Ivo Papazov in 2004 what bothered him most about the current state of wedding music, he answered: "That there are no longer gatherings of folk bands in Stambolovo. . . . Although abroad this is the most venerated festival, they can't find the finances. New talent was discovered there. People felt at ease there. Even now foreigners ask me about it" (Filipova 2004:17). Papazov stressed that it was a place for wedding musicians to socialize. The sponsors treated musicians well, not only offering prizes but hosting them with food and drink. In 1996 Ivo, cognizant of the financial woes of the festival, refused to accept any money for his performance. Despite the introduction of democracy, only Bulgarian music was permitted at the festival, illustrating the lasting power of socialist categories. Nevertheless, Ivo premiered his kyuchek composition "Celeste"[4] at the 1996 festival. Although the jury frowned on it, Ivo's fans went wild.

In April 1994, Payner sponsored a twelve-hour "megaconcert" in Sofia with thirty soloists and nine bands, but it was very poorly attended. In September 1994 Payner sponsored the first Trakiya Folk, a juried festival of wedding music with huge prizes. Payner invited bands to compete in two mutually exclusive categories: Thracian and Balkan, the latter meaning Romani, Turkish, Greek, and Serbian (Buchanan 2007:235). Payner produced cassettes and videotapes of the festival (video examples 7.2 and 7.9), and attendance was good. But the populace was too worried about their declining incomes to be active wedding music fans. In addition, new musical genres such as chalga (pop/folk) and new events such as the Stara Zagora Romfest were competing for listeners. In fact, in 1999 Payner changed the direction of Trakiya Folk toward chalga (see Chapter 9).[5]

Two magazines, *Folk Panair* (Folk Gathering) and *Folk Kalendar,* in the 1990s reported on folk music, wedding music, and Romani music. Contributors were well-respected academics and journalists; advertising and subscriptions supported the publications, but they faltered and folded. New radio programs debuted, including *Radio Signal Plyus* and *Radio Veselina* (founded by Veselina Kanaleva), offering a mixture of Bulgarian village music; wedding music; and Romani, Greek, Serbian, Turkish, and Macedonian music, funded by advertising and listeners' greetings and requests. These programs still exist today. A few television shows attempted to present wedding music in the 1990s, but they failed.

In the 1990s, weddings were a far cry from the three-day events of the 1980s. The economic crisis meant that Bulgarians could no longer afford lavish affairs with live music. True, there was freedom of repertoire, but few felt economically secure. A typical wedding lasted one afternoon or one evening, often with a DJ rather than live music. Weddings were bargained by the hour rather than the day. Moreover, rarely were musicians hired for transporting the bride from her home to the groom's home, which used to be an important musical moment; if instrumentalists were hired for this ritual, they tended to be lower-quality local musicians. All of this is still true today.

In 1994, Ivo remarked: "Now the businessmen rule Bulgaria, then [before 1989] the communists ruled. . . . Now there is no work for musicians in Bulgaria" (Dimitrova, Panayotova, and Dimov 1994:23). When a journalist asked him, "Has the great boom of wedding music passed?" he answered:

> Of course, such are the times. In the old days when I would play, twenty to thirty sheep would be slaughtered, 1,000–1,500 people invited under three to four huge tents. . . . Another 1,000 came to listen. But today times are such that a person can't relax. To make a wedding you need at least 50,000–60,000 leva, plus money for music. Look at the times—gasoline is 15–20 leva [per liter]. Sofia residents come and beg me [to play for weddings] but I can't take the soul of a person— tomorrow he won't have anything to eat. Categorically, I refuse them [Dimitrova et al. 1994:26].

In the 1990s, the families who put on relatively large weddings tended to be Roma and Turks, not because they were wealthier but because for these groups live music and dance were a necessary part of celebratory life (see Chapter 5). In 1994, an industrious Rom in the city of Septemvri told me: "We find a way to earn money, we manage. Bulgarians sit and complain. We still have big weddings, circumcisions, soldier send-off celebrations. Bulgarians don't do this any more—they invite just a few friends and family and use a disc jockey—that's it. Only Roma are having big events. We work and spend. Bulgarians are stingy. We spend money on our families."

In 1994 I attended a Turkish Romani wedding in the village of Tsar-atsovo, Plovdiv district, as a guest of Trakiya, which at that time included Yuri Yunakov's son Ahmed on saxophone (at that time, Yuri was in the United States). There were about 200 guests and the repertoire included Bulgarian, Romani, and Turkish music; as dawn approached, kyucheks dominated (video examples 7.5, 7.6, and 7.7). Trakiya played from about 9:00 P.M. to 3:00 A.M. and, in addition to Maria Karafezieva, who sang Bulgarian songs, a Turkish singer was hired. Afterward the assembled teenagers set up a disco and danced to rock music until dawn. Interest-ingly, their rock dancing incorporated stylistic moves from kyucheks.

In terms of compensation, Trakiya bargained for 8,000 leva (1,000 leva per hour) plus 8,000 leva in tips. Thus each performer received about $35, at a time when the average monthly salary for a factory worker was $80. Trakiya was therefore well paid; Ivo claimed he received 40,000 leva ($750) for one concert (Dimitrova et al., 1994:22). Other wedding musicians earned less and tried to supplement wedding work with additional jobs. Bulgarian singer Dinka Ruseva from Radnevo, Stara Zagora district, for example, explained that a well-paid five-to-six-hour wedding would gen-erate 6,000 leva plus 5,000 leva in tips, which came to $24 a person. Dinka also had a job singing in concerts for the House of Culture in Radnevo that paid $35 a month. Other wedding musicians had to take nonmusical work: they opened stores for car parts (Petŭr Ralchev), became administrators (Yuri Yunakov's brother was briefly a deputy major of Haskovo), or worked in sales.

In 1994 I attended the blessing of a new Romani house in a Sofia neigh-borhood (Hristo Botev district) as a guest of violinist Georgi Yanev; it was a one-evening event (see video excerpt 6.1 for solo kyuchek dancing at this event). Yanev's band Orfei includes both Romani and Bulgarian members; at this event the Romani members included Orlin Pamukov on clarinet and Paicho on drums, and the Bulgarian members included Yanev and his wife Pepa, accordionist Petŭr Ralchev, and guitarist Nikolai Georgiev. Orfei is well liked among Roma, and their kyuchek repertoire is quite var-ied. Several famous ensemble musicians came to hear the music, but the crowd was significantly smaller than at events in the 1980s.

Yanev described Orfei's typical weekly summer schedule in 1994: Thursday, drive from Plovdiv to Sofia (two hours each way) to play for six hours; Friday, drive to Vratsa (five hours each way) to play for seven hours; Saturday, drive to Stara Zagora (ninety minutes each way) to play for six hours; Sunday, drive to Dimitrovgrad (ninety minutes each way) to play for seven hours; Monday, Tuesday, and Wednesday, no work. Note that in comparison to the 1980, musicians played for shorter gigs, drove more, and suffered from more unengaged days. Because weddings were only one evening long, musicians had to play more weddings per week to make a decent income. This was more stressful and involved more driving and higher expenses. Even famous musicians could no longer earn enough to support their families. Many secured other jobs; Yanev struggled to create his own music studio. Yet, comparatively speaking, wedding musicians were lucky because at least they had some work, clustered in the summer months. Many Bulgarians and most Roma had no work at all.

A new genre of personal experience narrative arose in the 1990s among wedding musicians, illustrating the insecure times: the crime story. Georgi Janev and Orfei members, for example, were driving home from a large Romani wedding. A car passed them, swerving close to make them stop. Men emerged with guns and stockings over their heads and took all of their money. Dinka Ruseva's musician husband had a similar experience. He bluffed the thieves by pretending he was reaching for a gun; another time Dinka's son pretended his clarinet was a gun. Obviously, thieves were targeting wedding musicians. Also in the 1990s, Ivo Papazov and his family were robbed at gunpoint inside their own home in the village of Bogomilovo, Stara Zagora region. This happened in spite of his numerous guard dogs and watchmen. Other singers were tied up by mafia bosses and forced to perform in the back room of clubs. Indeed, the mafia emerged as a force in Bulgaria in the 1990s and had its finger in music, especially chalga (see Chapter 9).

An important concern of musicians during postsocialism became copyright and exploitation by record companies. For example, at the 1994 Trakiya Folk festival Payner required participating bands to be taped for a cassette release. Orfei refused to sign because they wanted to produce their own cassette. According to Petŭr Ralchev, "weaker groups were glad for the exposure." Producing an independent cassette required Orfei to overcome huge obstacles in financing, marketing, and distribution. Orfei's leader, Yanev, struggled to set up his own high-quality recording studio in Plovdiv and was eventually successful. For years he produced his band's recordings, but in 2006 Orfei signed up with Payner.

Musicians were especially worried about the widespread practice of pirating. Theoretically, a company like Payner would pay a royalty fee for every album it sold; musicians, however, complained that companies deliberately underreported the number. Petŭr Ralchev asserted: "The companies lie and say they sold 50,000 when they really sold 300,000. It's a big business." In addition, in the 1990s every city boasted a huge open-air market for pirated copies of albums, and Bulgaria was cited as one of the worst-offending countries (Kurkela 1997; Buchanan 2007:245). Recently, the situation has improved in terms of copyright laws; however, many problems remain.

Bulgarian Wedding Music in the Twenty-First Century

The current situation is challenging for wedding musicians, and some are nostalgic for the socialist period. According to Ivo Papazov, "I had more work back then. People were happier and had a lot of money. I don't think anything good has come of the new democratic Bulgaria. Now it is a place of corruption and everyone is fighting to get into the ruling party" (Cartwright 2006c:38). Nostalgia, however, should be seen not only as longing for socialism but also as a critique of capitalism and a desire for order and security.[6] It turns out that the free market is not so "free" after all; whatever sells receives the most media playtime, and in 2000 pop/folk was the best selling genre, not wedding music (see Chapter 9).

Moreover, wedding musicians now configure themselves as champions of Bulgarian folk music (of course, they mean the Bulgarian genres of wedding music). In some senses they are correct, if we conceive of folk music outside the narrow authentic socialist box, and if we see wedding music as opposed to chalga. When I asked Papazov what is Bulgarian about his style, he answered, "The foundation of wedding music is Bulgarian." He remarked that today, when few people are interested in Bulgarian music, "we wedding musicians play it. Ironically, I have preserved Bulgarian music. . . . We played pure Bulgarian folklore in spite of the fact that is wasn't really pure, but it was Bulgarian and it was beautifully embellished!"

Papazov complained that ethnic Bulgarian patrons request mostly kyucheks: "Recently I've played for several Bulgarian weddings, on purpose . . . they pay well. I opened with a Bulgarian horo and from then on it was all kyucheks" (Dimitrova et al. 1994:26). He and Yunakov have both proclaimed on television that Bulgarians should be ashamed that Roma are preserving their heritage: "Now we Roma are touring around playing Bulgarian music, while in Bulgaria, Bulgarians are playing Romani music." Here Ivo and Yuri are alluding to the popularity of chalga among Bulgarians, and the fact that wedding singers are collaborating with Romani bands. For example, ethnic Bulgarian vocalist Radostina Kŭneva appears on albums with the Romani band Kristal.

Wedding musicians blame chalga for the decline in popularity of wedding music; they assert that chalga is more pop than folk and that it is technically inferior to wedding music. Ivo exclaimed proudly: "Our music is not pop!" But aside from stylistic differences between wedding music and chalga, their respective positions vis-à-vis the state and capitalism need to be examined. In the socialist period the competitors of wedding music were the ensembles that were the purveyors of "authentic folk music"; the latter were supported by the state but, for the most part, rejected by the populace. Wedding music received some of its cachet by being countercultural, that is, oppositional to the state. More specifically, it represented capitalism in the midst of socialism. Now the competitor to wedding music is chalga, supported by unbridled capitalism. The state has withered and wedding music has lost its antistate oppositional positioning; it is emerging, however, as a force of nationalism (more on this later).

Today, however, wedding musicians are not totally negative. Although Ivo claimed that "it is sad to me that no one pays attention to wedding music," he also pointed out that wedding music still has many fans in Bulgaria:

In 2004 in Plovdiv we celebrated the [thirtieth] anniversary of Nikola Iliev and the Konushenska band. There was an audience of 6,000 people. . . . Wedding bands continue to exist and to have their fans. . . . Twenty-eight bands appeared. . . . The audience booed the lip-synched performers but the viewers stood up when we played live. That made Professor Radev [classical clarinetist] repeat with teary eyes: "We won't perish, we won't perish. If, from time to time, we, the elite of wedding music don't gather to play some kind of concert, the young generation will forget us. And for the rich music companies, it is unpleasant for us to appear in public because the people will realize they are being cheated with these lip-synchings" [Filipova 2004:17].

Similarly, in 2005 hundreds of wedding musicians attended the commemoration of Bulgarian wedding singer Dinka Ruseva's thirty-year career.

Wedding musicians have had to make compromises in the postsocialist period; one is strategically incorporating chalga singers, and another involves forgiving (but not forgetting) past detractors. Papazov recalled the past criticism of Nikolai Kaufman but admitted, "Now I'm going to play for his gala eightieth birthday. We will play some pieces he wrote for Maria and me!" In 1994 he elaborated: "I make compromises. . . . The other night . . . we were at Manol Todorov's sixtieth birthday celebration. Isn't that a gesture? For when one makes gestures, one makes money. After all, I have two children" (Dimitrova et al. 1994:26). Petŭr Ralchev bitterly criticized a televised birthday interview with Manol Todorov where the latter claimed he was glad he had the opportunity to help establish a place for wedding music. To the contrary, Ralchev remembered all the times Todorov called wedding music kitsch and impure.

Surveying the landscape of wedding music in 2010, immediately one notices that many of the hundreds of groups that existed in the late 1980s and early 1990s have simply disbanded, but several new ones have emerged. A solid group of bands have survived, including Vievska Grupa, Trŭstenik, Kanarite, Orfei, Konushenska Grupa, and Brestovica.[7] Vievska Grupa owes its popularity to its Rhodope regional focus and its backing by Payner. Yet the Vievska Grupa has also incorporated chalga and Macedonian and Serbian music to cater to current tastes.

The success of the Konushenska Grupa derives from its legendary clarinetist, Nikola Iliev, one of the founders of Bulgarian wedding style. Excelling in the Bulgarian repertoire and not emphasizing Romani and jazz elements, he has a regular following in the Plovdiv region, especially among the older generation. Orfei, with mastery of both Romani and Bulgarian repertoires, also has a steady output of albums and constant wedding work. In 1994 Orfei's singer Pepa Yaneva told me she would never sing chalga, but a year later Orfei albums included chalga; obviously, the market required it, and in 2006 Orfei signed with Payner, a company associated with chalga. In fact, Georgi and Pepa Yanev groomed their daughter Tsvetelina to be a chalga star; in 2009 she made a successful debut as the youngest singer with the Payner company.

Under the direction of ethnic Bulgarian Atanas Stoev, who arranges much of their material, Kanarite has emerged as perhaps the most prolific wedding band, producing an album every year with Payner. Their arrangements are sweet-sounding and pleasant, and their instrumental improvisations are short and do not veer toward jazz. Their sound is thus tamer and less aggressive than other bands, and this has struck a chord with a wide fan base; their 2003 album proclaims that it is "the tenth album in a row with typical Kanarite sound—composed music and texts distinguished by tradition and new authorship." Furthermore, they target a Bulgarian audience rather than Roma and Turks. Although they established their reputation in the 1980s with well-known Romani clarinetists Nesho Neshev and Delcho Mitev, now they underplay Romani associations and emphasize their Bulgarian affiliations.[8]

The trajectory of the Kanarite repertoire of the last fifteen years shows that they have moved away from kyucheks and chalga toward Bulgarian folk music. The *Kanarite '98* album, for example, contains several 2/4 and 9/8 kyuchek songs. "Biznesmen" (Businessman) has a typical chalga text (and Romani-style kaval solo): "I want to become a businessman, to drop a million every day, to buy a villa and two cars. . . . Bars, taverns, modern girlfriends." By 2000, however, the band was releasing fewer kyucheks and had veered away from texts about materialism, sex, and capitalism— instead embracing texts about love, family, friends, and village life. As early as 2000, they also cleverly converted chalga to something more ethnically Bulgarian and less Romani by inviting chalga singers to record Bulgarian folk songs with them as guests. Stoev could accomplish this because he is a good businessman; in addition, many chalga singers are also folk singers who were pleased with the exposure.

The video *Nie Bŭlgarite, Kanarite 25 Godini* (We Bulgarians, the Canaries, 25 years; 2000) illustrates this trend (video examples 8.1 and 8.2). The show begins with the announcement, "On this album, the beauty of Bulgaria has been collected." Staged in the Plovdiv amphitheater (which dates from Roman times), the video provides a visual spectacle that links the band to antiquity (and also to high-placed officials who authorized use of the site). Throughout the concert, the Smolyan Dance Ensemble, dressed in folk costume, performs choreographies and comic skits of village life. The dancers start the show with the propitious ritual of offering bread and wine. These visuals emplace the band in the realm of village and folklore.

The regular band is augmented by guest brass and string sections, but the most important instrumental guest is Petko Radev, beloved by many Bulgarians because, as a classical clarinetist with La Scala in Italy, he championed Bulgarian folk music. Note that in addition to Kanarite's standard instruments, the gaida and kaval link the band to tradition. The vocal guests on the video include eight chalga stars: Nelina, Gloria (video example 8.2[9]), Ekstra Nina, Toni Dacheva, Tsvetelina, Vesela, Desi Slava, and Slavka Kalcheva (who started as a wedding singer but crossed over to chalga), who all sing Thracian wedding songs. The crass sexuality of chalga has been tamed; the outfits are subdued (gowns are cut low but tasteful). In short, on this album Kanarite has assimilated chalga into their more wholesome folk aesthetic.

The band Kanarite continued to develop its Bulgarian profile in the last decade. Their standard formula includes songs (mostly sung in thirds) and instrumentals in folk style (major keys predominate) with shorter improvisations, more Macedonian/Pirin songs in 7/8, more city songs, and fewer and tamer kyucheks. Their 2001 album *Ne Godini, A Dirya* (Not Just Years, But a Path), has one 9/8 kyuchek and one 2/4 kyuchek (a duet with Stoev and chalga star Ivana); it also features the Eva Quartet in polyphonic a cappella arrangements reminiscent of the socialist era. The 2003 album *Na Praznik i v Delnik* (On Holiday and Weekday) has no 2/4 kyucheks and only one 9/8 song, with no instrumental improvisation. The video visuals feature a costumed folk ensemble in a village setting, and the singers wear large Eastern Orthodox crosses on their necks.

The 2003–04 album *S Ritŭma Na Vremeto* (With the Rhythm of the Times) epitomizes the band's evocation of national pride, with emerging themes of church, family, and patriotism. The religious theme surfaces in the first piece, where the band is filmed playing in a monastery in front of Byzantine icons. The title of the tune captures the theme: "Pravoslavno Horo" (Eastern Orthodox dance). The song "Bŭlgarski Cheda" (Bulgarian children) develops the themes of patriotism and family in a 7/8 Pirin rhythm that evokes nostalgia by poignantly narrating the sacrifices of Bulgarian soldiers and the suffering of the populace. Filmed in a church, with band members in black clothing lighting candles in memory of Bulgarian soldiers killed in Iraq, the somber atmosphere is interspersed with footage of military training. This song links past sacrifices to contemporary Bulgarian politics.[10]

Chalga singers (Emilia, Daniela, Gloria, and Ivana) are guests on this video, and again they sing Bulgarian songs (all composed by Stoev). Ivana's 7/8 Macedonian song, "Ah Lyubov, Lyubov," (Oh love, love) narrates a story about the pain of love that ends with separation and the birth of a child.[11] The accompanying visuals are close-ups of historical Bulgarian paintings depicting peasant mothers holding and nursing children, and Ivana relaxing with Stoev. Ivana's chalga glitziness is thus assimilated into the safe framework of the Bulgarian family and home.

These Kanarite albums position the band in opposition to the values of chalga (money, alcohol, and sex), but they manage to recuperate the association of chalga with success, modernity, and technology. In recent performances Atanas and his wife Nadya are featured together more prominently (singing, and even touching), as a symbol of stable marriage. On their 2005 video *Traditsiya, Stil, Nastroeniye* (Tradition, Style, and Spirit), the opening song, "Nie Sme Kanarite" (We are the Canaries), introduces them as successful and happy, content with their families and friends; it implores the audience to "forget your woes." The band has come to stand for the Eastern Orthodox religion, family values, optimism, and the nation (i.e., the Slavic majority). The band's 2009 album *Muzika s Lyubov* (Music with Love) featured neither kyucheks nor chalga guests; in general, they have distanced themselves from Romani and Turkish musical motifs and cultural symbols. I do not think this is accidental. Especially at a time when anti-Romani and anti-Muslim sentiments are being openly expressed by various political parties, Kanarite have tapped into a nationalistic musical vein.

The musical trajectory of Trakiya, on the other hand, is starkly different from Kanarite. Trakiya is perhaps the least recorded band, which is not only surprising but also quite a loss, considering its quality and fame. After *Balkanology* was released in 1991, the band did not make another recording until 2003. Papazov claims he was waiting and hoping that Boyd would record another project, but he never did (Cartwright 2006c:37). Perhaps Papazov was also suspicious of the reputations of the new Bulgarian companies, some of which allegedly had mafia ties. As mentioned earlier, in the 1990s Trakiya found most of its work abroad. Their older guitar and bass players were replaced by keyboardist Vasil Denev, adding the possibility of varied textures. Some Trakiya members developed their own paths; for example, Ivo collaborated with Hungarian Romani cimbalomist Kalman Balogh on a pan-Romani project, accordionist Neshko Neshev released an album with his own band, Yuri Yunakov emigrated to New York in 1994 and formed his own wedding band (see Chapter 11), and kaval player Matyo Dobrev often joined the band. For the most part, however, in Bulgaria Trakiya was ignored by the media.

All this changed in 2003 with the release of *Fairground/Panair* (Kuker Music KM/R 07), produced in Bulgaria but distributed in Germany. The album is a tour-de-force of Papazov's newer style, which is more arranged, more polished, more textured, more technically ambitious, and more varied than his music of the 1990s. Because *Fairground* was made for Western audiences, it features concertized versions of wedding compositions that are not danceable. Added to Trakiya's regular line-up are Bulgarian Turkish musician Ateshhan Yuseinov on guitar, Stoyan Yankulov on tupan and percussion, jazz pianist Vasil Parmakov, and two bass players. The repertoire includes the standard Bulgarian slow songs and dance songs, beautifully performed by Maria Karafezieva; and instrumental horos, rŭchenitsas, and kyucheks. The solo improvisations by Ivo and Neshko are longer, wilder, and much more inflected with a jazz sensibility than earlier recordings. This album is clearly intended to present Trakiya to Western jazz audiences.

The album's visuals solidly evoke Bulgarian folklore by displaying band members in folk clothing with folk motifs (men in red vests), Karafezieva in a Stara Zagora costume (which she rarely wears in a live performance), and six dancers in full village costume. I believe this image reflects the general repositioning of wedding music during postsocialism as closer to folk, in opposition to chalga. It also reflects Papazov's genuine attachment to folklore. Although the visuals eschew anything Romani or Muslim, the repertoire includes a Turkish slow melody and three kyucheks, one of which is titled "Gypsy Heart."

The album received triumphant reviews and in 2005 Papazov won the BBC (British Broadcasting Corporation) Radio 3 audience award in the category of World Music. Ivo was especially pleased because he was competing with international stars and because this award is determined by the BBC public, composed of 150 million listeners, not by a jury: "Other prizes are decided by two or three people . . . but my prize depended on the entire nation—on the voice of the audience, whether it is Bulgarian or English. For me that is a real prize!" In an emotional ceremony, Joe Boyd delivered the statue to his old friend. According to Boyd, although the BBC did not let Trakiya perform at the actual ceremony Papazov sneaked his instrument on stage and "brought down the house with a clarinet solo."

As a result of the BBC award, Trakiya received dozens of invitations to perform around the world, and the musicians captured the limelight once again. Trakiya now has a busy European touring schedule, and magazine articles have appeared about Ivo with titles such as "The King Returns" (Cartwright 2006c). American audiences warmly received members of Trakiya during their 2003 and 2005 reunion tours with Yuri Yunakov, and Traditional Crossroads produced a reunion album, *Together Again: Legends of Bulgarian Wedding Music* (2005). This album, unlike *Panair*, contains primarily dance music. In 2008 the British label World Village released *Dance of the Falcon*, featuring Ivo accompanied by several jazz and classical musicians.

What is perhaps most striking about the last few years is the official attention and adoration Ivo is finally receiving in Bulgaria. Special concerts have been organized for Trakiya in Sofia; Ivo was made an honorary citizen of Stara Zagora in fall 2005; and he now appears in the "Alley of the Stars" in Sofia. In 2004 Trakiya played for NATO leaders and in 2005 they played for a meeting of the presidents of Balkan nations. Ivo narrated that this concert brought up unpleasant memories:

> It was very prestigious but I couldn't perform in front of a row of officers guarding the room. To this day I am afraid of the police, of guards. I remember in the old days in Kŭrdzhali the whole neighborhood would clear out as soon as the police arrived. We were all scared. Everyone would go inside and wait. The police would let their dogs run and those dogs could kill you. My heart starts beating fast when I see those uniforms. So at the meeting, I took Georgi Pŭrvanov [the Bulgarian president] aside and asked him if he could dispense with the officers. So he asked them to wait outside. The fear of communist police is inside me—I can't get rid of it.

Ivo cannot help but notice the irony of receiving all these government accolades after years of being harassed followed by years of being ignored. He emphatically stated: "Only in 2005 did I start playing for large audiences again in Bulgaria. At one of these concerts, I told them bitterly,

'Now? Now you give me these honors? Now—when I'm getting old? Why not in my younger years when I was at the top of my fame?'" Similarly, Yuri Yunakov answered a Chicago reporter's question: "How do we feel about the press attention? Where was the press in the 1980s and 1990s? Not one Bulgarian paper wrote about us even though we were household names. Where was the press then?"

Recently, there are indications that wedding music is making a significant comeback; it is attracting larger audiences in Bulgaria and it is being marketed as a nationalistic genre. Payner has signed several wedding bands such as Kanarite and Orfei, and in March 2007 it launched a new twenty-four-hour television cable channel, Planeta Folk; cable channels Folklore TV (2006) and Tyankov TV (2007) also feature wedding music. According to Payner, Planeta Folk features "traditional and modern folklore, films about notable events in Bulgaria and historical and cultural achievements." It is aimed at "Bulgarian viewers at home and in Europe . . . who love Bulgaria and want to learn more about their natal culture and traditions" (http://planetafolk.tv). For example, to coincide with the holiday St. George's Day, on May 6, 2007, the channel sponsored an inaugural concert in London featuring Kanarite and Ivana (the combination I analyzed earlier), and a week later it sponsored a gala concert in Sofia with Kanarite, Vievska Grupa, and Orfei, as well as folk dance ensembles. Payner now regularly sponsors concerts at home and in the Bulgarian diaspora featuring a combination of chalga and wedding performers.

A typical day on the Planeta Folk cable channel includes not only Payner-sponsored wedding bands but also programs on the history and folklore of various regions and a bit of Eastern Orthodox liturgical music. Songs predominate in wedding music clips while instrumental improvisations are rather short and tame; again, it is locating wedding music in the realm of folk, rather than chalga or Romani music. No instrumental kyucheks are played, but songs in kyuchek rhythms are performed, e.g., chalga star Poli Paskova's Bulgarian language song "Moiite Pesni" (My songs). Unlike in chalga videos, Poli is dressed demurely and does not dance; rather, a folk dance ensemble wearing stylized costumes does line choreographies to kyuchek in an outdoor village setting. This staging plus the text (which extols how her songs express wholesome emotions) distance it from Roma and chalga.

The 7/8 Macedonian/Pirin rhythm is very common on Planeta Folk; this rhythm has a nostalgic symbolism for many Bulgarians, referencing a cross-border sentimental remembrance of family and folklore. Virtually all wedding music is depicted with a folk dance ensemble in costume performing choreographed dances, often staged outdoors in a village. These visual cues squarely define wedding music as rural and folk. Recently Payner has recruited many of its chalga stars into wedding band performances; for example, Nelina issued an entire *folkloren album* (folklore album) in 2008, including several of Nedyalka Keranova's signature songs. In wedding music clips chalga singers wear revealing but not overtly sexual clothing and sway to the music rather than dancing in sexually explicit ways, as they do in chalga videos. This illustrates the trend of assimilating the allure of chalga into a wholesome folk image of wedding music, and simultaneously accomplishes the ideological work of nationalism.

The creation of Planeta Folk by Payner, a company that had previously promoted chalga almost exclusively, is a clear sign that wedding music audiences are growing. The Bulgarian public is starting to become fatigued by the superficial glitz and the artificial formulas of chalga. Simultaneously, wedding music is becoming an ideological symbol of patriotism in a period where the definition of Bulgarian identity seems precarious. Chalga is criticized as too Romani, too eastern, but simultaneously too western, too much like Europop. Ironically, wedding music received the very same criticism in the socialist period, but now it is hailed as quintessential folk music. Nationalist parties such as Attack rail against chalga as corrupting the historical core values of Bulgaria; they encourage patriotic Bulgarians to support folk music, and for Payner, folk music means wedding music. Thus the popularity of wedding music today, just as in socialist times, is informed by a highly politicized environment where the meaning of Bulgarian identity is debated. The genre remains vital but must be seen in relationship to competing genres such as chalga and to developments in Bulgaria regarding Romani music, such as festivals and contests (discussed below).

Stara Zagora Romfest

Since 1993 the *Natsionalen Festival Za Romska Muzika, Pesni i Tantsi* (National Festival of Romani Music, Songs, and Dances, known as Romfest) has been held almost annually in Stara Zagora, Thrace, with growing crowds and growing media attention. Awards are given in several categories, and many musicians start their careers as a result of exposure at this festival (Peycheva and Dimov 2005). Playing a strong role in Romani identity politics, the Romfest can be compared with other European Romani festivals that are run by Roma, such as the Khamoro festival in the Czech Republic (www.khamoro.cz), the Amala Festival in Ukraine (Helbig 2007, 2008), and Šutkafest in Macedonia (see below); it can also be distinguished from the more numerous festivals and tours that are run by non-Roma, such as the American Gypsy Caravan tour and the New York Gypsy Festival (see Chapter 12).[12]

According to festival director Aleksandŭr Karcholov, the motivation was "to preserve the authentic and develop the music written [composed] by different Gypsy authors [composers], which has to represent the image of Bulgaria and with this music to reach Europe" (N. Georgieva 2006:13). The festival program asserts that the aims are "to retrieve, get acquainted with, proliferate and enrich Bulgarian Romani music and song; to elevate [it] to a higher level. . . .; to turn it into a 'bridge' for reviving the self-confidence of the Romani community. . . ." (Georgieva 2006:14; Peycheva and Dimov 2005:18). Note that these aims echo the rhetoric of the socialist period, namely themes of authenticity and the elevation of art. Ironically, during socialism the state used this ideology to exclude Romani music from the rubric authentic, but now Romani leaders are using the same ideology to shape Romani music in official settings. These goals also resonate with the emerging nationalist ideology of Roma to declare themselves a distinct and legitimate nation (see Chapter 3).

Anzhelo Malikov, who was a composer, arranger, cimbalom player, and graduate of the Sofia Academy, regularly served as president of the jury for the festival and was one of the founders. Before his death in 2009, he strongly defended purification of Romani music: "Foreign elements should be cleansed from Gypsy music, including the texts. Let's create one style called Bulgarian Romani music" (Peycheva and Dimov 2005:18). Aside from the problem of what his uniform style would sound like, Malikov's statement comes dangerously close to advocating the same sort of official control that the socialist state imposed on Muslims.

What Malikov and the festival directorship mean by purification is removing Turkish and Bulgarian elements from Romani music. Songs in Turkish and Bulgarian, for example, are prohibited; only songs in Romani can compete, despite the fact that many Bulgarian Roma speak only Turkish or Bulgarian. Purification of dance is also attempted. M. Angelov states: "it is not correct to say Romani dance is kyuchek, because it is a Turkism, imported into the country. Kyuchek is a type of Turkish dance. In Romani dance there is more lyricism . . . and temperament" (Peycheva 1999a:248).[13] When I interviewed him in 1994, Malikov emphatically explained to me that Ivo Papazov plays not Gypsy music but Turkish music. Sometimes Malikov uses Thracian as a euphemism for Turkish, as in this statement: "Take Ivo Papazov, he plays, let's say Gypsy or Turkish music but it all is in Thracian style. . . . But it should not be that way, however, it is loved by the wider public. Wherever you go in the rural areas Gypsy music is played in a Thracian manner" (Peycheva 1994a:17).

These leaders are trying to privilege a supposedly "unique" Romani music style with no neighboring influences. Malikov thinks that only in the Balkans does pure Gypsy music exist; everywhere else Roma merely play regional musics: "I believe in Bulgaria, Yugoslavia, and Greece, Gypsy music is the best preserved, as it was. No matter that in the west people think that Gypsy music is Hungarian or Russian. That music is Russian and Hungarian played in a Gypsy manner. The real Gypsy music is in the Balkans. There is no influence" from Bulgarian folk music (17). Here he falls prey to ethnocentrism, thinking his brand of Romani music is more pure than other brands. Festival leaders are vague as to what constitutes pure Romani style, and when asked how they judge pure Romani music (e.g., for prizes) they employ general terms such as lyricism and beauty (N. Georgieva 2006).[14] They seem to devalue the hybrid quality of Bulgarian Romani music. In spite of this official stance, audience members and Romani musicians in general value Turkish influences, Bulgarian influences, and the concepts of innovation and hybridity.

One important aim of the festival is to mainstream Romani culture "into Bulgarian national culture" (Peycheva and Dimov 2005:18). In speeches at the 2007 festival, for example, audience members were constantly reminded that Romani culture is part of Bulgarian culture. The 2007 festival also included a performance of a fourteen-year-old Romani boy singing a Bulgarian folk song with accordion accompaniment; he was introduced with the statement that Roma such as Boris Karlov and Ibro Lolov have expertly played Bulgarian music. This performance may have made a political point, but the audience was not appreciative.

Integration of Bulgarian music may be an admirable goal if it leads to acceptance of Roma, but unfortunately the festival is sometimes used by Bulgarian politicians for their own agendas. Indeed, politicians have regularly appeared at the festival to offer approval and recruit votes. One such agenda was Bulgarian accession to the European Union in 2007. An important criterion of membership in the EU was treatment of Roma, who were identified as a vulnerable group. The EU closely watched not only Romani unemployment statistics but also the visibility of Romani culture. In spite of the fact that Bulgarian government has only irregularly supported Romfest (through the Ministry of Culture and the National Advisor for Ethnic and Demographic Questions), it has become an ideal site for politicians to publicly affirm their commitment to Roma. In 2004, for example, Bulgarian President Georgi Pŭrvanov attended the festival and delivered this message: "Now, when Bulgaria is more intensively tying itself to Europe, we can proudly show that one of our strong points is peace and understanding between ethnic groups. Especially important here is the role of art. It doesn't know borders and restrictions" (Peycheva and Dimov 2005:19). Aside from the obvious romanticization of art, what is also glossed over is the real tension between ethnic groups in Bulgaria. Unfortunately, photo opportunities and speeches do not readily translate into tangible help for Roma. As I will discuss later in reference to pop music contests, that the state recognizes Romani art does not automatically mean progress in human rights; the state often recognizes a few talented Romani artists as tokens while ignoring the rest.

In addition to Bulgarian politicians, Romani activists also use the festival for educational purposes in the service of building Romani nationalism (N. Georgieva 2006). I have already mentioned promotion of the Romani language. For a few years, musical groups from India were invited, enacting the "homeland" idea. Although audience members appeared uninterested in the Indian groups, they serve as a visual and aural symbol of legitimate origins (2006:26–27). Other international Romani groups have been invited, and in 2005 the festival declared itself to be officially international. The act of gathering Romani musicians from disparate places gives legitimacy to the idea of a diasporic nation, with the message: we are a real people and we exist in many states (see Chapter 3).

Related to the nationalistic theme of the festival organizers is rejection of commercialism. According to Malikov, "As long as I have the strength, I will try to preserve pure Gypsy texts and music. But now many people are interested in Gypsy music for commercial reasons. It is very sad because they only want to profit" (Peycheva 1994a:16). Malikov implies that the profit motive of fusions taints them. On the other hand, virtually all Bulgarian Romani music is and historically has been in the commercial realm. However, Malikov is correct that not all commercial relations are equivalent. Privatization has indeed led to appropriation of Romani music into the pop/folk realm, with little financial benefit to most Roma. The few Romani chalga stars who have benefited financially, such as Azis (who got his start at the 1999 Romfest; see Chapter 9), do not compete in the official part of the festival. In 2004 he performed as "a guest" in a gala concert that was better attended than the official competition. Several Turkish groups that also performed "as guests" were very well received (Peycheva and Dimov 2005:20). And in 2007 the band Kristali, the Romani singer Sasho Roman, and the keyboardist Amza from Macedonia all performed as guests, to huge acclaim. In more recent years, other stars, such as clarinetist Sali Okka, have performed as guests.

The festival hopes to capture the attention of the Bulgarian state in order to interest it in cultural projects, and in some ways it is succeeding. On the other hand, state funding has declined, prize amounts have recently decreased, and Lozanka Peycheva and Ventsislav Dimov (Bulgarian ethnomusicologists) have not taken their honoraria. In 2010 Romfest was canceled because of insufficient funds. According to Malikov, the problem "boils down to the fact that the state does not pay attention to Gypsy culture. . . . It is necessary to have very strong state interference for the preservation and development of Romani culture" (N. Georgieva 2006:20). Malikov favored creation of state-supported Romani theaters and dance ensembles; this may be a nostalgic view of the socialist welfare state as paternalistic provider. Yet he is correct in that a crop of dance ensembles composed of Romani youth has emerged since 2005, and these ensembles are fostering pride (see Chapter 6). For example, in Sofia the Elit Center for Romani Culture has successfully sponsored many programs.[15]

However, I do not foresee the state fully embracing Romani music in its official categorization of folk music. Yes, Romani music may be embraced at Romani festivals for the audience of the EU, but little has changed in the realm of official folklore. Moreover, Romfest illustrates the problematic relationship of Roma to the state. As Imre suggests, Roma occupy a delicate position where they are suspect both because they can never be true representatives of the nation and because they are too closely allied to forces of commercialism and consumption (2006). For example, in 2008, a controversy erupted about the party atmosphere at the festival. Eran Livni reported that although the mayor of Stara Zagora opened the festival with a laudatory speech and the Open Society Institute funded a prefestival conference, the national government reduced funding. The head of the National Committee on Ethnic and Demographic Affairs announced she would not support a celebration of "kyuchetsi and kebap" (minced meat balls), symbolic of "boorish Gypsy music." Some Romani activists agreed with this sentiment, agreeing that it is shameful that the only public event in which Roma participate involves "boorish" music, eating kebap, and dancing kyuchetsi. Livni perceptively noted that the "stigmatic image of Roma as self-indulgent creators and consumers of 'boorish' music is so powerful (among both Roma and non-Roma) that any event that advances Roma recognition through music ends up marginalizing the music, the performers, as well as the audience" (Livni, personal communication; also see Livni 2011).

Official Postsocialist Bulgarian Views of Romani Music

Given the fundamental questioning of the past that has occurred in Bulgaria, the postsocialist period might be expected to reveal a grand shift in state folklore policy. Quite the opposite; little has changed.[16] Neither Romani music nor the music of other minorities is integrated into the curricula of folk music schools at the high school and college level.[17] No zurla or tŭpan is taught, no kyucheks are included, and few Romani children study at folk music schools. In fact no "modern" instruments (clarinet, saxophone, accordion) are taught.

Despite its current association of wedding music with folk music, it is often ignored in the folk music high schools and the Plovdiv Academy.[18] Only a few wedding musicians (including ethnic Bulgarians singer Ivan Handzhiev, gaida player Maria Stoyanova, and a few kaval players) have taught at the folk music high schools or the Plovdiv Academy. In the 1990s, Stoyanova invited Romani kaval player Matyo Dobrev to be a guest teacher. Students reported that he seemed nervous and out of place, and he was never invited back. Similarly, at folk festivals such as Koprivshtitsa, Pirin Pee, and Rozhen, wedding music and Romani music are not found in the official program. As during socialism, Roma do participate as individuals providing music for Bulgarian dance groups; zurna and tŭpan players perform for Pirin village dancers and gaida players perform for village dance groups. However, no Romani groups perform specifically Romani music.[19] Note that in the unofficial sphere of festivals, Roma are very visible. Just as during socialism, Roma zurna and tŭpan players and wedding musicians show up at folk festivals to play offstage for tips from Bulgarians of various ethnicities. The repertoire in these unofficial contexts includes wedding music, songs in the Romani language, and kyucheks, all of which fall outside the categories of official folk music.

Ditchev calls the Bulgarian situation "monoculturalism as prevailing culture." He points out that although there are numerous ethnographic villages devoted to Bulgarian folklore, there are none devoted to Romani, Pomak, and Turkish culture: "When travelling around the country, one discovers that what is thought and presented as folklore is without exception ethnically Bulgarian" (2004). Multicultural support comes only from NGOs, labeled "project culture" by Ditchev; "Any time you hear that a minority culture is being supported here or there, it means that there is project money behind it, with the backing of a Western donor. Take the initiative to write an all-Balkan history textbook, take trans-border cooperation, take the deliberate enrolling of Bulgarian and Roma kids together in school" (2004).[20]

Although the plight of Roma has received much attention from international organizations, in the realm of culture Roma receive little support from Bulgarian NGOs and even less support from the Bulgarian government. Most NGO aid is funneled to projects to build civil society, train leaders, and assist in social welfare. I am not criticizing these initiatives, but I note that music is assumed to be an area where Roma excel and thus need no help. For example, in the 1990s the Open Society Institute gave large grants that supported Romani "high culture," which they defined as "history, art, oral or written literature, cultural anthropology, and musicology." Specifically excluded were proposals for "pop music or folkloric music festivals." Recently, however, the Open Society Institute (now Open Society Foundation) began offering grants for East European Romani arts and culture, including CD production and festivals (see http://www.soros.org/initiatives/arts).

Since 1989 there have been several efforts by Bulgarian NGOs to introduce Romani culture to school-age children. The Interethnic Initiative for Human Rights Foundation (with money from the European Union) funded publication of several supplemental textbooks on Romani music, history, and folktales. The Romani history book for grades nine through eleven, for example, is an introduction to basic folklore genres. The song and folktale books feature collections illustrated with colorful designs and photographs, and the music book for grades five through eight presents contextual, regional, and historical material (Peycheva, Dimov, and Krŭsteva 1996 and 1997). All of these books were prepared by scholars and funded by NGOs. In addition, a few music textbooks issued for grades one to nine feature ethnic groups other than Bulgarians.

In 2005, however, a scandal erupted regarding inclusion of Romani music in elementary school curricula. A team of ethnomusicologists led by Gencho Gaytandzhiev prepared an educational music text, *Sharena Muzika* (Colorful music; Bulgarian), for preschool children in the town of Stara Zagora that included one children's song (among thirty-six songs) sung half in Romani and half in Bulgarian by Romani vocalist Sofi Marinova (discussed shortly; also see Chapter 9) and three photographs of Romani performers. The anti-Roma public outcry against this textbook was ferocious. Media headlines included "People . . . revolt against Roma textbooks; parents . . . sent a petition to the Ministry of Education," "What are our children exposed to? A wave of protest" and "Who must be integrated—us or them?" In addition, mothers from Stara Zagora were invited to appear on a prominent national television show where they declared "that they would never let their children listen to even a single Roma song" (Gaytandzhiev 2008:206). It is clear that racism surfaces in debates about music.

On the other hand, a successful project focusing on Romani folklore was initiated by the NGO Center for Interethnic Dialogue and Tolerance Amalipe (Friendship), based in Veliko Tŭrnovo, jointly funded by the Open Society Institute and the Bulgarian Ministry of Education. Begun in 2002 in thirteen schools with a group of 500 students, by 2007 the program attracted more than 5,500 children (53 percent of whom are Romani) in 230 schools, who were enrolled in classes on Romani folklore taught by Bulgarian teachers who took a training course and received an honorarium (http://amalipe.com).[21] In 2003 the "Open Heart" children's festival was organized by Amalipe for the children enrolled in the school culture program; and it has been held annually in Veliko Tŭrnovo and includes stories, music, and dance. In 2007 it received patronage from the city of Veliko Tŭrnovo, but no national funding. Ironically, this was the same year that Bulgaria became the leading nation in the Decade of Roma Inclusion. By 2009 the festival included 1,000 children and received some EU funding (http://amalipe.com/index.php?nav=news&id=332&lang=2); in 2010 it attracted 700 children and received funding from the America for Bulgaria Foundation.

Šutkafest

In postsocialist Macedonia, there are mixed signs of the legacy of Tito. Macedonia still prides itself on its multicultural fabric; Roma are now a "nationality" (Petrovski 2009). The Macedonian music school curriculum tends to omit Romani music, as in Bulgaria; however, there is a great deal of visibility of Romani music in the marketplace, and Romani political events often feature music. The situation of Roma is becoming more public as Macedonia lines up for European accession. Against this background, the first Macedonian Romani music festival that took place in Skopje in 1993 held great cultural and political importance. Šutkafest (named after the municipality of Šuto Orizari, outside Skopje) was sponsored by the NGO Union for Romani Culture, Macedonian National Radio and Television, Aura (a tourist agency), and Esma Redžepova and Stevo Teodosievski (see Chapter 10). It was a large event with several competitions and a jury composed of Macedonian performers and composers plus Romani clarinetist Medo Čun (see Chapter 2). Video example 3.2 features Esma Redžepova at the gala concert singing "Dželem Dželem," known as the Romani anthem (see Chapter 3 for analysis and lyrics); this song expresses the historical suffering of Roma (especially during the Holocaust) and their desire to rise up and unite.

Video example 6.7 features the festival's opening music and dance sequence; note that the festival concert featured a full classical orchestra (whose members used music stands and printed notation) with strings and woodwinds (including Ferus Mustafov on clarinet; see Chapter 2), plus a separate brass orchestra led by Stevo Teodosievski. The classical orchestra served as a symbol of elite "high art," which boldly legitimized

Romani culture. The dancers performed line and solo versions of the characteristic Romani dance čoček (the line dance pattern is *Bitolska gaida*) and were costumed with tasteful versions of šalvari. Note the omission of exposed skin as in belly dancing. The men wore identical red or white shirts, reminiscent of the Yugoslav KUDs that required folk costume. Several generations were represented.

Esma and Stevo delivered opening speeches that emphasized the theme of patriotism, lauding the fact that as far back as 1957 Esma sang publicly in the Romani language (see Chapter 10), and that the country was peaceful (this festival took place at the time of the Yugoslav wars). This rhetoric positioned Roma as true defenders of the country. Several young journalists spoke in Macedonian and Romani (symbolically equalizing the two languages in public space), and Faik Abdi, the leader of the Party for the Full Emancipation of Roma and the president of the festival organizing committee, greeted the European guests. The audience included numerous Macedonian politicians. Like the Bulgarian Romfest, Šutkafest featured Romani groups from other countries (Yugoslavia, Bulgaria, and Albania), legitimating the pan-Balkan public face of Roma.

Šutkafest continued successfully in 1994 and 1995 under the direction of Esma and Stevo, but then their sponsorship ended; funds eventually dried up and the last festival was held in 1999, when KUD Phralipe celebrated its fiftieth anniversary. Two commercial albums and a videotape were made of the 1993 festival. In addition to Šutkafest, other platforms and events in Macedonia have served as vehicles for Romani cultural visibility. They include the newspaper *Romano Sumnal*, several television stations, commemoration of International Romani April 8, beauty contests (see Chapter 6), and the very successful Golden Wheel Film Festival, held annually since 2002 for documentaries, fictional films, and radio programs. The latter is sponsored by the Romani television station BTR, which also sponsors the annual Miss Roma International contest. All these events serve the functions of both political and cultural mobilization and often feature Romani music.

Macedonia, World Heritage and UNESCO

The postsocialist state dramatically interacted with Romani music in the context of Macedonia's 2002 application to UNESCO to have a wedding from the village of Galičnik declared a "Masterpiece of Oral and Intangible Heritage of Humanity."[22] This UNESCO competition responds to the 1989 initiative titled Recommendations on the Safeguarding of Traditional Culture, which advocates "preserving cultural heritage which is in danger of disappearing due to cultural standardization, armed conflicts, tourism, industrialization, the rural exodus, migrations, and the degradation of the environment" (UNESCO 2001:3). Although I do not have the space here to interrogate all the problematic notions underpinning this UNESCO project and similar proclamations by the World Intellectual

Property Organization, I wish to point out that heritage is assumed to be coterminous with bounded territorial groups, so-called communities, and rural culture (see Chapter 7). Cultural heritage comes from "living communities with a sense of continuity" (5). These agencies have resurrected narratives of the impending loss or survival of selected items of authentic folklore (M. Brown 2004; Kurin 2004; Kirshenblatt-Gimblett 2004) that have rejected "unwanted hybridization" and "alien cultural forms."

Note that only nation/states can submit applications in the competition for Masterpieces. Thus the "humanity" designation elides into the nation/state, which chooses selected aspects of its culture to be masterpieces. Needless to say, minority cultures can be problematic. On the other hand, UNESCO specifically advocates "the preservation of cultural diversity" and "the tolerance and harmonious interaction between cultures," so one might expect cultural communication between ethnic groups to surface in applications. Not so for Macedonia's application for the Galičnik Wedding. Although the entire wedding is too complicated to describe here, at the turn of the twentieth century up to fifty weddings took place simultaneously on Petrovden, July 12, among the families of returning migrants. What is relevant here is the fact that whereas Galičnik is an exclusively ethnic Macedonian village with no Roma, all the musicians who provide music for the weeklong ritual are Muslim Roma from the nearest city, Debar.[23] These zurla and tapan players (all from the Majovci clan based in Debar) know the Slavic wedding rituals and dance repertoire intimately, and they signal every important ritual moment with appropriate music. There is even a proverb that says "no wedding will take place in Galičnik unless the Majovtsi family plays" (Kličkova and Georgieva 1951/1996). Thus not only are Roma integrated into the Galičnik wedding, but the villagers are dependent on them for their ritual, dance, and processional music.

Despite these facts, the UNESCO application from Macedonia hardly mentions Roma and omits them in relation to the goals of affirming cultural identity and preserving traditions. Roma are merely described in a few sentences as musicians.[24] The great potential in this project for recognizing and promoting cultural exchange between Roma and Macedonians is ignored.[25] Similarly, Roma are omitted from the section on training the next generation in folk practices. For example, one tangible way Roma could benefit is by UNESCO facilitating the learning of the ritual repertoire by young Romani zurla and tapan players, many of whom have few professional opportunities.

For the past few years the zurla and tapan players at the Galičnik wedding have not been from the Majovci clan but rather have been Roma from the capital city of Skopje who are employed by national dance ensembles. Furthermore, since the village of Galičnik was depopulated in the 1970s (for economic reasons), the ritual has been enacted in a two-day condensed version by summer returnees to the village and by members of the Skopje-based Kočo Racin dance ensemble in a specially built amphitheater. Thus the wedding is a revival staged by ensemble members.[26] The UNESCO application was submitted by the Union of Macedonian Folklore Ensembles, whose stated aim is "to preserve, protect, support and present Macedonian folklore which reflect [sic] . . . the heritage and traditions of the Macedonian people and the nationalities who live in the Republic of Macedonia." The submitted list of "custodians of the know how," however, omits Roma and is dominated by ensemble leaders and folklorists. The UNESCO application consists of florid language lauding the Galičnik wedding as "a masterpiece of human creative genius" embodying authentic folklore and national heritage. Referencing the organic tropes of romantic nationalism, the application implies that the wedding embodies the soul of the nation that finds expression in rural folklore. All this is quite paradoxical considering that the wedding is a re-creation. Ironically, Romani living traditions are excluded or minimized by the state, but the folklore of the majority ethnic Macedonians is coded as authentic even though it is staged. Finally, the UNESCO application needs to be seen in the context of the postsocialist economic crisis, with the possible motivation of increased tourism.

Popular Music Contests: Can Roma Represent the Nation?

In the last decade, several controversies have arisen around "music idol" contests in Eastern Europe over the role of Romani contestants; these controversies have exposed discriminatory tendencies that underlie the reluctance of states and majority citizens to accept Roma as representatives of the nation/state. Sofi Marinova, one of the few female Romani stars in mainstream Bulgarian pop/folk (see Chapter 9), was thrown into the middle of a huge scandal regarding the finals for the Eurovision (European popular music) contest in 2005 when Bulgaria participated for the first time. Singer Slavi Trifonov, the host of the most popular Bulgarian television show, produced a vocal duet, "Edinstveni" (The unique ones; Bulgarian), for Marinova and himself, which became a hit. It is no accident that Trifonov invited Marinova to collaborate with him. Trifonov was a seminal figure in the mainstreaming of Romani music in the 1990s; with his Ku-Ku (cuckoo) band he released an album of his show, *Roma TV,* that featured Romani music and comedy skits. Trifonov is also an intensely political figure who embraces biting political satire, often framed in music. For over a decade he has produced "The Slavi Show," broadcast on the BTV cable channel every evening Monday through Friday and watched by more than a million Bulgarians, combining live music, political comedy, and interviews with a variety of guests from Bulgaria and all over the world.[27]

Why did Slavi write a song for Sofi? Perhaps he wanted to highlight her talents (he called the song "a present for Sofi"), or perhaps he was motivated by politics. Remember that European Union conditions for Bulgarian accession required visible efforts at Romani visibility and integration. In any case, as soon as the song was nominated for Eurovision, immediately a virulent anti-Romani backlash was unleashed via the print media and the internet. Many Bulgarians were outraged that a Gypsy would be allowed to represent Bulgaria at Eurovision. Not referring to Sofi by name, critics called her "the dark girl" or "the Gypsy" and brought up the issue of the lower level of civilization of Roma.[28]

The song did not win in the audience voter call-in, and Slavi claimed that the votes were fixed.[29] In protest, the duo refused to sing in the final performance and instead Slavi read a speech denouncing the "fixed victory." He stated that it wasn't fair to any of the contestants, and especially to "the Gypsy woman Sofi Marinova" and to the country: "This is the selling of the country. . . . The country is not for sale, even for $50,000 leva. Bulgaria . . . is unique." During this speech, the atmosphere was very serious; Slavi wore a black suit and Sofi wore a black formal gown, and an instrumental version of the song played quietly behind them; she held his arm while he spoke.[30] Instead of singing, they broadcast a video of the national anthem, a symbol of their alignment with patriotism rather than corruption. Slavi sometimes reenacts this incident of betrayal in his concerts (almost like a memorial). Thus a song that could have been a national symbol of Romani integration turned into an example of the failure of multiculturalism.

Similar incidents with Roma in music contests have occurred in the Czech Republic, Hungary, Turkey, and Serbia. In 2004 in the Czech Republic, a Romani singer, Magda Balgova, was expected to win the national Česko Hleda Superstar (Pop Idol) vocal competition; when she was suddenly voted out in the final rounds, critics claimed it was due to racism. One newspaper wrote: "Did anyone . . . actually believe that the Superstar contest would be won by a girl who, without a shadow of a doubt, most deserved to win, but who is Romani?" Another Romani contestant was attacked in her local newspaper, which labeled her "the shame of the town" and claimed "people were surprised she reached the final forty because she was a Gypsy." In 2005, a male Romani singer fairly won the same contest, but activists said that this did not signal real acceptance of Roma in everyday life. The Romani organization Dženo pointed out that "a Romani man who can sing beautifully fails to challenge stereotypes [and] . . . actually reinforces the idea that Roma can do little else with success." This resonates with activist Ian Hancock's comment that success in music may actually harm the Romani rights movement by upholding stereotypes (see Chapter 12). At the same time that the singer won the Czech contest, a poll revealed that only 13 percent of Czechs consider Roma capable of being good neighbors. Dženo's web headline at the time read "Czech Superstar Can Sing But Not Move In" (Dženo 2005; also see Imre 2006:663).

A similar situation took place in Turkey in December 2006 regarding the Popstar Alaturka contest that is broadcast on national television. A Romani singer, Erkan, captured much of the audience vote, but one of the members of the jury seemed to be prejudiced against him.[31] In Serbia's first Pop Idol competition, in 2004, a similar pattern emerged. Romani singer Tanja Savić was defeated by two Serbian male singers despite her being by far the most talented artist. She told the Serbian media it was obvious that she didn't win the competition because she was Romani.[32]

On the other hand, there are several signs that Romani artists are becoming more acceptable in mainstream media. In 2007, a Romani singer was chosen to represent the Czech Republic in a project called the European Year of Equal Opportunities,[33] and in 2009 the Romani rap group Gipsy.cz was chosen to represent the Czech Republic at Eurovision. Most notably, Marija Šerifović, a Romani singer, won the 2007 Eurovision contest for Serbia. Similarly, in Hungary, the first two seasons of Megasztar (Hungarian Idol) in 2004 and 2005 featured Romani winners and runners-up. Romani singer Ibolya Olah, who won second place in 2004, was chosen to represent Hungary in her performance for the European Union Parliament. In 2005, many political leaders attended the Megasztar finals, when the Romani singer Caramel won, and three Romani finalists received Roma Civil Rights Foundation Awards for their outstanding service to the cause of Roma rights (Imre 2006:663).

As Roma begin to win a rightful place in pop music contests, it is questionable "whether the rise of Roma stars will elevate the status of the entire minority" (Imre 2008:333). Imre notes that "embracing selected Roma musicians has long been a strategy employed by the state . . . to handpick and isolate from their communities 'model' representatives of the minority, most of whom remain all the more excluded from the national community" (334). She reminds us that singers themselves are not anxious to be seen as Romani activists; they are "eager to shed the burden of representation" (333). For example, Hungarian Romani winners Olah and Caramel reveal nothing of their ethnicity; in the eyes of the global media world they are Hungarian. According to Imre, their images have been "whitewashed and nationalized' by their association with patriotic songs and stagings (334). Romani performers are, then, sometimes recruited for nationalistic projects of the state; Marija Šerifović has been involved in nationalistic projects; and the Romani festivals discussed earlier show this tendency. In Hungary, for example, Olah's performance of a patriotic song was used as a backdrop for fireworks on a national holiday. On the other hand, Olah has simultaneously been used "to exemplify the state's programmatic multicultural outreach and Europe's generosity towards minorities" (334).

Thus Romani stars can fulfill contradictory ideological discursive functions for the state: they can reinforce nationalism, or they can display the nation's commitment to diversity. But on the ground, they may do little to solve the problems of Roma. Imre argues that pop contests offer "rich case studies of the ambivalent relationship between Roma musicians and their nation states. . . . They provide the best illustration of the minefields that Roma entertainers have to negotiate, easily exploited as they are by both commercial media and state politicians for the economic and political capital they represent" (333). I further explore the relationship between Romani artists and anti-Romani sentiment in the next chapter.

These examples, including Bulgarian wedding music, Romani festivals, pop music contests, and the Galičnik wedding, all illustrate the dilemmas of Roma in the postsocialist period. Balkan Roma are squeezed between a weakening state and an expanding exploitative market. The state is not tangibly interested in Romani culture or the well-being of Roma except insofar as the state might reap certain rewards such as European Union membership, acknowledgment of supposed multicultural goals, or the stamp of UNESCO approval. Roma are disadvantaged in the realm of the free market in that they are poor, are despised, and lack start-up resources and connections to officials in high places. Commercial interests appropriate their music and their images, repackage them, and reap financial gain. In spite of this bleak picture, Roma are managing to survive as musicians; they are finding limited government support and some recognition for their talents at Romani festivals and in pop music contests, and outside of the Balkans in world music contexts, which I explore in the chapters ahead.

Bear trainer, Bulgaria, 1980

Monkey trainer, Bulgaria, 1980

Ferus Mustafov plays at a celebration for the birth of
Muamet Čun's granddaughter, Šutka, 1990

Esma Redžepova singing Dželem Dželem (Serbian text)
to American Kalderash Roma, private party sponsored by
Macedonian Roma, New York City, 1996

Women making stuffed grape leaves, Šutka wedding, 1994

Gifts on trays, Šutka wedding, 1990

Bride's female relatives dance at henna ceremony, Šutka, 1990

Groom's female relatives with a tray of henna with candle for the second henna ceremony, Šutka, 1990

Grooms' mother leads the dance line at double wedding,
New York, 1995

Solo dancer on top of a car, circumcision procession, Šutka, 1990

Solo dancer receives tips, Bulgaria, 1984

Tŭpan players, Pomak wedding, Avramovo, 1980

Mancho Kamburov, Pirin Pee, 1985

Yuri Yunakov (saxophone), Sunaj Saraçi (violin),
Severdžan Azirov (drums), Ilhan Rahmanovski (guitar),
Kujtim Ismaili (guitar), Trajče Džemaloski (keyboard),
wedding, New York, 1997

9

c/⊙ ⊙◯↩

Bulgarian Pop/Folk

Chalga

Chalga arose in the early 1990s as a fusion of pan-Balkan folk styles with pop Romani, Turkish, and wedding music; it has become a huge phenomenon in Bulgaria, with thousands of fans.[1] According to a sociological study, between 44 and 70 percent of the Bulgarian populace listen to chalga (Peev 2005:52), but the numbers may be higher because many upper-class educated Bulgarians conceal their affinity given its controversial status. On account of its low-class connotations and reputation as "uncivilized," the genre tends to be absent from national radio and television and the elite media (Ranova 2006:33). Nevertheless, chalga is widely disseminated via commercial CDs, radio programs (several twenty-four-hour), concerts, videos (via DVDs and a number of cable television stations, some twenty-four-hour; see www.fantv.bg and www.planeta.tv), and online (see www.chalgatube.com, www.planetaplay.com, and YouTube.com).

Early influences on chalga were Greek folk music (*laika*, e.g., use of bouzoukis), Turkish *arabesk* (use of Arabic melodic ornamentation, string orchestras, and instrumental fillers at the end of vocal phrases[2]), Pirin folk/pop (songs in 7/8, promulgated at the festivals Pirin Folk and Pirin Fest[3]), and, most important, *novokomponovana narodna muzika* (newly composed folk music). This last one is Serbian pop/folk that arose in the 1970s as an urban-based, "oriental identified," Romani-influenced genre and developed in the 1990s into *turbofolk* (Rasmussen 1995, 1996, 2002, and 2007). The word *chalga* comes from the Turkish *çalgı*, instrumental music or a musical instrument.[4] After 1989, in Bulgarian the word took on the designation of folk/pop or ethno-pop, vocal rather than instrumental, heavily influenced by Romani styles.

Kyuchek is the predominant rhythm, in varieties of 2/4 and 4/4, although there are some 7/8 and 9/8 pieces (see Figure 2.1); chalga also uses standard 2/4 pop rhythms. Kyuchek, shared by Turks and Roma, symbolically marks the genre as eastern or "oriental" (Said 1989; also see Chapter 1 of this volume). I argue that this easternness is often visualized as an oriental fantasy of sensuality, neither a real place nor a real ethnicity. As Kurkela pointed out, in chalga there are few specific references to Turkey or the Middle East in text or place (2007:156); rather, there are symbolic allusions in terms of rhythm, melody, texture, and imagery. As discussed earlier, Roma are coded as free, sexual, and musical; all three themes contribute to the "production of the oriental" (Kurkela 2007; Buchanan 2007).[5] For Bulgarians, kyuchek as a dance is a female Romani solo genre involving sensuous movements of the hips, shoulders, torso, and hands. In Chapter 6, I analyzed the diasporic manifestations of kyuchek, but here again I note the contrast between the demure style of dancing kyuchek at Romani family events with the sexualized, eroticized belly dance of chalga videos. Roma, however, sometimes participate in their own sterotypification, a point discussed throughout this book. Not only do stereotypes sell, but also music videos are manufactured by private music production companies, none of which are owned or operated by Roma; thus Roma are usually not participating in decisions about their representations. Romanian Roma, however, control their pop/folk industry.

Terminological Issues

Whereas most Bulgarians and scholars accept the designation pop/folk, I suggest we interrogate it more closely. What exactly is "pop" about chalga? Actually a great deal. Most texts have a pop or rock sensibility; they are about the dilemmas and emotions of modern life: sex, love, and money. Furthermore, chalga texts (which are usually in Bulgarian) rhyme, a characteristic present in Western pop music but absent from Bulgarian folk music. Much of the musical texture is pop, featuring synthesizers with a rock, techno, or rap texture. Dance moves and choreographies are also influenced by jazz dance and hip hop, and stagings reflect MTV aesthetics (when the mood of a song is sad, for example, the setting is often the beach). Finally chalga's use of dramatically overt emotion in the voice is characteristic of both pop music and Romani music, but not folk music or wedding music.

Conversely, we need to ask, What is "folk" about chalga? First we need to remember that *folk* in the Bulgarian language is not the same as *narodno*, which is defined as traditional or authentic by Bulgarian official institutions. In Chapter 7, I described how Romani music is excluded from the rubric narodno, and surely we can see why chalga is excluded. Very few chalga texts are in traditional style (traditional texts deal with village life and do not rhyme); traditional texts are sometimes featured in Pirin-based 7/8 chalga songs with nostalgic themes, but very rarely in songs with kyuchek rhythms. In addition, traditional instruments, such as gaida and kaval, are rarely used in chalga. So why is chalga called pop/folk?

To answer this question, we must consider the history of the term *folk* in Serbia, where, as in Bulgaria, it is distinguished from narodno. The term *pop/folk* in Bulgaria was inspired by the parallel term *turbofolk* in Serbia (which replaced the term *novokomponovana narodna muzika*). In Serbia folk was used to connote newness, as contrasted to the connotation of tradition for narodno; pairing turbo with folk insured its evocation of novelty, modernity, and ties to rock music. But still the term does retain ties to the local, something more homegrown than pop, which is obviously Western. Thus the hybrid terms *turbofolk* and *pop/folk* can escape the rigidity of tradition but keep a tie to the local while simultaneously locate themselves in a modern sensibility.

We may now ask, What elements of pop/folk are "folk" rather than "pop"? I posit that the kyuchek rhythm is the quintessential folk element. Remember, however, that kyuchek has always been excluded from the category narodno by official Bulgarian institutions; thus what is currently called folk/pop could be more accurately called Romani/pop. It is ironic that folk has become a gloss for Romani. Romani is now appropriated under the label *folk* in the world of commercial music, whereas it never was allowed to be narodno in the realm of state-sponsored official music. Chart 9.1 shows three intersecting circles: wedding music, Romani music, and chalga. They all intersect with the genre kyuchek, underlining its powerful influence.

Also note that chalga in Bulgaria has such a negative connotation that many performers and marketers avoid it; for example, music companies like Payner (www.payner.bg) and Ara/Diapason (www.ara-bg.com) and various websites use the marketing categories pop/folk, folk/pop, new folk, contemporary folk, but rarely chalga. In addition, singer Gloria said in a 2003 television interview that she sings pop/folk as opposed to chalga because she has a more Western approach and her lyrics are not gross. Similarly, Serbian singers Ceca Raznatović and Indira Radić and Macedonian singer Tatjana Lazarevska deny they sing turbo folk, preferring to emphasize that their music is "pop" or "European."

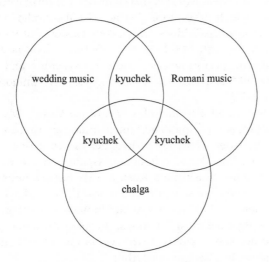

Chart 9.1. Intersecting Circles: Chalga, Wedding Music, Romani Music

Style, Text, and Imagery

The orient is evoked in chalga via symbolic Eastern instrumental styles plus Eastern references in texts, such as sheiks and harems (Dimov 2001; Buchanan 2007; Kurkela 1997; 2007); certain scales are sometimes used, for example phrygian (similar to makam kurd) and hicaz; and synthesized flutes and zurnas and arabesk-like instrumental fillers signal "easternness." Most important is the taksim or mane, the improvised free-rhythm solo, which is the hallmark of kyuchek. In videos a full range of eastern images are added: women belly dance wearing skimpy šalvari (with much exposed skin), sometimes in scenarios featuring palm readers, sultans, gongs, and horses.

In the 1998 video *Vsichko e Lyubov* (All is Love), featuring songs in the Bulgarian language by the pioneering chalga band Kristal (Crystal) with Toni Dacheva, "Chudesen Sŭn" (Wonderful dream[6]) displays an oriental fantasy. Female dancers wear veils and Arabic-style gold lamé and black belly dance outfits (with their buttocks exposed through netting). Women stroke a male sultan with an exposed chest; they fan him with a palm leaf and feed him (see video example 9.1 with text supplement). Although "Chudesen Sŭn" alludes to an eastern dream, other songs on the same video specifically document local Roma. "Svatba" (Wedding), in hicaz, includes footage of an actual Romani wedding where Roma modestly dance kyuchek. The text extols the music and lavish gifts at a Romani wedding. This ties the viewer to real Roma in the "ethnographic present" (video example 9.2 with text supplement).

In the song "Karavana Chayka" (the name of an entertainment venue) and other cuts on this album, the viewer sees the actual Romani musicians.[7] This video format, which Kurkela calls "concert documentation" (2007:151) is nonnarrative; for example, the performers are filmed on the beach at the Black Sea. Kurkela distinguishes the format from "music-based" videos, which have more abstract visuals unrelated to the text, and also from "narrative" videos, which depict a story (although the story may not follow the text; 148–152). Note that all videos as well as most concerts and club performances feature lip synching and instrument synching—it is all "playback." These video styles and techniques are drawn from western MTV.[8]

Kristal's 1998 video also includes "Zvezditse Moya" (My little star), depicting guitarists and Spanish/Flamenco dancers wearing black and red dresses with roses. There are no kyuchek moves, rather can-can type dances with swirling skirts, referencing Spain and the Romani diaspora. Another song, "Bashtinata Kŭshta" (My father's house), is a typical Pirin 7/8 melody whose text is about how the years have taken their toll. Visuals include old black-and-white footage of a village house and an elderly couple. Like most 7/8 songs, it evokes the nostalgic realm of folklore. Another song, "Dobro Utro" (Good morning), features Greek bouzouki playing in a 2/4 *syrtos* rhythm.

My final example from this album is a song with political commentary. Many chalga songs in the 1990s offered pointed critiques of social conditions, targeting local politicians, the mafia, and the banks (Kurkela 2007). "Sladka Rabota" (Sweet work) chronicles the ills of contemporary life; it incisively portrays how hard-working people are unemployed or poorly paid while swindlers have an easy life (audio example 9.1 with text supplement).

Thus on this one album we see how Kristal interpreted the major features of chalga of the 1990s (which is now known as "retro chalga"): kyuchek, 7/8, oriental imagery, Greek and Macedonian style, and texts about love, nostalgia, and politics. Note that Kristal (from Yambol and headed by keyboardist Krasimir Hristov) was one of the pioneering 1990s Romani bands that helped define the genre; in fact, the genre was briefly known as Kristal in the early 1990s (Buchanan 2007). Kristal remains one of the strongest bands today perhaps because it has become so adaptable Folk and is promoted by Payner. It currently records with different singers in many languages and styles, among them Turkish, Romani, and Bulgarian (including famous wedding singers).

Like pop music texts, chalga texts deal mostly with love and deception, but many lyrics are graphically sexual. As discussed earlier, texts and accompanying videos often depict the moneyed life filled with cigarettes, cell phones, fancy clothing, gambling, cars (especially Mercedes-Benz), sunglasses (Ray-Ban brand), alcohol, and sex. It is precisely these texts, coupled with erotic dance moves and skimpy clothing, that engendered a veritable backlash against chalga by both folk musicians and intellectuals (discussed later in this chapter). Some specifically Balkan themes, however, have emerged, such as emigration. Indeed, the "brain drain" of thousands of Bulgarians was of concern to many. In "Bŭlgarina v Evropa" (The Bulgarian in Europe; 2001), Mariana Kalcheva and Kristal satirically chronicle the journey of Bulgarians to Western Europe (audio example 9.2 with text supplement). Note that the instruments include gaida and kaval, referencing village life and provincialism as opposed to urbanity and cosmopolitanism.

Another text dealing with emigration is "Belgiiskite Vecheri" (Belgian evenings) sung by Amet. Aside from Azis (who is discussed extensively in this chapter), Amet is one of a handful of Romani singers to enter the chalga mainstream; most chalga stars are non-Romani women who capitalize on sex.[9] Male singers, unlike females, are not required to look sexy, or to dance. In his videos, Amet simply stands, sways, or sits, often wearing his signature hat, gazing at the undulating women surrounding him. Amet is also one of the few mainstream chalga stars to sing in many languages: Bulgarian, Romani, Greek, Serbian, and Turkish. Dual- or multiple-language songs are more often performed by those who are multilingual, that is, Roma and Turks.[10] "Belgiiskite Vecheri" is about the lure of Western material goods and gambling (see video example 9.3 with text supplement). Note that the text uses four languages (English, Bulgarian, French, and Romani) and satirically captures the immigrant's meager existence amidst plenty.

Post-2000 Trends in Chalga

Drawing from western videos by Madonna, Shakira, and Britney Spears, most chalga videos feature the partially unclothed female body as an object of male desire. As a Bulgarian musician in Chicago said to me, "The chalga crowd listens with their eyes." Sex was prohibited from the official media during socialism, so it is easy to comprehend why it exploded after 1989. This trend can be linked to the sexualization of the female body, the rise in pornography, and growing prostitution. After more than forty years of images of women as socialist neutered peasant workers, today many women have embraced femininity in its most commercial form: beauty products, cosmetic surgery, and chalga videos (Ranova 2006; Daskalova 2000:348–350).[11]

Chalga always had an erotic thread, but by 2005 the female star had eclipsed the musicians and all other elements. Now the female voice—and even more, the star's image—reigns. Stars are known by their first names (as with Madonna), wear designer wardrobes, have bodyguards, pose for pin-ups and men's magazines, and endorse products such as beer and telephones. Like movie stars, they have fan clubs and websites with interactive chat rooms.[12] Thousands of Bulgarians (mostly young) know their songs by heart and sing along at megaconcerts. Young girls who see no hope in the future cling to chalga not only as escape but also as a career goal. After seeing chalga stars interviewed on television (and seeing that some of them are not too well spoken), these girls think they too can make it. On the other hand, some stars are very intelligent, and a few are shrewd entrepreneurs, owning music clubs and hotels.

Note that although some chalga stars cannot sing well, others display a high level of technique. Tsvetelina, for example, was a wedding singer who switched to chalga for the income (Rice 2002), Ekstra Nina and Nelina are graduates of the folk music high school in Shiroka Lŭka, Slavka Kalcheva is a masterful wedding singer who also performs chalga, and Tsvetelina Yaneva was trained by her mother, the wedding singer Pepa Yaneva. Even when chalga began, Gloria, the oldest chalga star, sang wedding songs to display her mastery and link herself to tradition. In studio recordings, pitches are corrected electronically, and many singers never sing live. A few years ago, the trend of singing live developed. Some singers wear tiny monitors in their ears and can deactivate playback if they want to sing live, or activate it if they become breathless or tired. Sometimes they sing along with themselves. In the fan magazine *Nov Folk*, chalga star Maria boasted in 2006 that she performed live for the fifth birthday concert of the Payner television station. Ivana, on the other hand, is known to always sing live in concert.

The media hype about chalga stars is carefully orchestrated by production companies. In the last fifteen years, the Payner Company helped shape (and now dominates) the industry, causing Ivo Papazov to comment in 2005: "Payner is stronger than the government. They run the pop/folk empire." Since its debut in 2001, Payner's cable station *Planeta* has set the trends in chalga. Payner now records artists; produces and distributes CDs and DVDs; orchestrates promotions; sponsors tours, festivals, and contests; publishes calendars, pinups, and fan magazines; and runs a radio station, two cable television stations, a cosmetic surgery business, a party-planning service, an amusement park and hotel, and many music stores and chalga clubs. Chalga has also received journalistic attention in western travel magazines.[13] Stars regularly tour to other Balkan countries and to the West, especially where there are large Bulgarian émigré populations as in Spain and England; in the United States the tour circuit includes Chicago, Las Vegas, New York, and Seattle.[14]

I argue that since the early 2000s we can observe the development of "mainstream chalga," defined by the female sex star and orchestration of large media promotions by the companies. Events now tend to have the same formula: high production values and a bevy of female stars in skimpy outfits. In addition to the mainstream, other branches of chalga exist, among them Romani, Turkish, and wedding music collaborations; but they receive less media attention and sometimes the production values are inferior. For example, Sunrise Marinov's video *Nay-Dobri Kyuchetsi ot Mahalata* (The Best Kyucheks from the Neighborhood; Bulgarian) features excellent Romani musicians but amateurish stagings, similar to the 1998 Kristal video. Musicians are depicted playing in restaurants or on the street, and there is an absence of actors, choreography, and narrative. This is a far cry from the slick, polished videos of mainstream chalga.

Trends in mainstream chalga in the last five years include more sophisticated computer simulations, animation, complicated narrative stagings, better dancing, and a more pronounced pop aesthetic, specifically collaboration with DJs and the use of hip hop music, dance, and clothing styles. Political texts have dropped out almost entirely (emphasizing its entertainment function), though they are still found in other branches of chalga.[15] Recently more male singers have entered the scene, but women still predominate. Another trend is collaboration with pop/folk singers from other Balkan countries. For example, in the last few years, Emilia released a duet with the Greek Sakis Coucos, and Andreya and Maria recorded duets with Romanian manele singer Costi. Yanitsa also recorded a song with the Romanian manele singer Vali; note that it is staged on the Romanian-Bulgarian border, invoking Balkan connections.

In Chapter 8, I discussed how Bulgarian audiences have recently shown signs of fatigue with mainstream chalga and how cable channels have broadened their offerings. In 2007 Planeta (Payner's twenty-four-hour cable channel), for example, added to its previously all-chalga line up some wedding music, more Pirin and *starogradski pesni* (old city songs), Serbian music, and more western pop—in sum, more variety.[16] Similarly, the new cable channel K88 offers a large variety of genres, and *Balkanika*, a cable channel offering pop/folk in ten Balkan languages across eleven Balkan states, is gaining listeners. Thus variety is being introduced in the pan-Balkanization of pop/folk.

Payner's cable channel, Planeta Folk, which debuted in 2007, offers an interesting window to see how chalga and wedding music are moving closer to each other. In the last chapter I discussed how Payner now features some of its chalga stars singing wedding songs with bands and folk dance ensembles in village stagings. I described assimilation of the allure of chalga into a more wholesome folk image of folk and how this is tied to an emerging ideological strain of nationalism. It also reveals a conscious marketing strategy; Payner is not only legitimating chalga by allying it with wedding music but also putting glitz into folk and wedding music by using chalga stars. In the process, Roma tend to be erased as identifiably Romani.

The Romani elements are still visible and audible in mainstream chalga, but they have become part of a more stylized and abstract "orient" and absorbed into formulaic narratives enacted by larger casts of dancers and actors. For example, Emilia's 2005 song "Zabravi" (Forget! [Bulgarian]) features a text about failed love plus a bare-chested man striking a gong, with a haremlike group of women in sheer veils dancing synchronous steps that are closer to Hollywood or Bollywood than to belly dance. The dancers are then transformed into hip hop performers with a DJ, and the video concludes with the gong (video example 9.4 with text supplement).[17]

Kyuchek rhythms are still very common, but no longer are musicians depicted in mainstream chalga; there are fewer and shorter solo taksims and the synthesizer has taken over. Rather, as I have mentioned, the emphasis is on the star: the typical female chalga star is a non-Romani bombshell with fair skin and, often, blonde hair. Roma are now less visible in mainstream chalga. However, there are several Romani female singers: Boni, Yuliya Bikova, Ana-Maria, Toni Dacheva, and Sofi Marinova. In 2001 Payner's compilation hit mix CD included Turkish and Romani songs, but by 2004 the hit mix had only Bulgarian language songs sung by predominantly female stars. Romani and Turkish music is certainly released by Payner and other companies, but on separate CDs and videos, and often labeled "oriental rhythms." Mainstream chalga has become less ethnic precisely at a time when nationalism is on the rise.

As I mentioned in Chapters 1, and 8, the xenophobic Attack party achieved a stunning victory in 2005–06, when it captured 8 percent of parliamentary votes. Its anti-Romani platform (i.e., Roma are dangerous) and its anti-Turkish, anti-Muslim platform (Turks are fanatics) embrace cultural issues such as supporting the Eastern Orthodox Church and protesting construction of mosques (Ghodsee 2008); more recently, some of its tenets have been adopted by other parties. I don't think it is an accident that some artists, like the singers in the wedding band Kanarite (Chapter 8), started wearing large crosses at about the same time (crosses are also a symbol of wealth) and indentifying with nationalistic issues. For example, the last song in the cut titled "Ballads MegaMix" by DJ Jerry on Payner's 2004 DVD Collection 5 depicts the finale from a megaconcert (video example 9.5); after views of Bachkovo Monastery, we view the chalga stars on stage holding hands and performing a song about the Virgin

Mary (Bogoroditsa) with the refrain "Thank God we have such a clean and good land." This text is not so much about religion as about patriotism. In this concert, children are brought on stage, emphasizing the "family values" of chalga. There are several other recent chalga songs with patriotic texts.

These trends may have exclusionary consequences for Roma. By contrast, chalga star Maria released a song in 2007 in the Romani language, a news item so amazing it was announced in the fan magazine *Nov Folk* with the headline "Maria Sang in Romani."[18] Chalga singers look for new attention-getting motifs, and Romani language was a new frontier. Desislava sang a few lines of Romani in her duet with Azis, "Kazvash che me Obichash" (You tell me that you love me); however, many commentators criticized her for sullying herself by doing so. Maria's song was recorded with the Romani band Kristali (from the city of Montana, a different band from Kristal of Yambol) and is a remake/remix of "Telefoni" (Telephones), one of their hits. A question we may raise is, What happens when Maria collaborates with Kristali? Can Kristali now enter mainstream chalga because of Maria, or does she displace Roma? Indeed, Kristali's 2010 song "Ne Smenyai Kanala" (Don't change the channel) is a tribute to Payner's channel and lists its stars by name. Payner seems to have appropriated Kristali into its marketing program. The issue of appropriation is discussed in Chapter 13, but here I note that although Maria is clearly the star in Telefoni both of Kristali's singers still sing and get some screen time. On the other hand, a critical reading posits that once their music has been appropriated, Roma are not needed anymore. Chalga sometimes appropriates the exotic image and the oriental rhythm without Romani participation. In fact, with Roma being depicted as "dangerous" by some political parties, it is safer to take their music and exclude them. This presents special challenges for Romani chalga singers such as Sofi Marinova.

Sofi Marinova

Sofi Marinova is one of the most talented chalga vocalists, but being Romani she does not fall into the category of the standard bombshell, and thus her career has not been standard. She was born in 1975 in Sofia, speaks Romani, is self-taught, and has performed since she was seventeen years old in bars and at Romani weddings (Cartwright 2005c:42–43). Called the "Romska Perla" (Gypsy Pearl), she recorded for several years with the Romani band Super Ekspres, and in 1996 they won the grand prize at the Stara Zagora Romfest. Sofi's masterful technique can be heard in her Romani songs, where she executes exquisite descending runs and repeated mordents. It is also showcased in "Danyova Mama," a Bulgarian wedding song. Chalga singers usually do not attempt to sing slow wedding songs since they require such a high level of technical mastery (see Chapter 7). Audio example 9.3 with text supplement features the last verse of

"Danyova Mama," whose text depicts a mother speaking to her sons who have come home from the mountains, where they were *haidutsi* (guerrilla fighters).

Sofi adds a Romani sensibility to this standard wedding hit song: she includes more emotional phrasing, more dynamic contrasts (soft and loud), more exaggerated ornamentation (many repetitions), ascending slides, notes held for a long period, and glottals or breathy "ahs" at the ends of phrases. Note also how she ends the song by jumping up an octave (not typical of wedding songs). Some music journalists compared her emotional style to Esma Redžepova (see Chapter 10), but Marinova claimed she was more indebted to Džansever (see Chapter 2; Lozanova 2006).

In the past Sofi had many fans, but she never achieved the visibility of mainstream chalga, probably because of her ethnicity. In 2005, however, she became part of a huge public scandal involving television host Slavi Trifonov and the Eurovision pop music contest (see Chapter 8). As I described, she was the butt of cruel comments and racist remarks, and the polarization of audience members surfaced. A virulent anti-Romani campaign, focused on whether a Romani artist could properly represent Bulgaria in a Europe-wide contest. Sofi's song, which was derailed from Eurovision in 2005, has extremely subtle Romani elements. The rhythm of "Edinstveni" is the type of kyuchek typically used to invoke India (see Chapter 3 and number 4 in Figure 2.1); it is accented by *darabuka* (hand drum) and dajre, two typically "eastern" instruments. There are also strong pop elements in the song: the rhymed text is about love, the melody is in minor key, and the accompaniment consists primarily of strings, swelling in dynamics in the emotional parts. There is a guitar solo, neither flashy nor improvised; and a short emotional vocal solo by Sofi on the syllable "ah" that is not improvised and stays within pop style.[19]

After the Eurovision scandal, Sofi's career mushroomed; she has toured in the west and transformed her image form Middle Eastern playgirl to elegant star. On her 2004 album, titled *5 Oktava Lyubov* (5 Octaves of Love) referring to her five-octave range, she displays her versatility: two songs are in Serbian, two in Greek, and two in Turkish, in addition to several songs in her native Bulgarian and Romani languages. She explained: "I've been traveling to Turkey, Greece, and Serbia . . . so I wanted to learn some songs in their language. . . . In the recording studio I had musicians from those nations and tutors to make sure I got the phrasing" (Cartwright 2005c:42–43). On this album Sofi shows her mastery of multiple Balkan vocal styles as well as languages. For example, she sings "Ušest" (Serbian line dance) with regional ornamentation in a Serbian dialect of Romani. She also sings a duet with the popular Serbian Romani singer Zvonko Demirović. Finally, she uses talented guest musicians on this album rather than the bland synthesized arrangements of most chalga singers.

For the past five years, Marinova has collaborated with rap star Ustata on several songs, among them "Moy si Dyavole" (You are my devil), "Tochno Ti" (Exactly you), "Buryata v Sŭrtseto Mi" (The storm in my heart), "Bate Shefe" (Boss), and "Lyubov li Be" (Was it love?).[20] Some of these rap songs appeared on her 2006 album, *Ostani* (Stay) on Sunny Music, which also features "Vyatŭr" (Wind), a remake of pop diva Lili Ivanova's hit song resung in the Romani language with brass band participation. In an interview in *Nov Fok*, Sofi stated it was her manager's idea to rerecord Lili's song, and he secured permission from Lili. Indeed, Sofi used to be called the Romani Lili Ivanova, and in some of her songs she imitates Lili's style. Sofi explained that the wind metaphor in "Vyatŭr" is related to Roma via "the theme of travel, like in the hymn *Dželem Dželem*. . . . We live for our children . . . we want to make them big weddings" (Lozanova 2006:17). She also described her affinity for Indian films and claimed the timbre of her voice "resembles a bit an Indian voice" when she sings in Romani (Lozanova 2006:18). To the point, in her song "V Drug Svyat Zhiveya" (I live in another world) on the album *Studen Plamŭk* (Cold Flame) she uses a high-pitched Indian voice timbre and sings one verse in Hindi. The song also features a kyuchek rhythm associated with India (number 3 in Figure 2.1), synthesized drums (*dhol*), and string and flute fillers reminiscent of Indian film music.

Ostani also features a second duet with Slavi, "Lyubovta e Otrova" (Love is poison) that has an overtly political theme: war between Muslims and Eastern Orthodox people.[21] The black-and-white video opens with scenes of a snowy Balkan village, shots of a church and a mosque, and the text: "This is the Balkans. Over 300 wars have begun here. Here it is as if every wind brings sadness and the land smells of blood. But sometimes love is born from blood." The video depicts (somewhat abstractly) Muslim versus Slavic soldiers, with Slavi as a Slavic soldier and with Sofi, a conservative Muslim woman, as his love. She wears no makeup and is dressed in a long Muslim black coat and a white headscarf (covering her forehead and neck). The style of the song and its lyrics are squarely pop;[22] the visuals, however, clearly link the song to the Balkans. The text refers to ill-fated love between two people of warring religions, but the video ends on a hopeful note. Slavi is no longer in soldier's clothes, and the pair walk off holding hands.

I think it is interesting that although Sofi is not Muslim she is willing to portray a veiled Muslim in the clip. It is hard to picture another chalga star in this role; images of mainstream chalga stars are fixed in sexuality, whereas Marinova is a more adventurous and flexible artist. However, being a practical artist she does not avoid sexuality in her more mainstream videos. In 2008 Slavi wrote another song for Sofi, "Vinovni Sme" (We are guilty) that became an instant hit. He sang it with her on his show in a moving performance where they held hands; this was reminiscent of their defiant duet performing "Edinstveni" during the Eurovision scandal in 2005. Illustrating their commitment to social justice, Sofi and Ustata released "Lyubov li Be" (Was it love) in 2010, including a bold message against (and statistics about) sex trafficking.

The fact that Sofi clearly identifies as a Romani pop/folk singer was narrated by Nick Nasev, an astute fan:

Sofi performed at a concert in Gotse Delchev to a sold-out crowd of non-Romani elite businessmen and middle-class Roma; poor Roma who couldn't afford the tickets gathered outside and climbed on an adjacent building to see and hear her; they knew every word of her songs. When Sofi realized this, she went outside to sing directly to them. She said to them (in Romani): "This is where I was as a little girl." The management was annoyed but they knew the huge Romani crowd was a tribute to her, so they gave out free Cokes to the Roma.

In the last five years, Sofi has reemphasized her personal brand of pop/folk with many songs in Romani. In 2005 she released "Vasilica" (St. Basil's day), which describes the Romani customs of this winter holiday, and "Ah Lele" (Oh dear), a Bulgarian-language remake of a talava-style song (see Chapter 2) by Muharem Ahmeti, a masterful Albanian Romani singer from Tetovo, Macedonia. On her 2008 album *Vreme Spri* (Time stops), she sings the Romani anthem "Dželem Dželem" (Chapter 3), "Mik Mik" (Wink, wink; a remake of a popular older Romani song), and "Bubamara" (Ladybug), taken from the Serbian Romani singer Šaban Bajramović. Her ties to Romani music have been further cemented through collaboration with the Serbian brass band of Boban Marković (on the CD *Devla*, God [Romani], Piranha 2009) and with Azis.

Azis

A notable exception to my earlier observation about tame male chalga singers is Azis, who has emerged in the last decade as a megastar. Indeed, Azis is an exception to many of the rules of chalga. A Romani male who is ambiguous sexually, transgendered, and transvestite, he breaks every Balkan gender code of behavior. In his videos, he dances erotic kyucheks; loves fancy gowns, makeup, feathers, sequins, and high-heeled boots; and has sex with men, women, himself, or several people at once or watches others engage in sexual acts. He can be supermacho or superfeminine, or both simultaneously. The public fascination with him draws on his transgressive behavior, which is tolerable and even expected because he is Romani; if he were a Bulgarian man he would be despised.

Azis is by far the most radical Romani performer in Bulgaria today. He even has an extensive write-up in Wikipedia: "Azis has caused some controversy in Bulgaria with his queer-like ways and his campaigning on behalf of the somewhat downtrodden Roma gypsy minority." He is one of the "most famous people in Bulgaria," according to Wikipedia. In 2005, he was a candidate for a parliamentary position in the Evroroma political party but did not win the election; he did, however, serve as honorary party president. Bulgarians either love him or hate him, and consequently he has amplified the debate about the crassness of chalga. In 2006 his production company, Sunny Records, published his autobiography (with pin-up photographs), basically a guide to his sex life, including genital cosmetic surgeries (Azis 2006). Music journalist Garth Cartwright, who did an extensive interview with him, wrote: "You don't count Azis's press cuttings, you weigh them. . . ."[23] His metamorphosis into the most controversial entertainer in Bulgarian history involved a demonic appearance-shift and videos so lurid, so hallucinated with desire, they leave efforts by The Prodigy and Marilyn Manson gathering MTV dust" (2005b:262–263).

A Kalderash Rom, Azis was born Vasil Boyanov in 1978 in Sliven and started singing at an early age. He said: "Although my father was a professional accordion player, he didn't like the idea of me being a musician. . . . I started singing in the church choir in Romani. . . . We . . . formed a family ensemble and . . . I would perform every night impersonating Michael Jackson" (266). His family is Pentecostal, and his first recordings were of Christian Romani songs with his family. His family lived in Germany and Spain, but he returned to Bulgaria to sing (270). He performed in a bar where he worked as a waiter, and sang at weddings, eventually winning the best singer award at the 1999 Stara Zagora Romfest. "I'm for the big stage with professional sound and lighting," he says. "The small party, the small business, it doesn't interest me. So in a way I'm not the typical Romani performer. My parents did not make a big deal about being Gypsy and while I don't hide my Gypsy heritage I wouldn't say I was very proud of it" (268).

Azis realized from a young age that he was an outsider in multiple ways (in terms of ethnicity, gender, and sexuality) and that he could either suffer from this situation or capitalize on it. When British journalist Michael Palin interviewed him in 2008, Azis said "Because of my Gypsy ancestry, everywhere doors closed on me. . . . My mother took me to film castings but no one chose me because of the color of my skin."[24] His autobiography begins by describing the 1996 Bulgarian National Television pop music contest for young talent, which he says he deserved to win. He wore blue contact lenses and a great deal of hair gel:

They stopped me in the middle [of my song]. They told me thank you. By their tone I realized that I lost. And I knew why. Because I am a Gypsy. I was ashamed of this. That's how they lost a male pop singer. But Azis was born. Even Gypsies hate me. . . . Because I am fair and blue-eyed. They believe that I look like that, like in my photographs. They don't know about the existence of Photoshop. I wear a lot of make-up. This scares people. And no matter how good I am, for those close to me I'll never be good [enough]" [Azis 2006:12–13].

As a child Azis played with dolls and dressed in his mother's clothes. As a teenager he cleaned offices, walked dogs for rich people, and performed as a transvestite (Cartwright 2005b:260, 266). When he was interviewed on Slavi Trifonov's TV show in 2005, he admitted that as a young man he couldn't make a living as a wedding singer, so he and his agent invented the persona Azis. On the show he refused to define his sexuality; part of his mystique comes from audiences guessing. In October 2006, he married a man at a huge wedding in a Sofia nightclub in front of an audience of chalga stars and 200 journalists. Still, he wouldn't pigeonhole his sexuality.[25] He explained, "If I marry a woman, they will say it's only for show— he is homosexual. If I marry a man, they'll say it's only for show, he's a man!" (*Nov Folk* 2007:15). His wedding was the first public homosexual union in Bulgaria, a country with very traditional values. He and his partner now are raising their daughter, Raya, whose biological mother is a friend of Azis's. Although he is not overtly political, Azis underscores that part of being in the European Union is being tolerant toward homosexuals. He is now embraced by international gay artists and will be documented in the forthcoming *Encyclopedia of Gay Folklife*, to be published by M. E. Sharpe.

Like Madonna and Lady Gaga, Azis capitalizes on shock value in his shows. At a concert with Desislava in 2005 (where she was carried out on a palanquin, like a goddess, and where she referred to him as "her"), Azis was in makeup, a blue sequined leotard, and a feathered skirt. On stage he reenacted an incident that happened in Sofia regarding a billboard with his image on it. The billboard was sexual in nature and was met with protestors who complained not only about the image but also about the location (it was in front of Vasil Levski's monument, a sacred spot for many Bulgarians, as Levski was a nineteenth-century fighter for freedom). However, Azis claims he is also a fighter for freedom. On stage, actors displayed the billboard that replaced the controversial one. The audience heard the sound of police cars arriving and sirens screeching, and then a nearly nude dancer burst through the billboard—with Azis's name painted on her nude back and buttocks. The music for this sequence was the Bulgarian national anthem. This scene was followed by one in which two women kiss and simulate sex on a bed in the center of the stage, breaking the taboo against lesbianism.

Azis has thus breached numerous social codes, and the list is growing. Not only does he refuse to be categorized by sex (male, female), gender (masculine, feminine), or sexuality (homosexual, heterosexual, bisexual), but his clothing and image shift constantly. In 2001 there was a hysterical reaction when he released "Hvani Me" (Catch me), where he is dressed in Bollywood drag, toys with a python, and licks milk from the chest of two almost naked black men (Cartwright 2005b:271). He has also included the theme of Christianity in several videos. In "Obicham Te" (I love you, a Bulgarian cover of Sotis Volanis's Greek song *Poso Mou Lipis*, How much I miss you), he is in a white billowing suit and scarf on a rocky cliff, caressing himself; the video ends with lightning coming out of his outstretched arms, perhaps a reference to Jesus on the cross.[26] "Hajde Pochvay Me" (Come on, let's begin) takes place in a Catholic church and has scenes referencing sex by the clergy, flogging, and crucifixion.

Musically, Azis is a versatile and talented vocalist. He was invited by the world-renowned classical flautist Kristian Koev to sing "Ave Maria" at a classical Christmas concert in 2006; Koev claimed that Azis has an angelic voice. In 2005, he sang a slow Bulgarian wedding song live on Slavi Trifonov's TV show, which is a mark of a masterful singer; Azis recorded the song "Ne Kazvai Ljube Leka Nosht" (Don't tell me good night, love) on his 2004 album *Kralyat* (The King).[27] Ironically, this same album includes a poster-sized pin-up that exaggerates the contrast between his brown skin and bleached white hair. The album also includes a Bollywood song, and its inside cover shows Azis masturbating; in another shot he wears red thigh-high platform boots and a pink fur hat. Thus the traditional, the erotic, the exotic, and the transgressive are all juxtaposed.

Azis took his name from a Turkish movie, and indeed he can be fruitfully compared to two transgressive Turkish (non-Romani) singers: Zeki Müren and Bülent Ersoy. Müren had multiple gendered personas, sometimes wore female clothing, and played with a Liberace-type flamboyant male style. Ersoy is a transsexual, that is, he declared himself a woman after surgery, but she keeps much about her personal life hidden.[28] Both artists were influenced by the female singer Müzeyyen Senaras (Stokes 2003:319). Azis has similarly drawn one of his primary personas from the hypersexuality of mainstream chalga singers; the trick is that he is a man enacting the hypersexualized female role. Azis can also be compared to male Romani singers in Kosovo in the 1980s whose sexuality was assumed to be homosexual; they were respected for their mastery of the talava genre (see Chapter 2 and Pettan 2003). In historical perspective, Azis evokes professional Ottoman dancers who staged elaborate pageants and assumed various sexualities (Chapter 6).

Indeed, the comic and the playful have an important place in Azis's style. In 2007 Azis's *Night Show*, a television talk and music show, premiered on a cable channel, to mixed reviews. The program features musicians, actors, and media stars who perform and are interviewed by Azis, plus comedy skits and parodies of the evening news. Most viewers agree he is less adept as an interviewer than he is in presenting multiple personas via clothing and performative modes.

Azis has found affinities in the Romani homeland, India. He told Cartwright that when he was a child "a friend . . . gave me a cassette of Indian music and . . . I listened to it day and night. . . . Whenever they showed Bollywood movies . . . hundreds of Gypsies would be waiting and when the movie started we would all begin to cry" (2005b:266). Embracing India, "Antigeroi" (Antihero; video example 9.6), is filmed in grainy black and white and depicts kaleidoscopes of Azis wearing animalistic claws and green, yellow, and orange body paint while dancing in front of a Hindu temple that has erotic sculptures.[29] This is obviously not a typical mainstream chalga video. The song is taken from the popular 1993 Bollywood movie *Khalnayak* (released in Bulgaria as *Antigeroi*). Azis sings the original song (in mangled Hindi), and then we hear the original female film singer. The lyrics caused a scandal in India because the Hindi refrain *Choli ke peeche kya hai* (asking a woman "What is behind your shirt?") could be interpreted sexually instead of inviting the more acceptable answer, "my heart."[30] The music, in a rag similar to hicaz, features synthesized Indian instruments; the song opens with *pungi*, an instrument with two single reeds, associated with snake charming, followed by *shahnai*, similar to a zurna, and dhol (drums). Finally there is a short Bulgarian women's unison vocal section with glottals and yelps, characteristic of folk music.

Azis's song "Nyama" (I won't) illustrates the principle that many chalga videos are more sexual than their texts. Chalga songs are played on the radio for general audiences, while videos are watched by a more specialized group of fans. Nyama is a 2002 Bulgarian cover of the Serbian song "Sama" (Alone), by turbofolk star Dragana Mirković, a cover of a song by the Greek band Zig Zag. It has a harmless poetic text about failed (heterosexual, monogamous) love (video example 9.7 with text supplement).

The video, unlike the text, is extremely explicit. It opens with the sound of a heartbeat at night on a sleazy street; Azis is in the middle of a male pick-up scene, young men with their shirts off looking for customers. A transvestite walks by and Azis baits her. A limousine pulls up, the window rolls down, and a beautiful woman motions to him. The song begins when the chauffeur lets him into the back of the car and Azis and the woman start having sex. She winks at the chauffeur and he joins them in the back of the car for a threesome. The scene shifts to the woman's apartment, where the threesome continues, but the woman withdraws to facilitate the men kissing (blindfolded) as she watches and drugs them. The scene shifts again and the threesome are in a bed with black satin sheets located outside on a busy street corner in Sofia known as "five crossings." The woman is in the middle, but again she retreats and the two men are left embracing, finally sleeping in the bed as she grabs her coat and walks away, leaving them a generous tip.

This video, and other videos of Azis, can be analyzed via Judith Butler's theories of performativity, which remind us that we are dealing with discursive formations (in word and image) of gender, not immutable biological realities.[31] In her discussion of drag, Butler admits that drag may rework, mime, and resignify heteronormative gender categories (to which they always refer) but underscores that "there is no necessary relation between drag and subversion"; she "calls into question whether parodying the dominant norms is enough to displace them" (1993:125). Some have argued that drag is related to misogyny;[32] some have claimed that it reinscribes gender norms, and others have claimed it destabilizes norms. Butler states that drag both "appropriates gender norms and subverts them"; "it remains caught in an irresolvable tension" (128). Azis's drag shows are staged precisely as performances; the viewer is always reminded of the conscious display. In addition, he juxtaposes his drag persona with quite a number of heterosexual personas, so we are never really sure of his "true identity." He does not perform "classic" drag but rather combines masculine and feminine (for example, his bleached white beard is often juxtaposed to his makeup). His point, and Butler's, however, is that there is no stable core identity, neither for transgenders nor for heterosexuals.

Returning to "Nyama," I note that the viewer is led through a number of gazes or points of view.[33] When the video opens we think we are watching gay men waiting for other gay men to hire them; we are led to believe we are in a "gay low-class world." But the arrival of the limousine with the woman destabilizes the category and the label under which we assign her and Azis's sexuality. The tableau becomes more complicated when the chauffer joins them. The viewer then begins to identify with the woman, and we become the voyeur watching the two men. "The female gaze" overtakes the male gaze, but it is not simply the gaze of a female heterosexual desiring one male; rather she blindfolds the men (literally blocking their gaze) and watches male-male relations. Note the class transgression here as well as the gender transgression, as the chauffeur and Azis are markedly working-class while the woman is rich. The closing image of her leaving money for the two men underlines the transactional quality of the sex depicted.

Azis portrays the fluidity of categories not only through his sexual encounters but also through his musical voicings and the shifts in audience point of view in his videos. In "Dnevnik I Praznik" (Weekday and holiday), for example, he is without makeup and hair dye; in fact, he is a rather ordinary looking photographer, who embodies the heteronormative male gaze. He photographs beautiful models, making them into sexual objects; but he also inverts this gaze in the same video by singing as the persona of one of the models. In "Kak Boli" (How it hurts), Azis is a man rejected by his lover, a sexy transgendered male-to-female singer; at the same time, however, he sings her part as well as his own. In "No Kazvam Ti, Stiga" (But I'm telling you, enough), he is first pictured as a businessman/intellectual at his desk with his gaze on male bodybuilders; the latter become construction workers as Azis is morphed into a veiled, sequined, crowned apparition at the construction site. The video ends back at the desk but with a whole world of ambiguity introduced.

In the 2008 clip "Nakarai Me" (Force me), Azis introduces (by name) three buff men (sometimes wearing women's accessories, one noticeably dark-skinned) and one practically nude female, and suggestively offers variable sexual combinations of the assembled five actors. Finally, in "Teb Obicham" (I love you, 2008), he appears in a dress, earrings, and makeup and presents shifting sexualities in a black-and-white pageant; he also sings a vocal mane on the syllable ah, reminiscent of Romani instrumental improvisations.

Finally, let us turn to the duet "Edin Zhivot Ne Stiga" (One life isn't enough) because it pairs the two most prominent Romani chalga stars, Azis and Sofi Marinova. The vocal style is typically Romani, with florid ornamentation and emotional cries, glottals, and gasps. In a dramatic moment, Azis even switches to falsetto briefly at the end. The text is a love poem in Bulgarian, but it switches to Romani for the last two lines (see video example 9.8 with text supplement). The video depicts a male patron (Azis) watching Azis and Sofi belly-dance on stage (with a reference to pole dancing). The client snorts cocaine. Sofi and Azis are both wearing makeup and are similarly dressed in belly dance outfits. Sofi shows her midriff and Azis wears a skirt over pants, but the pants are cut off exposing one buttock cheek. This video equalizes males and females as sex objects, and the patron seems equally interested in Azis and Sofi. For their part, they seem to vie for his attention; they hardly sing to each other but rather each sings to the patron. The males seem interested in one another just as much as they are in Sofi, if not more. Thus instead of the standard heterosexual triangle (two males fighting over one woman), the video suggests other permutations.

I suggest that Azis is playing with stereotypes; sometimes he even gets other heterosexual singers to break heteronormative rules with him. For example, DJ Ustata appears in a homoerotic sequence with Azis in "Tochno Sega" (Exactly now; see http://www.youtube.com/watch?v=2Infm8wIJRA). I argue that Azis also paved the way for other chalga stars to explore non-heteronormative permutations of sexuality. For example, in the 2008 video "Ne Se Sramuvam" (I am not ashamed) Malina suggests she is not ashamed of lesbianism; in the 2008 video "Gubya Kontrol Kogato" (I lose control when), Miro and his cast display lesbianism, homosexuality, and heterosexuality via bondage and Goth costumes.

Azis often adopts the standard oriental stereotype but overlays it with a gendered stereotype of the superfeminine. In many of his videos, however, *he* is the superfeminine, which exposes the stereotype as constructed (in Butler's terms, as "performative"). I am not suggesting that he is critical of eastern stereotypes; rather, he loves play acting; the oriental is a fantasy world for him. But the oriental is a different type of fantasy for a Romani transvestite man than for a mainstream chalga star. To phrase it differently, if he can be as oriental as any chalga star, he can also be as feminine; and if he can be as feminine, he has destabilized the categories. Despite his elaborate stagings, Azis is actually much more grounded in Romani music than most mainstream chalga stars; he frequently sings live, he consistently uses real guest musicians and gives them solos, and he even has instrumental kyucheks on his albums—quite unheard of on the albums of the female mainstream stars.

Chalga, Morality, and Ethnic Politics

We may now return to an analysis of the culture wars over chalga, which have polarized Bulgarian society. Critics, composed of the intelligentsia, nationalists, and some folk musicians, accuse chalga of being crass, low-class, pornographic, banal, and kitsch, and of using bad or formulaic music and too many eastern elements. The debates have become so virulent that proposals to ban chalga have been suggested, and hundreds of articles and books such as *Chalga: Za ili Protiv* (Chalga: Pro and Cons; Bulgarian; Kraev 1999) and *The Seven Sins of Chalga* (Statelova 2005) have appeared.[34]

Defenders of chalga come from all social groups but are clustered in the working classes and in the under-thirty population. They often see chalga as a bridge between East and West, or as pan-Balkan feel-good entertainment, and they emphasize musical unity with Balkan neighbors. Indeed, chalga has both drawn from pan-Balkan styles and been exported to many countries, most notably Romania in the form of *manele* (see Chapter 12 and Beissinger 2007); pop/folk is now the most widely shared music in the Balkans. In fact, since 2005 the cable television channel Balkanika has broadcast pop/folk twenty-four hours a day in eleven countries. The channel emphasizes the unity of this new style despite the ten languages used in the songs.[35] Among scholars, chalga's Ottoman legacy in the form of inclusiveness, "symbiosis," or "cosmopolitanism" has been discussed as a strength and possibly as a counteraction to ethnic nationalism (Rice 2002:41; Buchanan 2007:260; Dimov 2001). I realize the idealistic potential of this sentiment, but I think the situation on the ground is more complicated.

I concur with Jane Sugarman's observation that the various recent manifestations of pop/folk across the Balkans are actually quite different from each other stylistically (2007:270). More important, each version of pop/folk does specific ideological work in its own locality, some of it even nationalist in nature.[36] For example, earlier I suggested that despite the oriental style of chalga, one recent strain emphasizes patriotism to the majority Eastern-Orthodox Bulgarian culture. Not surprisingly, debates about pop/folk in various countries have centered on what it means to be Balkan, often contrasted to what it means to be European.[37] Historian Maria Todorova, for example, has written eloquently on the ambivalent attitude of Bulgarians toward the Ottoman past (1997), and Kiossev reminds us that for Bulgarians *Balkan* can be coded as either positive or negative. It can mean uncivilized, oriental, and backward, or familiar and intimate,[38] or "tricksterlike" (2002:183). For Muslims, on the other hand, the Turkish legacy is often coded as positive in reference to the high Ottoman culture of urban Muslims (Ellis 2003).

Note that the figure of the Gypsy looms rather prominently in the imagery of the backward/oriental Balkans, or in Kiossev's terms "the stigma" (2002:189). In all Balkan languages (in fact in most European languages), Gypsy is used as a slur, meaning thief, and in Bulgarian *tsiganska rabota* (Gypsy work) means a job poorly done or a deceitful business move. The concept of "nesting orientalisms" (Bakić-Hayden 1995) can be helpful in teasing out who is perceived as more Balkan than whom. Bulgarians may be Balkan/oriental to Western Europeans (or Croatians) but Gypsies are Balkan/oriental to Bulgarians. Sugarman reminds us that not only are Roma the most marginalized group but they are precisely the group from which pop/folk appropriated its style: "Within this dynamic of musical 'nesting Orientalisms,' Roma are of course in a class by themselves, both as the group which all others have stigmatized and as the musicians who once dominated the spheres in which the majority of the new regional genres arose" (2007:303).

In the debate about chalga in Bulgaria, it is, then, not surprising that the criticism about eastern elements is often phrased specifically against Roma; I frequently heard the phrase "It is a shame that now Bulgarians only want to hear Gypsy music." Levy cites slogans from newspapers such as "Down with kyuchek" and "It wouldn't be surprising if soon the national anthem sounded oriental," and she describes a 1999 petition to parliament signed by prominent cultural figures that pleaded for a "cleansing" of the national soundscape, where the petitioners referred to chalga as "bad," "vulgar," and "strange" sounds coming from the "uncivilized experiences of the local Gypsies and Turks." The petition expressed concern "about an invasion by their music which might result in the 'gypsification' and 'turkification' of the nation" (2002:224). Note that chalga, and thus Roma, are associated with low morals and lack of civilization. For some opponents, then, chalga has become the enemy of the nation, and the Roma are to blame. Levy points out that these nationalists see heritage as threatened, and they personify the threat in Roma.

Imre uses the phrase "double cooptation" (2008:335) to refer to the untenable position Roma artists occupy as they are caught between the state and the market:

> The Roma are twice rendered abject in the negotiation between nation-states and corporate agents of globalization and Europeanization. First because they are perceived as unable and unwilling to assimilate to the national project, and thus are universally judged to be an impediment to full and furious EU accession. . . . Second, the Roma are also demonized because of their inherently transnational identity affiliations which in turn turns them into convenient suspects for allying themselves with the dreaded forces of globalization [2006:661].

Ditchev points out that chalga, as "low class music," is totally excluded from the rubric of culture; this is reserved for the high arts and folklore, which "instill love for the homeland." Bulgaria is not conceived of as a place where different ethnicities live together but rather a "form of kinship, based upon pure and direct (imagined, of course) filiation." Roma are, of course left out of this equation of place with monoethnicity. Furthermore, culture is opposed to pleasure and consumption (2004). Thus Roma are twice erased, first in terms of being outside the nation and second in terms of being too tied to consumption.

Ideological statements about music need to be placed in a larger political framework, specifically the rise in nationalism. Consider a 2005 Gallup poll on interethnic relations conduced by the Bulgarian Helsinki Committee. The results show that one-fifth of Bulgarians are so anti-Romani (and also anti-Turk, and to a slightly lesser extent anti-Semitic) that they do not recognize the right of these minorities to live in the same country as "pure Bulgarians." Twenty-seven percent of Bulgarians would not want to live in the same country as Roma. In answering the question, "Would you personally accept a Roma as a local police chief?" 82 percent said no; similar figures were obtained when the question asked about a government minister or an army officer. With the statement that Roma are lazy, irresponsible, and untrustworthy, 85–86 percent of respondents agreed (57 percent agreed that "Turks are religious fanatics," and 29 percent agreed that "Jews are taking up many leadership positions"—in spite of their total absence from state leadership; Cohen 2005). The poll found that these attitudes are clustered neither by age nor region, nor educational level, nor income level, but rather are spread out among all Bulgarians, indicating "deeply rooted prejudices, carried over . . . from the entire child-rearing and educational system" (Cohen 2005; also see Ghodsee 2008).

It is not surprising, then, that the racist themes of parties such as Attack are attractive to some Bulgarians. In the 2005 parliamentary elections, Attack received 8.14 percent of the vote and became the fourth largest party in parliament. In October 2006, Volen Siderov, head of Attack, received 21 percent of the votes for president. Since 2009 the party has held two of Bulgaria's seventeen seats in the European Parliament. Attack is against the European Union membership of Turkey, and one of Attack's campaign mottos was precisely the phrase used in the petition discussed above: "No to Gypsification! No to Turkification!" In 2006, on its cable channel SKAT, Ataka broadcast seven programs on criminality and "Gypsy terror" in which Siderov suggested that Bulgarians "were being murdered, robbed, beaten, and raped daily by an alien minority, and were not getting any protection from the law enforcement authorities who had united with the Roma against the Bulgarians because they are the employees of a corrupt ruling class" (Kanev 2005).

Attack claims that there is reverse discrimination and that Bulgarians are now the victims and Roma are the perpetrators. They have managed to take Roma, the most vulnerable citizens of Bulgaria, and constructed them as a criminal race that "sows terror against Bulgarians unhindered by the state." In fact, in one TV broadcast Roma were called cockroaches (Kanev 2005). Attack supporters use the slogan "Gypsies into soap," and a rap song with this phrase is being circulated on private channels.[39] Attack is currently finding allies in the European parliament and is part of an anti-immigration group called Identity, Tradition, and Sovereignty, The leader of the group was fined by the French Court for remarks made in 2004 questioning the Holocaust.

Thus—returning to the topic of music—we can see that as Romani music has been appropriated into chalga, Roma themselves have not become more integrated socially, economically, and politically into Bulgarian society. True, Romani musicians found work as chalga musicians in the 1990s, yet more recently the Romani presence is declining in mainstream chalga. But how can we explain the growing popularity of Azis, a Rom?

Azis breaks taboos and gets away with it, and he is even admired by some. His Romani ethnicity is the key, because a minority can function as the clown, the trickster, as society's other. He proves he is "other" by being Romani and gay, reinforcing society's opinion that Roma are doubly marginal. For a Bulgarian man to do what he does would be considered unthinkable by most Bulgarians. His minority status gives him the freedom of marginality; Azis can transgress more easily because he is already marginal by virtue of his being Romani.

PART IV

MUSICIANS IN TRANSIT

10

⌒⌒⌒

Esma Redžepova

"Queen of Gypsy Music"

All over the Balkans, Romani men are known as expert musicians; Romani women and their participation in musical arts have only recently been the focus of scholarly attention (Seeman 2002, 2007; Sugarman 2003; Pettan 1996a, b, and c, and 2003; Potuoğlu-Cook 2007; Silverman 1996b, 2000b, 2003). Building on Chapter 6, which examined dance in terms of gender and sexuality, this chapter looks at the history of Romani female singers and concretizes one performer's strategy. Esma Redžepova is perhaps the most famous Romani singer in the world. Though atypical, Esma's life illuminates how and why she resisted norms and became a star.[1] Esma was proclaimed "Queen of Romani Music" in India in 1976, European Primadonna in 1995, and Romani Millennium Singer in 2000; she has toured internationally for more than fifty years, has given some 10,000 concerts, and has recorded hundreds of albums (www.esma.com.mk).

I argue that Esma's success was built on a number of paradoxes: she succeeded in part because of her non-Romani mentor/husband's marketing ability; her image drew on sanitized stereotypes of Romani women as exotic, nubile, emotional, and musical on the one hand, yet rooted in families on the other; and finally, she bridged the ambivalent Romani attitude of requiring, aestheticizing, and respecting female musical performances in nonprofessional realms while stigmatizing them in professional settings. Professional music has been an important medium of exchange between Roma and non-Roma, and the musical marketplace has been the site where gendered images are exchanged. As I emphasized in Chapter 6, the association of women with sexuality is symbolic capital to use in the marketplace and negotiate in Romani contexts.

Romani Female Music Making in Historical Perspective

The history of Romani female musicians and the relationship between singing and sexuality are discussed in text supplement 10.1. Given the stigma of loss of modesty and reputation associated with singing in public for strangers, it is not surprising that among Balkan Muslim Roma there are very few female professional vocalists in comparison with male professional vocalists. A cursory review of Šutkafest 1993 (see Chapter 8) reveals that out of some seventeen Romani groups, only four or five singers were females, and of those, two had husbands also performing. Similarly, at the 1998 Romska Vasilica singing contest held in Macedonia, out of nine singers there was only one woman, Esma Redžepova. According to the 1997 Albanian Romani CD *Rromano Dives* (Romani Day), in the past women did not sing professionally at weddings. This situation also seems to be true for non-Romani Balkan Muslims and for Roma from other areas of Europe.[2]

Those women who defy convention are subject to ridicule and charges of immorality.[3] Salif Ali, a Bulgarian Romani drummer, explained that it was totally unacceptable for his daughter to become a singer. When the rare set of parents do agree to a daughter's singing, the career often ends with marriage if the husband is not a musician. Pettan writes of a young Kosovo Romani woman who was a wedding singer and became a recording artist: "After she married, her husband strongly opposed the continuation of her musical career, so now she sings only in a private setting for family and friends" (1996a:316–317).[4] One way to circumvent public disapproval is to marry a musician. This mitigates the professional's immodesty because one's husband (or father or brother) serves as the protector of the wife's honor. Indeed, many female Balkan Romani vocalists today perform with family members.[5] Given these restrictions, Esma Redžepova's life is quite extraordinary.

Esma Redžepova: Early Years

Esma was born in 1943 in Skopje, Macedonia, to a poor Muslim family. Her mother, Čanija, was from a village near Skopje and a seamstress who sewed šalvari for Romani patrons. Her father, Ibrahim, was a bootblack; as a child, Esma carried his shoe shining supplies for him. He lost a leg during World War II, was too poor to buy a prosthesis, and used crutches. In 1941 Ibrahim was wounded in a Nazi bombardment of Skopje; of about 400 people injured, only four survived. Esma recalled: "I was also a hardworking child. I delivered milk to households, and I cleaned windows in a four-story house for pocket money. I liked to go to the movies, the puppet theater; I was in love with all the arts. We all went to school and learned to read and write; one of my brothers went on to higher education."

Esma's father had a good voice and knew many songs but never performed professionally, and her older brother was a founding member of
the Phralipe KUD (see Chapter 6). As a child, Esma sang and danced in
school productions, and her talent was noticed by Pece Atanasovski, who
worked for Radio Skopje; he invited her to sing for an amateur program
called "The Microphone Is Yours." She feared her parents' wrath when
they found out that she had sung over the radio:

> This was in 1956. We were all sitting at home—we would listen to this
> program every Sunday. I knew I was going to be on the radio . . . so I
> suggested we take a nap, and I covered my head! On the radio [we
> heard]: "What's your name?" "Esma Redžepova." "How old are you?"
> "Eleven years old." "What will you sing?" "*A bre babi so kerdžan* [What
> did you do, father?]." I sang, and everybody hid their heads under the
> covers. Father said: "Is that our Esma? No it can't be, because our
> Esma is asleep under the covers . . . it must be another Esma. There
> are many Esma Redžepovas in Skopje. . . ." The next day when father
> went to work shining shoes, all his friends gathered around him and
> congratulated him for my performance. But he said, "No that was
> someone else, it wasn't my Esma. She was at home sleeping." His
> friends responded, "Don't you understand? That [program] was
> recorded earlier. It was Esma." I got a big slap when he got home!

Esma explained how her community was suspicious of a female singing
in public: "A Gypsy girl, beautiful, who also sang—that would have been
really dangerous. The family decided that I, like all other girls, I should
marry early, and have children, and obey my husband without question,
and work" (Teodosievski and Redžepova 1984:89).[6] Esma remembers, "I
was a girl at the time, I wasn't yet married. . . . According to our tradition
it was a shame to sing publicly." Singing was an especially sensitive topic
in the Redžepova household because of the disgrace Esma's sister Sajka
had brought on the family:

> For Ibrahim, my father, himself a wonderful singer, really hated singing!
> Or at least singing in public. For him singing in public meant singing in
> low grade restaurants (kafanas), it meant drinking and carousing. And
> he had every reason to think that way. My sister Sajka, a pretty talented
> girl had brought disgrace on the family and become a singer in a kaf-
> ana. Ibrahim couldn't get over it: his lovely Sajka singing to drunks who
> smashed glasses for kicks. For him, Sajka was "dead. . . ." I believe that
> had she kept on and had more luck, she would have become a great
> singer. . . . How beautifully she sang! I listened to her in wonder. My
> father and mother cursed her. If only Sajka had someone to lead her, to
> show her the way. But the kafana "ate her up. . . ." I remembered Sajka's
> fate, because something similar awaited me too. And it also helped me
> to understand why my parents would so bitterly resent me even
> thinking of becoming a singer [90–91].

Despite parental disapproval, Esma's brothers supported her first steps toward a singing career: "My brothers . . . never mentioned in front of my parents where and when they had seen me in town. . . . My brothers would say to our parents: 'Why do you worry so much about Esma, she is not Sajka! She has a will of her own and if she decides to sing she will sing! But she will be a real singer, an artist!'" (91). Similarly, Esma's teacher told her father: "Don't spoil your daughter's chances, Ibrahim! She is a great talent. Singing does not necessarily mean singing in a kafana" (93).

Esma was indeed strong-willed: "I became emancipated and stopped wearing dimije, which I thought clumsy and impractical, so I wore my shabby flowered dresses, handed down from my sister, but still, 'city-style'" (89). Esma also resisted her parents' marrying her off in her teens. When her mother mentioned marriage, Esma replied, "I tell you, I'll hang myself in the little square in front of the school, on the monument. . . . I don't know if my mother really believed my threats, but any way, they didn't manage to marry me off at the age of thirteen!" (92). She even had to fight off taunts from relatives. Her sister-in-law Veba often taunted her: "Ha! You want to be singer, do you? You'll wash windows and scrub floors as a married woman." "Hey, I will not, you know. I don't want to be a servant, I want to be an artist." "You can want all you want when your parents marry you off. They've already had offers" (91–92).

When she was eleven years old, Esma was brought to the attention of Stevo Teodosievski (1934–1997), an Eastern Orthodox ethnic Macedonian accordionist and folk music arranger who worked for Radio Skopje and later became her husband and mentor. The introduction was made by Medo Čun, a Romani clarinet player in Stevo's orchestra and a friend of Esma's brother's (see Chapter 2). According to Esma, Medo said to Stevo: "I have to show you this little girl because she is incredible when she sings and dances at weddings." Stevo was a self-taught musician from a poor Kočani family.

Esma was very intimidated during her first meeting with Stevo. His initial question to her was, "Do you smoke?" She answered negatively. He was struck by her talent and sparkle, and remarked: "You have some talent, but you really will have to work." Stevo wanted to take her on as a pupil and train her, but Esma's parents said no: "My father said, 'What? A singer? No, she's ready for marriage; people are already asking; she'll be married in a year or two. Why should singing break up my family?'" Her parents strongly opposed her singing career. They said: "She will not sing. She will listen to her mother and father." But Stevo managed to convince them that he would make her into an artist, not a cafe singer, if they would not marry her off until she was eighteen years old. "When Stevo promised him faithfully that he would help me to become a good and famous singer—not a singer in any old nightclub—when my father had reassured himself that Stevo's intentions were honest, that he would look after me, he finally agreed" (95).

However, the stigma of singing in public subtly undermined Esma's morality, and her parents faced many challenges. Esma asserted: "At that time it was the easiest thing to offend my girlish pride, my purity. Especially as we Skopje Roma were very sensitive about such things. Some busybody would go up to my father . . . and tell him there was 'something going on' between me and Stevo, always together on trips, in hotels. 'Poor Ibrahim' they would say and my father would wish the ground could swallow him" (96–97). The couple eventually decided to marry, but because Esma's father had passed away they had to wait a respectable mourning period.

At that time, it was virtually unknown for Roma and Macedonians to intermarry; neither group desired it. Esma narrated: "We were the first mixed marriage! That was a big deal! Can you imagine how many people were at our wedding in 1968. Ten to fifteen thousand people came to see if it were true that the two of us were getting married." They first celebrated in Dračevo, a suburb of Skopje, and then provided free buses to transport people to the celebration in Belgrade (Cartwright 2005b:105–106). She recalled:

> Even though Stevo was poor, the wedding arrangement was that he should provide new clothing for my mother and every single aunt— this was a great expense. We did all the Romani customs—henna, etc. Since my father had passed away, my brother defended me when Stevo came to get me. My brother demanded 10,000 dinars ($10) for me. Stevo said, "I can't possibly pay that much—I have to drive Esma around to perform, and pay for gas, food, lodging. I can only give 1,000." So I was bought for $1!

Esma's early career soared among Macedonian fans, but her relationship to Romani audiences was more ambivalent. According to Esma's cousin Šani Rifati, Roma at first rejected Esma not so much because she was a professional singer but because she spent time with, and married, a non-Romani man. For Roma, Stevo's Macedonian ancestry was even more important than any alleged indecent relationship. Eventually, after marriage and international stardom, Esma was accepted and embraced by her own Romani community.

Esma's Style and Image

Under the banner *Esma—Ansambl Teodosievski*, Esma and Stevo launched a career in the 1960s characterized by instantaneous success and daring innovations. Esma was the first Balkan Romani musician (male or female) to achieve commercial success in the non-Romani world; she was the first openly identified Romani singer to perform in the Romani and Macedonian languages for non-Roma; she was the first female Romani artist to record in Yugoslavia; and she was the first Macedonian woman (Romani or non-Romani) to perform on television. Esma claims her success is due to Stevo: "What I am singing is only what Stevo taught me. He was wise, about twenty years ahead of his time; He taught me how to understand music. . . . Whatever he promised to me came true."

Modesty aside, Esma herself composed many of her songs, choreographed her performances, and provided the talent propelling her success. On the other hand, Stevo planned Esma's career very carefully. One early strategy was not to allow Esma to perform at kafanas and weddings, but only at concerts and for radio and television recordings. In effect, Stevo created a new category of female concert artist that didn't have the stigma of cafe or wedding singer.[7] Today, Esma is very proud of the fact that she has not engaged in restaurant work and only sings at weddings of friends, for free.

Even before meeting Esma, Stevo was promoting Romani music at Radio Skopje, a radical move for which he was severely criticized. In 1956 he taught two Macedonian singers, Dragica and Dafinka Mavrovska, songs in the Romani language and arranged a performance in Belgrade (for which they wore dimije); later the songs were broadcast on Radio Skopje and recorded on Jugoton. Audiences were fascinated, and according to Stevo: "I knew that we had broken through a barrier" (Teodosievski and Redžepova 1984:30). Stevo remarked: "At that time it would have been impossible for a Romani woman to perform due to the racism." Esma, a child at the time, remembers thinking, "I can do better than that—Why don't I sing?" When Radio Skopje decided to make its first Romani records, director Blago Ivanovski substituted his girlfriend Anka Gieva for Dafinka. Thus the 1957 recording of Stevo's song "Bašal Seljadin" (Play Seljadin; Jugoton SY 1090) was sung by Anka Gieva and Dragica Mavrovska. This version seems tame and mild-mannered, hardly like a čoček, with a bland 2/4 rhythm and no syncopation (audio example 10.1 with text supplement). By contrast, Esma's version, recorded several years later, has a driving rhythm and gutsy vocals. The visuals in video example 10.1 with text supplement were filmed in 1988 (MP 31005), but the audio is Esma's recording from the 1960s (RTB SF 13085; I analyze the staging later in this chapter). The Romani text reflects the importance of music in the life of Roma.

Singing in the Romani language was Esma's statement of pride in her heritage: "I was the first Rom to sing in the Romani language. It was actually historical, that Yugoslavia was the first place to broadcast Romani songs on radio. It was kind of a shame to sing in Romani in my time; many singers hid the fact that they were Romani. When I came out singing my own songs in Romani, many came out after me." Note that Esma uses the phrase *come out*, to characterize the bravery a Romani artist needed to confront the prejudicial attitude of Yugoslav music production. For example, around 1967 two sisters, Živka and Jordana Runjaić, recorded singles in the Romani language. They said they were Serbs, but it was revealed they were Roma. According to Esma "many singers passed [as other ethnic groups] because there was an embargo on Romani singers. There was discrimination against them as performers. I risked a great deal when I said I was Romani and I want to sing in my own language." She continued: "Our Romani women were afraid at the time to say they were Roma—they said they were Turkish, Macedonian, Albanian, anything but Roma. . . . After the cleansings of World War II, Roma were afraid for their lives and at no time would admit they were Roma." Stevo commented: "Esma was the first leader with the flag! All the other people looked at her to see if she was accepted. . . . On her first record . . . 'Gypsy music' was written. It was very clear! For the first time 'Gypsy Music' was written on a label." Esma: "I opened the way for Roma, in the first place, to admit that they are Roma, and not to be ashamed they are Roma."

Esma and Stevo endured the racism of Macedonian institutions and the gossip of the public. At Radio Skopje, Stevo was repeatedly told: "Take Vaska Ilieva, take other singers—why a Gypsy?" His colleagues said cruelly, "Stevo why have you brought this Gypsy to disgrace us?" (Teodosievski and Redžepova 1984:95). In the beginning of her career, they deliberately denied Esma opportunities. Stevo recalled: "They took from her the song she knew and did best and gave it to another girl" (38); later, one of their films, *Zapej Makedonijo* (1968), won prizes, but it was rarely shown in Macedonia (54).

Stevo commented: "Some of the top officials from Radio Skopje's communist party leadership thought they needed to let me know that it would be better for the show to have a participant of Macedonian nationality. I was then, just like now, devoid of any nationalistic preconceptions. I consider myself a cosmopolitan" (Mamut 1993:3). Stevo recalled: "They chased me out of Macedonia because of Esma—we had to move to Belgrade. They said 'Why do you play that Romani music? Let it go—you are not Romani.' I was a member of the communist party through Radio Skopje. The party objected, they threw me out. . . . The secretary of the party said, 'Why do you bother with Esma? Vaska Ilieva, Anka Gieva, they are Macedonians, Esma isn't!' From then on I had nothing to do with the party—it didn't interest me any more. I played what they told me at work, period." The taunts became so stifling that in the 1960s Stevo and Esma decided to move to Belgrade, the capital of Yugoslavia, where they would have more opportunities. Esma said: "People knew me too well, they were talking too much about us in Skopje, and we had to get out of that environment."

Stevo was very conscious about creating a specific Romani niche for Esma in the commercial world. Part of his genius was to craft a trademark image and staging for Esma that evoked the historical stereotypes of Gypsy women as sensual and fiery but that kept the pageant tasteful (see photographs 10.1, 10.2, and 10.3). A survey of Yugoslav press reviews during Esma's early years reveals that critics focused on her Romani heritage in stereotypical prose: she was described as dark-skinned, hot-blooded, happy-go-lucky, and genetically talented; she was even hailed as the new "Koštana," referring to the Serbian 1902 opera about a seductive Romani songstress.[8] Stevo and Esma cultivated these stereotypes as long as they were positive. This resonates with a point I make throughout this book:

that Roma orientalize themselves when necessary for marketing purposes. Furthermore, historically Roma have had few opportunities to alter their imagery and discourse because they have never been in control of their representations.[9]

The 1988 staging of the song "Bašal Seljadin" (from 1957, discussed earlier, video example 10.1 with text supplement) includes several stereotypic elements: a barefoot man in a Hungarian Gypsy costume strutting, female dancers in belly dance outfits with flimsy veils doing modern dance choreographies, and a background of tents. Other songs in this video series are set in a pseudo-Gypsy camp near a "stream" (actually a swimming pool) with a fire, a setting sun, and pseudo-Russian Gypsy dancers. These videos feature the ballet troupe of Macedonian National Television. When I asked Esma what she thought of these stagings, she said she thought they were artistic. I discuss this point again in Chapters 12 and 13.

Romantic stereotypes do sometimes help break barriers. Esma, for example, may have reinforced the female Gypsy sensual image, but she herself never wore immodest belly dance outfits. Rather, she was the first Romani performer to appear in Romani-style dimije for non-Romani audiences. "I was the only one, with Teodosievski's help, to jump up publicly on stage and wear dimije, and I wear them to this day. I am not ashamed to wear them and I am not ashamed to say I am Romani." Dimije, which emphasize hip movements, linked Esma specifically to Roma, to other Muslims all over the Balkans, and to tradition. Esma's dimije were fashioned from modern fabrics and colors, and she further innovated with accessories and headpieces, some evoking Eastern themes.

Emotion is perhaps Esma's trademark affect. I have pointed out that emotion is iconically associated with Roma in terms of unbridled passion (Silverman 2011 and in press). Esma capitalized on emotion in both her voice and her stagings. Her iconic song in terms of emotion is "Hajri Ma Te, Dikhe, Daje" (May you see no good, mother), where she enacts the lament of a young girl being married off to an older man. In the 1970s video of the song she is dressed in a white veil and virtually cries while she sings the song (video example 10.2 with text supplement). The sobs become part of the unmetered melody. This staging depicts an older Muslim husband who is served by a young wife. In later concert stagings of this song, Esma sings from beneath a black veil with her face totally obscured and her accompanying musicians bowing their heads in sympathy. At the end of the song, one of them lifts her veil, and in a dramatic shift the musicians begin a new and lively rhythmic song; she dramatically plays on the emotional shift from despair to joy.

Another trademark feature introduced by Stevo was that all the performers stood up, giving them unprecedented freedom of movement on stage. Typically, they swayed right and left with their instruments in rhythm, evoking the back-up singers in Western pop groups of the 1960s. And most daring, Esma danced during musical interludes (see Chapter 6). She explained:

I am a traditional woman—growing up, the women gathered inside. I adopted all the old ways. Stevo told me when I was young, "you will dance exactly how you danced inside with the women at a wedding." I answered "but that is shameful." He said "it is not shameful—it is your tradition—it is your national dance. Others dance differently, but you are dressed in dimije, it is not a shame, you have something to dance about! You aren't bare, you don't dance (with your hips) in a circle, you dance with your stomach." And he persuaded me that I don't need to feel ashamed of that—I have to show my culture—it is our national dance. I have accepted, embraced, exactly what Stevo taught me. After I got on the stage and danced, it was easier for other Romani women and girls.

Stevo staged Esma's performances as miniature dramatic scenes in which she enacted the story of the song. Her voice showcased emotions evoked in the text (often using cries and yelps), and her hand gestures referred to story themes. Similar to professional female Ottoman dancers (see Chapter 6) and to generations of male musicians, she masterfully played to audience sentiment. Esma continues stagings of this type to the present, even though some have criticized them as too clichéd. Stevo also introduced the tarabuka (hand drum) to concert performances; associated with Muslims, it had never before been used on a concert stage. Furthermore, he engaged uninhibited young Romani boys to play the tarabuka dramatically while they playfully danced with Esma.[10] Not only did young boys provide visual and emotional interest but their participation also created a wholesome family image for Ansambl Teodosievski, with Esma as a maternal symbol. Indeed, she did serve as a "mother" to many of Stevo's pupils.

Esma's trademark song of the mid-1960s, "Čhaje Šukarije" (Beautiful girl), showcased both her voice and Stevo's arrangements. Although Esma claims she wrote the melody and text, her clarinetist at the time, Medo Čun, also claims credit for the melody; Čun displays his masterful playing in the opening slow section and the instrumental solo (see Chapter 6). Video example 10.3 with text supplement is excerpted from the 1968 film *Zapej Makedonijo*. Note the pastoral setting, the text about love, and the fact that Esma is barefoot and wearing šalvari, the čoček rhythm (pattern number 1 in Figure 2.1), and the melody in phrygian mode. Esma's voice is focused and emotional, featuring delicate ornamentation, yelps and glottals, and a wide rage of dynamics. The male instrumentalists engage in a question-and-answer dialogue with her and harmonize with her in the chorus, reminiscent of the "doo-wop" style of popular music of the 1960s.

Esma and Stevo were pioneers in producing music videos. They appeared on the Yugoslav music scene just when television was making inroads, and they correctly predicted that visuals would capture the public. Esma was involved in making four long films[11] and many short music videos. Video example 10.4, "Ciganski Čoček" is also from the film *Zapej Makedonijo* and features members of Esma's natal family dancing. This clip stands in marked contrast to the video example 10.1 of Bašal

Seljadin. Whereas Ciganski Čoček shows Roma of several generations dancing informally with no choreography, Bašal Seljadin show a non-Romani ballet troupe performing choreographies influenced by modern dance.

Video example 10.5 (with text supplement) of Čhaje Šukarije is part of a landmark 1965 Austrian television show. The staging embodies the classic Esma trademarks: dressed in šalvari, she emerges from behind the musicians' heads in an overhead shot; she playfully flirts with them, and she dances seductively but modestly. Another hit of this era, "Romano Horo" (Romani dance), about dancing among Roma, also appears on this show (video example 10.6 with text supplement). The song (in phrygian mode) features a male chorus and fade-out at the end, both reminiscent of pop music. Esma's voice demonstrates emotional variation, for example a breathy quality alternating with a throaty intensity. The rhythm is pseudo-Latin: a pseudo-clave (wooden sticks) and cow bell rhythmic pattern reflect the popularity of Cuban music in the 1960s.

This Austrian show encapsulates how Esma bridged the divide between East and West via music, language, costuming, and staging. In the first part of the show she wears a Romani costume, dances čoček, and stages her scenes in a "village." By contrast, in the second part of the show she wears a western cocktail dress and high heels, has short bobbed hair, and peeks through a curtain with a modern art design. To "Romano Horo" she dances the twist, the most popular dance in the west at the time. Esma ends her show with "Makedo," a pop song entirely in German, arranged by Stevo, and she encourages Germans to try Macedonian dances and songs. Esma composed a line dance in 7/8 (2+2+3) to this song and hoped it would catch on in Germany.

Stevo wanted Esma to appeal to wider Macedonian, Yugoslav, and international audiences, and so early in her career he broadened her repertoire and arranged tours. In 1960 Tito, the president of Yugoslavia, invited her to perform for a gathering of world leaders, and subsequently he sent her abroad to represent Yugoslavia (Cartwright 2003a). Her early repertoire included Macedonian folk songs, for which she dressed in traditional village costumes. For example, the video of the Macedonian song "Kolku e Mačno em Žalno" (How painful and sad it is) was filmed at the Sveti Naum monastery and shows close-ups of icons. This appealed to the visual and aural sense of nationalist pride (tied to religion and rural folklore) for Macedonian audiences. In addition, Stevo arranged concerts and recording sessions of duets with some of Macedonia's most famous vocalists, legitimating Esma's talent beyond the Romani sphere.

Vocal repertoire in other Yugoslav languages was added, including songs in Serbian, Croatian, Bosnian, Slovenian, Turkish, and Albanian. She embodied Tito's principle of *bratstvo i edinstvo* (brotherhood and unity) by performing the music of all the ethnic groups in Yugoslavia (see photograph 10.4). The video *Pesmom i Igrom Kroz Jugoslaviju* (With song and dance from across Yugoslavia; Serbian) features songs from all the republics with traditional regional costumes. Eventually songs of neighboring Balkan countries were incorporated (e.g., Bulgarian, Romanian, and

Greek), and then songs of far-flung ethnicities: Russian, Hebrew, German, and Hindi; again costumes reflected the region (see photograph 10.5). In publicity shots she is depicted as a performer of many ethnic musics and also a modern worldly citizen (photographs 10.6 and 10.7).

Perhaps the most important international tie in Esma's career was her link to India. In the early 1970s, Roma in Macedonia were beginning to develop a sense of their historical ties to India as part of a larger politicization process and a movement to define their identity.[12] Ensemble Teodosievski made its first (uninvited) trip to India in 1969, followed by two invited trips in 1976 and 1983. In 1976 Esma and Stevo were crowned "King and Queen of Romani Music" at the First World Festival of Romani Songs and Music in Chandigarh. A video of her 1983 trip documents Esma giving Indira Gandhi šalvari and showing her how to tie a Romani head scarf. As a result of her growing awareness of India and the pan-Romani identity movement, Esma incorporated a Hindi song into her repertoire; she also continued to perform the song "Dželem Dželem," which developed into the Romani anthem (see Chapter 3). Although Esma first recorded the song in Serbian (*Čerga Mala Luta Preko Sveta*, A small Gypsy tent wanders through the world; see video example 3.3), on video examples 3.2 and 3.4 she sings it in Romani (at Šutkafest in 1993 and in New Jersey at a Macedonian church in 2004; similar texts are in text supplement 3.1). This link between Romani identity politics and music helped to facilitate Esma's relationship to Roma.

Stevo's School

In the late 1960s, Stevo and Esma founded a music school in their home to train young boys from disadvantaged homes. Virtually all of the members of the Teodosievski ensemble throughout the last fifty years have come from Stevo's school.[13] Many, such as Medo Čun, Enver Rasimov, Sami Zekirovski, Pero Teodosievski (Stevo's nephew), Zahir Ramadanov, Eljam Rašidov, Simeon Atanasov, Bilhan Mačev, Tunan Kurtiš, Saško Velkov, and Šadan Sakip, went on to become famous musicians in their own right (Teodosievski and Redžepova1984:187). Sakip, one of the only vocalists, developed a singing career and won first prize at Šutkafest 1993; as a child he appeared in Esma's video playing tarabuka for "Kec" Ibro Demir's song *"Aj Leno, Lenorije Čhaje"* (Hey Lena, girl, 1979; video example 10.7 with text supplement). The song affirms that goodness and beauty exist in spite of poverty.

Many of Stevo's pupils came to the school at a young age from impoverished families; one was even rescued from abandonment. Although most of the boys were Romani, a few were not; Simeon, for example, became Romani by virtue of his upbringing with Esma and Stevo from the age of five. He was later officially adopted by them and became Esma's music arranger after Stevo's death in 1997. Both Esma and Stevo believed that anyone of any ethnicity could play Romani music well; Stevo said he was proof of this. During two trips to the United States, Esma was thrilled to teach Americans at the East European Folklife Center's Balkan Music and Dance Workshops in 1997 and 1998.

In Stevo's school, all the children received instruction, lodging, meals, and clothing free of charge. Esma served as an adopted mother and vocal coach. Because she never had children of her own, she achieved the role of motherhood through these boys, who to this day call her "mama." Stevo was a strict teacher, notorious for rigor and sternness. Neither Esma nor Stevo believed in talent; they believed only in hard work. All the boys began with tarabuka in order to master Balkan and Romani rhythms; then they switched to various instruments. Simeon, for example, was, according to Esma, sickly as a boy and could not blow hard enough for a wind instrument, so he was given an accordion. Zahir was already playing the trumpet when he was recruited, after Stevo heard him play at a wedding in Kočani. Zahir narrated:

I was a young student of twelve years when I came to Stevo's school. I played the trumpet incorrectly, on the side of my mouth, and Stevo taught me to play correctly. He was very strict—we were up at 6:00 AM, then we went to school, then we played; on weekends it was eight to fourteen hours a day. Mama had the watch, and we had to practice a certain *oro* [dance], for example, twenty-five times! It was a great deal of work. There were no excuses like "I can't do it." If we played something wrong we had to stand on one foot and play it!

Musicians received not only a musical education but also valuable exposure to a wider world. As a twelve-year-old Romani boy living in a Romani neighborhood, Zahir spoke Turkish well and Macedonian poorly; when he moved to Belgrade he learned Serbo-Croatian in school and Macedonian and Romani from Esma and the other boys. He also learned the ropes of the music industry and had a chance to travel to many foreign countries. Note that Esma and Stevo had only boys in their school. They did not accept girls because of the close living quarters; she asserted: "Stevo and I realized that it would be asking for trouble to put boys and girls together at that age, at puberty." Since the 1990s, Esma has trained several female singers, including her protégée (Eleonora Mustafovska, discussed below). Esma considers these forty-nine protégés her living legacy.

Esma, Politics, and Humanitarianism

Esma has always been vocal about her patriotism for Yugoslavia and Macedonia. These are her true personal beliefs, but this ideology also positions her as an ally of the nation/state rather than as an oppositional activist for a minority. She sees herself as an ambassador for Macedonia more than for Roma, and some Romani activists object to this. She and Stevo moved back to Macedonia in 1989, just before the outbreak of war.

In 2006 she said, "I represent Macedonia everywhere in the world and my ambassador mission is to present my country to my best [sic]" (www.culture.in.mk). The fact that this quote appeared on the country's website shows Esma's vision has a nationalist dimension rather than an ethnic one. Indeed she is an icon for many Macedonians. In 2007 she was awarded a diplomatic passport that allows her to travel without visas as a "cultural ambassador."[14]

Esma's patriotism extends to a defense of Macedonia as a haven for Roma. When I asked her in 1996 about problems Roma face in Macedonia, she answered:

Macedonia is the least oppressive place for Roma; it was one of the first countries in the world that early on had a radio show in the Romani language, with singing and music. One of the first Romani leaders was a mayor [of Šutka]. We have Romani members of parliament, we have two private Roma channels on TV and several radio stations in the Romani language, and on national TV, there are two half-hour weekly shows so all of Macedonia can watch us. Macedonia is definitely one of the most democratic and accepting places for Romani people.

As patriots, Esma and Simeon have often argued with her cousin Šani Rifati about their defense of Macedonia. Whereas Simeon pointed to his beautiful apartment and middle-class life as evidence that there is no prejudice in Macedonia, Šani pointed to the health crisis, police brutality, high unemployment, and squalor in Romani refugee shantytowns (European Roma Rights Centre 1998, 2006). Note that although Esma sponsored more than 2,000 benefit concerts for various causes throughout her career, it wasn't until 2002 that she sponsored a benefit specifically for Roma. This concert was organized by Šani as head of the NGO Voice of Roma and took place in Kosovo among refugees. I maintain that Esma has crafted a somewhat unthreatening profile. She stresses Macedonian patriotism in the realm of politics and Romani music in the realm of entertainment.

Proud of being middle-class, Simeon and Esma were critical of the documentary film on Romani music *When the Road Bends: Tales of a Gypsy Caravan* (2006) because it graphically showed the poverty of the native villages in Romania of the Romani bands Taraf de Haidouks and Fanfare Ciocarlia. They were afraid audiences would think all Roma lived in mud. Macedonian Roma in New York have expressed similar sentiments that international images of Roma focus on rural poverty and do not represent them. On the other hand, some activists thought that the film did not deal enough with prejudice. We may observe that both activists and musicians engage in "strategic essentialism" (Spivak 1988); the former essentialize Roma as victims and latter essentialize Roma as entertainers (see Chapter 3). This postcolonial concept helps us understand how subaltern activists reject some essentialized concepts of themselves as codified by their oppressors (all Roma are musical) while they embrace other essentialized concepts of themselves for identity politics (they are all oppressed). Musicians often confound the stance of activists, and vice versa.

When Šani, for example, introduced concerts in Esma's 2004 American tour with lectures on discrimination, Simeon objected, saying it would alienate the audience. Šani also asked Esma to open her concert in Sebastopol, California, with a Romani song, because the city was the home of the sponsoring organization, Voice of Roma. Esma refused, insisting that she open the show with a Macedonian song and reminding Šani that her artistic decisions were paramount. In general, Esma resists artistic advice from activists, claiming they aren't performers. Simeon and Esma, then, are more interested in the artistic, entertainment, and commercial value of music whereas activists are more interested in the educational aspects of music—or are hostile to it. I explore this topic further in Chapter 12.

This brings up the question of resistance. Esma certainly resisted the exclusionary categories of institutions by her pioneering use of Romani language, dance, music, and costume, but she also resisted political agendas that might hurt her commercial success and infringe on her artistic decisions. In addition, she collaborated with the commercial establishment by endorsing positive Romani stereotypes; finally, she embraced a broadly humanitarian stance rather than a narrow Romani activist stance. Esma's image is that of a universal humanist; her public statements repeatedly stress pacificism and cross-cultural understanding. When she discusses Roma, she points out that they have never started a war: "We are born naked and we die naked and we don't carry anything with us to the next world. So fighting doesn't make sense. The greatest barrier to all people is war." When interviewed by the Serbian newspaper *Blic*, she said "We Roma don't like war. . . . It doesn't matter what nationality you are. What matters is if you are a good person."[15] Esma's commentary brings up a point that I stress in this book: Romani musicians have selectively resisted, on the basis of strategic decisions about what they could actually accomplish and how resistance would affect their careers; furthermore, resistance is always paired with collaboration (Ortner 1995, 1999).

Related to Esma's nationalist stance vis-à-vis Macedonia is her elevation of Macedonian Romani music to being the most "authentic" Romani music. In Chapter 12 I discuss how she defended her use of the synthesizer even though European audiences rejected it as nontraditional. Although she embraces modern instrumentation, Esma insists that her style of singing is old, classic, and traditional: "It's all traditional. I try to keep the style pure so I don't mix cultural influences" (Cartwright 2005:103). Similarly, Stevo elevated Macedonian Romani music by claiming that "in India in 1976, all of the presenters from twenty-three countries agreed that Macedonian Romani music would be the music of all Roma." He interpreted the fact that he and Esma were crowned King and Queen at the Chandingarh festival as an implicit affirmation that Macedonian Romani was superior to other Romani musics. Esma concurred: "In 1976 twenty-three representatives of Roma from many countries gathered. They wanted to see who had the most traditional Romani way of singing. From the twenty-three representatives, I won first place!" Stevo elaborated: "In India we were crowned because only we played true Romani music. The other Roma played Turkish, Spanish, etc., music."

Esma continues to criticize the hybrid nature of Romani music in other countries, such as Spain and Hungary, where she claims Romani music sounds like the local non-Romani music: "Macedonia did not persecute the Romani language, and therefore the language, music, and the culture and traditions have been best preserved." She does not see her defense of her "authenticity" as contradictory to innovations in her music. She readily admits that her vocal style has become technically advanced, with more dramatic timing and numerous, complicated ornamentation; and she also defends the use of synthesizer and her collaborations with pop stars (to be discussed shortly). I believe she valorizes the concept of tradition in part because the concept was so venerated in the Yugoslav period. She feels Romani music is worthy of that veneration, and so her patriotism and elevation of Romani music are intertwined.

Esma became directly involved in Macedonian politics in the 1990s after the formation of Romani political parties (see Chapter 1). For a period of time, she was aligned with the Romani leader Amdi Bajram, and her performance at his son's wedding was filmed for Macedonian Television (*Romska Svadba*, discussed in Chapter 5). She was also aligned with Macedonian politician Vasil Tupurkovski's Democratic Alternative, a multiethnic party. Now she is aligned with Prime Minister Nikola Gruevski's ruling party VMRO-DPMNE and still maintains a strong public profile, which includes her extraordinary commitment to causes of the needy. She has given thousands of benefit concerts for hospitals, orphanages, disaster victims, poor children, and so on; she continues to generously donate her time and talents to charity. She was nominated for the Nobel Peace Prize in 2003 and 2005, is an Honorary president of the Red Cross, and received the 2000 Medal of Honor from the American Biographical Institute, the 2002 Mother Teresa Award, and several awards from UNICEF as well as from Tito. On International Roma Day in April 2010, she was awarded a Medal of Honor from the ruling government. She has a special interest in women's issues, and in 1995 the Macedonian Association of Romani Women took her name as its title. In 2002 she won the Woman of the Year from the Macedonian magazine *Žena* (Woman) and in 2010 she took part in a United Nations conference on women as part of the Macedonian delegation.

After Esma and Stevo returned to Skopje in 1989, they started work on a humanitarian and documentation project entitled the Home of Humanity and Museum of Music. This ambitious project includes construction of an outpatient clinic for underserved people, a recording studio, a performance space, and a museum and archive of Romani music. The economic crisis of the war years plus Stevo's death in 1997 have considerably slowed work on this project, but Esma is still committed to it. In 2010 the city of Skopje granted her land for the museum.

Esma briefly considered retiring after her husband's death, but instead she resumed her career, with renewed energy. She toured fairly regularly and has sought new international and local collaborations. She was offered a recording contract with the German label Asphalt Tango about ten years ago but declined in order to have more control over options in her career; in 2008, however, she accepted a contract with them and thus performs as a soloist in their "Queens and Kings" tours.[16] Her adopted son Simeon has assumed the role of her arranger, and several albums under his direction have been released.[17]

Collaborations and Current Directions

Throughout her career, Esma has collaborated with many non-Romani musicians, which broadened her appeal. Her early album and video *Legendi na Makedonska Narodna Pesna* (Legends of Macedonian Songs), for example, featured her in solos and in duets with non-Romani Macedonian singers. After Stevo's death, she widened her circle of collaborators. In 2001, for example, she concertized with the Italian guitarist/mandolinist Aco Bocina, and in 2000 she collaborated with American klezmer trumpetist Frank London on the album *Chaje Shukarije* (Times Square), which features new versions of older songs. In 2000 she also collaborated with Macedonian composer Duke Bojadžiev (who lives in New York City and graduated from the Berklee College of Music) on the album *Esma's Dream: Esma and Duke*. The recording features her vocals over an electronic mix of synthesized arrangements, sometimes augmented by added bass lines and Indian drums. (This album prefigures the electronic remixes I discuss in Chapter 13.) In 2007 she collaborated with the French guitarist Thierry (Titi) Robin on three tracks of her album *Mon Histoire* (Accord-Croises).

In the realm of pop music, Esma collaborated with the Serbian rock group Magazin on the 2002 song "Dani Su Bez Broja" (Days are endless; Serbian), where she sings an introductory verse in Romani and a melismatic passage on the syllable "ah."[18] She sang with the Macedonian pop star Kaliopi on her 2004 hit "Bel Den" (Fair day; Macedonian) and was a guest at Kaliopi's thirtieth anniversary megaconcert in 2006. She also collaborated with the ethnic Albanian pop singer Adrian Gaxha on the song "Ljubov e" (Love is; Macedonian). The song was entered in the preselection competition for the 2006 Eurovision contest and came in second by approximately 100 votes, which caused some controversy. Although it was not discussed openly in the media, the national mood was that the country and Macedonian national television (which sponsored the contest) were not ready to have a Rom and an Albanian represent them at a prestigious pan-European event such as Eurovision.[19]

In 2009 Esma was featured in electronic and film music composer Kiril Džajkovski's (formerly of the band *Leb I Sol*) fusion song "Raise Up Your Hand" with the Jamaican rap/dub artist Ras Tweed.[20] Ras Tweed sings in English creole and Esma sings in Romani; the video was shot in Macedonian

Romani neighborhoods. In 2010 she began her retirement with the premiere of her female vocal protégée Eleonora Mustafovska and with a new name: Esma's Band (see www.myspace.com/esmasbandskopje). The band's song "Džipsi Denz" qualified for Macedonia's 2010 Eurovision finals in tenth place. The band continues Esma's international humanitarian mission, according to leader Simeon Atanasov: "Our message is to fill our music as a bridge which connects the differences between the nations, because Roma (Gypsies) are living in every European country and feel themselves as cosmopolitans."[21]

Perhaps Esma's most famous, and most commercially successful, collaboration was with Toše Proeski in 2002, via the song titled "Magija" (Magic) in Macedonian and "Čini" (Spells) in Serbian. Toše (who died in an automobile crash in 2007) was one of the top young Macedonian pop singers and songwriters, with a huge following among Macedonian and Serbian youths; in 2004 he represented Macedonia at Eurovision (http://www.myspace.com/inmemoryoftodorproeski). The song won awards for best song and best video of 2002 at the 12 Veličanstveni (12 greatest) ceremony, Macedonia's version of the Grammy's. "Magija" is actually a combination of two songs, Toše's in Macedonian (or Serbian) and Esma's (in Romani). Esma's song "Naktareja mo Ilo Phanlja" (He closed my heart with a key) is a preexisting cut[22] that is inserted into Toše's song. Toše's song is about a love affair gone sour because of a magic spell, and Esma's is about a woman whose boyfriend deceived her and married her best friend. The Serbian version appears on video example 10.8 with text supplement.[23]

Although each of the songs has its own internal narrative, the pair seem to have few textual connections; they are combined for musical reasons rather than for narrative logic. Also remember that the intended audience is non-Romani, so the Romani text is irrelevant. The viewer, however, would immediately pick up on aural and visual cultural clues. The two songs contrast markedly in musical style: Toše's is in 2/4 rhythm in pop style, and Esma's is in 7/8 rhythm (number 10 in Figure 2.1), has a drone-based harmony, and has synthesized zurla and tapan accompaniment, this last a symbol of Roma.

In addition, the visuals portray two contrasting worlds: Toše's sunny daytime world of upper-class love and conviviality, and Esma's nighttime world of Gypsy magic, abandon, and the occult. Indeed, the text seems to suggest that Esma (and by extension all Gypsies) can cure Toše's despair (for perhaps she sent it as a spell) with music and dance. Esma is pictured in flowing dimije in the middle of a wild party on the beach, amidst tents, rusty cars, fire dancers, and couples who wear revealing clothes and sensuously belly-dance and kiss. All the elements of the standard Gypsy stereotype are here: sex, music, the occult—even a crystal ball, into which Toše gazes at the end of the video.

Why would Esma engage in such a stereotypical treatment of Roma? When her cousin, activist Šani Rifat of Voice of Roma, asked her this question, she replied, "It is an artistic staging. It is art." And when Garth Cartwright asked her about these images, she "insisted she liked the video and enjoyed the pop spotlight" (2005:110). Esma's reasons for collaboration with Toše are complex: she may have had an affinity for him because he was from a minority ethnicity (Aromun/Vlach) or because, like her, he was involved in humanitarian work and received several humanitarian awards. Toše seems to have had an affinity for Romani music and can be seen on Skopje television station BTR in several YouTube clips performing Bregović's song "Erdelezi" (see Chapter 13) with the Macedonian Romani singer Erdžan (see Chapter 2).

Collaboration with a pop star was certainly one way of increasing Esma's visibility and expanding her audience. The song was clearly listed as his, and he was definitely a rising star; however, she claims that she was helping him with his career. In the end, Esma probably chooses to collaborate whenever a good opportunity presents itself. For a decade after Stevo's death, she refused to sign an exclusive Western contract in order to manage her own career (with Simeon as arranger); but in 2008 she signed a contract with Asphalt Tango in Berlin. Does she have choices in her artistic products? Theoretically yes, but in a tight musical market she has fewer choices. Esma has carved a viable musical niche, but as an aging Romani star she is vulnerable.

In surveying Esma's life, we can see just show innovative she was. Under the tutelage of non-Romani Stevo, she created an unprecedented niche for Romani music and dance. Moreover, she raised female arts to a level of respectability by playing with images of emotionality and sexuality in the framework of the elite concert and recording stage. By achieving success among non-Roma first, she legitimated her role as a professional among Roma. By displaying her patriotism to Yugoslavia and Macedonia and by supporting international rather than Romani humanitarian causes, she achieved an unprecedented level of legitimacy. Today Esma is a living legend for many Roma, and many Macedonians.

The constellation of Romani female performers, including professional singers such as Esma and professional and nonprofessional dancers (whom I discussed in Chapter 6), points to a delicate convergence of a set of historical, economic, political, social, and aesthetic factors. Within Romani communities, female musicality and dance, although tinged with sexuality, is valued, prized, and encouraged to flower in appropriate settings. Moreover, female artistry as an occupation has a long history, as witnessed by Ottoman professional dancers and early-twentieth-century frame drum players and singers. In spite of the economic necessity propelling professionals, female singers and dancers are still scrutinized as immoral, but at the same time they are in demand by non-Roma and Roma. Sexuality is dangerous, but necessary. The position of female Romani performers to Roma structurally mirrors the position of Romani male performers to non-Roma: they are marginal, sexual, and dangerous, yet they are necessary for celebrations because they embody artistry and musicality and they bring out the "soul of the patrons." Okely makes a parallel point about British Romani fortune tellers who mingle freely with non-Roma for work but are ideally supposed to preserve modesty and reputation (1975).

Both fortune tellers and dancer/musicians have been stereotyped by non-Roma as quintessential images of Romani women. The marginal position of Roma, their lack of control over image making, and their role as service workers all contribute to the trafficking of their arts in the realm of the market. Females have a significant role in this market, as their talents, images, and bodies are a saleable commodity. Images of Romani women are rarely designed by women themselves; rather, they rely on patron fantasies that may be mimetically sold back. In all of these processes, female performers are not passive. Although they are rarely in charge of the institutions that shape their performances, Romani women have managed to exert control over certain realms of artistry and carve out new domains of performance. As Esma's case shows, they tailor their talents and sexuality to varying contexts. The nexus between in-group ideals of female modesty and the economic and aesthetic requirements of the marketplace has created a space for a variety of female performers. These women, like Esma, strategize to maximize both their commercial success and their reputation.

11

⌖

Yuri Yunakov

Saxophonist, Refugee, Citizen

This chapter explores identity in transnational contexts via a Bulgarian-Turkish-Romani-American male musician who has performed for both Romani communities (on several continents) and the world music market. This case study serves as a bridge between local and global sites and between Chapters 3, 4, and 5 (about Romani transnational communities) and Chapters 12, and 13 (about global marketing of Romani music). It is clear that the supposedly recent distinguishing characteristics of the global age, such as border-crossings, hybridity, multiplicity of identity, and interconnectedness of economic systems, have been operable for Roma for centuries. Furthermore, we must interrogate the local and national with as much vigor as we interrogate the global (Shuman 1993), for all three arenas reveal hierarchies and representational conflicts. Through examination of the life history of the saxophonist Yuri Yunakov, I illustrate how musical performances are strategies in personal identity politics. With music, Yunakov mediated the tension between supposed binaries such as official-unofficial, traditional-modern, authentic-hybrid, socialism-postsocialism, inclusion-exclusion, and local-global. But rather than a merely celebratory tale, this chapter also reveals the disjunctures and challenges in Romani identity making.[1]

Early Years: Border Crossings

Yuri Yunakov was born in 1958 in the Muslim Romani Turkish-speaking neighborhood of Haskovo, Bulgaria, about thirty miles from the Turkish and Greek borders:

> YY: My mother was born in Bulgaria, but my mother's father was born in Turkey, around the 1920s. There was border at that time, but it wasn't a secure border like with the Communists. My grandfather,

a butcher, came to Bulgaria and married my grandmother; after their daughters were married off, my grandfather went back to Turkey, Izmir.

CS: Why did he go back?

YY: I don't know why; I've never seen him. He died before I came to America. He married another woman back in Izmir. And there's a whole family there. I know an aunt from there. When I was a soldier in the army, my mother's sister came from Turkey for the first time to see her relatives, that is, the aunt I had never seen. My relatives couldn't go to Turkey. It would have been impossible during the Communist period. I was in the army and I had been given a leave; I was waiting for my mother and father to come see me. They [the officials] called me to the gate. I was looking at someone and I thought it was my mother. I said to her, "What's going on, you came so late?" I got closer. My mother was hiding behind this woman, because they wanted to see what I would do when I saw this aunt. She looked like my mother, but she had a younger face. And then my mother came out. I couldn't understand what was happening. And she started crying, and my aunt whom I'd never seen, she embraced me, and they started explaining, "This is my sister from Turkey."

CS: Did anyone from your family ever go to Turkey?

YY: When the Communist government fell, my mother went to Turkey to see her other brothers and sisters, her family from her father and his other wife. . . . There are some relatives in Istanbul, Edirne, and Izmir.

CS: Did you ever go to Turkey?

YY: No. I lived only thirty miles from the Turkish border, but I never went. . . . My father's family is more interesting. My father's mother was Greek. I don't know why, but a long time ago people came from Greece to Bulgaria, they were very poor. I'm just learning this from family stories. In those years, Bulgaria was somewhere to go, it was a good place for work. I think my father's mother was Muslim Greek Gypsy.

CS: Did they speak Romani?

YY: My grandparents on both sides didn't speak Romani. My mother's mother spoke Turkish, and maybe some Greek. . . . I spoke Turkish with all of my relatives from a very early age. It was my first language. . . . My father's father was from Sliven. He spoke Romani but not at home; at home, he spoke only Turkish. He married a woman who didn't speak Romani, so they spoke Turkish at home.

These comments reveal that Yunakov grew up with a strong connection to Turkey despite the fact that Bulgarians could not travel to Turkey during the communist period.

Note that Yunakov, like Ivo Papazov (see Chapter 7), identified as a Bulgarian Turk until quite recently:

YY: We identified as Turks growing up, not as Gypsies.

CS: If you didn't identify as Romani and you called yourselves Turks, then what was the difference between you and the other Turks who weren't Roma?

YY: Well, we called them *dahli*,[2] meaning a lower class. . . . We felt we were the real Turks. . . . Dahli are blond, with a different shape of the head, flat in the back! Our type of Turk was dark. . . . The dahli didn't like us, we didn't intermarry. We had nothing to do with them, we weren't like them at all. We never said we were Roma. Even today, my relatives don't identify as Roma because they don't know the Romani language. . . . To be Romani you had to know the Romani language. There wasn't a person in our group who didn't have relatives in Turkey, even though they didn't know them.

Yunakov's relatives identified as Turks; according to historians they were Roma who abandoned the Romani language and adopted Turkish during the Ottoman period in their effort to move up the social and economic hierarchy (see Marushiakova and Popov 1997 and Chapters 1, and 3). More important, most Bulgarians saw Yunakov as a Gypsy (despite his not speaking Romani). Yunakov apologetically noted that in his younger years he actually looked down on some Romani-speaking Roma: "There are some dirty Gypsies you know, so when we [the wedding band Trakiya] bargained for a wedding fee, we required a roast lamb just for the band so we wouldn't have to eat the cooked food from their plates." His attitude shifted in later years, as I shall describe.

"I'll tell you my life history in short form: all my male relatives are musicians. If a male child was born, he had to become a musician. . . . I have two sons and they both play," proclaimed Yunakov. "The neighborhood was my school" (*mahalata mi beshe uchilishte*; Bulgarian), Yunakov insists, meaning that informal music instruction was the rule. Yunakov's great-grandfather, grandfather, and three uncles were violinists, and his father was a prominent clarinetist: "My grandfather on my father's side, from Sliven, nicknamed 'Kemenche [violinist] Ali,' was a really good violinist. He was the best in the area and he even sang very well. He played only Turkish music. He didn't go to music school; nobody went to music school. There were no saxophones, guitars, accordions in Turkish music at that time. It was traditional music—*ud* [short necked fretless plucked lute], *kanun* [plucked zither], violin, clarinet, and tarabuka, tŭpan, sometimes, but mostly tarabuka. Clarinet was very important."[3]

Like many Romani musicians, Yunakov learned to play the tŭpan first to learn the rhythms; he also needed to accompany his father and older brother at weddings: "My father neither went to school nor learned notes, but his brother went to school and learned notes and so they were hired to play drums in the circus." One uncle played in first-class restaurants, and another joined Yunakov's father's band. Before sound systems were introduced in the early 1970s, multiple clarinets were used for outdoor performance. "One would play, another would play, we would take turns with the clarinet. You had to have stamina because you might lead a dance for seven hours nonstop, just one horo [line dance]. The clarinet players stood all night next to the person leading the dance line; they went home with swollen legs. My father played weddings every weekend, and one wedding would be five days long. . . . I was six or seven and the big drum was taller than me!" This commentary reveals that midcentury Romani musicians had to be versatile and adjustable, and they had to have stamina (see Peycheva 1999a). Yunakov's male relatives played light popular music in the circus as well as folk music for weddings.

Yunakov's second instrument was the kaval (end-blown flute), which he played in school: "I was already in first grade when a teacher, Mitko Angelov, came to teach us music. I was the first person to play kaval in my school; it was the first time they introduced folk music to the school, and I was the first person to sign up. In the first two hours I learned a whole song, I'll never forget that. The teacher said 'What are you doing—you don't even have the embouchure and you already have the notes!'" Yunakov, the only Romani member of the school ensemble, played with them in Haskovo performances for about five years.[4]

Simultaneously, Yunakov started playing the clarinet surreptitiously at home when his older brother wasn't around: "My elementary school was closer than my brother's school, so there was some time free time when I came home. I would grab my brother's clarinet before he got home and play a little. He would beat me up if he saw me touching it. . . . When my father realized that I was already fooling around on the clarinet, he was just amazed."

Regarding his childhood, Yunakov states that although his own community preferred Turkish music, some Bulgarian music was required for patrons. "In my family contexts, there was only Turkish music, but we liked Bulgarian music. It was necessary to know Bulgarian music—hora—for weddings—it was impossible not to know how to play it. Even for a Romani wedding it was necessary." Being a professional musician necessitated knowing multiple repertoires; in the Haskovo region the ethnic groups that Yunakov's family served included Turks, Turkish Roma, Romani-speaking Roma, Bulgarians, and Pomaks (Bulgarian-speaking Muslims). Pomaks migrated from the Rhodope mountain region to Haskovo after the 1950s; their celebrations require extensive knowledge of Rhodope songs. Regional singers were required because they received paid requests from patrons; thus a knowledgeable singer was essential to a band. Yunakov continued: "Yes, we played for Bulgarians as well. For them we wouldn't take the Turkish singers, rather we would take two to three Bulgarian singers. We'd play completely different music."

Yunakov's older brother Ahmed (Mecho) was his model because he had extensive ties with Bulgarian musicians as well as Turkish music and operated successfully in both spheres. Over the years, the membership of the family band changed. Yuri remembers his uncles and his father (on B-flat Albert system clarinet) in his grandfather's band; then his father took over, switched to saxophone, and he and his brother played clarinets

(Boehm system) in the band Aida. Then: "My father started getting ill; Mecho and I decided to relieve him from playing. We told him, you're getting older, we can take over, you've suffered so much. So my brother left the Bulgarian bands in which he was playing and he came back to Turkish music." Their father died when Yunakov was only seventeen.

Although Yunakov knew some Bulgarian music as a young man, he came to master it under the auspices of Ivan Milev. Milev is a legendary Bulgarian accordionist also from the Haskovo region; he was the leader of the prize-winning band Mladost (youth; Bulgarian; see Chapter 7). In 1982 Yunakov was playing clarinet in a small restaurant when Ivan Milev walked in:

> At that time Milev had his own wedding band and everyone knew him, he was really famous. . . . He listened, he drank and drank, and he said to me, "I want you in my band on saxophone." I said, "What do you mean, I have work here in this restaurant. I have my brother's band. No! What are you talking about? . . . I can't even play the instrument you want me to play—the saxophone." Milev said, "You will be able to. . . . You can do better. I'll make you the best saxophonist." I had actually started playing saxophone a little earlier with my brother. . . . So Ivan took me and convinced me. I hardly played Bulgarian music before that. I played when I had to, but not much. . . . Bulgarian music wasn't very clear to me then—then I played mostly Turkish and Romani music. . . . Ivan gave me the foundation of Bulgarian music. At the beginning he showed me the most unbelievable things—he was a virtuoso player with incredible technique. At first I refused—I was scared of his music. But Ivan said: "Now I'll show you that you can do it." So we started slowly, with easy dances; the ornaments were the most important. . . . It was very gratifying. . . . I found strength in myself. . . . Ivan didn't play other musicians' repertoires; he composed his own music, so I had to learn his repertoire. . . . I needed a great deal of time to master his repertoire—maybe a month and a half. . . . Ivan was up at seven or eight in the morning and we would play for twelve or thirteen hours. People were still sleeping but Ivan was ready to play. He would come in the morning, he didn't care if people were sleeping—he was ready to play. The women were running around making food for him all day long because he has a big appetite—more food, more food!

Yunakov recalls that Milev spent so much time at Yunakov's house that it was Milev who took Yunakov's wife to the hospital to give birth to their son Danko. Yunakov compares this incident with the fact that a few years later Ivo Papazov brought his wife to the hospital for the birth of his daughter Ani. His point is that these musicians were totally integrated into his life.

I dwell on Yunakov's acquisition and mastery of Bulgarian music because this was one of the first borders he crossed as he ventured outside his community to become adept in many musical styles. Another bridge to "being Bulgarian" was his boxing career, because boxing was a prestigious state-sponsored public sport. In fact, when Yunakov made his public debut with Milev's band at the Stambolovo festival of wedding music in 1985 (the band won third prize; see Chapter 7), the public knew him only as a boxer. "When I stepped on the large stage at Stambolovo the audience was amazed because they knew me as a boxer. They wondered, when did he become a musician, and start playing with the best of them?" Yunakov trained as a boxer in his teenage years and won three national middleweight championships, but he left the sport disappointed and returned to music. He realized that "you couldn't make a living from boxing. You couldn't support your family. All the best sportsmen from Bulgaria had emigrated. There was always discrimination against Turks and Roma."

Yunakov experienced prejudice in the sports arena during the late 1970s when the socialist government forced all Muslims to change their names to Bulgarian ones (see Chapter 7). Those who resisted were fined, beaten, harassed, or jailed. "I had to change my name to become a boxer. My trainer told me, 'If you want to succeed as a boxer, you have to make your name Bulgarian.' And my father was so angry with me for that that he wouldn't let me into the house for years. He hit me." Yunakov's birth name was Husein Huseinov Aliev, after his father Husein and his grandfather Ali. His government-issued name, Yunakov, came from the legend that his grandfather was a *yunak* (hero; Bulgarian) because he played for *haidutsi* (anti-Ottoman guerilla fighters in the mountains; Bulgarian). His new first name was chosen because Russian astronaut Yuri Gagarin had recently become famous. His brother's name was changed from Ahmed to Andrei, and his wife's name was changed from Nusret to Lidia. Yunakov often emphasized that despite living in Bulgaria for thirty years he never felt wholly Bulgarian because of racism and exclusionary practices. He was lauded for his talent in music and boxing, but there were always strings attached.

Wedding Music: Creativity, Resistance, and Accommodation

Yunakov's life presents insights into selective state-sponsored representations of folk music during the socialist period. Growing up, Yunakov not only never heard his community's music on the radio, television, on a recording, at a folk festival, in school, or played by an ensemble, but also he was bombarded by media that lauded only the socialist brand of Bulgarian music and vilified all others. Socialist officials and music scholars alike claimed Turkish and Romani music was "foreign" to Bulgaria and was corrupting folk music, to which no one listened anymore. The new genre of "wedding music" was seen as the culprit for the decline in folk music (see Chapter 7).

Precisely during this time, Yunakov was charting his path as a musician. He recalls the introduction of electrification and drum sets in the 1970s and his own switch from clarinet to saxophone: "I was developing a really different style on the saxophone. . . . I was used to playing very fast stuff on the clarinet, so I transferred that to the saxophone, . . . long runs and a richer tone." In 1985, after hearing Yunakov play with Milev, Ivo Papazov, the legendary star of wedding music, invited Yunakov to join Trakiya. Yunakov narrates: "I didn't refuse. I was ready, prepared. At that time, every musician's dream was to play with Ivo Papazov. We played together for ten years. . . . I spent more time with Ivo Papazov than with my wife!" As I described in Chapter 7, wedding music reached its apex of popularity in the mid-1980s, with fans crowding wedding tents to glimpse the super-stars. Video examples 7.3 and 7.4 show a rare television performance of Trakiya from 1987. Yunakov recalls: "Hundreds of uninvited onlookers would arrive from miles around." People booked Trakiya years in advance and married in the middle of the week to accommodate their busy sched-ule. Yunakov was earning a typical month's salary in a weekend and was able to build a large house. "Everyone wanted us to play at their celebra-tions—weddings, engagements. Everyone wanted us but we just couldn't travel everywhere. One day we would be at one end of Bulgaria, the next day at the other end, sometimes two weddings in one day, or even three. It was very hard but we needed the money." At this time virtually all the mu-sicians in Trakiya were investing in houses and cars.

The 1980s was also the era of socialist attempts to harass Roma, espe-cially those who were successful. Sometimes Yunakov was harassed just because he was Romani. He recalls a bank robbery in Haskovo: "I was playing in a restaurant and the police made me stop; they took me as a suspect. I was a boxer, Romani. We were in jail all night. My wife was pregnant; she asked them, 'Why are you holding my husband?' but they lied and said, 'We don't have anybody here.' The true culprit was the son of the Secretary General of the Communist party in Haskovo. That is the kind of corruption we suffered." This was the era of socialist attempts to suppress and control wedding musicians. Yunakov and his colleagues were jailed twice for playing the Muslim genre kyuchek; their heads were shaven, they had to break rocks, and their cars were confiscated by the police. Yunakov narrated:

In the early days, we didn't add many new musical elements because we were afraid of the authorities. Those were very difficult years. Our orchestra was the most well known in all of Bulgaria. We were so well known that there were ministers who weren't as well known as we were. Every kid knew us! But the most significant part of this story is that Romani and Turkish music was forbidden. I was in prison for fifteen days twice; also Ivo, Neshko, Sashko, and many other col-leagues. Even Petŭr Ralchev [a Bulgarian], one of the youngest and finest accordionists, was in prison. This was a shameful thing, all because of music! We could stir the poorest and richest with our music. But unfortunately, Bulgarian politicians mixed music with politics. According to me, music has nothing to do with politics; I think music remains music. Our politicians made music political. . . . Imagine yourself in a big field, in a tent where we hold our weddings, and you see fifteen police cars coming. We run away. Imagine Ivo Papazov with his weight, running, because he had been in prison already and he didn't want to go back. They arrested the sponsor of the wedding also, and if we were in a restaurant, the owner too. But in spite of this, we played Romani and Turkish music anyway. Jailing us was the most shameful thing for our country, and everyone learned about it via newspaper and radio. They put us, the most famous, in jail, so other musicians would see. They made examples of us so others would be afraid.

Yunakov's comments suggest that that his performance of kyuchek was not a deliberate antigovernment move, not conscious resistance, but rather a strategic, tactical, and subjective life choice based on his beliefs (see Chapter 7). Yunakov vividly remembered strategies for avoiding arrest, such as posting a lookout on the roof to scout for the police, stylistically morphing a kyuchek into a *pravo horo*, and developing intuition for approaching police officers. Many times Trakiya musicians ran away even before the police arrived. According to Ortner (1995), the literature on resistance unfortunately tends to be "thin" because it is not grounded in thick ethnography. Ortner calls for fieldwork that moves beyond the binary of domination vs. resistance in political terms and investigates cultural ramifications.[5] Scott's explication of "everyday forms of resistance" (1985, 1990) opens up the question of what can be counted as resistance, and his attention to performance as power is useful for music (Ebron 2002:117). Ortner highlights the ambiguity of resistance and the need to ground it in the subjectivity of actors who are all individuals with unique motives and histories (1995, 1999). Thus Yunakov's strategy of performing kyuchek in the face of sanctions made sense to him in aesthetic, cultural, and economic terms. He was intimately involved with Turkish and Romani music; it was the music of his community, and he was making a good living from it.

Yunakov's resistance, however, should neither be romanticized (Abu-Lughod 1990) nor elevated to heroic defiance, because in several arenas he (as well as other wedding musicians) accommodated the socialist government. For example, he ran away from the police, he did not resist the name changes even though his father ostracized him,[6] and he recorded sanitized, censored versions of his music so it could be disseminated via the state media. (Audio examples 7.1 and 7.2 compare a wedding version and a sanitized commercial version of the same piece; see Chapter 7.) Even in Stambolovo in the 1980s, he abided by regulations not to include kyucheks and to "clean up" Bulgarian music. Scott suggests that in public spaces "public transcripts" are performed in order to flatter elites, while backstage "hidden transcripts" express grievances (1990; Ebron 2002:117–118).

Indeed, wedding musicians courted favors with communist officials so they wouldn't be driven out of business. Yunakov recalls private parties where socialist officials requested kyucheks: "These ministers, they were our fans!"

It is difficult, however, to fit weddings into Scott's rubric "hidden." Family celebrations take place in public space (the street, the village square) but still should be coded as unofficial as opposed to official. Furthermore, they are located in the free market realm, one reason the socialist government was trying to regulate them. Precisely here, wedding musicians staged their resistance. They felt they were making an economic point rather than a political one. So then, is Yunakov naïve when he states that music is apolitical? Doesn't he know that prohibiting kyuchek was an anti-ethnic move? His statement may be either a utopian sentiment or a strategic defense of his resistance. Given the range of social acts I have considered, I underscore that collaborations with the dominating order exist side-by-side with acts of resistance; these are the performative contradictions that musicians enact.

Becoming Romani

As I discussed in Chapter 8, Trakiya morphed from a Bulgarian legend to an international touring phenomenon in the early 1990s thanks to the efforts of British impresario Joe Boyd (photograph 8.1). On an American tour, Yunakov made contact with the Macedonian Roma living in the Belmont neighborhood of Bronx, New York (described in Chapters 4 and 5). A group of Macedonian Roma went to hear Trakiya, warmly welcomed them, called them "brothers" in Romani, offered them hospitality, and invited them to play at a private dance party. They accepted the invitation and, as the legend is told, Romani community members were so excited at this event that the musicians received $5,000–$6,000 in tips. After this dance party, Yunakov was invited back to the United States to play at the wedding of a community member, who then sponsored him for a work permit. He connected with Macedonian Romani musicians and started playing for Macedonian Romani events.

At this time, Yunakov decided to stay in the United Sates and try to emigrate because the situation of Roma in Bulgaria was declining rapidly (see Chapter 1). In the mid-1990s, one of his closest friends was permanently maimed by a racially motivated attack. He had earned enough money in Germany to open a nice restaurant in downtown Haskovo, and Yunakov often played there. The mafia, in cahoots with the local police, ordered the business closed. According to Yunakov: "They said 'We can't have a thriving Gypsy business on main street. A Gypsy can't have a successful business in the middle of this town.' When my friend refused to close it down, the mafia, which was tied to the police, brutally beat him up and destroyed the restaurant. He is brain-damaged from the beating. They also beat up my cousin, who is a drummer." In addition, Yunakov claimed that his children, living in his hometown of Haskovo, were threatened with abduction by the local mafia in conjunction with the police, who were former communist officials. Yunakov decided to apply for political asylum, on the grounds of a well-founded fear of persecution if he returned to Bulgaria. I helped him assemble his file and was his translator for his hearing. Asylum was granted in 1995, and he received a green card several years later. In answer to the question, Why did you come to America?" Yunakov answered:

I wanted to live a calmer life. For Roma in Bulgaria, there weren't any possibilities. In my time, there were many terrible things. They wanted to abduct my children—I went to the police—they didn't want to help me—everything was getting worse at that time. I wanted a normal life. I wanted to play—I want to share my knowledge with others. Whatever I have in my soul I want to show people—I want to give my music. I am very uninvolved in politics—I don't even understand politics.

In the United States, Yunakov came to feel he was Romani on a deep personal level. In 2001 he reflected:

I realized I was a Rom when I was already grown. Only here in America, seven years ago, did I feel Romani. I'm speaking here of my inner feelings. Lidia and I have been together for twenty years. I never had the feeling that I was a Gypsy. Occasionally, people have told me I'm a Gypsy but my inner feeling was that I wasn't a Gypsy. I knew who Gypsies were—those that speak that language. I've played at every kind of Gypsy wedding; they are different people, and we are different. I felt that we were Turkish. But here, I understood. I thought about it and understood things, how it is, and why it is. In Bulgaria, the history wasn't clear, who you are, why. Even now, the exact history isn't clear. No one ever told us the true history, from where we came.

I believe one reason Yunakov came to identify as Romani was his immersion in the Macedonian Romani neighborhood in New York. He lived there for seven years and his closest friends were all Roma; he socialized with them and attended and performed at their celebrations; they helped him interface with American institutions such as hospitals, schools, lawyers, and immigration. As discussed in Chapter 4, these Macedonian Roma are Muslim and speak either Romani or Turkish as well as Macedonian (which is fully intelligible to Bulgarians) and English. They are quite clear in private about their Romani identity, although in public they may pass as other ethnic groups (see Chapter 4). Yunakov and his family felt comfortable among these Roma in large part because of their shared music, languages, religion, and culture. He played regularly for Macedonian Romani events. I am not implying that Yunakov admired everything about this community; to the contrary, he often criticized some aspects of the "neighborhood mentality, such as its inwardness. But I believe a cultural tie resonated within him; in addition, this was the only Balkan community in American that accepted him fully (there are very few Bulgarian Roma in New York City). Several thousand Bulgarians live in New York, but they do not constitute an organized community and do not have regular musical events. In addition, most Bulgarians are prejudiced against Roma and do not socialize with them. A second reason for Yunakov rethinking his identity may be his exposure to Romani history via conversations with me. He often asked me what historians have written about Roma, and we discussed the newest theories. A third reason may have been the public attention to Gypsy music, to be discussed here and in the next chapter.

As soon as he arrived in New York, Yunakov widened his musical niche, partly from necessity and partly because of his talent in quickly learning new genres. For several years he was the only immigrant musician I knew from Bulgaria who was able to support himself solely from music.[7] This was possible because he was so versatile. His instrument and his improvisatory style could be adapted to pan-Balkan and even Middle Eastern music. From Macedonian Romani performers, he learned Macedonian styles and repertoire, including the Romani dance gaida and rhythms such as berance (often 3+2+2+3+2) and the myriad types of 7/8. He also learned specific melodies that are popular among Roma, such as "Kjuperlika" and the many songs that are requested at events.

His first regular restaurant job, at the Turkish Kitchen restaurant in Manhattan (where he played for ten years), facilitated contact with several prominent Turkish, Armenian, and other Middle Eastern musicians, and they began to invite him to gigs. On the CD Gypsy Fire he plays Middle Eastern music; on this album the song "Fincan" is sung in Turkish and features the typical Armenian meter 10/8, which Yunakov learned in America. He has been flown to Los Angeles many times by wealthy Armenians for whom he mastered specific Armenian songs and rhythms.

Yunakov also performed with Avram Pengas, who introduced him to Greek, Israeli, and Arabic styles. Turkish music has become his mainstay—not the Turkish music he grew up with, but rather Turkish-American and belly dance music. This scene supports many restaurants, clubs, and family celebrations in New York. Several professional belly dancers began to rely on Yunakov for accompaniment. He feels very indebted to the many musicians who generously shared their knowledge with him: "The person who helped me learn Turkish repertoire was Hasan Isakut [a Turkish Romani kanun player and singer]—his repertoire is huge. Ara Dinkjian [a prominent Armenian keyboardist, guitarist, composer, and arranger] helped me a lot. The more technical aspects of classical Turkish music I learned from Tamer Pinarbasi [a kanun player]."

Yunakov further expanded his repertoire and contacts when he met the legendary Albanian singer Merita Halili and her husband and arranger, the famous Kosovar accordionist Raif Hyseni. They played together in clubs and weddings for several years, and Yunakov learned a large repertoire of

Albanian songs and musical genres.[8] Even before he met Merita and Raif, Yunakov was familiar with some Albanian repertoire through the Albanian-speaking Romani musicians from Macedonia in New York, including Ramiz Islami and Kujtim Ismaili (see Chapter 5). Yunakov collaborated with Kujtim for years and recorded with him in 1993 on an album that Kujtim produced.

Yunakov has also played with the noted Albanian singer Vera Oruçi and her husband, violinist Sunaj Saraçi, and also with the famous Albanian singer Haxhi Maqellara. He recorded a cassette in New York with Sunaj and Vera that was produced by the Cani company, which then distributed the album through its offices in Germany and Kosovo. On this album, Yunakov's name is listed as Juri (no last name), and on Kujtim's album his name is "Albanianized" to Nuri Hysein. Yunakov himself does not pass as Albanian (because he does not speak the language), though other Balkan Roma in New York can do so (in Chapter 5 I discussed the Umer musical family, who speak Albanian).

Not only was Yunakov invited to gigs by Turkish and Albanian musicians, but he also invited them to his Macedonian Romani gigs. I attended several Macedonian Romani weddings where Yunakov brought Hasan or Sunaj, and they were both very warmly received (photograph 11.1). Because both are excellent musicians and both speak Turkish, there was a shared language and plenty of musical repertoire in common. Video examples 6.4 and 11.1 features Hasan and Yunakov at a Bronx wedding, playing with Kujtim Ismaili, Trajče Džemaloski, and Severdžan Azirov. Yunakov has also been engaged to record improvisations in the middle of set musical pieces.

Yunakov has collaborated with New York musician Erhan Umer for many years. As I discussed in Chapter 5, Erhan is a New York keyboardist and singer from Bitola. Since he is talented in programming drum and harmonic arrangements for his instrument, he can work easily with a solo instrumentalist like Yunakov. They are a practical duet in tight economic times because they can sound like a full band; they can thus earn more money than they would have earned with a full band. Yunakov has also exposed Erhan to American audiences. He facilitated inviting Erhan to the several California Herdeljezi celebrations sponsored by the nongovernmental organization Voice of Roma, and they have also collaborated with Seido Salifoski (see Chapter 5).

Most recently, Yunakov has mentored a young Belmont clarinetist, Sal Mamudoski (see Chapter 5). Sal was only a child when he first heard Yunakov at community events and when his parents lived upstairs from Yunakov. Sal became very serious about music and devoted many years to teaching himself. Yunakov has shared repertoire and styling with Sal, but most important he has taught him the sense of how to communicate with an audience. Now Sal regularly performs with Yunakov and toured nationally with him in 2007, sponsored by Voice of Roma. In 2008 Sal and Yunakov, joined by Albanian keyboardist Alfred Popaj, began performing regularly at the club Mehanata under the name Grand Masters of Gypsy

Music. Thanks to Yunakov, Sal has connected with wider audiences, including Americans. Video example 5.51 from the 2008 California Voice of Roma Herdeljezi celebration features Sal, Erhan, and Rumen Sali Shopov, an excellent Turkish Romani drummer who emigrated to California from Gotse Delchev, Bulgaria.

In part because he missed playing the Bulgarian genres of wedding music, Yunakov formed the Yuri Yunakov Ensemble in 1995, a mixture of Roma, Bulgarians, and Americans. The ensemble changed personnel over the years, but there was always a mixture of ethnicities. The Americans were strategically included. I believe I was invited because Yunakov felt indebted to me for my help with his asylum case, plus I could translate (see Chapter 1 and Silverman 2000c). Catherine Foster was his first American protégée; as a female clarinetist, she was a rarity. Lauren Brody, who performed in the ensemble from 1996 to 2002, was a versatile performer who could translate and help with immigration documents. Other Americans who have played with the ensemble are Jerry Kisslinger (tŭpan) and Morgan Clark (accordion). The Yunakov Ensemble recorded four albums on Traditional Crossroads, notably *New Colors in Bulgarian Wedding Music* and *Balada* (recorded on an Australian tour).

From 1999 to 2002, the ensemble showcased Ivan Milev, who emigrated to New York from Bulgaria after he received a green card. This represented a historic reunion on American soil because Yunakov had not played with Milev since the mid-1980s. Milev is featured on the 2001 ensemble CD *Roma Variations*. In numerous cuts from this album, the pair display their improvisatory skills. The ensemble was successful in performing at festivals, universities, and clubs throughout the United States, but its identification with the global aspect of Gypsy music was cemented when it was invited to take part in the Gypsy Caravan tour in 1999 (which will be discussed shortly).

For years, Yunakov dreamed of inviting Ivo Papazov and his other former colleagues from Trakiya to the United States, but the logistics were difficult to arrange. In 1998, Neshko Neshev (accordion) joined the Yunakov Ensemble for a tour to Australia, and in 1999 Neshko and Salif Ali (drums) joined up for the Gypsy Caravan tour. In 2003, I helped arrange a national tour including Ivo, Neshko, and Salif, in addition to newcomer Kalin Kirilov (guitar and keyboard), who was already in the Untied States. The tour was very successful and a disk, *Together Again: Legends of Bulgarian Wedding Music*, was recorded and released on Traditional Crossroads. The label sponsored a second tour in 2005 in conjunction with the release of the album (see publicity photograph 11.2).

The excitement of these reunion tours was palpable. Papazov remarked:

Since Yuri left, we've constantly missed him. He's one of the best soloists; now we are so happy we are together—it doesn't matter if he lives in America, or Europe, it is still as if we are at a wedding. Because we all made . . . our recordings together. Even after ten years, we are very precise together—Yuri quickly gets up to par. We corresponded—we exchanged material ahead of time. With computers it is easier now. With Bulgarians and Americans going back and forth Yuri had continuous information about our music.

Indeed, the musicians had only one short rehearsal in Bulgaria and two days of rehearsals in New York before the first tour began. Moreover, it was Yunakov who had most of the catching up to do because Ivo, Neshko, and Salif had been expanding their Bulgarian wedding repertoire for the recording of their 2003 album *Fairground/Panair* (see Chapter 8). Yunakov, on the other hand, had been playing mostly Romani and Turkish music since he emigrated, and when he played Bulgarian wedding music it was to teach it to Americans. Their rehearsals were basically run-throughs (the concept of slowing music down was unknown to them). Yunakov learned the complicated new arrangements very quickly, but they also decided to revive several older pieces from the 1980s. In addition, Yunakov introduced Macedonian Romani pieces he had learned in America to the others. Aside from music, there were years of stories about weddings and about fellow musicians to recount. Ivo, for example, claimed that because Yuri had such a good appetite, they still request a portion of lamb for him at every Bulgarian wedding, in spite of his having emigrated to America!

Because of my respect for Yunakov's musicianship and my confidence that he could teach well, I tried to facilitate his connection to the world of Americans playing Balkan music. All over the United States (clustered on the two coasts), there is a network of Americans who are involved with Balkan music as dancers, instrumentalists, and singers (Laušević 2007). Yunakov taught Americans saxophone and clarinet at the Balkan Music and Dance Workshop for the first time in 1995, and was asked back several times. He has also been hired as a staff musician at several Turkish music camps and workshops. Many Balkan musicians, though they are excellent performers, lack teaching skills. Yunakov, however, is a gifted teacher and cultivates his relationships with Americans.

In addition, Yunakov is one of the few Balkan Romani musicians (along with Seido Salifoski) to become involved in activist Romani projects. Yunakov himself organized a benefit concert in New York for a Bulgarian Romani orphanage; he has also performed in several other benefit events. In 2005, 2006, and 2008 he played at the Herdeljezi festival in California sponsored by NGO Voice of Roma. His involvement in educational activities has been facilitated by Šani Rifati, head of Voice of Roma, and by me. As soon as he arrived in New York in 1994, I encouraged him to participate in panel discussions, lectures, newspaper interviews, and other informative events about Roma that were associated with concerts and dance parties. Both Rifati and I believe that music can be combined with education, but not all musicians agree and cooperate; for example, in Chapter 10 I described conflicts Rifati had with Esma Redžepova's musicians over the issue of activism, and in Chapter 13 I discuss managers' views of this issue. I believe that Yunakov agreed to participate in these projects not only because he believed in them but also because he wanted to cooperate with Rifati and me and facilitate future connections.

It is clear that Yunakov is not only a versatile musician but also a practical strategist. He is a consummate collaborator and, unlike most Balkan musicians, initiates diverse musical contacts for possible future business. One such contact was Frank London of the Klezmer All Stars; Yunakov performed with London's group at the National Folk Festival in Richmond, Virginia, in 2005. Another important contact is Eugene Hutz, founder of the Gypsy Punk band Gogol Bordello. In Chapter 12, I explore the Gypsy Punk movement, but here I want to underscore that Yunakov responded to Eugene's ideas for joint projects. Yunakov has performed with Gogol Bordello several times and has recorded with them (the recording has not been released). At the 2005 New York Gypsy Festival, Yunakov participated in the circuslike atmosphere of Gogol Bordello by stripping off his shirt, as Eugene does; he even convinced Ivo Papazov to perform with Gogol Bordello. As I discuss in the next chapter, Yunakov does not seem to object to the stereotypes Gogol Bordello portrays in its shows; he is willing to go along with anything that is "good for business," whether it is a stereotypical show or an activist panel.

Yunakov has also cultivated a relationship with the directors of the first New York Gypsy Festival, Alex Dimitrov (a Bulgarian) and Serdar Ilhan (a Turk); he communicates well with them because he speaks both Bulgarian and Turkish. He continued his relationship with Serdar, who sponsored subsequent Gypsy festivals. Yunakov performed regularly in his old downtown club Maia and continues to perform in Dimitrov's newer club, Mehanata on Ludlow Street. He was also involved in helping to lobby the community board for a liquor license for Mehanata, and he even recruited me into that endeavor (see Chapter 13). Another project Yunakov enjoyed was the "Clarinet All-Stars," which took place at several New York Gypsy Festivals; at the premiere of this event in 2005, Husnu Senlendirici (from Turkey; see Seeman 2007), Ismail Lumanovski (from Macedonia; see Chapter 5), Yunakov, and Ivo Papazov took turns in a dazzling showcase of solo reed playing. In 2009 Yunakov was honored at a concert sponsored by the Center for Traditional Music and Dance, and in 2010 his ensemble performed in the New York Black Sea Roma Festival. The most recent configuration includes Mamudoski, two keyboardists (Popaj and Erhan Umer), Erhan's brother Sevim on drums, Ali Ceyhan on dumbek, and Turkish singer Gamze Ordulu. In 2011, Yunakov was honored with a National Endowment for the Arts Heritage Fellowship Award.

The North American Gypsy Caravan tour of 1999 proved to be fertile ground to examine how Yunakov interacted with Roma from other regions. In the next chapter, I analyze the representational dilemmas surrounding the six groups in the Gypsy Caravan tour, including Musafir from Rajasthan, India; the Kolpakov Trio from Russia; Kalji Jag from Hungary; Taraf de Haiduk from Romania; Antonio El Pipa from Spain; and the Yunakov Ensemble. Here I want mention that despite journalists constantly asking him about what unified the groups, Yunakov underscored their diversity and hybridity, calling them "six different kinds of music." He said, "I haven't heard some of these kinds of music before—they are excellent musicians. The Indian music is perhaps closest to me. Of course the Hungarians and Romanians have taken a lot from Bulgarian Romani music." Here Yunakov is commenting on some Rajasthani tunes that sounded like kyucheks to him.

His comment that Taraf de Haidouks and Kalji Jag "have taken a lot" from us refers to manele, the post-1989 Romanian and Hungarian genre that is similar to the Bulgarian kyuchek (see Chapters 9, and 12). Yunakov was both annoyed and proud that his music was imitated, alternately using the words "they stole from us" and "they admire us." In the next chapter, I describe how manele was not allowed to be played by Taraf de Haidouks and Kalji Jag in the 1999 Caravan show because tour producer Robert Browning thought it was not distinctive enough to Romania and Hungary. Yet the genre kyuchek/manele was precisely what unified three of the six groups, and they enjoyed jamming backstage by playing kyucheks. Interestingly, Browning changed his mind by 2001, and in the second Gypsy Caravan tour Fanfare Ciocarlia was allowed to play manele.

The urban-rural dichotomy surfaced on the Caravan tour several times. As I discuss in the next chapter, Yunakov objected to the tattered peasant image of the Taraf de Haidouks and thought it would do damage to the professional image of Roma on the tour. When he personally volunteered to take the members of the Taraf shopping for new clothes and for instrument cases, he did so as an American urbanite helping less-fortunate Balkan villagers (he also gave bags of clothing to elder Nicolae Neascu to take home to his village). But when he learned that the Taraf managers actually cultivated the peasant image on stage, he became very angry with the managers and called the Taraf's appearance a "shame." This was amplified by his outrage that the managers did not "control" the behavior of Taraf members, allowing them to busk in the street. To Yunakov, busking was a low-class activity that professionals should disdain. He was also upset by surreptitious peddling on the part of Taraf members of pirated versions of their own CDs and other Taraf activities such as asking for gifts (gold jewelry) from fans. Yunakov perceived these actions as fulfilling the Gypsy stereotype, and therefore objectionable.

Note that Yunakov resisted certain Gypsy stereotypes (such as dirtiness, poverty, and dishonesty) but not others (exoticism and authenticity). Like many other Romani musicians, Yunakov was neither interested in nor surprised at how Roma were pictured and narrated in advertisements for the Caravan tour. When I showed him what I thought were objectionable images, such as photograph 12.1 (discussed in Chapter 12), he did not think it was worth our effort to complain. Perhaps he knew this was a battle already lost. Moreover, he accommodated to exoticism because it helped to sell tickets. In working with him in his ensemble, I have adopted some of that attitude. When Harold Hagopian (of Traditional Crossroads) contacted me about the publicity he was generating for the 2005 Yunakov/ Papasov tour, he said:

We have to think about marketing. In all these publicity packets, every group claims they are virtuosic. Our musicians really are virtuosic, but how do we convey that? Since no one will recognize Ivo's or Yuri's name or even the term Bulgarian wedding music, we need to figure out a catchy title; the subtitle can be "Gypsy wedding music." How about "Bulgarian Bebop"? Remember these guys were inspired by bebop in the 1970s—by the energy, the improvisation, the creativity, the speed? They passed around underground tapes when it was illegal to listen to jazz. I think this could be a way to sell them to the public.

I thought to myself that a few years ago I would have been appalled at this "misrepresentation." I would have thought, "These musicians don't play bebop, they play Bulgarian and Romani wedding music." But I found myself reacting exactly how Yunakov reacted: "Whatever will help the tour, the sales, the marketing is fine with me, unless of course it impinges on what I play."

Yunakov perceives himself as an urban modern performer whose clothing and music complement his image. He insists his band members wear dressy outfits, and he irons his clothes before every performance. A Bulgarian folk costume would be as foreign to him as it would to an American rock group. Musically, in his own view he is as modern as any jazz or rock performer, even if audiences interpret his music or his ethnicity differently. A controversy over instrumentation arose in 2000 when Yunakov toured Western Europe. His European tour managers, Henry Ernst and Helmut Neumann (from the Asphalt Tango production company), wanted him to replace the synthesizer in his band with a kanun because the latter is an acoustic instrument, and thus more authentic. As I discuss in Chapter 12, European audiences perceive Roma as the last bastion of tradition in a modern Europe devoid of authenticity. Ernst and Neumann told me that the Yunakov Ensemble had been rejected from some European festivals because their music was not viewed as authentic. But Yunakov was adamant in his decision to continue using the synthesizer (see Chapter 12 for similarities to Esma Redžepova's stance about the synthesizer). Resistance, then, is always paired with collaboration (Ortner 1995, 1999). "Neither submitting to power, nor 'resisting' it in any simple sense," Yunakov works through both resistance and collaboration and turns them as much as he can toward his goals (Ortner 1999:158).

On Multiple Identities

Much recent scholarship attends to the global forces affecting music; but rather than focusing only on the dichotomies local versus global or Romani musicians versus non-Romani marketers and mangers, the categories Roma and local need to be interrogated. This echoes Ortner's call to examine the internal conflicts within marginal groups, not just the politics between resisters and dominant forces (1995). All the local contexts of

Yunakov's life evince conflicts over representational practices; we do not have to wait for him to become a transnational musician to see this. Through Yunakov's life, we have also seen the myriad divisions and conflicts within the category Roma; in Bulgaria this label did not include him (he thought he was a Turk) and in transnational musical contexts it includes widely disparate groups with whom he does not identify (such as Rajasthanis). Note that the pan-Romani human rights movement is facing precisely this challenge in using the label *Roma* (see Chapter 3).

Throughout his life Yunakov accommodated to some representations of himself imposed by dominant structures, such as his name (imposed by socialists) and the Gypsy authenticity of his music (imposed by a capitalist market), but he also resisted in select arenas, such as repertoire and instrumentation. Yunakov flatly rejected certain stereotypes of Gypsies he found offensive, such as the dirty thief, but accommodated other stereotypes such as genetic talent and exoticism. Although his music is eclectic and amplified, it is viewed by many in Europe as the last bastion of tradition. Similar to Senegalese musician Youssou N'Dour, Yunakov faces "constant pressure from westerners to remain musically and otherwise pre-modern—that is culturally 'natural'—because of racism and western demands for authenticity" (Taylor 1997:126). According to Taylor, musicians such as N'Dour "are concerned with becoming global citizens and do this . . . by making cultural forms as (post)modern as the west's" (143). Yunakov favors an "alternative modernity," which rejects purity and embraces hybridity. According to Hall, "the aesthetic of modern popular music is the aesthetic of the hybrid . . . the crossover . . . the diaspora . . . creolization" (in Taylor 1997:xxi). Further, "the diaspora experience is defined not by essence or purity but by the recognition of a necessary heterogeneity and diversity; it is a conception of identity which lives through, not despite difference, by hybridity" (Hall 1989:80).

Yunakov is a diasporic hybrid musician; he is open and flexible, learns quickly, and can fit into a wide range of ethnic musical groups. Perhaps this heterogeneity describes a specifically Romani sense of adaptation, or perhaps it is his personal style. He may sound like a free-spirit hybrid, but part of Yunakov's strategy comes from necessity—being excluded—from being an outsider and having to fit in. He arrived in American with neither a stable band nor a saleable music product. His mainstay, wedding music, was not viable in America, so it had to be broken into parts and expanded. His fluency with Romani and Turkish music served as entry points into Albanian, Armenian, and Middle Eastern styles. He was a soloist who needed to find several musical niches because no one musical niche was reliable enough. His role as solo performer (rather than composer, arranger, and organizer) necessitated fitting into other musicians' groups.[9]

Music is Yunakov's language of artistry, commerce, and socialization. Music allowed him to cross borders, but barriers remained. Through his style of music and his dark-skinned physical appearance, he was known by various labels—Muslim, Turk, Gypsy, Bulgarian—which implied alternately inclusion and exclusion. Among Westerners, he could be seen as exotic; among Bulgarians he might be suspicious. He could never be fully accepted by Bulgarians because he is Muslim, Turkish-speaking, and Romani.[10] Even Macedonian Roma, with whom he felt most comfortable, often reminded him he is Bulgarian. In Chapter 3, I discussed how we should resist the urge to romanticize and valorize hybridity as creativity because celebration of hybridity often obscures its economic and political implications (Hutnyk 2000). Embracing hybridity might even suppress a critique of the world music market; in fact, the world music market promotes a depoliticized, consumption-oriented passive hybridity. Striving to essentialize neither capitalism nor hybridity, in this chapter I have instead focused on the negotiating practices within the market that musicians such as Yunakov have fashioned.

12

‿⁀⁀⊙ ⊙⁀

Romani Music as World Music

"Gypsy music" has become both a commodity and a discursive symbol in the trafficking of "authenticity" and "exoticism" in contexts such as world music festivals and tours.[1] Gypsy music, as a participatory, artistic, and processual means of commerce, encodes multiple meanings for performers, as well as for producers, marketers, and audience members. In this chapter, I examine the marketing and consumption of Gypsy music as it charts the relationships among festival producers and managers of Balkan Romani music acts (who provide a saleable item), audience members (who claim to support a liberal, multicultural agenda), the press (eager to create a catchy story), and Romani musicians (trying to make a living). Significant here is that the first three groups are elites with cash to invest, while most performers are members of a marginalized group. However, rather than viewing Roma simply as victims of manipulation, I explore how Roma manage to actively negotiate their representations, albeit within limited options. Sometimes engaging in a type of "self-orientalizing" (Ong 1997) that sells the product, Romani musicians, as well as their managers and producers, are all "cultural brokers" with ideological and economic agendas (Kurin 1997).

How did Romani music become the hip commodity labeled Gypsy music that is now found in world music festivals, in urban clubs, and endorsed by movie stars such as Johnny Depp and Madonna? First, we must remember that historically music is a positive romantic stereotype associated with Roma; second, for hundreds of years music has been a viable Romani occupation; and third, Roma have intimately interacted with non-Romani patrons via music making. Music is currently one of the few arenas for positive articulation of a public identity for Roma; this illustrates the paradox I raised in Chapter 1, namely that Roma are powerless politically and powerful musically. Indeed, music is one of the only bright spots for Eastern European Roma in an otherwise bleak picture. The few musical groups who travel abroad are truly lucky; many successful Romani musicians are supporting whole villages or extended families at home.

As discussed in Chapter 8, numerous festivals and tours of Romani music have been organized all over Europe since 1989 that serve various political and cultural functions. In 1999, the first Gypsy festival/tour was organized in North America. This tour constituted a rich opportunity to analyze the interaction of the American public and press with Balkan Romani musicians in the public commercial sphere. Why did an interest in Gypsy music suddenly arise in the 1990s in Western Europe and North America? One answer is that the end of socialism opened up a new vista for enterprising promoters (Gočić 2000). More important, the French documentary *Latcho Drom* (1993, The Good Road, discussed in this chapter) plus Emir Kusturica's fictional feature films[2] initiated a veritable craze for Gypsy music in the world music scene. These films became cult classics, and audiences began to flock to festivals and concerts to see and hear the performers. These films catapulted previously unknown performers into the world music scene; as an example, the ensemble Taraf de Haidouks went from villagers to stars almost overnight.

Gypsy music festivals can be divided into two broad categories, those sponsored by Roma and non-Roma, the latter usually Western European or North American impresarios.[3] The materials for this chapter fall into the latter category; I focus on two tours of the North American Gypsy Caravan (1999 and 2001) with comparative materials from Western European Gypsy music festivals and the New York Gypsy Festival. The first Western European Gypsy festivals were held in Berlin in 1992 (Musik and Kultureage der Cinti und Roma) and Paris (Les Tsiganes a l'Opéra); these were followed by a number of spectacles under the title Magneten, organized by German impresario Andre Heller. Other festivals followed, among them le Vie dei Gitani (Ravenna, Italy, 2000), Barbican: The 1000 Year Journey (London, 2000), the Time of the Gypsies (a tour to several countries, 2001), and the 1995 Lucerne festival, which featured a Gypsy component. More recent examples are the Iagori festival in Norway in 2005 (www.iagori.com); the annual Khamoro festival in Prague; the annual Gipsy Festival in Holland, started in 1997 (www.gipsyfestival.nl); the Festival Internazionale di Musica Romani (held in Italy since 1993); Barbican: The 1000 Year Journey, held in London for second time in 2007; and the Festival Tzigane (www.festival.tzigart.com), held in France annually since 2000. In addition, Roma often appear in general world music festivals such as WOMAD, the annual Balkan Trafik Festival in Belgium (www.1001valises.com) and Balkan Fever first held in Vienna in 2004 (www.balkanfever.at). In 2007 the Tenth Mediterranean Youth Festival at Akdeniz University in Turkey advertised a Gypsy focus, and the annual Athe Sam Romani festival began in Hungary.

North American interest in Gypsy music grew in the mid-1990s when recordings became available and Balkan groups such as Taraf de Haidouks toured. The 1997 festival Herdeljezi, sponsored by the NGO Voice of Roma, was the first American festival to feature Balkan Romani music in the context of activism. Voice of Roma (www.voiceofroma.com) was founded by a Kosovo Rom, Šani Rifati, and has successfully combined the goal of music programming with direct aid to Kosovo Romani refugees.

Its website states: "Voice of Roma is working domestically, primarily in the San Francisco and Northern Bay Area of California, presenting authentic Romani culture, music and art, counteracting the hype of the romanticized 'Gypsy,' and educating the public about history, current events and the plight of the Kosovo Roma." Since 1997, the California festival has grown to attract hundreds of Americans, and VOR has received funding from state and national folk arts agencies. VOR also presents an annual April International Roma Day celebration and has sponsored several tours of Balkan Romani musicians. A Herdelezi celebration was also held 2007–2009 in Maryland by the nonprofit World Music Folklife Center. VOR strives to present Romani artists; their audiences, however, are mostly Americans. VOR has Roma on its board of directors, has invested in significant educational programming, and has sponsored activist projects in Europe. This distinguishes VOR from most of the festivals mentioned above that are run by non-Roma.

The first multigroup North American tour was the Gypsy Caravan: A Celebration of Roma Music and Dance, sponsored by the World Music Institute of New York for six weeks in 1999; thanks to its success, a second tour followed in 2001. The 1999 tour featured six groups: Musafir from Rajasthan, Trio Kolpakov from Russia, Taraf de Haidouks from Romania, Kalyi Jag from Hungary, the Yuri Yunakov Ensemble from Bulgaria and the United States, and Antonio El Pipa Flamenco Ensemble from Spain. The 1991 tour featured four groups: Antonio El Pipa and Musafir again (under the name Maharaja), Fanfare Ciocarlia from Romania, and Esma Redžepova from Macedonia. (Note that the documentary Gypsy Caravan combines both tours.) In 2004 Gypsy Spirit: Journey of the Roma, performed by the Budapest Ensemble, toured North America.[4] In 2005 and 2007 the Romani Iag (fire) festival took place in Montreal. The annual New York Gypsy Festival began in 2005 (see Chapter 13).

The film Latcho Drom provided both the performers and the structural model for early Gypsy festivals. The model articulates a linear diaspora of Romani music, starting in India, the homeland, and ending in Western Europe, including groups from Rajasthan, Egypt, Turkey, Romania, Hungary, Czech Republic, France, and Spain (Malvinni 2004). Although the linear diaspora model is quite problematic, it serves as the unifying trope, as I critiqued in Chapter 3 and discuss later in this chapter. Latcho Drom is a staged documentary: performers are filmed in local settings prearranged by the French half-Romani director Tony Gatlif. Stunning musical performances and stark visuals accompanied by few words evocatively display artistry and marginality, but the filmic viewpoint is of an outsider looking into a world of supposed "authenticity." There are no naturally occurring contexts, and there is little attention to music as a profession. The film perpetuates several essentialist notions: all Roma are "natural" musicians; Roma constitute a bounded, unified ethnic group; and finally, there was a linear path of migration from India to Western Europe. Although it is beyond the scope of this chapter to analyze this documentary (Silverman 2000a), I emphasize that the film created an iconic sequence in which to present Gypsy music: from India to Spain. Furthermore, the groups featured in the film were the first to travel in world music circuits (e.g., Taraf de Haidouks, Musafir, and Musiciens du Nil from Egypt); finally, the film helped create the viability of the marketing category Gypsy music.

Marketing Exoticism and Authenticity

How is Gypsy music presented to the wider public in world music festivals and tours? Drawing on stereotypes, promoters and marketers emphasize exoticism, which is indeed a theme in much world music marketing (Taylor 1997, 2007). As I discussed in Chapter 3, world music thrives on heightened ethnic and racial difference. Most representations of Roma (like other oppressed groups) have been produced by outsiders, because historically Roma have had little control over hegemonic discourse and symbol systems (Hancock 1997). Recalling Edward Said's "Orientalism" (1978), Roma are pictured as located on the (eastern) margins of Western civilization, furnishing a figure of fantasy, escape, and danger for the imagination.

In the process of exoticization, the most eastern Gypsy groups are the most "orientalized" by marketers and producers. For example, the Rajasthani group Musafir's promotional packet reads:

> Classical and mystical musicians, unexpected instruments played by virtuosos, whirling desert drag queen, devotional and frantic folk dances, hypnotizing snake charmers, and dangerous fakirs, including fire eating, balancing acts, sword swallowing, and walking on crushed glass—a fantastic entertainment! . . . Sufi desert trance music by elegant gipsy wizards. . . . A music of ecstasy, whirlwinded of climaxes punctuated by the gentle gesture of a breathtaking tune. An authentic magical experience [Maharaja, email promotion, July 11, 2001].

The exotic trope also extends to Europe's margins. The poster for Fort Worth's Bass Hall concert (photograph 12.1) on the 1999 Gypsy Caravan tour reads: "Get in touch with your inner gypsy. Join in this impassioned celebration of Gypsy traditions. . . . The elders supply soul and experience, the young speed and energy. Come feel the heat of a Gypsy fire." The imagery includes eight photographs, only three of which feature groups from the actual show that the poster is advertising. The other five are stereotypical pictures of generic Gypsies: a dark-skinned man with a bare chest playing the violin, three women in seductive poses, and a red rose. Clearly, all Gypsy images are interchangeable, for Gypsies are merely a placeholder for the premodern, the exotic "other." Similarly, Zirbel reports that the campfire and caravan imagery used in marketing for the 1995 Lucerne festival heightened differences between the Swiss audience and "others": "such marketing reinforced the belief that the Gypsies were freshly imported, authentic exotics" (1999:38).

Music is an especially fruitful medium for trafficking in exoticism. As I discussed in Chapter 9, in the Balkans exoticism is coded as "oriental" or eastern (Turkish and Middle Eastern), and marked by scales and rhythmic patterns that are associated with the East, Gypsies, sex, and passion. These elements of musical style and text have been appropriated by non-Roma and are now widespread in pop and fusion styles such as chalga in Bulgaria and manele in Romania. On the other hand, otherness is sometimes tied it to inner truths: "The pattern whereby society's Others are recruited from the periphery in order to articulate musically the 'soul' of the more settled members is not an oddity from Serbia" (Van de Port 1999:292). Indeed, African Americans have historically served this role in Anglo-American society. Van de Port's research shows that devoted Serbian fans of Romani cafe music in Novi Sad, Vojvodina, Serbia, need Gypsy musicians to bring out their souls; the "stranger within" brings out "implicit social knowledge" (1999:292).

Exotic others, however, may conflict with local Roma, who are often less fortunate than touring musicians. The 2000 British Gypsy festival Barbican, for example, had an uneasy relationship with local Roma and Travelers while simultaneously capitalizing on "foreign" Roma. Traveler activist Jake Bowers pointed out that no British Travelers performed in prominent locations:

Call me a purist, but surely a Gypsy festival should predominantly feature Gypsies, especially those from the country hosting the event. . . . After receiving a few concessions from the organizers such as extra performances featuring British Travelers and cut-price tickets for Romani refugees, British traveler organizations gave the festival their reluctant stamp of approval. Even crumbs from a table are better than nothing at all in a time of starvation. . . . The trap they [festival organizers] fell into was one of exoticism where "real" Gypsies belong to some other place and time. They didn't consult any British Traveler organizations during the planning but used a world music consultant who wouldn't recognize a genuine Traveler if one slapped him with a hedgehog.[5] Musicians in the here and now were turned down in favor of people whose dress and music represented the there and then. Turks in tuxedos and Rajasthanis in turbans are a world apart from the average British Gypsy site or squalid refugee hotel (Patrin listserv, May 25 and 27, 2000).

Bowers's phrase "a time of starvation" refers to the current hostile climate in the United Kingdom (and elsewhere in Western Europe) for Roma, related to the fear of incoming waves of Romani refugees from the east.[6] Bowers writes: "Outside in the streets, Romani women from Romania were causing hysteria . . . by daring to beg for money. . . . Armed with nothing more threatening than children, the women were being vilified by the national press for threatening the shoppers of Chelsea. A housebreaker from a British Romani family had just been shot dead by a racist farmer causing even greater hysteria about Gypsy intimidation" (Patrin listserv, May 25, 2000). Indeed, we must remember that music festivals take place amidst growing xenophobia and anti-Romani violence. University of Texas Romani activist Ian Hancock took a stance squarely against Gypsy music festivals, finding them a poor substitute for real activism fighting discrimination: "It's not unusual for concerts to be funded for Roma to distract from the real issues. The money could be far better spent" (Patrin listserv, May 24, 2000).

In addition to displaying the exotic, festivals also cleanse, tame, appropriate, and colonize the exotic (Zirbel 1999:72). The structure of the festival is, in fact, a microcosm of colonialism: the Romani "darkies" wait at the margins of Europe (or in Western European ghettos) to be discovered by white promoters; they are then escorted to the west, briefly put on stage, and escorted home afterward. Kathryn Zirbel notes that at the Lucerne 1995 festival there was uneasiness among organizers that the Romani performers would overstep their place; indeed, when they were simply out on the street as nonperformers, they were met with hostility and suspicion (1999:84): "In response to community concerns, it was rumored that the festival organizers had to sign an affidavit promising to reimburse all goods stolen or damaged by the visiting 'Gypsies'" (Zirbel 2000:137).

Racism often lurks beneath artistic adoration. The connection between artistic adoration and colonialism has been noted by Paul Gilroy (1993); in fact, nostalgia for the lost authenticity of the past is often intertwined with domination (Taussig 1987). Rosaldo's phrase "imperialist nostalgia" similarly invokes how colonialists yearn for the very markers of non-Western life they have destroyed (1989). Susan Stewart reminds us that nostalgia is a representational practice (1984), a strategy of representation according to Lisa Rofel (1999:137). For Roma and their producers alike, nostalgia for the premodern authentic is a strategy not only of marketing but also of representing identity via music.

The concept of authenticity is evoked by sponsors and the media to convey the message that Gypsy music festivals are "the real thing." Statements such as "experience the true arts of the Gypsies," "authentic music," "authentic ensembles," "authentic culture," and "Take a ride on the Gypsy Caravan and discover the power and joy of traditional Gypsy culture" (Kennedy Center Performance Calendar, 1999) peppered Gypsy Caravan advertisements. Michel Winter, manager of the Taraf de Haïdouks, commented: "People are moved because they feel they are seeing something that they thought no longer existed" (quoted from the DVD *No Man Is a Prophet in His Own Land*). Similarly, Zirbel reports that for the Lucerne festival Gypsies were depicted as "nostalgic throwbacks in the midst of modern Western Europe, persisting embodiments of older values and customs" (1999:38). Live performances reinforce authenticity even more than the actual music; the musicians perform their Gypsy identities on stage. "Live music performance . . . explicitly and publicly encourages and directs audiences to imagine lives and subjectivities of the performers they see before them" (Zirbel 1999:45; also see Kirshenblatt-Gimblett 1998).

Unlike classical music festivals, where the interpretive and technical abilities of the individual artist are paramount, Gypsy festivals display the amazing fact that Gypsies are on stage: "The 'Gypsiness' of Gypsy music is a construct on the perceivers' part, but which is elaborated and commodified by Gypsy musicians. . . . Gypsy musicians return what is projected onto them" (Van de Port 1999:292). A good illustration of this is the theatrical framing used by the Romanian band Fanfare Ciocarlia: one of the trumpet players yells "Gypsies!" to the audience to initiate the show, ensuring that they are authentic.[7]

Similarly, Robert Browning, director of the World Music Institute, explained to me the challenges of producing Gypsy music: "The dilemma is how to market Romani musicians in an ethical manner given the fact that the Gypsy label has a host of stereotypes associated with it. If we present a Hungarian band or a Rajasthani group, audiences want to know if this is 'authentic' Gypsy music, and, I confess, the answer is very complicated. I don't know how to handle the situation when Gulabi Sapera calls herself 'Queen of the Gypsies.'" Here Browning highlights the fact that marketing relies on audience recognition, which in turn relies on historical stereotypes. Sapera, a Kalbelia dancer from Rajasthan, adopted the label of Gypsy when *Latcho Drom* made Kalbelia dancers famous (see Girgis 2007).

Western European audiences seem to be especially receptive to the trope of authenticity of Gypsy music, perhaps because they feel that they have lost their own authenticity and folklore. Suspicion of the homogenizing effects of the European Union is related to this fear of loss of local culture; this may cause Western Europeans to categorize Gypsy music as traditional and contrast it with their own pan-Europop. Ionitsa, the lead arranger for the band Taraf de Haidouks, commented in the liner notes to the CD *Honourable Brigands, Magic Horses and Evil Eye*: "At last I understand why Taraf de Haidouks is so successful in the West. The West has lost its own folklore and people are saturated with electronic music; they want something more natural." One manager concurred: "I think there is a . . . desire to keep something very pure and very traditional because we lost it—most of the Western audience, Western civilization, they lost stuff like this. . . . Music like Fanfare Ciocarlia, a huge brass band, it seems very rootsy, it hasn't been performed in Europe before. . . . For a world music audience, it can't be too electric, too modern, it has to be old time, roots." This current association of Gypsy music with tradition and authenticity is, however, ironic considering the historic Eastern European exclusion of Gypsy music from the category of traditional (see Chapter 7).

I concur with Paul Sant Cassia (2000) that European "modernity" is increasingly pursued through the celebration of "traditionalism" (282). Tradition and authenticity, however, are not self-evident categories; rather, they must be defined and narrated in discourse. "'Tradition' thus becomes not just something invented in an identifiable (recent) past (as Hobsbawn's contributors suggest), but a way of talking about the past and the present through the identification of certain practices that require preservation"

(289). Romani music becomes a "symbol of marginality . . . not so much power *from* the past, but power that has survived *in spite of* the past, and which is likely to 'disappear' because of the onslaught of the 'modern world'" (293). According to Kirshenblatt-Gimblett, heritage is a mode of cultural production in the present that has discursive recourse to the past (1998). She shows how authenticity oscillates between concealment and discovery of the marginal or authentic.

Sant Cassia elaborates on how the category marginal is "confabulated": the marginal is represented as exotic, as a "unique experience" that is discovered, then the narrating subject confers authenticity, and then authenticity is reproduced on a mass scale (2000:293). Oppression may also confer authenticity, as I discussed in Chapter 3. This advertisement narrates the tie between music and persecution: "Whirling wedding dances. Flamboyant fiddle and cymbalom [sic] music. Passionate lamentations born of centuries of persecution. . . . The Roma have kept alive their history, tradition and religions solely through oral and musical communication" (Dartmouth University concert advertisement, 1999). Similarly, Imre points out that Romani rappers in Hungary capitalize on the authenticity of the ghetto as fertile territory for the artistry of the marginal; they have turned the ghetto into a site of profitable entertainment (2006:663).

Roma themselves, however, do not usually buy into the dichotomy of tradition-modernity. For example, Esma Redžepova rejected the idea of recording an acoustic, more traditional album in her home; this idea was proposed by her cousin Šani Rifati, the sponsor of her 2004 American tour. Similarly, she argued with her Dutch manager, Anton Verdonk, when he suggested that audiences prefer the traditional sound of a tarabuka (hand drum) to a drum set. She insisted that the drum set provides a fuller, more modern sound for her "traditional music." Similarly, a controversy over authentic instrumentation arose when Esma's ensemble toured Western Europe in the early 2000s; they were met with hostile reactions, even booing in Spain, because they used a synthesizer, which is perceived as a nontraditional, modern intrusion. According to one of her managers, audiences want acoustic Gypsy music:

> The controversy is that many people say, "That is a great band, but it is a shame that the synthesizer is there. . . ." The crowds in Europe have this kind of purist view that it should be authentic; the management in Europe has been trying to talk Esma out of using the synthesizer. . . . It has to do with the image people have of a certain kind of music—they want to see that image on stage. They see something modern and they think it is not the real thing. . . . They might be looking for a certain stereotype of what people think Gypsy music is about.

Another manager claimed: "Now it is the fashion to hear real acoustic Gypsy music. . . . If it's amplified, electrified, audiences think it is not authentic. . . . It is quite ironical since Gypsies are very open, very influenced, very open to influences. But it is not accepted if they change something." Another manager concurred: "We can't convince audiences that the Gypsy community in the Balkans uses electric instruments. They want the acoustic way. . . . People want to keep music like it is. They don't see that Gypsies are in flux. They want to keep it so as not to change it. . . . I don't like this kind of purism because it is not so different from colonialism. You like it, it is so sweet, but you don't recognize their reality."

Indeed, labeling the synthesizer as nonauthentic is ironic considering the open and eclectic attitude musicians have toward styles and instrumentation (see Chapter 2). They have historically adopted and adapted both Western and Eastern elements, including rap, rock, jazz, rumba, and Indian motifs. They were among the first musicians in the region to use amplification, capitalizing its association of electrification with the West and with modernity. Also recall Yuri Yunakov's decision to reject his managers' advice to substitute a kanun for a synthesizer in order to make his music more traditional (see Chapter 11). Similarly, Esma was adamant in her decision to continue using the synthesizer: "Yes, there was this argument. My manger wanted us to use the contrabass, not the synthesizer— they wanted an older sound. This is stupidity—Romani music has used modern instruments for a long time. I insisted on having my way and we now use the synthesizer."

According to Simeon Atanasov, Esma's accordionist, "We had an argument at WOMEX [European booking conference]. Now there's a new fashion in Europe to do it the old way, to use older instruments. It is stupid. One time we agreed to use the contrabass—it was a total waste. We were very upset; it doesn't go with the music. The synthesizer fills out the sound." When I told Simeon that the managers claimed that the audiences didn't like the synthesizer, he answered: "The audience likes what you give them if you play well. You train an audience what to expect." Trumpet player Zahir Ramadanov concurred: "These managers are not musicians. They don't understand music, they don't play music. They shouldn't tell us what to play. We are musicians—we know what sound we want."

To these Roma, it was more important for them to control the music than the marketing images that are controlled by outsiders. Esma not only defied her managers' directives about the synthesizer but she fashioned her show as she pleased; for example, in Berkeley, California, in 2001, in front of more than 2,000 spectators, and in defiance of the union rules of the stage crew, she invited her cousin Šani Rifati on stage to dance with her. "What was important here was that culture was an object of self-conscious display and hence control" (Schein 1999:380). Performers, then, manage to exert artistic control, albeit within constraints.

Another marketing trope is depiction of Romani musicians as authentic peasant villagers. Among the groups in the North American tours of the Gypsy Caravan, the two Romanian groups Taraf de Haidouks and Fanfare Ciocarlia are composed of villagers, and the marketing imagery for these groups emphasizes dirt roads, frayed clothing, broken-down fences, old village houses, mud, and farm animals. The imagery of Taraf's village of Clejani is so iconic that Marc Hollander wrote: "All those who take an interest in the Taraf have seen it so often in magazines or in film: the wide dirt roads lined with long row houses—it feels as if you've been there before" (2001). According to one manager, "European audiences admire folklore (but not too staged)—they want naturalism—nothing styled up— just how they normally are. . . . People like groups such as Fanfare Ciocarlia and Taraf de Haidouks because it seems as if they just came in from the agricultural fields—they just washed their hands and picked up their instruments, and then they will go back to the fields."

Peasantry can always be staged if necessary. The Fanfare Ciocarlia album *Baro Biao* features a photograph of a trumpeter aiming his instrument at a chicken in its nest inside a village house; another photograph features a horn player on a horse, with a cow watching. Similarly, the Taraf album *Band of Gypsies* features the band in the back of a farm truck (photograph 12.2). Furthermore, when this album was recorded in 2001 in Bucharest, Taraf managers arranged a mandatory visit to the Taraf village of Clejani for the assembled two dozen journalists and photographers. They set up a huge media event that was filmed by director Elsa Gatlif.[8] She put the Taraf members in an old cart pulled by a tractor and had them driven through the village while playing, with views of the countryside passing by; indeed the album cover features this scene. Another photography shoot was arranged by Masataka Ishida, who had previously photographed Taraf members in designer Yohji Yanamoto's couture. Ishida's idea was to place the musicians in a farmyard, "surrounded by the entire population of the farm, chickens, pigs, and goats, huge bundles of firewood, assorted farming tools" (Hollander 2001).

While I am not denying that these musicians reside in a village, I underscore that peasantry has become cultural capital to market. The 2005 film *Iag Bari*, for example, chronicles the remoteness of the village of Fanfare Ciocarlia to illustrate how the musicians have been transformed into international travelers and cosmopolitan stars. Similarly, in the Taraf video *No Man is Prophet in his Own Land,* singer and violinist Pasalan is filmed standing awkwardly in front of his huge new house playing his violin and doing his famous animal vocalizations; off camera, farm animals echo his imitations. Indeed, manager Stephane Karo commented in the video about Taraf: "Wherever they've been they always end up in the same place." These images implicitly try to convince viewers that there is an authentic village mentality that never changes.

Clothing plays a significant role in the audience perception of authenticity. For example, both Fanfare Ciocarlia and Taraf de Haidouks perform in their everyday clothing, that is, shirts (often T-shirts) and pants. According to Fanfare manager Henry Ernst: "Fanfare never uses costumes at home in Romania for wedding or ceremonies. They just dress normally. On tour in Western Europe they just kept this practice, and afterwards we saw this is what audiences like." His partner Helmut Neumann concurred: "This creates, ironically, authentic Gypsy culture, because Europeans like to see a band which can create a really good party and they came on stage in absolutely normal clothing—not like folklore ensembles." Journalists often write about the appearance of these two Romanian groups: how everyday it is. As record producer Harold Hagopian of Traditional Crossroads comments, "Audiences and promoters want Gypsies to look and act like Gypsies." Similarly, Dušan Ristić, founder of the Serbian band Kal, said, "After the audience pays its money, it wants to see and hear what it expects. They have stereotypes, and the promoters and the managers have to satisfy the audience's taste."

Roma, however, are certainly not all villagers, and among themselves there is often disagreement about presentation styles. When Taraf members showed up at the start of the 1999 Gypsy Caravan tour carrying suitcases with holes, having no cases for their musical instruments, and wearing tattered clothing in which they performed, many of the other performers were horrified. They pointed out that this image would reinforce stereotypes of poor, dirty "Gypsies" and do a disservice to Roma all over the world. Bulgarian saxophonist Yuri Yunakov, who is from an urban clothes-conscious tradition, offered to personally take Taraf members shopping at his expense (see Chapter 11). Some performers spoke directly to Taraf manager Michel Winter about this "disgrace," but Winter replied that audiences actually like the tattered image; it is good for business.

Yunakov's group, by contrast, created an urban sophisticated image, wearing suits and ties and fashionable styles. Similar to Yunakov, Turkish Romani clarinetist Husnu Senlendirici views himself as a modern musician. Seeman relates how Senlendirici conceived the fusion album *Laço Tayfa* in response to his exposure to jazz and his assessment of the Western market. Only with Seeman's insistence did any distinctly Romani music get included. Ironically, Hagopian, who marketed the album in the United States, said it did not sell well perhaps because it wasn't perceived as "authentic" (Seeman 2002:363). By contrast, Winter markets Taraf as village peasants, part of the past, part of tradition, even though they play contemporary eclectic styles.

Esma Redžepova has yet another attitude toward clothing: her male band members wear identical costumes with Slavic folkloric elements (such as embroidery), while she wears traditional Muslim women's clothing (*šalvari*) with modern touches. Her choices reflect the specific musical history of Yugoslavia: in the socialist period, Romani musicians formed amateur collectives and appeared at state-sponsored folk festivals where costumes were obligatory (see Chapter 2). In socialist Bulgaria, by contrast, Romani music was not allowed in state-sponsored festivals. In short, with these examples it is clear that imagery accomplishes ideological work, whether dictated by managers or conceived by Roma.

Whereas scholars have questioned the concept of authenticity because it conveys a static view of history, marketers seek to promote it. Not surprisingly, some Romani performers themselves have begun to use the vocabulary of authenticity and tradition. Recall that in Chapter 8 I described how the organizers of Bulgarian Romfest strive for purity and authenticity. They may be tapping into nationalist discourse as well as responding to commercial forces. Not only have Romani musicians picked up on desirable marketing terminology, but their identities also accommodate authenticity as well as modernity, not finding them contradictory. Similarly, Esma Redžepova sees the diversity represented in Gypsy festivals as due to all the other Romani groups being assimilated, but not hers. "The Roma from Spain are assimilated to the Spaniards—they play Spanish music. The Indians play Indian music. They have all been assimilated except me. My music is authentic Romani music. We Roma in Macedonia aren't assimilated—we keep our language, we keep our traditions."[9] Remember, this is the same performer who refused to remove the synthesizer from her band even though European audiences found it too modern. Esma clearly has a stake in being "authentic"; she created a performance niche in Yugoslavia that displayed a consciously authentic Gypsy identity (see Chapter 10).

Education or Entertainment?

At the same time that promoters produce exoticism and authenticity, they also appeal to audiences to engage with diversity and multiculturalism. North American and Western European audiences for Romani concerts tend to be middle- or upper-class, from eighteen to forty years of age, well educated, with liberal leanings. Although Roma are familiar to them from popular and elite literature and art, and from the current refugee crisis, few have ever met or socialized with Roma. Suspicion is the main emotion in Europe, according to one manager: "Gypsies are present in European countries—like the begging of refugees at train stations. But Europeans know nothing about the culture—only that it could be dangerous. The concert is a window for people to learn something about this culture." Some European Gypsy festivals (for example, le Vie dei Gitani in Ravenna and Barbican in London) include educational components in the form of booklets, museum exhibitions, panels, lectures, and film showings with discussions. These events cover history, discrimination, and diversity, but in locations and times that are separate from the musical program. On the 1999 North American Gypsy Caravan tour, my role as education coordinator included lecturing and writing extensive concert notes. Lectures were always well attended, and many audience members appreciated the historical, political, and cultural information in the notes, but in general only a select portion of the audience is interested in education.

World music events are often assumed to have a progressive agenda. As Hutnyk's work shows, the type of multiculturalism produced by music promoters often turns into a bland form of liberal feel-good politics (2000, and see Chapter 3). Zirbel writes that Gypsy festival "audiences appeared to believe . . . that participating as audiences in such performances . . . constituted an act of progressive solidarity with whatever historical or current oppressions the performers' people were believed to face" (1999:80). In the xenophobic atmosphere of Western Europe, attending a

Gypsy concert or buying a Gypsy CD may appear to be a brave public statement of liberalism, but it can hardly be called activism. It does, however, make audience members feel good about their role. British activist Jake Bowers notes that "multiculturalism is fine and dandy when it is at an acceptable distance" (Patrin listserv, May 27, 2000). Gočić, a commentator on Balkan politics, remarked: "It is sad that the current fashion for Gypsy music, interest in Gypsy folklore, and dramatic depiction of the Gypsy soul has not translated into some kind of concern for Roma suffering. Beyond rousing applause for their musicians, Gypsies need substantial support from the West. . . . Of course, 'the art of the oppressed' is nothing new. Cultural adoration and political discrimination have often walked hand in hand. . . . Renewed interest in the culture of some ethnic group . . . often means it's in deep trouble" (Gočić 2000).

The uneasy relationship between education and entertainment was illustrated by the 2004 dance show Gypsy Spirit, sponsored by Columbia Artists. The show tried to educate the audience about Romani history and diversity by using slides of Roma as a backdrop and by incorporating Indian and Balkan music and dance, as well as its main focus, Hungarian dance. The program notes included commentary by professor and activist Ian Hancock on Romani history and language, and a reference to the Romani Archives and Documentation Center at the University of Texas; in addition, Hancock's 2002 introductory book on Romani history and culture was for sale at every performance. But many venues chose not to print the background information, opting to emphasize entertainment over education. The show was sponsored by the Hungarian Governmental Office of Equal Opportunity, Directorate of Romani Integration, and the program included testimonials from government officials affirming their commitment to Roma. Coming exactly at the time of Hungarian membership in the European Union, these statements rhetorically legitimated the government's involvement in Romani culture.

Ironically, there were very few Roma in the Gypsy Spirit show; the dancers were overwhelmingly Hungarian and the musicians were a mixed group.[10] Although the projected slides were of Roma, they depicted stereotypical poses and iconic occupations (such as fortune telling); along with a recurring motif of fire in the show, a myth was narrated about how Roma were once birds. This staging suggested that Roma were a mystical wandering people from a distant place. Audience members loved the high energy and impressive talent, but I doubt many were educated. When I asked one audience member what he learned about Romani culture, he responded, "It is all mixed up." Another said, "It reminds me of Fiddler on the Roof." A simplistic view of Romani psychology pervaded press releases, and the title Gypsy Spirit morphed into "free spirit." One sponsor said: "Gypsy Spirit is open, free, and very, very creative. They want to experience life to its fullest, and they are still very close to nature. They still believe in things that other people don't, like fortune-telling" (Peterson 2004). The New York Times reviewer, Brian Seibert, picked up on the superficial nod to education:

Like many folkloric shows, Gypsy Spirit . . . purports to correct history, then abandons that ambition in the effort of putting on a good show. The program notes glance at the centuries of oppression the Gypsies or Roma have endured, and the stereotypes with which they have been branded. And then it goes on to celebrate what it calls the "fiery" Gypsy spirit. . . . Except for projected slides of actual elderly Roma, the group is treated less as an ethnicity than as a mythic people [2004:12].

Seibert remarks that he is actually glad that the nod to education is superficial because "the performers and their art are the story here" (12).

Romani musicians have recently become hip images for Hollywood stars and the fashion industry. As mentioned above, Japanese fashion designer Yohji Yanamoto dressed the Taraf in his clothing, achieving a look of distressed chic (Hollander 2001). Johnny Depp has regularly hired the Taraf for his private parties in Los Angeles; Depp states he is "a fan of Taraf as musicians, as artists, as people, as human beings." He made this statement on the video about the Taraf, *No Man Is a Prophet in His Own Land,* and he is now part of Taraf's marketing strategy. Depp's testimonials and his publicity photographs with Taraf are by now iconic, and he gave Taraf its 2002 BBC prize for World Music in front of an audience of millions of television viewers. Depp met the Taraf on the set of the movie *The Man Who Cried*, where they played his family.

Because of Depp, other Hollywood figures have hired Romani musicians for their parties. Hollywood film composer Danny Elfman said: "Fanfare Ciocarlia . . . performed at my birthday last year on a rooftop in Hollywood. I was fortunate to catch them on a world tour, and hired them to perform at my party for the night" (Oseary 2004:443). When Taraf or Fanfare are unavailable, Hollywood stars sometimes hire Americans to play Romanian Gypsy music, and, according to these musicians, the stars expect them to dress up and act like Gypsies, and put on a "wild show." These stars, however, rarely "mention the general plight of the Gypsies. Neither do Roma artists themselves—attaching oneself to an already lost cause is not exactly a good career move" (Gočić 2000).

Roma know that they are paid to entertain, not educate, so they learn not to raise political issues on stage. In fact, several Romani performers, such as Macedonian accordionist Simeon Atanasov (of Esma Redžepova's ensemble), sincerely believe they do not face discrimination. Atanasov is a successful middle-class professional who thinks (as Esma does; see Chapter 10) that Macedonia affords Roma full rights; in fact, he blames poverty on the laziness of Roma themselves. His attitude, however, was in direct conflict with the American tour in which he appeared in 2004. Sponsored by the American NGO Voice of Roma, the tour had a strong educational component, featuring a short lecture on the history of exclusion delivered from the stage before the musical performance by tour manager and Voice of Roma president Šani Rifati. Because Rifati believes strongly in combining music with information, Voice of Roma founded the booking company Romani Routes precisely to foster educational music events and also to help Roma manage their own marketing. (The title of my book comes from this effort.) Rifati is very passionate about using the label *Roma* as opposed to Gypsy, and he has often confronted journalists, managers, and musicians. As discussed in Chapter 10, Rifati engaged in arguments in which Atanasov accused him of being too political and lecturing too much, and thus ruining the music; Rifati accused Atanasov of being a nationalist apologist for Macedonia.

Two Romani musicians who ardently believe in activism and seek to combine education and entertainment are Dušan and Dragan Ristić from the Serbian band Kal. Dragan stated, "We are not living in the past. . . . I'm an urban person, belong to the modern world, [and] go to rave parties . . . so mixing traditional and urban elements is the best way of presenting our culture. . . . And I'm a Roma. Gypsy is pejorative, a misnomer. We've always called ourselves Roma, so I find it distasteful to be called a Gypsy" (http://nygypsyfest.com). On Kal's 2006 tour Dragan actively participated in preconcert lectures and panel discussions organized by Rifati. In addition to the Ristić brothers, Yuri Yunakov has also willingly participated in numerous educational events (see Chapter 11). To educate non-Roma about Romani music, the Ristić brothers started the Amala (friends) Summer School in 2001 in their hometown of Valjevo, Serbia (www.galbeno.com).[11]

On the other hand, virtually all the producers and promoters I interviewed felt that education was not the main purpose of performance. One manager said, "It is not my idea to lead lessons with this music—it is entertainment"; another remarked, "I don't think it is very important for the public to be educated"; and a third said, "I don't like the very open educational style. I don't like to play with education . . . the best way is to take the music and give it to the audience, let them listen to it. They will like it and for those that are interested, there are lectures." One promoter expressed the dilemma between entertainment and education: "I think it is important to give the political background of the countries where the Roma are living. Well, I also understand that people go to a concert for experience and they don't care about politics—just to enjoy music for a couple of hours and forget about real life. So I wouldn't emphasize politics too much in the program—but they could also find time to read about what they are going to see. . . . You can't deny where these musicians are coming from. You shouldn't separate their suffering from their music."

Discrimination became a contested topic when the program notes that I wrote for the 1999 Gypsy Caravan tour were scrutinized by Michel Winter, the non-Romani manager of the Taraf de Haidouks. I had courteously asked Winter for feedback on my notes because the organizers were willing to make changes for the final New York City concerts. Winter insisted that I remove this paragraph:

In the 1970s, Ceausescu's policy of homogenization became more oppressive and Rom culture was targeted. Some Roma were removed from large government ensembles, where they made up 90% of professional musicians. The Rom ethnicity of musicians was frequently covered up and Roma were not allowed to perform in-group music, such as songs in Romani. Since the 1989 revolution, life has considerably worsened for Romania's approximately 2.5 million Roma. While they can now organize their own cultural and political organizations, they suffer numerous attacks on their homes, possessions, and persons. Groups like Taraf de Haïdouks salute the resilience of Rom music under trying conditions.

Winter claimed there was no discrimination against Roma in Romania and Roma could do anything they wanted, "even become president." Trying to mediate between Winter and me, Robert Browning, director of the sponsoring organization, the World Music Institute, reduced my entire paragraph to its last sentence.

The very artists Winter represented, the Taraf members, contradicted his absurd claim that there is no persecution. In conversations with Taraf members I learned of systematic abuse, taunting, and discrimination in schooling, employment, and health care. For me, the most moving moment of the 1999 Gypsy Caravan tour occurred during a panel discussion at Dartmouth University that I led on music and politics. Nicolae Neacsu, the elderly Romanian fiddler in Taraf (1924–2002), recounted his life history: how he left school in the fourth grade to work to support his mother and siblings, how he barely survived the Holocaust, how he was neglected during the Socialist period, and how in the 1990s, only because of Western European acclaim, did he have any respect in Romania.[12] Russian musician Sasha Kolpakov and Hungarian musician Gusztav Varga also shared experiences of prejudice and discrimination. Winter was not present to hear them. In fact, he forbade Ionitsa, an articulate leading younger member of Taraf, to take part in the panel even though he was slated to appear.

My interpretation of Winter's 1999 stance is that he believed Taraf's reputation would be tarnished by the intrusion of politics, specifically discrimination. Perhaps he felt it would hurt ticket sales. However, after the 1999 tour this stance changed; in the liner notes for the 2001 CD *Band of Gypsies*, the author, Marc Hollander, describes discrimination to illustrate how Taraf's talent was ignored by Romanians until it was discovered in the West by Winter. Similarly, on the 2006 Taraf DVD *The Continuing Adventures of Taraf de Haidouks*, manager Stephane Karo states, "There is no respect, in the cities, at least, for Gypsy musicians." The narrative strategy on these albums is to market members of Taraf as unsung heroes who are despised in their own country while exalted in the West. This relates to my point above and in Chapter 3 that discrimination confers authenticity. With Taraf, then, discrimination was either downplayed or, more recently, underlined as the key to Gypsy talent.

On the 1999 tour, Winter exhibited a patronizing attitude toward Taraf members.[13] He implied that Taraf members simply didn't know any better about how to dress and act; after all, they are just Romanian villagers.

For example, Winter made decisions for Taraf members without consulting them and without informing them. For their part, the Taraf members were not critical of Winter; on the contrary they were thankful to him as the person who had discovered them and made them famous. On the other hand, they often ignored Winter and did what they wanted, sometimes selling pirated version of their own albums or privately produced recordings instead of their official albums. Taraf musicians, then, had neither a role in creating their international image nor a desire to modify their image; they perceived themselves as powerless in this arena, and dependent on non-Romani mediators. Winter cultivated this dependency, but Taraf members asserted their independence in other ways.

Self-Stereotyping

Romani musicians seem not to resent the use of the exotic/authentic stereotype; the majority are neither interested in nor surprised at how they are pictured and narrated.[14] Most agree that exoticism helps to sell tickets. Several of Esma Redžepova's films made for Yugoslav audiences in the 1970s and 1980s feature campfires, tents, and caravans, which are totally foreign to her urban, sedentary Balkan culture (see Chapter 10). Similarly, Bulgarian Romani bands feature half-naked belly dancers even though Bulgarian Romani in-group dancing is subtler and clothed (see Chapter 6). Imre's research on Hungarian Romani pop music singers likewise shows how they participate in their own stereotyping; she labels this process "double cooptation, by both state discourses and by commercial media" (2008:336; also see 2009).[15]

Lemon's perceptive research in Russia deals precisely with the interplay of historical stereotypes of Roma and their constructed identities. She shows how non-Romani discourse has molded Romani perceptions of themselves; the Romen Theater (a professional Moscow company composed of Roma, founded in 1931) was significant in this process. For example, on a documentary film shoot in a Kelderara neighborhood the crew insisted on building a campfire in the snow and ordering all the young girls to dance simultaneously, behaviors which were totally foreign to the Roma; yet she learned that "the Kelderara did not criticize how they had been filmed. . . . In fact . . . Kelderara themselves shared and valued some of the same forms of stereotypic representation valued by the crew" (2000:156–157). This supports my point that Roma pragmatically essentialize themselves.

Van de Port similarly points out that Serbian Roma musicians in Novi Sad enact the stereotypes expected of them; Roma are embedded in a hierarchal patron-client relationship that depends on fulfilling dramatic roles (1998). Acton remarks that this is "shown to be as false and demeaning a relationship as that between southern aristocrats and nigger minstrels in the ante-bellum United States" (2004:110). He points to a parallel paradox in films about Roma that is "infuriating" to Romani activists: ". . . in many cases the more authentic the Roma involved in performance, the more powerfully dangerous is the stereotyping, a stereotyping all the more persuasive and damaging because of the authenticity of the actors and the backgrounds, and the fact that Gypsies will be bowled over by the rare privilege of hearing Romani spoken on screen in any context at all" (2004:112–113).

This hearkens back to my example in Chapter 6 of the Macedonian Romani journalist who embraced the stereotypical Romani dance performance of a Dutch group costumed with whips for the men and bare shoulders for the women; what is important to him was that songs were sung in Romani by an international group at a national folk festival. Similarly, Bosnian director Emir Kusturica's film *Time of the Gypsies*, although portraying stereotypes, was hailed by Roma for its use of Romani actors speaking Romani. Acton further explains that the "artistic collusion of the oppressed and the oppressor" is not unique to Roma; it has similarities, for instance, to "blaxploitation" movies, which were in opposition to but could not escape the stereotyping of African Americans in early cinema (113).

Romani musicians, then, do not actively resist stereotyping; they also often employ it fruitfully. Historically, Roma have sometimes believed and transmitted stereotypes (both positive and negative) about themselves, such as their "genetic" gift for music (Peycheva 1999a). Fortune tellers often presented themselves as exotic and powerful to their clients, and Ottoman female dancers capitalized on their perceived sexuality (see Chapter 6). Ong reminds us that "speaking subjects are not unproblematic representers of their own culture" (1997:194). Everyone speaks from a point of view with various motives. "Self-orientalizing" moves should not be taken at face value but should be examined within the webs of power in which they are located. "Self-orientalization" displays the predicaments of marginal "others" in the face of Western hegemony, but it also points to their "agency to maneuver and manipulate meanings within different power domains" (Ong 1997:195). Savigliano, writing about tango, coins the term "autoexoticism," defined as "exotic others laboriously cultivat[ing] passionate-ness in order to be desired, and thus recognized" (1995:212). Romani musicians, who have never been in control of their own imagery and reputations, are quite used to being made (and making themselves) into "exotic others" or "authentic originals." These tropes may be good for business, but more important they are just one of many labels and identities that Roma embrace.

Caravans, Nomadism, and Dilemmas of Romani Unity

The most ubiquitous trope of marketing is the caravan concept itself: the theme of linear nomadic migration, starting in India and ending in Spain. The World Music Institute labeled its two tours the "Gypsy Caravan," and their 1999 press packet described the festival as "a musical journey following the Romany trail from Asia to Europe" comparable to Tony Gat-lif's film *Latcho Drom*. Sponsors for the 1999 tour embellished this idea with "The 1000 Year Journey" (Barbican, London 2000 and 2007); "Take a journey in sound along the winding road followed over the span of centuries by Gypsies" (University Musical Society, Ann Arbor, Michigan); "Roma ensembles take you on a century-spanning journey of authentic Gypsy culture" (Hopkins Center, Dartmouth College); "thirty-five musicians and dancers will lead you on a nomadic musical journey through the traditions of the Roma people (gypsies) from their origins in Rajasthan India to Bulgaria, Romania, Hungary, Russia, and Spain" (Barclay Theater, Irvine, California); "The Gypsy caravan features authentic ensembles representing a sweeping scope of Gypsy migration, beginning with Asian and Indian influences from a thousand years ago and moving westward to contemporary Romanian fiddle music and Andalusian flamenco music and dance" (University Musical Society, Ann Arbor). Building on the migration theme, the World Music Institute arranged for sale of the compact disc *The Gypsy Road* (Alula) at concert venues. In addition, Jasmine Delall's 2006 documentary *When the Road Bends: Tales of a Gypsy Caravan* picks up this theme.

The problem with the theme of linear migration is that it distorts Romani history. As I discussed in Chapters 1, and 3, Roma were probably not a unified ethnic group who left Northwest India at one time, moving westward together; rather, Roma may have left India in several waves and coalesced as an ethnic and linguistic group outside of India (Hancock 1998, 2002). Furthermore, not all Roma are nomadic: Roma in Eastern Europe have tended historically to be more sedentary than Roma of Western Europe. In fact, most Romani groups currently performing in the Gypsy festival circuit have been sedentary since their arrival in Europe centuries ago.

As discussed earlier in this chapter, most festivals begin with Rajasthani music and end with Flamenco. Inclusion of Indian music conveys the simplistic linear message that Rajasthani music today represents what Romani music sounded like a thousand years ago. In truth we do not even know which specific groups in present-day Rajasthan are related to European Roma. In his insightful M.A. thesis, Girgis chronicles how *Latcho Drom* director Tony Gatlif and his music consultant Alain Weber rather arbitrarily selected performers in Rajasthan for the film, and how they were elevated to the category "Gypsies" (2007:87–90).

In my conversations with Caravan organizer Robert Browning, I tried to problematize the message that might be conveyed by including an Indian group: I was afraid audiences would assume that the Rajasthani group Musafir/Maharaja performed the music of European Roma of a thousand years ago. My program notes for the Gypsy Caravan tours therefore dwelt on the symbolic role of a contemporary Rajasthani group, the likelihood that it has no direct relationship to European Romani groups, and the fact that its music has changed a great deal during the last thousand years, as has the music of all the groups.

The market, however, has a life of its own; hence Musafir/Maharaja's promotional materials label them Indian "Gypsies." This label, however, is not merely a current invention; the British also mistakenly applied the term to virtually all professional musicians in India. The market dictates that whatever sells is adopted. Zirbel reports a similar ethnic marketing strategy for the Egyptian group Musiciens du Nil, who also performed in *Latcho Drom*. Although they had never used the label, after the film they were marketed as "Gypsy" (Zirbel 1999:60–64).

Related to the idea of a common Indian homeland is the question of the unity of all Romani musics.[16] The multiple cultural brokers (marketers, managers, audiences, journalists, musicians, and intermediaries like me) have varying views on the topic of the unity of Gypsy music. In general, non-Roma seek a bounded unit to label, describe, admire, or hate; they implicitly reason that if there is nothing that unifies Gypsies, then why bother with a festival? Reporters and audience members alike constantly want to know what unifies all these musicians. Cultural features among the performers are extremely varied: any one concert typically embraces several religions and linguistic groups. Consider the six groups that participated in the 1999 Gypsy Caravan: the Romanians and Russians were Eastern Orthodox, the Hungarians and Spaniards were Catholic, and the Bulgarians and half the Rajasthanis were Muslim, with the other half of the Rajasthanis being Hindu.

In terms of language, communication among the 1999 Caravan participants was almost impossible: Romani was spoken by two of the five Hungarians, a few of the Romanians, and all three Russians—but none of the Bulgarians, Spaniards, or Rajasthanis spoke Romani. I do not wish to give the impression that Romani performers at festivals do not get along with each other. On the contrary, Gusztav Varga, the director of Kalyi Jag, told interviewers in 1999 that all the performers in the Gypsy Caravan seemed like brothers; there was something familiar, perhaps a shared historical sense of discrimination. Indeed, at Gypsy concerts there is definitely a feeling of group camaraderie, but it is derived neither from musical specifics nor from lengthy conversations since few of the performers can communicate with one another. The performers, however, do carefully listen to each other's musics, clearly respect talent, and sometimes jam.

Focusing on music, the media insisted on knowing what the groups had in common and what made Romani music unique. In the last chapter, I mentioned that Bulgarian saxophonist Yuri Yunakov admitted in interviews that the groups in the 1999 tour shared nothing musically, except that the Romanian and Bulgarian groups knew Balkan rhythmic patterns. In general, Romani performers do not find the subject of unity worth discussing, and they often remark that the diversity is a strength of the festivals, a point with which I agree wholeheartedly. On the other hand, it seems as if it is intellectually impossible or morally wrong for journalists to accept the concept of diversity. One review was titled "Gypsy Show Offers a Lesson in Universality." If journalists could not readily find a common musical thread, they groped for one: passion, talent, soul, improvisation, which of course are not unique to Romani music.

Audience members too are caught up in the detective work of figuring out what the groups have in common. According to Zirbel's Lucerne research, one audience member suggested "Gypsies get a certain look in their eyes," while another "suggested they were linked by how they moved" (1999:53). Zirbel asserts that audiences assumed that amid the different musical cultures, "there was a sense that such elements formed a thin veneer over a similar underlying, originary identity. . . . In the case of the Gypsies, cultural difference provides a brilliant surface, but part of the curiosity and excitement for audiences and for scholars . . . has been the alleged underlying unity of racial and geographical origins" (52–53). She shows that Swiss audiences were engaged in a kind of safe nation building (55).

Typical audience comments include this observation of the 2001 tour:

One of the people I went with had JUST finished telling me how she could see/hear the commonalities between the Romanian and Macedonian Rom music but she just COULD NOT see any connection between those and the Flamenco or Rajasthani music. Then the three singers [Macedonian, Spanish, Rajasthani] did their three little bits [a cappella vocal phrases in the finale]—ALL strikingly similar but each done completely within their own styles. They were clearly put together to show EXACTLY that connection, and it was a little obvious but educational nonetheless for those who might still not have figured out what these four groups had to do with each other.

After hearing this appraisal, I still am puzzled as to how to evaluate what the groups have in common musically; clearly what they share is highly ornamented, unmetered (free rhythm, parlando rubato) singing; but neither the ornamentation nor the unmetered singing is unique to Roma—it is all shared with regional musics. One can just as plausibly posit that Muslim influence historically caused the similarities!

When Caravan producer Robert Browning suggested that the 1999 Gypsy Caravan performers arrange a finale, it was quite a challenge since the groups had no single tune, style, or language in common. Yuri Yunakov suggested performing "Dželem Dželem," the Balkan Romani song adopted as the Romani national anthem in 1971 at the first Romani International Congress (see Chapter 3), but the Spaniards and Rajasthanis had never heard of the song and the Romanians hardly knew it. What finally emerged was this (video example 12.1): after an introductory, unmetered a cappella section (performed first by a Rajasthani vocalist and then by a Spanish vocalist), the Hungarians began an instrumental tune, then the Russians joined, then the Romanians joined, and then the Bulgarians joined. The Indians joined, embellishing the rhythm and adding a vocal improvisation, and finally the Spaniards joined, dancing Flamenco to the group's rhythm. The Hungarians and the Russians shared the first tune; everyone else learned it by ear during rehearsal. Audiences remarked that the finale worked precisely because it was so unpolished and allowed spontaneous personal interactions between the groups. Indeed, in the finale Sasha Kolpakov, the elder Russian dancer, flirted with Tia Juana, the elder Flamenco dancer. Personal connections between performers happened despite lack of a common language and musical style.

The question of unity is further complicated in terms of repertoire. In festivals the mandate is to perform Gypsy music, but each group interprets what that means on its own. For example, before the 1999 Caravan tour the Bulgarians had prepared a program entirely composed of kyucheks and songs in Romani, even though at a typical Bulgarian Romani wedding the band would also perform Bulgarian music. When the Bulgarians heard Taraf from Romania perform, they realized that Taraf members sang in Romanian only and played Romanian village dance music; the Bulgarians then adjusted their program to include Bulgarian music. Ironically, the Romanians originally included a manele piece that was very similar to a Bulgarian kyuchek, but Browning cut it because it didn't sound distinctly Romanian. Similarly, the Hungarians also had a kyuchek-like instrumental in their performance, but Browning cut it for similar reasons.

Although we may think Browning was too rash or narrow in cutting these new pan-Balkan Romani styles, we should also realize that Roma themselves can be very possessive and essentialist about their supposedly "unique styles." As discussed in Chapter 3, essentialism helps define distinctive musical symbols of regional and ethnic identity in a competitive and politicized playing field. In other words, if you don't define and defend your own music, you can't sell it as your own. The band members in Esma Redžepova's ensemble were at first quite upset that Fanfare Ciocarlia performed manele at festivals. Accordionist Simeon Atanasov said: "I told the Romanians that they shouldn't play orientala on stage—they should play their own music—Romanian." Esma and Yuri Yunakov both agreed that the Romanians and the Hungarians "stole it from us, from the Balkans." On the other hand, their band members tremendously enjoyed jamming backstage with the younger Romanians precisely because they had the genre čoček/kyuchek/manele in common; the genre is a dynamic means of communication across borders. Also note that for the 2001 tour Robert Browning changed his mind and allowed Fanfare Ciocarlia to play manele; Fanfare even accompanied Esma for one song in 2001, and she sang with this band in the Queens and Kings show for several years (see Chapter 13).

As a result of festivals, many performers have been exposed to new, wider-ranging Romani musics and some have started to collaborate (see Chapter 13). Taraf de Haidouks invited Bulgarian wedding musician Filip Simeonov (see Chapter 2) to perform with them on several tours, to record on the album *Band of Gypsies*, and to play at a gala concert in Romania. As a result of this collaboration, Simeonov performed as a guest on the album of another Romanian Romani group, Mahala Rai Banda. Although Simeonov has no language in common with Taraf members (Simeonov speaks Bulgarian and Turkish, Taraf members speak Romanian and some

Romani), they share the musical genre of kyuchek/manele. Simeonov has now learned some Romanian and is a regular member of Taraf.

Taraf also worked with the Kočani Orkestar for a gala concert and in 2011 toured and recorded with them. Stephane Karo, the Belgian manager of both Kočani and Taraf, orchestrated this 2011 project. Other projects, such as one led by the Hungarian Romani cimbalom player Kalman Balogh, are based on collaborative compositions whereby each artist contributes a portion. All of these projects are fueled by audiences who enjoy watching different groups of Gypsies interact; Balogh's project was filmed as it progressed, and the film was issued under the title *Ušte Opre* (Rise Up).

These examples illustrate some paradoxes. First, Gypsy music means different things to different performers; some groups define it as the music that is distinctly Romani, while others define it as the entire range of music Roma perform. Second, although media critics and audience members look for unifying musical factors that might be indexical to older layers, the one genre in fact shared by the Bulgarians, Macedonians, Romanians, and Hungarians is not the oldest but rather the newest: kyuchek, a post-1989 phenomenon in Romania and Hungary, influenced by Bulgaria and Macedonia. Indeed, the Romanian performers are all avid listeners of Balkan wedding music and pop/folk, and they know the names of the famous Bulgarian and Macedonian Romani stars.

Gypsy Punk and the New York Gypsy Festival

In comparison to Europeans, Americans seem less overtly xenophobic about Roma; negative stereotypes certainly exist, but there is less pernicious violence (see Chapter 4). The reasons for this contrast include that there are far fewer poor Romani refugees in the United States than in Europe, and additionally in the United States there is less contact between non-Roma and Roma than in Europe, so knowledge about Roma is less available in America. Another factor is that the role of the American state is more circumscribed than in Europe. Whereas in Europe the state plays a greater role in citizens' lives in terms of social welfare and regulation, in the United States it does not track citizens as closely as in Europe; thus Roma are more invisible and mobile. Roma are more remote to most Americans than Europeans, so festivals and concerts serve as a window into an inaccessible foreign world.

For some Americans, music serves as the ticket to a fantasy realm of wild Gypsy music, exotic costumes, and freedom. A 2006 concert by the Portland, Oregon, band Vagabond Opera was titled by its sponsors at Reed College "Gypsies, Tramps, and Thieves" (from the song by Cher) despite the band insisting that the word *Gypsy* be omitted from its marketing. The American penchant for imitation (or simulation) of the exotic was illustrated during several 1999 Gypsy Caravan concerts when audience members dressed up in their idea of Gypsy clothing. Some men looked like Johnny Depp in the film *Chocolat*, and some women had colorful skirts, shawls, and gold earrings. This points to the American tendency to participate in Gypsy music, not only to patronize it. In fact, many non-Romani Americans perform versions of Gypsy music (see Chapter 13 and Laušević 2007).

In the last decade, Gypsy music has become a major factor in the New York punk fusion scene, almost entirely comprising non-Romani musicians. In 2002, the *New York Times* featured an article titled "'Gypsy Punk Cabaret,' A Multinational" (Sisario 2002:25 and 36), and in 2005 the paper featured a color photography display titled "The Rise of Gypsy Punkers: A Home-Grown Eastern European Hybrid Catches On" (Sisario 2005:A15); the first sentence of the latter article reads, "How many Gypsy punk groups does it take to start a movement?" (A15). The "movement" started in 1998 with the formation of the band Gogol Bordello and has expanded to include various groups and styles. What is this genre, what is Gypsy about it, what is its relationship to Roma, and why is it so popular?

There is no unified musical style in the Gypsy Punk movement; in fact, we need to ask what is punk and what is Gypsy in the groups clustered under this rubric. Slavic Soul Party, sometimes categorized as part of this movement, plays traditional Balkan Romani tunes as well as original compositions in Gypsy style, jazz style, and Latin style. The Hungry March Band draws repertoire "from their multi-cultural world community . . . influenced and inspired . . . with Latin flavor, Klezmer sounds, polish [sic] jigs, punk rock noise, hip-hop beats and music of the streets" (www.hungrymarchband.com). On rare occasions they feature a recognizable Balkan Romani melody. The bands Gogol Bordello on the east coast (www.gogolbordello.com) and Kultur Shock on the west coast (www.kulturshock.com) are the most popular groups in the movement. According to the latter's web page: "Kultur Shock isn't just the name of our band. It's Balkan punk rock gypsy metal wedding-meets-riot music from Bulgaria, the US, Japan, and Bosnia. Six members, and no two of us *really* speak the same language. You may wonder what brought such an unrelated, mixed-up group of people together, and you can read our biographies to find out. Call our music whatever the fuck you want—we'll still play every song of every performance as if it were our last." Kultur Shock derives its power precisely from the shock and clash of cultures. It was performing Balkan fusion pieces for many years before it was swept up in the Gypsy Punk scene.

With shows at the Whitney Museum and several successful albums and international tours (in 2007 to Japan, recently to Australia, and annual European tours), Gogol Bordello has set the standard for the Gypsy Punk movement. Founded by Eugene Hutz and composed of immigrants from Eastern Europe and Israel, the band combines "the passionate rage of punk with the ragtag theatricality of traditional Gypsy music" (Sisario 2005:A15). Hutz is an eccentric, larger-than-life character actor who had a leading comic role in the film *Everything Is Illuminated*. His acrobatic charisma on stage as a vocalist and guitarist is unmistakable, and his philosophical ideas on the revolutionary potential of music inform his song lyrics. When asked if he plays Gypsy music, Hutz remarked, "It's music of a traveling mind. We are doing music that is authentic even though it departs from its roots and from what is normally known as authentic. But it is authentic because it is the true immigrant experience music" (36). A typical song theme, often performed in multiple languages (Russian; English, sometimes mangled; and Romani, sometimes mangled), describes the refugee/immigrant caught between two cultures and inventing his own. Hutz portrays music as a cacophonous, unstable hybrid that is countercultural: the "chaotic clash of musical cultures. . . . Culture is a living being. The minute the culture is not challenged, it dies" (25). *Multi kontra culti vs. Irony*, the title of Gogol's 2002 album and Whitney show, highlights the satiric quality of the music, almost making fun of itself.

For Gogol Bordello and the Hungry Marching Band, the musical rubric Gypsy seems to signify a loud sound, brass instruments, and the aura of Eastern Europe. The tunes are reminiscent of pop or folk music, sometimes with Eastern European or Balkan motifs. Instrumental improvisation, though prominently featured in Slavic Soul Party, does not play a central role in these two bands; rather, they capitalize less on technique and more on punk texture, the feverish climax of volume and emotion. Perhaps what is most characteristic of their version of Gypsy style has little to do with music but is instead defined by a circuslike atmosphere on stage. A live show features provocative female dancers, cross-dressing, circus costumes, clownlike makeup, and dramatic scenarios, often depicting Eastern European military figures. The distinguishing profile of this music is its edgy, visceral quality rather than its technique.

The Gypsy punk scene was prominently featured in the first weeklong New York Gypsy Festival (www.nygypsyfest.com) in 2005, which had as its logo an image of the Statue of Liberty belly dancing with *zils* (finger cymbals) and with a red rose in her ear. This image combined iconic symbols of Middle Eastern female sexuality and Flamenco; thus Gypsies were simultaneously orientalized (located in the east) but also located in the west. In 2006 the logo of the festival featured the same Statue of Liberty with a red rose, this time playing a tapan. The 2007 festival image featured a trumpet, the 2008 festival featured a violin and the 2009 featured a guitar, and the 2010 festival featured a drum. According to its website, "The program of the festival was done very carefully to allow the mix of many genre-bending acts from punk-rock to jazz, hip-hop, global beats, funk and cabaret music with an underlying gypsy aesthetic."

Until 2007, the festival organizers were two club owners/entrepreneurs, Serdar Ilhan and Alex Dimitrov, a Turk and a Bulgarian who operated the clubs, Maia Meyhane and Mehanata, respectively (see Chapter 11). In 2008, Ilhan (sometimes in conjunction with the World Music Institute) took over programming. Marketing materials have emphasized the diversity of Gypsy music, but very few Roma actually participate in the festivals. In 2005, for example, participating groups were heavily drawn from the Gypsy Punk scene, and Gogol Bordello received top billing. In addition, there were Flamenco dancers, belly dancers, Eastern European folk/pop DJs (see Chapter 13), and Zlatne Uste (Golden Lips; Serbian), a brass band composed of Americans. Two Roma performed in the Russian Romani group Via Romen, and the Clarinet All-Stars were also Roma: Husnu Senlendirici from Turkey, Ismail Lumanovski from Macedonia (see Chapter 5), and Ivo Papazov and Yuri Yunakov (see Chapters 7, 8, and 11) from Bulgaria. More recent festivals have featured fewer Roma and more fusion "Gypsy inspired" genres.

Gypsy Punk fusion is a good example of the phenomenon of "appropriation"[17]: non-Roma pick and choose elements of "the Gypsy" to enact musically and theatrically for audiences of adoring fans. The New York Gypsy Festival differs from European Gypsy music festivals and the North American Gypsy Caravan tours in several ways: there is no reference to India, no diaspora model, no politics, no educational component, no concert notes. It is all about experiential simulation and visceral music. The festival does share themes with European festivals, among them displaying the exotic, the authentic, and the marginal outsider. But rather than dismiss this scene as pure fantasy disconnected from Roma, I think it is important to analyze how the scene works and how Roma are mobilized within it, even though they are not in charge of it and do not reap much profit from it.

I attended the 2005 festival with several Macedonian Roma from the Belmont community (see Chapters 4 and 5), and they were totally baffled at any connection Gogol Bordello and the other Gypsy Punk groups might have with Romani music. They saw Gogol's show as a circuslike parody of their own culture and were insulted and bored. But note that they had no problem with the phenomenon of Americans playing Gypsy music. They liked and danced to the Zlatne Uste Brass band. Thus tasteful, recognizable appropriations are fine with them. The Roma who participated as musicians in the festival had contrasting views of the event. Vadim Kolpakov, a member of Via Romen, felt it was misleading to use the label Gypsy for punk groups like Gogol Bordello; he told me "There's no Gypsy in it!" However, in later years he has collaborated with Gogol Bordello, and most recently with Madonna. Ivo Papazov viewed the Gypsy Punk scene with scorn (remember, it does not exist in Bulgaria), commenting that true musicianship was replaced by "a shallow show." He was, however, proud to play in the Clarinet All-Stars because the quality of the music was high. Yuri Yunakov had the most open and accommodating attitude; he proactively arranged with the organizers to participate in the clarinet extravaganza, and he made sure that he and Ivo Papazov could fit the festival into their tour schedule. He felt the festival offered good public exposure for his music, despite the low pay. During the event, he cultivated future collaborative possibilities with festival musicians. In fact he had already worked with Eugene Hutz and previously played as a guest with Gogol Bordello (see Chapter 11).

When Yuri was invited to join Gogol Bordello on stage at the festival during its climactic final evening, he readily agreed, motivated by the fact that hundreds of fans were ecstatically jumping to their music (Ivo also eventually joined them). His musical contribution could hardly be heard because the volume was so loud, but the visual spectacle was significant.

Yuri surprised his family and Balkan colleagues when he imitated Eugene Hutz's trademark of stripping off his clothes on stage: Yuri took off his shirt and played the saxophone. When I asked him later what motivated him to strip, he said, "It's good for business!"

Thus, although some Romani musicians were critical of the festival, none refused to perform, and some even embraced the future professional and economic possibilities of fusion. Indeed, the New York Gypsy Festivals have all been successful. In addition, Alex Dimitrov opened a new club in 2006 originally called House of Gypsy, and later renamed Mehanata, whose website at one time featured a photograph of a seductive belly dancer on a New York stoop. In the next chapter, I describe my interactions with the sponsors regarding the name and the image, but here I want to underscore that the stereotypes bothered me more than they bothered the Romani musicians.

Considering that Romani musicians depend financially on obtaining gigs and patrons, they have little choice as to where they play and how they are depicted. As I show in the next chapter, appropriations are exploitive of performers at the same time they are good for business (Feld 2000a, 2000b; Stokes 2004; Keil and Feld 1994; Meintjes 1990). Non-Roma appropriate from Roma with neither proper credit nor compensation, but as a result wider audiences listen to Gypsy music and more people buy albums and concert tickets. Roma, in the end, cannot object to the structure of the market because they are dependent on it.

Beyond Caravans: Concluding Comments

Festivals are instructive for investigating the motivations and choices of images and musical styles involved in cultural brokering of Gypsy music for Western audiences. In the world music scene, one can find a huge array of Gypsy musics: Romani music played by Roma (e.g., Esma Redžepova performing čočeks), Roma playing co-territorial music (Taraf de Haidouks playing Romanian village music), Roma playing fusion musics with pseudo-Gypsy elements (Yuri Yunakov playing Gypsy Punk), non-Roma playing Romani music (Zlatne Uste playing čočeks), and non-Gypsies playing fusion musics with pseudo-Gypsy elements (Gypsy Punk bands). In addition, there are a huge group of non-Romani performers who use Gypsy as one element in a long list of fusion styles. Here is a sample of the publicity for the groups featuring Americans that performed in the 2005 New York Gypsy Festival: "The Luminescent Orchestrii is a gypsy tango klezmer punk acoustic string band from Brooklyn. . . . An explosive union of Romanian Gypsy melodies, punk frenzy, salty tangos, hard-rocking klezmer, haunting Balkan harmony, hip-hop beats and Appalachian fiddle, all eaten and spit out by two violins, resophonic guitar, bullhorn harmonica and bass" (www.lumii.org). The band Romashka quotes the words of Bulgarian DJ Joro-Boro: "This lethal dose of gypsy fire water distilled from the Carpati to Canal street, Romashka will kick you in the ass and sing about it." Similarly, the band Cafe Antarsia claims: "This is high passion folk with streaks of Greek blues, Balkan goth/gypsy and sheer working-class fire from the NYC rebels of American opera. Rife with sudden transformation, disaster and ecstasy" (www.nygypsyfest.com). The stereotype of the passionate Gypsy has clearly been appropriated by non-Romani musicians.

We can also see that performers, whether they are Romani or not, make strategic artistic choices. Economics informs most choices about style and image; however, there are varied interpretations as to what sells. Whereas some promoters do not want politics to spoil the entertainment, others believe audiences need to know about persecution and that historical and political information augments the multicultural agenda of world music festivals. Romani performers themselves have varying artistic and historical interpretations of what Gypsy music is, which they then must negotiate with promoters and managers, who often have other interpretations. Whereas some Romani performers actively take a stake in creating their own images, such as urban sophisticates, others passively collaborate with their promoters to create other images, such as backward peasants.

Historically, Romani musicians are used to performing for varied audiences with varying expectations; they are also used to hostility. These skills are useful in the European festival circuit, where their exoticism and authenticity are displayed on stage while xenophobic sentiments rage outside. In the United States, where audiences know less and interact less with Roma than in Europe, Gypsy music has become a major factor in the punk fusion scene. In all these contexts, Roma negotiate their identities performatively on stage and off, sometimes rejecting, sometimes ignoring, and sometimes embracing stereotypes. Following Appadurai (1996), my approach has emphasized the sphere of the artistic and the imaginary, but always embedded in the political economy of inequality. As Nonini and Ong have written, "The concept of imaginaries therefore conveys the agency of diaspora subjects, who while being made by state and capitalist regimes of truth, can play with different cultural fragments in a way that allows them to segue from one discourse to another, experiment with alternative forms of identification, shrug in and out of identities, or evade imposed forms of identification" (Nonini and Ong 1997:26). Romani musicians excel precisely in this fluidity of cultural fragments. Whereas non-Romani audiences seek unity in culture and authenticity in music, Roma play with hybridity and with novel combinations, honing their adaptability. As Lemon remarks, "It is not Roma who find 'hybridity' problematic, but non-Roma who see it as shifty" (2000:212). Roma embrace a surprisingly modern cosmopolitan sensibility while dutifully fulfilling their multiple roles: either as Europe's last bastion of tradition or as New York's vanguard of punk fusion.

13

⌒⊙⊙⌒

Collaboration, Appropriation, and Transnational Flows

The global musical landscape of Balkan Romani music has expanded dramatically in the last two decades; in 2002 *Time* magazine's music section proclaimed that "Roma Rule" (Purvis 2002:70–71), and in 2007 *Newsweek* wrote "The World Embraces Gypsy Culture" (Brownell and Haq 2007; see also Bax 2007). With multiple BBC Planet awards from 2002 to 2008,[1] Balkan Romani music became increasingly visible in Western Europe and the United States. In addition, Bulgaria's Jony Iliev, Romania's Fanfare Ciocarlia and Mahala Rai Banda, Serbia's Kal, and several remix albums were all high on the European pop music charts in the years 2003–2010.[2]

In addition to buying albums, audiences heard Balkan Romani music on the *Borat* movie soundtrack and can currently dance to Gypsy remixes played by DJs in clubs in New York, San Francisco, London, Frankfurt, Brussels, Berlin, Vienna, Paris, and Amsterdam, often under the banner "Balkan Beats." Finally, Madonna's 2008–09 Sticky and Sweet tour included a section titled *Gypsy*, featuring the Russian Romani Kolpakov Trio.

What does all this mean for Romani music and musicians? Before we celebrate too glibly, we need to investigate not only the transmission of musical styles but also the flow of international capital and media attention. This chapter ties together previous threads to discuss issues of collaboration, appropriation, and the transnational movement of music in relation to the political and economic matrix. I address how Roma historically have appropriated from non-Roma and from other Roma, and how non-Roma are currently appropriating from Roma; moreover, I interrogate who is producing and marketing Romani music and how power relationships are implicated in these exchanges. I further examine issues of ownership and compensation through preliminary case studies of DJ remixes and the movie *Borat*.

Collaboration

Collaboration, at first glance, seems to be less messy than appropriation. In several chapters, I discussed collaboration between Romani and non-Romani musicians, examples being Esma Redžepova's duets with folk and pop singers and Yuri Yunakov's, Sofi Marinova's, and Azis's projects, along with the widened exposure that resulted. Turning to projects among Roma, I note that some collaborations are more artistically and commercially successful than others.[3]

The 2007 album *Queens and Kings* (Asphalt Tango) features collaborations between the Romanian brass band Fanfare Ciocarlia and guest Romani vocalists not only from the Balkans (Šaban Bajramović from Serbia, Esma Redžepova from Macedonia, Jony Iliev from Bulgaria, Ljiljana Butler from Bosnia, and Dan Armeanca and Florentina Sandu from Romania) but also from Hungary (Mitsou) and France (from the band Kaloome). The Serbian band Kal is also featured as a guest. According to Fanfare manager and producer Henry Ernst, the idea behind the disk was generated by the band, and management facilitated it; in one bold move they complemented Fanfare's instrumental virtuosity with top vocalists. This collaboration was feasible precisely because many of the artists are managed and produced by Henry Ernst and Helmut Neumann of Asphalt Tango; the others willingly cooperated.[4] Non-Romani producers routinely do the behind-the-scenes work paving the path to collaboration; they often envision and orchestrate artistic products. The album was very successful and has led to numerous tours; it reached the number two spot in the European world music charts only a few weeks after its release.

Asphalt Tango has pioneered in forging imaginative and tasteful collaborations that also seem to be fair to Romani artists in terms of financial compensation. Fanfare Ciocarlia's huge success is based on the band's talent and versatility, facilitated by its managers. In addition to the artists already mentioned, Fanfare has recorded with the Bulgarian women's choir Angelite (on the song "Lume Lume") and with the Croatian guitarist/mandolinist Aco Bocina (on the album *Aco Bocina and Fanfare Ciocarlia*, Ponderosa 2003), and it performed with Kodo drummers at the 2004 Japanese Earth Festival. Fanfare also appears briefly in the German film *Head On* (2004, directed by Fatih Akin), and it figures prominently in the soundtrack for the movie *Borat*.[5] In fact, Asphalt Tango deftly coordinates collaboration among its Romani artists and facilitates their recording together. Similarly, Piranha facilitates collaborations between its artists: the Serbian brass band of Boban Marković has collaborated with Frank London's Klezmer Brass All-Stars.

Reflecting on Fanfare's collaboration with Jony Iliev on the song "Godzila" on the album Gili Garabdi (audio example 13.1 with text supplement), Henry Ernst explained: "The band met Jony at a festival years ago. They started to have a good relationship, and I proposed that they play together. During the arrangement stage for their album I proposed to invite Jony to join them and the band loved this idea." The song, however, has a more complicated history. Most non-Romani fans in the west associate "Godzila" with Iliev from his own album (Cartwright 2005b:258; audio example 13.2 with text supplement) or from the Fanfare remake, both widely available in the west from Asphalt Tango. More recently, however, wider audiences know the song from its reissue on DJ Shantel's album *Bucovina Club 2* (Essay, 2006), discussed later in this chapter.

Bulgarian and Macedonian Roma, on the other hand, including those in the diaspora, associate "Godzila" with the band Kristali, from the Bulgarian city of Montana (several parties claim copyright). One of the premier Romani bands in the diaspora with a steady output of albums through the Payner label, Kristali is led by bassist Kiril Dimitrov and featured the masterful vocals and clarinet of Aleksei Atanasov Stefanov (Alyosha), who now performs with Orkestŭr Universal. Although it doesn't have the media visibility of "mainstream chalga" (see Chapter 9), Kristali's albums sell well, especially among Roma. Its performances feature songs in Bulgarian, Romani, and Turkish languages, but it is especially prolific in Romani songs.

Kristali's version of "Godzila" (video example 13.1 with text supplement) shows a clever combination of migration themes and popular culture images. The text is a love song from the point of view of a man who is lamenting the emigration of his girlfriend. The song also turns on the double meaning of Godzila: it is the girlfriend's name and also equates with the name of the movie monster. The video depicts the band entering a movie house in Bulgaria, watching the film *Godzilla*, and eating popcorn. Amidst gory scenes of the monster, Alyosha mounts the movie stage and, as his shadow overlays the film, sings about his girlfriend.

Audio example 13.3 features a live recording of "Godzila" by Kristali with guest vocalist Džansever (see Chapter 2) singing harmony at a Romani wedding in Bujanovac, in southern Serbia, near the border with Macedonia and Kosovo; you can hear shouts of *jaša* . . . (long live . . . and then names of musicians and wedding guests are inserted). This recording illustrates the transnational reputation of Kristali; the band is regularly hired for Romani events not only in Bulgaria but also in Macedonia, Serbia, and the Romani diaspora in Western Europe (this example features a Romani wedding in Serbia with a Bulgarian band and a guest vocalist from Macedonia). Many videos of Kristali and Alyosha performing at private Romani parties in Western Europe as well as concert videos from Bulgarian and Macedonia can be found on YouTube.[6] In addition, Macedonian Roma in New York regularly listen to Kristali.

These versions of "Godzila" illustrate the collaborations among Roma that are regularly occurring outside the recording industry. Famous artists such as Džansever, Kristali, Husnu Senlendirici, Amza, Ćita, Erdžan, Ševčet, and many others are hired in various combinations for family events in the Romani diaspora, where they are exposed to one another's repertoire and styles. It is a very rich context for fertilization, but the musical products are rarely on commercial albums; on YouTube, however, they receive thousands of hits. Parallel but rarely intersecting with this in-group Romani universe is the universe of non-Romani fans and commercial products. Non-Roma know "Godzila" from Jony Iliev on Western labels; Roma, on the other hand, don't buy Western labels because they can't afford them and seek alternative sources. Thus Jony, although famous in his hometown region of Sofia and Kyustendil (Cartwright 2003b, 2005b), is hardly known among Roma in Macedonia and in the diaspora. Roma know "Godzila" from Alyosha and Kristali, not from Jony; they rely on duplicated recordings obtained from relatives and friends. More young Roma are trading digital recordings and videos via the internet; as I mentioned, some young Macedonian Roma have thousands of songs on their computers and iPods.

To further illustrate the issue of collaboration, I turn to the popular Romanian band Taraf de Haidouks,[7] that invited the Bulgarian wedding clarinetist/saxophonist Filip Simeonov to perform and record on their albums *Band of Gypsies* and *The Continuing Adventures of Taraf de Haïdouks*. Taraf and Simeonov share the manele/kyuchek genre; in addition, because he is from north Bulgaria, Simeonov was familiar with Romanian village folk music, and he could learn tunes by ear very quickly. Conversely, Taraf members kept abreast of Bulgarian trends and followed respected performers such as Ivo Papazov, Yuri Yunakov, Neshko Neshev,[8] and Filip Simeonov. Yet collaboration is very delicate. In 2005, Simeonov complained about how little he played in a Taraf concert (in one ninety-minute concert he was featured in two numbers), and about how little he was paid; he also complained about lack of communication (he spoke Bulgarian and Turkish, and Taraf members spoke Romanian and Romani; he has since learned some Romanian); on the other hand, he was immensely grateful because of the economic crisis in Bulgaria.[9]

Like the Asphalt Tango managers/producers, Belgian managers Michel Winter and Stephane Karo of Divano Productions (together with the label Crammed Discs) arranged the terms of Taraf's collaborations.[10] For example, one reason they invited guest musicians Simeonov, Turkish darabuka player Tayik Tuysuzoglu, and the Macedonian Kočani brass band to perform in Taraf's 2000 Bucharest concert and for the subsequent recording *Band of Gypsies* was to draw the attention of the press. Indeed, before 2000 Taraf had been ignored in Romania despite their international success. In a wise marketing maneuver, Karo "decided to invite foreign journalists to Bucharest, to the village to see the reactions of the Romanians. . . . We would give them the works" (DVD *No Man Is a Prophet in His Own Land*). They also invited a European film crew to document the whole event, as described in the liner notes to the CD and on the DVD.

Through collaboration, Taraf's managers also wanted to show musical connections across borders. Winter explained: "It's all regional music. All these elements have always intermingled between Turkey, the south of Yugoslavia, Bulgaria, Romania, and Greece. These types of musics all have common elements, so this combination is in no way artificial. . . . They have a lot in common in terms of music, languages. The music of Taraf has a very Turkish sound to it" (*No Man Is a Prophet in His Own*

Land liner notes). Although Winter's comments emphasize unity, scholarly work shows that before the 1980s Taraf's music had little in common with Turkish music and the Romani musics of Bulgaria, Turkey, and Macedonia. Archival recordings, Speranta Radulescu's recordings of the early 1990s (released in 1996), and conversations with older Taraf members all reveal that before 1989 Taraf played mostly regional Romanian music and some Romani songs. Romanian manele in its current form is a recent phenomenon.[11] But as I discussed in Chapter 12, both audiences and the media want to see connections, not diversity; they then assume all connections are deeply historical when in fact some are quite recent.

Reinforcing the theme of unity, Cartwright's liner notes for Fanfare Ciocarlia's album *Queens and Kings* read in part:

> Casual observers may wonder how Fanfare Ciocarlia's roaring Balkan funk could possibly fuse with the flamenco guitars of French Gitans Kaloome or Macedonian legend Esma Redžepova's accordion driven music? Zece Prajini's musical magicians shrug off such concerns, noting that they share elements of language, experience, and an almost indescribable yet very Gypsy musical synergy with their guests. Hungarian music has permeated northern Romania for centuries, while Yugoslav and Bulgarian music came from encounters with travelling Gypsy communities or on pirate cassettes. Spain and France existed in pre-war memories, lost yet not forgotten Latin connections; as did jazz and pop flavours long filtered through closed borders. From these sources and their own ancient Gypsy roots, Zece Prajini's musicians built Fanfare Ciocarlia. Here, accompanied by some of Europe's finest singers, Romania's brass dervishes share tales of life, love and loss. "Queens and Kings" celebrates unity in diversity. . . .

Note that the terms "unity in diversity," "ancient Gypsy roots," and "Gypsy musical synergy" allude to a timeless mystical connection that these groups somehow magically possess (see Chapter 12). In actuality, the Balkan performers on the album currently share the Romani language and the new manele/kyuchek genre; even Kaloome's song "Que Dolor" works well as a kyuchek because its melody is so similar to Fanfare's and Dan Armeanca's song "Iag Bari"; in fact, the band uses the exact same instrumental break.

Appropriation

Collaboration is a concept used by many non-Romani musicians and producers when describing their relationship to Romani musicians, as if the term itself guarantees equal participation and equal benefits. Most non-Romani musicians subscribe to the belief that hybridity and fusions are inherently liberating (see Chapter 3), and some Roma would agree. Other Roma, by contrast, are aware of the slippery slope from collaboration to appropriation to exploitation. By appropriation I mean taking music from one group and using it in other musical projects, usually for profit.[12] I am aware of the underlying essentialism in the concept of appropriation; music cannot be ultimately assigned to unitary "sources." Postmodernists would argue that neither music nor any other part of culture is owned by individuals or groups, and I would agree that music cannot be ultimately owned; intermingling has always occurred.[13] Notwithstanding this observation, certain musics are associated with certain groups or individuals and do get used in new contexts outside the group.

Drawing from Murray Schafer, Steven Feld uses the term "schizophonia" to mean a split between source and use: "By 'schizophonic mimesis' I want to question how sonic copies, echoes, resonances, traces, memories, resemblances, imitations, duplications, all proliferate histories and possibilities. This is to ask how sound recordings, split from their source through the chain of audio production, circulation and consumption, stimulate and license renegotiations of identity" (2000a:263, 1994:258). Although Feld seems to presume one source, I would prefer avoiding the terms *source* and *origin* since for Roma they are often irrelevant. Rather, I focus on the process of transmission, that is, giving and taking, and ask who orchestrates and who benefits from these exchanges?

Historically, Roma have been characterized as the ultimate music appropriators. They have been accused of neither having nor creating music and merely appropriating the music of other ethnicities.[14] Although it remains true that Roma have taken numerous musical elements from co-territorial peoples (as well as from India and Western classical and pop music), it must also be remembered that they do not take indiscriminately but instead borrow selectively and then creatively rework what they take.[15] Throughout this book I have stressed that Roma contributed to many musical styles and genres in the Balkans: čoček/kyuchek, Bulgarian wedding music, and chalga. When Roma appropriate, however, their class relationship is rarely altered; no matter how powerful their music, they do not become powerful politically. They may supply a desirable commodity, but they have not lost their stigma. Furthermore, they still need patrons; even the most famous performers are managed by non-Romani producers. Appropriations by the powerful are different from appropriations by the marginal; when the powerful appropriate, the marginal often lose in the process because they can't fight back in terms of ownership or copyright.

Musical appropriations by non-Roma from Roma thus need to be investigated in terms of motivation, profit, and artistry. In the category of appropriators, we can find a diverse group: nonprofessional performers of Romani music, belly dancers who liberally use the term Gypsy, composers and arrangers who produce Romani music under their own name, Gypsy Punk musicians, DJs who remix Romani music, and celebrities such as Madonna. I will briefly discuss these groups in turn (but will leave Madonna and DJs for my next research project). Note that access to money and resources is not uniform in this group of appropriators.

Non-Roma who play Romani music as amateurs typically do not earn much money and are not internationally known. Balkan Roma tend to applaud these groups, have trained some of them, and sometimes collaborate with them.[16] The same benign attitude is often found toward non-Romani belly dancers for whom Romani musicians provide accompaniment. The term *Gypsy* has been widely appropriated into tribal belly dance, an American genre that features group choreographies (unlike Oriental or Egyptian belly dance), hand-made costumes (usually of natural fibers) reminiscent of Central and South Asia, a discourse about empowerment of women, and a focus on the tribe (group cohesion) that is imputed to be a Gypsy trait. The troupe name Gypsy Caravan is used in many locales, including Portland, Oregon (www.gypsycaravan.us); New Jersey (www.gypsycaravanenterprises.com); and Chicago; other popular names are Ultra Gypsy and Romani (Sellers-Young 2006:296). A founder of the genre describes Tribal as an "American fusion of elements from many countries along the Romani trial" (Sellers-Young 2006:285). In addition, tribal dancers often reproduce images from the film *Latcho Drom* as a source of inspiration (292). Some belly dancers have a superficial stereotypical grasp of Romani arts; others have adopted an educational and even activist role in relation to Romani dance.[17]

Goran Bregović

Whereas belly dancers are not routinely criticized by Balkan Roma, Goran Bregović is an object of wrath and is even labeled by some Roma as a thief and robber. At the same time, he is perhaps the most widely known performer/arranger of Gypsy music in the world, getting top billing at Gypsy and world music festivals (Marković 2008 and 2009). Why do Roma speak of him in these condemning terms, whereas they speak more kindly of other "collaborators"? How has Bregović's history pulled him squarely out of the category of collaborator and into the category of appropriator? Born in Bosnia of mixed Serbian and Croatian heritage, he was a rock guitarist in the 1970s band Bijelo Dugme (White Button; Bosnian/Croatian/Serbian), which pioneered in performing rock-folk fusions of all the ethnic groups in Yugoslavia. In the 1980s and 1990s, he became internationally famous for his musical scores for Bosnian film director Emir Kusturica, whose films deal with Romani themes and employ Romani actors.[18] The movie *Underground* (1995), for example, prominently features the Boban Marković Serbian brass band.

Although *Underground* helped launch Marković's career, he was critical of Bregović for not giving him proper credit. Indeed, according to the 1995 Polygram CD all the music in *Underground* is composed, directed, produced, and copyrighted by Bregović. This includes an instrumental version of "Čhaje Šukarije," Esma Redžepova's hit, which he rerecorded with Polish vocalist Kayah in 2000 with the credit line "Gypsy folklore." Esma commented: "His music is not original. Those records he makes, they use a lot of my songs. . . . I am not happy about this" (Cartwright 2003:98). She expanded: "Goran Bregović . . . took something from everybody. . . . He took 30% of my music and then some of Šaban Bajramović and other Roma musicians. So he made music for business. There's no quality in it" (Cartwright 2005b:109).

Marković claims that many tunes from *Underground* are his; other Roma say they are "traditional." According to Marković: "Bregović . . . well, we worked together and the music is my idea. One part is from a Šaban song and the other part is mine. I took the winning tune from that year's Guča [Serbian brass band festival] and played it and Bregović added his things and when the soundtrack came out he did not credit me for writing the music. This made me very angry. Bregović has asked me to do more work with him, but I've established myself and don't need to work [with him]. If we do work [together] in the future it'll be on my conditions" (2005b:76–77).

One *Underground* hit song can be traced directly to Šaban Bajramovič, one of the greatest Serbian Romani singers (Cahn 2001b; Cartwright 2005b:52–67). "Mesečina" (Moon; Serbian) is based on Šaban's song "Djeli Mara" (Mara left); Bregović transforms it from a soulful ballad with piano accompaniment into a brisk brass arrangement. Šaban had an ambivalent attitude; he has been quoted as saying that Bregović and Kusturica stole his songs, but when French Romani activist Cahn asked him whether he signed away his rights, he said he didn't remember (Cahn 2001b). "They took my song. . . . I probably signed things away. I was going to sue Bregović but taking him to court in Serbia—what a mess. So I don't bother, I forgive him" (Cartwright 2005b:62). He went on to work with Bregović in 2002 when he contributed three songs to his album *Tales and Songs from Weddings and Funerals*. This illustrates my point that Romani artists who don't have stable Western management have few options for egalitarian collaboration; they usually take whatever is offered to them because the market is so uncertain.

The case of Bregović's dubious ethics can be compared to other cases of world music collaboration/appropriations such as *Graceland* (Feld 1988; Meintjes 1990)[19] and *Deep Forest* (Feld 2000b). Feld discusses how Deep Forest producers sampled a Solomon Islands lullaby and used it in a series of musical moves involving questionable ethics.[20] Feld asks, "Is world music a form of artistic humiliation, the price primitives pay for attracting the attention of moderns . . .?" (166). Similarly, do Roma need appropriators like Bregović to achieve popularity in the modern world? Feld points out that collaborations with famous artists such as Paul Simon, Sting, and Peter Gabriel are often presented as part of "a politically progressive and artistically avant-garde movement. . . . This process has the positive effect of validating musicians and musics that have been historically marginalized, but it simultaneously reproduces the institutions of patronage" (Feld 2000a:270).[21]

The issue is how music moves between multiple contexts and levels of commercial power. For many non-Roma, Bregović has come to stand for all Balkan Romani music; for example, the program notes for his 2006 Lincoln Center concert state he has developed "a reputation as an eloquent spokesperson for Gypsy culture in eastern Europe." Furthermore, at the 2000 British Barbican Gypsy Music Festival, Bregović received top billing; he was featured in a prime Saturday night slot, while such artists as Kočani Orkestar and Fanfare Ciocarlia appeared in small print on advertisements and played at odd times on the free stage. He also had the prime slot at the 2010 Guča Serbian brass band festival and was received like a God. He is actually taken to be Romani by many fans; indeed, founder of the Serbian band Kal activist Dušan Ristić cynically called him part of the "Gypsy music industry."

Bregović's "reworked" Romani materials are sometimes reintegrated into the world of Romani musicians; the song "Erdelezi" moved from oral circulation to Bregović copyright and then out to oral circulation again. It was first released by Bijelo Dugme in 1986 in Serbian as "Djurdjevdan," but its fame was secured by its prominent place in Kusturica's film *Time of the Gypsies* (1989). The title refers to the spring holiday (known as Erdelezi/Herdelezi/Herdeljezi among Muslims and Djurdjevdan/Gjurgjovden among Eastern Orthodox) when families slaughter sheep (decorated with greenery to ensure fertility), clean their homes, and gather to dance and feast; there are many examples on YouTube (see text supplement 13.1). Many Roma in Šuto Orizari, Macedonia (where the film was shot), claim that they composed the song, but it was probably in oral circulation. Thanks to the popularity of the film and the haunting quality of the song, it has been reclaimed by Roma back into oral circulation, regardless of copyright. It has also been recorded by numerous Romani and non-Romani artists, most of the time with credit to Bregović.[22] A version in Hungarian and Serbian was entered by Hungary into the 2006 Eurovision contest, performed by Ruzsa Magdolna (a winner of the national Megasztar contest).[23]

Certainly Bregović deserves credit for a signature style of arrangement that makes Balkan Romani music more palatable to non-Roma. *Time* considered him "a pioneer in the gypsy music revival" (Purvis 2002:70), and the German *Financial Times* claims that he "found a way to shape Balkan music appealingly for a global audience."[24] Jane Sugarman points out: "He's making a whole career now of slightly arranged music of the former Yugoslavia, most of it heavily in Rom style, all of which gets packaged in his name. The arrangements do make a difference. Folks in Balkan countries who wouldn't be caught dead listening to a Rom band nevertheless love Bregović's music" (East European Folklife Center listserv posting, May 27, 1998). Thus he has widened the audience for Romani music, but he has clearly profited at the expense of others in the process.

Bregović positions himself as an antinationalist hybrid musician, mining the ethnicities of the Balkans as his repository; he implies that he can't be accused of taking other people's music since he is of mixed ethnic heritage himself. His concert projects have featured more than a hundred performers from various Balkan ethnicities, and his press reports often note his masterful "synthesis of the Balkans" (Alvaro Feto, *El Mundo*, April 23, 2001, www.goranbregovic.co.yu). One "composition" is "Kustino Oro," performed by the Athens Symphony Orchestra; this is a Romani čoček in hicaz embellished with West African "talking drums." His touring group, the Wedding and Funeral Orchestra, includes a symphony orchestra, a male choir, several Bulgarian and Serbian female vocalists, and a Gypsy brass band.[25] Recently he presented *Karmen with a Happy Ending*, an opera about Gypsies and sex trafficking. It is clear that generic, ethnic, and national border crossings are his forte.

In this chapter, however, I focus on Bregović's specific affinity for Roma. Indeed, the biography on his official website ends with the phase "Gypsy life full to the brim continues for this eclectic composer figure" (www.goranbregovic.rs). He seems to allude to the perceived hybridity, wildness, and unruly quality of Gypsies that he seeks to embrace. In a 2005 interview[26] he said:

> The Roma are those who are the first to suffer in any group, their life is difficult and tragic. Living such a life according to the principle of the wide smile, the gold tooth—isn't easy. But it's really true that the Roma are the cowboys of Europe; it's difficult to adapt to modern times and world views; it's hard to reach a compromise with the accoutrements of modern civilization and that's why I like them. . . . We'd all like to be Roma at least for one day just so that the rules of gravity don't apply to us, so that our system of values is a little different, a little old-fashioned, not of this world. Those are the Roma! [Jovanović 2005:44].

Bregović sees Gypsies as a different kind of human: a fun-loving, gold-toothed, music making tribe that has forsaken the modern world. His desire "to be Roma at least for one day" centers on their supposed freedom. Even though he mentions Romani poverty, it seems to melt into joy. His stereotypes and fantasies extend to Romani music:

> And as a composer I've always been impressed by the fact that the Roma treat music the same way as they treat nature. They don't understand music as something made up, but rather as something given by God, held in common. With unbelievable ease they take a Spanish harmony and lay a Turkish rhythm and an Arabic melody over it. This is the old, ancient way of making music. That's why the music of the Roma is so fascinating on all levels. Honestly, when I feel like drinking, I grab a bottle and go to some hotel, but when I want to write music I go to a Gypsy café. I'm telling you this in all honesty [44].

Bregović asserts Roma mix styles because they don't recognize belonging; perhaps this belief gives him license to appropriate from Roma—they wouldn't mind anyway. He ignores, however, their professional history of astutely serving patrons and instead imputes to them a childlike, ancient, static, romantic worldview. Van de Port writes of similar stereotypical scenarios involving sex, nature, violence, and music that Serbian patrons in Gypsy bars in Vojvodina impute to Romani musicians (1998).

Noting that his birth language no longer exists, Bregović turns to the Romani language as salvation: "These days I write in Romani and in a made-up language. In the place where I grew up only the Roma speak the same language they always spoke" (44). "I don't feel comfortable anymore writing in Serbo-Croatian. . . . Not long ago I discovered the Gypsy language, which I'm comfortable with. This is a language with very few words and it is simply swimming in rhymes. . . . The Gypsy language serves as a means of communication between the East and West" (Becković 2002). I doubt if Bregović actually composes in Romani because he doesn't speak it; perhaps he collaborates with Roma to write lyrics for him or, more likely he takes existing lyrics. Furthermore, linguists would disagree that it has very few words. Bregović posits the Romani language, and by extension Roma themselves, as an unchanging phenomenon amidst the pernicious conflicts of the Balkans. This fossilizes Roma as premodern relics.

We may clearly place Bregović in the "celebratory camp" of fusion artists. Feld and others have noted the divide between "anxious" and "celebratory" narratives of world music appropriation. Celebratory narratives valorize hybridity, feature hopeful scenarios about economic fairness, and "even have romantic equations of hybridity with overt resistance" (Feld 2000b:152).[27] Anxious narratives fret over purity and underline the economics of exploitation. I believe that we need to interrogate both narratives. Celebratory scholars and musicians eschew ownership and valorize the fertile artistic exchange of musical styles. Lipsitz, for example, shows that appropriations create cultural zones of contact where intercultural dialogue between ethnic groups can happen; he says hybridity "produces an immanent critique of contemporary social relations" (1994).

On the other hand, Lipsitz may "overstate the relative cultural power of these musics" (Born and Hesmondhalgh 2000:27) to effect change.[28] Celebratory tales tend to naturalize globalization, emphasizing its inevitability (Feld 2000b:152). They espouse a "democratic vision for world music," which then becomes part of the marketing scheme. When audiences observe the incredible diversity of music available, they see it "as some kind of sign that democracy prevails, that every voice can be heard, every style can be purchased, everything will be available to everybody" (167). But in celebrating diversity, we shouldn't confuse the flow of musical contents with the flow of power relations (Feld 1994:263). Often too much attention is paid to the sound aspect of hybrid musics and not enough to the social, political, and economic relationships that produce them. Anxious narratives often focus on the pitfalls of recorded music vs. live music[29] or claim that capitalism produces diluted, more commercial forms, less pure forms: "This fuels a kind of policing of . . . authenticity" (Feld 2000b:152).

Anxious accounts fear that world music erases musical diversity and focus on what is lost musically. Anxious scholars "want to calculate the kinds of loss and diminution of musical heterogeneity" (153).[30]

Regarding Roma, I am not concerned about authenticity; the music that is produced by Roma is not becoming more homogeneous. The fear of "damaging Western influence" belongs to purists; it does not illuminate the Romani case, historically or currently. As I have shown, Balkan Romani music has always been open to innovation. In addition, there is vital music making and dancing in the Balkan and diaspora Romani communities, tied to identity issues. I would locate myself in the celebratory camp in relation to artistic creativity and in the anxious camp in relation to political economy. Anxious narratives, however, need to focus less on the aesthetics of music and more on its production and consumption. A narrow aesthetic analysis ignores "who is doing the hybridity, from which position and with what intention and result" (Born and Hesmondhalgh 2000:19). Thus we need to focus more on questions of agency, profits, control, and the range of options available to performers. Along these lines I turn to Gypsy Punk and DJ remixes of Romani music.

Gypsy Punk and DJ Remixes

In Chapter 12, I discussed the New York Gypsy Festivals and the emergence of Gypsy Punk via the band Gogol Bordello. Here I update this thread to examine what some have labeled the "Gypsy rock movement" (http://www.crammed.be/index.php?id=34&art_id=10); I cover only a few key bands because my research is in progress. A number of bands draw from Balkan Gypsy materials but do not have Romani members: Kultur Shock (Seattle), Beirut (New Mexico), Balkan Beat Box (New York), A Hawk and a Hacksaw (New Mexico), and Basement Jaxx (England). Although Beirut prides itself on its naïve, fresh sound inspired by Gypsy brass filtered through Kusturica's films (Lynskey 2006), Kultur Shock and Gogol Bordello are distinguished by their overloaded punk sound and edgy circuslike shows. Balkan Beat Box overlaps with Gogol Bordello in its expression of the immigrant experience but has a more hip hop and electronic texture. Gogol Bordello consists of mostly East European immigrants (Eugene Hutz and Sergey Rjabtzev claim to be part Romani); the core of Balkan Beat Box are Israeli immigrants.[31]

Some bands use electronic as well as live music, and several band members notably Eugene Hutz of Gogol Bordello and Ori Kaplan of Balkan Beat Box, are DJs as well as live performers. In 2007, before Balkan Beat Box became internationally known, Kaplan served as DJ for a monthly party in Brooklyn in a loft where on another floor a live band such as Slavic Soul Party played. These dance parties were unadvertised; part of their cachet came from their underground, word-of-mouth feeling. Typically more than a hundred dancers, mostly under thirty years of age, gathered until dawn. Kaplan used the word "jump" to characterize the dancing of clubbers—and I would agree. He means literally jumping up and down rather than line dancing; there is some solo belly dancing, but most people "jump."

These bands overlap with the expanding world of DJs who remix Gypsy music in dance clubs to crowds of non-Roma in cities all over Western Europe and the United States (Szeman 2009). The online magazine *Exclaim* hailed "the rise of Balkan beats—Gypsy and other eastern European musics updated to a clubbing environment. . . . The anarchic and romantic sounds of Gypsy and Balkan music allow people of all ages to stomp to a different drummer"(Dacks 2005). And on the BBC website, Robert Jackman wrote: "It's impossible to ignore the soaring resurgence of Balkan music. Only a dwindling few could turn their backs on a movement which has persuaded European DJs—a breed famed for their unshakable faith in the synthesizer and the mullet—to swap their techno vinyls for fresh, gypsy-influenced flavours" (2007). A cursory glance at the touring schedules of DJs such as Shantel and Gaetano Fabri (from Belgium) reveals steady work in Western Europe and Turkey. Hutz used to DJ at Mehanata in New York on Thursdays, sometimes adding videos to his shows. At a 2006 show at Midway in New York (at the New York Gypsy Festival), he worked with videographer Al Jerrari (founder of the New York Gypsy Film Festival) to combine visuals from Kusturica's and Gatlif's films about Roma. DJs elicit an emotional response by employing these images of Gypsies; perhaps deliberately, they don't match specific Romani music to specific Romani communities in their video shows.

Shantel (Stefan Hantel) is perhaps the most famous Gypsy remix DJ/producer in Europe, having won the 2006 BBC Planet Club Global Award and produced four Balkan albums, *Bucovina Club 1 and 2* (Essay Recordings 2004 and 2005), *Disko Partizani* and *Planet Paprika* (Crammed Discs 2007, 2009); he has also started to perform live.[32] He launched the Bucovina Club in Frankfurt in 2001 after discovering his roots in this Ukrainian/Romanian/Moldavian borderland area, although some peg his Eastern European connection to media hype. According to music writer Garth Cartwright, Shantel was part of the enormous techno scene in Germany in the 1990s, and as it faded he picked up on the popularity of "Balkan Beats" nights in Berlin, popularized by Bosnian refugee DJ Robert Šoko.[33] A press releases states that the Bucovina Club is "a madhouse with scenes of drunkenness and fraternization, of anarchy and good vibrations that sometimes resemble the films of Emir Kusturica" (www.essayrecordings.com/press.htm). Shantel "describes his Bucovina Club nights as wild parties which absolutely destroy any sense of reserve amongst patrons" (Dacks 2005). He explains:

> You can play a party rocker, a wild Romanian belly dance tune, and the next one is a ballad, a very sad song, and there is no irritation. It's very tense these nights, people screaming and dancing. The audience is very diverse. We have young generation clubbers and then second, third generation immigrants from Yugoslavia, born in Germany with

parents from Serbia, Romania and they are exploring their own music traditions. Then there's the elder generation; its not a problem when you come with your parents to Bucovina Club (Dacks 2005).

Whereas New York Gypsy remix clubs attract Eastern European immigrants, clubs in Western European attract mostly non-Romani Europeans. Hutz's song texts reflect the alienation of being an "other," whether it is Gypsy, immigrant, or political refugee: "East European immigrants often reconstruct their 'home culture' by reaching out for their 'other'—Roma— identifying as 'nomads within the United States" (Budur 2007). Similarly, Ori Kaplan of Balkan Beat Box asserts: "We play an extension of Romani music. We are all immigrants; we are united in our fascination with immigrant cultures." And his colleague Tamir Muskat concurs: "Gypsy is the definition of a soul, not a color or place. It's a take on life" (http://www. myspace.com/balkanbeatbox, from an interview in SPIN, January 2007).

The Romani material that DJs use is markedly Balkan, mostly brass bands; Fanfare Ciocarlia, Kočani, Boban Marković, and Mahala Rai Banda are regularly sampled. In fact, brass bands have recently come to stand for all Balkan Gypsy music; for example, this is what appears on the website www.cocek.com: "There isn't an English word for Čoček although some refer to it as Gypsy brass." Of course, this frame obscures a great majority of Romani bands. As Feld says, "A region of musical variation gets reduced to one genre, a 'caricatured image'" (2000a:276).

One of the first remix albums, *Electric Gypsyland* was released in 2004 to rave reviews in the rock and electronic music world and was followed by a European tour featuring many of the DJs on the album. It was produced by Crammed Discs and sampled tracks from bands produced by that label: Fanfare Ciocarlia, Kočani, and Mahala Rai Banda, plus Taraf de Haïdouks. Thus the worlds of production and DJ remixes are tightly intertwined. *Electric Gypsyland 2* (2006) built on this roster of groups, and *Gypsy Beats and Balkan Bangers* (Atlantic Jaxx 2006, UK) reissued many remix hits. The list of DJs who are interested in Gypsy music is growing, and they come from and perform all over the world, including Mexico. According to the liners notes of *Electric Gypsyland 2*: "While some of these pieces stay close enough to the originals and can be described as remixes, most are poetic re-inventions, works of pure imagination." Indeed, one wonders what is Gypsy about some of the cuts.

The producers of Gypsy remixes appeal to the wider audience of rock fans rather than the smaller audience of ethnic music fans. Dacks's interview with Shantel asserts: "The sentiment of 'this is not world music' arises frequently. His aim is to portray this music as upfront, direct party music, as opposed to some half-baked fusion or the object of ethnomusicological study. . . . Essay Recordings' bio states 'it's not happy clappy multi-culti music'" (Dacks 2005). Marc Hollander, president of Crammed Discs and project director for Electric Gypsyland, said he welcomed the potential to "extend the fusion of Balkan energy with electronics under the banner of world music." He remarked that he was "pleasantly surprised that Gypsy music is gaining an unforeseen audience. . . . I never thought it would be a phenomenon. I was pretty much against the idea of remixing for a long time. But three years ago we got these requests from electronic music producers saying 'Please can we remix . . . Taraf de Haïdouks?' That's when we decided to do the Electric Gypsyland album. . . . It's a welcome change . . . for dance music" (Dacks 2005).

One song, "Red Bula," has taken a winding path to its remix by Balkan Beat Box on *Electric Gypsyland 2*. Although its origin is uncertain, by 2000 it was widely performed in Macedonia and Bulgaria. The title is a pun on the energy drink Red Bull; *bul* in Romani means "ass." Audio example 13.4 with text supplement features Ajgara, a band from Šutka, Macedonia, also known for comedy. It begins with an off-color parody of family greetings that are typically aired on Romani radio stations. Its refrain, "Red Bul, sexy bul, apogei" (Red Bull, sexy ass, the epitome) has migrated into many other songs.[34] Audio example 13.5 features a version by Kristali with Macedonian singer Džansever at a wedding in Bujanovac, Serbia; they sing nonsensical Bulgarian lyrics such as *mambo le, dupka do dupka* (hole to hole) and *adresa veche znai se, na Vitosha shestnaiset* (you already know my address: Vitosha 16). Finally, video examples 5.52 and 5.60 feature performances of Balkan Roma in America (see Chapters 5 and 11).

The "Red Bul," however, that is known by most non-Roma is the Balkan Beat Box remix of the Romanian band Mahala Rai Banda's 2004 version, mixed by Shantel, and produced by Michel Winter and Stephane Karo on Crammed Discs. "Red Bula" is listed in the public domain. Mahala Rai Banda's version is enhanced with a solo by Bulgarian saxophonist Filip Simeonov (audio example 13.6). Balkan Beat Box's remix separates several tracks and adds electronic beats (audio example 13.7). Ori Kaplan explained the process to me: "Marc Hollander offered us a few tracks. Red Bula had potential. We liked it—it has a dixieland vibe, and we thought we could give it a different kind of pulse. We separated out the tracks, took out the drums and repeated the vocals. We chopped it up and changed the structure. On a dead track hidden inside we found an accordion solo; like a phoenix we revived it from the dead! Tamir added keyboards, but there was no need to add much." Kaplan said that he realized what the song meant only after the remix was made.

Remixes raise questions of ownership and artistry; one journalist aptly asked: "Is this just another fad, or is it a sign of the times in New Europe. . . . And will Gypsies and the other disadvantaged citizens of Eastern Europe benefit from this attention?" (Dacks 2005). On the one hand, praise abounds; the online magazine *Know the Ledge* wrote: "DJ Shantel's Bucovina Club project is probably the most effective portrayal of the wildness and sheer ecstasy of traditional-modernized Gypsy party music" (Armstrong 2005). Similarly, music journalist Robert Christgau asserted that Gogol Bordello is "the most exciting new alt-rock band in the world" (2006). On the other hand, critics claim DJs like Shantel are opportunists; they similarly lament the fact that Gogol Bordello is now seen as the future of Gypsy music. German Popov, the founder of the band *OMFO* (Our Man from Odessa), which released *Trans Balkan Express* (Essay 2004), deliberately distances himself from "gypsyronica" (Dacks 2005).

The British music magazine *fRoots* published a scathing review of *Electric Gypsyland 2*: "The Balkans and 'gypsies' have become the new Cuba, and we're now drowning in opportunist 'Club Global' Romaxploitation compilations. Same bands, same tracks, different order. Buyer guide is to avoid anything . . . containing Gogol Bordello. Enough, already!" (fRoots 2006). Cartwright also weighed in against DJs: "I loathe remix albums. . . . The European remix album was a bad 90s phenomenon . . . and Electric Gypsyland's no exception. The Balkan grooves have been filtered out until all that's left is the drudgery of the EU dance floor. In Serbia, Vanjus of Modern Quartet (who remixed Kočani Orkestar), informed me 'I fucking hate folk music so I stripped everything but the clarinet' of MQ's effort. His comment provides a succinct review of the CD" (Cartwright 2005b:209–210). Cartwright also wrote: "Shantel bears obvious comparison with Bosnia's Goran Bregović—both are non-Romas who have imaginatively plundered Balkan Gypsy music to create a sound with popular appeal" (2006a).

But many other writers defend Shantel: "For the moment the modest and ebullient performer has managed to avoid the traps fallen into by the likes of Goran Bregović and Emir Kusturica. His plundering of Balkan Roma music to create popular dance music has been done with deep respect for the complex music forms" (D. Brown 2006). On the other hand, Cartwright puts Shantel and Gogol Bordello in the same category as Kusturica: all are appropriators who "skim the surface of Gypsy culture without dealing with the deeper artistry of the Romani people. Gogol Bordello are, then, a brown and white minstrel show" (Cartwright 2005a).[35]

Respect is something given freely by virtually all appropriators. Many DJs are effusive in their adoration of their "heroes," the Romani musicians from whom they draw: "Everyone, no matter how exoticizing, how patronizing . . . in their rhetoric . . ., declares their respect for the original music and its makers" (Feld 2000a:273). On several remix albums, bonus tracks are devoted to the unmixed sources, as homage or as a mark of "authenticity." DJ Russ Jones asserts:

> Alongside the young turks with their twisted interpretations the older guns still know how to rock it. Included here are some killer tracks from Fanfare Ciocărlia and Kočani Orchestra [sic]. See the guys live and it will be as riotous as any gig by Gogol or Shantel. The finest of musicians, . . . you . . . realize they could sit alongside the hottest Brazilian samba sound band or the most swinging jazz outfits and give them a good run for their money at any party or festival settings [2006].

For Jones, the mark of a good Balkan Romani band is how close it sounds to Gogol Bordello or Shantel. Ironically, the appropriators now set the standards, and it is their sources who must measure up to them. Furthermore, according to Jones, good Romani musicians must be fusionists according to the fashions of world music; they should be able to sit in with a Brazilian band or jazz band.[36]

Hutz claims a special relationship to Roma because of his Romani Ukrainian heritage (chronicled in the documentary *The Pied Piper of Hutzovina*): "Some have said we bastardized the culture. The thing is that we have roots in this culture. These are our personal heroes. This is the roots of my family—this is where we come from" (Budur 2007).[37] Reflecting on the 2006 New York Gypsy Festival, he asserted: "It's important to me that Roma came, not that the magazines thought it was happening. It would perhaps be the worst closure for me if Roma would give me the anathema" (2007). Some non-Romani musicians, including Bregović, legitimate their ties to Romani music through their regional roots.[38] Other musicians make a virtue of their naïveté; they admit they know little about Balkan Roma, but their genuine interest and admiration is enough. For example, Zach Condon of Beirut "makes a principle of naivety. His music may not be authentic—he never pretended it was—but it is never less than sincere." He said "I'm willing to take the music at face value. . . . I don't have to tie in historical and racial and political elements to make it mean anything more to me. Isn't melody enough?"[39] (Lynskey 2006).

For Condon, then, any music is "up for grabs." Critics might call him a "sonic scavenger."[40] Like Condon, Hutz "has no desire for border control. 'Gogol Bordello is about embracing other cultures because the more sources of joy you have, the better you are. There is no yes or no, no black and white here. I guess it all goes back to not where you came from but how deep in your bones does the music really reach. Does it reach that level of madness or not?'" (2006). Similarly, Hollander advocates "transculturalism—a genuine mash-up of cultures as opposed to multiculturalism where groups retreat into and protect their own enclaves." Hollander points out that Gypsy music has never been pure and is "by no means traditional music. They borrow bits and pieces of music right and left" (Dacks 2005).

These musicians disavow ownership theories of music; they believe that all music is available to everyone to use as they wish. They celebrate creativity and reject any hint that artistic or economic exploitation could exist. The opposite of this camp is the UNESCO and WIPO (World Intellectual Property Organization) approach, which sees discrete arts belonging to discrete cultural groups who should have control over them (see Chapter 8). In this model, it is supposedly clear who is taking from whom because ultimate sources can be identified. Historically, Roma test and ultimately refute the applicability of both of these two extremes. Their appropriations are from a marginal position; they have never officially had "rights" to any genre because they have never been in charge of institutions. Now that non-Roma are appropriating from them, they can't (and many wouldn't want to) fight back.

One strand in the celebratory camp asserts that appropriation is not problematic because there is no such thing as authentic Gypsy music.

Robert Christgau, for example, writes: "Purity is always a misleading ideal. With the gypsies, or Roma . . . it's an impossible chimera . . . real Gypsy music is a myth" (2006). Similarly, "There is no such thing as Gypsy music insists DJ Shantel . . . you can only talk about traditional music from different regions in southeastern Europe" (2006). Like Bregović, these artists claim that because Roma have appropriated, then appropriations from Roma are unproblematic. They confuse artistry with economics. As Lynskey writes: "There is no such thing as Gypsy music. From Basement Jaxx to Beirut to Gogol Bordello, bands are looking to the Balkans for inspiration, but . . . is this a genuine new musical hybrid or just cultural tourism? . . . Gypsy music has always been a hybrid, but for centuries the underdogs assimilated the music of the dominant society. Now they are the ones being assimilated" (2006).

Are DJs and Gypsy Punk bands putting Romani musicians out of work? Are they being hired instead of Romani musicians? It is certainly cheaper to hire one DJ or a local punk band than to bring a band from the Balkans. But DJs have argued with me on this very point. A few do invite guest Romani musicians. When I asked Ori Kaplan of Balkan Beat Box if he thought that remixes would take work away from Roma, he answered:

Not at all. We are not competing in any way—we are augmenting the scene. Audiences have grown. The whole scene is expanding. Would a Swedish hip hop band take work away from Eminem? Not at all! We in BBB are a completely different animal—we are Middle Eastern musicians. Would another band playing new Mediterranean music take work away from us? No, these bands would play an opening set for us—they help spread the word. We all spread the word, we tell audiences what albums to buy—Taraf, Ivo. I don't see the relevance of your question.

Kaplan views the market for Gypsy music as continuously growing, so everyone benefits. Similarly, Hutz says, "Does the bad reggae that's out there prevent the good reggae from existing?" Shantel goes even further, suggesting that the current wave is helping Balkan musicians: "In southeastern Europe these bands are totally vanishing away. The success of this sound helps the younger generation of musicians in the Balkans to continue the tradition" (Lynskey 2006). Postulating that Romani music is dying, he claims that he is inspiring a new generation in the Balkans. I would counter that not only is Balkan Romani music alive, but its vitality doesn't depend on him!

How do Roma feel about remixes? It is hard to ascertain their honest opinions because most are extremely practical and do not want to alienate possible "collaborators" from whom they may derive future revenues.[41] Also note that artistic opinions may diverge from economic decisions. Feld similarly observes that third-world musicians want more exposure, sales, and "a greater cut of the action. If their perception is that the same process that is screwing them over is the process that is eventually going to give them a larger cut, then how do you tell them to take a smaller cut?" (1994:315). In New York, for example, Yuri Yunakov is on excellent terms with Ori Kaplan (his former student) and Eugene Hutz.[42]

Eugene Hutz is such a polarizing figure in the United States that Romani activists find him hard to dismiss, even while they criticize his stereotypical displays. They hope that perhaps his fame can be recruited for activism. His biography from the website of the New York Gypsy Festival asserts:

> Eugene comes from mixed Russian-Ukrainian-Romany family, but it is the Romany side that became his biggest inspiration. Love for Romany people and music also lead [sic] him to become a Romany rights activist. Hutz constantly works with his native Ukrainian organization Romany Jag (Gypsy fire) and is currently translating works of his friend Gypsy writer and philosopher Vladimir Bambula. "Only through exploring my relationship with Gypsy culture I understood my place in the world," Hutz says.

When Hutz expressed his desire to become more involved in activism to Šani Rifati of the NGO Voice of Roma, Rifati invited him to perform at the 2008 Romani Herdeljezi festival in California. Several members of VOR's board of directors (especially the Romani members) were wary of this invitation, as they were of his biography. But Rifati paired him with Russian Romani guitarist Vadim Kolpakov rather than let him do his usual Gypsy Punk show. The event drew Gogol fans who would have never attended a Romani music festival; even so, Hutz's antics were offensive to many VOR volunteers and audience members.[43]

Cartwright "dismissed Gogol Bordello as a 'fiddle-driven Sham 69,'" and he reported that Fanfare Ciocarlia described Shantel's remixes as "dogshit" (Lynskey 2006).[44] Cartwright wrote that several Bucharest Romani musicians shrug off remixes as "for the west" (2005b:210). He also wrote that Henry Ernst (from Asphalt Tango) is "distrustful of most remix albums—Electric Gypsyland being an example of a less than satisfying attempt to turn Taraf de Haïdouks and Kočani Orkestar into café muzak" Cartwright 2004:31). In 2007 Ernst told me: "In the last ten years Gypsies built up a lobby through their artistic merits . . . and the next logical step is that non-Gypsy musicians start to pay attention. . . . Remixes are part of this observation and a result of the hungry music industry."

Remember that music is produced and copyrighted by a label, which decides whether to issue permission for remixing. The revenues go to the label, which may or may not have a revenue-sharing agreement with its artists. Asphalt Tango has clearly funneled profits to the musicians of Fanfare, but this is not the norm in the recording industry. As Feld writes, the music business is supported by three pillars: record companies, major contract artists who have control over their art, and musicians who "are laborers who sell their services for a direct fee and take the risk (with little expectation) that royalty percentages, spin off jobs, tours, and recording contracts might follow from the exposure and success of records with enormous sales" (2000a:245).

Another strand in the celebratory discourse of appropriators is the feel-good, peace-making "transcultural" aspect of the music. Bregović, for example, sees himself as transcending the conflicts of the Balkans through Gypsy music. Shantel similarly views his dance club experience as bringing people together: "It's only music, you know. It's to make people happy, not to fight against each other" (Lynskey 2006). The liner notes to *Bucovina Club 2* expound this ideal:

> The incomparable atmosphere and energy is tangible when people who have never set eyes on each other before end up dancing the Hora together, when they crash the stage to dance the Cocek along-side the musicians or to get a shot of vodka from the DJ. The Bucovina Club transforms every venue into an extraterritorial zone creating a dazzling euphoric event that transcends all generational and national boundaries. Ageism? No sign of it. Racism? No way. On the contrary: where else can you hear so many Gypsy sounds? How come Gypsy music is so popular with Shantel and his audience? The answer is simple: because it combines joy and sorrow so compellingly that only the most superficial or cold-hearted would fail to be swept along by it. . . . Tribes of émigré kids, Romanians, Serbians, Croats, Albanians, Ukrainians, and all those who know there is musical life beyond the increasingly cut-throat mainstream, gather here to celebrate their occasionally surrealistic boundary-crossing rituals, living out a sense of togetherness. . . . Forget the pain, forget the misery; celebrate this day as though it were your last [Trouillet 2006].

I applaud how music and dance can bridge ethnicities, but I wonder how Trouillet (who heads Essay records) can think that racism has disappeared in Western Europe. Perhaps inside some of these clubs Western Europeans and Eastern European immigrants dance together. But throughout Western Europe, there are other clubs where only émigrés gather. Furthermore, there are rarely any Romani patrons in remix clubs; not only can't they afford them, but also they can't relate to the scene.[45] Many Roma in Western Europe have a precarious legal, political, economic, and social status as refugees or underemployed workers. As discussed in Chapter 1, Roma face racism and deportation in Western Europe, and I doubt that the popularity of Gypsy remixes affects this fact. A romantic fantasy of harmony, however, overlays the club scene.

Among Americans, the fantasy element is especially prominent: Gypsies are associated with wildness and sexuality, and facts are scarce. Several years ago, when Alex Dimitrov moved his legendary New York club Mehanata (the tavern), he proposed renaming it House of Gypsy; this title was accompanied by a website photograph of a non-Romani woman in a seductive pose on the stoop of the building. When he contacted me about writing a letter to support a liquor license for the new location, I told him that I was offended at his use of a sexual stereotype. Perhaps he dropped the name and image when I said that the community board might think the bar was a brothel. His request to me to write about the beneficial multicultural musical exchanges among immigrants that happen at the club put me in an awkward position. On the one hand, I objected to the stereotypes, but on the other hand I knew the club provided employment to Roma, including Yuri Yunakov, Sal Mamudoski, and Seido Salifoski. After talking with these musicians, I told Dimitrov about my objections and wrote the letter. In the end, the stereotypes bothered me more than they bothered the Roma musicians, a point I have analyzed throughout this book.

Borat and Beyond

A huge public audience was exposed to Romani music via the record-breaking movie *Borat! Cultural Learnings of America for Make Benefit Glorious Nation of Kazakhstan* (2006; www.borat.com). Its soundtrack overwhelmingly features Balkan Romani music, with cuts from Esma Redžepova, Jony Iliev, Kočani Orkestar, Mahala Rai Banda, Fanfare Ciocarlia, Goran Bregović, and Ivo Papazov (with Maria Karafezieva). The Atlantic soundtrack album and *Borat* songs on iTunes are selling quite well; and DJ Shantel created a *Borat* tour and promoted the movie in a video clip (www.essayrecordings.com/vid_mtvtrailer.htm). Why would Romani music appear in this film, and what effect has this exposure had on Romani musicians?

The creation of British comedian Sasha Baron Cohen, *Borat* is a "mockumentary" about a Kazakh journalist documenting American life for the people in his Kazakh village. The film is a biting satire about Eastern Europe and the United States: both groups are portrayed as racists. But whereas Borat may be an equal opportunity offender, denigrating gays, Blacks, and Uzbeks, his specific targets are women, Jews, and Gypsies. Is Borat defending these views? I would contend that he is not defending them but rather attributing them to most East Europeans and some Americans. Many Western audience members have found the sexist, homophobic, and anti-Semitic remarks offensive, but at least they have a counter discourse readily available via vocal lobbies. On the other hand, most American audience members are ignorant about Roma; they have at best a vague sense of who Gypsies are, and that sense is shrouded in fantasy.

Note that references to Gypsies in the film center around theft (Borat assumes that a woman running a garage sale is a Gypsy who stole all the items), magic (he seeks a "jar of Gypsy tears" for protection against AIDS, and he assumes a garage sale proprietor can cast spells and shrink Barbie dolls), and violence against Gypsies that seems like fun (he seeks a car that goes fast enough to kill them). These are stock stereotypes, and because there is little counter-information available it is possible audience members might believe them as well as laugh at them.

One thing audiences won't notice is that, ironically, the Kazakh village depicted in the film is Glod, a Romani village in southern Romania. The primitiveness was staged (e.g., animals, such as cows, were brought inside homes), but the poverty and marginality were all too real. In fact, the villagers were outraged when they found out how the footage was used.[46] Although this backstage controversy is invisible to audiences, the soundtrack attracted global and sustained attention. All the reviewers on Amazon, for example, comment that the tunes are East European Gypsy, and one states: "The Borat soundtrack sounds just as you'd expect—imagine Beirut, Balkan Beat Box and a bit of Gogol Bordello, all with Sacha Baron Cohen as the frontman. God help us" (ea_solinas January 31, 2007, www. amazon.com). Actually none of these artists are featured, but this comment shows that virtually all Gypsy music is now associated with those groups. The BBC reported:

> The music from the film includes some of the most rollicking Balkan and gypsy music ever presented on such a wide scale to the American public. . . . For those who don't know much about Balkan or gypsy music, the soundtrack is an ideal tasting menu [Werman 2006].

Note that in the film, Gypsy music tends to be played when the scenes depict Kazakhstan, hence backwardness, illustrating the trope of Gypsy music as symbolic of the exotic other or the primitive marginal.[47] For example, the film opens with Esma Redžepova singing her signature song "Čhaje Šukarije" (see Chapter 10), while the viewer absorbs scenes of the muddy "Kazakh" village. Esma was upset at this use of her song and was seriously considering suing Cohen. She claimed her recording label did not know about the intended use of the song when it gave permission. However, the film properly credits Esma and the label, Times Square Records/World Connection (2000). In fact, Esma's album now proudly features an advertising sticker proclaiming "In Borat!"

The problem for artists is the way the music industry structures permissions. Cohen needed to ask the label, not Esma, for permission. The label may or may not inform the artist and may or may not pay her or him, depending on recording contracts; most do not give rights of distribution to artists. According to Esma, her Dutch managers negotiated the use of her song, and she was not consulted. She said she was angry, but royalties were paid. She decided not to sue perhaps because she did not have a case, but she received considerable compensation from World Connection. Contrary to Esma, however, Ivo Papazov did not complain about the use of his music, and Fanfare Ciocarlia's members were paid extremely well, according to their managers/producers. Villagers from Glod attempted to sue Cohen but failed (see the documentary film *Carmen Meets Borat*).

Fanfare's role in the film was unusual in that the band was commissioned to perform a piece new to them. "Born to Be Wild," whose chorus is sung in English, is a cover of the classic hit; it was hailed by the BBC as the "one standout number" (Werman 2006). According to the Asphalt Tango website: "Three months ago Hollywood knocked on our door and asked for an unmistakable Gypsy version of the Steppenwolf hit 'Born To Be Wild' to get the pictures of the film 'Borat' moving. So Fanfare Ciocarlia rushed into a studio and pepped up this ageless bikers' hymne [sic] from the 70's in their very own style" (June 2006, www.asphalt-tango.de/news. html). Henry Ernst explained the process: "The band created their own adaption of Born To Be Wild after listening for two hours to the original version. After two days rehearsal and final arrangements we went to a Berlin studio and we recorded the song. Afterwards we did some additional overdubs on tour (mostly in hotel rooms)." Fanfare was invited to perform at the film's premiere in London, but the trip was cancelled.

According to Ernst, Fanfare members "were compensated on a high level." In fact, a BBC report on the film's music assumes that "with this soundtrack, the royalties will now flow in for these Romanian gypsy musicians. They'll have Borat to thank, even though the ethno-musical connection between Romania and Kazakhstan is thin at best" (Werman 2006). Asphalt Tango had already funneled money into Fanfare's small village. According to the documentary film *Iag Bari: Brass on Fire* (2001), the first Romani church in Romania was built with this money. Fanfare's founder Ioan Ivancea readily admitted that "Henry Ernst's arrival had . . . saved Zeci Prajini and encouraged the younger generation to keep learning music" (Cartwright 2006b and 2005b:222). As opposed to Shantel's inflated claims about saving Gypsy music (discussed earlier), Alphalt Tango is an example of non-Romani producers investing in the music and the musicians.

Concluding Remarks

This book endeavors to combine several strands of inquiry: it analyzes the relationship of Balkan Romani music to Romani communities, to states, and to capitalist markets. Transnationalism ties together these three strands: Romani communities are mobile, music is part of a pan-Romani political articulation of identity, and Romani music has entered world music circuits. I have endeavored to confound the linear path that assumes music begins in bounded communities and is then changed by its use in state and market contexts. For Roma, the community is no longer, and never was, a bounded site; there are no original singular Romani homelands, neither in India nor in the Balkans. Rather, Romani communities are open-ended, transnational, and diasporic, with nodes in multiple sites such as New York and Dusseldorf, as well as Šutka. Music in Šutka or Belmont, then, may be as transnational as Shantel's remixes or even more so. Rather than looking at Romani communities as authentic cites of original music, I show how they are the sites of negotiation between economic and artistic diasporic forces.

Romani communities in the Balkans and the diaspora are dynamic centers of musical creativity where multiple generations of musicians interact for a knowledgeable patron base. But we need to be careful not to romanticize life in these communities. True, new waves of talent are developing, but the economic conditions are so dire that most musicians simply cannot make a living. Furthermore, the majority of Roma are not musicians, and their employment and life trajectories are even more precarious. When we hear of large weddings, we need to remember that families have saved for years for these events and they will sacrifice in the future because of their financial outlay. We should not judge the majority by the minority of famous artists who have steady incomes.

Emerging styles and genres also need to be seen in the political context of the growing pan-Romani human rights movement, which addresses past forms of injustice as well as new ones. Most Roma in Europe today lead a precarious existence thanks to widespread discrimination in all walks of life and threats of violence, but a select few have become musical stars. The rising tide of xenophobia in both Eastern and Western Europe exists side by side with the craze for Gypsy music among non-Roma. How do these spheres affect each other? Whereas most activists are not directly involved with music, and most musicians are not directly involved in activism, more Roma are mediating the gap. In the Balkans and the diaspora, they include Dušan and Dragan Ristić of Kal, Šani Rifati of Voice of Roma, Esma Redžepova, Yuri Yunakov, and the organizers of Romani music festivals. Music has always been an area of pride, but it is now tied to identity politics and festivals.

It is not surprising that musicians avoid the topics of history and discrimination; most non-Roma are fascinated by Gypsies precisely as "timeless" performers. They want to see Gypsies sing and dance, period; and some think that music is the totality of their lives. Many admirers presume that poverty doesn't bother Gypsies because they have music; as Kirshenblatt-Gimblett points out, "Spectacle suppresses conflict" (1998:72; see Chapter 12). For centuries, the marketing discourse of Romani music has promised exoticism, sexuality, passion, wildness, and abandonment; and precisely because Roma are in a marginal position, they have often promoted these stereotypes. Roma construct themselves in response to how non-Roma have constructed Gypsiness. Roma have never been in charge of their own representational discourse and imagery, and as a result they use the discourse available: the stereotypes that sell.[48] Thus sometimes they are marketed as authentic, sometimes as hybrid, and sometimes as authentic hybrids.

Historically, hybridity has been used by nation/states to discredit Romani music; official institutions claimed that Romani music was not pure enough to enter the national canon. But in the contemporary fusion music market, Romani hybridity is lauded. For hundreds of years, Roma have trafficked in border styles; long before the emigrant waves of postsocialism, Balkan Romani music was hybrid. But the new musical developments since the 1990s cannot merely be subsumed under old paradigms. When the Bulgarian socialist state banned kyuchek in the 1980s, something novel happened: Romani music was reconfigured as an antisocialist countercultural phenomenon, playing and dancing kyuchek became political, and resistance emerged. After 1989, Bulgarian Romani music was reconfigured in relation to more commercial media ventures such as chalga.

So what exactly is new since the 1990s? The rise in popularity of "world music" has opened up new possibilities for artists from every corner of the globe. Although this may sound like an inherently democratic move, in reality only a tiny fraction of Romani musicians benefit directly from the world music scene; most are ignored and the few that are picked up by Western production companies benefit less than their producers. Furthermore, their cultural capital may no longer be under their control. As I discussed in Chapter 3, world music thrives on marketing ethic and racial difference. Imre writes: "Global popular culture voraciously incorporates ethnic differences in the pursuit of selling and consuming non-stop entertainment. This process has two sides: it can be seen as liberating and democratic, empowering minorities whose voices would otherwise be missing or stereotyped. At the same time, it implies the appropriation of such voices and images by corporate multiculturalism and its cultures of simulation which re-trivializes racial difference on a commercial basis" (2006:661).

The transition from socialism to capitalism has opened up new arenas for expression and allowed the flowering of previously prohibited genres. At the same time, it has introduced contested forms of global appropriations. Whereas forty years ago, Hungarian Gypsy string bands stood for "Gypsy" music in the minds of many Americans and Western Europeans, today Balkan brass has taken its place. Are Gypsies merely a prop in the global vocabulary of sound, or are they active players? Ironically, the more popular Gypsy music becomes, the more non-Roma become involved in it, with both positive and negative consequences. With remixes, Gypsy music has become a set of sounds remote from Roma and available to all for appropriation and sampling.[49] Roma still operate primarily as musical providers to non-Romani patrons. The terms of commerce are very often out of their hands, and sometimes even artistic decisions are out of their hands.

The global market is mediated by record companies, managers, festivals, and clubs; these institutions and sites are all controlled by non-Roma.[50] Only a handful of artists have achieved world acclaim, while equally talented performers languish for lack of international ties. The famous bands support whole villages or extended families back home, but their managers and producers inevitably profit. The majority of Romani performers have little chance for international fame, but this situation is somewhat mitigated by a flourishing demand for quality musics in the Romani diaspora for in-group celebrations. In the parallel world of weddings, circumcisions, and baptisms, good musicians are in demand. The work is much harder, it may pay less, and is less stable. Some performers (Kristali, Ćita, Amza) are very well paid; some are not.

The contrast between these worlds is striking. In the international market, Roma perform for two hours at concerts, festivals, or clubs. Every detail, all the logistics (hotels, meals, transportation) are arranged by non-Romani managers. All this may be coordinated with a recording contract (meaning loyalty to one label), a steady stream of new albums, media events, and classy websites. Tours are arranged many months in advance. The chance of becoming passé and out of fashion, however, is very real. In this category, musicians do give up some autonomy for steady income. But few see it as a burden. For the most part, they are extremely grateful to non-Romani entrepreneurs for rescuing them from poverty.

By contrast, in the Romani world performers act as their own agents and have artistic independence but no financial stability. And because Romani audiences value novelty and are so discerning, there is fierce competition, which keeps the quality extremely high. Famous artists get hired, but for every famous artist there are hundreds of struggling musicians. Some performers, such as Yuri Yunakov, straddle these two worlds: they play Romani weddings and also have international reputations among non-Roma. There are also artists who no longer perform for the in-group world of weddings but rather rely on international work, such as Taraf de Haidouks; however, a few artists in this category, such as Ivo Papazov and Esma Redžepova, have sometimes lacked stable Western management and recording contacts.

It is clear that the contemporary Balkan Romani music scene should not be judged only by the international stars and their appearances of success. Neither should we ignore the tremendous impact the stars are making. The life histories of the mature stars remind us that they all emerged from communities (whether in Macedonia, New York, or Western Europe) where music is a vibrant part of life; furthermore, emerging new stars must pass the test of rigorous community standards. But these communities are not just places of ecstatic music making; in Europe they are sites of critical neglect and targets of discrimination, and in the United States they are the sites of social challenges. The arena of "world music" is not particularly relevant for many Roma, but for the successful musicians it is quite pertinent. The future could open up more possibilities for Romani musicians, or else Gypsy music could go out of style in dance clubs. Regarding the craze for remixes, Lynskey states: "Anyway, in the long history of Gypsy music this is just a blip. The music is strong enough to survive any number of reinterpretations" (2006). I hope and believe he is correct. Although Roma have sometimes been pictured as traditional, conservative, backward, and generally outside the framework of the modern, they are quintessentially cosmopolitan. Romani musicians embrace hybridity and eclecticism as a cultural resource, and thus they reveal to us a great deal about survival and cultural negotiation.

Notes

Chapter 1

1. In this book, I use *Gypsy* as an outsider term, although I acknowledge that it is sometimes used as an insider term. Along with its cognates *Gitan* (French), *Gitano* (Spanish), *Yiftos* (Greek), and *Gjupci* (Macedonian, plural), Gypsy connotes faulty history, i.e., Egyptian origins, and usually has strong negative connotations. Some groups, however, willingly embrace the term, e.g., the *Gitanos* in Spain (Gay y Blasco 2002:174–175) and the *Egjupkjani* in Macedonia and Kosovo (Marushiakova et al. 2001; V. Friedman 2005). Another common outsider term, *Tsigan* (and its cognates such as the German *Zigeuner,* Italian *Zingaro,* Turkish *Çingene*), derives from the Greek *atsingani,* a heretical sect in the Byzantine period (Soulis 1961:145). I use Roma as an umbrella ethnonym (singular Rom, adjective Romani) because it emerged as a unifying term in the last two decades, as political consciousness has been mobilized through political parties, conferences, and congresses (Petrova 2003:111–112). Note, however, that some groups, such as the *Sinti* in Germany, the *Rudari* and *Beyashi* in Hungary, the *Ashkalia* and *Egjupkjani* of Kosovo, and the Gypsies and Travelers of England and Scotland, distinguish themselves from Roma (Petrova 2003:111–112; Hancock 2002:34; Marushiakova 1992; Marushiakova et al. 2001). See Chapter 3 for a discussion of identity issues.

2. For example, see recent articles in *Time* and *Newsweek* magazines (Purvis 2002; Brownell and Haq 2007; Bax 2007).

3. The literature on folklore and performance is vast; overviews include Bauman and Briggs 1990; Fine 1984; Kapchan 1996a and b and 2007, Bendix 1997. Performance has also been discussed from the point of view of theater; see Schechner 2006 and Diamond 1996. Anthropologists Victor Turner (Turner and Bruner 1986) and Clifford Geertz (1973 and 1983) highlight the dramatic, performative quality of ritual.

4. Kapchan writes: "It has been the task of performance studies to understand what constitutes the differences between habitual practices and heightened performance, and how and why these differences function in society" (1996a:279).

5. Butler's notion of citationality derives from Derrida. Butler underscores the subversive potential that questions original identities: "The notion of gender parody . . . does not assume that there is an original which such parodic identities imitate. Indeed the parody is *of* the very notion of an original" (1990:138). See Chapter 9 for discussion of Bulgarian pop/folk star Azis's use of gender transgression in relation to Butler's theories.

6. The literature on blackface and minstrelsy is useful here (see Lott 1993).

7. Kapchan, for example, chronicles the transformation of Moroccan sacred Gnawa in world music markets, noting "how these changes spiral back to the local context and affect transformation there" (2007:235).

8. Herzfeld points out that the state is a prime reference point in post-socialist ethnography (2000). His work on "cultural intimacy" shows the subtle ways in which the state achieves rapport with citizens despite authoritarian regulations (1997).

9. For reportage on anti-Muslim, anti-immigration, anti-Semitic, and anti-Romani sentiments Europe, see Stracansky 2009; BBC 2008; Moore 2008; Donadio 2008; Kimmelman 2008a and b; Kulish 2007 and 2008; Sciolino 2007a and b; Sciolino and Bernard 2007; Erlanger 2008; Minchik 2007, Castle-Kanerova 2001; McCann 2007; Waringo 2004; Ghodsee 2008; on French deportations, see www.errc.org/cikk.php?cikk=3619; on Danish deportations, www.errc.org/cikk.php?cikk=3603; and on German deportations, http://www.dw-world.de/dw/article/0,6197201,00.html.

10. Because of space limitations, I do not focus on Serbia, Greece, and Albania. My specific focus on Macedonia and Bulgaria is a result of my historical and practical choices in fieldwork; see Silverman 2000c. The literature includes documentation of South Serbian Romani brass bands (Lovas 2003; Babić 2003 and 2004; Hedges 1996), Serbian Romani music from Vojvodina (Van de Port 1998 and 1999), Kosovo Romani music (Pettan 1996a, 1996b, and 1996c, 2001, 2002, 2003), Greek Romani music (Blau, Keil, Keil, and Feld 2002; Brandl 1996; Theodosiou 2003), and Romanian Romani music (Rădulescu 2004; Beissinger 1991, 2001, 2005, and 2007; Malvinni 2003 and 2004; Marian-Bălaşa 2002 and 2004; Szeman 2009). See Cartwright 2005b for a personal journalistic portrait of Balkan Romani musicians.

11. See Saul and Tebutt 2004. The cultural studies journal *Third Text* devoted a recent issue to interdisciplinary approaches to representations of Roma; see Imre 2008; Gay y Blasco 2008; Iordanova 2008; Hasdeu 2008.

12. There is an emerging field of genetic studies of Romani origins; see Kalaydjieva, Gresham, and Calafell 2005; Iovita and Schurr 2004.

13. The four major Romani dialects are Vlax, Balkan, Central, and Northern (Matras 2005:8); the word *Vlach* refers to Roma who lived in the southern Romanian principalities of Wallachia and Moldavia during the period of slavery (1300s to 1860s); see Fraser 1992 and Hancock 2002.

14. According to Fraser (1992) and Hancock (2002), nomadism was more prevalent in Western Europe than in Eastern Europe owing to regulations preventing settlement in the west. The binary division between nomadic Roma and settled Roma is somewhat artificial and is in fact partially an artifact of the scholarly disciplines that have studied Roma (Van de Port 1998). Whereas much of the literature contrasts nomadic groups, such as the Olah in Hungary, with sedentary groups such as Romungre, the on-the-ground situation was and is more fluid (Stewart 1997). The Roma discussed in this book from Macedonia and Bulgaria are all sedentary.

15. For similar reasons, other Balkan peoples such as Bulgarian Pomaks, Macedonian Torbeši, and a significant number of Albanians and Bosnians converted to Islam; see Hupchik 1994.

16. Pentecostalism among Roma has been documented by several scholars, including Gay y Blasco 1999 and Lange 2003.

17. For example, one Romani man born in 1930 in Skopje, Macedonia, spoke Turkish, Romani, Macedonian, Serbian, and Albanian (also see Lindemyer and Ramadonov 2004 for linguistic profiles of Bitola Roma). See Ellis 2003 for the fluidity of languages and identities of Skopje urban Muslims. See Victor Friedman's work (1995, 1999, 2005) for a thorough analysis of Balkan language interaction.

18. Todorova, discussing whether the Balkans are orientalized in reference to the rest of Europe, points out that we are not dealing with a colonial situation (1997); nevertheless, the Balkans are posed as "other" to Europe, and Roma are posed as "other" to the Balkans (also see Neuburger 2004). Ken Lee specifically extends Said's argument to Gypsies: "Whilst Orientalism is the discursive construction of the exotic Other *outside* Europe, Gypsylorism is the construction of the exotic Other *within* Europe—Romanies are the Orientals within" (2000:132).

19. The term *narodi* (nations) applied to the constituent peoples of Yugoslavia, those who were not a majority anywhere outside Yugoslavia: Serbs, Croats, Slovenians, Montenegrins, Bosnians (Muslims), and Macedonians. *Nardonosti* (nationalities) applied to those who were national minorities, i.e., they were majority populations outside Yugoslavia: Albanians, Turks, Hungarians, and others. *Etničeski grupi* (ethnic groups) were groups of distinctive people who had no nation/state outside of Yugoslavia, such as Roma. Privileges differed by category; for example, all of the narodnosti but not the etničeski grupi had the right to education in their native language; funding for the arts was also better for narodnosti (K. Brown 2000:137). Despite this schema, separatist ethnic ideologies were not tolerated in Yugoslavia; to the contrary, they were severely punished.

20. Petrovski, an Eastern Orthodox Rom from Skopje, received his Ph.D. in folklore in Belgrade and is employed by the Folklore Institute in Skopje. He has a number of publications (mostly collections) on Macedonian Romani customs and songs (1982, 1993, 2001, 2002, 2009) and has also been active in NGOs. Because money for publishing is so scarce Petrovski has secured funding for his books through Romani NGOs.

21. For detailed information on these topics, see Silverman 1995a, the World Bank Reports by Ringold 2000; Ringold, Orenstein, and Wilkins 2004; and Revenga, Ringold and Tracy 2002; and various issues of *Roma Rights*, the journal of the European Roma Rights Centre (www.errc.org). The center submitted this statement to the UN Committee on the Elimination of Racial Discrimination: "Roma remain to date the most persecuted people of Europe. Almost everywhere, their fundamental human rights are threatened. Disturbing cases of racist violence targeting Roma have occurred in recent years. Discrimination against Roma in employment, education, health care, and administrative and other services is common in many societies. Hate speech against Roma, also prevalent, deepens the negative stereotypes which pervade European public opinion" (ERRC 2002:5).

22. The twelve signatories are Bulgaria, Croatia, the Czech Republic, Macedonia, Serbia, Montenegro, Hungary, Romania, Slovenia, Albania, Bosnia and Hercegovina, and Spain; see www.romadecade.org.

23. The Macedonian census of 2002 listed 52,000 Roma, but scholars agree the actual number is much higher (Plaut and Memedova 2005:16; Petrovski 2009). Almost all Macedonian Roma are currently sedentary. See Chapter 8 for an overview of the legal and cultural status of Roma in Yugoslavia.

24. The oldest Romani party, the Party for the Complete Emancipation of Roma of Macedonia (PCER), led by Faik Abdi, was represented in parliament from 1990 to 1998; the Union of Roma, led by Amdi Bajram, was represented in parliament from 1998 to 2000; the Democratic Progressive Party of Roma is led by Bekir Arif; and the fourth is the United Party of Roma in Macedonia (see Plaut and Memedova 2005; and V. Friedman 1999). These Romani parties and NGOs have little effect on the larger political landscape. Strategically, they are neutral with regard to the main conflict of the nation: the ethnic tension between Macedonians and Albanians. An important Romani leader is Neždet Mustafa; in 1992 he helped establish *Bijandipe* (News), the Romani television program on the Macedonian national channel. He was the first general secretary of PCER and the leader of the United Party of Roma; he served in parliament 2006–2008, was elected mayor of Šutka in 1996 (when it became a municipality), and now serves in the national government as minister without portfolio for the Decade of Romani Inclusion.

25. The 2001 Bulgarian census officially listed only 371,000 (Rechel 2008:79 and 82; Pamporovo 2009). Roma affiliate with several major parties, including the Bulgarian Socialist Party, the United Democratic Front, and the Movement for Rights and Freedoms (MRF, the so-called Turkish party), which was already in existence at the time the ban was enacted; there have been only a few Romani representatives in government since 1989.

26. See *Monitoring the EU Accession Process 2002*. See also *European Union 2003*.

27. See http://www.sofiaecho.com/2009/02/25/681499_us-report-high lights-human-rights-problems-in-bulgaria, accessed June 19, 2011.

28. See http://bsanna-news.ukrinform.ua/newsitem.php?id=12811&;lang= en, accessed June 19, 2011.

29. See http://www.novinite.com/view_news.php?id=109779, accessed June 19, 2011

30. When Bulgaria joined the European Union in 2007, Ataka found allies in the European parliament. It joined forces with Western European xenophobic and anti-immigrant parties (such as the National Front in France) to establish the European Union platform "Identity, Tradition, Sovereignty," which defends "Christian values" and the "national identities of the countries." The platform has received €1 million because it has a sufficient number of members of parliament. In the 2009 European Parliament elections, Ataka won two of Bulgaria's seventeen seats.

31. Marushiakova et al. (2001) estimate that the pre-1999 war population of Roma in Kosovo was 120,000–150,000, and that only 30,000–35,000 remain; today there are far fewer. For accounts of Kosovo Roma, also see the website of Voice of Roma (www.vor.org), and European Roma Rights Centre reports at http://www.errc.org/cikk.php?cikk=2271.

32. See Castle-Kanerova 2001 and Graff 2002. Mark Landler's *New York Times* article (2004:4), for example, reported the hysteria of Germans in reference to the "human tidal wave" of incoming Roma. In January 2007 the *Los Angeles Times* published an article titled "EU's Ugly Little Challenge," which claimed that for many Western Europeans inclusion of Romania and Bulgaria in the EU "spells the inclusion of 3 million potential problems: yet more Gypsies. . . . Yet European newspaper editors are stumped by how they should address the largest minority on the continent. Town mayors all over Eastern Europe often avoid the term altogether and talk instead of 'whitening out' their inner cities" (McCann 2007).

33. I approximate that there are several hundred Macedonian Romani families in New York, numbering several thousand Roma, with half of them living in Belmont. In the last decade, more families have purchased private homes in the Pelham Parkway neighborhood. There are also a few families in the Philadelphia area and Roma from Struga in the Waterbury, Connecticut, area. Some Macedonian Roma reside in St Louis among a larger group of Bosnian Roma: http://stlouis.missouri.org/501c/gitana/roma.htm accessed June 15, 2010.

34. Macedonian Roma may be considered post-1965 "new immigrants" to New York, according to Foner 2001 and 2005.

35. See Lucassen 2005:143–150 for discussion of the German policy of recruiting Turkish guest workers.

36. The literature about roles and ethics in ethnographic fieldwork is vast; see Jackson 1987; Fluehr-Loban 2003; Lassiter 2005; Wolf 1996. For a discussion of fieldwork dilemmas in socialist and postsocialist states, see De Soto and Dudwick 2000.

37. This issue has been raised in the debates over representation in Marcus and Fischer 1986; Clifford 2004; Smith 1999; for the issue in terms of Roma, see Hancock 1997 and 2002; and Helbig 2007 and 2009.

38. My Balkan Romani fieldwork experiences until 1998 are reflexively analyzed in Silverman 2000c and 2008a; Silverman 1996c deals with fieldwork with American Kalderash.

39. The government policy of Bulgarization intended to turn all minorities into proper socialists. The policy began in the 1970s with the Pomaks and Roma and was extended in 1984–85 to the ethnic Turks, attracting international attention; see Poulton 1991. Chapter 7 discusses this policy in relation to Romani music and culture.

40. In the narrative of this book, all names are pseudonyms unless persons involved are well-known musicians.

Chapter 2

1. One of the earliest written sources about Balkan Romani music is found in the fifteenth-century archives of Dubrovnik (Gojković 1986:190).

2. For recent scholarly literature on East European and Balkan Romani music, see Seeman 1990a and 1990b (also see Seeman 2000, 2002, and 2007 for Turkish Thrace); Petrovski 2002; Kovalscik 1985, 1987; Sárosi 1978; Kertesz-Wilkinson 1992; Lange 1997a, 1997b, 1999, 2003; M. Stewart 1989, 1997; Dimov 2001; Peycheva 1993, 1994a, 1994b, 1994c, 1995, 1998, 1999a, 1999b, 2008a, 2008b, 2008c, 2009; Peycheva and Dimov 1994, 2002, 2005; Beckerman 2001; Van de Port 1998, 1999; Gojković 1986; Vukanović 1962, 1963, 1983; Pettan 1992, 1996a, 1996b, 1996c, 2001, 2002, 2003; Jakoski 1981; Blau, Keil, Keil, and Feld 2002; Brandl 1996; Radulescu 2004; Beissinger 1991, 2001, 2005, 2007; Malvinni 2003 and 2004; Helbig 2005, 2007, 2009; Rasmussen 1991, 1995, 1996, 2002, 2007; Marian-Bălaşa 2002, 2004; and the bibliographies in Silverman 2000b and Kertesz-Wilkinson 2001.

3. Reprinted in Djordjević 1984:38–39, cited in Pettan 2002:223.

4. In 1859 Franz Liszt claimed that what was called "Gypsy music" was in fact created by Roma, not Hungarians. Hungarians were, then, merely patrons for the Romani genius. Liszt wrote: "Hungarian songs as they are to be found in our villages . . ., being modest and imperfect, cannot command such respect as to be generally honored . . ., whereas the

instrumental music as it is performed and spread by Gypsy orchestras, is capable of competition with anything in the sublimity and daring of its emotion, as in the perfection of its form, and, we might say, the fineness of its development" (quoted in Sárosi 1978:142). Critics, most notably Bartók, countered, "What people (including Hungarians) call 'Gypsy music' is not Gypsy music but Hungarian music; it is not old folk music but a fairly recent type of Hungarian popular art music composed, practically without exception, by Hungarians of the upper middle class. But while a Hungarian gentleman may compose music, it is traditionally unbecoming to his social status to perform it 'for money'—only Gypsies are supposed to do that" (1931[1976]:206).

Liszt erringly dismissed the rich Hungarian rural peasant repertoire (which was collected a half-century later), romanticized about eastern survivals among Roma, and defended Roma as creative interpreters of urban songs. The opposite camp reproached Roma for inhibiting development of Hungarian creativity. For example, Spur wrote that Romani music "dammed the source of the far more valuable national folk song, and caused it to remain hidden in the isolated life of the village. . . . Such critics, the followers of Bartók and Kodály, turn off their radio set whenever a Gypsy orchestra is playing" (1947:130).

5. Note that there are female exceptions; see Chapter 10.

6. One common pitfall is attributing Indian scales to Roma with little documentation to substantiate the claim. Some scholars assert that the Romani scale is the Bhairava scale, which they define as "a 12-note Oriental chromatic scale" (see Acković 1989 and Fonseca 1995:106, the latter citing Hancock). On the contrary, the Bhairava rag (a hicaz pattern from the first degree repeated from the fifth degree) is not common in Romani music. In addition, Hancock (2002:72, as well as Fonseca 1995) attributes stick dancing among Roma in Hungary to the Romani migration from India. However, neither author accounts for the widespread performance of stick dancing among non-Roma in other locations (e.g., England and South Africa). Finally, Fonseca and Hancock attribute the vocal performance of rhythmic syllabics found in Hungary among Roma to Indian origins; again, this genre is also performed by non-Roma in Scotland and other locations. Indeed, it is difficult to attribute these far-flung musical styles and genres to Romani origins. These authors stretch musical interpretations to strengthen Indian ties (see Chapter 3). See below for further discussion of so-called Gypsy scales.

7. Bulgarian ethnomusicologist Vergilii Atanasov voiced this opinion to me and other colleagues (Buchanan 2006:266).

8. Although I do not discuss Serbian brass bands in this book, in Chapter 13 I do mention that in world music marketing Balkan music is often assumed to be synonymous with Romani brass band music; I am now researching this issue. See Babić 2003 and 2004 and Hedges 1996 for historical information.

9. For example, in Chapter 7 I discuss the fact that Bulgarian wedding musicians play for Bulgarians, Turks, Pomaks, and Roma. In Chapter 11 I illustrate this point with the life history of the Bulgarian wedding musician Yuri Yunakov, who added performing for Armenians, Albanians, and Macedonian Roma to the those groups when he emigrated to New York. Many other musicians, including non-Roma, can illustrate this principle.

10. For example, when Ivo Papazov (see Chapter 7) incorporated the theme from the movie *The Pink Panther* into his kyuchek "Pinko" in the

late 1970s, fellow musicians admired not only his whimsicality but also his skillful execution. (*Pinko* was recently rerecorded on Papazov's 2008 album *Song of the Falcon*. Kyuchek is discussed in the next section of this chapter.) Similarly, Boban Marković incorporated a theme from Mozart's Symphony No. 40 into his "Mundo Čoček," and Fanfare Ciocarlia released "007" on its 2005 CD *Gili Garabdi* (Hidden Songs), based on the theme from James Bond; this album includes a version of Duke Ellington's jazz classic "Caravan." (Although the liner notes for the CD claim thousands of former Romani slaves fled Romania for the American south, living in mostly black neighborhoods, the tie between jazz and Romani music is at best hypothetical.) Bulgarian wedding musician Ivan Milev is known for his quotes from classical and popular music; for example, his 2006 CD has quotes from Mozart's *Eine kleine Nachtmusik* and several genres of American popular music. A game that some wedding musicians play in jam sessions is to take a piece of classical or popular music (such as "Für Elise") and convert it into a kyuchek or a Bulgarian dance rhythm such as a rŭchenitsa.

11. Activist Mihail Georgiev of the Romani Baht (Romani Luck) Foundation was the organizer of the CD project. When he visited my home in 1994, he expressed interest in my commercial recordings of Romani music from the 1950s and 1960s, and I later gave him copies. He graciously thanked me in the CD notes.

12. "The Gypsy scale" has been defined by Central European writers as a scalar pattern with augmented seconds between the third and fourth degrees and sixth and seventh degrees. This scale is found in early *verbunkos* (a Hungarian recruiting dance) and is not common in the Balkans (Silverman 2000b). Another common mistake is to attribute Indian scales to Roma; see note 6 above.

13. The word *mane* refers to the Turkish and Greek vocal genre *amanes*, a free-rhythm vocal improvisation using makams where the singer vocalizes on the word *aman*, a pan-Balkan expression of emotion.

14. According to Seeman, the 9/8 *Roman oyun havası* rhythm became emblematic of Romani communities through Turkish marketing strategies of the 1960s (2002:276, 2007). Seeman discusses the types of 9/8 in 2002:278–279. Video example 5.12 shows a line dance in 9/8.

15. The tune is taken from a well-known kyuchek, "Leski Karuchka" (His Cart), and the text claims Barack Obama as "one of us": "You are so cool, you're from our tribe. Dark-skinned brother, you're our cousin." See www.youtube.com/watch?v=QXvgis02xwk, accessed February 15, 2009. Victor Friedman reminded me of a similar connection to African Americans when Muharem Serbezovski wrote a song in praise of Muhammad Ali several decades ago.

16. Patrons in one region often do not like the styles from another region. Papazov, for example, told me that he sometimes had a hard time pleasing north Bulgarian patrons.

17. Currently Serbian singers such as Mile Kitić, Šaban Šaulic, Dragana Mirković, and Marina Zivković are the source of many Bulgarian and Macedonian Romani songs. A website with a partial list of Balkan covers is www.bgpopfolk.free.fr.

18. In Chapter 6, I discuss the gendered dimensions of beauty contests. In 2008 the first festival of brass music was organized in Kumanovo.

19. Elsie Dunin, personal communication; also see Dunin 1997.

20. In 2002, Lolov was honored with an anniversary concert produced for Bulgarian National Television. Attention to Romani music is rare on national television, but Lolov regularly played Bulgarian as well as Romani

music, so he retained his reputation during socialism. Todor Kolev (a famous actor) hosted the event; classical and ethnic Bulgarian clarinetist Petko Radev spoke; Petŭr Ralchev (wedding music accordionist) performed with tŭpan accompaniment; and many other guests lauded Lolov's musical contributions. For a discussion of Lolov's life, see Peycheva 1999a and 2008.

21. See Chapter 7 for discussion of prohibitions during socialism, and how Roma sometimes resisted them. See Chapter 8 for an examination of the Stara Zagora festival. Peycheva refers to this period as the "media emancipation" of Roma (1995:13).

22. During this time, Romani song titles on album covers were routinely mangled by these Bulgarian-run record companies; Nikolai Gŭrdev's song "So Grešingjom" (What Did I Do Wrong?), which has the chorus *Da li Panda Roveja* (Do you still cry), was translated into Bulgarian as "Golyamata Panda" (The Large Panda; also see Peycheva 1995). For a discussion of piracy and the mafia, see Dimov 2001 and Kurkela 1997.

23. Many tavernas with Greek music opened in the early 1990s in Bulgaria. Kalin Kirilov reported that many of his fellow student tambura players in the Plovdiv Academy switched to bouzouki to find work; but since the hand position is so different for bouzouki, some of them had a hard time switching back to tambura to pass their examinations. See Chapter 8.

24. Using Hungarian examples, Anikó Imre discusses how Roma use rap to create a modern Romani sensibility and to enter the pop market. Furthermore, the representational tropes of rap tend to reinforce a ghetto image that may promote stereotypes (2006, 2008). She writes: "The most radical way in which global entertainment culture has mediated the post-socialist situation of Romany minorities in eastern Europe is by turning the ghetto, the place of the urban underclass, the very site of Romani segregation, into the site of profitable entertainment" (2006:663). Using the term *Ludic Ghetto*, Imre notes that some rappers in Hungary and the Czech Republic artfully play with stereotypes while criticizing racism (2006:664). Note that the Czech Romani rap band Gipsy.cz represented the Czech Republic at Eurovision in 2009 (see Chapter 8).

25. I would like to thank Sani Rifati and Rečko Ismail for insights into tel o vas. The oldest talava songs were sung in Albanian or Romani.

26. See the websites of the European Roma Rights Centre (www.errc.org) and Voice of Roma (www.vor.org) for more information about the displacement of Kosovo Roma as refugees in the aftermath of the Yugoslav wars and the plight of Roma who live in Kosovo. Džansever and Ćita improvise talava in Dusseldorf for Kosovo Roma, see: www.youtube.com/watch?v=NssFw766sJQ&;feature=mfu_in_order&list=UL accessed January 1, 2011.

27 See http://www.youtube.com/watch?v=sTFn3S-1vIA, accessed December 15, 2010.

Chapter 3

1. There is, however, significant literature on the African diaspora in music, including Gilroy 1993 and Garofalo 1994; the Caribbean musical diaspora has been documented by Ramnarine 2007a.

2. Pnina Werbner, drawing on Brah (1996), writes: "Conventionally, diasporas derive their imaginative unity from time-space chronotypes of shared genesis, homelands, sacred centers and cataclysmic events of

suffering (dispersion, genocide, slavery)" (2002b:11). These "homings" (Brah 1996) are articulated through commemorations and utopian visions. Thus ideologies of a shared past and a common destiny link diaspora communities (Werbner 2002b:11).

3. For example, Lemon demonstrates the patriotism of Russian Roma (2000), Theodosiou discusses the sense of local belonging of Epirot Greek Roma (2003), and Gay y Blasco emphasizes the Spanish emplacement of Gitanos (2002). Contrary to Safran, many Roma believe they are or will be accepted by their "host countries." See Toninato 2009 for a discussion of the tension between grassroots, scholarly and activist approaches to diaspora.

4. Caroline Brettel reminds us that "anthropologists have worked at both ends of the migration process, beginning in the country of origin and asking what prompts individuals to leave particular communities, and then what happens to them in their place of destination, including if and how they remain connected to their places of origin" (2003:1; see specific family trajectories in Chapters 4 and 5).

5. "*Because* of the fact that the very phenomenon of diaspora has produced a multiplicity of Chinese cultures, the affirmation of Chineseness may be sustained only through recourse to a common origin or descent that persists in spite of widely different historical trajectories, which results in the elevation of ethnicity and race over all of the other factors— often divisive-- that have gone into the shaping of Chinese populations and their cultures. Diasporic identity in its reification does not overcome the racial prejudices of earlier assumptions of national cultural homogeneity but in many ways follows a similar logic, now at the level not of nations but of off-ground 'transnations'" (Dirlik 2002:97).

6. Basch, Glick Schiller, and Blanc (1997) and others have distinguished among (1) transnational cultural studies that focus on the growth of global communications, media, consumerism, and public cultures; (2) globalization, which focuses on recent configurations of space such as the growth of global cities; and (3) transnational migration studies that examine the actual social interactions migrants construct across borders. My work focuses on the third area. Another useful term is "transborder," coined by Lynn Stephen (2007) to articulate processes on both sides of the U.S.-Mexican border.

7. In 1964 Milton Gordon wrote: "The individual who engages in frequent and sustained primary contacts across ethnic group lines, particularly racial and religious, runs the risk of becoming what, in standard sociological parlance, has been called 'the marginal man'" (56).

8. According to Werbner (2002b:120): "The powerful attraction of diasporas for postcolonial theorists has been that, as transnational social formations, diasporas challenge the hegemony and the boundedness of the nation-state, and indeed, of any pure imaginaries of nationhood."

9. Jonathan Friedman similarly writes: "Hybridity is founded on the myth of purity" (1997:82–83).

10. Dirlik writes of the danger in postcolonial and postmodern writings of focusing on identity at the expense of power. These writings focus on language as a discursive marker for registering and reaffirming difference, but they often fail to address broader networks of domination and exploitation (2000:188).

11. See Marushiakova et al. 2001 for a discussion of why Egjupkjani and Ashkalia in Kosovo do not employ the label *Roma*. See Gay y Blasco 2002

for a discussion of Gitanos and their identity labels in reference to religious mobilization. See Okely 1997 for a discussion of the politics of labels in the United Kingdom. Today, activists try to recruit all these groups into the Romani human rights movement; see Boscoboinik 2006.

12. Another excellent overview of postcolonial theory in relation to music is found in Born and Hesmondhalgh 2000.

13. Mirga and Gheorghe contrast a transnational minority that seeks rights within a nation/state framework to a transnational minority that seeks rights on a wider, i.e., European scale (1997). Some Romani groups, such as the Sinti of Germany, see themselves as a legitimate historic German minority; they reject the label *Roma*.

14. The International Romani Union (IRU) is a pan-Romani political organization that has sponsored congresses every few years since 1971 (see Acton and Klimova 2001; Klimova-Alexander 2005, and Barany 2002). Its legitimacy has been called into question by some Romani leaders. Since 2004 the European Roma and Travellers Forum (associated with the Council of Europe; www.ertf.org) has gained a respected place in Romani politics.

15. The few scholars who contest the Indian origin of Roma are often treated as traitors by activists. For example, Judith Okely (1983) and Leo Lucassen, Wim Willems, and Anna Marie Cottar (1998) suggest Gypsies might have an indigenous tie to Western Europe. Okely's historical claims are much more tentative than Lucassen; nevertheless she has been vilified by Romani and non-Romani activists (Okely 1997). From the point of view of linguistics, there is no question that Romani is an Indic language that separated from the rest of Indic at some point during the Middle Indic period and therefore indisputably arrived in Europe from India (Victor Friedman, personal communication).

16. Indian names were also popular in the 1970s among non-Roma. According to Garth Cartwright, Bosnian singer Indira Radić's father (a communist party member) named her deliberately after a strong Indian woman (personal communication). Indian music was also popular in Greece in the 1970s. Greek covers were released of songs from Hindi films; see Abadzi and Tasoulas 1998. In the 1990s the Bulgarian Romani band Kristal released *Maika India* (Mother India [Bulgarian]), and the wedding band Orfei released *Mafia ot India* (Mafia from India [Bulgarian]. Bulgarian chalga singers Sofi Marinova and Azis have also used Indian motifs; see Chapter 9.

17. Serbezovski's disk (RTB EP 16 306) cites the tune as "Indijska Narodna" (Indian folk; Serbian; arranged by B. Milivojević). Slobodan Ilić probably made the first recording; for an Ilić remake with the older vocals, see www.youtube.com/watch?v=2lqKhplhPNk, accessed October 20, 2009. Ethnomusicologist Jane Sugarman tried in vain to find the original tune, supposedly from the film *Dosti* (Hindi, friend, released with the English title *The Blind and the Lame*). She and Svanibor Pettan concurred that B. Milivojević wrote the song from the general sound of the Indian film score, with lyrics based on the plot of the film (Pettan 2002:240; East European Folklife Center listserv, November 29, 1997). The film is the story of two poor boys, one lame, one blind. Ramo Ramo continues to be covered, including Sasho Roman's "Oy Sashko" in Bulgarian and Keba's (Dragan Kojić) "Idem Idem Dušo Moja" (I'm going, my dear; Serbian, 2004).

18. Serbezovski's velvety vocal style captured audiences all over Yugoslavia. He was one of the first Yugoslav singers to adopt the practice of

making song "covers," for example taking the Turkish song "Allah Allah" and recording it in Serbian as "Bože Bože" (God, God; in both languages). He currently lives in Germany and has become a writer; his early life history was published in German (Serbezovski 1995, originally *Šareni Diamanti* [Colorful Diamonds], Sarajevo 1983). In 1997, he performed at a New Year's party for the New York Macedonian Romani community (see Chapter 5, photograph 5.22 and video examples 3.1 and 5.47), and in 2007 he performed in Bulgaria. His songs are still loved by the older generation from Macedonia and Serbia.

19. The 1970s recording opens with a dazzling unaccompanied clarinet solo by Medo Čun (see Chapter 2 for his family's history and other performances in video examples 10.3, 10.5, and 10.6), and the accompaniment is a rhythmic pattern that Balkan Roma associate with India (number 3 in Figure 2.1, discussed in Chapter 2). For Balkan Roma, pentatonic scales have come to symbolically represent India, perhaps because there seems to be a significant amount of pentatonicism in Indian film music. In Ramayana the scale is minor and five degrees of the scale are emphasized: 1, 3, 4, 5, and 7. In other pieces titled "Indiiski" (Indian), major and mixolydian (major with a lowered seventh degree) scales predominate, often with a pentatonic framework.

20. It has a 4/4 slow rhythm, number 5 in Figure 2.1, and is in the phrygian mode. Many other versions have been released.

21. Gelbart's version is from the German CD set *L'Epopee Tzigane/Road of the Gypsies*, (Network 24756, 1996). A similar version is referred to as "the canonical text" of the International Romani Union by Marushiakova and Popov (1995 2:13–14, 20).

22. Serbian and Macedonian versions include "Pilem Pilem" (I drank and drank; Rromano Centar, Opre CD 002, 1995) and "Djelem Djelem" (Kočani Orkestar, *L'Orient est Rouge*, Crammed Discs 1997/2006 CRAW 19), plus many YouTube versions. Šaban Bajramović recorded the song on *Mostar Sevdah Reunion* (Times Square TSQCD 9029) and other albums. Lyrics and mp3 files of the song can be found on the websites http://www.reocities.com/~patrin/gelem.htm and www.unionromani.org/gelem.htm (both accessed June 17, 2011). The version in the 1967 film can be found at www.youtube.com/watch?v=PQD6rWRiYVk, accessed December 13, 2009.

23. Dirlik continues: "so that it is often unclear whether the objection is to essentialism per se or to the politics, in which case essentialism serves as a straw target to discredit the politics" (2000:188).

Chapter 4

1. Gurbet, from Turkish, is a term and concept used in South Slavic languages, Albanian, and Romani.

2. I owe much insight to Jasmin and Aiše, who were very open with me in talking about the community and assessing traditional values.

3. Community members jokingly referred to her as one of "the original Mayflower people."

4. Being the super (superintendent) of a building means receiving free rent in exchange for doing maintenance and repairs. Many Romani families (as well as other Balkan families) seek this type of arrangement. Even middle-class families will hold on to their superintendent role to save money on rent.

5. See Foner 2000:127–141 for a discussion of immigrant women's work issues in New York.

6. Revitalization of Islam needs to be seen in a larger context; similar phenomena are happening in Macedonia and in other places in the Balkans. I am currently conducting research on this topic.

7. Unlike Balkan Roma, Kalderash typically do not work alongside non-Roma; they tend to be self-employed and marry at an earlier age (Sutherland 1975; Gropper 1975; Silverman 1988 and 1991). However, as Gropper and Miller point out, Kalderash also negotiate multiple identities, engaging in "selective multiculturalism" (2001).

8. Occasionally they call themselves *Horahane*, which means Muslim in Romani and Turkish.

9. According to Gropper and Miller, "a Rom may self-identify differently depending on the person to whom he is talking" (2001:87). I am currently doing research on identity issues among young Roma.

10. This comment arose when I distributed to Romani consultants a draft of my article on fieldwork with Roma (Silverman 2000c). All the other Roma I consulted didn't object to my using the real name of the neighborhood. However, from ethical concerns, I use pseudonyms except where Roma have public reputations as musicians.

11. Roma say that Americans in general are neither curious nor well informed about European geography. Before the breakup of Yugoslavia, many Belmont Roma told non-Romani Americans they were Yugoslavs. But in the 1990s, Yugoslavia became a synonym for violence, Macedonia declared independence, and thus the label *Yugoslav* faded. Roma sometimes strategically withhold speaking about their ethnicity and Americanize their names: for example, Severdžan is John, Tair is Tommy, Ferhan is Freddy; Nermin is Nancy, etc.

12. KUD stands for *Kulturno Umetničko Društvo*, Cultural Artistic Group (see Chapter 6).

13. Nevertheless, there are several community members who, for various reasons, have never married. Although homosexuality may be a factor, it is not discussed openly, as most Roma are homophobic; on the other hand, families do not ostracize someone even if he or she is perceived as outside the community norms. Divorced men and women usually remarry.

14. Much of what I am describing about family and community life is not unique to Roma and can be compared to other Balkan ethnic groups. There is a huge cross-cultural literature on the topic of negotiation of female power; for the Middle East, see Nelson 1974. See later in this chapter, the next chapter, Sugarman 2003 and Silverman 1996b for female knowledge about ritual.

15. Leila claimed that "ninety-nine percent of the time that's the only objective—to get to America. And then once they come here, it becomes an issue." The zet in one failed Belmont marriage could not adjust to America and became abusive to the bride's entire family. He was sent back within a few months, and the bride remarried a local man. On the other hand, many marriages of this type work out well.

16. The uneasy relationship of human rights to culture has been discussed widely in the anthropological literature; see Cowan, Dembour, and Wilson 2001; Abu-Lughod 2002; and Goodale 2006.

17. Eminova expanded her prototype virginity surveys to organizations in Novi Sad, Serbia; and Pecs, Hungary, with successful results. In addition to tackling the issue of virginity, a goal of the project is to teach educated Romani women computer skills for future research and networking;

see http://www.advocacynet.org/resource/492, accessed June 17, 2011 and Eminova 2005. Furthermore, the virginity issue is only one of many regarding female subordination that European Romani women activists are addressing; see European Roma Rights Centre 2000, Plaut and Memedova 2005 and http://www.advocacynet.org/page/irwn, accessed June 17, 2011.

18. I am exploring this topic in my current research on youth and education in the Romani community.

19. There is a huge literature on the domestic-public split, which emerged as a concept in the early anthropology of women (Lamphere and Rosaldo 1974; Lewin 2006; di Leonardo 1991). I am exploring this issue in my current research with Romani youth.

20. See *Roma Portraits* (1998), a series of short documentaries by young Bulgarian Roma, supervised by Asen Balicki. By contrast, many scholars have noted the focus on place and religion in Native American filmmaking (Leuthold 1998:183; Worth and Adair 1972).

Chapter 5

1. For example, in his memoir, Serbezovski recounts his circumcision celebration in Topana, Skopje, in the 1950s, where the absence of his father (who had abandoned the family for a mistress) almost ruined the event. His uncle filled the role of his father, but emotions were noticeable and raw (1995). Similarly, at a New York wedding there was almost a fight between the in-laws because of an old schism. At another wedding, a female guest was upset with the order in which families were called up to lead dance lines (see Chapter 6); to protest, she did not attend the blaga rakija.

2. Creative solutions are devised to save money on celebrations. Relatives who are musicians may be asked to play, and they often perform just for tips. Sometimes ceremonies are combined; a wedding in 1990 in Šutka, described later, was deliberately combined with a circumcision. The contracted musicians charged a combined fee, and the costumes (and even the meals) did double duty.

3. In the Bulgarian language there is a proverb that states "the Gypsy will throw all his money for a wedding and the next day he will not be able to buy bread for his children" (Marushiakova and Popov 1997:149).

4. Roma Portraits is the outcome of a 1997 video workshop in Sliven, Bulgaria, where Balicki trained young Roma to film subjects of interest to them in Romani communities.

5. Her precarious and transitional status is manifested in the myriad ways she is protected from the evil eye; for similarities to ethnic Macedonian weddings, see Silverman and Wixman 1983.

6. See Petrovski 2001 and 2002:18–20 for a discussion of songs texts. There is, of course, a distinction between the numerous songs that describe weddings and the few songs which are sung at weddings for ritual purposes, such as *Oj Borije* see below.

7. The role of Roma in keeping rituals active (or even newly adopting them when they have died out) among the majority populations has been noted by several scholars (see, e.g., Popov 1993). Sugarman (1997) documented multiday weddings among Prespa Albanian Macedonians in the early 1990s, but currently they have shorter weddings.

8. The literature on Balkan weddings includes Sugarman 1997; Ivanova 1984; and Kligman 1988.

9. For a description of Macedonian Romani weddings, see Petrovski 1993 and 2002; Dunin 1971 and 1984; Seeman 1990b; and the Macedonian television film "Romska Svadba" (1992; see next note). For Bulgarian Romani weddings see I. Georgieva 1966 and Marushiakova and Popov 1997:144–150.

10. Amdi Bajram, the founder of the textile factory Šuteks in Šutka, was a Romani representative to the national parliament during the 1990s. Because he was well known and wealthy, his son's 1992 wedding, which had hundreds of guests, was filmed for Macedonian television and broadcast as "Romska Svadba." Gjulizar Dželjadin, a respected family friend, narrated the wedding for the film; I interviewed her several times about ritual.

11. Gjulizar is referring to Kalderash Roma who practice the custom of brideprice, that is, the groom's family paying the bride's family a sum of several thousands of dollars. It is probable that the brideprice was once widespread among many groups of Roma but has survived only among the Kalderash (Marushiakova and Popov 1997:144). Most groups still have at least a symbolic buying (or bargaining) for the bride; see later discussion in this chapter. See Sutherland (1975) for brideprice among American Kalderash; Marushiakova and Popov 1997:144 for Bulgarian brideprice customs; and Brunwasser 2011 for a description of a Kalajdži bride fair in Stara Zagora, Bulgaria. Among Bulgarian Kalajdži the current average brideprice is $5,000. In the United States among Kalderash the price for a virgin can reach $10,000.

12. The days of the week are sometimes modified (e.g., henna ceremony starting on Thursday), but the sequence of the events remains the same in Macedonia; however, in New York, the sequence often changes (see below).

13. Henna is used throughout the Middle East and South Asia in Muslim and Hindu communities. Unlike in South Asia, where it is applied in intricate designs, in the Balkans it is smeared on the hands and feet of women merely for the color. In Macedonia and Bulgaria, henna is associated with Muslims, but in many parts of Greece henna is used in Eastern Orthodox weddings.

14. Michelle Rosaldo's classic article, which explains female subordination as a result of the domestic-public split, claims that women are identified with the domestic sphere (defined as the sphere of mothers and children) and men with the public, leading to hierarchy (1974). This argument has been challenged by scholars for many reasons, for example regarding slippage between domestic and private (MacCormack and Strathern 1980; Ortner 1996). The domestic, rather than being marginal and excluded from politics, is the site of important decision making regarding family budgets, marriage choices for children, and reputations. As for gender geography, the spatial segregation during celebrations has been described in terms of the "inside" women's world and the "outside" men's world. This concept of space is shared with non-Romani Muslims of the region (Sugarman 1997), but it must be noted that the inside female world spills out into public space during rituals (Silverman 2003).

15. According to zurla player Muzafer Mahmud (see the next note), there was a special melody (*alaj*, procession) for getting the bride, but it declined in Skopje around 1960. In 1990, Mahmud only knew three ritual melodies, one for the kana and two for waking up the boy the morning after his circumcision.

16. Muzafer Mahmud was the resident zurla player at Radio Skopje from the early 1970s and made dozens of recordings. At his daughter's

henna ceremony, his brother, cousin, and brother's grandson played. He comes from a long line of zurla and tapan players. In Šutka there are about ten families of performers. For comparative information about Greek Macedonia, see Blau et al. 2002; for Bulgaria, see Peycheva and Dimov 2002.

17. After her wedding, Ramisa ended up living around the corner from her parents. For Romanian comparisons of how the bride is made to cry, see Kligman 1988.

18. See Dunin 1971 for a description of this ceremony in an old communal bathhouse.

19. In the 1980s in Macedonia, women started using hairdressers as a status symbol. In the 1990s upswept hairstyles were common, with white flowers and beads interwoven in the hair. Romani women patronized certain hairstylists who knew how to design these styles. Before styling, a woman had to put on her white underblouse (worn under the jacket of her šalvari) because the volume of the hairstyle would not allow the underblouse to be slipped over the head.

20. In the Balkans, written invitations are rarely used; rather, sponsoring families invite people face-to-face with formulaic language by visiting their homes. See later discussion for contrast to American weddings.

21. The bajraktar role is common in Skopje but not in Prilep. It seems to be more prevalent among the Džambazi subgroup of Roma than among other groups. The bajraktar is a very well-respected, trusted male member of the community, and his whole family is treated royally during the wedding both by the groom's and the bride's sides. He and his family give and receive lavish gifts. In fact, many of the gifts he receives (such as shirts) are pinned to the flag itself. The flag is decorated with fruit on the end of its pole, greenery, and red ribbons. At a double wedding I attended in Šutka in 1994, there were two bajraktari, one with the Macedonian flag and one with the Romani flag, signifying two levels of identity. According to Gjulizar Dželjadin: "He has to be from a good family, married only once." According to Aiše, the flag is so important that "if it falls to the ground, you have to sacrifice an animal immediately"; Aiše also explained that "the bride's family tried to steal it, so it must be guarded!"

22. Temana is a Turkish term for a symbolic hand gesture indicating respect. See Ellis 2003:128.

23. In the early 1990s in Bulgaria, it was fashionable among wealthy Macedonians, Turks, and Roma to rent a helicopter to transport the bride.

24. The mother-in-law and daughter-in-law relationship is the most important and most precarious in the time period after the wedding. The bride is subordinate to her mother-in-law as well as to the males of her new family. She has to get along with her mother-in-law, who may be very demanding. On the other hand, the mother-in-law may be her greatest ally, even against the males of the family. See Sugarman 1997, Kligman 1988, and Ellis 2003 for analyses of Prespa Albanian, Romanian, and Turkish/Albanian urban Macedonian female-to-female relationships, respectively.

25. Džumaluk comes from *Cuma*, which is the Turkish word for Friday. According to Seeman, before World War II the taking of the bride occurred on Thursday and the džumaluk occurred on Friday (1990b:42).

26. Interestingly, Kalderash Roma in the United States hold their weddings midweek and still get large attendance. One reason Kalderash have weddings midweek is that they do not plan them far in advance; banquet halls are simply not available on weekends at the last minute.

Kalderash can attend midweek because they tend to be self-employed and thus have more control over their schedule. Also, because their children are less integrated into the school system, they have no qualms about keeping a child out of school for several days because of a wedding. This is not true for Macedonian Roma, for whom school is often very important (see Chapter 4).

27. The parents of the groom entered first, followed by the parents of the bride, followed by close relatives.

28. The event typically begins with the mother of the bride leading the dance line with the decorated bottle of rakija.

29. The family does not speak Romani; rather, they speak Turkish and Macedonian in addition to their primary language, Albanian. In fact, they do not refer to themselves as Roma, but everyone else in the community accepts them as Roma, and they are intermarried with Roma. For a 2010 release of Ramiz's sons' band, see www.youtube.com/watch?v=uykObURJNns, accessed January 1, 2011.

30. See Leibman 1974 and Sugarman 1997 for a discussion of the Prespa song, instrumental, and dance repertoire.

31. Ilhan is referring to violent incidents that occurred in the 1990s in New York night clubs, where Bosnian Muslims, Albanians, and Serbs congregated. I witnessed several such fights at these clubs.

32. Severdžan's wife also loves to dance. The whole family, including the parents and both brothers' children, were invited to teach at a weekend dance workshop for Americans in Maryland. Yuri Yunakov and Seido Salifoski were among the invited musicians. In Chapter 4, I discussed how the two brothers tried unsuccessfully to organize a dance group with Roma from Belmont.

Chapter 6

1. Portions of this chapter are reprinted from Silverman 2008b with permission from McFarland & Company Inc.

2. Community members refer to dances by names that are not standardized. When a leader requests a song or dance, there is sometimes miscommunication, and the leader might refuse to dance until the musicians play the "right" melody. From the point of view of musicians, this can be very frustrating because they sometimes have to guess several times what the leader wants.

3. Mahala means neighborhood in Turkish and the Balkan languages, but its use implies that it is a low-class Turkish or Romani neighborhood. Manele, from the Turkish *amane*, means an instrumental or a vocal free-rhythm improvisation. These terms are also used for the accompanying music, which may also be referred to as *musică orientală* (oriental music); see Chapter 9.

4. The film *Iag Bari: Brass on Fire* (by Ralf Marschalleck, HS Media Consult, 2002) features several performances with these female dancers.

5. Markovic's Vranje suite (by Ensemble *Djido*, Bogatić, Serbia) can be seen at www.youtube.com/watch?v=rpiDsHdxKt8&;feature=related, accessed December 12, 2010. The poster remarks: "Boiling Gypsies temperament with the accompaniment of tambourine and drums contribute to the value of this spectacle (sic)." Kolo introduced a new Vranje choreography in the 1990s, but it retains many of the stereotypical movements (see www.youtube.com/watch?v=gKp8PbTR5hQ&;feature=related accessed December 11, 2010) (Alexander Markovic, personal communication).

6. Sugarman recalled a regional dance competition in the early 1980s where an amateur Romani group from Kumanovo performed, ending their choreography with a mock drunken fight. Neither the performers nor the jury nor the folklorists seemed concerned that thus was "feeding a negative stereotype" (Sugarman, personal communication).

7. See Hancock 2008 and Hasdeu 2008 for discussion of how the female Gypsy body is portrayed in museums and literature.

8. For example, the Bulgarian Romani wedding band *Trŭstenik's* 1990s cassettes are titled *Gol Kyuchek* (Naked Kyuchek) 1 and 2 and feature bare-breasted women.

9. See as examples the videos MP 31003: *Volim Te/U Zemlji Baro-Than* (I Love You/In the Great Land [refers to India; Serbian/ Romani]) and MP 31005 *Romano Horo/Čhaje Šukarije* (Romani Dance/Beautiful Girl). In Chapter 10, I discuss Esma's video collaboration with Toše Poeski, which features stereotypical dance scenes.

10. Sonneman reports that Roma sometimes defend romantic images of themselves because they elicit sympathy from non-Roma (1999:129).

11. I would like to thank Lozanka Peycheva for updating me on this phenomenon (personal communication). Clips from the 2007 Stara Zagora festival, including dance groups, are often aired on the Sofia cable television station Gypsy TV (www.gypsytv.tv).

12. See Helbig 2005, 2007, 2008, and 2009 for discussion of the role of Romani NGOs in music activism in Ukraine.

Chapter 7

1. Some ideas in this chapter and the next were first presented at the conference Cultural Circulations, at the Ohio State University, 2005. I would like to thank the participants, especially Amy Shuman, for their comments. Some concepts were further developed in Silverman 2007a (reprinted with permission from the publisher).

2. See Chapter 11 for a discussion of how and why Yuri Yunakov changed his name.

3. Although not widespread, there were some notable instances of resistance among Pomaks, for example, in the village of Ribnovo in the Pirin region.

4. Of course, the teaching and practice of the Qur'an was prohibited, as was the teaching of Eastern Orthodoxy, the majority religion; however, the ban against Islam was enforced more severely. For example, there were virtually no working mosques in villages, whereas there were a few working Eastern Orthodox churches in villages and quite a few in towns.

5. Verdery (1996) and Gal and Kligman (2000a and b) have written extensively about the private-public dichotomy in socialist societies.

6. *Tsiganski Pesni* (Gypsy Songs) BHA 11087 omits the Romani titles of songs; rather, songs are translated (often mistranslated) into Bulgarian. The same policy applied to Turkish music. Other early 1980s Balkanton releases of Romani music include *Tsiganski Pesni*, BHA 10183; *Ivo Barev /Asiba Kemalova: Tsiganski Pesni*, BHA 10645; *Ibro Lolov: Tsiganski Pesni*, BHA 10890. These albums often featured famous wedding musicians.

7. This celebration (*izprashtane na voinik*) is sponsored by the parents of the soldier and can be as elaborate as a wedding.

8. See N. Kaufman 1989; D. Kaufman 1990; Buchanan 1991:522–529. Non-Roma also played a major role in the history of wedding music. For

example, Atanas Milev, the father of Ivan Milev, was one of the founders of the influential band Pŭrvomayskata Grupa (May First Group, referring to the town named after International Workers' Day). I would like to thank Ivan Milev for many fruitful discussions about the history of wedding music.

9. In the 1970s there were often two accordions and no bass guitar; the bass was introduced a few years later. The drum set is sometimes modified to include bongos or *indiyanki* (roto-toms).

10. Ironically, tambura is a traditional village instrument only in the Pirin region. It was physically modified to become a chordal instrument and accepted into the canon of national traditional instruments by the ensembles in the 1960s (Buchanan 2006:151).

11. Pravo horo is in 2/4 meter (often 3+3), rŭchenitsa is in 7/8 meter (2+2+3), paidushko is in 5/16 meter (2+3), and krivo horo is in 11/16 meter (2+2+3+2+2). Another common rhythm is 7/8 (3+2+2), identified with Macedonia and Pirin (see rhythm number 10 in Figure 2.1).

12. Nikolai Kaufman observed, "The most important feature of this musical genre is improvisation. . . . Different from folk orchestras composed of traditional instruments which strictly play pieces of composed multi-part music, the groups . . . play more freely, often without knowing how long a piece will take, how it will be built, who will solo—how it was at the dance (horo)" (1987:79). Here Kaufman favorably compares the spirit of the wedding bands to the spirit of the horo, the traditional village dance event, where dancers and musicians communicated constantly; he distinguishes it from the more formal, stilted, and formulaic atmosphere of ensemble music. See Buchanan 1991 and 2006 for thorough discussion of ensemble style, and Rice 1994 and 1996 for comparison of wedding, ensemble, and village styles. Indeed, wedding music shares many characteristics with the horo: it is village-based, is open-ended in terms of length (with some horos lasting four to five hours), and thrives on dancer reaction. Todor Todorov, for example, describes Bulgarian accordionist Ivan Milev as an "artist who grasps the audience and leads them to react violently to every one of his gestures" (1986:7). There is, then, a great deal of performer-audience interaction in wedding music, and both dancers and listeners alike are energized, especially when the musicians improvise. In comparing weddings to concerts, clarinetist Ivo Papazov stated: "In truth, a wedding is equal to a dozen concerts. There a person can create. . . . A great deal of music is introduced into a wedding, and in a concert you lack this thrill." Saxophonist Yuri Yunakov concurred: "You can't compare a wedding with any other performance. At weddings people have gathered for joy, they know each other. On the concert stage it is more like an examination."

13. The full version is at www.youtube.com/watch?v=Q_FgpZ87R_M, accessed January 2, 2011.

14. Papazov is known as Ibryam by most fans; other wedding musicians are also known by their first names or nicknames (which are sometimes based on their village or town).

15. Art Pepper is an American jazz saxophone player of the bebop era. Alan Stivell is a harpist who played a large role in the Celtic revival. Joan Baez is a prominent singer in the folk revival.

16. This act of contagious magic, whether accidental or purposeful, bonded him to the instrument. If this incident ever happened, however, the thread could have been from a zurna, not a clarinet, since his father played both. Ivo also related that when a male child was born, the parents put a clarinet in the cradle, also illustrating contagious magic.

17. This quote is taken from the BBC (British Broadcasting Corporation) Bulgarian web site www.bbc.co.uk/bulgarian/news/story/2005/03/printable/050306_papazovbbc.shtml, accessed March 10, 2008.

18. See biographies of wedding musicians in Bakalov 1992, 1993, 1998; and Peycheva 1999. Note that Ivo's wife, Maria Karafezieva, is a well-known performer of the Bulgarian vocal repertoire of wedding music.

19. See the full version at www.youtube.com/watch?v=leA_UQ6GJUI&; feature=related accessed January 2, 2011.

20. Gaida player Maria Stoyanova is an exception to this pattern. See Rice 1994:268–271; also see Chapters 6 and 10.

21. In 1985, Trakiya charged 1,300 leva for a one-night soldier send-off party. Also see N. Kaufman 1989.

22. In the 1980s the category of wedding musician was somewhat motley, including hundreds of lesser-known performers of doubtful ability. These imitators contributed to the mass phenomenon by disseminating the core repertoire.

23. This sometimes causes conflict, for at the end of a wedding there are often drunken men demanding "one more song."

24. In the western Thracian area around Plovdiv and Pŭrvomai, a decorated box is used to collect tips. If the collected amount is less than the agreed sum, the sponsors pay the musicians the difference; if it is more, the musicians keep the difference.

25. The label was used by the socialist government for wealthy families who resisted collectivization of land.

26. For example, *Popularni Trakiiski Klarinetisti* (BHA 11188, Popular Thracian Clarinetists) includes Petko Radev, Nikola Iliev, Nikola Yankov, Hari Asenov, Ibryam Hapazov, and Yashko Argirov; the first three are Bulgarians, and the last three are Roma (although no ethnicity is revealed on the album).

27. Arrangers are indeed paid per arrangement; see Buchanan 2006 for a full discussion of obrabotki.

28. The first studio was established in 1980 in Plovdiv. By the mid-1980s every major city had one or several studios.

29. The proprietors of the studios often traveled to events to record, sometimes plugging into the amplification equipment and paying the musicians a modest state-set fee. Many fans also recorded at events, but musicians received no compensation. Proprietors were state employees who worked on a percentage system, which in 1985 was 50 percent. In one Sliven studio, the average monthly intake in 1985 was 1,000 leva, which means the proprietor received 500 leva or $250. He boasted he had even made 900 leva, or $450 a month, occasionally. This proprietor was a Romani drummer who gave up wedding work because studio work was easier. Thus the world of the studios and the world of wedding musicians intersected; the proprietors were often extremely knowledgeable about wedding music, and they had their fingers on the pulse of popular taste.

Chapter 8

1. The Plovdiv Folk Jazz Band, composed of jazz musicians, had a style much closer to jazz than to folk or wedding music, For connections between folk and jazz, see Levy 2009.)

2. This sentiment is still current in Bulgaria. Later in this chapter I discuss the 2005 controversy about Romani singer Sofi Marinova's role in the competition leading to Eurovision.

3. The 1994 festival cost 1 million leva ($17,000), and prizes were approximately $30 each, equivalent to about two weeks' salary in a factory. The Haskovo regional government contributed one-quarter of the 1994 festival funding, but the rest had to be raised from private firms. In 1994 there were 40,000 audience members, but by 1996 there were only 4,000. In 1996, for the first time, Bulgarian television and radio did not broadcast the event; they demanded a huge subsidy from the sponsors to defray their expenses, and when the sponsors refused they did not attend. The sponsors had a hard time raising even the prize funds. After 1996 the festival was abandoned.

4. Ivo composed "Celeste" earlier and named it after a popular television series. It was later recorded on the album *Panair/Fairground* (2003); see later discussion.

5. Trakiya Folk was held in 1994 (Dimitrovgrad), 1995 (Haskovo), 1999 (Stara Zagora), 2000 (Stara Zagora), and 2003 (Plovdiv).

6. There is am emerging literature on nostalgia for socialism; see Berdahl 1999.

7. I will deal with Trakiya in detail because its trajectory is unique. Some wedding performers have become active in the growing Romani music scene and in the chalga scene. Others have been featured as guests in international Romani productions; Yashko Argirov and Slavcho Lambov, for example, appeared in the Hungarian production Gypsy Spirit, which toured Europe and North America (see Chapters 6 and 12). Filip Simeonov appeared with the Romanian group Taraf de Haidouks and recorded with them on the album *Band of Gypsies* (Nonesuch; see Chapter 13). Note that there are newer wedding bands that command solid reputations, such as Orkestŭr Plovdiv and Folk Palitra (folk palette, Bulgarian).

8. As early as the 1980s Kanarite were known as a "well-behaved band." According to Rice, Stoev insisted that members arrive on time, wear identical white jackets, and refrain from smoking and drinking on the job. In 1988, their Romani clarinetist Nesho Neshev complained to Rice about how reserved the music was (Rice 1994:246).

9. See full version at www.youtube.com/watch?v=74NS_VpYbZs&; feature=related, accessed January 2, 2011.

10. In the Iraq war, Bulgaria was known as a staunch ally of the United States.

11. See www.youtube.com/watch?v=gMEFyAk6Y-0, accessed January 2, 2011.

12. Another type of Romani festival is emerging in the United States with the work of the NGO Voice of Roma (www.voiceofroma.com); their Herdeljezi festival is run by both Roma and non-Roma and tries to combine education and entertainment. The lack of an organized Romani community in the United States has been an impediment to connecting with and attracting more Roma. On the other hand, the festival has received state and national grant funding for its pioneering efforts to bring the culture of an invisible minority to the attention of the American public.

13. Despite the official rhetoric, kyuchek is the main dance genre found on stage and off stage at the Stara Zagora festival. See Chapter 6 for a discussion of kyuchek, and video example 2.5 for the music and dance at the festival.

14. Jury members have included, for example in 2007, Romani singers Nikolai Gŭrdev, Ivo Barev, and Sasho Roman along with Bulgarian folklorist Lozanka Peycheva, as well as Malikov.

15. The Elit Center is associated with the *chitalishte* (reading room or cultural center) of the Krasna Polyana district. Recent activities have included shows for the holiday of Christmas and Vasilitsa (St. Basil's Day, a Romani holiday after New Year's), dances, and political speeches. These activities have been led by Sali Ibrahim, a Romani poet who directs the center. The center also sponsored *Romane: International Magazine for Romani Culture, Literature, and Art*, whose first issue appeared in 2005. See www.chitalishteelit.piczo.com.

16. On the other hand, scholarly study now includes the culture, music, and history of Roma and other ethnic groups. Pioneering Bulgarian scholars writing on Roma include Lozanka Peycheva, Ventsislav Dimov, Antonina Zhelyaskova, Ilona Tomova, Elena Marushiakova, Vesselin Popov, Rosemary Statelova, Claire Levy, Ivalyo Ditchev, Alexey Pamporovo, etc. (see Valtchinova 2004 and Silverman 2008a).

17. In 2001, when an American visitor asked Prof. Slavchev of the Academy of Musical Arts in Plovdiv if the musics of Jews, Roma, Turks, and other minorities were included in the curriculum, he said no, but there were restaurants in town where one could hear these musics (Henry Goldberg, personal communication). For the past several years, a summer program for foreigners has been organized at the Plovdiv Academy (www.folkseminarplovdiv.net). Instruction on clarinet, violin, and accordion is featured in this program; however, it is not officially part of the regular curriculum. It is ironic that the rubric "folk music" includes these instruments for foreigners but not for Bulgarians.

18. There are some exceptions: at a 2005 concert celebrating European accession, students from the Shiroka Lŭka High School performed one wedding song; the rest of their program included solo and arranged village music. A new folk dance curriculum track has been introduced in Shiroka Lŭka, but this too excludes Romani and Turkish dance.

19. Note the contrast between the place of Romani music and the place of Pomak music in contemporary folk festivals. Unlike Romani music, Pomak music is now embraced at festivals and draws enthusiastic audiences. Pomaks now wear their Muslim costumes freely and sing texts that include Muslim names and references to Muslim celebrations. Why the difference between Romani and Pomak music? Pomaks are Bulgarian-speaking, and thus, their folklore is configured by scholars as purely Bulgarian with a Muslim overlay. Thus it can be embraced in the domain of folk. But note that when Pomaks do perform, their ethnic label is neither announced nor printed in written programs (Ditchev 2004). The label *Pomak* has become a contested term; Bulgarian Muslim (Bulgaro-mohamedanin) is preferred. Although the rise of Pomak ethnic consciousness is not the topic of this book, I note that escalation of Pomak Muslim religious identity receives financial support from countries such as Libya and Saudi Arabia (for mosque building, teaching of Arabic language, and distribution of Qur'ans; see Ghodsee 2009). Whereas rich Muslim nations have an interest in the Pomaks and Turks of Bulgaria as potential allies, they have less interest in Roma. Pentecostals, however, are interested in Roma.

20. Ditchev is referring to an innovative Romani educational integration project initiated by the NGO Drom in Vidin and funded by the Open Society Institute. Integrated educational projects now operate in several locations around the country.

21. The July 2006 training was supported by the Democracy Commission Small Grants program of the American Embassy. The participants were

exposed to Romani culture and folklore; an additional aim of the seminar was also to stimulate teachers to diversify their way of teaching and more actively engage both children and parents. The teachers dedicated a day of the week to the culture of each ethnicity and invited parents to present, see www.amalipe.com/index.php?nav=projects&;id=8&lang=2, accessed June 18, 2011. Textbooks were prepared for the classes, e.g., *Stories from the Fireplace* (for grades two through four) and *Roads Retold* (for grades five through eight), although the Bulgarian Ministry delayed in disbursing the necessary funds for their distribution. The folklore texts are organized according to classic generic categories such as fairy tales, calendrical and family feasts, and song texts, attributed to various groups of Bulgarian Roma. The books tend to treat folklore as a collection of items to be classified and categorized. In reality, Romani scholars and Roma alike agree that Roma in Bulgaria cannot be divided into neat groups (Marushiakova and Popov 2001). In addition, Romani folklore items are compared to Slavic Bulgarian variants and West European variants such as those from the Grimm brothers. I suggest this framework seeks to legitimize Romani folklore by showing that its structural features are similar to Bulgarian folklore. In spite of these small caveats, this project is a welcome sign of support of Romani folklore by the state.

22. Note that this designation is accompanied by a monetary award plus international prestige. See www.unesco.org/culture/ich/index.php?pg=00103, accessed June 10, 2010.

23. These Albanian-speaking Roma refer to themselves as Egjupci or Egjupkjani (Egyptians); see Chapter 1. Historically they are Roma who moved up the social scale by adopting the Albanian language and distancing themselves from the stigmatized label *Roma*. In a 1955 film of the wedding, these Romani musicians are featured prominently; see www.europafilmtreasures.eu/PY/262/see-the-film-galichnik_wedding, accessed October 25, 2010.

24. Note that the descriptive part of the application is based on Kličkova and Georgieva's 1951/1996 study, which also minimizes the role of Roma.

25. Compare this to Giguère's research, which deals with Spain's unsuccessful 2005 application to UNESCO to have Flamenco declared a Masterpiece of Intangible Cultural Heritage. She found that the ownership of Flamenco was contested and that the role of Gitanos was minimized (2008).

26. See http://www.youtube.com/watch?v=mfXNtcNG2SE, accessed December 15, 2010.

27. Trifonov's impact on Bulgarian cultural life is considerable. His show is watched by Bulgarian émigrés, by Macedonians and Serbs. He not only is a singer, song writer and arranger, but he produces concerts, tours, a reality TV show, and contests for talented singers. He is acutely aware of and promotes ethnic diversity in Bulgaria and is closely tied to Bulgarian folk music, wedding music, and Romani music. One of his projects was a televised dance contest (Dance with Me) in which one category of competition, Oriental Dance, included Gypsy Dance, Belly Dance, North Indian Dance, Arabic dance, and several more. The grand prize was a red Ferrari, and in February 2007 it was won by a Romani dancer. See www.slavishow.com.

28. Some writers said Sofi and Slavi looked static, like statues, and didn't dare hold hands; many others wrote racist comments on internet forums and in YouTube commentaries. Activist organizations responded; the NGO Romani Baht (Romani Luck) in Sofia called on public officials to denounce the anti-Romani backlash.

29. The votes were tallied by cell phone; Slavi accused the producers of the winning group, Kaffe, of purchasing 50,000 leva worth of SIM cards and distributing them to people who voted for Kaffe (see *Standart* February 15, 2005).

30. See www.youtube.com/watch?v=vvHF8SK6dHY, accessed March 20, 2010.

31. However, by June 2008 the Roman Star music contest, devoted exclusively to Roma, was launched in Turkey; see www.medyakafe.com/haber.php?haber_id=6366, accessed July 20, 2008.

32. I would like to thank Nick Nasev for this information; also see www.zvezdegranda.com accessed June 15, 2010.

33. This was a nonmusical ambassadorial post. See www.romea.cz/english/imdex.php?id=detail+detail+2007_517, accessed October 30, 2007.

Chapter 9

1. Audiences for chalga shows have reached 27,000 fans, for example, at the 2006 Planeta Prima show in Varna. Some authors use pop/folk as a broad category under which they place wedding music, pop music, chalga, and other contemporary fusions (Buchanan 2007; Dimov 1995). Several authors emphasize the continuity of chalga from the nineteenth century (D. Kaufman 1995; Levy 2002; Vŭlchinova-Chendova 2000). The scholarly literature on chalga is quite extensive, encompassing works in Bulgarian (Dimov 1995, 1996, 1997, 1998, 1999, 2001; Peycheva 1995, 1999a, 1999b; Peycheva and Dimov 1994; Kraev 1999; Ivanova 2001) and in English (Rice 2002; Statelova 2005; Levy 2002; Kurkela 1997 and 2007; Apostolov 2008; and Buchanan 2007). There are also hundreds of Bulgarian newspaper articles, some scholarly, some journalistic, and some merely descriptive. See bibliographies in Dimov 2001 (the definitive book of its era) and Statelova 2005.

2. For a discussion of arabesk as a controversial genre in Turkey, see Stokes 1992 and 2003.

3. These festivals started in the early 1990s; see Buchanan 1999, 2006, and 2007:436-452.

4. As mentioned in Chapter 2, čalgija in Macedonia refers to improvisational urban Turkish-influenced music that was prominent until World War II and performed mainly by Roma. In Bulgaria at the end of the nineteenth century the word *chalgadzhii* meant professional urban musicians (mostly Roma) who performed the repertoires of various ethnic groups in both urban and rural settings (Vŭlchinova-Chendova 2000). By the 1970s it referred to wedding musicians who could improvise. Peycheva writes that "among Romani musicians, chalga is used to mean our music, free, virtuosic, impressive, masterful, celebratory, beautiful" (1999b: 64). According to Seeman, professional wedding Romani musicians in Turkey call themselves *çalgici* (2002:264–266).

5. Bulgarians code Turks as more religious Muslims than Roma (only half of whom are Muslim). Some Bulgarians believe that Turks are fanatical Muslims, and are thus conservative in dress, dance, and treatment of women. The xenophobic Attack party reflects these racist views (see later discussion; Cohen 2005; and Kanev 2005).

6. All album and song titles in this chapter are in Bulgarian unless otherwise noted.

7. Toni Dacheva is a member of an Eastern Orthodox Romani group known as *sivi gŭlŭbi* (grey doves; Bulgarian; Marushiakova and Popov 1997:96); Slavka Kalcheva and Nedyalka Keranova are also alleged to be from this group.

8. Remember, however, that just as chalga looks toward MTV for models, MTV has also looked to world music for new ideas.

9. Other Romani male singers such as Valentin Valdes and Kondyo were popular in the 1990s. In 2005 Kondyo was arrested for sex-trafficking offenses and sentenced to three years in prison; after prison, his career continued. Newer male Romani singers are Erik and Iliyan.

10. They may constitute a marked form of in-group communication for those who can understand the multiple languages. This would privilege Roma and Turks over Bulgarians, inverting the usual power hierarchy. Linguist Traci Lindsey at the University of California, Berkeley, is doing research on this topic. Also see Azis and Sofi Marinova's duet, discussed in this chapter.

11. In fact, the chalga production company Payner also runs a cosmetic surgery business. Along with commercialization of the female body is the expectation of sexual services with many female jobs. Some women desire to become high-class prostitutes, secretaries report harassment at work, and some chalga stars provide sexual services for money. The image of the *mutresa* (well-kept woman) is rampant in the media; see Ranova 2006. In 2007 it was common on unmoderated websites (for example, those associated with the mainstream newspaper *Standart*) for recruiters to communicate openly with sex workers about services they offered.

12. See www.bg-fen.com and www.chalgatube.com for chalga gossip, news, music reviews, and interactive discussions.

13. In 2004 Matt Pointon wrote in a British travel magazine: "Listen not to the intelligent and educated Bulgarians who deride this peasant/Tsigani/stupid form of entertainment. Instead gather some friends, a fine carafe of rakiya, a mouth-watering salad and turn up the CD player. Get up on your table, click your fingers, move every part of your body, feel proud of that beer belly and then kiss the person next to you, be it a scantily-clad, bad-perm-sporting young maiden, or an overweight, transvestite Gypsy. It's a pleasure that's divine and one that can only be had in Bulgaria," See http://travelmag.co.uk/?p=611, accessed June 19, 2011.

14. In Las Vegas in 2005 for example, Gloria performed lip synching to her own CD. There was no live music, and the show was basically a visual spectacle. In 2006 in Chicago, Azis performed several songs live to a sold-out crowd. In 2010 I heard Poli Paskova in Portland, Oregon, singing live for four hours to her CDs (karaoke style).

15. Exceptions include Neilna's 2008 song *Nyama Nashi, Nyama Vashi* (Neither mine nor yours) in which she playfully urges listeners to forget their political differences and have a good time (http://www.youtube.com/watch?v=yGOtNGhLAmo, accessed June 18, 2011). Other exceptions are Slavi Trifonov's texts and Sofi Marinova and Ustata's 2010 song (discussed later in this chapter).

16. It is true that late at night chalga predominates on Planeta TV, but on their website a greater variety of genres can be heard and seen.

17. Emilia, one Payner's top stars, has been featured on the cover of the Bulgarian edition of *FHM* (For Him Magazine) and in several revealing photo spreads in the fan magazine *Nov Folk*. The gong in the video "Zabravi" is engraved with E for Emilia, but it also looks like the sign for the Euro.

Note that the narrative is not literal, but an abstract oriental flavor is very clear. The full version is at www.youtube.com/watch?v=MqBbAtRyJTc, accessed January 2, 2011.

18. See www.bg-fen.com, accessed June 10, 2010.

19. The text of "Edinstveni" is by Georgi Milchev-Godzhi; he wrote the music with Evgeni Dimitrov and Slavi Trifonov. (See www.youtube.com/watch?v=M8u6XqhExts, accessed June 20, 2010). In the (nonnarrative) video of the song, Sofi and Slavi are dressed in business suits, pictured (separately) in an airplane and in winter coats in a snowstorm.

20. Rap singers sometimes satirically comment on chalga in their texts; in "Tochno Ti" Ustata sings about Sofi's duet with Slavi in the song "Edinsteveni." For discussion of why rap's ties to African-American culture resonate with Roma, see Chapter 2 and Imre 2006 and 2008. In 2009 Sofi and Ustata were awarded a prize for general quality in music by Fen TV.

21. "Ljubovta e Otrova" was produced by Slavi Trifonov. He had used the theme of war in his earlier videos, for example in staging the clip for the old Macedonian favorite 7/8 song "Yovano Yovanke" (Oh Yovana).

22. Kalin Kirilov called my attention to the use of rock-style guitar chords with a bit of distortion, also characteristic of Serbian turbofolk.

23. Cartwright explained that this is a quote from an interview in *Nov Folk* magazine that was released to coincide with his huge 2003 stadium concert.

24. www.youtube.com/watch?v=whi99J4B_2Q, accessed August 25, 2010.

25. He said: "On my first wedding night Desislava [chalga star] participated. . . . I wanted a little bit of that female happiness that all women try to get. Because I'm a girl in my soul—I didn't choose that—I was born that way" (*Nov Folk* 2007:32).

26. Volanis is a Greek Romani singer from Thessaloniki. The same song became very popular in Israel in 2002. One version is in Hebrew and Greek, performed by Moshik Afia and Shlomi Saranga; another version is performed by Lebanese vocalist Fadl Shaker in Arabic; another is in Turkish disco style, performed by Serdar Ortac; finally, a Romanian version is performed by Romani manele star Adrian Minune, or Adrian the Wonder Boy (Eva Broman, East European Folklife Center listserv, March 10, 2007).

27. Azis also sang the Macedonian slow song "Zajdi Zajdi Jasno Sonce" (Set, bright sun) at the end of his 2003 Sofia stadium concert; Ceca and other Serbian turbofolk singers routinely perform this song to prove they can really sing (Garth Cartwright, personal communication).

28. Stokes analyzes these two singers in relation to concepts of modernity and reminds us that they cannot be assumed to be critical of existing categories (2003). In Pakistan today, a transvestite occupies an acceptable place as a TV host, a similar role to Azis's role as a singer. The Pakistani host dares to bring up taboo subjects, and somehow these topics become more acceptable because a nonmainstream person brings them up (Masood 2007).

29. The temple pictured is Khajuraho, in North India, and it is indeed known for its erotic Hindu sculptures. There are other temples pictured in this video as well as shots of Tibetans praying at a Buddhist sacred site. I would like to thank Ron Wixman for help identifying this temple.

30. For references to the lawsuit over the immorality of this song, see http://planetbollywood.com/Film/Khalnayak/ accessed June 20, 2010. I would like to thank Francis Fung and Farrukh Raza for help in researching this song.

31. Butler writes: "Sexual difference . . . is never simply a function of material differences which are not in some way both marked and formed by discursive practices" (1993:1).

32. See bell hooks 1991.

33. I am using the word *gaze* in the sense in which it has been used by cultural and film studies. The normative gaze is assumed to be a male, that is, looking at females as sexual objects; this replicates the unequal power relationship between the sexes (see Gamman and Marshment 1988).

34. For cogent analyses in English, see Rice 2002; Statelova 2005; Levy 2002; Kurkela 1997, 2007; Apostolov 2008; and Buchanan 2007. Nick Nasev suggested to me that to many members of the middle class, who saw their status drop after the fall of communism, chalga represents the evils of capitalism.

35. Balkanika was started by Victor Kasamov, the owner of Bulgarian Ara Audio-Video; advertising brings in "65% of Balkanika's income, with the rest coming from on-screen sales of ring tones or fortune-telling and romance-forecasting services" (Brunwasser 2007).

36. Sugarman shows how the Kosovar commercial folk/pop industry is involved in the ideological work of defining a specific Albanian national modernity (2007).

37. Ditchev writes "the new identity debate in the 1990s was largely dominated by the question of whether or not to be Balkan" (2002:235). The issue of European Union membership has heightened these issues.

38. Kiossev writes of the "dark intimacy" of acts of identification, as in "we're all just Balkan shit" (2002:182 and 189). Herzfeld's concept of cultural intimacy can be fruitfully applied here (1997). Kiossev writes: "Balkan culture domesticates the official codes of national representation . . . through the multiple uses, misuses, and flexible appropriations performed by social actors in everyday life. Popular amusements in the Balkans produce ironic self-images and display them in semi-public spaces of insiders' 'collective privacy'. . . . It also often scandalously perverts these negative auto-stereotypes into positive ones, with a peculiar emotional ambivalence" (2002:189–190).

39. Here is a post on an internet forum: "The success of Attack doesn't come from anywhere else than the fact that they behave like strong men. Nobody likes the *mangali* [derogatory name for Roma, meaning a black pot], everyone thinks they should be 'neutralized' but nobody's doing anything about it, they just keep watching them reproduce. Then suddenly along comes this guy . . . and says "let's wipe them out" and you think, "that's easy, all I have to do is put a check in the little box, and I've solved the problem." The high Romani birthrate (in comparison to the low Bulgarian birthrate) is indeed causing hysteria in some circles. On October 9, 2006, the Minister of Health Radoslav Gaidarski announced a proposed legislative change to ban births by women under the age of eighteen. He further said that the measure would be directed mainly to girls of minority origin. According to the newspaper *Sega* (Now), Gaidarski told journalists that if the birthrate among Roma is not limited, then the mortality rate in Bulgaria would become among the highest in Europe, since many of these children do not survive to adulthood. Gaidarski also suggested that a meeting of the health ministers of Bulgaria, Hungary, Romania, and Slovakia

(countries where large populations of Roma live) should be held to tackle the social problems of these groups. Although the health minister feebly tried to frame the demographic issue in terms of the welfare of Roma, the underlying message was that there are simply too many of them. Another example of racism surfacing in high places occurred in October 2006, when a Bulgarian observer in the European Parliament, from the Attack party, commented on the nomination of a Hungarian Romani woman for a human rights prize by writing an e-mail to all MEPs: "In my country there are tens of thousands of gypsy girls way more beautiful than this honorable one. In fact, if you're in the right place at the right time, you even can buy one (around twelve or thirteen-years-old) to be your loving wife. The best of them are very expensive—up to 5000 euros a piece. Wow!"

Chapter 10

1. A number of ideas in this chapter appear in Silverman 2003 (reprinted with permission, ©2003 University of Chicago) and Silverman 2011c and 2011b, this last volume being from the conference Interpreting Emotions in Eastern Europe, University of Illinois, Fisher Forum, 2008.

2. Sugarman writes that "until recently, no south Albanian women from Macedonia had ever performed at an event as a professional singer" (1997:369). On the 1999 Gypsy Caravan Tour, of thirty Romani musicians only one was female, and she was the wife of a participant. The 1999 CD *Gypsy Queens* (Network 32843) was an attempt to highlight the contribution of women to Romani music.

3. Sugarman reports that an ethnic Albanian female singer in Chicago from Kosovo "endured a few years of gossip from community members" (1997:342).

4. This ideology exists among non-Roma of the Balkans as well. I collected a number of stories of Bulgarian women whose parents, mothers-in-law, or husbands prohibited them from joining professional ensembles in the 1960s because it was shameful.

5. They include Lisa Angelova and Zlatka Chinchirova from Bulgaria, who performed with their fathers; and Natalia Borisova from Bulgaria and Ramiza Dalipova and Esma Redžepova from Macedonia, who performed with their husbands. When Zlatka's father, Hasan Chinchiri, and Esma Redžepova's husband died, their careers were already launched. A similar pattern exists for Bulgarian Eastern Orthodox Slavic vocalists. Most Bulgarian wedding singers, notably Maria Karafezieva, Ruska Kalcheva, Binka Dobreva, and Pepa Yaneva, are in bands with their husbands. The same pattern can be found among Hungarian Roma; in the group Kalyi Jag, the only female participant, Agnes Kunstler, is the wife of male participant Jozsef Balogh. Sugarman also reports that the few female Prespa Albanian singers are in bands with their husbands (1997:342). For Middle Eastern parallels, see Van Nieuwkerk 1995:68 and 128.

6. Teodosievski and Redžepova1984 is an autobiographical book with photographs, newspaper clippings, and testimonials.

7. Van Nieuwkerk claims that in Egypt there is a hierarchy, with nightclub entertainers at the bottom, wedding entertainers a little higher, and concertizing entertainers at the top (1995:122–132).

8. See Chapter 6 and Teodosievski and Redžepova 1984:137 and 194. Essentializing and racist press quotes from the 1960s and 1970s include: "She is a Gypsy girl, hot blooded, happy as a bird! For her money means a new hat,

a ticket for the movies, a new dress, nothing more" (138); "Esma has a lovely dark complexion, it would be a wonderful advertisement for suntan creams and lotion; it has the shade of well-baked bread" (141); "this music reveals the Gypsy philosophy, the simple philosophy and wisdom close to all colors and tongues" (143); "Gypsies are a strange people—they live in their own way from their very birth. . . . Music is the soul and philosophy of Gypsies, simple, clear and deep. Music is their first and eternal occupation" (141).

9. Acton termed this the "artistic collusion between the oppressed and the oppressor" and illustrated his point with materials from Lemon's book on Russian Romani performers (2000) and Van de Port's book on Serbian Romani musicians (1998) (Thomas Acton, Patrin listserv, April 18, 2001).

10. Stevo wrote that the idea of using young drummers arose almost by accident when their usual drummer could not attend a concert; he was visiting the home of a possible replacement drummer and spotted his young son, Enver Rasimov, playing the tarabuka. Enver became their trademark young performer (Teodosievski and Redžepova1984:39-40). When he was married off at a young age, he was replaced with another young drummer because a precedent was firmly established.

11. The longer films are *Krst Rakoc* (Rakoc's Cross; Macedonian), *Skopje '63, So Sila Tatko* (How are you, father?), *and Zapej Makedonijo* (Sing, Macedonia! Macedonian); her shorter music videos were compiled and released by RTS (Radio Television Skopje) under the series titled *Putevima Pesme: Esma Ansambl Teodosievski* (Song Paths; MT 31001-5). There are many recent video clips of Esma performing Čhaje Šukarije on YouTube; for example, a 2006 video of her performing a lip-synched version of Čhaje Šukarije on BTR Skopje (one of the Romani TV channels) is at www.youtube.com/watch?v=IglS8eJayUY, accessed June 10, 2010. The video includes greetings in Romani and Macedonian, from Roma in Germany to Roma in Macedonia and vice versa, and it gives a telephone number and price to order greetings.

12. In Chapter 3 I chronicled this process. Yugoslav Roma, for example, gave their children Hindu names, and musicians used Indian themes in their čočeks.

13. An exception is keyboardist Elvis Huna, who met Esma through her adopted son, Simeon Atanasov, with whom he served in the army.

14. Two other performers were awarded diplomatic passports several years ago, pop singer Toše Proeski and rock guitarist Vlatko Stefanovski. Nick Nasev (a long-term fan of Esma) pointed out that of the three artists, Esma had the most solid international reputation, but the government delayed her passport for years. He felt that the state preferred certain ambassadors to others (personal communication).

15. http://www.blic.rs/stara_arhiva/kultura/22531/Pozitivne-emocije-i-cisto-srce, March 27, 2002, accessed July 19, 2011; also see Cartwright 2005b:111.

16. In Chapter 13 I discuss how western recording contracts have had an impact on Balkan Romani musicians.

17. They include *Nasvali me so Iljum* (I fell ill), available on her website, and *Pomegu Dva Života* (Between two lives [Macedonian], KMP: KA005).

18. http://www.youtube.com/watch?v=K2ewdetNASI, accessed December 15, 2010.

19. I thank Nick Nasev for these observations. In Chapter 8 I discussed similar controversies related to contests in Bulgaria, Turkey, Hungary, and the Czech Republic.

20. See www.youtube.com/watch?v=2aDYAfA_plQ, accessed May 30, 2010.

21. http://www.esctoday.com/news/read/15114, accessed May 30, 2010.

22. This song can be found on the albums mentioned in note 17. The text is by Oskar Mahmut, the arrangement by Simeon Atanasov, and the melody by Esma.

23. The whole song can be found at http://www.youtube.com/watch?v=ZzQukXu0ARE, accessed June 19, 2011.

Chapter 11

1. Fieldwork with Yunakov spanned the mid-1980s to the present and took place in Bulgaria and New York City, and on multiple tours. Portions of this chapter are reprinted from Silverman 2009, with permission of the publishers.

2. According to Seeman, dahli may come from *dagli,* the label used in Erdine, Turkey, for zurna and tŭpan families who came to Turkey from the Yambol, Bulgaria, region in the nineteenth century. In Turkish, the word means mountain folk (2002: 260).

3. The ensemble Yunakov describes is similar to the urban Macedonian čalgija ensemble discussed in Chapter 2.

4. See www.ctmd.org/pages/enews0509yunakov.html, accessed December 15, 2010.

5. See John and Jean Comaroff 1993:34 and Ortner 1995:174. As noted in Chapter 7, domination as well as resistance needs to be interrogated and its pluralities revealed.

6. I know of no wedding musicians who resisted the name changes. Whereas many Turks and a small number of Roma resisted the name changes, Roma in general did not resist.

7. In 2001 Yunakov purchased a condominium and, because of financial pressure, took a nonmusical job; he registered his own corporation and purchased a van. For the last decade, he has worked as a driver but continues to perform music at night and on weekends. Yunakov has always been the primary income producer in his family; he is extremely hard-working, and at one time he was supporting at least eight people.

8. He also introduced Merita and Raif to Americans involved in the Balkan Music and Dance Workshops sponsored by the East European Folklife Center, which facilitated their being asked to teach at the workshops.

9. Like many Balkan musicians, Yunakov cannot afford a stable booking agent; the Center for Traditional Music and Dance (www.ctmd.org) and Harold Hagopian of Traditional Crossroads (www.traditionalcrossroads.com) have served in that role, and when he first emigrated I did so informally as well. For the most part, Yunakov handles his own bookings, which has disadvantages and advantages.

10. Turks also notice that he is from Bulgaria, has not lived in Anatolia, and is Romani. It is obvious that Yunakov could never be fully accepted by Armenians and Albanians because of language and religious differences, not to mention racism.

Chapter 12

1. As mentioned in Chapter 8, world music emerged in the late 1980s in Europe and America as a marketing category (Taylor 1997; Feld 1994). Portions of this chapter are reprinted from Silverman 2007b with permission from the publishers.

2. Kusturica's films include *Time of the Gypsies*; *Underground*; and *Black Cat, White Cat*. They all prominently feature Romani music, and many of the Romani performers became famous as a result (see Iordanova 2001, 2002, 2003a, 2003b). See Chapter 13 for analysis of the role of Goran Bregović, the music collaborator with Kusturica for *Time of the Gypsies* and *Underground*.

3. Roma-sponsored festivals such as the Khamoro festival in the Czech Republic, Šutkafest in Skopje, and Romfest in Bulgaria (discussed in Chapter 8) serve overt political functions, but their cultural displays are sometimes just as stereotypical as non-Romani-sponsored events. The Guča brass band festival in Serbia is sponsored by the regional administration and by private Serbian sponsors, all non-Roma. Although south Serbia is not the focus of this book, I underscore that the issue of representation deserves to be examined at Guča. Serbian music garners international fame, but several participating Romani musicians have complained about discrimination at Guča in the 1990s.

4. Gypsy Spirit, directed by Hungarian choreographer Zoltan Zsurafski, featured the Budapest Dance Ensemble (composed of non-Roma) performing dances from India, the Balkans, and Hungary (http://centrummanagement.org/gypsy-spirit/, accessed June 19, 2011; see Chapter 6 for a discussion of Gypsy dance suites). Romani musicians participated in the band.

5. Hedgehogs are traditional food for Travelers.

6. See Chapter 1; see Clark and Campbell 2000 for media coverage in England.

7. This can be seen a number of times in the film *Iag Bari*.

8. Tony Gatlif, director of Latcho Drom, was scheduled to film the events but he took ill; he sent his daughter instead. The resulting film *No Man Is a Prophet in His Own Land* is found on the DVD *The Continuing Adventures of the Taraf de Haidouks* (2006); also see Chapter 13.

9. At this point, Esma launched into a proud narrative of how tolerant Macedonia is toward Roma. As mentioned in Chapter 10, Esma is very patriotic; other Roma are similarly loyal even to states that have not protected their freedoms (see Lemon 2000).

10. The masterful Kalman Balogh Gypsy Cimbalom Band was prominently featured, with guest Romani musicians from Bulgaria Yashko Argirov and Slavcho Lambov. Yashko and Slavcho were extremely grateful for the opportunity to tour, but they complained of the low salary; indeed, several Bulgarian musicians turned down the offer to tour with Gypsy Spirit because the compensation was so poor.

11. According to Dušan, the purpose of the Amala school is "to promote Romani music and culture, to teach it to non-Roma, and to provide jobs for local Romani musicians. When they see non-Roma, especially western people, interested in their music, they start to feel more pride in their culture." Dušan served as a board member of Voice of Roma and also published the collection *Rromani Songs from Central Serbia and Beyond* (Ristić and Leonora 2004).

12. Cartwright documented the widespread discrimination of Romanian musicians in his report for Free Muse (Cartwright 2001); also see Cartwright 2007.

13. Note that condescension was not typical of all the managers I met. Winter's relationship to Taraf also needs to be seen in the light of the fact that his managerial partner, Stephane Karo, is married to a relative of a Taraf member, and is thus considered adopted family. The managers of

Fanfare Ciocarlia, Henry Ernst and Helmut Neumann, are also married to Romani women from the Taraf village of Clejani; these managers are very respectful of their clients.

14. Zirbel's research with "Gypsies" from Egypt who perform at European festivals supports this claim: "Most groups either did not realize or were just not interested in what they . . . signified for audiences" (1999:86). Exceptions to this observation are Dušan and Dragan Ristić, discussed earlier in this chapter, who are also activists.

15. Imre discusses Hungarian Romani pop singer Gyozo Gaspar, leader of the band Romantic. His reality television show portrays his family as stereotypically childish, loud, argumentative, materialistic, and unable to lead a civilized life: "The show seems to confirm nothing but Gypsies' inability to function as hard-working citizens" (2006:335). Gaspar seeks to fulfill expectations that non-Roma have of Roma, thereby making him unoffensive.

16. As discussed in Chapter 3, the question of unity is also an important political issue; to build a human rights movement, Roma have to establish unity based on something tangible.

17. For discussion of appropriation and music, see Born and Hesmondhalgh 2000 and Chapter 13. I am currently researching this topic.

Chapter 13

1. In 2002, Taraf de Haidouks won the BBC Radio 3 Planet Europe Award; in 2006 Fanfare Ciocarlia won the award and Gypsy-inspired DJ Shantel won the BBC Club Global Award; in 2005 Ivo Papazov won the BBC Planet Audience Award; in 2007 Gogol Bordello won the BBC Planet Americas award and Balkan Beat Box was nominated in the Club Global category; in 2008 French guitarist Thierry (Titi) Robin, known for Gypsy fusions, was nominated in the Europe category, and Balkan Beat Box was nominated in three categories (Newcomer, Club Global, and Culture Crossing); in 2006 Taraf and its label Crammed Discs won the Edison Award in Holland (equivalent to a Grammy).

2. For example, Jony Iliev's album *Ma Maren Ma* (Don't Beat Me) was on the European world music charts for two months in 2003; Fanfare Ciocarlia's *Gili Garabdi* (Secret Songs) was in the top twenty for two months in 2005 and in April it was number one; the Serbian Romani band Kal's (Black) album *Kal* was on the charts for four months in 2006 and was number three in the annual list; its *Radio Romanista* was number two in March 2009; Mahala Rai Banda's *Ghetto Blasters* was number two in November 2009, and the remix album *Electric Gypsyland 2* was on the top of the charts in December 2006 after two months in the top twenty. In 2007 Fanfare Ciocarlia's *Queens and Kings* was and was voted among the top ten world albums of 2007 by the British magazines *Songlines, fRoots,* and *Mojo,* and the French magazine *Mondomix* (www.asphalt-tango.de/news.html). Here are the albums in the top twenty of the European world music charts at some point in 2008 to 2010: Kal's *Radio Romanista*, Shantel's *Disko Partizani* and *Planet Paprika*, Kočani Brass Band's *The Ravished Bride*, the DJ compilations *Balkan Beats 3* and *Balkan Grooves*, Balkan Beat Box's *Blue-Eyed Black Boy*, Boban and Marko Marković's *Devla*, and the CD compilation accompanying Cartwright's book *Princes Among Men* (www.wmce.de). Note that although sales were good for these albums, they never approximated the sales of the top pop and rock albums. Henry Ernst remarked that Joni Iliev's album sold

poorly, and that although *Gili Garabdi* sold 41,000 albums in two years this was very low in comparison to pop acts (personal communication). Appropriation of Romani music into the popular music realm is the subject of my current research project.

3. For example, after the 1999 Gypsy Caravan tour, Robert Browning of the World Music Institute wanted to develop the potential musical relationship between the Rajasthani and the Flamenco artists in preparation for the 2001 tour. He secured a commissioning grant from the Rockefeller Foundation for the piece *Maharaja Flamenco*; it was performed in the 2001 tour, but the collaborative project seemed to neither excite audiences nor lead to future work.

4. The creativity of Asphalt Tango was recognized by their WOMEX and World Music Charts Europe Top Label Award in 2006; in 2009 they were honored as one of WOMEX's top twenty labels. Asphalt Tango's roster of Romani artists includes Fanfare Ciocarlia, Kal, Jony Iliev, Esma Redžepova and as of 2009 Mahala Rai Banda (see www.asphalt-tango.de). The Western European companies Asphalt Tango, Piranha, Essay, Divano Productions, and Crammed Discs currently dominate the Romani music market.

5. The Macedonian brass band Kočani has also collaborated with several groups as a result of its managers' efforts. Kočani played with two Italian jazz combos: *Kočani Orkestar meets Paola Fresu and Salis Antonello: Live* (Manifesto 2005); and *Harmana Ensemble and Kočani Orkestar: Ulixes* (Materiali Sonori 2002). Kočani, unfortunately, has suffered because of an ugly split between its founder Naat Veliov and Crammed Discs (more specifically, Stephane Karo and Michel Winter of Divano Productions), regarding money (Cartwright 2005:136–138). Now there are two bands: Kočani Orkestar whose name is registered in Belgium by Crammed Discs and which is managed by Divano Productions, and King Naat Veliov and the Original Kočani Orkestar which records on the small German label Plane.

6. One video (accessed January 2, 2011), for example, shows Alyosha performing Godzila with Orkestar Universal on Veselina TV in Bulgaria.

7. The popularity of Taraf and Fanfare may be due to the Romanian meters (mostly 2/4 and 4/4) being more accessible to non-Romani audiences than southern Balkan meters (such as 11/16, 7/16, and 9/16); however, all Balkan Romani groups now perform 2/4 or 4/4 kyuchek/čoček/manele. These two groups achieved their popularity in part thanks to the efforts of their managers and record producers. Note that Ernst, Neumann, and Karo are all married to women from Clejani, Taraf's village. This cements their ties (and also their obligations) to Romania and the musicians.

8. On the 1999 North American Gypsy Caravan tour, Taraf accordionist Ionitsa asked Neshko Neshev to play Ionitsa's accordion "for good luck." He was clearly in awe of Neshko's playing.

9. Similarly, the two Bulgarian musicians on the 2004 Gypsy Spirit tour, Yasko Argirov and Slavcho Lambov, were grateful for the work but complained of low pay and short performances (see Chapter 12). We might think that all Roma-Roma relationships would be less problematic than non-Roma–Roma ones, but this is not the case. Although Romani Routes, a division of the NGO Voice of Roma that arranges tours in the United States, aims to avoid the exploitation that non-Romani managers might impose on Romani musicians, sometimes it is accused of the very same exploitation.

10. Taraf's most famous collaborators are the Kronos Quartet (*Caravan* album from 2000; see Broughton 1999). I too had a role in this collaboration: after several phone conversations with violinist David Harrington

about Romani bands that might be suitable for collaboration with the quartet, I suggested Taraf because of its string base.

11. Beissinger asserts that earlier manele (called *manea* and performed in the eighteenth-century Ottoman Phanariot or Greek-run court in Romania) possibly resembled the contemporary genre in rhythm but differed from it in many ways (also see Garfias 1981 and 1984). She writes (2005, 2007) that Romanians in the 1980s forged the genre from musics covertly imported from Serbia and Bulgaria (also see Voiculescu 2005). I posit that Romanians themselves may overemphasize the linear unbroken connections from the eighteenth century so that manele seems more home-grown, and thus legitimate. The same recasting occurred in Bulgaria, where scholars emphasized the historical roots of wedding music to legitimize it.

12. Ziff and Rao define cultural appropriation as "the taking—from a culture that is not one's own—of intellectual property, cultural expressions or artifacts, history and ways of knowledge and profiting at the expense of the people of that culture" (1997:1 and their footnote 1).

13. The literature on ownership of culture is vast; recent scholarship on UNESCO initiatives to copyright culture examines the legal and theoretical frameworks of this debate (M. Brown 2004; Kurin 2004, Kirshenblatt-Gimblett 2004).

14. See Gojković 1977 and Djordjević 1984 (1910). In his discussion of culture and appropriation, Samson states "almost axiomatically, Rom musics from all over the region" can be subsumed in the category synthetic (2005:44).

15. This is related to the professional role and the requirement to provide music that the patron knows and desires. Samson and Pettan remind us that Kosovo Roma have appropriated to remain neutral in a war. Samson writes: "Kosovo Rom musicians deliberately adopted transnational idioms, including Western popular music, if not to promote a universalist ideology then at least to maintain ethnic neutrality at a time of prevailing ethnic tension and dispute" (2005:46; also see Pettan 1996b and 1996c).

16. In the United States, for example, Macedonian Romani drummer Seido Salifoski (see Chapter 5) performs with the Zlatne Uste Brass Band, Serbian Romani accordionist Peter Stan performs with Slavic Soul Party, and clarinetist Catherine Foster was trained by and performed for many years with Yuri Yunakov. Zlatne Uste has been particularly sensitive to ethical issues. They routinely credit the sources of their music, they encourage local Balkan Roma to attend their events, and the proceeds of their annual Golden Festival have been donated to NGOs working for peace and justice in the Balkans.

17. At the Portland showing of the film *When the Road Bends: Tales from a Gypsy Caravan*, a tribal dance representative proudly announced that his troupe was "keeping the Gypsy spirit alive." On the other hand, dancers such as Artemis (Elizabeth Mourat) and Helene Ericksen have done fieldwork in Romani communities and educate their students about the political situation of Roma (www.serpentine.org/artemis/artemis.htm; www.helene-eriksen.de/). Tribal dancer Kajira Djoumahna has interviews with Esma and activist Šani Rifati on her website (www.blacksheepbellydance.com/writings/files/rom.html).

18. Bregović collaborated on *Time of the Gypsies* and *Underground*. See Iordanova 2002, Gočić 2001, and Malvinni 2004 for a discussion of Kusturica.

19. *Graceland* (1986), a collaboration among Paul Simon, Ladysmith Black Mambazo, and others, won awards, sold millions of copies, and even figured in the anti-apartheid movement. Simon's lyrics contributed to the project, and he clearly respected his collaborators, paid them well, toured with them, and donated to political causes. But in terms of ownership, Simon's name appeared above the title and he copyrighted the music (Feld 1988). In the end, perhaps "musicians fill the role of wage laborers" (2000a:242).

20. *Deep Forest*'s 1992 multimillion-dollar-selling album features digitally sampled and manipulated African sounds mixed with synthesized tracks. The liner notes are stereotypical romanticizations, but they ask listeners to contribute to a Pygmy Fund (which has received little money; Feld 2000a:271). As mentioned in Chapters 3 and 12, world music is often tied to soft social altruism, in part to make fans feel good. But the real money being made is in record sales, and "hardly any of this money circulation returns to or benefits the originators of the cultural or intellectual property in question" (274).

21. Feld asks if we should believe Peter Gabriel when he says he wants to make third-world artists as famous as he is, and when Youssou N'Dour gives him special thanks (1994:271). Feld asked Charles Keil: "How do you respond to Joseph Shabalala when he says that without Paul Simon Ladysmith Black Mambazo would have never gotten a record contract with Warner Brothers?" CK: "I would tell Joseph to be content with Shanachie Records. . . . If that is the price to pay for keeping Warner Brothers and Paul Simon from having the copyright and ownership rights to those grooves, it is worth it." SF: "I don't think you can say that to third-world musicians. You just can't. . . ." CK: "Everybody is hoping that they are going to make money because of this ownership principle of music, but they never do. All the musicians with the exception of Michael Jackson end up poor" (Keil and Feld 1994:315).

22. The list of Romani artists and groups who have recorded Erdelezi includes Serbian singer Džej, Macedonian performers Muharem Serbezovski and Kočani, the Bulgarian band Džipsi Aver, the French band Bratsch, and the Hungarian singer Mitsou. Non-Romani singers of Erdelezi include Albanian vocalist Merita Halili. Bregović rerecorded Erdelezi in collaborations with Greek performers Giorgos Dalaras and Alkisti Protopsalti, with Polish singer Kayah, and with Turkish pop star Sezen Aksu (who recorded a whole album of his songs; see Stokes 2003).

23. The Magdolna version can be viewed at http://www.youtube.com/watch?v=t-H47c3xnZo, accessed June 19, 2011.

24. The paper wrote: "Balkan music is a volatile concoction. Though instantly identifiable, it can also be difficult to define. Selling it outside the region is even harder" (www.ftd.de/karriere_management/business_english/149236.html). Bregović's website contains this press excerpt: "The fashion of reviving all sorts of popular music in most spectacular manners has recently set its heart on the musical world of the Gypsies. It's been sixty years since the Belgian guitarist Django Reinhardt conducted, for the first time, an orchestra, becoming thus the greatest reviver of Gypsy music. Today a number of names join him in this enterprise, but no one does it with the intelligence of the Yugoslav Goran Bregović. The result of his inventions resembles no other. He is light years from them, both by means but even more by his ingenuity" (Luis Martin, ABC, April 25, 2001, www.goranbregovic.co.yu, accessed January 2, 2011).

25. Note that despite his notorious reputation among Roma, Bregović has always been able to hire Romani musicians. We can benignly suggest that perhaps he pays musicians well, or else, cynically, that Romani musicians are desperate for work. In either case, his ownership and copyright practices have not changed. In his 2009–2011 tours, however, he seemed to employ fewer Romani musicians. The lineup included a string quartet, a male vocal sextet, a five-piece brass band, two Bulgarian female singers, and lead singer and percussionist Alen Ademović (whose father, Ninoslav, does many of Bregović's arrangements).

26. I would like to thank the late Mirjana Laušević for several fruitful conversations about Bregović and for drawing my attention to several published interviews with him.

27. For example, "Hybridity can rebound from its discursive origins in colonial fantasies and oppressions and can become instead a practical and creative means of cultural rearticulation and resurgence from the margins" (Born and Hesmondhalgh 2000:19). Postmodernists tend to see a "resolution of issues of appropriation into unproblematic notions of crossover and pluralism." Aesthetic pluralism is then divorced from extant socioeconomic differences and "held to be an autonomous and effective force for transforming those differences. The aesthetic is held to portend social change; it can stand . . . for wider social change" (21). Hutnyk critiques this stance of postmodernists (2000; see Chapter 3).

28. African Americans are a useful comparison as a marginal group with musical power. Monson's statement that "African-Americans invert the expected relationship between hegemonic superculture and subculture" (1994:286) could apply equally to Roma. But is this another from of exploitation? Has the socioeconomic position of blacks improved as a result of their music becoming popular throughout the world? Simon Jones (a celebratory scholar) asserts that when white British youths adopt black musical styles they are implicitly rejecting racism (1988). Others, however, focus on how black music never lost its imputed exoticism and primitiveness even when taken into white commercial forms (Born and Hesmondhalgh 2000:22–23). In jazz, white musicians have tended to receive greater rewards; similarly, in rock, "its white stars have generally been paid much more attention than black innovators" (23).

29. Keil, for example, is concerned that mediated musics, because of their frozen electronic form, are separated from communities (Keil and Feld 1994).

30. Perhaps the earliest critic of commercial forms of recording and appropriation was Alan Lomax, who warned of a cultural gray-out, a homogenization of the world's music toward Western forms. Similarly, the *New York Times* music critic Jon Pareles wrote in his 1988 article "Pop Passports—At a Price": "When Paul Simon, Peter Gabriel and Talking Heads sell millions of records using Jamaican reggae and South African mbaqanga, the sources deserve a piece of the action. But to reach the world audience how much will these musicians have to change—and for better or worse?" (cited in Feld 1994:267).

31. Although some writers have grouped the bands together, there is also dissention among them. When Beirut band leader Zach Condon told New York magazine that "half of what makes that band [Gogol Bordello] work is the fact that the singer [Eugene Hutz] dresses crazy," Hutz retorted: "To me, that's digging your open grave. For us, the whole movement was about getting people to think about authenticity rather than the ironic plastic crap we've been force-fed for generations. Then, of course,

there's people who are simply in it for fashion" (Lynskey 2006). On the other hand, Ori Kaplan of Balkan Beat Box claims to be inspired by Hutz, with whom he performed for three years in Gogol Bordello, and Kaplan and Tamir Muskat of Balkan Beat Box collaborated with Hutz and Oren Kaplan on the JUF (Jewish Ukrainian Friendship project).

32. Although Shantel himself has never claimed to be Romani, the liner notes to the album *Gypsy Beats and Balkan Bangers* by DJ Russ Jones clearly state "he was captivated by his Gypsy heritage" (2006). Gypsy wannabes do liberally populate the club scene. Non-Romani dancers at the New York Gypsy Festivals often dress in versions of the fantasy Gypsy. The outfits of musicians and dancers in Gypsy Punk bands also reinforce the image of a circus (see Chapter 12).

33. I would like to thank Garth Cartwright for many comments on this chapter. On his website, www.bealkanbeats.de, accessed January 2, 2011, Šoko reports that as early as 1990 he was DJ'ing using Balkan music at the Berlin Mudd Club; his April 2007 performance in Paris featured "Gypsy Punk, electrogypsy, hiphop, Klezmer, Balkan and Gypsy, Reggae, traditional, gypsy, hungary, brass bands of the Balkans [sic]." Šoko also has several albums titled *Balkan Beats* (Dimova 2007). I am currently conducting research on the DJ scene in Western Europe.

34. For example, a video in Greek and Romani was released by Payner in Bulgaria by Sakis Coucos. The refrain is also used in the Turkish/Bulgarian/Romani song "Yak Motoru" (Light the motor, Turkish), sung by Habibi and Malki Kristalcheta (The Young Crystals) on Payner. Versions by Paultalia of Kyustendil and Kristali of Montana, Bulgaria, are on www.cocek.com. A Romanian manele version was released by Brandy.

35. Indeed, parallels can be drawn with American minstrel shows where whites enacted stereotypes of blacks, and sometimes even blacks wore blackface (see Johnson 2003 and Lott 1993).

36. For example, at the 2000 British Barbican Gypsy festival, Fanfare Ciocarlia was paired with Transglobal Underground, a multi-ethnic London electronic fusion group that has its own issues of appropriation (see Hesmondhalgh 2002).

37. Similarly, Feld reports that when he asked African-American jazzman Herbie Hancock if he had any moral concerns when he copied a central African phrase on a remake of his hit "Watermelon Man," Hancock answered, "It's just a brothers kind of thing" (Feld 2000a:257). With Hancock and the fusion group Zap Mama (composed of urban Africans and hyphenated African-European musicians), there is the issue of "the place of condescension, even subjugation within a sphere overtly marked by inspiration and homage." Hancock and Zap Mama are not critiqued for cultural theft the same way that Europeans and Americans are; they are even hailed as "cultural ambassadors." "Nonetheless, the power differentials separating cosmopolitan African Americans, Afropeans, and Africans from their forest pygmy muses cannot be elided" (270).

38. For example, Tamir Muskat of Balkan Beat Box explains: "Our connection to the Balkans, blood wise, is both of our families came from Eastern Europe. . . . The beginning of our Balkan Beat Box experience was falling in love with bands like Taraf de Haïdouks, Fanfare Ciocărlia, so many others. That just started the whole thing for us. . . . And then incorporated into what we do and mixing it with all this beautiful Mediterranean music we grew up on, from Turkey to Greece to Egypt, Morocco, and tons of other places. That would be part of why we use the name. The other part is just so much love

to our music, kind of a nonsense approach, not to take it too serious. We are not only necessarily dealing with Balkan music. We are and we will deal with music from all around the world. So don't take it too serious there." http://www.essayrecordings.com/essay_bbb.htm, accessed January 2, 2011.

39. *Spin* magazine wrote: "Condon may hail from Albuquerque, call Brooklyn his home, and choose Lebanon's capitol for his nom-de-plume, but this teenager sounds straight up Balkan. His orchestral gypsy dirges feature a string of somber horns that sound fueled on the tears from a torn Soviet Bloc" (www.Spin.com/articles/beirut June 23, 2006, accessed June 10, 2010). In fact, Condon has recently abandoned the Balkans and has more recently collaborated with a Zapotec Mexican brass band.

40. I would like to thank my colleague John Fenn for this term.

41. A few Roma collaborated in Shukar Collective's album of electronic remixes, *Urban Gypsy* (2005, Riverboat Records). Producer Marc Hollander realizes that in all remixes "electronics are added by a producer rather than a member of the band, which distinguishes it from fusion from within, though this too can be very satisfying for all participants" (Dacks 2005). He, as well as Dacks, still seems to emphasize the fairness of the process: "In the end, call it world fusion or party music, all parties are concerned to make sure each project is an equitable work situation and generates goodwill amongst the participants" (Dacks 2005).

42. As I mentioned in Chapters 11 and 12, Yuri appeared with Eugene at the 2005 New York Gypsy Festival and the 2008 Herdeljezi festival. Yuri was projected to appear with J.U.F. on the album *Balkan Gypsy Reggaeton* (www.myspace.com/jewishukrainianfreundschaft, accessed June 10, 2010), although it appears that this project has been shelved as Balkan Beat Box and Gogol Bordello have become more famous. Yuri told me this relationship with Hutz was "good for business."

43. I am currently researching new manifestations of and reactions to of Gypsy Punk and the DJ scene. Also see Stankova 2009 and Szeman 2009.

44. In response to these accusations, Shantel said of Cartwright: "I think he is racist. Who is he to judge this is wrong and this is right?" (Lynskey 2006). Actually Cartwright is by no means against fusions; he admires musicians "who listen with open ears and wish to share music" and differentiates them from "those . . . who simply lust after pop fame." For example, he thinks the Beirut album is "very much a student effort . . . to bring Balkan brass into a pop-rock setting;" he admires the Hawk and a Hacksaw album *The Way the Wind Blows*, which was recorded in Zece Prajini with members of Fanfare laying down brass; "it's eerie sounding and not imitation, more using the Rom musicians to colour a folk rock canvas" (personal communication).

45. Roma also can't afford to buy western albums. As I discussed in Chapter 12, Macedonian Roma attending Gogol Bordello performances at the 2005 New York Gypsy Festival were baffled by the music and offended at the visual spectacle. They stood on the side and either laughed or sneered.

46. According to a British *Daily Mail* article, the Roma of Glod:

. . . eke out meagre livings peddling scrap iron or working patches of land. . . . Just four villagers [out of 1,000] have permanent employment in the nearby towns . . . while the rest live off what little welfare benefits they get. . . . But now the villagers of this tiny, close-knit community have angrily accused the comedian of exploiting them, after discovering his new blockbuster film portrays them as a backward group of rapists, abortionists and prostitutes, who happily engage in casual incest. They claim film-makers lied to them about the true nature of the project, which they believed would be a documentary about their hardship, rather than a comedy mocking their poverty and isolation. Villagers say they were paid just £3 each for this humiliation, for a film that took around £27 million at the worldwide box office in its first week of release. Now they are planning to scrape together whatever modest sums they can muster to sue Baron Cohen and fellow film-makers, claiming they never gave their consent to be so cruelly misrepresented [Pancevski and Ionescu 2006].

Villager testimonies show just how marginal these Roma were: "'Our region is very poor, and everyone is trying hard to get out of this misery. It is outrageous to exploit people's misfortune like this to laugh at them.' When a Hollywood film crew descended on a nearby run-down motel last September, with their flashy cars and expensive equipment, locals thought their lowly community might finally be getting some of the investment it so desperately needs." The filmmaking process replicated the very prejudice it seeks to mock. Indeed, when the non-Romani local vice-mayor was asked whether the villagers felt offended, he replied: "They got paid so I am sure they are happy. These gipsies will even kill their own father for money." Moreover, Sasha Cohen "insisted on traveling everywhere with bulky bodyguards, because, as one local said: 'He seemed to think there were crooks among us.'" Finally, no villagers have seen the film since they can't afford a trip to the nearest movie theater, 20 miles away (Pancevski and Ionescu 2006). This echoes the situation with expensive western albums that feature Balkan Romani music: Balkan Roma can't afford to buy them. The Mail article reported that one actor from Glod said: "It was very uncomfortable at the end and there was animal manure all over our home. We endured it because we are poor and badly needed the money, but now we realise we were cheated and taken advantage of in the worst way." Another said: "All those things they said about us in the film are terribly humiliating. They said we drink horse urine and sleep with our own kin. You say it's comedy, but how can someone laugh at that?" Another actor said: "What I saw looks disgusting. Even if we are uneducated and poor, it is not fair that someone does this to us." A local official helped the crew but he claims he was never told what sort of movie it was, and that the crew failed to get a proper permit: "I was happy they came and I thought it would be useful for our country, but they never bothered to ask for a permit, let alone pay the official fees. I realise I should have taken some legal steps but I was simply naive enough to believe that they actually wanted to do something good for the community here. They came with bodyguards and expensive cars and just went on with their job, so we assumed someone official in the capital Bucharest had let them film." The production company that facilitated the filming claims the crew donated computers and TV sets to the local school and the villagers, but villagers have denied this. "The school got some notebooks, but that was it. People are angry now, they feel cheated" (Pancevski and Ionescu 2006).

47. According to Borenstein (2006), *Borat* is both indebted to and a parody of the films of Emir Kusturica, and this is why the song "Erdelezi" is featured so prominently. Borenstein sees the "over-the-top squalor and old-country festivities" as similar to the film *Time of the Gypsies*.

48. This situation is similar to performers from other marginal groups; see Taylor 1997, 2007. As Feld observes, Central African Pygmies "are disempowered precisely because they have never gained control over how they are discursively represented" (2000a:262). Similarly Johnson discusses the construction of "blackness" and its appropriation by whites (2003).

49. Taussig aptly writes: "Once the mimetic has sprung into being, a terrifically ambiguous power is established: there is born the power to represent the world, yet that same power is a power to falsify, mask and pose. The two powers are inseparable" (1993:42–43).

50. In commenting on third-world musicians, Keil claims corporations and privileged people profit the most from collaborations: "With the high-quality recording and distribution and all the rest, ninety percent of the money winds up going to white people . . . who are already in positions of power: the gatekeepers, the copyright holders, and the distributors" (Keil and Feld 1994:317). But Feld says the same moves that are being read as "cannibalizing" from an ethical point of view are also read as "empowering in various third-world locations" (315).

References

Abadzi, Helen, and Emmanuel Tasoulas. 1998. *Indoprepon Apokalypsi* (Hindi-style Songs Revealed). Athens: Atrapos.

Abrahams, Roger. 1977. Toward an Enactment-Centered Theory of Folklore. In *Frontiers of Folklore,* ed. William Bascom, 79–120. Washington, DC: American Association for the Advancement of Science.

———. 1993. Phantoms of Romantic Nationalism in Folkloristics. *Journal of American Folklore* 106:3–37.

Abu-Lughod, Lila. 1990. The Romance of Resistance: Tracing Transformations of Power Through Bedouin Women. *American Ethnologist* 17:41–55.

———. 2002. Do Muslim Women Really Need Saving? Anthropological Reflections on Cultural Relativism and Its Others. *American Anthropologist* 104(3):738–790.

Acković, Vesna, 1989. Harmonske i Tonalne Osnove Ciganske Lestice i Njena Sličnost s Bhairava Ragom (Harmonic and Tonal Basis of the Romani Scale and Its Similarity to the Bhairava Rag). In *Jezik i Kultura Roma* (Romani Language and Culture), ed. Sait Balič et al., 421–428. Sarajevo: Institut za Proučvanje Nacionalnih Odnosa.

Acton, Thomas. 2004. Modernity, Culture, and "Gypsies": Is There a Meta-Scientific Method for Understanding the Representation of "Gypsies" and Do the Dutch Really Exist? In *The Role of the Romanies: Images and Counter-Images of "Gypsies"/Romanies in European Cultures,* eds. Nicholas Saul and Susan Tebbutt, 98–116. Liverpool: Liverpool University Press.

———. 2006. Romani Politics, Scholarship and the Discourse of Nation-building: Romani Studies in 2003. In *Gypsies and the Problem of Identities: Contextual, Constructed and Contested,* eds. Adrian Marsh and Elin Strand, 27–37. Swedish Research Institute in Istanbul, Transactions vol. 17.

———, and Ilona Klimova. 2001. The International Romani Union: An East European Answer to West European Questions? In *Between Past and Future: The Roma of Central and Eastern Europe,* ed. Will Guy, 157–219. Hatfield: University of Hertfordshire Press.

Al-Ali, Nadje, and Khalid Koser. 2002. Transnationalism, International Migration and Home. In *New Approaches to Migration,* eds. Nadje Al-Ali and Khalid Koser, 1– 14. London: Routledge.

And, Metin. 1959. *Dances of Anatolian Turkey.* New York: Dance Perspectives 3, Summer.

————. 1963–64. *A History of Theatre and Popular Entertainment in Turkey*. Ankara: Forum Yayinlari.

————.1976. *A Pictorial History of Turkish Dancing*. Ankara: Dost Yayinlari.

Andrews, Walter, and Mehmet Kalpaklı 2005. *The Age of Beloveds: Love and the Beloved in Early-Modern Ottoman and European Culture and Society*. Durham, NC: Duke University Press.

Apostolov, Apostol. 2008. The Highs and Lows of Ethno-cultural Diversity: Young People's Experiences of Chalga Culture in Bulgaria. *Anthropology of East Europe Review* 26(1):85–97.

Appadurai, Arjun. 1996. *Modernity at Large: Cultural Dimensions of Globalization*. Minneapolis: University of Minnesota Press.

Appiah, Kwame Anthony. 2006a. *Cosmopolitanism: Ethnics in a World of Strangers*. New York: Norton.

————2006b. Toward a New Cosmopolitanism: The Case for Contamination. *New York Times* January 1:30–37, 52.

————, and Henry Gates, Jr., eds. 1995. *Identities*. Chicago: University of Chicago Press.

Aretxaga, Begoña. 2003. Maddening States. *Annual Review of Anthropology* 32:393–410.

Armstrong, John. 2005. Shantel/Bucovina Club Volume 2. *Know the Ledge*, August 23, www.knowtheledge.net/bucovinaclub2.htm.

Azis. 2006. *Az, Azis* (I, Azis). Sofia: Sunny Music.

Babić, Dragan. 2003. A Century and a Half of the Trumpet in Serbia. *Večernje Novosti* August 3–9, 2003 (7 parts). www.VečernjeNovosti.co.yu, accessed September 9, 2003.

————. 2004. *Priča o Srpskoj Trubi* (Stories of Serbian Trumpets). Belgrade, Serbia: Beogradska Knjiga.

Bajraktarović, Mirko. 1970. Ciganu u Jugoslaviji Danas (Gypsies in Yugoslavia Today). *Zbornik Filozofskog Fakuteta u Beogradu* 11(1):743–748.

Bakalov, Todor. 1992. *Svatbarskite Orkestri: Maystori na Narodna Muzika* (Wedding Orchestras: Masters of Folk Music). Tom II. Sofia: Muzika.

————. 1993. *Maystori na Narodna Muzika* (Masters of Folk Music). Tom I. Sofia: Voennoizdatleski Kompleks Sv. Georgi Pobednosets.

————. 1998. *Anthology of Bulgarian Folk Musicians*. Sofia: St. Kliment Ohridski University Press.

Bakić-Hayden, Milica. 1995. Nesting Orientalisms: The Case of the Former Yugoslavia. *Slavic Review* 5:917–931.

Balicki, Asen. 1998. *Roma Portraits* (documentary film).

Barany, Zoltan. 2002. *The East European Gypsies: Regime Change, Marginality, and Ethnopolitics*. London: Cambridge University Press.

Barth, Fredrik, ed. 1969. *Ethnic Groups and Boundaries*. London: Allen and Unwin.

Bartók, Béla. 1931 (1976). Gypsy Music or Hungarian Music? In *Béla Bartók Essays*, ed. Benjamin Suchoff, ed., 206–223. New York: St. Martin's.

Basch, Linda, Nina Glick Schiller, and Cristina Blanc. 1997. *Nations Unbound: Transnational Projects, Postcolonial Predicaments and Deterritorialized Nation- States*. Amsterdam: Gordon Breach.

Bauman, Richard. 1975. Verbal Art as Performance. *American Anthropologist* 77:290–311.

————, and Charles Briggs. 1990. Poetics and Performance as Critical Perspectives on Language and Social Life. *Annual Review of Anthropology* 19:59–88.

Bax, Daniel. 2007. Grooving the Gypsy Way: The Music of Eastern European Roma Captures Hearts in the West. *Atlantic Times*, Life Section, May 2007, p. 22.

BBC. 2008. EC Warns on Roma Finger Printing. June 28, 2008. http://news.bbc.co.uk/go/em/fr/-/2/hi/europe/7479298.stm.

Becerra, Hector. 2006. Gypsies: The Usual Suspects. *Los Angeles Times*, January 30, A1.

Beckerman, Michael. 2001. Pushing Gypsiness, Roma or Otherwise. *New York Times*, April 1(sec 2):32 and 34.

Bećković, Olja. 2002. Ima i Gorih od Mene. *Nin*, October 24, 2002, issue 2704. http://www.nin.co.rs/2002-10/24/25596.html, accessed June 20, 2011.

Beissinger, Margaret. 1991. *The Art of the Lautar: The Epic Tradition of Romania*. New York: Garland.

———. 2001. Occupation and Ethnicity: Constructing Identity Among Professional Romani (Gypsy) Musicians in Romania. *Slavic Review* 60(1):24–49.

———. 2005. Romani (Gypsy) Music-Making at Weddings in Post-Communist Romania: Political Transitions and Cultural Adaptions. *Folklorica* X(1):39–51.

———. 2007. *Muzică Orientală*: Identity and Popular Culture in Post-Communist Romania. In *Balkan Popular Culture and the Balkan Ecumene: Music Image and Regional Political Discourse*, ed. Donna Buchanan, 91–141. Lanham, MD: Scarecrow Press.

Bendix, Regina. 1997. *In Search of Authenticity: The Formation of Folklore Studies*. Madison: University of Wisconsin Press.

Berdahl, Daphne. 1999. Nostalgie for the Present: Memory, Longing, and East German Things. *Ethnos* 64(2):192–211.

Berzatnik, Hugo. 1930. *Europas Vergessenes Land*. Vienna: Seidel.

Bhabha, Homi. 1994. *The Location of Culture*. London: Routledge.

Bhattacharya, Deben. 1965. *The Gypsies*. London: Record Books.

Bilefsky, Dan, and Ian Fisher. 2006. Across Europe, Worries on Islam Spread to Center. *New York Times*, October 11: A1 and A12.

Blau, Dick, Charles Keil, Angeliki Keil, and Steven Feld. 2002. *Bright Balkan Morning: Romani Lives and the Power of Music in Greek Macedonia*. Middletown, CT: Wesleyan University Press.

Borenstein, Eliot. 2006. The Cosmopolitan and the Yokel. Unpublished manuscript. New York University, Department of Slavic Languages and Literature.

Born, Georgina, and David Hesmondhalgh, eds. 2000. Introduction: On Difference, Representation, and Appropriation in Music. In *Western Music and Its Others: Difference, Representation, and Appropriation in Music*, eds. Georgina Born and David Hesmondhalgh, 1–58. Berkeley: University of California Press.

Boscoboinik, Andrea. 2006. Becoming Rom: Ethnic Development Among Roma Communities in Bulgarian and Macedonia. In *Ethnic Identity: Problems and Prospects for the Twenty-first Century*, eds., Lola Romanucci-Ross, George De Vos, and Takeyuki Tsuda, 295–307. Walnut Creek, CA: Alta Mira Press.

Bourdieu, Pierre. 1984. *Distinction: A Social Critique of the Judgement of Taste*. Cambridge, MA: Harvard University Press.

Boyarin, Daniel, and Jonathan Boyarin. 1993. Diaspora: Generational Ground of Jewish Identity. *Critical Inquiry* 19(4):693–725.

Boyd, Joe. 1989. Liner Notes to *Orpheus Ascending: Ivo Papasov and His Bulgarian Wedding Band*. London: Hannibal.

Brah, Avtar. 1996. *Cartographies of Diaspora: Contesting Identities*. London: Routledge.

Brandl, Rudolf. 1996. The "Yiftoi" and the Music of Greece, Role and Function. *World of Music* 38(1):7–32.

Brettel, Caroline. 2003. Anthropology and Migration: *Essays on Transnationalism, Ethnicity and Identity*. Walnut Creek, CA: Alta Mira Press.

Briggs, Charles. 1996. The Politics of Discursive Authority in Research on the "Invention of Tradition." *Cultural Anthropology* 11(4):435–469.

Broughton, Simon 1999. Gypsy Music: Kings and Queens of the Road. In *World Music: The Rough Guide,* vol. 1, eds. Mark Ellington and Richard Trillo, 146–158. London: Rough Guides.

Brown, Catherine. 1933. Gypsy Wedding at Skoplje. *Folk-Lore* 49: 305–309.

Brown, Daniel. 2006. DJ Shantel. *Mondomix World Music*, September, http://shantel.mondomix.com/en/portrait3261.htm.

Brown, Julie. 2000. Bartok, the Gypsies and Hybridity in Music. In *Western Music and Its Others: Difference, Representation, and Appropriation in Music*, eds. Georgina Born and David Hesmondhalgh, 119–142. Berkeley: University of California Press.

Brown, Keith. 2000. In the Realm of the Double-Headed Eagle: Para-Politics in Macedonia 1994–9. *In Macedonia: The Politics of Identity and Difference*, ed. Jane Cowan, 122–139. Sterling, VA: Pluto Press.

Brown, Michael. 2004. *Who Owns Native Culture?* Cambridge, MA: Harvard University Press.

Brownell, Ginanne, and Amber Haq. 2007. The World Embraces Gypsy Culture. *Newsweek International*, July 30, 2007, http://www.newsweek.com/2007/07/29/living-like-gypsies.html, accessed June 22, 2011.

Brunwasser, Matthew. 2007. Bridging the Great Balkan Divide with Music. *International Herald Tribune*, February 25. http://www.nytimes.com/2007/02/25/technology/25iht-music26.4713036.html?scp=1&sq=Bridging%20the%20Great%20Balkan%20Divide%20with%20Music&st=cse, accessed June 18, 2011.

———. 2011. Subtle Shift at the Gypsy Bride Market. *New York Times*, May 3, http://www.nytimes.com/2011/05/04/world/europe/04iht-letter04.html, accessed June 18, 2011.

Bryceson, Deborah, and Ulla Vuorela. 2002. *The Transnational Family: New European Frontiers and Global Networks*. Oxford: Berg.

Buchanan, Donna. 1991. The Bulgarian Folk Orchestra: Cultural Performance, Symbol, and the Construction of National Identity in Socialist Bulgaria. Ph.D. dissertation, University of Texas, Austin.

———. 1996. Wedding Musicians, Political Transition and National Consciousness in Bulgaria. In *Retuning Culture: Musical Change in Eastern Europe,* ed. Mark Slobin, 200–230. Durham, NC: Duke University Press.

———. 1999. Democracy or "Crazyocracy"? Pirin Folk Music and Sociocultural Change in Bulgaria. In *New Countries, Old Sounds? Cultural Identity and Social Change in Southeastern Europe*, ed. Bruno Reuer, 164–177. Munich: Verlag Südostdeutsches Kulturwerk.

———. 2006. *Performing Democracy: Bulgarian Music and Musicians in Transition*. Chicago: University of Chicago Press.

——— 2007. Bulgarian Ethnopop Along the Old Via Militaris: Ottomanism, Orientalism, or Balkan Cosmopolitanism? In *Balkan Popular Culture and the Balkan Ecumene: Music Image and Regional Political Discourse*, ed. Donna Buchanan, 225–267. Lanham, MD: Scarecrow Press.

Budur, Diana. 2007. *Gypsies/Roma People Imagined in Films and Festivals*. Self-produced film.

Buonaventura, Wendy. 1994. *Serpent of the Nile: Women and Dance in the Arab World*. New York: Interlink Books.

Burton, Kim. 1995. Notes to *King Ferus: Ferus Mustafov Macedonian Soul Cooking*. Ace CDORBD 089.

Butler, Judith. 1990. *Gender Trouble: Feminism and the Subversion of Identity*. New York: Routledge.

———. 1993. *Bodies That Matter: On the Discursive Limits of "Sex."* New York: Routledge.

Cahn, Claude, ed. 2001a. *Roma Rights: Race, Justice and Strategies for Equality*. New York: International Debate Education Association.

———. 2001b. Šaban Bajramović: The Maximum King of Romani Pop Music. Amala School website. www.galbeno.com/saban-bajramovicthe-maximum-king-of-yugoslav-romani-pop-music/, accessed December 15, 2010.

Cartwright, Garth. 2001. "A Little Bit Special" Censorship and the Gypsy Musicians of Romania. www.freemuse.org/sw1174.asp, accessed February 23, 2009.

———. 2003a. Esma Redzepova. *fRoots* 25(244):47–49.

———. 2003b. Jony Iliev. *fRoots* 25(242/243):32–35.

———. 2004. Hard as Brass. *fRoots* 26(257):26–31.

———. 2005a. Gogol Bordello. *Independent*, December 5.

———. 2005b. *Princes Among Men: Journeys with Gypsy Musicians*. London: Serpent's Tail.

———. 2005c. Romani Pearl, Sofi Marinova. *fRoots* 27(6):42–43.

———. 2006a. Artist Profile: Winner 2006 DJ Shantel (Germany). www.bbc.co.uk/radio3/worldmusic/a4wm2006/a4wm_shantel.shtml, accessed February 23, 2010.

———. 2006b. Ioan Ivancea: Sad News: Gypsy Leader of Fanfare Ciocarlia, the Romanian Village Band That Found International Fame. *Guardian*, November 14, 2006.

———. 2006c. The King Returns. *fRoots* 28(5, 281):37–38.

———. 2007. Wild at Heart? *New Statesman*. February 14. www.newstatesman.com/arts-and-culture/2007/05/gypsy-life-musicians-reality, accessed February 23, 2009.

Castle-Kanerova, Mita. 2001. Romani Refugees: The EU Dimension. In *Between Past and Future: the Roma of Central and Eastern Europe*, ed. Will Guy, 117–133. Hatfield: University of Hertfordshire Press.

Cheah, Pheng. 1998. Given Culture: Rethinking Cosmopolitical Freedom in Transnationalism. In *Cosmopolitics: Thinking and Feeling Beyond the Nation*, ed. Pheng Cheah and Bruce Robbins, 290–328. Minneapolis: University of Minnesota Press.

Chow, Rey. 1998. *Ethics After Idealism: Theory, Culture, Ethnicity, Reading*. Bloomington: Indiana University Press.

Christgau, Robert. 2006. The New Bohemians. *Salon*, October 26, 2006. http://www.salon.com/entertainment/feature/2006/10/29/christgau, accessed June 22, 2011.

Ciobanu, Claudia. 2008. Bulgaria: Far Right Goes out of Fashion. Inter Press Service News Agency, March 28, 2008. www.ipsnews.net, accessed June 22, 2011.

Clark, Colin, and Elaine Campbell. 2000. "Gypsy Invasion": A Critical Analysis of Newspaper Reaction to Czech and Slovak Romani Asylum Seekers in Britain, 1997. *Romani Studies* 10(1):23–47.

Clifford, James. 1988. *The Predicament of Culture: Twentieth Century Ethnography, Literature and Art*. Berkeley: University of California Press.

———. 1992. Traveling Cultures. In *Cultural Studies*, ed. Lawrence Grossberg, 96–116. London: Routledge.

———. 1994. Diasporas. *Cultural Anthropology* 9(3):302–338.

———. 1997. *Routes: Travel and Translation in the Late Twentieth Century*. Cambridge, MA: Harvard University Press.

———. 2003. *On the Edges of Anthropology (Interviews)*. Chicago: Prickly Paradigm Press.

———. 2004. Looking Several Ways: Anthropology and Native Heritage in Alaska. *Current Anthropology* 45(1):5–30.

———, and George Marcus. 1986. *Writing Culture: The Poetics and Politics of Ethnography*. Berkeley: University of California Press.

Cohen, Emil. 2005. The Data Indicate: Our Society Is Ill from Racism. *Obektiv* 123 (April–July). www.bghelsinki.org, accessed July 10, 2010.

Comaroff, John, and Jean Comaroff. 1993. *Modernity and Its Malcontents: Ritual and Power in Postcolonial Africa*. Chicago: University of Chicago Press.

Cowan, Jane. 1990. *Dance and the Body Politic in Northern Greece*. Princeton, NJ: Princeton University Press.

———,M-B. Dembour, and R. A. Wilson, eds. 2001. *Culture and Rights: Anthropological Perspectives*. Cambridge, UK: Cambridge University Press.

Currid, Brian. 2000. "Ain't I People?" Voicing National Fantasy. In *Music and the Racial Imagination*, eds. Ronald Radano and Philip Bohlman, 113–144. Chicago: University of Chicago.

Dacks, David. 2005. Balkan Beats: East European Fusions Burn up the Dance Floor. *Exclaim*, September 1, http://exclaim.ca/Features/Research/balkan_beats-eastern_european_fusions_burn_up_dance, accessed June 23, 2011.

Daskalova, Krassimira. 2000. Women's Problems, Women's Discourses in Bulgaria. In *Reproducing Gender: Politics, Publics, and Everyday Life After Socialism*, eds. Susan Gal and Gail Kligman, 337–369. Princeton, NJ: Princeton University Press.

De Soto, Hermine, and Nora Dudwick, eds. 2000. *Fieldwork Dilemmas: Anthropologists in Postsocialist States*. Madison: University of Wisconsin Press.

Diamond, Elin, ed. 1996. *Performance and Cultural Politics*. New York: Routledge.

di Leonardo, Micaela. 1987. The Female World of Cards and Holidays: Women, Families, and the Work of Kinship. *Signs* 12(3):440–454.

———. 1991. *Gender at the Crossroads of Knowledge: Feminist Anthropology in the Postmodern Era*. Berkeley: University of California Press.

Dimitrova, Ilka, Rumyana Panayotova, and Ventsislav Dimov. 1994. Az Vinagi Sŭm Si Az (I Am Always Me). *Folk Panair* 4:23–26.

Dimov, Ventsislav. 1995. Folkbumŭt i Popharakteristikite mu (Kŭm Sotsiokulturniya Portret na Sŭvremennata Etnopopmuzika v Bŭlgaria)

(The Folk Boom and Its Pop Characteristics [Towards a Sociocultural Portrait of Contemporary Ethnopop Music in Bulgaria). *Bŭlgarski Folklor* 21 (6):4–19.

———. 1996. V Presa Na Presa (Pogled kŭm Folka na Masmediite i Negovata Auditoriya (In the Press of the Press [A View Towards Folk in the Mass Media and Its Audience]. *Bŭlgarski Folklor* 22 (5–6):35–45.

———. 1997. Krŭchmata v Kasetnata Kultura (Nabludeniya vŭrhu Temata na Krŭchmata v 100 Pesni) (The Bar and Cassette Culture [Observations on the Subject of the Bar in 100 Songs]). *Bŭlgarski Folklor* 23(1–2):91–102.

———. 1998. "Boretsŭt"—Geroi v Sŭvremennata Mediina Kultura v Bŭlgaria (Kŭm Otnoshenieto Identichnost-sŭvremenna Etnopopmuzika (The Thug—A Hero in Contemporary Media Culture in Bulgaria [Towards the Relationship of Identity and Contemporary Ethnopop Music]. *Bŭlgarski Folklor* 24(1–2):142–149.

———. 1999. Vŭrhu Nyakoi Ideologemni Aspekti na Bŭlgarskata Etnopopmuzika (On Some Ideological Aspects of Bulgarian Ethnopop Music). *Bŭlgarsko Muzikoznanie* 23(1):45–58.

———. 2001. *Etnopopbumŭt* (Ethnopop Boom). Sofia: Zvezdan.

———. 2009a. Romani Musicians in the Bulgarian Music Industry (1944–1989). In *Voices of the Weak: Music and Minorities*, ed. Zuzana Jurkova and Lee Bidgood, 179–184. Prague: Slovo 21.

———. 2009b. Roma Musicians in the Music of Post-1989 Bulgarian Media. In *Struga Musical Autumn: First Symposium of ICTM Study Group for Music and Dance in Southeastern Europe*, ed. Velika Serafimovska, 205–212. Skopje, Macedonia: Sokom.

Dimova, Rozita. 2007. BalkanBeats Berlin: Producing Cosmopolitanism, Consuming Primitivism. *Ethnologia Balkanica* 11:221–235.

Dirlik, Arif. 1997. *The Postcolonial Aura: Third World Criticism in the Age of Global Capitalism*. Boulder, CO: Westview.

———. 2000. *Postmodernity's Histories: The Past as Legacy and Project*. Lanham, MD: Rowman and Littlefield.

———. 2002. Bringing History Back In: Of Diasporas, Hybridities, Places, and Histories. In *Beyond Dichotomies: Histories, Identities, Cultures, and the Challenge of Globalization*, ed. Elisabeth Mudimbe-Boyi, 93–127. Albany: State University of New York Press.

Ditchev, Ivaylo. 2002. The Eros of Identity. In *Balkan as Metaphor: Between Globalization and Fragmentation*, eds. Dušan Bjelić and Obrad Savić, 235–250. Cambridge, MA: MIT Press.

——— 2004. Monoculturalism as Prevailing Culture. *Eurozine*, February 5. http://www.eurozine.com/articles/2004–02–05–ditchev-en.html, accessed July 10, 2010.

Djordjević, Tihomir. 1984 (1910). Cigani I Muzika u Srbiji (Gypsies and Music un Serbia), *Naš Narodni Život* 7:32–40.

Donadio, Rachel. 2008. Italy's Attacks on Migrants Fuel Debate on Racism. *New York Times*, October 13, 2008, A5.

Doubleday, Veronica. 1999. The Frame Drum in the Middle East: Women, Musical Instruments and Power: *Ethnomusicology* 43(1): 101–134.

Dunin, Elsie. 1971. Gypsy Wedding: Dance and Customs. *Makedonski Folklor* IV(7–8):317–326.

———. 1973. Čoček as a Ritual Dance Among Gypsy Women. *Makedonski Folklor* VI(12):193–197.

―――. 1977. The Newest Changes in Rom Dance (Serbia and Macedonia). *Journal of the Association of Graduate Dance Ethnologists, University of California, Los Angeles,* 1(Spring):12–17.

―――. 1984. A Gypsy Celebration of St. George's Day in Skopje. In *Dance Occasions and Festive Dress in Yugoslavia,* ed. Elsie Dunin, 24–27. UCLA Monograph Series 23. University of California at Los Angeles: Museum of Cultural History.

―――. 1985. Dance Change in the Context of the Gypsy St. George's Day, Skopje, Yugoslavia, 1967–1977. In *Papers from the Fourth and Fifth Annual Meetings, Gypsy Lore Society, North American Chapter,* ed. Joanne Grumet, 110–120. New York: Gypsy Lore Society.

―――. 1997. Miss Roma of 1997 is 15 year old Tirana, Albanian High School Student Eriona Gjonaj. *Newsletter of the Gypsy Lore Society* 20 (3August):1, 3–4.

―――. 1998. *Gypsy St. George's Day—Coming of Summer: Romski Gjurgjuvden, Romano Gjurgjovdani—Erdelezi, Skopje, Macedonia 1967–1997.* Skopje: Association of Admirers of Rom Folklore Art: Gypsy Heart.

―――. 2000. Dancing at the Crossroads by the Skopje Roma During St. George's Day. In *ICTM Study Group on Ethnochoreology 20th Symposium Proceedings.* August 19–26, 1998, 244–254. Istanbul Turkey: Danzmuzik Kultur Folklora Dogru.

―――. 2006. Romani Dance Event in Skopje, Macedonia: Research Strategies, Cultural Identities, and Technologies. In *Dancing from Past to Present: Nation, Culture, Identities,* ed. Theresa Buckland, 175–198. Madison: University of Wisconsin Press.

―――. 2008. Čoček in Macedonia: A Forty Year Overview. In *The Balkan Peninsula as a Musical Crossroad: Struga, Macedonia, September 2007,* ed. Velika Serafimovska, 115–125. Skopje, Macedonia: Sokom.

―――. 2009. The "Cloning" of Čoček in Macedonia: Media Affecting Globalization as Well as Localization of Belly Dancing. In *Struga Musical Autumn: First Symposium of ICTM Study Group for Music and Dance in Southeastern Europe,* ed. Velika Serafimovska, 213–225. Skopje, Macedonia: Sokom.

Dženo Association. 2005. Czech Superstar—Roma Can Sing But Not Move In. June 6, http://www.dzeno.cz/?c_id+7787, accessed July 10, 2010.

Džimrevski, Borovoje. 1983. *Vie se Oro Makedonsko* (The Macedonian Dance Line Winds Around). Skopje: Nova Makedonia.

―――. 1985. *Čalgiskata Tradicija vo Makedonija* (The Čalgija Tradition in Macedonia). Skopje: Makedonska Kniga.

Ebron, Paula. 2002. *Performing Africa.* Princeton, NJ: Princeton University Press.

Economist, The. 2008. Briefing: Europe's Roma: Bottom of the Heap. June 21, 2008, 35–38.

Ellis, Burcu Akan. 2003. *Shadow Genealogies: Memory and Identity Among Urban Muslims in Macedonia.* New York: Columbia University Press.

Eminov, Ali. 1997. *Turks and Other Muslim Minorities in Bulgaria.* New York: Routledge.

Eminova, Enisa. 2005. Raising New Questions About an Old Tradition. *Open Society News,* Summer/Fall:13.

Engelbrecht, Kjell. 1993. Bulgaria: The Weakening of Postcommunist Illusions. *RFE/RL Research Report* 2(1):78–83.

Erlanger, Steven. 2008. After U.S. Breakthrough, Europe Looks in Mirror. *New York Times*, November 12, 2008, A1, A12.

European Roma Rights Centre. 1998. *A Pleasant Fiction: The Human Rights Situation of Roma in Macedonia*. Budapest.

———. 2000. "Romani Women in Romani and Majority Societies." *Roma Rights*, no. 1, 28–46.

———. 2002. *Biannual Report 2001–2002*. Budapest.

European Roma Rights Centre/National Roma Centrum. 2006. *Written Comments of the European Roma Rights Centre and the National Roma Centrum Concerning the Former Yugoslav Republic of Macedonia* (presented to the United Nations Committee on Economic, Social, and Cultural Rights, September 19, 2006), 1–14.

"European Union Support for Roma for Roma Communities in Central and Eastern Europe." ec.europa.eu/enlargement/pdf/brochure_roma_oct2003_en.pdf, accessed June 17, 2011.

Feld, Steven. 1988. Notes on World Beat. *Public Culture* 1(1):31–37.

———. 1994. From Schizophonia to Schismogenesis: On the Discourses and Commodification Practices of "World Music" and "World Beat." In *Music Grooves*, eds. Charles Keil and Steven Feld, 257–289. Chicago: University of Chicago Press.

———. 2000a. The Poetics and Politics of Pygmy Pop. In *Western Music and Its Others: Difference, Representation, and Appropriation in Music*, eds. Georgina Born and David Hesmondhalgh, 254–279. Berkeley: University of California Press.

———. 2000b. A Sweet Lullaby for World Music. *Public Culture* 12(1):145–171.

Filipova, Nedyalka. 2004. Virtuozŭt Ivo Papazov: I Vseki Den da me Bombardirat, Shte si Ostana Tuka (Even If They Bomb Me Every Day, I'll Stay Here). *Trud*, April 27, 115:17.

Fine, Elizabeth. 1984. *The Folklore Text: From Performance to Print*. Bloomington: Indiana University Press.

Fluehr-Loban, Carolyn. 2003. *Ethics and the Profession of Anthropology: Dialogue for an Ethically Conscious Debate*. Walnut Creek, CA: AltaMira Press.

Foner, Nancy. 2000. *From Ellis Island to JFK: New York's Two Great Waves of Immigration*. New Haven: Yale University Press.

———. 2001. *New Immigrants in New York*. New York: Columbia University Press.

———. 2005. *In a New Land: A Comparative View of Immigration*. New York: New York University Press.

Fonseca, Isabel. 1995. *Bury Me Standing: The Gypsies and Their Journey*. New York: Vintage.

Fordham, Signithia. 1996. *Blacked Out*. Chicago: University of Chicago Press.

Fraser, Angus. 1992. *The Gypsies*. Oxford: Blackwell.

Friedman, Jonathan. 1997. Global Crises, the Struggle for Cultural Identity and Porkbarrelling: Cosmopolitans Versus Locals, Ethnics, and Nationals in an Era of Global De-Hegemonisation. In *The Dialectics of Hybridity*, ed. Pnina Werbner, 70–89. Cambridge: Zed Books.

Friedman, Victor. 1985. Problems in the Codification of a Standard Romani Literary Language. In *Papers from the Fourth and Fifth Meetings*,

Gypsy Lore Society, North American Chapter, ed. J. Grumet, 56–75. Cheverley, MD: Gypsy Lore Society.

———. 1995. Romani Standardization and Status in the Republic of Macedonia. In *Romani in Contact: The History, Structure, and Sociology of a Language*, ed. Yaron Matras, 177–188. Amsterdam: Benjamins.

———. 1999. The Romani Language in the Republic of Macedonia: Status, Usage and Sociolinguistic Perspectives. *Acta Linguistica Hungarica* 46(3–4):317–339.

———. 2005. The Romani Language in Macedonia in the Third Millennium: Progress and Problems. In *General and Applied Romani Linguistics: Proceedings from the 6th International Conference on Romani Linguistics*, eds. Barbara Schrammel, D. Halwachs, and G. Ambrosch, 163–173. Munich: Lincom Europa.

Frigyesi, Judit. 1994. Bela Bartok and the Concept of Nation and Volk in Modern Hungary. *Musical Quarterly* 78(2):255–287.

fRoots (anonymous). 2006. And the Rest Reviews: Electric Gypsyland 2. *fRoots* December, issue 282.

Gadjev, Vladimir. 1987. En Avant La Zizque! *Nouvelles de Sofia*, November 12:10.

Gal, Susan, and Gail Kligman. 2000a. *The Politics of Gender After Socialism*. Princeton, NJ: Princeton University Press.

———, eds. 2000b. *Reproducing Gender: Politics, Publics, and Everyday Life After Socialism*. Princeton, NJ: Princeton University Press.

Gamman, Lorraine, and Margaret Marshment. 1988. *The Female Gaze: Women as Viewers of Popular Culture*. London: Women's Press.

Garfias, Robert. 1981. Survivals of Turkish Characteristics in Romanian Musica Lautaresca. *Yearbook for Traditional Music* 13:97:107.

———. 1984. Dance Among the Urban Gypsies of Romania. *Yearbook for Traditional Music* 16:84–96.

Garofalo, Reebee. 1994. Culture vs. Commerce: The Marketing of Black Popular Music. *Public Culture* 7:275–287.

Gay y Blasco, Paloma. 1999. *Gypsies in Madrid: Sex, Gender and the Performance of Identity*. Oxford: Berg.

———. 2002. Gypsy/Roma Diasporas: A Comparative Perspective. *Social Anthropology* 10(2):173–188.

———. 2008. Picturing "Gypsies": Interdisciplinary Approaches to Roma Representation. *Third Text* 22(3):297–303.

Gaytandzhiev, Gencho. 2008. Roma Children in Bulgarian Schools: Have the Internal Obstacles Been Surmounted? In *The Human World and Musical Diversity: Proceedings from the Fourth Meeting of the ICTM Study Group Music and Minorities*, eds. R. Statelova, A. Rodel, L. Peycheva, I. Veleva, and V. Dimov, 204–206. Sofia: Institute of Art Studies.

Geertz, Clifford. 1973. *The Interpretation of Cultures*. New York: Basic Books.

———. 1983. *Local Knowledge*. New York: Basic Books.

Gelbart, Petra. 2004. Music as Territory: The Romani National Anthem, Representation and Transnational Sociopolitical Spaces. Paper presented at the Society for Ethnomusicology Annual Meeting, Tucson, AZ.

Georgiev, Krum 1986. Pŭrva Natsionalna Sreshta na Svatbarskite Orkestri (The First National Gathering of Wedding Orchestras). *Bŭlgarski Folklor* 3:90–91.

Georgieva, Ivanichka. 1966. Izsledvaniya vŭrhu Bita i Kulturata na Bŭlgarskite Tsigani v Sliven (Studies on the Life and Culture of the Bulgarian Gypsies of Sliven). *Izvestiya na Etnografskiya Institut i Muzei* IX:25–48.

Georgieva, Nadezhda. 2006. Contestation and Negotiation of Romani Identity and Nationalism Through Musical Standardization. *Romani Studies* (series 6) 16(1):1–30.

Gheorghe, Nicolae, and Thomas Acton, 2001. Citizens of the World and Nowhere: Minority, Ethnic, and Human Rights for Roma. In *Between Past and Future: the Roma of Central and Eastern Europe*, ed. Will Guy, 54–70. Hatfield: University of Hertfordshire Press.

Ghodsee, Kristen. 2008. Left Wing, Right Wing, Everything: Xenophobia, Neo-totalitarianism and Populist Politics in Bulgaria. *Problems of Post-Communism* 55(3):26–39.

———. 2009. Identity Shift. *Transitions on Line*. January 21, 2009 http://www.tol.org/client/article/20319-identity-shift.html, accessed June 22, 2011.

Giguère, Hélène. 2008. Musique Ethnique ou Musique Internationale? Diversité et Unicité dans le Patrimoine "Flamenco." In *Il Patrimonio Culturale Immateriale: Analisi e Prospettive*, ed. C. Bortollo, 129–143. Milan: Instituto Poligrafico e Zecca dello Stato.

Gilliat-Smith, Bernard. 1910. Notes and Queries. *Journal of the Gypsy Lore Society* IV(1):79.

Gilroy, Paul. 1993. *The Black Atlantic: Modernity and Double Consciousness*. Cambridge, MA: Harvard University Press.

———. 1994. Black Cultural Politics: An Interview with Paul Gilroy by Timmy Lott. *Found Object* 4:46–81.

———. 2004. Migrancy, Culture, and a New Map of Europe. In *Blackening Europe: The African American Presence*, ed. Heike Raphael Hernandez, xi–xxii. New York: Routledge.

Girgis, Mina. 2007. *Latcho Drom* for a *Gadjo Dilo*: The Problem with the Gypsy's Indian Origin in World Music. M.A. thesis, Music. University of California, Santa Barbara.

Glick Schiller, Nina, Linda Basch, and Cristina Blanc Szanton. 1992. *Towards a Transnational Perspective on Migration: Race, Class, Ethnicity and Nationalism Reconsidered*. New York: New York Academy of Sciences.

———. 1995. From Immigrant to Transmigrant: Theorizing Transnational Migration. *Anthropological Quarterly* 68(1):48–63.

Gočić, Goran. 2000. Gypsies Sing the Blues. Institute for War and Peace Reporting. *Balkan Crisis Report* no. 141. www.iwpr.net, accessed July 10, 2010.

———. 2001. *The Cinema of Emir Kusturica: Notes from the Underground*. London: Wallflower Press.

Goffman, Erving. 1974. *Frame Analysis: An Essay on the Organization of Experience*. Boston: Northeastern University Press.

Gojković, Adrijana. 1977. Romi u Muzičkom Životu Naših Naroda (Roma in the Musical Life of Our People). *Zvuk* 3:4 5–50.

———. 1986. Music of Yugoslav Gypsies. *Traditional Music of Ethnic Groups—Minorities, Proceedings of the Meeting of Ethnomusicologists on the Occasion of the European Year of Music 1985* 7(7):187–194.

Goodale, Mark ed. 2006. Anthropology and Human Rights in a New Key. *American Anthropologist* 108(1):1–83.

Gordon, Milton. 1964. *Assimilation in American life: The Role of Race, Religion, and National Origins*. New York: Oxford University Press.

Graff, James. 2002. Across the New Frontier. *Time*, June 24:24–33.

Gropper, Rena. 1975. *Gypsies in the City*. Princeton, NJ: Princeton University Press.

———, and Carol Miller. 2001. Exploring New Worlds in American Romani Studies: Social and Cultural Attitudes Among the American Mačwaia. *Romani Studies* 11(2):81–110.

Guy, Will, ed. 2001. *Between Past and Future: The Roma of Central and Eastern Europe*. Hatfield: University of Hertfordshire Press.

Hall, Stuart. 1989. Cultural Identity and Cinematic Representation. *Framework* 36: 68–81.

———. 1996a. Introduction: Who Needs Identity? In *Questions of Cultural Identity*, eds. Stuart Hall and Paul DuGay, 1–17. London: Sage.

———. 1996b. When Was "The Post-Colonial?" Thinking at the Limit. In *The Post-Colonial Question: Common Skies, Divided Horizons*, eds. Ian Chambers and Lidia Curti, 242–260. London: Routledge.

Hancock, Ian. 1987. *The Pariah Syndrome*. Ann Arbor, MI: Karoma.

———. 1997. The Struggle for the Control of Identity. *Transitions* 4(4):36–44.

———. 1998. Introduction. In *Roads of the Roma*, ed. Ian Hancock, Siobhan Down and Rajko Djurić, 9–21. Hatfield: University of Hertfordshire Press.

———. 2002. *We Are the Romani People: Ame Sam e Rromane Džene*. Hatfield: University of Hertfordshire Press.

———. 2008. The "Gypsy" Stereotype and the Sterilization of Romani Women. In *"Gypsies" in European Literature and Culture*, eds. Valentina Glajar and Domnica Radulescu, 181–192. New York: Palgrave Macmillan.

Handler, Richard, and Jocelyn Linnekin. 1984. Tradition: Genuine or Spurious. *Journal of American Folklore* 97:273–290.

Hanson, Allan. 1989. The Making of the Maori: Culture Invention and Its Logic. *American Anthropologist* 91:890–902.

Hasdeu, Julia. 2008. Imagining the Gypsy Woman: Representation of Roma in a Romanian Museum. *Third Text* 22(3):347–357.

Hasluck, Margaret. 1938. The Gypsies of Albania. *Journal of the Gypsy Lore Society*, (series 3) 17(2):49–61; 17(3):18–30; 17(4):110–122.

Hedges, Chris. 1996. Gypsy Folk Bands Play Loudly on for Serbs. *New York Times*, July 15, 1996, A4.

Helbig, Adriana. 2005. "Play for Me Old Gypsy": Music as Political Resource in the Roma Rights Movement in Ukraine. Ph.D. dissertation, Columbia University Department of Music.

———. 2007. Ethnomusicology and Advocacy Research: Theory in Action Among Romani NGOs in Ukraine. *Anthropology of East Europe Review* 25(2):78–83.

———. 2008. Managing Musical Diversity Within Frameworks of Western Development Aid: Views from Ukraine, Georgia, and Bosnia and Hercegovina. *Yearbook for Traditional Music* 40:46–59.

———. 2009. Representation and Intracultural Dynamics: Romani Musicians and Cultural Rights Discourse in Ukraine. In *Music and Cultural Rights*, eds. Andrew Weintraub and Bell Yung, 169–186. Chicago: University of Illinois Press.

Herzfeld, Michael. 1982. *Ours Once More: Folklore, Ideology, and the Making of Modern Greece*. Austin: University of Texas Press.

———. 1997. *Cultural Intimacy: Social Poetics in the Nation-State*. London: Routledge.

———. 2000. Afterword: Intimations from an Uncertain Place. In *Fieldwork Dilemmas: Anthropologists in Postsocialist States*, eds. Hermine De Soto and Nora Dudwick, 219–236. Madison: University of Wisconsin Press.

———. 2007. Small-Mindedness Writ Large: On the Migrations and Manners of Prejudice. *Journal of Ethnic and Migration Studies* 33(2):255–274.

Hesmondhalgh, David. 2000. International Times: Fusions, Exoticism, and Anti-Racism in Electronic Dance Music. In *Western Music and Its Others: Difference, Representation, and Appropriation in Music*, eds. Georgina Born and David Hesmondhalgh, 280–304. Berkeley: University of California Press.

Hobsbawm, Eric, and Terence Ranger, eds. 1983. *The Invention of Tradition*. Cambridge, UK: Cambridge University Press.

Hollander, Marc. 2001. Liner Notes to the CD. *Band of Gypsies: Taraf de Haidouks*. Nonesuch.

Hooker, Lynn. 2007. Controlling the Liminal Power of Performance: Hungarian Scholars and Romani Musicians in the Hungarian Folk Revival. *Twentieth Century Music* 3 (1):51–72.

hooks, bell. 1990. *Yearning: Race, Gender, and Cultural Politics*. Boston: South End.

———. 1991. "Is Paris Burning?" *Z* (Sisters of the Yam Column):61.

Hristov. Hristo. 2007. Dzhansever: Iskam da sŭm Zdrava, Kakto Predi, No Imam Nuzhda ot Pomosht (I Want to Be Healthy, Like Before, But I Need Help). *Drom Dromendar* XIII(5, March):1,6, and 12.

Hunt, Yvonne. 1995. Ta Kechekia—A Greek Gypsy Carnival Event. In *Dance and Ritual: Proceedings of the 18th Symposium of the ICTM Study Group on Ethnochoreology*, 97–103. Warsaw: Institute of Art.

———. 2009. Crossing the Border: The Case of Zurnaci-Tapan Ensembles of Bulgaria and the Daoulia of the Serres Prefecture of Greece. In *Struga Musical Autumn: First Symposium of ICTM Study Group for Music and Dance in Southeastern Europe*, ed. Velika Serafimovska, 153–158. Skopje, Macedonia: Sokom.

Hupchick, Dennis. 1994. "Nation or Millet? Contrasting Western European and Islamic Political Cultures in the Balkans." In *Culture and History in Eastern Europe*, 121–155. New York: St. Martin's Press.

Hutnyk, John. 2000. *Critique of Exotica: Music, Politics, and the Culture Industry*. London: Pluto Press.

Hymes, Dell. 1975. Folklore's Nature and the Sun's Myth. *Journal of American Folklore*. 88:345–369.

Ilnitchi, Gabriela. 2007. Ottoman Echoes, Byzantine Frescoes, and Musical Instruments in the Balkans. In *Balkan Popular Culture and the Balkan Ecumene: Music Image and Regional Political Discourse*, ed. Donna Buchanan, 193–223. Lanham, MD: Scarecrow Press.

Imre, Anikó. 2006. Play in the Ghetto: Global Entertainment and the Roma "Problem." *Third Text* 20(6):659–670.

———. 2008. Roma Music and Transnational Homelessness. *Third Text* 22(3):325–336.

———. 2009. *Identity Games: Globalization and the Transformation of Media Cultures in the New Europe*. Cambridge, MA: MIT Press.

Iordanova, Dina. 2001. *Cinema of Flames: Balkan Film, Culture, and the Media*. London: British Film Institute.

———. 2002. *Emir Kusturica: The Films*. London: British Film Institute.

———. 2003a. *The Cinema of the Other Europe: The Industry and the Artistry of East Central European Film*. London: Wallflower Press.

———. 2003b. Introduction. *Framework* (Special issue of Romanies and Cinematic Representation 44(3).

———. 2008. Mimicry and Plagiarism: Reconciling Actual and Metaphoric Gypsies. *Third Text* 22(3):305–310.

Ioviţă, Radu, and Theodore Schurr. 2004. Reconstructing the Origins and Migrations of Diasporic Populations: The Case of European Gypsies. *American Anthropologist* 106(2):267–281.

Ivanova, Radost. 1984. *Bŭlgarskata Folklorna Svatba* (The Traditional Bulgarian Wedding). Sofia: Bŭlgarskata Akademiya na Naukite.

———. 2001. Pop Folk (Čalga): Pros and Cons. *Ethnologia Bulgarica* 2:88–99.

Ivy, Marilyn. 1995. *Discourses of the Vanishing: Modernity, Phantasm, Japan*. Chicago: University of Chicago Press.

Jackman, Robert. 2007. Review, Shantel: Disko Partizani. http://www.bbc.co.uk/music/release/bbvr/.

Jackson, Bruce. 1987. *Fieldwork*. Urbana: University of Illinois Press.

Jakoski, Voislav. 1981. Pesnite na Gjurgjovden na Romite vo Skopje (St. George's Day Songs of the Roma of Skopje). *Etnološki Pregled* 17:293–302.

Janković, Ljubica, and Danica Janković. 1939. *Narodne Igre* (Folk Dances), vol. 3. Beograd: Stamparija Drag.

Johnson, E. Patrick. 2003. *Appropriating Blackness*. Durham, NC: Duke University Press.

Jones, Russ, 2006. Notes to the CD *Gypsy Beats and Balkan Bangers*. Atlantic Jaxx JAXXCD004.

Jones, Simon. 1988. *Black Youth, White Culture: The Reggae Tradition from JA to UK*. Basingstoke: Macmillan.

Jovanović, Boris. 2005. Interview with Goran Bregović. *Max Magazine*, June 29, 2005, No. 123, p. 44.

Kalaydjieva, Luba, David Gresham, and Fracesc Calafell. 2001. Genetic Studies of the Roma (Gypsies): A Review. *BMC Medical Genetics* 2:5–18.

Kanev, Krassimir. 1996. Dynamics of Inter-ethnic Tensions in Bulgaria and the Balkans. *Balkan Forum* 4(2):213–252.

———. 2005. How Should We Think of "Attack"? *Obektiv* 123 (May 9). http://www.bghelsinki.org/en/publications/obektiv/by-krassimir-kanev/2005-05/how-should-we-think-attack/, accessed June 22, 2011.

Kapchan, Deborah. 1996a. *Gender on the Market: Moroccan Women and the Revoicing of Tradition*. Philadelphia: University of Pennsylvania Press.

———. 1996b. Performance. *Journal of American Folklore* 108 (430): 479–508.

———. 2007. *Traveling Spirit Masters: Moroccan Gnawa Trance and Music in the Global Marketplace*. Middletown, CT: Wesleyan University Press.

Katzarova, Raina and Kiril Djenev. 1976. *Bulgarian Folk Dances*. Ann Arbor, MI: Slavica.

Kaufman, Dimitrina. 1989. Kŭm Problemite na Istrumentalna Improvizatsiya vŭv Sŭvremennite Svatbarski Orkestri. *Muzikalni Horizonti* 12/13:235–237.

———. 1990. Ot Vŭzrozhdenskata Chalga kŭm Sŭvremennite Svatbarski Orkestri (From the Chalga of the National Period to Contemporary Wedding Orchestras). *Bŭlgarski Folklor* 16(3):23–32.

———. 1995. Sŭvremennite Svatbarski Orkestri kato "Disidentski" Formatsii (Contemporary Wedding Orchestras as Dissident Formations). *Bŭlgarski Folklor* 21(6):49–57.

Kaufman, Nikolai. 1968. *Bŭlgarski Gradski Pesni* (Bulgarian Urban Songs). Sofia: Bŭlgarskata Akademia na Naukite.

———. 1987. Vtora Natsionalna Sreshta na Instrumentalnite Grupi za Bŭlgarska Narodna Muzika (Second National Gathering of Instrumental Groups of Bulgarian Folk Music). *Bŭlgarski Folklor* 13(1): 78–79.

———. 1989. Nyakoi Predshestvenitsa na Sŭvremennite Orkestri na Narodna Muzika (Some Precursors of Contemporary Folk Music Orchestras). *Muzikalni Horizonti* 12/13:220–228.

———, and Todor Todorov. 1967. *Narodni Pesni ot Yugozapadna Bŭlgaria: Pirinski Krai* (Folk Songs from Southwestern Bulgaria: Pirin Region). Sofia: Bŭlgarskata Akademiya na Naukite.

Kavaldzhiev, Lyubomir. 1988. Te Sa Profesorite! (They Are Professors) *Narodna Kultura* 40 (September 30):5.

Keil, Charles, and Steven Feld. 1994. *Music Grooves*. Chicago: University of Chicago Press.

Kelsky, Karen. 2001. *Women on the Verge: Japanese Women, Western Dreams*. Durham, NC: Duke University Press.

Kenrick, Donald, and Grattan Puxon. 1972. *The Destiny of Europe's Gypsies*. New York: Heinemann.

Kertesz-Wilkinson, Iren. 1992. Genuine and Adopted Songs in the Vlach Gypsy Repertoire: A Controversy Re-examined. *British Journal of Ethnomusicology* 1:111–133.

———. 2001. Gypsy Music. In *The New Grove Dictionary of Music and Musicians*, ed. Stanley Sadie, 613–620. London: Macmillan.

Kidikov, Nikola. 1988. Zavŭrshi Stambolovo '88 (Stambolovo '88 is Over). *Rabotnichesko Delo*, September 19:10.

Kimmelman, Michael. 2008a. In Hungary, Roma Get Art Show, Not a Hug. *New York Times*, February 6, 2008, B1, B6.

———. 2008b. Italy Gives Cultural Diversity a Lukewarm Embrace. *New York Times*, June 25, 2008, B1, B7.

Kiossev, Alexander. 2002. The Dark Intimacy: Maps, Identities, Acts of Identifications. In *Balkan as Metaphor: Between Globalization and Fragmentation*, eds. Dušan Bjelić and Obrad Savić, 165–190. Cambridge, MA: MIT Press.

Kirshenblatt-Gimblett, Barbara. 1994. Spaces of Dispersal. *Cultural Anthropology* 9(3):339–344.

———. 1998. *Destination Culture: Tourism, Museums, and Heritage*. Berkeley: University of California Press.

———. 2004. Intangible Heritage as Cultural Production. *Museum International* 56(1–2):52–65.

Kličkova, Vera, and Milica Georgieva. 1996 (1951). *Wedding Customs in Galičnik*. Skopje: NIP.

Kligman, Gail. 1988. *The Wedding of the Dead: Ritual Poetics, and Popular Culture in Transylvania*. Berkeley: University of California.

Klimova-Alexander, Ilona. 2005. *The Romani Voice in World Politics: The United Nations and Non-State Actors*. London: Ashgate.

Kossev, Dimitur, H. Hristov, and D, Angelov. 1963. *A Short History of Bulgaria*. Sofia: Foreign Language Press.

Kovalcsik, Katalin. 1985. *Vlach Gypsy Folk Songs in Slovakia*. Budapest: Institute for Music of the Hungarian Academy of Sciences.

———. 1987. Popular Dance Music Elements in the Folk Music of Gypsies in Hungary. *Popular Music* 6(1):45–65.

Kraev, Georg. 1988. Folklorŭt—Zanayat, Profesia ili? (Folklore—Craft. Profession or?) *Hudozhestvena Samodeinost* 38(January):33.

———. 1999, ed. *Chalagata: Za I Protiv* (Chalga: Pros and Cons). Sofia: Nov Bŭlgarski Universitet.

Kulish, Nicholas. 2007. Hungarian Extremists Reflect Discontent and Add to It. *New York Times*, October 24, 2007, A3.

———. 2008. Far-Right, Anti-Immigrant Parties Make Gains in Austrian Elections. *New York Times*, September 29, 2008, A7.

Kurin, Richard. 1997. *Reflections of a Culture Broker: A View from the Smithsonian*. Washington, DC: Smithsonian Press.

———. 2004 Safeguarding Intangible Cultural Heritage in the 2003 Convention: A Critical Appraisal. *Museum International* 56(1–2):66–75.

Kurkela, Vesa. 1997. Music Media in the Eastern Balkans: Privatized, Deregulated, and Neo-Traditional. *Cultural Policy* 3(2):177–205.

———. 2007. Bulgarian *Chalga* on Video: Oriental Stereotypes, Mafia Exoticism and Politics. In *Balkan Popular Culture and the Balkan Ecumene: Music Image and Regional Political Discourse*, ed. Donna Buchanan, 143–173. Lanham, MD: Scarecrow Press.

Lamphere, Louise, and Michelle Rosaldo, eds. 1974. *Woman, Culture, Society*. Stanford: Stanford University Press.

Landler, Mark. 2004. A Human Tidal Wave, or a Ripple of Hysteria? *New York Times*, May 5, 2004, A4.

Lange, Rose. 1997a. Hungarian Rom (Gypsy) Political Activism and the Development of *Folklor* Ensemble Music. *World of Music* 39(3):5–30.

———. 1997b. "What Was That Conquering Magic . . ." The Power of Discontinuity in Hungarian Gypsy *Nota*. Ethnomusicology 41(3): 517–537.

———. 1999. Political Consciousness in the Vernacular: Versions and Variants of the Hungarian *Ciganyhimnusz* (Gypsy Anthem). Journal of the Gypsy Lore Society (series 5) 9(1):29–54.

———. 2003. *Holy Brotherhood: Romani Music in a Hungarian Pentecostal Church*. New York: Oxford University Press.

Lassiter, Luke. 2005. *The Chicago Guide to Collaborative Ethnography*. Chicago: University of Chicago.

Laušević, Mirjana. 2007. *Balkan Fascination: Creating an Alternative Music Culture in America*. New York: Oxford University Press.

Lawless, Elaine. 1992. "I Was Afraid Someone Like You . . . an Outsider . . . Would Misunderstand": Negotiating Interpretive Differences Between Ethnographers and Subjects. *Journal of American Folklore* 105(417):302–314.

Leblon, Bernard. 1994. *Gypsies and Flamenco: The Emergence of the Art of Flamenco in Andalusia*. Hatfield: University of Hertfordshire Press.

Lee, Ken. 2000. Orientalism and Gypsylorism. *Social Analysis* 44(2): 129–156.

Leibman, Robert. 1974. *Traditional Tosk (South Albanian) Song and Dances from the Lake Prespa Area*. Booklet to accompany Selo LP.

Lemon, Alaina. 2000. *Between Two Fires: Gypsy Performance and Romani Memory from Pushkin to Postsocialism*. Durham, NC: Duke University Press.

Leuthold, Steven. 1998. *Indigenous Aesthetics: Native Art, Media and Identity*. Austin: University of Texas Press.

Levy, Claire. 2002. Who Is the "Other" in the Balkans: Local Ethnic Music as a *Different* Source of Identities in Bulgaria. In *Music, Popular Culture, Identities*, ed. Richard Young, 215–229. Amsterdam: Rodopi.

———. 2009. Folk in Opposition? Wedding Bands and New Developments in Bulgarian Popular Music. *Music and Politics* 3(1): 1–12.

Lewin, Ellen. 2006. *Feminist Anthropology*. New York: Blackwell.

Lewy, Guenter. 2000. *The Nazi Persecution of the Gypsies*. New York: Oxford University Press.

Lindemyer, Jeff, and Sotir Ramadonov. 2004. *History of the Roma in Bitola: Conversations with Older Roma in Bitola*. Bitola: Bairska Svetlina.

Lipsitz, George. 1994. *Dangerous Crossroads: Popular Music, Postmodernism and the Poetics of Place*. London: Verso.

Liszt, Franz. 1859. *Des Bohémiens at de Leur Musique en Hongrie*. Paris: Bourdillat.

Livni, Eran. 2011. Why Was Iordan Not Interested in Pictures of Dancing Gypsies: A Bulgarian-Romani Festival and the Discourse of European "Civility." In *Audiovisual Media and Identity in Southeastern Europe*, eds. Eckehard Pistrick, Nicola Scaldaferri, and Gretel Schwörer, 273–291. London: Cambridge Scholarly Publishers.

Lockwood, William, and Sheila Salo. 1994. *Gypsies and Travelers in North America: An Annotated Bibliography*. New York: Gypsy Lore Society.

Lopez, Nancy. 2003. *Hopeful Girls, Troubled Boys: Race and Gender Disparity in Urban Education*. New York: Routledge.

Lott, Eric. 1993. *Love and Theft: Blackface Minstrelsy and the American Workplace*. New York: Oxford University Press.

Lovas, Lemez. 2003. Gucaholic. *fRoots* 25(238):49–51.

Lozanova, Margarita. 2006. Sofi Marinova: Ne Davai na Horata Poveche Otkolkoto Iskat (Don't Give People More Than They Ask For). *Nov Folk* 39(October 30):17–19.

Lubkemann, Stephen. 2002. Refugees: Worldwide Displacement and International Response. *AnthroNotes* 23(2):1–10 and 19.

Lucassen, Leo. 2005. *The Immigrant Threat: The Integration of Old and New Migrants in Western Europe Since 1850*. Urbana: University of Illinois Press.

———, Wim Willems, and Anna Marie Cottar. 1998. *Gypsies and Other Itinerant Groups: A Socio-historical Approach*. London: Macmillan.

Lynskey, Dorian. 2006. Dorian Lynskey on Bands Looking to the Balkans for Inspiration. *Guardian*, November 24, 2006, http://www.guardian.co.uk/music/2006/nov/24/worldmusic?INTCMP=SRCH, accessed June 22, 2011.

MacCormack, Carol, and Marilyn Strathern. 1980. *Nature, Culture, Gender*. Cambridge, UK: Cambridge University Press.

Magrini, Tullia, 2003. *Music and Gender: Perspectives from the Mediterranean*. Chicago: University of Chicago Press.

Mahler, Sarah, and Patricia Pessar. 2006. Gender Matters: Ethnographers Bring Gender from the Periphery to the Core of Migration Studies. *International Migration Review* 40(1):27–63.

Malkki, Lisa. 1995. Purity and Exile: *Violence, Memory and National Cosmology Among Hutu Refugees in Tanzania*. Chicago: University of Chicago Press.

Malvinni, David. 2003. Gypsy Music and Film Music. *European Meetings in Ethnomusicology* 10:45–76.

———. 2004. *The Gypsy Caravan: From Real Roma to Imaginary Gypsies in Western Music and Film*. New York: Routledge.

Mamut, Oskar. 1993. Brak po Merak I od Merak (A Marriage of Necessity and Desire). *Romano Sumnal* (Skopje), November 17:3.

Maners, Lynn. 2006. Utopia, Eutopia, and E.U.–topia: Performance and Memory in the Former Yugoslavia. In *Dancing from Past to Present: Nation, Culture, Identities*, ed. Theresa Buckland, 75–96. Madison: University of Wisconsin Press.

Marcus, George, and Michael Fischer. 1986. *Anthropology as Cultural Critique: An Experimental Moment in the Human Sciences*. Chicago: University of Chicago Press.

Marian-Bălaşa, Marin. 2002. Birds in Cages Still Sing Well: An Introduction to the Musical Anthropology of the Romanian Jail. *European Meetings in Ethnomusicology* 2:250–264.

———. 2004. Romani Music and Gypsy Criminality. *Ethnologica Balkanica*. 8:195–225.

Marković, Aleksandra. 2008. Goran Bregović, the Balkan Music Composer. *Ethnologia Balkanica* 12(2008):9–23.

———. 2009. Sampling Artists: Gypsy Images in Goran Bregović's Music. In *Voices of the Weak: Music and Minorities*, eds. Zuzana Jurkova and Lee Bidgood, 108–121. Prague: Slovo 21.

Marushiakova, Elena. 1992. Ethnic Identity Among Gypsy Groups in Bulgaria. *Journal of the Gypsy Lore Society* (series 5) 2(2):95–116.

Marushiakova, Elena and Vesselin Popov. 1994. *Studii Romani*, vol. I. Sofia: Club' 90.

———. 1995. *Studii Romani*, vol. II. Sofia: Club' 90.

———. 1997. *Gypsies (Roma) in Bulgaria*. Frankfurt: Peter Lang.

———. 2000. *Tsiganite v Osmanskata Imperia* (Gypsies in the Ottoman Empire). Sofia: Litavra.

———. 2001. Bulgaria: Ethnic Diversity—A Common Struggle for Equality. In *Between Past and Future: The Roma of Central and Eastern Europe*, ed. Will Guy, 370–388. Hatfield: University of Hertfordshire Press.

———. 2005. The Roma—A Nation Without a State? Historical Background and Contemporary Tendencies. In *Nationalismus Across the Globe: An Overview of the Nationalism of State-endowed and Stateless Nations*, ed. W. Burszta, T. Kamusella and S. Wojciechedowski, 433–455. Poznan: School of Humanities and Journalism.

———, Herbert Heuss, Ivan Boev, Jan Rychlik, Nadege Ragary, Rubin Zemon, Vesselin Popov, and Victor Friedman. 2001. *Identity Formation Among Minorities in the Balkans: The Cases of Roms, Egyptians and Asjkali in Kosovo*. Sofia: Minority Studies Society Studii Romani.

Masood, Salman. 2007. When She Speaks, He's Breaking All of Islam's Taboos. *New York Times*, January 3:A4.

Massey, Douglas, Joaquín Arango, Graeme Hugo, Ali Kouaouci, Adela Pelligrino, and J. Edward Taylor. 1993. Theories of International Migration: A Review and Appraisal. *Population and Development Review* 19(3):431–466.

Matras, Yaron. 2002. *Romani: A Linguistic Introduction*. Cambridge: Cambridge University Press.

———. 2004. The Role of Language in Mystifying and Demystifying Gypsy Identity. In *The Role of the Romanies: Images and Counter-Images of "Gypsies"/Romanies in European Cultures*, eds. Nicholas Saul and Susan Tebbutt, 53–78. Liverpool: Liverpool University Press.

———. 2005. The Future of Romani: Toward a Policy of Linguistic Pluralism. *Roma Rights* 1:31–44.

McCann, Colum. 2007. EU's Ugly Little Challenge. *Los Angeles Times*, January 6, 2007, http://articles.latimes.com/2007/jan/06/opinion/oe-mccann6, accessed June 22, 2011.

Meintjes, Louise. 1990. Paul Simon's Graceland, South Africa, and the Mediation of Musical Meaning. *Ethnomusicology* 34(1):37–73.

Mernissi, Fatima. 1975. *Beyond the Veil: Male-Female Dynamics in a Modern Muslim Society*. Rochester, VT: Schenkman.

Mihaylova, Milena, 2005. The Effects of Neo-Liberal Reforms on the Roma in Bulgaria. Honors thesis, International Studies, University of North Carolina, Chapel Hill, NC.

Miller, Daniel. 1994. Modernity: An Ethnographic Approach: Dualism and Mass Consumption in Trinidad. Provincetown, RI: Oxford University Press.

Minchick, Adam. 2007. Waiting for Freedom, Messing It Up. *New York Times*, March 25, 2007, section 4, 13.

Mirga, Andrzej, and Nicolae Gheorghe. 1997. The Roma in the Twenty-first Century: A Policy Paper. Princeton, NJ: Project on Ethnic Relations.

Monitoring the EU Accession Process: Minority Protection. Open Society Foundations 2002. http://www.soros.org/resources/articles_publications/publications/euminority_20021125, accessed June 17, 2011.

Monson, Ingrid. 1994. Doubleness and Jazz Improvisation: Irony, Parody, and Ethnomusicology. *Critical Inquiry* 20(2):283–313.

Moore, Malcolm. 2008. Italian Children Back Burning of Gypsy Camps. *Telegraph,* May 28, http://www.telegraph.co.uk/news/worldnews/europe/2043851/Italy-Italian-children-back-burning-of-gypsy-camps.html, accessed June 22, 2011.

Nelson, Cynthia. 1974. Public and Private Politics: Women in the Middle Eastern World. *American Ethnologist* 1(3):551–563.

Neuburger, Mary. 2004. *The Orient Within: Muslim Minorities and the Negotiation of Nationhood in Modern Bulgaria*. Ithaca: Cornell University Press.

Nieuwkerk, Karin van. 1995. *"A Trade Like Any Other:" Female Singers and Dancers in Egypt*. Austin: University of Texas Press.

———. 2003. On Religion, Gender and Performing: Female Performers and Repentance in Egypt. In *Music and Gender: Perspectives from the Mediterranean*, ed. Tullia Magrini, 267–286. Chicago: University of Chicago Press.

No Man Is a Prophet in His Own Land. 2001. Film directed by Elsa Dahmani (Princes Film). On DVD *The Continuing Adventures of Taraf de Haïdouks*. Crammed Discs 2006.

Nonini, Donald, and Aihwa Ong, eds. 1997. *Ungrounded Empires: The Cultural Politics of Modern Chinese Transnationalism*. London: Routledge.

Nov Folk. 2007. *Azis: Fabtazii I Realnost* (Azis: Fantasy and Reality). Broi 18, May 21.

Okely, Judith. 1975. Gypsy Women: Models in Conflict. In *Perceiving Women*, ed. Shirley Ardener, 55–86. London: Malaby.

———. 1983. *The Traveler-Gypsies*. Cambridge, UK: Cambridge University Press.

———. 1996. Trading Stereotypes. In *Own or Other Culture*, pp. 45–61. London: Routledge.

———. 1997 Cultural Ingenuity and Travelling Autonomy: Not Copying, Just Choosing. In *Romany Culture and Gypsy Identity*, eds. Thomas Acton and Gary Mundy, 188–203. Hatfield: University of Hertfordshire Press.

Oldenburg, Veena Talwar. 1990. Lifestyle as Resistance: The Case of the Courtesans of Lucknow, India. *Feminist Studies* 16(2):259–287.

Ong, Aihwa. 1996. Anthropology, China, Modernities: The Geopolitics of Cultural Knowledge. In *The Future of Anthropological Knowledge,* ed. by Henrietta Moore, 60–92. London: Routledge.

———. 1997. Chinese Modernities: Narratives of Nation and of Capitalism. In *Ungrounded Empires: The Cultural Politics of Modern Chinese Transnationalism,* ed. Aihwa Ong and Donald Nonini, 171–202. London: Routledge.

———. 1999. *Flexible Citizenship: The Cultural Logics of Transnationalism*. Durham, NC: Duke University Press.

———. 2003. *Buddha Is Hiding: Refugees, Citizenship, and the New America*. Berkeley: University of California Press.

———, and Donald Nonini. 1997. Towards a Cultural Politics of Diaspora and Transnationalism. In *Ungrounded Empires: The Cultural Politics of Modern Chinese Transnationalism*, eds. Aihwa Ong and Donald Nonini, 323–332. London: Routledge.

Ortner, Sherry. 1995. Resistance and the Problem of Ethnographic Refusal. *Comparative Studies in Society and History* 37(1):173–193.

———. 1996. *Making Gender: The Politics and Erotics of Culture*. Boston: Beacon.

———. 1999. Thick Resistance: Death and the Cultural Construction of Agency in Himalayan Mountaineering. In *The Fate of Culture: Geertz and Beyond*, ed. Sherry Ortner, 136–163. Berkeley: University of California Press.

Oseary, Guy. 2004. *On the Record*. New York: Penguin.

Öztürkmen, Arzu. 2001. Politics of National Dance in Turkey: A Historical Appraisal. *Yearbook for Traditional Music* 33:139–143.

Pamporovo, Alexey. 2009. Roma in Bulgaria. *Gesis: Leibniz Institute for the Social Sciences (Roma in Central and Eastern Europe)* 2:27–32.

Pancevski, Bojan, and Carmiola Ionescu. 2006. Borat Film "Tricked" Poor Village Actors. *Daily Mail* (London), November 11, 2006. http://www.dailymail.co.uk/news/article-415871/Borat-film-tricked-poor-village-actors.html, accessed June 22, 2011.

Peev, Todor. 2005. Popfolk Forever. *Edno* 3(37):52.

Peters, Karen. 2005. Contemplating Music and the Boundaries of Identity: Attitudes and Opinions Regarding the Effect of Ottoman Turkish Contact on Bulgarian and Macedonian Folk Musics. *Folklorica* X(2): 5–25.

Peterson, Diane. 2004. "Gypsy Spirit" Tracks Roma People's History. *Press Democrat*, February 8, p. Q11.

Petrova, Dimitrina. 2003. The Roma: Between a Myth and a Future. *Social Research* 70(1):111–161.

Petrovski, Trajko. 1982. Običajot Vasilica Kaj Skopskite Romi, *Etnološki Pregled* 17.

———. 1987. Nekoi Karakteristiki na Romsata Ljubovna Narodna Poezija od Makedonija (Some Characteristics of Romani Love Folk Poetry from Macedonia). *Zbornik Radova XXXIV Kongresa Saveza Udruženja Folklorista Jugoslavija, Tuzla:*397–401.

———. 1993. *Kalendarskite Običai Kaj Romite vo Skopje I Okolinata* (Calendrical Customs Among the Roma of Skopje and Environs). Skopje: Feniks.

———. 2001. *Romski Narodni Pesni* (Romani Folk Songs). Skopje: Bis-Grafik.

———. 2002. *Romski Folklor* (Romani Folklore). Skopje: Romani Ilo.

———. 2009. Roma in Macedonia. *Gesis: Leibniz Institute for the Social Sciences (Roma in Central and Eastern Europe)* 2:41–43.

Pettan, Svanibor. 1992. Lambada in Kosovo: A Profile of Gypsy Creativity. *Journal of the Gypsy Lore Society* 2:117–130.

———. 1996a. Female to Male—Male to Female: Third Gender in the Musical Life of the Gypsies in Kosovo. *Narodna Umjetnost* 33(2):311–324.

———. 1996b. Gypsy Music and Politics in the Balkans. *World of Music* 38:33–61.

———. 1996c. Selling Music: Rom Musicians and the Music Market in Kosovo. In *Echo der Vielfalt: Traditionelle Musik von Minderheitenethnischen Gruppen/Echoes of Diversity: Traditional Music of Ethnic Groups*, ed. Ursula Hemetek, 233–245. Wien: Bohlau Verlag.

———. 2001. Encounter with "The Others from Within": The Case of Gypsy Musicians from the Former Yugoslavia. *World of Music* 43(2–3):119–137.

———. 2002. *Rom Musicians in Kosovo: Interaction and Creativity*. Budapest: Institute for Musicology of the Hungarian Academy of Sciences.

———. 2003. Male, Female and Beyond in the Culture and Music of Roma in Kosovo. In *Music and Gender: Perspectives from the Mediterranean*, ed. Tullia Magrini, 287–305. Chicago: University of Chicago Press.

———. 2007. Balkan Boundaries and How to Cross them: A Postlude. In *Balkan Popular Culture and the Ottoman Ecumene: Music, Image, and Regional Political Discourse*, ed. Donna Buchanan, 365–383. Lanham MD: Scarecrow Press.

Peycheva, Lozanka. 1993. Nablyudenia Vŭrhu Zurnadzhiiskata Traditsia v Yugozapadna Bŭlgaria (Notes on the Zurna Tradition in Southwest Bulgaria). *Bŭlgarski Folklor* 19(2):48–58.

———. 1994a. Na Balkanite e Istinskata Tsiganska Muzika (In the Balkans Is True Gypsy Music). *Folk Panair* 3:16–17.

———. 1994b. V Bŭlgaria mezhdu Kalkuta i Viena: Muzikata na Bŭlgarskite Roma s Gidove Hasan Chinchiri I Kiril Lambov (In Bulgaria Between Calcutta and Vienna: The Music of Bulgarian Roma Guided by Hasan Chinchiri and Kiril Lambov). *Bŭlgarski Folklor* 20(2):83–91.

———. 1994c. Vtori Festival na Romskata Muzika I Pesen—Stara Zagora '94 (The Second Festival of Romani Music and Dance—Stara Zagora '94). *Folk Panair* 1(7–8):29–30.

———. 1995. Muzikalniyat Polilingvizŭm na Tsiganite v Bŭlgaria (Nabyudenia vŭrhu 84 Audiokaseti) (Musical Polylingualism Among the Gypsies of Bulgaria [Notes on 84 Audiocassettes]). *Bŭlgarski Folklor* 21(6):58–72.

———. 1998. "Tsigania" v Bŭlgarska Identichnost (Aksiologichni Aspekti) ("Gypsiness" in Bulgarian Identity [Axiologic Aspects). *Bŭlgarski Folklor* 24(1–2):132–141.

———. 1999a. *Dushata Plache, Pesen Izliza* (The Soul Cries and a Song Comes Out). Sofia: Terart.

———. 1999b. Sŭblaznenata. Muzika (Chalgata spored Romskite Muzikanti) (Alluring Music [Pop/folk according to Romani Musicians]) *Bŭlgarsko Muzikoznanie* 23(1):59–63.

———. 2008a. Gypsy Musicians from Bulgaria: Trans-boundary and Trans-ethnic Mediators. In *The Balkan Peninsula as a Musical Crossroad: Struga, Macedonia, September 2007*, ed. Velika Serafimovska, 127–134. Skopje, Macedonia: Sokom.

———. 2008b. The Hybridization of Local Music from Bulgaria: The Role of Gypsy Clarinetists. In *The Human World and Musical Diversity: Proceedings from the Fourth Meeting of the ICTM Study Group Music and Minorities*, eds. R. Statelova, A. Rodel, L. Peycheva, I. Veleva, and V. Dimov, 124–133. Sofia: Institute of Art Studies.

———. 2008c. *Mezhdu Seloto i Vselenata: Starata Folklorna Muzika ot Bŭlgaria v Novite Vremena* (Between the Village and the Universe: Old Folk Music from Bulgaria in Contemporary Times). Sofia: Marin Drinov.

———. 2009. A Romani Musician from Bulgaria: Between the Local and the Global. In *Voices of the Weak: Music and Minorities*, eds. Zuzana Jurkova and Lee Bidgood, 59–66. Prague: Slovo 21.

———, and Ventsislav Dimov. 1994. Demokasetite: Za Edin Neizsledvan Fact ot Sofisiyat Muzikalen Pazar (Democassettes: On an Unstudied Fact of the Sofia Music Market). *Bŭlgarski Folklor* 19(2):25–34.

———. 2002. *Zurnadzhiiskata Traditsiya v Yugozapadna Bŭlgaria* (The Zurna Tradition in Southwestern Bulgaria). Sofia: Bŭlgarsko Muzikoznanie Izsledvania.

———. 2005. *Muzika, Romi, Medii* (Music, Roma, Media). Sofia: Zvezdan.

Peycheva, Lozanka, Ventsislav Dimov, and Svetla Krŭsteva. 1996. *Romska Muzika: Priturka kŭm Učebnitsite po Muzika, 1 Klas* (Romani Music: Supplement to Textbooks of Music, 1st Grade). Sofia: Papagal.

———. 1997. *Romska Muzika: Priturka kŭm Učebnitsite po Muzika za 5–8 Klas* (Romani Music: Supplement to Textbooks of Music for 5–8 Grades). Sofia: Papagal.

Plaut, Shayna, and Azbija Memedova. 2005. Blank Face, Private Strength: Romani Identity as Represented in the Public and Private Sphere. In *Roma's Identities in Southeast Europe: Macedonia*, eds. Azbija Memedova et al., Ethnobarometer Working Paper No. 9. http://www.ethnobarometer.org, accessed June 22, 2011.

Pollack, Sheldon, Homi Bhabha, Carol Breckenridge, and D. Chakrabarty, eds. 2000. Cosmopolitanisms. *Public Culture* 12(3):577–589.

Popov, Veselin. 1993. The Gypsies and Traditional Bulgarian Culture. *Journal of the Gypsy Lore Society* (series 5) 3(1):21–33.

Potuoğlu-Cook, Öyku. 2006. Beyond the Glitter: Belly Dance and Neoliberal Gentrification in Istanbul. *Cultural Anthropology* 21(4):633–660.

————. 2007. Sweat, Power, and Art: Situating Belly Dancers and Musicians in Contemporary Istanbul. *Music and Anthropology* 11. http://www.umbc.edu/MA/index/number11/potuoglu/pot_0.htm, accessed June 22, 2011.

Poulton, Hugh. 1991. *The Balkans: Minorities and States in Conflict*. London: Minority Rights Publications.

Povinelli, Elizabeth. 2002. *The Cunning of Recognition: Indigenous Alterities and the Making of Australian Multiculturalism*. Durham, NC: Duke University Press.

Purvis. Andrew. 2002. Roma Rule. *Time* 159(25, June 24):70–71.

Radano, Ronald, and Philip Bohlman, eds. 2000. *Music and the Racial Imagination*. Chicago: University of Chicago.

Rădulescu, Speranta. 2004. *Chats About Gypsy Music*. Bucharest: Paideia.

Ramnarine, Tina. 2007a. *Beautiful Cosmos: Performance and Belonging in the Caribbean Diaspora*. London: Pluto Press.

————. 2007b. Musical Performance in the Diaspora. *Ethnomusicology Forum* 16(1):1–17.

Ranova, Elitza. 2006. Of Gloss, Glitter and Lipstick: Fashion, Femininity and Wealth in Post-socialist Urban Bulgaria. *East European Anthropology Review* 24(2):25–34.

Rasmussen, Ljerka Vidić. 1991. Gypsy Music in Yugoslavia: Inside the Popular Music Tradition. *Journal of the Gypsy Lore Society* (series 5) 1(2):127–139.

————. 1995. From Source to Commodity: Newly-Composed Folk Music of Yugoslavia. *Popular Music* 14(2):241–256.

————. 1996. The Southern Wind of Change: Style and the Politics of Identity in Prewar Yugoslavia. In *Retuning Culture: Musical Change in Eastern Europe*, ed. Mark Slobin, 99–116. Durham, NC: Duke University Press.

————. 2002. *Newly Composed Folk Music of Yugoslavia*. New York: Routledge.

————. 2007. Bosnian and Serbian Popular Music in the 1990s: Divergent Paths, Conflicting Meanings, and Shared Sentiments. In *Balkan Popular Culture and the Ottoman Ecumene: Music, Image, and Regional Political Discourse*, ed. Donna Buchanan, 57–93. Lanham, MD: Scarecrow Press.

Rechel. Bernd. 2008. *The Long Way Back to Europe: Minority Protection in Bulgaria*. Stuttgart: Ibidem Verlag.

Revenga, Ana, Dena Ringold, and William Tracy. 2002. *Poverty and Ethnicity: A Cross-Cultural Study of Roma Poverty in Europe*. Washington, DC: World Bank.

Rice, Timothy. 1982. The Surla and Tapan Tradition in Yugoslav Macedonia. *Galpin Society Journal* 35:122–137.

————. 1994. *May It Fill Your Soul: Experiencing Bulgarian Music*. Chicago: University of Chicago Press.

————. 1996. The Dialectic of Economics and Aesthetics in Bulgarian Folk Music. In *Retuning Culture: Musical Change in Eastern Europe*, ed. Mark Slobin, 176–199. Durham, NC: Duke University Press.

————. 2002. Bulgaria or Chalgaria: The Attenuation of Bulgarian Nationalism in Mass-Mediated Popular Music. *Yearbook for Traditional Music* 34:25–47.

Ringold, Dena. 2000. *Roma and the Transition in Central and Eastern Europe: Trends and Challenges*. Washington, DC: World Bank.

———, Mitchell Orenstein, and Erika Wilkins. 2004. *Roma in an Expanding Europe: Breaking the Poverty Cycle*. Washington, DC: World Bank.

Ristić, Dušan, and Suzanne Leonora. 2004. *Rromani Songs from Central Serbia and Beyond*. Beograd: Galbeno.

Robbins, Bruce. 1992. Comparative Cosmopolitanism. *Social Text* 31/32:169–186.

———. 1999. *Feeling Global: Internationalism in Distress*. New York: New York University Press.

Rofel, Lisa. 1992. Rethinking Modernity: Space and Factory Discipline in China. *Cultural Anthropology* 7(1):93–114.

———. 1999. *Other Modernities: Gendered Yearnings in China After Socialism*. Berkeley: University of California Press.

Romanucci-Ross, Lola, and George De Vos, eds. 1995. *Ethnic Identity: Creation, Conflict, and Accommodation*. Walnut Creek, CA: AltaMira Press.

———, and Takeyuki Tsuda, eds. *Ethnic Identity: Problems and Prospects for the Twenty-first Century*. Walnut Creek, CA: AltaMira Press.

Rosaldo, Michelle. 1974. Woman, Culture and Society: A Theoretical Overview. In *Woman, Culture, Society*, eds. M. Rosaldo and L. Lamphere, 17–42. Palo Alto: Stanford University Press.

Rosaldo, Renato. 1989. *Culture and Truth: The Remaking of Social Analysis*. Boston: Beacon Press.

Rouse, Roger. 1991. Mexican Migration and the Space of Postmodernism. *Diaspora* 1(1):8–23.

Safran, William. 1991. Diasporas in Modern Societies: Myths of Homeland and Return. *Diaspora* 1(1):83–99.

Said, Edward. 1978. *Orientalism*. New York: Vintage.

———. 1989. Representing the Colonized: Anthropology's Others. *Critical Inquiry* 15(2):202–225.

Samson, Jim. 2005. Borders and Bridges: Preliminary Thoughts on Balkan Music. *Musicology* (Belgrade) 5:37–65.

Sant Cassia, Paul. 2000. Exoticizing Discoveries and Extraordinary Experiences: "Traditional" Music, Modernity, and Malta and other Mediterranean Societies. *Ethnomusicology* 44(2):281–301.

Sárosi, Bálint. 1978. *Gypsy Music*. Budapest: Corvina.

Saul, Nicholas, and Susan Tebbutt. 2004. *The Role of the Romanies: Images and Counter-Images of "Gypsies"/Romanies in European Cultures*. Liverpool: Liverpool University Press.

Savigliano, Marta. 1995. *Tango and the Political Economy of Passion*. Boulder, CO: Westview.

Schechner, Richard. 2006. *Performance Studies: An Introduction*. London: Routledge.

Schein, Louisa. 1999. Performing Modernity. *Cultural Anthropology* 14(3):361–395.

———. 2000. *Minority Rules: The Miao and the Feminine in China's Cultural Policy*. Durham: Duke University Press.

Sciolino, Elaine. 2007a. Identity, Staple of Right, Moves to Center of French Campaigns. *New York Times*, March 30, 2007, A1, A6.

———. 2007b. Immigration, Black Sheep and Swiss Rage. *New York Times*, October 8, 2007, A1, A7.

Sciolino, Elaine, and Bernard Ariane. 2007. On France's Far Right, Tweaking an Image. *New York Times*, April 8, 2007, section 1, 13.

Scott, James. 1985. *Weapons of the Weak: Everyday Forms of Peasant Resistance*. New Haven, CT: Yale University Press.

———. 1990. Domination and the Arts of Resistance: Hidden Transcripts. New Haven, CT: Yale University Press.

Seeman, Sonia Tamar. 1990a. Continuity and Transformation in the Macedonian Genre of Chalgija: Past Perfect and Present Imperfective. M.A. thesis, University of Washington.

———. 1990b. Music in the Service of Prestation: The Case of the Rom of Skopje. M.A. thesis, University of Washington.

———. 1998. Notes to *Sulukule: Rom Music of Istanbul*. Traditional Crossroads CD 4289.

———. 2000. Liner Notes. *The Road to Keşan: Turkish Rom and Regional Music of Thrace*. Traditional Crossroads CD 6001.

———. 2002. "You're Roman!" Music and Identity in Turkish Roman Communities. Ph.D. dissertation, University of California, Los Angeles, Department of Ethnomusicology.

———. 2007. Presenting "Gypsy," Re-presenting Roman: Towards as Archeology of Aesthetic Production and Social Identity. *Music and Anthropology* http://www.umbc.edu/MA/index/number11/seeman/see_0.htm accessed June 22, 2011.

Seibert, Brian. 2004. Legacies and the Artists Who Escape Them. *New York Times*, March 21, section 2, p. 12.

Sellers-Young, Barbara. 2006. Body, Image, Identity: American Tribal Belly Dance. In *Belly Dance: Orientalism, Transnationalism and Harem Fantasy*, eds. Anthony Shay and Barbara Sellers-Young, 277–303. Costa Mesa, CA: Mazda.

Serbezovski, Muharem. 1995. *Bunte Diamanten*. Zurich: Schweizer Verlaghaus.

Shay, Anthony. 2002. *Choreographic Politics: State Folk Dance Companies, Representation, and Power*. Middletown, CT: Wesleyan University Press.

———. 2005. The Male Oriental Dancer. In *Belly Dance: Orientalism, Transnationalism and Harem Fantasy*, eds. Anthony Shay and Barbara Sellers-Young, 85–113. Costa Mesa CA: Mazda Publishers.

———. 2008. Choreographing the Other: The Serbian State Folk Dance Ensemble, Gypsies, Muslims, and Albanians. In *Balkan Dance: Essays on Characteristics, Performing, and Teaching*, ed. Anthony Shay, 161–175. Jefferson, NC: McFarland Press.

———, and Barbara Sellers-Young. 2006. Introduction. In *Belly Dance: Orientalism, Transnationalism and Harem Fantasy*, eds. Anthony Shay and Barbara Sellers-Young, 1–27. Costa Mesa, CA: Mazda.

Shuman, Amy. 1993. Dismantling Local Culture. *Western Folklore* 52(1):345–364.

Silverman, Carol. 1981. Pollution and Power: Gypsy Women in America. In *The American Kalderash: Gypsies in the New World*, ed. M. Salo, 55–70. Centenary College, NJ: Gypsy Lore Society.

———. 1982. Everyday Drama: Impression Management of Urban Gypsies. *Urban Anthropology* 11(3–4):377–398.

———. 1986. Bulgarian Gypsies: Adaptation in a Socialist Context. *Nomadic People* 21–22:51–62.

———. 1988. Negotiating Gypsiness: Strategy in Context. *Journal of American Folklore* 101(401):261–275.

———. 1989. Reconstructing Folklore: Media and Cultural Policy in Eastern Europe. *Communication* 11(2):141–160.

———. 1991. Strategies of Ethnic Adaptation: The Case of Gypsies in the United States. In *Creative Ethnicity: Symbols and Strategies of Contemporary Ethnic Life*, eds. Stephen Stern and John Cicala, 107–121. Logan: Utah State University Press.

———. 1992. The Contemporary Bulgarian Village Wedding: The 1970's. *Indiana Slavic Studies* 6 (Balkanistica 8, special issue: Bulgaria Past and Present, ed. John Treadway):240–251.

———. 1995a. Persecution and Politicization: Roma (Gypsies) of Eastern Europe. *Cultural Survival* 19(2):43–49.

———. 1995b. Roma of Shuto Orizari, Macedonia: Class, Politics, and Community. In *East-Central European Communities: The Struggle for Balance in Turbulent Times*, ed. David Kideckel, 197–216. Boulder, CO: Westview Press.

———. 1996a. Music and Marginality: The Roma (Gypsies) of Bulgaria and Macedonia. In *Retuning Culture: Musical Change in Eastern Europe*, ed. Mark Slobin, 231–253. Durham, NC: Duke University Press.

———. 1996b. Music and Power: Gender and Performance Among Roma (Gypsies) of Skopje, Macedonia. *World of Music* 38(1):63–76.

———. 1996c. Who's Gypsy Here? Reflections at a Rom Burial. In *The World Observed: Reflections on the Fieldwork Process*, eds. Bruce Jackson and Edward Ives, 193–205. Champaign: University of Illinois Press.

———. 1999. Rom (Gypsy) Music. In *Garland Encyclopedia of World Music*, Europe volume, eds. Timothy Rice, James Porter, and Christopher Goertzen, 270–293. New York: Garland.

———. 2000a. *Latcho Drom* (Film Review). *Ethnomusicology* 44(2):262–264.

———. 2000b. Macedonian and Bulgarian Muslim Romani Women: Power, Politics, and Creativity in Ritual. *Roma Rights*, 1 (April):38–41.

———. 2000c. Researcher, Advocate, Friend: An American Fieldworker Among Balkan Roma 1980–1996. In *Fieldwork Dilemmas: Anthropologists in Postsocialist States*, eds. Hermine De Soto and Nora Dudwick, 195–217. Madison: University of Wisconsin Press.

———. 2003. The Gender of the Profession: Music, Dance and Reputation Among Balkan Muslim Romani (Gypsy) Women. In *Gender and Music in the Mediterranean*, ed. Tullia Magrini, 119–145. Chicago: University of Chicago Press.

———. 2007a. Bulgarian Wedding Music Between Folk and Chalga: Politics, Markets, and Current Directions. *Musicology* (Belgrade) 7:69–97.

———. 2007b. Trafficking in the Exotic with "Gypsy" Music: Balkan Roma, Cosmopolitanism, and "World Music" Festivals. In *Balkan Popular Culture and the Ottoman Ecumene: Music, Image, and Regional Political Discourse*, ed. Donna Buchanan, 335–361. Lanham, MD: Scarecrow Press.

———. 2008a. Fieldwork in Bulgaria. An Interview with Chris Hann. In *Studying Peoples in the People's Democracies II; Socialist Era Anthropology in South-East Europe*, eds. Vintila Miailescu, Ilia Iliev, and Slobodan Naumović, 397–404. Berlin: Lit Verlag.

———. 2008b. Transnational Chochek: The Politics and Poetics of Balkan Romani Dance. In *Balkan Dance: Essays on Characteristics, Performing, and Teaching*, ed. Anthony Shay, 37–68. Jefferson, NC: McFarland Press.

———. 2009. Music and Transnational Identity: The Life of Romani Saxophonist Yuri Yunakov. *Džaniben* Winter:59–84.

————. 2010. Queen of Gypsy Music. *City* 14(6):51–55.

————. 2011. Music, Emotion, and the "Other": Balkan Roma and the Negotiation of Exoticism. In *Interpreting Emotions in Russia and Eastern Europe*, ed. Mark Steinberg and Valeria Sobol, 224–247. DeKalb, IL: Northern Illinois University Press.

————. submitted. Balkan Romani Culture, Humans Rights, and the State: Whose Heritage? In *Intangible Rights in Transit*, ed. Deborah Kapchan. Philadelphia: University of Pennsylvania Press.

————. in press. Producing Music, Sexuality, and Emotion: Gendered Balkan Romani Dilemmas. *Etudes Tsiganes*.

————, and Ron Wixman. 1983. *Macedonian Bridal Costumes*. Museum of Natural History, Eugene: University of Oregon.

Sisario, Ben. 2002. "Gypsy Punk Cabaret," A Multinational. *New York Times*, April 14, 25 and 36.

————. 2005. The Rise of Gypsy Punkers. *New York Times*, July 2, p. A15.

Sivaramakrishnan, K. 2005. Some Intellectual Genealogies for the Concept of Everyday Resistance. *American Anthropologist* 107(3): 346–355.

Slobin, Mark. 2003. The Destiny of "Diaspora" in Ethnomusicology. In *The Cultural Study of Music*, eds. Martin Clayton, Trevor Herbert, and Richard Middleton, 284–296. New York: Routledge.

Smith, Linda. 1999. *Decolonizing Methodologies: Research and Indigenous Peoples*. London: Zed Press.

Soja, Edward. 1996, *Thirdspace: Journeys to Los Angeles and Other Real-and-Imagined Places*. Oxford: Blackwell-Wiley.

Sonneman, Toby. 1999. Dark Mysterious Wanderers: The Migrating Metaphor of the Gypsy. *Journal of Popular Culture* 32(4):119–139.

Soulis, George. 1961. *The Gypsies in the Byzantine Empire and the Balkans in the Late Middle Ages*. Washington, DC: Dumbarton Oak Papers No. 161.

Spivak, Gayatri. 1988. Can the Subaltern Speak? In *Marxism and the Interpretation of Culture*, eds. Carey Nelson and Lawrence Grossberg, 271–313. London: Macmillan.

Spur, Endre. 1947–1949. Jozsi the Second: The Problem of a Gypsy Musician's Career. Journal of the Gypsy Lore Society (series 3) 25(4–3):132–145; 27(3–4):117–132; 28(3–4):97–115.

Stankova, Maria. 2009. *Gogol Bordello and Kultur Shock: Genre, Representation, and Musical Transformation in Gypsy Punk*. M.A. thesis, Ethnomusicology, Wesleyan University.

Starkie, Walter. 1933. *Raggle-Taggle: Adventures with a Fiddle in Hungary and Romania*. London: John Murray.

Statelova, Rosemary. 2005. *The Seven Sins of Chalga: Toward an Anthropology of Ethnopop Music*. Sofia: Prosveta.

Stephen, Lynn. 2007. *Transborder Lives: Indigenous Oaxacans in Mexico, California, and Oregon*. Durham, NC: Duke University Press.

Stewart, Michael. 1989. "True Speech": Song and the Moral Order of a Hungarian Vlach Gypsy Community. *Man* 24(1):79–102.

————. 1997. *The Time of the Gypsies*. Boulder, CO: Westview.

Stewart, Susan. 1984. *On Longing: Narratives of the Miniature, the Gigantic, the Souvenir, the Collection*. Baltimore, MD: Johns Hopkins Press.

Stokes, Martin. 1992. *The Arabesk Debate*. Oxford: Clarendon.

————. 1994. *Ethnicity, Identity, Music: The Musical Construction of Place*. Oxford: Berg.

————. 2000. East, West and Arabesk. In *Western Music and Its Others: Difference, Representation, and Appropriation in Music*, eds. Georgina Born and David Hesmondhalgh, 213–233. Berkeley: University of California Press.

————. 2003. The Tearful Public Sphere: Turkey's "Sun of Art," Zeki Muren. In *Gender and Music in the Mediterranean*, ed. Tullia Magrini, 307–328. Chicago: University of Chicago Press.

————. 2004. Music and the Global Order. *Annual Review of Anthropology* 33:47–72.

————. 2007. On Musical Cosmopolitanism. *The Macalester International Roundtable 2007*. http://digitalcommons.macalester.edu/intlrdtable/3.

Stracansky, Pavol. 2009. Europe: the Right Rises. Inter Press Service, May 5. www.ipsnews/net, accessed June 22, 2011.

Sugarman, Jane. 1997. *Engendering Song: Singing and Subjectivity at Prespa Albanian Weddings*. Chicago: University of Chicago Press.

————. 2003. Those "Other Women": Dance and Femininity Among Prespa Albanians. In *Music and Gender: Perspectives from the Mediterranean*, ed. Tullia Magrini, 87–118. Chicago: University of Chicago Press.

————. 2007. "The Criminals of Albanian Music:" Albanian Commercial Folk Music and Issues of Identity Since 1990. In *Balkan Popular Culture and the Ottoman Ecumene: Music, Image, and Regional Political Discourse*, ed. Donna Buchanan, 269–307. Lanham, MD: Scarecrow Press.

Sŭrnev, Tsano. 1988. Nenadsvireniyat Ivo Papazov: Kontsertŭt (The Unsurpassed Ivo Papazov: The Concert). *Otchestvo* 3(301, April 26):23–25.

Sutherland, Anne. 1975. *Gypsies: The Hidden Americans*. Prospect Heights, IL: Waveland.

Szeman, Ioana. 2009. "Gypsy Music" and DeeJays: Orientalism, Balkanism and Romani Music. *TDR: The Drama Review* 53(3):98–116.

Taussig, Michael. 1987. *Shamanism, Colonialism, and the Wild Man: A Study in Terror and Healing*. Chicago: University of Chicago Press.

————. 1993. *Mimesis and Alterity*. New York: Routledge.

Taylor, Timothy, 1997. *Global Pop: World Music, World Markets*. London: Routledge.

————. 2007. *Beyond Exoticism: Western Music and the World*. Durham, NC: Duke University Press.

Teodosievski, Stevo, and Esma Redžepova. 1984. *On the Wings of Song*. Kočani: Dom Kulture Beli Mugri.

Theodosiou, Aspasia. 2003. "Be-longing" in a "Doubly Occupied Place": The Parakalamos Gypsy Musicians. *Romani Studies* 14(3):25–58.

Thomas, Nicholas. 1994. *Colonialism's Culture: Anthropology, Travel and Government*. Princeton, NJ: Princeton University Press.

Todorov, Manol. 1985. Pŭrva Reshitelna Krachka: Natsionalna Sreshta na Orkestrite za Narodna Muzika—Stambolovo '85 (The First Decisive Step: National Gathering of the Orchestras for Folk Music). *Hudozhestvena Samodeinost* 12:30–31.

Todorov, Todor. 1986. Na Svatba i na Kontsertniya Podium (At a Wedding and on the Concert Stage). *Kultura* 49(December 5):7.

Todorova, Maria. 1997. *Imagining the Balkans*. New York: Oxford University Press.

Toloyan, Khachig. 1996. Rethinking Diasporas: Stateless Power in the Transnational Moment. *Diaspora* 5(1):3–36.

Toninato, Paola. 2009. The Making of Gypsy Diasporas. *Translocations: Migration and Social Change* 5(1). http://www.dcu.ie/imrstr/volume_5_issue_1/index.shtml, accessed June 22, 2011.

Trask, Haunani-Kay. 1991. Natives and Anthropologists: The Colonial Struggle. *Contemporary Pacific* 3:159–167.

Trouillet, Jean. 2006. Notes to the CD *Bucovina Club 2*. Essay Records.

Trumpener, Katie. 1992. The Time of the Gypsies: A "People Without History" in the Narratives of the West. *Critical Inquiry* 18:843–884.

Turino, Thomas. 2000. *Nationalists, Cosmopolitans, and Popular Music in Zimbabwe*. Chicago: University of Chicago Press.

Turner, Victor, and Edward Bruner, eds. 1986. *The Anthropology of Experience*. Urbana: University of Illinois Press.

UNESCO. 2001. *Proclamation of Masterpieces of the Oral and Intangible Heritage of Humanity*. Paris: UNESCO.

Valtchinova, Galia. 2004 Folkloristic, Ethnography or Anthropology: Bulgarian Ethnology at the Crossroads. *Journal of the Society for the Anthropology of Europe* 4(2):2–18.

Van de Port, Mattjis. 1998. *Gypsies, Wars, and Other Instances of the Wild: Civilisation and Its Discontents in a Serbian Town*. Amsterdam: Amsterdam University Press.

———. 1999. The Articulation of the Soul: Gypsy Musicians and the Serbian Other. *Popular Music* 18(3):291–308.

Van Gennep, Arnold. 1961. *Rites of Passage*. Chicago: University of Chicago Press.

Van Nieuwkerk, Karin. 1995. *"A Trade Like Any Other": Female Singers and Dancers in Egypt*. Austin: University of Texas Press.

———. 2003. On Religion, Gender and Performing: Female Performers and Repentance in Egypt. In *Music and Gender: Perspectives from the Mediterranean*, ed. Tullia Magrini, 267–286. Chicago: University of Chicago Press.

Vekerdi, Joszef. 1976. The Gypsy's Role in the Preservation of Non-Gypsy Folklore. *Journal of the Gypsy Lore Society* 4/1(2):49–86.

Verdery, Katherine. 1996. *What Was Socialism and What Comes Next?* Princeton, NJ: Princeton University Press.

Vermeersch, Peter. 2006. *The Romani Movement: Minority Politics and Ethnic Mobilization in Contemporary Europe*. New York: Berghahn Books.

Višinski, Stanimir, and Elsie Dunin. 1995. *Dances in Macedonia: Performance, Genre, Tanec*. Prilep:11 Oktomvri.

Visweswaran, Kamala. 1988. Defining Feminist Ethnography. *Inscriptions* (special issue on feminism and the critique of colonial discourse) 3/4:27–46.

Voiculescu, Cerasela. 2005. Production and Consumption of Folk-Pop Music in Post Socialist Romania: Discourse and Practice. *Ethnologia Balkanica* 9:261–283.

Vukanović, Tatomir. 1962. Musical Culture Among the Gypsies of Yugoslavia. *Journal of the Gypsy Lore Society*, Series 3, 41:41–61.

———. 1963. The Gypsy Population of Yugoslavia. *Journal of the Gypsy Lore Society* 42:10–27.

———. 1983. *Romi (Cigani) u Jugoslaviji* (Roma in Yugoslavia). Vranje: Nova Jugoslavija.

Vŭlchinova, Elisaveta. 1989. Kŭm Sotsialnite Izmerenia na Edno Yavlenie (Towards the Social Dimensions of a Phenomenon). *Bŭlgarska Muzika* XL(1):31–43.

Vŭlchinova-Chendova, Elisaveta. 2000. *Gradskata Tradisionna Instrumentalna Praktika i Orkestrovata Kultura v Bŭlgaria (Sredata na XIX–Kraya na XX Vek)* (Traditional Urban Instrumental Practice and Orchestral Culture [Mid 19th to End of 20th century]). Sofia: Poni.

Wagner, Roy. 1979. *The Invention of Culture*. Chicago: University of Chicago Press.

Waringo, Karin. 2004. Who Is Afraid of Migrating Roma? *EU Map Journal*. http://www.warwithoutend.co.uk/uk-and-europe/2004/09/04/xenophobic-europe.php, accessed June 22, 2011.

Werbner, Pnina. 1997. Essentialising Essentialism, Essentialising Silence: Ambivalence and Multiplicity in the Constructions of Racism and Ethnicity. In *Debating Cultural Hybridity: Multi-Cultural Identities and the Politics of Anti-Racism*, eds. Pnina Werbner and Tariq Modood, 226–254. London: Zed.

———. 2002a. *Imagined Diasporas Among Manchester Muslims*. Oxford, UK, and Santa Fe, NM: School of American Research.

———. 2002b. The Place Which Is Diaspora: Citizenship, Religion and Gender in the Making of Chaordic Transnationalism. *Journal of Ethnic and Migration Studies* 28(1):11–123.

Werman, Marco. Global Hit—Borat Soundtrack. *The World*, November 9, 2006. http://www.pri.org/theworld/?q=node/5713, accessed June 21, 2011.

Wilson, William. 1973. Herder, Folklore, and Romantic Nationalism. *Journal of Popular Culture* 6:819–835.

Wolf, Diane. 1996. *Feminist Dilemmas in Fieldwork*. Boulder, CO: Westview Press.

Worth, Sol, and John Adair. 1972. *Through Navajo Eyes*. Bloomington: Indiana University Press.

Xenos, Nicholas. 1996. "Refugees: The Modern Political Situation." In *Challenging Boundaries: Global Flows, Territorial Identities*, eds. Michael Shapiro and Hayward Alker, 233–246. Minneapolis: University of Minneapolis Press.

Zaharieva, Svetlana. 1988. Bŭlgarska Vyara Hubava . . . Za Biteto na Folklora Dnes (Beautiful Bulgarian Faith . . . The Life of Folklore Today). *Bŭlgarska Muzika* 39(3):3–5.

Ziff, Bruce, and Pratima Rao. 1997. Introduction to Cultural Appropriation: A Framework for Analysis. In *Borrowed Power: Essays on Cultural Appropriation*, eds. Bruce Ziff and Pratima Rao, 1–27. New Brunswick, NJ: Rutgers University Press.

Zirbel, Kathryn. 1999. Musical Discursions: Spectacle, Experience, and Political Economy Among Egyptian Performers in Globalizing Markets. Ph.D. dissertation, University of Michigan, Department of Anthropology.

———. 2000. Playing It Both Ways: Local Egyptian Performers Between Regional Identity and International Markets. In *Mass Mediations: New Approaches to Popular Culture in the Middle East and Beyond*, ed. Walter Armbrust, 120–145. Berkeley: University of California Press.

Žižek, Slavoj. 1997. Multiculturalism, or, the Cultural Logic of Multinational Capitalism. *New Left Review* 225 (September/October):28–51.

Index

Page references in italics indicate illustrations.

Abdi, Faik, 170, 298n24
accordion, 131, 132
Acton, Thomas, 47, 50, 257–58,
    322n9
Ademović, Alen, 329n25
Ademović, Ninoslav, 329n25
adet (custom of), 76, 77
aesthetic pluralism,
    329n27
African Americans
  identity of, 55–56
  as a marginal group with
    musical power, 329n28
  and minstrel shows,
    330n35
  and Roma, 35, 45, 122
Ahmeti, Muharem, 36, 188
Aida, 224–25
Ajvazi, Nešo, 64, 101, 104
Akana ka khela (now So-and-So
    will dance), 86–87
Aksu, Sezen, 328n22
Albania
  Albanian influence in
    Skopje, 28
  Albanian style in Romani
    music, 35
  as Decade of Roma Inclusion
    signatory, 297n22
  Macedonians vs. Albanians,
    298n24
  modernity in, 320n36
Albanian music, 231–32
Ali, Muhammad, 301n15
Ali, Salif, 16, 17, 135, 136, 142,
    202, 233–34
"Allah Allah," 304–5n18

"Alo Taxi," 30, 133
Alyosha (Aleksei Atanasov
    Stefanov), 34, 36–37,
    271–72, 326n6
Amala Festival (Ukraine), 163
Amala Summer School (Valjevo,
    Serbia), 255, 324n11
Amalipe (Veliko Tŭrnovo,
    Bulgaria), 169
amanes, 36, 301n13
Amanet, 102
America for Bulgaria Foundation,
    169
Amet: "Belgiiskite Vecheri," 181
Amza Tairov, 35, 37, 166, 271
Amzoski, Severdžan, 100
Ana-Maria, 184
Andreya, 183
Angelite, 270
Angelov, Mitko, 224
Angelova, Lisa, 321n5
animal trainers, 25
Ansambl Teodosievski. *See*
    Esma—Ansambl
    Teodosievski
Ansel, 122
anthems, 48, 50, 51, 169, 173, 188,
    190, 211, 261
Anti-Discrimination Act
    (Bulgaria), 12
Antonio El Pipa Flamenco
    Ensemble, 235, 243
anxious account of appropriation.
    See celebratory vs. anxious
    account of appropriation
Appadurai, Arjun, 42–43,
    81, 268

appropriation, 273–75, 279, 327n12, 327nn14–17, 329n27, 329n30
arabesk, 177
Arabic music, 231
"Arabski," 30
Ara/Diapason, 34, 179
Ara TV (Bulgaria), 34
Aretxaga, Begoña, 129, 147
Argirov, Yashko, 136, 313n26, 314n7, 324n10, 326n9
Arif, Bekir, 298n24
Armeanca, Dan, 270
  "Iag Bari," 273
Armenian diaspora, 39
Armenian music, 231
arranged marriages, 71, 87
Artemis (Elizabeth Mourat), 327n17
artistic collusion between oppressed and oppressor, 258, 322n9
artistic/cultural adoration, 246, 253
Asenov, Hari, 313n26
Ashkalia, 295n1
Asphalt Tango, 216, 218, 270–71, 287, 290–91, 326n4
assimilation, 8, 66, 252
"Astargja o horo," 85
Ataka party. See Attack
Atanasov, Simeon, 211, 212–14, 216–17, 249, 254, 322n13, 323n22
Atanasov, Vergilii, 300n7
Atanasovski, Pece, 203
atsingani, 295n1
Attack (Ataka; Bulgaria), 12–13, 163, 184, 197, 298n30, 317n5, 320n39
Aura, 169
authenticity
  clothing as conveying, 250–51
  conceptions of, 54, 237
  of the ghetto, 248
  of Gypsy music festivals, 246–49
  and hybridity, 46, 215, 292
  in instrumentation, 248
  marketing of, 241, 244–52, 250–52, 257

and modernity, 54, 248, 251–52
and nationalism, 163–64, 251
of the past, nostalgia for, 246
search for, 22
as a social construction, 53
and tradition, 53, 247–48
See also heritage and the Bulgarian socialist state; purity
autoexoticism, 258
Azirov, Menderes, 68, 103–4, 310n32
Azirov, Sabuhan, 104
Azirov, Severdžan, 64, 68–69, 103–4, 119, 232, 310n32
Azis (Vasil Boyanov), 188–94, 270
  "Antigeroi," 191–92, 319n29
  autobiography of, 188–89
  clothing/image of, 190
  discrimination experienced by, 189
  "Dnevnik I Praznik," 193
  "Edin Zhivot Ne Stiga," 193–94
  gender ambiguity/transgression by, 188, 189–90, 191, 192–93, 197, 295n5, 319n25
  "Hajde Pochvay Me," 190
  homosexual wedding of, 189–90, 319n25
  "Hvani Me," 190
  Indian motifs used by, 304n16
  influence of, 194
  influences on, 191
  "Kak Boli," 193
  "Kazvash che me Obichash," 185
  Kralyat, 191
  live performances in Chicago by, 318n14
  and Marinova, 188, 193–94
  musical background/upbringing of, 189
  "Nakarai Me," 193
  "Ne Kazvai Ljube Leka Nosht," 191
  Night Show, 191
  "No Kazvam Ti, Stiga," 193
  "Nyama," 192–93
  "Obicham Te," 190
  oriental stereotype used by, 194
  popularity of, 188–89, 197, 319n23

at Romfest, 166
shock value in his
    performances, 190
"Teb Obicham," 193
"Tochno Sega," 194
versatility/talent as a singer,
    190–91
"Zajdi Zajdi Jasno Sonce,"
    319n27

babina (party for a newborn), 83
Babuš, 36
Baez, Joan, 134, 312n15
bajraktari (flag bearers), 81, 93,
    309n21
Bajraktarović, Mirko, 10
Bajram, Amdi, 88, 93–94, 215,
    298n24, 308n10
Bajramović, Šaban, 51, 188, 270,
    305n22
    "Djeli Mara," 276
bakshish (tips), 138
Balgova, Magda, 173
Balicki, Asen: *Roma Portraits*, 84,
    307n4, 307n20
Balkana, 150
Balkan Beat Box, 280, 325n1,
    330n31, 330–31n38
    Blue-Eyed Black Boy, 325n2
    "Red Bula," 283
Balkan Beats 3, 325n2
Balkan Beats music, 3, 20, 43,
    269, 281, 330n33
Balkan culture/identity, 195,
    320nn37–38
Balkan Fever (Vienna), 242
Balkan Grooves, 325n2
Balkanika (cable television
    channel), 183, 195,
    320n35
Balkan Music and Dance
    Workshops, 104, 105–6,
    212, 234
Balkan Roma. *See* Roma, Balkan
Balkan Romani dialect, 296n13
Balkanton, 130, 146, 147–48, 152,
    311n6
Balkan Traffic Festival (Belgium),
    242
Balkanturist (Bulgaria), 139
Ballet Troupe of Macedonian
    Television, 32–33, 119–20,
    208

Balogh, Jozsef, 321n5
Balogh, Kalman, 160, 263,
    324n10
banja (bath ceremony), 91
"Barack Obama Kyuchek," 30,
    301n15
Barbican: The 1000 Year Journey
    (London, 2000 and 2005),
    242, 245, 259, 330n36
Barev, Ivo, 314n14
Baro Bajrami or Šeker Bajram
    (big or sweet festival), 83
Bartók, Béla, 300n4
Basch, Linda, 41, 303n6
Basement Jaxx, 280
bass guitar, 312n9
bath ceremony, 91
bathhouses, 91
BBC (British Broadcasting
    Corporation), 290
BBC (British Broadcasting
    Corporation) Radio
    3 award, 161
bear leaders, 25
beauty contests, 33, 120–21, 170,
    301n18
Beirut (band), 280, 329n31,
    331n44
Beissinger, Margaret, 327n11
belly dancers
    association with prostitutes, 108
    in Bulgaria, 121, 122–23
    and çengi dancing, 108
    and čoček/kyuchek dance,
        98, 112, 116, 118, 119–20,
        121–23
    commercial success of, 123
    as exotic/oriental others, 116,
        121, 123
    exposed skin of, 32, 98, 112
    non-Romani, 275
    stomach movements of, 32
    tribal, 275, 327n17
    Turkish, 231
    wedding guests' reactions to, 98
Belmont (Bronx, New York)
    Balkan/Macedonian Roma in,
        13–14, 299n33
    education in, 66
    identity/ethnicity in, 66–71
    as an Italian neighborhood,
        13–14
    male/females roles in, 79

Belmont (*continued*)
  as multigenerational, 64–65
  music/dance as emblematic of
    Romani identity, 4
  musicians in, 99–106, 310n32
  Muslim practices in, 65–66
  Roma's contact with relatives in
    Macedonia, 59
  weddings in, 95–99,
    309–10nn26–28
  work and family life in, 63–66
  young male dancers in, 112–13
  Yuri Yunakov in, 229–31
Beranče (a line dance), 114
Beyashi, 295n1
Bhabha, Homi, 42–43
Bhairava scale/rag, 300n6
Bijandipe (Macedonian television
    program), 33, 298n24
bijav (celebration), 86
  See also weddings
Bijelo Dugme, 275
  "Djurdjevdan"/"Erdelezi," 277
Bikov, Sasho, 36–37
Bikova, Yuliya, 184
"Bingo," 30
Bistijani, 33
Bitola (Macedonia), 14, 62, 70,
    102, 114
Bitolska Gaida (a line dance),
    114, 120
Black Cat, White Cat (Kusturica),
    324n2
black market, 146, 147
blackness, white appropriation of,
    332n48
blaga rakija (sweet brandy) ritual,
    75–76, 95, 99, 310n28
Blagica Pavlovska, 101
Blanc, Cristina, 41, 303n6
blaxploitation movies, 258
blue color, 92
Bocina, Aco, 216, 270
Bojadžiev, Duke, 216
Boni, 184
Borat (Cohen), 269, 270, 289–91,
    331–32nn46–47
Borenstein, Eliot, 332n47
bori (bride), 85
  See also weddings
Borisova, Natalia, 321n5
Bosnia and Hercegovina, 297n22
bouzouki, 302n23

Bowers, Jake, 245, 253
Boyarin, Daniel and Jonathan,
    39, 40
Boyd, Joe, 149–51, 160, 161, 229
Brah, Avtar, 302–3n2
Brandy, 330n34
brass bands, 25, 300n8
  See also Boban Marković
brass music festival (Kumanovo,
    Macedonia, 2008), 301n18,
    (Guča, Serbia), 324
Bratsch: "Erdelezi," 328n22
bratstvo i jedinstvo (brotherhood
    and unity), 10, 210
Bregović, Goran, 275–80,
    324n2
  Balkan music revived by, 277,
    328n24
  "Čhaje Šukarije" 275
  criticized as an appropriator,
    275–78, 284, 329n25
  "Erdelezi," 218, 277
  Karmen with a Happy Ending,
    278
  "Kustino Oro," 278
  "Mesečina," 276
  musical background of, 275
  on the peace-making aspect of
    music, 288
  reputation/success of, 276–77
  on the Roma, 278–79
  Romani language used by, 279
  Roman musicians employed by,
    329n25
  *Tales and Songs from Weddings
    and Funerals*, 276
  Time of the Gypsies, 327n18
  Underground, 275–76, 327n18
Brestovica, 157
Brettel, Caroline, 79, 303n4
Briggs, Charles, 54, 55–56
Brody, Lauren, 233
Brown, Catherine, 111
Browning, Robert, 236, 247, 256,
    259, 261–62, 326n3
Bryceson, Deborah, 59
BTR (Macedonian television
    station), 33, 170, 218
Buchanan, Donna, 129
Bucovina Club (Frankfurt),
    281–82
Budapest Dance Ensemble, 118,
    243, 324n4

Bugarsko (a line dance), 114
Bulgaria
  belly dancers in, 121, 122–23
  brideprice customs in,
    308n11
  communism's fall in, 151
  as Decade of Roma Inclusion
    signatory, 297n22
  democracy in, 152
  economic crisis in (1990s),
    151–55
  European Union membership
    of, 6, 12, 165, 172,
    298n30
  in the Iraq war, 159, 314n10
  recording industry in, 33–34,
    146–48, 152, 155, 302n22,
    313nn26–29
  Romani population in, 12,
    297n25
  Romani rights in, 12–13, 196
  socialist, 10, 16–17, 33, 119,
    164, 227, 292–93, 299n39,
    302n21 (see also heritage
    and the Bulgarian socialist
    state)
  western currency/goods in, 146
  See also Sofia
Bulgarian Helsinki Committee,
  196
Bulgarian language, xiii, xiv
Bulgarian Ministry of Education,
  169, 316n21
Bulgarian Socialist Party,
  298n25
busking, 236
Butler, Judith, 5, 192, 194, 295n5,
  320n31
Butler, Ljiljana, 270

cable television, 34, 311n12
čačak (a line dance), 114
Cafe Antarsia, 268
çalgı (instrumental music or a
  musical instrument), 177
çalgici (wedding Romani
  musicians in Turkey),
  317n4
čalgija, 156, 160, 162, 179
  See also chalga
Cani company, 232
capitalism, 44–45, 151–52, 153,
  155–56, 293

Cappadocia, Rufus, 105
Caramel, 174
Cartwright, Garth, 189, 217–18,
    273, 281, 284, 287, 304n16,
    319n23, 331n44
  Princes Among Men,
    325n2
čeiz (bride's trousseau), 90
celebrations, transnational,
    19, 83–106
  babina (party for a newborn),
    83
  Baro Bajrami or Šeker Bajram
    (big or sweet festival), 83
  circumcision, 83, 89, 307nn1–2,
    308n15
  combined, 307n2
  engagement parties, 88, 96
  getting the bride, 93–95, 97,
    309n21, 309nn23–25
  guest musicians at, 93, 96
  henna and bath ceremonies,
    88–91, 96–97, 308n13,
    308–9n16
  Herdelezi (St. George's Day)
    (Herdeljezi, Erdelezi),
    32, 33, 83–84
  home videos of, 80
  igranka (dance party), 91–93,
    309nn19–20
  Kurban Bajrami (festival of
    sacrifice), 83
  of life-cycle vs. calendrical
    events, 83
  money spent on, 84,
    307nn2–3
  musicians in Belmont, 99–106,
    310n32
  overview of, 83–85
  patriarchal values in rituals, 94
  Ramazan (fasting month), 83
  tensions surrounding, 84,
    307n1
  Vasilica (St. Basil's Day), 83
  weddings, generally, 83–85,
    307nn2–3 (see also wedding
    music; weddings)
  women's vs. men's roles in,
    86–87, 98, 308n14
celebratory vs. anxious account
  of appropriation, 279–80,
  285–88, 329n28, 329n30
"Celo Dive Mangasa," 49

çengis (professional female
dancers), 107–8, 112, 123
Center for Interethnic Dialogue
and Tolerance Amalipe
(Veliko Tŭrnovo, Bulgaria),
169
Center for Traditional Music and
Dance, 235, 323n9
Central Romani dialect, 296n13
čest (honor) vs. sram (shame),
109–10
Ceyhan, Ali, 235
Chaje Shukarije (album), 216
See also Redžepova, Esma
Čhaje Šukarije, 31, 35, 209–10,
216, 275–76, 290, 322n11
See also Bregović, Goran and
Redžepova, Esma)
chalga (Bulgarian pop/folk
music), 19, 177–97
"Ballads MegaMix," 184–85
Bulgarian folk songs sung by
chalga singers, 158–59,
162–63
costumes for performers of,
159, 162–63, 180
critics of, 115, 163, 177, 181,
194, 195–96, 320n34
defenders of, 194–95
definition of, 317n4
erotic elements of, 182, 191,
318n11
folk elements of, 178–79
history/origins of, 31–32,
177–78, 317n1, 317nn3–4
influences on, 177, 195
kyuchek rhythms/style in,
177–78, 184, 194
and kyuchek's association with
belly dancing, 121, 178, 180
live vs. studio-recorded, 182,
318n14
mainstream, 183–84,
194, 271
male singers of, 181, 188, 318n9
and morality/ethnic politics,
194–97, 320–21nn34–39
and MTV, 180, 318n8
nationalistic/patriotic elements
of, 184–85, 195
orient evoked in, 177–78, 180,
184, 195–96
pop elements of, 178–79

popularity of, 156, 177, 183,
195, 317n1
and rap, 122, 186–87, 319n20
retro, 181
in Romani, 185, 187
Romani elements in, 184–85,
195–96
as social critique, 181
stars of, 182–83, 184, 318n14
style, text, and imagery of,
180–81, 318nn7–10
taksim/mane in, 180
terminological issues regarding,
178–79, *179*
at the Trakiya Folk festival, 152
travel magazines on, 183, 318n13
trends in (2000–), 182–85,
318n11, 318nn13–14,
318–19nn16–17
variety in, 183, 195
videos vs. texts of songs, 192
See also Azis; Kristal;
Marinova, Sofi
chalgadzhii (professional urban
musicians), 317n4
chetvŭrtŭk pazar (Thursday
market), 137
children of adulterous
relationships, 110
Chinchiri, Hasan, 27, 33, 137–38,
321n5
Chinchiri, Tome, 137–38
Chinchirova, Zlatka, 321n5
Chinese diaspora, 40, 303n5
chitalishte (reading room or
cultural center), 315n15
Chow, Rey, 43
Christgau, Robert, 283, 286
Cibula, Jan, 50
čifteteli, 28
"Ciganski Čoček," 209–10
Ciganski Pesni, 311n6
"Ciganyhimnusz," 50
čintijani (wide, billowing pants), 87
See also dimije, šalvari
circumcision, 83, 89, 307nn1–2,
308n15
Ćita, 35–36, 271
clarinet, 131, 132, 135, 223–24
Clarinet All-Stars, 105, 235, 266
Clark, Morgan, 233
class distinctions, 152
Clifford, James, 52, 53, 54–55

The Cobra (film), 49
čoček/kyuchek dance, 27
    as artistic/civilized, 115
    and belly dancing, 112, 116,
        118, 119–20, 121–23
    body movements in, 112
    Bulgarian, 119, 121–22
    in diaspora Balkan Romani
        communities, 112–15,
        310n2
    line dances, 114
    manele/mahala (dance), 116,
        310nn3–4
    by men, 112–13
    names of dances, 310n2
    oro, 114, 115
    Ottoman roots of, 107–8, 116, 123
    and pravo horo, 144
    as professional dance, 116–23,
        310–11nn5–6, 311nn8–9,
        311n11
    recordings of, 148
    rhythms of, 114
    for ritual vs. entertainment, 113
    as social dance among
        non-Roma, 115–16,
        310nn3–4
    as a solo dance, 112–13, 115–16,
        120
    at the Stara Zagora festival,
        314n13
    stomach flick in, 113–14, 119
    at Šutkafest, 170
    as Turkish vs. Romani, 164
    at weddings, 115
    by women, 112–15
čoček/kyuchek music, 98
    Bulgarian ban on, 119
    eclecticism in, 133
    "Leski Karuchka," 301n15
    meters used in, *29,* 112, 114
    scales used in, 27–28
    after socialism, 151
    spread of, 27
    styles of, 30
    tunes for, 30, 49, 133
    See also chalga; manele (music)
Cohen, Sasha Baron: *Borat,* 269,
    270, 289–91, 331–32nn46–
    47
collaboration, appropriation, and
    transnational flows, 20,
    269–94

appropriation, 273–75, 279,
    327n12, 327nn14–17,
    329n27, 329n30
Borat, 269, 270, 289–91,
    331–32nn46–47
collaboration, 270–73,
    326–27nn3–11, 333n50
    (*see also individual
    musicians*)
Gypsy Punk and DJ remixes,
    5–6, 280–89, 294,
    329–30nn31–42, 331nn44–
    45
overview of, 269, 325–26nn1–2
transnationalism, 291
See also Bregović, Goran
Columbia Artists, 118, 253
Comité Internationale Tsigane
    (London, 1971), 50
Condon, Zach, 285, 329n31,
    3301n39
copyright/pirating, 155, 287
    See also appropriation;
    DJ remixes
corruption, 152, 155–56, 173, 227
cosmopolitanism, 44, 181, 195
Costi, 183
Cottar, Anna Marie, 304n15
Coucos, Sakis, 183, 330n34
Council of Europe, 11, 50, 304n14
Cowan, Jane, 109
Crammed Discs, 282, 283, 325n1,
    326nn4–5
creativity, interpretations of, 22
Croatia, 297n22
crosses, 184
cultural/artistic adoration,
    246, 253
cultural intimacy, 127, 129, 145,
    147, 296n8, 320n38
cultural politics of postsocialist
    markets/festivals, 19,
    149–75
    Bulgarian wedding music
        (1990s), 149–55, 313n1,
        314n3, 314n5
    Bulgarian wedding music
        (2000–), 155–63, 313n2,
        314nn7–8
    copyright/pirating issues, 155
    Macedonian UNESCO
        application, 170–72,
        316n22, 316nn24–25

cultural politics (*continued*)
  music idol contests, 172–75,
    316–17nn27–29, 317n31,
    317n33
  official views of Romani music,
    167–69, 315–16n21
  overview of, 175
  Romani representation of
    Bulgaria, 150, 166, 172–75
  Romfest, 34, 122–23, 152,
    163–67, 185, 189, 311n11,
    314nn13–14
  Šutkafest, 50, 120, 169–70
cultural reification, 40
Čun, Medo, 169, 211, 305n19
  "Čhaje Šukarije", 31, 35, 209
Čun, Muamet, 31, 32, 93
Čun family, 31, 32
Čuperlika/Kjuperlika, 118
Czechoslovakia, 10
Czech Republic, 173, 174, 297n22

Dacheva, Toni, 158–59, 180, 184,
  318n7
Dacks, David, 282, 331n41
dahli, 223, 323n2
Daily Mail, 331–32n46
dajre (type of drum), 31
Dalaras, Giorgos, 328n22
Dalipova, Ramiza, 321n5
dance, transnational, 19, 107–23
  at bath ceremonies, 91
  Čuperlika/Kjuperlika, 118
  dance contests, 316n27
  igranka (dance party), 91–93,
    309nn19–20
  order of dance leaders,
    86–87, 98
  Ottoman roots of professional
    Balkan dancers, 107–8,
    116, 123
  segregated dancing, 110,
    112, 120
  self-stereotyping in, 118,
    121–22, 123, 311nn9–10
  sexuality and dance,
    109–12, 115
  social relations displayed via, 80
  stigma of professional female
    dancers, 109, 115, 119,
    123, 218
  at weddings (*see under*
    weddings)

  women's vs. men's, 87, 98
  See also belly dancers; čoček/
    kyuchek dance; Phralipe
dance parties (Brooklyn), 280–81
Daniela, 159
Darriau, Matt, 105
Decade of Roma Inclusion, 11,
  169, 297n22
Declaration of a Nation, 47
Deep Forest, 276, 328n20
Delall, Jasmine: *When the Road
  Bends: Tales of a Gypsy
  Caravan*, 213, 259, 327n17
Demir, Ibro: "Aj Leno, Lenorije
  Čhaje," 211
Demirov, Rifat, 100
Demirović, Zvonko, 186
  "Stranci," 49–50
Democracy Commission Small
  Grants program (American
  Embassy), 315–16n21
Democratic Alternative
  (Macedonia), 215
Democratic Progressive Party of
  Roma (Macedonia), 298n24
Denev, Vasil, 136, 160
Depp, Johnny, 254
Derrida, Jacques, 295n5
Dervisoski, Šaban, 103, 105
Desislava, 158–59, 185, 190,
  319n25
Diamond, Elin, 5
diaspora, hybridity, and identity
  conceptions of diaspora,
    39, 302–3n2
  diasporic identity in reification,
    40, 303n5
  and displacement/emplacement,
    40, 41–42
  essentialized identities, 51–53
    (*see also* essentialism)
  and exoticism, 46
  and ghettos/marginalization,
    41–42, 45–46, 303n7
  Hall on, 238
  and homelands, 8, 39–41,
    66–67, 302–3nn2–3 (*see also*
    Indian homeland)
  identity politics and the Romani
    rights movement, 44,
    47–48, 304n11, 304nn13–14
  Indian origins, 8
  and migration, 40, 303n4

music in the Romani rights movement, 49–51, 304–5nn16–22
overview of, 39–40, 221
and the Roma label, 44, 303–4n11
tensions/modalities within diasporas, 41
transnationalism and hybridity, 41–44, 303n6, 303–4 nn8–11
world music and hybridity, 44–47, 151, 239 (*see also* Romani music as world music)
diasporic Romani communities. *See* Belmont; Šutka
di Leonardo, Micaela, 86
dimije (wide, billowing pants), 87, 204, 208–9, 217
Dimitrov, Alex, 235, 265, 267, 288–89
Dimitrov, Evgeni: "Edinstveni," 172–73, 186, 319n19
Dimitrov, Kiril, 271
Dimitrovgrad (Bulgaria), 314n5
Dimov, Ventsislav, 26, 33, 166, 315n16
Dinkjian, Ara, 231
Dirlik, Arif, 40, 42–44, 46–47, 52–54, 303n5, 303n10, 305n23
Ditchev, Ivaylo, 167–68, 196, 315n16, 315n20, 320n37
Divano Productions, 272, 326n4
Djordjević, Tihomir, 21
Djoumahna, Kajira, 327n17
DJ remixes, 5–6, 280–89, 294, 329–30nn31–42, 331 nn44–45
Dobrev, Matyo, 16, 136, 145, 160, 167
Dobreva, Binka, 321n5
domazet (live-in son-in-law), 74, 306n15
domestic-public split, 78, 89, 109, 111, 307n19, 308n14
Dosta, 50
Dosti (film), 304n17
double cooptation, 196, 257
drag and gender, 192
Drom, 105, 315n20
drum set, 312n9

Dule, 101
Dunin, Elsie, 32, 110, 112, 113, 116–17, 119
Dǔrzhavna Atestatsionna Komisiya (State Certifying Commission; Bulgaria), 139
Dǔrzhavno Obedinenie Muzika (State Music Society; Bulgaria), 139
Džafer, 36–37
Džajkovski, Kiril: "Raise Up Your Hand," 216
Džambazi, 309n21
Džansever, 186, 271
  "Astargja o horo," 85
  "Romani Čhaj Sijum," 113
Džej, 328n22
"Dželem Dželem," 50–51, 169, 188, 211, 261, 305nn21–22
Dželjadin, Gjulizar, 75, 88, 308n11, 309n21
Džemaloski, Trajče, 98, 232
Dženo, 173
Dzhambazov, Marin, 30
Dzhamgyoz, Halil, 135
Džipsi Aver, 34–35, 122
  "Erdelezi," 328n22
džumaluk (first visit of bride's family to groom's home), 95, 309n25
džumbuš (type of lute), 31, 49

Eastern Orthodoxy, 311n4
East European Folklife Center, 104, 105–6, 212
The Economist, 6
Egjupci or Egjupkjani, 295n1, 316n23, *see also* Gjupci
Egypt, hierarchy of entertainers in, 321n7
Ekstra Nina, 158–59, 182
Electric Gypsyland/Electric Gypsyland 2, 282–83, 284, 287
Eleno Mome (a line dance), 114
Elfman, Danny, 254
Elit Center for Romani Culture (Sofia, Bulgaria), 166, 315n15
Ellington, Duke: "Caravan," 301n10

elopement, 74, 77
Emilia, 159, 183
 "Zabravi," 184, 318–19n17
Eminova, Enisa, 76,
 306n17
*Encyclopedia of Gay
 Folklife*, 190
Enev, Dragiya, 144
engagement parties, 88, 96
ensemble dance groups,
 166, 171
 See also čoček/kyuchek dance,
 as professional dance; *and
 individual groups*
Erdžan, 36, 218, 271
Erickson, Helene, 327n17
Erik, 318n9
ERIO (European Roma
 Information Office), 11
Erkan, 173
Ernst, Henry, 237, 250, 270, 287,
 291, 324–25n13, 325–26n2,
 326n7
ERRC (European Roma Rights
 Centre), 11, 297n21
Ersoy, Bülent, 191, 319n28
Esma—Ansambl Teodosievski,
 31, 205, 209, 211. *See also*
 Redžepova, Esma
Esma's Band: "Džipsi Denz," 217
essentialism
 of appropriation, 274
 Clifford on, 54–55
 and collective identity, 53
 demonization of, 52, 305n23
 and ethnic/cultural identity, 54,
 243, 262
 and multiculturalism, 18
 strategic, 51–52, 213–14
ethics, individualistic vs.
 collective, 66
ethnogenesis, 47
ethnography, 15–17
etničeski grupi (ethnic groups), 10,
 297n19
European Gypsy Festivals, 242,
 252, 266
European Parliament, 13, 197,
 298n30, 321n39
European Roma and Travellers
 Forum, 11, 304n14
European Roma Information
 Office (ERIO), 11

European Roma Rights Centre
 (ERRC), 11, 297n21
European Union
 and Balkan identity, 320n37
 Bulgaria's membership in, 6, 12,
 165, 172, 298n30
 platform of, 298n30
 and the Romani human rights
 movement, 11
European Year of Equal
 Opportunities (Czech
 Republic), 174, 317n33
Eurovision, 172–74, 186, 187,
 216–17, 302n24, 313n2,
 316–17nn28–29
Eva Quartet, 159
evil eye, 92, 94, 307n5
"Evro," 30
Evroroma, 188
Exclaim, 281

Fabri, Gaetano, 281
families, transnational, 19,
 59–81
 bride's reputation/virginity,
 74–77, 95, 306–7n17
 definition of transnational,
 59
 early emigration stories, 61–62,
 305nn3–4
 identity issues surrounding,
 66–71, 306nn7–11
 marriage in, 71–74,
 306nn13–15
 men's role in, 77–78
 migration narrated in song,
 59–60
 Muslim practices by, 65–66
 superintendent roles for free
 rent, 305n4
 travel and keeping in touch with
 relatives, 60–61
 videos used in the diaspora,
 79–81, 307n20
 women's power/knowledge in,
 71, 77–79, 88, 94, 109–10,
 306n14, 307n19
 work and family life, 63–66
Fanfare Ciocarlia, 287, 330n36
 "007," 301n10
 awards received by, 325n1
 Baro Biao, 250
 in *Borat*, 290–91

"Born To Be Wild," 290–91
clothing of, 250–51
collaborations by, 270
Gili Garabdi, 270–71, 301n10, 325–26n2
"Godzila," 270–71
in the Gypsy Caravan tour, 236, 243, 262
Hollywood party performances by, 254
managers of, 324–25n13
manele dancers who tour with, 116, 310n4
peasant image of, 249–50
popularity of, 269, 326n7
poverty of, 213
Queens and Kings, 270, 273, 325n2
success of, 270
theatrical framing by, 247
Feld, Steven, 274, 276, 279, 282, 286–88, 328n21, 330n37, 332n48, 333n50
Fen (Bulgaria), 34
Festival Internazionale di Musica Romani (Italy), 242
festivals. *See* Gypsy festivals *and individual festivals*
festivals, dance groups at. *See* čoček/kyuchek dance, as professional dance; *and individual groups*
Festival Tzigane (France), 242
Filipovska, Jagoda, 93
First World Festival of Romani Songs and Music (Chandigarh, India),211, 214
First World Romani Congress (London, 1971), 48, 50
See also International Romani Union
flags, 48, 51, 106, 309n21
Flamenco
dance, 180, 261–62, 265–66
as heritage, 316n25
music, 35, 243, 259, 326n3
Folk Kalendar, 153
Folklore TV, 162
Folklorna Grupa Trabotviše, 119
folk music
Greek, 177

instruments, 131–32, 312n10 (*see also specific instruments*)
schools for, 167, 315nn17–18
turbofolk, 177, 178–79, 319n22, 319n27
and wedding music, 132–33, 140, 156, 160, 162, 167, 226, 312n12, 315n18
See also chalga
Folk Palitra, 314n7
Folk Panair, 153
folk (new) vs. narodno (traditional) music, 178–79
Foner, Nancy, 299n33
Fonseca, Isabel, 300n6
fortune tellers, 218–19, 258
Foster, Catherine, 233, 327n16
Framework Program for Equal Integration of the Roma in Bulgarian Society, 12
Fraser, Angus, 296n14
Friedman, Jonathan, 303n9
Friedman, Victor, 48, 301n15
Frigyesi, Judit, 22
fRoots, 284
Frula, 117–18
fusion genres, 5–6, 18, 166. *See* chalga; Gypsy Punk

Gabriel, Peter, 276, 328n21, 329n30
gaida (bagpipe), 23, 31, 128
Galičnik Wedding (Macedonia), 170–72, 175, 316n23
Gandhi, Indira, 211
Gardjian, 100
Gaši, Džemailj, 36, 37
Gaši, Nehat, 36
Gaspar, Gyozo, 325n15
Gatlif, Elsa, 250, 324n8
Gatlif, Tony, 281, 324n8
Latcho Drom, 49, 151, 242, 243–44, 247, 258–59, 260, 275
Gaxha, Adrian, 216
Gaytandzhiev, Gencho, 168
Gay y Blasco, Paloma, 303n3
gaze, 192–93, 320n33
Gazoza, 33

Gelbart, Petra, 50, 51, 305n21
gender
　and dance, 87, 98, 109–15, 120
　　(see čoček/kyuchek dance)
　and drag, 192
　inside world of women vs.
　　outside world of men,
　　111, 308n14
　parody of, 5, 295n5
　and performance, 5, 192, 295n5
　See also Azis; domestic-public
　　split; women
Georgiev, Lyudmil, 136
Georgiev, Mihail, 301n11
Georgiev, Nikolai, 154
Georgieva, Milica, 171, 316n24
"Germaniya," 30
Gheorghe, Nicolae, 47, 304n13
Gieva, Anka, 206
Giguère, Hélène, 316n25
Gilroy, Paul, 42, 43, 46–47,
　　55–56, 246
Gio Style, 26
Gipsy.cz, 174, 302n24
Gipsy Festival (Holland), 242
Gipsy Kings, 26, 34–35
Girgis, Mina, 259
Gitan/Gitanos, 48, 295n1, 316n25
Gjupci, 66, 295n1
Glick Schiller, Nina, 41, 303n6
global imaginaries, 42–43
Glod (Romania), 290, 331–32n46
Gloria, 158–59, 179, 182, 318n14
Gnawa, 296n7
Gočić, Goran, 253
Gogol Bordello, 235, 264, 266,
　　280, 283–84, 325n1,
　　330n31, 331n45
　Multi kontra culti vs. Irony, 265
Gojković, Adrijana, 21
Golden Wheel Film Festival, 170
Goodman, Benny, 135
Gordon, Milton, 303n7
Gostivar (Macedonia), 94
Grand Masters of Gypsy Music,
　　232–33
Greek music, 34, 231, 302n23
Gropper, Rena, 71, 306n7, 306n9
Gruevski, Nikola, 215
Grupi Sazet E Ohrit, 101
Guča brass band festival (Serbia),
　　324n3
gŭdulka/kemene, 25

gurbet/pečalba (working abroad),
　　59, 103, 305n1
Gŭrdev, Nikolai, 314n14
　"So Grešingjom," 302n22
Guy, Will, 47
Gypsies
　authenticity of, 46
　conceptions/stereotypes of,
　　3, 67–69, 117–18, 123, 195,
　　207–8, 217, 292, 311n6,
　　321–22n8
　fictional Gypsy musicians in
　　Western culture, 3, 7
　hypersexualized female Gypsy
　　body, 121, 123
　orientalization/exoticization of,
　　116, 292, 297n18
　as refugees, 14–15
　vs. Roma, 295n1
　use of term "Gypsy," 195, 255,
　　295n1
　See also Roma, Balkan
Gypsy Caravan tour (1999)
　American response to, 263–64
　caravan concept in, 258–59
　communication/camaraderie
　　among participants in,
　　260, 262
　diversity vs. unity in, 260–62
　"Dželem Dželem" rejected as
　　finale piece, 51
　educational component of, 252
　exoticized marketing of, 244
　female musicians in, 321n2
　finale for, 261–62
　groups in, 235–36, 243, 260
　marketing/publicity for,
　　236–37, 249–50
　program notes for, 255–56, 259
　reception of, 260–62
　vs. Romfest, 163
　sponsorship of, 243
　stereotypes used in, 236
　Together Again: Legends of
　　Bulgarian Wedding Music,
　　161, 233
　urban-rural dichotomy in, 236
　Yuri Yunakov Ensemble in,
　　233–34
Gypsy Caravan troupe, 275
Gypsy festivals, 105, 235, 242,
　　247, 252, 266, 330n32, see
　　Romfest, Šutkafest

Gypsy Punk music, 3, 20, 43, 235
  and DJ remixes, 5–6, 280–89, 294, 329–30nn31–42, 331nn44–45
  and the New York Gypsy Festival, 265–67
  popularity of, 264
  Romani music appropriated by, 5–6
Gypsy Queens, 321n2
The Gypsy Road, 259
Gypsy scale, 27–28, 301n12
Gypsy Spirit (2004), 118, 243, 253–54, 314n7, 324n4, 324n10, 326n9
Gypsy TV (www.gypsytv.tv; Sofia, Bulgaria), 34, 311n12

Habibi and Malki Kristalcheta: "Yak Motoru," 330n34
Hagopian, Harold, 236–37, 251, 323n9
hairstyles/hairdressers, 309n19
Hajgara, 283
Halili, Merita, 102, 231–32, 323n8
  "Erdelezi," 328n22
Hall, Stuart, 7, 55–56, 238
Hancock, Ian, 8, 15, 106, 173, 246, 253, 296n14, 300n6
Handler, Richard, 53
Handzhiev, Ivan, 167
Hannibal records, 149
Hanson, Allan, 53
Hapazov, Ibryam. *See* Papazov, Ivo
Harrington, David, 326–27n10
Haskovo (Bulgaria), 224, 227, 229–30, 314n3, 314n5
hate speech, 11, 297n21
A Hawk and a Hacksaw, 280
  The Way the Wind Blows, 331n44
"Hazart," 30
Head On (film), 270
hedgehogs, 324n5
Heller, Andre, 242
henna
  ceremonies/rituals involving, 88–91, 96–97, 111, 308n13, 308–9n16
  use in different cultures, 308n13

Herdelezi (Herdeljezi, Erdelezi) (St. George's Day), 32, 33, 83–84
Herdelezi festivals, VOR-sponsored, 102–3, 105, 232–34, 242–43, 287, 314n12, 331n42
heritage, meanings of, 127, 171
heritage and the Bulgarian socialist state, 19, 127–48
  Bulgarization, 16–17, 127, 129, 299n39
  category system for musicians, 139–40, 152
  "fascist" families, 139, 313n25
  the free market and state control, 137–40, 148, 313nn21–25
  in inclusion vs. exclusion in the nation/state, 127
  meanings of heritage, tradition, and folk, 127, 171
  Muslim emigration to Turkey, 130
  name-changing campaign, 129–30, 142, 226, 228, 323n6
  official rhetoric of purity, 129, 140–45
  overview of, 127
  resistance to state policy, 127, 128–30, 143–45, 147, 148, 228, 292–93, 311n3
  state ambivalence, 145–48, 313nn26–29
  Turkish resistance to state policy, 129–30
  wedding music (1970s–1989), 131–33, 311–12nn7–12
  zurna and the anti-Muslim campaign (1980s), 128–30, 141, 311nn3–4, 311n6 (ch. 7)
  See also Papazov, Ivo; wedding music, Bulgarian; Yunakov, Yuri
heritage movements, indigenous, 54
Herzfeld, Michael, 51–52, 127, 129, 145, 147, 296n8, 320n38
hicaz, 27–28, 180, 191–92, 278, 300n6
hidden vs. public transcripts, 228–29

Hindi language, 49
hip-hop music, 122
Hleda Superstar competition, 173
Hobsbawm, Eric, 53
Hollander, Marc, 249–50, 256,
    282–83, 285, 331n41
Holocaust, 39, 48, 51, 169, 197
homelands, 39–41, 55, 66–67, 196,
    291, 302–3nn2–3
    See also Indian homeland
Home of Humanity and
    Museum of Music, 215
honor, 109, 202
hooks, bell, 52
Horahane (Muslim), 306n8
horo (village dance event),
    160, 312n12
Hristov, Krasimir, 181
Hristovski, Jonče, 93
Huna, Elvis, 76, 77, 322n13
Hungarian Governmental Office
    of Equal Opportunity,
    Directorate of Romani
    Integration, 253
Hungary, 130, 253, 297n22
Hungry March Band, 264, 265
Husein, Husein. See Yunakov, Yuri
Hutnyk, John, 43, 44–46, 252, 329n27
Hutz, Eugene, 235, 264–65,
    266–67, 280, 281, 285, 287,
    329–30n31, 331n42
hybridity
    transnationalism and, 41–44,
        303n6, 304n8–11
    world music and, 44–47, 151, 239
    See also diaspora, hybridity, and
        identity
Hyseni, Raif, 102, 231–32, 323n8

Iag Bari: Brass on Fire, 250, 291,
    310n4, 324n7
Iagori festival (Norway, 2005), 242
Ibrahim, Sali, 315n15
Ibraim Odža (a line dance), 114
Ibraimov, Muren, 101
Ibro Lolov: Ciganski Pesni, 311n6
identity
    African American, 55–56
    Balkan, 195, 320nn37–38
    Balkan vs. European, 195, 320n37
    in Belmont, 66–71
    essentialized identities, 51–53
    ethnic/cultural, 54, 243, 262
    female, 86

issues surrounding
    transnational families,
    66–71, 306nn7–11
and modernity, 44, 81
Native American, 56
and performance, 4–5, 41
Romani rights movement and
    the politics of, 44, 47–48,
    304n11, 304nn13–14
of Yuri Yunakov (see under
    Yunakov, Yuri)
See also diaspora, hybridity, and
    identity
Identity, Tradition, and
    Sovereignty, 197
igranka (dance party), 91–93,
    309nn19–20
Ilhan, Serdar, 235, 265
Ilić, Slobodan, 304n17
Iliev, Boril, 136
Iliev, Jony, 269,
    272, 289
    "Godzila," 270–71, 272
    Ma Maren Ma, 325–26n2
Iliev, Nikola, 134, 136, 139, 157,
    313n26
Iliyan, 318n9
Imam li Dobŭr Kŭsmet (Džipsi
    Aver), 35
immigration, 14, 15
imperialist nostalgia, 45, 246
Imre, Anikó, 166, 174,
    196, 248, 257, 293,
    302n24, 325n15
inbetweenness, 42–43
Indian homeland, 23, 39–40, 165,
    191, 243, 259–60
Indian music and films, 49,
    304nn16–17, 305n19
indigenous rights movements, 48
    See also Romani rights
        movement
"Indiiski" scales, 305n19
interethnic relations, Gallup poll
    on, 196
intermarriage, 71–72, 205
International Roma Day
    (April 8), 33, 50, 119, 170,
    243
International Romani Union
    (IRU), 11, 47, 48, 50,
    304n14, 305n21
invented traditions, 53
Ionitsa, 247, 256, 326n8

IRU. *See* International Romani Union

Isakut, Hassan 231, 232

Ishida, Masataka, 250

Islam
 Bulgarian anti-Muslim campaign (1980s), 128–30, 141, 311nn3–4, 311n6 (ch. 7)
 conversions to, 8–9, 296n15
 revitalization of, 66, 306n6
 on women as sexual, 110

Islami, Kurte, 99, 101

Islami, Ramiz, 98, 101, 102, 104, 105, 232

Islami, Redžep, 101

Islami, Romeo, 101

Islamic Center (Belmont, Bronx, New York), 66

Ismail, Ferhan, 102
 "Gurbeti," 59

Ismaili, Kujtim, 98, 101, 232

Ismaili, Muamed, 101

Israeli music, 231

"Isuara," 30

Ivana, 159, 162, 182

Ivancea, Ioan, 291

Ivanova, Lili: "Vyatŭr," 187

Ivanovski, Blago, 206

Ivo Barev/Asiba Kemalova: Tsiganski Pesni, 311n6

Jackman, Robert, 281

Jackson, Michael, 328n21

Jackson, Michael, imitators of, 122

Janev, Georgi, 155

Janković sisters, 118

Jankuloski, Toni, 104, 105

Jašarov, Ilmi, 32

jazz, 135, 136, 329n28

Jeni Jol (a line dance), 114

Jerrari, Al, 281

Jewish diaspora, 39

Johnson, E. Patrick, 332n48

Jones, Russ, 284–85
 Balkan Bangers, 282, 330n32
 Gypsy Beats, 282, 330n32

Jones, Simon, 329n28

Joro-Boro, 267–68

Jovanović, Jarko, 50

Julliard Conservatory, 104

Južni Kovači, 37

K88 (cable television channel), 183

kaba zurna, 28

Kaffe, 317n29

Kal, 251, 255, 269, 270, 277, 326n4
 Kal, 325n2
 Radio Romanista, 325n2

Kalbelia dancers, 247

Kalcheva, Mariana, 181

Kalcheva, Ruska, 321n5

Kalcheva, Slavka, 158–59, 182, 318n7

Kalderash Roma. *See* Roma, Kalderash

Kaliopi, 216

Kalji Jag, 235–36

Kalman Balogh Gypsy Cimbalom Band, 324n10
 Ušte Opre, 263

Kaloome: "Que Dolor," 273

Kalyi Jag, 243, 321n5

Kamburov, Mancho, 128

kana. *See* henna

Kanarite, 184, 314n8
 "Ah Lyubov, Lyubov," 159
 "Biznesmen," 158
 "Bŭlgarski Cheda," 159
 Kanarite, 98, 158
 Muzika s Lyubov, 160
 Na Praznik i v Delnik, 159
 Ne Godini, A Dirya, 159
 Nie Bŭlgarite, Kanarite 25 Godini, 158–59
 "Nie Sme Kanarite," 159
 popularity of, 157
 "Pravoslavno Horo," 159
 recordings by/repertoire of, 158
 S Ritŭma Na Vremeto, 159
 St. George's Day concert, 162
 Traditsiya, Stil, Nastroeniye, 159
 vs. Trakiya, 160

kanun (type of zither), 31

Kapchan, Deborah, 4–5, 295n4, 296n7

Kaplan, Ori, 280–81, 282, 283, 286–87, 330n31

Karafezieva, Maria, 143, 154, 160, 289, 313n18, 321n5

Karlov, Boris, 165
Karo, Stephane, 250, 256, 272, 283, 324n13, 326n5, 326n7
Kasamov, Victor, 320n35
Kaufman, Nikolai, 140, 157, 312n12
kaval (type of flute), 23, 31
Kayah, 328n22
Kazakov, Radi, 136
Keba (Dragan Koyć): "Idem Idem Dušo Moja," 304n17
Keil, Charles, 328n21, 329n29, 333n50
"Kemano Bašal," 50, 305n20
kemene/gŭdulka, 25
Kenrick, Donald, 50
Keranova, Nedyalka, 133, 162, 318n7
kerava bijav (putting on a wedding), 86
Kertesz-Wilkinson, Iren, 26
Khajuraho (North India), 191, 319n29
Khalnayak (film), 191
Khamoro festival (Prague), 163, 242, 324n3
King of the Gypsies (Maas), 67
kinwork, 86
Kiossev, Alexander, 195, 320n38
Kirilov, Kalin, 132, 136, 145, 233, 302n23, 319n22
Kirshenblatt-Gimblett, Barbara, 5, 41–42, 248, 292
kissing hands of elders, 91, 99
Kisslinger, Jerry, 233
Kitić, Mile, 301n17
Kjuperlika, 118
Klezmer All Stars, 235, 270
Kličkova, Vera, 171, 316n24
Know the Ledge, 283
Kočani Brass Band/Orkestar, 263, 272, 277, 282, 284, 287, 289
"Erdelezi," 328n22
Harmana Ensemble and Kočani Orkestar, 326n5
Kočani Orkestar meets Paola Fresu and Salis Antonello, 326n5
The Ravished Bride, 325n2
köçek (professional male dancer), 107, 112
Kočo Racin, 68, 103, 171
Kodály, Zoltán, 300n4

Koev, Kristian, 190
Kolev, Nikolai, 144–45
Kolev, Todor, 302n20
Kolo (Serbian State Ensemble): *Vranje*, 116–17, 310n5
Kolpakov, Sasha, 256, 262
Kolpakov, Vadim, 266, 287
Kolpakov Trio, 235, 243, 269
Kondyo, 318n9
Konushenska Grupa, 134, 139, 157
Kopanari, 25
Koprivshtitsa festival, 167
Kosovo Roma, 13, 242–43, 298n31, 327n15
"Koštana," 207
"Kote Isi Amalalen," 94, 95
Kotel school, 30, 144
Kracholov, Aleksandŭr, 163
Kristal, 34, 156
"Bashtinata Kŭshta," 180
"Bŭlgarina v Evropa," 181
"Chudesen Sŭn," 180
"Dobro Utro," 180
"Karavana Chayka," 180
Maika India, 304n16
Maria's collaboration with, 185
"Ne Smenyai Kanala," 185
"Sladka Rabota," 181
"Svatba," 180
Vsichko e Lyubov, 180–81
"Zvezditse Moya," 180
Kristali, 34, 166, 330n34
"Godzila," 271–72, 326n6
"Ne Smenyai Kanala," 185
"Red Bula," 283
"Telefoni," 185
transnational reputation of, 271
krivo horo, 132, 312n11
Kronos Quartet, 326–27n10
Krst Rakoc (film), 322n11
Krŭstev, Angel, 128
KUD (Kulturno Umetničko Društvo), 68, 101, 118–19, 170, 306n12
See also Phralipe
Ku-Ku, 172
Kulturno Umjetnčko Društvo. *See* KUDs
Kultur Shock, 264, 280

Kŭneva, Radostina, 156
Kunstler, Agnes, 321n5
Kurban Bajrami (festival of
    sacrifice), 83
"Kŭrdzhaliisko Horo," 146
Kŭrdzhali school, 30, 135
Kurkela, Vesa, 178, 180
Kurtiš, Tunan, 100, 211
Kurtov, Samir, 24
Kusturica, Emir, 242
    Black Cat, White Cat, 324n2
    Time of the Gypsies, 258, 277,
        324n2, 332n47
    Underground, 275–76, 324n2
Kutev, Filip, 129
Kuti, 30
kyuchek. See čoček/kyuchek

Laço Tayfa, 36
Lady Gaga, 190
Ladysmith Black Mambazo,
    328n21
    Graceland, 276, 328n19
Lambov, Slavcho, 324n10, 326n9
Landler, Mark, 298n32
Lassiter, Luke, 15
Latcho Drom (T. Gatlif), 49, 151,
    242, 243–44, 247, 258–59,
    260, 275
Lawless, Elaine, 15
Lazarevska, Tatjana, 179
Lazarov Records (Bulgaria),
    34, 152
Lee, Ken, 297n18
Lemon, Alaina, 122, 257, 268,
    303n3
Lenovska Grupa, 139, 141
lesbianism, taboo against, 190
Lesno (a line dance), 114, 115
Leviev, Milcho, 135
Levski, Vasil, 190
Levy, Claire, 195–96, 315n16
Lindsey, Traci, 318n10
Linnekin, Jocelyn, 53
Lipsitz, George, 279
lip synching, 157, 189
Liszt, Franz, 299–300n4
Livni, Eran, 166–67
Lolov, Ibro, 27, 33, 165,
    301–2n20
Lomax, Alan, 329n30
London, Frank, 216, 235, 270
Lucassen, Leo, 304n15

Lucerne festival (1995), 242, 244,
    246, 261
Lumanovski, Ismail ("Smajko"),
    102, 103, 104–5, 235, 266
Lumanovski, Remzi, 104
Luminescent Orchestrii, 267
Lynskey, Dorian, 286, 294

Maas, Peter: King of the Gypsies,
    67
Macedonia
    as Decade of Roma Inclusion
        signatory, 297n22
    European Union membership
        of, 6, 169
    independence of, 14, 306n11
    Indian culture in, 49, 304n16
    Macedonians vs. Albanians,
        298n24
    multiculturalism in, 169
    prejudice in, 213
    recording industry in, 33
    Romani political parties in,
        215
    Romani population in, 11,
        297n23
    Romani rights in, 11–12, 254
    Romani television stations in, 33
    UNESCO application of,
        170–72, 316n22, 316nn24–
        25
    virginity test in, 76
    See also Šutka
Macedonian Association of
    Romani Women, 215
Macedonian language, xiii–xiv
Macedonian National Radio and
    Television, 169
    See also Ballet Troupe of
        Macedonian Television
Mačev, Bilhan, 36, 100, 211
Madonna, 3, 182, 190, 266, 269,
    274
Maestro (television program),
    120
mafia, 34, 155, 160, 181, 229–30
Magazin, 216
Magdolna, Ruzsa: "Erdelezi," 277
Magneten, 242
Mahala Rai Banda, 262, 269
    Electric Gypsyland 2, 325n2
    Ghetto Blasters, 325n2
    "Red Bul," 283

Mahmud, Muzafer, 89–90, 308–9nn15–16
Mahmut, Oskar, 323n22
Maia Meyhane (East Village, New York City), 103, 235, 265
Majovci clan, 171
makams (Turkish-derived modes/scales), 27–28, 31, 301n13
Malakov, Mladen, 136
Malikov, Anzhelo, 27, 33–34, 130, 138, 151, 164–66, 314n14
Malikov, Yashar, 27, 33, 130
Malina: "Ne Se Sramuvan," 194
Malkki, Lisa, 41
Malvinni, David, 27
Mamudoski, Sal, 103, 232–33, 235, 289
manele (music), 194–95, 236, 245, 262–63, 272, 273, 326n7, 327n11
manele/mahala (dance), 116, 310nn3–4
mane/taksim, 27–28, 180, 209, 210
manglardi čhaj (asked-for girl), 74
Maqellara, Haxhi, 232
the marginal as exotic, 248
Maria (chalga star), 182, 183, 185
Maria Theresa, Empress, 8
Marinova, Sofi ("Romska Perla"), 168, 172–73, 184, 185–188, 270, 304n16, 316–17nn28–29
    5 Oktava Lyubov, 186
    "Ah Lele," 188
    "Bate Shefe," 186–87
    "Buryata v Sŭrtseto Mi," 186–87
    "Danyova Mama," 172–3, 185–86
    "Dželem Dželem," 188
    "Edinstveni," 172–73, 186, 319nn19–20
    "Edin Zhivot Ne Stiga," 193–94
    and the Eurovision scandal, 186, 187
    "Lyubov li Be," 186–87
    "Lyubovta e Otrova," 187, 319nn21–22
    "Mik Mik", 188
    "Moy si Dyavole," 186–87
    Ostani, 187
    and Slavi Trifonov, 172–73, 187
    Studen Plamŭk, 187
    "Tochno Ti," 186–87
    "Ušest," 186
    "Vasilica," 188

"V Drug Svyat Zhiveya," 187
"Vinovni Sme," 187
Vreme Spri, 188
"Vyatŭr," 187
markets, postsocialist, 149–163, 286
Marković, Boban, 25, 188, 270, 275, 282, 301n10
    on Bregović, 276
    Devla, 325n2
Marković, Branko: Vranje, 116–17, 119, 310n5
Marković, Marko: Devla, 325n2
marriage, 71–74, 306nn13–15
    See also intermarriage, weddings
Marushiakova, Elena, 12, 48, 298n31, 305n21, 307n3, 315n16
Massey, Douglas, 40
Masterpiece of Oral and Intangible Heritage of Humanity, 170–72, 316n22, 316nn24–25
Matras, Yaron, 8, 48
Mavrovska, Dafinka and Dragica, 206
Mefailov, Vehbi, 100
Megasztar (Hungary), 174
Mehanata (New York City), 103, 232–33, 235, 265, 267, 281, 288–89
meter
    2/4 meter, 29, 32, 112, 114, 136, 158–59, 177, 180, 217, 312n11, 326n7
    4/4 meter, 29, 49–50, 177, 305n20, 326n7
    5/16 meter, 312n11
    7/8 meter
        for čoček, 112, 114
        fast, light forms of, (lesno), 28, 114
        folklore evoked by, 180
        on Planeta Folk, 162
        rŭchenitsa, 132, 312n11
        slower forms of, 28
        variations of, 29
    9/8 meter, 28, 29, 112, 114, 136, 158–59, 177, 301n14
    10/8 meter, 231
    11/16 meter, 312n11
migration
    difficulties of, 13, 298n32

linear, 243, 258–59, 291
narrated in song, 59–60
to the United States, 13–15,
    299nn33–34 (*see also*
    Belmont)
See also diaspora, hybridity, and
    identity; *Latcho Drom*
Milchev-Godzhi, Georgi:
    "Edinstveni," 172–73, 186,
    319n19
Milena (Bulgaria), 34
Milev, Atanas, 311–12n8
Milev, Ivan, 136, 225–26, 227, 233,
    301n10, 312n12
Milivojević, B., 304n17
Miller, Carol, 71, 306n7, 306n9
millet system, 8
minstrel shows, 330n35
Minune, Adrian ("Adrian the
    Wonder Boy"), 319n26
Mirga, Andrzej, 47, 304n13
Mirković, Dragana, 301n17
    "Sama," 192
Miro: "Gubya Kontrol Kogato," 194
Miss Roma International contest,
    170
Mitev, Delcho, 158
Mitsou: "Erdelezi," 328n22
Mixolydian scales, 305n19
Mladeshki Tants (young person's
    dance), 150–51
Mladi Talenti, 33
Mladost, 136, 146, 225
modern dance, 32–33, 119–20, 210
modernity
    alternative, 238
    amplified music associated
        with, 131, 248–49
    and authenticity, 54, 251–52
    and chalga, 159
    and cosmopolitanism, 44
    and identity, 44, 81
    and tradition, 55, 247–48
Modern Quartet, 284
modesty, 109–10, 118, 120, 123,
    202, 218–19
monkey leaders, 25
Monson, Ingrid, 329n28
Montenegro, 297n22
Movement for Rights and
    Freedoms (MRF; Turkish
    party; Bulgaria), 298n25
MTV, 180, 318n8

MTV 2 (Macedonia), 33
Muabet Bez Parsa (radio
    program), 34
multiculturalism
    commercial manipulation of, 43
    Disney version of, 45
    by music promoters, 252–53
    Romani diversity, 47
    selective, 71, 306n7
Müren, Zeki, 191, 319n28
Musafir/Maharaja, 235, 243–44,
    259–60
Musa Mosque (Belmont, Bronx,
    New York), 65–66
music idol contests, 172–75,
    316–17nn27–29, 317n31,
    317n33
Musiciens du Nil, 260
Muskat, Tamir, 282, 330n31,
    330–31n38
Mustafa, Neždet, 298n24
Mustafov, Ferus, 100, 119, 120,
    133
    "Tikino," 32
Mustafov, Ilmi, 32
Mustafovska, Eleonora, 217
mutresa (well-kept woman),
    318n11
Muzafer, Abas: "To Phurano
    Bunari," 60
Le Mystère des Voix Bulgares,
    150

"Naktareja Mo Ilo Phanlja," 28,
    217, 323n22
nardonosti (nationalities), 10,
    297n19
narodi (nations), 10, 297n19
Nasev, Nick, 187–88, 320n34,
    322n14
našli čhaj (runaway girl), 74
Nasmi'ler, 34
National Committee on Ethnic
    and Demographic Affairs
    (Bulgaria), 166
National Council for
    Cooperation on Ethnic
    and Demographic Issues
    (NCCEDI; Bulgaria), 12
National Ensemble of Folk Music
    and Dance (Bulgaria), 129
Native American filmmaking,
    307n20

Native American identity, 56
Nay-Dobri Kyuchetsi ot Mahalata, 183
NCCEDI (National Council for Cooperation on Ethnic and Demographic Issues; Bulgaria), 12
N'Dour, Youssou, 238
Neascu, Nicolae, 236, 256
Nelina, 158–59, 162, 182
Neshev, Neshko, 30
    on Gypsy Caravan tour, 233–34, 326n8
    and Papazov, 135
    recordings of, 160
    stature of, 136
    in Trakiya, 134, 136
    and the Yuri Yunakov Ensemble, 233
Neshev, Neshko, 30
    in Kanarite, 158, 314n8
    Kŭrdzhali style of, 30
Neumann, Helmut, 237, 250, 270, 324–25n13, 326n7
Newsweek, 269
New York Black Sea Roma Festival, 235
New York City
    blaga rakija banquets in, 99, 310n28
    Herdelezi celebrations in, 84
    Macedonian Roma in, 13–15, 299nn33–34 (see also Belmont)
    videotaping in, 79–80
    weddings in, 99, 308n12
New York Gypsy Festivals, 105, 163, 235, 243, 265–67, 330n32
    See Belmont
New York Times, 264
NGOs (nongovernmental organizations), 11, 12, 167–69, 168
Nieuwkerk, Karin van, 321n7
nihavent, 27–28
nomadism, 8, 296n14
nongovernmental organizations. See NGOs
Nonini, Donald, 42, 268
Northern Romani dialect, 296n13
northern style of Romani clarinet playing
nostalgia 30, 45, 246

Novi Sad (Vojvodina, Serbia), 115, 245
novokomponovana narodna muzika (newly composed folk music), 25–26, 177, 178–79
Novoselsky, Valery, 51
NY Gypsy All-Stars, 105

Obama Kyucheck, Barack, 30, 301n15
O Clone (television program), 119
Ogneni Ritmi (Sofia, Bulgaria), 130
"Oj Borije," 28, 94, 97
Okely, Judith, 5, 218–19, 304n15
Okka, Sali, 34, 166
Olah, Ibolya, 174
"Olimpiada," 30, 133
OMFO: Trans Balkan Express, 283–84
Ong, Aihwa, 42, 258, 268
Ongeni Momčinja, 33
"Open Heart" children's festival, 169
Open Society Institute, 11, 76, 166, 168–69, 315n20
Orce Nikolov, 68, 103
Ordulu, Gamze, 235
Orfei, 136, 154–55, 157–58, 162
    Mafia ot India, 304n16
Organization for Security and Cooperation in Europe, 11
Orkestŭr Plovdiv, 314n7
Orkestŭr Universal, 34, 36–37, 271, 326n6
oro, 114, 115
Ortner, Sherry, 7, 145, 228, 237
Oruçi, Vera, 232
otherness, 116, 121, 123, 245, 258, 282
Ottoman dancers, 107, 108, 112, 116, 123, 191, 209, 218, 327n211
Ottoman Empire, 8, 24, 31, 44, 107, 108, 195, 223, 226, 258

Paicho, 154
paidushko (a line dance), 132, 312n11
Pamporovo, Alexey, 315n16
Pamukov, Orlin, 136, 154
Panair/Fairground, 314n4

pan-Romani human rights movement. *See* Romani rights movement
Papazov, Ivo (*formerly* Ibryam Hapazov), 16, 17
arrest/imprisonment of, 142–43, 145
awards received by, 325n1
BBC Radio 3 award won by, 161
bitterness of, 161–62
and Boyd, 149–50, 161
on capitalism and the economic crisis, 153
"Celeste," 152, 314n4
on chalga, 156
clarinet played by, 235, 266
on corruption, 155–56
on critics of wedding music, 157
Dance of the Falcon, 161
on democracy vs. socialism, 155–56
ethnicity of, 141–42
ethnic variety/tastes among patrons of, 301n16
gaida imitations on clarinet, 132
and Gogol Bordello, 235
on Gypsy Caravan tour, 233–34
on Gypsy Punk music, 266
influence of, 101
Kŭrdzhali style of, 30
legends surrounding, 135–36
Anzhelo Malikov on, 164
musical family/upbringing of, 135, 141, 312n16
"A Musical Stroll Around Bulgaria," 132
and Ferus Mustafov, 32, 133
name change of, 142
on the NATO concert, 161
in the New York Gypsy Festival, 266
on Payner, 182
"Pinko," 30, 133, 300–301n10
police evasion by, 143–44, 161
popularity in Bulgaria, 161–62
popularity/stature of, 134–36
recordings of, 146–47, 160, 313n26
reunion tours with Yuri Yunakov, 161
robbed at gunpoint, 155
and Salieva, 108
and Salifoski, 105

self-censorship by, 147
Song of the Falcon, 301n10
on Stambolovo festivals, 152
on state constraints on music, 141
Together Again: Legends of Bulgarian Wedding Music, 161, 233
on tour, 151, 161
in Trakiya, 134, 136, 160
wage labor resisted by, 139
on wedding music, 141, 143, 156–57, 312n12
and Yuri Yunakov, 225, 227, 233–34
Paradox Trio, 105
Pareles, Jon, 329n30
parsa (tip collection), 138
Party for the Complete Emancipation of Roma (PCER; Macedonia), 170, 298n24
Pasalan, 250
Paskov, Dimitŭr, 30, 136
Paskova, Poli, 318n14
"Moiite Pesni," 162
pativ (respect) vs. ladž (shame), 109–10
patriarchy, 74, 77, 85, 88, 94, 109–10
Payner (Bulgaria), 34, 152, 155, 157–58, 162–63, 179, 182–83, 318n11
See also Planeta Folk, Planeta Prima, and Planeta TV
pazari za muzikanti (musicians' markets), 137–38
PCER (Party for the Complete Emancipation of Roma; Macedonia), 170, 298n24
peasantry, marketing of, 249–50, 324n8
pečalba/gurbet (working abroad), 59, 103, 305n1
Pengas, Avram, 231
pentatonic scales, 305n19
Pepper, Art, 134–35, 312n15
performance
Judith Butler on, 5, 192, 194, 295n5
definition of, 4
framework of, 4–5
and gender, 5, 192, 295n5

performance (*continued*)
and identity, 4–5, 41
Kapchan on, 4–5, 295n4
as power, 228
subjectivities created via, 5,
296n7
Petrova, Dimitrina, 9
Petrović, Aleksandar: *Skupljači
Perja*, 50
Petrovski, Trajko, 10, 297n20
Pettan, Svanibor, 26, 31, 108, 111,
119, 202, 304n17, 327n15
Peycheva, Lozanka, 26, 30, 33,
114, 166, 302n21, 314n14,
315n16, 317n4
"Phirava Daje," 30
Phralipe, 32, 68, 103–4, 118–19,
170, 203
Phrygian scales, 27–28, 180, 209,
210
"Pinko," 30, 133, 300–301n10
pipiza. *See* zurna/zurla
Piranha, 270, 326n4
Pirin Ensemble, 128
Pirin Folk/Fest, 177, 317n3
Pirin Pee folk festival, 128,
129, 167
Planeta Folk (cable television
channel), 162–63
Planeta Prima, 317n1
Planeta TV (Bulgaria), 34, 182–84,
318n16
Plovdiv (Bulgaria), 137, 139,
143–44, 157, 158, 313n28,
314n5
Plovdiv Academy, 132, 144–45,
167, 302n23, 315n17
Plovdiv Folk Jazz Band, 149,
313n1
Pointon, Matt, 318n13
political parties, Roma, 11,
298nn24–25
Pomaks (Bulgarian Muslims), 170,
215, 224, 311n3, 315n19
Popaj, Alfred, 103, 232–33, 235
pop/folk music. *See* chalga
Pop Idol competition, 173–74
Popov, German, 283–84
Popov, Vesselin, 12, 48, 305n21,
307n3, 315n16
Popstar Alaturka contest, 173
Popularni Trakiiski Klarinetisti,
313n26

pornography, 121, 182, 311n8
porrajamos (Holocaust), 48
See also Holocaust
postmodernists, 274, 329n27
postsocialist markets/festivals, 19,
149–75
Povinelli, Elizabeth, 5
Prashtakov, Todor, 146
pravo horo, 132, 144, 312n11
Prespa Albanian Macedonians,
101, 307n7, 321n5
Prilep (Macedonia), 14, 62, 69–70,
309n21
privatization, 5–6, 166
Proeski, Toše, 322n14
"Erdelezi," 218
"Magija"/"Čini," 217
prostitution, 182, 318n11
Protestant work ethic, 9
prvić (first visit of bride to her
natal home), 95
public vs. hidden transcripts,
228–29
purity
ghettoization of, 46
of Gypsy music, 164–66,
285–86
vs. hybridity, 42
myth of, 303n9
rhetoric of, 129, 140–45
See also authenticity
Pŭrvanov, Georgi, 161, 165
Pŭrvomayskata Grupa, 312n8
*Putevima Pesme: Esma Ansambl
Teodosievski*, 322n11
Pygmies, 332n48

Qur'an, 311n4

racial profiling, 11
racism toward Roma, 6, 69–70,
168, 173, 196–97, 207,
320–21n39
Radev, Petko, 157, 158, 302n20,
313n26
Radić, Indira, 179,
304n16
Radio Signal Plyus, 34, 153
Radio Skopje, 31, 93, 203, 204,
206–7, 308n16
Radio Veselina, 34, 153
Radulescu, Speranta, 273
Rahmanovski, Džengis, 103

Rahmanovski, Ilhan, 102, 103, 105
Rajasthani music, 49, 236, 243, 259–61, 326n3
Rajasthani Roma, 247, 259–61, 326n3
Ralchev, Petŭr, 136, 143, 154, 155, 157, 227, 302n20
Ramadanov, Zahir, 211, 212, 249
Ramazan (fasting month), 83
Ramko (Ramadan Bislim): "O Gurbetluko," 59–60
"Ramo Ramo," 49, 304n17
Ranger, Terence, 53
Rao, Pratima, 327n12
rap music
  and the authenticity of the ghetto, 248
  and chalga, 186–87, 319n20
  in pop/folk videos, 122
  in Romani, 26, 34–35
  stereotypes used in, 302n24
Rašidov, Eljam, 211
Rasimov, Enver, 211, 322n10
Rasmussen, Ljerka Vidić, 25–26
Raznatović, Ceca, 179
reciprocity, 79, 93
Recommendations on the Safeguarding of Traditional Culture (UNESCO), 170
"Red Bul," 283, 330n34
red color, 92
Redžepova, Esma, 20, 32, 201–19
  artistic control by, 214, 249
  on assimilation, 252
  on authenticity, 214–15, 248
  awards received by, 215
  "Bašal Seljadin," 206, 208, 209–10
  "Bel Den," 216
  on belly dancing, 120
  on Bregović, 275–76
  "Čhaje Šukarije" 31, 35, 209–10, 275–76, 290, 322n11
  Chaje Shukarije, 216
  "Ciganski Čoček," 209–10
  collaborations by/current directions of, 216–19, 270, 323n22
  criticism of, 248
  dancing by, 119, 120, 208–9, 210

"Dani Su Bez Broja," 216
dimije worn by, 204, 208–9, 217, 251
"Dželem Dželem," 50, 51, 169, 211
early years of, 202–5
emotion used by, 208–9
in Esma—Ansambl Teodosievski, 31, 205, 209, 211
Esma's Band, 217
Esma's Dream: Esma and Duke, 216
family of, 202–5
and Fanfare Ciocarlia, 262
in films, 209–10, 257, 322n11
in the Gypsy Caravan tour, 243, 262
Gypsy persona of, 46
"Hajri Ma Te, Dikhe, Daje," 208
Indian ties of, 211
"Kolku e Mačno em Žalno," 210
Legendi na Makedonska Narodna Pesna, 216
"Ljubov e," 216
"Magija"/"Čini," 217
"Makedo," 210
and Marinova, 186
Mon Histoire, 216
multiethnic socialist agenda embodied in her music, 46
"Naktareja Mo Ilo Phanlja," 28, 217, 323n22
overview of success of, 201
patriotism of, 324n9
performances with her husband, 321n5
Pesmom i Igrom Kroz Jugoslaviju, 210
as a pioneer, 207, 209, 214, 218
and Proeski, 311n9
politics/humanitarianism of, 212–16, 217, 218, 322n14
as Queen of Romani Music, 201, 211, 214
racism/prejudice faced by, 46, 321–22n8
"Raise Up Your Hand," 216
relationship with Stevo, 204–6, 209–10
respectability of, 109

Redžepova (*continued*)
  retirement of, 217
  Romani identity of/songs sung
    in Romani by, 206–7
  Romani reception of, 205
  "Romano Horo," 210
  at Romska Vasilica, 202
  on segregated dancing, 110
  stereotypes in her videos,
    121–22, 311n9
  stereotypes of her by the press,
    207, 321–22n8
  and Stevo's music school,
    211–12, 322n13
  style and image of, 205–11,
    214–15, 321–22n8, 322n11
  Šutkafest involvement of,
    169–70
  synthesizer used by, 248, 252
  as visiting artist in New York,
    100, 101
  wedding performances by, 93
  on women's čoček dancing,
    113
  Yugoslavian and other Balkan
    songs sung by, 210–11
Reinhardt, Django, 328n24
representation in fieldwork and
    writing, 7, 15–17
resistance, everyday forms of, 147,
    228–29
Rice, Timothy, 132–33, 314n8
The Riches (television program),
    67
Rifati, Šani, 205, 213–14, 217,
    234–35, 242, 248–49,
    254–55, 287, 292
  See also Voice of Roma
Ristić, Dragan, 255, 292, 325n14
Ristić, Dušan, 251, 255, 277, 292,
    325n14
  *Rromani Songs from Central
    Serbia and Beyond*, 324n11
ritual, 86. *See* celebrations,
    circumcision, Herdelezi,
    weddings
Rjabtzev, Sergey, 280
Robert Browning, 236
Robin, Thierry ("Titi"), 216, 325n1
Rofel, Lisa, 246
Roma, Balkan, 3–18
  and African Americans, 35,
    45, 122

conceptions/stereotypes of, 3,
    9, 207–8 (*see also under*
    Gypsies)
diasporic communities of (*see*
    Belmont; Šutka)
dichotomous conceptions of, 3
discrimination against, 8, 10–
    11, 47, 69–70, 152, 206–7,
    255–56, 292, 297n21
Eastern Orthodox, 9
educational integration of
    Romani children, 168,
    315n20
vs. Gypsies, 295n1
historical and political overview
    of, 7–13, 296nn13–15,
    297nn18–19, 297–98nn21–
    25, 298nn31–32
homophobia among, 306n13
Indian origins of, 39, 40, 49–50,
    55, 211, 304nn15–17,
    305n19
vs. Kalderash Roma, 66, 306n7
marginalization/poverty of, 11,
    45–46, 218–19, 254, 274,
    292, 331–32n46
media emancipation of, 302n21
migration of, difficulties of, 13,
    298n32
migration to the United States,
    13–15, 298n32, 299nn33–34
    (*see also* Belmont)
multilingual, 9, 297n17
Muslim, 8–9, 13
nomadic vs. sedentary, 8, 10, 22,
    47, 111, 259, 296n14
oppression of/xenophobia
    toward, 6, 8, 9–11, 13,
    297n21 (*see also* Holocaust)
orientalization/exoticization of,
    9, 10, 244, 292, 297n18
overview of, 3–7
Pentecostals' interest in, 315n19
population figures for,
    European, 11
poverty-stricken majority vs.
    successful musicians,
    overview of, 4, 45
racism toward, 6, 69–70, 168,
    173, 196–97, 207, 320–
    21n39
representation in fieldwork and
    writing, 7, 15–17

romanticization of, 9, 291–92, 300n4

scholarship on, 10, 315n16

self-identification of, 306n9

as slaves, 8

under socialism, 10

tensions with other Balkan ethnic groups, 101

use of term "Roma," 255, 295n1

violence against, 6, 10–11, 152, 229–30, 245–46

See also Gypsies

Roma, Bosnian, 14

Roma, Bulgarian, 12–14. *See also* heritage, Romfest, wedding music, Yunakov, Yuri

Roma, Greek, 303n3

Roma, Kalderash, 13, 16, 51, 66–67, 306n7, 308n11, 309–10n26

Roma, Kelderara, 122, 257

Roma, Russian, 122, 303n3

Roma Civil Rights Foundation Awards, 174

Roman, Sasho, 166, 314n14

"Oy Sashko," 304n17

Roma National Congress, 11

Romane: International Magazine for Romani Culture, Literature, and Art, 315n15

Romane Merikle/Roma Beads, 27

Romania, 6, 8, 297n22

Romani Baht Foundation, 301n11, 316n28

Romani culture

and Bulgarian culture, 165

and folklore, 5

folklore taught in schools, 169, 315–16n21

and high art, 169–70

NGO support for, 168–69

and postsocialist agendas, 5–6

Romani Iag festival (Montreal), 243

Romani language

chalga in, 185, 187

dialects of, 7–8, 48, 296n13

dictionary of, 48

in films, 258

as an Indo-Aryan language, xiii, 7–8, 304n15

literary, 48

in Macedonia, 11

transliteration of, xiii–xiv

Romani music, 18, 21–37

2/4 meter, *29*, 32, 112, 114, 136, 158–59, 177, 180, 217, 312n11, 326n7

4/4 meter, *29*, 49–50, 177, 305n20, 326n7

9/8 meter, 28, *29*, 112, 114, 136, 158–59, 177, 301n14

27, 138

Albanian style in, 35

Americans who play Balkan music, 105–6

amplified, 32, 36

as appropriated, 274, 286, 327nn14–15 (*see also under* collaboration, appropriation, and transnational flows)

and Balkan historical threads, 23–27

in beauty contests, 120

brass bands, 25, 300n8 (*see also* Boban Marković)

Bulgarian, history of, 33–34, 301–2n20, 302nn22–23

in cafes, 115–16, 245

čoček/kyuchek, 27–28, *29*, 30, 301n15

concerts after the fall of socialism, 33–34, 302n21

in elementary schools, scandal regarding, 168

ethnic variety among patrons of, 26, 300n9, 301n16

in fusion genres, 5–6, 18, 166

at the Galičnik wedding, 170–72, 175, 316n23

history of Romani female musicians, 202, 218, 321nn2–5

homelands evoked in, 40

Hungarian, 22, 299–300n4

improvisation/innovation in, 26–27

and Indian scales, 49, 300n6, 305n19

innovation/hybridity valued by audiences, 164, 292

instruments used in, 23, 31–32

Kalderash, 27

Kosovo-style, 36, 49

in Macedonia, 50, 120, 169–70

makams, 27–28, 31, 301n13

Romani music (*continued*)
  mane/taksim, 25, 28, 180, 184,
      193, 301n13
  by non-Roma, 275, 327n16
  *novokomponovana narodna
      muzika*, 25–26, 177, 178–79
  oral, 30
  overview of, 21–23
  piracy of, 34
  as polyethnic/-lingual, 26
  popular, 25–26
  popularity in the West, 269,
      325n2
  pure Gypsy music, 164–66
  on radio, 34, 153
  recordings of, generally,
      32–34, 302n22 (*see also
      specific recordings*)
  resistance by musicians, 7
  resistance songs, 50
  scales used in, 27–28, 49,
      301n12
  scholarship on, 21–23
  Sofia-based, 33, 151
  song variants/versions, 30–31,
      301n17
  stereotypes used strategically
      by, 7, 292 (*see also under*
      Romani music as world
      music)
  stylistic trends in, 31–37,
      302nn23–24
  talava songs, 36–37, 60, 191,
      302n25
  technique/passion in, 26–27,
      300–301n10
  on television, 301–2n20
  themes in song texts, 50
  tours, 293–94 (*see also
      specific tours*)
  as traditional, 214–15
  turbofolk, 177, 178–79, 319n22,
      319n27
  urban folk, 22, 25–26
  U.S. reception of Bulgarian and
      Macedonian musicians, 7
  videos, 34
  village folk, 25–26, 131, 272
  vitality of, 36–37
  See also meter; chalga;
      collaboration,
      appropriation, and
      transnational flows;
      pop/folk music; wedding
      music; *and individual
      instruments and musicians*
Romani music as world music,
    20, 241–68
  caravans, nomadism, and
      Romani unity, 258–63, 268,
      325n16
  and fashion, 254
  festivals/tours, generally, 242–
      43, 262, 267–68, 293–94,
      324n3 (*see also specific
      festivals and tours*)
  marketing exoticism and
      authenticity, 241, 244–52,
      257, 293
  and New York Gypsy Festivals,
      265–67
  overview of, 241–44, 324n2
  and self-stereotyping, 257–58,
      325nn14–15
  world music events as education
      vs. entertainment, 252–57,
      324n11
  See also Gypsy Punk music
Romani Music Festivals. *See*
    Romfest, Šutkafest
Romani rights movement
  emergence of, 15
  and the European Union, 11
  growth of, 292
  and identity politics, 44, 47–48,
      304n11, 304nn13–14
  music in, 49–51, 304–5nn16–22
  national symbols of, 48
  and NGOs, 11
  and the Roma label, 238,
      304n11
  and stereotypes, 173
  and unity, 325n16
Romani Routes (booking
    company), 254–55, 326n9
Romaniya, 122
Romano Sumnal, 170
Romano Suno 2, 21
Romanov, Manush, 129
Roman oyun havası, 301n14
Roman Star contest, 317n31
Roma Portraits (Balicki), 84,
    307n4, 307n20
Roma Reggaeton Hip Hop,
    26
Roma Rights, 297n21

Romashka, 267–68
Roma TV, 172
Roma Variations, 233
Romen Theater (Moscow), 257
Romfest (Stara Zagora,
    Bulgaria), 34, 122–23, 152,
    163–67, 185, 189, 311n11,
    314nn13–14
Romska Gaida (a line dance),
    114
Romska Ubavica (Most
    Beautiful Romani Woman
    contest), 33
Romska Vasilica (Romani
    St. Basil's Day), 33
Romska Veseliya, 122
Romski Boji, 106
Rosaldo, Michelle, 308n14
Rosaldo, Renato, 45, 246
Rouse, Roger, 41
Rozhen festival, 167
R Point, 50
Rromani Dives, 50–51, 202
rŭchenitsa, 132,
    312n11
Rudari, 25, 295n1
Runjaić, Živka and Jordana,
    206
Ruseva, Dinka, 139, 144, 154,
    155, 157

Safran, William, 39, 303n3
Said, Edward, 9, 244, 297n18
Sakip, Šadan, 100, 211
Šakir, Feta, 33, 92
Saleas, Vasillis, 36
Salieva, Zvezda, 108
Salifoski, Seido
    activism of, 106, 234
    family life of, 64–65, 71, 72,
        73, 105
    as a musician, 101, 102, 103,
        104, 105–6, 232, 289,
        310n32, 327n16
Salijević, Slobodan, 93
šalvari (wide, billowing pants), 87,
    91, 96
    See also dimije
Sambolovo festivals (1985–1988),
    148
Samson, Jim, 327nn14–15
Sandu, Florentina, 270
Sant Cassia, Paul, 247, 248

Sapera, Gulabi, 247
"Sapeskiri Čoček," 49
Saraçi, Sunaj, 232
"Sarajevo, 84," 30, 133
Šaulic, Šaban, 301n17
Savić, Tanja, 173–74
Savigliano, Marta, 258
Schafer, Murray, 274
schizophonia/schizophonic
    mimesis, 274
Scott, James, 147, 228–29
Seeman, Sonia Tamar, 108, 114,
    251, 301n14, 309n25,
    317n4, 323n2
Seibert, Brian, 253–54
self-orientalization, 258
Sellers-Young, Barbara, 116
Senaras, Müzeyyen, 191
Senlendirici, Husnu, 36, 235, 266
    Laço Tayfa, 251
Serbezovski, Muharem, 100,
    301n15, 304n17, 307n1,
    328n22
    "Bože Bože," 304–5n18
    popularity of, 305n18
    "Ramajana," 31, 49, 305n19
    "Ramo Ramo," 49, 304n17
    "Sine Moj," 94
Serbia, 25, 115–16, 297n22
Serbian language, xiii–xiv
Šerifović, Marija, 174
Serbian-Albanian conflict, 13
Ševćet, 26, 36
sexual consummation, 74, 75, 95
sexuality and dance, 109–12, 115
    See also belly dancers
sexualized female body, 121, 123,
    182, 191
shahnai (zurna-like instrument),
    191–92
Shakira, 182
shame, 109–10
Shantel (Stefan Hantel), 281–82,
    286, 325n1, 330n32
    Borat tour, 289
    Bucovina Club 1 and 2, 271,
        281, 283, 288
    on dance clubs as bringing
        people together, 288
    Disko Partizani, 281, 325n2
    Planet Paprika, 281, 325n2
    reception of, 284, 287
    response to critics, 331n44

Sharena Muzika, 168
Shay, Anthony, 116–17
Shepik, Brad, 105
Shiroka Lŭka, 144–45, 182, 315n18
Shopov, Rumen Sali, 103, 233
Shukur Collective: *Urban Gypsy*, 331n41
Siderov, Volen, 197
sieve (sita), ritual function of, 28, 91
Simeonov, Filip "Fekata,"
    northern style of, 30
    "Shalvar Kyuchek," 28
    stature of, 136
    and Taraf de Haidouks, 262–63, 272
    in Taraf de Haidouks, 314n7
Simon, Paul, 328n21, 329n30
    Graceland, 276, 328n19
Sinapov, Traicho, 136
"Sine Moj," 94
Sinti, 48, 295n1, 304n13
"Sitakoro Oro," 28
sivi gŭlŭbi, 133, 318n7
Skopje (Macedonia), 115, 215, 308n15
    Albanian influence in, 28
    bajraktari in, 309n21
    celebrations/festivals in, 32–33
    racism toward Roma in, 207
    Romani immigrants from, 14
    Romani spoken in, 62
    temana in, 94
    See also Šutka
Skopje, 63, So Sila Tatko (film), 322n11
Skupljači Perja (Petrović), 50
Slavchev, Prof., 315n17
Slavcho Lambov, 314n7, 324n10, 326n9
Slavic Soul Party, 264, 265, 327n16
"The Slavi Show" (television program), 172
Sliven (Bulgaria), 139–40
Slobin, Mark, 40
Slovenia, 297n22
smiruvanje (reconciliation) ceremony, 74
Smolyan Dance Ensemble, 158
socialist state, Bulgarian, 10, 16–17, 33, 119, 164, 227,

292–93, 296n8, 299n39, 302n21 (*see also* heritage and the Bulgarian socialist state)
Sofia (Bulgaria)
    Alley of Stars in, 161
    megaconcert in, 152
    musicians' market in, 137–38
    Romani music based in, 33, 151
    Romani NGOs in, 12
    Romani rights/living conditions in, 12
Soja, Edward, 42–43
Šoko, Robert, 281, 330n33
soldier send-off celebrations, 131, 311n7, 313n21
Sonneman, Toby, 311n9
S Orkestŭr Na Kanarite Na Svatba, 146
Spain, 297n22
Spears, Britney, 182
Spin, 331n39
Spivak, Gayatri, 51–52
Spur, Endre, 300n4
Stambolovo festivals, 146–47, 148, 152, 228, 314n3
Stan, Peter, 327n16
Stara Zagora (Bulgaria), 139–40, 154, 161, 168, 308n11, 314n5
Stara Zagora festivals. *See* Romani Music Festival; Romfest
starogradski pesni (old city songs), 183
state
    Bulgarian socialist, 10, 16–17, 33, 119, 164, 227, 292–93, 296n4, 296n8, 299n39, 302n21 (*see also* heritage and the Bulgarian socialist state)
    post-socialist, 296n8
Statelova, Rosemary, 315n16
Stefanovski, Vlatko, 322n14
Stephen, Lynn, 303n6
stereo zapis studio (tape recording studio), 147–48, 152, 313nn28–29
Stewart, Susan, 246
stick dancing, 300n6
Sting, 276
Stivell, Alan, 134, 312n15

Stoev, Atanas, 158–59, 314n8
Stokes, Martin, 7, 319n28
Stoyanova, Maria, 167, 313n20
strategic essentialism, 51–52,
    213–14
Struškite Svadbari, 101
Sugarman, Jane, 108, 115, 195,
    277, 304n17, 311n6,
    320n36, 321nn2–3, 321n5
Šukaripe, 122
sunet. See circumcision
Sunny Music/Sunny Records
    (Bulgaria), 34, 188–89
Sunrise Marinov, 183
Super Ekspres, 185
surla. See zurna/zurla
Šuteks (Šutka, Skopje,
    Macedonia), 308n10
Šutel (Macedonia), 33
Šutka (Šuto Orizari, Skopje,
    Macedonia)
    families of musicians in, 309n16
    music/dance as emblematic of
        Romani identity, 4
    Romani population in, 12
    settlement of, 12
    weddings in, 86–88, 308n10
Šutkafest (Skopje, Macedonia,
    1993), 50, 120, 163, 169–70,
    202, 324n3
Šuto Orizari. See Šutka
svatbarska muzika. See wedding
    music
svirki (flutes), 128
synthesizer, 35, 131, 237,
    248–49, 252

Takev, Ventsislav, 138
Takev brothers, 33
taksim. See mane/taksim
talava songs, 36–37, 60, 191,
    302n25
Talking Heads, 329n30
tambura, 132, 302n23, 312n10
Tanec, 101–2, 118
tapan/tŭpan
    in ensembles, history of, 23–24,
        128
    as exclusively male, 24
    at folk festivals, 167
    at the Galičnik wedding, 171
    in henna processions, 89
    improvisation on, 25

multipart playing of, 25
ritual function of, 24, 36, 89,
    308n15
tarabuka (hand drum), 31, 209
Taraf de Haidouks, 242, 247,
    263–64, 325n1, 326n7
    awards received by, 325n1
    Band of Gypsies, 250, 256, 262,
        272, 314n7, 324n8
    clothing/image of, 213, 236,
        249–50, 251, 256–57
    The Continuing Adventures of
        Taraf de Haidouks, 256, 272,
        324n8
    on the Gypsy Caravan tour, 213,
        235–36, 243, 255–56, 262
    and the Kočani Orkestar, 263, 272
    and the Kronos Quartet,
        326–27n10
    No Man Is a Prophet in His Own
        Land, 250, 254, 272–73,
        324n8
    and Simeonov, 262–63, 272
    and Winter, 246, 251, 255–57,
        272–73, 283, 324n13
Tarkan, 37, 105
Taussig, Michael, 333n49
Tavče Gravče, 93
Taylor, Timothy, 238
Tekbilek, Omar Faruk, 105
tel, 89
temana (hand gesture of respect),
    94, 97, 98, 309n22
Tenth Mediterranean Youth
    Festival (Akdeniz
    University, Turkey), 242
Teodosievski, Pero, 211
Teodosievski, Stevo
    on authentic Romani music,
        214–15
    "Bašal Seljadin," 206, 208,
        209–10
    death of, 211, 215
    in Esma—Ansambl
        Teodosievski, 31, 205, 209,
        211
    on Esma's leadership, 207
    as King of Romani Music,
        211, 214
    music school of, 211–12,
        322n13
    relationship with Esma, 204–6,
        209–10

Teodosievski (*continued*)
  Romani music promoted by,
    206, 214
  Šutkafest involvement of,
    169–70
  young drummers used by, 209,
    322n10
Ternipe, 119
Tetovo (a Romani KUD), 119
Tetovo (Macedonia), 94
Theodosiou, Aspasia, 303n3
thirdspace, 42–43
Thomas, Nicholas, 52–53
Thrace (Bulgaria), 33, 132, 133,
    137, 164
  See also Stara Zagora
Tia Juana, 262
Time, 269, 277
Time of the Gypsies (2001), 242
Time of the Gypsies (Kusturica),
    258, 277, 324n2, 332n47
Times Square Records/World
    Connection, 290
Titanik, 28, 33
Tito, Josip Broz, 10, 32, 116, 169,
    210, 215
Todorov, Manol, 139, 140, 146,
    157
Todorov, Todor, 312n12
Todorova, Maria, 195,
    297n18
Tomova, Ilona, 315n16
tradition
  and authenticity, 53, 247–48
  conceptions of, 54–55
  invented, 53
  and modernity, 55, 247–48
  oral, 137
Traditional Crossroads, 161, 233,
    323n9
Trakiya, 313n21
  Balkanology, 150–51, 160
  and Boyd, 149, 161, 229
  Fairground/Panair, 160–61, 234
  fees for weddings, 137
  founding of, 134
  "Gypsy Heart," 160–61
  improvisation by, 136
  vs. Kanarite, 160
  NATO concert by, 161
  Orpheus Ascending, 150
  police evasion by, 228
  popularity in Bulgaria, 161

popularity of, 134, 227, 229
Stambolovo festival success of,
    146–47
targeted by Bulgarian officials,
    142
on tour, 151, 161, 229
See also Papazov, Ivo
Trakiya Folk festival, 133, 152,
    314n5
transborder processes, 303n6
Transglobal Underground, 330n36
transliteration, xiii–xiv
transnational flows. *See*
    collaboration,
    appropriation, and
    transnational flows
transnationalism and hybridity,
    41–44, 303n6, *see also*
    hybridity
transnational minority, 47, 304n13
Trask, Haunani-Kay, 53
Travelers, 245, 295n1, 324n5
Traykov, Dzhago, 35
Trifonov, Dimitǔr, 146
Trifonov, Slavi, 189, 190–91
  "Edinstveni," 172–73, 186,
    319nn19–20
  and the Eurovision scandal,
    172–73, 316–17nn28–29
  influence of, 316n27
  "Ljubovta e Otrova," 187,
    319nn21–22
  Roma TV, 172
  "The Slavi Show," 172
  "Vinovni Sme," 187
  "Yovano Yovanke," 319n20
truba (trumpet or flugelhorn), 31
Trumpener, Katie, 9
Trǔstenik, 30, 157, 311n8
Tsigan, 48, 295n1, 318n13
Tsiganska Muzika, 122
Tsiganska Muzika (Gypsy Music),
    122
tsiganska rabota (Gypsy work),
    195
Tsintsarska, Rumyana, 149
Tupurkovski, Vasil, 215
turbofolk music, 177, 178–79,
    319n22, 319n27
Turkey, 173–74, 197, 317n31
Turkish culture
  as a mark of civilization, 8,
    44, 115

music and belly dancing, 231
translations of songs into Bulgarian, 311n6
Turks seen as Muslim fanatics, 196–97, 317n5
Turkish language, xiii, xiv, 8, 44
Tuysuzoglu, Tayik, 272
Tweed, Ras, 216–17
Tyankov TV, 162

ud (type of lute), 31
Ultra Gypsy, 275
Umer, Dževat, 102
Umer, Erhan ("Rambo"), 70, 98, 101–2, 232–33, 235
Umer, Husamedin ("Uska"; Erhan's son), 102–3
Umer, Husamedin (Erhan's father), 98, 101–2
Umer, Jusuf, 102
Umer, Sevim, 102, 235
Umer, Turan, 102
Umer, Vebi, 102
Underground (Kusturica), 275–76, 324n2
unemployment, 152, 155
UNESCO, 285
UNESCO world heritage applications, 149, 170–72, 316n22, 316nn24–25
Union for Romani Culture, 169
Union of Macedonian Folklore Ensembles, 171–72
Union of Roma (Macedonia), 298n24
Unison Stars (Bulgaria), 34, 152
United Democratic Front (Bulgaria), 298n25
United Kingdom, Roma in, 245–46
United Party of Roma (Macedonia), 298n24
U.S. State Department, 12
"Ustaj Kato," 94
Ustata
    and Azis, 194
    "Bate Shefe," 186–87
    "Buryata v Sŭrtseto Mi," 186–87
    "Lyubov li Be," 186–87
    "Moy si Dyavole," 186–87
    "Tochno Ti," 186–87, 319n20

Vagabond Opera: "Gypsies, Tramps, and Thieves" concert, 263
Valdes, Valentin, 318n9
Vali, 183
Van de Port, Mattjis, 115, 245, 247, 257–58, 279
Van Gennep, Arnold, 85–86
Vanjus, 284
Varga, Gusztav, 256, 260
Vasilica (St. Basil's Day), 83
Veliov, Naat, 326n5
Velkov, Saško, 211
verbunkos (Hungarian recruiting dance), 301n12
Verdery, Katherine, 146, 147
Verdonk, Anton, 248
Versace (Romani band), 33
Vesela, 158–59
Veseli Momci, 33
Veselina TV (Bulgaria), 34
Via Romen, 266
Vie dei Gitani (Ravenna, Italy, 2000), 242
Vievska Grupa, 157, 162
Visweswaran, Kamala, 16
Vlax Romani dialect, 296n13
VMRO-DPMNE (Macedonia), 215
Voice of Roma (VOR), 16, 213–14, 242–43, 254–55, 324n11, 326n9
    See also Herdeljezi festivals, VOR-sponsored
Volanis, Sotis: *Poso Mou Lipis*, 190, 319n26
VOR. *See* Voice of Roma
Vranje (V. Marković), 116–17, 119, 310n5
Vuorela, Ulla, 59

Wagner, Roy, 53
Weber, Alain, 259
Wedding and Funeral Orchestra, 278
wedding music, Bulgarian
    amplified, 32, 131
    Bulgarian ban on, 16–17, 19, 119, 131, 142–43, 146, 227
    Bulgarian instrumentation/ style/repertoire (1970s– 1989), 131–33, 141, 311– 12nn7–12

wedding music (*continued*)
  Bulgarian vs. Romani, 132
  vs. chalga, 156, 160, 162, *179*:
    See also čoček/kyuchek
    music; Chalga; cultural
    politics of postsocialist
  category system for musicians,
    139–40, 152
  comeback of, 162–63
  crimes against musicians,
    155
  criticism of, 131, 140, 157
  development of, 141
  eclecticism in, 132
  economic crisis's effects on
    musicians, 153–55
  economic framework of,
    137–40, 148, 313nn21–24
  vs. ensemble music, 138, 156,
    312n12
  fees for musicians, 99–100,
    154
  female singers for, 137
  and folk music, 132–33, 140,
    156, 160, 162, 167, 226,
    312n12, 315n18
  guest musicians, 93, 96
  history of, 131, 311–12n8,
    327n11
  improvisation/innovation in, 27,
    132, 301n10, 312n12
  learned in secret, 144–45
  as a male realm, 86, 137
  and nationalism/patriotism,
    156, 159–60, 162–63
  as an oral tradition, 137
  popularity of, 151, 156–57,
    227, 293
  on radio and television, 146–47
  recordings of, 146–48,
    313nn26–29
  self-censorship by musicians,
    147
  stars of, 16, 36–37, 136, 137, 294
  structure of, 132
  tipping for, 138, 313n24
  uses of, 131
  versatility in, 132
  See also čoček/kyuchek
    music; cultural politics
    of postsocialist markets/
    festivals; Papazov, Ivo;
    Stambolovo festivals

weddings
  American, average cost of, 99
  American customs at, 97–98
  in Belmont, 95–99, 309–
    10nn26–28
  brideprice, 88, 308n11
  bride's clothing, 90
  bride's crying, 90, 92
  bride's importance and
    transitional status, 85, 90,
    307n5
  bride's transfer to groom's
    home, 93–95, 97, 153,
    309n21, 309nn23–25
  čoček/kyuchek danced at, 115
  dance's role in, 86–87, 96–98,
    310n28
  economic crisis's effects on,
    153
  Esma and Stevo's wedding,
    205
  food preparation for, 87
  gift giving, 87–88, 92–93, 99,
    309n21
  homosexual, 189–90
  igranka (dance party), 91–93,
    309nn19–20
  invitations to, 92, 95, 309n20
  length of, 85, 96, 153, 307n7
  Macedonian, 85, 307n7
  midweek, 134, 309–10n26
  mother-in-law and
    daughter-in-law
    relationship after, 94,
    309n24
  Muslim vs. Eastern Orthodox,
    86
  order of the wedding week, 88,
    308n12, 309–10n26
  pan-Balkan structure of,
    85–86
  pregnant brides at, 84
  Romani and Turkish vs.
    Bulgarian, 153–54
  season for, 89
  segregated dancing/parties,
    110–11
  as status symbols, 133
  in Šutka, 86–88, 308n10
  women's clothing at, 87–88, 91
  women's vs. men's roles in,
    86–87, 98
  See also henna

Werbner, Pnina, 52, 302–3n2, 303n8
Western European Gypsy festivals, 242
When the Road Bends: Tales of a Gypsy Caravan (Delall), 213, 259, 327n17
Willems, Wim, 304n15
Winter, Michel, 246, 251, 255–57, 283, 326n5
  See also under Taraf de Haidouks
WIPO (World Intellectual Property Organization), 285
WOMAD (Belgium), 242
women
  freedom of movement of, 111
  hairstyles of, 309n19
  inside world of, vs. men's outside world, 111, 308n14
  kinwork by, 86
  modesty of, 109–10, 118, 120, 123, 202, 218–19
  objectification/commodification of, 121
  power/knowledge of, 71, 77–79, 88, 94, 109–10, 306n14, 307n19
  role in weddings (*see under* weddings)
  sexuality of, 109–10, 120–21
  subjugation of, 76, 306–7n17, 308n14
work ethic, 9, 63
World Bank, 11, 12
World Intellectual Property Organization, 170–71
world music and hybridity, 44–47, 151, 239
  See also hybridity; Romani music as world music
World Music Folklife Center, 243
World Music Institute (New York), 243, 258–59
World Village, 161
World War II Nazi extermination of Roma, 9–10, 48
  See also Holocaust

xenophobia, 3, 13, 246, 252–53, 268
Xenos, Nicholas, 14–15

Yambol Ensemble, 128
Yanamoto, Yohji, 250, 254
Yanev, Georgi, 136, 154–55, 157–58
Yaneva, Pepa, 154, 157–58, 182, 321n5
Yaneva, Tsvetelina, 157–59, 182
Yanitsa, 183
Yankov, Nikola, 139, 313n26
Yankulov, Stoyan, 160
Yiftos, 295n1
YouTube, 17, 34, 81, 271
Yugoslavia
  dance/music in, 116–17, 119
  economic crisis in, 14
  guest worker policy of, 14
  multiculturalism in, 10, 32, 116, 297n19
  socialist, 10
  violence associated with, 306n11
  wars in (1991–1995), 14
Yunakov, Ahmed, 136, 224–25
Yunakov, Danko, 104, 225
Yunakov, Yuri (*formerly* Husein Huseinov Aliev), 16, 20 221–39
  activist projects of, 234–35
  Americans taught by, 234
  arrests/imprisonment of, 227
  awards received by, 235
  bitterness toward the press, 162
  boxing career of, 226
  Bulgarian Turkish identity of, 222–23
  clarinet played by, 223–25, 235
  clothing/image of, 237, 251
  club performances by, 235
  collaborations by/contacts of, 231–35, 270, 323n8
  at dance workshops, 310n32
  early years of, 221–26
  educational events, participation in, 255
  emigration to the United States, 17, 20, 160, 229–30
  ethnicity of, 142
  ethnic variety among patrons of, 100, 300n9
  "Fincan," 231

Yunakov, Yuri (*continued*)
  and Foster, 327n16
  on Gypsy Caravan tour, 233–34,
    235–36, 260, 262
  Gypsy Fire, 231
  at Herdeljezi festival, 234
  and Hutz, 287, 331n42
    identities of, 44, 337n39,
      323n10
  and Kaplan, 287
  kaval played by, 224
  kyuchek played by, 227–28
  and Mamudoski, 232–33,
    235
  as mentor to young musicians,
    103, 104
  and Milev, 225, 233
  multiple identities of, 237–39,
    323n10
  musical background/training of,
    223–26
  name change of, 142, 226,
    228
  and the New York Gypsy
    Festival, 235, 266–67
  on oppression of musicians,
    227–28
  and Papazov, 225, 227, 233–34
  on Papazov, 134
  police evasion by, 143–44, 228
  on professional female dancers,
    109
  on Rajasthani music, 236
  repertoire/versatility of, 231–32,
    235
  reunion tours with Papazov, 161
  Romani identity of, 223, 230–37
  and Salifoski, 105
  saxophone played by, 227
  synthesizer used by, 237, 249
  on Taraf de Haidouks, 236
  *Together Again: Legends of
    Bulgarian Wedding Music*,
    161, 233
  on tour, 237
  in Trakiya, 136, 227
  Turkish music played by, 231
  and Erhan and Sevim Umer,
    102
  on wedding music, 312n12
  wedding music by, 226, 227–29

Yuri Yunakov Ensemble, 17, 243
  Balada, 233
  clothing/image of, 237, 251
  in the Gypsy Caravan tour, 243
  *New Colors in Bulgarian
    Wedding Music*, 233
  in the New York Black Sea
    Roma Festival, 235
Yuseinov, Ateshhan, 160

zadruga (patrilineal familial unit),
  64
"Zajdi Zajdi Jasno Sonce,"
  319n27
Zapej Makedonijo (film), 207,
  209, 322n11
"Zapevala Sojka Ptica," 28
Zap Mama, 330n37
Zekirovski, Sami, 211
Žekov, Osman, 136
Zhekov, Osman, 30
Zhelyaskova, Antonina, 315n16
Ziff, Bruce, 327n12
Zig Zag, 192
Zirbel, Kathryn, 244, 246, 252,
  260, 261, 325n14
Zivković, Marina, 301n17
Žižek, Slavoj, 43
Zlatne Uste, 105, 265–66,
  327n16
Zsurafski, Zoltan, 324n4
zurna/zurla
  and the Bulgarian anti-Muslim
    campaign (1980s), 128–30,
    141, 311nn3–4, 311n6 (ch.
    7)
  Bulgarian ban on, 24,
    128–29, 148
  vs. clarinet, 135
  in ensembles, history of, 23–24
  as exclusively male, 24
  at folk festivals, 167
  at the Galičnik wedding, 171
  in henna ceremonies, 89–90,
    308–9n16
  mane (free rhythmic
    improvisations) on, 25
  multipart playing of, 24–25
  of Pirin, 24
  recordings of, 148
  ritual function of, 24, 36, 89
  training in, 24